2008 SEISMIC ENGINEERING CONFERENCE

COMMEMORATING THE

1908 MESSINA AND REGGIO CALABRIA EARTHQUAKE

Reggio Calabria, Italy   8 – 11 July 2008

PART ONE

EDITORS
Adolfo Santini
Nicola Moraci
University of Reggio Calabria, Italy

SPONSORING ORGANIZATIONS
Municipality of Reggio Calabria
Regional Council of Reggio Calabria
Regional Province of Reggio Calabria
Mediterranean University of Reggio Calabria
Faculty of Engineering of Reggio Calabria
Department of Mechanics and Materials (University of Reggio Calabria)
Department of Computer Science, Mathematics. Electronics and Transportations
  (University of Reggio Calabria)
University of Catania
CARICAL Foundation
Chamber of Commerce of Reggio Calabria
Regional Province of Messina
Association of Architects of Reggio Calabria
Association of Architects of Messina

AMERICAN INSTITUTE OF PHYSICS

Melville, New York, 2008
AIP CONFERENCE PROCEEDINGS ■ VOLUME 1020

Editors:

Adolfo Santini
Nicola Moraci

Dipartimento di Meccanica e Materiali
Facoltà di Ingegneria
Via Graziella - Feo di Vito
I-89122 Reggio Calabria
Italy

E-mail: adolfo.santini@unirc.it
           nicola.moraci@unirc.it

L.C. Catalog Card No. 2008928042
ISBN 978-0-7354-0542-4
ISSN 0094-243X
Printed in the United States of America

CONTENTS

SPECIAL SESSIONS

STEEL SHEAR WALLS FOR SEISMIC APPLICATIONS

Chair: Prof. Abolhassan Astaneh-Asl – University of California at Berkeley, USA
Chair: Prof. Federico Mazzolani – University of Naples "Federico II", Italy

SEISMIC RETROFITTING FOR THE MONUMENTAL CULTURAL HERITAGE
IN THE MEDITERRANEAN BASIN

Chair: Prof. Fabio Casciati – University of Pavia, Italy

STRUCTURAL CONTROL AND MONITORING

Chair: Prof. Lucia Faravelli – University of Pavia, Italy

CONFERENCE TOPICS

SITE CHARACTERIZATION, MICROZONATION AND SITE EFFECTS

vii

SOIL LIQUEFACTION AND LIQUEFACTION COUNTERMEASURES

SLOPE, EMBANKMENTS, DAMS AND WASTE FILLS

FOUNDATIONS AND SOIL-STRUCTURE INTERACTION

EARTH RETAINING STRUCTURES AND GEOSYNTHETICS

CODES AND GUIDELINES

STRUCTURAL ENGINEERING

NUMERICAL METHODS

ADVANCED TECHNOLOGIES IN CONSTRUCTION
AND RETROFIT OF STRUCTURES

STRUCTURAL SAFETY AND RELIABILITY

STRUCTURAL DYNAMICS

URBAN PLANNING AND POLICIES FOR SEISMIC RISK REDUCTION

PREFACE

On 28 December 1908, at dawn, a strong earthquake shook the area of the Strait of Messina in Southern Italy, which separates Sicily from Calabria. This was the most powerful and destructive earthquake recorded in Europe. Its magnitude was about a 7.5 according to the Richter scale, and in the epicentral zone it reached the 10th degree of the Mercalli scale. After the earthquake, a tsunami with about ten meter high waves struck the coasts, causing even more destruction. The two major cities on either side of the Messina Strait - Reggio Calabria and Messina - suffered extensive, heavy damages, as well as all the other nearby coastal towns. Due to poor construction practice at that time, about 90 percent of the buildings could not withstand the shaking of the ground and collapsed. Estimates of casualties vary, but may be more than 70,000.

In the years following 1908, new seismic provisions were established and safety measures were taken when reconstruction began, aiming at a new building architecture able to resist earthquakes. The lesson learned from this earthquake can be considered as the starting point of the *seismic engineering* in Italy. Nowadays, one century after, our scientific knowledge and preparedness are considerably improved, including the awareness that another strong quake could hit, in the future, the area of the Strait of Messina. However, though we are better prepared than before, we are not yet ready for the next great earthquake. The seismic vulnerability degree of a relevant part of our buildings and infrastructures, in fact, is still too high and should be reduced.

Our "2008 Seismic Engineering International Conference commemorating the 1908 Messina and Reggio Calabria Earthquake" (MERCEA'08) had the aim to provide a forum to discuss the state-of-the-art, the best practices and the new research results in the field of earthquake engineering and geotechnics. The success of the Conference ensured that this purpose was achieved.

This volume contains almost all the papers which were presented at the Conference, held at the University of Reggio Calabria, Italy, from 8 July to 11 July 2008. There are over 210 papers written by authors from approximately 30 different countries, making the Conference a true international meeting. The papers form a broad set covering the main seismic engineering and geotechnics issues. The following topics are addressed: Seismic retrofitting for the monumental cultural heritage in the Mediterranean basin; Structural control and monitoring; Site characterization, microzonation and site effects; Soil liquefaction and liquefaction countermeasures; Slopes, embankments, dams and waste fills; Foundations, and soil-structure interaction; Earth retaining structures and geosynthetics; Codes and guidelines; Structural engineering; Numerical methods; Passive protection devices and seismic isolation; Advanced technologies in construction and retrofit of structures; Seismic risk; Structural safety and reliability; Structural dynamics; Urban planning and policies for seismic risk reduction.

The Editors are grateful to all authors and participants for their support and to the members of the International Scientific Committee for their help. They are particularly indebted to the Institutions sponsoring the Conference: Municipality of Reggio Calabria, Regional Council of Reggio Calabria, Regional Province of Reggio Calabria, Mediterranean University of Reggio Calabria, Faculty of Engineering of Reggio Calabria, Department of Mechanics and Materials (University of Reggio Calabria), Department of Computer Science, Mathematics, Electronics and Transportations (University of Reggio Calabria), University of Catania, CARICAL Foundation, Chamber of Commerce of Reggio Calabria, Regional Province of Messina, Association of Architects of Messina. Finally, a special thank to the publishers, the American Institute of Physics, for their advice throughout the preparation of this volume.

R. Pietropaolo
President of the Organizing Committee of MERCEA'08

A. Santini and N. Moraci
Book Editors

Local Organizing Committee

President

Rosario Pietropaolo	*(University of Reggio Calabria)*

Vice Presidents

Nicola Moraci	*(University of Reggio Calabria)*
Giuseppe Muscolino	*(University of Messina)*
Adolfo Santini	*(University of Reggio Calabria)*

Members

Pier Luigi Antonucci	*(University of Reggio Calabria)*
Giordano-Bruno Arato	*(ENEA)*
Michele Buonsanti	*(University of Reggio Calabria)*
Ernesto Cascone	*(University of Messina)*
Piero Colajanni	*(University of Messina)*
Paolo Clemente	*(ENEA)*
Enzo D'Amore	*(University of Reggio Calabria)*
Giuseppe Failla	*(University of Reggio Calabria)*
Giovanni Falsone	*(University of Messina)*
Giuseppe Fera	*(University of Reggio Calabria)*
Massimo Forni	*(ENEA)*
Pasquale Giovine	*(University of Reggio Calabria)*
Lidia La Mendola	*(University of Palermo)*
Maria Rossella Massimino	*(University of Catania)*
Massimiliano Mattei	*(University of Reggio Calabria)*
Giuseppe Mortara	*(University of Reggio Calabria)*
Antonina Pirrotta	*(University of Palermo)*
Daniela Porcino	*(University of Reggio Calabria)*
Giuseppe Ricciardi	*(University of Messina)*
Michelangelo Savino	*(University of Messina)*

International Scientific Commitee

Mustafa Aktar, Turkey
Jalal Al Dabbeek, Palestine
David Alexander, Italy
Nicholas N. Ambraseys, UK
Atilla Ansal, Turkey
Farhad Ansari, USA
Abolhassan Astaneh-Asl, USA
Alessandro Baratta, Italy
Mohamed Belazougui, Algeria
Gianmario Benzoni, USA
Lawrence Bergman, USA
Raimondo Betti, USA
Franco Bontempi, Italy
Antonio Borri, Italy
Franco Braga, Italy
Alberto Burghignoli, Italy
Gianmichele Calvi, Italy
Paolo Carrubba, Italy
Fabio Casciati, Italy
Daniele Cazzuffi, Italy
Mehmet Celebi, USA
Anil K. Chopra, USA
Edoardo Cosenza, Italy
Luis Decanini, Italy
Armen Der Kiureghian, USA
Antonello De Luca, Italy
Giovanni Dente, Italy
Mario Di Paola, Italy
Mauro Dolce, Italy
Amr S. Elnashai, USA
Peter Fajfar, Slovenia

Lucia Faravelli, Italy
Michael Fardis, Greece
Filip C. Filippou, USA
Takafumi Fujita, Japan
Gorge Gazetas, Greece
Mohsen Ghafory-Ashtiany, Iran
Vinay K. Gupta, India
Mahmood Hosseini, Iran
Michele Jamiolkowsky, Italy
James Kelly, USA
Takaji Kokusho, Japan
Stephen Kramer, USA
Helmut Krawinkler, USA
Sergio Lagomarsino, Italy
Renato Lancellotta, Italy
Cinna Lomnitz, Mexico
Paulo B. Lourenco, Portugal
Guido Magenes, Italy
George Manos, Greece
Alessandro Martelli, Italy
Nicos Makris, Greece
Gaetano Manfredi, Italy
Michele Maugeri, Italy
Marcello Mauro, Italy
Federico Mazzolani, Italy
Alberto Mazzucato, Italy
Mikayel Melkumyan, Armenia
Claudio Modena, Italy
Nicola Moraci, Italy
Roberto Nova, Italy
Shunsuke Otani, Japan

Giuliano Panza, Italy
Manolis Papadrakakis, Greece
Maurizio Papia, Italy
Manolo Pastor, Spain
Kiriazis Pitilakis, Greece
Shamsher Prakash, USA
Giuseppe Ricceri, Italy
Silvana Rizzo, Italy
Rodolfo Saragoni, Chile
Abdoreza Sarvghad-Moghadam, Iran
Pedro Seco y Pinto, Portugal
Francesco Silvestri, Italy
Enzo Siviero, Italy
Pierre Sollogoub, France
Pol D. Spanos, USA
Bill Spencer, USA
Fumio Tatsuoka, Japan
Colin Taylor, UK
Stephen O. Tobriner, USA
Mihailo D. Trifunac, USA
Alexander Vakakis, Greece
Calogero Valore, Italy
Giovanni Vannucchi, Italy
Filippo Vinale, Italy
Alfonso Vulcano, Italy
Martin S. Williams, UK
David M. Wood, UK
David Whittaker, New Zealand
Fu Lin Zhou, China

Sponsors

Financial support has been received by:

Municipality of Reggio Calabria

Regional Council of Reggio Calabria

Regional Province of Reggio Calabria

Mediterranean University of Reggio Calabria

Faculty of Engineering of Reggio Calabria

Department of Mechanics and Materials (University of Reggio Calabria)

DIMET

Department of Computer Science, Mathematics, Electronics and Transportations (University of Reggio Calabria)

University of Catania

CARICAL Foundation

Chamber of Commerce of Reggio Calabria

Regional Province of Messina
Association of Architects of
Messina

Furthermore, the Conference has been held under the auspices of:

 Ministry of Cultural Heritage

 Region Calabria

 University of Messina

 University of Palermo

 ENEA – Italian National Agency for New Technologies, Energy and the Environment

 Italian Geotechnics Association

 Anti-Seismic System International Society

 International Geosynthetics Society

 **Isolation and other antiseismic design strategies**

 **Network of Seismic Engineering University Laboratories**

 Association of Engineers of Reggio Calabria

 Association of Architects of Reggio Calabria

SPECIAL SESSIONS

STEEL SHEAR WALLS
FOR SEISMIC APPLICATIONS

Chair: Prof. Abolhassan Astaneh-Asl

University of California at Berkeley, USA

Chair: Prof. Federico Mazzolani

University of Naples "Federico II", Italy

STEEL SHEAR WALLS,

BEHAVIOR, MODELING AND DESIGN

Abolhassan Astaneh-Asl[a]

[a]University of California, Berkeley, 781 Davis Hall, Berkeley, CA, 94720-1710, USA

Abstract. In recent years steel shear walls have become one of the more efficient lateral load resisting systems in tall buildings. The basic steel shear wall system consists of a steel plate welded to boundary steel columns and boundary steel beams. In some cases the boundary columns have been concrete-filled steel tubes. Seismic behavior of steel shear wall systems during actual earthquakes and based on laboratory cyclic tests indicates that the systems are quite ductile and can be designed in an economical way to have sufficient stiffness, strength, ductility and energy dissipation capacity to resist seismic effects of strong earthquakes. This paper, after summarizing the past research, presents the results of two tests of an innovative steel shear wall system where the boundary elements are concrete-filled tubes. Then, a review of currently available analytical models of steel shear walls is provided with a discussion of capabilities and limitations of each model. We have observed that the tension only "strip model", forming the basis of the current AISC seismic design provisions for steel shear walls, is not capable of predicting the behavior of steel shear walls with length-to-thickness ratio less than about 600 which is the range most common in buildings. The main reasons for such shortcomings of the AISC seismic design provisions for steel shear walls is that it ignores the compression field in the shear walls , which can be significant in typical shear walls. The AISC method also is not capable of incorporating stresses in the shear wall due to overturning moments. A more rational seismic design procedure for design of shear walls proposed in 2000 by the author is summarized in the paper. The design method, based on procedures used for design of steel plate girders, takes into account both tension and compression stress fields and is applicable to all values of length-to-thickness ratios of steel shear walls. The method is also capable of including the effect of overturning moments and any normal forces that might act on the steel shear wall.

Keywords: Steel Shear Walls, Seismic Design, AISC Seismic Provisions, Structural Modeling

INTRODUCTION

There are two types of steel shear walls; stiffened and unstiffened. This paper focuses on unstiffened steel shear walls. Figure 1 shows typical configurations used for steel shear walls. The most common type of steel shear wall system is one- bay shear wall, Figure 1(a), where the steel plate is welded to boundary columns and beams and the beam-to-column connections are special ductile moment connections. In narrow and tall structures, where the width of shear wall is relatively small, in order to resist the overturning moment, outrigger steel shear wall system, shown in Figure 1(b) can provide the necessary stiffness and strength to resist such overturning moments. In this case also, the wall panel is usually welded to the boundary beams and columns and beam-to-connections of the wall framing are special ductile moment

CP1020, *2008 Seismic Engineering Conference Commemorating the 1908 Messina and Reggio Calabria Earthquake*,
edited by A. Santini and N. Moraci
© 2008 American Institute of Physics 978-0-7354-0542-4/08/$23.00

connections. Depending on the stiffness and strength demand, we can also make the connections of the side girders to columns moment connections.

Figure 1(c) shows coupled steel shear wall system where two single bay shear walls of Figure 1(a) are connected to each other by girders with end moment connections. This system is very suitable for cases where the steel shear wall is used in the core of the building and there is a need to have a corridor opening in the wall practically dividing the single bay wall into two separated but shorter walls.

Figure 1(d) shows a steel shear wall system developed by the Magnusson Klemencic and Associates which was used successfully in a few buildings in the United States. In this system, which generally is used in the core of tall buildings, there are two large concrete-filled tubes that act as boundary columns. These concrete-filled tubes, because of their size, carry a significant amount of shear due to lateral load effects. In some cases, the shear carried directly by the four concrete-filled tubes can be up to 50% of the total story shear. The concrete-filled tubes also carry a significant amount of floor gravity load, in some cases up to 70% of the total gravity load. In this system, the steel columns and beams located between the two concrete-filled tubes are used only for lateral load effects with almost no gravity load transferred to them. The shear walls are welded to the steel beams and columns as well as the boundary concrete-filled tubes. The beam-to-column connections located between the two concrete-filled tubes in this system are all special ductile moment resisting connections. The beam-to-column connections outside the shear wall system are simple connections.

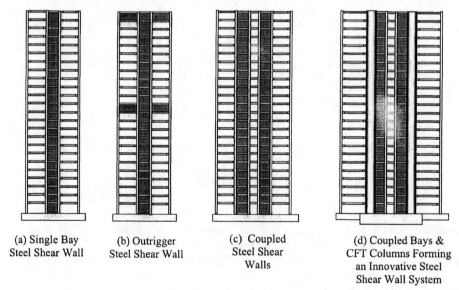

| (a) Single Bay Steel Shear Wall | (b) Outrigger Steel Shear Wall | (c) Coupled Steel Shear Walls | (d) Coupled Bays & CFT Columns Forming an Innovative Steel Shear Wall System |

FIGURE 1. Typical configurations of steel shear walls

Development and use of steel shear walls are relatively recent phenomena. Traditionally, reinforced concrete (RC) shear walls were used as an efficient lateral load resisting system in buildings. During the 1970's, a number of existing Veterans Administration Hospitals in California were seismically retrofitted by adding steel shear walls. In these early applications, the steel shear walls had horizontal and vertical stiffeners with relatively close spacing between the stiffeners to prevent local buckling of steel shear wall plates prior to yielding of the wall under shear. The unstiffened steel plate shear walls have been used in modern steel structures as an efficient seismic lateral load resisting system in California while most steel shear walls used in Japan have been stiffened.

Each of the reinforced concrete (R/C) and steel shear walls has its advantages and disadvantages. Based on the information available in the literature on seismic behavior of R/C and steel shear walls, the main disadvantage of R/C shear walls is the development of tension cracks in tension zones while the main disadvantage of the steel shear walls is buckling of steel plate panel under relatively small load. Both tension cracking of R/C walls and compression buckling of steel plate shear walls result in pinching of the hysteretic loops and reduced stiffness, strength and energy dissipation capacity of the system. In addition in R/C shear walls, under large cyclic deformations, the tension cracking and compression crushing can result in spalling and severe damage that may require extensive repairs after an earthquake. In steel shear walls, the buckling of plate that take place out-of-plane of the wall, can result in damage to lifeline pipes placed next to the walls. In steel shear walls lateral deformations, due to out-of-plane buckling of the steel plate, can be significant and may require steel shear walls to be replaced after a large earthquake. In addition, tests of steel shear walls as summarized in the following section, have shown that during the large inelastic deformation cycles, the boundary columns of steel plate shear walls develop plastic hinges and undergo severe inelastic deformations. Also, thin steel shear walls are not expected to provide much of noise isolation compared to R/C walls.

Another disadvantage of both steel and R/C shear walls is that these systems show relatively large initial stiffness when used in low rises. Such a large stiffness is in most cases far more than necessary to control and reduce the lateral drift. The large initial stiffness can attract large amount of seismic energy which in turn results in large seismic forces. Such large seismic forces in turn need to be resisted by shear wall, boundary columns and foundations resulting in increased and unnecessary construction costs. More importantly, such large seismic forces can result in damage to non-structural elements.

Combining advantages of both R/C and steel shear walls results in a composite shear wall. The composite shear walls, which are the combination of steel plate and R/C wall, have the potential of performing much better than either of the steel plate or R/C shear walls. More on the composite shear walls can be found in Astaneh-Asl [1] and Zhao and Astaneh-Asl [2] including design procedures and cyclic test results.

PAST STUDIES OF STEEL SHEAR WALLS

Seismic behavior of steel shear walls have been studied in recent years in the United States, Canada, United Kingdom, Italy, Iran, Japan, China and elsewhere. For a list of references on these see Astaneh-Asl [3]. Following paragraphs provide summaries of some of the major experimental and analytical studies on this subject.

In the U.K., Sabouri-Ghomi and Roberts [4] have reported the results of 16 tests of steel shear panels diagonally loaded. They have also developed an analytical model of hysteretic behavior of the shear panel.

In Canada Timler and Kulak [5], Driver et al. [6] and Rezai et al [7] have performed cyclic testing of steel shear walls as well as development of the so-called tension only "strip model" for shear wall design. The strip model and its limitations are discussed later in this paper under Modeling of Steel Shear Walls.

In the United States, Elgaaly and Caccese [8] have conducted a number of studies of steel plate shear walls including cyclic tests of small scale steel frames in-filled with steel plate shear walls as well as analytical studies. The specimens in these tests were relatively small scale and the length-to-thickness ratio of the walls was very large compared to practical values. In recent years, at the University of California, Berkeley, Zhao and Astaneh-Asl [2] have conducted cyclic tests of two ½ scale realistic steel shear wall systems similar to those shown in Case (d) of Figure 1 as well as have performed analytical studies of steel shear walls. The studies, summarized in the next section included seismic design recommendations as well.

In Japan, a number of valuable studies on steel plate shear walls have also been conducted and steel shear walls have been used in a number of modern and tall steel structures. Sugii and Yamada [9] have reported the results of tests on 14 steel plate shear walls. Their studies in all specimens show pinching of hysteretic loops due to buckling of compression field. Yamada and Tamakaji [10] have conducted tests of single panel steel shear walls and center core type steel shear walls on frames similar to case (a) in Figure 1 above, to study cyclic behavior of steel shear walls. One of the important conclusions of their research was that the behavior of steel shear walls can be modeled as plate girder with diagonal compression and tension fields. The width of tension field was established by assuming that the length of the anchored tension field was equal to 1/3 of the length of the beam length and column height at early stages of loading, and later the 1/3 ratio becoming 1/4. They also concluded that the angle of buckled waves was near 45° even though the span to height ratio of tested wall systems was 2.0:1.0

In Italy, Mazzolani et al. [11] have reported on tests conducted on steel and aluminum shear panels for seismic upgrading of new and exiting structures. Combined with analytical studies, the tests indicated that use of metal shear panels can be a very attractive strategy to reduce the seismic vulnerability of existing structures.

In Iran, recently, Sabouri-Ghomi and Gholhaki [12] reported on tests of two ½-scales, three-story steel shear wall systems subjected to cyclic shear. The specimens had high strength steel for boundary beams and columns and low yield steel in shear wall plates. The beam-to-column connections in one specimen were simple and in the other rigid. Both specimens showed good ductility although behavior and energy dissipation capacity of the specimen with rigid connections was better than the specimen with simple connections.

TESTS OF STEEL SHEAR WALLS AT U.C. BERKELEY

One of the most recent projects on steel shear walls was done at the University of California, Berkeley by Zhao and Astaneh-Asl [2]. A brief summary of this project follows. Figure 1(d) shows the system that was tested. In the system, coupled steel shear walls are used to resist lateral forces. The boundary columns are concrete filled composite columns while the interior columns are wide flange shapes. The system was developed by the Magnusson Klemencic and Associates of Seattle and successfully used in tall buildings. The main objective of studies conducted by Zhao and Astaneh-Asl [2]was to investigate this type of steel plate shear wall and to develop seismic design recommendations. Figures 2 shows Specimens 1 and 2 which were taken out of the global system for testing. Figure 3(a) shows Specimen 2 inside test set-up while Figure 3(b) shows main components of the system tested. Figure 4 shows cyclic hysteretic loops in terms of shear force- lateral drift for two specimens. The two specimens were similar with one difference that the length/height ratio for Specimen 1 and 2 were 1.5 and 1.0 respectively. More information on the specimens and test results can be found in Zhao and Astaneh-Asl [2]. Figure 5 shows Specimens 1 and 2 at the end of the tests. Initially, the diagonal compression zone of the shear wall specimens buckled and tension field developed. As the loading of the specimens continued plastic hinges formed in the boundary beams and columns. Eventually, one of the plastic hinges in the coupling beams fractured and the load dropped below 80% of the maximum load achieved during the tests, which was considered to be the failure point. It must be mentioned that in the tested system the steel columns are not carrying any gravity and are only for seismic lateral resistance. Therefore, their yielding not only is acceptable but helps to increase the energy dissipation capacity of the system.

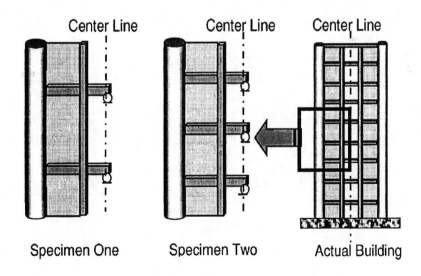

FIGURE 2. Specimen 1 and 2 representing part of the building on the right [2]

The connections in the specimens behaved in a very ductile manner. The connections of girders to the concrete filled tube (CFT) column were moment connections where the web of the girders were filed welded to the CFT and the flanges of the girders were welded to four horizontal rebars that were embedded in the concrete-filled tube with sufficient length to develop their cross section, Figure 5. During construction, after placing the shear wall unit in its proper position, the four rebars for each flange were welded to the girder flanges. In order to provide sufficient ductility for these moment connections, about 50 mm of the length of the rebars between the CFT and the girders were not welded and were left free to elongate inelastically, providing ductility in rotation to the girder connections to CFT. The

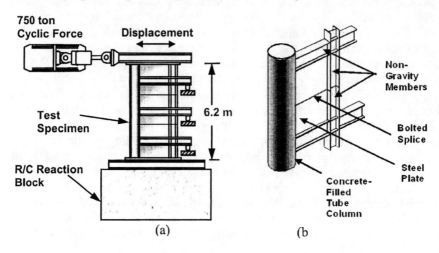

(a) (b

FIGURE 3. Specimen 2 inside the set-up (left) and main components of the system [2]

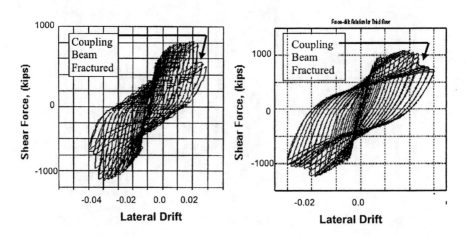

FIGURE 4. Test results for Specimen 1 (left) and Specimen 2 (right) [2]

50 mm fuses worked well during testing and the connections survived cyclic rotations of more than 0.03 radians during the testing with no fracture. The panel zones also performed well and had distributed yielding with no buckling or fracture of the web or flanges in the panel zone areas. The specimens had a horizontal field-bolted splice at mid-height of the floors, similar to actual buildings. These splices performed well and initially did not slip. However, during later cycles, they developed slippage and contributed to energy dissipation of the system.

As mentioned earlier, the steel columns in this system are only for lateral loads and are expected to yield and even buckle in later cycles. An interesting test observation was that due to slippage of splices at the mid-height of columns, the axial strains in the columns were released during large cycles and the columns never buckled. Figure 5 shows appearance of two specimens at the end of the tests.

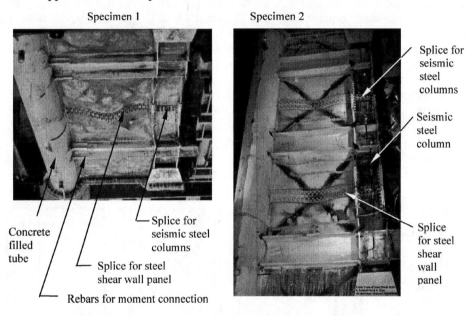

FIGURE 5. Specimen 1 (left) and Specimen 2 (right) at the end of the tests [2]

MODELING OF STEEL SHEAR WALLS

Past test results and analytical studies of steel shear walls indicate that they behave more or less like a plate girder with boundary columns acting as flanges of the plate girder and the boundary beams acting as the stiffeners in the plate girders. In the literature there are several models proposed to be used in the analysis to represent stiffness and strength of the shear walls. The simplest model shown in Figure 6(a) is to replace the steel shear wall with an X brace where the areas of tension and compression diagonal braces are selected to provide the same stiffness as the tension and compression diagonal fields in the shear wall. Another model proposed by Kulak [13] is to replace the steel shear wall by a series of usually 10-11 parallel tension only

11

bars diagonally positioned as shown in Figure 6(b). This model does not have any compression element to represent the stiffness and strength of the diagonal compression field. Also, the model is not capable of properly transferring the stresses, due to overturning moment to the steel shear walls. The model called tension only "strip model" is used in the ANSI/AISC-341 [14] standard, currently the governing seismic design standard in the United States. Tests and recent realistic finite element analytical studies by Shi and Astaneh-Asl [15] indicate that this model can only be appropriate to represent the behavior of *very slender* steel shear walls where the length-to-thickness ratio of the wall panel is greater than 1500 which is inappropriate to be used to represent steel shear walls that have length-to-thickness ratios less than 600. In very slender shear wall panels the compression diagonal buckles under very small lateral drift values leaving the wall to act as tension only element justifying the tension only strip model of the ANSI/AISC-341 [14]. However, the tension-only strip model is incapable of representing steel shear walls with length-to-thickness ratio less than about 600 where not only there can be significant compression field but also the tension side of steel plate shear wall participates with boundary columns in resisting overturning moments. Such participation for typical steel shear walls can be significant.

In almost all actual application of steel shear walls in buildings the length-to-thickness ratio of steel shear wall is in the range of 100-600 and more or less in the narrow range of 200-400. Therefore, using the tension only strip model to represent the steel shear walls for almost all cases of practical applications is questionable at best and will result in erroneous drift as well as unrealistic force values applied to elements of the system in particular the boundary columns.

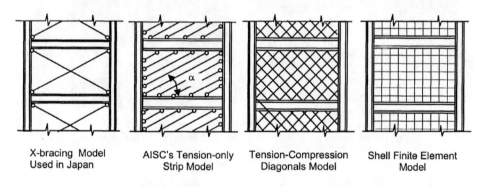

FIGURE 6. Models Proposed to Represent Steel Shear Wall in Analysis

A more appropriate model for steel shear walls with length-to-depth ratio of less than 1500 is to replace the shear wall with a series of diagonal tension and compression members connected to each other to act as a grid, Figure 6(c). This model was used in modeling the shear walls of the 74 –story Jinta building in Shanghai by Skidmore Owing & Merril engineers, Mathias et al.[16].

Currently, with the availability of structural analysis software there may be no need to model the shear wall using stick elements as discussed in previous paragraph. The shear wall simply can be modeled as suggested by the author , A. Astaneh-Asl, [3] as orthotropic shell elements shown in Figure 6(d). Such shell elements can be assigned two different stiffness values in principle directions, one representing the stiffness of tension field and the other representing the stiffness of compression field. Then the properties of orthotropic shell elements can be rotated 45 degrees to be aligned with the actual diagonal direction of tension and compression fields.

SEISMIC DESIGN OF STEEL SHEAR WALLS

In seismic design of structural systems three types of information are needed: (1) information for the load side of the design equation, which consists of values of Response Modification Factor, R, the Over-strength Factor Ω and the Deflection Amplification Factor C_d, (2) information on how to model the steel shear wall in the analysis to obtain realistic forces for design of components of the system such as members, shear wall itself and connections; and (3) information on how to establish strength of components to ensure that the load effects are less than resistance for all possible failure modes. Of course the design procedures should ensure the high ductility for the system to justify the high value of Response Modification Factor of 8.

In 2000, for the first time, the author, A. Astaneh-Asl [3] proposed the values of R, Ω and C_d, given in Table 1, for a number of steel shear wall systems. The value of R equal to 8, given in the table, is currently adapted by the ANSI/AISC-341[14], seismic provisions for seismic design of steel plate shear walls.

Table 2. Values of Seismic Design Parameters for Steel Shear Walls [3]

Basic Seismic Force resisting System	Response Modification Factor, R	System Over-Strength Factor, Ω	Deflection Amplification Factor, C_d
1. Unstiffened steel plate shear walls inside a gravity carrying steel frame with simple beam to column connections	6.5	2	5
2. Stiffened steel plate shear walls inside a gravity carrying steel frame with simple beam-to-column connections	7.0	2	5
3. Dual system with special steel moment frames and unstiffened steel plate shear walls	8	2.5	4
4. Dual system with special steel moment frames and stiffened steel plate shear walls	8.5	2.5	4

13

The second information needed for seismic design of steel shear walls is a rational model of steel shear wall that can be used in the analysis to establish the internal forces and stresses in the components of the system such as shear wall, boundary elements and connections. As mentioned earlier, the current ANSI/AISC-341 [14] standard has the strip model as the representative model of the steel shear wall. Again, the use of tension only strip model to represent steel shear walls can only be justified for very thin and slender shear walls where the length-to-thickness ratio of the wall is in the range of 600-1500 and more. Most practical cases of steel shear walls have a length-to-thickness ratio of 100-600. As mentioned earlier, the most rational model for steel shear walls appears to be the use of orthotropic shells with material properties defined such that the principle direction of the shells represent the stiffness of the tension field and compression field.

Finally, the most important information needed in design of steel shear walls is design criteria that establish the philosophy of design as well as provide rational procedure to establish design strength and stiffness of the components of the system. The most rational philosophy for design of steel shear walls is capacity design such that the designed shear wall system has the ductility to justify the R factor of 8 used in establishing the seismic forces applied to it. In order to ensure that steel shear wall has the necessary ductility in applying capacity design, one has to ensure that the capacities associated with brittle failure modes are greater than the capacities associated with ductile ones. For this, in 2000 the author proposed design procedures in [3] as summarized below.

PROPOSED SEISMIC DESIGN PROCEDURES

1. Introduction

Steel shear wall systems can be divided into two categories of: (a) *singular steel shear wall system* where the steel shear wall is the only lateral load resisting system and; (b) *dual steel shear wall system* where the steel shear wall is placed parallel to moment frames or within the moment frames and together the steel shear wall and moment frame resist the lateral load. The main elements of a steel shear wall system are the steel shear wall boundary columns, horizontal floor beams and connections.

Steel shear walls can be stiffened or unstiffened. Stiffened shear walls have sufficient horizontal and vertical stiffeners welded to wall plate that buckling of the wall plate will not occur prior to the shear yielding. In the unstiffened shear walls, no stiffeners are used on the wall plate. As a result under applied shear the steel plate wall buckles along the compression diagonal in a multiple wave form prior to shear yielding of steel plate. After buckling of the compression diagonal, the shear stiffness and strength of the wall is primarily due to the development of tension and compression field along the tension diagonal. Total shear capacity of the wall *plate* is the sum of shear capacity due to compression field buckling and tension field yielding. The total shear capacity of steel shear wall *system* is sum of the shear capacity of the wall plate and the boundary moment frame.

2. Provisions for Design of Shear Wall Steel Plate Panel

2. a. The material of shear wall should be selected such that the $R_y F_y$ of the steel shear wall is less than or equal to the $R_y F_y$ of the boundary columns and horizontal beams connected to the wall. This will result in yielding of wall plate while boundary elements remain elastic.

2. b. The design shear strength of steel shear wall is established using procedures for design of steel plate girders in the AISC Specifications [17]. Other rational design procedures based on test results or realistic inelastic analyses can also be used.

2. c. At the top floor, if tension field action is used in design, the horizontal beams and boundary columns shall be designed to be strong enough to resist the horizontal and vertical components of the diagonal tension field. Alternatively, by using stiffened shear walls or thicker unstiffened shear walls, the story shear is resisted without utilizing tension field action.

2. d. In stiffened shear walls, horizontal as well as vertical stiffeners shall be spaced such that the maximum h/t_w of all steel panels bounded by the stiffeners complies with the following:

$$\frac{L_w}{t_w} \leq 1.1\sqrt{k_v\, E / F_{yw}} \tag{1}$$

The plate buckling coefficient, k_v, is given as:

$$k_v = 5 + \frac{5}{(h_w / L_w)^2} \tag{2}$$

In the above equations, L_w is the length of steel shear wall panel, t_w is the thickness of the wall, E is the modulus of elasticity of steel, F_{yw} is the specified yield stress of the steel and h_w is the height of the steel shear wall panel.

2. e. At the bottom floor, where shear wall is attached to the foundation, special arrangements shall be made to ensure proper transfer of horizontal and vertical components of tension field action to the foundation.

2. f. Slip-critical bolts or continuous welds can be used to connect the steel shear wall to boundary columns and horizontal beams. The connections shall be designed to develop expected shear strength of the wall plate.

2.g. It is important that in design of steel shear walls, provisions be made to ensure that steel plate part of the shear wall system is not carrying significant amount of gravity loads. This is due to the fact that such gravity loads can result in buckling of the steel shear wall under gravity load which can adversely affect its performance during future earthquakes. Another issue is that in tall buildings shortening of the boundary columns due to addition of gravity loads as the upper floors are added, can result in buckling of the steel plate panel in vertical direction. Provisions should also be made in design and construction of steel shear walls such that the wall plates can accommodate shortening of the boundary columns as the structure rises and more and more gravity load is applied to boundary columns.

3. Provisions for Design of Boundary Columns in Steel Shear Wall Systems

3. a. The web of boundary columns shall be in plane of the steel shear wall. Otherwise, the connection of wall plate to the column web perpendicular to it should be such that out-of-plane bending of column web is prevented.

3. b. If boundary columns of steel shear walls are carrying gravity loads, the columns should be designed to remain elastic under the Design Earthquake.

3. c. In steel shear wall systems where boundary columns are not carrying gravity load, the columns may be designed to undergo yielding and cyclic local buckling provided that their width thickness ratios are limited to values given in Table I-9.1 of the ANSI/AISC-341 [14].

3. d. The web thickness of boundary columns should be greater than the thickness of the steel plate walls connected to them.

3. e. Base connections of boundary columns to foundations shall be designed to develop tension yield capacity of the boundary columns. The governing failure mode of a boundary column base connection shall be a ductile failure mode, such as yielding of base plate or limited yielding of anchor bolts, but not a fracture mode.

4. Provisions for Design of Boundary Beams in Steel Shear Wall Systems

4. a. Horizontal beams in a steel shear wall system shall be designed to carry the gravity loads without participation of the steel shear wall.

4. b. Web thickness of horizontal beam shall be greater than the thickness of steel plate walls above and below the beam.

4. c. The shear connection of horizontal beams to boundary columns should be designed to develop shear strength of the beam web. Yielding of the shear plate shall be the governing failure mode of the connection. Opening and sharp corners are not allowed in the connections.

4. d. In steel shear wall systems where horizontal beams are not carrying gravity load, they may be permitted to undergo significant yielding and some inelastic local buckling. Their width-thickness ratios should satisfy the limits given in Table I-9.1 of the AISC/ANSI -341 [14] for preventing premature local buckling during seismic events.

5. Design of Boundary Frame

5. a. In shear wall systems where special moment frame(s) are used parallel to steel shear wall or in the same plane as steel shear wall and as its boundary frame(s), the design of special moment frame shall comply with the seismic design provisions for the Special Ductile Moment Resisting frames in the ANSI/AISC -341 [14].

5. b. In dual systems it is preferred that the steel shear wall be an infill to the special moment frame instead of being outside the moment frame and parallel to it.

6. Provisions for Design of Coupling Beams in Steel Shear Wall Systems

6. a. Steel shear walls can be connected to each other to act as a coupled shear wall system.

6. b. Coupling beams shall be connected to boundary columns with special ductile moment resisting connections designed in compliance with the applicable provisions of the ANSI/AISC-341 [14].

6. c. Coupling beams shall be compact sections satisfying the width-thickness ratios of Table I-9.1 of the AISC/ANSI -341 [14] for preventing premature local buckling during seismic events.

CONCLUSIONS

The provisions currently in the ANSI/AISC-341 [14] for design of steel shear walls may be applicable to *very slender* shear walls with a length-to-thickness ratio of more than 1000 or so but their applicability to *all* shear walls including those with length-to-thickness ratio less than 600 is questionable. There is no test data or analytical studies to justify the use of tension-only strip model which is the basis of the ANSI/AISC-341 [14] procedures in shear walls with length-to-thickness ratio less than 600, the most common range of applications.

A procedure for design of steel shear walls which was proposed in 2000 by the author, A. Astaneh-Asl, [3] is included in this paper. The procedure is applicable to all ranges of length-to-thickness ratios of steel shear walls.

ACKNOWLEDGMENTS

The author would like to thank the Structural Steel Educational Council for providing support in development of the author's Steel TIPS report on "Behavior and Design of Steel Shear Walls". The test results summarized in this paper on the work of Zhao and Astaneh-Asl were part of the dissertation work of Professor Qiuhong Zhao when she was a doctoral student at the University of California, Berkeley during 2000-2006 . The tests were supported in part by the General Services Administration of the U.S. Government and by the Skilling Ward Magnusson Barkshire (currently

Magnusson Klemencic Associates). The opinions expressed in this paper are solely those of the author.

REFERENCES

1. A. Astaneh-Asl, "Seismic behavior and design of composite shear walls ", *A Steel Technical Information and Product Services (TIPS) report,* Structural Steel Educational Council (www.steeltips.org), Moraga, California, (2001).
2. Q. Zhao and A. Astaneh-Asl, "Experimental and analytical studies on cyclic behavior of steel and composite shear wall systems", *Report Number UCB/CE-Steel-2006/01,* Department of Civil and Env. Engineering, Univ. of California (www.ce.berkeley.edu/~astaneh), Berkeley (2006).
3. A. Astaneh-Asl, "Seismic behavior and design of steel shear walls" *A Steel Technical Information and Product Services (TIPS) report,* Structural Steel Educational Council (www.steeltips.org), Moraga, California, (2000).
4. S. Sabouri-Ghomi, and T.M. Roberts, "Nonlinear dynamic analysis of steel plate shear walls including shear and bending deformations", *Engineering Structures*, Vol. 14, no.5, pp. 309-317 (1992).
5. P.A. Timler, and G.L. Kulak, "Experimental study of steel plate shear walls", *Structural Engineering Report No. 114,* University of Alberta, Canada (1983).
6. R.G. Driver, G.L. Kulak, A.E. Elwi, and D.J.L. Kennedy "Cyclic tests of four-story steel plate shear wall", *Journal of Structural Engineering,,* ASCE, Vol. 124, No. 2, Feb., pp. 112-120 (1998).
7. M. Rezai, C. E. Ventura, and H.G.L. Prion, "Numerical investigation of thin unstiffened steel plate shear walls", *Proceedings,* 12th World Conf. on Earthquake Engineering, New Zealand, (2000).
8. M. Elgaaly and V. Caccese "Post-buckling behavior of steel- plate shear walls under cyclic loads", Journal *of Structural Engineering,* ASCE, Vol. 119, n. 2, pp. 588-605 (1993).
9. K. Sugii and M. Yamada "Steel panel shear walls with and without concrete covering", *Proceedings,* 11th World Conference on Earthquake Engrg., Acapulco, Mexico (1996).
10. M. Yamada and T. Yamakaji, "Steel panel shear wall-as single and center core type aseismic element," Proceedings, Second International Conference, STESSA, Kyoto, Japan (1997).
11. F. M. Mazzolani, G. De Matteis, S. Panico, A. Formisano and G. Brando, "Shear panels for seismic upgradeing of new and existing structures", *Prceedings* , Urban Habitat Construction under Catastrophic Events Workshop, COST, Prague, Czech Republic, (2007).
12. S. Sabouri-Ghomi and M. Gholhaki, "Tests of two three-story ductile steel plate shear walls", *Proceedings,* Structures Congress, American Society of Civil Engineers, Vancouver, Canada, (2008).
13. G.L. Kulak, "Behavior of steel plate shear walls", *Proceedings,* AISC International Engineering Symposium on Structural Steel, American Institute of Steel Construction, Chicago, (1985).
14. ANSI/AISC-341, "Seismic provisions for structural steel buildings", *A National Standard,* American National Standards Institute, New York, and American Institute of Steel Construction (www.aisc.org), Chicago (2005).
15. Y. Shi, and A. Astaneh-Asl, "Lateral stiffness of steel shear wall systems", *Proceedings,* Structures Congress, American Society of Civil Engineers (www.asce.org), Vancouver, Canada, (2008).
16. N. Mathias, N. , M. Sarkisian, E. Long, and Z. Huang "Steel plate shear walls: efficient structural solution for slender high-rise in China", *Proceedings,* The 2008 Seismic Engineering Int. Conference commemorating the 1908 Messina and Reggio Calabria Earthquake (MERCEA'08), Reggio Calabria and Messina, Italy (2008).
17. ANSI/AISC-360, "Specifications for structural steel buildings", *A National Standard,* American National Standards Institute, New York, and American Institute of Steel Construction (www.aisc.org), Chicago (2005).

Aluminum Shear Panels for Seismic Protection of Framed Structures: Review of Recent Experimental Studies

G. De Matteis[a], G. Brando[a], S. Panico[b] and F.M. Mazzolani[b]

a Dept. of Design, Rehabilitation and Control of Architectural Structures,
University of Chieti-Pescara "G. D'Annunzio", V.le Pindaro, 42 - 65127 Pescara (I)
b Dept. of Structural Engineering, University of Naples "Federico II",
P.le Tecchio, 80 - 80125 Naples (I)

Abstract. An important experimental campaign on pure aluminum shear panels, to develop new devices for the seismic passive protection of buildings, has been recently carried out at the University of Naples "Federico II" in cooperation with the University "G. d'Annunzio" of Chieti/Pescara. In particular, several pure aluminum shear panels, suitably reinforced by ribs in order to delay shear buckling in the plastic deformation field, have been tested under cyclic loads. The choice pure aluminium, which is really innovative in the field of civil engineering, is justified by both the nominal low yield strength and the high ductility of such a material, which have been further improved through a proper heat treatment. Two different testing layouts have been adopted. In the former, six "full bay" pure aluminum shear panels, having in-plane dimensions 1500x1000 mm and thickness of 5 mm, have been taken in consideration. In the latter, four 5 mm thick stiffened bracing type pure aluminum shear panels (BTPASPs) with a square shape of 500 mm side length have been cyclically tested under diagonal load. In the whole several plate slenderness ratios have been considered, allowing the evaluation of the most influential factors on the cyclic performance of system. In the current paper a review of the most important results of these recent experimental activities is provided and discussed.

Keywords: aluminum alloy, cyclic tests, dissipative devices, low yield stress material, shear panels.

INTRODUCTION

Since few decades, thin metal shear panels (MSPs) have been established as an effective device for the seismic protection of new and existing buildings, constituting a valid alternative to other valuable systems, namely concentric, eccentric and buckling restrained braces [1]. They may provide very satisfactory energy dissipation capabilities [2], as a result of the exhibited post-buckling behaviour, which consists in a pure shear deformation mechanism followed by a tension field resisting mechanism, the latter, in case adequate stiffeners are applied, being activated for large shear deformation levels. Furthermore, it has to be underlined that these systems are able to provide also a suitable initial stiffness, which can be helpful for satisfying the serviceability limit state requirements of the whole structure. In addition, it is worthy of note that the large ductility of the base material allows a useful employment also when large seismic events provoke a large excursion of the system in the plastic field.

CP1020, *2008 Seismic Engineering Conference Commemorating the 1908 Messina and Reggio Calabria Earthquake,*
edited by A. Santini and N. Moraci
© 2008 American Institute of Physics 978-0-7354-0542-4/08/$23.00

The above features may result particularly attractive when MSPs are used for the retrofitting of existing buildings [3].

The actual hysteretic behavior exhibited by the system is strongly conditioned by the mechanical features of the adopted material. Therefore, in recent years, aiming at improving the energy dissipation capability of the system, the use of low yielding strength materials has promoted. Although the first studies were carried out in Japan, where a low yield steel characterized by a low carbon content has been proposed and then largely adopted [4], in Europe a special attention has been devoted to pure aluminum MSPs, which have been recently investigated by the authors. In particular, a wide experimental campaign on both full-bay multi-stiffened and bracing type pure aluminum shear panels has been undertaken at the Department of Structural Engineering of the University of Naples "Federico II", in cooperation with the Department of Design, Rehabilitation and Control of Architectural Structures of the University "G. d'Annunzio" of Chieti/Pescara, in order to investigate the potentiality of these devices, which should be adopted to improve the seismic performance of framed structures. In the present paper the main results achieved are summarized and discussed.

THE EXPERIMENTAL ACTIVITY

General

Two different typologies of shear panels have been considered, namely full bay type and bracing type. In the first case, tested shear panels have a dimension of 1000x1500x5 mm and are intended to be used as an infill of the whole bay of the base frame structure. In the second case, tested specimens have a dimension of 500x500x5 mm, since shear panels have to be intended as a plate to be allocated within a bracing system. For the first typology of MSP, different types of ribs (either welded or bolted) for the basic aluminum plate have been considered, whereas for the second only welded ribs have been adopted. A key aspect of the study is represented by the employment of the pure aluminum (AW1054A) as base material, which is an innovative proposal for civil engineering applications. In fact, this material is characterized by very convenient mechanical features, namely a very low yielding stress point (about 20 MPa), a high hardening ratio (about 4), a large ductility (over 40%) and lastly it is also easily available on the market [5]. In addition, in order to allow a useful comparison with other materials, tested full-bay type shear panels have been also fabricated by using a different aluminium alloy, namely the AW5154A, which is commonly adopted in the field of aluminum structures. In both cases, in order to improve the mechanical features of the material, proper heat treatments, consisting in a re-crystallization which provokes a smoothing of the material imperfections, have been applied. In Figure 1, the stress-strain curves for the adopted materials before and after the heat treatment are depicted. It is worth noticing that the same heat treatment produces different results for the two tested alloys. In particular it resulted less important in case of AW 5154A, where while the conventional yield strength f_{02} of heat treated alloy was halved, the ultimate strain ε_u remained almost unchanged owing to the higher content of allowing elements. On the other hand, it is apparent that the heat treated aluminum alloy AW 1050A is very suitable for the application under consideration. In fact, it is characterized by a higher value of the E/f_{02} ratio, which means that is less susceptible to incur into buckling before undergoing plastic deformations.

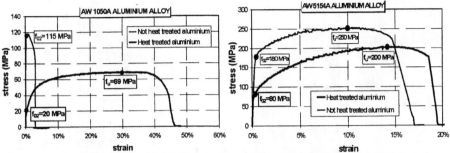

FIGURE 1. Behavior of the adopted alloys before and after the applied heat treatment process.

The cyclic response of the material has been also investigated by means of specific tensile-compression cyclic tests carried out on specimens equipped with a steel "jacket" able to inhibit out-of-plane deformations due to buckling (Figure 2a). The main purposes of these tests were (1) to define the behavior both in compression and in tension, (2) to characterize the hysteretic behavior and (3) to verify the hardening features. The obtained results shown a good dissipative behavior of the material, characterized by full hysteretic cycles, a substantial iso-resistance for each displacement level and the existence of a isotropic hardening component (Figure 2b).

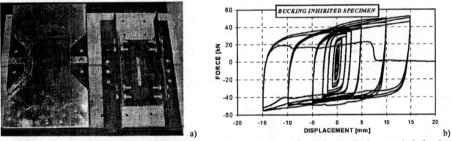

FIGURE 2. Buckling inhibited pure aluminum specimen: (a) tested specimen; (b) hysteretic behavior.

Full-bay Type Shear Panels

The structural responses of pure aluminum stiffened shear panels has been investigated by performing cyclic tests under shear loads considering six different shear panel configurations. For the basic configurations (shear panels type B, F and G), stiffeners made of longitudinal and transversal ribs with a rectangular cross section (depth of 60 mm and thickness of 5 mm) were used. Such stiffeners were connected to the base shear plate by means of welding. In addition, in one case, steel channel shaped stiffeners, connected to the basic aluminum plate by means of bolted joints, were adopted (shear panel type H). Finally, in order to assess the influence of the mechanical properties of the applied aluminum alloy, panel type B and panel type F made of both the AW 1050A and the AW 5154A aluminum alloy were considered. The selected configurations of the tested panels, together with the adopted testing apparatus, are shown in Figure 3.

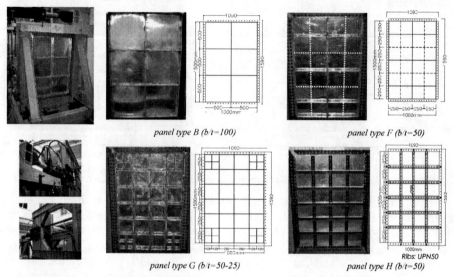

panel type B (b/t=100)

panel type F (b/t=50)

panel type G (b/t=50-25)

panel type H (b/t=50)

FIGURE 3. Adopted testing apparatus and geometrical configuration of tested specimens.

The results (see Figure 4) of the experimental tests are provided in terms of obtained cyclic response, i.e. as relationship between the applied shear force (F) and the corresponding shear deformation (γ). The response of tested specimens has been also characterized and compared in terms of maximum hardening ratio (τ_{max}/τ_{02} with $\tau_{02}=f_{02}/\sqrt{3}$), secant shear stiffness (G_{sec}) and equivalent viscous damping factor (ζ_{eq}), as defined in Figure 4d.

The examination of test results allows the identification of two different classes of shear panels according to the adopted base material, namely dissipative shear panels (AW 1050A) and stiffening shear panels (AW 5154A). Based on the obtained results, tested specimens pointed out a very high ductility and a good structural performance in terms of strength, stiffness and dissipative capacity, proving that the considered system can be usefully adopted as passive seismic protection device [6-8].

Bracing Type Shear Panels (BTPASP)

The obtained results of the above mentioned activity related to full-bay aluminum shear panels showed that their structural response, in terms of energy dissipation capability, is significant for medium-large lateral displacements, while some slipping phenomena were observed for small deformation levels. In order to improve the dissipative behavior of tested pure aluminum shear panel also for limited inter-story drift values, a different prototype, with reduced side length, has been also tested, so to establish a favorable ratio between the inter-story drift displacement of the relevant primary structure and the shear deformations of the applied panels [9].

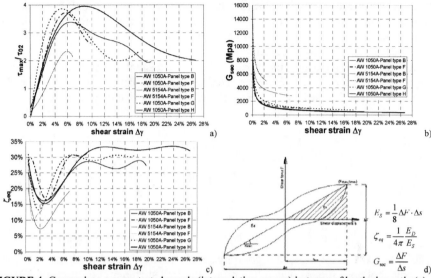

FIGURE 4. Comparison among tested panels (interpolation curves) in terms of hardening ratio (a), secant global stiffness (b) and equivalent viscous damping ratio(c); definition of parameters (d).

A suitable experimental procedure has been set up in order to characterize the main behavioral parameters of four specimens with different aspect ratio values (Figure 5).

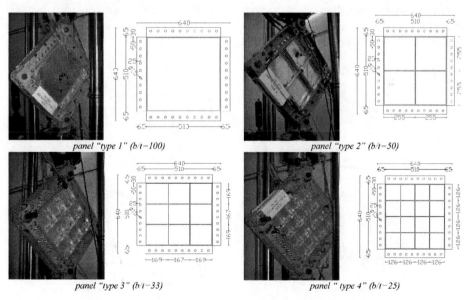

FIGURE 5. Tested specimens: bracing type shear panels (BTPASP).

The tested specimens are characterized by global dimensions of 500 mm by 500 mm and thickness of 5 mm. They present slenderness ratio values a_w/t_w equal to 100 (BTPASP "type 1"), 25 (BTPASP "type 2"), 33 (BTPASP "type 3") and 25 (BTPASP "type 4") respectively. These values have been obtained through the application of welded rectangular-shaped ribs. Such ribs are equally spaced on the two faces of the panels, have a depth of 60 mm and are made of the same material and thickness of the base plates. The testing apparatus is composed by a pin jointed steel framework and linked to the panel edges by means of tightened steel bolts. Tested specimens are subjected to diagonal cyclic forces, according to the displacement history shown in Figure 6, where the ordinate indicates the applied diagonal displacement.

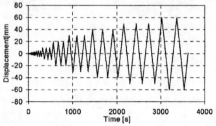

FIGURE 6. The displacement history on the tested specimens

The obtained results showed a good hysteretic performance which was influenced by different collapse modes depending on the applied stiffener configuration [10]. In fact, the internal rib system acts as a sort of framework axially stressed, providing a resistant contribution to the panel and, therefore, a larger stress in the connecting system, which usually represents the weakest component of the studied devices, which could provoke their anticipated failure. In Fig. 7, the obtained hysteretic cycles for the tested specimens, together with the relevant collapse modes are provided, while in Figure 8, consistently with the definitions given in Figure 4d, the cumulated dissipated energy and the equivalent viscous damping ratio are illustrated.

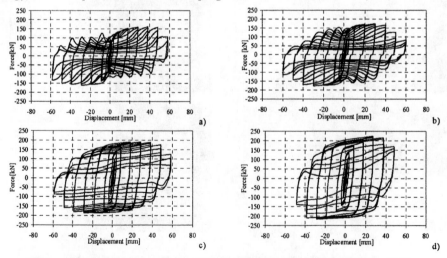

Figure 7. Hysteretic cycles: "type 1" (a), "type 2" (b), "type 3" (a) and "type 4" (b).

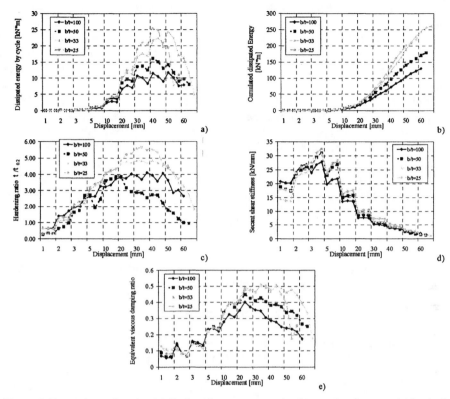

Figure 8. Comparison of results: (a) dissipated energy per cycle; (b) cumulated energy; (c) hardening ratio; (d) secant shear stiffness; (e) equivalent viscous damping factor.

These results clearly emphasize that the panel configurations provide a good hysteretic performance, with large hysteretic cycles also for high deformation levels. It is worth noticing the higher values of the equivalent viscous damping factor (about 50%) was achieved for large shear strains.

It has also to be observed that the two tested BTPASPs "type 3" and "type 4" provided substantially the same dissipative behaviour even though the configuration "type 4" exhibited a larger peak strength. On the contrary, it presents a lower ductility due to a premature collapse of the connection system, which is also highlighted by a quick reduction of the equivalent viscous damping factor starting from a diagonal displacement value of 50 mm. Hence, shear panel "type 3" seems to represent the optimum configuration as a good compromise between reduced fabrication costs and good hysteretic behaviour.

CONCLUDING REMARKS

In the current paper a wide experimental campaign carried out for the assessment of the structural performance of stiffened aluminum shear panels has been reviewed. Two different typologies of shear panels, namely "full bay" and "bracing type" shear

25

panels, have been considered in order to evaluate their structural response for different shear strain demands. In particular four "full bay" shear panel made of the heat treated alloy AW 1050A and two "full bay" shear panel made of the alloy AW 5154A installed in a pin-jointed steel frame have been tested under cyclic shear loads. In particular, the former panel typology evidence a suitable energy dissipation capability, up to a peak value of shear strain demand equal to 10%.

In order to favor a larger hysteretic performance of the system also for shear deformation lower than 10%, pure aluminium shear panels characterized by a reduced side length, to be employed as a "bracing type" system have been proposed. The obtained results allowed a optimum slenderness ratio of the plate, namely $a_w/t_w=33$, to be detected.

In the whole, based on the above outcomes, aluminum shear panels appear to be an effective passive device to be applied for both steel and RC moment resisting frames.

ACKNOWLEDGMENTS

The experimental studies presented in this paper has been developed in the framework of theme n.5 "Development of innovative approaches for the design of steel and composite structures" of the Italian research project RELUIS.

REFERENCES

1. F. M. Mazzolani, "Innovative metal systems for seismic upgrading of rc structures" (keynote lecture). Proc. of the Int. Conference on "Stability and Ductility of Steel Structures" (SDSS '06), Lisbon, Portugal, September 6-8, pp. 29-48.
2. G. De Matteis, F.M. Mazzolani, S. Panico, "Pure aluminium shear panels as dissipative devices in moment-resisting steel frames". *Journal of Earthquake Engineering and Structural Dynamics.* June 2007; vol. 36, n. 7: pp. 841-859.
3. G. De Matteis, A. Formisano, S. Panico, B. Calderoni, F. M. Mazzolani, "Metal shear panels". Seismic upgrading of RC buildings by advanced techniques – The ILVA-IDEM Research Project. Mazzolani, F. M. co-ordinator & editor, Polimetrica International Scientific Publisher, Monza, pp. 361-449, 2006.
4. M. Nakashima, S. Iwai, M. Iwata, T. Takeuchi, S. Konomi, T. Akazawa, K. Saburi, "Energy Dissipation Behaviour of Shear Panels Made of Low Yield Steel". *Earthquake Engineering and Structural Dynamics,* 1994, 23: 1299-1313.
5. G. De Matteis, F.M. Mazzolani, S. Panico, "Pure aluminium shear panels as passive control system for seismic protection of steel moment resisting frames". Proc. of the 4th STESSA Conference (Behaviour of Steel Structures in Seismic Areas), Naples, Italy, pp. 599-607, 2003.
6. G. De Matteis, S. Panico, F.M. Mazzolani, "Experimental tests on stiffened aluminium shear panels". Proc. of XI International Conference on Metal Structures, Rzeszów, Poland, 2006.
7. G. De Matteis, F.M. Mazzolani, S. Panico, "Experimental tests on pure aluminium stiffened shear panels". *Engineering Structures.* DOI:10.1016/j.engstruct.2007.11.015.
8. G. De Matteis, A. Formisano, S. Panico, F.M. Mazzolani, "Numerical and experimental analysis of pure aluminium shear panels with welded stiffeners". *Computers and Structures.* 2008; n. 86: pp. 545-555.
9. G. De Matteis, G. Brando, S. Panico, F.M. Mazzolani, "Experimental study on stiffened bracing type pure aluminium shear panels (BTPASPs)", Proceeding of the third International Conference on Structural Engineering, Mechanics and Computation (SEMC '07). Cape Town, South Africa, 10-12 September, 2007.
10. G. De Matteis, G. Brando, S. Panico, F.M. Mazzolani "Bracing type pure aluminium stiffened shear panels: an experimental study", submitted for publication in International Journal of Advanced Steel Construction (IJASC), S.L. Chan, W.F. Chen and R. Zandonini (editors-in-chief), ISSN 1816-112X, Vol. 4, No. 1 March 2008.

RC structures strengthened by metal shear panels: experimental and numerical analysis

G. De Matteis[a], A. Formisano[b] and F. M. Mazzolani[b]

[a] Dept. of Design, Rehabilitation and Control of Architectural Structures,
University of Chieti-Pescara "G. D'Annunzio", V.le Pindaro, 42 - 65127 Pescara (I)
[b] Dept. of Structural Engineering, University of Naples "Federico II",
P.le Tecchio, 80 - 80125 Naples (I)

Abstract. Metal shear panels (MSPs) may be effectively used as a lateral load resisting system for framed structures. In the present paper, such a technique is applied for the seismic protection of existing RC buildings, by setting up a specific design procedure, which has been developed on the basis of preliminary full-scale experimental tests. The obtained results allowed the development of both simplified and advanced numerical models of both the upgraded structure and the applied shear panels. Also, the proposed design methodology, which is framed in the performance base design philosophy, has been implemented for the structural upgrading of a real Greek existing multi-storey RC building. The results of the numerical analysis confirmed the effectiveness of the proposed technique, also emphasising the efficiency of the implemented design methodology.

Keywords: Metal shear panels, RC structures, experimental tests, numerical analyses, structural upgrading.

INTRODUCTION

In the past years, Metal Shear Panels (MSPs) have been effectively used as a lateral load resisting system for new and existing buildings. When compared with other traditional solutions based on the use of concrete shear walls, they put into evidence some advantages, such as less weight transferred to the foundations, less space occupied into the building, less seismic forces (due to the reduced structure mass) and increased speed of erection [1].

In recent years, applications of such devices have been mainly devoted to increase the seismic performance of new buildings, whereas the potential applicability to existing buildings for seismic retrofitting purposes, with particular reference to RC structures, has not been deeply investigated. On the contrary, the authors' opinion is that light-gauge metal plate shear walls could actually represent a useful system for strengthening week RC frames, due to the large strength and stiffness they possess, the versatility of the collocation in the structure and also the possibility to avoid a strong reinforcement of the existing structural members of the frame.

In the present paper, such a retrofitting strategy is investigated, assuming as starting point the results of two full-scale experimental tests previously carried out on an existing two-storey RC structure endowed with steel and pure aluminium slender shear plates. On the basis of such results, both simplified and sophisticated numerical

CP1020, *2008 Seismic Engineering Conference Commemorating the 1908 Messina and Reggio Calabria Earthquake*,
edited by A. Santini and N. Moraci
© 2008 American Institute of Physics 978-0-7354-0542-4/08/$23.00

models of both the single applied device and the whole upgraded structure have been implemented. The evaluation of the performance improvement due to the applied metal shear panels allowed a general design methodology for seismic retrofitting of RC structures to be set up. Therefore, the efficiency of such a method has been checked through the development of a significant study case, i.e. the design application referred to an existing irregular multi-storey RC building located in Athens, whose dynamic response has been evaluated by means of detailed numerical dynamic analysis.

STATE OF THE ART

MSPs can be considered as an attractive option for lateral load resisting systems of new and existing constructions [2]. The applied shear panels can be made up of either simple unstiffened or stiffened metal plates. The latter have a major cost and could be able to develop a pure shear collapse mechanism with large hysteretic cycles, whereas the former are characterised by a poor hysteretic behaviour, which is conditioned by the activation of a tension field resisting mechanism developing after the plate buckling in shear [3]. Several studies have been addressed to investigate the behaviour of unstiffened steel plate shear walls, quantifying the effect of many significant factors, namely the effect of simple versus rigid beam-to-column connections on the overall behaviour of the system, the advantage of using a low yield point steel, the effects of bolted versus welded infill connections, the influence of the flexural and axial stiffness of the surrounding members [4]. Based on such studies, unstiffened metal shear panels have been largely applied within either rigid or pinned steel frame buildings, while few numerical studies and none practical applications related to RC structures equipped with metal shear plates have been carried out.

In the above field, the first application was due to Mo and Perng [5], who tested RC frames equipped with steel trapezoidal sheeting having different thickness and connected to the frame members by means of steel bolts. In all tests, the revealed small amount of dissipated energy due to large relative displacements between RC members and the panel put into evidence that this system was not appropriate for practical design purposes, although its high potentiality was demonstrated. For this reason, the examined system was revised by Kono et al. [6] in order to improve the resistance against seismic forces. Four half-scale specimens, composed by corrugated steel shear panels connected with different location of diameter studs to a RC portal frame (Figure 1a), were tested under cyclic loads. In the whole, all hybrid specimens showed a lateral strength increase of more than 30% with respect to the original one. The panel sheeting tearing and the collapse mechanism of the RC frame was the final failure mode of the system.

Wang analysed under numerical way the behaviour of a non-ductile RC concrete hospital building retrofitted with four different techniques, namely steel braces, RC jackets, RC shear walls (RCSWs) and Steel Plate Shear Walls (SPSWs), the latter depicted in Figure 1b [7]. The maximum inter-story drift of the retrofitted structure was only slightly reduced (about 9%) with respect to the original one, whereas the number and maximum rotation of plastic hinges in the frame members were significantly reduced when SPSWs were adopted. The comparison among the responses of the retrofitted structure and the original one showed that, after RCSWs, SPSWs produces the best improvement of both the strength and stiffness features of the bare building.

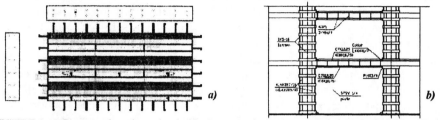

FIGURE 1. a) Corrugated steel panel tested by Kono et al. (2007) and system proved by Wang (2006).

The above research studies have been carried out by developing numerical analyses; only two cyclic experimental tests on a real two-story RC structure retrofitted with metal shear panels are provided in the technical literature [8], which are framed in the ILVA-IDEM research project [9]. Such tests concerned the use of two different shear panels, which were separately applied on the first storey level of the structure, made of a common steel and a very low-strength pure aluminium. The adopted shear panels, which were divided into six sub-parts having a width b=600 mm and a height h=400 mm, had a thickness of 1.15 mm and 5 mm when steel and aluminium were used, respectively. They were inserted within an external steel frame composed by UPN 220 profiles. For the sake of example, the global view of aluminium panels is depicted in Figure 2a. The results of the experimental tests are depicted in Figure 2b and 2c in terms of envelope curves and hysteretic loops, respectively, and may be summarized as follows: 1) in both cases a very large improvement of the retrofitted structure in terms of strength (10 and 11.5 times with steel and aluminium panels, respectively) and stiffness (about 2.5 and 2 times with steel and aluminium panels, respectively) was achieved; 2) the hysteretic cycles obtained when aluminium panels were used are decidedly larger than those related to steel panels, due to the reduced slenderness of the applied aluminium plates. Therefore, on the basis of such results, it was concluded that aluminium and steel plates could be profitably used as dissipative and strengthening devices of RC structures, respectively.

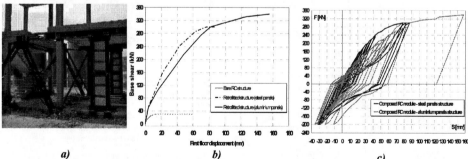

FIGURE 2. The experimental tests in Bagnoli, Naples: a) the applied aluminium shear walls; b) comparison among test results and the response of the bare RC structure; c) comparison between the cyclic behaviour of applied metal shear panels.

29

THE NUMERICAL STUDY

Starting from the results of the above experimental tests [9], both a refined numerical model of the adopted shear panel and a simplified FEM model of the reinforced structure have been set up. Firstly, a sophisticated FEM model has been implemented through the ABAQUS non linear calculation software [10]. The shear plates, which have been modelled by using S4R shell elements, have been placed within a supporting steel frame realised with beam elements made of S275 steel (Figure 3). A tie constraint has been assumed between panel sides and frame members and a mesh of 20 mm side length has been used to model the plate. The effect of geometrical imperfections have been also considered, their amplitude being assumed equal to 1/1000 of the panel depth. Additional details on the implemented numerical model are available in [11]. In Figure 3, in addition to the deformed shape of the panel related to the final phase of the loading test, the numerical result reported in the shear-displacement plane are also depicted, showing that a very good agreement between the experimental and the numerical results is obtained.

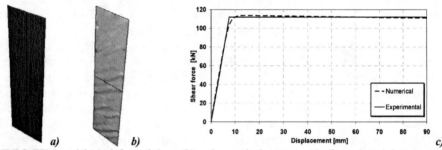

a)   *b)*   *c)*

FIGURE 3. FEM model (a), deformed shape (b) and numerical response (c) of tested steel shear panel.

Aiming at evaluating the interaction between the seismic devices and the base RC module, the global analysis of the upgraded structure has been performed by using the SAP2000 analysis software [12]. In such a case, the metal shear panels have been directly inserted within an external steel frame on both sides of the structure at the first floor (Figure 4a).

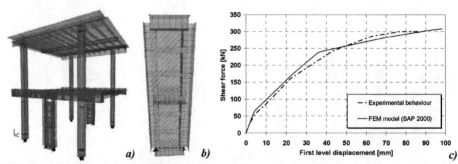

a)   *b)*   *c)*

FIGURE 4. The RC structure strengthened with steel shear panels: the FEM model (a), the shear wall (b) and comparison between numerical and experimental results (c).

Frame members have been modelled as beam elements, while ten trusses connected to the steel beams and columns through pinned joints and inclined of 45° in the same direction of the main tensile stresses shear panels have been used to represent each shear wall field, according to the strip model theory (Figure 4b) [3]. The numerical behaviour of the retrofitted structure is represented in Figure 4c, showing a very good agreement with the obtained experimental results.

THE DESIGN PROCEDURE

The encouraging results of the numerical study presented in the previous Section allowed to set up a general design methodology for the design of steel shear panels for seismic retrofitting purposes. The procedure has been applied according to the ATC-40 American guidelines [13]. In particular, in the spectral acceleration – spectral displacement plane (ADRS format), the target displacement of the retrofitted structure is set equal to the one assumed for the original building. Then, based on the simplified assumption of "equal displacements", the initial stiffness for the retrofitted structure is calculated starting from the knowledge of the corresponding structural period, leading to the following relationship:

$$K_{ret} = K_{ini} \left(\frac{T_{ini}}{T_{ret}} \right)^2$$

(1)

where K_{ini} and T_{ini} are the initial stiffness and the period, respectively, of the original structure and K_{ret} is the lateral stiffness of the retrofitted structure.

Therefore, assuming that the retrofitted structure is able at least to provide the same damping of the initial structure, the "desired performance point" can be defined and the required ultimate base shear capacity for the retrofitted structure can be obtained from the following equation:

$$V_{ret} = \frac{S_{a_{ret}}}{S_{a_{ini}}} V_{ini}$$

(2)

where V_{ini} is the ultimate base shear capacity of the initial structure, V_{ret} is the required ultimate shear capacity of the retrofitted structure, $S_{a_{ini}}$ and $S_{a_{ret}}$ are the ultimate spectral acceleration for the initial and retrofitted structures, respectively.

Accordingly, since the behaviour of the bare structure in terms of both strength and stiffness is available, the knowledge of both K_{ret} and V_{ret} allows the shear panel characteristics V_p and K_p to be determined. In particular, knowing K_{ret} and V_{ret} the design curve of the strengthening intervention can be plotted. As a sake of example, in Figure 5 the design curve related to the tested structure previously illustrated is compared with the ones achieved by the application of the above procedure, showing a good agreement among these results, therefore proving the effectiveness of the implemented theoretical procedure, which can be usefully applied for designing metal shear panels to be applied for seismic retrofitting of existing RC buildings.

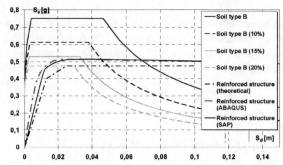

FIGURE 5. Comparison between the theoretical design curve and the numerical ones for the tested RC building located in Bagnoli and strengthened with steel shear panels.

THE STUDY CASES

To validate the achieved results, the proposed design methodology has been applied to two existing Greek RC buildings, the first one characterised by two storeys and the latter being a multi-story irregular structure. In the first case, the original structure was composed by ten bays (width 3.20 m) and developed on two levels having depth of 4.00 m and 3.00 m at the first and the second floor, respectively [14]. The cross-section of beams and columns was of 20x50 cm and 30x40 cm, respectively. The non-linear analysis of this structure gave the response diagram shown in Figure 6a. The performance point of this structure was equal to 6.2 cm, while the initial stiffness and the base shear capacity were K_{ini} =15.500 kN/m and V_{ini} = 258 kN, respectively. In order to achieve a "Life safety" performance level, a maximum displacement of about 3 cm was required for the upgraded structure. This corresponded to a required period T_{ret}=0.55 sec. Assuming that the structural damping could be equal to the one of the original structure (about 25%), a required spectral acceleration capacity $S_{a,ret}$ =0.16 g can be determined. Consequently, by eqs. (1) and (2), K_{ret} = 40610 kN/m and V_{ret} = 497.9 kN and therefore, since K_{ini} and V_{ini} are known, the minimum required stiffness K_p and strength V_p for shear panels may be calculated, in the case under investigation they being equal to 25110 N/mm and 239.9 kN, respectively.

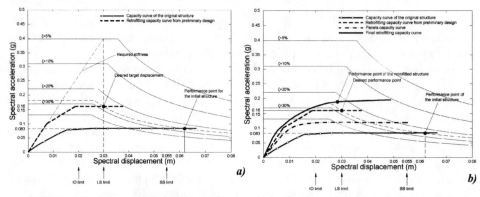

FIGURE 6. Design curve (a) and theoretical-numerical comparison (b) of the seismic retrofitting procedure for the analysed two-storey RC building.

Steel shear panels were installed in the middle of the central bay of the existing frame, they having at the first and the second floor cross-section of 1200xx4 mm and 1200x1.7 mm, respectively. Shear panels were made of a low-yield steel having the yield stress $f_{0.2}$ = 85.8 MPa and the ultimate stress f_u= 236.20 MPa. They were installed within an external steel frame composed by HEB180 columns and HEB160 beams made of Fe430 steel. In order to validate the proposed theoretical procedure, a numerical analysis of the upgraded structure was also carried out. The obtained results are shown in Figure 6b, where it appears that the response of the structure is correctly interpret by the initial retrofitting design. The second analysed structure is the so-called Koletti building, which is a multi-story irregular RC building located in the centre of Athens (Figure 7a) [15]. The building, which was conformed into a rectangular shape of 41x33 m, consisted of two storeys (total height of 8.25 m) placed below the ground floor and six levels (total height of 26.50 m) above the ground floor. C16/20 concrete and steel bars with yield stress of 420 MPa represented the base materials for reinforced concrete. Aiming at evaluating the structure behaviour under earthquake loading, pushover analyses have been carried out by using a FEM model of the structure implemented by the SAP 2000 software (Figure 7b). In the current case, the retrofitting design methodology provided K_p = 167 kN/m and 110 kN/m and V_p = 3708 kN and 3497 kN in direction x and y, respectively. The corresponding curves of the retrofitted structure in both directions are provided in Figure 7c. Starting from these values, the geometry of steel shear panels have been determined, leading to plates having a width b = 1.65 m and 1.30 m and a thickness t ranging between 4 and 8.4 mm and between 4 and 7 mm in direction x and y, respectively. Hence, a FEM model of the retrofitted structure has been developed (see Figure 8a). The behavior of a frame about the y axis is shown in Fig. 8b, where it appears that a global collapse mechanism develops with a plastic engagement of all applied shear panels. In addition, as shown in [15] the obtained results evidence that the proposed design methodology is fully capable to correctly interpret the actual response of the structure.

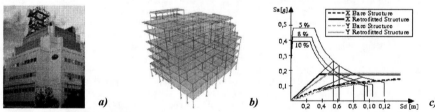

FIGURE 7. The Koletti building in Athens: a) real view; b) FEM model of the bare structure; c) retrofitting design curves

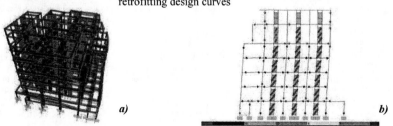

FIGURE 8. The Koletti building in Athens: a) FEM model of the retrofitted structure and b) developed collapse mechanism of the structure.

CONCLUDING REMARKS

In this paper an innovative design procedure for seismic retrofitting of existing RC buildings by means of MSPs has been presented. Starting from an experimental activity carried out on a real building and according to the results of numerical analyses carried out on both single devices and the whole structure, a theoretical procedure to determine the response of the strengthened structure has been set up, it allowing the characteristics of the shear panels to be identified in terms of strength and stiffness. This design procedure has been successfully validated by analysing two different study cases, which are referred to two RC buildings located in Greece. In both cases the obtained results prove that the proposed design procedure may be effectively applied to finalize the design of metal shear panels to be applied for retrofitting existing RC buildings.

ACKNOWLEDGMENTS

The present study is framed within both the European PROHITECH and the Italian RELUIS research projects.

REFERENCES

1. A. Astaneh-Asl, "Seismic Behavior and Design of Steel Shear Walls", *Steel Tips Report*, 2001.
2. M. Bruneau, J. Berman and D. Vian, "Steel Plate Shear Walls - From Research to Codification", *ASCE 2005 Structures Congress*, New York, 2005.
3. L. J. Thorburn, G. L. Kulak and C. J. Montgomery, "Analysis of Steel Plate Shear Walls", *Structural Engineering Report No. 107*, University of Alberta, Edmonton, Alberta, Canada, 1983.
4. G. De Matteis, "Effect of lightweight cladding panels on the seismic performance of moment resisting steel frames", *Engineering Structures - The Journal of Earthquake Wind and Ocean Engineering*, Elsevier, Vol. 27/11, pp 1662-1676, Elsevier, 2005
5. Y.L. Mo and S. F. Perng, "Seismic Behavior of Reinforced Concrete Frames with Corrugated Steel Walls", *Advances in Structural Engineering*, Vol.3, No. 3, 2000, pp. 255-262.
6. S. Kono, Y. Ichioka, Y. Ohta, F. Watanabe, "Seismic Performance of Hybrid System with Corrugated Steel Shear Panel and RC Frame", *Proc. of the 1st Int. Workshop on Performance, Protection & Strengthening of Structures under Extreme Loading*, Whistler, Canada, 2007.
7. H. Wang, "Seismic Retrofit of a Reinforced Concrete Hospital Building in the Eastern US", *Annual Meeting of the Multidisciplinary Center for Earthquake Engineering Research*, 2006.
8. A. Formisano, "Seismic upgrading of existing RC buildings by means of metal shear panels: design models and full-scale tests, Ph.D. Thesis, University of Naples "Federico II", 2007.
9. F.M. Mazzolani (co-ordinator & editor), *Seismic upgrading of RC buildings by advanced techniques – The ILVA-IDEM Research Project*, Polimetrica Publisher, Monza, 2006.
10. Hibbitt, Karlsson, Sorensen, Inc., *ABAQUS/Standard, version 6.4*, Patwtucket, RI, USA, 2004.
11. A. Formisano, G. De Matteis, S. Panico, F.M. Mazzolani, "Metal shear panels as innovative system for seismic upgrading of existing RC buildings: from numerical analyses to full-scale experimental tests", *Proc. of the 15th UK Conference of the Association of Computational Mechanics in Engineering*, Glasgow, UK, CD-ROM, paper no. 73, 2007.
12. CSI, "*SAP 2000 Non linear version 10.01*", Berkeley, California, USA, 2006.
13. ATC 40, "Seismic evaluation and retrofit of concrete buildings", *Report No. SSC 96-01*, 1996.
14. E. S. Mistakidis, G. De Matteis and A. Formisano, "Low yield metal shear panels as an alternative for the seismic upgrading of concrete structures", *Advances in Engineering Software*, 38, 2007, pp.626-636.
15. A. Formisano, G. De Matteis, F. M. Mazzolani, "Seismic retrofitting of existing RC buildings by means of steel shear panels: the Koletti buidings in Athens", *Proc. of the Italian Conference "Evaluation and reduction of the seismic vulnerability of existing RC buildings"* (accepted for publication), Rome, 29-30 May, 2008.

Development of Partially Encased Composite Columns for use in Steel Shear Walls for Seismic Applications

Robert G. Driver

Dept. of Civil and Environmental Engineering, University of Alberta,
Edmonton, Alberta, T6G 2W2, Canada.

Abstract. Partially encased composite columns consist of a welded thin-plate H-shaped steel section with concrete cast between the flanges. Transverse steel links are welded between the flanges, spaced at regular intervals, to enhance the resistance of the flanges to local buckling. Although they were developed originally to resist gravity loading in mid- and high-rise buildings, they are currently being investigated for applications as the vertical boundary elements in steel shear walls in both low and high seismic zones. The paper provides an overview of several developments from a major ongoing investigation leading up to a recent large-scale test of a steel shear wall that incorporates partially encased composite columns. Key results from the various components of the research program are presented.

Keywords: Composite column, experimental, finite element, seismic, steel shear wall.

INTRODUCTION

Partially encased composite (PEC) columns consist of a welded thin-plate H-shaped steel section with transverse steel links welded between the flanges, spaced at regular intervals, to enhance the resistance of the flanges to local buckling. After the column is erected, concrete is cast between the flanges to provide the strength required under service loads. Construction loads are temporarily carried by the steel section alone and the columns are then cast at the same time as the floor above. The system capitalizes on both the speed of erection of steel construction and the relatively economical strength contribution of concrete. Moreover, since the column flanges act as formwork on two sides, formwork construction is simple. A typical PEC column cross-section and three different stages in the construction process are depicted in Fig. 1.

The PEC column concept, using sections built-up from relatively thin plates to reduce crane loads in high-rise construction, was originally developed by the Canam Group as an economical means of resisting gravity loads. However, to expand the applicability of these columns, a major ongoing collaborative research program has been initiated and is being carried out at the University of Alberta and Ecole Polytechnique. The objective is to develop the use of PEC columns for a variety of new applications where they are subjected to both axial and flexural loads. One of these applications is as the vertical boundary elements in steel shear walls. The paper summarizes the various components of the research program that were conducted at the University of Alberta. These include an experimental investigation into the

CP1020, *2008 Seismic Engineering Conference Commemorating the 1908 Messina and Reggio Calabria Earthquake*,
edited by A. Santini and N. Moraci
© 2008 American Institute of Physics 978-0-7354-0542-4/08/$23.00

behaviour of these members under combined axial compressive load and flexure (i.e., as beam-columns) using either regular or high performance concrete, a finite element study wherein a numerical PEC column model was developed to simulate their full response under axial load and flexure including the post-peak behaviour, and a large-scale test on a steel shear wall with PEC columns as the vertical boundary elements to establish the overall system behaviour. Two additional large-scale steel shear wall tests designed to investigate uses specifically intended for economical applications in low and high seismic regions are also discussed briefly.

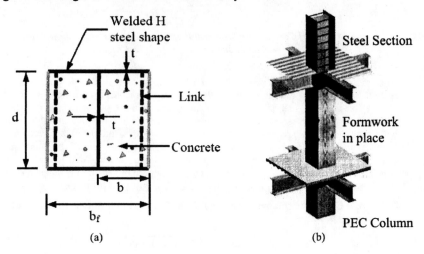

FIGURE 1. Partially encased composite (PEC) column: (a) cross-section; (b) sequence of construction (3-D illustration courtesy of R. Vincent, Canam)

RESEARCH PROGRAM COMPONENTS

PEC Columns Under Combined Axial Compression and Flexure

To investigate the behaviour of PEC columns under combined compression and flexure, eleven full-scale specimens were tested under either concentric or eccentric axial loading. Since no data were available for PEC columns with high performance concrete, seven tests were conducted with concentric loading to achieve a good understanding of the effect of using concrete strengths higher than those used previously. In addition, four test specimens were loaded eccentrically, examining both strong and weak axis bending. A typical eccentrically loaded specimen is shown in Fig. 2. Additional information on these tests can be found elsewhere [1].

The test specimens measured 400 mm×400 mm×2000 mm and the nominal plate thickness for the steel shapes was 7.9 mm. The link diameter and spacing were 12.7 mm and 240 mm (0.6d), respectively. The concrete in the eccentrically loaded specimens was nominally 60 MPa, although in the end zones, 80 MPa was used to force the failure region away from the loading fixtures. Two columns were oriented so that bending took place about the strong axis, with eccentricities of 23 mm and

100 mm, and two were oriented for weak axis bending, with eccentricities of 25 mm and 74 mm. The eccentricities were chosen so that the failure load would be close to either 85% or 55% of the maximum cross-sectional capacity in pure compression.

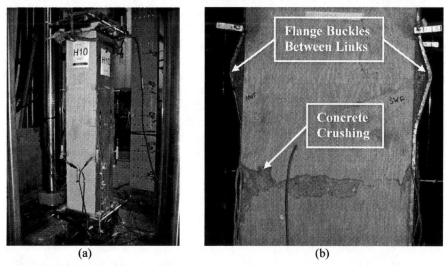

(a)   (b)

FIGURE 2. Eccentrically loaded PEC column (weak axis): (a) prior to test; (b) failure mode

Interaction diagrams, based on a strain compatibility approach assuming composite behaviour, were developed as a means for predicting the capacities of the eccentrically loaded PEC columns, as shown in Fig. 3. Confinement of the concrete by the steel section is small and therefore neglected. Although the initiation of local buckling was not detected in all tests prior to achieving the peak load, in constructing the diagrams for design, local buckling of the steel flanges in compression was accounted for by using an effective width according to the method of standard CAN/CSA-S16-01 [2]. The design diagrams resulted in average test-to-predicted ratios for column load and moment of 1.12 and 1.15, respectively, with minimum values of 1.04 for both load effects, providing a conservative strength prediction in all cases.

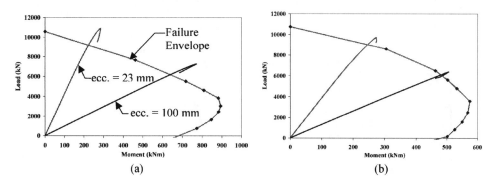

(a)   (b)

FIGURE 3. PEC column strength predictions: (a) strong axis bending; (b) weak axis bending

Numerical Modelling of PEC Columns

A finite element model was developed with the goal of accurately simulating the full response of PEC columns loaded to failure. The use of a dynamic explicit formulation and a concrete damage plasticity model resulted in good predictions of both concentrically and eccentrically loaded PEC column tests reported in the literature. The model accounts for local buckling of the thin steel flanges and the rapid volumetric expansion of the concrete near the ultimate load. The model provides good representations of the axial deformation at the peak load, the post-peak behaviour, and the failure mode. The model was then used to conduct a parametric study to gain insight into the effects of various parameters including link spacing, concrete strength, and loading conditions, as well as local fabrication imperfections. Additional information on this component of the research program can be found elsewhere [3].

Four-node finite-strain, reduced integration shell elements were used to model the steel shape. The mesh was optimized to produce proper representations of local buckling of the steel flange, while maintaining reasonable computing economies. The concrete infill was modelled with eight-node reduced integration brick elements using a 6×6 mesh over each quarter cross-section. The transverse steel links were modelled using two-node beam elements. The interfaces of the steel flange and the encased concrete were modelled using a friction-type master-slave contact with no bond between the surfaces. Geometric nonlinearities caused by large rotations in some of the analyses were accounted for in the model. A continuum, plasticity-based damage plasticity model that is capable of predicting both compressive and tensile response under low confining pressures was used to simulate the concrete material behaviour.

The finite element model gave good predictions of the behaviour of PEC columns tested by a variety of researchers and with many different geometric, material, and loading parameter combinations. The failure modes under strong and weak axis bending and a comparison of experimental and numerical load vs. strain responses are shown in Fig. 4. The mean test-to-predicted ratio for the columns tested in the first part of this research program was 1.01, with a coefficient of variation of 0.03.

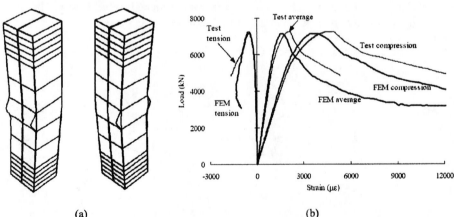

(a)   (b)

FIGURE 4. Finite element results: (a) failure mode predictions (strong and weak axis);
(b) test and finite element load vs. strain curves (strong axis bending; eccentricity = 100 mm)

Large-scale Test of a Steel Shear Wall With PEC Columns

With the understanding of the behaviour of PEC columns under combined axial and flexural loads gained from previous research, the first stage of exploring the efficacy of their use as the vertical boundary elements in steel shear walls could be carried out. A large-scale two-storey steel shear wall with PEC columns was tested under gravity load and gradually increasing cyclic lateral loads to study the behaviour of the PEC columns themselves, as well as of the system as a whole. The specimen tested, shown in Fig. 5 along with the loading system, had an overall height of 4.09 m and an overall width of 2.69 m, excluding the base plate. Storey heights were 1.9 m and the column centreline spacing was 2.44 m. Plates of 3 mm thickness were used for the infill panels. Gravity loads of 600 kN were applied to the top of each column using gravity load simulators, which kept the gravity loads vertical when the specimen experienced large lateral displacements. Lateral loads were applied to each floor equally according to the loading procedure outlined in Report 24 of the Applied Technology Council [4].

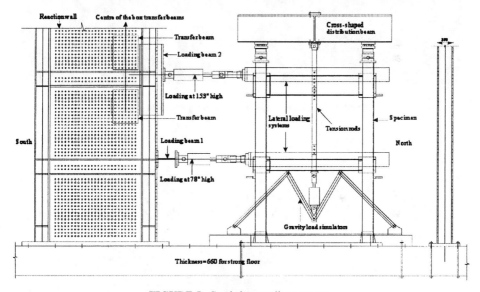

FIGURE 5. Steel shear wall test set-up

The test specimen prior to casting the columns and during the test near the peak load are shown in Fig. 6. The hysteresis curves for the first storey response are shown in Fig. 7. The first floor deflection was used as the deformation control parameter to control the lateral loading starting from cycle 10. Concrete cracks developed in the PEC columns during cycles 10 to 15, causing a decrease in the stiffness of the system. The fist sign of local buckling occurred in the outside flanges at the bottoms of the columns in cycle 19, which reached a load very near to the ultimate value eventually reached in cycle 21. The peak load coincided with the initiation of column flange tears near their bases and the full development of plastic hinges at the bottoms of the columns. The loading capacity of the specimen dropped gradually after the ultimate capacity was reached as damage accumulated in the form of concrete spalling, flange buckling and tearing, and eventual rupture of a few of the link welds that were exposed at the column bases by concrete spalling.

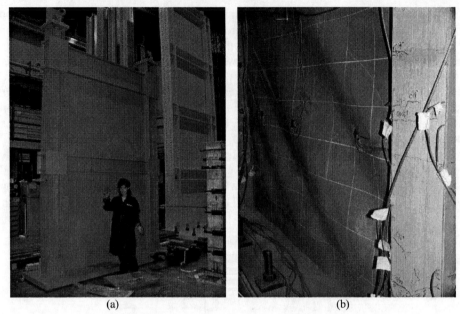

(a)   (b)

FIGURE 6. Test specimen: (a) prior to casting concrete;
(b) first storey panel under load (cycle 20)

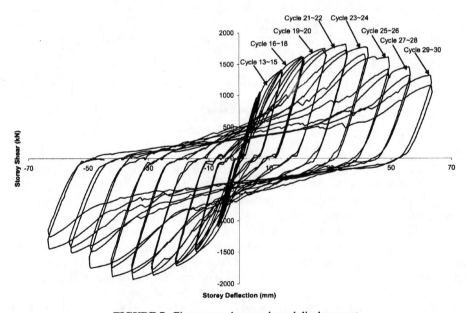

FIGURE 7. First storey shear vs. lateral displacement

Overall, the steel shear wall system with PEC columns performed well in the test and these columns are considered appropriate for use in this application. The existing PEC column finite element model is currently being expanded to predict the behaviour of the whole system. Experience from this first shear wall test has provided input for the remaining two tests in this research program that are described in the next section.

Shear Walls With PEC Columns for Low and High Seismic Regions

Although steel shear walls are usually associated with applications for high seismic regions, they can also be economical in zones of low or moderate seismicity. With economics in mind, a bolted modular steel shear wall with PEC columns that completely eliminates the need for field welding will be tested to assess its performance under cyclic loading. Each module consists of a floor beam and half-storey infill panels connected to the top and bottom of the beam, as shown in Fig. 8(a). A fish plate is pre-attached to the PEC columns for connection to the infill plate. The individual modules are fabricated under controlled shop conditions and are bolted together at the site. Lap plates are used at mid-height in the storey to create a bolted splice of the infill plate segments of adjacent modules. Wall segments could be pre-fabricated in two- or three-storey modules to maximize the assembly efficiency in the field. Since this system is not intended for use in high seismic regions, simple beam-to-column connections are used for economy. This effectively eliminates any contribution of frame action to the performance of the wall, so the modular system will likely qualify for a somewhat lower seismic force modification factor for design than the tested shear wall with moment-resisting beam-to-column connections.

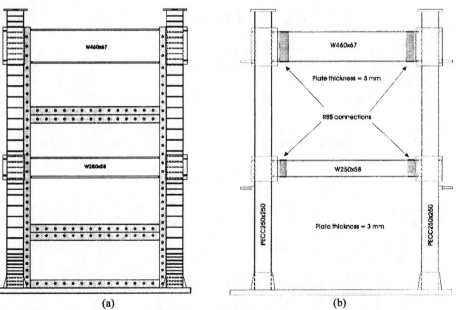

(a)   (b)

FIGURE 8. Shear wall specimens for upcoming tests: (a) modular construction for low seismic zones; (b) ductile frame with RBS connections for high seismic zones

With maximum ductility and performance in mind, the third test in the steel shear wall series will incorporate reduced beam sections (RBS)—or dogbones—where a portion of each beam flange is trimmed in the region adjacent to the beam-to-column connections, as indicated in Fig. 8(b). The RBS detailing reduces the beam capacity locally to reduce the demand on the columns and the joints. As a result, somewhat improved performance over the test described in the previous section, where plastic hinges developed at both the top and the bottom of the first storey column, is anticipated. Instead, plastic hinges are expected to form at the reduced section in the beams under large lateral displacements of the shear wall.

SUMMARY

An ongoing research program with the goal of extending the use of PEC columns to applications that involve significant flexural loads in addition to axial loads is described. One of these applications is as the vertical boundary elements of steel shear walls. Initial experimental and numerical studies have enhanced understanding of the behaviour of PEC columns under combined loads and have produced a sophisticated finite element modelling technique for predicting their response to these loads. Three large-scale, two-storey steel shear wall specimens with PEC columns are being tested to study the overall behaviour of the system under cyclic loading. The first test showed that the system behaves in a ductile and stable manner. Two additional specimens are in the final stages of design: a modular wall designed with economics in mind for low and moderate seismic regions and a shear wall that incorporates reduced beam sections for high seismic regions.

ACKNOWLEDGMENTS

Funding for this research program is provided by the Le Groupe Canam and the Natural Sciences and Engineering Research Council of Canada. The four research program components described in the paper were conducted by graduate students Brent Prickett, Mahbuba Begum, Xiaoyan Deng, and Mehdi Dastfan. Their major contributions to the success of the research are acknowledged. Valuable interactions with Bruno Massicotte and Robert Tremblay of Ecole Polytechnique de Montréal and Richard Vincent and Pièrre Gignac of Le Groupe Canam are also acknowledged.

REFERENCES

1. Prickett, B.S. and Driver, R.G. "Behaviour of Partially Encased Composite Columns Made With High Performance Concrete", Structural Engineering Report No. 262, Dept. of Civil and Environmental Engineering, University of Alberta, 2006.
2. "Limit States Design of Steel Structures", CAN/CSA-S16-01, Canadian Standards Association, Toronto, ON, 2001.
3. Begum, M., Driver, R.G., and Elwi, A.E. "Finite Element Modeling of Partially Encased Composite Columns Using the Dynamic Explicit Method", Journal of Structural Engineering, American Society of Civil Engineers, 133(3): 326-334, 2007.
4. "Guidelines for Cyclic Seismic Testing of Components of Steel Structures", Report No. 24, Applied Technology Council, Redwood City, CA, 1992.

Steel Plate Shear Walls: Efficient Structural Solution for Slender High-Rise in China

Neville Mathias, PE, SE [a]; Mark Sarkisian, PE, SE [b] Eric Long, PE[c] and Zhihui Huang PhD, PE[d]

[a]*Associate Director, Skidmore, Owings & Merrill LLP, One Front Street, San Francisco, CA 94111*
[b]*Structural Director, Skidmore, Owings & Merrill LLP, One Front Street, San Francisco, CA 94111*
[c]*Associate Director, Skidmore, Owings & Merrill LLP, One Front Street, San Francisco, CA 94111*
[d]*Senior Engineer, Skidmore, Owings & Merrill LLP, One Front Street, San Francisco, CA 94111*

Abstract. The 329.6 meter tall 74-story Jinta Tower in Tianjin, China, is expected, when complete, to be the tallest building in the world with slender steel plate shear walls used as the primary lateral load resisting system. The tower has an overall aspect ratio close to 1:8, and the main design challenge was to develop an efficient lateral system capable of resisting significant wind and seismic lateral loads, while simultaneously keeping wind induced oscillations under acceptable perception limits. This paper describes the process of selection of steel plate shear walls as the structural system, and presents the design philosophy, criteria and procedures that were arrived at by integrating the relevant requirements and recommendations of US and Chinese codes and standards, and current on-going research.

Keywords: Slender steel plate shear wall; SPSW; tension field action.

INTRODUCTION

The 329.6 meter tall 74-story Jinta Tower (Tower) in Tianjin, China, is expected, when complete, to be the tallest building in the world with slender steel plate shear walls (SPSW) used as the primary lateral load resisting system. The Tower has four stories of parking space below existing grade and 74 stories of office space above grade (Figure 1). It has an elliptical footprint approximately 42m by 81m at the base which changes with height to create an "entasis" effect. The total framed area of the tower is approximately 205,000 sq. m. The building is intended for office use.

Because of the tower's slender form, it has an overall aspect ratio close to 1:8, a key design challenge was to develop an efficient lateral system capable of resisting significant wind and seismic lateral loads while simultaneously keeping wind-induced oscillations under acceptable perception limits.

Several structural system options were considered in the concept and early schematic design phases including a concrete dual system with perimeter ductile moment resisting frames and core shear walls; composite systems with perimeter steel ductile moment resisting frames, steel floor framing and composite metal deck slabs, and composite concrete and steel plate shear walls; and steel systems with perimeter ductile moment resisting frames and braced or SPSW cores.

Concrete systems were eliminated primarily on account of the large sizes of the members required that had a significant impact on rentable area. Composite steel and concrete shearwall systems were eliminated after detailed investigation showed that there was insufficient precedent and research / testing data available (considering the specific features of the project such as the CFT columns) to convincingly demonstrate

CP1020, *2008 Seismic Engineering Conference Commemorating the 1908 Messina and Reggio Calabria Earthquake*,
edited by A. Santini and N. Moraci
© 2008 American Institute of Physics 978-0-7354-0542-4/08/$23.00

Figure 1. Architectural Rendering of Tower

the feasibility of these systems to the authorities without very significant research, testing, cost and, most significantly, time. Steel dual systems with braced cores were found to require as much as 20 - 25% more steel to satisfy structural performance requirements than SPSWs. This fact, taken together with the minimal dimensions of the steel plates, the availability of substantial code provisions and design guides, research and testing data that highlighted the superior ductility of SPSWs, and excellent predicted structural performance resulted in a decision to use SPSWs over steel braces in the tower core.

Because of the relative newness of the structural system as well as its significant height that exceeded code limits, at the end of the Design Development Phase the project was subjected to review by panels of seismic and wind experts in accordance with the regulations in China. The experts reviewed the seismic and wind performance of the proposed structure and imposed requirements additional to the codes to address the unique nature of the project and ensure its safety.

STRUCTURAL SYSTEM DESCRIPTION

The main lateral force resisting system for the tower consists of a frame-shear wall system, with perimeter and core ductile moment-resisting frames, and core SPSWs linked together with outrigger and belt trusses (Figure 2).

The ductile moment-resisting frames consist of concrete filled steel pipe composite (CFT) columns and structural steel wide flange beams. The SPSWs consist of the CFT columns and structural steel wide flange beams in-filled with structural steel plates. Outrigger trusses, which are placed in the short direction of the tower plan, are used to engage the perimeter columns in resisting overturning. (Figure 3).

Belt trusses (Figure 4) are used at the outrigger levels to better distribute the axial loads resulting from outrigger action among the perimeter columns.

Four sets of outrigger and belt trusses are provided and located at the mechanical levels of the tower. Strengthened diaphragm slabs are used at the outrigger levels.

The gravity system for the tower consists of conventional rolled structural steel wide-flange framing and composite metal deck slabs. Metal decking is typically 65mm deep with 55mm normal weight concrete topping for a total slab depth of 120mm. Composite structural steel wide-flange beams are typically 450mm deep and span from the interior shear wall core to the perimeter ductile moment-resisting frame. Beams are typically spaced at 3.25m on center. CFT columns at the interior shear wall core and perimeter ductile moment-resisting frame are also used to resist gravity loads. The frame CFT columns typically vary from 1700mm to 700mm in diameter over the tower height. Concrete in the CFT columns varies in grade from C80 (80 MPa 28 day cube strength) at the base to C60 at the top of the tower.

The lateral and gravity systems of the superstructure are typically continued down into the substructure to the top of the foundations.

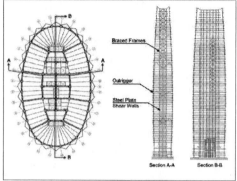

Figure 2. Typical Structural Plan, Sections

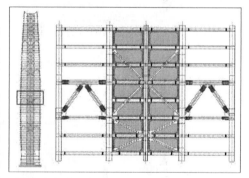

Figure 3. Outrigger Truss

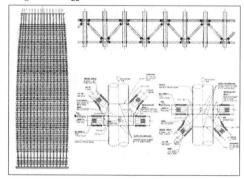

Figure 4: Belt Truss

The tower foundation system consists of a 4000mm conventionally reinforced concrete mat foundation supported by augured cast-in-place caissons. The caissons are typically 800mm in diameter and extend down approximately 60m below the bottom of mat foundation to local soil layer 11. All foundation concrete is C45. Strengthened reinforced concrete diaphragm slabs at and below the ground level transfer lateral shear forces to the perimeter reinforced concrete foundation walls.

LOADING AND PERFORMANCE REQUIREMENTS

The design was required to meet the Chinese code loading and performance requirements including those for gravity, wind and seismic loads, as well as those for strength and stiffness. Code design procedures typically utilize 50 year return wind and seismic loads (63.5% probability of exceedance in 50 years). The seismic event corresponding to this recurrence period is termed the "frequent earthquake".

Based on the project's size and importance, the codes required that the tower be designed to meet drift requirements under the 50 year wind (basic wind pressure 0.5 kN/m^2 in Tianjin) and strength requirements under the 100 year wind (basic wind pressure 0.6 kN/m^2 in Tianjin). Damping was set at 3.5% considering the composite effect of the CFT columns. Inter-story drifts under the 50 year wind load were limited to 1/400. Code wind acceleration perception requirements were based on a 10 year return event (basic wind pressure 0.3 kN/m2 in Tianjin) with damping set at 1.5% by code, once again considering the CFT columns. Acceleration was limited by code to 0.28 m/s^2 at the highest occupied floor. Wind tunnel testing was required. In accordance with local practice, wind speeds used in the tests were at least as high as those stipulated in the codes in the predominant wind direction, but directional effects were permitted to be considered.

Tianjin is located in seismic intensity zone 7 and the peak ground acceleration corresponding to this seismic event, per local codes, is 0.15g (147cm/sec^2). Inter-story drifts are limited to 1/300 in this event, with damping set at 3.5%.

The codes also required that the tower be analyzed dynamically using two measured and one simulated site specific time histories (length of record at least eight times the fundamental period), and further required that results be checked using two different analysis programs.

Wind and seismic experts who reviewed the project made several recommendations additional to the codes that are intended to ensure the safety and sound performance of the tower structure considering that it is not typical of the structures the codes were written to address. They recommended, among other things, that the design team:

- Satisfy code drift requirements using the code static 100 year wind loads (instead of 50 year static wind loads) in addition to 50 year wind tunnel loads. Strength requirements were to be satisfied using 100 year loads determined using code procedures and wind tunnel tests.
- Perform scaled testing of the typical proposed SPSW assembly.
- Perform non-linear time-history analysis to evaluate the behavior of the structure in the code rare (2% in 50 year) earthquake. Two measured and one simulated record were to be used. Damping was to be 5%.
- Design the columns and outriggers to typically not yield in a moderate (10% in 50 year) earthquake.
- Design the columns in the lower 16 floors (below the level of the lowest outriggers) to typically not yield in the rare (2% in 50 year) earthquake.

46

The computed peak accelerations at the highest occupied floor using code 10 year winds were found to be 0.20 m/s^2 as compared to 0.214 m/s^2 from the wind tunnel studies. These were less than the 0.28 m/s^2 limiting criteria in the code.

Rotational velocities were also checked in the wind tunnel studies and found to be 1.9 milli-rads/sec based on the code 10 year winds. This value is less than the 3 milli-rads/ sec criteria recommended by the CTUBH/Isyumov. The Chinese code does not currently have any acceptance criteria relating to rotational velocities.

SPSW DESIGN PHILOSOPHY AND PROCEDURES

The philosophy and procedures for the design of the SPSWs was based on the integration of the US and Chinese code requirements. A "slender" SPSW design approach was adopted which means that the lateral strength (and stiffness) of the shear walls results from tension field action in the steel plates. The relevant requirements following US codes and references were utilized:

- 2005 AISC – 341 "Seismic Provisions for Structural Steel Buildings"[1]
- FEMA 450 " Recommended Provisions for New Buildings and Other Structures"[2]
- AISC "Steel Design Guide 20: Steel Plate Shear Walls"[3]

Chinese codes currently only have a few requirements that pertain to the design of SPSWs. A key feature of the Chinese code requirements, and one that is not mirrored in the US codes and references, is that the SPSW's not buckle in the code frequent earthquake (50 year return). The requirements of the following Chinese codes were utilized:

- JGJ 99-98 Technical Specification for Steel Structure of Tall Buildings, Appendix 4[4].

Based on the documents listed above as well as a review of he Canadian codes and other pertinent papers and references[5]-[8], the design philosophy and procedures outlined below were developed.

General Design Philosophy

- The building frames are designed to carry gravity loads while neglecting the contribution of the SPSW plates, which ensures that the building fames have sufficient capacity to support the gravity loads during seismic events, when the plates could experience buckling due to the development of the tension-field action.
- SPSW plates are sized to respond elastically without tension-field action or buckling under frequent earthquake loads and design wind loads as required by Chinese code JGJ 99-98.
- Tension field action is expected to be the primary lateral load resisting mechanism in the SPSW plates in the event of moderate or rare earthquakes.
- The beams (horizontal boundary elements – HBEs) and columns (vertical boundary elements - VBEs) bounding the SPSW plates are designed for the forces determined from elastic analyses to meet the requirements of the Chinese code. The strength design forces include the component forces from the steel plates.
- Plastic hinging (but no failure / significant strength loss) is permitted at the ends of HBEs at moderate earthquake demand levels as well as at rare

- As per the requirement of the seismic experts, some minor yielding but no plastic hinging is permitted in the VBE's at moderate earthquake levels, and, in the lower 16 stories, some minor yielding but no plastic hinging is permitted in the VBEs at rare earthquake levels.

Selection of SPSW plate thickness

The determination of the required thickness of SPSW plates is based on satisfaction of the requirements described below; to not be less than any of the following:

- Thickness required to satisfy frequent earthquake / design code wind drift limits.
- Thickness required to satisfy formulas from the Chinese code JGJ 99-98 that assure no buckling under frequent earthquake and design wind loads.
- Thickness required to satisfy equations from FEMA 450 and AISC 341-05 that assure nominal strength (based on tension-field action) is greater than the code factored load demands.

SPSW MODELLING FOR ANALYSIS

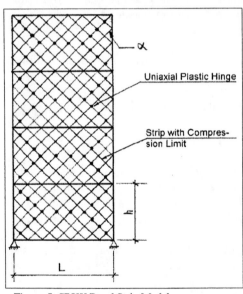

Figure 5: SPSW Panel Strip Model

In accordance with the design philosophy detailed earlier, SPSW panels were designed to not buckle under frequent seismic and design wind loads. Therefore, SPSW panels were modelled for elastic analysis with code level design lateral loads using full shell elements and isotropic materials.

Steel panels of the SPSWs are expected to buckle along compressive diagonals when subjected to moderate and rare earthquake loads depending on their slenderness ratios. After buckling, tension field action of the tension diagonals becomes the primary mechanism to resist shear forces in the steel plates.

The SPSW plates are replaced with a series of truss members parallel to the tension field with the inclination angle α as suggested in AISC 341-05 (Figure 5). As non-linear time history analyses are performed, strips are provided in two directions with compression resisting capabilities depending on their effective un-braced lengths. The stiffness of the strips are adjusted to ensure that the dynamic properties of the non-linear model in the elastic range match those of the elastic model. The number of strips per panel in each direction shall be taken greater than or equal to 10. Each strip is assigned a non-linear uni-axial plastic hinge. (AISC 341-05, FEMA 450, CAN/CSA S16-01).

SPSW STEEL PLATE BUCKLING CHECK

As has been pointed out, a key requirement of the Chinese codes is that the SPSW panels not buckle under frequent seismic and design wind loads.

A buckling check of the SPSW steel panels under frequent seismic and design wind loads in addition to that recommended in the Chinese codes and described in Section 5.2.2 was performed based on procedures suggested in "Guide to Stability Design Criteria for Meetal Structures" Edited by Theodore V. Galambos[9]. For plates supported along four edges while under combined bending and axial stresses at the ends, along with shear, an approximate evaluation of the critical combined loads can be obtained using a three-part interaction formula suggested by Gerard and Becker, 1957/1958).

CONSTRUCTION CONSIDERATIONS

On the Jinta project, the contractors recommended the use of slip-critical bolted connections between the steel panels and the boundary elements (Figure 6). In order to ensure that the SPSW panels did not buckle under the effects of frequent seismic and design wind loads, it was deemed necessary to minimize axial gravity loads transmitted to the steel panels. This is accomplished by providing vertically slotted or oversized holes for the bolted connections along the top and sides of the steel plates. The steel panels are installed as the tower construction proceeds, but the slip-critical bolts are only tightened after the tower has reached its full height and most of the dead loads have been imposed. The stability of the structure in the temporary condition is provided by the core and perimeter ductile moment resisting frames. The buckling checks described in the preceding section consider only the gravity loads applied after the bolts are tightened.

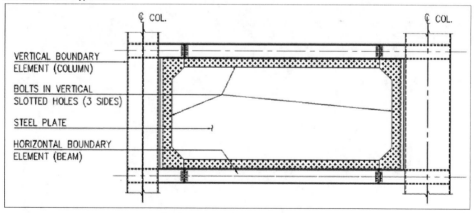

Figure 6: Typical SPSW elevation

CONCLUSIONS AND ACKNOWLEDGEMENTS

Steel plate shear walls acting in tandem with ductile moment resisting frames were determined to be the optimal solution to structuring the 329.6m tall Tianjin Jinta Tower with an aspect ratio of 1:8. The design solution was subjected to rigorous review by panels of seismic and wind experts in China at the end of the "Design Development" phase. This review resulted in the adoption of enhanced analysis and design procedures and performance goals; and testing. The project is nearing completion of the "Construction Documents" phase. Excavation of the site is complete and construction of the foundations is under way.

The authors wish to acknowledge the efforts of the Skidmore, Owings and Merrill LLP, San Francisco (SOM) design team, particularly those of Shihua Nie PhD, PE, who was involved in developing the SPSW design philosophy and procedures, and responsible for performing the analysis and design computations for the lateral system of the tower. Design responsibilities for the project were shared by SOM and the East China Architectural Design Institute (ECADI) of Shanghai, China with SOM in the primary role through the Design Development phase and ECADI in the primary role thereafter. Thanks are given to Prof. A. Astaneh-Asl of the University of California, Berkeley and Prof. Y. J. Shi of Tsinghua University, Beijing, China for their invaluable advice during the various design phases of the project.

Finally, and most importantly, thanks are due to the client, Finance Street Holdings of China together with their local expert consultants Prof. Baisheng Rong, Prof. D. X. Wang and Dr. W. B. Yang for their vision in supporting this innovative structural solution.

REFERENCES

1. AISC-341, *Seismic Provisions for Structural Steel Buildings*, 2005.
2. NEHRP *Recommended Provisions for New Buildings and Other Structures*, FEMA 450, 2003.
3. R. Sabelli , M. Bruneau, AISC *Steel Design Guide 20: Steel Plate Shear Walls*, 2007
4. JGJ 99-98, *Technical Specification for Steel Structure of Tall Buildings*, 1998.
5. ASTANEH-ASL A., *Seismic Behavior and Design of Steel Plate Shear Walls*, 2nd edition, 2006
6. CAN/CSA S16-01, *Limit States Design of Steel Structures*, 2001.
7. ATC 40, *Seismic Evaluation and Retrofit of Concrete Buildings*, Vol. 1, 1996.
8. FEMA 356, *Prestandard and Commentary for the Seismic Rehabilitation of Buildings*, 2000
9. T. V. Galambos (Editor), *Guide to Stability Design Criteria for Meetal Structures*, 5th edition, pp. 124-145

SEISMIC RETROFITTING FOR THE MONUMENTAL CULTURAL HERITAGE IN THE MEDITERRANEAN BASIN

Chair: Prof. Fabio Casciati – University of Pavia, Italy

Retrofitting Heritage Buildings by Strengthening or Using Seismic Isolation

Moshe Danieli, Jacob Bloch and Yuri Ribakov

Department of Civil Engineering, Ariel University Center of Samaria, Ariel, Israel

Abstract. Many heritage buildings in the Mediterranean area include stone domes as a structural and architectural element. Present stage of these buildings often requires strengthening or retrofitting in order to increase their seismic resistance. Strengthening is possible by casting above existing dome a thin reinforced concrete shell with a support ring. It yields reduction of stresses and strains in the dome. This paper deals with examples of actual restoration and strengthening of three structures in Georgia, two of them damaged by an earthquake in 1991, (a temple in Nikortzminda and a synagogue in Oni, built in 11th and 19th century, respectively) and a mosque in Akhaltzikhe, built in 18th century. Retrofitting of these structures was aimed at preservation of initial geometry and appearance by creating composite (stone – reinforced concrete, or stone – shotcrete) structures, which were partially or fully hidden. Further improving of seismic response may be achieved by using hybrid seismic isolation decreasing the seismic forces and adding damping. A brief description of the design procedure for such cases is presented.

Keywords: ancient stone domes, reinforced concrete, strengthening, stress, concentration, earthquake resistance, seismic isolation.

INTRODUCTION

Ancient stone domes are usually heritage architecture and have utilitarian value. Currently there is great number of stone domes in the Mediterranean countries, at the Caucasus and in the East (India, Iran, etc.). In the past these structures passed severe strong earthquakes, accompanied by significant destruction and human victims. The state of ancient stone domes after earthquakes proves their relatively high seismic resistance caused mostly by axisymmetry, continuity and closed form in perimeter.

In cases of cracks or other damages it is necessary to receive seismic loading during severe earthquakes without significant damages, hence the question of strengthening or conservation of domes rises. For strengthening of stone domes traditionally metal rings are applied, straining beams, doubling constructions, masonry injection, shotcreting, reinforced concrete one- or two-sided thin shells, etc.

With appearance of reinforced concrete (RC) the heavy stone domes were actually replaced by the lighter thin RC domes. Use of RC elements allows strengthening of stone domes, as RC is well compatible with stone masonry. It is possible to create almost invisible elements, in order to keep the original view of the monument. There are certain examples of conservation and restoration of the stone domes in the seismic regions (Aoki *et al.*, 1998; Emerson and Van Nice,1943; Heyman, 1995; Danieli *et al.*, 1998; 2002 etc.).

CP1020, *2008 Seismic Engineering Conference Commemorating the 1908 Messina and Reggio Calabria Earthquake*,
edited by A. Santini and N. Moraci
© 2008 American Institute of Physics 978-0-7354-0542-4/08/$23.00

SIGNIFICANCE AND METHODS

Conservation is aimed at preventing damage by maintaining and improving loading capacity to fit a required level. Thus, it is based on proper structural design in each individual case. One should distinguish between temporary and permanent conservation (Danieli et al., 1998). Restoration is primary aimed at architectural aspects, such as external and internal appearance, and functional roles of internal spaces, courtyards, etc. Maintenance refers mostly to utilization of buildings. It implies large variety of checkouts and repair works carried out periodically and based on a detailed program. The major purpose of maintenance is to prevent malfunctioning of engineering systems as well as loss of structural integrity and problems of durability. Repairs are done only if needed, according to the results of periodical checkouts.

Three building preservation projects are reported further on. All three refer to domes made of natural stone and bricks, situated in mountainous seismic areas of Caucasus.

SEISMIC SITUATION

Several historical buildings, like a temple in Nikortzminda and a synagogue in Oni, faced an earthquake with the Richter magnitude M = 6.9, which occurred in Racha, Georgia, on April 29, 1991 (Engineering... 1996). This magnitude corresponds to the intensity of I_0=9.5 on the MSK-64 twelve-step scale. Several thousands of aftershocks were registered later on, during four months. Their magnitudes varied between M = 6.2 and M = 5.9, being sometimes almost as powerful as the primary shock. The following chart illustrates this situation.

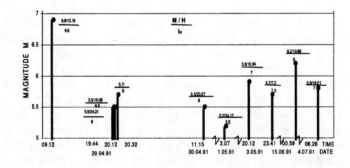

FIGURE 1. Seismic situation caused by the Racha earthquake, Georgia, between April 29 and July 04, 1991, in the epicenter. **M** – Richter magnitude; **H** – focus depth, km; **I_0** – intensity on MSK-64 scale

NIKORTZMINDA TEMPLE

The temple was built in the 11th century. It has a cross – shaped structure with a drum and a brick dome (Fig. 2). The overall height of the structure is about 26 m, the dome span is 6.4 m, and its rise is 2.8 m. The dome shell thickness is 0.6÷0.8 m. The drum has three-layer walls with thickness of 1.0÷1.5 m, weakened by windows. The distance from Nikortzminda and the earthquake epicenter was about 95 km.

The estimated intensity of the primary shock was 7 on the seismic scale MSK-64 ($0.5÷1.0$ m/sec^2 at the foundation). It caused cracking of the drum, and actually divided it into a group of vertically separated segments. Series of radial cracks and a closed-loop horizontal crack at the shell top of the dome developed. The central disk of the shell moved downwards by 60÷100 mm, but was wedged by surrounding parts of the dome, and so did not collapse (Fig. 2). Temporary conservation of the dome was carried out in order to prevent its collapse by possible aftershocks (Danieli et al.,1998;2002). The following measures were taken.

 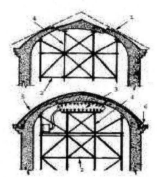

FIGURE 2. The temple in Nikortzminda, damage of the dome and repair works: 1-dome; 2- scaffolding; 3- air pillow; 4- pillow support; 5- reinforced shell; 6- ring beam .

External reinforcement of the dome ring beam at the top of the drum included a ring bundle of ø10 mm steel wires, designed for taking a tensile force of 150 kN. Horizontal force in the ring due to the dome weight was determined following Timoshenko's theory of plates and shells. Safety coefficient of 2.5 was applied. External reinforcing "hoops", consisting of ø10 mm wires, mounted atop vertical timber boards. Each "hoop" was designed to take a tensile force of 200 kN. The encountered resistance of the drum to axial tension was 0.20 MPa, higher than up-to-date standard requirement of 0.18 MPa. Internal scaffolding, designed to support damaged areas of the structure, including the drum and the dome. The scaffolding was used later for the dome repair (Fig. 2). These temporary measures proved safety along the aftershock period. The temple performed at no further damage. The permanent conservation project was completed two years later (Danieli et al., 1998, 2002). Its major steps were as follows:
1. Construction of a steel belt around the ring beam of the dome.
2. Filling all cracks in the drum and dome by a lime-based mortar under pressure.
3. Construction of a reinforced concrete shell atop the dome: the reinforcement mesh was tied to ø10 mm steel anchors, embedded in the shell and in the cracks prior to their filling. The shell thickness was 80÷90 mm. Concrete ($f_{ck} = 30$ MPa) was used.
4. Lifting of the central disk, weighing about 4 t, back to its original position using air pillows situated on the scaffolding (Fig. 2). The gap between the disk and the

surrounding shell, which remained after lifting, was filled with a lime-based mortar under pressure.

5. Completing the reinforced concrete shell in the central zone, atop the lifted disk.

As a result, the dome became a two-layer composite structure. Its original shape was restored and strengthened by an external RC shell. Therefore the internal view of the dome did not change, which allowed further restoration of ancient paintings.

SYNAGOGUE IN ONI

The synagogue in Oni, Georgia was completed in 1895; it is an architectural and historical monument of Georgia. A general view of the building and its cross section are presented in Fig. 3. It is a rectangular symmetrical 18.5 x 14.9 m structure, built of local stone, with a maximum height of 15 m. The dome has a 6.7 m span and a 3.0 m rise and it is situated in the center of the structure, atop a drum supported by arches. The rectangular inner structure, on its four sides has a flat ceiling on wooden beams. The beams, as the vaults, are atop outer walls and inner arches. Cement-lime mortar was used for construction. The arches in the plan are located on mutually perpendicular directions. They are situated atop four stone columns symmetrically located inside the building and inner pilasters of the outer walls. The arches are tied at their bottom by steel rods with square cross section (25 x 25mm), for taking horizontal tensile forces. The stele and sculptures are made of separate stones connected to each other by metal staples. Portico of the building is made of stone columns covered by stone vault. The columns are made of separate stones. Their bottoms are tied on the edges by metal tie bars, and rear columns of the portico are connected with the building by metal staples, located in the body of the vault and preventing free horizontal shift of the portico's structures (Danieli et al., 1998; 2002). .

The 1991 earthquake caused substantial damage to the building, which did not collapse due to symmetrical structure, rigid walls, steel ties, and small arch spans The crack opening in the arches was 5÷10 mm. Many architectural elements and sculptures were damaged. From the falling stele blow the stone portico vault collapsed. The front portico columns stones shifted. Some of them collapsed (Fig. 3).

FIGURE 3. Overall view of building and its cross section: 1- arch; 2- ceiling; 3- dome; 4-drum; 5- collapsed sculpture; 6- collapsed facade element.

A method for estimation of structural seismic resistance (Danieli and Bloch., 2006) was applied to this building. Relative earthquake resistance coefficient was estimated based on the design data and the results of the inspection. Temporary conservation was not carried out, but it was no further damage during the aftershock. Permanent conservation was aimed primarily at strengthening of bearing elements. The drum has been reinforced as well. The project included the following major steps (Danieli et al., 1998; 2002):

- Removal of plaster, filling of cracks by cement-lime mortar. Steel wedges were used to for filling of wide cracks (> 6 mm).

- Reinforced plastering of arch and shell surfaces in damaged areas. One steel meshes layer (150x150 mm of $\varnothing$6 mm wires) served as reinforcement. The meshes were tied to $\varnothing$10 mm steel anchors, each situated in a pre-drilled hole at a 30° angle to the surface. Cement-lime mortar was applied (f_{ake} = 10 Map), the layer thickness was 40 mm.

- Drum strengthening. A 10 x 10 mm mesh of $\varnothing$1 mm wires was used for external reinforcement. Cement lime mortar (f_{ake} = 10 Map) was used to provide a 30 mm layer.

- Restoration of non – bearing external parts and architectural elements.

Fragments of strengthening are shown in Fig. 5.

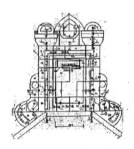 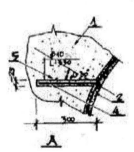

FIGURE 4. Fragments of strengthening (dimensions in mm). A - Details of arches and vaults reinforcement 1- arch; 2- reinforced plastering layer; 3- steel mesh; 4- drilling; 5- steel anchor B - Strengthening of the frontal stele. 1-frontal façade; 2-lateral facade.

THE MOSQUE IN AKHALTZIKHE

The mosque is situated in seismic areas of Georgia. Its dome was built of locally manufactured thin bricks (240 x 240 x 40 mm), using a lime - clay mortar. The dome dimensions are: span - 16.2 m, rise – 8.0 m, shell thickness – 0.6 ÷ 0.8 m (Fig. 5).

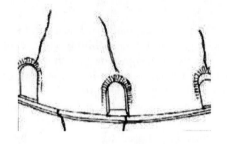

FIGURE 5. The mosque in Akhaltzikhe: general view, radial cracking of the dome.

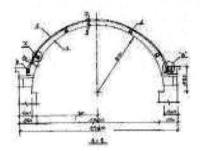
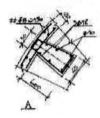
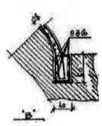

Figure 6. Strengthening of the dome in Akhaltzikhe(dimensions in mm).
1- dome; 2- reinforcing shell; 3- ring beam (detail B); 4- connection element between the shell and the dome   (detail A)(dimensions in cm)

The expected Richter magnitude for the Akhaltzikhe area is M=7. The mosque, built in 18th century, was damaged by a few major earthquakes: Akhalkalaki, Georgia, 1899, (45 km, M=5.4); Spitak, Armenia, 1988, (125 km, M=6.9); and Racha, Georgia, 1991, (100 km, M=6.9). Although the structure did not loose stability, the dome was damaged. Wide radial cracks appear on its internal surface, mainly in relatively weak window areas (Fig. 7). At the bottom part, they were about $15 \div 20$ mm width. Tensile stresses in this area could be 0.045 MPa due to the dome weight

The conservation project was aimed at the dome's repair and at increasing the seismic resistance of the whole structure (Figure 6) (Danieli et al., 1998;2002).. Preservation of interior appearance was considered. For these purposes, an external, 120 mm thick, RC shell cast atop the existing dome was designed. The shell is to be connected to the dome, by means of connecting elements (detail A in Fig. 6) forming a composite two – layer structure. The calculations of dead loads were based on Timoshenko's shells theory and on RC shells limit analysis (Akhvlediani and Danieli, 1990). Stress-strain analysis under seismic and gravity loads was done by Finite Element Method (Danieli et al., 2003).

The following material parameters were used in calculations: concrete compressive strength f_{ck}=30 MPa, reinforcing steel design tensile strength f_{sk}=340 MPa, Young modulus for brick walls E_b=1,500 MPa, and for reinforced concrete E_c=30,000 MPa. Horizontal seismic stresses in the ring beam of the dome, calculated for MSK-64 earthquake intensities 7, 8, and 9, were 0.075 MPa, 0.105 MPa, and 0.135 MPa accordingly. Considering the current condition of the dome, severe damage, and even collapse, is likely to occur. Therefore, the RC shell is designed, with two layers of ø8 mm steel mesh 150x150 mm and the ring beam reinforcement by eight ø20 mm steel bars. The ring beam cross section is 600 x 400 mm, and it is designed with an additional safety factor of 1.5÷2.0 to the calculation results. This approach allows preservation of the dome internal appearance, even without filling the existing cracks.

BASE ISOLATION AND HYBRID SYSTEMS

Base isolation is a well known method for increasing seismic resistance of structures. It has been implemented in hundreds of buildings and bridges all over the world in the last decades. Parducci1 et al. (2005) reported on base isolated buildings of the new Emergency Management Centre of Foligno (Umbria) in Italy shown in Fig. 7. The structure has been designed to overcome without damage an earthquake having an average return period of about 1000 years.

FIGURE 7. Base isolated Emergency Management Centre of Foligno building (Parducci1 et al., 2005).

The building is a circular false-dome with a diameter of 31 meters. A set of 10 reinforced concrete semi-arches, supported by 10 isolators Ø=1000 mm, are the main structural elements. The radial semi-arches stand from a peripheral ring situated at the first floor and sustain, at their top, a suspended prestressed cylindrical core containing lifts and stairs. A rigid peripheral shell system connects the ring to the rubber isolators situated at the base of the building, upon the soil. The central core is a suspended element, and the whole structure is supported just by 10 isolators.

Using supplemental dampers as a part of a base isolation system allows further improvement of structural seismic response and also decreases the base displacement. An original idea for selecting the optimal properties of passive friction dampers being a pert of a hybrid base isolation system was proposed by Ribakov et al. (2007). The method includes modeling an artificial earthquake excitation having desired peak ground acceleration, spectrum bandwidth and duration. and it is aimed to select the properties of passive friction dampers so that the seismic response of a base isolated structure with these dampers would be as close as possible to that of a structure with optimally controlled active damping devices. Such behavior is achieved if the energy, dissipated in passive friction dampers, will be equal to that in active controlled devices.

Addition to the base isolation system passive friction dampers, designed according to this method, yields reduction in base displacements up to 25%, compared to those obtained in the structure with a base isolation system without supplemental dampers. Maximum values of floors' displacements in a base isolated structure with passive friction dampers are up to 30% lower, compared to those in a base isolated stricture without dampers. No increase in floors' accelerations was observed for the selected earthquakes.

Conclusions

Wherever possible, external reinforced concrete shells are a preferable permanent solution. Design of such shells is aimed at creating dual – layer composite structures able of taking expected seismic loads. The Reinforcement by Request approach to design of dome reinforcement was applied to external concrete shells. Although reinforcing shell is part of a composite structure, it takes loads, others than its own weight, only if the original part, i.e. the dome itself, needs "help". This can make lighter shells. Construction technology of such external shells is relatively simple, and they can be more economical than internal shells. They also may take higher loads and, at last, they allow preservation of internal appearance.

REFERENCES

1. T. Aoki, S.Kato, K. Ishikawa, K. Hidaka, M. Yorulmaz and F. Gili, "Principle of Structural Restoration for Hagia Sophia Dome", *STREMAH International Symposium*, San Sebastian, 467-476 (1998).
2. W. Emerson and R.L. Van Nice, "Hagia Sophia, Istanbul: Preliminary Report of a Recent Examinatioh of the Structure", *Supplement to American Journal of Archaeology*, **47** (1943).
3. J. Heyman, *The Stone Skeleton – Structured Engineering of Masonry Architecture*, Cambridge University Press, England, 1995
4. M. Danieli, G. Gabrichidze, Y. Melashvili and O. Sulaberidze, "Study of Some Reinforced and Metal Spatial Structures, in Seismic Regions of Georgia", *ICSS-98, Spatial Structures in New and Renovation Project of Buildings and Construction Theory, Investigation, Design, Erection*. **I**, Moscow, 1998, pp. 396-403
5. M. Danieli, G. Gabrichidze, A. Goldman and O. Sulaberidze, "Experience in Restoration and Strengthening of Ancient Stone Made Domes in Seismic Regions", *Seventh US Conference on Earthquake Engineering*, Boston, 2002.
6. *Engineering Analysis of the Racha Earthquake Consequences, In Georgia, 1991*, edited by Gabrichidze G., Metsnereba, Tbilisi, Georgia (1996), (in Russian).
7. M. Danieli (Danielashvili), J. Bloch, *Evaluation of Earthquake Resistance and the Strengthening of Buildings Damaged by Earthquake*. In: First European Conference on Earthquake Engineering and Seismology.Geneva, Switzerland, 3-8 September 2006, Paper No: 673 (SD-R).
8. M. Danieli, A. Goldman, A. Aronchik and J. Bloch, "Method of Increasing Seismic Resistance in Ancient Stone and Brick Domes", *Fifth National Conference on Earthquake Engineering*, 26-30 May 2003, Istanbul, Turkey.
9. N.A. Akhvlediani and M.A. Danieli, "Limit Analysis of Reinforced Concrete Shells", *Arkhivum Inzynezii Ladowey*, **36**, Poland, 187-205 (1990).
10 . A. Parducci, S. Costantini, A. Marimpietri, M. Mezzi, R. Radicchia and G. Tommesani, "Base Isolation and Structural Configuration the New Emergency Management Centre in Umbria", *9th World Seminar on Seismic Isolation, Energy Dissipation and Active Vibration Control of Structures*, Kobe, Japan, June 13-16 2005.
11. Y. Ribakov and G. Agranovich, "Design of Base Isolation Systems with Passive Friction Dampers", *European Earthquake Engineering*, 2007.

Seismic Protection of an Ancient Aqueduct Using SMA Devices

Christis Z. Chrysostomou[a], Themos Demetriou[b], Andreas Stassis[c], and Karim Hamdaoui[d]

[a]Department of Civil Engineering and Geomatics, Cyprus University of Technology, P.O.Box 50329 3603 Limassol, Cyprus
[b]Themos Demetriou and Associates, 13 Hera str., 1061 Nicosia, Cyrpus
[c]Department of Mechanical Engineering, Higher Technical Institute, P.O.Box 20423, 2152 Aglantzia, Cyprus
[d]Karim Hamdaoui, Department of Structural Mechanics, University of Pavia, Via Ferrata, No. 1, 27100 Pavia, Italy

Abstract. The effectiveness of the use of Cu-based shape memory alloy (SMA) prestressing devices on an ancient aqueduct is examined in this paper. The dynamic characteristics of the aqueduct were measured within the span of three years and computational models were developed that matched very closely its dynamic behaviour. Using this as a bench mark, SMA prestressing devices were applied on the structure and the effects on its dynamic characteristics were assessed. It was noted that the SMA prestressing devices have a significant effect on the dynamic response of the structure. This is attributed to the stiffening of the structure due to the increase in contact between the masonry units and hence the increase of its stiffness through the increase of the modulus of elasticity of the masonry matrix. It can be concluded that the SMA prestressing devices can provide an inconspicuous means of stiffening masonry structures and increase their resistance to earthquake loads.

Keywords: Conservation, shape-memory-alloy, passive, monuments, seismic-protection.

INTRODUCTION

Monuments are structures that managed to survive through centuries and stand their ground as a testament of the achievements of the people who have constructed them and their civilization. Many times people in their effort to protect these monuments are changing their structural system in harmful rather than protective way. A code of practice is evolving that states that any intervention in the structural system of an ancient monument should be such that it neither violates its form, nor changes drastically its structural behaviour. Another important issue is that the materials used should be compatible with those of the monument and any intervention should be reversible.

While most of the traditional retrofitting methods violate the above conditions innovative seismic-protection techniques, such as shape memory alloy (SMA) based devices, can be made very inconspicuous and therefore not violating the form of the monuments and in addition are removable which means that they satisfy the condition of reversibility of the retrofitting method. At the same time these devices can be very

CP1020, *2008 Seismic Engineering Conference Commemorating the 1908 Messina and Reggio Calabria Earthquake*,
edited by A. Santini and N. Moraci

effective in dissipating the energy generated by earthquakes and hence protect the monuments. The application of SMA and other innovative devices in protecting monuments are reported by Biritognolo et al. [1], Croci [2], and Chrysostomou et al. [3], [4] and [5].

In this paper a literature review on shape memory alloys is presented along with the characteristics of the SMA used in this research. In addition, the results of the system identification study, which includes measurements and computational models and is described by Chrysostomou et al. ([4] and [6]), are presented briefly. Finally, the experimental results of the application of SMA prestressing devices on the monument are presented along with conclusions drawn.

SHAPE MEMORY ALLOYS

The series of Nickel/Titanium alloys developed by Buehler and Wiley [7] in the 60's, exhibited a special property allowing them to regain and remember their original shape, although being severely deformed, upon a thermal cycle, which became known as the shape memory effect. Later, it was found that at sufficiently high temperatures, such materials also have the super-elasticity property: the recovery of large deformations during mechanical loading-unloading cycles performed at constant temperature [8]. Subsequently, SMAs lend themselves to innovative applications.

A detailed references list on the use of SMA metals in different engineering applications is provided in [9]. More recently, investigations are focused on the possibility of using the pre-stressed SMA wires to sew ancient buildings made by blocks ([10] and [11]) and on the possibility of using SMA based base-isolators ([12]).

Because of their high cost and their limited range of potential transformation temperatures, the application of the NiTi alloy might not be the best candidate for the seismic retrofit of cultural heritage structures. Therefore, Casciati and Faravelli [13] checked the suitability of an alloy different from NiTi for the application in monumental retrofitting. The interest was focused on Copper-based SMA that consists of Cu, Al and Be chemical composites.

The thermo-mechanical characteristics of the 3.5 mm diameter AH140 SMA wires used in this research are described by Casciati F. and Faravelli [14]. This type of alloy is selected on the basis of its typical temperature window, in which it maintains its super-elastic properties.

METHODOLOGY

The methodology used in order to establish the effects of the SMA devices was first to obtain the dynamic characteristics of the monument using measurements and develop a finite element model to match those characteristics. Then, apply the SMA devices and measure the dynamic characteristics again in order to establish any changes. In this section the above methodology is described.

System Identification Without SMA Devices

Chrysostomou et al. ([4] and [6]), describe the methodology used for obtaining the dynamic characteristics of the monument and the development of a computational model. Figure 1 shows the FFT obtained from the signal analysis and the first five frequencies that were identified. The signals were obtained from two triaxial accelerometers (EpiSensors) positioned at 2 different locations on the aqueduct indicated with numbers 1 and 2 in Fig. 2.

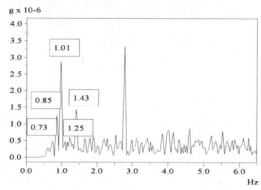

FIGURE 1. Measured frequencies for aqueduct without SMA devices

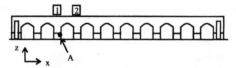

FIGURE 2. Positions of accelerometers and location of impact (point A)

Experimental Setup for the SMA Devices

Twenty 60 cm long SMAs were prepared by adding fixing devices at each of their ends. These devices were then connected to steel strands that extended from the top to the bottom of a pier of the aqueduct.

In order to be able to apply a prestressing force, prestressing devices were used by placing them along the length of the steel strand. Two such devices were used for each strand, in order to be able to provide the required strain, taking into account the elongation of the SMA wire and that of the steel strand. The use of 10 wires on each side of the pier would require the drilling of 20 holes. This was considered excessive damage to the monument; therefore it was decided to connect two wires on a rigid pipe, which was in turn connected to one steel strand. This resulted in a decrease of the required holes from 20 to only 10. Figure 3 shows 10 of the wires hanging from the rigid base at the top of the pier. At the bottom of the pier the strands were anchored on bolts.

Having setup all twenty wires, tensioning of the wires took place in a symmetric manner. First the four central wires were tensioned. At first the prestressing devices were used to remove any slack that existed in the assembly. When that was done, the initial length of the SMA devices was measured, as shown in Fig. 3. This length was used to calculate the strain in the wire as the tensile force was increased by turning the prestressing devices. Based on the stress- strain data for the SMA it was decided that a 2% strain would be applied in the SMA wires since that constituted the center of the plateau.

FIGURE 3. SMA wires supported at the rigid base at the top of the pier of the aqueduct

As it can be observed from Fig. 3 only two of the SMA wires (on each side of the pier) appear to be stretch since the experiment started by applying a prestressing force to only 4 of the 20 wires (wire no. 5, 6, 15 and 16 in Fig. 4). Two more measurements were taken and in each case 8 more wires (4 on each side) were prestressed giving a total of 12 and 20 wires, respectively, by stressing first wires no. 1, 2, 9, 10, 11, 12, 19 and 20, and then wires no. 3, 4, 7, 8, 13, 14, 17 and 18. Then an unloading cycle took place by removing first the central 4 wires (wires no. 5, 6, 15 and 16) resulting in a 16 wire loading and then removing 8 more wires (wires no. 1, 2, 9, 10, 11, 12, 19 and 20) resulting in an 8 wire loading.

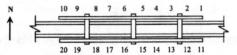

FIGURE 4. Plan view of the top support showing the numbering of the SMA wires

TABLE 1. Strains in the loaded wires

Wire No.	Strain (%)	Wire No.	Strain (%)
1	2.80	11	1.30
2	3.88	12	1.85
3	2.16	13	2.11
4	1.54	14	1.69
5	1.90	15	2.45
6	2.67	16	2.96
7	0.94	17	1.92
8	1.34	18	2.45
9	1.51	19	1.31
10	2.43	20	2.04
Average strain	2.12	Average strain	2.01

The strains for the loading phase are shown in Table 1. It should be noted that during unloading no measurements were taken since from checks that were made, there was insignificant change in the length of the loaded wires when either the number of loaded wires was increased or decreased.

It can be observed from the table that wire no. 11 has a small strain. This is due to the fact that the thread of the prestressing device failed and no additional load could be applied. Although it was very difficult to reach the 2% strain in all the wires, an attempt was made to have the strains between 1% and 3%. As it can be seen from the table on the average the strain in the wires at each side of the pier was very close to 2%.

System Identification With SMA Devices

Two EpiSensors were positioned at 2 different locations on the aqueduct indicated with numbers 1 and 2 (see Fig. 2), and a rubber impact hammer was used to induce vibrations in the aqueduct in addition to ambient vibrations. One impact location was used indicated in Fig. 2 with the letter A. The x-axis, y-axis and z-axis of the EpiSensors were aligned with the longitudinal, perpendicular to its plane and vertical directions of the aqueduct, respectively.

As it was explained in the previous section, five loading conditions were applied on the structure by prestressing 4 wires, 12 wires and 20 wires in a loading cycle, and then by removing the load in 4 wires, hence leaving 16 wires loaded, and finally removing 8 wires leaving 8 wires loaded. The results of the FFT analysis of the signal at position 1 are shown in Fig. 5 (a) to (e). It should be noted that the same results were obtained by the sensor which was placed at position 2.

DISCUSSION

In comparing Fig. 5 (a) to (e) with Fig. 1 it is obvious that the application of the SMA wires have a significant effect on the dynamic behaviour of the aqueduct. While in the case that there are no wires on the structure the obtained signal is complex and only a few distinct frequencies appear, as soon as the SMA wires are applied, the signal in all the cases clears-up and four distinct frequencies appear.

The results presented in Fig. 5 (a) to (e) are summarized in Table 2. In the same table the total force in the loaded wires is shown per loading case, which was obtained by converting the strain into stress using the stress-strain curve of the material and hence calculating the force in the wire. These results are plotted in Fig. 6. It is obvious from the results that as the number of wires increases from 4 to 20, there is a shift of the fundamental frequency from 0.85 Hz to 1.01 Hz (period shift from 1.18 s to 0.99 s). The same is observed for the second recorded frequency. This indicates a stiffening of the structure due to the application of a prestressing force. This stiffening can be explained by the increase in contact between the masonry units and hence the increase of its stiffness through the increase of the modulus of elasticity of the masonry matrix. For the third and the forth frequencies though a slight decrease is recorded with the increase of the number of wires, from 2.93 Hz to 2.75 Hz and from 4.76 Hz to 4.70 Hz, respectively.

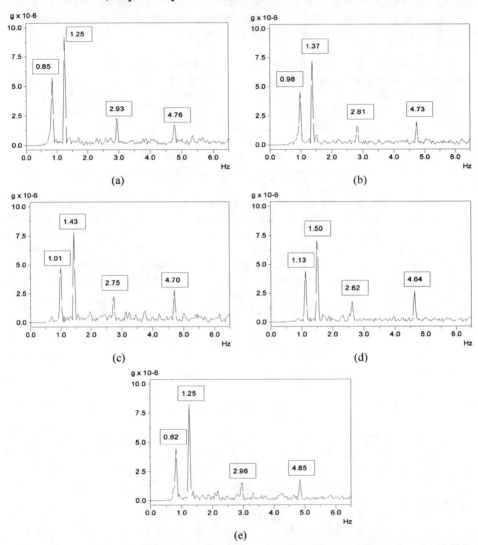

FIGURE 5. Frequencies obtained from position 1 for (a) 4 SMA wires, (b) 12 SMA wires, (c) 20 SMA wires, (d) 16 SMA wires, and (e) 8 SMA wires

TABLE 2. Load in the wires and corresponding frequencies

	Loading			Unloading	
No. of wires	4	12	20	16	8
Load in wire cluster (kN)	6.44	18.28	29.99	23.55	11.72
Frequency 1 (Hz)	0.85	0.98	1.01	1.13	0.82
Frequency 2 (Hz)	1.25	1.37	1.43	1.50	1.25
Frequency 3 (Hz)	2.93	2.81	2.75	2.62	2.96
Frequency 4 (Hz)	4.76	4.73	4.70	4.64	4.85

66

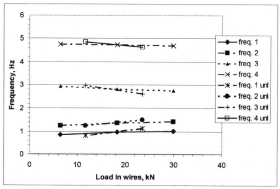

FIGURE 6. Variation of the measured frequencies with the load in clusters of SMA wires for the loading (4, 12, 20 wires) and unloading (unl., 16, 8 wires) cycles.

Using now the results shown in Fig. 6 for the loading cycle, a second order polynomial was fitted to the results for the fundamental frequency which is shown in Fig. 7. From the equation for the fundamental frequency, it can be predicted that the fundamental frequency of the structure when no wires are applied is 0.74 Hz. In comparing with the measured fundamental frequency of the structure (Fig. 1), it is obvious that there is a close match and therefore this equation can be used for the prediction of the modification of the fundamental frequency of the structure as a function of the applied load in the SMA wires.

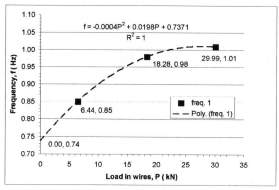

FIGURE 7. Extrapolation of the fitted curve of the 1st frequency to zero-load for the loading (4, 12, 20 wires) cycle.

CONCLUSIONS

In this paper the results of the application of copper-based shape-memory-alloy prestressing devices on an ancient aqueduct are presented. The dynamic characteristics of the monument without the application of SMA are reported by Chrysostomou et al. [6].

Based on the above findings it can be concluded that the application of the SMA wires on a real structure has shown that they significantly change the dynamic characteristics of the structure. This is a matter that should be further investigated by laboratory tests in order to establish the relationship between the variation of the frequencies of the structure and the modulus of elasticity of the masonry matrix. This relationship should then be tested by the development of a computational model that will be able to predict the behaviour of the aqueduct when the SMA prestressing devices are applied, and hence investigate numerically the effectiveness of the SMA wires in the protection of monuments from earthquakes.

ACKNOWLEDGMENTS

The authors acknowledge the financial contribution of the European Commission and the cooperation of ing. Mauro Mottini for the design of the tie anchorage. The authors are also thankful to Professor Fabio Casciati, coordinator of the project. We would also like to express our thanks to the Director of the Department of Antiquities of the Republic of Cyprus for giving us permission to apply the SMA devices on the monument, in order to perform the experiment.

REFERENCES

1. Biritognolo, M., Bonci, A., and Viskovic, A, "Numerical models of masonry façade walls with and without SMADs," *Proc. Final Workshop of ISTECH Project – Shape Memory Alloy Devices for Seismic Protection of Cultural Heritage Structures*, Joint Research Centre, Ispra, Italy, June 2000, pp. 117-140.
2. Croci, G., "General methodology for the structural restoration of historic buildings: the cases of the Tower of Pisa and the Basilica of Assisi," *Journal of Cultural Heritage*, 1, 7–18 (2000).
3. C. Z. Chrysostomou, Th. Demetriou, and M. Pittas, "Conservation of historical Mediterranean sites by innovative seismic-protection techniques," *Proceedings 3rd World Conference on Structural Control*, Como, Italy, April 7-12 2002, v. 2, pp. 947-954.
4. C. Z. Chrysostomou, T. Demetriou, and A. Stassis, "Seismic protection of an aqueduct by innovative techniques," *Proceedings 3rd European Conference on Structural Control*, Vienna, July 2004.
5. C. Z. Chrysostomou, T. Demetriou, M. Pittas and A. Stassis (2005), "Retrofit of a church with linear viscous dampers," *Journal of Structural Control and Health Monitoring*, 12, No. 2, 197-212 (2005).
6. Chrysostomou C. Z., Demetriou Th. and Stassis A., "Health-Monitoring and System-Identification of an Ancient Aqueduct," *Smart Structures and Systems*, v. 4, No. 2, pp. 183-194 (2008).
7. Buehler W.J. and Wiley R.C., "Nickel-based Alloys," *Technical report*, US-Patent 3174851 (1965).
8. Auricchio F., "Shape Memory Alloys: Applications, Micromechanics, Macromodelling and Numerical Simulations," PhD Thesis, University of California at Berkeley, USA, 1995
9. Auricchio F., Faravelli L., Magonette G. and Torra V. (eds.), "Shape Memory Alloys: Advances in Modelling and Applications," *CIMNE*, Barcelona, 2001
10. Casciati S. and Faravelli L., "Fastening Cracked Blocks by SMA Devices," *Proceedings of the 3rd European Conference on Structural Control*, Vienna, Austria, 2004, M1-1/M1-4.
11. Casciati S., Faravelli L, and Domaneschi M., "Dynamic tests on Cu- based shape memory alloys toward seismic retrofit of cracked stone monuments," *Workshop on smart structures and advanced sensor technologies*, June 26-28 2005, Santorini Greece.
12. Casciati F., Faravelli L. and Hamdaoui K., "Numerical Modelling of a Shape Memory Alloy Base Isolator", *Proceedings of the 3rd International Conference on Structural Engineering Mechanics and Computation*, September 10-12 2007, pp. 223-228, Cape Town, South Africa.
13. Casciati S. and Faravelli L., "Thermo-mechanic Characterization of a Cu-Based Shape Memory Alloy," *Proceedings SE04*, Osaka, Japan, 2004, pp. 377-382.
14. Casciati, F. and Faravelli, L., "Experimental Characterisation of a Cu-based Shape Memory Alloy toward its Exploitation in Passive Control Devices," *Journal de Physique IV*, 115, 299-306 (2004).

Dynamic and Fatigue Analysis of an 18th Century Steel Arch Bridge

Nadir Boumechra[a] and Karim Hamdaoui[b]

[a]Department of Civil Engineering, Faculty of Engineering, University of Abou Bekr Belkaid, Chetouane B.P. 230, 13000, Tlemcen, Algeria.
[b]Department of Structural Mechanics, University of Pavia, Via Ferrata 1, 27100, Pavia, Italy.

Abstract. Within the "Oran-Tlemcen" railway line realization project (159km), several bridges were built by the Railroads Algerian West Company. 7km from the east of Tlemcen city, this railway line must cross a very broken mountainous collar, that's why the French engineer "Gustave Eiffel" was solicited to construct a 68m length bridge. In 1890, an arch steel truss bridge was realized. The bridge presents 300m of apron curvature radius and, currently, is considered as one of the most important monuments of the Algerian historical heritage. Considering the age of the bridge and the evolution of the railway loads in time, it was essential to check the good behavior of the studied structure. For that, analyses to verify the physical and mechanical properties of the growth iron members are made. A finite element model of the bridge was built and numerical simulations were drawn. The structural vibration conducted analysis permit to understand the behavior of this particular structure, then to evaluate (in detail) the rate of the structure fatigue.

Keywords: Steel Bridge, Railway, Arch, Vibration, Seismic, Fatigue Analysis.

INTRODUCTION

The construction of the railway line between the two towns of Oran and Tlemcen was started with the railway segment Tabia to Tlemcen (64 km), conceded by a law of July 16, 1885. It was carried out by the "Compagnie des Chemins de fer de l'Ouest Algérien" (Railroads Algerian West Company) and finished in 1889 [1] & [2].

The bridge, placed between five tunnels, had been built in a way to avoid the digging of cliff. It was built in 1889 by the very known Engineer "Gustave Eiffel" and usually known as "El-Ourit bridge". The bridge is an arch riveted steel structure. It is composed of two beams, stiffened between them by the higher apron in addition to the horizontal and vertical wind-bracings disposed in Saint-André cross'. Its opening length is of 68m and the overall length of the higher apron is of 70m. The apron presents a small curvature, of a radius of 300m, in conformity with the supported railway line curvature. In the same year of its construction, a test of loading was carried out by the Railroads Company using a train of 250 Tons [3]. The railway line entered in service since 1890.

The regular and rigorous maintenance of the bridge limited the metal's corrosion problems. Because its old age (118 year) and the current evolution of the rail traffic, it was essential for the Railroad services to check its static and dynamic resistance and compare it to the actual tests in addition to its fatigue rate.

CP1020, *2008 Seismic Engineering Conference Commemorating the 1908 Messina and Reggio Calabria Earthquake*, edited by A. Santini and N. Moraci
© 2008 American Institute of Physics 978-0-7354-0542-4/08/$23.00

The bridge overpass by the Mefrouche wadi (river). The very mountainous environment of the bridge as its history confer it a role of tourist attraction (Figure 1).

(a)   (b)

FIGURE 1. El-Ourit arched steel bridge. (a) in the past and (b) presently.

DESCRIPTION OF THE STRUCTURE

The bridge is 68m length with a height of 8m. At mid-span, it presents a joint and an articulation connecting the two half spans. In addition to the articulation, the two parts of the bridge are connected by rods of pre-stressed ties placed in diagonals. This system was conceived to partially relieve the structure, therefore to increase its flexibility, considering the supports that are fixed on the abutments.

This choice was made to limit the problems of temperature variation and any possible geological movement.

The principal carrying element is the arched girder with lattice made up of two soles of variable thickness. This frame is simply supported on the abutments. The structure consists of steel elements riveted out of wrought iron. The principal carrying element is the bottom arc beam of a curvature radius of 300m. This arc supports the apron via amounts and strengthened by horizontal and vertical wind-bracings. The transverse distance between the two arched girders is of 4.50m. The physical and mechanical tests on metallic samples of the bridge were made and gave the following results: Density = 7.30, E = 1,785.108 kN/m², υ = 0.30 and f_e = 245N/mm².

The finite element model of the structure was made by using the SAP2000 computer software and by considering the exact geometric description of all its elements. The model is composed of 1548 frame elements and 170 Shell elements (Figure 2).

FIGURE 2. The finite element model of the studied structure.

STATIC AND SEISMIC ANALYSIS OF THE BRIDGE

Because of the evolution on time of the trains and, especially, the weight of the locomotives, it was necessary to check the resistance of the structure. For this reasons, a research was made on the different type of trains circulating since the construction, as well as on the intensity of the traffic.

The initial trains operating on this railroad are schematically defined in the Figure 3a. For the current trains, used since the sixties, the model defined by the UIC71 (Figure 3b), is a correct average of the railway loads in operation [4].

The study of the structural behavior is concentrated on the follow-up of the maximum stresses on 3 sample-elements located on the bottom beam of the arched girder. These elements are located near the support (elem.462), in the quarter of span (elem.481), near the mid-span (elem.489) and that of the amount near to the other support in its symmetric position (elem.1126). The stresses are checked with the limit states of service.

For the case of the dead load, the maximum stress is 22.02 N/mm². Considering the trains operating in the past, the maximum stress is 68.92 N/mm², therefore a resistance safety-factor of F=3.55 was reached. For an actual train loading, the stress increase to 120.62 N/mm², so, a safety-factor decreases to the value of F=2.03 (see Table 1).

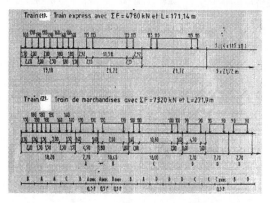 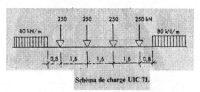

(a) Trains du passé types 1 & 2 (19ème siècle)   (b) Train actuel (Train 3)
FIGURE 3. The train types defined in the bridge analyses.

The modal dynamic analysis shows that the bridge is more flexible transversely than vertically, with the fundamental free period of vibration of T1=0.336s. The first six free periods of vibration are T1=0.336s, T2=0.315s, T3=0.250s, T4=0.180s, T5=0.155s and T6=0.135s (Figure 4).

Two codes were applied to perform the seismic analysis of El-Ourit bridge: i) the Algerian para-seismic code for bridges, abbreviated RPOA 2006 [5] and ii) the European para-seismic code Eurocode 8 [6]. A peak ground acceleration of 0.20g on the rock was considered and the classification of the bridge as of "great importance". The seismic response spectrum takes account of the horizontal and vertical components of the first 60 free modes of vibration. The gathered results from the two para-seismic codes are very similar. The maximum stresses do not exceed 48.69 N/mm².

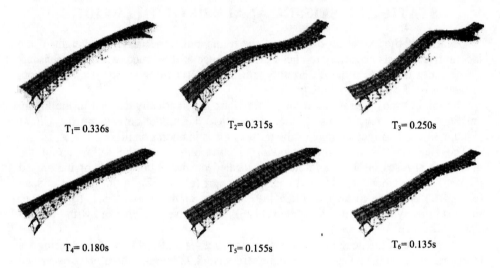

T_1= 0.336s T_2= 0.315s T_3= 0.250s

T_4= 0.180s T_5= 0.155s T_6= 0.135s

FIGURE 4. The first six mode shapes.

Loading Case	Elem. Beam (462)	Elem. Beam (481)	Elem. Beam (489)	Elem. Column (1126)
G	-22.02	-13.34	-7.94	-1.69
G + Past Train (1)	-11.40 / -68.92	+5.86 / -54.81	-1.26 / -26.16	-0.49 / -19.08
G + Past Train (2)	-14.41 / -54.55	+2.28 / -43.54	-3.32 / -22.03	-0.70 / -10.79
G + Present Train (3)	+2.64 / -120.62	+30.95 / -103.53	+7.31 / -47.12	+0.65 / -35.46
G+0.30 Present Train + Earthqu. RPOA	-33.54 / -48.69	-19.81 / -29.59	-11.25 / -17.90	-5.36 / -6.21
G+0.30 Present Train + Earthqu. Eurocode 8	-34.40 / -47.83	-20.54 / -28.84	-10.46 / -18.69	-5.30 / -6.27

TABLE 1. Maximum and minimum stresses for the different loading cases [N/mm²] ((-) for a compression and (+) for a tension).

It is seen that the forces and stresses corresponding to railway moving loads are more severe than those obtained by the seismic analysis and, in general, the bridge resistance is largely sufficient.

FATIGUE ANALYSIS OF THE BRIDGE

The puddeled steel is obtained in the original puddling technique, molten iron in a reverberate furnace was stirred with rods, which were consumed in the process. Then the Wrought iron is obtained. This product is more flexible than the cast iron. For these reasons, the coefficient of resistance to the fatigue is almost one half the values generally obtained for steel structures.

Following the computation results of the structure's stresses for the different cases of railway loads, it was useful to know the state of fatigue of the bridge. The study of the influence lines for the elements of the bottom arched beam, defined previously, shows that the stresses vary from a maximum in compression to a minimum in tension at the time of the train passing (Figure 5).

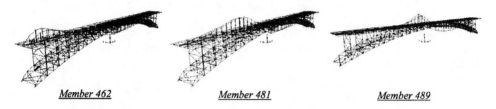

Member 462   *Member 481*   *Member 489*

FIGURE 5. Influence lines of the bottom beam elements (elements462, 481 & 489).

The method to estimate the state of fatigue of an old structure and its residual life time is based on the Wohler curve specific to wrought iron, of the partial damages resulting from the past and future loads and on the accumulation of the various damages according to the Palmgren-Miner assumption [4] & [7]. It is worth noticing how, with a number of $N_i=2.10^6$ cycles of a stress amplitude of $\sigma_i=70$ N/mm², the rupture by fatigue is reached (Figure 6).

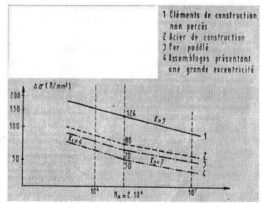

FIGURE 6. Wohler curve specific to iron [4]

Traffic Periods	Year	Train	
		Type	Nbr/day
1890-1940	50	1	4
		2	4
1940-1990	50	3	4
1990-2007	17	3	8
Future	-	3	12

TABLE 2. The railway traffic spectrum

Loading Case	Elem. Beam (462)	Elem. Beam (481)	Elem. Beam (489)	Elem. Column (1126)
Train 19rd (I)	+10.58 / -46.93	+19.20 / -41.47	+6.68 / -18.22	+1.20 / -17.39
Train 19rd (II)	+0.76 / -32.56	+15.62 / -30.20	+4.63 / -14.08	+0.99 / -9.09
Train Actual (UIC)	+24.63 / -98.63	+44.29 / -90.19	+15.26 / -39.17	+2.34 / -33.77

TABLE 3. Maximum and minimum stresses for the different railway loading cases [N/mm²] ((-) for a compression and (+) for a tension).

The damage in the current state is calculated from the moving load spectrum presented in Table 2. The analysis of the more stressed element of the bridge (elem.462) is presented in (Table 3):

$$d = \sum d_i = \sum \frac{n_i}{N_i} = d_{(1890-1940)} + d_{(1940-1990)} + d_{(1990-2007)}$$

$$d = \sum d_i = \sum \frac{n_i}{N_i} = (\frac{73000}{1,50.10^7} + \frac{73000}{1.10^8}) + (\frac{73000}{7.10^5}) + (\frac{49640}{7.10^5}) = 0.0765$$

For the more stressed metallic element of the bridge, one notices that its fatigue ratio now is rather weak, thus the capacity of residual fatigue resistance is appreciable. For a future frequency of loading of 12 cycles per day with a standard train UIC71, the remaining theoretical life time to the bridge is evaluated at 147 years. This result is extremely positive for a 118 years old bridge!

CONCLUSION

El-Ourit bridge has a structural, historical and patrimonial importance for Tlemcen region. It was necessary to check the resistance of its elements for the service limits state. To this end and for the current railway loads, an interesting resistance reserve was evaluated for the bridge. Concerning the seismic action it is worth noticing how the induced stress level is quite moderate.

Considering the wrought iron bridge age, the estimating of the fatigue ratio is an obligatory step in this study. The result of the current cumulated damages is rather appreciable. This process, which is analytical and theoretical, remains limited because of the fact that several specific problems were not taken into account in the performed analysis. Among them the dynamic bridge– train interaction, the influence of the riveting connections and thermal fatigue effect should be included in a further analysis.

REFERENCES

1. Maurice Antoine BERNARD, 1913, *Les chemins de fer algériens*, éditions A. JOURDAN. (In french).
2. Pierre Morton, 2003, *Le développement des chemins de fer en Algérie*, Extrait de la revue du Gamt, n°71/2003/3 et 72. (In french).
3. Louis ABADIE, *Tlemcen au Passé Retrouvé*, Editions Jacques Gandini, 1994. (In french).
4. ITBTP, 1989, *Etude de la fatigue dans les ouvrages d'art*, Annales de l'ITBTP. (In french).
5. CTTP, 2006, Projet du règlement parasismique des ouvrages d'art, Direction des Routes, Ministère des Travaux Publics. (In french).
6. Afnor 2000, Eurocode 8 : Design of structures for earthquake resistance- Part 2 : Bridges.
7. Schijve J., 2003, Fatigue of structures and materials in the 20th century and the state of the art, Materials Science, Vol.39, N°3.

Assessment of the Structural Conditions of the San Clemente a Vomano Abbey

Francesco Benedettini, Rocco Alaggio and Felice Fusco

DISAT University of L'Aquila, 67040 Monteluco di Roio, L'aquila, Italy

Abstract. The simultaneous use of a Finite Element (*FE*) accurate modeling, dynamical tests, model updating and nonlinear analysis are used to describe the integrated approach used by the authors to assess the structural conditions and the seismic vulnerability of an historical masonry structure: the Abbey Church of San Clemente al Vomano, situated in the Notaresco territory (TE, Italy) commissioned by Ermengarda, daughter of the Emperor Ludovico II, and built at the end of IX century together with a monastery to host a monastic community. Dynamical tests "in operational conditions" and modal identification have been used to perform the *FE* model validation. Both a simple and direct method as the kinematic analysis applied on meaningful sub-structures and a nonlinear 3D dynamic analysis conducted by using the *FE* model have been used to forecast the seismic performance of the Church.

Keywords: Historical Buildings, Dynamical Tests, *FE* modeling, Seismic Vulnerability.

OUTLINE

Reading from the "Chronicon Casauriense", a collection of transcripted deeds dated back to the years from 866 to 1182, we get to know that the Abbey Church of San Clemente al Vomano, situated in the Notaresco territory, was commissioned by Ermengarda, daughter of the Emperor Ludovico II, and built at the end of IX century together with a monastery to host a monastic community. In 1108 the abbey underwent a process of deep reconstruction which determined its current architectural and structural configuration. As the years go by, the church of San Clemente al Vomano, in spite of the sudden collapse of the adjoining monastery, was able to remain stable and stand up though in the need of maintenance.

The church, as it appears nowadays, has a nave and two side aisles covered with a wooden truss-beam roof and divided by pillars and columns that support round arches. Opposite to the front, three different apses close the church, which has no transept and shows a presbytery set directly above the crypt.

The outside walls show some re-used stony carved decorations (as the portal and the window in the central apse) that are attributed to a branch of the School who worked in San Liberatore in Maiella. The wall themselves, made up of a rough juxtaposition of blocks of stone and bricks, have an irregular texture which suggests a systematic re-use of materials; such a building procedure appears to be in perfect agreement with the strait circumstances of the local community and with the need to recycle parts of other buildings in the second reconstruction. Wide re-use of constructive elements is evident also in the inside of the church, as columns and

CP1020, *2008 Seismic Engineering Conference Commemorating the 1908 Messina and Reggio Calabria Earthquake,*
edited by A. Santini and N. Moraci
© 2008 American Institute of Physics 978-0-7354-0542-4/08/$23.00

capitals of different shapes and materials offer a magnificent and harmonious view to the visitor but, simultaneously, do create a lack of structural harmony that led to continuous repairs over the centuries.

To this purpose, several restoration works have been conducted on the ancient structure. The first documented works were conducted during the XVIII century when a new roof of the church was erected. More recent actions were conducted on 1926, 1936, 1971 and, eventually during the 80's. During the last restoration a quite complete reinforcement of the structural components has been conducted. The assessment of the present structural conditions after the last restoration phase, is conducted in this study. In fact, even if the Church is now in an excellent conservation state for what is concerned the outward appearance, a series of symptoms suggested a deeper analysis from the structural point of view (Fig.1_a).

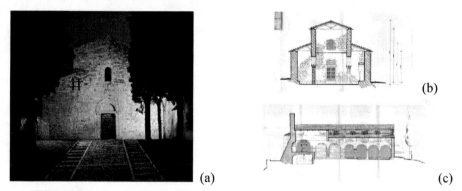

(a)   (b)   (c)

FIGURE 1. The *San Clemente a Vomano* Abbey (a), technical drawings (b) and (c)

To this end an accurate *FE* model of the church has been developed on the basis of existing technical documents and on an accurate geometrical survey. Even though, the numerical model, even if based on the field survey and mechanic characterization of the materials, always involve simplifying assumptions and several uncertainties in the material, geometric properties and boundary conditions; hence, the model needs to be validated by full-scale tests. Dynamical tests and a modal analysis using output only identification methods, were recently conducted and the relevant results have been used to validate the numerical model.

The importance of dynamical tests and the related optimization analysis on the *FE* model is crucial at least for two different aspects: they can help in constructing a realistic numerical model having, after the validation, a "realistic performance" and, secondly, it may be used as a starting point to put into evidence anomalous local behaviors and possible damage occurrences that can be even localized by using a structural (parametric) identification technique. Some comments on the use of the modal analysis conducted under linearity hypothesis on structures whose behavior is often (if not always) non linear, are added to clarify if, when and way such an analysis could be conducted and meaningfully used [1,2].

After the validation, the structural analysis has been performed to assess the tensional level due to dead loads and to assess the seismic vulnerability of the Church with respect to the actual state of conservation. In particular, the seismic analysis conducted with the *FE* model, taking into account finite displacement and nonlinear materials behavior, has been compared with a simple kinematic analysis [3-5] . The quite high level of local stresses under the dead loads and the comparison of the demand versus capacity under lateral loads, confirms the critical state of the Church to extensive damage and collapse, in case of a seismic event.

The analysis of possible repairing actions and strengthening techniques put into evidence the effectiveness of classical structural reinforcement in term of global behavior and increased seismic capacity.

THE *FE* MODEL

The *FE* modeling was based on technical drawings available from the archives of the "Soprintendenza ai beni Architettonici e Paesaggistici" of Abruzzi region (Fig.1$_b$ and 1$_c$) and on *in situ* survey that permitted a detailed 3D modeling. The model is based on the use of 3D-solids elements generally having 8 nodes for element; in case of local structural complexity, 20-nodes elements have been adopted (Fig.2$_a$-2$_c$). Both a linear and a non-linear constitutive laws have been adopted for the composing materials. The linear model has been used to describe the stress level under gravity loads and to perform a modal analysis, while the non-linear model has been used to obtain the Peak Ground Acceleration corresponding to the structural collapse. Both fixed boundary conditions and elastic springs have been considered in the relevant analysis.

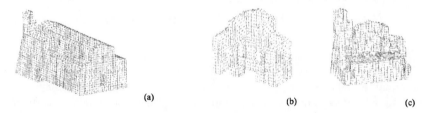

(a)   (b)   (c)

FIGURE 2. The *FE* model: general views (a), front segment of Abbey (b), apses segment (c)

THE DYNAMICAL TESTS IN OPERATIONAL CONDITIONS

To validate the *FE* model, dynamical tests in operational conditions and modal identification have been used to obtain a measure of the global structural behavior to be used in a validation/updating process of the numerical model. In particular, the identified frequencies and modal shape of the Abbey, have been compared, in such validation process, with the companion quantities forecasted by the *FE* model. Considering such a validation procedure, usually necessary because the models often are very far from a correct description of the structural behavior, the first addressed question is if, when and why, the modal identification (obtained under hypothesis of linearity) is applicable to the analysis of old stone/masonry structures, whose behavior is often (if not always) non linear. However, during the analysis of an historical structures, always an interdisciplinary approach is needed and the modal testing is only one of the possible means to be applied to understand and, when possible, to accurately model every analyzed case. The deep knowledge of the materials, the composition of structural elements and the accurate geometric description, play a key role in the interpretation of tests-results.

Concerning the dynamical tests used in this study, among other considerations, when dealing with monuments, the excitation devices are generally responsible of interferences with the normal use of the structure and could be responsible of activation of non linear regimes; on the contrary, when using the ambient excitation, excluding particular cases, the linearity assumption is more easily maintained because the vibrations used in the analysis are not externally induced and exactly correspond to the usual ambient level of vibration induced by natural causes. Nevertheless, if this is the case, a very sensitive acquisition chain is needed, able to capture the meaningful structural dynamics embedded in the noise.

In the analyzed case, to identify frequencies and modal shapes to be used in the subsequent model updating, there was the possibility to put only 4 accelerometers for each side (Fig. 3$_a$), located at the position of the windows (Fig. 3$_b$) of the main nave (long terms to have permission from the Authority "in charge", to attach the instruments on the walls).

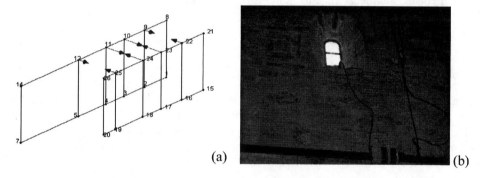

(a)   (b)

FIGURE 3. The instruments lay-out (a), an instrumented window (b)

In Fig. 4 the obtained results are reported together with the companion numerical ones forecasted by the *FE* model after a preliminary calibration of the material parameters. The experimental identified modes clearly enlighten the circumstance that is not guaranteed an efficient link between the wooden-roof and the corresponding walls as evidenced by the part (c) and (g) of the figure showing a clear independence of movement between the left and right walls of the main nave. In the following collapse analysis, such a circumstance will clearly identify the kind of mechanism responsible of the structural failure and, accordingly, the coherent *FE* model has been adopted in the analysis (roof not connecting the opposite walls).

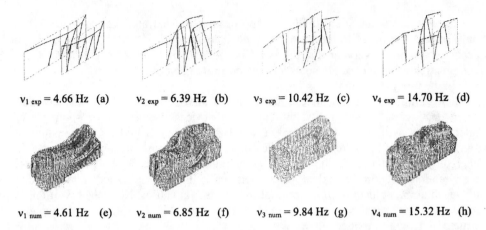

$v_{1\ exp} = 4.66$ Hz (a) $v_{2\ exp} = 6.39$ Hz (b) $v_{3\ exp} = 10.42$ Hz (c) $v_{4\ exp} = 14.70$ Hz (d)

$v_{1\ num} = 4.61$ Hz (e) $v_{2\ num} = 6.85$ Hz (f) $v_{3\ num} = 9.84$ Hz (g) $v_{4\ num} = 15.32$ Hz (h)

FIGURE 4. Experimental [(a), (b), (d) walls in phase, (c) walls not in phase] and numerical [(e), (f), (h) walls in phase, (g) walls not in phase] identified modes

The agreement between the "real results" and the numerical prediction obtained by the calibrated model is quite good and the updated model can be thought as a reliable model to be used with confidence in the structural analysis.

THE STRUCTURAL ANALYSIS

As a first step a linear static analysis under dead loads has been performed by using the updated *FE* model. Concerning the stress field, the worst situation is localized in correspondence of the basis of the third column on the right side of the main nave (Fig. 5), where the maximum value of the compressive stress reaches the value of 1.95 MPa. It is worth noticing that at the basis of the other columns, the compressive stress is bounded between 1.3 and 1.6 MPa while the stress level at the basis of the continuous walls is less than 0.2 MPa.

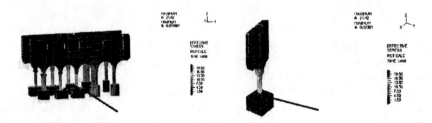

FIGURE 5. The stress at the basis of the columns reach considerable levels even only under the action of the dead loads.

To evaluate the minimum Peak Ground Acceleration (PGA) producing the collapse, a static nonlinear approach was first followed: the obtained PGA will be compared with the expected value for the site located in the Notaresco municipality, III seismic zone according to Italian technical rules, where the PGA for an 'A' class soil is $a_g = .15 \, g$.

The aim of the analysis is to identify the actual collapse mechanism among all the (global and local) possible ones. In the examined case, it is necessary to remove any simplifying hypothesis on global structural behavior usually done in standard masonry buildings; the assumed constitutive laws in the *FE* model is based on a full nonlinear characterization and a 3D failure criterion. The relevant analysis was performed taking also into account a finite displacement regime.

The adopted constitutive law is characterized by a brittle tensile failure for small value of principal tensile stress, ductile-softening compression failure for an high value of principal compression stress, softening strain on the compression branch and collapse due to reach of the ultimate strain. According with the tests conducted on the materials by a different research group, the model parameters have been chosen as follows: tangent modulus at zero strain 4000 MPa, uniaxial cut-off tensile stress .3 MPa, uniaxial maximum compressive stress -3. MPa, uniaxial ultimate compressive stress -1. MPa, uniaxial ultimate compressive strain -0.08.

According to the results of the dynamical tests already discussed, the link between the masonry walls and the wooden truss-beam roof is judged unable to transfer actions when their values are of the order of magnitude expected in a SLC seismic analysis. A push-over analysis in the transversal direction is performed in large displacement regime and the capacity curve requested by OPCM 3362/2004 is obtained.

Among all the possible failure mechanisms for the analyzed Abbey, the maximum vulnerability is found in correspondence of a transversal seismic action. Such a result is due to the presence of a long span lateral masonry walls without any intermediate transversal constraint. In particular, the analysis select the right wall as the most critical element of the whole building. In the evaluation of the capacity curve, the seismic action is modeled through a mass-proportional force distribution practically coincident with the first modal shape and kept constant during the analysis.

After a nonlinear dead loads solution the push-over analysis is executed, by using the restart option, according to a load-displacement-control (LDC) algorithm [6]. The analysis result is a curve ground-acceleration vs. a control point displacement. The control point is located at the top of the wall mid-span. This curve, representing the out-of-plane inflection until the collapse, is then transformed in an equivalent capacity curve for a one degree of freedom equivalent system.

The response curve (Fig. 6) exhibit a substantially fragile behavior, coherently with the specified local collapse mechanism.

A sensitivity analysis performed on the material law and on collapse surfaces shows no significant sensitivity to maximum compression failure parameters, a very low sensitivity to tensile failure stress and an appreciable sensitivity to post-critical tensile behavior. The obtained capacity curve considering a 1 d.o.f. oscillator having the frequency of the first mode when the roof is assumed non connecting the two walls of the main nave (2.24 Hz), is compared with the response spectrum in the ADSR form (acceleration-displacement) in Fig. 7. In such a figure the structural performance of a safe structure would be identified locating the so called performance point (P.P.) at the intersection between the capacity and the demanding curves, unachievable in this case.

The evaluated collapse peak ground acceleration on rock is $PGA_{collapse}.(rock)=0.09$ g considerably lower than the $PGA(rock)=0.15g$ requested by the Italian code for the considered location.

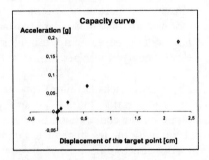

FIGURE 6 Acceleration Vs displacement curve at the target point

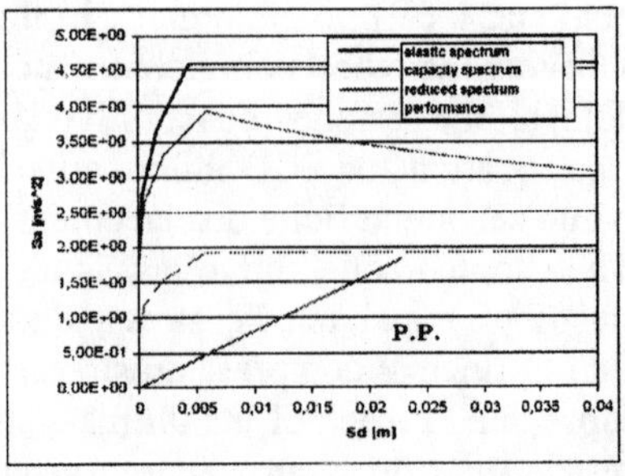

FIGURE 7 Comparison between the capacity and the demanding curves (type B soil)

The global analysis puts into evidence the lacking of intermediate transversal elements and, consequently, the expected collapse is a local out of plane instability mechanism of the more vulnerable wall (Fig. 8).

Because the Italian seismic code (OPCM 3431/2005) suggests to evaluate the PGA on global and local mechanisms by using, among the other, the equilibrium limit state analysis, for the sake of comparison and validation, the alternative approach to capacity evaluation furnished by such an approach on local mechanisms, is applied to the already evidenced out-of-plane collapse mechanism of the right side wall of the main nave (Fig. 9).

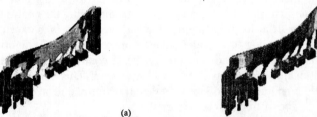

(a)   (b)

FIGURE 8. Crack distribution (a) and max principal tension (b) in the pre-collapse configuration

The local failure mechanism is activated if the masonry quality is sufficiently good to avoid distributed failure and a monolithic behavior of the wall is guaranteed. The assumed hypotheses are: rigid walls, no tensile resistance, unlimited compressive stress. The collapse multiplier of horizontal actions is evaluated by means of a kinematic linear analysis and the spectral acceleration that activate the assumed mechanism is determined.

Among all the admissible local mechanisms, the one exhibiting the lowest PGA$_{collapse}$ is, as expected, in correspondence of the overturn of the right wall of the nave, inadequately linked to the wooden truss-beam roof.

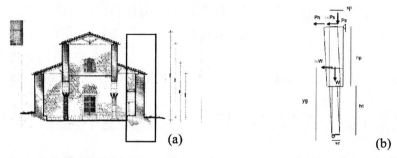

FIGURE 9. Transversal section of the church (a) and equilibrium scheme (b)

The resulting peak ground acceleration on rock is PGA$_{(collapse, rock)}$ =0.08g resulting in good agreement with the value obtained by using the 3D *FE* analysis.

CONCLUSIONS

The analysis put into evidence how the thin columns, particularly on the right side of the analyzed Church, are vulnerable elements even under the dead loads only. The thin columns and the lacking in the transversal connection between the longitudinal walls, is responsible of the activation of a local instability mechanism of the right wall of the main nave under the action of horizontal excitation and, consequently, the PGA$_{collapse}$ of the Church is of about the 50% of the requested value. New connections (dissipative frames) between such walls could dramatically improve the global behavior, permitting a box response, instead of local response, to horizontal loads avoiding in such a way the independent response of opposite walls.

REFERENCES

1. C. Gentile, A. Saisi, "Ambient vibration testing of historic masonry towers for structural identification and damage assessment", *Constr. and Buildings Materials*, 21, 1311-1321, (2007)
2. A. De Stefano, R. Ceravolo, P. Zanon, "Seismic assessment of bell towers through modal testing" *Proceedings of SPIE - The International Society for Optical Engineering*, 4753 II, 1265-1271, (2002).
3. M. Betti, A. Vignoli, "Assessment of seismic resistance of a basilica-type church under earthquake loading: Modelling and analysis" *Advances in Engineering Software*, 39 (4), 258-283, (2008).
4. P. Fajfar, "*Capacity spectrum method based on inelastic demand spectra*, Earthquake Engng. Struct. Dyn., 28, 979-993, (1999).
5. G. Augusti, M. Ciampoli, P. Giovenale, *Seismic vulnerability of monumental buildings*, Structural Safety, 23, 253-274, (2001).
6. K.J. Bathe and E.N. Dvorkin, "On the Automatic solution of nonlinear Finite Element Equations", *J. Computers and Structures*, 17, 5-6, 871- 879, (1983).

Integrated Approach to Repair and Seismic Strengthening of Mustafa Pasha Mosque in Skopje

Veronika Sendova, Predrag Gavrilovic, Blagojce Stojanoski

Institute of Earthquake Engineering and Engineering Seismology, IZIIS
University "Ss. Cyril and Methodius", P. O. Box 101, 1000 Skopje, Republic of Macedonia

Abstract. Mustafa Pasha's Mosque is one of the biggest and the best preserved monuments of the Ottoman sacral architecture in Skopje and the Balkan. As a cultural historic monument of an extraordinary importance for the city of Skopje and Republic of Macedonia, it is under protection of the Law on Protection of Cultural Heritage. IZIIS is currently elaborating a project on repair and strengthening of Mustafa Pasha's mosque. Respecting the modern requirements in protection of historical monuments, as is the main principles of seismic strengthening are: application of new technologies and materials, reversibility and invisibility of the applied technique. The concept of structural strengthening and repair aimed at reaching the designed level of earthquake protection has been selected based on: (i) investigations of the soil conditions, (ii) investigations of the characteristics of the built-in materials, (iii) investigation of the main dynamic characteristics, as well as (iv) previous experimental investigation of the mosque model.

Key words: historic monument, repair, strengthening, earthquake protection, innovative materials and techniques.

INTRODUCTION

The repair and/or strengthening of historical monuments in seismic regions is highly dependent on the earthquake conditions to which they have been exposed in their past history and the ground motion to which they are expected to be frequently exposed in future, as well as the materials and methods used for their construction. Due to these reasons, it will be of importance that repair and/or strengthening, as a part of preservation, conservation and restoration of historical monuments be planned based on detailed studies of the following factors: seismic hazard; local soil conditions and its dynamic behaviour under earthquake loading; dynamic properties of the structural systems and their strength and deformability and the dynamic response of the structure under the expected ground motions.

Considering that the above factors are of main importance for the determination of the earthquake response of historical monuments, as well as the fact that seismic analysis cannot be performed using seismic design codes for modern buildings, determination of the criteria, methods and techniques for strengthening and the process of restoration and preservation should be based on detailed studies with consideration of the cost effectiveness of the alternative solutions.

CP1020, *2008 Seismic Engineering Conference Commemorating the 1908 Messina and Reggio Calabria Earthquake*,
edited by A. Santini and N. Moraci
© 2008 American Institute of Physics 978-0-7354-0542-4/08/$23.00

IZIIS' EXPERIENCE IN SEISMIC PROTECTION
OF CULTURAL HERITAGE

Within the frames of the IZIIS' research activities, in addition to seismic design and protection of modern structures, particularly noteworthy is also the experience gathered in the field of protection of structures pertaining to the cultural historic heritage. During a period of more than 30 years of activities in this field, the Institute has realized important scientific research projects involving experimental and analytical research, field surveys of historic structures and application of knowledge during earthquake protection of important cultural historic structures and monuments.

Within the frameworks of the scientific research projects realized at the Institute in the period 1990-2000 for the purpose of development of appropriate methods for repair and strengthening of the Byzantine monuments in general, and particularly the Byzantine churches located within Macedonia, shaking table testing of a church model in a realistic geometrical scale was performed for the first time in the world, [1]. 1:2.75 scaled model of St. Nikita church was constructed and tested on the seismic shaking table in the IZIIS laboratory in its original state, strengthened state by use of "ties and injection", and as a base isolated model. The knowledge gained in this way is unique and incomparable and hence necessary for seismic strengthening of individual important cultural-historic structures where it is important to have an insight into the effect of the interventions upon the authenticity of the monument.

After realization of these projects, IZIIS became partner of the Republic Institute for Protection of Cultural Historic Monuments which enabled direct application of the gained knowledge in actual conditions and for specific historic monuments. Presented further are the two most characteristic examples of application of the developed methodology by implementation of vertical and horizontal strengthening elements.

1. Reconstruction and seismic strengthening of the St. Athanasius church, [3]: On August 21, 2001, during the armed conflict in R. Macedonia, the monastic church of St. Athanasius in Leshok experienced strong detonation, which resulted in its almost complete demolition, (Fig. 1). From structural aspects, there have been two approaches taken in the attempt to renovate and reconstruct the structure, (Fig.2). Based on the performed detailed analysis of the structure, (i) solution for repair and strengthening of the existing damaged part of the monastic church and (ii) solution for seismic strengthening of the ruined part of the church to be reconstructed were made.

FIGURE 1. The Church after detonation   **FIGURE2.** The church after reconstruction

2. Reconstruction of the St. Pantelymon church in Ohrid, [2]: In the process of conservation and rebuilding of the St. Panteleymon church, Ohrid, (Fig. 3) having in mind the importance and specific nature of the structure representing a historic monument classified in the first category and a structure of a particular national interest, it was necessary to design a building structure that will satisfy the stability conditions in the process of application of the conservation principles regarding shape, system and identification of materials. Seismic strengthening was provided in accordance with the previously developed and verified methodology, (Fig.4).

FIGURE 3. The church during reconstruction   FIGURE 4. The rebuilt St. Panteleymon church

Integrated Approach to Repair And Strengthening of Historic Monuments

The decisions for repair and strengthening of a historic structure should be thoroughly clarified and justified in advance because of the existence of some unusual aspects (compared to modern structures) that have to be taken into account. The characteristic structural entity, the variety of the built-in material, the complex history of successful modifications done in the past, as well as the degree of deterioration makes each historic structure a case for itself. Only a team of experts of different profiles, (architects, archaeologists, art historians, conservators and other profiles), who are completely competent in their fields but sufficiently flexible to accept the arguments of the others can successfully protect a historic structure. Nevertheless, protection of structures in seismically active regions is mainly a task to be done by a civil engineer - a structural engineer.

Extensive research activities have been performed by IZIIS for the purpose of evaluation of a procedure for repair and strengthening of valuable historic monuments. Such a procedure is based on conventional understanding of retrofitting, although, in our concepts, there are also techniques, which are based on the idea of structural control. As a result of several decades of gathering of experience, it can be said that an integral approach to seismic protection of extraordinarily important cultural historic structures has been adopted by the Institute. This approach, first of all, complies with all the restoration and conservation requirements set in a number of international documents and declarations, as well as procedures and legislative regulations for high category structures. This integrated approach to repair and seismic strengthening of historic monuments should encompass the following:

- Definition of expected seismic hazard;
- Definition of soil conditions and dynamic behaviour of soil media;
- Determination of structural characteristics along with the bearing and deformability capacity of existing structures;
- Definition of criteria and development of a concept for repair and/or strengthening;
- Design of structural methods, techniques, materials and types of excitation;
- Determination of the response of repaired and/or strengthened structures and verification of their seismic stability;
- Definition of field works, execution and inspection.

Although the above stated seems to be the "normal procedure", it is the only way of providing high quality in protection of cultural heritage. This task is certainly much more than simply listing of what is to be done since it requires a lot of knowledge and efforts.

REPAIR AND SEISMIC STRENGTHENING OF MUSTAFA PASHA MOSQUE IN SKOPJE

Mustafa Pasha's Mosque is one of the biggest and the best preserved monuments of the Ottoman sacral architecture in Skopje and the Balkan. The building style belongs to the early Constantinople period at the beginning of the second half of the 15th century. The structural system of the mosque consists of massive peripheral walls in both orthogonal directions with a thickness of about 170 cm constructed partially of hewn stone and brick. The walls of the building are original in principle. They are constructed with two faces between which there is an infill of lime mortar and pieces of bricks and stone.

The catastrophic Skopje earthquake of 1963 inflicted damage to the mosque structure that dominantly affected the central dome and the domes of the porch, the east facade and the minaret. In 1968, these damages were repaired by injection of cement mortar based mixtures as well as incorporation of RC belt courses, (Fig.5).

FIGURE 5. Mustafa Pasha Mosque after the 1963 Skopje earthquake

Today, Mustafa Pasha's Mosque represents a cultural historic monument of an extraordinary importance for the city of Skopje and Republic of Macedonia. As such, it is under the protection of the Law on Protection of Cultural Heritage and it is categorized as a structure belonging to the first category.

The authors of this paper had the opportunity and the challenge to design a system for seismic strengthening of the structure, (financed by the Turkish foundation TIKA) for the needs of the conservation project on repair of Mustafa Pasha Mosque prepared by the Foundation of the University of Gazi in cooperation with the Ministry of Culture and Tourism of Turkey, the Ministry of Culture of Republic of Macedonia and the National Conservation Centre in Skopje and realized in 2006.

Respecting the modern requirements in the field of protection of historical monuments, as is the application of new technologies and materials, reversibility and invisibility of the applied technique, the authors have decided to choose a concept of repair and strengthening involving the use of composite materials. In realization of this project, the established integrated approach has thoroughly been respected. The concept of structural strengthening and repair aimed at reaching the designed level of earthquake protection has been selected based on:

    (i)   investigations of the characteristics of the built-in materials,
    (ii)   investigation of the main dynamic characteristics,
    (iii)   shaking table testing of the mosque model;
    (iv)   investigations of the soil conditions;
    (v)   detailed geophysical surveys for definition of geotechnical and geodynamic models of the site.

Investigation of the characteristics of the built-in materials and the main dynamic characteristics,[4] : In the course of 2006, within the bilateral scientific project with the Yildiz University, Turkey, the dynamic characteristics of the Mustafa Pasha Mosque structure were investigated by use of the ambient vibration technique. The obtained natural periods of the structure were Tx=3.0s and Ty=3.2s, whereas the fundamental period of the minaret structure is T=1.04s. Within this project, the mechanical characteristics of the built in materials were also investigated through testing of samples of material taken from the structure. These data are particularly important and necessary for correct analysis of the seismic stability of the monument.

FIGURE 6. Damages to the mosque model   FIGURE 7. Strengthening of the model

88

Shaking table testing of the Model of Mustafa Pasha Mosque: Experimental shaking table tests on a model of the Mustafa Pasha Mosque were carried out in IZIIS within the frames of PROHITECH project "Earthquake Protection of Historical Buildings by Reversible Mixed Technologies", [5]. The 1:6 scaled model has been subjected to the effect of a series of earthquakes that caused damage (Fig. 6). Then the model was strengthened by application of CFRP elements and subjected to iterative tests (Fig. 7). The results from the investigation, although obtained for small geometrical scale, have shown that the system is efficient, which was the starting point in making the decision about the concept of strengthening of the prototype.

Investigations of the soil conditions: To define the bearing capacity of the local soil and identify the foundation level of the structure, geomechanical drilling (3 sondages down to the depth of 8.00 m) was carried out in June 2007. These geomechanical investigations enabled sufficient definition of the lithological composition of the soil, definition of the geotechnical soil profiles below the structure as well as definition of the physical-mechanical characteristics of the present materials. It has been also concluded that the structure is founded at level –4.00 m on a layer of semi-bound to well compacted sandstone of high bearing capacity.

Detailed geophysical surveys for definition of geotechnical and geodynamic models of the site: The main purpose of these investigations has been to define the seismic parameters for evaluation of the seismic stability of the structure. Namely, for structures of extraordinary importance according to the valid technical regulations, it is necessary to define the seismic input concretely for the site of the structure in order to perform correct dynamic analysis. The investigations have been carried out in compliance with the latest achievements in the field of earthquake engineering, [6]. The main concept of the applied procedure is to consider the expected earthquake effect through a probabilistic approach, including also the local soil effects through nonlinear dynamic analysis of a representative geotechnical model. Table 1 shows the expected values of maximum acceleration for different return periods, whereas Table 2 shows the defined seismic design parameters through input acceleration and characteristic spectra for the design and maximal earthquake. For such defined input parameters, the characteristic records whose frequency content covers the frequency range of interest have been selected in the dynamic analysis.

TABLE 1. Maximal acceleration (in g) for different return periods						
	DAF	Return period (years)				
		50	100	200	500	1000
Bedrock	1.00	0.13	0.19	0.25	0.27	0.36
Foundation level	1.35	0.176	0.257	0.338	0.365	0.480

TABLE 2. Seismic design parameters			
Serviceability life (years)	Level of seismic risk (%)	Earthquake	Max. acceleration $a_{max}(g)$
100 and more	30-40	Design	0.34
	10-20	Maximum	0.39

Concept for seismic strengthening of Mustafa Pasha Mosque: Based on these investigations and the defined seismic parameters, as well as detailed analysis of the seismic stability of the structure, the solution of structural strengthening has been accepted, (Fig. 8, 9), that complies thoroughly with the conservation principles for repair and strengthening of cultural historic monuments, [7]. It consists of incorporation of strengthening elements in the process of conservation of the architecture of the structure and in accordance with the existing project on conservation of the structure, with the main purpose of providing integrity to the structure as well as simultaneous behaviour of the bearing walls at the corresponding levels.

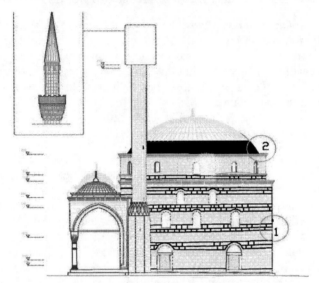

FIGURE 8. Strengthening of Mustafa Pasha Mosque, (facade)

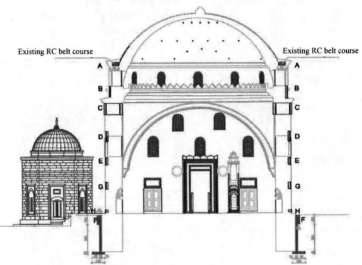

FIGURE 9. Strengthening of Mustafa Pasha Mosque, (cross section)

- DOME: After removal of the cement mortar layer over the dome placed in 1968, the following is anticipated to be carried out: (1) coating of the formerly constructed reinforced concrete ring in the base of the main dome with an injection mixture based on lime mortar and (2) placement of a CFRP strip in a layer of epoxy glue along the perimeter of the dome base within a width of 2.9 m. Then the entire dome is anticipated to be externally coated with a protective layer of lime mortar and covered in accordance with the project on the conservation of the architecture, (Fig.9-A).

- TAMBOUR, BEARING WALLS: After cleaning of all the joints on the outside with a depth of 8 – 10 cm, it is anticipated to place CFRP bars of defined mechanical characteristics (tensile strength of ft = 1800 – 2000 MPa) in an epoxy mortar layer and connect them in the vertical joints. Then, it is planned to fill the joints with pointing lime mortar in accordance with the project on conservation of the architecture, (Fig.9-B,C,D,E,G,H).

- FOUNDATION STRUCTURE: The solution consists of construction of a reinforced concrete wall with a thickness of d=25 cm along the perimeter of the foundation walls, on the external side, below the terrain level and down to the foundation level. From conservation reasons, this reinforced concrete wall will be physically separated from the existing foundation walls by a polyurethane coating for the purpose of separating the concrete from the existing stone masonry. To ensure interaction between the newly designed RC wall and the existing foundation structure, the solution anticipates placement of anchors made of chrome steel to an alternating length according to the reinforcement details, in previously formed openings filled with epoxy mortar (mixture of quartzite sand and epoxy) in accordance with the enclosed scheme, (Fig.9-F).

In the entire part of the structure extending above the terrain level, it is anticipated to remove the cement mortar injected in the cracks after the 1963 earthquake. With the project on conservation of the architecture, it is anticipated that these cracks as well as all the cracks detected after the opening of the external joints be injected with lime mortar with defined mechanical characteristics. Strengthening of the mosque structure in accordance with the designed system started in the fall of 2007.

REFERENCES

1. P. Gavrilovic, V. Sendova and W.S.Ginell, "Seismic Strengthening and Repair of Byzantine Churches", *Journal of Earthquake Engineering*, Imperial College Press, Vol. 3 No. 2, 1999
2. P. Gavrilovic, G. Necevska –Cvetanvska, R. Apostolska, "Consolidation and Reconstruction of St. Panteleymon Church in Ohrid", *IZIIS Report 2001*, IZIIS - Skopje, 2001.
3. V. Sendova, B. Stojanoski, "Main Project on Repair, Strengthening and Reconstruction of St. Ahtanasius Church in Leshok", *IZIIS Report 2004*, IZIIS – Skopje, 2004.
4. Lj. Tashkov, L. Krstevska, "Experimental testing of the historical monuments from ", *Bilateral scientific research project*, Yildiz Technical University, Tureky – IZIIS Skopje, 2006.
5. L. Krstevska, Lj. Tashkov, K. Gramatikov, "Shaking Table Testing of Mustafa – Pasha Mosque Model", *EU-FP6 Programme, Project PROHITECH, WP7*, 2007.
6. V. Sesov et al., "Definition of the seismic parameters for the evaluation of seismic stability of Mustafa Pasha Mosque in Skopje", *IZIIS Report 2007-47*, Skopje 2007.
7. V. Sendova, B. Stojanoski, P. Gavrilovic, "Main Project on Repair and Strengthening of the Mustafa Pasha Mosque in Skopje", *IZIIS Report 2007-41, Vol.1- 3*, Skopje 2007.

Vulnerability assessment of medieval civic towers as a tool for retrofitting design

Sara Casciati[a] and Lucia Faravelli[b]

[a]ASTRA Department, University of Catania, Siracusa, Italy

[b]Department of Structural Mechanics, University of Pavia, Pavia, Italy.

Abstract. The seismic vulnerability of an ancient civic bell-tower is studied. Rather than seeing it as an intermediate stage toward a risk analysis, the assessment of vulnerability is here pursued for the purpose of optimizing the retrofit design. The vulnerability curves are drawn by carrying out a single time history analysis of a model calibrated on the basis of experimental data. From the results of this analysis, the medians of three selected performance parameters are estimated, and they are used to compute, for each of them, the probability of exceeding or attaining the three corresponding levels of light, moderate and severe damage. The same numerical model is then used to incorporate the effects of several retrofitting solutions and to re-estimate the associated vulnerability curves. The ultimate goal is to provide a numerical tool able to drive the optimization process of a retrofit design by the comparison of the vulnerability estimates associated with the different retrofitting solutions.

Keywords: ancient masonry tower; damage assessment; seismic vulnerability; dynamic measurements; modal parameters; time history analysis.

INTRODUCTION

At the beginning of the twentieth century, the failure of the San Marco bell-tower in Venice represented a general warning of the durability of these slender masonry structures, which are characteristic of many urban nuclei in the Mediterranean basin. Table 1 provides a non-exhaustive synthesis of the studies and interventions which were performed in Italy on towers during the last twenty years [1-11]. Similar studies were also conducted in Islamic areas to preserve mosque minarets [12-14]. Within all these works, three stages can usually be distinguished: monitoring, diagnostics, and retrofitting.

The authors were responsible of organizing the instrumental monitoring of the Soncino civic tower [15-19]. The recorded signals were used to localize the damage with structural effects and to clearly distinguish it from architectonic damage (diagnostics) [20]. The third stage, retrofitting, is approached in this paper by numerically investigating the effects of applying to the structures devices formed by pre-stressed shape memory alloy (SMA) wires, following the technique proposed in [21]. Since different spatial distributions of the retrofitting devices can be conceived, the design is driven by comparing the associated simplified vulnerability curves, as originally suggested in [22-24].

CP1020, *2008 Seismic Engineering Conference Commemorating the 1908 Messina and Reggio Calabria Earthquake*,
edited by A. Santini and N. Moraci
© 2008 American Institute of Physics 978-0-7354-0542-4/08/$23.00

TABLE 1. Summary of the main studies conducted in Italy on medieval civic and/or bell-towers.

Group	Location	Name	Height	Construction period (in centuries)	Tipology	Historical	Materials	Visual	Instrumental	Numerical	DIAGNOSTICS	RETROFITTING	Reference
ISOLATED	Pisa		58.36 m	XII-XIV	Belfry	x	Borehole tests, petrographic, chemical and biological analyses	Geometry and cracking survey	Laser, photogrammetry and geotechnical investigation	Deterioration mapping	Structural stabilisation	Architectonic retrofit, confining, geotechnical consolidation (balancing with lead and excavation with forestakes)	[1]
	Roma	Capecci	36.2 m	XIII	Tower	x		Geometry and cracking survey	Ambient vibrations (accelerometers, velocimeters, seismometers)	FEM	Soil-structure interaction	Geotechnical stabilisation	[2]
	Pavia	Civica	60 m	XI	Tower	x	Laboratory compression tests on material samples	Geometry and cracking survey			Soil-structure interaction	*** The monitoring and the diagnostics phases were performed after the collapse, occurred on march 1989	[3]
	Pavia	Maimo	51 m	VIII	Tower	x		Geometry and cracking survey				Structural rehabilitation, geotechnical stabilisation (micro-poles), application of metallic tensile members	[4]
	Pavia	Fraccaro			Tower	x		Geometry and cracking survey				Implacement of a reinforcement cage	[5]
	Cremona	Torrazzo	112 m	VIII-XIII	Belfry	x	Borehole tests and laboratory compression tests on material samples	Geometry and cracking survey	Static loading, ambient vibrations (accelerometers, velocimeters), temperature measurements	FEM			[6]
EMBEDDED	Teramo		48 m	XII-XV	Belfry	x	Borehole tests	Geometry and cracking survey	Ambient vibrations (accelerometers, velocimeters, seismometers)	FEM	Non-destructive (ND) techniques and sensitivity analyses		[7]
	Alba	Artesano	36 m	XI-XII	Tower	x	Borehole tests and endoscopic investigations	Geometry, topography and cracking survey	Acoustic emissions, topography survey, laser, endoscope, thermography, jacks	FEM	Cracking, tilting, material deterioration, subsoil consolidation	Retrofit of exterior degradation and large cracks, reconstruction of removed masonry parts, subsoil consolidation	[8]
	Alba	Bonino	35 m	XI-XII	Tower	x	Borehole tests and endoscopic investigations	Geometry, topography and cracking survey	Acoustic emissions, topography survey, laser, endoscope, thermography, jacks	FEM	Cracking, tilting, material deterioration, subsoil consolidation	Retrofit of exterior degradation and large cracks, reconstruction of removed masonry parts, subsoil consolidation	[8]
	Alba	Sineo	39 m	XI-XII	Tower	x	Borehole tests, endoscopic investigation and topographic survey	Geometry, topography and cracking survey	Acoustic emissions, topography survey, laser, endoscope, thermography, jacks	FEM	Cracking, tilting, material deterioration, subsoil consolidation	Retrofit of exterior degradation and large cracks, reconstruction of removed masonry parts, subsoil consolidation	[8]
	Monza		70 m		Belfry	x	Fatigue compression tests	Geometry and cracking survey	Tests with flat jacks	FEM	Structural stabilisation	Implacement of reinforcement bars and reconstruction of removed masonry parts	[9]
	Padova	S.Giustina	70 m	XIII	Belfry	x					Structural stabilisation	Implacement of reinforcement bars in the interior joints	[9]
	Trignano	S.Giorgio	18.5 m	XIII	Belfry	x		Geometry and cracking survey	Ambient vibrations (accelerometers, velocimeters, seismometers)	FEM	Structural stabilisation	Architectonic retrofit and implacement of devices in shape memory alloy	[10]
	Positano	S.Maria Assunta		XII	Belfry	x	-	Geometry and cracking survey	Foundations			Architectonic retrofit	[11]

THE SONCINO TOWER

The Soncino Civic Tower (Figure 1) is a slender masonry building with hollow square section of side 5.95 m. Its original height (in 1128 A.D.) was 31.5 m, but, in 1575, it was elevated up to 39.5 m by adding the bell room at the top. The thickness of the walls decreases from the bottom to the top as follows:
- 1.55 m is the thickness of both the part of the tower embedded in the surrounding buildings, and the above walls up to the height of 19.05 m. At level 8.1 m, there is a masonry vault ceiling of thickness 1.4 m;
- 1.40 m is the walls thickness within 19.05 m and 22.55 m of height;
- 1.25 m is the walls thickness within 22.55 m and 26.25 m of height;;
- 1.18 m is the walls thickness within 26.25 m and 31.35 m of height;;
- 1.08 m is the walls thickness within 31.35 m and 34,95 m of height; here there is a second horizontal element made of brick and concrete of thickness 0.15 m;
- four corner pillars support the roof of the bell room at the height of 39.50 m.

The perimeter walls stand directly on their foundations, but along three of the four sides are surrounding buildings (up to 10 m high) which increase the boundary conditions of the structure. The fourth unbounded side will be denoted as side A, and the other sides will be named consecutively in a counter clockwise manner.

An experimental measurements campaign was conducted [16, 17] by mounting along the height of the tower four uni-axial accelerometers (Kinemetrics FBA-11) working in parallel with four velocimeters (Lennarz, Le-3D/5s). Ambient vibrations were recorded, among which there are also signals collected when the bell was operating, or when heavy trucks were driven along side A. From the elaboration of the records, the dynamic signature of the tower was captured and its main features are summarized in Table 2. In particular, the detected experimental frequencies are compared with the numerical ones as calculated from the finite element model of Figure 1a, which is characterized by a constant masonry equivalent Young modulus equal to 2400 MPa. The masonry mass density is assumed to be 1800 kg/m^3, and the Poisson's ratio is assigned equal to 0.3.

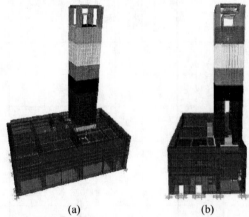

(a)   (b)

FIGURE 1. Finite element model of the Soncino civic tower: a) homogeneous equivalent model; b) model in which the damage is accounted by removing masonry blocks.

TABLE 2. Experimental and calculated frequencies.

Mode		Frequency [Hz]		
		Experimental	Numerical	
			Fig. 1a	Fig. 1b
I	Bending along A-C (Y axis)	1.05	1.0793	1.0469
II	Bending along B-D (X axis)	1.15	1.3073	1.1568
III	Torsion	2.5	3.9263	2.6927
IV	Bending along A-C (Y axis)	4.1	4.1925	3.9314
V	Bending along B-D (X axis)	4.3	4.6431	4.1999

As shown in Table 2, the torsion frequency is not well estimated by using a homogeneous equivalent model. When attempting to account for the damage by assigning to identified areas a lower value of the masonry equivalent Young modulus [25], one observes that the numerical estimates of the first five frequencies are most sensitive to changes in the material properties of the elements forming the tower ring located immediately above the roof of the surrounding buildings.

However, modifications of the Young modulus in this area only alter the frequencies associated to the bending modes, but not the one associated with the torsion mode. In order to achieve a convergence also on the experimental frequency associated to the torsion mode, it is necessary to modify the geometry, by decreasing the torsion stiffness of the ring over-standing the surrounding building. After removing a suitable cluster of masonry blocks in this area, the numerical frequencies reported in the last column of Table 2 were found by adopting a Young modulus value of 3400 everywhere along the tower height, except for the pillars which resulted from the blocks removal. In these pillars, the Young modulus is set equal to 3900 MPa [20].

RETROFITTING CONFIGURATIONS

Once the damage has been localized, the next step consists of investigating the possible retrofit solutions. Following the strategy proposed in [21], the retrofit is pursued by adding ties made of pre-tensioned shape memory alloy wires. They are mounted using the device shown in Figure 2, with the two steel oblique plates placed across the entire wall thickness. This device can be mounted inside the tower and, as such, it has the advantage of not being invasive. Furthermore, this type of intervention also offers the advantage of reversibility, since it can be removed without leaving any permanent effect on the structure.

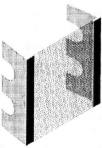

FIGURE 2. The retrofitting device in SMA wires, to be mounted on the internal part of the walls.

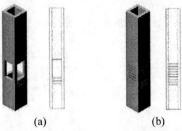

(a)   (b)

FIGURE 3. Different extensions of the retrofitting, named as: (a) solution 1, and (b) solution 2.

The task consists of selecting which is the most effective spatial distribution of the retrofitting devices along the tower height. Several different configurations were actually considered, but, for sake of conciseness, only the two in Figure 3 were selected for analysis in this paper. Namely, the first configuration consists of applying the retrofit device of Figure 2 along a limited strip of length 6.68 m within the damaged area identified after the block removal; it is denoted as solution 1 in Figure 3a. Alternatively, in solution 2 of Figure 3b, the retrofitting covers the entire height of the damaged area. As suggested in [24], the effectiveness of these two different solutions can be compared by referring to the associated vulnerability curves.

VULNEARBILITY CURVES AND THEIR COMPARISON

The assessment of structural vulnerability requires the a priori selection of a structural performance scalar, R, which may represent the maximum inter-story drift, the size of the crack opening, the number of failed connections, etc.. For the specific case study under consideration, the following two quantities are studied:
- the drift ratio, defined as the ratio between the top displacement and the height of the tower over-standing the surrounding buildings;
- the maximum equivalent von Mises stress detected in specifically located finite elements.

A segment of the El Centro 1940 N-S accelerogram is selected as excitation and it is applied in either the A-C (north-south), or B-D (east-west) direction by performing two separate analyses (one for each direction).

For the two considered performance parameters, the damage thresholds are assumed as follows:
- for the drift ratio, the three levels of damage are fixed at 1/500, 1/300, and 1/200 of the height above the surrounding buildings;
- for the maximum von Mises stress detected in the rings of finite element at the connection of the tower with the surrounding buildings, the three damage levels are selected as 75%, 100%, and 125% of the material reference value (2 MPa).

A response coefficient of variation of 0.25 is assigned in order to account of all the uncertainties.

The vulnerability curves obtained for the drift ratio limit state are very similar in the considered cases of homogenized model, damaged model, and two retrofit solutions. Therefore, only those relative to the first case are reported in Figure 4.

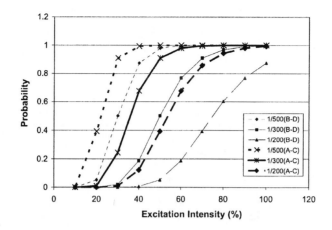

FIGURE 4. Vulnerability curves obtained by using the initial homogeneous equivalent numerical model with excitation in one of the two directions (either B-D or A-C), and the drift ratio as limit state.

By contrast, the comparison of the vulnerability curves in Figure 5, which refers to the von Mises stress limit state, shows that a significant improvement is achieved by the retrofit. In particular, using the amount of translation to the right and the clockwise rotation of the curves as indicators of the retrofit benefit, solution 1 seems to be more effective than solution 2.

CONCLUSIONS

The seismic vulnerability of a civic bell-tower is studied. Rather than seeing it as an intermediate stage toward a risk analysis, the assessment of vulnerability is here pursued for the purpose of optimizing the retrofitting design. The tower vulnerability curve is drawn by carrying out numerical analyses of a model calibrated on the basis of experimental data. The models including the damaged situation and several possible retrofit solutions are also analysed, and the resulting vulnerability curves are compared to evaluate the effectiveness of the retrofit alternatives.

ACKNOWLEDGMENTS

The research was supported by a grant from the University of Pavia (FAR 2007) and by a grant from the University of Catania (2007). The authors are grateful to Ing. Stefania Camerini who carried most of the numerical analyses as part of her duties toward MS graduation.

REFERENCES

1. R. Bartelletti, G. Berardi, M. Jamiolkowski, et al. "Stabilization of the Leaning Tower of Pisa", in *Proc. 13th Int. Congress of Int. Ass. for Bridges and Structural Eng. (IABSE)*, Helsinki, 1988.

2. D. Abruzzese, A. Vari , "Comportamento dinamico di torri in muratura attraverso misurazioni di vibrazioni ambientali" (in Italian), *Proceedings AIAS03*, Salerno, 2003.
3. A. Carpinteri, R. Cerioni, I. Iori, "Alcuni pensieri in merito al crollo della torre civica di Pavia" (in Italian), *Studi e Ricerche – Polytechnic of Milan*, **17**, 491-511 (1996).
4. A. Pavese, "Diagnosi di torri medioevali mediante identificazione dinamica" (in Italian), in *Monitoraggio delle strutture dell'ingegneria civile*, edited by P.G. Malerba, CISM Proceedings, Udine, Italy, 1995.
5. L. Binda , G. Gatti , G. Mangano, C. Poggi, G. Sacchi Landriani, "The Collapse of the Civic Tower of Pavia: a Survey of the Materials and Structure, *Masonry International*, **6** (1), 11-20 (1992).
5. G. Ballio, "Structural Preservation of the Fraccaro tower in Pavia", *Structural Engineering International*, **3** (1), 9-11 (1993).
6. L. Binda, G. Mirabella Roberti, C. Poggi, R. Tondini Folli, M. Falco, R. Corradi, "Static and Dynamic Studies on the Torrazzo in Cremona (Italy): the Highest Masonry Bell Tower in Europe", *Proceedings IASS – MSU Int. Symp. Bridging Large Spans from the Antiquity to the Present*, Istanbul, Turkey, 2002, pp. 100-110.
7. F. Benedettini, C. Gentile, "Ambient Vibration Testing and Operational Modal Analysis of a Masonry Tower", *Proceedings 3rd IOMAC Conference*, Copenhagen, Denmark, 2007.
8. A. Carpinteri, G., Lacidogna G., "Damage Evaluation of Three Masonry Towers by Acoustic Emission", *Engineering Structures*, **29**, 1569-1579 (2007).
9. M. R. Valuzzi, L. Binda, C. Modena, "Mechanical Behavior of Historic Masonry Structures Strengthened by Bed Joints Structural Repointing", *Construction and Building Materials*, **19**, 63-73 (2005).
10. F. Auricchio, L. Faravelli, G. Magonette, V. Torra, *Shape Memory Alloy. Advances in Modelling and Applications*, CIMNE, Barcelona, Spain, 2001.
11. www.comunedipositano.it
12. A. G. El Attar, A. M. Saleh, A. H. Zaghw, Conservation of Slender Historical Mamluk-style Minaret by Passive Control Techniques, *Structural Control &Health Monitoring*, **12** (2), 157-177 (2005).
13. C. Syrmakezis, Seismic Protection of Historical Structures and Monuments, *Structural Control & Health Monitoring*, **13** (6), 958-979 (2006).
14. A. El-Attar, A. Saleh, I. El-Habbal, A.H. Zaghw, and A. Osman , "The Use of SMA Wire Dampers to Enhance the Seismic Performance of Two Historical Islamic Minarets", *Smart Structures & Systems*, **4** (2), 221-232 (2008).
15. S. Camesasca, S. Candusso, "Diagnosi dello stato di conservazione della Torre Civica di Soncino. Proposta di restauro" (in Italian), Polytechnic of Milan, Master Thesis, 2005.
16. A. Marcellini, A. Tento, R. Daminelli, V. Poggi, "Tecniche sperimentali per la valutazione simultanea delle caratteristiche dinamiche di suoli e strutture", Institute for the Dynamics of Environment Processes, National Research Council (CNR), Milan, 2005.
17. M. Beccari, "Monitoraggio e identificazione della Torre Civica di Soncino" (in Italian), Dept. Of Structural Mechanics, University of Pavia, Master Thesis, 2006.
18. S. Persano, "La Torre Civica di Soncino: modellazione e diagnostica" (in Italian), Dept. Of Structural Mechanics, University of Pavia, Master Thesis, 2006.
19. S. Casciati, K. Hamdaoui, "Optimal Robust Design via SMA Dissipative Devices", *Proceedings ICASP10, 10th International Conference on Applications of Statistics and Probability in Civil Engineering.* July 31-August 3, Tokyo, Japan, 2007.
20. S. Casciati , "Monitoring Data for the Structural Assessment of Historical Buildings", *Proceedings of SHM2008*, Cracow, 2008.
21. S. Casciati, K. Hamdaoui, Experimental and numerical studies toward the implementation of shape memory alloy ties in masonry structures, *Smart Structures & Systems*, **4** (2), 153-170, 2008
22. F. Casciati, L. Faravelli, *Fragility Analysis of Complex Structural Systems*, Research Studies Press, Taunton, UK, 1991.
23. V. Sepe, E. Speranza, A. Viskovic A., "A Method for Large-Scale Vulnerability Assessment of Historic Towers, *Structural Control & Health Monitoring*, **15** (3), 389-415 (2008)
24. L. Faravelli, S. Casciati, "Vulnerability Assessment for Medieval Civic Towers", accepted for publication in *Structure and Infrastructure Engineering*.
25. Camerini S., "Adeguamento strutturale della torre civica di Soncino e curve di fragilità", (in Italian) Dept. of Structural Mechanics, University of Pavia, Master Thesis, 2008.

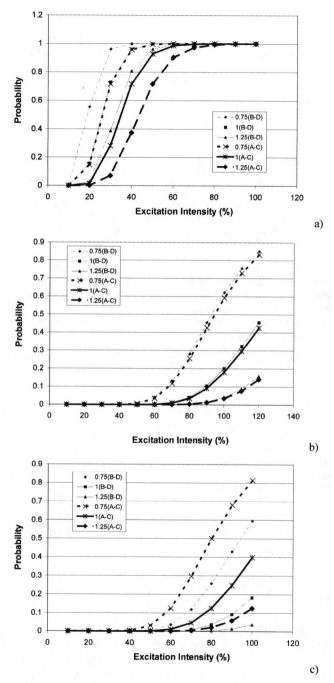

FIGURE 5. Vulnerability curves obtained with the excitation in one of the two directions (either B-D or A-C) and the von Mises stress as limit state: (a) damaged model; (b) retrofit solution 1; and (c) retrofit solution 2.

Advanced Techniques for Seismic Protection of Historical Buildings: Experimental and Numerical Approach

Federico M. Mazzolani

Department of Structural Engineering, University of Naples "Federico II"
Piazzale Tecchio, 80, 80125, Naples, Italy

Abstract. The seismic protection of historical and monumental buildings, namely dating back from the ancient age up to the 20th Century, is being looked at with greater and greater interest, above all in the Euro-Mediterranean area, its cultural heritage being strongly susceptible to undergo severe damage or even collapse due to earthquake. The cultural importance of historical and monumental constructions limits, in many cases, the possibility to upgrade them from the seismic point of view, due to the fear of using intervention techniques which could have detrimental effects on their cultural value. Consequently, a great interest is growing in the development of sustainable methodologies for the use of Reversible Mixed Technologies (RMTs) in the seismic protection of the existing constructions. RMTs, in fact, are conceived for exploiting the peculiarities of innovative materials and special devices, and they allow ease of removal when necessary. This paper deals with the experimental and numerical studies, framed within the EC PROHITECH research project, on the application of RMTs to the historical and monumental constructions mainly belonging to the cultural heritage of the Euro-Mediterranean area. The experimental tests and the numerical analyses are carried out at five different levels, namely full scale models, large scale models, sub-systems, devices, materials and elements.

Keywords: Seismic protection, historical buildings, reversible systems, mixed technologies.

INTRODUCTION

This paper illustrates the ongoing experimental and numerical parallel campaigns aimed at studying and developing Reversible Mixed Technologies (RMTs) for the seismic protection of historical and monumental buildings. RMTs are based on the integration of structural members of different materials and/or construction methods into a single constructional organism. The basic feature of RMTs is that their application should be always completely recoverable, that is reversible, if required. This is considered as an essential design requirement in order to prevent historical and monumental buildings from unsuitable rehabilitation operations. The main aim of RMTs is the best exploitation of material and technology features, in order to optimize the structural behaviour under any condition, including very severe limit states produced by strong seismic actions.

The activity described in this paper is carried out within the PROHITECH ("Earthquake protection of historical buildings by reversible mixed technologies") project, a scientific research project involving sixteen academic institutions coming from twelve Countries belonging to the Euro-Mediterranean area (Italy, Algeria,

CP1020, *2008 Seismic Engineering Conference Commemorating the 1908 Messina and Reggio Calabria Earthquake,*
edited by A. Santini and N. Moraci

Belgium, Egypt, Macedonia, Greece, Israel, Morocco, Portugal, Romania, Slovenia, Turkey). The PROHITECH research project is framed within the INCO thematic areas, devoted to "Protection and conservation of cultural heritage", and its duration is of four years, since October 1, 2004 to September 30, 2008. The scientific activity is subdivided into four parts, and the object of this paper mainly belongs to the third part of the research. Details on the whole PROHITECH project can be found in Mazzolani [1-5], and on the website www.prohitech.com.

EXPERIMENTAL ACTIVITY

General

The experimental analyses described in this section are carried out with the objective of giving a suitable experimental contribution to the assessment and set-up of new mixed techniques for repairing and strengthening of historical buildings and monuments belonging to the Cultural Heritage of the Mediterranean basin. The experimental activity is developed at five different levels, namely full scale building, large scale models, sub-systems, full devices, materials and elements, and it is dealt with in the following sub-sections.

Full Scale Tests

The full scale experimental tests are referred to the following constructions: a reinforced concrete building located in the Bagnoli area in Naples (Italy); the Mustafa Pasha Mosque in Skopje (Macedonia); the Gothic Cathedral in Fossanova (Italy); the Byzantine St. Nikola Church in Psacha, Kriva Palanka (Macedonia); the Beylerbeyi Palace in Istanbul (Turkey).

The experimental studies carried out at the University of Naples "Federico II" on the Bagnoli r.c. building (Fig. 1a) have been extremely exhaustive and detailed. This building, in fact, is not an "ad hoc" built model but it is a "real" construction, actually representative of a large part of the building stock present in many Countries during the 20th Century, and so it represents a unique occasion of knowledge of wide interest.

(a)   (b)

FIGURE 1. The r.c. building in the Bagnoli area, Naples, Italy: (a) the original building; (b) the reaction structure for carrying out the push-pull cyclic tests on the building.

After preliminary tests on the materials, aimed at characterizing them from the mechanical point of view, the dynamic identification of the structure has been carried out [6]. In order to perform inelastic cyclic tests under lateral loading conditions, a special steel structure has been designed and realized (Fig. 1b), which allows to alternately push and pull the construction up to reach pre-fixed horizontal displacement values. The experimental tests have been carried out in three phases. In the first phase, the original structure has been strongly damaged by applying a seismic input corresponding to a return period of more than three thousand years (Fig. 2a). In the second phase, it has been repaired by means of FRP bars placed in the mortar joints of the external walls (Fig. 2b) and damaged again [7]. At last, in the third phase, an intervention by means of buckling restrained braces (BRBs) has been carried out (Fig. 2c), with subsequent further tests [8].

 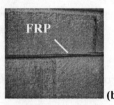 

(a) (b) (c)

FIGURE 2. (a) Damage in the full scale building after the test on the original structure; (b) first repair intervention, by means of FRP bars; (c) second repair intervention, by means of BRBs.

The full scale experimental work on the other above mentioned buildings, say the Mustafa Pasha mosque, the Gothic Cathedral in Fossanova, the St. Nikola Church in Psacha, Kriva Palanka, and the Beylerbeyi Palace in Istanbul, have been non-destructive tests, mainly focused on the characterization of the structural materials and on the dynamic identification of the constructions.

Large Scale Tests

The programme of large scale tests is now under development. It includes experiments on the following models: Mustafa Pasha Mosque in Skopje; Fossanova Gothic Cathedral; Greek Temple; St. Nikola Byzantine Church in Psacha.

The large scale model of the Mustafa Pasha Mosque (scale 1:5) has been tested on shaking table (Fig. 3) at the IZIIS Laboratory in Skopje, Macedonia [9], [10]. The seismic shaking table testing was performed in three main phases: 1) Testing of the original model under low intensity level, with the aim to induce damage in the minaret only; 2) Testing of the model with strengthened minaret under intensive earthquakes, with the aim to provoke the collapse of the minaret and damage to the mosque; 3) Testing of strengthened mosque model until reaching heavy damage.

(a) (b) (c)

FIGURE 3. Shaking table tests on the Mustafa Pasha Mosque large scale model (Skopje, Macedonia): (a) before testing; (b) after the consolidation of the minaret; (c) after the consolidation of the mosque.

With regard to the tests on the large scale models of a Greek temple (scale 1:3 with respect to the columns of Parthenon), two series of experiments have been conducted on the shaking table of the Earthquake Engineering Laboratory of the National Technical University of Athens (NTUA), namely three experiments on three freestanding columns in a row (Fig. 4a) and three experiments on three columns in a row with architraves (Fig. 4b); in addition, tests on columns with architrave in angle configuration are planned.

In addition to the above tests, which have already been carried out, large scale models of the Fossanova Gothic Cathedral (scale 1:5.5 – Fig. 4c) and the St. Nikola Byzantine Church (scale 1:3.5 – Fig. 4d) are going to be tested at the Skopje IZIIS Laboratory.

   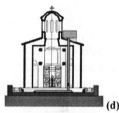

(a)   (b)   (c)   (d)

FIGURE 4. Experimental set-up in Athens (Greece) of three columns in a row: (a) freestanding columns, (b) with architraves; (c) large scale model of Fossanova Cathedral in Skopje, Macedonia; (d) design of the experimental model of St. Nikola Church.

Tests on Sub-Systems

A large number of experimental tests on sub-systems has been carried out or is in phase of development within the project. The tests are mainly related to the application of RMTs to masonry, timber, reinforced concrete and iron structures.

With regard to masonry panels, two main groups of experiments have been carried out. The first group, at the "Politehnica" University of Timisoara (Romania), deals with the behaviour of masonry panels consolidated by means of metal (steel or aluminium) sheeting plates or steel wire mesh, which are applied at the external faces of the panel. In this case, the connection of the metal sheet plates to the masonry wall can be realised in two ways, namely by means of chemical anchors or pre-stressed ties; the wire mesh is glued to the masonry wall by using epoxy resin. The second group, at the University of Naples "Federico II" (Italy), deals with the behaviour of masonry walls strengthened by FRP bars, which are located in the mortar joints. These tests are aimed at investigating the behaviour of such masonry walls in three conditions, namely in absence of the retrofitting system, in presence on the FRP bars at one side of the wall only, and in presence of the FRP bars placed at both sides of the masonry wall.

Also to the experimental activity on timber sub-systems is developed in two main groups of tests. The first group is related to the tests carried out on timber composite beams and floors, at the University of Naples "Federico II" (Italy) and at the Istituto Superiòr Tecnico of Lisbon (Portugal), both systems being based on an innovative technological system useful for connecting timber elements and concrete slabs (Fig.

5a) [11]. The second group of tests is carried out at the Boğaziçi University of Istanbul and it is related to the study of the behaviour of timber frames equipped by means of metal shear panels (Fig. 5b).

(a)

(b)

FIGURE 5. Tests on timber sub-systems: (a) composite timber-concrete beam (Lisbon, Portugal); (b) timber frame retrofitted by a metal shear panel (Istanbul, Turkey).

At last, experimental tests on reinforced concrete columns retrofitted by means of three techniques, namely r.c. jacketing, the FRP jacketing, and steel jacketing, have been carried out at the Technical University of Bucharest (Istanbul), whereas tests on iron elements retrofitted by means of FRPs are in phase of development at the University of Liège (Belgium) and at the University of Naples – Architecture Faculty (Italy).

Tests on Devices

The experimental investigation of the innovative devices is aimed at characterizing the cyclic performances of the Reversible Mixed Technologies developed within the project, in order to optimize their use in the seismic protection of historical and monumental buildings.

A wide campaign on the cyclic behaviour of pure aluminium shear panels is currently in progress at the University of Naples "Federico II" (Italy) and at the University of Chieti-Pescara "G. D'Annunzio" (Italy) [12], [13]. The experimental tests have been carried out on both full bay and bracing type pure aluminium shear panels (Figs. 6a, b). In particular, four full bay and four bracing type specimens have been considered. For both groups of experiments, the main differences among the tested systems are related to the presence of adequate stiffening ribs on the panels and to the connection (bolted or welded) between the ribs and the panels.

The basic innovative devices used for the realization of composite timber-steel-concrete elements, have been subjected to extensive experimental investigations (Figs. 6c, d) at the University of Naples "Federico II" (Italy) and at the Istituto Superiòr Tecnico of Lisbon (Portugal).

A special dissipative beam-to-column node has been conceived at the University of Naples "Federico II" (Italy), and it has been subjected to experimental investigations devoted to evaluate the node capability to dissipate the input seismic energy by a torque mechanism in metal elements placed in the nodal area (Fig. 6e).

For the connection of marble elements, special steel anchors in marble have been studied at the Technical University of Athens (Greece), by performing pull-out tests on threaded reinforcement bars which are installed in drilled holes and connected to the marble by means of a suitable cementitius material. Moreover, special metallic devices for the connection of marble architraves have been tested at the University of Ljubljana (Slovenia), in order to assess the effectiveness of such innovative system in linking marble blocks each other.

Further experiments on devices are referred to iron connections (University of Liège, Belgium, together with University of Naples-Architecture Faculty, Italy), to magneto-rheological devices at the Second University of Naples (Aversa, Italy), and to DC90 dampers at the University of Ljubljana (Slovenia).

(a) (b) (c) (d) (e)

FIGURE 6. Tests on devices: (a) full bay type and (b) bracing type pure aluminium shear panels (Naples, Italy); timber-steel-concrete connections (c) in presence of concrete elements (Lisbon, Portugal), and (d) in absence of concrete (Naples, Italy); (e) dissipative beam-to-column wooden connection (Naples, Italy).

Tests on Materials and Elements

The tests on materials and elements have represented the basis for all the experimental analyses carried out at different scales, as previously described. Simple elements and materials have been characterized from the mechanical point of view, so allowing the correct interpretation of the experimental results coming from the tests at larger scales.

Experimental campaigns have been performed, in particular, on elements made of: adobe (Rabat, Morocco), bricks with mortar (Skopje, Macedonia), stone (Rabat, Morocco and Algiers, Algeria), marble and limestone (Ljubljana, Slovenia and Athens, Greece), iron (Liège, Belgium and Naples, Italy), aluminium (Naples, Italy), timber (Istanbul, Turkey and Naples, Italy), and concrete (Bucharest, Romania).

NUMERICAL ACTIVITY

General

The numerical analyses represent the counterpart of the experimental tests described in the above section, since most of them are focused on models of the experimented test specimens. Consequently, also for the numerical analyses, the activity is developed at five levels, from full scale building to materials and elements. The clear aim of this activity is the set up of reliable numerical investigation tools, useful for both studying aspects difficult to catch in the experimental tests and providing the basis for the set up of calculation models adequate for historical buildings retrofitted by Reversible Mixed Techniques.

Full Scale Model

The whole Bagnoli r.c. building, already described in the section on full scale experimental tests, has been modeled by means of the non-linear finite element program SAP2000 at the University of Naples "Federico II" [8]. In the numerical model (Fig. 7a) the presence of the innovative BRB retrofitting system has been taken into account, and the numerical results, from the static non-linear analysis of the building, have well matched the experimental ones.

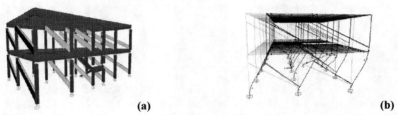

(a)   (b)

FIGURE 7. The numerical model of the full scale building: (a) geometry; (b) column-sway mechanism.

Large Scale Models

Pre- and post-experimental numerical analyses, devoted to support the development of advanced analytical models, have been performed for the large scale models interested also by the experimental tests. The Mustafa Pasha Mosque has been modelled in cooperation between the University of Naples "Federico II"-Architecture Faculty and the University of Skopje "Sts. Cyril and Methodius" [14] (Fig. 8a). The Gothic Cathedral of Fossanova has been modelled at the University of Chieti-Pescara "G. d'Annunzio" (Fig. 8b). The model of the St. Nikola Church in V. Psacha has been set up at the University of Skopje "Sts. Cyril and Methodius" (Fig. 8c) and the Greek Temple has been modelled at the National Technical University of Athens (Fig. 8d).

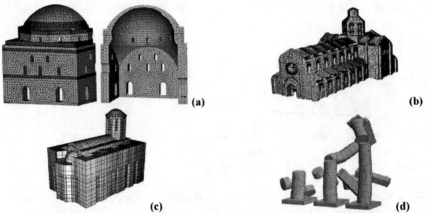

(a)   (b)

(c)   (d)

FIGURE 8. FEM models: (a) Mustafa Pasha Mosque; (b) Fossanova Gothic Cathedral; (c) St. Nikola Church; (d) Greek Temple.

Models of Sub-Systems

Numerical models aimed at investigating the behaviour of sub-systems endowed with RMTs have been set up or are in phase of development. Several retrofitted sub-systems have been considered in the study, namely: masonry walls and metal panels (Fig. 9a) ("Politehnica" University of Timisoara – Romania and University of Chieti-Pescara "G. d'Annunzio" – Italy), masonry walls and FRPs (University of Naples "Federico II" – Italy), timber frames and metal shear panels (Boğaziçi University of Istanbul – Turkey and University of Chieti-Pescara "G. d'Annunzio" – Italy), timber composite floors (Fig. 9b) (University of Naples "Federico II" – Italy and Istituto Superiòr Tecnico of Lisbon – Portugal), iron elements and FRPs (University of Liège – Belgium).

In all cases, pre- and post-experimental analyses are foreseen. The pre-experimental analyses are used for setting up the models and carrying out preliminary studies. The post-experimental analyses are developed in five phases, namely: the numerical simulation of the original specimens, the modelling of the strengthening devices, the simulation of the strengthened specimens, the comparison with the experimental results, and the calibration of numerical procedures for the analysis of the strengthened structural systems.

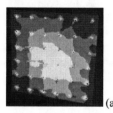 (a)  (b)

FIGURE 9. Numerical models: (a) masonry wall reinforced by metal panels; (b) timber composite floor.

Models of Devices

The activity concerned with the numerical modelling of innovative devices has been subdivided in several steps, following the same approach of the numerical analyses on sub-systems. In fact, after the pre-experimental analyses aimed at setting up adequate numerical models and carrying out preliminary investigations, the post-experimental analyses have been focused on the modelling of the devices, on the comparison of the numerical results with the experimental ones and, as a final step, on the calibration of numerical procedures for the analysis and the design of the devices.

The devices modelled numerically are: iron connections (University of Naples "Federico II" – Italy), architrave connections (University of Ljubljana – Slovenia and National Technical University of Athens – Greece), wood-to-concrete connectors (University of Naples "Federico II" – Italy and Istituto Superiòr Tecnico of Lisbon – Portugal), wooden node dissipative device (University of Naples "Federico II" – Italy) (Fig. 10), pure aluminium shear panels (University of Chieti-Pescara "G. d'Annunzio" – Italy) [15], [16], magnetorheological devices (Second University of Naples – Italy), and DC90 dampers (University of Ljubljana – Slovenia).

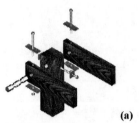

(a)   (b)

FIGURE 10. Dissipative beam-to-column wooden connection: (a) geometry ; (b) FEM model of the nodal metal element.

Models of Materials and Elements

The characterization of the behavioural features of the materials and the elements used for the construction of the models of devices, sub-systems, large scale and full scale systems has represented the basic issue for carrying out the numerical activities. In fact, the calibration of material and element simple numerical models is essential for the set up of more complicated models, from the modelling of the single devices to the numerical investigation of the complex behaviour of the full scale building retrofitted by reversible mixed systems.

In this perspective, numerical models of adobe, stone, marble, iron and aluminium have been set up. The models have preliminarily been used for carrying out pre-experimental numerical analyses; then, after the experimental tests, the numerical models have been calibrated, and numerical post-experimental analyses have been carried out with the aim of calibrating the numerical constitutive laws to be used in the models at larger scales.

CONCLUSIVE REMARKS

This paper presents a comprehensive overview of the experimental and numerical activities carried out within the PROHITECH research project. This project, which involves sixteen academic institutions coming from twelve Countries belonging to the European and Mediterranean area, is focused on the seismic protection of historical and monumental buildings by means of Reversible Mixed Technologies. The experimental and numerical investigations obviously represent the core part of the PROHITECH research project. Two preliminary parts already produced are related to the traditional intervention strategies in the field of seismic protection [17] and to the collection of information referred to innovative materials and devices [18], these two parts being preliminary to the experimental and numerical ones. The final part will deal with validation criteria illustrated also by selected study cases and a proposal of codification for the use of RMTs in the seismic protection of historical buildings.

The experimental and numerical analyses summarized in this paper have been developed at different levels and they represent an important contribution in the field of the seismic protection of historical and monumental constructions by RMTs, due to both the encouraging obtained results and the comprehensiveness of the whole activity, which will probably lead towards the set up of a proposal of codification well framed into the international standards for the seismic protection of constructions.

REFERENCES

1. F. M. Mazzolani, "Earthquake protection of historical buildings by reversible mixed technologies: the PROHITECH project", Proceedings of the Symposium on Damage and repair of historical and monumental buildings, Venice, Italy, 2005.
2. F. M. Mazzolani, "Earthquake protection of historical buildings by reversible mixed technologies", Keynote lecture in Proceedings of the 5th International Conference on the Behaviour of Steel Structures in Seismic Areas (STESSA 2006), Yokohama, Japan, 2006.
3. F. M. Mazzolani, "Earthquake protection of historical buildings by reversible mixed technologies: the PROHITECH project", Proceedings of the 7th European Conference "SAUVEUR" Safeguarded Cultural Heritage, Prague, Czech Republic, 2006.
4. F. M. Mazzolani, "Earthquake protection of historical buildings", Invited lecture, Reluis workshop, Salerno, Italy, 2007.
5. F. M. Mazzolani, "The PROHITECH research project", Proceedings of SAHC 2008 Conference, Bath, UK, 2008.
6. F. M. Mazzolani, M. D'Aniello and G. Della Corte, "Modal testing and dynamic identification of a two-story RC building", International Conference on Earthquake Engineering to mark 40 years of IZIIS, Skopje-Ohrid, Macedonia, 2005.
7. G. Della Corte, L. Fiorino and F. M. Mazzolani, "Lateral-loading tests on a real RC building including masonry infill panels with and without FRP strengthening", ASCE Journal of Materials in Civil Engineering, 2008, accepted for publication.
8. M. D'Aniello, G. Della Corte and F. M. Mazzolani, "A special type of buckling-restrained brace for seismic retrofitting of RC buildings: design and testing", Proceedings of XXI C.T.A. Conference, Catania, Italy, 2007.
9. L. Krstevska, L. Taskov, K. Gramatikov, R. Landolfo, O. Mammana, F. Portioli and F. M. Mazzolani, "Experimental and numerical investigations on the Mustafa Pasha Mosque large scale model", COST C26 Workshop, Prague, Czech Republic, 2007.
10. L. Krstevska, L. Taskov, K. Gramatikov, R. Landolfo, O. Mammana, F. Portioli and F. M. Mazzolani, "Shaking table tests on the large scale model of Mustafa Pasha Mosque without and with FRP", Proceedings of SAHC 2008 Conference, Bath, UK, 2008.
11. L. Calado, J. M. Proença, A. Panão, F. M. Mazzolani, B. Faggiano and A. Marzo, "Experimental analysis of rectangular shaped sleeve connectors for composite timber-steel-concrete floors: bending tests", Proceedings of SAHC 2008 Conference, Bath, UK, 2008.
12. G. De Matteis, F. M. Mazzolani and S. Panico, "Pure aluminium shear panels as dissipative devices in moment-resisting steel frames", Earthquake Engineering and Structural Dynamics, Wiley InterScience, DOI: 10.21002/eqe, vol. 36: 841-859, 2007.
13. G. De Matteis, G. Brando, S. Panico, F. M. Mazzolani, "Bracing type pure aluminium stiffened shear panels: an experimental study", International Journal of Advanced Steel Construction, accepted for publication, 2008.
14. R. Landolfo, F. Portioli, O. Mammana, and F. M. Mazzolani, "Finite element and limit analysis of the large scale model of Mustafa Pasha Mosque in Skopje strengthened with FRP", Proceedings of APFIS 2007 Conference, Hong Kong, China, 2007.
15. A. Formisano, F. M. Mazzolani, G. Brando, G. De Matteis, "Numerical evaluation of the hysteretic performance of pure aluminium shear panels", Proceedings of the 5th International Conference on the Behaviour of Steel Structures in Seismic Areas (STESSA 2006), Yokohama, Japan, 2006.
16. G. Brando, G. De Matteis, S. Panico, F. M. Mazzolani, "Prove cicliche su pannelli a taglio di alluminio puro: Modelli numerici", Proc. of the XXI Congresso C.T.A., Catania, Italy, 2007.
17. F.M. Mazzolani, G. De Matteis, A. Mandara, G. Altay Askar, D. Lungu. (eds). PROHITECH first main deliverable D-I "Assessment of intervention strategies for the seismic protection of historical building heritage in the Mediterranean basin", 2005. www.prohitech.com.
18. F.M. Mazzolani, G. De Matteis, L. Calado, D. Beg (eds). PROHITECH second main deliverable D-II "Reversible mixed technologies for seismic protection: guide to material and technology selection", 2006. www.prohitech.com.

Seismic Risk Mitigation of Historical Minarets Using SMA Wire Dampers

Adel G. El-Attar[a], Ahmed M. Saleh[b], and Islam R. El-Habbal[c]

[a]Structural Engineering Department, Cairo University, Giza, Egypt, adelattar291258@link.net
[b] Structural Engineering Department, Cairo University, Giza, Egypt, amsaleh001@yahoo.com
[c] Structural Engineering Department, Cairo University, Giza, Egypt, islam_elhabbal@yahoo.com

Abstract. This paper presents the results of a research program sponsored by the European Commission through project **WIND-CHIME** (**Wi**de Range **N**on-**IN**trusive **D**evices toward **C**onservation of **HI**storical Monuments in the **ME**diterranean Area), in which the possibility of using advanced seismic protection technologies to preserve historical monuments in the Mediterranean area is investigated. In the current research, two outstanding Egyptian Mamluk-Style minarets, are investigated. The first is the southern minaret of Al-Sultaniya (1340 A.D, 739 Hijri Date (H.D.)), the second is the minaret of Qusun minaret (1337 A.D, 736 H.D.), both located within the city of Cairo. Based on previous studies on the minarets by the authors, a seismic retrofit technique is proposed. The technique utilizes shape memory alloy (SMA) wires as dampers for the upper, more flexible, parts of the minarets in addition to vertical pre-stressing of the lower parts found to be prone to tensile cracking under ground excitation. The effectiveness of the proposed technique is numerically evaluated via non-linear transient dynamic analyses. The results indicate the effectiveness of the technique in mitigating the seismic hazard, demonstrated by the effective reduction in stresses and in dynamic response.

Keywords: Seismic risk; historical monuments; Shape-memory alloy; dampers; retrofit.

INTRODUCTION

Historically, Cairo city used to be named the city of one thousand minarets [1]. It possesses large inventory of historical Islamic minarets that date back to the early Islamic period (641A.D). Following the 1992 Dahshur earthquake, large numbers of these minarets were recorded to experience damage. Examining damage records, indicated that minarets built during the Mamluk period were the most severely hit. Irregular mass and stiffness distribution along their heights with large displayed stalactite carving made them more vulnerable to damage during earthquakes compared to other minaret styles.

Earlier work by the authors [2, 3] constituted the dynamic evaluation of two such historic minarets. Field measurements of the material and soil mechanical properties in addition to the ambient vibration measurements were made. Seismic performance of the two minarets was evaluated via the capacity spectrum method [2]. Also, a 1/16th scale model of one of the minarets was constructed at Cairo University Concrete Research Laboratory. Lab tests as well as numerical simulations were

CP1020, *2008 Seismic Engineering Conference Commemorating the 1908 Messina and Reggio Calabria Earthquake*,
edited by A. Santini and N. Moraci
© 2008 American Institute of Physics 978-0-7354-0542-4/08/$23.00

conducted under free vibration with and without SMA wire dampers. Indications on the effectiveness of using SMA wires as dampers to enhance the seismic performance of the minarets were drawn based on those tests [3].

In this work, numerical analyses of the two minarets are performed to explore the proper use of SMA wire dampers on the chosen minarets and to assess the effectiveness of such retrofitting technique.

STUDIED MINARETS

The first minaret is the southern minaret of Al-Sultaniya (1340 A.D, 739 H.D) located in El-Suyuti cemetery on the southern side of the Salah El-Din citadel in Cairo. It is now standing alone but seems to have been, previously, attached to a n unidentified structure. The minaret total height is 36.69 meters and has a 4.48 × 4.48 m rectangular base. The second minaret is the Qusun minaret (1337 A.D, 736 H.D). It is located about 30.0 meters from Al-Sultaniya minaret. The minaret is currently separated from the surrounding building and is directly resting on the ground. The total height of the minaret is 40.28 meters with a base rectangular shaft of about 5.54 × 5.20 m. Figure 1. shows recent photos of the two minarets.

(a) Al-Sultaniya Minaret   (b) Qusun Minaret

FIGURE 1. Recent photos of the chosen minarets

Field investigations were conducted to obtain the geometrical description of the minarets, actual material properties, soil conditions at the minarets' location and ambient vibration measurements. Details of the field and lab measurements as well as the dynamic characteristics of the minarets, can all be found in [2] and [3].

NON-LINEAR FINITE ELEMENT ANALYSIS

Three dimensional finite element models are built for each minaret via ANSYS 8.1 commercial package [4]. Care is taken to capture the architectural details of the minarets (wall construction, balconies, openings, etc.). Solid elements are used to model stone, filling material within walls, steel and SMA dampers. The stone and fill material are assigned non-linear material properties based on measured values [3]. Tables 1. and 2. describe the adopted material parameters.

The material model available in ANSYS 8.1 for SMA is utilized. Figure 2. shows the model, where the assigned material parameters are as follows:

1. σ_s^{AS}: Starting stress value for the forward phase transformation, taken 140 MPa;
2. σ_F^{AS}: Final stress value for the forward phase transformation, taken 270 MPa;
3. σ_s^{SA}: Starting stress value for the reverse phase transformation, taken 200 MPa;
4. σ_F^{SA}: Final stress value for the reverse phase transformation, taken 70 MPa;
5. ε_L: Maximum residual strain, taken 0.03 in the analysis;
6. α: Parameter accounting for material responses in tension and under compression, taken 0.27.

TABLE 1. Elastic material properties

Zone	Young's modulus (MPa)	Density (Kg/m³)	Poisson's ratio	Damping multiplier
Al-Sultaniya limestone in Zone-1	3.35×10^3	2172	0.21	0.02
Al-Sultaniya limestone in Zone-2	2.35×10^3	2172	0.21	0.02
Al-Sultaniya limestone in Zone-3	3.05×10^3	2172	0.21	0.02
Al-Sultaniya limestone in Zone-4	3.35×10^3	2172	0.21	0.02
Qusun limestone	3.35×10^3	2172	0.21	0.02
Fill material for both minarets	50	2000	0.2	0.005

TABLE 2. Non-Linear Material Parameters

Material	Shear Coeff. (Open/ Closed) crack	Tensile strength (MPa)	Comp. Strength (MPa)	Tension Cracking factor	Cohesion (MPa)	Angle of Friction (Deg.)	Flow Angle (Deg.)
Lime-stone	0.50/ 0.70	0.55	16.70	0.60	5.0	85	85
Fill	0.20/ 0.30	0.05	0.15	0.20	0.1	60	60

The minaret models were analyzed under the effect of three different earthquake records. In selecting these earthquake records, two main aspects were considered, namely, the frequency content of the record and the peak ground acceleration to peak ground velocity ratio (a/v). As such, the selected records were picked to cover a wide spectrum of frequencies and a/v ratios. The three records selected were: (a) the N-S component of 1940 Imperial valley earthquake recorded at El-Centro with high a/v ratio (a/v >1.2); (b) the N-S component of 1966 Parkfield earthquake recorded at Temblor with intermediate a/v ratio (1.2> a/v >0.8; and (c) the N-S component of 1985 Mexico earthquake recorded at Zihuatenejo with low a/v ratio (a/v <0.8). All three records were scaled down to a peak ground acceleration of 0.15g that corresponds to the maximum expected earthquake acceleration within Cairo city for a return period of 475 years [5].

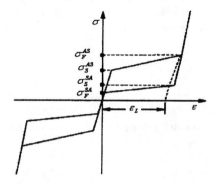

FIGURE 2. Idealization of super-elastic behavior of SMA at room temperature.

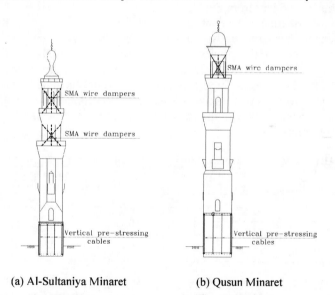

(a) Al-Sultaniya Minaret   (b) Qusun Minaret

FIGURE 3. Schematic of the proposed retrofitting

Two sets of analyses were performed, the first was to analyze the minarets at their current conditions then the second was to re-analyze after introducing the proposed retrofitting technique utilizing the SMA wire dampers. Different retrofitting options were considered using SMA wire dampers with or without vertical prestressing of the minaret body. The wire dampers served the purpose of reducing the structural

113

response while the prestressing helped alleviate the tensile stresses in the stone. It is to be noted that other researchers [7] have also employed combinations of SMA wires for seismic retrofitting of bell towers. In the following section, the results of analyzing the minarets in their current conditions as well as those of one retrofitting option are reported. The details of the proposed retrofit are shown in Fig. 3, where SMA wire-dampers are used in the upper lighter and more flexible parts of the minarets. For the lower bulky parts, it is found that an unrealistic amount of SMA material would be required to be effective in reducing the dynamic response. It is found sufficient to use vertical prestressing to reduce the tensile stresses in these parts.

RESULTS

The results are given in Fig. 4. for the cracking pattern, in Figs. 5. through 10. for the dynamic response and in Tables 3. through 6. for the maximum calculated stresses. As seen from Fig. 4, the minarets experienced many cracks along their heights. Al-Sultaniya minaret is seen to be more susceptible to cracking throughout its height, while in the minaret of Qusun, the cracking is more localized at the lower part, at changes of the minaret cross section and at the upper columns carrying the top dome. However, after introducing the proposed retrofitting, cracks only exist in the lower part of the minaret body in Al-Sultaniya minaret and in the upper columns of the Qusun minaret, as evident from Fig. 4.

In terms of maximum calculated stresses, Tables 3. to 6. give the maximum calculated tensile and compressive stresses in the stone. As seen, earthquakes with medium to high a/v ratio produce much higher stresses in the minarets than earthquakes dominated by low frequency. The proposed retrofit has an effect of reducing the tensile stresses in the minarets. It is to be noted that the vertical prestressing adds to the compressive stresses and accordingly no reduction in these stresses is observed, however the compressive stresses remain within the stone strength in both minarets.

TABLE 3. Maximum stresses (MPa) in Al-Sultaniya minaret before retrofitting

Level (m)	Maximum Tensile			Maximum Compressive		
	Elcentro	Parkfield	Mexico	Elcentro	Parkfield	Mexico
Bottom	0.55	0.55	0.55	11.13	12.69	10.71
	0.55	0.55	0.55	1.46	1.66	0.70
	0.55	0.55	0.55	0.51	0.58	0.48
	0.55	0.55	0.55	9.80	11.17	8.67
Top	0.55	0.55	0.55	12.96	14.77	11.48

TABLE 4. Maximum stresses (MPa) in Al-Sultaniya minaret after retrofitting

Level (m)	Maximum Tensile			Maximum Compressive		
	Elcentro	Parkfield	Mexico	Elcentro	Parkfield	Mexico
Bottom	0.43	0.33	0.30	11.34	12.90	10.91
	0.14	0.12	0.12	1.69	1.70	0.73
	0.12	0.10	0.07	0.55	0.61	0.51
	0.32	0.28	0.20	9.90	10.89	7.34
Top	0.44	0.39	0.35	10.00	11.00	8.65

TABLE 5. Maximum stresses (MPa) in Qusun minaret before retrofitting

Level	Maximum Tensile			Maximum Compressive		
(m)	Elcentro	Parkfield	Mexico	Elcentro	Parkfield	Mexico
Bottom	0.55	0.39	0.34	5.17	2.41	1.04
	0.55	0.29	0.27	3.35	1.56	1.10
	0.55	0.40	0.35	3.62	1.69	2.50
Top	0.55	0.55	0.55	8.90	4.16	4.00

TABLE 6. Maximum stresses (MPa) in Qusun minaret after retrofitting

Level	Maximum Tensile			Maximum Compressive		
(m)	Elcentro	Parkfield	Mexico	Elcentro	Parkfield	Mexico
Bottom	0.38	0.27	0.19	5.30	2.76	1.40
	0.29	0.21	0.20	3.10	1.61	1.50
	0.32	0.24	0.22	3.45	1.79	2.40
Top	0.41	0.45	0.42	8.50	4.42	3.85

In terms of the dynamic response of the minarets, Figs. 5. to 10. illustrate the acceleration of the top of the minarets for each ground motion. All the figures are scaled by the gravitational acceleration (g). As seen, the proposed retrofit results in considerably reducing the top most point acceleration.

(a) Al-Sultaniya Minaret (before/after)

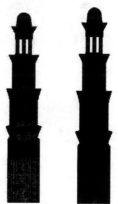
(b) Qusun Minaret (before/after)

FIGURE 4. Cracking Patterns in the minarets' bodies at current conditions (before) and after introducing the proposed retrofitting (after).

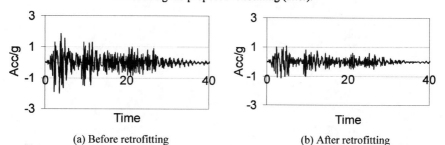

(a) Before retrofitting   (b) After retrofitting

FIGURE 5. Acceleration of Al-Sultaniya minaret top most point under Elcentro earthquake

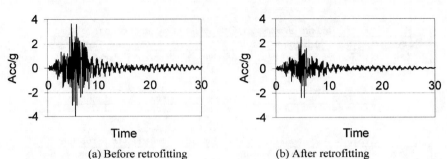

(a) Before retrofitting   (b) After retrofitting

FIGURE 6. Acceleration of Al-Sultaniya minaret top most point under Parkfield earthquake

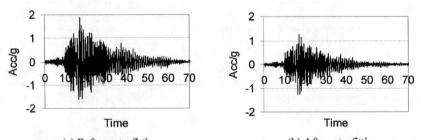

(a) Before retrofitting   (b) After retrofitting

FIGURE 7. Acceleration of Al-Sultaniya minaret top most point under Mexico earthquake

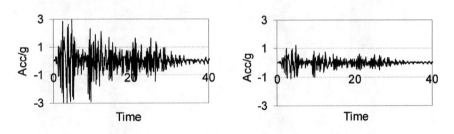

(a) Before retrofitting   (b) After retrofitting

FIGURE 8. Acceleration of Qusun minaret top most point under Elcentro earthquake

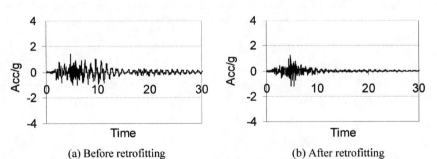

(a) Before retrofitting   (b) After retrofitting

FIGURE 9. Acceleration of Qusun minaret top most point under Parkfield earthquake

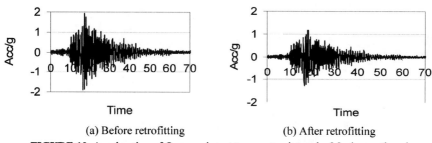

(a) Before retrofitting   (b) After retrofitting

FIGURE 10. Acceleration of Qusun minaret top most point under Mexico earthquake

CONCLUSIONS

A new retrofitting technique is proposed for mitigating the seismic risk of historical minarets belonging to the Mamluk style of the 14th century in Egypt. Two minarets were chosen for the study of the applicability of the technique. Numerical simulations are utilized for the study. The technique is based on the use of SMA dampers in the form of wires in combination with vertical prestressing. This application resulted in effectively reducing both the tensile stresses in the stone body as well as the dynamic response of the structures.

ACKNOWLEDGEMENT

This work is jointly funded by the European Commission through project **WIND-CHIME** (**W**ide Range Non-**IN**trusive **D**evices toward **C**onservation of **HI**storical Monuments in the **ME**diterranean Area).

REFERENCES

1. Abouseif, D. B. , The Minarets of Cairo, The American University in Cairo Press, 1987.
2. Saleh, A. M. and Zaghw, A. I., "Seismic Performance Evaluation of Two Historical Minarets Using the Capacity Spectrum Method," in Proceedings of the Third European Conference on Structural Control (3ECSC) 12-15 July 2004, Vienna University of Technology, Vienna, Austria.
3. El-Attar, A., Saleh, A., El-Habbal, I., Zaghw, A. and Osman, A., " The use of SMA wire dampers to enhance the seismic performance of two historical Islamic minarets," in Smart Structures and Systems, Vol. 4, No. 2-3, 2008.
4. Ansys 8.1 , 'Static and Dynamic Finite Element Analysis of Structures', 2003.
5. Egyptian Code for Calculating Loads and Forces in Structural and Building Works, Ministry of Housing, Infrastructure and Urban Communities, 2001.
6. Uniform Building Code, Vol. 2, International Conference of Building, Whittier, CA, 2002.
7. Castellano, M. G., Indirli, M., Azevedo, J. J., Tirclli, D., & Croci, G., "Seismic Protection of Cultural Heritage Using Shape Memory Alloy Devices", in Internationa! Posi-SMiRT Conference Seminar on Seismic Isolation, Passive Energy Dissipation and Active Control of Vibrations of Structures, Cheju, Korea, August 1999.

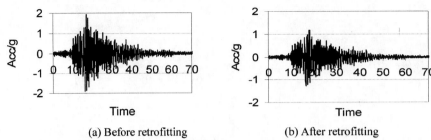

(a) Before retrofitting   (b) After retrofitting

FIGURE 10. Acceleration of Qusun minaret top most point under Mexico earthquake

CONCLUSIONS

A new retrofitting technique is proposed for mitigating the seismic risk of historical minarets belonging to the Mamluk style of the 14th century in Egypt. Two minarets were chosen for the study of the applicability of the technique. Numerical simulations are utilized for the study. The technique is based on the use of SMA dampers in the form of wires in combination with vertical prestressing. This application resulted in effectively reducing both the tensile stresses in the stone body as well as the dynamic response of the structures.

ACKNOWLEDGEMENT

This work is jointly funded by the European Commission through project **WIND-CHIME (W**ide Range Non-**IN**trusive **D**evices toward **C**onservation of **HI**storical Monuments in the **ME**diterranean Area).

REFERENCES

1. Abouseif, D. B. , The Minarets of Cairo, The American University in Cairo Press, 1987.
2. Saleh, A. M. and Zaghw, A. I., "Seismic Performance Evaluation of Two Historical Minarets Using the Capacity Spectrum Method," in Proceedings of the Third European Conference on Structural Control (3ECSC) 12-15 July 2004, Vienna University of Technology, Vienna, Austria.
3. El-Attar, A., Saleh, A., El-Habbal, I., Zaghw, A. and Osman, A., " The use of SMA wire dampers to enhance the seismic performance of two historical Islamic minarets," in Smart Structures and Systems, Vol. 4, No. 2-3, 2008.
4. Ansys 8.1 , 'Static and Dynamic Finite Element Analysis of Structures', 2003.
5. Egyptian Code for Calculating Loads and Forces in Structural and Building Works, Ministry of Housing, Infrastructure and Urban Communities, 2001.
6. Uniform Building Code, Vol. 2, International Conference of Building, Whittier, CA, 2002.
7. Castellano, M. G., Indirli, M., Azevedo, J. J., Tirclli, D., & Croci, G., "Seismic Protection of Cultural Heritage Using Shape Memory Alloy Devices", in Internationa! Posi-SMiRT Conference Seminar on Seismic Isolation, Passive Energy Dissipation and Active Control of Vibrations of Structures, Cheju, Korea, August 1999.

STRUCTURAL CONTROL AND MONITORING

Chair: Prof. Lucia Faravelli – University of Pavia, Italy

A System Identification and Change Detection Methodology for Stochastic Nonlinear Dynamic Systems

Hae-Bum Yun, Sami F. Masri and John P. Caffrey

University of Southern California. Los Angeles, CA, U.S.A.

Abstract. In this paper, a component-level detection methodology for system identification and change detection is discussed. The methodology is based on non-parametric, data-driven, stochastic system identification and classifications using statistical pattern recognition techniques. In order to validate the methodology discussed in this paper, an experimental study was performed using a complex nonlinear magneto-rheological (MR) damper. The results of this study show that the proposed methodology is very promising to detect and interpret changes in critical structural components, such as nonlinear springs and joints, as well as various types of dampers.

Keywords: structural health monitoring, magneto-rheological damper, non-parametric system identification, Restoring Force Method, support vector classification, error analysis

INTRODUCTION

The development of an effective Structural Health Monitoring (SHM) methodology is performed to avoid catastrophic structural failure by detecting structural deterioration and to reduce maintenance cost by establishing effective means and time schedules for structural maintenance and rehabilitation with predicted deterioration, based on detected structural changes over time. However, current methodologies of SHM have many limitations to be practical to deal with complicated structures often found in the fields of aerospace, mechanical and civil engineering.

In this paper, a component-level detection methodology for system identification and change detection is discussed. The methodology is based on non-parametric, data-driven, stochastic system identification and classification methods for complex nonlinear dynamic systems. The methodology was developed to achieve the following three goals: (1) it should be able to detect various types of structural changes. The changes include not only the changes in system parameter values but also changes into different classes of nonlinearity. The system changes are often relatively "small"; (2) it should be able to interpret the physical meanings of detected changes; and (3) it should be able to quantify the uncertainty of detected changes.

In order to validate the methodology discussed in this paper, an experimental study was performed using a complex nonlinear magneto-rheological (MR) damper.

CP1020, *2008 Seismic Engineering Conference Commemorating the 1908 Messina and Reggio Calabria Earthquake*,
edited by A. Santini and N. Moraci
© 2008 American Institute of Physics 978-0-7354-0542-4/08/$23.00

121

Employing the semi-active MR damper, a precise control in the statistics of the damper characteristics with user-controlled input current could be obtained. Although the MR damper was used for the experimental validation of the proposed methodology, the methodology is applicable to different types of nonlinear systems, such as nonlinear springs and joints, since the methodology is based on data-driven techniques. Using the Restoring Force Method, a non-parametric identification method based on two-dimensional Chebyshev polynomial series expansion, the complex nonlinear MR damper could be accurately identified. Once the damper was identified, the changes of the stochastic system characteristics could be detected and interpreted, using the statistical distributions of the identified Chebyshev coefficients. Statistical pattern recognition techniques, such as supervised support vector classification, were used to classify the different levels of system changes. An error analysis was also performed on the classification results to find the optimal strategies for the classifier design.

EXPERIMENTAL STUDIES

In order to perform experimental studies for change detection in uncertain nonlinear systems, a single degree-of-freedom (SDOF) magneto-rheological (MR) damper was used. MR dampers are semi-active energy dissipating devices [1-4]. The MR dampers typically consist of a piston rod, electromagnet, damper cylinder filled with MR fluid, accumulator, bearing and seal. The magnetic field generated with the electromagnet changes the characteristics of the MR fluid, which consists of small magnetic particles and fluid base. Consequently, the strength of the electromagnet's input current determines the physical characteristics of MR dampers.

To investigate the effects of system characteristics uncertainty using this semi-active device, performing a numerous series of tests is necessary. The MR damper was mounted on the actuator, controlling the damper displacement with a PID controller. The MR damper was fully instrumented with various sensors, including an LVDT, LVT, accelerometer, load cell (force), and thermocouple to measure damper surface temperature to prevent overheating, as shown in Figure 1.

The MR damper used in this study had very complicated nonlinearities: a hysteretic nonlinearity due to the viscous action of the MR fluid, combined with a dead-space nonlinearity due to a mechanical gap in the damper (Figure 1).

A series of tests were performed with eight different combination sets of mean (μ_I) and standard deviation (σ_I) of the MR damper input current (I): μ_I = 1.0 A, 0.8 A, 0.6 A and 0.4 A and σ_I = 0.1 A and 0.15 A. Therefore, with the MR damper input current, the effective (nominal) damper characteristics can be precisely controlled through μ_I and the uncertainty of the damper characteristics through σ_I. For each set of tests, 500 tests were performed. Consequently, a total of 4000 tests were conducted. The distributions of MR damper input currents used in the experimental studies are shown in Figure 2. The MR damper was subjected to broadband random excitation with the cutoff frequencies of 0.1 ~3.0 Hz.

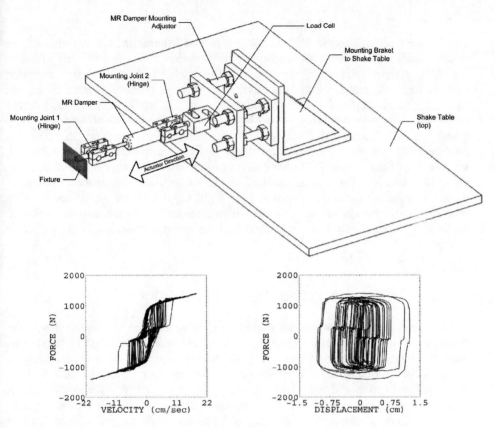

FIGURE 1. Magneto-rheological damper test apparatus phase diagrams of the damper subjected to sinusoidal excitation.

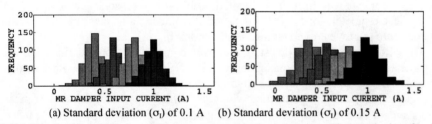

(a) Standard deviation (σ_I) of 0.1 A   (b) Standard deviation (σ_I) of 0.15 A

FIGURE 2. Distributions of MR damper input current used in the experiments. The sample size of each distribution is 500.

NON-PARAMETRIC RESTORING FORCE METHOD

The Restoring Force Method (RFM) is a non-parametric, data-driven system identification method for nonlinear dynamic systems [5]. Using two-dimensional normalized Chebyshev polynomial series expansion, the restoring force of a single-degree-of-freedom nonlinear system can be expressed as

$$r(x,\dot{x}) \approx \sum_{i=0}^{P}\sum_{j=0}^{Q} \overline{C}_{ij} T_i(\overline{x}) T_j(\overline{\dot{x}}) = \sum_{i=0}^{P}\sum_{j=0}^{Q} \overline{a}_{ij} \overline{x}^i \overline{\dot{x}}^j , \qquad (1)$$

where $r(x,\dot{x})$ is the nonlinear restoring force, $\overline{C}_{ij}$ is the orthogonal Chebyshev coefficient, $\overline{a}_{ij}$ is the non-orthogonal power-series coefficients, P and Q are the series expansion orders, and $\overline{x}$ and $\overline{\dot{x}}$ are the normalized displacement and velocity, respectively, in the range of [-1, 1]. For SHM purposes, Yun *et al.* showed in their experimental studies that the RFM has many advantages over other parametric and non-parametric identification methods [6]. Three different identification approaches, including parametric, non-parametric RFM, and non-parametric artificial neural networks (ANN) are compared in Table 1.

TABLE 1. Comparison of different identification approaches for structural health monitoring purposes.

	Advantages	Disadvantages
Parametric	- Most accurate if the exact system model is known - Direct physical interpretation is possible using identified parameters	- A priori knowledge of system required - Significant error in the parameter identification when the model is incorrect
Non-parametric RFM	- No assumption on the system - The same model is usable even if the system changes over time - Applicable to wide range of nonlinearity (polynomial type) - Physical interpretation is possible - Both Chebyshev and PS coefficients are identified	- Not applicable to some types of nonlinearity
Non-parametric ANN	- No assumption on the system - Applicable to wide range of nonlinearity - System change could be detected through comparison of the fitting error	- Change detection possible, physical interpretation not possible

SYSTEM IDENTIFICATION AND STATISTICAL CHANGE DETECTION FOR NONLINER MR DAMPER

The complex nonlinear MR damper was identified with different orders of series expansion of the RFM, and sample identification results with orders 5 and 20 are shown in Figure 3.

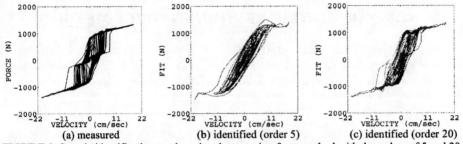

| (a) measured | (b) identified (order 5) | (c) identified (order 20) |

FIGURE 3. Sample identification results using the restoring force method with the orders of 5 and 20.

Once the MR damper was identified using 4000 data sets, the probability density functions of the identified normalized Chebyshev coefficients for eight test cases were obtained. Sample bivariate distributions of the identified Chebyshev coefficients for two dominant terms (the first-order damping and first-order stiffness terms) are shown in Figure 4. The figure illustrates that the identified Chebyshev coefficients can be used as excellent *features* to detect changes in nonlinear systems. In addition, the RFM is more advantageous than other non-parametric identification approaches since some level of physical interpretation is possible by looking at the types of identified coefficients. For example, Figure 4 shows that the decrease of MR damper restoring force with the decreasing input current (i.e., from $I = 1.0$ A to 0.4 A) is due to the decrease of both stiffness and damping-related characteristics of the damper.

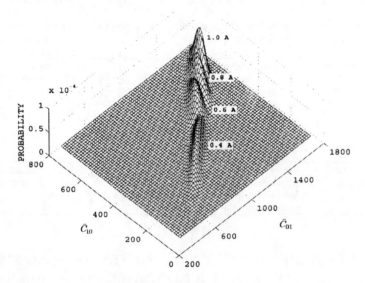

FIGURE 4. The probability density functions of the normalized Chebyshev coefficients.

SUPPORT VECTOR CLASSIFICATION FOR CHANGE DETECTION IN NONLINEAR SYSTEMS

The Support Vector algorithms are statistical learning techniques for various classification and regression problems. For the classification problems, the Support Vector Classifiers (SVC) have been successfully used for various system identification and damage-detection-related applications [7-13]. An excellent overview of SVC can be found in Scholkopf and Smola [14].

Figure 5 shows that the classification precision increases as the number of features increases for both Chebyshev and power series coefficients. The classification precision with Chebyshev coefficients, however, is always greater than that with power series coefficients for the same number of features. Consequently, using the orthogonal coefficients is more advantageous for change detection in nonlinear systems, due to better classification precision as well as reduced influence of model complexity.

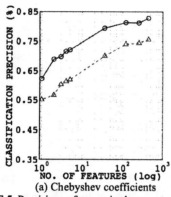
(a) Chebyshev coefficients

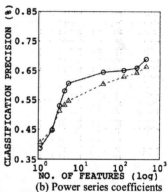
(b) Power series coefficients

FIGURE 5. Precisions of supervised support vector classification for different number of features. In the figure, the solid line with circular symbols is for the normalized Chebyshev coefficients, and the dashed line with triangular symbols for the normalized power series coefficients.

In order to detect "small" changes in the complicated nonlinear MR dampers with system parameter uncertainty, a series of analyses were conducted on the classification results of SVC to find the optimal strategy of designing classifiers to detect changes in nonlinear systems. Using the detection theory [15, 16], an error analysis was performed with the following null and alternative hypotheses:

$$H_0 : \text{The MR damper does NOT belong to this class} \qquad (2)$$
$$H_a : \text{The MR damper belongs to this class} \qquad (3)$$

In the MR damper change detection, two types of classification errors can be considered:

$$\text{Type-I error} : H_0 \text{ is rejected when } H_0 \text{ is true ("false alarm")} \qquad (5)$$
$$\text{Type-II error} : H_0 \text{ is accepted when } H_0 \text{ is false ("missed")} \qquad (6$$

Figure 6 shows that both Type-I and Type-II errors decrease with a larger number of features in the classification. For a given number of features, in general, there is a "trade-off" between Type-I and Type-II errors (i.e., if Type-I error increases, Type-II error decreases, and vice versa). For the purpose of SHM, minimizing Type-II errors would be more appropriate since the probability of "missing" significant damages can be reduced, and increasing "false-alarm" errors is more conservative in practical damage monitoring applications.

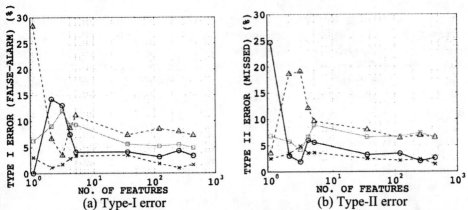

(a) Type-I error   (b) Type-II error

FIGURE 6. Type-I and Type-II errors of the support vector classification for different numbers of the normalized Chebyshev coefficients (features). In the classification, four classes of data are classified, including $\mu_l = 1.0$ A ($\circ$), 0.8 A ($\triangle$), 0.6 A ($\square$) and 0.4 A (X) with $\sigma_l = 0.1$ A.

CONCLUSIONS

A stochastic change detection methodology for complex nonlinear system was studied using a semi-active magneto-rheological (MR) damper. Using a non-parametric, data-driven restoring force method, the complicated nonlinear MR damper was accurately identified without a priori knowledge of the identified system. The support vector classification, a supervised statistical pattern recognition technique, was successfully applied to detect changes in the physical characteristics of the MR damper. The analysis results of change detection errors showed that the change detection classifiers should be designed on the basis of minimizing Type-II error ("missed" error) for reliable damage monitoring applications.

ACKNOWLEDGMENTS

This study was supported in part by grants from the U.S. National Science Foundation (NSF), the National Aeronautics and Space Administration (NASA), the Air Force Office of Scientific Research (AFOSR), and the California Department of Transportation (Caltrans).

REFERENCES

1. R. C. Ehrgott and S. F. Masri. Modelling of oscillatory dynamic behavior of electrorheological materials in shear. *Journal of Smart materials and structures*, 4:275-285, 1992.
2. R. C. Ehrgott and S. F. Masri. Structural control applications of an electrorheological device. *In Proceedings of the International Workshop on Structural Control*, pages 115-129, Los Angeles, California, 1994. USC Publication.
3. C.-W. Zhang, J.-P. Ou and J.-Q. Zhang, Parameter optimization and analysis of a vehicle suspension system controlled by magnetorheological fluid dampers. *Structural Control and Health Monitoring*, Volume 13, Issue 5, pp. 885-896, 2006.
4. P. Y. Lin, P. N. Roschke and C. H. Loh, Hybrid base-isolation with magnetorheological damper and fuzzy control, *Structural Control and Health Monitoring*, Volume 14, Issue 3, pp. 384-405, 2007.
5. S. F. Masri and T. K. Caughey. A nonparametric identification technique for nonlinear dynamic problems. Journal of Applied Mechanics, *Trans. ASME.*, 46(2):433-447, June 1979.
6. Hae-Bum Yun, Farzad Tasbihgoo, Sami F. Masri, John P. Caffrey, Raymond W. Wolfe, Nicos Makris, and Cameron Black. Comparison of modeling approaches for full-scale nonlinear viscous dampers. *Journal of Vibration and Control* Vol. 14, No. 1-2, 51-76 (2008).
7. K.Worden and A. J. Lane. Damage identification using support vector machines. *Smart Materials and Structures*, 10:540-547, 2001.
8. Junfeng Gao, Wengang Shi, Jianxun Tan, and Fengjin Zhong. Support Vector Machines based approach for fault diagnosis of valves in reciprocating pumps. *In Proceedings of the 2002 IEEE Canadian Conference on Electrical and Computer Engineering*, 2002.
9. Akira Mita and Hiromi Hagiwara. Damage diagnosis of a building structure using support vector machine and modal frequency patterns. In Shih-Chi Liu, editor, *Smart Systems and Nondestructive Evaluation for Civil Infrastructures*, 5057: 118-125, 2003.
10. S-H Park, C-B Yun, and Y. Roh. PZT-induced Lamb waves and pattern recognitions for on-line health monitoring of jointed steel plates. *In Proceedings of SPIE International Symposium on Smart Structures and Materials*, pages 364-375, 2005.
11. C. K. Oh and J. L. Beck. Sparse Baysian learning for structural health monitoring. *In 4th World Conference on Structural Control and Monitoring*, San Diego, 2006.
12. C-B Yun, S. Park, and S. Inman. Health monitoring of railroad tracks using PZT active sensors associated with Support Vector Machines. *In Proceedings of 4th China-Japan-US Symposium on Structural Control and Monitoring*, Hangzhou, China, 2006.
13. J. Zhang, T. Sato, and S. Iai. Incremental Support Vector Regression for non-linear hysteretic structural identification. *In 4th World Conference on Structural Control and Monitoring*, San Diego, CA., 2006.
14. Bernhard Scholkopf and Alexander J. Smola. *Learning with Kernels*. The MIT Press, 2002.
15. William Mendenhall and Terry Sincich. *Statistics for Engineering and the Sciences*, 4th Edition. Prentice Hall, New York, 1995.
16. Robert V. Hogg and Elliot A. Tanis. *Probability and Statistical Inference*, 5th Edition. Prentice Hall, New York, 1997.

Discrete-Time ARMAv Model-Based Optimal Sensor Placement

Wei Song and Shirley J. Dyke

Washington University in St. Louis, St. Louis, MO 63130, USA
weisong@wustl.edu, sdyke@seas.wustl.edu.

Abstract. This paper concentrates on the optimal sensor placement problem in ambient vibration based structural health monitoring. More specifically, the paper examines the covariance of estimated parameters during system identification using auto-regressive and moving average vector (ARMAv) model. By utilizing the discrete-time steady state Kalman filter, this paper realizes the structure's finite element (FE) model under broad-band white noise excitations using an ARMAv model. Based on the asymptotic distribution of the parameter estimates of the ARMAv model, both a theoretical closed form and a numerical estimate form of the covariance of the estimates are obtained. Introducing the information entropy (differential entropy) measure, as well as various matrix norms, this paper attempts to find a reasonable measure to the uncertainties embedded in the ARMAv model estimates. Thus, it is possible to select the optimal sensor placement that would lead to the smallest uncertainties during the ARMAv identification process. Two numerical examples are provided to demonstrate the methodology and compare the sensor placement results upon various measures.

Keywords: Structural health monitoring ARMAv optimal sensor placement Fisher information matrix   steady state Kalman filter

INTRODUCTION

Reliable estimates of the structural parameters are crucial to the success of structural health monitoring techniques. To estimate the current structural status as well as update the previous mathematical structural model, the estimates of the model parameters certainly require good quality. For ambient vibration based system identification techniques, the uncertainties involved will depend on several factors. With limited resources, manipulating the placement (location) of the sensors would have an effect on the variance of the estimates, and thus on the quality of the results. This paper is intended to answer the question, if the identification algorithm is ARMAv, what is the optimal sensor placement scheme so that the estimates have the smallest uncertainties.

The optimal sensor placement problem has been studied by many researchers [7-8, 12-13]. The approaches have evolved from case studies on the sensitivity of perturbing the modeling parameters, to rigorous application of multivariate estimation theory. Estimation theory indeed provides a sound explanation to the placement problem, that the covariance of parameter estimates has a lower bounded defined in terms of the inverse of Fisher information matrix [12]. Another feature of all the above approaches

CP1020, *2008 Seismic Engineering Conference Commemorating the 1908 Messina and Reggio Calabria Earthquake,*
edited by A. Santini and N. Moraci
© 2008 American Institute of Physics 978-0-7354-0542-4/08/$23.00

is that a specific formulation of the identification model and associated estimation method must be determined if one is to quantify the uncertainties.

In this paper, a different approach to solve the problem of optimal sensor placement is proposed based on the use of ARMAv model identification. As a widely used discrete-time dynamic system identification method, the ARMAv model has been applied in time-series analysis, signal processing, and structural identification [1, 9]. Using the prediction error method (PEM), the ARMAv parameter estimate is asymptotically normal and the covariances may be obtained analytically. Various matrix norms and measures are provided to interpret the covariance, and by minimizing certain measures, the optimal sensor placement scheme is selected. Genetic algorithms (GA) are used to solve the resulting discrete minimization problem.

Though the closed form of estimate covariance is obtained and used in the optimal sensor selection, it requires full knowledge prior to the identification process. So the scenarios to apply the analytical form are limited (Type II problem in next section). For structural health monitoring, only limited information are typically available for selecting the optimal sensor placement (Type I problem in next section). Therefore, the use of estimated form of the covariance is also proposed in this paper. In the numerical examples, both the analytical and estimated forms are applied to demonstrate the methodology.

BASIC ASSUMPTIONS

Before presenting the theory, we first discuss the assumptions required in the subsequent derivations, namely: the structure is assumed to be linear and time-invariant; the ambient excitation is considered to be a stationary, Gaussian white-noise, or colored noise which can be generated by a stationary Gaussian white-noise passing through a linear filter (in the latter case, we can obtain a new augmented system with white-noise as input); both the process noise ξ and the measurement noise η are present (see Eq. 3); the final ARMAv model has to be put in a predictor form, such as [5]

$$M(\theta): \hat{y}(k \mid \theta) = g\left(k, Z^{k-1}; \theta\right), \text{ with } \varepsilon(k, \theta) = y(k) - \hat{y}(k) \text{ are independent r.v.} \quad (1)$$

with Z^{k-1} representing all the information up to time instant $k-1$ and θ denoting the parameter set.

In practice, there are two types of sensor placement problems: Type I is structural damage diagnosis, and does not usually have a proper mathematical model of the structure to begin with; all structure properties are unknown, as well as the variance of the noise and input. In this case, the optimal sensor placement is obtained to get a more reliable estimate of structure model. The second type, Type II, is structural damage prognosis, which is based on a prior or current structural model and noise/ambient input information and optimal sensor placement is needed for better future monitoring. Numerical examples with Type II are the focus here and are shown in a later section.

MODEL FORMULATION

Even though civil engineering structures are, in general, temporally and spatially continuous, discretized models are still preferred when analyzing the digitized data obtained from distributed sensors. The finite element method (FEM) is usually used to formulate spatially discrete models. For discrete-time models, there are several popular techniques, such as zero-order hold, first-order hold of transfer function, bilinear approximation, and sample impulse response technique [6]. This paper adopts a covariance equivalence technique, because the purpose of this study is aimed at civil structures under broad-band ambient excitation, and the covariance equivalent model would guarantee that the covariance equals to the original system at each time sampling instant. If the governing differential equation of motion is given as

$$M\ddot{x} + C\dot{x} + Kx = f(t) \tag{2}$$

The corresponding covariance equivalent state space model is formulated as [4]

$$
\begin{aligned}
X(k+1) &= AX(k) + Bu(k) + \xi(k) \\
Y(k+1) &= CX(k) + \eta(k)
\end{aligned} \tag{3}
$$

with $X(k) = \begin{bmatrix} x(kT) \\ \dot{x}(kT) \end{bmatrix}$, $A = e^{A_c T}$, $B = \int_0^T e^{A_c \tau} d\tau B_c$, T is the sampling period and

$$
A_c = \begin{bmatrix} 0 & I \\ -M^{-1}K & -M^{-1}C \end{bmatrix} \text{ and } B_c = \begin{bmatrix} 0 \\ M^{-1} \end{bmatrix}. \tag{4}
$$

Here, matrix C is special, because it relates to sensor placement. It is an upper triangular matrix with each row containing all zeros but for one, which is unity. For example, $C_{m,n} = 1$ would imply that the m-th channel outputs the n-th states. However, because of the requirement in Eq. 1, a predictor form of the discrete-time state space model (Eq. 3) must be sought. Here the steady-state Kalman filter is applied to obtain such a predictor form [2], called the innovation state space model, expressed as

$$
\begin{aligned}
\hat{X}(k+1|k) &= A\hat{X}(k|k-1) + Le(k) \\
Y(k) &= C\hat{X}(k|k-1) + e(k)
\end{aligned} \tag{5}
$$

with $L = (A\Pi C^T + \Omega)\Sigma^{-1}$ as the Kalman gain, and e as white noise with covariance $\Sigma = C\Pi C^T + \Theta$. Matrices Λ, Ξ, Θ, and Ω represent the covariance of uu^T, $\xi\xi^T$, $\eta\eta^T$ and $\xi\eta^T$, respectively. Π is the solution to the algebraic Riccati equation given by

$$A\Pi A^T - \Pi + B\Lambda B^T + \Xi - L\Sigma L^T = 0 \tag{6}$$

131

Now we are ready to estimate the model.

ASYMPTOTIC DISTRIBUTION OF ARMAV ESTIMATES

The probabilistic model in (Eq. 1) can be estimated by maximum likelihood estimation (MLE), which is an asymptotic unbiased and efficient estimate. This approach implies that the model class M in (Eq. 1) can be estimated with a covariance that is at least equal to its Cramer-Rao lower bound [5, 12]. This bound can be written as (when $\varepsilon(k,\theta)$ is normally distributed with covariance Σ)

$$\text{COV}\left[\sqrt{N}\left(\hat{\theta}-\theta_0\right)\right] \geq \left[E\psi(\theta_0)\Sigma^{-1}\psi^T(\theta_0)\right]^{-1} = I_\theta^{-1} \tag{7}$$

with $\psi(\theta_0) = \dfrac{d}{d\theta}\hat{y}(\theta)|_{\theta=\theta_0}$, and the right hand side is actually the inverse of Fisher information matrix I_θ. However, for the multivariate case, the MLE is very difficult to calculate, and sometimes does not even exist. Therefore, the ARMAv model estimate is used as alternative. A general order (*na,nc*) ARMAv model is expressed as

$$y(k)+A_1 y(k-1)+\ldots+A_{na}y(k-na)=e(k)+C_1 e(k-1)+\ldots C_{nc}e(k-nc) \tag{8}$$

It is shown that if the ARMAv is globally identifiable, then with the prediction error method (PEM), the asymptotic distribution of its estimates converges to a Gaussian random variable [5]

$$\sqrt{N}\left(\hat{\theta}_{ARMAv}-\theta_0\right)\xrightarrow[N\to\infty]{D} N\left(0,I_\theta^{-1}\right) \tag{9}$$

where $\xrightarrow{D}$ means "converge by distribution."

To satisfy global identifiability, the following conditions should be justified:
1. A proper prediction error function has to be chosen.
2. Model class M in (Eq. 1) contains the 'true' model T with parameter θ_0.
3. The matrix polynomials $A_0(z)$ and $C_0(z)$ of the 'true' model T are left coprime with $A_0(z)=Iz^{na}+A_1^0 z^{na-1}+\ldots+A_{na}^0$, and $C_0(z)=Iz^{nc}+C_1^0 z^{nc-1}+\ldots+C_{nc}^0$.

In the next section we will present the condition to obtain the 'true' model based on the steady-state Kalman filter, and hence build the model class M.

ARMAV REALIZATION OF STATE SPACE MODEL

To build an ARMAv realization of the obtained steady state Kalman filter, complete observability is required. As shown by [3], if the number of states in the model is *n*, the number of sensors (outputs) is *m*, there always exists ARMAv realization that is restricted to an order with an integer value *p=n/m*, dimension *m*.

Upon obtaining the ARMAv realization of the steady state Kalman filter, which is considered as the 'true' system, we are ready to build a model class M, which can assume the similar structure of the realization. Note that throughout this paper, all the sensor placement cases discussed satisfy the above globally identifiability condition.

CALCULATE THE COVARIANCE OF ARMAV ESTIMATES

After obtaining the 'true' ARMAv model, it is possible to calculate the covariance of the parameter estimates. Based on (Eq. 7-9), define the parameter vector as

$$\theta = \text{vec}[A_1 \quad \cdots \quad A_{na} \quad C_1 \quad \cdots \quad C_{nc}] \tag{10}$$

where $\text{vec}[X]$ represents the vector formed by concatenating all the columns of X. Then the Fisher information matrix can be derived as [11]

$$I_\theta = E\psi(\theta_0)\Sigma^{-1}\psi^T(\theta_0) = E\phi\phi^T \otimes \Sigma^{-1} \tag{11}$$

where $\phi = \left[-y^T(k-1) \quad \cdots \quad -y^T(k-na) \quad e^T(k-1) \quad \cdots \quad e^T(k-nc)\right]^T$, and $\otimes$ is denoted as Kronecker product.

When solving the Type I problem, implying the covariances of the parameters in building the structure are unknown, then Eq. 11 can be approximated by the following estimate form

$$E\phi\phi^T \cong \frac{1}{N}\sum_{k=1}^{N}\phi(k)\phi^T(k), \text{ and } \Sigma \cong \frac{1}{N}\sum_{k=1}^{N}e(k)e^T(k) \tag{12}$$

When solving the Type II problem, Eq. 11 can be directly calculated by invoking the discrete-time Lyapunov equation of the steady-state Kalman filter shown in Eq. 5. Hence the closed form solution is obtained.

NUMERICAL EXAMPLES

Two numerical examples are designed to demonstrate the proposed methodology, and utilizing various matrix norms, to interpret the uncertainties from the covariance obtained in the previous section. It is a challenge to compare the covariances directly. The authors tried to apply the "positive semi-definite ordering"--- "$A \geq B$ if and only if $A - B$ is positive semi-definite", to identify the optimal sensor placement scheme. However, this approach was not successful. In the end, the information entropy (differential entropy) and some matrix norms are used to do the selection. Instead of measuring the covariance directly in Eq. 7, all the measures are applied to its inverse, the Fisher information matrix I_θ. Generally speaking, the larger the measure of I_θ is, the more information embedded in the estimate, and hence the less uncertainty. The below two examples are modified from the work by [7].

133

Shear Building Model

A nine story shear building model shown in Figure 1 is used to demonstrate the proposed methodology. The stiffness and mass are the same for each floor, 1.45E6 (N/m) and 1E3 (Kg), respectively. 5% modal damping for each mode is applied. The covariance matrices are $\Lambda = 2e4 \cdot I \,(\mathrm{N}^2)$, $\Xi = 1e\text{-}6 \cdot I \,(\mathrm{m}^2)$, $\Theta = 4e\text{-}10 \cdot I$ (m^2) and $\Omega = 0$. The sampling frequency is 200 Hz. The optimal sensor placement results calculated analytically from Eq. 11 are summarized in Table 1, and the results calculated by the estimated form (Eq. 12) are shown in Table 2. Only the placements related to identifiable ARMAv models are shown here. The "fully instrumented" case (all the nine displacements are measured) is also shown for comparison purpose.

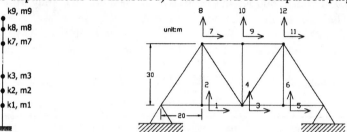

FIGURE 1. Shear Building   **FIGURE 2.** Truss Model (12 DOFs)

TABLE 1. Optimal Sensor Placement for Shear Building (based on analytical result)
The number in the bracket indicates the measure value of the corresponding placement, same hereafter

Number of sensors	Different measures of I_θ				
	Entropy	Trace	Frobenius	$\|A\|_1 = \|A\|_\infty$	$\|A\|_2$
2	4 8 (7.71E22)	7 9 (1684.85)	7 9 (835.86)	7 9 (711.88)	7 9 (616.07)
3	4 7 9 (3.59E29)	6 8 9 (2572.99)	6 8 9 (1017.49)	6 8 9 (703.58)	6 8 9 (622.51)
6	3 4 5 6 8 9 (5.74E35)	3 4 6 7 8 9 (4754.41)	2 5 6 7 8 9 (1109.74)	2 5 6 7 8 9 (511.52)	2 5 6 7 8 9 (446.35)
9	1 2 3 4 5 6 7 8 9 (8.21E36)	1 2 3 4 5 6 7 8 9 (6662.06)	1 2 3 4 5 6 7 8 9 (1092.78)	1 2 3 4 5 6 7 8 9 (390.64)	1 2 3 4 5 6 7 8 9 (332.62)

TABLE 2. Optimal Sensor Placement for Shear Building (based on estimated result)

Number of sensors	Different measures on I_θ				
	entropy	Trace	Frobenius	$\|A\|_1 = \|A\|_\infty$	$\|A\|_2$
2	4 8 (1.75E19)	7 9 (2135.28)	7 9 (1171.17)	7 9 (1000.44)	7 9 (881.83)
3	4 7 9 (4.43E23)	6 8 9 (3300.59)	6 8 9 (1441.55)	6 8 9 (997.20)	6 8 9 (93.30)
6	3 4 5 6 7 8 9 (1.862E??)	3 4 6 7 8 9 (5979.75)	2 5 6 7 8 9 (1566.22)	2 5 6 7 8 9 (794.86)	2 5 6 7 8 9 (669.63)
9	1 2 3 4 5 6 7 8 9 (2.369E28)	1 2 3 4 5 6 7 8 9 (8195.51)	1 2 3 4 5 6 7 8 9 (1528.28)	1 2 3 4 5 6 7 8 9 (622.81)	1 2 3 4 5 6 7 8 9 (509.66)

From Table 1 and 2, it is observed that the analytical results (Table 1) agree well with the estimated ones (Table 2). The difference lies in the entropy measure (shaded cell in Table 2). The reason is the condition of the Fisher information matrix obtained from estimate is not always very good, and thus affects the results on the quality of estimated entropy. Other matrix measures are very stable even by using estimated form. However, the entropy measure and trace norm indicate that when the number of sensors increases, the value of this measure increases, which implies that more information is involved. This coincides with common sense. Another observation is that, all the matrix norms yield the same optimal sensor placement, except the case with 6 sensors measured under trace norm.

Truss Model

The second example considers a truss model built with 13 elements, and 12 DOFs (Figure 2). The elastic modulus, cross sectional area and density for each element are 200 GPa, 0.05m^2, and 7800 Kg/m^3, respectively. 5% modal damping for each mode is applied. The covariance matrices are $\Lambda = 4e12 \cdot I$ (N^2), $\Xi = 1e-6 \cdot I$ (m^2), $\Theta = 4e-6 \cdot I$ (m^2), and $\Omega = 0$. The sampling frequency is 500 Hz. As in the previous example, both the analytical and estimated results are shown in Tables 3 and 4.

TABLE 3. Optimal Sensor Placements for Truss (based on analytical result)

Number of sensors	Different measures on I_θ				
	entropy	Trace	Frobenius	$\|A\|_1 = \|A\|_\infty$	$\|A\|_2$
3	9 10 11 (7.69E13)	2 8 10 (2265.16)	2 8 10 (891.84)	2 8 10 (789.34)	2 8 10 (600.88)
4	2 4 8 10 (1.36E15)	2 4 8 10 (3329.36)	2 4 8 10 (1217.72)	2 4 10 12 (883.46)	2 4 8 10 (679.11)
6	3 4 6 8 10 12 (8.54E15)	2 4 6 7 8 10 (4431.72)	2 4 7 8 10 12 (1151.79)	2 4 7 8 10 12 (785.32)	2 4 7 8 10 12 (552.15)
8	2 4 6 8 9 10 11 12 (5.76E16)	2 4 6 8 9 10 11 12 (5690.25)	2 4 6 8 9 10 11 12 (1205.92)	2 4 6 8 9 10 11 12 (707.04)	2 4 6 8 9 10 11 12 (480.01)
12	1~12 (1.27E22)	1~12 (7027.02)	1~12 (1105.81)	1~12 (545.15)	1~12 (348.04)

TABLE 4. Optimal Sensor Placement for Truss (based on estimated result)

Number of sensors	Different measures on I_θ				
	entropy	Trace	Frobenius	$\|A\|_1 = \|A\|_\infty$	$\|A\|_2$
3	2 8 10 (1.97E14)	2 8 10 (2.18E3)	2 8 10 (873.03)	6 10 12 (755.93)	2 8 10 (579.96)
4	2 4 8 10 (1.66E15)	2 4 8 10 (3.19E3)	2 4 8 10 (1.17E3)	4 6 8 10 (843.31)	2 4 8 10 (655.57)
6	2 4 7 8 10 12 (6.06E15)	2 4 6 8 10 11 (4.25E3)	2 4 7 8 10 12 (1.10E3)	2 4 7 8 10 12 (765.94)	2 4 6 7 8 10 (529.53)
8	2 4 6 8 9 10 11 12 (6.44E15)	2 4 6 8 9 10 11 12 (5.46E3)	2 4 6 8 9 10 11 12 (1.14E3)	2 4 6 7 8 9 10 12 (695.79)	2 4 6 7 8 9 10 12 (463.97)
12	1~12 (7.00E20)	1~12 (6.80E3)	1~12 (1.05E3)	1~12 (534.90)	1~12 (330.07)

It is noted that, because the structure is symmetric, symmetric placement of the obtained optimal sensor placement in the above tables are actually found to be optimal. For example, with 4 sensors, the optimal placement under the $\|A\|_1 = \|A\|_\infty$ measure is (2, 4, 10, 12) which is the same as (2, 4, 8, 10) due to symmetry. After taking into account the symmetry, by comparing Tables 1 and 2, it is also observed that there are several discrepancies in the shaded cells, which can be attributed to entropy sensitivity to the ill-conditioning of I_θ and to numerical errors. In reading the values computed for the various measures, the entropy and trace norm still possess the property that, the larger the number of sensors, the more information is involved.

CONCLUSIONS

The results presented in this paper are an extension of the work in [10]. By realizing the mathematical FE model of the structure as an ARMAv model, this paper investigates the analytical form of the Fisher information matrix for the estimate of model parameters. Thus, this paper proposes a closed form relationship to determine the covariance of the ARMAv parameters, as well as the estimated form. The optimal sensor placement scheme is obtained by selecting the smallest resulting uncertainty with certain measures. Through two numerical examples, various measures are applied to examine the information matrix obtained. The estimated results agree well with the analytical results. Also, the entropy and trace measures are found to be more meaningful measures of the information matrix through comparisons with other choices.

ACKNOWLEDGEMENTS

The financial support from the National Science Foundation grant nos. CMS-0625640 and 0245402 is gratefully acknowledged.

REFERENCES

1. G. Box, G. Jenkins and G. Reinsel, *Time Series Analysis: Forecasting & Control*, 3rd Edition, 1994.
2. J. Durbin, and S. J. Koopman, *Time Series Analysis by State Space Methods*, Oxford University Press, 2001.
3. W. Gawronski, and H. G. Natke, "On ARMA Models for Vibrating Systems", *Probabilistic Engineering Mechanics*, **1**(3), 150-156, (1986).
4. J.-N. Juang, *Applied System Identification*, Prentice Hall, 1993.
5. L. Ljung, *System Identification-Theory for The User*, 2nd ed., Prentice-Hall Inc, 1999.
6. A. V. Oppenheim, R. W. Schafer, and J. R. Buck, *Discrete-Time Signal Processing*, 2nd Ed., Pearson Education, 1999.
7. C. Papadimitriou, J. Beck, and S. Au, "Entropy-based optimal sensor location for structural model updating," *Journal of Vibration and Control*, **6** (5), 781-800, (2000).
8. Z. H. Qureshi, T. S. Ng, and G. C. Goodwin, "Optimum Experimental Design for Identification of Distributed Parameter Systems," *International Journal of Control*, **31**(1), 21-29, (1980).
9. W. Song, D. Giraldo, E. H. Clayton, S. J. Dyke, and J. Caicedo, "Application of ARMAV for modal identification of the Emerson bridge" *Proc. 3rd Intl. Conf. on Bridge Main., Safety and Manag.*, 2006.
10. W. Song, S. J. Dyke, "Entropy Based Optimal Sensor Placement for Discrete Time ARMAv Model Identification", *Proc. of Inaugural International Conference of the Engineering Mechanics Institute* (EM08), ASCE, MN, May 2008.
11. W. Song, S. J. Dyke, "Sensor Optimal Placement of Discrete-Time ARMAv Model", *In preparation*.
12. F. Udwadia, "Methodology for Optimum Sensor Locations for Parameter Identification in Dynamic Systems," *Journal of Engineering Mechanics*, **120**(2), 368-390, (1994).
13. F. E. Udwadia, and J. A. Garba, "Optimal Sensors Locations for Structural Identification," JPL *Proceedings of the Workshop on Identification and Control of Flexible Space Structures*, 247-261, (1985).

Recent Advances In Structural Vibration And Failure Mode Control In Mainland China: Theory, Experiments And Applications

Hui Li[a] and Jinping Ou[a,b]

[a] School of Civil Engineering, Harbin Institute of Technology ,Harbin, 150090, China
[b] School of Civil and Hydraulic Engineering, Dalian University of Technology, Dalian,116024, China

Abstract. A number of researchers have been focused on structural vibration control in the past three decades over the world and fruit achievements have been made. This paper introduces the recent advances in structural vibration control including passive, active and semiactive control in mainland China. Additionally, the co-author extends the structural vibration control to failure mode control. The research on the failure mode control is also involved in this paper. For passive control, this paper introduces full scale tests of buckling-restrained braces conducted to investigate the performance of the dampers and the second-editor of the Code of Seismic Design for Buildings. For active control, this paper introduces the HMD system for wind-induced vibration control of the Guangzhou TV tower. For semiactive control, the smart damping devices, algorithms for semi-active control, design methods and applications of semi-active control for structures are introduced in this paper. The failure mode control for bridges is also introduced.

Keywords: structural vibration control, failure mode control, recoverable structure; smart damping.

INTRODUCTION

In 1972, two-story buildings were constructed equipped with isolators, which is the first example to use structural vibration control technology in mainland China. However, more researchers focused on their interest in structural vibration control since 1981 after Wang (1981) introduced the concept of structural vibration control in mainland China [1]. A number of achievements associated with structural vibration control have been made in mainland China, which include passive energy dissipation technologies, active structural vibration control, semi-active control and smart control. Viscouselastic dampers, viscous dampers, metallic dampers, friction dampers, tuned mass dampers (TMD) and tuned liquid dampers (TLD) have been comprehensively studied during past two decades. Design methods of buildings incorporated with passive dampers have been proposed by Ou [2], which are adopted by the Code (GB5001-2001). In 2001, the new version of Code for Seismic Design of Buildings (GB 50011-2001) was published and the passive energy dissipation technology was the first time to be involved in the Code. After that, more and more new and existing buildings are incorporated with passive dampers for improvement or retrofit of seismic

CP1020, *2008 Seismic Engineering Conference Commemorating the 1908 Messina and Reggio Calabria Earthquake*,
edited by A. Santini and N. Moraci
© 2008 American Institute of Physics 978-0-7354-0542-4/08/$23.00

performance. Recent years, more and more full-scale tests of passive dampers have been conducted to investigate the damper performance.

Active control algorithms for linear and nonlinear structures have been studied by Chinese researchers [3,4]. Considering that the power system is more complicated for driving active mass damper, Ou (2003) proposed an electro-magnetic driving active mass damper (EAMD) and the performance of EAMD has been systematically studied [3]. Recently, an HMD (TMD combination with AMD) is designing for the Guangzhou TV tower to suppress the wind-induced vibration, which is the first time of active structural vibration control system to be used in the practical structure in mainland China.

Magneto-rheological fluid (MR) smart dampers, PZT-based actuators, shape memory alloy (SMA)-based dampers and magnetostrictive smart dampers have been studied and MR dampers have been implemented into bridges to suppress the wind-induced vibration.

This paper summarizes recent advances mentioned above.

PASSIVE ENERGY DISSIPATION TECHNLOGIES

Full scale performance tests of all-steel buckling-restrained brace

The buckling-restrained brace (BRB) consists of an inner steel core surrounded by an outer encasing member, in which the former carries the axial load, while the later provides lateral support to the core and prevents it from buckling in compression at the target lateral displacement. Therefore, BRB is able to provide stable hysteretic energy dissipation through the hysteretic yielding of the steel core.

Due to its outstanding seismic behavior, BRB has been extensively used in Japan, USA and Taiwan. However, in mainland China, research related with BRB is so insufficient.

Most of BRB consist of a conventional brace encased in a steel tube filled with mortar or concrete. This configuration is little complicated and bored. All-steel BRB with cruciform cross section encased in a square steel tube is relatively simply fabricated. This kind of BRBs is designed to be incorporated into the Beijing Tonghui Jiayuan for improvement of its seismic resistance. The Beijing Tonghui Jiayuan is a six-story office building located over the subway station and may suffer from strong earthquake attacks. Because the performance of all-steel BRBs remains insufficiently examined, full-scale performance tests of all-steel BRBs are carried out before they are implemented into the building.

First, two full-scale specimens are tested under uniaxial quasi-static cyclic loading by the MTS system at Harbin Institute of Technology. It is observed from the test that the buckling appears at the top end of the specimen because the core become slimmer due to deformation and is off from the outer square steel tube, the out square steel tube drops down and thus the core at the top end out of the outer square steel tube becomes longer and relatively readily buckling. To avoid this fault, one bolt is welded with the core and outer steel tube at the middle part along the height. However, weld generates fault that results in the abrupt rupture of the core at the weld cross-section. Finally, a barrier is welded at bottom end to prevent outer steel tube from dropping down and thus the buckling of the core at the top end can disappear. Another full-scale BRB is implemented into a frame for investigation of its performance and the reliability of the connection. The test is carried out at the Shenyang Architecture University.

The picture and the detailed configuration of the BRB are shown in **FIGURE** 1 and **FIGURE** 2, respectively. **FIGURE** 3 and **FIGURE** 4 show the test scenario and the detailed dimension of the test system. The configuration of loading system is intended to represent the geometry and loading conditions for braces used in the practical structure. Two hydraulic actuators, one end connected to the react wall and the other end connected with a column at its free top end. The BRB-frame system is loaded horizontally by these two hydraulic actuators. To prevent the out-plane buckling of the frame, two lateral supports are setted at each side of the column. Each of the actuators had a capacity of 960kN in tension and 1500kN in compression, thus the total horizontal tension force and compression force exerted by them are 1920kN in tension and 3000kN, respectively.

FIGURE 1 Picture of the BRB

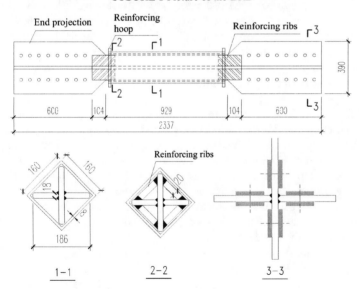

FIGURE 2 Detailed configuration of the BRB(the black area represents weld)

FIGURE 3 Scenario of the BRB –frame test system

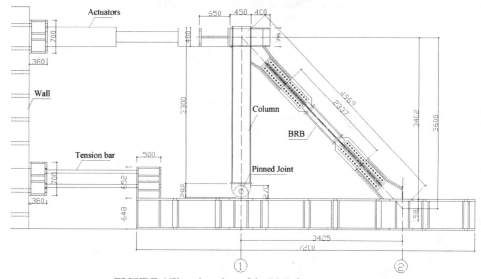

FIGURE 4 Elevation view of the BRB-frame test system

Taking the horizontal displacement of the actuators as the control loading-parameter, the full-scale BRB is tested under a cyclic loading. The loading history is shown in FIGURE 5.

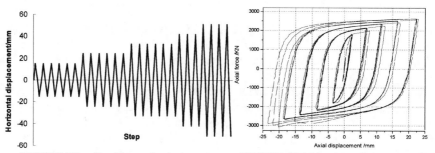

FIGURE 5 Loading history for the test   **FIGURE 6** Displacement-force curve of BRB

FIGURE 6 shows the relationship between axial displacement and axial force of the BRB. The maximum strain is 3.2%, which indicates that the ductility of the brace is excellent. The over-strength factor β defined as the ratio of the maximum compression force to the maximum tension force in the same cycle is less than 1.1, indicating the friction between the core and the outer tube is not significant. The cumulative inelastic ductility factor is 720, indicating that the brace has excellent energy dissipation capacity.

Full scale performance tests of X-shaped metallic damper

The Capital Stadium, constructed in 1950s, is needed to be retrofitted for the 2008 Beijing Olympic Game. 67 X-shaped metallic dampers were designed and implemented into this building in 2007. The yield displacement of this building is very small, while the ultimate deformation is larger to meet the requirement of the New Code. Therefore, the ductility of the metallic damper is required to be large. The X-shaped metallic damper with a special configuration is proposed by authors, as shown in **FIGURE** 7. The yield displacement and ultimate displacement of the damper are about 2mm and 45mm, respectively. Full-scale performance tests are carried out by using MTS system at Harbin Institute of Technology, and the damping force versus displacement of the damper is shown in **FIGURE** 7.

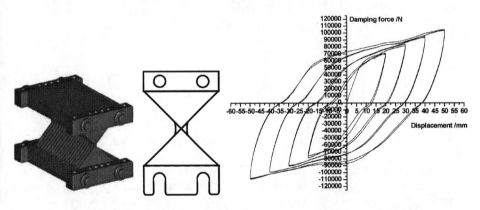

FIGURE 7 X-shaped metallic damper and its damping force versus displacement

ACTIVE CONTROL FOR THE GUANGZHOU TV TOWER

The Guangzhou TV tower is a steel-concrete composites structure with a height of 618m, a weight of 200,000t and an area of 102,000m^2, as shown in **FIGURE** 8. Guangzhou suffers from strong Typhoon and the wind-induced vibration of this tower is very dramatic and should be suppressed. First, finite element model of this tower is established and the modal parameters are calculated. The first period of this tower is 10.0135s. Two hybrid mass dampers, which is two TMDs combining with two AMDs are preliminarily designed by Professor Zhou's group and Professor Ou's group together, as shown in **FIGURE** 8. The wind-induced vibration of the tower without/with HMD is shown in **FIGURE** 9. It can be seen from **FIGURE** 9 that the HMD can effectively suppress the wind-induced vibration of this tower. However, further analysis indicates that the AMD can only increase the control efficiency in a very small extent because the stroke of TMD is limited and AMD cannot push the mass of TMD to increase its stroke. Additionally, the mass of the TMD is the water tank and water for fire resistance and it is not necessary to be decreased. Therefore, AMD seems no useful for this system.

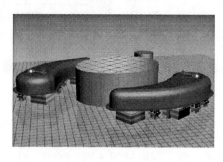

FIGURE 8 Guangzhou TV Tower and its HMD system

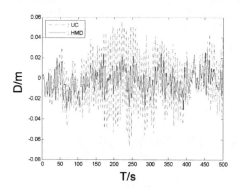

FIGURE 9 Wind-induced vibration of the Guangzhou TV Tower without/with HMD system

SMART DAMPING CONTROL FOR STRUCTURES

Smart dampers, including MR dampers, PZT damper, SMA dampers and magnetostrictive dampers, have been comprehensively studied by authors. At present, only MR dampers are commercially produced, so this paper mainly introduces the MR dampers and their applications.

MR dampers with various capabilities and their performance have been studied, as shown in **FIGURE** 10. The minimum capability of the MR damper can be 100N (the adjust times may be 10, meaning that the damping force of MR damper without current is 10 N), while the maximum capability of the MR damper can be 40t.

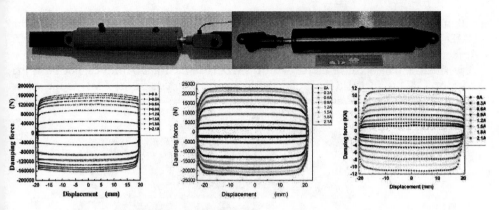

FIGURE 10 MR dampers and their performance

Due to flexible and small inherent damping, stay cables are prone to vibrate under wind-rain. The MR dampers have been incorporated into cables to suppress wind-rain induced vibration. The tests of a stay cable with a MR damper have been conducted at Harbin Institute of Technology, as shown in **FIGURE** 11. An algorithm based on limited observable locations is proposed for the cable vibration control using MR damper [5]. The response of the cable without/with MR damper is shown in **FIGURE** 11. It can be seen that the MR damper can effectively mitigate the vibration of the cable and can achieve further reduction than passive-on MR damper due to negative stiffness provided by MR damper [6]. The MR dampers have been implemented into the Binzhou Yellow River Highway Bridge to suppress the wind-rain induced vibration of cables, as shown in **FIGURE** 11.

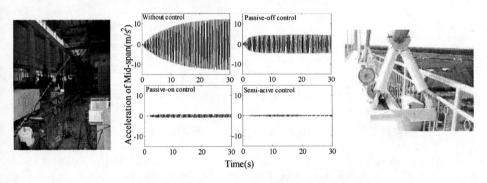

FIGURE 11 Cable-MR damper system and response

The bridges are readily to be collapse due to pounding and unseating under an earthquake event. MR dampers are used to control the pounding and unseating of bridges. The shake table tests are carried out, as shown in **FIGURE** 12. The results are shown in **FIGURE** 12. It can be seen from **FIGURE** 12 that the pounding and unseating can be avoided using MR damper and thus seismic behavior of the bridge can be improved.

143

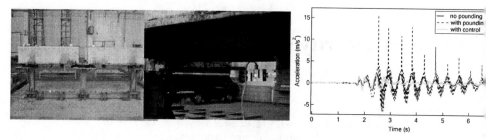

FIGURE 12 Control of pounding and unseating of bridges using MR dampers

Finally, the MR damper is used to mitigate the ice-induced vibration and earthquake response of offshore platforms. 8 MR dampers have been implemented into a offshore platform located in Bohai Bay, China, as shown in **FIGURE** 13.

FIGURE 13 Offshore platform with 8 MR dampers

ACKNOWLEDGMENTS

This study is financially supported by the Ministry of Science and Technology with grant No. 2007CB714204 and 2006BAJ03B06.

REFERENCES

1. G. Y. Wang, "Vibration control for tower structures", Proceedings of the National Conference of Tall Buildings, Wuhan,1981
2. J. P. Ou and F. L. Zhou, "Isolation and Passive Energy Dissipation Technologies (Chapter 12)" in *Code of Seismic Design for Buildings* (GB 10011-2001), Beijing, Press of Architecture and Buildings, 2001
3. J. P. Ou, *Structural Vibration Control*, Beijing, Press of Science, 2003.
4. L. Y. Li, " Vibration Control for Nonlinear Structures: Simulation and experiments", PhG. Thesis, Harbin Institute of Technology, 2008
5. H. Li, M. Liu, J. P. Ou, "Vibration control of stay cables of Shandong Binzhou Yellow River highway bridge by using magnetorheological fluid dampers", ASCE *Journal of Bridge Engineering*, 2007,12(4), pp 401-409,
6. H. Li, M. Liu and J. P. Ou, "Nagative stiffness characteristics of active and semi-active control systems for stay cables", *Structural Control and Health Monitoring*, 2008, 15:120-142

Seismic Integrity of a Long Span Bridge with Tower-Deck Buffers

Fabio Casciati[a] and Fabio Giuliano[b]

[a]University of Pavia, Department of Structural Mechanics, Via Ferrata 1, Pavia
[b]Univerity of Rome "La Sapienza"

Abstract. The wind action is acknowledged to be the primary environmental load in the design of suspension bridges. It affects both the performance of safety and serviceability. Nevertheless, the impact of earthquakes cannot be neglected especially when the construction site is located in a high seismicity region. The evaluation of the vulnerability of a suspension bridge requires a hazard analysis on the site, the definition of the seismic intensity related to the desired return periods, the generation of artificial signals for the reference site and the collection of critical records for integrity analyses,. The signals will then be used in the structural analyses for the verification of structural safety and structural robustness. This paper is focused on the generation of these critical signals, ,with their own a-synchronicity, and the associated numerical nonlinear analyses, which must be conducted in the time domain.

Keywords: Suspension bridges, seismic loads modeling, critical scenarios, analysis and response, the Messina Strait bridge.

INTRODUCTION

It is well known that long span bridges have a low seismic vulnerability, thus allowing the transportation authorities to construct large bridges in seismic areas such as Japan and California. Major evidence went from the earthquake in Kobe in 1995, with the almost absence of damage in the Akashi Kaikyo bridge, in construction at that time. The damage was limited to some residual displacements at the tower location after the event.

On the other side the design of a suspension bridge comes with safety and robustness verifications for actions at different return period, in different constructive stages, at different stages of its lifetime. For the seismic action this requires the collection of recorded signals, and the assemblage of a database calibrated on the geo-morphologic characteristics of the reference site, for the safety verifications (Serviceability Limit State – SLE - and Ultimate Limit State - SLU) and for the robustness verifications (Structural Integrity Limit State - SLIS).

A numerical model of the bridge gives then the designer the chance to evaluate the safety performance of the main sub-structures (cables, towers, anchorages), which must resist to extreme events proper of the SLIS verification, and of the secondary substructures (hangers, deck) whose damage is allowed to produce confined and temporary out of service.

Control devices with finite free gap can also be introduced to allow finite transversal and longitudinal displacements of the deck, to absorb the thermal

CP1020, *2008 Seismic Engineering Conference Commemorating the 1908 Messina and Reggio Calabria Earthquake*,
edited by A. Santini and N. Moraci

displacements without coactions and to reduce the peak of horizontal curvature of the deck runnability. Such devices are shown not to provide compressive reactions even in the case of severe seismic scenarios, confirming the lower impact of earthquake in the rank of the environmental actions.

THE REFERENCE CASE STUDY

The reference case study adopted in this paper is the design configuration approved in 1992 for the Messina Strait bridge. It is a suspension bridge whose main span is 3300m long, while the total length of the deck, 60m wide, is 3666m (including the side spans). The deck is formed by three box sections, the outer ones being used in the two directions by the roadway and the central by the railway. The two towers are 383m high and the bridge suspension system relies on two pairs of steel cables, North and South, each with a diameter of 1,24m and a total length, between the anchor blocks, of around 5000m. The main characteristics of the structure are summarized in Figure 1.

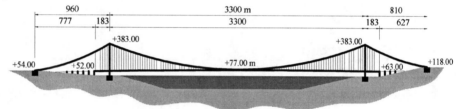

FIGURE 1. The Messina Strait bridge.

In order to increase the railway runnability, some adjustments of the design were also introduced, modifying its kinematics under the towers by inserting free-gap buffers between the deck and towers in both the longitudinal and transversal directions (Figure 2). The former buffers allow thermal and braking displacements and avoid too large expansion joints; the buffers in the latter direction reduce the peak of horizontal curvature of the track in case of transversal wind.

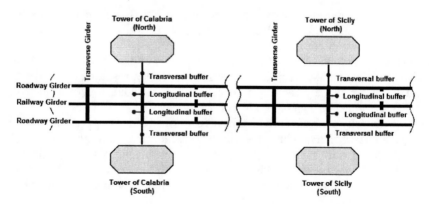

FIGURE 2. Plan of the deck with buffers under the towers of the Messina Strait bridge.

THE SEISMIC ACTION: EVOLUTION OF THE MODELING

The analyses conducted before 1992 for the Messina Strait bridge considered the seismic action by scaling the design spectrum for the design seismic intensities, assumed as 5.7 m/s^2 for the SLIS level. The Project of 1992 had been verified considering only four spectrum-coherent time-histories, as represented in Figure 3.

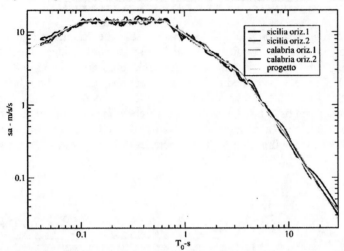

FIGURE 3. Response Spectra of the four accelerograms selected as seismic inputs for the 1992 Project

During the activities of the Scientific Committee, the following evidences emerged:
- Hazard analyses (with Professor Prestininzi serving as princiapal investigator) demonstrated that the spectrum used in the 1992 Project underestimated the seismic action for the SLIS level;
- There was scientific consensus in regarding as incorrect to scale seismic spectra or time-histories to get signals of different intensity and return period;
- The duration of the signals used in the 1992 Project (10, 15, 25 s) were not consistent with the dynamic properties of the bridge, whose fundamental period is about 30 seconds;
- Each considered event was composed of two signals, one per side (Sicily and Calabria) without distinguishing the signals of towers and anchorages, which are several hundreds meters far each from the other..

As a consequence, the Scientific Committee promoted the update the definition of the seismic action on the bridge. The seismic intensity resulting from the hazard analyses, in terms of the peak ground accelerations, resulted in 1.2-2.6 m/s^2 for the SLE, 5.7 m/s^2 for the SLU and 6.3 m/s^2 for the SLIS.

For high intensity analyses, and so for the SLIS verifications, it is difficult to extend the results of hazard analyses based on the classical framework originally proposed by Cornell. So the use of asynchronous spectrum-coherent signals was limited to the (SLE and SLU) verifications, while for the SLIS level an adequate number of accelerograms was generated through a different approach, without scaling the signals to different intensities.

THE SEISMIC ACTION: DATABASE OF CRITICAL SIGNALS

A study was committed by the responsible authority, Stretto di Messina SpA, with the first author serving as interface. The goal was to generate a database of signals, potentially critical for the structural integrity of the bridge, likely to happen at the return period resulting from the hazard analysis, and keeping into account the geometry and the geo-mechanics of the site.

The database of signals was generated by a numerical algorithm which uses as input a selected signal (source record), the position of the source of the earthquake and solve the equations of the propagation of the seismic waves in the multilayered formations. Seismic records of the past were led to the reference earthquakes in the region, accounting for the magnitudes and the possible sources of the most severe events expected all around., Their association with the desired return periods was supported by probabilistic methods.

The source records selected for the SLIS verifications cover different cases:
- Accelerograms with relevant high frequency content;
- Accelerograms with relevant low frequency content;
- Signals with high displacements;
- Signals with high out-of-phases between towers and anchorages.

Examples of the records considered are Chi-Chi (20/09/1999, TCU046 St., Mw=7.6, Ms=7.6, Ml=7.3), Santa Cruz (17/10/1989, Capitola&SanJose St., Ms=7.1, Ml=7.0), Landers (28/06/1992, Joshua Tree St., Mw=7.4 , Ms=7.5), Duzce (C0375, D0531, C1062 St., 12/11/1999, Mw=7.22 , Ms=7.3), Tabas (16/09/1978, Tabas&Dayhook St., Mw=7.35 , Ms=7.4). The three symbols Mw, Ms and Ml denote the magnitude moment, the magnitude and the local magnitude, respectively. Many of these records exceed the response spectrum specified in the design specifications (Figure 4).

Among them Chi-Chi record has a high content of energy at low frequencies, and the Tabas at the high, and higher values of the peak ground acceleration, about 1g for the horizontal component.

Each source record has been then manipulated in such a way to obtain groups of signals, expressed in terms of time-histories of horizontal and vertical displacements, velocities and accelerations, for the four bridge boundaries (2 towers + 2 anchorages).

Groups of realistic vertical and horizontal signals have been determined for the four boundaries of the bridge, in such a way to configure critical scenarios.

The output signals inside each group are distinguished for different features:
- out of phase in the input motion;
- incongruence of amplitudes and exciting frequencies.

These aspects depend on the position of the hypocenter, on the magnitude, on the mechanical properties of the soils and the surface layers, on the propagation path of the waves, and the incidence angles between source and site. Random phases are not physically acceptable, and the expected seismic motion must be determined on the base of the real knowledge of the seismic-tectonic of the site.

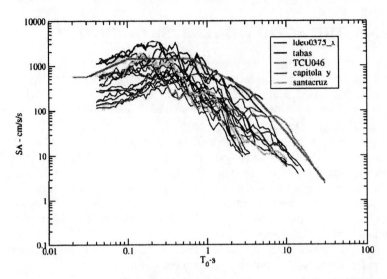

FIGURE 4. Response spectra of the horizontal components of the selected as critical source signals and SLIS Design Spectrum.

The size of the bridge, i.e. its main span and the distance between towers and anchorages suggests that the delay of the four signals is generated in such a way to keep into account the different mechanical characteristics of the rock and soils and the azimuth of the seismic wave incidence.

In fact the complete synchronicity and congruence of the signal between towers (or between anchorages) is possible only if the soil is homogeneous and the epicenter is in the middle of the central span. However the finite speed of the seismic waves produces a delay in the signal of towers and anchorages. The a-synchronicity of the signals can result in different intensities and versus of displacements, producing possible critical scenarios for the bridge as with opposite motion between towers and anchorages, both in the horizontal and vertical directions (Figure 5).

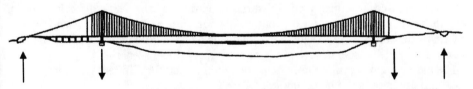

FIGURE 5. Scenario of vertical excitation.

The response spectra of the resulting database, for the longitudinal horizontal component of the motion at the Calabria anchorage, is represented in Figure 6.

149

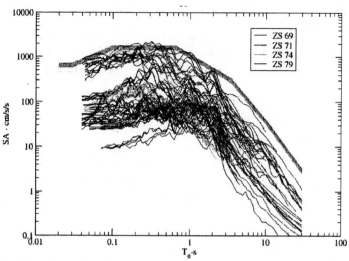

FIGURE 6. Response spectra (damping= 5%) of the horizontal signals of Calabria anchorage (SLIS).

STRUCTURAL INTEGRITY: RESPONSES AND SYNTHESIS

The dynamic response of a long span suspension bridge is evaluated by nonlinear time-domain simulations. For ordinary structures it is computed applying the same ground acceleration on all structural masses as inertial terms; for long span bridges such procedure is incorrect because the acceleration in the edges is in general delayed, because of the finite speed of propagation of the seismic waves. The seismic motion must be expressed in terms of time-histories of displacements, obtained by the accelerograms through a double integration through the trapezium method and imposed to the edges of the numerical model of the bridge.

The database of critical signals and the manipulation of the records allowed the evaluation of the seismic performance of the Messina Strait bridge in the years 2002-2004. For the SLIS verification, the attention was focused on the response of the main structural system: towers, cables and anchorages.

The results are re-elaborated in this paper in terms of envelopes and statistics of the computed quantities: displacements, deformation, stresses and forces in the structural members.

Figures 7 and 8 summarize some results of the numerical analyses conducted for the SLIS verifications. They are expressed in terms of statistical frequency of the maximum stress detected in cables and towers.

The effects on the cables are expressed at the anchorage and at the midspan sections, while those on the towers are computed at the basement and at the third transverse (i.e., at the height of 250 meters).

The diagram shows that the maximum increase in the cable stress for very extreme events can be evaluated as only a 20% of the value computed under the permanent loading condition. But the increase in the tower members is much higher.

Figure 9 provides the statistical frequency of the values achieved by the deck displacements under the tower, in the longitudinal and transversal direction. They are

150

smaller than the free gap of the buffer, introduced against the effects of wind and temperature loading.

It is worth noting that the bridge deck remains completely suspended on the hangers and unbounded at the edges, and receives the seismic input as filtered by the series system made by the sequence foundation-tower-cables-hangers.

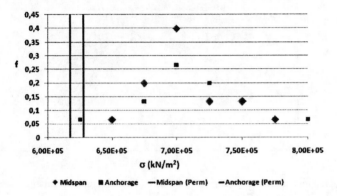

FIGURE 7. Frequency of the maximum cable stress and permanent values.

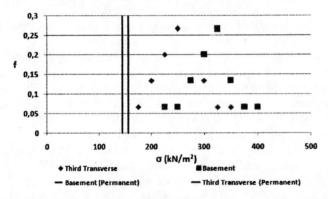

FIGURE 8. Frequency of the maximum tower stress and permanent values.

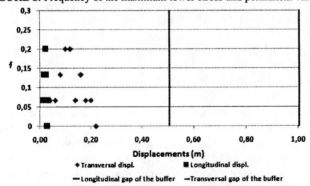

FIGURE 9. Frequency of the deck displacements and free gaps of the buffers.

CONCLUSIONS

Among the initiatives of the Scientific Committee of the Messina Strait Bridge, active in the period 2003-2006, there was the development of a database of seismic signals, consistent with the reference return period specified in the Design Prescriptions.

The signals have been generated on the basis of the geomorphology of the site, and manipulated in the perspective of the structural analysis of a long span bridge, whose peculiarities have been briefly discussed in this paper.

The analyses confirmed the expected low vulnerability of the Messina Strait bridge (as in the described configuration) to the seismic excitation, even in the presence of the extreme scenarios associated to a return period 10000 years.

ACKNOWLEDGMENTS

The authors wish to thank the President of the Scientific Committee for the Messina Strait bridge, Professor R. Calzona, and Professor F. Bontempi who coordinated the structural analyses during the activity (2003-2006) of the same Committee, for which the first author served as Member and Executive Secretary.

REFERENCES

1. Ambraseys N., Douglas J., Sigbjornsson R., Berge-Thierry C., Suhadolc P., Costa, G., Smit P. , 2004, CD-ROM European Strong-Motion Database – Volume 2.
2. Bontempi, F. 2005. Frameworks for Structural Analysis, in Innovation in Civil and Structural Engineering Computing, B.H.V. Topping (Editor), Saxe-Coburg Publications, Stirling, Scotland, 1-24.
3. Calzona, R., Casciati, F. 2005. "Provisions for Robustness and the Italian Code. Robustness of Structures", Workshop JCSS&IABSE WC 1 BRE, Garston, Watford, UK, Nov. 28-29, 2005.
4. Camassi R., Stucchi M. 1997, NT4.1.1, "Un catalogo parametrico di terremoti di area italiana al di sopra della soglia di danno. Rapporto tecnico", GNDT, Milano.
5. Casciati, F. 2006. The Challenge of Long-Span Suspended Bridges. Keynote lecture at the 3rd European Workshop on Structural Health Monitoring. Granada, Spain, 5-7 July, 2006.
6. Casciati F., Faravelli L., 1991, *Fragility Analysis of Complex Structural Systems*, Research Studies Press, Taunton, UK.
7. Levine, Marie-Bernard P. 1990. "Accelerogram processing using reliability bounds and optimal correction methods. Tech. Rep. CaltechEERL:1990.EERL-90-02", California Inst. of Technology.
8. Meletti C., Patacca E., and Scandone P. 2000 "Construction of a seismotectonic model:the case of Italy", , Pure and Apll. Geoph.,157,11-35.
9. Rey J., Faccioli E., Bomber J.J. 2002, "Derivation of design soil coefficients and response spectra shapes for Eurocode 8 using the European Strong Motion Database", Jour. Of Seism. 6: 547-555.
10. Stretto di Messina Spa, 2004, *Fondamenti progettuali e prestazioni attese per l'Opera di attraversamento*.
11. Stucchi (coord.). 2004 "Redazione della mappa di pericolosità sismica: Rapporto conclusivo.", (http://zonesismiche.mi.ingv.it/).
12. Trifunac, Mihailo D. 1970 "Low frequency digitization errors and a new method for zero baseline correction of strong-motion accelerograms. Technical Report: CaltechEERL:.EERL-70-07", 1970, California Institute of Technology.
13. Valensise G., Pantosti D., 2001 "Database of potential sources for earthquakes lager than M 5.5 in Italy", Ann.Geofis. suppl. Vol. 44(4),180 pp.
14. Vannucci G., GasperiniP. "A database of revised fault plane solutions for Italy and surrounding regions", 2003, Computer Geosciences, 29-7-pp 903-909.

Semiactive Control Using MR Dampers of a Frame Structure under Seismic Excitation

Vincenzo Gattulli[a], Marco Lepidi[a], Francesco Potenza[a], Rùbia Carneiro[b]

[a]DISAT, University of L'Aquila, Monteluco di Roio, Italy
[b]ENC, University of Brasilia, Brazil

Abstract. The paper approaches the multifaceted task of semiactively controlling the seismic response of a prototypal building model, through interstorey bracings embedding magnetorheological dampers. The control strategy is based on a synthetic discrete model, purposely formulated in a reduced space of significant dynamic variables, and consistently updated to match the modal properties identified from the experimental response of the modeled physical structure. The occurrence of a known eccentricity in the mass distribution, breaking the structural symmetry, is also considered. The dissipative action of two magnetorheological dampers is governed by a clipped-optimal control strategy. The dampers are positioned in order to deliver two eccentric and independent forces, acting on the first-storey displacements. This set-up allows the mitigation of the three-dimensional motion arising when monodirectional ground motion is imposed on the non-symmetric structure. Numerical investigations on the model response to natural accelerograms are presented. The effectiveness of the control strategy is discussed through synthetic performance indexes.

Keywords: three-dimensional building, seismic response, semiactive control, magnetorheological dampers, clipped-optimal strategy.

INTRODUCTION

Great research effort has been focused over the last years on reducing the seismic response of engineering structures through dissipative systems. Presently, an increasing attention is being paid to combine the reliable and cost-saving passive technology with the highly performant active strategies, by means of hybrid and semiactive solutions. In this field, magnetorheological dampers are considered the most promising devices to mitigate the structural vibrations, due to their mechanical simplicity, high dynamical range, low power requirements, large force capacity and mechanical robustness [1]. Experimental testing on large scale models show that this technology can be effectively implemented to control the structural dynamical response, and is suited for applications on civil structures [2]. To date, the experimental verification of magnetorheological dampers in reducing the three-dimensional response of asymmetric buildings subjected to seismic excitations is confined to light models [3].

CP1020, *2008 Seismic Engineering Conference Commemorating the 1908 Messina and Reggio Calabria Earthquake*,
edited by A. Santini and N. Moraci
© 2008 American Institute of Physics 978-0-7354-0542-4/08/$23.00

The paper approaches the multifaceted task of semiactively controlling the three-dimensional seismic response of a prototypal building model, through interstorey bracings embedding magnetorheological dampers. The model simulates the mechanical behaviour of a two-storey, laboratory-scaled (2:3), steel-made structure [4], realized at the Structural Laboratory of the DiSGG (University of Basilicata - Italy). As a benchmark in the framework of Research Line n.7 of the Italian DPC-ReLUIS Research Project, it is presently under investigation for the experimental assessment of different passive and semiactive strategies in the mitigation of seismic-induced vibrations.

In particular the effectiveness of the semiactive control strategy strongly depends on the availability of reliable models for the damper hysteretic behaviour, as well as on the formulation of precise dynamical models for the structure to be controlled. Therefore, the control strategy is based on a synthetic discrete model, purposely formulated in a reduced space of significant dynamical variables, consistently updated to match the modal properties identified from the experimental response of the benchmark structure. The efficacy of the semiactive strategy, implemented according to the clipped-optimal control law, is discussed through synthetic performance indexes.

ANALYTICAL MODEL

The three-dimensional motion of the prototypal building is described by a discrete dynamical model, formulated according to the direct displacement method, assuming Euler-Bernoulli frame elements and lumped masses. Starting from the cumbersome model including 72 degree-of-freedom (dofs) presented in Figure 1a, which considers complete flexibility of all the beams and columns, some reasonable simplifying hypotheses are introduced to suitably reduce the model dimension without compromising its reliability.

The hypotheses concerning the high axial stiffness of the frame elements and the rigid behaviour of the two storeys are treated as internal geometric constraints and implemented imposing master-slave relationships between the involved nodes. Differently, the assumptions regarding negligible rotational inertia of the beam-column nodes are considered to determine quasi-static dynamics of the associated displacements, which are still taken into account, even if condensed through a classical Guyan reduction [5]. Therefore, the model is reduced into a six degree-of-freedom space, including the only in-plane components of the barycentrical displacement for each storey (Figure 1b).

The obtained model is completed adding the structural damping and the supplementary dissipation due to the magnetorheological dampers. Finally, an imposed mono-directional ground acceleration history $u_g(t)$ is considered as external excitation, acting in the direction of the column minimum flexibility (see Figure 1b). For sake of brevity, the two translational components of displacement, which are respectively collinear or transversal to the ground acceleration direction, are distinguished as *along excitation* (*ae*) or *cross excitation* (*ce*) in the following.

Once the total displacement vector $\mathbf{u}_t = \{u_{b1}, u_{b2}, \varphi_{b3}, u_{t1}, u_{t2}, \varphi_{t3}\}^\top$ is suitably decomposed to separate the rigid translation due to the ground motion $\mathbf{u}_t = \mathbf{u} + \mathbf{r}\, u_g$, the equations governing the forced damped (controlled) oscillations of the model read

$$\mathbf{M\ddot{u}} + \mathbf{C\dot{u}} + \mathbf{Ku} + \mathbf{f(\dot{u}, u, v)} = -\mathbf{M r}\ddot{u}_g \qquad (1)$$

where $\mathbf{M}$ and $\mathbf{K}$ are the mass and stiffness matrix, and the rigid translation vector is $\mathbf{r} = \{1, 0, 0, 1, 0, 0\}^\top$. The structural damping is considered of viscous nature, and conventionally introduced by a mass-proportional damping matrix $\mathbf{C} = \alpha\mathbf{M}$. The control action vector is $\mathbf{f(\dot{u}, u, v)} = \{(F_{d1} + F_{d2}), 0, \frac{1}{2}(F_{d1} - F_{d2})\ell_y, 0, 0, 0\}^\top$, where $F_{d1,2}$ are the horizontal forces applied by the two magnetorheological dampers placed on the top of the first storey bracings (Figure 1b). The eccentricity of the application points lets the damper forces work on both the ae-translation u_{b1} and rotation φ_{b3} of the first storey.

It is worth noting that each damper force depends on the displacement u_d and velocity $\dot{u}_d$ of the corresponding application point, which can be expressed itself through the structure displacement and velocity vectors ($\mathbf{u}$ and $\dot{\mathbf{u}}$, recalling the assumption of rigid storey motion). Moreover, since the hysteretic properties of the magnetorheological fluids can be modified by an external magnetic field induced by a pair of potential-differences $\mathbf{v} = \{V_{d1}, V_{d2}\}^\top$, the vector $\mathbf{v}$ contains the command signals.

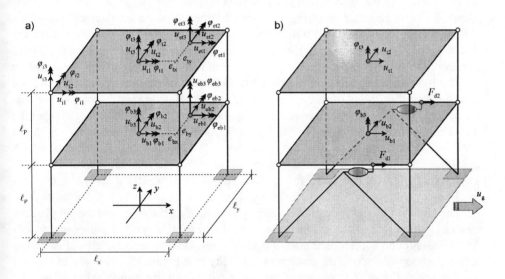

FIGURE 1: Two-storey frame building: (a) wholly-flexible model, (b) dynamically active dofs, control forces and ground acceleration of the condensed model.

Different rheological models are commonly used to describe the constitutive relationship $f(\dot{u}, u, v)$ for magnetorheological dampers. To the aim of the present paper a suitable description of the highly nonlinear damper behaviour is given by the 9-parameter model proposed by Spencer et al. [6], in which the *Bouc-Wen* block [7] is combined with a series dashpot and a parallel spring (Figure 2b). The equation governing the relationship between the damper force and the application point displacement/velocity

$$F_d(u_d, \dot{v}_d) = c_1 \dot{v}_d + k_1(u_d - u_{d0}) \tag{2}$$

is based on the evolution of the variables v_d and ζ, which is governed by the equations

$$\dot{v}_d = (c_0 + c_1)^{-1} [k_0(u_d - v_d) + c_0 \dot{u}_d + \alpha\zeta] \tag{3}$$

$$\dot{\zeta} = A(\dot{u}_d - \dot{v}_d) - \beta(\dot{u}_d - \dot{v}_d)|\zeta|^n - \gamma\zeta|\dot{u}_d - \dot{v}_d||\zeta|^{n-1} \tag{4}$$

where the coefficients k_0 and c_0 in the Bouc-Wen block assess the stiffness and damping at higher velocities, the stiffness k_1 of the parallel spring accounts for the damper accumulator, while the series dashpot with viscosity c_1 reproduces the roll-off phenomenon.

The parameters defining the Bouc-Wen model of the magnetorheological dampers in absence of semiactive control ($v = 0$, passive control) are properly tuned to simulate the experimental behaviour of the commercial device Lord RD1005-3 (Figure 2a), as experimentally identified by dynamical tests at different voltage amplitudes [8].

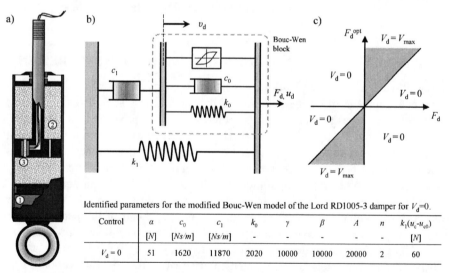

Identified parameters for the modified Bouc-Wen model of the Lord RD1005-3 damper for $V_d=0$.

Control	α	c_0	c_1	k_0	γ	β	A	n	$k_1(u_c-u_{c0})$
	[N]	[Ns/m]	[Ns/m]	-	-	-	-	-	[N]
$V_d=0$	51	1620	11870	2020	10000	10000	20000	2	60

FIGURE 2: Magnetorheological damper: (a) Lord RD1005-3 commercial device, (b) modified Bouc-Wen rheological model, (c) clipped-optimal control scheme.

Different control strategies have been recently proposed for the use of semiactive devices in preventing large oscillations of frame structures under seismic excitation and a comparative study has been conducted for evaluating the performance of different proposed algorithms through a numerical model of a six-storey planar frame [9]. The authors' conclusions indicate that *clipped-optimal controllers* posses the flexibility needed by the control designer to achieve a large range of control objectives.

Accordingly to these results, in the present study a clipped-optimal control, based on the Linear-Quadratic Regulator (LQR), has been designed and its performance tested through the seismic response of the model representing the prototype structure. In particular, the reference optimal control $\mathbf{f}_d^{opt} = \{F_{d1}^{opt}, F_{d2}^{opt}\}^\top$ has been designed based on a full-state LQR feedback law; if a monotonically increasing force-voltage relationship is assumed, inducing the i-th magnetorheological damper to generate approximately the desired optimal force is obtained letting the command signals in the vector $\mathbf{v}$ obey

$$V_{di} = V_{max} H\left[\left(F_{di}^{opt} - F_{di}\right) F_{di}\right] \tag{5}$$

where V_{max} is the voltage associated with the damper saturation, and $H(\cdot)$ is the Heaviside step function. The algorithm selecting the command signal produces the following effects: when the magnetorheological damper is providing the desired optimal force, the voltage applied remains at the present level; if the magnitude of the force produced by the damper is smaller than the optimal required force and the two forces have the same sign, the applied voltage is increased to the maximum allowed; otherwise the commanded voltage is set to zero. Figure 2c represents graphically the control law.

MODEL IDENTIFICATION THROUGH DYNAMICAL MEASUREMENTS

Aiming to achieve the best performances in the semiactive control of the structure, a key point is represented by the updating of the model mass and stiffness, in order to reproduce, as close as possible, the real behaviour of the prototype, as measured during dynamical the tests performed at the DiSGG Structural Laboratory in November 2007 [4]. The structure was tested under different configurations: a basic configuration (named CB in the following), characterized by double symmetry of both stiffness and mass distribution; a second configuration (CS) is obtained adding four equal and eccentric lumped masses on each storey, without destroying the structural double symmetry; finally a third configuration (CN) is obtained removing a pair of lumped masses from the CS configuration, destroying the structure mass-symmetry with respect to the *ae* direction. Therefore, three-dimensional motion, involving coupling between the structure *ae* displacement and rotation, is expected to arise when the mono-directional ground motion is applied to the CN configuration, due to the mass eccentricity.

TABLE 1: Experimental frequencies [Hz] compared with those of the updated physical model.

	EFDD identification			Updated model			Differences		
	f_{CB}	f_{CS}	f_{CN}	f_{CB}	f_{CS}	f_{CN}	Δf_{CB}	Δf_{CS}	Δf_{CN}
Mode 1	3.3762	2.8505	3.0813	3.3732	2.8505	3.0776	-0.089	0.000	-0.120
Mode 2	4.2335	3.5773	3.8417	4.2335	3.5773	3.8408	0.000	0.000	-0.023
Mode 3	5.8901	5.1069	5.5098	5.9126	5.1272	5.5546	0.382	0.398	0.813
Mode 4	9.4096	8.4202	8.9092	9.9583	8.4202	9.0895	5.831	0.000	2.024
Mode 5	14.6418	12.3802	12.9691	14.6410	12.3802	13.2628	-0.005	0.000	2.265
Mode 6	18.7417	16.2237	17.6379	19.0385	16.4744	17.9049	1.584	1.545	1.514

The entire identification process has been driven through output-only procedures, based on the experimental response (known-output) of the structure to environmental excitation (unknown-input). First, the frequencies and modal shapes were estimated applying the Enhanced Frequency Domain Decomposition (EFDD) procedure [10], in order to assess a representative modal model. The experimental frequencies are collected in Table 1. The first (forth) mode is characterized by a dominant translation component in the *ce* direction, with in-phase (counter-phase) motion of the two storeys. The second (fifth) and the third (sixth) modes are instead characterized by dominant translation in the *ae* direction and rotation components, respectively, again with in-phase (counter-phase) motion of the two storeys. The asymptotic method proposed in [11], based on the sensitivity of the modal properties to slight changes in the structural mass configuration, was then used to orthonormalise the experimental modes. Finally, the modal properties allowed the complete identification of the structure mass and stiffness (Tables 2,3), since the physical model balances the dimension of the modal information available. The physical model, as obtained from separate procedures for the mass and stiffness identification, matches the experimental modal properties with excellent agreement (see Table 1). An exhaustive description of the whole identification process can be found in [12].

TABLE 2: Non-null mass matrix coefficients of the updated physical model.

	M_{11}	M_{22}	M_{33}	M_{44}	M_{55}	M_{66}	$M_{13,31}$	$M_{46,64}$
	[kg]	[kg]	[kg m²]	[kg]	[kg]	[kg m²]	[kg m]	[kg m]
Configuration CS	4741.7	4741.7	10915.1	4695.0	4695.0	11093.7	4741.7	4695.0
Configuration CN	4069.7	4069.7	9488.3	4027.0	4027.0	9679.1	4069.7	4069.7

TABLE 3: Non-null stiffness matrix coefficients (multiplier 10^6) of the updated physical model.

	K_{11}	K_{22}	K_{33}	K_{44}	K_{55}	K_{66}	$K_{14,41}$	$K_{25,52}$	$K_{36,63}$
	[N/m]	[N/m]	[Nm]	[N/m]	[N/m]	[Nm]	[N/m]	[N/m]	[Nm]
Configuration CS,CN	24.180	10.719	97.282	6.9251	4.0883	31.935	-9.9959	-4.8959	-42.074

PASSIVE AND SEMIACTIVE CONTROL OF THE SEISMIC RESPONSE

The seismic response of the structure has been numerically simulated using a set of natural accelerograms, which provide mono-directional ground acceleration histories consistent with the spectrum of the B-type soil, according to the Italian Standard. A nondimensional factor $a = 0.15$ is used to linearly scale the acceleration signals.

The effectiveness of the control system in mitigating the structural response is measured by means of four performance indexes, defined on the purpose

$$\mathcal{J}_j^u = \frac{u_{j1,\max}^c}{u_{j1,\max}^u}, \qquad \mathcal{J}_j^a = \frac{\ddot{u}_{j1,\max}^c}{\ddot{u}_{j1,\max}^u}, \qquad \mathcal{J}_j^\varphi = \frac{\varphi_{j3,\max}^c}{\varphi_{j3,\max}^u}, \qquad \mathcal{J}_j^F = \frac{F_{di,\max}^c}{F^u} \qquad (6)$$

Six indexes represent the ratio between the maxima of the controlled (c) and uncontrolled (u) response, in terms of ae displacement $(\mathcal{J}_i^u)$, ae acceleration $(\mathcal{J}_i^a)$ and rotation $(\mathcal{J}_i^\varphi)$ of the j-th storey (j=b,t). Two indexes $(\mathcal{J}_i^F)$ represent the ratio between the maximum force in the i-th damper of the controlled structure and a conventional force $F^{uc}=1000N$.

The seismic response of the structure in the configuration CS is characterized by mono-dimensional motion in the ae direction, whereas in the configuration CN the coupling between the ae displacement and rotation gives three-dimensional motion (Figure 3). The performance indexes in Table 4 show that the passive control is effective in mitigating the peak response, with a mean reduction of about 14% for the displacement, and about 17% for the acceleration of the second storey, in CS configuration. In CN configuration the peak reduction is lower, but still ranging around 11% for the displacement and 12% for the acceleration, while a satisfying 17% reduction of the rotation peak is also obtained. The benefit of introducing the semiactive control is evident from Table 5. With the only exception of the earthquake 3, characterized by highly non-stationary content [5], the response peaks are further reduced, up to 30% for both the displacement and the acceleration of the second storey, in CS configuration. The semiactive control in CN configuration still gives good results for the displacement (29%). Differently, the results in terms of rotation (17%-reduction) are similar to those of the passive control, while even negative performances affect some of the acceleration peaks.

TABLE 4: Performance indexes of the passive control for the structure seismic response.

	Configuration CS						Configuration CN							
	$\mathcal{J}_b^u$	$\mathcal{J}_t^u$	$\mathcal{J}_b^a$	$\mathcal{J}_t^a$	$\mathcal{J}_1^F$	$\mathcal{J}_2^F$	$\mathcal{J}_b^u$	$\mathcal{J}_t^u$	$\mathcal{J}_b^a$	$\mathcal{J}_t^a$	$\mathcal{J}_b^\varphi$	$\mathcal{J}_t^\varphi$	$\mathcal{J}_1^F$	$\mathcal{J}_2^F$
Earthquake 1	0.939	0.854	0.717	0.814	0.228	0.228	0.772	0.893	0.711	0.952	0.835	0.808	0.236	0.201
Earthquake 2	0.813	0.821	0.736	0.778	0.374	0.374	0.880	0.876	0.717	0.823	0.732	0.731	0.334	0.282
Earthquake 3	0.843	0.850	0.789	0.832	0.568	0.568	0.812	0.813	0.804	0.814	0.772	0.785	0.497	0.425
Earthquake 4	0.849	0.854	0.880	0.846	0.338	0.338	0.913	0.913	0.865	0.905	0.862	0.860	0.325	0.294
Earthquake 5	0.881	0.884	0.788	0.841	0.369	0.369	0.887	0.890	0.852	0.908	0.907	0.903	0.241	0.220
Earthquake 6	0.791	0.790	0.826	0.788	0.343	0.343	0.913	0.910	0.792	0.887	0.879	0.857	0.357	0.294
Earthquake 7	0.878	0.952	0.835	0.921	0.204	0.204	0.873	0.909	0.744	0.863	0.829	0.840	0.312	0.270

TABLE 5: Performance indexes of the semiactive control for the structure seismic response.

	Configuration CS						Configuration CN							
	$\mathcal{J}_b^u$	$\mathcal{J}_t^u$	$\mathcal{J}_b^a$	$\mathcal{J}_t^a$	$\mathcal{J}_1^F$	$\mathcal{J}_2^F$	$\mathcal{J}_b^u$	$\mathcal{J}_t^u$	$\mathcal{J}_b^a$	$\mathcal{J}_t^a$	$\mathcal{J}_b^\varphi$	$\mathcal{J}_t^\varphi$	$\mathcal{J}_1^F$	$\mathcal{J}_2^F$
Earthquake 1	0.797	0.706	0.821	0.653	2.104	2.104	0.673	0.725	1.130	1.176	0.888	0.697	2.115	2.122
Earthquake 2	0.551	0.546	0.805	0.584	2.114	2.114	0.685	0.669	1.327	1.640	0.784	0.641	2.102	2.113
Earthquake 3	0.798	0.794	1.242	0.849	2.140	2.140	0.625	0.632	1.364	1.577	0.956	0.867	2.121	2.126
Earthquake 4	0.709	0.690	1.026	0.731	2.124	2.124	0.767	0.770	0.848	1.216	0.995	0.862	2.127	2.140
Earthquake 5	0.681	0.659	0.915	0.650	2.123	2.123	0.657	0.587	0.728	1.420	0.798	0.813	2.070	2.046
Earthquake 6	0.501	0.502	0.982	0.585	2.155	2.155	0.888	0.884	0.883	1.093	1.185	0.996	2.143	2.127
Earthquake 7	0.803	1.004	0.834	0.841	2.103	2.103	0.705	0.737	1.301	1.444	1.111	0.907	2.074	2.088

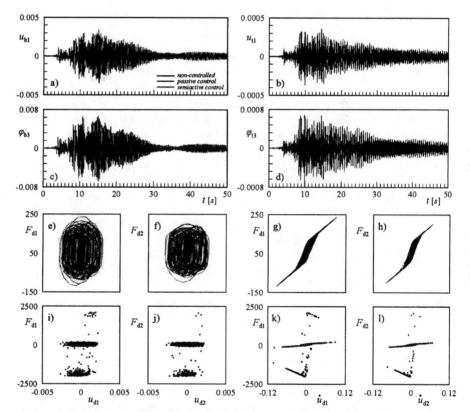

FIGURE 3: Non-controlled and controlled response of the CN structure to earthquake 1: time histories of the (a),(c) *ae* displacement u_{b1} and u_{t1} [*m*] and (b),(d) rotation φ_{b3} and φ_{t3}; cycles of the (e),(f),(g),(h) passive and (i),(j),(h),(k) semi-active control forces $F_{d1,2}$ [*N*] versus displacement $u_{d1,2}$ [*m*] and velocity $\dot{u}_{d1,2}$ [*m/s*] of the application point.

160

CONCLUSIONS

A reduced order analytical model of a prototypal frame structure is formulated and updated according to the spectral identification of a physical model. Passive and semiactive strategies for mitigating the structure seismic response through magnetorheological dampers are analyzed, and their performance in controlling the three-dimensional dynamics due to monodimensional ground acceleration history in presence of mass eccentricity are compared. A clipped-optimal semiactive control law is demonstrated to efficiently mitigate the peak response, enhancing the passive control performance.

ACKNOWLEDGMENTS

The present paper has been supported within the Project *DPC-ReLUIS 2005-2008, RL n.7 "Technologies for the isolation and control of structures and infrastructures"*.

REFERENCES

[1] Yang G., Spencer B., Carlson J., Sain M. (2002), Large-scale MR fluid dampers: modeling and dynamic performance considerations, *Engineering Structures* 24, 309-323.

[2] Occhiuzzi A., Spizzuoco M., Serino G. (2003), Experimental analysis of magnetorheological dampers for structural control, *Smart Materials & Structures* 12, 703-711.

[3] Yoshida O., Dyke S.J., Giocosa L.M., Truman K.Z. (2003), Experimental verification of torsional response control of asymmetric buildings using MR dampers, *Earthquake Engineering & Structural Dynamics* 32, 2085-2105.

[4] Ponzo F., Cardone D., Di Cesare A., Moroni C., Nigro D., Vigoriti G., Dynamic tests on JETPACS steel frame: experimental model set up. *DPC-ReLUIS 2005-2008 Report* 3/2007.

[5] Carneiro R., Gattulli V., Lepidi M., Potenza F., Mitigation of three-dimensional vibrations of a frame structure using MR dampers. 11th *Int. Conference on Civil, Structural & Environmental Engineering Computing* CC2007, St. Julians (Malta), September 2007.

[6] Spencer B.F., Dyke S.J., Sain M.K., Carlson J.D. (1997), Phenomenological model for magnetorheological dampers. *J. Engineering Mechanics* 123(3), 230-238.

[7] Wen Y.K. (1976), Method of random vibration of hysteretic systems. *ASCE J. Engineering Mechanical Division* 102 (2), 249-263.

[8] Basili M. (2006), Controllo semi attivo di strutture adiacenti mediante dispositivi magnetoreologici: teoria, sperimentazione e modellazione. *DSc.Thesis*, University of Rome 'La Sapienza' (in Italian).

[9] Jansen L.M., Dyke S.J. (2000), Semi-active control strategies for MR dampers: a comparative study. *ASCE J. Engineering Mechanics* 126(8), 795-803.

[10] Peeters B., De Roeck G. (2001), Stochastic System Identification for Operational Modal Analysis: A Review. *J. Dynamic Systems Measurement & Control* 123(4), 659-667.

[11] Parloo E., Verboven P., Guillaume P., VanOvermeire M. (2002), Sensitivity-based operational mode shape normalisation. *Mechanical Systems & Signal Processing* 16(5), 757-767.

[12] Gattulli V., Lepidi M., Potenza F., Identification of analytical and finite element models for the JETPACS three-dimensional frame. *DPC-ReLUIS 2005-2008 Report* 2/2007.

Smart Structures with Fibre-Optic Technologies

Andrea Del Grosso[a], Donato Zangani[b], and Thomas Messervey[b]

[a]Department of Civil, Environmental and Architectural Engineering, University of Genoa,
Via Montallegro 1, 16145 Genova, Italy.
[b]D'Appolonia S.p.A.,
Via San Nazaro 19, Genova, Italy

Abstract. A number of smart structures have been proposed, and some of them realized, to reduce the effect that seismic motions induce on the structure themselves. In particular, active and semi-active control devices have been studied for being applied to buildings and bridges in seismic prone regions. The heart of the application for these devices consists of a network of sensors and computational nodes that produces the input to the actuating mechanisms. Despite the initial enthusiasm for these developments, only a few practical applications involving active devices have been implemented to-date, the main reason residing in questions concerning the reliability of active systems over time. Nevertheless, the allocation of sensory systems and computational intelligence in structures subjected to earthquakes can provide very important information on the real structural behavior, provide self-diagnosis functions after events, and allow for reliability estimates of critical components. The paper reviews several recently developed sensory devices and diagnostic algorithms that may be applied to existing structures or embedded in new ones for the above purpose. Special emphasis will be given to fibre optic technology and its applications.

Keywords: Smart Structure, SHM, Multi-Functional Textiles, Fiber Optic Sensor.

INTRODUCTION

A smart structure can be defined as any engineered device that can react to an external input by modifying some of its features like stiffness, damping properties, shape, characteristics of its surface, etc. as the result of a built-in sense-decide-act capability. In the structural engineering field, this concept is aimed at taking control over the mechanical behaviour of a structural system. The implementation of the smart structure concept, originally developed for application to aerospace, terrestrial and marine vehicles, into civil engineering structures dates back to some twenty years ago, when the conception of actively controlled buildings was first realized in Japan [1] to help tall buildings and bridges withstanding earthquake and wind forces.

Since then, a great research effort has been devoted to the study of actively controlled civil structures, resulting in a number of proposals to realize fully active or semi-active systems aimed at reducing the effect of seismic motions. However, very few of these ideas have been implemented in practice. Several reasons explain this gap between research and applications. Apart from cost of smart structures and the fact that most of the risk in seismic prone areas is associated to the response of existing structures, where the complete implementation of the smart structure concept

CP1020, *2008 Seismic Engineering Conference Commemorating the 1908 Messina and Reggio Calabria Earthquake*,
edited by A. Santini and N. Moraci
© 2008 American Institute of Physics 978-0-7354-0542-4/08/$23.00

is obviously out of question, skepticism towards the usefulness of this concept mainly resides in ensuring the durability and reliability of such systems. Indeed, it should not be expected that design codes allow the use of such systems as a general tool until their reliability (functionality over time) can be demonstrated in practice. To date, very few studies have appeared in the literature on this subject. The related concepts to conduct such a study are discussed in the following section. Afterwards, the subject of monitored structures that implement only a part of the smart structure paradigm (the sense-decide chain) will be presented as part of an ongoing European Commission funded project named POLYTECT [2] with emphasis on fibre optic technologies. Finally, the topic of damage detection using the collected information is introduced.

RELIABILITY CONCEPTS FOR SMART SYSTEMS

To transfer into practice the smart structure concept, a civil structure must be equipped with a network of sensors able to detect the structural response. Then, a computerized system must log data and perform various logic functions on the data sending instructions to a driving a set of actuators that modify some of the structural features and therefore the structural response in a closed-loop configuration.

These components must perform for the entire life of the structure, i.e. at least for several decades. It is worth noting that this requirement is very hard to be satisfied by currently available products. Electronic components, for example, may be very quickly outdated because of technological evolution and their maintenance could become impossible even a few years after installation, thus requiring a complete substitution. Mechanical components can have a longer life but require continuous maintenance. On the other hand, modern sensor technologies offer products that may really remain in service for the entire life of the structure, practically without requiring maintenance.

These considerations must be specifically taken into account if the structural safety against limit states is made dependent upon smart systems. In a previous work [3], it was noted that the introduction of smart devices raises the level of complexity of civil structures. In addition, smart devices are generally heterogeneous in nature with respect to conventional structural components. To better discuss the problem, let us consider the following example. Suppose that the safety of a structure subjected to given load conditions for a reference period (e.g., the design life of the structure) is measured by the probability of failure with respect to a limit state L: $P_f = P(L)$. If a smart device potentially able to prevent the limit state under the hypothesis of correct operation during the entire reference period is introduced, the probability of failure becomes

$$P_f = P(L * O)P(O) + P(L * \tilde{O})P(\tilde{O}). \tag{1}$$

where $P(L*O)$ is the probability of failure for the limit state L, given that the smart device operates properly, $P(O)$ is the probability that the device is operating properly, $P(L*\tilde{O})$ is the probability of failure given that the smart device is not operating properly, and $P(\tilde{O})$ is the probability that the device is not operating properly. In this example, it is assumed that the device has only two states of functioning: operational and not operational; therefore $P(\tilde{O}) = 1 - P(O)$.

Computation of the terms appearing in Equation (1) requires different approaches. Performance of reliability, availability and maintainability analysis of the device over time is needed to compute $P(O)$. This analysis shall take into account events that may be internal as well as external (e.g. fault of energy supply, etc.) to the device. Complexity and dependability of the analysis will be a function of the nature of the device and of the availability of stochastic models. The term $P(L*\tilde{O})$ can be derived from a reliability analysis of the original structure, while $P(L*O)$ requires the development of a stochastic mechanical model that includes the interaction between structure and device. Unless the probability of malfunction of the device can be rendered as small as to neglect the second addend, reliability analyses for smart structures may require performance of system reliability analyses based on the development of complex and heterogeneous sequences of events.

Therefore, it appears that simple safety measures based on conventional partial safety factors for resistance and loads related to an acceptable probability of failure or on a reliability index, as implemented in current design practices, could not be adequate to cope with the generality of the situations encountered in smart structures. Consequently, it is unlikely that, at the present stage and apart from other considerations, a smart structure in which the introduction of smart devices is aimed at preventing severe consequences like the attainment of ultimate limit states will be accepted in the civil engineering field without clear restrictions.

The problem is completely different when considering monitored structures, in which smart technologies are applied to provide structural health monitoring (SHM) functions. In this case, sensing equipments and the associated hw/sw logic have the aim of assessing the integrity state of a given structure in order to support a continuous reliability estimate of the structural system over its entire operational life. The addition of monitoring technologies to a conventional structure cannot negatively affect the structural reliability. Although the safety index may remain unchanged, it is reasonable to assume a higher level of safety through reduced uncertainty of the resistance and load parameters. However, a reduction of the partial safety factors must also include the consideration of the probability for false positive/negative conclusions from the sensor data acquisition (DAQ) logic processing chain. To this end, reliability studies for monitored structures have not been performed yet for civil engineering applications, but the implementation of SHM to perform self-diagnosis functions, reduce design uncertainties, and to realize highly efficient light-weight structures is already a common practice in aeronautical engineering. In addition, such procedures are admitted for the safety evaluation of existing offshore platforms and recent standards even open the door for interesting developments in the field of marine vehicles.

The exploitation of the SHM concept in civil engineering structures is therefore the implementation of the smart structures concept that has the highest potential for practical applications in the short term and may actually lead to a breakthrough in the design/construction and life-cycle management of civil structures [4].

MONITORED STRUCTURES

The application of SHM to civil infrastructure has intensified in recent years as technological advancements have made such applications smaller, cheaper, and less intrusive to the normal functionality of the structure. Coupled with advances in the assessment techniques that utilize SHM data, designers, engineers, and infrastructure managers alike are becoming more aware of and accepting to the use of these technologies. For some applications, such as the assessment of an intact structure after an earthquake without disrupting its structural integrity or use, smart systems provide compelling advantages. Although most common to the structural health assessment (SHA) of highway bridges, the use of optical sensors for monitoring also extends to offshore structures and pipelines, historical buildings, and geotechnical applications.

Although monitoring requirements may vary from case to case, many needs can be met by providing information on displacement, acceleration, temperature, and means to test for the presence of a desired agent or chemical. Under the Polytect research project funded by the European Commission [2], new monitoring and sensing technologies are being developed for application in construction by embedding sensors of various types into textile materials. These new smart multi-functional textiles are intended to both provide reinforcing strength and monitoring information for masonry construction and geotechnical structures including embankments, dikes, retaining structures, and encased columns. By integrating sensors directly into these materials during their fabrication, direct measurements of stress/strain are possible during all phases of construction and during the service life. Obtaining such information after the fact can be both difficult and expensive.

The sensing elements under consideration for integration into textiles include optical fibres and sensitive fibres, such as those characterized by piezoelectric properties. The fabrication process to integrate these sensors into the textiles requires special attention to the stresses placed upon the sensors during production. In contrast, application to masonry structures requires special attention to the adhesives, coatings, and bonding agents used both during textile fabrication and during application to the structure of interest. These aspects are under investigation as part of the POLYTECT project. Typical fabric architectures considered in the study are shown in Figure 1 and include net-like structures that integrate high strength fibres and multi-axial fabrics.

 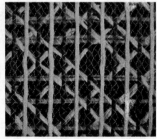

FIGURE 1: Textile architectures used for reinforcement of masonry structures include net-like structures integrating high strength fibres (left, courtesy of STFI, Germany) and multi-axial fabrics (right, courtesy of SELCOM, Italy).

FIBRE OPTIC SENSING TECHNOLOGY

Fibre optic sensors act on the principle of transmitting a light signal through the fibre and measuring the status of the returning or transmitted signal. The change in signal properties is then translated into appropriate quantities that allow the measurement of a wide range of mechanical, physical, and chemical quantities. The core of the fibre is generally made of silica glass and the cladding is made of plastic materials which give the fibre protection and mechanical strength. Typical core diameters are ~10 microns for single mode fibres and 50, 62.5, and 100 microns for multimode fibres. Their small size and the ability to operate in harsh environments are notable advantages of these sensors. A complete description of fibre optic sensors types and capabilities, data acquisition systems, and case studies of their application can be found in the newly published book Fibre Optic Methods for Structural Health Monitoring [5].

Most common fibre optic sensors used for short and long based temperature and/or strain sensing in civil structures include Fibre Bragg Gratings, Optical Time Domain Reflectometry (OTDR), Fabric Perot, and the SOFO system. In order to realize distributed fibre sensors for mechanical and physical parameters, fibre sensor technologies based on non-linear effects in optical fibres, like stimulated Brillouin scattering as well as OTDR sensor techniques, are used. The use of Fibre Bragg Grating sensors is suitable to be applied as point-wise or quasi-distributed sensors. Some of these fibre sensors can be realized not only based on glass fibres but also based on polymer optical fibres (POF) - a new and very promising technology - which has a number of advantages, including a higher deformability than glass fibres and therefore a higher suitability to be processed by textile machines.

Fibre Bragg Grating (FBG) is a short segment of an optical fibre with a periodically varying refractive index in the core of the fibre. Such segments then act as a mirror reflecting only a specific wavelength of light propagating through the fibre and which is incident on this segment. The rest of the spectrum is transmitted. The wavelength of the reflected light then only depends on the grating period of the FBG. When the FBG is exposed to the influence of the ambient environment (stress, temperature change, pressure) it may be elongated or shrink causing the change of reflected light wavelength. This wavelength change is detected by means of optical spectroscopy and is interpreted as a change of corresponding physical quantity measured. As one FBG can provide sensing only in one point, it is possible and necessary to inscribe more FBGs into a single fibre if there is a need for distributed measurements.

The Optical Time Domain Reflectometry (OTDR) uses Rayleigh light scattering in optical fibres to analyze local attenuations along the fibre. For these purposes, a short light pulse is sent into the fibre and the backscattered light is recorded as a function of time or distance, respectively. Any mechanical deformations like strain applied to any location of the fibre will change its physical properties (e.g. the refractive index) at this location and will result in a change of the scattered light. Experimental investigations show that strain up to more than 40 % can be measured using this sensor technique with a strain resolution of about 1 %.

Extrinsic Fabry-Perot Interferometers (EFPIs) are constituted by a capillary silica tube containing two cleaved optical fibres facing each other's, but leaving an air gap of a few microns or tens of microns between them. When light is launched into one of the fibres, a back-reflected interference signal is obtained. This is due to the reflection of the incoming light on the glass-to-air and on air-to-glass interfaces. This interference can be demodulated using coherent or low-coherence techniques to reconstruct the changes in the fibre spacing. Since the two fibres are attached to the capillary tube near its two extremities (with a typical spacing of 10 mm), the gap change will correspond to the average strain variation between the two attachment points.

The SOFO system is a fibre optic displacement sensor with a resolution in the micrometer range and an excellent long-term stability. It was developed at the Swiss Federal Institute of Technology in Lausanne (EPFL) and is now commercialized by SMARTEC in Switzerland. The measurement setup uses low-coherence interferometry to measure the length difference between two optical fibres installed on the structure to be monitored. The measurement fibre is pre-tensioned and mechanically coupled to the structure at two anchorage points in order to follow its deformations while the reference fibre is free and acts as the temperature reference. Both fibres are installed inside the same tube and the measurement basis can be chosen between 200mm and 10m. The resolution of the system is of 2 μm independently from the measurement basis and its precision of 0.2% of the measured deformation is maintained even over years of operation.

Brillouin scattering is an intrinsic property related to the propagation of light in the silica material from which the sensing fibre is made. The Brillouin scattering effect exhibits a well-known and reproducible response to external measurands such as temperature and strain and this method is well suited to distributed measurements.

Currently, sensor integrated textiles are being tested for the two applications considered within the POLYTECT project (soil structures/embankments and masonry walls). In particular, Figure 2 shows field testing of the geotextiles and the laboratory set-up for the testing of a masonry wall subjected to lateral loading. Different sensor types and sensor configurations are under investigation for comparison.

FIGURE 2: Testing of technical textiles integrating sensors for geotechnical applications (left, courtesy of STFI, Germany) and reinforcement/monitoring of masonry structures (right, courtesy of Karlsruhe University).

DAMAGE DETECTION ALGORITHMS

Data acquisition systems and the software that interfaces with smart structures must be able to perform various functions including data fusion, data analysis and interpretation, and eventually decision support. The core task of the envisioned software system is the processing of the data that enables the identification, quantification, and localization of damage. Indeed, the reliability of an SHM system is dependent upon the uncertainties associated with how well these functions are performed. It is clear that an efficient and fully functional SHM system in post-earthquake conditions could dramatically reduce time and costs required for the evaluation of building usability and speed up the design of repairs.

Approaches based on the measurement of the dynamic response have been more extensively applied and the corresponding algorithms are better known and understood than for the approaches based on the measurement of static responses. This is because the measurement of the dynamic response of a structure for ambient as well as forced excitations can be performed with standard accelerometric instrumentation, more familiar to civil and industrial engineers. Identification of the relevant dynamic properties from dynamic measurements is also well established. Reliable software systems are commercially available to perform this task.

Damage detection and localization are possible by comparing the dynamic properties of the structure at different times, and analyzing the changes in the eigenfrequencies and eigenmodes. However, the procedures that are used to interpret these differences are not simple. In addition, the deterioration of the structural resistance is not the sole reason that may induce changes in the dynamic response, thus rendering the identification of damages of small intensity very difficult.

Damage identification by means of a static approach requires the availability of sensors that are stable over a very long period of time (typically years) and automatic continuous data collection. Recent advances in fibre-optic technologies has made such approaches both possible and cost effective. A program using these technologies to monitor the columns of high rise buildings in Singapore is found in [6].

In the continuous static approach the identification of the structural system is not necessary and damage identification can be performed solely by comparing signal features at different time-windows. Variations in these features may be associated with structural degradation and the location of the damage can also be identified. Several processing tools have been made available to perform damage identification from static monitoring signals, each one possessing advantages and disadvantages.

Data analysis and interpretation is the subsystem that is responsible for feature extraction and damage identification. The selection of the appropriate algorithms is however application dependent. An extensive critical review of available methods can be found in [7]. Basically, model-based or non-model based schemes can be used.

Model-based techniques utilize changes in the response functions to identify damage location and levels, while non-model based techniques determine direct changes in the sensor output signals to identify the presence of damage in the structure. Non-model based techniques are also capable of localizing and quantifying damage without making use of structural models.

Recent studies on data analysis and interpretation for continuous static and periodic dynamic monitoring using fiber optic sensors have shown that algorithms based on proper orthogonal decomposition methods [8] and wavelet packet decomposition [9] have been successfully applied to detect relatively small damages produced in computer simulated experiments on r.c. beams. Laboratory experiments are presently under way.

It has to be pointed out that damages produced by earthquakes induce permanent changes in the structural behavior and can therefore be detected by continuous static as well as periodic dynamic monitoring. Simple analysis of distributed sensing data from Brilloin type sensors has also been successfully used to detect distresses and leakage in pipelines [10]. Ground movements due to earthquakes very often cause such events.

CONCLUSIONS

The smart structure concept offers great potential to the design, management, and use of civil infrastructure to include the considerations surrounding extreme events such as earthquakes. Although the development of fully operational smart systems are not yet complete or implemented in practice, promising developments treating parts of the end product are emerging. After reviewing the probabilistic requirements necessary to describe the reliability of such systems, recent advances in the development of sensor-embedded multifunctional textiles was presented as part of an ongoing research project. Such materials are envisioned to provide strength, reinforcement, and the monitoring data required by the data processing module of the smart system. Recent progress in the damage detection algorithms that use this data, their capabilities, and limitations were also discussed. Through such developments and continued research efforts on these fronts, the realization and implementation of the smart structure concept will be feasible in the foreseeable future.

REFERENCES

1. M Sakamoto and T Kobori, "Research, development and practical applications on structural response control of buildings", *Smart Mater. Struct.* **4** A58 (1995).
2. POLYTECT, 2006, "Polyfunctional Technical Textiles against Natural Hazards", EC contract number NMP2-CT-2006-026789A.
3. Del Grosso, "Safety concepts and safety measures in smart civil structures", *Proc. 1st European Conference on Structural Control*, A Baratta and J Rodellar Eds., World Scientific (1996).
4. Frangopol, D.M., Messervey, T.B. Integrated Maintenance-Monitoring-Management for Optimal Decision Making in Bridge Life-Cycle Performance. *Proceedings* of the NSF Civil, Mechanical and Manufacturing Innovation (CMMI) Engineering Research and Innovation Conference, Knoxville, Tennessee (2008).
5. Glisic, B. and Inaudi, D., *Fibre Optic Methods for Structural Health Monitoring*. John Wiley & Sons, LTD, England, (2007).
6. Glisic, B., Inaudi, D., Lau, J.M. Long-term monitoring of high-rise buildings using long-gage fiber optic sensors, *Proceedings IFHS2005*, Dubia, United Arab Emirates, (2005)
7. F Lanata, *Damage detection algorithms for continuous static monitoring of structures*, PhD Thesis, University of Genoa, Dep. of Struct. & Geot. Eng. (2005).
8. Lanate, F., Del Grosso, A., "Damage detection and localization for continuous static monitoring of structures using a Proper Orthogonal Decomposition (POD) of signals", *Smart Mater. Struct.* Vol. 15 (2006).
9. Lanata, F., Del Grosso, A., "Corrosion detection in reinforced concrete structures using static and dynamic response measurements," *Proceedings, IALCCE'08*, Como, Italy (2008).
10. Inaudi D., A. Del Grosso, "Fiber Optic Sensing Technologies for Smart Materials and Structures", *World Forum on Smart Materials and Structures*, Chongquing & Nanjing (2007).

On the effectiveness of smart technologies in the seismic protection of existing buildings
Part I : Masonry structures

A. Mandara, F. Ramundo and G. Spina

Department of Civil Engineering, Second University of Naples
Real Casa dell'Annunziata, via Roma 29 – Aversa (CE), Italy.

Abstract. The first part of a study concerning innovative intervention techniques for dissipate a share of the input seismic energy compatible with the preservation of existing buildings, including historical and monumental constructions, is presented in this paper. The case of a typical scheme of a long-bay box-like masonry building fitted with a dissipative floating roof is analyzed. In the examined building a wide simulation analysis has shown the achievement of a very satisfying performance. Furthermore, the effectiveness of the system can be maximized by means of active or semi-active devices implemented in the floating roof and a significant reduction of the seismic impact on the building can be obtained compared with non-controlled or simply passively controlled structure. The results prove the remarkable increase of the energy dissipation capability of the system, as well as the reduction of structural damage, independently of any specific strengthening intervention.

Keywords: Dynamic control, masonry structures, seismic upgrading, smart technologies.

INTRODUCTION

The use of structural control systems for seismic protection of structures represents a relative new area of research that is growing rapidly. The dynamic response control techniques in structural engineering has been applied in the last decades to preserve both comfort and structural integrity of relevant buildings, like towers, hospitals and financial headquarters. Innovative intervention techniques based on the use of special devices, able to dissipate a share of the input seismic energy, have been demonstrated to be effective and compatible with the preservation of the features of historical constructions [1]. A recent, natural evolution of these techniques involves the use of smart materials, which can be used to create controllable devices, whose performances are much more suitable to the aim of a reduction of the seismic response.

The terms *smart structures, intelligent structures, adaptive structures* all belong to the same field, namely that of *smart technologies*. All these terms refer to the integration of actuators, sensors in structural components, together with some kind of control unit or enhanced signal processing, with a material or structural component. The goal of this integration is the creation of a mechanical system having optimized structural performance, but without adding too much mass or consuming too much power. Because of its inherent nature and contrary to simply passive systems, the field of smart structures relies upon inter-disciplinary research, since numerous disciplines (e.g. material science, applied mechanics, control theory, etc.) are involved in the design of any structure making

CP1020, *2008 Seismic Engineering Conference Commemorating the 1908 Messina and Reggio Calabria Earthquake,*
edited by A. Santini and N. Moraci

use of these solutions [2]. In comparison with passive systems, there are some significant advantages associated with smart systems: enhanced effectiveness in dynamic control, relative insensitivity to site conditions and ground motion, applicability to multi-hazard mitigation situations, selectivity of control objectives. Among smart control systems, the semi-actives are the most attractive for structural protection, because of the very low energy consumption requirements, the effectiveness comparable to active systems and the reliability when working as passive systems. Magnetorheological Fluid Dampers (MRD), an application of which is presented herein, represent one of the best examples of smart devices, due to their ability to dissipate energy and their low power requirements.

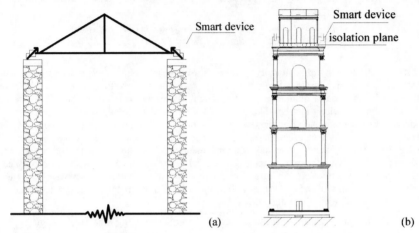

FIGURE 1. Examples of application for the MD system in a box-like building (a) and a tower(b).

INTERVENTION APPROACH

A possible implementation of MRD in structural control is within the strategy of mass damping (MD). Structural control systems based on mass dampers demonstrated their effectiveness in application on flexible structures, such as towers and long span bridges, especially for the reduction of wind induced vibrations. Mass damping system represents a simple and effective solution for the control of seismically induced vibrations, applicable when it is possible to couple the main structure with an auxiliary mass, so as to make such a mass floating respect to the rest of the construction. Since in this case the energy dissipated by MD does not depend on the relative motion of internal parts of the structure, the system can be relatively easily implemented on existing structures, including historical and monumental buildings [3,4]. The disconnection of a part of the construction itself to obtain a significant mass for the MD system represents a profitable strategy of intervention, two possibilities of which are represented in Fig. 1.

The seismic response of structural systems representing typical constructions of Italian architectural heritage, like long-bay, box-like masonry building and masonry towers can be approximately described by means of simplified mechanical models. In such a way, the reference models of Fig. 1 can be represented by the equivalent dynamic scheme of Fig. 2, where the mass of the top is connected to the masonry structure by equivalent links arranged in parallel, corresponding to a spring, a viscous

171

damper and a smart device. In the case of active control, an actuator is considered instead of the MRD present in case of semi-active control.

ANALYTICAL MODEL

The aim of active and semi-active structural control is to give a "sense of equilibrium" to the structure, so as to provide it with the capability to self-regulate instantaneously its properties as a function of the structural response [5]. The response control concept can be easily understood referring to the dynamic equilibrium equation of a generic n-degree-of-freedom structural system subjected to external excitations and control forces:

$$M\ddot{x}(t) + C\dot{x}(t) + Kx(t) = L_e f_e(t) + L_c f_c(t) \qquad (1)$$

in which M, C and K are the $n \times n$ mass, viscous and stiffness matrices of the system, respectively, $x(t)$ is the n-dimensional displacement vector, $f_e(t)$ is the r-dimensional external excitation vector and $f_c(t)$ is the m-dimensional control force vector. L_e and L_c are the $n \times m$ and $n \times r$ matrices that define the locations of the excitation and control force vector, respectively. In the more general feedforward-feedback configuration, the applied control force f_c is normally a linear function of structural displacements and velocities as well as of external forces. In a general form, it is given by:

$$f_c(t) = K_c x(t) + C_c \dot{x}(t) + E_c f_e(t) \qquad (2)$$

where K_c, C_c and E_c are the control gain matrices, the choice of which depends on the control algorithm designed. By substituting the general form of the control force vector into the dynamic equilibrium equation (1) one may obtain:

$$M\ddot{x}(t) + (C - L_c C_c)\dot{x}(t) + (K - L_c K_c)x(t) = (L_e + L_c E_c)f_e(t) \qquad (3)$$

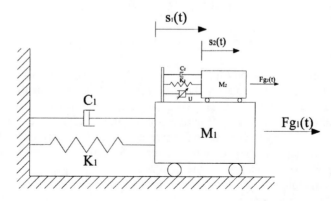

FIGURE 2. Scheme of the dynamic system considered in the numerical analysis.

172

Observing the new equation of the controlled structure, it can be seen that the effect of the feedback control is to modify the mechanical parameters (stiffness and damping) so that the structure can respond more favorably to the external excitation. On the contrary, the effect of the feedforward control is a modification (reduction or total elimination) of the excitation.

The analyzed control system is based on MD as shown in Fig. 2, where a simplified model of the controlled structure is represented in the form of a SDOF system, whose natural frequency vibration is tuned to the first mode frequency of the real structure, evaluated according to its relevant dynamic parameters (Table 1). When a either active or semi-active control strategy is applied, the equations of motion for the structure provided with the control system are:

$$\begin{cases} m_1\ddot{s}_1 + c_1\dot{s}_1 + k_1 s_1 = -m_1\ddot{s}_g - u(t) + k_2(s_2 - s_1) + c_2(\dot{s}_2 - \dot{s}_1) \\ m_2\ddot{s}_2 + c_2(\dot{s}_2 - \dot{s}_1) + k_2(s_2 - s_1) = -m_2\ddot{s}_g + u(t) \end{cases}$$

(4)

In the state space the equations of motion take the form:

$$\dot{x} = Ax + B_1 f_e + B_2 f_c$$
$$y = Cx + D_1 f_e + D_2 f_c$$

(5)

where the matrices that define the system are the following:

$$A = \begin{bmatrix} 0 & 0 & 1 & 0 \\ 0 & 0 & 0 & 1 \\ -\dfrac{k_1 + k_2}{m_1} & \dfrac{k_2}{m_1} & -\dfrac{c_1 + c_2}{m_1} & \dfrac{c_2}{m_1} \\ \dfrac{k_2}{m_2} & -\dfrac{k_2}{m_2} & \dfrac{c_2}{m_2} & \dfrac{c_2}{m_2} \end{bmatrix} \quad B_1 = \begin{bmatrix} 0 \\ 0 \\ \dfrac{1}{m_1} \\ \dfrac{1}{m_2} \end{bmatrix} \quad B_2 = \begin{bmatrix} 0 \\ 0 \\ -\dfrac{1}{m_1} \\ \dfrac{1}{m_2} \end{bmatrix}$$

(6)

$$C = I_{4x4}; \quad D_1 = O_{4x1}; \quad D_2 = O_{4x1}$$

CONTROL STRATEGIES

The controlled system of Fig. 2 represents the equivalent SDOF system of the reference structures, the secondary mass M_2 representing the mass of the floating top, which is connected to the principal mass M_1 by a spring, a viscous damper and a smart device. As already said, in case of active control, the MRD allowed for in the case of semi-active control is replaced by an active actuator. A Linear Quadratic Regulator (LQR) methodology is applied with reference to actively controlled structure, the control strategy being focused on the displacement of the principal mass (M_1). Two different methodologies of semi-active control are then considered, in order to compare the results obtained.

TABLE 1. Dynamic parameters of the equivalent SDOF dynamic model considered in the analysis.

Principal Mass M_1	Mass ratio (M_2/M_1)	Vibration period	Link device threshold
50 t	5 %	0,45 s	5 kN

The optimal control strategy based on the LQR performance criteria is a simple and effective way to design a control law able to reach the highest performance at the lowest cost (limit the control force). In this case, the dynamic equilibrium equation (1) can be expressed in the state space as:

$$\dot{x}(t) = Ax(t) + B_1 f_e(t) + B_2 f_c(t) \qquad (7)$$

The LQR performance index is given by:

$$J = \frac{1}{2} \int_{t_0}^{t_f} \left[x(t)^T Q x(t) + f_c(t)^T R f_c(t) \right] dt \qquad (8)$$

where Q and R are the time invariant weights of the state and of the control force, respectively, $u(t)$ is a linear function of the state and the gain matrix G depends on the solution of the algebraic Riccati equation:

$$-PA - A^T P + PBR^{-1}B^T P - Q = 0; \quad G = R^{-1}B^T P; \quad f_c(t) = -Gx(t) \qquad (9)$$

By substituting the second of Equations (9) into Equation (7), the state equation of the optimal controlled system becomes:

$$\dot{x}(t) = \left(A - GB_2 \right) x(t) + B_1 f_e(t) \qquad (10)$$

In the case of semi-active system the control strategy is defined by a relationship between the relative structural motion (i.e. velocity) and the control force. An MRD-MD system requires a robust control to account for the possible nonlinear behaviour of the damper, nonlinear excitation and the potential hysteretic behaviour of the structure. The simplest method of control is based on an *on-off* behaviour for the damper. In a few words, a continuous decision loop evaluates the motion within the structure adjusting the damper state based on the following criteria, defined by energy dissipation considerations:

$$\text{if } F_d \cdot \dot{x}_1 \leq 0 \rightarrow F_d = F_{d\max} \qquad \text{if } F_d \cdot \dot{x}_1 > 0 \rightarrow F_d = F_{d\min} \qquad (11)$$

This algorithm works so as to activate the MRD device only when its action is effective on the reduction of dynamic response of the structure. Dyke and Spencer [6] proposed an improved *on-off* clipped optimal algorithm for structural control based on MRD. The basic concept of the approach proposed in [6] is to create a feedback loop in order to drive the MRD to react with a force F_{opt} approximately equal to that of an ideal active actuator,

174

which is determined with the LQR optimal control strategy. The MRD is brought to generate the desired optimal control force by means of a command signal (i.e. the voltage V necessary to induce the required magnetic field) set out in a such a way to satisfy the following conditions:

$$\left(\left|F_{opt}\right|-\left|F_d\right|\right)\cdot sgn\left(F_{opt}\cdot F_d\right)\geq 0 \rightarrow F_d = F_{d\max}\left(V = V_{\max}\right)$$

$$\left(\left|F_{opt}\right|-\left|F_d\right|\right)\cdot sgn\left(F_{opt}\cdot F_d\right)< 0 \rightarrow F_d = F_{d\min}\left(V = 0\right)$$

$$(12)$$

RESULTS OF THE NUMERICAL ANALYSIS

The evaluation of the proposed control strategy effectiveness has been assessed carrying out a set of non-linear dynamic analyses of the system shown in Fig. 2 by means of the MatLab-Simulink toolbox, considering 8 seismic input recordings scaled at different values of PGA. For the sake of comparison, the system is considered in the not-controlled configuration as well as in passive, semi-active and active controlled configurations as described before. A synopsis of results is provided in Table 2.

The active controlled structure represented a target point for the semi-active system, whereas the passive control corresponds to the least performance level which can be obtained. The semi-active control, thanks to its adaptability to input characteristics and low power requirements, represents a reasonable and affordable compromise between performance and costs. In particular, the parametric analysis carried out put into evidence the good performance of Clipped-Optimal algorithm when applied to semi-active control systems. This result is related to the possibility of this control law to adjust the reaction force of MRD, keeping it quite close to the optimal control force value. Moreover, contrary to passive systems, the effect of both semi-active and active control, do not depend on the input intensity, as one may observe from the mean value of the maximum response reduction coming from each time history analysis (Table 2).

The reduction of maximum response in case of semi-active control is averagely 15% as respect to the solution with passive control and 36% as respect to the uncontrolled structure. The reduction of response in case of active control (Active Mass Driver) showed even better results, and this would represent a new frontier in the seismic protection target of historical heritage.

The benefit of structural control is more evident observing the global reduction of the structural response (Fig. 3). This aspect is strictly related to a lower elastic strain energy transmitted to the system, which is a fundamental aspect from the safeguard point of view. Observing the time history diagrams in term of elastic strain energy (Fig. 4) of the main structure (M_1) it is clear as control acts to drive the system toward the stable position, corresponding to a zero value of elastic strain energy (E_s). This consideration can be better understood observing the results summarized in Table 3, where the maximum values of E_s and the global elastic work of the internal force W of the main system are reported, for each time history analysis carried out with Calitri 1980 seismic input scaled at 0.15-0.25-0.35g. The reduced values of the elastic work of internal forces, in fact, for the semi-active controlled system respect to the passive and uncontrolled system, proves the significant increase of the structural system capacity to dissipate input energy without undergoing structural damage.

175

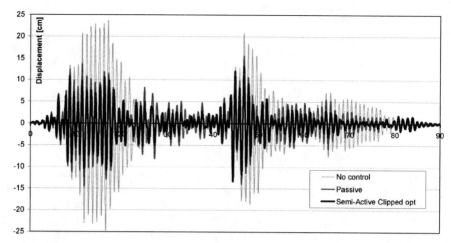

FIGURE 3. Dynamic model response time history under Calitri seismic input scaled to PGA 0.25 g.

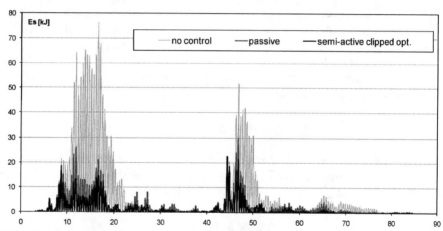

FIGURE 4. Dynamic model elastic strain energy time history under Calitri seismic input scaled to PGA 0.25 g.

TABLE 2. Percentage reduction of the peak values of the dynamic response under different seismic inputs (for P.G.A. 0.35-0.25-0.15 g) of the controlled systems with reference to the not-controlled case.

Seismic input	Passive Control	Semi-active Control ON-OFF	Semi-active Clipped Optimal	Active Control
	0.15 / 0.25 / 0.35g	0.15 / 0.25 / 0.35g	0.15 / 0.25 / 0.35g	0.15 / 0.25 / 0.35g
Bevagna	36 / 30 / 16	41 / 40 / 38	48 / 48 / 47	56 / 56 / 56
Colfiorito	35 / 31 / 24	41 / 41 / 38	35 / 42 / 42	69 / 69 / 69
Calitri	38 / 34 / 27	44 / 43 / 41	44 / 44 / 45	57 / 57 / 71
El Centro	28 / 23 / 12	37 / 36 / 35	41 / 42 / 42	64 / 64 / 64
Gubbio	16 / 14 / 12	30 / 28 / 24	30 / 30 / 31	37 / 37 / 37
M. S. Severino	26 / 21 / 16	29 / 27 / 23	39 / 39 / 39	63 / 63 / 63
Taiwan	25 / 25 / 14	24 / 23 / 20	21 / 22 / 20	49 / 49 / 49
Thessaloniki	23 / 18 / 9	33 / 30 / 25	40 / 40 / 38	63 / 63 / 63

TABLE 3. Elastic Strain Energy E_s (peak) and global elastic work W of internal force for the analyzed system under Calitri seismic input scaled to PGA = 0.15-0.25-0.35g.

P.G.A. (g)	No Control		Passive Control		Semi-active Control	
	E_s max (kJ)	W (kJ)	E_s max (kJ)	W (kJ)	E_s max (kJ)	W (kJ)
0.35	150	3759	53	1179	44	871
0.25	77	1918	30	669	23	439
0.15	28	961	13	284	8	155

CONCLUSIONS

Mass damping system represents a simple and effective solution for the control of seismically induced vibrations. The system turns to be applicable in all cases when it is possible to add to or disconnect a mass from the main structure, so as to make such a mass floating respect to the rest of the construction. This may happen in many categories of historical and monumental buildings. In the examined case a relatively wide simulation analysis has shown that a very satisfying performance can be achieved by means of this technique. In particular, the effectiveness of the system can be maximized by means of active or semi-active devices implemented in the floating roof. Moreover, the presented structural control system produces a significant reducing effect on the elastic strain energy transmitted by the external perturbation to the structure, which is itself an important target in the protection perspective. These results prove the significant increase of the structural system capacity to dissipate input energy without structural damage and this represents a significant aspect in the refurbishment of historical and monumental constructions.

ACKNOWLEDGMENTS

This paper has been developed within the research project "*Seismic protection of existing buildings by semi-active control techniques*" issued by Regione Campania - Italy (L.R. 5 - 28.03.2002) and coordinated by A. Mandara.

REFERENCES

1. G. Spina, "Seismic protection of historical buildings by response control systems" (in Italian), PhD Thesis, Second University of Naples, Naples, Italy, 2006.
2. A. Mandara, A. Durante, G. Spina, S. Ameduri, A. Concilio, Seismic protection of civil historical structures by MR dampers. Proceedings of the SPIE's 13th Annual International Symposium on Smart Structures and Materials. San Diego, CA, USA, 2006.
3. A. Mandara, A. Durante, F. Ramundo, G. Spina, Smart technologies for seismic protection of historical structures. Proceedings of the First European Conference on Earth-quake Engineering and Seismology. Geneva, Switzerland, 2006.
4. A. Mandara, A. Durante, F. Ramundo, G. Spina, Control of the seismic response of historical buildings by mass damping systems. Proceedings of the XII National Conference of ANIDIS. Pisa, Italy, 2007.
5. L. Meirovitch, *Elements of Vibration Analysis*. McGraw Hill, New York, 1986.
6. S.J. Dyke, B.F. Jr. Spencer, M.K. Sain, J.D. Carlson, An experimental study of MR dampers for seismic protection, in *Smart Structures and Materials*, 7, 1998, pp. 693-703.

On the effectiveness of smart technologies in the seismic protection of existing buildings
Part II : Reinforced concrete structures

A. Mandara, F. Ramundo and G. Spina

Department of Civil Engineering, Second University of Naples
Real Casa dell'Annunziata, via Roma 29 – Aversa (CE), Italy.

Abstract. The second part of a study concerning innovative intervention techniques for seismic protection of existing buildings is presented in this paper. The case of an existing framed r.c. structure, not designed for horizontal forces and extremely vulnerable to seismic action, is analyzed both in terms of maximum response reduction and energy dissipation. The proposed intervention approach, based on steel braces linked to the existing structure by passive or smart devices comes out appropriate and effective in the case of this type of buildings. The adopted control strategy produces a significant reducing effect on the elastic strain energy transmitted by the external perturbation to the structure, which is itself a fundamental safeguard aspect. The results prove the significantly improved capability of the system to dissipate input energy without structural damage, regardless of the specific seismic input.

Keywords: Dynamic control, r. c. structures, seismic upgrading, smart technologies.

INTRODUCTION

The use of innovative techniques, able to maintain the structures into elastic range also under earthquakes of the highest intensity for the considered site, comes out appropriate in the case of existing framed r.c. structures. This type of buildings, most of times not designed against horizontal seismic forces, have been found extremely vulnerable to seismic action, in terms of both strength and deformability [1]. Structural control systems based on dynamic coupled structures demonstrated their effectiveness in the reduction of seismic induced vibrations and can be relatively easily implemented and installed on existing structures [2]. This is also the case for smart technologies such as those relying on either active or semi-active systems [3]. Contrary to passive protection techniques, which are at now widely experienced and tested and are also codified in the form of guidelines and practice rules, the use of smart techniques is still far away from finding a definitive framing, especially when the seismic upgrading of existing constructions is concerned. In order to assess the effectiveness of such solutions, as an accomplishment of a similar investigation on masonry buildings [4], a study on the seismic improvement of existing r.c. buildings by means of integrative steel structures and energy dissipation devices is presented in this paper, dealing with a two-storey frame structure housing a primary school building complex. The presence of a gym at the first level involves a system characterized by large spans and considerable deformability in both directions, requiring the upgrading with a suitable bracing system.

CP1020, *2008 Seismic Engineering Conference Commemorating the 1908 Messina and Reggio Calabria Earthquake*,
edited by A. Santini and N. Moraci

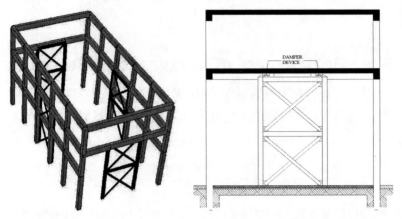

FIGURE 1. Possible implementation of dissipative braces in existing buildings

INTERVENTION APPROACH

The upgrading interventions on relatively deformable r. c. structures often involves the use of bracing or core systems aimed at reducing the lateral floor displacements. An evolution of this kind of traditional intervention consists of adding energy dissipation devices in the bracing elements or in the connection of braces to the existing structure (Fig. 1). The use of such devices as a link between the structure and the bracing elements gives way to a coupling system acting due to relative displacements between the two connected parts of the structure, with consequent large energy dissipation [5]. This system can be also implemented by means of smart devices, like magnetorheological dampers or actuators [6]. When the dynamic response of the existing structure can be well described by referring to the first vibration mode, a simple mechanical model representing the coupled system can be used for the study of the problem. The basic dynamic scheme of Fig. 2 represents the reference model, where M_1, K_1 and C_1 are the mass, the lateral stiffness and damping coefficient of the existing structure, respectively. Likewise, M_2- K_2 (K_d)- C_2 (C_d) are the same dynamic parameters for the braces (for the link), whereas f_d represents the controllable part of the force acting in the device. These two SDOF systems are coupled through either passive dampers or more complex smart devices.

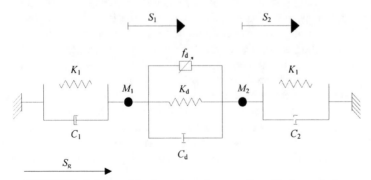

FIGURE 2. Scheme of the dynamic system considered in the analysis.

179

ANALYTICAL MODEL

An active or semi-active response control strategy based on coupled structures like the system shown in Fig. 2 is governed by the following equations of motion:

$$\begin{cases} M_1\ddot{s}_1 + C_1\dot{s}_1 + K_1 s_1 = -M_1\ddot{s}_g + f_d + K_d(s_1 - s_2) + C_d(\dot{s}_1 - \dot{s}_2) \\ M_2\ddot{s}_2 + C_2\dot{s}_2 + K_2 s_2 = -M_2\ddot{s}_g - f_d - K_d(s_1 - s_2) - C_d(\dot{s}_1 - \dot{s}_2) \end{cases} \qquad (1)$$

In the state space the equations of motion take the form:

$$\begin{cases} \dot{x} = Ax + B_1\ddot{s}_g + B_2 u \\ y = Cx + D_1\ddot{s}_g + D_3 u \end{cases} \qquad (2)$$

where the matrices that define the system are the following:

$$\begin{Bmatrix} \dot{x}_1 \\ \dot{x}_2 \\ \dot{x}_3 \\ \dot{x}_4 \end{Bmatrix} = \begin{bmatrix} 0 & 0 & 1 & 0 \\ 0 & 0 & 0 & 1 \\ \dfrac{K_d - K_1}{M_1} & -\dfrac{K_d}{M_1} & \dfrac{C_d - C_1}{M_1} & -\dfrac{C_d}{M_1} \\ -\dfrac{K_d}{M_2} & \dfrac{K_d - K_2}{M_2} & -\dfrac{C_d}{M_2} & \dfrac{C_d - C_2}{M_2} \end{bmatrix} \begin{Bmatrix} x_1 \\ x_2 \\ x_3 \\ x_4 \end{Bmatrix} + \begin{Bmatrix} 0 \\ 0 \\ -1 \\ -1 \end{Bmatrix} \ddot{s}_g + \begin{Bmatrix} 0 \\ 0 \\ \dfrac{1}{M_1} \\ -\dfrac{1}{M_2} \end{Bmatrix} u \qquad (3)$$

$$\begin{Bmatrix} y_1 \\ y_2 \\ y_3 \\ y_4 \end{Bmatrix} = \begin{bmatrix} 1 & 0 & 0 & 0 \\ 0 & 1 & 0 & 0 \\ 0 & 0 & 1 & 0 \\ 0 & 0 & 0 & 1 \end{bmatrix} \begin{Bmatrix} x_1 \\ x_2 \\ x_3 \\ x_4 \end{Bmatrix} + \begin{Bmatrix} 0 \\ 0 \\ 0 \\ 0 \end{Bmatrix} \ddot{s}_g + \begin{Bmatrix} 0 \\ 0 \\ 0 \\ 0 \end{Bmatrix} u$$

In the case of active control, the link between the coupled systems is created by an actuator, whose performance is governed by a control algorithm. An optimal control strategy, based on the Linear Quadratic Regulator (LQR) performance criteria, is considered to this purpose to govern the control law [7]. In the case of semi-active system, instead, the link is based on a magnetorheological damper (MRD) and the control strategy is defined by a relationship between the relative structural motion (i.e. velocity) and the control force. A system equipped with MRD requires a robust control to account for the possible nonlinear behaviour of the damper, nonlinear excitation and the possible hysteretic behaviour of the structure due to inelastic deformations [8]. The simplest method of control is based on an *on-off* criterion for the damper. In a few words, a continuous decision loop evaluates the motion within the structure adjusting the damper state based on the following criterion, based on energy dissipation considerations:

$$\text{if } F_d \cdot \dot{x}_1 \le 0 \rightarrow F_d = F_{d\max} \qquad \text{if } F_d \cdot \dot{x}_1 > 0 \rightarrow F_d = F_{d\min} \qquad (7)$$

This algorithm works so as to optimise the energy dissipation in the system by switching the damper off when its contribution to structural motion is detrimental to the response of the structure. An improvement of the *on-off* criterion has been proposed by Dyke and Spencer [9], who developed a clipped optimal algorithm for structural control based on MRD. The approach proposed is to append a feedback loop in order to drive the MRD to produce approximately a desired control force F_{opt}, which is determined with optimal control strategy as described before. To bring the MRD to generate approximately the desired optimal control force, the command signal (i.e. the voltage V necessary to induce the magnetic field) is applied so as to satisfy the relationships:

$$\left(\left|F_{opt}\right|-\left|F_d\right|\right)\cdot sgn\left(F_{opt}\cdot F_d\right)\geq 0 \rightarrow F_d = F_{d\max}(V=V_{\max})$$

$$\left(\left|F_{opt}\right|-\left|F_d\right|\right)\cdot sgn\left(F_{opt}\cdot F_d\right)< 0 \rightarrow F_d = F_{d\min}(V=0)$$

$$(8)$$

THE CASE STUDY

The analyzed structure is a two-storey RC building located in the town of Vico Equense near Naples (Italy) (Fig. 1). It has a rectangular plan whose dimensions are 10.80×20.40 m^2. The elevation of the first floor is 7.40 m, whereas the second one is at 11.10 m. The building is characterized by a reinforced concrete structure framed in the longitudinal direction only and is designed against vertical loads, without account for seismic action. Vertical structural elements are twelve columns supporting perimeter r.c. beams and two floors made of hollow tile r.c. slabs. Columns and beams have rectangular 40×50cm^2 and 40×70cm^2 cross-sections, respectively.

Some characterization tests have been carried out by core boring and steel bars drawing to find information about the quality of the structural materials. The tests have allowed to evaluate the mechanical resistance of both concrete and steel and the results revealed the presence of materials of satisfying quality: $f_{ck} \approx 20$ N/mm^2 for concrete and $f_{yk} \approx 380$ N/mm^2 for steel bars. On the first floor a direct load test has been carried out, showing adequate capacity of the r.c. slab against vertical loads, which allowed making the assumption of infinitely rigid floor in the seismic analysis.

Vulnerability assessment

The area of Vico Equense is classified as third category with an expected value of PGA of 0.15g by the recent Italian seismic code [10]. For the analysis carried out a PGA of 0.25g has been considered, considering the combination of site effect and the importance of the structure with regard to collapse (primary school). A three-dimensional FEM model of the structure has been developed considering floors like rigid diaphragms in the horizontal plane. Two nonlinear static analyses and a set of linear and nonlinear time history analyses have allowed to evaluate the vulnerability of the structure in the as-built condition and the effectiveness of the upgrading interventions. A calculation of the natural frequencies of the system has been carried out, evaluating the first mode period in each direction. The founded values are 0.482 s in x direction (longitudinal) and 1.256 s in y direction (transverse) (Table 1). The mass participation factors are higher than 95% for such modes, so that the structure can be assumed as a matter of fact as made of two mutually independent SDOF systems in both x and y directions. This consideration assumes relevance in the definition of the optimal control strategy as aforementioned.

In order to assess the seismic vulnerability of the existing structure two non linear static analyses have been carried out. The response of the as-built structure in the two principal directions has then been evaluated in terms of capacity curves F-d. These curves have been represented in approximate way by means of equivalent SDOF elastic-plastic relationships. The most significant parameters have been determined (Table 1), namely the force F_{el}, that represents the base shear at which the first element attains its strength capacity, and the corresponding lateral displacement of the control point d_{el}, the ultimate strength capacity F_u and the ultimate lateral displacement d_u. These parameters are also useful to determine the effectiveness of the different control strategies.

TABLE 1. Performance parameters of the as-built structure obtained by pushover analysis.

Force direction	Fundamental period (s)	F_{el} (kN)	F_u (kN)	d_{el} (cm)	d_u (cm)
longitudinal	0.482	1106	1200	2.10	8.88
transverse	1.256	320	373	6.40	7.68

Up-grading intervention

The analysis carried on the as-built structure has shown an excessive deformability of the original structure, not compatible with the structural safety and immediate post-earthquake occupancy requirement. The upgrading interventions are aimed at reducing the lateral floor displacements of the structure by means of steel braces fitted with additional special devices. Such devices connect the original structure at the first floor level with rigid steel braces and act due to relative displacements between the original structure and the steel braces. The study has been first carried out considering a rigid connection, then passive hysteretic devices and eventually smart devices and actuators. As shown in the analyses, the reduction of horizontal floor displacements obtained thanks to the addition of smart devices is significantly greater than the one obtained with a rigid connection of the original structure to the steel braces.

For the purposes of the analysis, the controlled system has been represented in the form of two coupled SDOF systems, whose natural vibration frequencies correspond to the first mode frequency of the respective real sub-structures, evaluated according to the 3-D model analysis. The evaluation of the proposed control strategy effectiveness has been assessed carrying out a set of non-linear dynamic analyses of the system shown in Fig. 2 by means of the MatLab-Simulink toolbox, considering 8 seismic input recordings scaled at different values of PGA. For the sake of comparison, the system is considered in the not-controlled configuration as well as in passive, semi-active and active controlled configurations as described before. Percentage of the reduction of maximum top displacements are summed up in Tables 2-3, for each seismic input. Comparing these maximum values with the control point displacement at the elastic limit, obtained from pushover analysis, it is clear that the upgraded structure displacements are much lower than this limit. Some top displacement time histories, obtained from the analysis of both as-built and upgraded structure, are plotted in Fig. 3.

TABLE 2. Percentage difference in the peak values of the dynamic response in the longitudinal direction under different seismic inputs (for P.G.A. 0,35-0,25-0,15 g) of the upgraded structure with reference to the not-controlled case.

Seismic input	Rigid link	Passive link	Smart link	Active link
	0.15 / 0.25 / 0.35g	0.15 / 0.25 / 0.35g	0.15 / 0.25 / 0.35g	0.15 / 0.25 / 0.35g
Bevagna	-61 / -61 / -61	-61 / -61 / -61	-60 / -61 / -62	-73 / -73 / -73
Colfiorito	+9 / +9 / +9	+9 / +8 / +3	-11 / -12 / -14	-20 / -20 / -20
Calitri	-37 / -37 / -37	-37 / -37 / -37	-44 / -45 / -45	-74 / -74 / -74
El Centro	-44 / -44 / -44	-44 / -47 / -50	-59 / -58 / -58	-71 / -71 / -71
Gubbio	-23 / -23 / -23	-23 / -23 / -24	-35 / -37 / -38	-72 / -72 / -72
M. S. Severino	-45 / -45 / -45	-45 / -45 / -46	-59 / -58 / -58	-76 / -76 / -76
Taiwan	-17 / -17 / -17	-17 / -22 / -25	-31 / -33 / -33	-71 / -71 / -71
Thessaloniki	-30 / -30 / -30	-30 / -30 / -31	-38 / -40 / -40	-70 / -70 / -70

TABLE 3. Percentage difference in the peak values of the dynamic response in the transverse direction under different seismic inputs (for P.G.A. 0,35-0,25-0,15 g) of the upgraded structure with reference to the not-controlled case.

Seismic input	Rigid link	Passive link	Smart link	Active link
	0.15 / 0.25 / 0.35g	0.15 / 0.25 / 0.35g	0.15 / 0.25 / 0.35g	0.15 / 0.25 / 0.35g
Bevagna	+53 / +53 / +53	- 6 / -18 / -23	-23 / -25 / -25	-65 / -65 / -65
Colfiorito	-25 / -25 / -25	-22 / -30 / -35	-41 / -43 / -44	-82 / -82 / -82
Calitri	-36 / -36 / -36	-41 / -46 / -47	-49 / -49 / -50	-84 / -84 / -84
El Centro	-58 / -58 / -58	-58 / -58 / -58	-36 / -36 / -36	-39 / -39 / -39
Gubbio	-60 / -60 / -60	-64 / -65 / -64	-65 / -65 / -64	-88 / -88 / -88
M. S. Severino	+102/ +102/ +102	- 1 / -18 / -25	-34 / -35 / -36	-72 / -72 / -72
Taiwan	-22 / -22 / -22	-38 / -40 / -41	-42 / -44 / -44	-80 / -80 / -80
Thessaloniki	- 3 / - 3 / - 3	-16 / -16 / -18	-27 / -29 / -31	-77 / -77 / -77

The results of the analysis as a whole demonstrate the effectiveness of both active and semi-active systems and confirm the aptitude of this strategy to the use in retrofitting and upgrading cases. In particular, the active controlled structure represented a target point for the semi-active system, whereas the passive control corresponds to the least performance level which can be obtained. The semi-active control, thanks to its adaptability to input characteristics and low power requirements, represents a reasonable compromise between performance and costs. The parametric analysis put into evidence the good performance of Clipped-Optimal algorithm when applied to semi-active control systems. This result is related to the possibility of this control law to adjust the reaction force of MRD, keeping it close to the optimal control force value. Moreover, Clipped-Optimal semi-active control, as well as active control, do not depend on the input intensity, as one may observe from the mean value of the maximum response coming from each seismic input (Tables 2-3).

The benefit of structural control is more evident observing the global reduction of the structural response (Fig. 3). This aspect is strictly related to a lower elastic strain energy transmitted to the system. Observing the time history diagrams in terms of elastic strain energy (E_s) of the existing structure (Fig. 4), it is evident that control effectively acts to drive the system toward the stable position, corresponding to a zero value of elastic strain energy. The reduced values of E_s for the semi-actively controlled system respect to passive and rigid link, proves the significant increase of the structural system capacity to dissipate input energy, thus limiting the possibility to undergo structural damage.

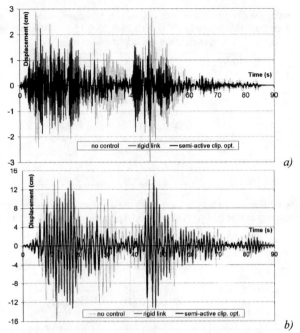

FIGURE 3. Response time history under Calitri seismic input scaled to PGA 0.25 g.
a) longitudinal, *b)* transverse

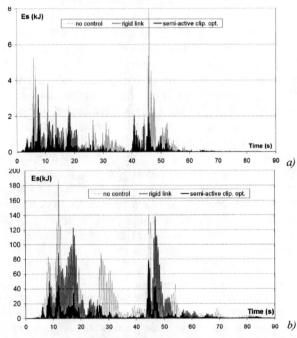

FIGURE 4. Elastic strain energy time history under Calitri seismic input scaled to PGA 0.25 g.
a) longitudinal, *b)* transverse

CONCLUSIONS

The application of steel bracing elements equipped with structural control devices to a real case has put into evidence the beneficial effect of such technique. As clearly shown by the results of the analysis, thanks to this kind of intervention it is possible to reach a highly dissipative global behavior. In such a way, the existing structure can remain in the elastic range even under seismic inputs of intensity higher than the maximum expected for the site. This technique, therefore, confirms to be one of the most appropriate for the seismic upgrading of existing r.c. buildings, the majority of which, even though characterized by acceptable material quality, do not meet most basic anti-seismic requirements and do not posses adequate resistance to earthquake actions. Eventually, smart systems have also demonstrated significant advantages in comparison with passive ones: enhanced effectiveness in dynamic control, relative insensitivity to site conditions and ground motion, significant increase of the capacity of the structural system to dissipate seismic input energy without undergoing structural damage.

ACKNOWLEDGMENTS

This paper has been developed within the research project "Seismic protection of existing buildings by semi-active control techniques" issued by Regione Campania - Italy (L.R. 5 - 28.03.2002) and coordinated by A. Mandara. The helpful cooperation of Eng. Vincenzo Garofalo in the execution of the dynamic numerical analysis is gratefully acknowledged.

REFERENCES

1. G. Spina, "Seismic protection of historical buildings by response control systems" (in Italian), PhD Thesis, Second University of Naples, Naples, Italy, 2006.
2. Mandara, A. Durante, F. Ramundo, G. Spina, Control of the seismic response of historical buildings by mass damping systems. Proceedings of the XII National Conference of ANIDIS. Pisa, Italy, 2007.
3. Mandara, A. Durante, F. Ramundo, G. Spina, Smart technologies for seismic protection of historical structures. Proceedings of the First European Conference on Earthquake Engineering and Seismology. Geneva, Switzerland, 2006.
4. Mandara, F. Ramundo, G. Spina, On the effectiveness of smart technologies in the seismic protection of existing buildings. Part I : Masonry structures. Proceedings of the MERCEA'08 International Conference. Messina and R. Calabria, Italy, 2008 (submitted).
5. Mandara, F. Ramundo, G. Spina: Seismic up-grading of an existing r.c. building by steel braces and energy dissipation devices. XXI National Congress of CTA. Catania, Italy, 2007.
6. T.T. Soong, M.C. Constantinou, *Passive and Active Structural Vibration Control in Civil Engineering.* Springer, New York, 1994.
7. L. Meirovitch, *Elements of Vibration Analysis.* McGraw Hill, New York, 1986.
8. Mandara, A. Durante, G. Spina, S. Ameduri, A. Concilio, Seismic protection of civil historical structures by MR dampers. Proceedings of the SPIE's 13th Annual International Symposium on Smart Structures and Materials. San Diego, CA, USA, 2006.
9. S.J. Dyke, B.F. Jr. Spencer, M.K. Sain, J.D. Carlson, An experimental study of MR dampers for seismic protection, in *Smart Structures and Materials,* 7, 1998, pp. 693-703.
10. Ordinanza P.C.M. n. 3274 e successive modifiche e integrazioni, 2003.

CONFERENCE TOPICS

SITE CHARACTERIZATION, MICROZONATION AND SITE EFFECTS

Site Effect and Expected Seismic Performance of Buildings in Palestine- Case Study: Nablus City

Jalal N. Al-Dabbeek and Radwan J. El-Kelani

Earth Sciences and Seismic Engineering Center (ESSEC)
An-Najah National University, P.O. Box 7, Nablus, Palestine
E-mail address: seiscen@najah.edu

Abstract. The effects of local geology on ground-motion amplification and building damage were studied in Palestine-West Bank. Nakamura's method of microtremor analysis was applied in this study. The measurements showed significantly higher amplification in the frequency range of building vulnerability in different parts of Nablus city. This finding is consistent with the distribution of the earthquake damage grades in the urban areas struck by the 11 February 2004 earthquake (ML= 5.2) with a focal depth of 17 km beneath the northeastern part of the Dead Sea Basin. Quite large differences in amplification between around 1 and 9 were computed between the eastern and western rims of the city. The downtown built in the central part of the city on soft clay, marl and valley deposits, whereas the northern and southern parts of urban areas in Nablus city lying on mountains consist of consolidated carbonates bedrock. In the central part of the city and at the rims, where the thickness of fluvial deposits and soft formations is about 15 m, amplifications between 6.74 and 8.67 for dominant natural period range of 0.8 - 1.1 sec were obtained. On the southern and northern mountains, which are located on limestone rocks covered with a thin layer of soil, the amplification in the same frequency range was low. Calculating the natural period of the existing common buildings (T_b) in the studied area (buildings with 10-12 stories), by using the dynamic analysis method. The values of T_b obtained were much closed to the site dominant natural period (Ts).The findings of this study indicate that the expected differences in damage grades for urban areas in Nablus city could be attributed to variations in the thickness and physical properties of Tertiary-Quaternary sediments, which appear to be rather heterogeneous.

Keywords: site effect, amplification, buildings, vulnerability, Palestine.

INTRODUCTION

Local site effect (landslides, liquefaction, amplification and faulting systems) plays an important role on the intensity of earthquakes. Thus, Earthquake- resistant design of new structures and evaluating the seismic vulnerability of existing buildings are taking into account their response to site ground motions. Geophysical studies of seismic activity in Palestine, deep seismic sounding, paleoseismic excavation, and instrumental earthquake studies of half a century [1-14] demonstrate that the damaging earthquakes were located along the Dead Sea Rift/Transform fault (Fig. 1). These damaging earthquakes caused in several cases severe devastation and many hundreds and sometimes thousands of fatal casualties.

Recent studies of large destructive earthquakes have shown that damages during the earthquakes are often caused by the amplification of seismic waves in near-surface geology [15-20], where the post disaster damage assessment showed that the local site effect may have a dominant contribution to the intensity of damage and destruction.

CP1020, *2008 Seismic Engineering Conference Commemorating the 1908 Messina and Reggio Calabria Earthquake,*
edited by A. Santini and N. Moraci
© 2008 American Institute of Physics 978-0-7354-0542-4/08/$23.00

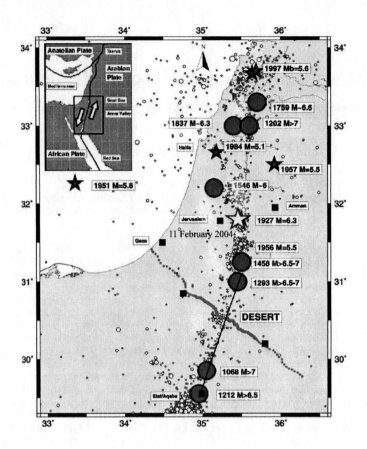

FIGURE 1. Seismic activity in the Dead Sea Transform region; the map shows locations of historical earthquakes [11-14]. Also shown is the most recent earthquake of 11 February 2004, ML 5.2.

The estimation of site response is therefore critical, in order to evaluate the seismic hazard potential of a given area. Since sedimentary deposits are often the prime locations for development of urban areas, local amplification is a major concern in earthquake-prone regions, but also in moderate seismicity areas where the mid-size cities developed could be struggle with future damaging events due to the combination of site effects and urban development. In the present study, the local site effect on ground-motion amplification and buildings vulnerability were studied in Nablus city/ West Bank (Fig. 2). The regions in Palestine suffer, in general, from planning absence, random urban expansion, and multiple land use with lack of seismic information and shortage of studies on the soil and ground response due to earthquakes [21-23].

The primary research goal is to construct the theory of urban planning for disaster reduction in Palestine. It is necessary that the planning contain both hardware and software in order to raise the force of the urban areas for the disaster such a way of reducing the impact of potential earthquakes (Fig. 2).

Local Geology

The exposed sequence of rocks in Nablus region mainly consists of carbonates; limestone, dolomite, marl and chalk and it includes other sediments as chert, clay, gravel and some sandstone, with ages ranging from upper Cretaceous to Recent (Fig. 2). The limestone formation is massively bedded at base and becomes increasingly thin towards the top [24].the formation outcrops and reported lithology from groundwater boreholes indicate that the limestone thickness ranges from about 170-280 m containing occasionally dolomite. Besides, some chalk, chalky marl and marl appear in thin laminations throughout the formation. Tertiary and Quaternary sediments as gravel deposits and clay sediments cover the central part of Nablus city and form the main soil layer at the eastern and western sides of the city.

METHODOLOGY AND DATA ANALYSIS

Nakamura's Technique

One of the most appealing techniques for estimating site response is Nakamura's method [25] since it only requires records from a single three-component station deployed at the site of interest an does not need a reference seismogram measured at the substratum bedrock. As introduced by Nakamura [25], the technique was intended to assess S-wave amplification from microtremor measurements. There are four components of spectral amplitudes involved in this one- layer problem, namely, the horizontal components of motion at the surface and bottom of the sedimentary layer, referred as $H_s(f)$ and $H_b(f)$, respectively; and the vertical components of motion at surface and bottom, correspondingly denoted as $V_s(f)$ and $V_b(f)$.

The prime objective of Nakamura's technique is to isolate the amplification effect suffered by horizontal components of substratum motion. In order to do this, he first constructs the theoretical borehole ratios that are widely regarded as the most reliable transfer function estimates for horizontal and vertical components, as given below, respectively:

$$S_h = \frac{H_s}{H_b} \, and \tag{1}$$

$$S_v = \frac{V_s}{V_b} \tag{2}$$

With these two ratios Nakamura constructs an additional transfer function S_t which gives formally the factor by which the horizontal ratio exceeds the vertical one:

$$S_t = \frac{S_h}{S_v} = \frac{H_s / H_b}{V_s / V_b} \, or \tag{3}$$

$$S_t = \frac{H_x / V_s}{H_b / V_b} \tag{4}$$

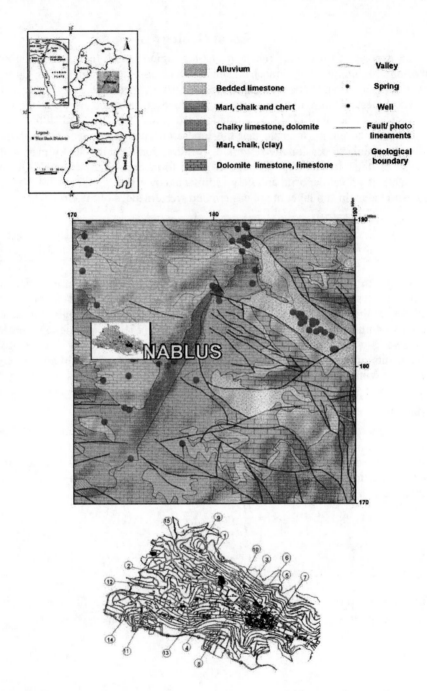

FIGURE 2. Geological map shows the main outcropping formations in Nablus region, and a sketch site map shows the distribution of the microtremor measurements in Nablus city.

Microtremor Measurements

An attempt was made to understand the site effect conditions on damage distribution during a probable strong ground motion in Nablus City [22]. Several points have been selected in different sites of the city. The criterion for each point is the variations in the geology as well in the topography, where points have been measured on rock cover of mainly limestone, chalk and marl (mountain areas), while other points selected to be on soil and soft sediments (valley areas of caly and gravels). Microtremor measurements were carried out at 15 selected sites to represent different geological formations and to cover a comprehensive area in order to give a macrozonation general idea about the dominant frequency in the study area, see Fig 2.

The site effects have been investigated by taking measurements of ambient noise collected by short period seismic station and making the spectral analysis using the packages programs [26]: SDA software for data acquisition and SEISPECT for data analysis. The measurements were made during the daytime, when the contaminating effects of traffic and industrial noise were significant.

Spectral Analaysis Results

The analysis of ambient vibration measurements developed a spectral ratio site response for each site, an example of average Fourier spectra for selected windows is presented in Fig. 3, and also the relation between dominant frequency at the site and the amplification factor (spectral ratio) is presented in Fig. 4 for the sites 3 and 4. The dominant frequency (or period) and its amplification factor of all measured locations are presented in Table 1. The results showed obvious difference among the dominant frequencies even the studied area is small.

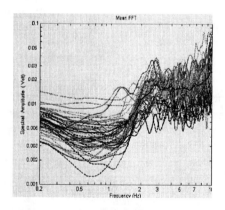 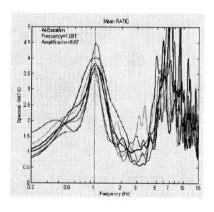

FIGURE 3. Spectral amplitude for site 4.　　　　　　**FIGURE 4**. Spectral ratio for site 3.

The distribution of the dominant frequencies values of the measured sites varied from eastern part to the western part. Quite large differences in amplification between around 1 and 9 were computed between the eastern and western rims of the city. The downtown built in the central part of the city on soft clay, marl and valley deposits, whereas the northern and southern parts of urban areas in Nablus city lying on mountains consist of consolidated carbonates bedrock. In the central part of the city and at the rims (see table 1, sites 1-8), where the thickness of fluvial deposits and soft formations varies between 15-25 m, amplifications between 2.6 and 8.66 for dominant natural period range of 0.8 - 1.4 sec were obtained. On the southern and northern mountains, which are located on limestone rocks covered with a thin layer of soil (see Table 1, sites 9-15), the amplification was relatively low.

merical Methods Using Seismic-Geotechnical Data

The period of vibration corresponding to the fundamental frequency is called the characteristic site period:

$$T_s = \frac{4H}{V_s} \tag{5}$$

The characteristic site period which depends only on the thickness (H) and shear wave velocity (V_s) of the soil provides a very useful indication of the period of vibration at which the most significant amplification can be expected [27]. The selected sites (sites 3, 5, 6 and 8) underlain by 18 to 25m of soft soil with an average s-wave velocity of about 100 m/sec, where the characteristic site periods were estimated by using equation 5 at 0.72 to 1 sec. The value of the characteristic site period obtained by using the numerical method was closed to the values obtains by using microtremor measurements indicated in Table 1.

TABLE 1. Results list of dominant frequencies, amplifications and natural periods.

	Microtremor Measurements			Numerical method
Site	Dominant Frequency Hz	Amplification Factor	Natural period sec	Natural period sec
1	0.700	3.55	1.429	-
2	0.688	3.35	1.453	-
3	1.001	7.44	0.999	0.9
4	0.846	2.634	1.182	-
5	0.952	3.405	1.050	0.88
6	0.906	8.66	1.103	1
7	0.797	6.736	1.255	-
8	1.2	2.6	0.833	0.72
9	1.306	1.393	0.766	-
10	1.074	1.416	0.931	-
11	1.251	1.894	0.799	-
12	1.453	2.337	0.688	-
13	1.492	2.952	0.670	-
14	1.13	1.937	0.885	-
15	1.7701	1.104	0.565	-

The Fundamental Natural Period of Common Buildings in Nablus City

The fundamental period of few buildings (buildings between 8 and 12 levels) in selected sites were calculated by using building code empirical formula (UBC 97 and IBC Codes) and dynamic analysis (see Table 2 for the common buildings with 10 levels), most of these buildings had fundamental periods equal to or some what less than the characteristic site period. Accounting for the period-lengthening effect of soil-structure interaction and the tendency for the fundamental period of a structure shows that they increase during a strong earthquake, due to the reduction in stiffness caused by cumulative architectural and structural damage [27].

This double-resonance conditions (amplification of bedrock motion by the soil deposit and amplification of the soil motion by the structure) combined with structural design and construction deficiencies are expected to increase the seismic vulnerability of the common buildings in the studied area.

TABLE 2. Structural analysis results, cracked and non-cracked sections, show building natural periods using UBC97.

Building type	Period, T For Exterior frame		Period, T For Interior frame		Period, T For Space frame		Period, T
	Uncr.	Cracked	Uncr.	Cracked	Uncr.	Cracked	Using UBC97
10-story frames (h=3.25 m)	0.85	1.25	1.2	1.78	1.04	1.5	1.0
10-story perimeter walls (h=3.25 m)	0.31	0.47	1.29	2.92	0.0.47	0.50	0.664

CONCLUSIONS

Based on the effects of local geology, there is good correlation between the different values of the amplification factor and the changes in the lithology. Where in the southern and northern mountains, consist of consolidated carbonates bedrock, slight amplification is expected in comparison with quite larger amplification factor was computed at the eastern and western rims of the city and the downtown built on soft clay, marl and fluvial deposits. This is also in good agreement with the reported obvious difference of intensity grades felt in the different parts of Nablus city during the past and recent seismic activity in Palestine.

Regarding the spectral ratio analyses of the selected sites in the study area, the high spectral ratio range of 6.7-8.7 is shown in the sites 3, 6 and 7 in the frequency range of about 0.8 - 1. 0 Hz. This ratio range shows good agreement with the reported obvious differences in damaging levels reported at the different urban areas in Nablus city after 11 July 1927 and 11 February 2004 earthquakes (ML 6.3 and 5.2, respectively) that have been attributed to site effect variations related to the physical properties of

Tertiary-Cretaceous sediments, which appear to be rather heterogeneous in the lithology [22].

The natural frequency of many common buildings in the studied area are very close to the site dominant frequency, therefore this will increase their seismic vulnerability.

REFERENCES

1. A. M. Quennell, *Tectonic of the Dead Sea Transform*, International Geological Congress, Maxico City, 1959, pp. 385-405.
2. R. Fruend, I. Zak and Z. Garfunkel,. Nature, **220**, 253-255 (1968).
3. A. Ginzburg, J. Makris, K. Fuchs and C. Prodehl, Tectonophysics, **80**, 109-119 (1981).
4. Z. Garfunkel, Tectonophysics, **80**, 81-108 (1981).
5. Z. El-Isa, J. Mechie, C. Prodehl, J. Makris and R. Rihm, Tectonophysics, **138**, 235-253 (1987).
6. E. A. Al-Tarazi, Natural Hazards, **10**, 79-96 (1994).
7. E. A. Al-Tarazi, J. African Earth Sci., **28**, 743-750 (1999).
8. Ch. Haberland, A. Agnon, R. El-Kelani, N. Maercklin, I. Qabbani, G. Rümpker, T. Ryberg, F. Scherbaum and M. Weber, J. Geophys. Res.,**108**, 1-11 (2003).
9. G. Ruepker, , T. Ryberg, G. Bock, M. Weber, K. Abu-Ayyash, Z. Ben-Avraham, R. El-Kelani, Z. Garfunkel, C. Haberland, A. Hofstetter, R. Kind, J. Mechie, A. Mohsen, I. Qabbani and K. Wylegalla,. Nature, **425**, 497-501 (2003).
10. R. J. El-Kelani, An-Najah Uni J. for Research-A (Natural Sciences), **19**, 185-208 (2006).
11. N. Abou Karaki"Synthese et Carte Seismotectonique des pays de la Borrdure Orientate de la Mediterranee: Sismicite du System de Failles du Jourdain-MerMorte," Ph.D. Thesis, Univ. Strasbourg, France, 1987.
12. A. Hofstetter, T. van Eck, and A. Shapira, Tectonophysics, **267**, 317-330 (1996).
13. N. Ambraseys and J. Jacson, Geophys. J. Int., **133**, 390-406 (1998).
14. J. Mechie, K. Abu-Ayyash, Z. Ben-Avraham, R. El-Kelani, A. Mohsen, G. Rümpker, J. Saul, M. Weber, Geophys. J. Int., **156**, 655-681 (2005).
15. K. Aki, Tectonophysics, 218, 93-111 (1993).
16. K. Atakan, B. Brandsdottir and g. Fridleifsson, Natural Hazards, **15**, 139-164 (1997).
17. C. Gutierrez and S. Singh, Bull. Seismol. Soc. Am., **82**, 642-659 (1992).
18. S. E. Hough, R. Borcherdt, P. Friberg, R. Busby, E. Field and K. Jacob, Nature, **344**, 853-855 (1990).
19. J. Lermo, M. Rodriguez and S. Singh, Earthquake spectra **4**, 805-814 (1988).
20. W. S. Phillips and K. Aki, Bull. Seismol. Soc. Am., **76**, 627-648 (1986).
21. I. S. Jerdaneh, Journal of Applied Sciences **3**, 364-368 (2004).
22. J. Al-Dabbeek, and R. El-Kelani, *Dead Sea Earthquake of 11 February 2004, ML 5.2: post earthquake damage assessment*, The Int. Earthquake Conf. (TINEE), Amman, 2004, pp. 1-8.
23. J. Al-Dabbeek, The Islamic Uni J. (Se. of Natural. Studies. and Eng.), **15**, 193-217 (2007).
24. Rofe and Raffely consulting engineers, Geological and hydrological report, Jordan Central Water Authority, 1963, pp. 79.
25. Y. Nakamura, Quarterly Report of Railway Tech. Res. Inst. of Japan, **30**, 25-33 (1989).
26. U. Peled, V. Avirav and Hofstetter, GII, Report No. 550/019/04, 2004, 1-23.
27. S. Kramer, Geotechnical Earthquake Engineering, New Jersey: Prentice Hall, 1996, pp. 308-315.

Seismic Response of Alluvial Valleys to SH Waves

Ernesto Ausilio, Enrico Conte and Giovanni Dente[a]

[a] *Dipartimento di Difesa del Suolo, Università della Calabria*

Abstract. This paper presents a theoretical study on the seismic response of alluvial valleys. The considered model consists of a two-dimensional elastic inclusion of arbitrary shape embedded in a stiffer half-plane excited by vertically or obliquely incident SH waves. Computations are conducted using a procedure based on the boundary element method. As known, this numerical technique is well suited to deal with wave propagation in infinite media as it avoids the introduction of fictitious boundaries and reduces by one the dimensions of the problem. This provides significant advantages from a computational point of view. A one-dimensional closed form solution is also used for comparison, and the most significant differences between the results obtained using the two methods are highlighted.

Keyword: microzonation studies, site effects, one and two-dimensional analyses.

INTRODUCTION

Macroseismic observations during various historical and recent earthquakes have shown that the local geological conditions can generate large amplifications and important spatial variations in the ground motion. Consequently, prediction of the local site effects is of great importance for the microzonation studies and the analysis of the seismic response of engineering works. To this purpose, it is necessary to understand the physical phenomena associated with the seismic wave propagation, and at the same time to develop methods capable of predicting reasonably the ground motion at a given site. In many situations, the simple shear beam model is not completely appropriate, thus use of two and three dimensional solutions is generally mandatory. In this context, both analytical and numerical methods are available.

The analytical solutions deal with simple geometric situations, such as semi-cylindrical or semi-elliptical alluvial valleys subjected to incident SH-waves [1-2]. Their application has allowed the role of the parameters involved to be highlighted. In addition, they represent unfailing references against which numerical solutions can be tested.

The numerical methods that are widely used for the analysis of seismic wave propagation can be classified into domain, boundary and asymptotic methods. The finite difference method [3] and the finite element method [4] fall within the first class as these numerical techniques require that the entire domain be discretized. On the

CP1020, *2008 Seismic Engineering Conference Commemorating the 1908 Messina and Reggio Calabria Earthquake*,
edited by A. Santini and N. Moraci
© 2008 American Institute of Physics 978-0-7354-0542-4/08/$23.00

other hand, when use is made of the methods falling within the second class such as the boundary element method, discretization of the boundaries needs only to be performed. In this context, two main approaches can be distinguished: one is based on the use of complete systems of solutions [5], and the other on the boundary integral equations [6]. Finally, the asymptotic methods are useful when the solution has an interest in the high frequency range and the diffraction effects may be ignored [7]. These methods solve the differential equations governing wave propagation under two or three-dimensional conditions [8-12]. However, engineering applications usually rely on the one-dimensional analysis to predict the surface motion at a site. As a consequence, the effects of the geometric shape and limited lateral extent of the soil deposit under consideration are completely ignored in the analyses.

In this paper, the boundary element method is used to analyse SH wave scattering by alluvial valley of arbitrary shape under conditions of plane strains. This numerical technique is well suited to deal with wave propagation problems, because it avoids the introduction of fictitious boundaries and reduces by one the dimensions of the problem. This provides significant advantages from a computational point of view. However, the involved materials are assumed to behave as linear elastic media. To assess the accuracy of the proposed method, the results in terms of surface displacement amplitude are compared with those calculated using the analytical solution derived by Trifunac [1] for semi-cylindrical alluvial valleys. In addition, a simple one-dimensional solution is also considered with the purpose of highlighting the main differences between the results.

PROBLEM FORMULATION

The problem considered in this study concerns a two-dimensional alluvial valley of arbitrary shape embedded in a half-plane excited by incident harmonic SH waves with frequency ω, and angle of incidence γ (Fig. 1). The material of the valley and that of the half-plane is assumed to be homogeneous, isotropic and linearly elastic. It is also assumed that the valley is perfectly bonded to the half-plane at the interface Γ_j indicated in Fig. 1. Under conditions of plane strain, the differential equation governing the propagation of SH waves is:

$$\frac{\delta^2 \bar{u}_j}{\delta x^2} + \frac{\delta^2 \bar{u}_j}{\delta y^2} = \frac{1}{\beta_j^2} \frac{\delta^2 \bar{u}_j}{\delta t^2} \qquad (1)$$

where x and y are the spatial coordinates, t is the time, $\bar{u}_j$ indicates the out-plane displacement field of the half-plane (when j=1) or the valley (when j=2), $\beta_j = (G_j/\rho_j)^{1/2}$ is the shear wave velocity, G_j is the shear modulus and ρ_j is the mass density of the half-plane (j=1) or the valley (j=2). For harmonic motion (i.e. $\bar{u}_j = u_j e^{i\omega t}$, with $i = \sqrt{-1}$), Eq. 1 reduces to the Helmholtz equation, that is:

200

$$\frac{\delta^2 u_j}{\delta x^2} + \frac{\delta^2 u_j}{\delta y^2} + k_j^2 u_j = 0 \qquad (2)$$

in which $k_j = \omega/\beta_j$ and ω is the excitation frequency. Owing to the linearity of the problem, the displacement field of the half-plane can be cast in the form:

$$u_1 = u_o + u_d \qquad (3)$$

where u_o is the amplitude of the displacement field due to the incident and reflected waves, and u_d is the displacement amplitude due to the diffracted waves caused by the presence of the valley. The former represents the free-field motion amplitude and is provided by the following equation, under the assumption that the displacement amplitude due to the incident waves is equal to 1:

$$u_o = 2\cos(k_1 y \cos\gamma)e^{i k_1 x \sin\gamma} \qquad (4)$$

The displacement of the valley, u_2, is due to the diffracted waves at the boundary Γj. In the case under consideration, the unknown fields are u_d and u_2. Both these fields satisfy Eq. 2.

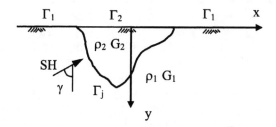

FIGURE 1. The problem considered

In the present study, the solution to Eq. 2 is achieved using the boundary element method. Following Brebbia et al. [6], this method is based on the integral equation:

$$c(P)u(P) = \int_\Gamma \left[u^*(P,Q)\frac{\delta u(Q)}{\delta n} - u(Q)\frac{\delta u^*(P,Q)}{\delta n} \right] d\Gamma \qquad (5)$$

where P is the point under consideration and Q is a point located on the boundary Γ (where $\Gamma = \Gamma_1 \cup \Gamma_2 \cup \Gamma_j$, as shown in Fig. 1), c is a coefficient depending on the position of P, n indicates the normal direction to Γ, and u^* is the fundamental solution of the Helmholtz equation, which for two dimensions results in

$$u^* = \frac{1}{4}i H_o^1(k_j r) \qquad (6)$$

in which H_o^1 is the Hankel function of first kind and zero order, and r is the distance between P and Q. It is worth noting that Eq. 5 only involves boundary integrals. In addition, it directly accounts for the radiation condition for infinite media owing to the presence of the fundamental solution. This avoids the introduction of fictitious boundaries, unlike other numerical techniques such as the finite element method or the

201

finite difference method. As suggested by Kobayashi [13] and Conte et al. [14], it is more convenient, from a computational point of view, to use the following equation instead of Eq. 6:

$$u^* = \frac{1}{4}i\left[H_o^1(k_j r) + H_o^1(k_j r')\right]$$ (7)

where r' is the distance between P and the image point of Q with respect to the horizontal ground surface (Γ_1 and Γ_2 in Fig. 1). Unlike Eq. (6), this latter directly satisfies the boundary condition on Γ_1 and Γ_2 (i.e., $\partial u_d/\partial n = 0$ at Γ_1, and $\partial u_2/\partial n = 0$ at Γ_2). As a consequence, only the interface Γ_j needs to be considered in Eq. 5.

The basic integral equation (Eq.5) is first applied separately to the valley (in terms of u_2) and the half-plane (in terms of u_d), and then the compatibility and equilibrium conditions are enforced at the interface Γ_j. These conditions are expressed by the equations:

$$u_1 = u_2$$ (8a)

$$G_1\left(\frac{\delta u_1}{\delta n}\right) = -G_2\left(\frac{\delta u_2}{\delta n}\right)$$ (8b)

To achieve a solution to Eq. 5, the interface Γ_j is divided into a finite number of one-dimensional elements. For simplicity, the values of the unknowns u and $\partial u/\partial n$ are assumed to be constant over each element and equal to the value at the mid-element node. In this type of element, the boundary is always smooth as the node is at the centre of the element, hence the coefficient c in Eq. 5 is 0.5 [6]. After discretizing Eq. 5 for each node, an algebraic system of equations is obtained, the solution of which provides the nodal values of the diffracted displacement and its normal derivative at the interface Γ_j. Once these quantities are known, it is possible to calculate the displacement field of the valley or that of the half-plane using Eq. 2 in which c is 1, together with Eqs. 3 and 8. It is worth noting that the resulting values are complex. They provide the normalized amplitude of the out-plane displacements due to wave propagation with respect to that due to the incident waves. As already said, this latter amplitude is assumed equal to 1. The variation of these quantities with time is obtained multiplying them by the factor $e^{i\omega t}$.

To assess the accuracy of the method, comparisons were performed with the analytical solution derived by Trifunac [1] for a semi-cylindrical alluvial valley excited by incident SH waves. Some results are presented in Fig. 2, in terms of the displacement amplitude at the ground surface for different values of the incidence angle and two values of the dimensionless frequency (η=0.5 and 1). This latter is defined as

$$\eta = \frac{\omega a}{\pi \beta_1}$$ (9)

where a is the radius of the valley. The other data assumed in the calculations are ρ_1/ρ_2=1.5 and G_1/G_2=6. As can be seen from Fig. 2, the results are in very close agreement.

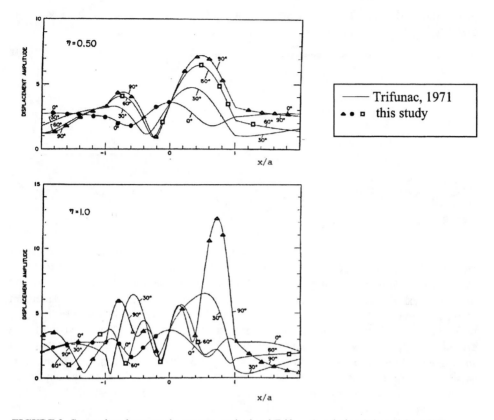

FIGURE 2. Comparison between the present method and Trifunac's solution (adapted from [1])

APPLICATIONS

In the engineering applications, the local site response is often evaluated using a one-dimensional model consisting of a soil deposit resting on a deformable or rigid bedrock excited by vertically incident shear waves. This implies that the effects of wave scattering due to the actual shape and the limited lateral extent of the soil deposit are ignored. Consequently, it is of interest to compare the results obtained by such a simple model with those calculated using the two-dimensional solution presented in the previous section. To this purpose, a semi-cylindrical valley embedded in a homogeneous half-plane is considered, and the results are compared with those obtained for a soil layer with thickness equal to the radius of the valley, resting on a deformable bedrock. In addition, a valley with triangular cross-section is also considered (Fig. 3). This latter is characterized by a height equal to the radius of the semi-cylindrical valley, and a slope of ½ with respect to the horizontal direction. For all these soil systems, the excitation consists of a train of SH harmonic waves with unit amplitude and vertical incidence. Using the same notation for the one and two-dimensional models, ρ_2 and β_2 are the mass density and the shear wave velocity of the valley and soil layer, whereas ρ_1 and β_1 are the above material properties for the half-plane and bedrock. The displacement amplitude at the surface of the layer was evaluated by the equation [15]:

$$u = \frac{2}{\sqrt{\left[\cos^2\left(\frac{\omega a}{\beta_2}\right) + \left(\frac{\mu_2}{\mu_1}\frac{\beta_1}{\beta_2}\right)^2 \sin^2\left(\frac{\omega a}{\beta_2}\right)\right]}} \tag{10}$$

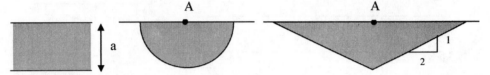

FIGURE 3. Schemes considered

The calculations were carried out assuming $\rho_1/\rho_2 = 1.5$ and two different values of the ratio β_1/β_2 (i.e., 2 and 3). The results are presented in Figs. 4, 5 and 6, in terms of the surface displacement amplitude of the soil layer and the valleys, versus the dimensionless frequency, η. In the two-dimensional models, the surface displacement amplitude was calculated at the middle point of the valley (point A in Fig. 3). As can be seen, the one-dimensional response generally results more attenuated than that provided by the two-dimensional models, with the exception of the triangular valley at some frequencies. The ratio β_1/β_2 is higher, larger is this attenuation. In addition, the peak values of the displacement amplitude at the upper surface of the layer are attained at lower frequencies than those calculated under two-dimensional conditions. In other words, using the one-dimensional model leads to a reduction of the surface displacement amplitude accomplished by a shifting in the fundamental frequencies with respect to the two-dimensional case. This is due to the wave scattering effects that are ignored when a one-dimensional solution is employed.

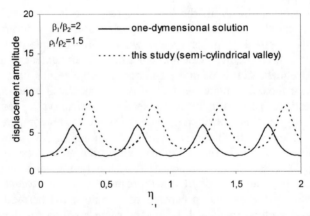

FIGURE 4. Displacement amplitude at the surface of a soil layer and at the middle point of a semi-cylindrical valley versus the dimensionless frequency η.

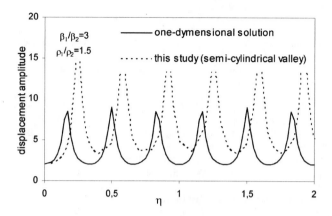

FIGURE 5. Displacement amplitude at the surface of a soil layer and at the middle point of a semi-cylindrical valley versus the dimensionless frequency η.

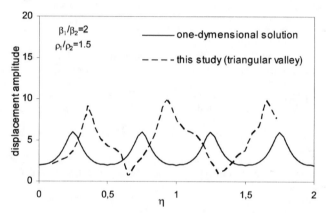

FIGURE 6. Displacement amplitude at the surface of a soil layer and at the middle point of a triangular valley versus the dimensionless frequency η.

CONCLUDING REMARKS

A numerical solution based on the boundary element method has been presented for the dynamic response of alluvial valleys to SH waves. Taking advantage of a special fundamental solution of the Helmholtz equation, the interface between the valley and the underlying half-plane needs only to be considered for solving the integral equation on which the method is based. Comparisons have been made with a well-known one-dimensional solution to highlight the most significant differences between the results. It has been shown that the one-dimensional model in principle leads to a reduction of the surface displacement amplitude and a shifting in the fundamental frequencies towards lower values than those obtained using the two-dimensional method presented in this paper.

REFERENCES

1. M. D. Trifunac "Surface motion of a semi-cylindrical alluvial valley for incident plane SH-waves", *Bull. Seism. Soc. Am.*, **61**, 1755–1770 (1971).
2. H. L. Wong and M. D. Trifunac "Surface motion of a semi-elliptical alluvial valley for incident plane SH waves", *Bull. Seism. Soc. Am.*, **64**, 1389–1408 (1974).
3. Z. Alterman and F.C. Karal "Propagation of elastic waves in layered media by finite difference methods ", *Bull. Seism. Soc. Am.*, **58**, 367–398 (1968).
4. J. Lysmer and L.A. Drake "Afinite element method for seismology", in: B.A. Bolt (Ed.), Methods in Computational Physics, vol. 11, Seismology,Academic Press, New York, 1972.
5. I. Herrera, "Variational principles for problems with linear constraints, prescribed jumps and continuation type restrictions", *J. Inst. Math.Appl.*, **25**, 67–96 (1980).
6. C.A. Brebbia, J.C.F. Telles and L. C. Wrobel, Boundary Element Techniques, Spring-Verlag Ed., 1984.
7. J.A. Rial, "Caustics and focusing produced by sedimentary basins. Application of catastrophe theory to earthquake seismology, Geophys.", *J. Roy. Astr. Soc.*, **79**, 923–938 (1984).
8. J.E. Luco, H.L. Wong, F.C.P. de Barros.A. Rial, "Three-dimensional response of a cylindrical canyon in a layered half-space ", *Earthquake Eng. Struct. Dyn.* 419, 799–817 (1990).
9. F. Luzón, S. Aoi, D. Fäh, F.J. Sánchez-Sesma "Simulation of the seismic response of a 2D sedimentary basin: a comparison between the indirect boundary element method and a hybrid technique three-dimensional response of a cylindrical canyon in a layered half-space ", *Earthquake Eng. Struct. Dyn.* 419, 799–817 (1990).
10. M. A. Bravo, F. J. Sànchez-Sesma and F. J. Chávez-Garcia "Ground motion on stratified alluvial deposits for incident SH waves", *Bull. Seism. Soc. Am.*, **78**(2), 436–450 (1988).
11. J. T. Chen, P.Y. Chen and C.T. Chen "Surface motion of multiple alluvial valleys for incident plane SH-waves by using a semi-analytical approach", *Soil Dynamics and Earthquake Engineering*, **28**, 58–72 (2008).
12. F. J. Sànchez-Sesma, F. J. Chàvez-Garcia and M. A. Bravo, "Simulation of the seismic response of a 2D sedimentary basin: a comparison between the indirect boundary element method and a hybrid technique", *Bull. Seism. Soc. Am.*, **85**, 1501–1506 (1995).
13. S. Kobayashi "Some problems of the boundary equation method in elastodynamics", Proc. 5th Int. Conf. BEM in Eng., 775-784 (1983).
14. E. Conte, G. Dente and I. Guerra "Calcolo dell'effetto sismico locale mediante il metodo degli elementi di contorno", Proc. 7th Convegno Gruppo Nazionale di Geofisica della Terra Solida 341-354 (1988).
15. F. J. Sànchez-Sesma and J. A. Esquivel "Ground motion on alluvial valley under incident plane SH waves", *Bull. Seism. Soc. Am.*, **69**, 1107–1120 (1979).

The Resonance of the Surface Waves. The H/V Ratio in the Metropolitan Area of Bucharest

Stefan F. Balan, Carmen O. Cioflan, Bogdan F. Apostol,

Dragos Tataru, Bogdan Grecu

National Institute for Earth Physics, Calugareni 12, POBox MG2, Bucharest-Magurele, Romania

Abstract. The purpose of this work is to evaluate the natural period of oscillation T_0 for soils in Bucharest city area. We will start by examine the elastic waves excited at the surface of an isotropic body by an oscillatory, localized force (Rayleigh waves). We define the "H/V"-ratio as the ratio of the intensity of the in-plane waves (horizontal waves) to the intensity of the perpendicular-to-the-plane waves (vertical waves). It is shown that this ratio exhibits a resonance at a frequency which is close to the frequency of the transverse waves. It may serve to determine Poison's ratio of the body. We consider the ratio H/V of the horizontal to the vertical component of the Fourier spectrum for the seismic events recorded at 34 locations during the period October 2003 to August 2004. The method gives reliable data regarding the fundamental frequencies for soil deposits and the results of this experiment allows us to improve the known distribution of T_0 – regularly calculated with the approximate formula $T = 4h/v_s$. The earthquakes with $M_w > 4$ that occurred on 21.01.2004, 07.02.2004, 17.03.2004 and 04.04.2004 will be used as input to compute H/V ratios for each site of a URS stations in the area of Bucharest city. The H/V ratio is also calculated from noise recordings in the same areas. Computation of H/V spectral ratios are performed by means of the SeismicHandler and J-SESAME software showing the reliability of the method used for the sites located in Bucharest. The fundamental period obtained for the majority of sites is in accordance with already known results. By obtaining the fundamental period for much more and different spots situated in the Bucharest area we covered the zones where these data did not exist before. This study is significant in seismic risk mitigation for the Bucharest city area, for a safer seismic design and for the improvement of microzonation efforts.

Keywords: surface (Rayleigh) waves, resonance, H/V - ratio.

INTRODUCTION

Recently there is a great deal of interest in the method of the "H/V"- ratio for assessing the elastic properties of soils by means of their response to external excitations.[1]-[8] We analyze here the surface (Rayleigh) waves excited in an elastic isotropic body by an oscillatory, localized force. The "H/V"- ratio, defined as the ratio of the intensity of the horizontal waves (in-plane waves) to the intensity of the vertical waves (perpendicular-to-the-plane waves), exhibits a resonance peak at a frequency which is close to the frequency of the transverse waves.

Continuous recordings (24 hours/day) were performed during almost 10 months (from October 2003 to August 2004) in the framework of Urban Seismology Project

CP1020, *2008 Seismic Engineering Conference Commemorating the 1908 Messina and Reggio Calabria Earthquake,*
edited by A. Santini and N. Moraci

(URS), a collaboration between the National Institute for Earth Physics and the University of Karlsruhe – Collaborative Research Center 461 (CRC 461).

The recordings were done at 34 sites with 31 seismic stations that were equipped with broadband velocity sensors (Karlsruhe BroadBand Array - KABBA) uniformly spread over Bucharest city and its adjacent zones (Magurele, Voluntari, Otopeni, Popesti-Leordeni, Buftea, Surlari, Ciorogarla). During the time interval indicated above several earthquakes occurred in the Vrancea seismic zone: 4 seismic events with $M_W>4$, 48 seismic events with $M_W>3$ and 67 seismic events with $M_W>2$. In Fig. 1 it is shown the distributions of the seismic stations over the whole area. The number of the seismic stations has supported modifications in time, some of them being disaffected, some moved or shut down. The seismic stations are of last generation technology. The sensors are of type STS-2, LE-3D, 4OT, 3ESP and KS2000. The performance of carrying out recordings for such a long period of time was possible by using for each seismic stations a self-data storage system consisting of a hard-drive of 120 Gb which gives a maximum 3 month/station autonomy. The seismic stations being placed in an urban area or in public institutions, schools, parks, industrial areas they are exposed to shock-type ambient noise or accidents. Therefore a maintenance & service were conducted monthly for the proper functioning of the stations, for the recordings' collection and for diminishing the decalibration risks. We obtained a very good quality of the recordings and a minimum of data loss.

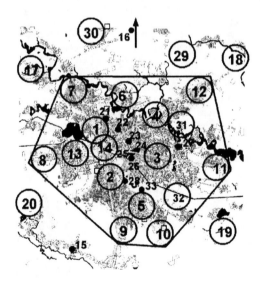

FIGURE 1. The distributions of the seismic stations over the whole area of interest

GEOLOGY

Bucharest city is located in the North-Eastern part of the Moesian Subplate (age: Precambrian and Paleozoic) in the Romanian Plain. After a Cretaceous deposit and a Miocene one (having the top at about 1000m of depth) a Pliocene shallow water deposit (about 700m thick) was settled. The surface geology consists mainly of Quaternary alluvial deposits. Starting from the surface, seven lithological formations are identified [9]: (i)-Backfill (thickness h up to 3m); (ii)-Sandy-clay superior deposits (loess and sand, h=3-16m) from Holocene, and the others from Pleistocene; (iii)-„Colentina" gravel (gravel and sand, h=2-20m); (iv)-Intermediate cohesive deposits of lacustral origin (clay and sand, h=0-25m); (v)-„Mostistea" banks of sands (mainly sand, sometimes lenses of clay included, h=10-15m); (vi) Lacustral deposits from clay and sands (h=10-60m) and (vii)-„Fratesti" gravel (gravel and sands separated by clay, h=100-180m).

A considerable number of geotechnical drillings are available for the central part of the city. The majority of these holes penetrate the upper 30-40m, and some of them extend to depths between 70m and 180m [10]. Strong lateral variations in depth and thickness of these 7 layers can be observed everywhere in Bucharest. From the synthetic seismograms, computed for the 1986 and 1990 strong Vrancea earthquakes in the 0.05-1.0 Hz frequency range, it can be concluded that the thickness of the Quaternary and Tertiary sediments strongly affects the seismic ground motion in the Bucharest area. An important role in the evaluation of the site effects in Bucharest must be associated with the numerous aquifers in the underground of the city. There are three main aquifer systems: (i) „Colentina" located about 8m deep, (ii) „Mostistea" situated at about 25-30m depth, and (iii) „Fratesti" - the deepest aquifer - consisting of three layers situated between 120 and 200m of depth. The horizont of the "Fratesti" gravels has a general slope of 7-8% from South to the Northern part of the city, while the layers above it are less tilted [11].

THE MODEL

In a simplified model we can consider an isotropic elastic body. The equation of elastic waves in this case [12] is given by

$$\ddot{\mathbf{u}} = c_t^2 \Delta \mathbf{u} + \left(c_l^2 - c_t^2\right) grad \cdot div \mathbf{u} + \mathbf{F} \tag{1}$$

where $\mathbf{u}$ is the local displacement, $c_{t,l}$ are the velocities of sound for transversal and, respectively, longitudinal waves and $\mathbf{F}$ is an external force (per unit mass). The sound velocities are given by

$$c_t^2 = \frac{E}{2\rho(1+\sigma)}, \, c_l^2 = \frac{E(1-\sigma)}{\rho(1+\sigma)(1-2\sigma)} \tag{2}$$

where E is Young's modulus, σ is Poison's ratio ($0<\sigma<1/2$) and ρ is the density of the body.

We consider surface waves in a half-space z<0 excited by a localized external force

$$\mathbf{F} = -fe^{-i\Omega t}\delta(\mathbf{r})e^{\kappa z} \qquad (3)$$

where $\mathbf{f}$ is a force per unit superficial mass, Ω is the frequency of the force, κ is an attenuation coefficient and $\mathbf{r} = (x,y)$ are in-plane coordinates. This would correspond to surface waves excited at Earth's surface by seismic waves or other external perturbations. The localization of the force means that we detect the surface waves far away form the source of excitation.

We look for solutions of equation (1) of the form $\mathbf{u} \sim e^{ikr}e^{\kappa z}$ and introduce the notation $\mathbf{u}=(u_l, u_t, u_v)$ and $\mathbf{k}=(k,0)$. In addition we assume $\mathbf{f}=(f_l, 0, f_v)$. Equation (1) becomes

$$
\begin{aligned}
\ddot{u}_l &= \left(-c_t^2 k^2 + c_l^2 \kappa^2\right)u_l + i\kappa k\left(c_l^2 - c_t^2\right)u_v - f_l e^{-i\Omega t}\\
\ddot{u}_v &= \left(-c_l^2 k^2 + c_t^2 \kappa^2\right)u_v + i\kappa k\left(c_l^2 - c_t^2\right)u_l - f_v e^{-i\Omega t}\\
\ddot{u}_t &= c_t^2\left(-k^2 - \kappa^2\right)u_t
\end{aligned}
\qquad (4)
$$

We pass now to solving equations (4) with the force term. The solution is of the form $\mathbf{u} \sim e^{ikr}e^{\kappa z}e^{-i\Omega t}$, where κ is the attenuation coefficient in the force. We get easily

$$
\begin{aligned}
u_l &= \frac{\left(\Omega^2 - c_t^2 k^2 + c_l^2 \kappa^2\right)f_l - i\kappa k\left(c_l^2 - c_t^2\right)f_v}{\Delta}\\
u_v &= \frac{\left(\Omega^2 - c_l^2 k^2 + c_t^2 \kappa^2\right)f_v - i\kappa k\left(c_l^2 - c_t^2\right)f_l}{\Delta}
\end{aligned}
\qquad (5)
$$

where $\Delta = \left[\Omega^2 - c_l^2\left(k^2 - \kappa^2\right)\right]\left[\Omega^2 - c_t^2\left(k^2 - \kappa^2\right)\right]$. We define the "H/V"- ratio as $H/V = \frac{|u_l|^2}{|u_v|^2}$. It is convenient to introduce the notation $s = \frac{f_l^2}{f_v^2}$. We get

$$
\frac{H}{V} = \frac{\left(\Omega^2 - c_t^2 k^2 + c_l^2 \kappa^2\right)^2 s + \kappa^2 k^2\left(c_l^2 - c_t^2\right)^2}{\left(\Omega^2 - c_l^2 k^2 + c_t^2 \kappa^2\right)^2 + \kappa^2 k^2\left(c_l^2 - c_t^2\right)^2 s}
\qquad (6)
$$

We introduce $\omega = c_t \xi k$ for $\xi \cong 1$ and $r = \frac{c_l}{c_t}$. It is natural to assume $\kappa \cong 0$, comparable with κ_t given by $\kappa_t^2 = \left(1 - \xi^2\right)k^2$ for $\xi \cong 1$. Equation (6) becomes then approximately

$$
\frac{H}{V} \cong \frac{\left(\Omega^2 - \omega^2\right)^2 s}{\left(\Omega^2 - r^2\omega^2\right)^2}
\qquad (7)
$$

We can see that the "H/V" - ratio exhibits a resonance at $\omega = \omega_0 \cong \frac{\Omega}{r} = \left(\frac{c_t}{c_l}\right)\Omega$. If we take $\Omega = c_l k$ this resonance is in the vicinity of the S-wave frequency $\omega \cong c_t k$, in agreement with previous results. [1] For $s = 0$ the H/V - ratio is given by

$$\frac{H}{V} \cong \frac{\left(1-\xi^2\right)\left(r^2-1\right)^2 \omega^4}{\left(\Omega^2 - r^2\omega^2\right)^2} \tag{8}$$

and one can see that the resonance is rather sharp. For $s \to \infty$ the resonance disappears.

We may also use $\kappa^2 = \kappa_t^2 = \left(1 - \frac{\xi^2}{r^2}\right)k^2$, and equation (6) becomes

$$\frac{H}{V} \cong \frac{\left[\Omega^2 + \left(r^2-2\right)\omega^2\right]^2 s + \left(r^2-1\right)^3 \dfrac{\omega^4}{r^2}}{\left[\Omega^2 - \left(r^2 + \dfrac{1}{r^2} - 1\right)\omega^2\right]^2 + s\left(r^2-1\right)^3 \dfrac{\omega^4}{r^2}} \tag{9}$$

This expression has a rather broad maximum. For $s = 0$ equation (9) exhibits a resonance at $\omega \cong \left(r^2 + \dfrac{1}{r^2} - 1\right)^{-\frac{1}{2}} \Omega = \left(1 + \dfrac{1}{r^4} - \dfrac{1}{r^2}\right)^{-\frac{1}{2}} \omega_0$ which is greater than ω_0 ($r^2 > 2$). For $s \to \infty$ the maximum of (9) disappears.

It is likely that the attenuation coefficient κ in the expression of the force is very small. The surface waves given by (5) (e.g., Rayleigh wave, [13]) and the *H/V* - ratio from (6) acquire then a very simple expression. A small but finite value of κ shifts the resonance frequency and smooth out the resonance, giving it a small width. The frequency Ω is related to the frequency of the in-depth waves of excitations, or it may have other sources. If the force is a superposition of various frequencies Ω then the resonance is smoothed out and gets a finite width.

The model presented here can be extended to include damping effects and various other distributions of external forces.

DATA PROCESSING

The seismic events chosen for the H/V spectral ratios computation are presented in Table 1 (source: „Romanian Earthquake Catalogue ROMPLUS"). The selection was made by taking into account the recordings quality and the magnitude higher than 3.5. They belong to a data base which consists in events recorded in the time period of interest. The H/V ratios computations were performed with J-SESAME software program for the noise and with "H/V-ratio" software [14] for the seismic events. This J-SESAME software program was developed in the frame of European SESAME Project (Site Effects assessment using Ambient Excitations). We used here all the seismic events recorded during the URS experiment with magnitudes M_W between 3.5 and 4.5. The spectral ratios are computed for each seismic station as shown in the Figs. 2 which are denoted by the name of the station. The spectral ratio curve is period-dependent. The dashed lines are the root mean square (rms) of the H/V ratio. The average shear wave velocity for the layers above Fratesti has values between 340m/s (in the South-East) and 390m/s at TIT station, in the North-Western part of the city. Due to this velocity structure dangerous amplifications in the long period range could be expected in this city area in case of strong Vrancea earthquakes.

Table 1. Characteristics of the seismic events

No.	Data of the event	Time of Occurrence	Lat. (N)	Long. (E)	Depth [km]	Mw
1	21.01.2004	5:49	45.52	26.46	117.70	4.10
2	07.02.2004	11.58	45.67	26.62	143.50	4.40
3	13.02.2004	17.48	45.68	26.69	129.50	3.80
4	17.03.2004	23.42	45.69	26.53	157.50	4.10
5	18.03.2004	20.51	45.63	26.57	140.80	3.70
6	04.04.2004	6.41	45.64	26.48	149.70	4.30
7	06.04.2004	22.35	45.63	26.51	150.50	3.90
8	21.04.2004	1.03	45.70	26.50	165.70	3.50
9	22.04.2004	16.08	45.48	26.40	131.60	3.70
10	24.04.2004	13.00	45.57	26.58	129.30	3.80

In Table 2, a comparison is given between the fundamental period of the lithological columns for boreholes F1 (East of the city), F2 (centre of the city) and F3 (South of the city) (NATO SfP Project 981882, 2005) computed with empirical equation $T=4h/v_s$ and the period evaluated with H/V ratio method. The H/V ratio is also calculated from noise recordings in the same areas. In the previous formula h stands for thickness of the layers above Fratesti horizont and v_s represents the corresponding average shear wave velocity. These boreholes are in the vicinity of the URS05, URS08, URS13 and URS23 seismic stations for which we have computed the spectral ratios H/V presented in Figs. 2.

Table 2. The fundamental period at resonance of the lithological columns (in [s])

Borehole	Fundamental period evaluated with H/V ratio method	Fundamental period computed with empirical equation $T=4h/v_s$	Fundamental frequency [Hz] computed with empirical equation	Fundamental period corresponding to the noise
F1	URS08= 1.45 URS13= 1.5	1.42	0.703	1.38
F2	URS23= 1.45	1.523	0.656	1.56
F3	URS05= 1.48	1.39	0.717	1.42

We considered the ratio H/V of the horizontal to the vertical component of the Fourier spectrum for low-medium intensity seismic events. The scientific relevance of this ratio and its connection to the elliptical trajectory of the Rayleigh waves [15], [16] have been emphasized. This ratio can be used in the identification of the fundamental frequency of soft soils by noticing that the vertical component of the Rayleigh waves

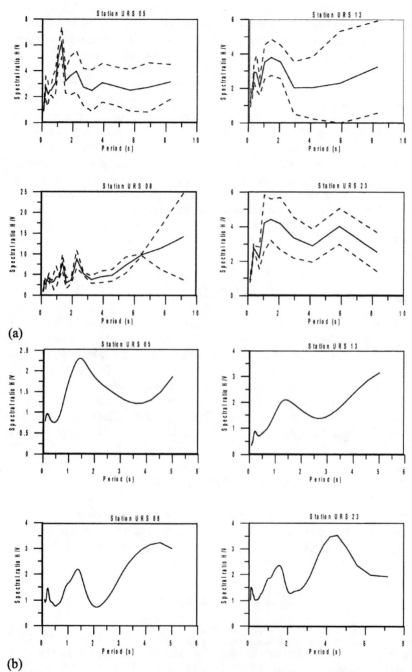

(a)

(b)

Figs.2 The spectral ratios H/V computed for the URS05, URS08, URS13 and URS23 seismic stations: (a) for seismic events and (b) for noise

vanishes in the vicinity of the fundamental frequency of the S waves. Making use of qualitative arguments Nakamura [17] showed that the H/V ratio is a reliable tool for estimating the seismic response at a given site. According to this author, this ratio gives us not only information about resonance period and frequency but also on the corresponding amplification. In his opinion, due to the vertical component, the effects of the source and Rayeigh waves are diminished. In a subsequent paper [18] he assigned a much more important role to the surface waves. There is a consensual conclusion that the H/V ratio for the soft soils exhibits a peak (maximum) value correlated with resonance frequency. These observations are maintained by numerous theoretical studies [6] showing that the synthetic signals obtained for aleatory distributed surface seismic sources lead to H/V ratios with peaks around fundamental frequency everywhere the surface strata have an emphasized impedance different from more rigid strata situated below. At the same time the experimental observations support the H/V spectral ratio method. The comparison between fundamental frequencies obtained from the H/V ratio method allows us to conclude that H/V ratios even for the low intensity earthquakes gives us reliable data regarding fundamental frequencies for soil deposits [7], [19].

CONCLUSIONS

The spectral ratios obtained for the 3 locations confirm the correlation of the periods of H/V peaks – from medium seismicity and noise – to fundamental period of the soils from the drills. The computation of H/V spectral ratios shows the reliability of the method applied for the sites located in the Bucharest area. The fundamental period obtained is in accordance to data from literature about Bucharest [20].

By obtaining the fundamental period for 34 sites situated in the Bucharest area we covered the zones where these data don't exist. This study is significant in seismic risk mitigation for the Bucharest city area, for a safer seismic design and for microzonation improvement.

ACKNOWLEDGMENTS

The authors are grateful to the institutions involved in the funding of the URS experiment: the Deutsche Forschungsgemeinschaft and the National Institute for Earth Physics at Bucharest, Magurele. Microzonation and drilling activities are funded by NATO through Science for Peace Project grant SfP 981882.

REFERENCES

1. M. Nogoshi and T. Igarashi, *J. Seism. Soc. Japan*, **24**, 26, 1971.
2. Y. Nakamura, *QR of Railway Technical Research Institute*, **30**, 1, 1989.
3. Y. Nakamura, *Proc. of the X-th World Conf. Earthquake Eng.*, Mexico, Elsevier, pp. 2134, 1996.
4. E. H. Field and K. H. Jacob, *Geophys. Res. Lett.*, **20**, 2925, 1993.
5. J. F. Lermo, S. Francisco and J. Chavez-Garcia, *Bull. Seism. Soc. Am.*, **83**, 1574, 1993.
6. C. Lachette and P.-Y. Bard, *J. Phys. Earth*, **42**, 377, 1994.

7. H. Yamanaka, M. Dravinski and H. Kagami, *Bull. Seism. Soc. Am.*, **63**, 1227, 1993.

8. E. Field and K. Jacob, *Bull. Seism. Soc. Am.*, **85**, 1127, 1995.

9. N. Mandrescu, *Rev. Roum. Geol., Geophysique et Geographie, serie de Gheophysique,* **10**, 1972, pp. 103, Acad. Roumaine.

10. F. Wenzel, V. Ciugudean, W. Wirth, A. Kienzle, D. Hnnich, K.-P. Bonjer and T. Moldoveanu, V. Socolov, in *Earthquake Hazard and Countermeasures for Existing Praagile Bulidings*, edited by D. Lungu, T. Saito, 2000, pp. 81, Independent Film, Bucharest.

11. N. Mandrescu, M. Radulian and G. Marmureanu, *First European Conference on Earthquake Engineering and Seismology (a joint event of the 13th ECEE&39TH General Assembly of the ESC)*, Geneva, Switzerland, 3-8 September, 2006.

12. L. Landau and E. Lifshitz, *Theory of Elasticity*, Elsevier, 2005.

13. Rayleigh, Lord, *Proc. London Math. Soc.*, **17**, 4, 1885.

14. L. Oncescu, M .Rizescu, *Spectral Ratio*, User's Guide, July, 1997.

15. K. Kobayashi, *Proc. of the 7th WCEE*, 1980, **1**, pp. 237, September 8-13, Istanbul, Turkey.

16. M. Nogoshi and T. Igarashi, *Journal Seism. Soc. Japan*, 1971, **24**, pp. 26.

17. Y. Nakamura, 1989, QR of R.T.R, pp. 30.

18. Y. Nakamura, 1996, *Xth World Conf. Earthquake Engineering*, **2134**, Elsevier Science Ltd., Acapulco, Mexico.

19. E. Field and K. Jacob, *Bull. Seism. Soc. Am.*, **85**, pp. 1127, 1995.

20. A. Aldea and C. Arion, 2001, *2nd National Conference of Seismic Engineering*, 8-9 November, Bucharest.

Probabilistic Simulation of Territorial Seismic Scenarios

Alessandro Baratta[a] and Ileana Corbi[a]

[a] *Department of Structural Engineering, University of Naples "Federico II"*
Via Claudio 21, 80125 Naples, Italy.

Abstract. The paper is focused on a stochastic process for the prevision of seismic scenarios on the territory and developed by means of some basic assumptions in the procedure and by elaborating the fundamental parameters recorded during some ground motions occurred in a seismic area.

Keywords: Seismic simulation, site characterization, historical earthquakes.

INTRODUCTION

The prevision of the damages deriving from a seismic motion, as an instantaneous and violent event, is the first important approach to the problem of a building's seismic protection. A large number of specialist and multidisciplinary technicians (engineers, geologists, seismologists and others) collaborate by many years in the resolution of this problem. It is evident that the abstract prevision of the damage, due to the effect of the single event or to the combination of more than one event, is on the basis of the seismic provision and of the planning of the defense organs' activities [1].

So it is essential the rule taken by the method adopted to have this result. Moreover, because the test of the resisting capacity of a building at the seismic action is extremely randomly and consists in undergoing a quake without any possible intervention in the course of the loading process, the only possible check of congruity of the adopted method consists in the control of the innermost logical coherence of the argumentations assumed as support of the taken choices and in the coordination between the multi-disciplinary contributions.

In the last years a lot of methods have been developed to simulate the propagation of the seismic input by a point-wise source through a semi-infinite deformable and multi-layered space. It is evident by the observation of the earthquake records that they are affected by a pronounced random character due to many different circumstances as, for example, that seismic waves originating at the hypocenter undergo a very large number of reflections and refractions before reaching the earth surface, influenced by the deep geological stratifications [9], [10], [14] and by the site amplification [12]. Moreover, the earthquake recording instruments are very sensitive to the local conditions at the site where they are placed and to the own setting of the

CP1020, *2008 Seismic Engineering Conference Commemorating the 1908 Messina and Reggio Calabria Earthquake,*
edited by A. Santini and N. Moraci

record instruments, and many other errors are added when record are digitized or manipulated [4] [13], [15]. Nevertheless, the random character of the recorded signal presents many analogy with a sample function deriving from a stochastic process. By means of the stochastic process, after fixing some simple parameters solved by the recording data, a population of functions can be defined in a number proportional to the probability that a quake with known structural response and damaging occurs in the site.

THE ASSUMPTION OF SEISMIC POINT-WISE SOURCES

Some basics are assumed in the building of the stochastic process. One of these is the assumption that the seismic source is a point-wise font (for the extended procedure see [7]). Usually the epicenter of the historical earthquakes is recorded as a point source, by contrast instead in the real situation the source is constituted by a fracture along a fault plane. That means that the energy emission from the source is inhomogeneous along the fault line and so the energy contribution on the site, which depends on the distance from the source, is different in the space.

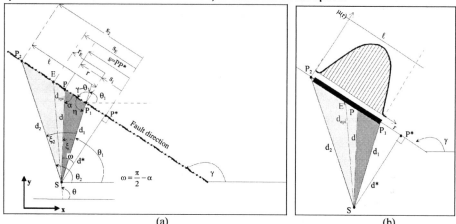

(a)   (b)

FIGURE 1. Schemes of: (a) epicenter and site in plane for the geometrical solution of the methodology developed for a point source moving along a fault line, (b) energy distribution along the segment P_1P_2.

So if one assumes that $E(x_e, y_e)$ is the *nominal* epicenter of a considered historical earthquake, having coordinates x_e and y_e with respect to an axes system <xy> and lying on a fault plane with length ℓ, the epicenter E represents only the average position of an instantaneous point source P which moves along the fault line from the initial position P_1 to the final position P_2

$$P_1 \equiv (x_1, y_1) = \begin{cases} x_1 = x_e - \dfrac{\ell \cos\gamma}{2} \\ y_1 = y_e - \dfrac{\ell \sin\gamma}{2} \end{cases} \quad ; \quad P_2 \equiv (x_2, y_2) = \begin{cases} x_2 = x_e + \dfrac{\ell \cos\gamma}{2} \\ y_2 = y_e + \dfrac{\ell \sin\gamma}{2} \end{cases} \quad (1)$$

Assume that d_{epi} is the distance between the epicenter E and the site S, analogously d_1, d_2 and d are respectively the distance between the site S and the fixed points P_1 and P_2 and the moving point P

$$d_1 = \sqrt{\Delta x_1^2 + \Delta y_1^2} \quad ; \quad d_2 = \sqrt{\Delta x_2^2 + \Delta y_2^2} \quad ; \quad d^* = d_1 \sin(\gamma - \theta_1)$$

$$(2)$$

with $\quad \Delta x_1 = x_1 - x_s \quad ; \quad \Delta x_2 = x_2 - x_s \quad ; \quad \Delta y_1 = y_1 - y_s \quad ; \quad \Delta y_2 = y_2 - y_s$

and d* represents the length of the perpendicular from the site to the fault line having the foot in P*.

Consider that (Fig. 1.a):
$- \theta_1$, θ_2 and θ are the absolute angles respect to the axis x of d_1, d_2 and d;
$- \xi$, ξ_2 and ω are the relative angles between d-d_1, d_1-d_2 and d-d*;
$- \gamma$, α and η are respectively the angle of the fault line with respect to the axis x and the angles between the fault line and the directions d and d_1.
From the geometrical construction one gets

$$\theta_1 = \operatorname{arctg}\left(\frac{\Delta x_1}{\Delta y_1}\right)$$

$$\xi_2 = \arcsin\left(\frac{\ell \sin \eta}{d_2}\right) \quad ; \quad \omega = \frac{\pi}{2} - \alpha \qquad (3)$$

$$\eta = \pi - \gamma + \theta_1$$

after which

$$\alpha = \gamma - \xi - \theta$$

$$\xi = \arcsin\left(\frac{d^*}{d}\right) - \eta \qquad \text{with } \xi \in (0, \xi_2) \qquad (4)$$

By assuming that, during the propagation of the fracture, the source point moves along the segment P_1P_2, the abscissa r is expressed by

$$r = s - s_1 = d\cos(\gamma - \theta_1 - \xi) - d_1 \cos(\gamma - \theta_1) \qquad \text{with } r \in (0, \ell) \qquad (5)$$

where s, s_1 and s_2 are the distances between the points respectively P, P_1 and P_2 to P* , so geometrically $s = d*/tg\alpha$ (Fig. 1.a).
Then the total magnitude M can be assumed as distributed in function of the source point position with the distribution $\mu(r) \geq 0 \ \forall r$ (Fig. 1.b).
The attenuation law which expresses the intensity at the site I(S) as a function of magnitude M and of distance d(r) between instantaneous point source P and site S is

$$I(S) = \int_0^{\ell} F[M, d(r)]\mu(r)dr = F[M, d_{eq}] \qquad (6)$$

By combining the Richter's relationship between the epicentral intensity I_o and the magnitude value M (where the coefficients a, b and c are experimentally estimeted as a= −0.0238, b= 1.8095, c= −1.7857)

$$I_o = aM^2 + bM + c \qquad (7)$$

that corresponds to the probability function of the magnitude and of the *equivalent distance* d_{eq}, which is a function of the distribution $\mu(r)$ with the attenuation law in according with [5] (where L is a characteristic length of the order 10 Km and d is the epicentral distance expressed in km, with Q = 4,38)

$$I_o - I = Q \log_{10} \left(1 + d^2/L^2\right)^{1/2} \qquad (8)$$

one gets

$$F[M, d_{eq}] = I_o - Q \int_0^\ell Log_{10} \sqrt{1 + [d(r)/L]^2} \, \mu(r) dr \qquad (9)$$

Then by introducing the *nominal epicentral distance*, as the distance between site and epicenter d_{epi}=d(E,S), and posing $d_{eq} = \varepsilon \, d_{epi}$, after some algebra one gets

$$\varepsilon = \frac{1}{d_{epi}} 10^{\left(1 + \int_0^\ell Log_{10} \sqrt{1 + [d(r)/L]^2} \, \mu(r) dr\right)} \qquad (10)$$

where the factor ε depends on the distribution $\mu(r)$, and by the relative position of site with respect to fault. In practice the coefficient ε should be introduced in the modeling process, and, in order to simplify the solution and to assess the importance of the source-length effect, the maximum and minimum values of ε can be evaluated as function of the nominal epicentral distance (Fig. 2).

If one assumes that the distribution of magnitude has a bell-form in the interval $(0,\ell)$, symmetrical with respect to the nominal epicenter, a possible choice is

$$\mu(r) = \begin{cases} \rho \dfrac{2}{\ell} \exp\left\{-1 \Big/ \left[1 - \dfrac{4}{\ell^2}\left(\dfrac{\ell}{2} - r\right)^2\right]\right\} & \text{for } 0 \leq r \leq \ell \\ 0 & \text{for } r < 0 \text{ and } r > \ell \end{cases} \qquad (11)$$

with $\quad \rho = 2.2522828$

Then by referring to an approximate evaluation of the integral of Eq. (10)

219

$$\int_0^\ell \text{Log}_{10}\sqrt{1+[d(r)/L]^2}\,\mu(r)dr \tag{12}$$

After some algebra and putting

$$r_0 = \int_0^\ell r\mu(r)dr \qquad ; \qquad R_0^2 = \int_0^\ell (r-r_0)^2\mu(r)dr \tag{13}$$

which are functions of $\mu(r)$ and λ, so they are known as a datum because they are solved during the procedure.

Then, by considering the simplified case with $r_0=r_E=\ell/2$ (i.e. where the distribution $\mu(r)$ is symmetrical with respect to the centre E of the fracture), and the condition $\varepsilon^2\approx1$, after Taylor's truncation, one gets

$$\varepsilon^2 \approx 1 + \frac{R_0^2}{d_{epi}^2}\frac{\left(1+[d_{epi}/L]^2\right)}{[d_{epi}/L]^2}\cdot$$

$$\cdot\frac{\left(2[s_E/L]^2-[d_{epi}/L]^2\right)-[d_{epi}/L]^4+[d_{epi}/L]^2[s_E/L]^2-[s_E/L]^2[d_{epi}/L]^2\sqrt{1+[d_{epi}/L]^2}}{\sqrt{1+[d_{epi}/L]^2}\left(1+[d_{epi}/L]^2\right)^2} \tag{14}$$

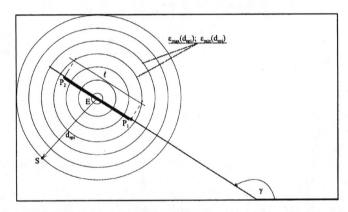

FIGURE 2. Fault line, epicenter and site positions in plane over the earth-surface.

In order to evaluate quantitatively the entity of the coefficient ε, one can consider some concentric circles with centre E on which the possible localization of the site is taken (Fig. 2). The maximum and minimum values of ε are evaluated as function of the radius of the circle and it is evident that they result higher for small distance epicenter-site and quickly converge to $\varepsilon=1$ for a distance equal to 10 km, which is about the value of the length of the fault (fixed to 5 km) (Fig. 3).

Also varying the length of fault for various distances from the epicenter the convergence of the coefficient ε to the value 1 is in correspondence of a distance epicentre-site approximately equal to the length of the fault. However the coefficient ε solved as in Eq. (14) works well if it is about 1, but not well when it assumes different values. In order to solve this problem, the coefficient ε has been evaluated with more accuracy as in the complete form of Eq. (10) for various length of the fault (from 1 to 100 km). The convergence of the coefficient ε to the value 1 is still in correspondence of a distance epicenter-site approximately equal to the length of the fault.

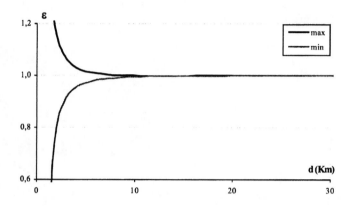

FIGURE 3. Maximum and minimum values of the fault coefficient function of the nominal epicentral distance d_{epi} (Length of the fault=5 Km).

That means that for distances epicentre-site smaller than the length of fault, the intensity at the site can be considered as coincident with the epicentral intensity; on the contrary, for distance epicentre-site larger than the length of fault, the approximation of linear source (as a fracture along a plane of fault) with punctual source (epicentre) does not influence the distribution of intensity function over territory, so the evaluation of the seismic hazard of a region.

EPICENTRAL DISTRIBUTION OF SEISMIC SOURCES OVER THE TERRITORY

In order to build a stochastic process for the simulation of seismic scenarios, some earthquakes, occurred in the last century in a large seismic area of the Centre and South Italy, have been selected where the fundamental macro-seismic parameters, as epicentral intensity, magnitude, intensity at the site, distance between the site and the epicenter. In particular some earthquakes with varying intensities and magnitude values larger than 4 and occurred in the area in the time period between the years 1000-1997, have been collected from the principal seismic catalogues, i.e. the Catalogue of the strong motions in Italy [5] and the Macro-seismic Bulletins of Italy [10] and selected on the basis of the availability and reliability of the data (Fig. 4.a).

If a seismic event with magnitude M has occurred, the geographical location of the epicenter for an expected earthquake can be looked at as a two-dimensional random vector ruled by a Joint Probability Density Function (JPDF) of the epicentral location E

$$P(E|M) = \text{Prob}\{\tilde{E} = E\} = \sum_{j=1}^{n} k_j p_j\left(x_E, y_E, \overline{x}_{Ej}, \overline{y}_{Ej}, \sigma_{xj}, \sigma_{yj}, \rho_j\right) \qquad (15)$$

where k_j and p_j are respectively some combination coefficients which depend on the magnitude, and some probability functions which depend on the geographical position of the possible epicenters in the area (with x_E and y_E the geographical coordinates), and on some other probabilistic parameters.

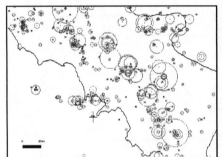 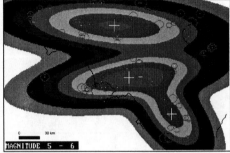

FIGURE 4. (a) Distribution of the 542 selected historical epicenters (the circles show the magnitude values relevant to the event); (b) sample of a macro-seismic map obtained for 3-Gaussian distributions for the recurrence of earthquakes' sources with a magnitude of 5-6.

Such unknown parameters (medium values $\overline{x}_{Ej}, \overline{y}_{Ej}$, standard deviations σ_{xj}, σ_{yj}, correlation coefficients ρ_j and combination parameters k_j), which are necessary to solve the probability functions of epicentral location, can be deduced by the statistical elaboration of the recorded data of historical earthquakes occurred in the sample area over a fixed time period.

By fixing a number of magnitude intervals and distributing the earthquakes in some classes with a fixed value of magnitude, the epicentral density is modeled by means of a linear convex combination of a number of bi-variate Gaussian distributions, whose basic variables are the two coordinates of the epicenter. The parameters that identify the Gaussian distributions (mean values, standard deviations, correlation coefficients) and the relevant combination coefficients are calibrated by means of two different procedures: the first procedures (conventionally called "local fitting") fixes a grid and solves the various JPDFs for each grid's quadrant, then the JPDF of the total area is given by the sum of these distributions (Fig.4.b); the second procedure (conventionally called "global fitting"), which is alternative to the first one, solves the unknown parameters of the Gaussian distribution by minimizing the difference between the simulated moments and the moments calculated by elaboration of the data of historical earthquakes already occurred in the sample area. Both procedures are validate instruments for the elaboration of the macro-seismic maps, where the "local fitting" produces in a very simple way the macro-seismic maps of an area which are more detailed of the maps obtained by the "global fitting". On the other hand the "global fitting", by using a moment procedure, gets macro-scale results of the epicentral distribution which have no constrains due to the selected grid (for the extended procedure see [2], [3], [7]).

Anyway the advantage of both methodologies, that can be viewed at as a generalization of the well established Cornell method [8], is that the maps showing the epicentral probabilistic distribution over the territory (in practice the "seismogenetic areas" for any given value of magnitude) are obtained automatically by simply elaborating in a proper way the known seismic parameters of the historically occurred earthquakes.

CONCLUSIONS

The stability of both procedures is demonstrated by checking the variation of the epicentral JPDF's results for different time ranges; which also demonstrates the robustness of the data used in conjunction with the proposed procedure. With respect to other statistical methodologies, both procedures for the elaboration of seismic maps, although simplified, present the advantage to be prompt and to be based on the fundamental macroseismic data, which are easily available. Some disadvantage can be found in the necessity of elaborating a large quantity of basic data, which, anyway is a common characteristic of any probabilistic study. On the other side, an important difference with respect to the classical models so far available, is that the proposed procedures do not let the choice of the seismogenetic area to the subjectivity of the operator.

REFERENCES

1. A. Baratta, T. Colletta, and G. Zuccaro, *Seismic Risk of Historic Centres*, La Città del Sole, 1996.
2. A. Baratta an I. Corbi, *Epicentral Distribution of seismic sources over the territory*, Int. Jr of Adv. in Eng. Software, **35**(10-11), 663-667 (2004).
3. A. Baratta an I. Corbi, *Evaluation of the Hazard Density Function at the Site*, Int. Jr of Comp. & Struct., **83**(28-30), 2503-2512 (2005).
4. G.V. Berg and G.W. Housner, *Integrated velocity and displacement of strong earthquake ground motion*, Bull. Seism. Soc. Am., **51**, 175-189 (1961).
5. A. Blake, *On the estimation of local depth from macro-seismic data*, Bull. Seism. Soc. Am., **31**, 225-231 (1941).
6. E. Boschi, G. Ferrari, P. Gasperini, E. Guidoboni, G. Smriglio and G. Valensise, *Catalogo dei forti terremoti in Italia dal 461 a.C. al 1980*, ING-SGA, Bologna, 1995 (in Italian).
7. I. Corbi, *Stochastic Models for the Regional Attenuation and the Simulation of Seismic Ground Motions at the Site*, Ph.D. Thesis, University of Naples, 2003.
8. C. A. Cornell, *Engineering seismic analysis*, Bull. Seism. Soc. Am., **58**(5), 1583-1606 (1968).
9. M. Erdik and E. Durukal, *A Hybrid Procedure for the Assessment of Design Basis earthquake Ground Motion for Near-Fault Conditions*, Soil. Dyn. Earthq. Eng., **21**(5), 431-443 (2001).
10. G.W. Housner, *Characteristics of strong-motion earthquakes*, Bull. Seism. Soc. Am. , **45**, 197-218 (1947).
11. INGV. Macro-seismic Bulletin. 1981-1997 (in Italian).
12. K.B. Olsen, *Site Amplification in the Los Angeles Basin from Three-dimensional Modeling of Ground Motion*, Bull. Seism. Soc. Am., **90**(63), 77-94 (2000).
13. A. Schiff and J.L. Borganoff, *Analysis of current methods of interpreting strong-motion accelerograms*, Bull. Seism. Soc. Am., **57**, 857-874 (1967).
14. A. Theoharis and G. Deodatis, *Seismic Ground Motion in a Layered Half-Space due to a Haskell-Type Source. I. Theory*, Soil Dyn. Earthq. Eng., **13**(4), 281-292 (1994).
15. M.D. Trifunac, *Zero baseline correction for strong-motion accelerograms*, Bull. Seism. Soc. A., **61**, 1201-1211 (1971).

Interaction Between the Himalaya and the Flexed Indian Plate - Spatial Fluctuations in Seismic Hazard in India in the Past Millennium?

Roger Bilham and Walter Szeliga

CIRES & Dept. of Geological Sciences, University of Colorado,
Boulder CO 80309-0399 USA

Abstract. Between the tenth and early 16th centuries three megaquakes allowed most of the northern edge of the Indian plate to slip 20-24 m northward relative to the overlying Himalaya. Although the renewal time for earthquakes with this large amount of slip is less than 1300 years given a geodetic convergence rate of 16-20 mm/yr, recently developed scaling laws for the Himalaya suggest that the past 200 years of great earthquakes may be associated with slip of less than 10 m and renewal times of approximately 500 years. These same theoretical models show that the rupture lengths of the Himalaya's Medieval earthquakes (300-600 km) are too short to permit 24 m of slip given the relationships demonstrated by recent events. There is thus reason to suppose that recent earthquakes may have responded to different elastic driving forces from those that drove the megaquakes of Medieval times.

An alternative source of energy to drive Himalayan earthquakes exists in the form of the elastic and gravitational energy stored in flexure of the Indian plate. The flexure is manifest in the form of a 200-450 m high bulge in central India, which is sustained by the forces of collision and by the end-loading of the plate by the Himalaya and southern Tibet. These flexural stresses are responsible for earthquakes in the sub-continent. The abrupt release of stress associated with the northward translation of the northern edge of the Indian plate by 24 m, were the process entirely elastic, would result in a deflation of the crest of the bulge by roughly 0.8 m. Geometrical changes, however, would be moderated by viscous rheologies in the plate and by viscous flow in the mantle in the following centuries.

The hypothesized relaxation of flexural geometry following the Himalayan megaquake sequence would have the effect of backing-off stresses throughout central India resulting in quiescence both in the Himalaya and the Indian plate. The historical record shows an absence of great Himalayan earthquakes in the late 16th to early 19th centuries, and colonial records for this period contain few records of earthquakes in central India. Although this may be an artifact caused by a poor recorded history, it is unlikely that Mw>8.2 earthquakes have escaped notice in the Mughal or early colonial histories.

Recent mid-plate earthquakes in India may thus represent a redevelopment of flexural stressing of the Indian plate. Their return also signifies the development of stresses in the Himalaya that will eventually be released in great Himalayan earthquakes.

Keywords: Earthquakes, Plate flexure. Himalaya

INTRODUCTION

GPS geodesy in the past two decades indicates that convergence between the India and EuroAsian plates cause the Himalaya to contract at a rate of 16-18 mm/year [1-3]. The rate is not precisely defined because the velocity field across the Himalaya merges with the velocity field of Tibet [4], which itself contracts at a rate of more than 1 cm/yr [5-7]. Additional measurements in southern Tibet may permit the local

CP1020, *2008 Seismic Engineering Conference Commemorating the 1908 Messina and Reggio Calabria Earthquake,*
edited by A. Santini and N. Moraci
© 2008 American Institute of Physics 978-0-7354-0542-4/08/$23.00

contraction rate across the Himalaya to be further refined, however, for the purposes of this article a maximum convergence rate of 18 mm/yr will be assumed. This convergence rate, if entirely elastic and released in earthquakes can renew the 5-9 m of slip associated with recent great Himalayan earthquakes in approximately 300-500 years.

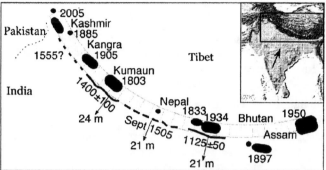

FIGURE 1. Earthquake rupture zones in the Himalaya are shown where inferred from historical and instrumental records. Slip in these earthquakes is believed to be less than 10 m, and only for the Kashmir 2005 earthquake has a surface rupture been identified. In contrast, palæoseismic investigations have identified surface ruptures exceeding 20 m (solid lines). Six hundred km of the central Himalaya slipped >20 m in 1505, and the western Himalaya may have slipped in 1555.

Although less than one third of the Himalaya has slipped in the past 200 years, palæoseismic investigations reveal that three large earthquakes occurred in Medieval India (Figure 1): in c.1100, [8], c1400 [9,10] and in 1505 [Yule, personal communication 2007], the last corresponding to a known earthquake that destroyed Agra and numerous buildings in Tibet and Nepal [11]. Each of these earthquakes was associated with slip exceeding 20 m requiring a elapsed time of more than 1100-1300 years for the slip released in these earthquakes to have developed at current rates. This suggests that the predecessors to these megaquakes would have occurred 200BC to 400 AD, a time during which few historical data, and no palæoseismic data currently exist. A large earthquake is supposed to have occurred at the time of the birth of Buddha, however, no details of this event are known [12].

The difference between the recent great earthquakes and the Medieval megaquakes is striking. In each of these megaquakes surface rupture occurred, raising and back-tilting structures at the front of the Himalayan foothills. In contrast, with the exception of the Kashmir earthquake of 2005, no primary surface ruptures were associated with earthquakes in the past two centuries, and although it is possible that surface rupture, had it occurred, may have escaped the notice of contemporary observers, palæoseismic trench investigations have not recorded offsets at the times of these known events.

The Medieval megaquakes, due to their large slip and long rupture lengths also dominate the cumulative seismic-moment released in the past 1000 years. A previously published summation of seismic-moment for known M>7.5 earthquakes since 1800 concluded that a slip deficit in the Himalaya currently amounts to four M>8 earthquakes [13]. However, when the three megaquakes 1125-1505 are included, and the moment-summation extended to 1000 years the moment deficit is much reduced, or entirely eliminated, depending on the rupture lengths and widths assumed for these events (Figure 2). The 1505 rupture length is estimated at ≤600 km from the reports of damage to monasteries in southern Tibet [11], which when combined with its estimated 21 m of slip (Yule, personal communication, 2007) and an assumed 80-100 km wide rupture, yields a magnitude of **Mw8.9-9**. The minimum rupture length of the c.1125 earthquake is 180 km (**Mw 8.4**), and that for the c.1400 earthquake is ≈250 km (**Mw8.7**).

For the entire Himalayan plate boundary length L_h, and width W_h, the mean slip rate v_h mm/yr can be calculated from the sum of the seismic moments of all the earthquakes ΣMo for a given interval of time t,

$$v_H = \Sigma Mo / L_h W_h t\sigma \text{ mm/yr} \tag{1}$$

Where the length of the Himalaya plate boundary, L_h, is here taken to be 1800 km and the width is 100 km, t=1000 years and σ

The mean slip rate, assuming minimum rupture lengths for the pre-17th century earthquakes is 14.2 mm/yr. Were we to increase this by 20% to account for the large numbers of small and moderate earthquakes missing from the historical record we would obtain a slip rate close to that observed geodetically. The slip rate increases to 25 mm/yr if the plate boundary is narrowed from 100 km to 80 km. If the rupture lengths of the c.1125 and c.1400 earthquakes were each 600 km, the calculated mean slip rate would exceed the 18 mm/yr geodetic convergence rate by 1.5-5 mm/yr, suggesting that the rupture lengths of these earthquakes are not significantly underestimated.

The range of calculated slip rates is thus close to the observed geodetic rate and it suggests that 1000 years is sufficiently long interval to observe a complete earthquake cycle at the Himalayan plate boundary. An inspection of Figure 1 shows that part of the plate boundary has no clear record of rupture in this interval. For example, the region from Bhutan eastwards has no known earthquake, except for some suggestion of an earthquake of unknown size in 1713 [11]. Similarly an earthquake in Kashmir in 1555 may have ruptured a segment of the plate boundary west of the 1905 Kangra earthquake [14,11]. Earthquakes in these regions may have contributed to the 20% of slip considered missing from the historical data base described above.

Our assumption of a 100 km width for the plate boundary may also be inappropriate. The width of the plate boundary is commonly taken to lie between the zone of microseismicity that defines the transition between the locked Indian plate and the zone of aseismic slip beneath southern Tibet, a zone that Avouac [15] notes follows the 3.5 km contour rather closely. This points to a locked width of approximately 90-100 km between here and the start of the foothills of the Himalaya. The transition, however, is rather diffuse and in practice it cannot be an abrupt step function, but a zone of tapered slip. It is not clear how much of this transition zone participates in coseismic slip. An analytic solution in an elastic half-space suggests that it may narrow the effective locked zone to 83 km [4] with a region to the north participating in coseismic slip or immediate post-seismic afterslip.

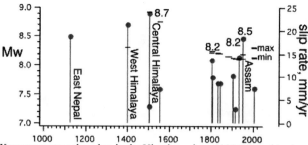

FIGURE 2. Known great earthquakes in the Himalaya since 1000 AD and horizontal tick marks indicating instantaneous convergence velocities calculated from the moment release. The calculated slip rate (min), if adjusted upwards by 20% for the numerous smaller earthquakes that have escaped historical documentation (max), is close to the observed geodetic convergence rate of ≈18 mm/yr.

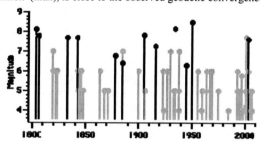

FIGURE 3. Detailed earthquake data are available only for the past 200 years in the Himalaya (black) and mid-plate (grey).

A further complication in the Himalaya that may render our current data on earthquakes heterogeneous, is caused by the high rates of erosion particularly in the eastern Himalaya. Monsoon rains result in the removal of mass from the Himalayan accretionary wedge, a critical taper whose wedge shape is maintained by a balance between thrust faulting, internal friction and gravitational forces [16]. Were it not for this erosion the Himalayan wedge would taper like those in submarine environments, rising northward at approximately 2°. Erosion removes on average 3 mm/yr of material from the frontal slopes of the Himalaya, destabilizing the accretionary wedge, and promoting activity on high level thrusts. Out-of-sequence thrusts have been reported recently active in the Sikkim Himalaya [17]. Should these thrust faults break through to the surface they will temporarily short-circuit slip on the basal thrusts, delaying or circumventing basal thrust earthquakes that would otherwise have occurred. It is possible, for example, that some of the recent earthquakes in the historical record for which no surface breaks have been recognized (eg. Kangra 1905, Nepal 1833) may have occurred on out-of sequence thrust faults. It is also possible that slip on other historical earthquakes may have been reduced by previous high level thrusts of which we have no record. Thus the historical record of the past thousand years may be quite heterogeneous.

227

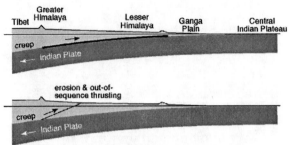

FIGURE 4. Great earthquakes occur on a decollement between the Indian plate and the base of the Himalaya (upper figure). Erosion of the Himalaya has destabilized the 6° tapered accretionary wedge, which in recovering its shape by the mobilization of slip on steep, high-level thrust faults north of the Lesser Himalaya, does so at the expense of slip on the main decollement.

Notwithstanding the uncertainties in the Himalayan earthquake record, it appears that the Medieval earthquake sequence was significantly more energetic than the earthquakes that have occurred in the past 200 years. There is also a suggestion that there may have been an absence of great earthquakes in the 17th and 18th centuries.

Listings of earthquakes in the Indian peninsular reveal numerous earthquakes in post 1800 colonial times but very few in the centuries prior to 1800. While this is also true of the Himalaya, there were few colonial centers of population in the Himalaya, whereas Portuguese, Dutch, French and British colonies flourished along the coasts and navigable rivers in India from the mid 16th century onward.

One possibility is that these apparent periods of low seismicity are real. The other is that they are artifacts of the historical record. and that earthquakes in the 16th and 17th centuries were simply not recorded or that records of events in these times have not survived. Neither of these alternatives can be tested easily.

A physical link between Himalayan megaquakes and mid-plate seismicity, however, appears plausible. In the next section we discuss a possible causal connection.

Flexure of the Indian plate and the Himalayan Earthquake Cycle

The earthquake cycle in the Himalaya is driven by elastic energy resulting from India's convergence with Asia. In simple models of the process [18], strain is concentrated at the transition between the creeping Indian plate and the locked decollement at approximately 15 -18 km depth and decays from this point (the locking line) towards the surface within the Himalaya and Tibet, and downwards into the descending Indian plate [4]. More complex models of the process incorporate tapered slip, more realistic geometries and depth-dependent rheologies [19-21].

Elastic models of the earthquake process in the Himalaya have yielded a scaling law linking the length of earthquake rupture to the amount of slip anticipated in a Himalayan earthquake [4]. The strain at failure, and hence the amount of slip, depends on the recurrence interval between earthquakes. For those earthquakes that have occurred in the past 200 years a recurrence interval of 500 years appears appropriate, yet the Medieval megaquakes require a significantly longer renewal time. One possibility is that the megaquakes (Mw>8.3) release a different reservoir of elastic energy that that available to great earthquakes where Mw<8.2. A possibly suggested in [4] is that a large volume of southern Tibet can participate in driving these great earthquakes due to the longer rupture zones with which they are associated. We discuss here an alternative reservoir of elastic energy - the gravitational and elastic energy of India's flexural bulge

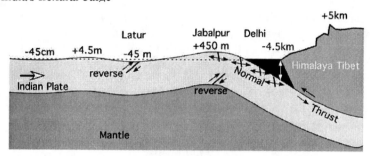

FIGURE 5. Flexure of the Indian plate results in a region of downwarping near Latur (depressed ≈45 m) bordered to its north by a 450-m-high, ≈650-km-wide bulge, which is itself bordered by the downwarped plate to the north. The plate is found at a depth of 4-6 km near the Lesser Himalaya. The flexural forebulge has weakened the upper surface of the plate where it descends beneath the Ganges foredeep (shaded black).

The weight of the Himalaya, and the end loading stress caused by India's northward motion result in the downwarping of the Indian plate and the generation of a bulge to its south [22]. The bulge is responsible for the central Indian plateau beneath which a gentle rise in the Moho and the underlying high density mantle causes a positive gravity anomaly. The anomaly was recognized during the late 19th century Survey of India and named the "hidden range". An outer trough lies south of the flexural bulge causing compressive stresses near the surface of the plate, and it has been suggested that earthquakes of peninsular India are to a large degree caused by the flexural stresses of the Indian plate as it streams northwards toward the Indian Plate at a rate of ≈18 mm/yr [23].

It is important to note that stresses throughout the plate, in the absence of additional stressing from topographic or tectonic loads, are in many places approaching failure, and that the thickness of the plate result in tensile and compressional fiber stresses far exceeding those needed to fracture rock. The release of a 1000 years of accumulated stress along India's northern edge in the Medieval earthquake sequence would have considerably reduced the in-plane stress that supports the flexed Indian plate. Had this occurred it would have reduced stresses throughout the plate. In the subsequent few hundred years stresses would gradually be returned to their previously high values resulting in the resumption of earthquakes in the Himalaya and in the Indian plate. In this view the great earthquakes that have occurred in the past 200 years mark the end of a period of quiescence and the beginning of a period of increasing seismic severity that will culminate in a another sequence of megaquakes 300-500 years hence.

Calculations show that the geometric change in shape of the bulge is likely to be very modest - an elevation reduction of a few cm at most - moderated by viscous flow in the mantle. However, the accompanying in-plane stress changes are likely to exceed a few bars in places, a value that elsewhere is known to retard aftershocks, and which would require several hundred years for their re-development.

However, it is not clear that the Indian plate would in fact slip 10 m northward as a result of 20 m of slip in an earthquake rupture in the Himalaya. For example, were the Indian plate significantly higher modulus than the overlying Himalayan materials we could imagine that the slip in these large earthquakes, in an absolute sense, would occur mainly in the hanging wall. This would result in a relatively minor change in in-plane stress , but an increased incremental end load on the Indian plate, equivalent to adding a layer of rock 1-2 m thick (assuming a 6° dipping decollement) causing an elevation of the flexural bulge. This would act in an opposite sense to releasing the in-plane stress that supports the flexural bulge.

We recognise that elastic models of the interseismic process are consistent with the hanging wall and footwall being effectively the same modulus, supporting a simple interpretation that cosiesimic slip is antisymmetric.

Discussion

Earthquake related stresses in the Himalaya are unable to cause significant stress changes deep within the Indian plate to the south. However, changes in end-loading stress, almost certainly accompany long ruptures, especially those that permit a significant fraction of the northern edge of Indian plate to slip. When this occurs, as is speculated to have occurred between 1125 and 1505, it will tend to change the geometry of the flexural bulge. A reduction in the amplitude of the flexural bulge would retard the timing of earthquakes in central India, and these earthquakes would not be brought back to their former state of imminence until the flexural geometry is re-established by further plate motion.

The historical earthquake data available to us, though consistent with such a scenario occurring in the 16th and 17th centuries, are equally consistent with the data for this time period having been lost. Similarly too little is know of the rheology of northern edge of the buried Indian plate to constrain the response of the flexed Indian plate to a sequence of large contiguous earthquakes. Although the historical record of Indian earthquakes is unlikely to improve in the next century, there is little doubt that improved rheological constraints may be available following the next large earthquake in the Himalaya, that may enable us to address this interesting question analytically.

If in fact the flexural bulge is beginning to tighten following a historial period of partial relaxation, we may anticipate that it will be accompanied by a corresponding increase in mid-plate seismicity in central India and the Himalaya.

ACKNOWLEDGMENTS

This study was supported by the National Science Foundation. We thank Stacey Martin for the use of his catalog of mid-plate Indian earthquakes which was used to develop Figure 3.

REFERENCES

1. Bettinelli, P., J.P. Avouac, M. Flouzat, F. Jouanne, L. Bollinger, P. Willis, and G.R. Chitraker, Plate motion of India and interseismic strain in the Nepal Himalaya from GPS and DORIS measurements, Journal of Geodesy, doi: 10.1007//s00190-006-0030-3, 2006.
2. Banerjee, P. & Burgmann, R. Convergence across the northwest Himalaya from GPS measurements. *Geophysical Research Letters* **29** (2002).
3. Jouanne, F. et al. Current shortening across the Himalayas of Nepal. *Geophysical Journal International* **157**, 1-14 (2004).
4. Feldl, N and R. Bilham, Great Himalayan Earthquakes and the Tibetan Plateau, *Nature* **444**, 165-170. doi:10.1038/nature05199;
5. Wang, Q. et al. Present-Day Crustal Deformation in China Constrained by Global Positioning System Measurements. *Science* **294**, 574-577 (2001).
6. Zhang, Pei-Zhen, Shen, Zhengkang, Wang, Min, Gan, Weijun, Bürgmann, R, Molnar, P, Wang, Qi, Niu, Zhijun, Sun, Jianzhong, Wu, Jianchun, Hanrong, Sun, Xinzhao, You, Continuous deformation of the Tibetan Plateau from global positioning system data, Geology 2004 32: 809-812
7. Chen, Q., J. T. Freymueller, Z. Yang, C. Xu, W. Jiang, Q. Wang, and J. Liu (2004), Spatially variable extension in southern Tibet based on GPS measurements, J. Geophys. Res., 109, B09401, doi:10.1029/2002JB002350.
8. Lave, J. & Avouac, J. P. Active folding of fluvial terraces across the Siwaliks Hills, Himalayas of central Nepal. *Journal of Geophysical Research* **105**, 5735-5770 (2000).
9. Kumar, S., S. G. Wesnousky, T. K. Rockwell, R. W. Briggs, V. C. Thakur, and R. Jayangondaperumal (2006), Palæoseismic evidence of great surface rupture earthquakes along the Indian Himalaya, *J. Geophys. Res.*, *111*, B03304, doi:10.1029/2004JB003309.
10. Wesnousky, S. G., Kumar, S., Mohindra, R. & Thakur, V. C. Uplift and convergence along the Himalayan Frontal Thrust of India. *Tectonics* **18**, 967-976 (1999).
11. Ambraseys, N. & Jackson, D. A note on early earthquakes in northern India and southern Tibet. *Current Science* **84**, 570-582 (2003).
12. Bilham, R. Earthquakes in India and the Himalaya: Tectonics, geodesy, and history. *Annals of Geophysics* **74**, 839-858 (2004).
13 Bilham, R. & Ambraseys, N. Apparent Himalayan slip deficit from the summation of seismic moments for Himalayan earthquakes, 1500-2000. *Current Science*, **88**, 1658-1663, (2005).
14. Iyengar R.N., Sharma D. (1998) Earthquake history of India in medieval times, *Rep. Central Build. Res. Inst.*, 124pp, Roorkee.
15. Avouac, J. P. Mountain Building, Erosion, and the Seismic Cycle in the Nepal Himalaya. *Advances in Geophysics* **46**, 1-80 (2003).
16. Davis, Dan; Suppe, John; Dahlen, F. A. Mechanics of fold-and-thrust belts and accretionary wedges, J. Geophys. Res., 88, Issue B2, p. 1153-1172
17. Mukul, M., Jaiswal, M., Singhvi, A. K., 2007. Timing of recent out-of-sequence active deformation in the frontal Himalayan wedge: Insights from the Darjiling sub-Himalaya, India: Geology, 35 (11), p.999-1002
18. Savage, J. C. A dislocation model of strain accumulation and release at a subduction zone. *Journal of Geophysical Research* **88**, 4984-4996 (1983).
19. Vergne J., R. Cattin and J.P. Avouac, On the use of dislocations to model interseismic strain and stress build-up at intracontinental thrust faults, *Geophys. J. Int.*,147, 155-162, 2001.
20. Perfettini, H. & Avouac, J. P. Stress transfer and strain rate variations during the seismic cycle. *Journal of Geophysical Research* **109** (2004).
21. Cattin, R. & Avouac, J. P. Modeling mountain building and the seismic cycle in the Himalaya of Nepal. *Journal of Geophysical Research* **105**, 13389-13407 (2000).
22. Watts, A. B., Isostasy and Flexure of the Lithosphere, Cambridge University Press pp.458.
23. Bilham, R., R. Bendick, and K. Wallace, (2003). Flexure of the Indian Plate and intraplate earthquakes, *Proc. Indian Acad. Sci. (Earth Planet Sci.)*,112(3) 1-14

Geotechnical Seismic Hazard Evaluation At Sellano (Umbria, Italy) Using The GIS Technique

P. Capilleri[a] and M. Maugeri[b]

[a] PhD. in Geotechnical Engineering, Department of Civil and Environmental Engineering, University of Catania, viale A. Doria n° 6, 95125 Catania (Italy), e-mail: pcapilleri@dica.unict.it.
[b] Full Professor of Geotechnical Engineering, Department of Civil and Environmental Engineering, University of Catania, viale A. Doria n° 6, 95125 Catania (Italy), e-mail: mmaugeri@dica.unict.it

Abstract. A tool that has been widely-used in civil engineering in recent years is the geographic information system (GIS) [1]. Geographic Information systems (GIS) are powerful tools for organizing, analyzing, and presenting spatial data. The GIS can be used by geotechnical engineers to aid preliminary assessment through to the final geotechnical design. The aim of this work is to provide some indications for the use of the GIS technique in the field of seismic geotechnical engineering, particularly as regards the problems of seismic hazard zonation maps. The study area is the village of Sellano located in the Umbrian Apennines in central Italy, about 45 km east of Perugia and 120 km north-east of Rome The increasing importance attributed to microzonation derives from the spatial variability of ground motion due to particular local conditions. The use of GIS tools can lead to an early identification of potential barriers to project completion during the design process that may help avoid later costly redesign.

Keywords: Microzonation maps, amplification, response analysis, laboratory test, risk, earthquake .

INTRODUCTION

Computer-based tools, such as geoinformation systems and decision support systems, are expected to be very helpful in realizing natural hazard and risk zonation mapping [2]. One important tool for this purpose is the geographic information system (GIS). The scope of the tool is to supplement abilities and knowledge that have already been acquired, not to serve as a replacement for those which are lacking.

By using GIS it is possible to organize spatial data, visualize complex spatial relationships of a particular location, make a spatial investigation of the characteristics of a particular location, combine data sets from different sources, analyze the data to understand its meaning, and make predictions by combining data layers in accordance with purposely developed rules [2].

Hazard mapping related to seismic activity is a useful application of the above-listed potentialities of the GIS.

Geotechnical seismic hazards are studied systematically, in order to evaluate damage to human life, buildings, and infrastructure, such as roads, dams, bridges and so on.

CP1020, *2008 Seismic Engineering Conference Commemorating the 1908 Messina and Reggio Calabria Earthquake,*
edited by A. Santini and N. Moraci

To achieve this objective, data from geotechnical borings, geophysical investigations and geotechnical laboratory tests have been collected, reprocessed and organized in a digital database.

The analysis of such hazards implied the creation and management of a geotechnical zoning of the urban area of Sellano using GIS techniques, and the adopted procedure for the final creation of the Arc View GIS database of the town of Sellano [3]. Efforts have been directed to the creation of a GIS compatible database of all available geotechnical borings and to using all available data to draw up a geotechnical zonation map.

THE GIS APPLICATION IN SEISMIC MICROZONATION MAPPING

A GIS is defined as a powerful set of tools for collecting, storing, retrieving at will, displaying, and transforming spatial data [2]. One of the main advantages of the use of this technology is the possibility of improving hazard occurrence models, by evaluating their results and adjusting the input variables.

This paper briefly introduces GIS, how the technology can be applied, and discusses the benefits of its use in design. To illustrate these applications actual projects utilizing GIS are presented and discussed. GIS tools can be used to integrate existing data such as soil survey maps and aerial photos with project specific data; identify potential geological and geotechnical hazards; plan and track field work; catalog and review sampling and laboratory testing results; create maps and figures for reports; and improve communication between office and field staff, consultants and their clients, and project team members.

The GIS can combine spatial data of any kind for the purpose of ordering and presenting them. It also enables data related to the territory to be processed by storing the cartographic elements in digital size, and for these to be combined with descriptive alphanumeric information or photos [3].

The basis of the geographical data is similar to a collection of layers: each of them is a separate category of information; territorial objects and complex phenomena are the result of the combination of various layers.

A large database has been created to store the values of the geotechnical parameters referring to each geological formation to be found throughout the study area. Each geological unit is described in the database according to certain attributes. The great variability of geotechnical parameters, such as cohesion, angle of internal friction, thickness of layers or depth of ground-water level, is inconsistent with the homogeneity of data required in deterministic models; this is the reason why only average values have been used. Both a database, which contains the geological information, and a digital terrain model (DTM) with the geomorphologic information have been built.

GEOGRAPHICAL AND GEOTECHNICAL SETTING

The city of Sellano was founded in 81 a.c., by Lucio Silla, from who it took its name. It is located 640 m above sea level in the Umbrian Apennines in central Italy, about 45 km east of Perugia and 120 km north-east of Rome (Fig. 1). The village stands on a hill crest which elongates in an approximately WE direction. The ancient town is located on the eastern side of the hill while the most recent area of expansion (from 1800 onwards) is to be found on the western part. From a geological point of view, the study area is made up of the "Scaglia" unit constituted by alternating layers of limestones and marly limestones embedded in a sandy-silt or clayey-silt matrix [4]. The "Scaglia" unit may be further differentiated into four sub-units, i.e. the "Scaglia Bianca", "Scaglia Rossa", "Scaglia Variegata" and "Scaglia Cinerea". The calcareous levels are prevalent in the oldest units ("Scaglia Bianca" and "Rossa") while the marly-clayey levels predominate in the youngest ("Scaglia Variaegata" and "Cinerea"). Sellano is mainly constituted by the "Scaglia Cinerea" formation, which is weathered in the upper part and frequently covered by fill material of variable thickness [4].

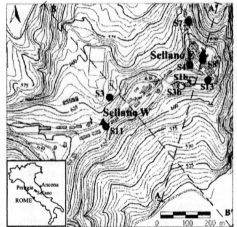

FIGURE 1. Plain view of the Sellano area [4].

Historical seismicity and accelerometer data

The Sellano area is characterized by a significant historical seismicity that produced extensive damage to the village. The seismic catalogues show numerous earthquakes with intensities ranging between the VII and X degree on the Mercalli-Cancani-Sieberg (MCS) scale [5].

The most important recent events have been the 19 September, 1979, Valnerina earthquake (M_s=5.5) and the 29 April, 1984, Gubbio earthquake (M_s=5.3-5.5). Table1 shows the other historical earthquakes in the Sellano area.

The Umbria-Marche seismic sequence occurred between September 1997 and April 1998 along a seismogenetic structure constituted by several faults running essentially in a NW-SE direction [5]. The sequence started on September 3, 1997, at 22.07 GMT, with an earthquake of magnitude M_L=4.4 localized in the Colfiorito area, about 20 km north of Sellano. The first strong event of magnitude M_L=5.5 occurred on September 26, at 00:33 GMT, followed at 09:40 GMT by another strong quake with magnitude M_L=5.4. The epicenters of both earthquakes were located around Colfiorito. The successive seismic activity was characterized by smaller events with epicenters migrating towards the southern part of the tectonic structure, i.e. the area of Sellano. On October 14, 1997, another strong event (M_L=5.4) occurred, with its epicenter at about 2.5 km north of Sellano, which produced severe damage (I=VIII-IX MCS) to the historical buildings of Sellano.

This seismic sequence was recorded by several permanent S.S.N. stations. [4]. After, in addition to the RAN (National Network Acceleration) permanent Station a mobile digital network was installed, in order to increase the density of strong motion instruments in the epicentral area. In particular, two mobile stations were established in the town of Sellano respectively, Sellano East and Sellano West. The localization the new mobile stations has been integrated into the GIS project.

TABLE 1. Historical earthquakes

Epicentral area	Richter Magnitude	Annex
Norcia	5.9	-99
Spoleto	5.5	-63
S.Sepolcro	6.7	1328
S.Sepolcro	7.6	1349
Norcia- Cascia	6.5	1707
Mugello	5.9	1919
Marche	5.7	1930
Valfabbrica	4.5	1978
Alta Valnerina	5.9	1979
Colfiorito	5.5	1997
Colfiorito	5.8	1997
Sellano	5.4	1997

Site characterization

Geological and physical information regarding the land (Fig 2), and, in particular, its geotechnical features, has been obtained by incorporating borehole investigations into the GIS so as to check their stratigraphic view through active links in the Arc View GIS.

The location of the boreholes is reported in Fig 3

Geotechnical investigations were performed in the most damaged area, between November 1998 and November 1999, in addition to other field tests carried out in January 2003.

In some cases the borehole was accompanied by the results of the geotechnical in situ test or laboratory test.

The determination of the dynamic characteristics of the soil was obtained through precise analyses performed on the test site and in laboratory tests.

Dynamic laboratory tests were performed on the undisturbed samples, with the aim of evaluating, the shear modulus (G) and the damping ratio (D) linked to the current shear deformation γ [6].

FIGURE 2. Geological map of the Sellano area.

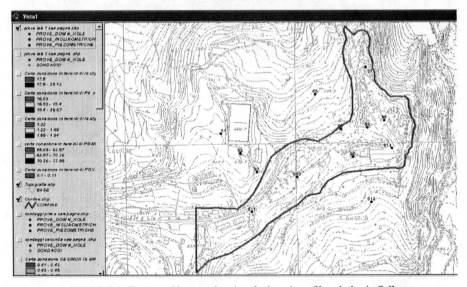

FIGURE 3. Topographic map showing the location of boreholes in Sellano.

The shear modulus G and damping ratio D for the Sellano soil were obtained in the laboratory from the Resonant Column test (RCT), cyclic loading torsional shear test (CLTST) and monotonic loading torsional shear test (MLTST) and using a double specimen direct simple shear device (DSDSS).

Figure 4a shows the results of RCT, CLTST and MLTST in terms of degradation of shear modulus with strain; while Figure 4b shows the results obtained by RCT, CLTST and MLTST in terms of increase of damping with strain.

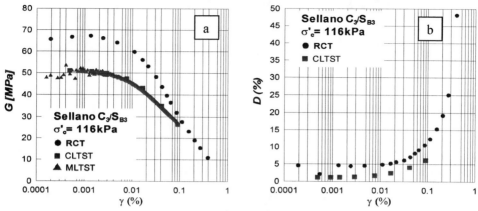

FIGURE 4. a)Curve G-γ; b)Curve D-γ

The GIS includes the boreholes, the down-hole test and the results from laboratory tests, mixing the layers of the schematic geo-settled geological map of the urban area of the town of Sellano (Fig. 2) with the layers of borehole and geotechnical data.

SEISMIC HAZARD ZONATION MAP

The accuracy of zoning based on local site investigations can be further enhanced using ground motion modeling. Computer codes which are now available include one-dimensional linear anon linear analysis and two – dimensional analysis, which are necessary to achieve zoning [7].

According to the Manual for the zonation of seismic geotechnical hazard [7], the soil response spectra for each zone must be evaluated by means of a computer code modeling the soil response [4,8].

To investigate the role of soil stratigraphy on the amplification of seismic motion, a 1-D analysis with Proshake [4,9] was carried out. To further explore the influence of the hill morphology on the surface motion, a 2-D numerical investigation was also performed using the FEM code QUAD4M [4,10].

The data from all the analyses were collected and organized in a digital data base. To achieve the hazard seismic map 1:5000 topographic maps were utilized.

The area of Sellano to be mapped was divided into a mesh, the grids having 10x10 meter long sides. The data from borings or measurements of shear wave velocity were used to build up the surface ground model for each mesh. The specific parameter required for such an analysis is the shear - wave velocity. When boring data is not available in all the elements of the mesh, the elements are grouped according to the generic ground condition.

In implementing the maps, the degree of risk was associated to a chromatic scale consisting of the three basic traffic-light colors, green, yellow and red (Fig. 5).

Finally, Sellano suffered great damage during the 1997-1998 seismic sequence due to the greater vulnerability of its buildings (Fig. 6)and the greater seismic site amplification [11]. On the basis of local surveys of different buildings in Sellano the damage was inserted into the GIS project (fig.7).

237

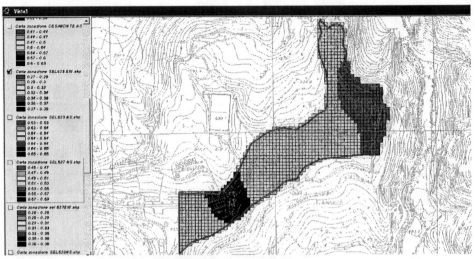

FIGURE 5. Microzonation map of Sellano

FIGURE 6. Example of structural damage

CONCLUSIONS

During the last decade, the rapid development of the GIS has proved to be a very helpful tool in hazard and risk zonation mapping.

In this case the GIS tools were used to finalize the Sellano microzonation project. Microzonation maps were created to identify and demarcate areas of homogeneous response.

Therefore, the use of the GIS was important to deal with the vast amounts of information including spatial (geographical) and geotechnical data. The GIS is capable of managing spatial and non spatial data, and can perform a large number of operations analyses and cartographic representations.

The GIS system can be updated with new data and new information. It is not a static system, because the various information levels can be constantly updated within the existing layers and implemented to obtain a new layer.

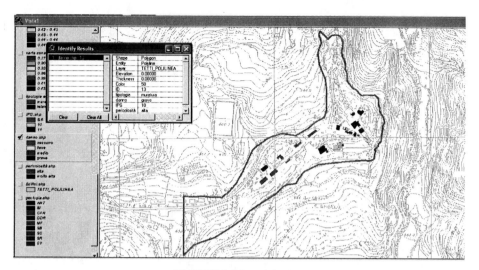

FIGURE 7. Map of the damage

REFERENCES

1. Scott B. Miles and Carlton L. Ho. "Application and issue of GIS as tool for civil engineering modeling". *Journal of computing in civil engineering ASCE* Vol 13 No 3, July 1999.
2. M. G. Sakellariou and M. D. Ferentinou. "GIS- base estimation of slope stability". *Natuaral Hazards Review ASCE* Vol 2 No 1, February 2001.
3. S. Grasso and M. Maugeri. "Vulnerabilty of physical environment of the city of Catania using GIS technique". Seismic prevention of damage: a case study in a Mediterranean city. Edited by: M. Maugeri, Università di Catania , Italy. Wit Press (2005).
4. Capilleri P., Lanzo G., Maugeri, M. Pagliaroli A. "Site Effects Evaluation In Sellano (Italy) By Means Of 1-D And 2-D Numerical Analyses (Italy)" *Fifth International Conference on Earthquake Resistant Engineering Structures - ERES 2005* 30 May-01 June 2005, Skiathos, Greece.
5. Decanini L., Mollaioli F. & Oliveto G., Structural and seismological implications of the 1997 seismic sequence in Umbria and Marche, Italy, in *Innovative Approaches to Earthquake Engineering*, Edited by G. Oliveto, WIT Press, Southampton , Boston, pp. 229-323, (2002).
6. Capilleri, P. "Caratterizzazione Geotecnica e Microzonazione Sismica della Città di Sellano", Ph.D. Thesis, Università degli Studi di Catania, (2003).
7. TC4. Manual for Zonation on Seismic Geotechnical Hazards, Technical Committee for Earthquake Geotechnical Engineering, TC4, ISSMFE, Japanese Society of Soil Mechanics and Foundation Engineering, Tokyo, Japan(1999).
8. Capilleri P., Massimino M.R., Maugeri M., "The ground-motion grade-3 microzonation of Sellano", *Italian Geotechnical Journal*, Vol. XXXV, No. 4, 97-111 (2001).
9. EduPro Civil System, Inc., ProShake–Ground Response Analysis Program, EduPro Civil System, Inc., Redmond, Washington, (1998).
10. Hudson, M., Idriss, I.M. & Beikae, M., QUAD4M: a computer program to evaluate the seismic response of soil structures using finite element procedures and incorporating a compliant base, Dpt. of Civil and Environmental Eng., Univ. of California Davis, Davis California, (1994).
11. Massimino M. R., Maugeri M., Zuccarello S. "The grade-2 microzonation of Sellano". *Italian Geotechnical Journal*, Vol. XXXV, No. 4, 97-111 (2001).

Site Response Analysis of the Monte Po Hill in the City of Catania

A. Cavallaro[a], A. Ferraro, S. Grasso and M. Maugeri[b]

[a] CNR-IBAM, Via A. di Sangiuliano n. 162, 95124 Catania, Italy.
[b] Department of Civil and Environmental Engineering, University of Catania, Viale A. Doria 6, 95125 Catania, Italy

Abstract. The Monte Po Hill is located in the North-eastern part of the city of Catania; this area is prone to high seismic risk due to the presence of several constructions, including a school, in the vicinity of a toe of a slope, characterized by precarious stability conditions. For site characterization of the soil, deep site investigations have been undertaken. Borings and dynamic in situ tests have been performed. Among them Down-Hole (D-H) and Seismic Dilatometer Marchetti Tests (SDMT) have been carried out, with the aim to evaluate the soil profile of shear waves velocity (Vs). The Seismic Dilatometer Marchetti Tests were performed up to a depth of 15 meters. The results show a very detailed and stable shear waves profile. The shear waves profiles obtained by SDMT compare well with other in situ tests. Synthetic seismograms have been drawn for the site long a set of receivers placed at different depths, starting from the surface up to almost 40 m. After evaluating the synthetic accelerograms at the bedrock, the ground response analysis at the surface, in terms of time history and response spectra, has been obtained by a 1-D non-linear model. In particular the study has regarded the evaluation of site effects in correspondence of the site, to which corresponds a different value of the Seismic Geotechnical Hazard. In the beginning of 2007 a seismic station has been also located into the school building, with the aim of recording seismic events. Seismograms obtained by the seismic station have been also used to evaluate the ground response analysis at the surface. In the beginning of 2007 a seismic station has been also located into the school building, with the aim of recording seismic events. Seismograms obtained by the seismic station have been also used to evaluate the ground response analysis at the surface. Finally the 1-D computer code EERA was also used to model the equivalent-linear earthquake site response analyses of layered soil deposits of the hill. The detail with which the hill has been studied has allowed the construction of a detailed 2-D model of its structure. It has been explored the differences between the computed ground motion for different Vs profiles using QUAD4M 2-D code. It has been also possible to compare the results from different 1-D models reflecting current approaches to the determination of site response.

Keywords: Seismic Response, Dilatometer, Shear Waves Velocity, 2-D model.

INTRODUCTION

Monte Po hill is located in the north-eastern part of the city of Catania. In the area a school and private constructions were built in the past in the nearness of the hill and sometimes near the toe. Then, high-risk conditions arose since the stability of the area is poor even under static loading condition. During the first construction times, instability phenomena occurred in the hill without determining an interruption in the works. In the same time damages in some of the buildings existing near the hill occurred and failure of several earth-retaining structures were observed in the area.

CP1020, *2008 Seismic Engineering Conference Commemorating the 1908 Messina and Reggio Calabria Earthquake,*
edited by A. Santini and N. Moraci
© 2008 American Institute of Physics 978-0-7354-0542-4/08/$23.00

Successively, damages and instability phenomena were observed as a consequence of meteoric events, during and after some excavations performed near the toe and, finally, after the December 13, 1990 Sicilian earthquake.

Due to the significant seismic geotechnical hazard related to this site an intensive research activity was performed in the framework of the research project "*Detailed scenarios and actions for seismic prevention of damage in the urban area of Catania*" [1]. The aim of the present research was the study of the site response of the Monte Po hill. The detail with which the hill has been studied has allowed the construction of a detailed 2-D model of its structure. In this paper it has been explored the differences between the computed ground motion for different Vs profiles. In order to obtain a reliable model of the subsoil, data concerning the soil geotechnical properties were collected using both in situ and laboratory test results. In particular results of the geotechnical characterization performed during the different building activities occurred in the area was considered. Moreover, data obtained during the research programme carried out by the geological office of the Catania municipality as a consequence of the instability phenomena occurred in the area were also analyzed.

In this paper the procedure adopted to detect the more reliable subsoil model to be used in the local seismic response analyses is described. The results were analyzed in the attempt of evaluating the possible occurrence of an earthquake-triggered landslide focusing the attention on the effects of the earthquake-induced permanent displacements and on the post-seismic serviceability of the structures involved in the area.

SEISMICITY OF THE AREA

The study area is located in one of the most seismically active zone of the Mediterranean. In the last 900 years, the east cost of Sicily has been struck by various disastrous earthquakes with MKS intensity varying in the range IX - XI, and estimated magnitude ranging from less than 5.0 to greater than 7.0. The most probable source of earthquakes in the area is the Malta Escarpment, a system of sub-vertical normal faults, NNW-SSE oriented which runs for about 70 - 100 km offshore along the Ionian coast of Sicily. This structure appears to be subdivided into different segments, the northernmost ones bordering the eastern Hyblean coast and extending inland as far as the Etna volcano area.

Using the historical seismicity data available for the area it is reasonable to assume as a maximum expected earthquake the repetition of the two M > 7 events which hit western Sicily in the past and destroyed Catania: the 1169 and the 1693 earthquakes. These catastrophic events were characterized by intensity equal to X MCS and XI MCS and estimated magnitudes equal to 7.0 - 7.4. According to Azzaro et al. [2] the first level scenario event for the Catania area may be reasonably assumed the January 11, 1693 earthquake that caused the largest seismic catastrophe in eastern Sicily. The most probable source of this event is located along the northern part of the Ibleo-Maltese fault and is commonly associated to rupture with a normal pure mechanism along the escarpment [3].

In particular at the beginning of 1991 during the construction of the school, a shallow instability phenomenon occurred causing a high-risk conditions for the workyard of the school and for the existent private constructions. As a consequence of the occurred instability, the Catania municipality carried out an intense site investigation activity and monitored of the hill with piezometers and inclinometers.

BASIC SOIL PROPERTIES

The investigated Monte Po hill area, located in the North-West zone of the city, has plane dimensions of 40,000 m^2 and a maximum depth of 20 m. The area pertaining to the investigation program and the locations of the boreholes and field tests are shown in Figure 1.

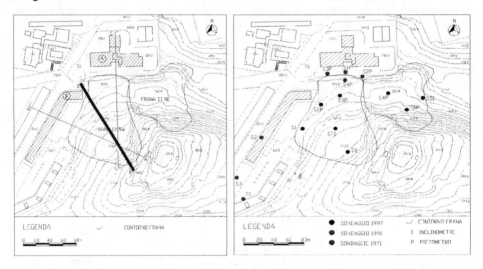

FIGURE 1. Lay-out of Monte Po hill area with landslide localization and indication of the three cross sections considered, localization of borings location.

Generally, a thin layer of altered soil can be observed in the area with thickness ranging from 0 to about 1 m. Then, four different unit can be recognized: a layer of medium-stiff alluvial silt of medium plasticity with thickness ranging from 50 cm to 4 m; a layer of very sandy gravel which was detected only in some boreholes with thickness ranging from 25 cm to about 4 m; a formation of conglomerate and sand with thickness ranging from 50 cm to about 3 m; finally a layer of clay of upper plasticity range locally representing the sub-grade.

In the Catania Monte Po hill area, the clay fraction (CF) is predominantly in the range of 28 - 44 %. This percentage decreases to 17 % at the depth of 7 m where a sand fraction of 42 % is observed. The gravel fraction is always zero. The silt fraction is in the range of about 3 - 42 %.

Typical range of physical characteristics, index properties and strength parameters of the deposit are reported in Table 1.

TABLE 1. Mechanical characteristics for Catania Monte Po hill area.

Depth (m)	0 - 5	5 – 10	10 – 15	15 – 20
γ [kN/m³]	18.9÷20.4	19.7÷20.1	20.6÷20.3	20.1÷20.7
γ_s [kN/m³]	26.8÷26.9	20.1÷27.2	27.0÷27.0	2.71÷27.1
γ_d [kN/m³]	15.8÷17.2	15.5÷16.2	16.3÷16.8	16.2÷17.6
γ_{sat} [kN/m³]	20.4÷20.8	19.7÷20.2	20.2÷20.6	20.2÷21.1
w_n [%]	17.5÷ 25.6	21.5÷26.9	20.2÷23.0	17.1÷23.9
e	0.56÷0.67	0.65÷0.72	0.59÷0.66	0.62÷0.67
S_r [%]	76.3÷91.4	87.9÷99.2	91.0÷96.2	87.7÷97.1
w_L [%]	45.3÷51.2	44.3÷49.8	47.1÷54.2	37.7-51.5
w_p [%]	19.5÷27.8	18.5÷19.3	21.3÷22.6	16.5÷20.5
PI	17.5÷31.7	25.8÷30.5	25.8÷31.6	21.2÷32.1
IC	1.03÷1.20	0.69÷0.86	1.02÷1.04	0.86÷0.97
c_u [kPa]	17.4÷19.6		46.5÷63.6	14.4÷20.1
c' [kPa]	12.4÷18.6		24.0÷35.1	10.6÷12.0
ϕ' [°]	21.8÷22.8		21.6÷23.7	21.3÷24.7

where: c_u (Undrained shear strength) was calculated from and U-U Triaxial Tests, c' (Cohesion) and ϕ' (Angle of shear resistance) were calculated from C-U Triaxial Tests.

SHEAR MODULUS

Shear-Wave Velocity V_s: In Situ Measurements

The small strain ($\gamma \leq 0.001$ %) shear modulus, G_o, was determined from SDMT and Down Hole (D-H) tests. The Seismic Dilatometer Marchetti (SDMT) is an instrument resulting from the combination of the DMT blade with a modulus measuring the shear wave velocity. The seismic modulus is an instrumented tube, located above the blade (see Figure 2), housing two receivers at a distance of 0.50 m. The test configuration "two receivers"/"true interval" avoids the problem connected with the possible inaccurate determination of the "first arrival" time sometimes met with the "pseudo interval" configuration (just one receiver). Also the pair of seismograms at the two receivers corresponds to the same blow, rather than at two successive blows - not necessarily identical. The adoption of the "true interval" configuration considerably enhances the repeatability in the V_s measurement. The SDMT provides a simple means for determining the initial elastic stiffness at very small strains and in situ shear strength parameters at high strains in natural soil deposits. Source waves are generated by striking a horizontal plank at the surface that is oriented parallel to the axis of a geophone connected to a co-axial cable with an oscilloscope [4], [5]. The measured arrival times at successive depths provide pseudo interval V_s profiles for horizontally polarized vertically propagating shear waves (Figure 2). V_s may be converted into the initial shear modulus G_o. The combined knowledge of G_o and of the one dimensional modulus M (from DMT) may be helpful in the construction of the G-γ modulus degradation curves [6]. The V_s determinations are executed at 0.50 m depth intervals. A summary of SDMT parameters are shown in Figure 2 where: I_d: Material Index; gives information on soil type (sand, silt, clay); M: Vertical Drained Constrained Modulus; C_u: Undrained Shear Strength; K_d: Horizontal Stress Index; V_s: Shear Waves Velocity.

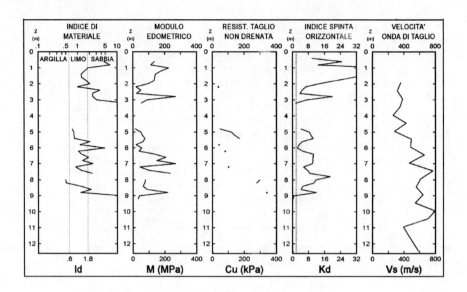

FIGURE 2. Summary of SDMT's in Monte Po hill area.

The profile of K_d is similar in shape to the profile of the overconsolidation ratio OCR. $K_d = 2$ indicates in clays OCR = 1, $K_d > 2$ indicates overconsolidation. A first glance at the K_d profile is helpful to "understand" the deposit. Figure 3 shows the values of V_s obtained in situ from a D-H test and SDMT. In the superficial strata V_s by SDMT is about 350 m/s. The V_s values increased with depth. At the depths of 6 and 10 m V_s is about 700 m/s in correspondence of sandy strata. These high V_s values by SDMT show the effect of soil disturbance during the test. The V_s values, experimentally determined during D-H tests, show an important variation in the transition zone at depths of 8 and 19 m, where thin layers of sandy soil exist.

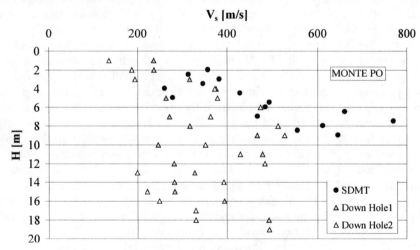

FIGURE 3. V_s from different in situ tests.

1-D LOCAL SITE RESPONSE ANALYSIS USING DIFFERENT V_S PROFILES

The Monte Po hill in the city of Catania, located in the South-Eastern Sicily (Italy), has been affected by several destroying earthquakes of about magnitude 7.0+ in past times. It is so reasonable to assume in Catania a maximum expected earthquake as a repetition of the January 11, 1693 event, with intensity XI MCS and estimated magnitude M = 7.3. Synthetic seismograms have been drawn for the site long a set of receivers placed at different depths, starting from the surface up to almost 40 m. After evaluating the synthetic accelerograms at the bedrock, the ground response analysis at the surface, in terms of time history and response spectra, has been obtained by some 1-D non-linear codes. The Sicilian earthquake of December 13, 1990 brought to an end a period of seismic dormancy lasted a long time thus reaching a local magnitude M_L = 5.4 with a focus depth of about 6-12 km. Even if it was internationally recognized as a "moderate" earthquake it provoked serious damages to many buildings.

By the beginning of 1991, during the building up of a public school situated at the bottom of the Monte Po hill, signs of gravitational motions were noticed, which proved to be dangerous both for the building in progress and for some council buildings some decades old. It is well-known that the slope is linked to rain phenomena of a certain intensity, which often cause a relevant rise in its piezometric surface, thus risking to jeopardize the stability of the whole slope. The heavy precipitations occurred in the days before the sliding, with the concurrence of the 1990 earthquake represented the cause of the hydrogeological breakdown of the Monte Po hill, which had lost its passive defenses that had previously and so far granted a good surface drainage; moreover, during the same period, cuttings and excavations for the marking of the external perimeter of the school were performed. The flowing and gliding landslide which came out and affected the north-western side of the hill at about 190 mt. upon sea level, shook huge masses of ground with a roto-traslational kinematic mechanism, causing in some cases a significant damage to buildings situated in the valley. Another detachment, occurred later in the north-eastern side, affected by collapses of the sub-vertical faces of the surfacing conglomerates, but that motion did not jeopardize any houses.

In the beginning of 2007 a seismic station has been also located into the school building, with the aim of recording seismic events. Seismograms obtained by the seismic station have been also used to evaluate the ground response analysis at the surface.

Local site response analyses have been brought for the Monte Po Hill by 1-D linear equivalent computer codes. The codes implement a one-dimensional simplified, hysteretic model for the non-linear soil response. The Seismic Dilatometer Marchetti Test (SDMT) was performed up to a depth of 15 meters. The results show a very detailed and stable shear waves profile. The S-wave propagation obtained by D-H and SDMT occurs on a 1-D column having shear behavior. The column is subdivided in several, horizontal, homogeneous and isotropic layers characterized by a non-linear spring stiffness $G(\gamma)$, a dashpot damping $D(\gamma)$ and a soil mass density ρ. Moreover, to take into account the soil non-linearity, laws of shear modulus and damping ratio against strain have been inserted in the code [7]. The 1-D columns have a height of 20 m and of 40 m and are excited at the base by accelerograms obtained from the seismograms and from the recordings of local earthquakes. The analysis provides the time-history response in terms of displacements, velocity and acceleration at the surface. Using this time history, response spectra concerning the investigated site have been deduced. The soil response at the surface was also modeled using the Equivalent –linear Earthquake site Response Analyses of Layered Soil Deposits computer code EERA [8] for calculus of amplitude ratios and spectral acceleration. An evaluation on buildings most directly affected or menaced by the landslide (see Figure 1) has been carried out, in order to focus on possible existing damage. As the NE landslide section did not affect any buildings, the attention has been directed on NW landslide section, which seriously jeopardized both a public school situated to the north of the landslide (building A), and some IACP buildings to the west of it (building B).

In Figure 4 is reported cross section B-B of the Monte Po hill (red line in Figure 1), plotted in the direction of maximum gradient just next to the point where the ground overlaps a restraint wall, with localized borehole points S2 (near the school), S7 and S8 (on the hill) along which local site response has been calculated. Figure 5 shows maximum shear stresses and maximum accelerations obtained using 1-D code EERA, considering as input the Sortino recording (E-W component) of the 1990 earthquake. Figure 6 shows maximum shear stresses and maximum accelerations obtained using 1-D code EERA, considering as input the synthetic accelerograms of the 1693 earthquake. Figure 7 shows amplitude ratios obtained using 1-D code EERA, while Figg. 8 and 9 show response spectra obtained using 1-D codes EERA and GEODIN.

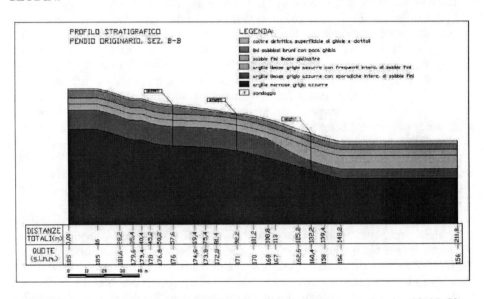

FIGURE 4. Section B-B of the Monte Po hill with localization of the borehole points S2, S7, S8.

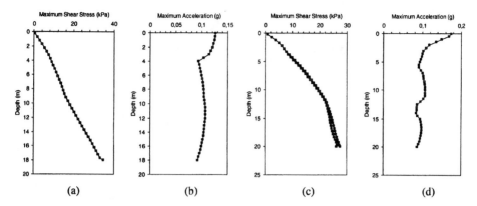

FIGURE 5. Maximum shear stress (kPa): (a) S2 point; (c) S7 point. Maximum accelerations (g): (b) S2 point; (d) S7 point, considering as input the December 13, 1990 earthquake.

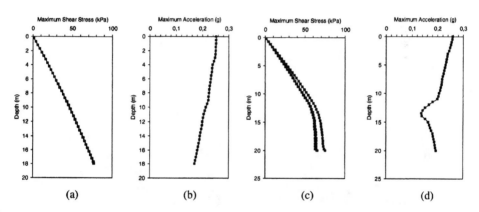

FIGURE 6. Maximum shear stress (kPa): (a) S2 point; (c) S7 point. Maximum accelerations (g): (b) S2 point; (d) S7 point, considering as input the January 11, 1693 earthquake.

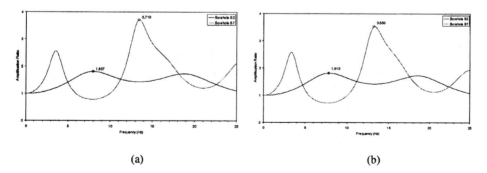

FIGURE 7. Amplitude ratios: (a): 1990 earthquake input; (b): 1693 earthquake input.

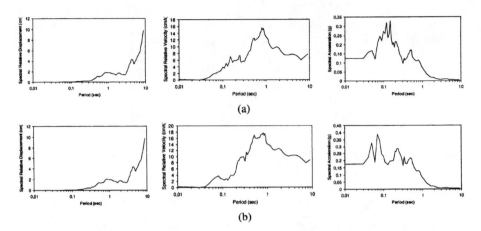

(a)

(b)

FIGURE 8. Response spectra using EERA code of relative displacement, relative velocity and acceleration, considering 1990 earthquake input: (a) S2 point; (b) S7 point.

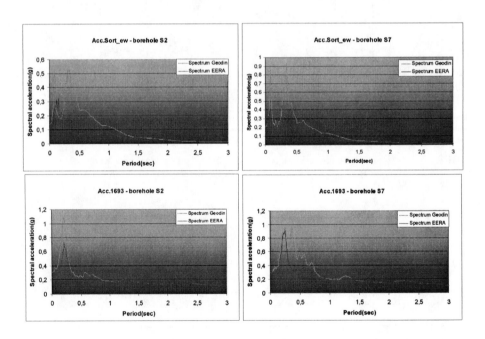

FIGURE 9. Comparisons between response spectra obtained by 1-D codes EERA and GEODIN.

248

2-D SITE RESPONSE ANALYSIS USING QUAD4M CODE

A detailed 2-D model has been constructed and validated for the Monte Po hill. The advantage of this model is to investigate the parameters that, in addition to surface soil conditions, can be used to correctly characterize site response in a 2-D structure. Through analyses using 2-D numerical simulations for SH waves, differences between the computed ground motion along some points of the Monte Po hill structure have been evaluated. It was also possible to compare the results from different 1-D models used reflecting current approaches to the determination of site response.

Figure 10 shows the cross section B-B modeled into the 2-D FEM code QUAD4M [9], modified from QUAD4 [10]. QUAD4M is a dynamic, time-domain, equivalent linear two dimensional computer program, as a modification to QUAD4 to implement a transmitting base and an improved time stepping algorithm. Figure 10 shows also the 2m x 2m mesh (1185 finite elements) of the cross section B-B adopted for calculations and the green points along which seismic response has been monitored.

Figg. 11 shows the results obtained using 2-D code QUAD4M in correspondence of points S2 and S7. Results compare well to EERA 1-D code.

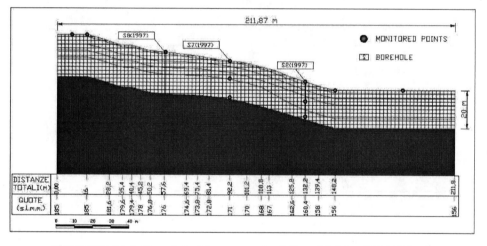

FIGURE 10. Mesh (2m x 2m) of the cross section B-B modeled into QUAD4M code.

CONCLUSIONS

It has been presented a study of the site response of the Monte Po hill, in the city of Catania. The detail with which the hill has been studied has allowed the construction of a detailed 2-D model of its structure. In the paper it has been explored the difference between the computed ground motion for different Vs profiles. It has been also possible to compare the results from different 1-D models reflecting current approaches to the determination of site response.

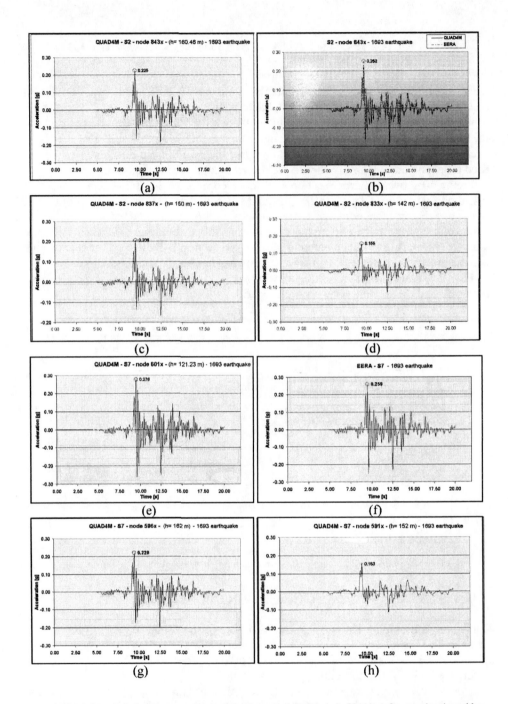

FIGURE 11. Seismic response analysis of the Monte Po hill. Borehole S2: (a) surface acceleration with QUAD4M; (b) comp. between QUAD4M and EERA surface accelerations; (c) acceleration at a depth of 10.5 m; (d) acceleration at a depth of 18.5 m. Borehole S7: (e) surface acceleration with QUAD4M; (f) comp. between QUAD4M and EERA surface accelerations; (g) acceleration at a depth of 9.5 m; (h) accel. at a depth of 19.5 m.

250

REFERENCES

1. Maugeri M. Detailed Scenarios and Action for Seismic Prevention of Damages in the Urban Area of Catania, *GNDT*, 2000.
2. Azzaro R., Barbano M.S., Moroni A., Mucciarelli M. Stucchi M. The Seismic History of Catania. *Journal of Seismology*, 3,3, 1999, pp. 235-262.
3. Postpischl D. Atlas of Isosismal Maps of Italian Earthquakes. Editors Postpischl. *CNR-Progetto finalizzato Geodinamica*, CNR (Italian Research Council), Rome, 1985, 164 pp..
4. Martin, G.K. and Mayne, P.W. Seismic Flat Dilatometers Tests in Connecticut Valley Vaeved Clay. *ASTM Geotechnical Testing Journal*, 20 (3), 1997: pp. 357-361.
5. Martin, G.K. and Mayne, P.W. Seismic Flat Dilatometers Tests in Piedmont Residual Soils. *Geotecnical Site Characterization*, Vol. 2, Balkema, Rotterdam, 1998, pp. 837-843.
6. Cavallaro A., Grasso S. and Maugeri M. Clay Soil Characterisation by the New Seismic Dilatometer Marchetti Test (SDMT). *Proceedings of the 2nd International Conference on the Flat Dilatometer*, Washington, 2 - 5 April 2006, pp. 261-268.
7. S. M. Frenna and M. Maugeri, (1995), "GEODIN: a Computer Code for Seismic Soil Response", *Proceeding of the. 9th Italian Conference of Computational Mechanics*, Catania, Italy, 20-22 June 1995, (in Italian): 145 - 148.
8. J. P. Bardet, K. Ichii, and C. H. Lin, „EERA: a computer program for equivalent-linear earthquake site response analyses of layered soil deposits, user manual", *University of Southern California*, 2000, 40 pp.
9. M. Hudson, I.M. Idriss, M. Beikae, "QUAD4M: a computer program to evaluate the seismic response of soil structures using finite element procedures and incorporating a compliant base", Center for Geotechnical Modeling, Department of Civil and Environmental Engineering, *University of California Davis*, 1994, Davis California.
10. I.M. Idriss, J. Lysmer, R. Hwang, H. B. Seed, "QUAD-4: A computer Program for Evaluating the Seismic Response of Soil Structures by Variable Damping Finite Element Procedures", *Earthquake Engineering Research Center, Report N. EERC 73-16*, 1973, University of California, Berkeley.

One-dimensional seismic response of two-layer soil deposits with shear wave velocity inversion

YuQin Ding [a], Alessandro Pagliaroli[b] and Giuseppe Lanzo[b]

[a]*College of Civil Engineering, Chongqing University, Chongqing 400045, P.R. China, visiting Ph.D. student at the Dipartimento di Ingegneria Strutturale e Geotecnica, Sapienza Università di Roma*
[b]*Dipartimento di Ingegneria Strutturale e Geotecnica, Sapienza Università di Roma, Via A. Gramsci 53, 00197, Rome, Italy*

Abstract. The paper presents the results of a parametric study with the purpose of investigating the 1D linear and equivalent linear seismic response of a 30 meters two-layer soil deposits characterized by a stiff layer overlying a soft layer. The thickness of the soft layer was assumed equal to 0.25, 0.5 and 0.75H, being H the total thickness of the deposit. The shear wave velocity of the soft layer was assumed equal to $V_s=90$ and 180 m/s while for the stiff layer $V_s=360$, 500 and 700 m/s were considered. Six accelerograms extracted by an Italian database characterized by different predominant periods ranging from 0.1 to 0.7s were used as input outcropping motion. For the equivalent liner analyses, the accelerograms were scaled at three different values of peak ground acceleration (PGA), namely 0.1, 0.3 and 0.5g. The numerical results show that the two-layer ground motion is generally deamplified in terms of PGA with respect to the outcrop PGA. This reduction is mainly controlled by the shear wave velocity of the soft layer, being larger for lower V_s values, by the amount of nonlinearity experienced by the soft soil during the seismic shaking and, to a minor extent, by the thickness of the soft soil layer.

Keywords: two-layer deposit, soft soil, stiff soil, shear wave velocity inversion, site response, linear and equivalent linear analyses.

INTRODUCTION

Site response is a critical component of geotechnical earthquake engineering studies. It is well established that amplification of seismic waves propagating up through a soil column is a function of the dynamic properties of soil layers, the characteristics of bedrock motion and the mechanical properties of the base rock. Site response of soil columns with either relatively homogeneous stratigraphy or heterogeneous stratigraphy with shear wave velocity continuously increasing with depth have been extensively investigated in the literature through experimental, numerical and analytical studies. These studies have led to the development of procedures and criteria for establishing the earthquake actions in several seismic codes worldwide as a function of different subsoil categories.

In contrast, sites characterized by shear wave velocity inversion, i.e. a stiff layer resting on top of a soft, low velocity layer has not received much attention (e.g., [1]). This peculiar stratigraphic feature can determine a complex dynamic response of a soil column significantly affecting the in-depth and at surface ground motion. This response may be of practical interest in many geotechnical earthquake engineering

CP1020, *2008 Seismic Engineering Conference Commemorating the 1908 Messina and Reggio Calabria Earthquake*,
edited by A. Santini and N. Moraci
© 2008 American Institute of Physics 978-0-7354-0542-4/08/$23.00

problems such as the liquefaction potential of soils, the design of pile foundations and of critical buried structures. Moreover in the Italian territory, especially in central and southern Italy, many historical centres are founded on large stiff soils/soft rock slabs overlying more deformable clay deposits [e.g., 2,3]. In these situations, a careful site response assessment may be essential for reliably estimating the magnitude of earthquake shaking intensity within the ground.

In this paper preliminary results of a parametric study on the seismic response of two-layer soil deposits with a stiff layer overlying a softer layer on elastic bedrock are presented. The main purpose is to assess which factors most influence the dynamic response of a two-layer soil column with shear wave velocity inversion and to which extent the seismic response of such layered systems differ from that of homogeneous soil deposits. Factors that were considered include the thickness and the shear wave velocity of both soft and stiff layers, and the level of input motion; this latter can be especially relevant in the soft layers which can exhibit pronounced soil nonlinearity for strong shaking. The effects of these factors on peak ground acceleration and maximum shear strain profiles have been investigated by means of linear and equivalent linear analyses.

TWO-LAYER SOIL DEPOSITS

Soil deposit consisting of two layers on top of an elastic bedrock was considered (Fig. 1). The upper layer was assumed stiffer than the lower one. The total thickness of the soil deposit is H=30 m. The thickness of the lower soft layer (h_{soft}) was varied such that the ratio h_{soft}/H had values of 0.25, 0.50 and 0.75 corresponding to h_{soft}=7.5, 15 and 22.5 m, respectively. For each h_{soft}/H, different V_s values were assumed, specifically for the top stiff layer V_{stiff} = 360, 500 and 700 m/s while the lower soft layer had V_{soft} = 90 and 180 m/s. The V_s values were selected according to the EC8 [4] subsoil categories. The limiting cases of homogeneous stiff (h_{soft}=0) and homogeneous soft (h_{soft}=H) soil deposits over elastic bedrock were also examined. All of the above combinations led to 23 configurations as illustrated in Table 1. In this table the geometric characteristics of soil deposits and the mechanical properties of soil layers are reported along with the fundamental linear period (T_1). Unit weight in both layers was assumed equal to 20 kN/m^3. For the bedrock, unit weight and shear wave velocity were assumed equal to 22 kN/m^3 and 800 m/s, respectively.

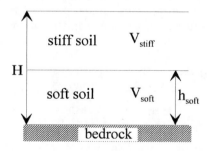

FIGURE 1. Layout of the two-layer soil deposit assumed for the analyses.

TABLE 1. Geometric characteristics of soil deposits and mechanical soil properties.

#	h_{soft} (m)	h_{soft}/H	V_{stiff} (m/s)	V_{soft} (m/s)	V_{stiff}/V_{soft}	Average V_S (m/s)	T_1 (s)
1	0	0	360	0	-	360	0.33
2	0	0	500	0	-	500	0.24
3	0	0	700	0	-	700	0.17
4	7.5	0.25	360	90	4	292.5	0.41
5	7.5	0.25	500	90	5.56	397.5	0.30
6	7.5	0.25	700	90	7.78	547.5	0.22
7	15	0.5	360	90	4	225	0.53
8	15	0.5	500	90	5.56	295	0.41
9	15	0.5	700	90	7.78	395	0.30
10	22.5	0.75	360	90	4	157.5	0.76
11	22.5	0.75	500	90	5.56	192.5	0.62
12	22.5	0.75	700	90	7.78	242.5	0.49
13	30	1	0	90	0	90	1.33
14	7.5	0.25	360	180	2	315	0.38
15	7.5	0.25	500	180	2.78	420	0.29
16	7.5	0.25	700	180	3.89	570	0.21
17	15	0.5	360	180	2	270	0.44
18	15	0.5	500	180	2.78	340	0.35
19	15	0.5	700	180	3.89	440	0.27
20	22.5	0.75	360	180	2	225	0.53
21	22.5	0.75	500	180	2.78	260	0.46
22	22.5	0.75	700	180	3.89	310	0.39
23	30	1	0	180	0	180	0.67

SEISMIC RESPONSE ANALYSES

The seismic response of two-layer systems was evaluated using the PROSHAKE code [5] through linear and equivalent linear analyses. For these latter, nonlinear soil behavior was taken into account iteratively by adjusting the values of shear modulus and damping ratio to the strain level induced by the shaking. Damping ratio for the linear case was set equal to 3%. For the nonlinear case the reduction of shear modulus and the increase of damping ratio with shear strain was modeled using Vucetic and Dobry curves [6] corresponding to PI=30 for both soft and stiff clays.

Six real rock outcropping accelerograms were selected from a recently developed Italian database [7]. For each recording the earthquake name, date and magnitude along with the peak ground acceleration (PGA) and the predominant period (T_p), are reported in Table 2. The recordings were selected to cover essentially the whole range of fundamental periods of soil deposits; in particular, 4 records exhibit T_p between 0.25 and 0.4s while the other 2 records are characterized respectively by T_p =0.1 and 0.6s (Fig. 2).

TABLE 2. Earthquake records used as input motion for site response analyses.

Recording station	Earthquake (dd/mm/yyyy)	M_w	PGA (g)	T_p (s)
Assisi Stallone - NS	Umbria Marche – 26/09/1997	6.0	0.19	0.32
Cascia - NS	Umbria Marche – 14/10/1997	5.6	0.05	0.26
Nocera Umbra Biscontini - NS	Umbria Marche – 14/10/1997	5.6	0.04	0.10
Pontecorvo - NS	Lazio Abruzzo – 07/05/1984	5.9	0.06	0.36
Tolmezzo Diga Ambiesta - NS	Friuli – 05/06/1976	6.5	0.36	0.26
Torre del Greco - NS	Irpinia – 23/11/1980	6.9	0.06	0.66

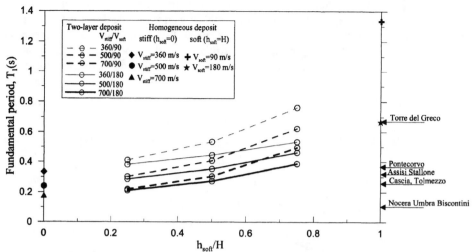

FIGURE 2. Fundamental periods of soil deposits vs. thickness ratio (predominant periods of input motions are also reported).

For each of the 23 assumed configurations and for all the 6 input motions selected, the dynamic response of the two-layer systems was calculated. The equivalent linear analyses were executed by scaling the accelerograms to 0.1, 0.3 and 0.5g.

RESULTS OF NUMERICAL ANALYSES

Representative results of the numerical analyses are plotted in Fig. 3 with reference to the Torre del Greco input outcropping motion in terms of peak acceleration ratio versus thickness ratio h_{soft}/H. The peak acceleration ratio is defined as the ratio of the peak surface acceleration of the two-layer system (PGA_{surf}) to the peak acceleration of the outcropping motion (PGA_{out}). The results, presented for the linear and equivalent linear analyses, allow the following considerations.

For the linear case (Fig. 3a) the peak acceleration at the surface of the homogeneous stiff deposit ($h_{soft}/H=0$) is slightly higher than the outcrop PGA ($PGA_{surf}/PGA_{out}=1-1.4$) while for the homogeneous soft deposit ($h_{soft}/H=1$) the surface PGA reaches values as high as 2.5 times the outcrop PGA. The two-layer systems show a different behavior depending on the soft layer shear wave velocity. In particular, for $V_{soft}=180$ m/s, the PGA ratio plots between the limiting values calculated for the homogeneous deposits; on the contrary for $V_{soft}=90$ m/s the PGA ratio is lower than 1 for $h_{soft}/H=0.25$ and 0.5, that is a deamplification of motion occurs with respect to outcrop PGA, and slightly higher than 1 for $h_{soft}/H=0.75$. Further, for a given V_{soft}, the PGA ratio is not much affected by the shear wave velocity of the stiff layer V_{stiff}.

The results of nonlinear analyses are presented for PGA_{out} scaled to 0.3g. They show, as expected, a strong reduction of surface PGA for the homogenous soft deposits (Fig. 3b). For the two-layer systems a marked reduction of the peak acceleration ratio can be noted: for both V_{soft} =90 m/s and V_{soft}=180 m/s the PGA ratio attains values lower than 1 and lower than those that would occur in respective homogenous deposits. The maximum deamplification with respect to PGA_{out}, of about 70%, can be observed for V_{soft} =90 m/s and h_{soft}/H=0.25. Like the linear case, for a given V_{soft}, the PGA ratio is rather insensitive to the shear wave velocity of the stiff layer V_{stiff}.

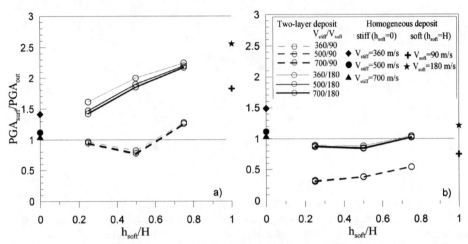

FIGURE 3. Peak acceleration ratio vs. thickness ratio for (a) linear and (b) equivalent linear analyses.

Peak ground acceleration and peak shear strain profiles corresponding to h_{soft}/H=0.25 are plotted for the linear and equivalent linear analyses in Fig. 4 and 5, respectively. For the linear case (Fig. 4a) the PGA profile for homogeneous soft deposit with V_{soft}=180 m/s shows a gradual increase from the bedrock up to the surface while for V_{soft}=90 m/s a larger variation of PGA with depth is observed according to the shorter wavelength involved. Peak shear strain profiles show a fairly uniform trend with depth, the highest values being about 0.1% and 0.15% for V_{soft}=180 m/s and V_{soft}=90 m/s, respectively (Fig. 4b). For the two-layer systems the variation of PGA with depth occurs essentially in the soft layer while the upper stiff layer experiences almost uniform PGA (Fig. 4a). The presence of a stiff layer overlying a soft layer leads to a PGA reduction with depth with respect to the homogenous deposits, especially pronounced close to the surface. The corresponding peak shear strain profiles show that comparable or higher strains ($\gamma_{max}\cong$0.3%) develop in the soft layer with respect to those occurring in the deposits totally constituted by soft soil while much lower strains develop in the upper stiff layer ($\gamma_{max}\cong$0.03%), as illustrated in Fig. 4b. The PGA profiles for the two-layers systems seems to be essentially insensitive to the shear wave velocity of the stiff layer.

For the nonlinear analyses, results are presented for PGA_{out}=0.3g in Fig. 5. In this case, the deposit behaves in a manner similar to what have occurred in the linear case, i.e. the presence of a stiff layer over a soft layer determines a reduction of PGA, especially at the ground surface, with respect to homogenous soft deposits.

Nonetheless, a larger deamplification occurs with respect to the linear case because of the nonlinearity effects in the soft soil. This fact is evident is Fig. 5b where large strains develop ($\gamma_{max} \cong 1\%$). This results in a substantial increase of damping ratio which consequently causes a reduction in the high frequency response of the soil profile and therefore a reduction of peak ground acceleration.

Therefore, the effect of the stiff layer on top of the soft layer is to impose a boundary condition on the soft layer response, i.e. large strains may be developed in the soft layer when subjected to higher levels of input motion which in turn can significantly affect the surface response. The upper stiff layer acts essentially as a rigid body in which uniform PGA occurs.

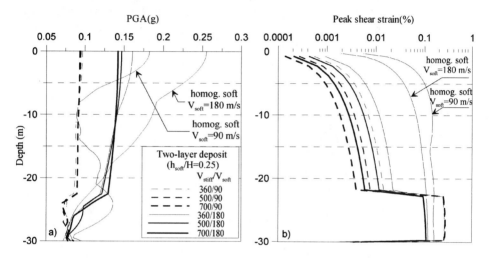

FIGURE 4. Peak acceleration (a) and peak shear strain (b) profiles for linear analyses.

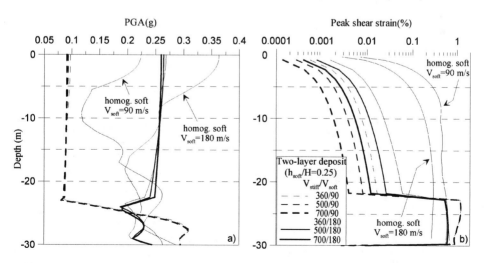

FIGURE 5. Peak acceleration (a) and peak shear strain (b) profiles for equivalent linear analyses.

In Fig. 6 peak acceleration ratio versus the thickness ratio h_{soft}/H is plotted for all the linear and equivalent linear analyses carried out for the three increasing levels of input motions PGA_{out}=0.1, 0.3 and 0.5g. It should be pointed out that the results obtained for PGA_{out}=0.5g must be considered as approximated because of the questionable adequacy of the equivalent linear method to reliably determine the seismic response of very soft soils at large strains.

Considering the negligible influence of V_{stiff} on PGA_{surf}/PGA_{out} ratio, an average value of this ratio was computed for all three V_{stiff} values (V_{stiff}=360, 500 and 700 m/s). Moreover, the ratio was averaged over the six selected accelerograms. The standard deviation was also computed for 18 values corresponding to each h_{soft}/H. The average (m) and average $\pm$ 1 standard deviation (σ) values are plotted in Fig. 6 as thick and thin lines, respectively. A slight amplification at the surface of the two-layer systems is observed only for the linear case with V_{soft}=180 m/s. Conversely, for the linear case with V_{soft}=90 m/s and for all the equivalent linear analyses a deamplification of motion occurs on average.

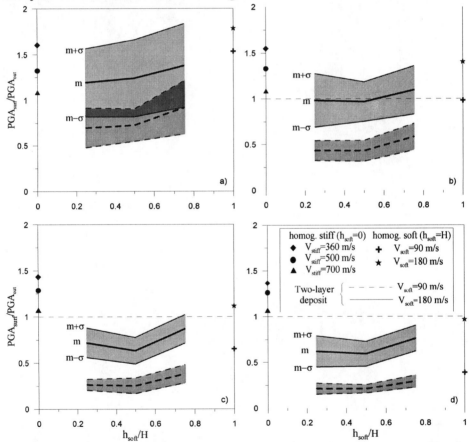

FIGURE 6. Peak acceleration ratio vs. thickness ratio for linear (a) and equivalent linear analyses for (b) PGA_{out}=0.1g, (c) PGA_{out}=0.3g and (d) PGA_{out}=0.5g.

258

The controlling factors governing the magnitude of PGA amplification ratio are the shear wave velocity V_{soft}, nonlinearity which develops in the soft layer especially marked for higher levels of input motion and, to a minor extent, the thickness ratio. It can also be noted that dispersion of calculated ratios significantly reduces with increasing level of input motion.

CONCLUSIONS

The paper illustrates preliminary results of a parametric study of two-layer soil deposits characterized by a stiff layer overlying a soft layer. On the whole the parametric study included 23 soil deposits and 6 real input accelerograms resulting in 138 linear and equivalent linear analyses.

The dynamic response of the layered system is essentially controlled by the geometric and mechanical characteristics of the underlying soft layer. In particular the results indicate that surface motion response in terms of PGA is generally attenuated with respect to the outcrop PGA and to the motion that would occur in homogenous soft deposits of equal depth. This attenuation is larger for lower V_s values of the soft layer, for higher levels of input motion because of soil nonlinearity and, to a minor extent, for lower values of the soft layer thickness. On the other hand, large peak shear strains tend to occur into the soft layer, especially for limited thickness, with respect to the overlying stiff layer. Therefore, for soil profiles with shear wave velocity inversion numerical site response analyses are required to reliably estimate the response of the entire stratum for design considerations.

ACKNOWLEDGMENTS

The study stage of Dr. YuQin Ding at the Dipartimento di Ingegneria Strutturale e Geotecnica was supported by a fellowship of the Sapienza University of Rome. This support is gratefully acknowledged.

REFERENCES

1. C. P. Aubeny, J. L. Von Thun and N. Y. Chang, "Response characteristics of soil deposits", 3th U.S. National Conference on Earthquake Engineering, Charleston, South California, vol. 1 (1986).
2. G. Lanzo, L. Olivares, F. Silvestri and P. Tommasi, "Seismic response analysis of historical towns rising on rock slabs overlying a clayey substratum", 5th International Conference on Case Histories in Geotechnical Engineering, New York, April 13-17 (2004).
3. A. Costanzo, F. Silvestri, S. Lampitiello, L. Olivares, G. Lanzo and P. Tommasi, "Vulnerabilità sismica di centri storici su rilievi: i casi di Bisaccia, Orvieto e Gerace", XI Congresso Nazionale "L'Ingegneria Sismica in Italia", Genova, January 25-29 (2004) (in italian).
4. Eurocode 8, "Design for structures for earthquakes resistance, Part 1: General rules, seismic actions and rules for buildings" Final Draft – prEN 1998-1 December 2003 Edition (2003).
5. EduPro Civil System, Inc., ProShake – Ground Response Analysis Program, Redmond, WA (1998).
6. M. Vucetic and R. Dobry, "Effect of soil plasticity on cyclic response", ASCE, *Journal of Geotech. Eng.*, **117**, 89-107 (1991).
7. G. Scasserra, G. Lanzo, J.P. Stewart and B. D'Elia, "SISMA (Site of Italian Strong Motion Accelerograms): a web-database of Italian records for engineering applications", (this conference) available at URL: http://sisma.dsg.uniroma1.it.

On the tsunami wave – submerged breakwater interaction

P. Filianoti and R. Piscopo

Department of Mechanics and Materials, University Mediterranean of Reggio Calabria, Italy.

Abstract. The tsunami wave loads on a submerged rigid breakwater are inertial. It is the result arising from the simple calculation method here proposed, and it is confirmed by the comparison with results obtained by other researchers. The method is based on the estimate of the speed drop of the tsunami wave passing over the breakwater. The calculation is rigorous for a sinusoidal wave interacting with a rigid submerged obstacle, in the framework of the linear wave theory. This new approach gives a useful and simple tool for estimating tsunami loads on submerged breakwaters.

An unexpected novelty come out from a worked example: assuming the same wave height, storm waves are more dangerous than tsunami waves, for the safety against sliding of submerged breakwaters.

Keywords: Tsunami; solitary wave; submerged breakwater; wave loads.

(1) INTRODUCTION

Submerged breakwaters are increasingly employed for the coast protection. In respect to different structures, like detached breakwaters or seawalls, submerged breakwaters have no visual impact. These structures are generally designed to withstand the strongest storms occurring during their life. In the past, many structures have been seriously damaged and even destroyed by tsunamis. Therefore, this kind of events must be necessarily considered during structure design in seismic area, where the probability of occurrence of tsunami is greater.

The interaction solitary wave – submerged breakwater was dealt with the aim to estimate the wave energy transmission behind the breakwater [2]; to study the deformation [3], [4], [5], [6] and the breaking [7], [8] of the wave passing over the breakwater; to determine vortex shedding near the breakwater corners [12].

In our knowledge, the study of loads exerted by a solitary wave on a submerged impermeable breakwater, has been dealt with exclusively in [13]. They solved the Navier-Stokes equations, for unsteady two-dimensional flows, to numerically simulate the flow field generated by the interaction of a solitary wave and a submerged rectangular breakwater. Following this approach they determined the time histories of the horizontal force exerted by tsunami waves with different heights, on breakwaters with different section ratios. In the present paper the above cited horizontal force has been calculated in a completely different manner, that is based on the speed drop of a periodic wave passing over the breakwater.

CP1020, *2008 Seismic Engineering Conference Commemorating the 1908 Messina and Reggio Calabria Earthquake,*
edited by A. Santini and N. Moraci
© 2008 American Institute of Physics 978-0-7354-0542-4/08/$23.00

The paper has the following articulation. In Sect. 2 a description of the REWEC1 breakwater and operational conditions is made. The mathematical representation of a tsunami wave in the light of the solitary wave theory is illustrated in Sect. 3. The new approach followed for estimating the solitary wave loads is described in Sect. 4. In the same Section the results obtained are put in comparison with those obtained in [13]. Finally, in Sect. 5 some practical calculations are carried out for comparing the effects produced by a solitary wave and a wind wave, on the overall stability of a submerged breakwater.

(2) REWEC 1: THE ACTIVE BREAKWATER

The REWEC1 is a caisson breakwater embodying a device able to absorb the wave energy. Such a device consists essentially in a U-conduit (see the scheme of Fig. 1), a branch of which is a vertical duct which extends along the wave-beaten wall of the caisson, and it is connected with the sea through an upper opening. The other branch is a box which extends along the whole caisson with an air pocket. Under the wave action the pressure on the outer opening of the vertical duct fluctuates and, as a consequence, water flows up and down in the vertical duct and in the box, and the air pocket acts as a gas spring. The air quantity must be regulated so that the eigenperiod of oscillations be close to wave periods of the sea state.

For simplifying operations and reducing both running and installation costs, we can fix the quantity of air inside the caisson, by tuning the plant to reach resonance with the most frequent seas during the season. In fact, in the fixed air regime, the plant is "active" (capable of energy absorption) only during summer seasons, whereas in the rest of the year it operates as an no-active breakwater, with the absorption box flooded. The running is very simple, since the plant needs only two operations in a year: before summer seasons air is pumped into the box; passed the summer, air is discharged from the plant. This working system presents disadvantage in terms of absorption performance, that decrease far from resonance conditions.

Varying the air mass inside the box, the plant can be tuned to reach resonance with sea states having different significant heights, optimizing performances in the whole significant wave heights spectrum. This working system is restricted during the most strong sea states, since there would need too much air in the box to achieve the resonance. Moreover, this working system involves bigger running costs in comparison with fixed air working, as each caisson must be equipped with a control system for the tuning, and must be connected to a pumping station.

(3) MATHEMATICAL REPRESENTATION OF A TSUNAMI WAVE

A tsunami is a series of ocean waves generated by rapid disturbances of the seawater. Usually, submarine earthquakes are the most common causes that generate the tsunamis; other causes are landslides, volcanic eruptions, underwater explosions or meteor impacts. Not all submarine earthquakes generate tsunamis. Tsunamis are generated by vertical deformations of the earth crust. An earthquake can generate a tsunami if it produces significant vertical displacement on the seafloor. Magnitude of earthquake has an important part as regards the intensity of tsunami.

Tsunami waves have a low steepness in the generation area such that they can pass unnoticed. In their propagation toward the coast, the more decrease depths, the more the height of tsunami increases and wave celerity and wave length decrease. Because tsunamis have wide wave lengths, their propagation can be modeled as a wave propagating in shallow waters.

In the open sea, tsunami wave has a sinusoidal shape. Therefore, it can be described by means of the linear wave theory. When tsunami waves approach the coasts, the wave through disappears and only the peak remains. Usually, to represent the dynamics of the interaction of tsunami waves with structures and coast, scientists make use of the "solitary wave theory" [15], [16], [17]. Recently, wave forms can be also described as N-waves. N-waves have crest and trough, while solitary wave has only the crest. The solitary wave model is more used than N-waves model owing to it is easier to deal with and to generate.

The first approximation of the solitary wave theory gives

$$\eta(y,t) = \frac{H}{\cosh^2[k(y-ct)]},$$ (1)

where η is the surface displacement, k has been defined as

$$k \equiv \sqrt{\frac{3H}{4d^3}},$$ (2)

and c is the propagation speed (wave celerity), which can be expressed as a function of the wave height H at the depth d, by means of equation

$$c = \sqrt{gd}\left(1 + \frac{H}{2d}\right).$$ (3)

Pressure fluctuations are equals to

$$\Delta p(y,t) = \rho g \frac{H}{\cosh^2 q}.$$ (4)

Horizontal components of velocity and acceleration are given by

$$v_y(y,t) = \frac{1}{\cosh^2 q} \frac{H}{d} \sqrt{gd},$$ (5)

$$a_y(y,t) = \sqrt{3}\frac{\sinh q}{\cosh^3 q}\left(\frac{H}{d}\right)^{1.5} g.$$ (6)

Solitary wave length is infinite in theory. But, for practical calculations we can define a wave length L, as a consequence of the rapid decrease to zero of surface displacement η leaving from the crest. Assuming $L = 2\pi/k$, at a distance $y = L/2$ from the crest, the wave elevation is 0.74% of its maximum value H, and the relationship between the relative depth and the solitary wave height assumes the form

$$\frac{d}{L} = \frac{\sqrt{3}}{4\pi}\sqrt{\frac{H}{d}}.$$ (7)

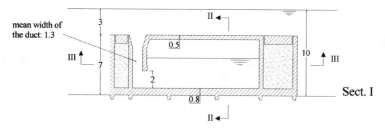

mean width of the duct: 1.3

Sect. I

FIGURE 1. Cross section of a REWEC1 breakwater [14].

(4) LOADS OF A SOLITARY WAVE ON A SUBMERGED BREAKWATER

In ideal flow conditions, surface waves propagating over an isolated body have a reduction of the propagation speed [18]. This is the physical reason why the amplitude of the horizontal wave force on the solid body is different from the amplitude of the horizontal wave force on the equivalent water mass, (the mass having the same volume and shape of the solid body), which is called the Froude-Krylov force. The reduction of the propagation speed of a sinusoidal wave passing over a submerged breakwater can be calculated solving the problem of the wave diffraction in the presence of the breakwater. In this way we arrive to determine pressure fluctuations on two vertical walls: the sea-wall and the lee-wall.

Here, the wave diffraction problem has been solved by means of the integral equations method. The velocity potential of diffracted waves is calculated numerically through the boundary element method (BEM).

When the breakwater is placed on a relative depth d/L less than 0.1, we have found that pressure fluctuations at different points on the sea-wall and on the lee-wall have practically the same shape and amplitude. We have also received confirmation that a larger phase angle exists between two points placed respectively on the sea- wall and on the lee-wall of the solid body, in respect to the same points placed on the equivalent water mass. The larger phase angle of pressure waves on the breakwater is due to the reduction of the propagation speed of these waves, this reduction generates a horizontal wave force on the breakwater greater than the horizontal Froude-Krylov force.

Fig. 2 shows the speed drop factor $\mathscr{F}_r$, as a function of the relative width b/d of the breakwater. The three curves are relevant to three different values of the relative height a/d. Upper abacus represents the case $d/L_0=0.05$ ($H/d=0.15$, according to eq. 7), lower abacus, the case $d/L_0=0.075$ ($H/d=0.30$). We can see the speed drop factor decreases monotonically to the increasing of the breakwater width, and it increases to the increasing of the breakwater relative height.

Usually, the ideal flow pattern is suitable to describe the wave field around a solid body until the Keulegan-Carpenter number K_E, is less or equal to about 6 [18]. In a first approximation, for a breakwater of width b, interacting with a solitary wave of height H, the K_E can be expressed in the form

$$K_E = \frac{2\pi}{b} \frac{v_{y\,max}^2}{a_{y\,max}}, \tag{8}$$

from which, replacing maximum values of the horizontal velocity and acceleration, obtained by means of eq. (5) and (6) respectively, we arrive to the form

$$K_E \cong \frac{2}{3}\sqrt{\frac{H}{d}} \bigg/ \frac{b}{d}. \tag{9}$$

The horizontal wave force per unit length exerted by a solitary wave of H height on a breakwater of a height and b width, placed on a floor of d depth, can be calculated by means of eq. (4), bearing in mind that the difference of phase angle of the pressure fluctuation between the points on the sea-wall and the points on the lee-wall is $\mathcal{F}_r$ times larger than the phase angle between the same points on the equivalent water mass:

$$F_y(t) = \rho g H a \left[\cosh^{-2}\left(\frac{\sqrt{3}}{2}\sqrt{\frac{H}{d}} \frac{-\mathcal{F}_r b/2 - ct}{d}\right) - \cosh^{-2}\left(\frac{\sqrt{3}}{2}\sqrt{\frac{H}{d}} \frac{\mathcal{F}_r b/2 - ct}{d}\right) \right]. \tag{10}$$

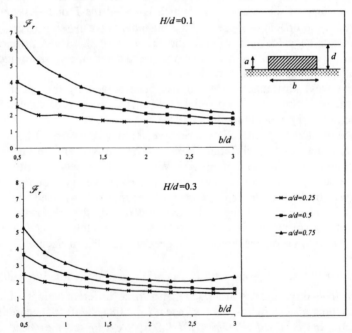

FIGURE 2. The speed drop factor $\mathcal{F}_r$ as a function of b/d, for a fixed value of the ratio a/d, in the cases of solitary waves with H/d equal to 0.1 (upper abacus) and 0.3 (lower abacus).

Fig. 3 shows the ratio F' between the F_y and the hydrostatic force $0.5\rho g a^2 (2d/a - 1)$ as a function of $t' \equiv t(g/d)^{0.5}$. Three cases are shown:

(a): $H/d = 0.15$, $b/d = 1$, $a/d = 0.5$;
(b): $H/d = 0.30$, $b/d = 1$, $a/d = 0.5$;

(*c*): $H/d = 0.15$, $b/d = 20$, $a/d = 0.5$.

For comparison, the values obtained in [13] are shown too. They were obtained by solving the two-dimensional unsteady Navier-Stokes equation, in the presence of a submerged rectangular breakwater. In order to simulate the generation of the solitary wave, a numerical piston-type wavemaker was incorporated in the model. The Navier-Stokes equations were linearized in a discrete computational domain, which included the wavemaker and the breakwater, and then solved using the analytic finite method.

We can see an essential agreement between two estimates of F'. In details, it appears that:

(i) the time history exhibited by two curves are quite similar. In all three cases (*a*), (*b*), (*c*) crests, troughs, and zeros occur at the same time instants in the two curves;

(ii) the F' maximum value calculated by eq. (10) is equal to that calculated by [13] for the wider breakwater (case *c*), while it is overestimated in the other two cases;

(iii) the trough heights of forces calculated by [13] are always lower than crest heights, while those calculated by eq. (10) are equal between them.

In synthesis, we can affirm that the horizontal forces obtained by [13] are typically inertial, despite of vortex generation in the proximity of breakwater corners. The asymmetric shape of the force is probably due to the deformation of the wave passing over the breakwater. Such a deformation is not provided by the simple inertial scheme represented with eq. (10).

(5) BREAKWATER GLOBAL STABILITY ANALYSIS UNDER THE ACTION OF A TSUNAMI AND A WIND WAVE

We consider a hypothesis of design of a REWEC1 for defending the Jonian coast of Calabria (south-east Italy). The structure winds parallel to the coast, and it is placed on a seabed 10 m deep below the mean sea level, the height of the water column above the breakwater roof is 3 m. Each caisson, whose sketch is represented in Fig. 1, is built in reinforced concrete. It is reasonable to design the REWEC1 assuming two running conditions: during summer seasons, the internal pocket is partially filled with air (active plant); during winter seasons, the internal pocket is totally filled with water (inactive breakwater). The air quantity in the pocket is regulated so that the absorber device enters in resonance with the most frequent seas during summer season.

In winter (flooded pocket), the weight in still water of the structure is 1086 kN/m. In summer (pocket partially filled with pressurized air), the weight in still water is 773 kN/m.

Considering a maximum tsunami wave height of 9 m (actually it is already broken at the 10 m depth), the maximum value of Keulegan-Carpenter number, calculated by means of eq. (9) is 3.97. Thus, we can assume the ideal flow pattern and apply the eq. (10) to estimate the wave force on the structure.

In fig. 4 the speed drop factor for the REWEC1 of fig. 1 is represented as a function of H/d. We notice that, for $H/d=0.7$, $\mathcal{F}_r$ tends asymptotically to 2.3. The safety factor against sliding $C_s = \mu P^* / F_{y\max}$, may be evaluated applying the eq. (10) for the calculation of $F_{y\,max}$, and fixing the friction coefficient μ (=0.6). Fig. 5 shows C_s as a function of height H of the solitary wave, for the cases of active breakwater and no-active breakwater. $C_s > 1$ means the structure is able to withstand, thanks to its weight, to the tsunami wave force without sliding. Calculations show that the structure is able to resist to tsunami waves up to 6.6 m high, without sliding, during active breakwater operations (summer season). During no-active operations (winter season), this height grows to 9.2 m.

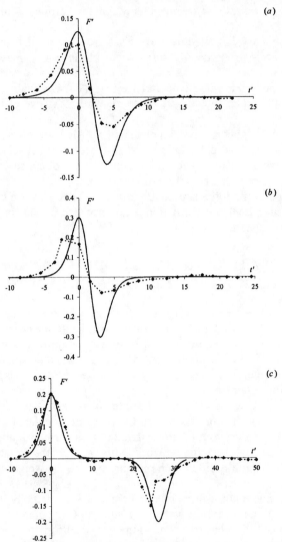

FIGURE 3. $F' \equiv F_y \,/[0.5\rho gd^2(1 - a^2/d^2)]$ as a function of $t' \equiv t(g/d)^{0.5}$. Continuous line: model with speed drop factor; dotted line: Huang & Dong [13] numerical solution.
(a): H/d=0.15, b/d=1, a/d=0.5; (b): H/d=0.30, b/d=1, a/d=0.5; (c): H/d=0.15, b/d=20, a/d=0.5.

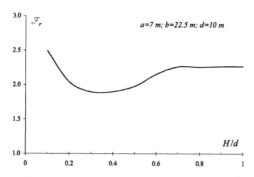

FIGURE 4. The speed drop factor as a function of H/d for the REWEC1 of Fig.1.

Fig. 6 shows the time histories of horizontal and vertical forces exerted by a wind wave with $H = 6$ m, on the same breakwater of Fig. 1, during no-active operations. Profiles of the forces have been calculated applying the quasi-determinism theory [18], and the boundary element method to determine the diffraction coefficients of wave forces on the breakwater [20]. During active operations, the height of the wind wave which starts to generate the instability against the breakwater sliding ($C_s \cong 1$) is 4.5 m.

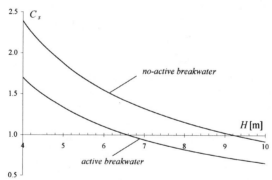

FIGURE 5. The sliding safety factor as a function of H for the REWEC1 of Fig.1.

Comparing safety factors showed in Figs. 5 and 6, we can conclude that before the occurrence of breakwater sliding, the REWEC1 is able to withstand a solitary wave 1.5 times higher than a wind wave. This result, surprising in a first sight, is explained by the fact that the vertical force exerted by the solitary wave on the submerged breakwater is zero, in that pressure fluctuations acting simultaneously on the roof and the basis of the breakwater are equal, being hydrostatical. On the contrary, a wind wave passing over the submerged breakwater produces a vertical force whose amplitude is even larger than the amplitude of the horizontal wave force (see Fig. 6). As a consequence, the lowest value of the safety factor C_s doesn't occur in the time instant of the maximum horizontal force ($T/T_p = 0$), but shortly before, when the vertical force is upward and assumes a value more than two times bigger than the corresponding horizontal force.

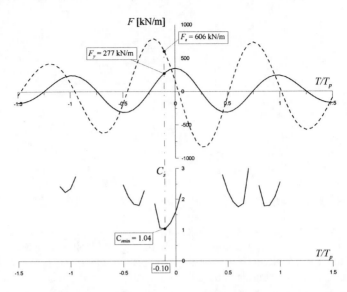

FIGURE 6. Time histories of the horizontal wave force (continuous line) and vertical wave force (dotted line) exerted by a wind wave of 6 m of height on the REWEC1 of Fig.1, and time variation of the sliding safety factor C_s, for the no-active plant (flooded pocket).

(6) CONCLUSIONS

The solitary wave is a mathematic theory frequently used in coastal engineering to represent certain characteristics of tsunamis. Systematic studies have been carried out by many authors on the interaction between a submerged breakwater and a solitary wave. A topic little investigated is the force produced on the breakwater and its effect on the overall stability.

This work proposes a new approach to calculate the loads exerted by a solitary wave on a submerged rectangular breakwater. The force is supposed inertial, like the one acting on the equivalent water mass (the mass of water having the same volume and shape as the breakwater), which is called the Froude-Krylov force. Nevertheless there is a variation of the flow field from the water rectangular barrier and the solid rectangular barrier, which consists in a drop of the propagation speed of the pressure head waves in the presence of the solid body. The reduction of the propagation speed means a larger phase angle, which implies a larger pressure difference between points on the sea-wall and points on the lee-wall of the breakwater. Increasing the phase angle between two points (where the pressure fluctuations have the same amplitude and frequency) implies increasing the pressure difference (i.e. the horizontal force) between this two points.

In the paper the propagation speed drop has been calculated with the boundary element method (BEM), assuming that a periodic sinusoidal wave and a solitary wave with nearly the same wavelength, would undergo the same slowing down passing over the breakwater.

Results have been compared with calculations [13] carried out by solving numerically the unsteady two-dimensional Navier-Stokes equations. The inertial characteristics of the horizontal force have been clearly confirmed. This is proved (cf. Fig. 3) by the matter that maximum, minimum and zero of horizontal force occur at the same time instants in the two curves obtained with different models.

The proposed methodology is a valid tool for design purposes in the coastal engineering. A worked example showed that, under the same wave height, wind waves are more pernicious than a solitary wave, for the safety against sliding of the considered submerged breakwater. The reason get readily apparent looking at Fig. 6, which shows that the vertical force produced by wind waves on the breakwater may be even larger than the horizontal force, leading to a significant reduction of the immersed weight of the structure.

REFERENCES

1. P. Boccotti, *On a new wave energy absorber*, Ocean Eng., 2003, (30), pp. 1191-1200.
2. P. Lin, *A numerical study of solitary wave interaction with rectangular obstacles*, Coastal Engineering, 2004, (51), pp. 35-51.
3. O. S. Madsen and C. C. Mei, *The transformation of a solitary wave over an uneven bottom*, J. Fluid Mech., 1969, (39), pp. 781-791.
4. F. J. Seabra-Santntos, D. P. Renouard and A. M. Temperville, *Numerical and experimental study of the transformation of a solitary wave over a shelf or isolated obstacle*, J. Fluid Mech., 1987, (176), pp. 117-134.
5. D. G. Goring and F. Raichlen, *Propagation of long waves onto shelf*, Proc. 17th Int. Conf. Coastal Eng., ASCE, 1980, pp. 763-783.
6. S. Beji and J. A. Battjes, *Numerical simulation of nonlinear wave propagation over a bar*, Coastal Eng., 1994, pp. 1-16.
7. A. Otta, I. A. Svendsen and S. T. Grilli, *The breaking and runup of solitary wave on beaches*, Proc. 23rd Int. Conf. Coastal Eng., ASCE, 1992, pp. 1461-1474.
8. S. T. Grilli, R. Subramanya, I. A. Svendsen, and J. Veeramony, *Shoaling of solitary wave on plane beaches*, J. Waterway, Port Coastal, Ocean Eng., 1993 (120), pp. 609-628.
9. S. T. Grilli, M. A. Losada and F. Martin, *Characteristics of solitary wave breaking induced by breakwaters*, J. Waterway, Port Coastal, Ocean Eng., 1994 (120), pp. 74-92.
10. F. Zhuang and J. J. Lee, *A viscous rotational model for wave overtopping over marine structure*, Proc. 25th Int. Conf. Coastal Eng., ASCE, 1996, pp. 2178-2191.
11. F. C. K. Ting and Y. K. Kim, *Vortex generation in water waves propagating over a submerged obstacle*, Coastal Eng., 1994 (24), pp. 23-49.
12. C. J. Huang and C. M. Dong, *Wave deformation and vortex generation in water waves propagating over a submerged dike*, Coastal Eng., 1999 (37), pp. 123-148.
13. C. J. Huang and C. M. Dong, *On the interaction of a solitary wave and a submerged dike*, Coastal Eng., 2001 (42), pp. 265-286.
14. P. Boccotti, *Gli impianti REWEC*, Ed. BIOS 2004, pp. 1-117 (in italian).
15. J. B. Keller, *The solitary wave and periodic waves in shallow water*, Comm. Applied Mathematics, 1948, (1), pp. 323-339.
16. E. V. Laitone, *The second approximation to cnoidal and solitary waves*, Journal of Fluid Mechanics, 1961, (9), pp. 430-444.
17. R. Grimshaw, *The solitary wave in water of variable depth*, Journal of Fluid Mechanics, 1970, (42), pp. 639-656.
18. P. Boccotti, *Wave Mechanics for Ocean Engineering*, Elsevier Amsterdam, 2000, pp. 1-496.
19. T. Sarpkaya and M. Isaacson, *Mechanics of wave forces on offshore structures*, Van Nostrand Reinhold Co., 1981.
20. P. Filianoti and R. Piscopo, *Sulla efficienza e stabilità di barriere sommerse REWEC1 per la difesa costiera*, Atti del XXX Conv. Naz. di Idraulica e Costr. Idrauliche, Roma, 2006, pp. 1-22 (in italian).

Seismic Hazard Assessment of Tehran Based on Arias Intensity

G. Ghodrati Amiri[a], H. Mahmoodi[b] and S.A. Razavian Amrei[c]

[a]*Professor, Center of Excellence for Fundamental Studies in Structural Engineering, College of Civil Engineering, Iran University of Science & Technology, Narmak, Tehran 16846, Iran*
E-mail: ghodrati@iust.ac.ir
[b]*MSc Student, College of Civil Engineering, Iran University of Science & Technology, Tehran, Iran*
[c]*PhD Student, College of Civil Engineering, Iran University of Science & Technology, Tehran, Iran*

Abstract. In this paper probabilistic seismic hazard assessment of Tehran for Arias intensity parameter is done. Tehran is capital and most populated city of Iran. From economical, political and social points of view, Tehran is the most significant city of Iran. Since in the previous centuries, catastrophic earthquakes have occurred in Tehran and its vicinity, probabilistic seismic hazard assessment of this city for Arias intensity parameter is useful. Iso-intensity contour lines maps of Tehran on the basis of different attenuation relationships for different earthquake periods are plotted. Maps of iso-intensity points in the Tehran region are presented using proportional attenuation relationships for rock and soil beds for 2 hazard levels of 10% and 2% in 50 years. Seismicity parameters on the basis of historical and instrumental earthquakes for a time period that initiate from 4th century BC and ends in the present time are calculated using Tow methods. For calculation of seismicity parameters, the earthquake catalogue with a radius of 200 km around Tehran has been used. SEISRISKIII Software has been employed. Effects of different parameters such as seismicity parameters, length of fault rupture relationships and attenuation relationships are considered using Logic Tree.

Keywords: Seismic Hazard Assessment, Arias Intensity, Tehran, Iran.

INTRODUCTION

Tehran as the capital of Iran besides its population which is over than 10 million people is known as an economic and political center. For this reason, destruction in this city due to its centric role has severe effects on the whole country. So, it seems to be necessary to evaluate the occurrence of earthquake intensity.

Existence active faults like north of Tehran, Mosha, north and south of Ray and the past strong earthquakes, indicate the great seismicity of this region and high probability of occurrence an earthquake with the magnitude of more than 7.

An earthquake is a natural phenomenon caused by a sudden rupture of a fault. When an earthquake occurs, energy stored in rocks is released and seismic waves radiate away from the seismic source and travel rapidly through the earth crust to the ground surface [1]. The seismic hazard resulting from an earthquake may include soil liquefaction, landslides and ground motion, that ground motion is considered the most critical seismic hazard. In engineering applications, the

CP1020, *2008 Seismic Engineering Conference Commemorating the 1908 Messina and Reggio Calabria Earthquake*,
edited by A. Santini and N. Moraci

ground motion is expressed usually in terms of amplitude, frequency content and the duration of ground motion.

Since the Arias intensity parameter includes all characteristics of ground motion and because of the importance of Tehran, it is intended in this paper to perform hazard analysis of Tehran using Arias intensity. For this purpose, two seismicity relationships, two relationships of fault rupture length and five attenuation relationships associated with Arias intensity will be used. Finally, according to the importance of these studied relationships, the desired coefficients will be applied in the logic tree.

ARIAS INTENSITY

A quantitative measure of earthquake-shaking intensity, often termed Arias intensity, is used to represent the total energy per unit weight absorbed by an idealized set of oscillators during earthquake motion [2].

Using the above definitions and based on a series of simplifications the Arias intensity formula could be expressed as follows:

$$I_{x-x} = \frac{\pi}{2g} \int_0^t a_x^2(t) dt \tag{1}$$

Where I_{x-x} is Arias intensity in x direction, $a_x(t)$ is acceleration time history in x direction and t is the total duration of ground motion.

The above definition applies only to a single component of ground acceleration and the total intensity should be represented as the sum of the two horizontal components of ground motion. The total Arias intensity is:

$$I_h = I_{x-x} + I_{y-y} \tag{2}$$

Some researchers such as Dobry et al. [3], Wilson and Keefer [4], believe that Arias intensity is related to the larger component of horizontal acceleration and not their sum.

SEISMOTECTONIC STRUCTURE OF TEHRAN

Tehran's extent is the northest depression of central Iran that in this region, Alborz mountains heights are forced to the Tehran plain. Tehran plain has southern slope and has been divided to the following different districts by mountains and eastern-western depressions:

High Alborz – Alborz Border Folds – Pedminent Zone – North Central Iranian depression (Tehran Plain)

For Tehran's case, the fault of Fasham-Mosha, North of Tehran, Kahrizak, North and South of Ray are the most susceptible faults which cause ground shaking. The way these faults have been chosen was that, a point was considered

271

as a center of Tehran and then a circle with a radius of 200 km was drawn. Those faults which fully or partially located in this circle were considered in this analysis.

Study Area

The network which hazard analysis has been done for it, is a square-shaped with 30×30 km. It should be noted that this network is divided into squares with dimensions of 1×1 km and in the four corners of these squares, probabilistic Arias intensity is obtained by software.

Earthquake Catalogue

Earthquake catalogue helps us to obtain comprehensive information about ground shaking happened in Tehran and its vicinity. The method of selecting the earthquake like previous section is to draw a circle with the radius of 200 km around the center of Tehran and to choose all earthquakes that their M_s are greater than 4 and are located inside the circle. Earthquake catalogue includes information like occurrence time, geographical latitude and longitude of the location of earthquake occurrence, type of magnitude, the value of magnitude, focal depth and distance between the location of earthquake occurrence and the center of Tehran.

There are 3 types of magnitude available in the earthquake catalogue as M_L, M_s and m_b. Since the appropriate magnitude used in seismic hazard analysis in Iran is M_s, all magnitude has been converted to M_s. In this paper, IRCOLD relationship [5] is used to convert m_b into M_s. The relation is expressed as following:

$$M_s = 1.2 * m_b - 1.29 \tag{3}$$

The correlation coefficient of this relationship is 0.87.

Seismicity Parameters

Seismic hazard analysis needs the determination of seismicity parameters. Parameters used in this paper are maximum expected magnitude (M_{max}), b value of Gutenberg-Richter relationship [6] and activity rate (λ).

Two approaches are used to determine these parameters: 1) Kijko [7] approach 2) Tavakoli's [8] method.

As an example, Table 1 shows the outputs of Kijko [7] software.

TABLE 1. Seismicity Parameters in Different Cases for Tehran

Parameter	Value	Data Contribution to the Parameters(%)		
		EXTREMES	COMPLETE	COMPLETE
Beta	1.63	34.4	32.7	33
Lambda(for Ms=4)	0.85	19.5	15.9	64.6

PROBABILISTIC SEISMIC HAZARD ANALYSIS (PSHA)

In this paper, by using the SEISRISK III software [9] the values of Arias intensity with 2% and 10 % probabilities of exceedence in 50-year lifetime of structure are obtained. The corresponding return periods with these hazard levels are 475 and 2475 years, respectively.

The values obtained from SEISRISK III are plotted as iso-intensity contours in desired periods.

Attenuation Relationships

One of the most important parts of seismic hazard assessment is attenuation relationship. Attenuation relationship describes decrease in the ground motion as a function of distance and magnitude. Many factors affect the attenuation relationships which are: the geology effects of the site, source specifications, magnitude, fault mechanism, reflection and refraction, etc. In this paper, relationships of Mahdavifar et al. [10], Travasarou et al. [11], Hwang et al. [1], Kayen and Mitchell [12] and Tselentis et al. [13] were employed using the logic tree method.

Relationships Between Maximum Expected Magnitude and Fault Rupture Length

The general form of relationship between maximum expected magnitude and fault rupture length is as follows:

$$\text{Log } L = a + b * M \tag{4}$$

Where L is the rupture length, M is the maximum expected magnitude and a and b are constant coefficients. The rupture length is percentage of the fault length where this percentage lies between 30 and 50. In this paper, two fault rupture length relationships are used. The first relationship comes from Nowroozi's [14] work that belongs to Iran and the second relationship comes from Wells and Coppersmith's [15] work that is obtained based on the collection of historical earthquake around the world.

Logic Tree

Logic tree is a popular tool used to compensate for the uncertainty in PSHA [16]. Figure 1 shows the logic tree considering the uncertainty of attenuation relationships, seismicity parameters and fault rupture length relationships.

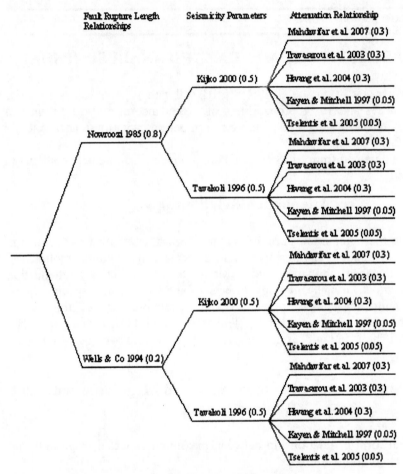

FIGURE 1. Applied logic tree

CONCLUSION

In this paper, the seismic hazard analysis for city of Tehran is performed by using Arias intensity parameter. Important results of this analysis are expressed as follows:
1) Developing a full and up-to-date catalogue
2) Determining seismicity parameters for city of Tehran
3) Drawing Iso-intensity maps according to the type of soil for city of Tehran and based on the different attenuation relationships, fault rupture length relationships and the methods of determining seismicity parameters (Figure 2).

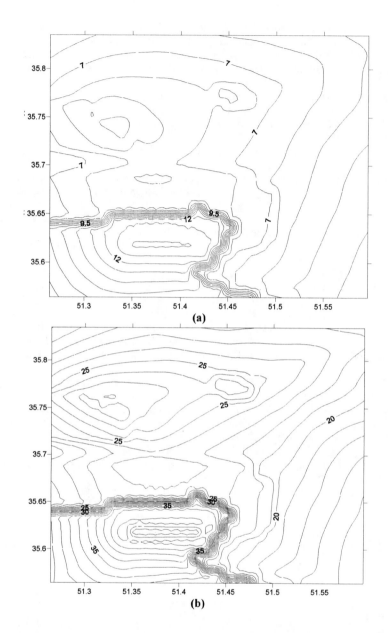

FIGURE 2. Seismic hazard maps in terms of Arias intensity in Tehran and its vicinity using Logic Tree for 475 and 2475 years return periods (a) two-dimensional zoning map showing Arias intensity with 475 year return period (b) two-dimensional zoning map showing Arias intensity with 2475 year return period

REFRENCES

1. H. Hwang, C. K. Lin, Y. T. Yeh, S. N. Cheng and K. C. Chen, "Attenuation relations of arias intensity based on the Chi-Chi Taiwan earthquake data", *Soil Dynamic and Earthquake Engineering* **24**, 509-517 (2004).

2. A. Arias, "A Measure of Earthquake Intensity," in *Hansen, R. J. (ED.), Seismic Design for Nuclear Power Plants*, MIT Press. Cambridge, MA, 1970, pp. 438-483.

3. R. Dobry, I. M. Idriss, and E. Ng, "Duration characteristics of horizontal component of strong-motion earthquake records", *Bulletin of the Seismological Society of America* **68**, 1487-1520 (1978).

4. R. C. Wilson and D. K. Keefer, "Predicting areal limits of earthquake-induced landsliding", *U.S.Geological Survey*, Prof. Paper **1360**, 317-345 (1985).

5. IRCOLD, *Iranian Committee of Large Dams Relationship between Ms and mb*, Internal Report (in Persian), 1994.

6. B. Gutenberg and C. F. Richter, *Seismicity of the Earth and Associated Phenomena*, Princeton University Press, New Jersey, 1954.

7. A. Kijko, "Statistical estimation of maximum regional earthquake magnitude Mmax" in *Workshop of Seismicity Modeling in Seismic Hazard Mapping*, Poljce, Slovenia, 2000, May 22-24.

8. B. Tavakoli, *Major Seismotectonic Provinces of Iran*, International Institute of Earthquake Engineering and Seismology, Internal Document (in Persian), 1996.

9. B. Bender and D.Perkins, *SEISRISK III, A Computer Program for Seismic Hazard Estimation*, US Geological Survey, 1987, Bulletin 1772.

10. M. Mahdavifar, M. K. Jafari and M. R. Zolfaghari, "The attenuation of arias intensity in Alborz and central Iran" in *Proceedings of the Fifth International Conference on Seismology and Earthquake Engineering*, Tehran, Iran, 2007.

11. T. Travasarou, J. D. Bray and N. A. Abrahamson, "Empirical attenuation relationship for arias intensity", *Earthquake Engineering and Structural Dynamic* **32**, 1133-1155 (2003).

12. R. E. Kayen and J. K. Mitchel, "Assessment of liquefaction potential during earthquakes by arias intensity", *Journal of Geotechnical and Geoenvironmental Engineering* **123**, 1162-1174 (1997).

13. G. A. Tselentis, L. Danciu and F. Gkika, "Empirical Arias Intensity Attenuation Relationships for the Seismic Hazard Analysis of Greece" *in Brebbia et al (eds.), Earthquake Resistant Engineering Structures, The Built Environment*, 81, 2005.

14. A. Nowroozi, "Empirical relations between magnitude and fault parameters for earthquakes in Iran", *Bulletin of the Seismological Society of America* **75(5)**, 1327-1338 (1985).

15. D. L. Wells and K. J. Coppersmith, "New empirical relationships among magnitude, rupture length, rupture width, rupture area and surface displacement", *Bulletin of the Seismological Society of America* **84(4)**, 974-1002 (1994).

16. G. Ghodrati Amiri, R. Motamed, H. R. ES-Haghi, "Seismic hazard assessment of metropolitan Tehran, Iran", *Journal of Earthquake Engineering* **7(3)**, 347-372 (2003).

Evaluation of Horizontal Seismic Hazard of Shahrekord, Iran

G. Ghodrati Amiri[a], M. Raeisi Dehkordi[b], S.A. Razavian Amrei[c] and M. Koohi Kamali[d]

[a]Professor, Iran University of Science & Technology - Islamic Azad University of Shahrekord, Narmak, Tehran 16846, Iran, E-mail: ghodrati@iust.ac.ir
[b]Assistant Professor, Department of Civil Engineering, Islamic Azad University of Shahrekord, Iran.
[c]PhD Student, College of Civil Engineering, Iran University of Science & Technology, Tehran, Iran
[d]MSc Student, Department of Civil Engineering, Islamic Azad University of Shahrekord, Iran

Abstract. This paper presents probabilistic horizontal seismic hazard assessment of Shahrekord, Iran. It displays the probabilistic estimate of Peak Ground Horizontal Acceleration (PGHA) for the return period of 75, 225, 475 and 2475 years. The output of the probabilistic seismic hazard analysis is based on peak ground acceleration (PGA), which is the most common criterion in designing of buildings. A catalogue of seismic events that includes both historical and instrumental events was developed and covers the period from 840 to 2007. The seismic sources that affect the hazard in Shahrekord were identified within the radius of 150 km and the recurrence relationships of these sources were generated. Finally four maps have been prepared to indicate the earthquake hazard of Shahrekord in the form of iso-acceleration contour lines for different hazard levels by using SEISRISK III software.

Keywords: Seismic Hazard Assessment, Seismicity Parameters, PGA, Shahrekord, Iran.

INTRODUCTION

Iran, one of the most seismic countries of the world, is situated over the Himalayan-Alpied seismic belt. Shahrekord with the exceeding 200 thousands people located in the west of Iran, very close to Isfahan. Due to be the province center it needs a very precise investigation of seismicity and seismic hazard. This paper presents probabilistic horizontal seismic hazard assessment of Shahrekord.

SEISMOTECTONIC STRUCTURE OF SHAHREKORD

Shahrekord city is situated on the west plateau of central Zagros mountain. In order to evaluate the seismic hazard of a region or zone, all the probable seismic sources must be detected and their potential to produce strong ground motion must be checked. The major faults in Shahrekord region are Zagros, Dena, Ardal, Cherou, Gazolk, South Rokh, Dopolan and Zardkooh. The location of these faults with respect to Shahrekord is shown in Figure 1.

CP1020, *2008 Seismic Engineering Conference Commemorating the 1908 Messina and Reggio Calabria Earthquake*, edited by A. Santini and N. Moraci

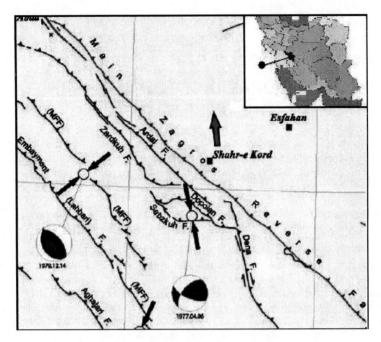

FIGURE 1. The active faults of Shahrekord [1]

SEISMICITY OF SHAHREKORD

The seismicity of each region is indicated by the past earthquakes occurred in that region and to obtain the seismotectonic properties, a thorough list of each region's earthquake events must be collected and studied.

Shahrekord Seismicity Parameters

The evaluation of seismicity parameters is performed based on the seismic data of earthquakes occurred in the region under study and employing probabilistic methods. The seismic catalogue has been collected, assuming that earthquakes follow Poisson distribution. The method which is used to eliminate the foreshocks and aftershocks is the variable windowing method in time and space domains [2].

Due to the very high importance of the seismicity parameters in seismic hazard evaluation, in this study the new Kijko [3] method has been employed which is based on double truncated Gutenberg-Richter [4] relationship and the maximum likelihood estimation method.

In the maximum likelihood estimation method, it is possible to use historic and instrumentally recorded data at the same time. The values of seismicity parameters β, λ resulting from this method was: 1.52 and 0.76.

The annual average occurrence rate of earthquake versus magnitude for earthquakes with magnitude greater than Ms=4.0 in the extent of 150 km around

Shahrekord is shown in Figure 2 based on these investigations and the performed calculations with Kijko method.

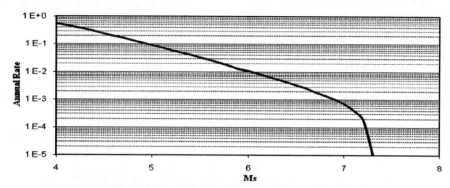

FIGURE 2. Annual rates estimated by Kijko method for Shahrekord

EVALUATION OF HORIZONTAL SEISMIC HAZARD

In order to evaluate Peak Ground Horizontal Acceleration (PGHA) for the return period of 75, 225, 475 and 2475 years, probabilistic seismic hazard analysis method has been used. In this method, seismicity parameters (β, λ) are given to the seismic sources based on the seismicity investigations, then based on earthquake magnitude, distance of epicenter or hypocenter from site and application of an appropriate attenuation relationship, Peak Ground Horizontal Acceleration (PGHA) at the corresponding site is evaluated.

Attenuation Relationship

Selection of appropriate attenuation relationship is very important in validity and reliability of the analysis results. Therefore, there are some important notes that must be paid attention for the selection of attenuation relationship. The most important ones are source specifications, magnitude, fault rupture type, distance to the seismogenic sources, geology and topology of site.

Based on the mentioned remarks, in this research six weighted horizontal attenuation relationships; Ramazi [5], 0.2; Ambraseys and Bommer [6], 0.15; Ghodrati et al. [7], 0.3; Zare et al. [8], 0.15; Sadigh et al. [9], 0.1 and Campbell [10], 0.1, in logic tree method were employed.

Probabilistic Seismic Hazard Analysis

For the seismic hazard probabilistic evaluation, the software SEISRISK III [11] was utilized to calculate the Peak Ground Horizontal Acceleration (PGHA) in the specific hazard level in the structure lifetime. The calculated values can be shown in the form of iso-acceleration lines for each specific hazard level in the structure lifetime. In this study the seismic hazard analysis carried out was based on the assumption of an ideal bedrock case and therefore no influence of local soil condition is taken into consideration. Based on the Iranian seismic rehabilitation code of the existing buildings [12], 4 hazard levels are most considered: 50%, 20%, 10% and 2% probabilities of exceedence in 50 years. Before the calculations, a grid of sites must be considered in the region where seismic hazard analysis is performed. For this purpose a grid of 8*7 is considered.

As a result, our outputs are Peak Ground Horizontal Acceleration (PGHA) with 50%, 20%, 10% and 2% probabilities of exceedence in 50-year lifetime of structure. The result of the seismic hazard analysis is graphically shown in Figures 3 to 6.

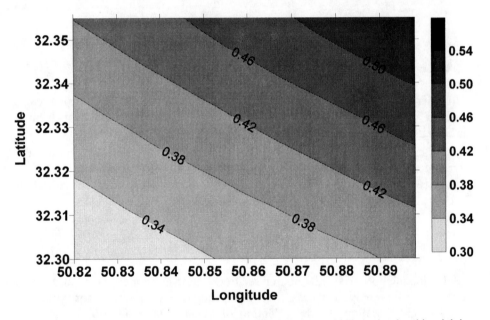

FIGURE 3. Horizontal seismic hazard (PGA over bedrock) map of Shahrekord and its vicinity using logic tree for 2475 year return period.

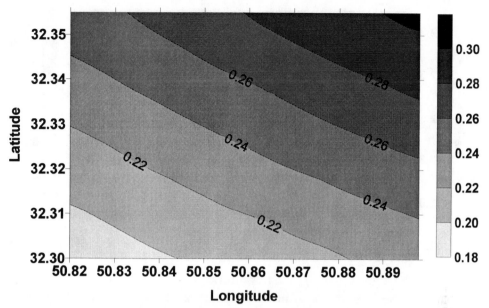

FIGURE 4. Horizontal seismic hazard (PGA over bedrock) map of Shahrekord and its vicinity using logic tree for 475 year return period.

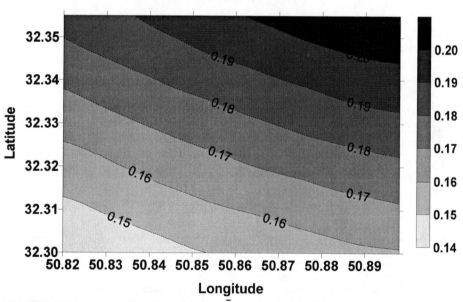

FIGURE 5. Horizontal seismic hazard (PGA over bedrock) map of Shahrekord and its vicinity using logic tree for 225 year return period.

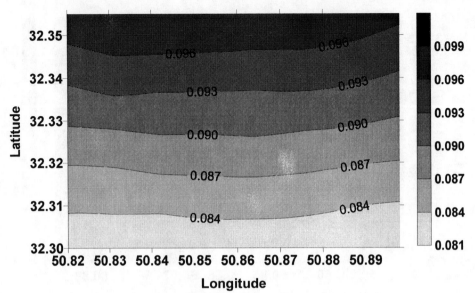

FIGURE 6. Horizontal seismic hazard (PGA over bedrock) map of Shahrekord and its vicinity using logic tree for 75 year return period.

CONCLUSIONS

This paper presents seismic hazard maps of Shahrekord and its vicinity based on Peak Ground Horizontal Acceleration (PGHA) for 50%, 20%, 10% and 2% probabilities of exceedence in a time span of 50 years. The significant results of this study can be summarized as:

(1) The contour levels of the horizontal acceleration hazard maps showed that the PGHA for 2% ranges from 0.3(g) to around 0.54(g), PGHA for 10% ranges from 0.19(g) to around 0.31(g), PGHA for 20% ranges from 0.14(g) to around 0.21(g) and PGHA for 50% ranges from 0.089(g) to around 0.099(g).

(2) The highest acceleration contours locate in the northeast parts of Shahrekord.

(3) The smallest accelerations are expected in the southwest of Shahrekord.

REFRENCES

1- IIEES, International Institute of Earthquake Engineering & Seismology, http://www.iiees.ac.ir/.
2- J. K. Gardner and L. Knopoff, "Is the sequence of earthquake in southern California, with aftershocks removed, Poissonian?", *Bulletin of the Seismological Society of America* **64(5)**, 1363-1367 (1974).
3. A. Kijko, "Statistical estimation of maximum regional earthquake magnitude Mmax" in *Workshop of Seismicity Modeling in Seismic Hazard Mapping*, Poljce, Slovenia, 2000, May 22-24.
4. B. Gutenberg and C. F. Richter, *Seismicity of the Earth and Associated Phenomena*, Princeton University Press, New Jersey, 1954.
5. H. R. Ramazi, "Attenuation laws of Iranian earthquakes" in *Proceedings of the 3rd International Conference on Seismology and Earthquake Engineering*, Tehran, Iran, 337-344, 1999.
6. N. N. Ambraseys and J. J. Bommer, "The attenuation of ground accelerations in Europe", *Earthquake Engineering & Structural Dynamics* **20(12)**, 1179-1202 (1991).

7. G. Ghodrati Amiri, A. Mahdavian, and F. Manouchehri Dana, "Attenuation relationships for Iran", *Journal of Earthquake Engineering* **11(4)**, 469-492 (2007).
8. M. Zare, M. Ghafory-Ashtiany and P. Y. Bard, "Attenuation law for the strong-motions in Iran" in *Proceedings of the 3rd International Conference on Seismology and Earthquake Engineering*, Tehran, Iran, 345-354, 1999.
9. K. Sadigh, C.Y. Chang, J. A. Egan, F. Makdisi and R. R. Youngs, "Attenuation relationships for shallow crustal earthquakes based on California strong motion data", *Seismological Research Letters*, **68(1)**, 180–189 (1997)
10. K. W. Campbell, "Empirical near-source attenuation relationships for horizontal and vertical components of peak ground acceleration, peak ground velocity, and pseudo-absolute acceleration response spectra", *Seismological Research Letters*, **68(1)**, 154–179 (1997).
11. B. Bender and D. Perkins, *SEISRISK III, A Computer Program for Seismic Hazard Estimation*, US Geological Survey, Bulletin 1772, 1987.
12 IIEES, *Seismic Rehabilitation Code For Existing Buildings in Iran*, International Institute of Earthquake Engineering and Seismology (In Persian), 2007.

Earthquake Risk Management of Underground Lifelines in the Urban Area of Catania

S. Grasso[a] and M. Maugeri[b]

[a]Ph. Doctor in Geotechnical Engineering, University of Catania, Department of Civil and Environmental Engineering, Viale A. Doria 6, 95125 Catania, Italy. e-mail: sgrasso@dica.unict.it
[b]Full Professor in Geotechnics, University of Catania, Department of Civil and Environmental Engineering, Viale A. Doria 6, 95125 Catania, Italy. e-mail: mmaugeri@dica.unict.it

Abstract. Lifelines typically include the following five utility networks: potable water, sewage natural gas, electric power, telecommunication and transportation system. The response of lifeline systems, like gas and water networks, during a strong earthquake, can be conveniently evaluated with the estimated average number of ruptures per km of pipe. These ruptures may be caused either by fault ruptures crossing, or by permanent deformations of the soil mass (landslides, liquefaction), or by transient soil deformations caused by seismic wave propagation. The possible consequences of damaging earthquakes on transportation systems may be the reduction or the interruption of traffic flow, as well as the impact on the emergency response and on the recovery assistance. A critical element in the emergency management is the closure of roads due to fallen obstacles and debris of collapsed buildings.

The earthquake-induced damage to buried pipes is expressed in terms of repair rate (RR), defined as the number of repairs divided by the pipe length (km) exposed to a particular level of seismic demand; this number is a function of the pipe material (and joint type), of the pipe diameter and of the ground shaking level, measured in terms of peak horizontal ground velocity (PGV) or permanent ground displacement (PGD). The development of damage algorithms for buried pipelines is primarily based on empirical evidence, tempered with engineering judgment and sometimes by analytical formulations.

For the city of Catania, in the present work use has been made of the correlation between RR and peak horizontal ground velocity by American Lifelines Alliance (ALA, 2001), for the verifications of main buried pipelines. The performance of the main buried distribution networks has been evaluated for the Level I earthquake scenario (January 11, 1693 event I = XI, M 7.3) and for the Level II earthquake scenario (February 20, 1818 event I = IX, M 6.2). Seismic damage scenario of main gas pipelines and water has been obtained, with PGV values calculated for level I and level II earthquake scenarios.

Keywords: Site characterisation, microzonation and site effects, lifelines vulnerability.

INTRODUCTION

The primary purpose of this paper is to provide basic tools for estimating the Earthquake Risk Management and vulnerability assessment (and some related effects) of Underground lifelines and essential facilities in the urban area of Catania, and for producing suitable map representations. Such estimations and maps are intended as a basic input for developing relatively detailed earthquake damage scenarios for the city of Catania. The paper identifies and briefly describes the methods considered suitable for tackling the task involved, and provides specific maps.

CP1020, 2008 Seismic Engineering Conference Commemorating the 1908 Messina and Reggio Calabria Earthquake,
edited by A. Santini and N. Moraci

South-Eastern Sicily has been affected by several destroying earthquakes in past times. The area to the south of Volcano Etna, on the east of the Ibleo-Maltese escarpment, to the south of the graben of the Sicilian channel and on the east of the overlapping front of Gela, known as Iblean Area, is placed on area of contact between the African and the Euro-asiatic plates, and it is therefore a seismogenic area.

The city of Catania located in the South-Eastern Sicily has been affected by several destroying earthquakes of about magnitude 7.0+ in past times. The area to the south of Volcano Etna, on the east of the Ibleo-Maltese escarpment shown in Figure 1, to the south of the graben of the Sicilian channel and on the east of the overlapping front of Gela, known as Iblean Area, is placed on area of contact between the African and the Euro-asiatic plates, and it is therefore a seismogenic area. According to the frequency and the importance of the seismic effects suffered in past times, Eastern Sicily must be considered one of the most high seismic risk areas in Italy. Today, on such a densely populated territory, a huge patrimony of historical and industrial buildings is placed.

It is so reasonable to assume in Catania a maximum expected earthquake as a repetition of the 1693 event, with intensity XI MCS and estimated magnitude M= 7.3 (Azzaro and Barbano, 2000). The 1693 earthquake may be selected as a first level scenario event. As second level scenario event may be chosen the 1818 earthquake with intensity IX MCS and estimated magnitude M = 6.2.

In the south-eastern Sicily there are two areas where seismicity is mainly distributed: long the Ionian coast (earthquakes of magnitude M>7.0) and in the hinterland area (earthquakes of magnitude lower than 5.5).

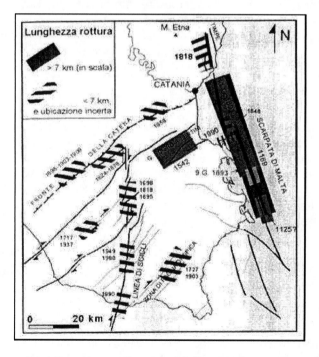

FIGURE 1. Map of the seismogenetic structures of south-eastern Sicily, modified.

285

There are evidences from the late Quaternary period that the Ibleo-Maltese fault system is the most probable source for the great earthquakes that struck the region (1693, 1818 earthquakes). This fault system is mainly made up of normal faults NNW-SSE oriented, divided into three segments of fault, the most northern of which continues on the ground up to the Etnean area (Timpa), the central segment reaches the Gulf of Catania while the most southern part lies at sea between Augusta and Siracusa (see Figure 1).

BASIC FEATURES OF LIFELINES AND CRUCIAL FACILITIES

Lifelines and essential facilities are vital elements at risk. As it is reported by Schiff and Buckle (1995), "Lifelines are those systems that are vital for the continued operating of communities in an industrialized society. They include power, communication, water, sewer, gas, liquid fuel and many type of transportation systems. Even without an earthquake, the disruption of any one of these systems, even for a day, would constitute a major disaster. In the aftermath of a damaging earthquake, many of these systems play a critical role for the emergency response community and the community in general to save lives and prevent additional damage to property." Lifelines and essential facilities are singular and complex elements at risk. The spatial distribution of lifelines usually widely exceeds the urban area. Thus, any lifeline is actually a unique and singular element at risk. All lifelines are composed by "lines" and "nodes" of different typology in each system.

There is a specific exposure of lifelines to earthquake. The wide extent of lifelines implies spatial variability of seismic motion (ground acceleration and velocity) and higher probability of exposure to permanent ground displacement induced by: Fault offset; Liquefaction phenomena; Landslides. The objective of the paper is to evaluate the vulnerability assessment of lifelines and essential facilities of the city of Catania.

Utility Systems

In urban areas, several utility systems may be closely located to each other. Thus, it is possible to have water, waste and gas pipes, telecommunication and electric power distribution lines etc. in the same location, but in different burial depths. An emergency repair (during crisis and recovery period) of one network can have a serious impact to the others. Additionally, the increased demand of one system (i.e. mobile telecommunication calls) due to the disruption or the reduced serviceability of another system (public telecommunication network) could be an indirect interaction between utility networks (Monge et al., 2003). Interaction means synergies in mutual or reciprocal actions or influence. High dependence among different lifeline systems and also with essential facilities is very important feature.

Water Pipes Utility System

Pipes can be free-flow or pressure conduits, buried or elevated. Several materials can be used. In order to avoid contamination of treated water, potable water pipes are most of the time pressurized. Waste water pipes are most of the time free flow conduits. Pile supports can be made of wood, concrete or concrete-encased steel. Buried pipes are buried 1 to 5 m or deeper in the ground. For detailed diagnostics of pipe failure, mechanical characteristics of material are required. Pipes are commonly made of: - Asbestos Cement (AC), - Concrete (C), - Cast Iron (CI), - Ductile Iron (DI), - High Density PolyEthylene (HDPE), etc. The diameter of distribution pipe is important both in terms of pipe damage algorithms and post-earthquake performance of the entire water system. Pipe diameters are generally greater than 4 inches and one should consider the following classes: - Small diameter means 4 to 12 inches (≈100 to 300 mm); - Large diameter mean 16 inches and more (>400 mm). A jointed pipeline consists of pipe segments coupled by relatively flexible (or weak) connections (e.g., a bell-and-spigot cast iron piping system). Continuous pipelines are those having rigid joints, such as continuous welded steel pipelines. Corrosion will accentuate damage, especially in segmented steel, threaded steel and cast iron pipes. Different modes of failures are possible (Table 1), namely well described by O'Rourke and Liu, (1999).

TABLE 1. Possible modes of failures for pipes subject to earthquake, (O'Rourke and Liu, 1999).

ATC-25	Continuous pipes (O'ROURKE and LIU, 1999)	Segmented pipes (O'ROURKE and LIU, 1999)	ALA, 2001a
Pipe crushing and cracking Joint breaking Joints pulling	Tensile failure Local buckling (wrinkling) Beam buckling Welded slip joint	Axial pull-out Crushing of bell and spigot joints Joint rotation failure/leakage Round flexural cracks	Pullout at a joint Excessive rotation at a joint Excessive tensile and bending deformations of the pipe barrel

The possible consequences for the pipes are described by a repair rate combining breaks and leaks. A pipe repair can be due either to a complete fracture of the pipe, a leak in the pipe or damage to an appurtenance of the pipe. In any case, these repairs require the water agency to perform a repair in the field. It is often assumed that damage due to seismic waves will consist of 80% leaks and 20% breaks while damage due to ground failure will consist of 20% leaks and 80% breaks. The translation of repair rates into damage state or serviceability is not evident. Radius project used classes of repair rates, possibly associated to damage states for water pipes (Table 2).

TABLE 2. Possible consequences for water pipes subject to earthquake.

Repair Rate (repair /km)	Serviceability	Damage states
≥0.60	≤10%	Complete
0.15 to 0.60	10 to 50%	Extensive
0.05 to 0.15	50 to 85%	Moderate
≤0.05	≥85%	Minor

A GIS database of the water pipes system of the city of Catania has been then created. All information about water pipes have been collected thanks to the collaboration with SIDRA water agency of the city of Catania. Information regards location and depth of the pipe, material of the pipe, diameter of the pipe, location of tanks. Data have been inserted in the database and a GIS map has been created (Figure 2).

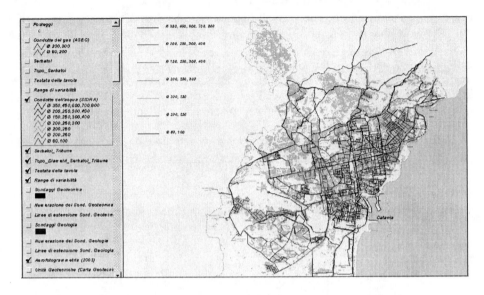

FIGURE 2. GIS map of the water pipes system of the city of Catania.

Gas Pipelines Utility System

Natural gas networks are operating at different pressures: supra-regional transmission pipelines operate at very high pressures. These pipelines have a maximum diameter of 1.40 m and are operating at pressure higher than 100 bars. Such gas pipelines can cover distances of up to 6 000 km (e.g. from west Siberia to western Europe). Regional networks range between from 1 to 70 bars, while local distribution systems are usually operates in the medium (0.1 - 4 bars) or low-pressure (<0.1 bars) range (Alexoudi and Pitilakis, 2003). Beyond classification, it is also possible to classify gas pipelines importance, according to one or several of the following points (Table 3). Transmission pipelines are typically large- diameter welded steel pipes that are expected to perform in earthquakes in a manner superior to that of typical underground pipelines (ATC-25, 1991). High damage to pipelines was caused by traveling ground waves during 1995 Hyogo-Ken Nanbu (Kobe) earthquake with local magnitude M=7.2. Almost all utilities, roadways, railways, the port, and other lifelines to the city center suffered severe damage, greatly delaying rescue efforts. The destruction of lifelines and utilities made it impossible for firefighters to reach fires started by broken gas lines. Large sections of the city burned, greatly contributing to the loss of life.

TABLE 3. Classification of gas pipelines.

GS3: Gas pipeline	1 (High)	2 (Medium)	3 (Low)	
Radiance	Regional and above	In between	Local	
Redundancy capability	Mainly single	In between	Mainly redundant	
Level of risk	Particular section[1]		Common section[2]	
Transmission versus distribution	Mainly transmission	In between	Mainly distribution	
Pressure	>60 bar	10 to 60 bar	4 to 10 bar	<4 bar
Valves	>5 km	2.5 to 5 km	<2.5 km	
Diameter (Transmission)	>200 mm[3]	100 to 200 mm	<100 mm	
Diameter (Distribution)	>90 mm	32 to 90 mm	<32 mm	

Different modes of failures are possible (Table 1), namely well described by O'Rourke and Liu, (1999). The possible consequences for the pipes are described by a repair rate combining breaks and leaks. A pipe repair can be due either to a complete fracture of the pipe, a leak in the pipe or damage to an appurtenance of the pipe. In any case, these repairs require the gas agency to perform a repair in the field. It is often assumed that damage due to seismic waves will consist of 80 % leaks and 20 % breaks, while damage due to ground failure will consist of 20 % leaks and 80 % breaks. Consequences of gas leakage are very serious. A GIS database of the gas buried pipes system of the city of Catania has been then created. All information about gas pipes have been collected thanks to the collaboration with ASEC gas agency of the city of Catania. Information regards location and depth of the pipe, material of the pipe, diameter of the pipe, location of tanks. Data have been inserted in the database and a GIS map has been created (Figure 3).

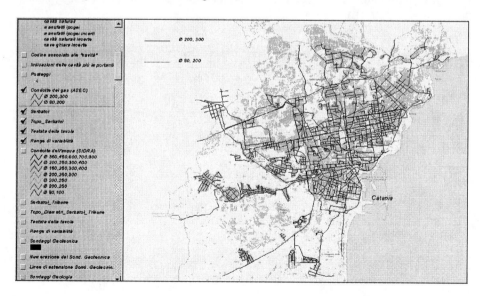

FIGURE 3. GIS map of the gas pipes system of the city of Catania.

VULNERABILITY OF UNDERGROUND LIFELINES

The damage algorithm for buried pipe is expressed as a repair rate (RR), defined as the number of repairs divided by the pipe length (km) exposed to a particular level of seismic demand; this number is a function of the pipe material (and joint type), of the pipe diameter and of the ground shaking level, measured in terms of peak horizontal ground velocity (PGV) or permanent ground displacement (PGD). The following equations have been used to evaluate peak horizontal ground velocity (PGV) of the 1693 and 1818 scenario earthquakes:

Sabetta and Pugliese (1996):

$$\log PGV = -0.710 + 0.455M - \log \sqrt{R^2 + 3.6^2} + 0.133S + 0.215P \tag{1}$$

Spudich et al. (1999):

$$\log PGV = b1 + b2(M - 6) + b3(M - 6)^2 + b4R + b5\log R + b6\Gamma \tag{2}$$

Bommer et al. (2002):

$$\log PGV = C1 + C2M + C4\log \sqrt{R^2 + h_0^2} + C_A S_A + C_S S_S + \sigma P \tag{3}$$

in which M is the Richter Magnitude, R is the Joyner-Boore distance, S and Γ are soil parameters, P is the standard deviation, b1-b6 and C1-C4-C_A-C_S are smoothed coefficients for geometric mean 5% damped psuedo-velocity response spectrum, function of the period T of the earthquake. For the 1693 and 1818 scenario earthquakes have been estimated respectively T=0.7 sec and T=0.4 sec.

The development of damage algorithms for buried pipe is primarily based on empirical evidence, tempered with engineering judgment and sometimes by analytical formulations. The damage algorithm is expressed as a repair rate per unit length of pipe, as a function of ground shaking (Peak Ground Velocity, according to Isoyama et al., 1998) or ground failure (Permanent Ground Displacement, according to ALA, 2001a and ALA, 2001b):

Isoyama et al. (1998):

Induced hazard	Vulnerability model
Wave propagation	Equation 1 : Repair Rate (repair/km) = $C_p \cdot C_d \cdot 0.00311 \left(\dfrac{PGV}{100} - 15\right)^{1.3}$

ALA (2001a):

Induced hazard	Vulnerability model	Lognormal standard deviation
Wave propagation	Equation 1 : Repair Rate (repair/km) = $\dfrac{0.00187}{0.3048} K_1 \cdot \left(\dfrac{PGV}{0.0254}\right)$	$\beta = 1.15$

ALA (2001b):

Ground failure	Equation 2 : Repair Rate (repair/km) = $\dfrac{1.06}{0.3048} K_2 \cdot \left(\dfrac{PGD}{0.0254}\right)^{0.319}$ Lognormal standard deviation ($\beta = 0.74$)

In which: C_P, K_1 and K_2 are coefficients of ground shaking vulnerability model, according to various pipe material, while C_D is the coefficient of ground shaking vulnerability model, according to various pipe diameter.

The performance of the main buried distribution networks has been evaluated for the level I earthquake scenario (1693 event) and for the level II scenario (1818 event), with the RR values shown respectively in table 4 and table 5. RR values distribution for gas pipelines and for the level I scenario is as an example shown in Fig. 4.

TABLE 4. Repair rate (RR) for the 1693 earthquake scenario.

Pipe	PGV	Bommer et al. (2002)	Sabetta and Pugliese (1996)	Spudich et al. (1999)
Water	Isoyama et al. 98	0.06-0.075	0.045-0.06	0.075-0.09
Water	ALA,2001a	0.09-0.105	0.09-0.105	0.135-0.15
Gas	Isoyama et al. 98	0.06-0.075	0.06-0.075	0.075-0.09
Gas	ALA,2001a	0.09-0.105	0.09-0.105	0.135-0.15

TABLE 5. Repair rate (RR) for the 1818 earthquake scenario.

Pipe	PGV	Bommer et al. (2002)	Sabetta and Pugliese (1996)	Spudich et al. (1999)
Water	Isoyama et al. 98	0.045-0.06	0.045-0.06	0.045-0.06
Water	ALA,2001a	0.06-0.075	0.045-0.06	0.075-0.09
Gas	Isoyama et al. 98	0.045-0.06	0.045-0.06	0.06-0.075
Gas	ALA,2001a	0.045-0.06	0.045-0.06	0.06-0.075

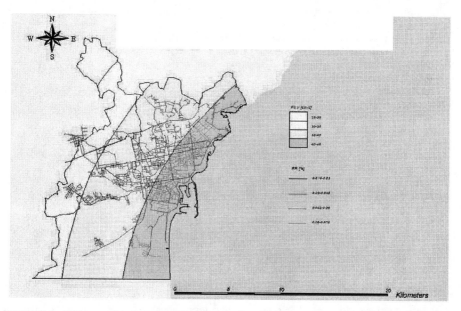

FIGURE 4. GIS map of the RR values distribution (from Isoyama et al., 1998) for gas pipelines and for the level I scenario (PGV from Spudich et al., 1999).

CONCLUSIONS

For the city of Catania, in the present work use has been made of the correlation between RR and peak horizontal ground velocity by American Lifelines Alliance (ALA, 2001), for the verifications of main buried pipelines. The performance of the main buried distribution networks has been evaluated for the Level I earthquake scenario and for the Level II earthquake scenario. Seismic damage scenario of main gas pipelines and water has been obtained, with PGV values calculated for level I and level II earthquake scenarios. Results shown in Tables 4 and 5 make clear that RR values, as number of repairs divided by the pipe length (km) exposed to seismic demand, are not very high. However it must be noted that there is a lack of information about location and depth of the pipe, joint types, material of the pipe, diameter of the pipe, location of tanks.

ACKNOWLEDGMENTS

The authors wish to thank the geotechnical engineer Sebastiano Pistone for his contribution to the work. The authors wish also to thank SIDRA water agency and ASEC gas agency for their contribution and for accessibility to data.

REFERENCES

1. R. Azzaro and M.S. Barbano, "Seismogenic features of SE Sicily and Scenario Earthquakes for Catania", in *The Catania Project: Earthquake Damage Scenarios for a high risk area in the Mediterranean*. Editors: Faccioli and Pessina. Roma 2000, pp. 9-13.
2. A.J. Schiff and I.G. Buckle, "Critical issues and state-of-the-art in Lifeline earthquake engineering", *TCLEE monograph n°7*. ASCE. 92 p.
3. O. Monge, M. Alexoudi, S. Argyroudis, C. Martin, K. Pitilakis, "Vulnerability assessment of lifelines and essential facilities (WP06): basic methodological handbook". In: *RISK-UE An advanced approach to earthquake risk scenarios with applications to different European towns*, Report n°GTR-RSK 0101-152av7, 2003, 71 pages.
4. M. J. O'Rourke, X. Liu, "Response of buried pipelines subject to earthquake effects". In: *MCEER. Monograph series* n°3, 1999, 249 p.
5. M. Alexoudi, K. Pitilakis, "Vulnerability assessment of lifelines and essential facilities (WP06): methodological handbook Appendix 7: Gas system". In: *RISK-UE An adv. approach to earthquake risk scenarios with appl. to different European towns*, Rep. n°GTR-RSK 0101-152av7, 2003, 35p.
6. Applied Technology Council, "Seismic Vulnerability and Impact of Disruption of Lifelines in the Conterminous United States". *ATC-25*, Redwood City, 1991, CA.
7. F. Sabetta and A. Pugliese, "Estimation of response spectra and simulation of nonstationary earthquake ground motion". *Bull. of the Seismol. Society of America*, 1996, vol. 86, pp. 337-352.
8. P. Spudich, , W. B. Joyner, , A. G. Lindh, B. M. Margaris, D. M. Boore, , J. B. Fletcher, "SEA99 - A revised ground motion predict. rel. for use in extens. tectonic regimes", in *BSSA*, 1999, v. 89, no. 5, 1156-1170.
9. J. Bommer, J. Rey, E. Faccioli, "Derivation of design soil coefficients (S) and response spectral shapes using the European strong motion database", Journ. of Seismology, 2002, 6, 4, pp. 547-555.
10. Isoyama R, Ishida E, Yune K, Shirozu T (1998) Seismic damage estimation procedure for water supply pipelines. In: Proceedings of water & earthquake '98 Tokyo, IWSA international workshop.
11. American Lifelines Alliance, "Seismic fragility formulations for water systems. Part 1 – Guideline". ASCE-FEMA, 2001a, 104 p.
12. American Lifelines Alliance. "Seismic fragility formulations for water systems. Part 2 – Appendices". ASCE-FEMA, 2001b 239 p.

Keeping the History in Historical Seismology: The 1872 Owens Valley, California Earthquake

Susan E. Hough

U.S. Geological Survey
525 South Wilson Avenue
Pasadena, California 91106

Abstract. The importance of historical earthquakes is being increasingly recognized. Careful investigations of key pre-instrumental earthquakes can provide critical information and insights for not only seismic hazard assessment but also for earthquake science. In recent years, with the explosive growth in computational sophistication in Earth sciences, researchers have developed increasingly sophisticated methods to analyze macroseismic data quantitatively. These methodological developments can be extremely useful to exploit fully the temporally and spatially rich information source that seismic intensities often represent. For example, the exhaustive and painstaking investigations done by Ambraseys and his colleagues of early Himalayan earthquakes provides information that can be used to map out site response in the Ganges basin. In any investigation of macroseismic data, however, one must stay mindful that intensity values are not data but rather interpretations. The results of any subsequent analysis, regardless of the degree of sophistication of the methodology, will be only as reliable as the interpretations of available accounts—and only as complete as the research done to ferret out, and in many cases translate, these accounts. When intensities are assigned without an appreciation of historical setting and context, seemingly careful subsequent analysis can yield grossly inaccurate results. As a case study, I report here on the results of a recent investigation of the 1872 Owen's Valley, California earthquake. Careful consideration of macroseismic observations reveals that this event was probably larger than the great San Francisco earthquake of 1906, and possibly the largest historical earthquake in California. The results suggest that some large earthquakes in California will generate significantly larger ground motions than San Andreas fault events of comparable magnitude.

Keywords: Historical earthquakes, earthquake hazard

PACS: 91.30.Bi, 91.30.mv, 91.30.Px.

INTRODUCTION

Instrumental seismic data is non-existent for earthquakes prior to the late 19th century, and strong motion data are sparse at best for earthquakes prior to the mid-20th century. Typically, especially in low-strain-rate areas, no instrumental data are available to investigate the largest historical earthquakes in a region. So-called macroseismic data have been shown to be of enormous value for determining source parameters as well as to investigate ground motions [e.g., 1, 2]. In recent years, sophisticated new methods have been developed to analyze intensity values quantitatively [e.g., 3, 4]. In these studies there is, or can be, a tendency to talk about

CP1020, *2008 Seismic Engineering Conference Commemorating the 1908 Messina and Reggio Calabria Earthquake*,
edited by A. Santini and N. Moraci
2008 American Institute of Physics 978-0-7354-0542-4/08/$23.00

"intensity data," a term that reflects a lack of appreciation for the nature of intensity values, and for key issues associated with their determination.

The interpretation of intensity values is moreover a critical step in its own right. Modern intensity scales take building vulnerability into account (see discussion in [2]), but values cannot be assigned without an appreciation of the historical context of each earthquake, including local building styles. Values also cannot be assigned without careful consideration of individual intensity indicators, some of which (for example, liquefaction and rockslides) have been recognized to be very poor indicators of overall shaking severity. Finally, any interpretation of macroseismic data must consider carefully the veracity and possible biases associated with archival sources, for example the fact that news accounts focus on dramatic rather than representative damage [e.g., 5].

Many early intensity assignments have been shown to be inappropriately high [e.g., 6] and to yield inflated estimates of magnitude. However, it is neither expected nor plausible that every historical earthquake is smaller than early studies suggest. I report here on a case in which careful reinterpretation reveals that the magnitude of an important earthquake, the Owen's Valley, California, earthquake of 26 March 1872, has generally been underestimated in earlier studies.

THE OWENS VALLEY EARTHQUAKE

The Owens Valley earthquake (hereinafter OV1872) occurred along the eastern flank of the Sierra Nevada range at approximately 2:30 in the morning, local time, on 26 March, 1872. By 1870 the population of the state of California had grown to over 560,000, and a number of mining settlements, including Independence, Lone Pine, and Bishop Creek, had been established along Owens Valley (Figure 1a.) Population was sparse in the Owens Valley region at the time of the earthquake. For example, there were only about 500 voters (e.g., male citizens of any race, 21 years old and older) scattered between a half-dozen principle settlements [7].

The Owens Valley earthquake generated a dramatic surface rupture that was described crudely by Josiah Whitney [8, 9]. Over a century later the surface rupture was mapped by Beanland and Clark [10] (hereinafter BC94), who identified a break 90-100 km in length, with an average right-lateral slip of 6 m and a total oblique slip of 6.1 m. Later investigations [e.g., 11] conclude that the break extended approximately 17 km further south. Revisiting the contemporary account of Whitney [8], one finds that he describes "frequent cracks in the earth" as far south as Haiwee (Figure 1), and further notes as much as 4-5 ft of subsidence along the edge of Haiwee Meadows.

No surface break has been identified to the north of the northern terminus identified by BC94. However, early reports describe that the ground was pervasively cracked between Independence and Bishop Creek [12].

Instrumentally recorded background seismicity along the Owens Valley corridor reveals a striking gap that extends between Haiwee to the south and Bishop Creek to the north (Figure 1b). Thus both geological and seismological observations point to a significantly longer rupture than that mapped by BC94, 130 rather than 90-100 km. A further note is that the rupture was clearly complex, with mapped breaks on multiple strands, and so the straight-line distance between the endpoints might underestimate the true length of the rupture.

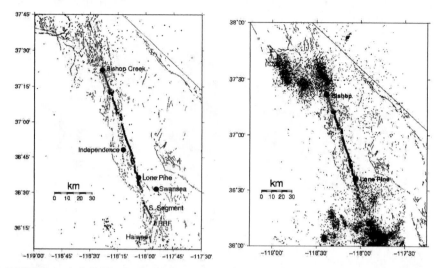

FIGURE 1. a) (left) The Owens Valley region and rupture of the 1872 earthquake, including break mapped by Beanland and Clark [10] (heavy black line) and possible extensions of the rupture (heavy grey lines.) b) (right) Instrumentally recorded earthquakes (small circles) between 1932 and 1990.

The depth extent of seismicity in the Owens Valley region is poorly constrained, with estimates ranging from 15 to 25 km. Taking a rupture length of 130 km, a depth of 20 km, and the average slip reported by BC94, one infers M_w 7.8. Considering the uncertainties in rupture parameters, one estimates a range of M_w 7.5-7.9.

Macroseismic Observations

As noted by authors in the early-to-mid 20th century [e.g., 12], the Owens Valley earthquake "has generally been considered the largest known in the entire California-Nevada region, thus placing it in magnitude above those of 1857 and 1906 on the San Andreas fault." Richter further noted that this assessment rests, "on the violence of effects over the large meizoseismal area, as well as perceptibility extending to great distances."

Archival accounts of OV1872 were compiled and interpreted in an earlier study [13]. Based on accounts that the earthquake stopped clocks and awakened many or most people throughout most of California, Toppozada *et al.* [13] assigns Modified Mercalli Intensity values of at least V throughout all but the northernmost 1/3 of the state. It is now recognized, however, that the long-period waves from large regional earthquakes can stop pendulum clocks at intensities much lower than V [*e.g.*, 14]. Further, data from the Community Internet Intensity Map website [15] reveal that during large earthquakes, intensities of III-IV are generally strong enough to awaken many or most people.

In the Owens Valley region, intensities are difficult to assign with precision because most of the damage was to buildings of weak masonry (generally adobe) construction. However, descriptions and a few photographs (e.g., Figure 3) suggest MMI values were generally no higher than VIII in the near-field region.

FIGURE 3. a) (left) Collapsed adobe structure flanked by two wod-framed structures that escaped serious damage. The location of the photograph is not identified, but is probably Independence. (Photo reprinted courtesy of Laws Railroad Museum, Laws, California). b) (right)The Edwards House in Independence, built in 1868, serves as an example of a well-built wood frame house of the era.

One intriguing account describes the chimney of a writing lamp being thrown straight up into the air and landing upright on the desk, while the lamp fell over, spilling hot oil (*Inyo Independent*, 6 April 1872.) While this suggests vertical acceleration in excess of 1g, the worst damage was concentrated in Lone Pine, where almost all structures were of weak adobe construction. One thus does not have a basis for assigning intensity values higher than VIII.

Although intensities in the near-field are typically difficult to estimate precisely, accounts from more distant locations can be equally if not more important for estimating magnitude. As noted, for example, the shaking was strong enough to awaken many or most sleepers at 2:30 in the morning throughout much of California, at nearest-fault distances of 300-500 km. At one location (Visalia) approximately 100 km west of the rupture, shaking was strong enough to throw items off of store shelves and damage some brick buildings. At another location (Chico) approximately 400 km northwest of the rupture, brick walls were cracked. Even interpreting such accounts conservatively, one cannot reasonably lower MMI estimates too much.

The accounts of OV1872 can furthermore be compared to those of the 1906 San Francisco earthquake, which also struck at a time of day when most people were asleep. Accounts of the 1906 earthquake were recently reinterpreted by Boatwright and Bundock [15]. As a first step, these assignments were confirmed to have been consistent with those of this study. Comparing the MMI values for the two earthquakes, one finds that intensities for OV1872 are systematically higher than those of the 1906 earthquake at regional distances (Figure 4.)

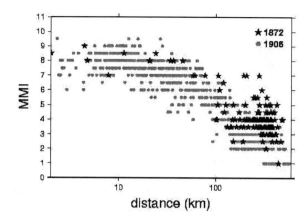

distance (km)

FIGURE 4. MMI values for the Owens Valley earthquake (black stars) and the 1906 San Francisco earthquake (grey circles) as estimated by Boatwright and Bundock [15].

I conclude that, while intensity assignments are inevitably uncertain to some degree, it is a robust conclusion that shaking intensities were systematically higher at regional distances in OV1872 than the 1906 earthquake. As G.K. Gilbert summarized in 1907, at a distance of only 20 miles from the San Andreas fault, "only an occasional chimney was over-turned." At 75 miles, the shock was observed by nearly all persons awake at the time; at 200 miles it was "perceived by only a few persons" [17]. The intensities assigned by Boatwright and Bundock [15] are consistent with Gilbert's descriptions. The rapid attenuation of strong shaking away from the fault is consistent with the new "Next-Generation Attenuation" (NGA) relations developed from strong motion data collected from large earthquakes around the world (see http://peer.berkeley.edu/products/nga_project.html for a description of the project and recent results.).

CONCLUSIONS

Reinterpretations of historical observations often yield lower magnitudes than earlier estimates, in large part because many (not all) early intensity assignments assigned higher values than what one would assign according to current practice. Indeed, the reinterpreted MMI values in this study are significantly lower than those assigned by Toppozada *et al.* [4]. However, these results confirm what was widely recognized early in the 20th century: the shaking effects of the Owens Valley earthquake were more dramatic at regional distances than those of the 1906 San Francisco earthquake. These results therefore suggest that, of the two, OV1872 was the larger event.

The magnitude of the 1906 San Francisco earthquake can be estimated from early instrumental seismic data and early geodetic data. Recent estimates have ranged from a low of M_s 7.7 [18] to as high as Mw 7.9 [19, 20]. A key question involves the length of the rupture, which geodetic observations suggest was longer than the break mapped following the earthquake, and the rupture length inferred from the seismic data. Song *et al.* [20] show that the seismic and geodetic data can be reconciled if the northernmost part of the earthquake had supershear rupture velocity. Clearly, however, the early seismic data have significant limitations, and so magnitude estimates remain uncertain.

The results of this study argue that OV1872 was larger than SF1906. If one assumes the magnitude is larger by at least 0.1 units, then, within the uncertainties of the magnitude estimates for both earthquakes, one is left with M_w 7.9 for OV1872 and M_w 7.8 for SF1906. An even larger magnitude for OV1872 is not inconsistent with available observations.

An alternative possibility is that the shaking from OV1872 was systematically higher for its magnitude than the shaking from SF1906. It is possible that OV1872 and SF1906 are representative of different classes of large earthquakes in California; respectively, they represent events on large, well-developed faults versus large events on relatively low-slip-rate faults. Sagy *et al.* [21] show that the surfaces of faults with low overall displacement are rougher than well-developed, high-slip faults. A number of earlier studies concluded that intraplate earthquakes, which generally occur on low-slip faults, have higher stress drops than interplate earthquakes [*e.g.*, 22]. The results of this study suggest that there might be important differences between stress drops of interplate earthquakes.

The results of this study further provide evidence that large earthquakes (i.e., close to M_w 8.0) in California, and indeed other regions, are not restricted to large, high-slip faults such as the San Andreas.

ACKNOWLEDGMENTS

I thank Karen Felzer, Jack Boatwright, Bill Bakun, Gary Fuis, Tousson Toppozada, Jim Evans, Andy Michael, and Brad Aagard for constructive feedback. I also gratefully acknowledge the hospitality of the Laws Railroad Museum and the Inyo County Historical Museum.

REFERENCES

1. N.N. Ambraseys and R. Bilham, Reevaluated intensities for the great Assam earthquake of 12 June 1897, Shillong, India, *Bull. Seismol. Soc. Am.*, **93**, 655-673 (2003)
2. N.N. Ambraseys and J. Douglas, Magnitude calibration of north Indian Earthquakes, *Geophys. J. Int.* **159**, 165-206 (2004).
3. W.H. Bakun and C.M. Wentworth, Estimating earthquake location and magnitude from seismic intensity data, *Bull. Seismol. Soc. Am.*, **87**, 1502-1521 (1997)
4. G.M. Atkinson and D.J. Wald, "Did You Feel It?" Intensity Data: A surprisingly good measure of earthquake ground motion, *Seismol. Res. Lett.* **78**, 362-368 (2007).

5. S.E. Hough and P. Pande, Quantifying the "media bias" in intensity surveys: lessons from the 2001 Bhuj, India earthquake, *Bull. Seismol. Soc. Am.* **97**, 637-645 (2007).
6. S.E. Hough, L. Seeber, J.G. Armbruster, and J.F. Hough, On the modified Mercalli intensities and magnitudes of the 1811-1812 New Madrid, central United States earthquakes, *J. Geophys. Res.*, **105**, 23,839-23,864 (2000).
7. W.A. Chalfant, *The Story of Inyo*, Bishop: Community Printing and Publishing, 1975, pp 239-249.
8. J.D. Whitney, *The Owens Valley earthquake*, Part I, Vol. **IX**, 130-140, *Overland Monthly* (1872a).
9. J.D. Whitney, *The Owens Valley earthquake*, Part II, Vol. **IX**, 266-278, *Overland Monthly* (1872b).
10. S. Beanland and M. Clark, The Owens Valley fault zone, eastern California, and surface faulting associated with the 1872 earthquake, *U.S. Geol. Surv. Bull.* **1982** (1994).
11. E. Vittori, G.A. Carver, S. Jayko, A.M. Michetti, and D.B. Slemmons, Quaternary fault map of Owens Valley, eastern California, 16th INQUA Conference Program with Abstracts, 106 (2003).
12. E.S. Holden, List of recorded earthquakes in California, Lower California, Oregon, and Washington Territory, Sacramento: State printing office, 1887, pp. 59-60.
13. C.F. Richter, *Elementary Seismology*, W.H. Freeman, San Francisco, 1958, pp. 499-501.
14. T.R. Toppozada, C.R. Real, and D.L. Parke, Preparation of isoseismal maps and summaries of reported effects of pre-1900 California earthquakes, *California Division of Mines and Geology Open-File Rep. 81-11* (1981).
15. J. Boatwright and H. Bundock, Modified Mercalli intensity maps for the 1906 San Francisco earthquake plotted in ShakeMap format, in *U.S. Geol. Surv. Open-File Rep. 2005-1135.*
16. D.J. Wald, V. Quitoriano, T.H. Heaton, and H. Kanamori, Relationships between peak ground acceleration, peak ground velocity, and modified Mercalli intensity in California, *Earthq. Spectra* **15**, 557-564 (1999).
17. G.K. Gilbert, The investigation of the California earthquake of 1906, in *The California Earthquake of 1906*, Jordan, D.S. ed., San Francisco: A.M. Robertson, 1907, pp. 215-256.
18. D.J. Wald, H. Kanamori, D.V. Helmberger, and T.H. Heaton, Source study of the 1906 San Francisco earthquake, *Bull. Seismol. Soc. Am.* **83**, 981-1019 (1993).
19. W. Thatcher, G. Marshall, and M. Lisowski, Resolution of fault slip along the 470-km-long rupture of the great 1906 San Francisco earthquake and its implications, *J. Geophys. Res.* **103**, 5353-5367, (1997).
20. S.G. Song,, G.C. Beroza, and P. Segall, A unified source model for the 1906 San Francisco earthquake, in press, *Bull. Seismol. Soc. Am.* **98** (2008).
21. A. Sagy, E.E. Brodsky, and G.J. Axen, Evolution of fault-surface roughness with slip, *Geology* **35**, 283-286, 2007.
22. C.H. Scholz, The mechanics of earthquakes and faulting, 2nd edition, New York: Cambridge Univ. Press, 2002, pp. 206-212.

Sensitivity of Base-Isolated Systems to Ground Motion Characteristics: A Stochastic Approach

Yavuz Kaya[a] *and* Erdal Safak[b]

[a,b] *Kandilli Observatory and Earthquake Research Institute, Cengelkoy Uskudar*
Bogazici University, Istanbul, Turkey

Abstract. Base isolators dissipate energy through their nonlinear behavior when subjected to earthquake-induced loads. A widely used base isolation system for structures involves installing lead-rubber bearings (LRB) at the foundation level. The force-deformation behavior of LRB isolators can be modeled by a bilinear hysteretic model. This paper investigates the effects of ground motion characteristics on the response of bilinear hysteretic oscillators by using a stochastic approach. Ground shaking is characterized by its power spectral density function (PSDF), which includes corner frequency, seismic moment, moment magnitude, and site effects as its parameters. The PSDF of the oscillator response is calculated by using the equivalent-linearization techniques of random vibration theory for hysteretic nonlinear systems. Knowing the PSDF of the response, we can calculate the mean square and the expected maximum response spectra for a range of natural periods and ductility values. The results show that moment magnitude is a critical factor determining the response. Site effects do not seem to have a significant influence.

Keywords: Base isolation, Ground Motion Characteristics, Random Vibration, Expected Maximum Response, Site effects

INTRODUCTION

Lead-Rubber-Bearing (LRB) type isolators are one of the most commonly used base isolators in practice. The force-deformation behavior of an LRB isolator under earthquake loads can be approximated by a bilinear hysteretic oscillator, as schematically shown in Fig.1. Dynamic response of such oscillators under base accelerations are sensitive to input characteristics, and can change significantly depending and the time and frequency domain parameters of the input. Rather than studying the response for specified input time histories, it is more convenient to investigate the statistical characteristics of the response directly for a given statistical description of ground shaking. This can be accomplished by using the methods of random vibration theory for nonlinear systems.

In this paper, we utilize a frequency-domain description of ground accelerations, and present equations to calculate inelastic response spectra by using equivalent-linearization techniques of random vibration theory for hysteretic nonlinear systems. Ground accelerations are characterized by its power spectral density function (PSDF), which includes corner frequency, seismic moment, moment magnitude, and site effects as its parameters. The PSDF of the oscillator response, as well as the mean square and the

CP1020, *2008 Seismic Engineering Conference Commemorating the 1908 Messina and Reggio Calabria Earthquake*,
edited by A. Santini and N. Moraci
© 2008 American Institute of Physics 978-0-7354-0542-4/08/$23.00

expected maximum response spectra are calculated for a range of natural frequencies and ductility values.

STOCHASTIC DESCRIPTION OF GROUND ACCELERATIONS

We will utilize frequency-domain descriptions of ground acceleration to characterize its stochastic description. One of the simple frequency-domain models available in the literature is the one suggested by Anderson and Hough (1984). It is defined by the following equation for the Fourier Amplitude Spectra (FAS) of ground accelerations:

$$F_a(f) = C \cdot \frac{(2\pi f)^2}{1 + \left(\dfrac{f}{f_c}\right)^2} \cdot e^{-\pi \kappa f} \qquad (2.1)$$

where:

$F_a(f)$: Fourier amplitude spectra of ground accelerations

κ: Attenuation factor

f_c: Corner frequency of the earthquake

f: Frequency in Hz

C: Scaling constant

The attenuation factor κ accounts for the fall-off in the shaking energy at high frequencies, as well as site amplification. It can be shown that the decay of resonant peaks due to site amplification can be modeled by an exponentially decaying function of frequency (Safak, 1995). For example, for a single soil layer over bedrock, the envelope of resonant peaks in the frequency response function of the soil layer can be approximated by the following equation:

$$E(f) \approx \frac{1+r}{1-r} e^{-\beta f}$$

with

$$\beta = \frac{\pi \tau}{Q} + 4\tau \ln\left(\frac{1 - re^{-\pi/2Q}}{1-r}\right) \qquad (2.2)$$

$$r = \frac{(\rho V)_{rock} - (\rho V)_{soil}}{(\rho V)_{rock} + (\rho V)_{soil}}$$

where ρ and V denote mass density and the shear wave velocity, respectively, r is the reflection coefficient for up-going waves at the bedrock-soil interface, and Q is the quality factor representing the damping in the soil layer. More detail on the derivations of the equations are given in Safak (1995).

The corner frequency f_c accounts for the magnitude of the earthquake. It is related to the seismic moment, M_0, and the moment magnitude, M, by the following equations:

$$f_c = 4.9 \times 10^6 \, \beta (\Delta\sigma / M_0)^{1/3} \quad \text{with} \quad M = \frac{2}{3} \log_{10} M_0 - 10.7$$

where

β = rupture velocity in km/s.   (2.3)

$\Delta\sigma$ = average stress drop in $bars$ in the fault plane

M_0 = seismic moment in $dyne\text{-}cm$

M = Moment magnitude of the earthquake.

The probabilistic description of ground accelerations in frequency domain can be represented by its Power Spectral Density Function (PSDF), $S_a(f)$. The frequency variation of $S_a(f)$ is correlated to the square of Fourier amplitude spectra, that is

$$S_a(f) \propto F_a^2(f) \tag{2.4}$$

Therefore, we can represent $S_a(f)$ by the following expression:

$$S_a(f) = S_0 \cdot \frac{(2\pi f)^4}{\left[1 + \left(\dfrac{f}{f_c}\right)^2\right]^2} \cdot e^{-2\pi\kappa f} \tag{2.5}$$

S_0 is the scaling constant and can be determined in terms of a specified measure of ground shaking. For example, if ground shaking is specified in terms of its *mean-square* acceleration value, σ_a, we can calculate S_0 from the condition that the area under the PSDF is equal to the mean-square value, that is

$$\sigma_a^2 = \int_{f=0}^{\infty} S_a(f) \cdot df \tag{2.6}$$

S_0 can also be determined by specifying other parameters of ground shaking (e.g., peak acceleration, peak velocity, etc.) and using their relation to $S_a(f)$ and its moments. These details can be found in Safak (1988).

STOCHASTIC RESPONSE OF BILINEAR HYSTERETIC OSCILLATOR

A bilinear hysteretic behavior is equivalent to the response of a linear spring-dashpot and a Coulomb slider connected in parallel. The system and its force-deformation relationship are schematically shown in Fig. 1. The equation for the dynamic response, $x(t)$, of the oscillator relative to its base, when subjected to base accelerations can be written as

$$m \cdot \ddot{x}(t) + c \cdot \dot{x}(t) + k_2 \cdot x(t) + g(x, \dot{x}) = -m \cdot a(t) \qquad (3.1)$$

where m, c, and k_2 are the mass, damping, and stiffness of the linear component of the oscillator, $g(x, \dot{x})$ is the nonlinear hysteretic restoring force, and $a(t)$ denote the ground accelerations. There is no exact solution of this equation for stochastic $a(t)$. One of the classical techniques used in random vibration theory to solve such equations is the equivalent linearization technique (e.g., Lutes and Sarkani, 1997). The equivalent linearization technique that will be used here is based on the equivalency of energy dissipation. The area under the hysteresis loop represents the energy dissipated by the bilinear system during one cycle of motion. We seek, by adjusting the stiffness and damping, an equivalent linear system that will dissipate the same amount of energy in one cycle. This can be accomplished by representing the nonlinear hysteretic force as a linear function of oscillator displacement and velocity, that is

$$g(x, \dot{x}) = b_0 + b_1 \cdot x(t) + b_2 \cdot \dot{x}(t) \qquad (3.2)$$

With the linearization, the equation of vibration becomes

$$m \cdot \ddot{x}(t) + (c + b_2) \cdot \dot{x}(t) + (k_2 + b_1) \cdot x(t) = -m \cdot a(t) \qquad (3.3)$$

Therefore, the frequency, f_a, and the damping, ξ_a, of the equivalent linear oscillator are

$$f_a = \frac{1}{2\pi} \sqrt{\frac{k_2 + b_1}{m}} \quad \text{and} \quad \xi_a = \frac{c + b_2}{2m\omega_a} \qquad (3.4)$$

The coefficients b_0, b_1, and b_2 are determined by minimizing the mean-square value of the energy difference between the bilinear system and the equivalent elastic system. Assuming that $x(t)$ and $\dot{x}(t)$ are zero-mean and jointly Gaussian random variables, the minimization results in the following expressions for b_0, b_1, and b_2:

$$b_0 = 0; \quad b_1 = \frac{E\left[x(t) \cdot g[x(t), \dot{x}(t)]\right]}{E\left[x^2(t)\right]}; \quad b_2 = \frac{E\left[\dot{x}(t) \cdot g[x(t), \dot{x}(t)]\right]}{E\left[\dot{x}(t)\right]} \qquad (3.5)$$

where $E[..]$ denotes ensemble average.

To determine b_1 and b_2, we assume that the response $x(t)$ is a narrow-band process of the following form:

$$x(t) = A(t) \cdot \cos\left[2\pi f_a \cdot t + \theta(t)\right] \tag{3.6}$$

where $A(t)$ and $\theta(t)$ are independent and slowly varying random processes, and $\theta(t)$ is distributed uniformly between 0 and 2π. With these assumptions, following equations can be developed to calculate b_1 and b_2 (the detail of the derivations can be found in Lutes and Sarkani, 1997):

$$b_1 = k_1 \cdot \left[1 - \frac{8}{\pi} \int\limits_{z=1}^{\infty} \left(\frac{1}{z^3} + \frac{x_y^2}{2\sigma_x^2 z} \right) (z-1)^{1/2} \exp\left(-\frac{x_y^2 z^2}{2\sigma_x^2} \right) dz \right] \tag{3.7}$$

$$b_2 = \left(\frac{2}{\pi}\right)^{1/2} \cdot \frac{k_1 x_y}{2\pi f_a \sigma_x} \cdot \left[1 - \Phi(x_y / \sigma_x) \right]$$

where x_y is the yield displacement of the oscillator as shown in Fig. 1, and $\Phi(\cdots)$ denotes the cumulative distribution function of a standardized Gaussian random variable.

Once b_1 and b_2 are determined, we can calculate the PSDF, $S_x(f)$, of the response from the random vibration theory as

$$S_x(f) = \left| H(f) \right|^2 \cdot S_a(f) \tag{3.8}$$

where $|H(f)|^2$ is the frequency-response function of the equivalent oscillator, defined by

$$\left| H(f) \right|^2 = \frac{1}{(2\pi)^4 \left[\left(f_a^2 - f^2 \right)^2 + \left(2\xi_a f_a f \right)^2 \right]} \tag{3.9}$$

The expected mean-square response, σ_x^2, and the peak response, x_{max} can be calculated from the following equations:

$$\sigma_x^2 = \int\limits_{f=0}^{\infty} S_x(f) \cdot df \quad \text{and} \quad x_{max} = g \cdot \sigma_x \tag{3.10}$$

where g denotes the peak factor and can be approximated for narrow-band processes as (Davenport, 1964):

$$g \approx \sqrt{2\ln(2f_a T)} + \frac{0.577}{\sqrt{2\ln(2f_a T)}} \qquad (3.11)$$

where T denotes the duration of the earthquake.

NUMERICAL EXAMPLE

As an application of the formulation given above, we calculate the response spectra of a bilinear oscillator and investigate their sensitivity to various ground motion and oscillator parameters. The nonlinear behavior is defined by assuming $k_2/k_1=0.1$. The initial (i.e., for the linear behavior) damping ratio is taken as 5%, and the response spectra is calculated by varying the initial frequency from 0.1 Hz to 10 Hz. The variation of inelastic response spectra with ductility is investigated for three values of the mean-square ductility, $\mu = 2, 5, 10$, where μ is defined as $\mu = \sigma_x / x_y$.

For the ground motion, four values of the corner frequency, corresponding moment magnitudes $M = 5.0, 5.5, 6.5$, and 7.0, and three values of the high-frequency attenuation / site amplification coefficient, $\kappa = 0.01, 0.04, 0.05, 0.07$, are considered. The transient characteristics of the ground motion is modeled by a box-car time window of 20 sec. duration.

The results are presented in Figs.2 through Fig.4. When varying a parameter, all other parameters are kept constant. We plot the response spectra against the elastic period, as commonly done, not against the frequency. Fig.2 shows the variation of response spectra with ductility. The specified ductility values are matched by changing the yield displacement. Fig.3 shows the variation of response spectra with moment magnitude, and Fig.4 with spectral decay factor κ. The results show that moment magnitude is the most critical factor determining the response. The parameter κ does not seem to have a significant influence.

CONCLUSIONS

The force-deformation behavior of lead-rubber bearing (LRB) isolators can be modeled by a bilinear hysteretic oscillator. Dynamic response of such oscillators under base accelerations are sensitive to input characteristics, and can change significantly depending and the time and frequency domain parameters of the input. Rather than studying the response for specified input time histories, it is more convenient to investigate the statistical characteristics of the response directly for given statistical descriptions of ground shaking. This can be accomplished by using the equivalent-linearization techniques of random vibration theory for hysteretic nonlinear systems. For a given stochastic description of ground motions, the stochastic description of the nonlinear response is obtained in terms of power spectral density functions, which are then used to calculate the mean-square and expected maximum responses. Numerical examples show that the response spectra of bilinear systems are sensitive to moment magnitude, but not the high-frequency decay and site amplification.

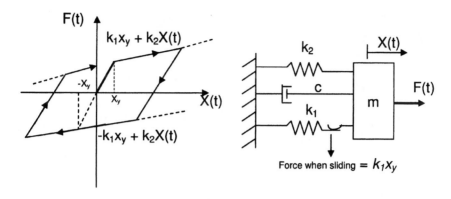

FIGURE 1. Schematic representation and the parameters of the bilinear hysteretic system considered to model rubber-lead bearing isolators

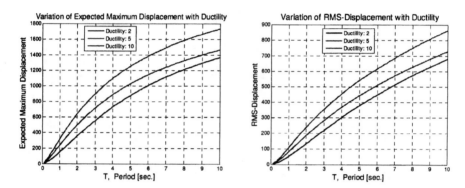

FIGURE 2. Variation of mean-square and expected maximum response spectra with mean-square ductility ($k_2 / k_1 = 0..01$, $M=7.0$, $\kappa=0.07$)

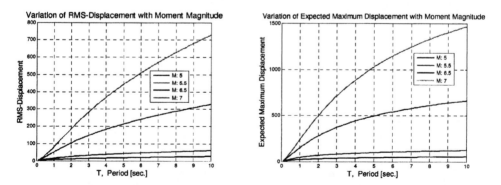

FIGURE 3. Variation of mean-square and expected maximum response spectra with moment magnitude ($k_2 / k_1 = 0..01$, $\mu=5$, $\kappa=0.07$)

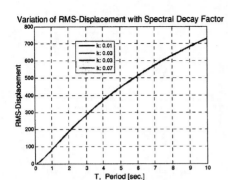

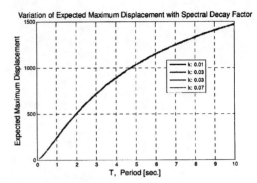

FIGURE 4. Variation of mean-square and expected maximum response spectra with spectral decay factor κ ($k_2 / k_1 = 0..01$, M=7.0, $\mu=5$)

REFERENCES

1. Anderson, J.G. and S. Hough (1984). A model for the shape of the Fourier amplitude spectrum of acceleration at high frequencies, *Bulletin of Seismological Society of America*, 74, 1969-1993.
2. Davenport, A.G. (1964). Note on the distribution of the largest value of a random function with application to gust loading, *Proc. Inst. Civil Eng.*, 28, 187-196.
3. Lutes, L.D. and S. Sarkani (1997). *Stochastic Analysis of Structural and Mechanical Vibrations*, Prentice-Hall Inc., Upper Saddle River, NJ.
4. Safak, E. (1988). Analytical approach to calculation of response spectra from seismological models of ground motion, *Earthquake Engineering & Structural Dynamics*, Wiley Inter-Science, Vol.16, No.1, January 1988, pp.121-134.
5. Safak, E. (1995). Discrete-time analysis of seismic site amplification, *Journal of Engineering Mechanics*, ASCE, Vol.121, No.7, July 1995, pp.801-809.

Predictive Equations to Estimate Arias Intensity and Cumulative Absolute Velocity As A Function of Housner Intensity

J. Enrique Martinez-Rueda[a] , Giannis Moutsokapas[a] and
Evdokia Tsantali[a]

[a] Civil Engineering and Geology Division, School of Environment and Technology,
University of Brighton, Lewes Rd, Cockcroft Bldg, Brighton BN24GJ, UK.
Email: jem11@bton.ac.uk

Abstract. This paper deals with the development of empirical relationships for the estimation of Arias Intensity and Cumulative Absolute Velocity as a function of Housner Intensity. These ground motion parameters are evaluated for a data set of natural accelerograms previously used in the calibration of attenuation equations in Europe and the Middle East. The family of the predictive equations proposed account for different combinations of seismic site (rock, stiff soil, or soft soil) and fault mechanism (normal, strike-slip, or thrust). Once the target Housner Intensity has been defined in frequency domain, the proposed equations can be applied to estimate the target instrumental intensity of the seismic input in time domain.

Keywords: Housner Intensity, Arias Intensity, Cumulative Absolute Velocity.

INTRODUCTION

In an attempt to define a numerical indicator of earthquake damage potential a number of ground motion parameters have been proposed in the past. Peak ground acceleration is now generally considered as a poor indicator of earthquake damage potential. Accordingly, the evolution of ground motion parameters has been motivated by the search of better predictors of damage. And this has lead to a collection of indicators of damage that account for duration, amplitude and frequency content of the seismic input. Because some of these parameters have practical application in the definition of seismic input or in the estimation of damage potential, studies on the derivation of attenuation relationships for ground motion parameters have began to gradually appear in the literature [*e.g.* 1,2].

A number of ground motion parameters belong to the time domain; hence they may have limited application, unless a direct link with seismic code design provisions can be established. Over the years the scaling of natural accelerograms has been adopted as a practical way to estimate the seismic input for nonlinear analysis [3-6]. Scaling criteria that can be directly associated with a design spectrum have a good balance between simplicity (in the definition of the scaling factor to match a property of the design spectrum) and applicability (favoring a rational use of scaled natural accelerograms).

CP1020, *2008 Seismic Engineering Conference Commemorating the 1908 Messina and Reggio Calabria Earthquake*,
edited by A. Santini and N. Moraci

The main objective of this paper is to derive empirical equations to estimate ground motion parameters of the time domain as a function of a ground motion parameter of the frequency domain. The proposed equations are assumed to be applicable for Europe and the Middle East as they are calibrated for ground motion recorded in this region.

DEVELOPMENT OF PREDICTIVE EQUATIONS TO ESTIMATE GROUND MOTION PARAMETERS

To visualize the interdependence between the frequency and the time domains in terms of ground motion parameters, a number of these were evaluated for an extensive dataset of natural accelerograms. Each pair of numerical values of the parameters under study is considered here as an observation. As explained in more detail below, empirical relationships were then obtained by identifying first trends in the observations, followed by the fitting of analytical expressions by nonlinear regression.

Ground Motion Parameters

The ground motion parameters selected for the study were the Housner Intensity SI_H, the Arias Intensity I_A and the Cumulative Absolute Velocity CAV. The adequacy of these parameters to assess both instrumental earthquake intensity and earthquake damage potential is well documented in the literature as summarized below.

Housner Intensity SI_H

This parameter is defined in frequency domain (period domain more precisely). It was introduced by Housner [7] to measure the instrumental intensity of an earthquake at a given site. Housner Intensity is defined as the mean value of the elastic velocity spectrum between the periods 0.1 and 2.5 secs, in mathematical form:

$$SI_H = \frac{1}{2.4} \int_{0.1}^{2.5} SV(T,\xi)\,dT \qquad (1)$$

where $SV(T,\xi)$ is the spectrum velocity curve, T is the natural period of a SDOF system and ξ is the damping ratio of the system.

It has been shown that there is a good correlation between Housner Intensity and displacement ductility demand [5]; particularly for structures with fundamental period greater than 0.2 secs [6]. This correlation is meaningful as ductility demand is an effective damage index that characterizes damage potential once ductility capacity is has been estimated.

Arias Intensity I_A

This is defined in time domain and it was introduced by Arias [8] to define the damage potential of earthquake records. In mathematical form I_A is given as:

$$I_A = \frac{\pi}{2g} \int_0^{t_d} \ddot{u}_g^2(t)\, dt \qquad (2)$$

where $\ddot{u}_g(t)$ is the ground acceleration at a given time t and t_d is the total duration of the ground motion.

In terms of damage potential assessment I_A appears to be of limited application for structural design as it correlates well with seismic demands but mainly for short period structures; however, I_A finds application in geotechnical problems such us the estimation of liquefaction potential and slope instability triggered by earthquake ground motion [1]

Cumulative Absolute Velocity (CAV)

Defined as the area under the absolute ground acceleration curve this parameter of the time domain is given as:

$$CAV = \int_0^{t_d} \left| \ddot{u}_g(t) \right| dt \qquad (3)$$

CAV was originally introduced by Kennedy and Reed [9] to set an exceedance criterion for the operating basis earthquake of nuclear power plants.

Earthquake Ground Motion

The strong motion data set used here consists of 866 accelerograms of horizontal ground motion and were downloaded from the *European Database of Strong Motion Data Site* [10]. Table 1 summarizes the main characteristics of the strong motion data set. The seismological parameters of the accelerograms including site-to-source distance d, moment magnitude M_w, seismic site and fault mechanism are those reported by Ambraseys *et al.* [11] in connection with the calibration of empirical equations to predict spectral acceleration in Europe and the Middle East.

TABLE 1. Summary of properties of the strong motion database used in this study

property	Range	Mean	S.D.
d [km]	[0; 100]	33.58	25.37
M_w	[5: 7.6]	5.90	0.69
PGA [m/sec^2]	[0.01; 10.81]	0.97	1.15
SI_H [m/sec]	[7E-4; 1.38]	0.11	0.16
I_A [m/sec]	[1E-5; 15.49]	0.24	0.93
CAV [m/sec]	[0.013; 34.82]	2.57	3.42

In terms of seismic site, the number of records classified according to the categories rock, stiff soil and soft soil are: 242, 376 and 248, respectively. However, for the current study the soils classified in [11] as soft soil and very soft soil were considered as a single category denoted here as soft soil. In terms of fault mechanism, the number of records classified according to the categories normal, strike-slip, thrust and odd are: 310, 162, 152 and 242, respectively. Note that the category odd corresponds to fault mechanisms reported in [11] as mixed or uncertain. Therefore, while studying the influence of fault mechanism all the accelerograms within the 'category' odd were ignored.

Correlation between ground motion parameters

The degree of association between ground motion parameters was first assessed graphically by producing scatter plots of the observations. These plots showed the observed relationships between Housner intensity and either Arias intensity or Cumulative absolute velocity. Different types of observed relationships were plotted for different combinations of seismic site and faulting mechanism. This allowed the study of the influence of seismological parameters on the degree of association between ground motion parameters. In general, it was observed that the overall trend of the observed relationships between ground motion parameters was nonlinear. To derive predictive equations to estimate either I_A or CAV as a function of SI_H the observations were fitted to the power equations:

$$I_A = A_1 \left(SI_H \right)^{A_2} \tag{4}$$

$$CAV = B_1 \left(SI_H \right)^{B_2} \tag{5}$$

where A_1, A_2, B_1 & B_2 are fitting constants.

The fitting model used in equations (4) and (5) has a good balance between simplicity and accuracy as it realistically predicts nil values of I_A or CAV when $SI_H = 0$. The goodness of fit of the predictive equations (4) and (5), as well as, the degree of correlation between pairs of ground motion parameters was assessed by the coefficient of determination R^2.

The power eqs. (4) & (5) were adopted as an alternative to the truncated second-degree polynomials ($y = C_1 x + C_2 x^2$) used by the authors in a previous study [12].

This past study relied on a reduced data set of 478 accelerograms recorded in the Greco-Italian region. For the current study, preliminary comparisons with linear as well as the truncated polynomials indicated that equations (4) and (5) fitted better the observations as indicated by consistently higher R^2 values. No attempt was made to fictitiously reduce the scatter of the observations by working with the logarithm of the ground motion parameters.

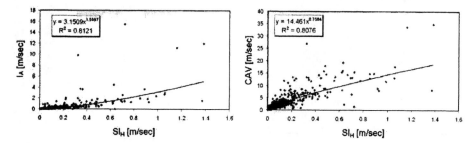

FIGURE 1. Observed relationships between ground motion parameters for the whole of the strong motion data set.

Figure 1 shows an example of observed relationships between ground motion parameters as well as the calibrated predictive equations ($x = SI_H$ & $y = I_A$ or CAV). In this example all accelerograms are being analyzed as a single group; hence the influence of seismic site or fault mechanism is neglected. Despite this limitation a good correlation is observed between the ground motion parameters. Figure 2 illustrates the correlation between ground motion parameters when the fault mechanism is accounted for. There are changes in the R^2 values but these do not appear to be drastic.

TABLE 2. Fitting constants and R^2 associated with equation (4)

Case No.	Seismic Site	Fault Mechanism	N	A_1	A_2	R^2
1	All combined	All combined	866	3.1509	1.5097	0.8121
2	Rock	All combined	242	5.4407	1.5961	0.7826
3	Stiff soil	All combined	376	3.6825	1.5630	0.8346
4	Soft	All combined	248	2.1075	1.4565	0.8353
5	All combined	Normal	310	3.7777	1.5497	0.7739
6	All combined	Strike-slip	162	2.6399	1.464	0.8703
7	All combined	Thrust	152	2.8177	1.3488	0.7610
8	Rock	Normal	102	4.9855	1.485	0.6792
9	Rock	Strike-slip	50	2.8049	1.5231	0.7489
10	Rock	Thrust	40	5.2619	1.608	0.9341
11	Stiff soil	Normal	136	4.6458	1.6595	0.8270
12	Stiff soil	Strike-slip	62	2.7753	1.4558	0.9091
13	Stiff soil	Thrust	86	2.9953	1.3493	0.7797
14	Soft soil	Normal	72	2.3441	1.5473	0.8918
15	Soft soil	Strike-slip	50	2.189	1.3669	0.9192
16	Soft soil	Thrust	26	0.8571	0.891	0.4165

313

TABLE 3. Summary of fitting constants and R^2 values associated with equation (5)

Case No.	Seismic Site	Fault Mechanism	N	B_1	B_2	R^2
1	All combined	All combined	866	14.461	0.7684	0.8076
2	Rock	All combined	242	16.236	0.7908	0.7484
3	Stiff soil	All combined	376	14.835	0.7819	0.8356
4	Soft	All combined	248	13.346	0.7468	0.8158
5	All combined	Normal	310	15.177	0.7682	0.8261
6	All combined	Strike-slip	162	15.832	0.7925	0.8426
7	All combined	Thrust	152	15.089	0.7922	0.7488
8	Rock	Normal	102	19.268	0.7905	0.7857
9	Rock	Strike-slip	50	7.8761	0.652	0.6591
10	Rock	Thrust	40	21.18	0.9117	0.8929
11	Stiff soil	Normal	136	14.108	0.7688	0.8251
12	Stiff soil	Strike-slip	62	16.678	0.803	0.8463
13	Stiff soil	Thrust	86	15.339	0.779	0.831
14	Soft soil	Normal	72	14.652	0.7908	0.9242
15	Soft soil	Strike-slip	50	18.699	0.7978	0.9298
16	Soft soil	Thrust	26	8.8044	0.6435	0.3872

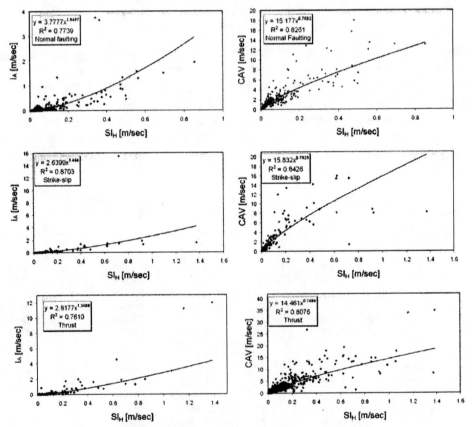

FIGURE 2. Observed relationships between ground motion parameters when the influence of the fault mechanism is taken into account.

Tables 2 and 3 summarize the coefficients of the predictive eqs. (4) and (5) for different combinations of seismic site and fault mechanism. In general there seems to be a good correlation between the parameters under study, except for case 16 (ground motion recorded on soft soil due to thrust faulting) where R^2 is rather low; however R^2 in this case relies on 26 records only (*i.e.* the smallest number of records from all the 16 case studies). Hence, the collection of a larger number of records for case 16 to arrive at more reliable predictive equations is recommended for further studies.

Each case study i in Tables 2 and 3 is associated with a data subset with a different number of observations N_i. Hence, to compare the influence of seismic site and fault mechanism on the correlation between ground motion parameters a weighted average value $\overline{R}^2$ of coefficient of determination was defined as:

$$\overline{R}^2 = \frac{\sum N_i R_i^2}{\sum N_i} \tag{6}$$

where R_i^2 is the coefficient of determination of the nonlinear regression under study.

TABLE 4. Summary of $\overline{R}^2$ values of the correlation study between ground motion parameters accounting for the influence of different seismological parameters

Seismological parameters considered	SI_H vs. I_A	SI_H vs. CAV
None	0.8121	0.8076
Seismic site	0.8027	0.8056
Fault mechanism	0.7958	0.8116
Seismic site and Fault mechanism	0.8028	0.7760

Table 4 summarizes the $\overline{R}^2$ values for different combinations of the seismic parameters considered in the study of the correlation between ground motion parameters. In general, this table suggests that on the average the goodness of fit of the empirical relationships is not significantly different when seismological parameters are taken into account. It is also evident that, in practical terms, the degree of correlation between SI_H & I_A and between SI_H & CAV is very similar. This suggests that one could use indistinctly either eq. (4) or eq. (5) to estimate the intensity in time domain of the seismic input for a given target Housner Intensity.

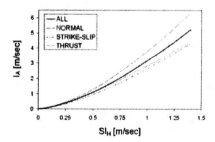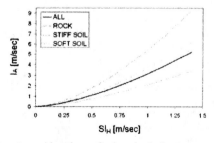

FIGURE 4. Predicted Arias Intensity for different considerations of seismological parameters.

Figure 4 shows the prediction of I_A when the seismic site or the fault mechanism is accounted for. The plots reveal the influence of seismological parameters on the estimation of instrumental intensity in time domain. For instance, if one neglects the seismic site influence, then the ground motion intensity generated by normal faulting can be underestimated. On the other hand, if one neglects the fault mechanism influence, then the intensity predicted on rock can be largely underestimated.

CONCLUDING REMARKS

This paper introduced empirical SI_H vs. I_A and SI_H vs. CAV relationships for ground motion recorded in Europe and the Middle East. Predictive equations for these relationships were proposed. The influence of seismological parameters on the fitting constants and on the degree of correlation between ground motion parameters was accounted for. Overall comparisons of the goodness of fit of the predictive equations revealed that the degree of correlation between the ground motion parameters is not significantly influenced by the seismological parameters. On the other hand, the predictive equations confirm the fact that a more reliable estimation of the seismic input in time domain is obtained when the seismological parameters are accounted for.

REFERENCES

1. T. Travasarou, D.J. Bray & N.A.Abrahamson, "Empirical attenuation relationship for Arias intensity", *Earthquake Engineering and Structural Dynamics*, **32**, 2003, pp. 1133-1155.
2. J.E. Martinez-Rueda, "Proposal of an attenuation relationship of Housner spectrum intensity in Europe", *Proceedings of the First European Conference on Earthquake Engineering and Seismology*, Geneva, 2006, paper 1193.
3. S.K. Ghosh and M. Fintel, "Explicit inelastic design procedure for aseismic structures", *ACI Journal*, **79**, 1982, p.p. 110-118.
4. A.J. Kappos, "Analytical prediction of the collapse earthquake for R/C buildings: suggested methodology", *Earthquake Engineering and Structural Dynamics*, **20**, 1991, p.p. 167-176.
5. J.E. Martinez-Rueda, "Scaling procedure for natural accelerograms based on a system of spectrum intensity scales", *Earthquake Spectra*, **14**, 1998, pp. 135-152.
6. J.E. Martinez-Rueda, "Proposal of a system of spectrum intensity scales for the scaling of natural accelerograms accounting for hysteretic behaviour and local site conditions. A new system and its application on displacement-based design", *Proceedings of the First European Conference on Earthquake Engineering and Seismology*, Geneva, 2006,paper 1196.
7. G.W. Housner, "Spectrum intensities of strong motion earthquakes", *Proceedings of Symposium on Earthquake and Blunt Effects on Structures*, 1952, pp. 20-36.
8. A. Arias, "A measure of earthquake intensity", in Hansen, R.J.(ed.), in *Seismic Design of Nuclear Reactors*, MIT Press, 1969, pp. 438-483.
9. EPRI, "A criterion for determining exceedance of the operating earthquake", EPRI NP-5930, Electrical Power Research Institute, Palo Alto California, 1998.
10. www.isesd.cv.ic.ac.uk, *The European Strong Motion Database*.
11. N.N.Ambraseys, J. Douglas, J., K. Sarma, & P.M. Smit, P.M., "Equations for the Estimation of Strong Ground Motions from Shallow Crustal Earthquakes Using Data from European and Middle East", *Bulletin of Earthquake Engineering*, **3**, 2005, pp. 1-53.
12. J.E. Martinez-Rueda and E. Tsantali, "Analysis of the correlation between instrumental intensities of strong earthquake ground motion", paper submitted to 7th *European Conference on Structural Dynamics*, 2008.

Effects of Wave Propagation Direction on the Evaluation of Torsional Motion

Gholam Reza Nouri[a] and Heiner Igel[b]

[a] Department of Civil Engineering, University of Mohaghegh Ardebili, Ardebil, Iran,
r.nouri@iiees.ac.ir
[b] Department of Earth and Environmental Sciences, Section Geophysics, Ludwig-Maximilians-
Universität, Munchen, Germany

Abstract. The benefits of the determination of rotational motion in seismology and engineering are still under investigation. Four main approaches have been developed to incorporate the rotational motions in engineering and seismology: one is numerical simulation of radiation field from source mechanism. The second approach is based on theoretical formulation of spatial distribution of ground motion. The third approach is the application of recorded strong motion data from seismic arrays and the last one is using instruments to direct measurement of torsional component. Due to the slow development of instruments for direct measurement of rotational motion, dense arrays are one of the unique sources of experimental information on rotational motion estimation by the spatial derivatives. The average rotational motions can be evaluated from difference of two translational records in an array of stations on the ground. Usually the conventional spatial derivative of two accelerograms, are employing to estimate torsional ground motion in the array. In some cases this method were applied without attention to the effect of wave propagation direction on the rotational motion. In this study assuming simple plane wave propagation and variation of propagation angle, it was shown that direction of wave propagation has principal role in the evaluation of torsional motion. Results showed that based on using two or four stations and the arrangement of those stations, torsional motion has different values. This is very important issue in the evaluation of torsional motions that must be considered.
Keywords: rotational motion, wave direction, array data.

INTRODUCTION

The rotational part of earthquake-induced ground motion has basically been ignored in the past decades, compared to the substantial research in observing, processing and inverting translational ground motions, although their importance was understood [1-2]. Rotational ground motions can be evaluated and measured by four main approaches: the first one is numerical simulation of radiation field from source mechanism. The second approach is based on theoretical formulation of spatial distribution of ground motion. Application of recorded strong motion data from seismic arrays is the third approach [3] and the last one is using instruments to direct measurement of torsional component [4]. The interest in the observation of rotational ground motions has increased in recent years due to development of sensors and optical instruments such as ring laser gyros. Because of the slow development of direct measurements of rotational motions dense arrays data are one of the unique

CP1020, *2008 Seismic Engineering Conference Commemorating the 1908 Messina and Reggio Calabria Earthquake*,
edited by A. Santini and N. Moraci
© 2008 American Institute of Physics 978-0-7354-0542-4/08/$23.00

sources of experimental information on rotational motion estimation by the spatial derivatives. The average rotational motions can be approximated from the difference of two translational records in an array of stations on the ground. In some cases this method has been applied without attention to the wave propagation direction.

In this paper by applying simple model for plane wave propagation, effect of wave direction and stations arrangement in the assumed array were studied. For this, four stations with distance of Δx and Δy were considered the angle of wave propagation is varied form zero to 360 degrees and torsional motion using 2 or 4 stations were evaluated.

TORSIONAL MOTION

Consider a Cartesian system of reference, whose coordinates along axes are x, y and z, the latter being along the vertical axis, and let u, v, and w be the translation along these axes. The rotation around the vertical axis can be obtained by Eq.1 [3]:

$$\psi_z = \frac{1}{2}(\frac{\partial u}{\partial y} - \frac{\partial v}{\partial x})$$ (1)

Based on Eq. 1, the torsional ground acceleration is calculated for different events and pairs of stations with different separation distances. As explained in the introduction it is assumed that the plane wave is propagating with wave number of k and frequency ω that is given by Eq. 2: The angle of propagation direction and vertical axis is α, as shown in Figure 1. As shown in Figure 2, four stations with distances of Δx and Δy are assumed to evaluate torsional motion using translational ones. For three cases torsional motion using Eq. 1 was evaluated with the variation of α from zero to 360 degrees.

$$U = \cos(k_x x + k_y y - \omega t)$$ (2)

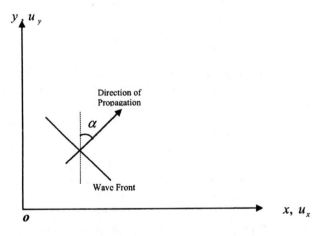

FIGURE 1. Simple Model of Plane Wave in the x, y Coordinates.

In the first case stations S1 and S2 were selected and torsional motion were evaluated with the variation of α as illustrated in Figure 3a. For the next step stations S3 and S4 have been used to estimation of torsional component that the result was shown in Figure 3b. Finally four stations of S1-S4 in the assumed array were used to estimation of torsional motion. As it is visible in the Figure 3, when two stations are used to evaluation of torsional motion the wave propagation direction affect torsional component strongly.

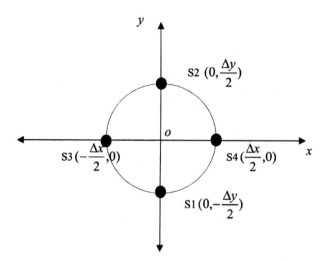

FIGURE 2. Assumed Array and Four Stations to Evaluation of Torsional Motion with Distance of Δx and Δy in $x - y$ Coordinates

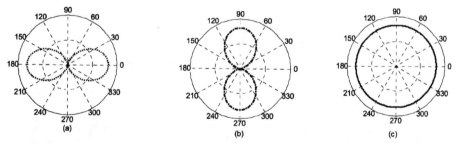

FIGURE 3. Evaluation of Torsional Motion with Variation of wave propagation direction α from 0 to 360 degrees (a) Using Stations S1 and S2, (b) Using Stations S3 and S4, (c)Using Stations S1-S4

CONCLUSION

Effect of wave propagation direction on the evaluation of torsional motion in an assumed array was studied. For this, simple plane wave propagation was applied and torsional motion with the variation of propagation angle from 0 to 360 degrees was computed. Assumed array has four stations that for three cases are used to computation of torsional motion. First stations S 1and S 2, second stations S 3 and S 4 and in last all of four stations selected to evaluation of torsional motion by Eq. 1. By comparing the results of three cases importance of wave propagation direction will be well-defined as shown if Figure 4.

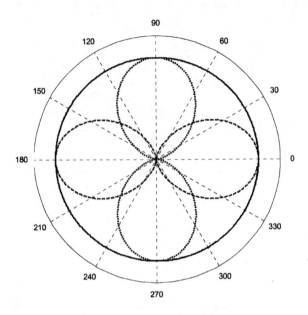

FIGURE 4. Torsional Motion with Variation of wave propagation direction α from 0 to 360 degrees Using Stations S1 and S2 (dash line), Stations S3 and S4 (doted line) and Stations S1-S4 (solid line).

REFERENCES

1. K. Aki and PG. Richards, *"Quantitative seismology"*, University Science Books, 2nd ed, Sausalito, CA (2002).
2. M. Takeo and HM. Ito, *"What can be learned from rotational motions excited by earthquakes"*, Geophys J Int 129: 319-329(1997)
3. M.R Ghayamghaminan and G.R Nouri, *"On the characteristics of ground motion rotational components using Chiba dense array data"*, Earthquake Engineering Structural Dynamics, DOI: 10.1002/eqe.687,(2007)
4. W. Suryanto, H. Igel, J.Wassermann, A. Cochard., B. Schuberth, D. Vollmer, F. Scherbaum, U. Schreiber, A. Velikoseltsev, *"First Comparison of Array-Derived Rotational Ground Motions with Direct Ring Laser Measurements"*, Bulletin of the Seismological Society of America, Vol. 96, No. 6,, December 2006, doi: 10.1785/0120060004

The Uncertainty in the Local Seismic Response Analysis

A. Pasculli[a], A. Pugliese[b], R.W. Romeo[c] and T. Sanò[d]

[a]*Dept. of Geotechnologies, University G. D'Annunzio, Chieti, Italy*
[b]*National Agency for the Environmental Protection, Rome, Italy*
[c]*GISLab, University Carlo Bo, Urbino, Italy*
[d]*Research Centre on Geological Risks, Valmontone, Italy*

Abstract. In the present paper is shown the influence on the local seismic response analysis exerted by considering dispersion and uncertainty in the seismic input as well as in the dynamic properties of soils. In a first attempt a 1D numerical model is developed accounting for both the aleatory nature of the input motion and the stochastic variability of the dynamic properties of soils. The seismic input is introduced in a non-conventional way through a power spectral density, for which an elastic response spectrum, derived – *for instance* – by a conventional seismic hazard analysis, is required with an appropriate level of reliability. The uncertainty in the geotechnical properties of soils are instead investigated through a well known simulation technique (Monte Carlo method) for the construction of statistical ensembles. The result of a conventional local seismic response analysis given by a deterministic elastic response spectrum is replaced, in our approach, by a set of statistical elastic response spectra, each one characterized by an appropriate level of probability to be reached or exceeded. The analyses have been carried out for a well documented real case-study. Lastly, we anticipate a 2D numerical analysis to investigate also the spatial variability of soil's properties.

Keywords: local seismic response; uncertainty; Monte Carlo method; response spectra.

INTRODUCTION

An old story says: your God is Jewish, your culture is Greek-Latin, your numbers are Arabic, ..., why is your neighbor foreign alone?.

Let's transpose the metaphor: earthquakes' sources are uncertain, their recurrence laws are also uncertain, waves' path and propagation are still uncertain, ..., may only local seismic response be certain?, sure no!.

Usually, in the local seismic response analysis (LSRA) the uncertainty is concentrated in the seismic hazard alone, as a summation of the aleatory and epistemic uncertainties in source and path effects. The same result, though intrinsically uncertain, is nevertheless used as a deterministic input (read *accelerometric time-history*) to be read by a soil column whose transfer function is itself deterministically defined in turn. As a result we forget the stochastic nature of the seismic action as well as the soil's properties uncertainty and the soil column seismo-stratigraphic variability. To account for these sources of uncertainty an approach and a new set of computer programs have been set forth. In a first approach they afford the problem to solve the wave equations using a one-dimensional equivalent linear-elastic analysis.

CP1020, *2008 Seismic Engineering Conference Commemorating the 1908 Messina and Reggio Calabria Earthquake,*
edited by A. Santini and N. Moraci
© 2008 American Institute of Physics 978-0-7354-0542-4/08/$23.00

Nevertheless, computational routines have been developed for the uncertainty treatment that can be used also in a two-dimensional full non-linear approach whose development is in progress. The paper synthetically shows this new approach and compares the results with those coming from a conventional deterministic LSRA.

REFERENCE SOIL PROFILE

Recently, many local seismic response studies have been performed in Italy due to a recent new uniform buildings code which now forces the designers to assess the influence of local site conditions on the seismic actions to which structures shall withstand. Among these a research project carried out in a small town in central Italy [1] allowed us to select a reference in-hole seismic profile and related soils' dynamic parameters and their variances. Figure 1 shows the seismic soil profile derived from a cross-hole investigation.

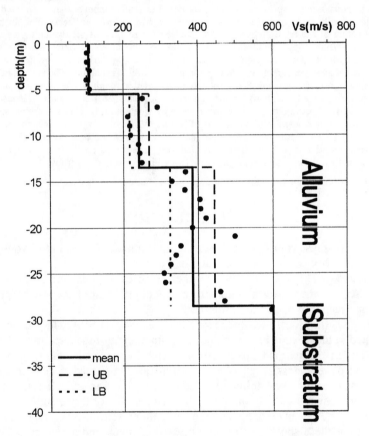

FIGURE 1. Shear waves seismic profile of the investigated soil column. Upper and lower bounds of shear waves refer to one standard deviation around mean values of shear waves. Alluvial soils consist of quaternary clays and silts; the substratum belongs to a Plio-Pleistocenic marly clays formation.

For the purpose of reducing hard seismic impedance contrasts between two successive seismic layers, a linear velocity rate has been imposed within each layer, preserving the variance observed in the shear waves velocity.

Geotechnical characterization

Dynamic soils properties needed for a 1D-LSRA reduce to stiffness and damping variations with shear strain, respectively, $G/Go(\gamma)$ and $D(\gamma)$ functions. Due to the availability of a statistical ensemble of such parameters for the alluvial soils, a Monte Carlo simulation technique following a normal distribution has been applied in order to generate normal variate distributions for both functions, whose trends are shown in Figure 2.

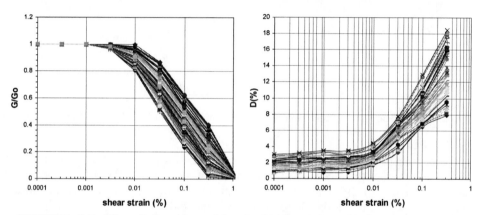

FIGURE 2. Soil stiffness (left) and internal damping (right) variations with shear strain. 100 out of 1,000 realizations following a normal distribution truncated at ±2σ. Each realization has been ordered to avoid physically inconsistent hardening of the soil.

The assumption of no vanishing standard deviation ($\sigma \neq 0$) for $G/Go(mean)\cong 1$, means to include the variability of Go in the normalized G/Go random variable. It seems to be more physically sound to consider the two random variables G/Go and Go not linked each other, at least for $G/Go(mean)\cong 1$. Therefore, it means that $\sigma=0$ for $\gamma(\%)\leq 0.001$. For the same reasons and to avoid a non-physically constrained null value of soil stiffness at high shear strains, the generated normal distributions have been truncated at ±2σ. Lastly, an inverse relationship has been assumed between damping and soil stiffness, namely, the lower the $G/Go(\gamma)$ curve, the higher the $D(\gamma)$ curve, and vice versa, according to the theory of soil dynamics [2], although without a formal correlation coefficient.

Seismic Input

The seismic input has been chosen according to a seismotectonic study carried out in the research project [1] and selecting a compatible motion among the accelerometric records available in the Italian strong-motion database [3]. The strong motion record, whose time-history is shown in Figure 3, is characterized by a peak ground acceleration of 0.25g, which is furthermore consistent with the seismic zoning of the investigated site.

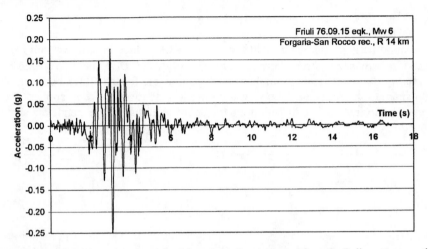

FIGURE 3. Time-history acceleration of the seismic input selected from the Italian strong-motion database [3] according to the regional seismotectonic setting.

According to the variation of soil's stiffness, the maximum shear strains induced by the seismic input reduce the G/Go modules not more than 70% of the initial shear modulus (Figure 4). In a first approximation this assures that an equivalent linear-elastic analysis is still reasonable for the case under study.

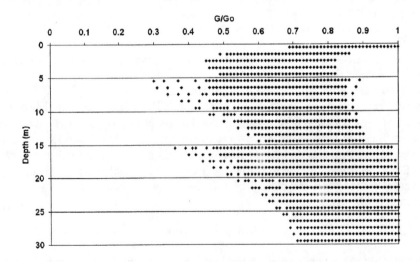

FIGURE 4. Soil's modules reduction with depth due to shear strains induced by the seismic input and given uncertainty in the soils' geotechnical parameters.

Local Seismic Response Analyses

Three statistical sets of LSRA were carried out plus one set according to a standard engineering practice based on a sensitivity study carried out with deterministic methods.

One out three of the statistical sets is based on the computation of the LSR through a standard computer program (SHAKE [4], case S), while two out of three were performed through a computer program which allows to account for the intrinsic variability of the input seismic motion (PSHAKE [5]). The latter implies computing the power spectral density function of the seismic input and attributing it a reliability level, that means, a probability the peak factor of the seismic input may be reached or exceeded. Two reliability levels have been considered: one kept to a constant 50% level (case A), and the second where the level ranges between 30 and 70% (case B).

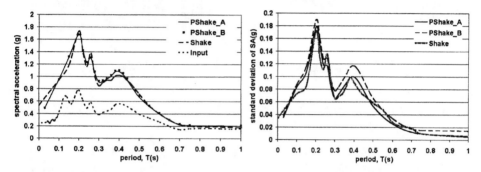

FIGURE 5. Response spectra (left) for the three examined cases incorporating uncertainty: cases A and B differ in the way they consider the aleatory nature of the input seismic motion (see text for details). Trend of standard deviations of spectral accelerations (right) for the same three cases: allowing for considering also the uncertainty in the seismic input leads to more scattered results especially in the period range where the soil column amplifies.

Figure 5 shows the comparison between the use of the two approaches carried out by means of the two computer programs, PShake vs. Shake, that means, considering or neglecting the stochastic nature of the input motion, respectively. Though the computed spectral accelerations (left figure) don't show an appreciable difference between the outputs, nevertheless accounting for the intrinsic variability of the seismic motion leads to more scattered results as can be inferred from looking at the standard deviations of the computed spectral accelerations (right figure). This seems to be more physically suitable with the intrinsic aleatory of the seismic motion, even if its influence on the results may be even negligible.

The differences among all the investigated cases slightly increase in the interval of the fundamental periods of vibration of the soil column (0.35÷0.55s), whose shifting toward longer periods is shown in Figure 6 (on the right). In the same figure the scattering of the first two modes of vibration of the soil's amplification function is shown on the left: the second fundamental frequency is more scattered than the first one, due to the presence of multiple peaks of the input spectrum at low frequencies.

Since there's a good agreement among the results of the three investigated cases, for the sake of comparison the statistical analysis carried out with the Shake program (case S) has been compared with a standard engineering approach based on a deterministic sensitivity analysis carried out with the same computer program. This comparison allows to highlight possible deviations of the standard practice from a rigorous uncertainty analysis.

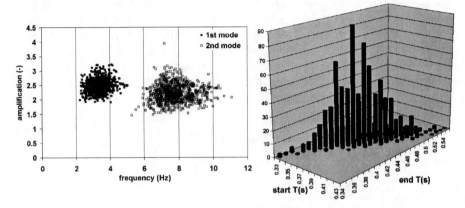

FIGURE 6. First two modes of vibration of the soil's amplification function (left). Starting vs. ending fundamental period of vibration of the soil column due to induced shear stresses (right).

The sensitivity analysis has been set forth considering the statistical variance of the three variables Go, G/Go(γ) and D(γ). In Table 1 the conditions under which the sensitivity analysis has been carried out is shown.

TABLE 1. Summary of the analyses carried out to compare a standard deterministic sensitivity analysis with a fully probabilistic one. All the analyses have been performed using the Shake computer program, modified to call – *when required* – a subroutine to compute the random generation of soils' properties (Monte Carlo method).

Case	Go	G/Go(γ)	D(γ)
S0	Mean	Mean	Mean
S1a	Mean + 1σ	Mean	Mean
S1b	Mean – 1σ	Mean	Mean
S2a	Mean	Mean + 1σ	Mean – 1σ
S2b	Mean	Mean – 1σ	Mean + 1σ
S	*Full uncertainty analysis (Monte Carlo technique)*		

Figure 7 summarizes our preliminary observations about considering the intrinsic variability of dynamic soil properties. The mean curve derived from a full uncertainty analysis (*S_mean*) overlaps well the deterministic case *S0*; this means that a standard approach, based on a good characterization of soil properties may capture the mean response of a soil column to seismic excitations. But, when soils' properties uncertainty is taken into account, the standard sensitivity analysis carried out considering the soils' properties varying within their statistical moments (mean ±1σ) may underestimate the overall soil response (compare *S2a,b* spectra with *S_mean+1sd* spectrum). It is worth noting the effect exerted by considering the uncertainty in the initial soil stiffness, which means to take into account the seismic soil profile variability; the *S1a* spectrum amplifies at lower periods than the *S1b* spectrum, according to the observation that the stronger the stiffness the lower the amplification period, and vice versa. On the other hand, when a full uncertainty analysis is carried out, the upper bound of the response (*S_mean+1sd* spectrum) allows taking into account all the sources of uncertainty, thus leading to a conservative but more realistic estimate of the response spectrum all over the entire period range.

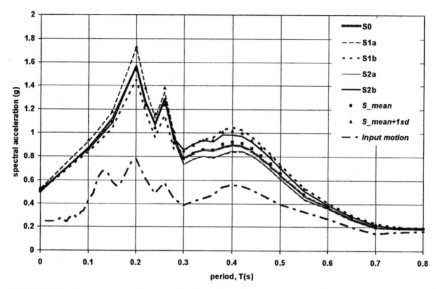

FIGURE 7. Comparison of results obtained through a sensitivity analysis commonly used in engineering practice with those deriving from a full uncertainty analysis of the LSR. The cases analyzed and displayed are detailed in Table 1.

CONCLUSIONS

A full probabilistic approach to the local seismic response analysis has been applied and its main results shown in the present paper. Both uncertainty in the seismic input motion as well as in the seismic soil profile and dynamic soil properties have been considered. Incorporating uncertainty in the seismic soil profile and/or taking into account the stochastic nature of the seismic motion provides more scattered results in the spectral period range where the soil column amplifies; this may lead, when it is taken into account, to more conservative results respect to an approach neglecting them or even considering only the uncertainty in the geotechnical properties.

The influence exerted by considering the uncertainty in the soil properties has been investigated comparing a full probabilistic approach, implementing a Monte Carlo simulation technique, with the results coming from a commonly adopted sensitivity analysis using a deterministic procedure. Incorporating uncertainty can better describe the scattering of the results in terms of expected spectral accelerations in the entire period range investigated by the local seismic response analysis. In fact, though a sensitivity analysis carried out with a deterministic approach may capture the scattering in turn, on the other hand it generally relates to the period range to which the soil column is more sensitive, disregarding the period range where the input motion, rather than the soil column, can give amplification.

Both these effects show how uncertainty cannot be neglected when a local seismic response analysis is carried out, in order to avoid unforeseen dynamic amplification effects due to our lack of knowledge regarding the soil's properties or profile and because of the stochastic nature of the seismic input motion.

Finally, further studies are anticipated to be in progress investigating in depth the influence exerted by uncertainty when strong non-linear effects can arise due to high shear stresses, as well as the influence exerted on the local seismic response by spatial variability of soil's properties which can be examined through 2D random finite element methods [6].

ACKNOWLEDGMENTS

The Authors wish to thank Teresa Crespellani and Claudia Madiai (Department of Civil and Environmental Engineering, University of Florence, Italy), to have kindly provided the data regarding the investigated soil profile.

REFERENCES

1. T. Crespellani, C. Madiai, J. Facciorusso, G. Simoni, R. Bardotti and M. Formelli, "Caratterizzazione dinamica dei terreni di Senigallia e analisi della risposta sismica locale", in *Studio delle sorgenti sismogenetiche lungo la fascia costiera marchigiana*, edited by M. Mucciarelli and P. Tiberi, Rome, INGV, 2007, pp. 150-186.
2. Braja M. Das, *Principles of Soil Dynamics*, Thomson-Engineering, London, 1992, pp. 592.
3. Working Group S6, "Data base of the Italian strong-motion data (1972-2004)", (2007) http://itaca.mi.ingv.it/.
4. P.B. Schnabel, H.B. Seed and J.B. Lysmer, "SHAKE -- A Computer Program for Earthquake Response Analysis of Horizontally Layered Sites", Report No. EERC 72-12, Earthquake Engineering Research Center, University of California, Berkeley, CA (1972).
5. T. Sanò and A. Pugliese, "PSHAKE, Analisi probabilistica della propagazione delle onde sismiche", RT/DISP/91/03, ENEA (1991).
6. D.V. Griffiths, G.A. Fenton and H.R. Ziemann, "Seeking out failure: the Random Finite Element Method (RFEM) in probabilistic geotechnical analysis", ASCE Proceedings on Probabilistic Modeling and Design, Atlanta, 2006, e-paper in CD-ROM.

Cerreto di Spoleto (Umbria-Italy): Seismic amplification at the ENEA local array stations.

Dario Rinaldis [a]

[a] ENEA, C.R. Casaccia, Via Anguillarese 301, 00060 S.M. Galeria (Rome), Italy..

Abstract. The Nerina valley, where Borgo Cerreto is located, is surrounded by the Apennine mount chain at the top of which lies the historical centre of Cerreto di Spoleto. The study is part of a research project aiming at analysing natural disasters and their impact on the Italian cultural heritage. Within the framework of this research project, local seismic records were analysed for both the carbonate ridge and the bordering alluvial valley. The choice of Cerreto di Spoleto as a test site derives from the analysis of Italian seismic hazard maps, obtained in terms of peak ground velocity and taking into account regional geology. The maps highlight the considerable seismic hazard which characterises the Apennine belt and its possible increase due to the effect of alluvial deposits. To this aim, ENEA installed in the 80's an accelerometric array (CODISMA up to 2000 and, in the following years, ETNA; for more detailed description see [1]. The 14 October event, was recorded both at the roof of CSM and at BCT stations. This is important to check the features observed comparing the FAS of acceleration at CSM and BCT during the 26 September events. Unfortunately the station at CSM basement did not record the above mentioned events but several aftershocks were recorded at each array station. Velocimetric records of both ambient noise and small-magnitude earthquakes were analysed in order to identify amplification conditions. The analysis was carried out in the time domain, through directional energy evaluation, and in the frequency domain, through H/V spectral ratios and spectral ratios with respect to a reference station.

Keywords: Seismic amplification, accelerometric records, trapped waves, beatings, azimuthal energy distribution

INTRODUCTION

Since many years ENEA has recognized the importance of installing strong motion instrumentation in areas prone to seismic risk. The Umbria-Marche portion of the Apennine chain has a long history of seismic crisis characterized by small to moderate events, lasting several months [2]. During the '80-s ENEA installed several instruments in the Umbria portion of the Apennine around the Nerina valley. At Cerreto di Spoleto, the array originally composed of 8 instruments, installed at 5 recording stations, was modified in such a way that only 4 instruments were surviving when the 26 September events struck the Umbria-Marche region. The instruments were CODISMA (Contraves Digital Strong Motion Accelerographs) with 12 bits acquisition system and PCB triaxial accelerometers. The 4 accelerographs were installed at 3 different sites: 1) Cerreto tower, BCT, rock site; 2) Cerreto football field, CCS, narrow alluvial valley, 200 m from CTO; 3) Cerreto City Hall basement, CMC, rock site, top hill, about 300 m from BCT; 4) Cerreto City Hall top floor, CMS. Starting from January 2000, ETNA 18 bits accelerographs, in all stations of the

CP1020, *2008 Seismic Engineering Conference Commemorating the 1908 Messina and Reggio Calabria Earthquake,*
edited by A. Santini and N. Moraci
© 2008 American Institute of Physics 978-0-7354-0542-4/08/$23.00

Cerreto di Spoleto permanent array, were deployed and CODISMA removed. Finally from May 2000, a temporary array of 3 K2 (12 accelerometric sensors each) was set up to perform a seismic instrumentation of the CEDRAV (Centre for Anthropological Documentation and Research of Nerina Valley) building of Cerreto di Spoleto, up to September 2001. The second array in Norcia, about 20 km far from Cerreto, is composed of two analog stations, with triaxial accelerograph Kinemetrics SMA-1: 1) Norcia Altavilla, NAL, firm soil site at the border of a debris plane; 2) Norcia industrial area, NZI, debris site, about 2 Km from NAL. During the days following the main shock the permanent arrays were extended installing a CODISMA accelerometer in Preci, PRE, a rock hill top site north-west of Cerreto di Spoleto, and an accelerometric station in Foligno, FOL, center of an alluvial basin, at the basement of the bell tower of the "Santa Maria Infraportas" basilica. The station FOL was composed of a FBA-23 Kinemetrics triaxial force-balance accelerometer connected to a MARS-88/fd Lennartz acquisition system.

RECORDS FROM THE 26TH SEPTEMBER 1997 EVENTS

The record obtained at Borgo Cerreto Torre (BCT) during the fore shock of the 26 September at 00: 33 show a larger PGA (but shorter duration) if compared with the record obtained at same station during the main shock of the 26 September at 09: 40. This is connected to the fault rupture propagation indicated in the NE-SW direction for the main shock and opposite for the fore shock [3]. Records at Municipio Soffitta of Cerreto di Spoleto (CMS) should show a clear influence of the building. Unfortunately the station at CMC did not record the events. So we did not be able to de-aggregate the contribution to the FAS of the acceleration obtained in CMS due to the building amplification of the ground motion at the basement from some other contribution (i.e. topographic effect).

RECENT RECORDS.

Records obtained at the Cerreto di Spoleto array stations, after the ETNA kinemetrics digital strong-motion accelerographs deployment, were analyzed. In particular the event of the 15th Decenber 2005 13:28:35 GMT, 4.2 Ml, was recorded at all the array stations. Fig 1 shows a comparison between the FAS of the recorded acceleration for the horizontal components obtained at BCT (rock site), at CMS and at CMC. It is interesting to note that not significant departures are noticeable, in both WE and NS FAS of acceleration horizontal components, in the frequency interval between 0 and 1 Hz. Furthermore, the plots of the WE components (fig. 1 b) seems to indicates that, for this event, the seismic energy in this frequency interval is equal at all the examined stations.

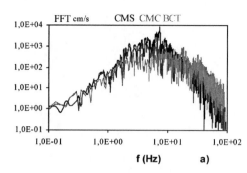 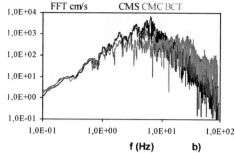

FIGURE 1. FAS of NS (a) and WE (b) component of acceleration.

At higher frequencies the FAS of the recorded accelerations at CMS and CMC stations, continued to be equivalent up to 7-8 Hz but departed respect to the FAS of the acceleration recorded at BCT. BCT station is in a different morphological condition respect to the CMS and CMC stations [1,4]. Then taking into account that some contribution to the value of the spectral ordinates is due to an increase in jointing of the rock, producing a low velocity layer between two active faults (see the red lines starting from Municipio and Faglia) that will amplify the seismic motion at frequencies between 6.5-7.0 Hz [4] then is possible an amplification of the seismic waves, in the 2-5 Hz frequency interval, may be due to the morphological conditions at the CMC and CMS stations. Records at CMS show a clear influence of the building. In fact if comparison is made between the plots of the FAS of the acceleration recorded at CMS and CMC , the 6-12 Hz frequency interval seems to contain the contribution of structural modes to the seismic amplification. It is very unlike the situation at frequencies close to7 Hz where to the seismic waves amplification is added some frequency content due to the building resonance frequency(municipio).

ENERGY AZIMUTHAL DISTRIBUTION

In this paragraph we examine the records by means of the graphic representation of the azimuthal distribution of the energy (AED) at the recording point. The energy is intended in the sense given by Arias [5]. Records of the prinipal shocks (26th September 1997 09:40:00 and 14th October 1997 15:23:00 UTM) at rock site BCT have most of the energy concentrated within 0°-15° that was erroneously interpreted [3] just consistent with the fault rupture process along the direction NW-SE (Figure 2, record 7043). For the opposite geological condition of the site, NZI, the energy was very dispersed in azimuthal direction (Figure 2, record 7038) according to some expected mechanical and geometrical effects of the soil condition on the seismic field.

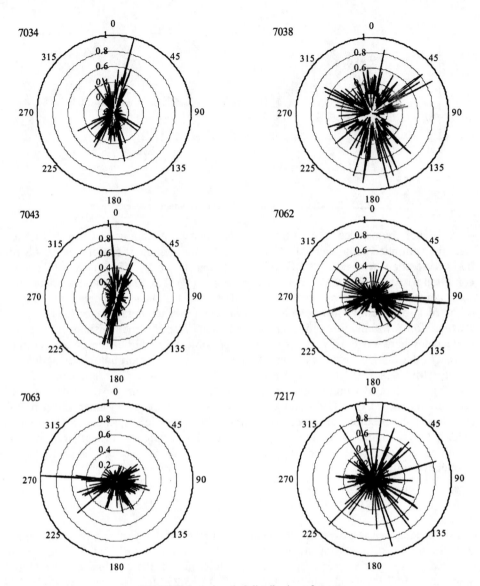

FIGURE 2: Azimuthal distribution of energy

At site NAL (Figure 2, record 7034) there was relevant energy concentration in direction consistent with the rupture process and the rest along a sharp direction. This direction could be related to some border effect. The AED of the record at FOL (Figure 2, record 7217) was spread over 360° denoting the absolute prevalence of the local effects, geometrical and mechanical, with respect to the source and path effects. The energy distribution for records at CCA of aftershocks individuates angles where there is almost absence of energy (Figure 2, record 7063). Since the events had very low magnitude, we attributed this behavior mainly to geometrical effects on the seismic field inside the valley(for more details see [3]). Any way, the main goal of this paper was to study records obtained at accelerometric stations of the Cerreto di Spoleto array. Figure 3 shows the Arias Intensity azimuthal distribution of several accelerometric records obtained BCT station. All records, if exception is made of the one obtained in 2005, were obtained during the main shocks (26th of September and 14th of October 1997) that stroke the Umbria-Marche Apennine centers. They were recorded by CODISMA strong-motion accelerograph, when the 2005 event was recorded by an ETNA. The ETNA instrument was installed as N40E, instead of the NS direction as usual. So data were rotated to be compared with the other records (CODISMA was NS oriented).

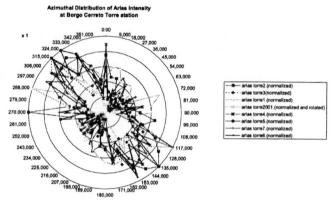

FIGURE 3 : Azimuthal distribution of energy at BCT

INVESTIGATIONS IN THE 5-7 HZ FREQUENCY BAND.

To investigate the above mentionned results, seismometric investigations aiming at the analysis of the seismic response in the carbonate ridge were studied [4]. In particular the velocimetric records obtained during the Umbria-Marche sequence, filtered at the 5-7 Hz frequency band, were analyzed at the temporary array stations of "Faglia" (F), "Antenne"(A) and the station installed at bottom of the slope. Stations A and F show an energy distribution at, more of less, 315°, roughly perpendicular to the direction of the ridge not observed at the others recording stations of the temporary velocimetric array [4]. Then because A and F are deployed not far from the top of the relief, the peak at 315° degrees is caused by a topographic effect. In the same paper has been observed that: "two main fault systems were identified: the first, transtensional, with an approximately NS ± 10° direction; the second, estensional, with an about N30°W direction". One fault of the second system is supposed to be not far from the BCT station. To investigate this peak on the azimuthal energy distribution at the BCT station we may analyze results of a measurements campaign in the Nerina valley. In particular a velocimetric array was set up at the sites of "Cacciatore" and "Case Popolari", that recorded some events of the Umbria-Marche sequence. As is possible to see in figure 4 the AED at "Case Popolari" and "Cacciatore", filtered with a pass-band filter of 5-7 Hz, shows a very interesting result: peaks at the same angle (330° or to be more clear at N30°W) recorded at the station BCT.

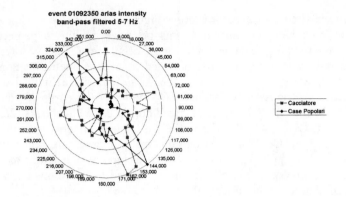

FIGURE 4 : Azimuthal energy distribution at Case Popolari and Cacciatore

Corrected versions (5-7 Hz band-pass filtered) of accelerometric records obtained at the BCT station during the 13 October 1997 at 13:05:46 UTM event (Md=3.9) and 14 October 1997 at 15:23:00(Ml=5.4), were analyzed and results of the AED are shown in fig. 5. It is interesting to note that the 13th October filtered and unfiltered accelerations show similar AED, when the 14th October one shows that the main contribution at the azimuthal distribution of energy, of the filtered acceleration, is in the NS direction.

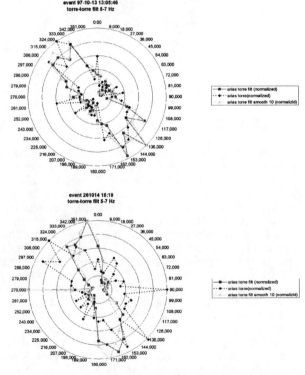

FIGURE 5 : Azimuthal energy distribution in the 5-7 Hz frequency band at BCT station.

Finally fig. 6 and 7 shows a very interesting result: the FAS of acceleration of the filtered version have two peaks at 5.7 and 6.7 Hz for the Ml=5.4 event record and three peaks at 5.3 Hz, 6.0 Hz and 6.8 Hz for the Md=3.9 event record, at the BCT station. The NS component, of the 14th October event, shows as well a stationary wave lasting at least 1 second and starting at 7.8 seconds.

CONCLUSIONS

The observation that AED is very spreaded for records at soft soil NZI and FOL in conjunction with spectral ratios having several small peaks, while records at soft soil CCA have relevant amplification at single frequencies, although for small energy event, underline the relative importance of geometrical and mechanical effects.

Moreover records obtained at the BCT station seems to indicate that energy is azimuthally distributed mainly at three angles:0°, 315°, 330°. The first two peaks, in energy distribution, were already discussed in previous papers [1], [4] and were attributed to trapped waves and topographic amplifications respectively. The third one is instead a new observation; the proposed interpretation of this peak, in the AED, is due to an estensional fault system, with an about N30°W direction[4].

The investigation in the 5-7 Hz frequency band carried out two main results: first of all, filterd and unfiltered acceleration gave similar AED for weak motion records. The frequency domain analysis evidenced a main peak at 7 Hz that correspond to the time-histories strong phases: the shape of the time-histories seems to indicate that seismic waves, traveling in this fault system, were amplified. The second result is obtained from the 14th October event record: filtered data (5-7 Hz) show a different AED if compared with the unfiltered ones, where the main peak correspond to the 0° angle (i.e. trapped waves amplification). Data from seismometric investigations aiming at the analysis of the seismic response in the carbonate ridge (velocimetric temporary array) seems to confirm the former results. In fact both filtered data at "Case Popolari" and "Cacciatore" stations show a main peak in the AED (filtered data) at 330° angle. Finally the comparison between FAS of acceleration at the BCT and CMC and CMS stations for the 15th Decenber 2005 weak motion event, shows how useful could be the AED analysis to evaluate the seismic risk for the Municipio building. In fact the record at the BCT do not shows peaks in the distribution at 0° , indicating that not trapped waves amplification was verified. Other events show instead an evident peak at 0° in the AED giving the idea of seismic waves amplification in a frequency interval were the main resonance frequencies of the Municipio (and other important buildings) are included. To have a better view of these studies ,records of municipio basement, from larger events, should be analyzed.

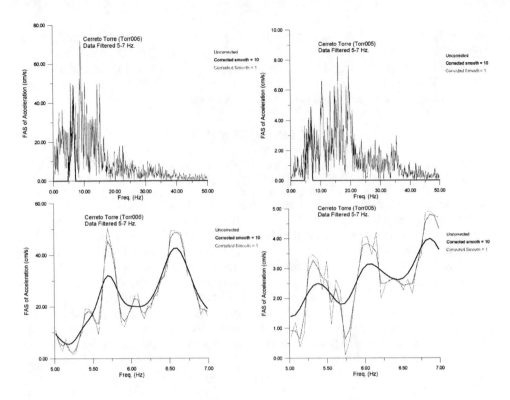

FIGURE 6: FAS of acceleration at BCT station : the frequency zoom shows a shift at lower frequency for the 14 October 1997 event respect to the 13 October 1997 event.

REFERENCES

1. Rinaldis, D. (2008), The Umbria-Marche sequence: digital recordings at ENEA stations. Accepted at IACMAG 12th International Conference, 1-6 October, Goa, India.
2. Boschi E. and Cocco M. (ed) (1997), Studi preliminari sulla sequenza sismica dell'Appennino Umbro-Marchigiano del settembre-ottobre 1997, ING, Roma (in Italian).
3. Clemente, P., Rinaldis, D. and Bongiovanni, G.: 2000, The 1997 Umbria-Marche earthquake: analysis of the records obtained at the ENEA array stations. Proc. XII World Conference on Earthquake Engineering (Auckland, NZ), paper n°1630.
4. Martino, S.,Minutolo A., Paciello, A. Rovelli, A. Scarascia Mugnozza G. and Verrubbi, V.: 2006, Evidence of Amplification Effects in Fault Zone Related to Rock Mass Jointing. Natural Hazards, 39: 419-449.
5. Arias A. (1970), «A measure of earthquake intensity», In Hansen R. (ed), Seismic Design of Nuclear PowerPlant, MIT Press, Cambridge.

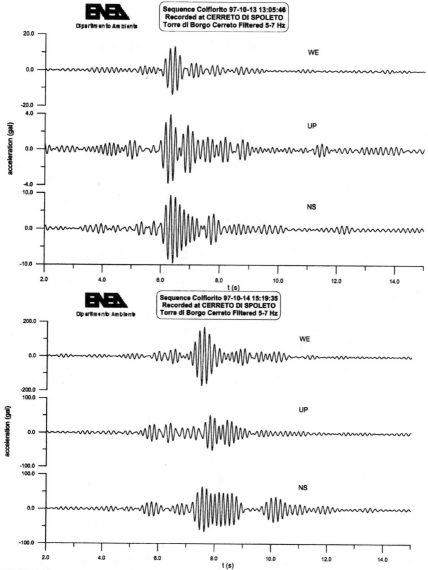

FIGURE 7: Time-histories of the acceleration at BCT: 14th October 1997 and 13th October 1997 event records.

Site Characterization of Italian Strong Motion Recording Stations

Giuseppe Scasserra[a], Jonathan P. Stewart[b], Robert E. Kayen[c] and
Giuseppe Lanzo[a]

[a] Dipartimento di Ingegneria Strutturale e Geotecnica, Sapienza Università di Roma, Via A. Gramsci
53, 00197, Rome, Italy
[b] Civil and Environmental Engineering Department, University of California, Los Angeles, 5731 Boelter
Hall, Los Angeles, CA 90095
[c] U.S. Geological Survey, Menlo Park, California, USA

Abstract. A dataset of site conditions at 101 Italian ground motion stations with recorded motions has been compiled that includes geologic characteristics and seismic velocities. Geologic characterization is derived principally from local geologic investigations by ENEL that include detailed mapping and cross sections. For sites lacking such detailed geologic characterization, the geology maps of the by Servizio Geologico d'Italia are used. Seismic velocities are extracted from the literature and the files of consulting engineers, geologists and public agencies for 33 sites. Data sources utilized include post earthquake site investigations (Friuli and Irpinia events), microzonation studies, and miscellaneous investigations performed by researchers or consulting engineers/geologists. Additional seismic velocities are measured by the authors using the controlled source spectral analysis of surface waves (SASW) method for 18 sites that recorded the 1997-1998 Umbria Marche earthquake sequence. The compiled velocity measurements provide data for 51 of the 101 sites. For the remaining sites, the average seismic velocity in the upper 30 m (V_{s30}) is estimated using a hybrid approach. For young Quaternary alluvium, V_{s30} an existing empirical relationship for California sites by Wills and Clahan (2006) is used, which we justify by validating this relationship against Italian data. For Tertiary Limestone and Italian Mesozoic rocks, empirical estimates of V_{s30} are developed using the available data. This work is also presented in Scasserra et al. (2008: JEE, in review).

Keywords: Italian accelerograms, recording stations, V_s profile, SASW, geology.

INTRODUCTION

A major effort has been undertaken by U.S. and Italian investigators to improve the quality of the Italian strong motion database and to use that database to evaluate the compatibility of Italian data with ground motion prediction equations derived for active tectonic regions by investigators with the Next Generation Attenuation (NGA) project. In this article, we summarize a critical aspect of this broader effort, which is the characterization of conditions at strong motion recording sites in Italy. The emphasis is on both the classification of surface geology and average shear wave velocity in the upper 30 m of the site (V_{s30}). The work described here has been previously presented by Scasserra et al. [1].

CP1020, *2008 Seismic Engineering Conference Commemorating the 1908 Messina and Reggio Calabria Earthquake*,
edited by A. Santini and N. Moraci

PARAMETRIZATION OF SITE CONDITIONS FOR GROUND MOTION STUDIES

The V_{s30} parameter is now widely used to represent site condition in engineering practice and was also adopted by four of five of the NGA ground motion prediction equations (GMPEs). In the development of the NGA database (described in detail by [2]), each site was assigned a V_{s30} value, with approximately 1/3 coming from on-site measurements and the remaining coming from correlations with other, more readily available site information.

In the development of the NGA database, protocols were followed for estimating V_{s30} when on-site measurements (extending to a depth of at least 20 m) are not available. Those protocols are as follows [3]:

✓ Velocity estimated based on nearby measurements on same geologic formation (site conditions verified based on site visit by geologist).
✓ Velocity estimated based on measurements on same geologic unit at site judged to have similar characteristics based on site visit by geologist.
✓ Velocity estimated based on average shear wave velocity for the local geologic unit; presence of the unit verified based on site visit by geologist.
✓ Velocity estimated based on average shear wave velocity for the geologic unit as evaluated from large-scale geologic map (1:24,000 to 1:100,000).
✓ Velocity estimated based on average shear wave velocity for the geologic unit as evaluated from small-scale geologic map (1:250,000 to 1:750,000).

We adopt similar procedures for estimation of V_{s30} values at Italian strong motion stations. Each site has been assigned a V_{s30} value along with an index pertaining to how the value was derived. Those indices are defined as:

✓ Category A: Velocity measured on-site using cross-hole, down-hole, or spectral analysis of surface wave methods;
✓ Category B: Velocity estimated based on nearby measurements on same geologic formation (site conditions verified based on site visit by geologist). This is similar to Categories (1)-(2) by [3].
✓ Category C: Velocity estimated based on measurements from the same geologic unit as that present at the site (based on local geologic map). This is similar to Categories (2)-(3) by [3].
✓ Category D: Velocity estimated based on general (non-local) correlation relationships between mean shear wave velocity and surface geology.

The following three sections describe how velocities were assigned to strong motion sites. As described in the next section, for 33 sites, velocity profiles from the literature and the files of practicing engineers, geologists, and public agencies are used to assign V_{s30} values that are assigned as Categories A, B, or C. We then describe velocity profiling performed for 18 additional sites as part of this study (Category A). Next, we describe how V_{s30} values are assigned on the basis of surface geology for the remaining 50 sites. A table presenting the site classifications for each recording station is given in Table 1 of Ref [1].

Data from the Literature

The previous site characterization for Italian strong motion stations can be grouped into the following major categories: (1) site investigations at selected instruments that recorded the 1976 Friuli earthquake [4] and 1980 Irpinia earthquake [5,6,7]; (2) microzonation and other studies for local municipalities such as Ancona [8], Tarcento [9] and Sant'Agapito [10]; and (3) individual site studies documented in the literature (e.g., Catania-Piana site [11]) and from the files of consulting engineers and geologists with local experience (e.g., Naso station [12]).

The Friuli and Irpinia site investigations were generally performed at the recording sites and are classified as Category A. The work in the Friuli and Irpinia regions examined 7 and 16 accelerograph sites, respectively. For each site, cross-hole measurements were made to evaluate shear wave velocity profiles. Additional in situ and geotechnical laboratory testing was also performed.

The microzonation and individual site studies were used to assign velocities to strong motion stations that are listed as Categories A-C depending on the proximity of the measurement to the strong motion station and the verification (or not) of similar geologic conditions at the two locations from a site visit by a geologist. Most of these velocity profiles are from cross-hole or down-hole measurements.

Velocity Measurements as Part of the Present Study

The 1997-98 Umbria-Marche earthquake sequence produced a significant number of recordings, but prior to this study velocity profiles had been evaluated and disseminated for relatively few of the recording sites in that region. Accordingly, on-site measurements were performed at numerous sites using a controlled sine wave source and the spectral analysis of surface waves (SASW) method [13]. The SASW method of testing is a portable, inexpensive, and efficient means of non-invasively estimating the stiffness properties of the ground. The equipment utilized in the present work can typically be used to profile velocities to depths of approximately 100 m.

The testing program investigated 17 sites in Umbria and Marche. Typically, the strong motion recording (SMR) stations are located in residential or light industrial sites outside the town center, in parks, or on private farm land. We located next to the SMR stations, or the GPS location of the site if we could not observe the SMR.

Details of the testing system and data analysis procedure are provided in [1]. The results of the testing are plots of surface wave velocity versus wavelength (also referred to as dispersion curves). Figure 1 presents a plot of a group of eight individual dispersion curves that together cover a range of wavelengths from 0.6-400 m for the Cascia site in Umbria. The averaged dispersion curve from these eight profiles is used to invert the velocity structure.

The inversion process is used to estimate a soil velocity model having a *theoretical* dispersion curve that fits the data. Typically, a 10-15 layer model was used for the inversion, with layer thicknesses geometrically expanding with depth. The increasing layer thicknesses correspond with decreasing dispersion information in the longer wavelength (deeper) portion of the dispersion curve. The profiles generally increase in stiffness with depth, though low velocity layers are present at depth in several profiles. Figure 2 shows the inverted shear wave velocity profile for the Cascia, Umbria site, in which velocity rapidly climbs from less than 300 m/s at the surface to >1900 m/s at 40 m. Values of V_{s30}, calculated as 30 m divided by shear wave travel time through the upper 30 m, range from 182 to 922 m/s (NEHRP categories B to D).

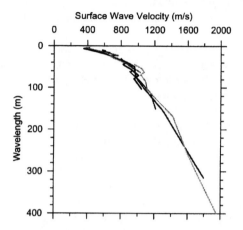

FIGURE 1. A group of eight dispersion curves covering a wavelength range of 1-400 m (Site 267CSC, Cascia, Umbria)

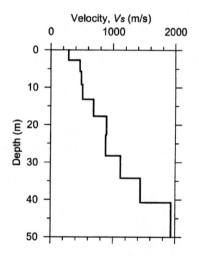

FIGURE 2. Shear wave velocity profile for Cascia, Umbria site 267CSC (V_{s30} =540 m/s, Site Class C).

Estimating Velocities for Sites without Measurements

For sites for which no local measurements of seismic velocities are available, we estimate V_{s30} values based on correlations with surface geology. Correlations to estimate V_{s30} from surface geology are not available in the literature for geologic units in Italy. Accordingly, we evaluate the effectiveness in Italy of correlations developed for California and develop preliminary additional correlations for geologic units not represented in the California models.

The geology maps available for Italy include large-scale maps (1:100,000) by Servizio Geologico d'Italia [14] that provide coverage of the entire country (and hence all recording stations) and local geologic maps/sections (typical scale 1:2000) by ENEL. The local maps/sections are derived from a site visit by an ENEL geologist and are available for 77 of 104 strong motion sites. Additional geologic information is available for a few sites from local microzonation reports or geologic reports for individual sites. The geologic classifications are based on the largest map scale that is available for the site. The map scale from which the classification was taken is indicated in the table, with "local" referring to the aforementioned microzonation studies or geologic reports.

We judge the best available correlations for California to be those of Wills and Clahan [15]. A number of the Wills-Clahan geologic categories are descriptive of conditions encountered at Italian sites. Among these are Quaternary alluvium categories segregated by sediment depth and material texture (Qal,thin; Qal,deep; Qal,coarse), older Quaternary alluvium (Qoa), Quaternary to Tertiary alluvial deposits (QT), and Tertiary sandstone formations (Tss). The relatively firm rock categories used by Wills-Clahan are generally not descriptive of Italian firm rock sites, which are often comprised of limestone, marls, and volcanic rocks.

Wills and Clahan [15] provide mean and standard deviation values of V_{s30} for each geologic category based on California data. We evaluate the applicability of those estimates to Italian sites by calculating V_{s30} residuals as:

$$R_i = \left(V_{s30}\right)_{m,i} - \left(V_{s30}\right)_{WC} \tag{3}$$

where R_i = V_{s30} residual for site i, $(V_{s30})_{m,i}$ = value of V_{s30} from measurement at Italian site i, and $(V_{s30})_{WC}$ = mean value of V_{s30} [15]. Due to the small number of sites falling in individual categories, we group sites into two general categories for analysis of residuals – Quaternary alluvium (combination of the thin, deep, and coarse sub-categories) and late Quaternary and Tertiary sediments (combination of Qoa, QT and Tss). Figure 3 shows histograms of residuals grouped in this manner. Also shown in Figure 3 is the range of velocities within ± two standard deviations of zero using average values of standard deviation [15] for the grouped categories (taken as $\sigma_{WC}=85$ m/s for the Qal categories and $\sigma_{WC}=170$ m/s for the Qoa/QT/Tss categories).

The histogram for Qal categories (Figure 3a) shows that the mean of residuals is nearly zero, but only 78% of the data fall within the ± $2\sigma_{WC}$ bands (approximately 95% should fall within this range if the Italian data shared the standard deviation of the California data). The histogram for the Qoa/QT/Tss categories (Figure 3b) similarly shows a nearly zero mean, and 85% of the data fall within the ± 2σ_WC bands. Similar results are obtained if the grouped categories are broken down to smaller sub-categories (e.g., Qal,deep from Qal). Hence, our preliminary conclusion is that the Wills-Clahan recommendations provide an unbiased estimate of V_{s30} for Italian alluvium sites of Quaternary to Tertiary age. However, the standard deviation of the Italian data is different, being larger for the Qal categories are perhaps slightly smaller for the older alluvium and Tertiary categories.

As mentioned above, many of the rock sites have conditions geologically dissimilar to California such as limestone, marls, and volcanic rocks. Since we are unaware of existing correlations to V_{s30} for these types of materials, we assembled rock categories descriptive of Italian conditions that seem to generally have similar seismic velocities. These categories are summarized as follows:

✓ Tm: This category consists of Tertiary Marl, often with surficial overconsolidated clays. It is common along the central-southern Apennines, and 13 sites in our database have this classification. A histogram of the Tm velocities is given in Figure 4a, showing a mean V_{s30} = 670 m/s and standard deviation = 190 m/s.

✓ Pc: This category consists of Pleistocene to Pliocene cemented conglomerate. Its occurrence is widespread in Sicily and the Apennines. Five sites in our database have this classification, two of which have velocity measurements with V_{s30} = 972 and 1156 m/s. We use V_{s30} =1000 m/s for sites without measurements.

✓ Ml, Mv, and Mg: This category comprises Mesozoic limestone (Ml), volcanic rocks (Mv), and gneiss (Mg). We group these three together for velocity characterization because the available data is inadequate to justify further discretization and the seismic velocities are generally high (> 1000 m/s). The Ml category includes 14 sites located in the Alps and Apennines. The Mv category applies to three sites located near the active volcanoes of Mt. Etna (Sicily) or Mt. Vesuvio (near Napoli). The Mg category is encountered only at the Messina and Milazzo Station in Sicily. A measured shear velocity of 1800 m/s is reported in Table 1 for Messina, but this measurement was made in a tunnel deep in the ground. Shallow velocities should be slower and hence the preferred V_{s30} value is given as 1000 m/s to be consistent with other the other Mesozoic categories.

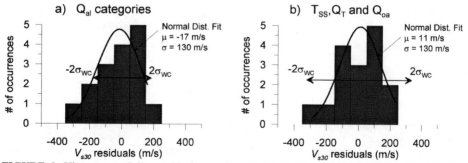

FIGURE 3. Histograms of V_{s30} residuals and normal distribution fits for (a) Quaternary alluvium categories and (b) older Quaternary, Quaternary-Tertiary, and Tertiary sandstone categories. The ±2σ_WC limits indicate ±2 standard deviations above and below zero from the Wills and Clahan [15] correlation.

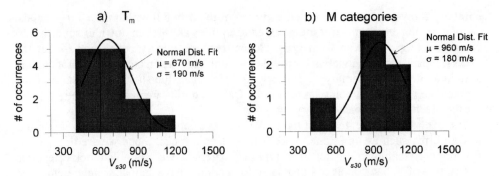

FIGURE 4. Histograms of V_{s30} values and normal distribution fit for (a) Tm category and (b) M categories

SUMMARY

In this article, we describe work directed towards the evaluation of site conditions at strong motion sites that have recorded past earthquakes in Italy. We have also developed a framework for estimation of V_{s30} for sites without measurements. We hope this procedure will be updated in the future as additional data becomes available.

ACKNOWLEDGMENTS

The SASW site characterization work was sponsored by the Pacific Earthquake Engineering Research Center's Program of Applied Earthquake Engineering Research of Lifeline Systems supported by the State Energy Resources Conservation and Development Commission and the Pacific Gas and Electric Company. This work made use of Earthquake Engineering Research Centers Shared Facilities supported by the National Science Foundation under Award #EEC-9701568. In addition, the support of the California Department of Transportation's PEARL program is acknowledged. The SASW site characterization work was also partially funded with the financial support of Ministero dell'Università e della Ricerca (MIUR). This support is gratefully acknowledged. We would like to thank Drs. Paolo Bazzurro, John Douglas, Julian Bommer, Roberto Paolucci, and Pierre-Yves Bard for their helpful suggestions about optimal locations for site characterization.

REFERENCES

1. G. Scasserra J.P. Stewart, R.E. Kayen and G. Lanzo, "Database for earthquake strong motion studies in Italy," Journal of Earthquake Engineering, Imperial College Press, in review (2008).
2. B.S.J. Chiou, R., Darragh, D. Dregor and W.J. Silva, "NGA project strong-motion database" Earthquake Spectra, 24 (S1), accepted for publication (2008).

3. R.D. Borcherdt, "Empirical evidence for acceleration-dependent amplification factors," Bull. Seism. Soc. Am., **92**, 761-782 (2002).

4. A. Fontanive, V. Gorelli, L. Zonetti, "Raccolta di informazioni sulle postazioni accelerometriche del Friuli," Commissione ENEA-ENEL per lo studio dei problemi sismici connessi con la realizzazione di impianti nucleari, Rome, Italy, (1985).

5. S. Palazzo, "Progetto Irpinia - Elaborazione dei risultati delle indagini geotecniche in sito ed in laboratorio eseguite nelle postazioni accelerometriche di Bagnoli Irpino, Calitri, Auletta, Bisaccia, Bovino, Brienza, Rionero in Vulture, Sturno, Benevento e Mercato S.Severino," Ente Nazionale Energia Elettrica (ENEL), Direzione delle Costruzioni, Rome, Italy (1991).

6. S. Palazzo, "Progetto Irpinia - Elaborazione dei risultati delle indagini geotecniche in sito ed in laboratorio eseguite nelle postazioni accelerometriche di Sannicandro, Tricarico, Vieste, Areinzo, S.Severo e Garigliano," Ente Nazionale Energia Elettrica (ENEL), Direzione delle Costruzioni, Rome, Italy (1991).

7. E. Faccioli "Selected aspects of the characterization of seismic site effects, including some recent European contributions," Proc. International Symposium on The Effects of Surface Geology on Seismic Motion (ESG1992), Odawara, Japan, **1**, 65-96 (1992).

8. Working Group, "Elementi di Microzonazione sismica dell'area Anconetana." Consiglio Nazionale delle Ricerche-Progetto Finalizzato Geodinamica, pub. n.430 (1981).

9. A. Brambati, E. Faccioli, G.B. Carulli, F. Cucchi, R. Onori, S. Stefanini and F. Ulcigrai, "Studio di Microzonizzazione delle'area di Tarcento," Regione autonoma Friuli-Venezia Giulia – Università degli studi di Trieste (1979).

10. Comune di Isernia, "Indagini geognostiche prospezioni geofisiche e prove di laboratorio finalizzate all'adozione della variante generale del piano regolatore comunale", GeoTrivell Teramo, (1998).

11. S.M. Frenna and M. Maugeri "Fenomeni di amplificazione sismica sulla piana di Catania durante il terremoto del 13\12\1990". Atti del VI convegno nazionale L'ingegneria sismica in Italia. 13-15 ott. 1993. Perugia, Italy (1993).

12. G. Copat (private communication)

13. S.Nazarian and K. Stokoe, "In situ shear wave velocities from spectral analysis of surface waves, Proc. Eighth World Conference on Earthquake Engineering, San Francisco, California, Vol. III, pp. 31-39 (1984).

14. Working Group, "Redazione della mappa di pericolosità sismica (Ordinanza PCM 20.03.03, n.3274) All.to 1 Rapporto Conclusivo" Istituto Nazionale di Geofisica e Vulcanologia (2004).

15. C.J. Wills and K.B. Clahan, "Developing a map of geologically defined site-condition categories for California," Bull. Seism. Soc. Am., **96**(4a), 1483-1501 (2006).

Engineering Seismic Base Layer for Defining Design Earthquake Motion

Nozomu Yoshida[a]

[a]Department of Civil and Environmental Engineering, Tohoku Gakuin University,
Tagajo 1-13-1, Miyagi, Japan

Abstract. Engineer's common sense that incident wave is common in a widespread area at the engineering seismic base layer is shown not to be correct. An exhibiting example is first shown, which indicates that earthquake motion at the ground surface evaluated by the analysis considering the ground from a seismic bedrock to a ground surface simultaneously (continuous analysis) is different from the one by the analysis in which the ground is separated at the engineering seismic base layer and analyzed separately (separate analysis). The reason is investigated by several approaches. Investigation based on eigen value problem indicates that the first predominant period in the continuous analysis cannot be found in the separate analysis, and predominant period at higher order does not match in the upper and lower ground in the separate analysis. The earthquake response analysis indicates that reflected wave at the engineering seismic base layer is not zero, which indicates that conventional engineering seismic base layer does not work as expected by the term "base". All these results indicate that wave that goes down to the deep depths after reflecting in the surface layer and again reflects at the seismic bedrock cannot be neglected in evaluating the response at the ground surface. In other words, interaction between the surface layer and/or layers between seismic bedrock and engineering seismic base layer cannot be neglected in evaluating the earthquake motion at the ground surface.

Keywords: surface ground, bedrock, engineering seismic base layer, reflection.

INTRODUCTION

Earthquake motion spreads from a fault to the site where an engineer is going to evaluate a seismic motion in order to design the structure. Ideally, an analysis considering the fault mechanism and path characteristics simultaneously is to be carried out in order to evaluate the earthquake motion at the ground surface. It is, however, very difficult and complicated, and far from the engineering practice. A chain method is frequently employed in practice, in which response of the ground surface is evaluated by the product of fault mechanism, attenuation characteristics from a fault to the seismic bedrock, and amplification characteristics from the seismic bedrock to the ground surface. Here, seismic bedrock is characterized by shear wave velocity V_s to be 3 km/s or larger, and engineering seismic base layer to be 300 to 400 m/s (maximum 700 m/s). The last path is usually separated into two parts: a path from the seismic bedrock to the engineering seismic base layer and a path in a surface ground, i.e., from the engineering seismic base layer to the ground surface. In this method, the outcrop motion at the engineering seismic base layer is first computed and

CP1020, *2008 Seismic Engineering Conference Commemorating the 1908 Messina and Reggio Calabria Earthquake*,
edited by A. Santini and N. Moraci
© 2008 American Institute of Physics 978-0-7354-0542-4/08/$23.00

a half of the outcrop motion is applied to the base of the surface ground as incident wave to the surface ground.

This concept is widely accepted. For example, when an engineer want to evaluate the seismic motion by an earthquake by using an earthquake motion obtained a little apart from the site, he first evaluates the incident wave at the engineering seismic base layer of the site where earthquake record was obtained by a deconvolution analysis, and treats it as an incident wave at the engineering seismic base layer at the interested site. Moreover, many design specifications specify design earthquake motion at the engineering seismic base layer when it outcrops. In the North American practice, rock or hard deposit outcrop is used instead of the engineering seismic base layer, but both are same meaning.

This kind of separation of the ground is justified when there is no interaction between two parts. However, validity of this concept has not been investigated in detail. In this paper, we will show that conventional method to separate the ground at the engineering seismic base layer is shown to lead incorrect evaluation of earthquake motion at the ground surface through the analysis.

EXHIBITING EXAMPLE THROUGH AMPLIFICATION CHARACTERISTICS

A site in Hachinohe-city, Japan, is investigated. Soil profiles at this site up to the seismic bedrock ware evaluated by Midorikawa and Kobayashi [1], and is shown in Table 1 and Fig. 1. Here, h denotes depth from the ground surface, V_s denotes shear wave velocity, ρ denotes mass density and Q denotes a quality factor by which damping ratio D is evaluated as $D=0.5/Q$. The engineering seismic base layer is set at GL-180 m whose shear wave velocity is 690 m/s. This velocity corresponds to the definition in design specification of nuclear power plant [2].

Two one-dimensional models are used in analyzing the ground in Table 1. The one is a continuous analysis in which the ground from the seismic bedrock to the ground surface is solved simultaneously, and will be referred as continuous analysis. The other is a separate analysis in which the whole ground is analyzed by dividing into two parts, i.e., from the seismic bedrock to the engineering seismic base layer and a surface layer. The analysis is carried out in two steps in the latter model. The outcrop motion at the engineering seismic base layer is first computed by the analyzing the lower part of the ground, and a half of the outcrop motion is applied at the bottom of the surface layer as incident wave. This analysis will be called separate analysis, and the lower part and upper part of the model will be referred as lower and upper grounds, respectively, in this paper. The upper ground is the same meaning of the surface ground.

A multi-reflection theory, which solves equation of motion in frequency domain, is employed in the following analyses in order to distinguish incident and reflected waves. In the case to analyze nonlinear behavior, improved equivalent linear method, DYNEQ [3] is used, which overcomes shortages that conventional equivalent linear method such SHAKE [4] as have.

Figure 2 shows amplification factor from the seismic bedrock to the ground surface. Here, E and F denote amplitude of incident and reflected waves, respectively, and subscripts S, E and G indicate the seismic bedrock, the engineering seismic base layer and the ground surface, respectively. For example, E_E/E_S is ratio of Fourier amplitudes at the engineering base layer to those at the seismic bedrock. The amplification in the separate analysis is computed by multiplying the amplification factor in the lower ground (E_E/E_S) and that in the upper ground ($(E+F)_G/E_E$). If $(E+F)_G/E_S$ by both continuous and separate methods coincides to each other, use of the separate analysis is justified. In other words, differences between them are error included in the separate analysis.

TABLE 1 Soil profiles
(modified from Midorikawa and Kobayashi, 1978)

Layer	h (m)	V_s (m/s)	ρ (t/m³)	Q
1	2.0	107	1.80	14
2	4.0	176	1.80	13
3	6.5	201	1.90	12
4	9.0	193	1.90	12
5	15.5	239	1.70	12
6	22.0	234	1.70	9
7	32.0	248	1.80	7
8	40.0	309	1.80	7
9	50.0	378	1.80	7
10	180.0	379	1.70	100
11	360.0	690	2.00	100
12	380.0	1100	2.10	100
13		2800	2.50	200

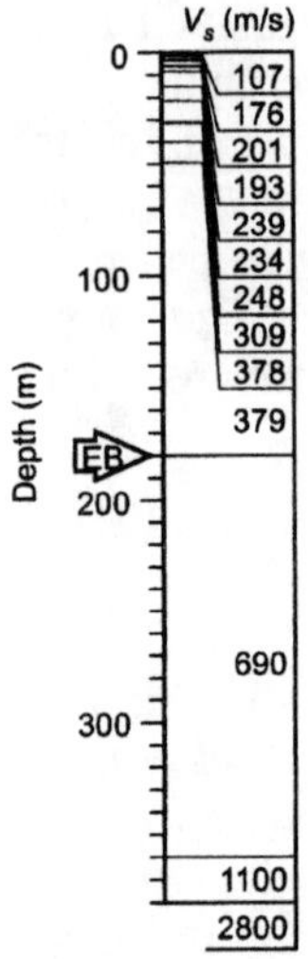

FIGURE 1. Soil Profiles

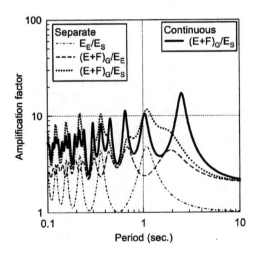

FIGURE 2. Comparison of amplification characteristics

348

As seen in Fig. 2; amplification characteristics by both methods do not agree with enough accuracy. For example, separate analysis does not have peak amplification at around 2.5 seconds and 0.65 second that the continuous analysis has. Since period in this range is important in many structures, design based on the separate analysis may underestimate the earthquake motion at the ground surface. Disagreement is also seen in short period, too. The reason why predominant period of 2.5 seconds does not appear seems clear; there is no peak amplification in the separate analysis in this period. Since amplification ratio at the ground surface is product of two amplifications in the separate analyses, it is impossible to produce large amplification without peak response in one or both amplification characteristics. The same discussion can be made for another peak. For example, since there is peak amplification at about 1 second in the lower ground, and amplification ratio of the surface layer is greater than two, large amplification around 1 second is produced. On the other hand, amplification at about 0.65 second is almost minimal in both the upper and lower grounds and it is the reason why peak of amplification does not appear in the separate analysis.

INVESTIGATION THROUGH EIGEN VALUE PROBLEM

Typical eigen values (predominant periods) and eigen vectors are summarized in Fig. 3. In the figure, circled number denotes mode number. Figure 3(a) shows result of the continuous analysis; eigen mode up to 5th order is shown in the figure. On the other hand, results of the separate analyses are shown in Fig. 3(b). Eigen modes are shown up to the mode whose shortest predominant period is less than that of fifth mode of the continuous analysis. Periods shown in the figure are same with the ones derived from the periods where amplification characteristics become local maximum.

The overall ground from the seismic base layer to the ground surface vibrates

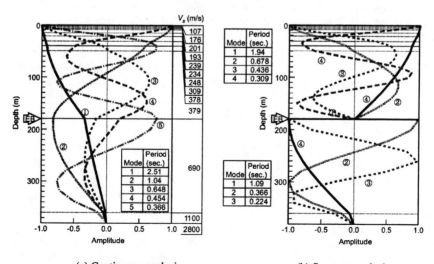

(a) Continuous analysis   (b) Separate analysis

FIGURE 3. Eigen vectors and predominant periods

without reflection point in the first mode in the continuous analysis. Since there is no such vibration mode in two separate analyses, it is natural that the separate analysis cannot produce this mode. The second mode period is 1.04 seconds in the continuous analysis. On the other hand, the period at the first mode of the lower ground in the separate analysis is 1.09 seconds. They are close to each other and this agreement is the reason why period at maximal amplification about 1 second appears in both continuous and separate analyses. In this sense, this agreement can be said to be accidental.

As can be seen, agreement and disagreement of the natural periods between the upper and lower layers determine the ground vibration characteristics of the overall ground by the separate analysis, and predominant periods are determined individually without interaction to each other, there is no guarantee that predominant period of the continuous analysis agrees with the ones by the separate analysis.

WAVE PROPAGATION: INCIDENT AND REFLECTED WAVES

The separate analysis is justified if interaction between the lower ground and the surface ground does not occur. This will occur in two cases. The one is that all reflected wave does not pass the engineering seismic base layer, and the other is that all the energy that the incident wave has is absorbed in the surface ground by such as nonlinear effect. The former case will be discussed in this section by comparing the earthquake motions at the engineering seismic base layer by two analyses.

The earthquake motion at Furofushi site obtained during the 1983 Nihonkai-Chubu earthquake is used in the investigation. The waveforms at the engineering seismic base layer are shown in Fig. 4. Here, response between 35 and 45 seconds are enlarged in order to sea the differences easily although analysis is made considering whole duration. Reflected wave is compared with incident wave in Fig. 4(a). Although magnitude of the reflected wave is smaller than that of the incident wave in general, they are the same order, and are sufficiently larger than zero. In other words, wave traveled downward is not completely reflected at the engineering seismic base layer; quite a bit waves enters into the base.

Incident and reflected waves by the separate analysis are compared with those by the continuous analysis in Fig. 4(b) and (c). It is clear that differences of incident wave caused difference of seismic motion at the ground surface by two analyses.

NONLINEAR ANALYSIS

In two cases that the separate analysis is justified described in the preceding, the possibility that all energy is absorbed by the hysteresis behavior of the surface ground is investigated. Nonlinear behavior is taken into account in the layers up to GL-50m. The ground below GL-50m is treated as elastic media with Q value of 10. Average of dynamic deformation characteristics compiled by Imazu and Fukutake [5] are employed. Accelerations at the ground surface by the continuous and separate analyses are compared in Fig. 5. Maximum acceleration and strain distributions are compared in Fig. 6. They cannot be said to be identical nor similar. Amplification characteristics and acceleration response spectrum with 5% damping ratio are compared in Fig. 7. Difference of amplification characteristics has the same tendency as Fig. 2; peaks of amplification at period of 2.5 and 0.65 seconds are not reproduced in the separate analysis, for example. Acceleration response is underestimated at periods longer than about 0.3 second and that is overestimated at shorter periods. Differences are especially large at period longer than about 1 second.

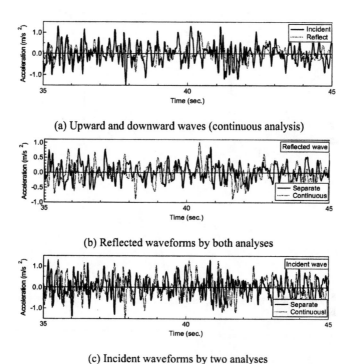

(a) Upward and downward waves (continuous analysis)

(b) Reflected waveforms by both analyses

(c) Incident waveforms by two analyses

FIGURE 4. Ground shaking at engineering seismic base layer

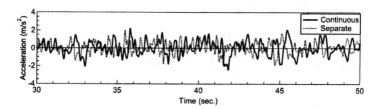

FIGURE 5. Waveforms at the ground surface by nonlinear analysis

351

CONCLUDING REMARKS

The earthquake motion at the ground surface is usually evaluated by the analysis of the surface ground by assuming that incident wave at the engieering seismic base layer is uniquely defined in widespread area. When incident wave at the seismic bedrock is specified, for example, the incidet wave at the engineering seismic base layer is first evaluated by assuming outcrop engineering seismic base layer, and the surface ground is analysized separately under the incident wave obtained in the previous analysis. Validity of this separate analysis is examined by comparing the continuous analysis in

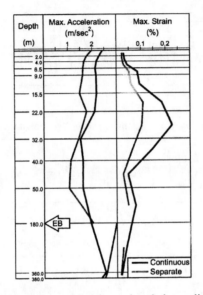

FIGURE 6. Maximum acceleration and strain by nonlinear analysis

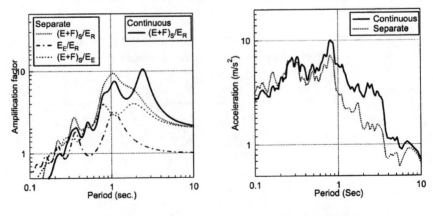

(a) Amplification factor   (b) Acceleration response spectrum
FIGURE 7. Comparison of response in nonlinear analysis

which ground from the seismic bedrock to the ground surface is analysized simultaneously. The following conclusions are obatined.

1) Vibration characteristics of overall ground cannot be obtained by the separate analysis.
2) Behavior at period longer than the predominat period of each separated ground is not reproduced by the separate analysis.
3) All these result indicates that engineering seismic base layer does not work as base, even if shear wave velocity is about 700 m/s which is the largest shear wave velocity used in desigh specificaitons in Japan.
4) The wave that goes down through the engineering seismic base layer and reflect at the seismic bedrock and/or between seismic bedrock and engineering seismic base layer cannot be neglected in evaluating the earthquake motion at the ground surface. In other words, interactive behavior of the grounds above and below the engineering seismic base layer cannot be negrected.
5) Difference appears in the important period range in the geotechnical engineering.

ACKNOWLEDGEMENTS

This study was supported by the Special Project for Earthquake Disaster Mitigation in Urban Areas from the Ministry of Education, Culture, Sports, Science and Technology of Japan.

REFERENCES

1. Midorikawa, S. and, Kobayashi, H. "Spectral characteristics of incident wave from seismic bedrock due to earthquake," Transactions of AIJ, No. 273, 1978, pp. 43-54 (in Japanese)
2. Electric Technical Guideline Committee. "Technical guidelines for aseismic design of nuclear power plants," Japan Electric Association, 1987
3. Yoshida, N., Kobayashi, S., Suetomi, I. and Miura, K. "Equivalent linear method considering frequency dependent characteristics of stiffness and damping," Soil Dynamics and Earthquake Engineering, Vol. 22, No. 3, 2002, pp. 205-222; published at http://boh0709.ld.infoseek.co.jp/
4. Schnabel, P. B., Lysmer, J. and Seed, H. B. (1972): SHAKE A Computer program for earthquake response analysis of horizontally layered sites, Report No. EERC72-12, University of California, Berkeley
5. Imazu, M. and Fukutake, K., "Dynamic shear modulus and damping of gravel materials," Proc., The 21st Japan National Conference of Soil Mechanics and Foundation Engineering, 1986, pp. 509-512

Shear wave velocities from noise correlation at local scale

G. De Nisco[a], C. Nunziata[a], F. Vaccari[b], G. F. Panza[b,c]

[a] Dipartimento di Scienze della Terra, Univ. Napoli Federico II, Italy. E-Mail: conunzia@unina.it
[b] Dipartimento di Scienze della Terra, Univ. Trieste, Italy
[c] The Abdus Salam International Center for Theoretical Physics, ESP-SAND Group, Trieste, Italy

Abstract. Cross correlations of ambient seismic noise recordings have been studied to infer shear seismic velocities with depth. Experiments have been done in the crowded and noisy historical centre of Napoli over inter-station distances from 50m to about 400m, whereas active seismic spreadings are prohibitive, even for just one receiver. Group velocity dispersion curves have been extracted with FTAN method from the noise cross correlations and then the non linear inversion of them has resulted in Vs profiles with depth. The information of near by stratigraphies and the range of Vs variability for samples of Neapolitan soils and rocks confirms the validity of results obtained with our expeditious procedure. Moreover, the good comparison of noise H/V frequency of the first main peak with 1D and 2D spectral amplifications encourages to continue experiments of noise cross-correlation. If confirmed in other geological settings, the proposed approach could reveal a low cost methodology to obtain reliable and detailed Vs velocity profiles.

Keywords: Noise, Rayleigh waves, shear-wave velocities

INTRODUCTION

One of the key parameters for the study of the effects of local site conditions is the S-wave velocity structure of the unconsolidated sediments and the S-wave contrast between bedrock and overlying sediments. These parameters together with the geometry of the bedrock-sediment interface mainly control the behaviour of seismic waves during earthquakes.

The S-wave velocity structure can be obtained through active in-situ measurements such as surface-wave measurements with single stations. A signal analysis performed by a multifilter technique in the frequency-time domain, called FTAN (Frequency Time Analysis) is able to extract the group velocity dispersion curve of the fundamental mode of Rayleigh waves from the recorded signal [1,2]. Typically a weight drop of 30 kg is used as source and one or more low frequency (1Hz) vertical geophones are used as receivers [3]. Only one receiver is requested, or, alternatively, a seismic refraction spreading, in order to evaluate an average group velocity dispersion curve from 4-5 receivers or 4-5 sources. Anyway, at urban sites, the prohibitive use of explosive sources or heavy masses blows, limits the penetration depth to the uppermost 20-30 m, depending upon the rock velocities.

Then, a non linear inversion of the Rayleigh waves group velocity average dispersion curve, with an optimized Monte Carlo method (hedgehog method), gives S-wave velocity profiles with depth and their uncertainty [4, 5, 6, 7]. Such an approach

CP1020, 2008 Seismic Engineering Conference Commemorating the 1908 Messina and Reggio Calabria Earthquake,
edited by A. Santini and N. Moraci
© 2008 American Institute of Physics 978-0-7354-0542-4/08/$23.00

has been successfully tested in different geological environments of the Italian territory, mainly in urban areas and archaeological sites [8].

Passive methods that are based on ambient vibrations or microtremors are attractive because of field simplicity. One method that is based on ambient vibrations is the H/V method of Nakamura [9], defined as the ratio between the mean of the Fourier spectra of the horizontal components and the spectrum of the vertical component, which has proven to be a convenient technique to estimate the fundamental frequency of soft deposits (e.g. [10]). An inversion scheme to retrieve the S-velocity structure from a single ambient vibration record has been proposed for observed H/V ratios from microtremors [11]. Moreover, if Vs models representative of average geological structures are available, the main peak of average H/V spectral ratios is in agreement with the ellipticity of the fundamental mode Rayleigh wave [12]. The ellipticity at each frequency is defined as the ratio between the horizontal and vertical displacement eigenfunctions in the P-SV case, at the free surface.

Recently, cross correlations of long time series of ambient seismic noise have been demonstrated to recover surface wave dispersion (Green function) over a broad range of distances, from a few hundred meters to a several hundred kilometres (e.g. [13, 14, 15]). This approach is based on the work after [16], who proved the identity between the time derivate of full Green's function of the medium structure, inclusive of all reflections and scatterings and propagation modes and the field correlation-function. The basic idea of the method is that a time-average cross correlation of a random, isotropic wavefield computed between a pair of receivers will result in a waveform that differs only by an amplitude factor from the Green function between the receivers. Ambient seismic noise can be considered as a random and isotropic wavefield both because the distribution of the ambient sources responsible for the noise randomizes when averaged over long times and because of scattering from heterogeneities that occur within the Earth. Several researchers have used the noise cross correlation instead of the time derivate (e.g. [13, 17]). This assumption is acceptable from the ambient noise recorded on broadband seismic stations, typically with relatively small bandwidth, being the difference between the cross correlation and its derivate a phase shift. Cross correlations of ambient seismic noise were used also to construct tomographic images of the principal geological units of California, improving the resolution and fidelity of crustal images obtained from surface-wave analysis [15].

In this paper we present the results of noise cross correlation experiments in the historical centre of Napoli, which is prohibitive for active seismic surveys. The quality of the results is checked against the well known stratigraphies and V_S velocity profiles with depth in analogous accessible sites of the urban area.

DATA PROCESSING AND EXPERIMENTAL RESULTS

Three experiments of noise cross correlation have been conducted in the urban area of Napoli over distances from 60 to about 400 m (Fig. 1). The geological setting of Napoli, reconstructed from numerous geotechnical and stratigraphic data [18], is mainly characterized by pyroclastic materials, soil (pozzolana) and rock (tuff), from Campi Flegrei different eruptive centres and Somma-Vesuvio volcano, that often underwent morphological changes in the past due to the influence of the meteoric and

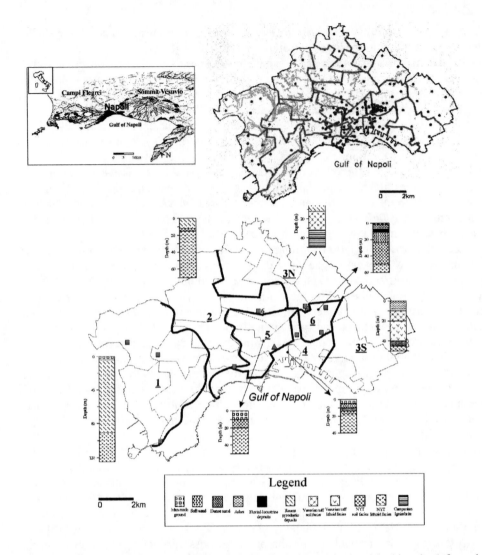

FIGURE 1. Urban map of Napoli with quarter limits and representative stratigraphic column for each of the six geological zones (bold underlined numbers). In the right top, dots represent drillings. Symbols indicate noise measurement sites: grey squares indicate noise and FTAN measurement sites; the grey triangle indicates the location of the noise cross correlation sites.

marine agents and to the urban settlement. Neapolitan soils include sands along the coast, and alternations of volcanic soils, alluvial soils and organic materials. Taking into account the stratigraphies, 6 geologically homogeneous zones have been recognized in Napoli. The tuff horizon is on average at 10-30 m of depth, but deepens to more than 90 m of depth in the western area and more than 40 m in the south-eastern area of Napoli. The compact tuff horizon represents the neapolitan seismic bedrock having Vs > 750 m/s [19].

For the evaluation of the seismic response, Vs data from hole tests have been collected and organized for lithotypes [19]. Wide ranges of variability have resulted, mostly due the pumiceous and scoriaceous nature of the Neapolitan pyroclastic soils. Instead, detailed average Vs profiles over distances of about 100 m have been obtained from the non linear inversion (hedgehog method) of group velocity dispersion curve of Rayleigh waves artificially generated ([19] and references therein). The Vs profiles so obtained have resulted in a very good agreement between noise H/V ratios and computed spectral amplifications, regarding not only the frequency of the main peak but also the peak amplitude. Instead, no correlation resulted with the assumption of down- and cross-hole point measurements [12].

The experiments have been conducted at the historical centre of Napoli (zone 5 in Fig. 1) with limited car traffic (Fig. 2). Noise recordings for at least 1 hour, using a 20-Hz sampling frequency, have been performed at 4 sites, being one, hereafter SMN station, permanently installed at the rock foundations of the SS. Marcellino and Festo monumental complex, an ancient monastery built in 1567. The other 2 sites are: S. Lorenzo Maggiore basilica (in front of the main entrance), the most important gothic church of Napoli, whose construction began in 1266 on an ancient (VI century AD) paleochristian basilica and finished in 1350; the De Laurentiis-Capano palace at the famous street S. Biagio dei Librai, built at the end of 1500, and nowadays hosting the Sociology Faculty of the University of Napoli Federico II.

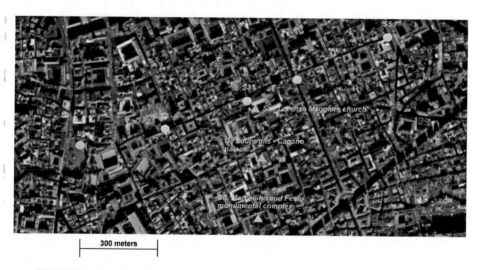

FIGURE 2. Location of the seismic stations (full triangle) employed in the cross correlation experiments and drillings (full circle) at the historical centre of Napoli.

Broadband Kinemetrics Quanterra Q330 stations equipped with 3 component Episensor broadband FBA (Force Balance Accelerometer) sensors have been employed. Another experiment has been conducted at the courtyard of SS. Marcellino and Festo complex by noise recordings with a 24 bit Geometrics StrataVisor

357

seismograph and 1 Hz vertical geophones (Geospace GS-1), with geophone distance of 60 m.

The subsoil of the historical centre of Napoli (Fig. 3) is characterized by a cover of man-made ground, up to about 10 m thick, and pyroclastic soils (pozzolana) overlying a Neapolitan Yellow tuff (NYT) horizon. The geological setting of the investigated area has been reconstructed from a drilling close to S. Lorenzo Maggiore basilica and the measured depth (17 m from the ground level) of NYT tuff at the SMN station (Fig. 4). In the light of the short distance between the end points of the experiment, a stratigraphy has been interpolated below the Sociology Faculty.

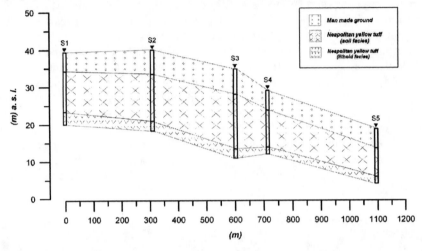

FIGURE 3. Geological cross section of the historical centre of Napoli along the alignment between Dante (S1) and Tribunali (S5) squares, located in Fig. 2.

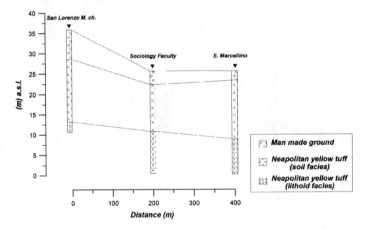

FIGURE 4. Geological cross section representative of the experiment (located in Fig. 2)

Cross correlations have been computed for 2 s windows, for a total duration of 1 hour (Fig. 5) with the Seismic Analysis Code (SAC) [20]. FTAN analysis [1, 2] has been employed on radial and vertical components of noise recordings and non-linear inversion with hedgehog method [4, 6] of the average group dispersion curve of Rayleigh waves has resulted in Vs models compatible with stratigraphies (Fig. 6). A representative solution has been chosen among the solution set with the criterion that the root mean square (rms) of the chosen solution is as close as possible to the average rms, computed from all solutions.

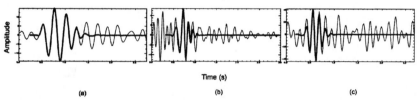

FIGURE 5. Noise cross correlations (thin) and fundamental mode extracted with FTAN method (bold) for the paths: (a) San Lorenzo Maggiore church-SMN station; (b) Sociology Faculty- SMN station; courtyard of S. Marcellino monumental complex.

The chosen solutions clearly show at 15-20 m of depth a V_S increase to 500-600 m/s. According to stratigraphies and velocity ranges for NYT tuff [19], it can be argued that such V_S values can be referred to NYT tuff. At greater depth, below the San Lorenzo Maggiore basilica-SMN station path, further increases of V_S are observed, to 1000 m/s and 1200 m/s at about 70 m and 150 m of depth, respectively. This V_S distribution versus depth is very consistent with the stratigraphy of a deep borehole (400 m) drilled in the Plebiscito square, in front of the Royal Palace, by order of king Ferdinando II in 1859, and close to the investigated area. In fact, in the shallower 150 m, layers of Whitish tuff, 90 m thick, and Campanian Ignimbrite, below a layer of NYT tuff, 50 m thick, were found [21].

At each investigated site the H/V function has been computed as the average from about 150 windows of noise data, each of 2 s length. A cosine tapering (10% of the total duration) has been applied to each window, then the Fourier spectra and H/V ratios have been computed and no smoothing has been applied to the spectra. The H/V ratios have been averaged over all windows and finally have been slightly smoothed (Fig. 7). The frequencies of the main peaks of the H/V ratios are in good agreement with the average 2D spectral amplifications computed, along a representative geological cross section of the zone 5, with a hybrid method consisting of mode summation technique in the 1D part of the model and finite difference technique in the 2D part of the model with heterogeneities ([22, 18] and references therein). A similar fit is not observed at the SS. Marcellino and Festo courtyard and can be explained as a singularity with respect to the average geological subsoil of the zone 5, being a man made refilling. Moreover, the agreement is good also with 1D amplifications computed with Shake program [23] for the chosen hedgehog solutions, meaning that the lateral geological heterogeneities are seismically homogeneities.

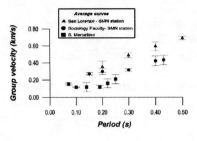

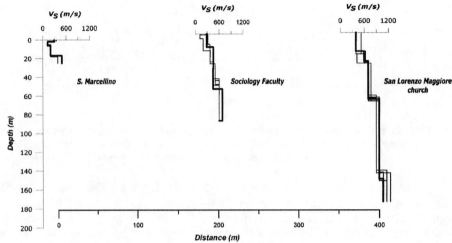

FIGURE 6. V$_S$ solution set obtained with non linear inversion, hedgehog method, of the average group velocity dispersion curves (top). The chosen solutions are indicated in bold.

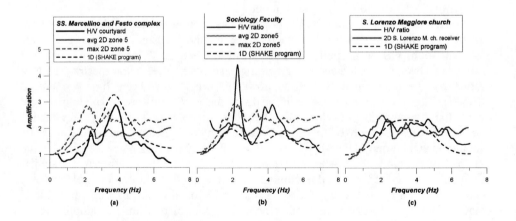

FIGURE 7. Comparison between H/V functions, 1D spectral amplification calculated by means SHAKE program [23] and 2D spectral amplification [18] for the three sites.

CONCLUSIONS

The successful results obtained at Napoli with the non linear inversion of Rayleigh wave group velocity dispersion curve extracted with FTAN method from seismic noise cross correlation are promising a powerful instrument to get Vs measurements in urban areas, almost always prohibitive for active seismic experiments. The proposed methodology is low cost and let us to define Vs measurements at great depths, depending on the stations distance. This is particularly interesting in places like Napoli where it is impossible to reconstruct the thickness of the different eruptions. Hence the methodology is an excellent candidate to perform reliable structural studies necessary for the sound assessment of both seismic and volcanic hazard.

REFERENCES

1. Dziewonski, S. Bloch and M. Landisman, *Bull. Seism. Soc. Am.* **59**, 427-444 (1969).
2. Levshin, L. Ratnikova and J. Berger, *Bull. Seism. Soc. Am.* **82**, 2464-2493 (1992).
3. Nunziata, G. Costa, M. Natale and G.F. Panza, "FTAN and SASW methods to evaluate Vs of neapolitan pyroclastic soils", in *Earthquake Geotechnical Engineering*, edited by P. S. Seco e Pinto, Rotterdam: Balkema, 1999, pp. 15-19.
4. V.P. Valyus, V.I Keilis-Borok and A.L Levshin, Doklady Akad. Nauk SSSR, **185** (8), 564-567 (1968).
5. L. Knopoff and G.F. Panza, A*nn. Geophys.* **30**, 491-505 (1977).
6. G.F. Panza, "The resolving power of seismic surface wave with respect to crust and upper mantle structural models," in *The solution of the inverse problem in Geophysical Interpretation* edited by R. Cassinis, New York: Plenum Press, 1981, pp. 39-77.
7. G.F. Panza, A. Peccerillo, A. Aoudia and B. Farina, *Earth-Science Reviews* **80**, 1-46 (2007).
8. M. Natale, C. Nunziata, G. F. Panza, 2004, "FTAN method for the detailed definition of Vs in urban areas" in *Proc. 13th World Conference on Earthquake Engineering,* Vancouver, B.C., Canada, August 1-6, 2004, in CD-Rom, pp. 11.
9. Y. Nakamura, *QR RTRI* **30**, 25-33 (1989).
10. J. Lermo and F. J. Chávez-García, *Bull. Seism. Soc. Am.* **84** (5), 1350-1364 (1994).
11. D. Fäh, F. Kind and D. Giardini, *Geophys. J. Int.* **145**, 535-549 (2001).
12. C. Nunziata, *Engineering Geology* **93**, 17-30 (2007).
13. M. Campillo and A. Paul, *Science* **299** (5606), 547-549 (2003).
14. P. Roux, K. G. Sabra, P. Gerstoft and W. A. Kuperman, *Geophys. Res. Lett.* **32**, L19303, doi:10.1029/2005GL023803 (2005).
15. N. M. Shapiro, M. Campillo, L. Stehly and M. H. Ritzwoller, *Science* **307**, 1615-1618 (2005)
16. R. L. Weaver and O. I. Lobkis, *J. Acoust. Soc. Am.* **110**, 3011-3017 (2001).
17. N.M Shapiro and M. Campillo, *Geophys. Res. Lett.* **31**, L07614, doi:10.1029/2004GL019491 (2004).
18. C. Nunziata, *Pageoph.* **161** (5/6), 1239-1264 (2004).
19. C. Nunziata, M. Natale, and G.F. Panza, *Pageoph*, **161** (5/6), 1285-1300 (2004).
20. P. Goldstein and L. Minner, *Seism. Res. Lett.* **67** (39), (1996).
21. Perrotta, C. Scarpati, G. Luongo and V. Morra, "The Campi Flegrei caldera boundary in the city of Naples" in *Volcanism in the Campania Plain: Vesuvius, Campi Flegrei and Ignimbrites*, edited by B. De Vivo, Amsterdam: Elsevier, 2006, pp. 85-96.
22. G.F. Panza, F. Romanelli, and F. Vaccari, *Advances in Geophysics* **43**, 1-95 (2001).
23. Schnabel, J. Lysmer and H. Seed, "Shake: a computer program for earthquake response analysis of horizontally layered sites", Rep. E.E.R.C. 70-10, Earthq. Eng. Research Center, Univ. California, Berkeley, 1972.

Low-Frequency Seismic Ground Motion At The Pier Positions Of The Planned Messina Straits Bridge For A Realistic Earthquake Scenario

A.A. Gusev[a,b], V. Pavlov[b], F.Romanelli[c], and G.Panza[c,d]

[a] *Institute of Volcanology and Seismology Petropavlovsk-Kamchatskii, Russia*
[b] *Kamchatka Branch, Geophysical Service Petropavlovsk-Kamchatskii, Russia*
[c] *Department of Earth Sciences, University of Trieste, Italy*
[d] *The Abdus Salam International Centre for Theoretical Physics, Trieste, Italy*

Abstract. We estimated longer-period (period $T > 0.5$ s) components of the ground motion at the piers of the planned Messina straits bridge. As the shortest fault-to-site distance is only 3-5 km, the kinematic earthquake rupture process has to be described in a realistic way and thus, the causative fault is represented by a dense grid of subfaults. To model the 1908 event, we assume a Mw=7 earthquake, with a 40×20 km rectangular fault, and pure reverse dip-slip. The horizontal upper side of the rectangle is at 3-km depth, and the N corner of the rectangle is just between the piers. For the fault nucleation point, the least favorable place is assumed and a randomized rupture velocity is used in a particular run. In a typical simulation, the fault motion is initially represented by the time history of slip in each of the subfaults and by the distribution of the final seismic moment among the subsources (forming "asperities"), both generated as lognormal random functions. The time histories are then filtered in order to fit a chosen source spectral model. The parameters that are conditioning the random functions can be based on the bulk of published fault inversions, or reproduced from an earlier successful attempt to simulate ground motions in the epicentral zone of the 1994, M=6.7 Northridge, California, earthquake. In the second step of calculations, the Green functions (for each subfault and pier combination) are calculated for a layered halfspace model of the pier foundation stratigraphy, using an advanced Green function calculator, that allows an accurate calculation over the entire relevant frequency band including static terms. Finally, the 3-components of the strong ground motion are obtained at the two piers through convolution and summation over the different subsources. We compare a set of response horizontal velocity spectra (PRV) obtained from our calculations with a reference PRV that is considered as a reasonable upper bound for the possible ground motion near the piers. Our results suggest that the seismic ground motion under Torre Sicilia dominates over this under Torre Calabria and that the median (average log) PRV is generally above the reference one, about 1.1-1.3 times for $T > 4$ s, and up to 2 times for $1 < T \leq 4$ s. The use of advanced fault and medium models, accounting also for the natural scatter of individual PRV due to events with the same gross source parameters, provides a sound basis for the deterministic engineering estimates of future seismic ground motion.

Keywords: Ground motion; Seismic source; Fault; Response spectra.

GROUND MOTION COMPUTATION

The Messina straits bridge might be an outstanding engineering achievement. The earthquake engineering challenge is the estimate of the seismic load that should be expected at the sites of the piers of the bridge. As the natural period of the construction should be longer than 2-3 s, only the longer-period component of the motion is estimated. From the seismological point of view, the largest difficulty in the problem is accounting for the earthquake fault motion with large detail, because the shortest distance from a pier to the nearest point on the fault surface is only about 3-5 km. Thus, a sufficiently dense grid of source nodes, from now on called subfaults is needed for the numerical representation of the earthquake fault slip history. As a general earthquake fault model, we adapted a known model of Valensise and Pantosti [1], assuming a $Mw=7$ earthquake, with a 40×20 km rectangular fault, with pure reverse dip-slip motion, with dip angle 29°, and strike N20°E. The upper side of the rectangle (see Figure 1) is buried at 3 km depth, the inclined NNE fault edge is located near to the bridge, and the N corner of the fault rectangle is located just between the piers. Fault nucleation is assumed to occur at the less favorable place, in the middle of the rectangle side farthest from the bridge, thus producing considerable directivity-related relative amplification of ground motion. In a particular simulation run, a random value of rupture velocity was drawn from the interval 2.38-2.72 km/s, or about 75-85% of the average ambient S-wave velocity [2]. We use 33×15=495-element subfault grid, each subfault represents a 1.33×1.25 km piece of the fault surface. The time step is selected as 0.05 s, resulting in sufficient time resolution for long-period (LP) motion.

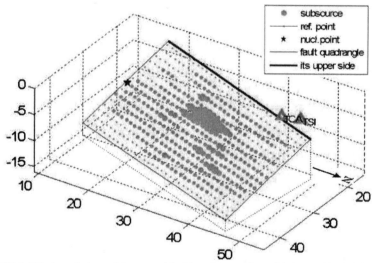

FIGURE 1. 3D view: fault model geographical location together with the position of the two piers: Torre Calabria, TCA and Torre Sicilia, TSI. Star is the rupture nucleation point.

The simulation is performed in three steps (see [3] for a more detailed description of the computational technique). First, fault motion is simulated: it is represented as the time history of the seismic moment rate (or equivalently, of slip) in each of the 495 subfaults. For each subfault, a preliminary version of this time history is first generated as a positive non-stationary random function, called 1D lognormal multifractal, with a duration that is about 10% of the total rupture. The distribution of the final seismic moment among subsources (forming "asperities" of slip function over the fault surface) is also random, and is called 2D lognormal multifractal. I.e., both 2D final slip and any of the 495 subsource time functions are constructed from white Gaussian noise by power-law coloring and exponentiation. As a final step, the preliminary subfault time functions are filtered in order to fit a certain far-field source spectral model: in our case we consider Gusev [4] spectral scaling law. The values of the parameters that conditioned the multifractal random functions are either selected on the basis of published fault inversions, or are reproduced from values used in our earlier study that successfully simulated the ground motion in the epicentral zone of 1994, M=6.7 Northridge, California, earthquake. As the second step we compute 990 Green functions for a layered halfspace, taking each pier, with its stratigraphy [5,6,7,8] as a receiver point, and each subfault as a source. The Green function calculator used employs a novel analytical approach that permits accurate calculation for the entire relevant frequency band including static (near-field) terms. As the final step, the source time functions are convolved with the Green functions and summed, to produce 3-component simulated strong motion at the foundation of each pier (see Figure 2 and 3).

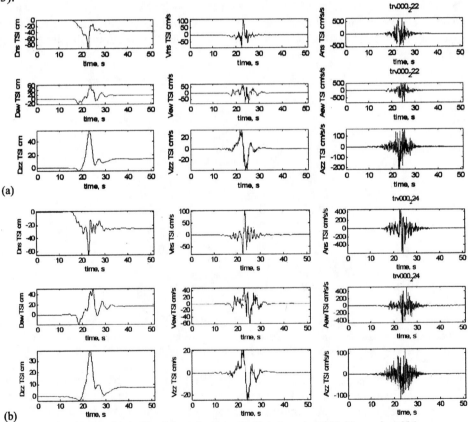

FIGURE 2. Ground motion at pier Torre Sicilia: a): variant 222, b) – variant 224; in each block, from left to right: displacement, velocity and acceleration; from top to bottom: NS, EW and Z component.

364

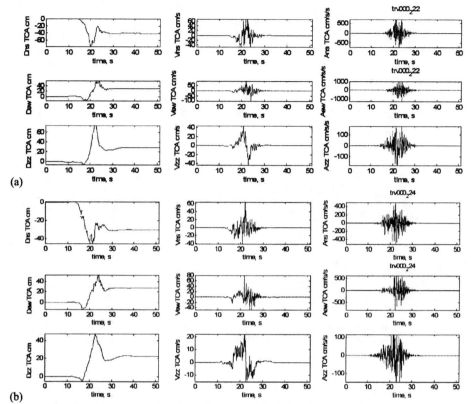

FIGURE 3. Ground motion at pier Torre Calabria: (a): variant 222, (b) – variant 224; in each block, from left to right: displacement, velocity and acceleration; from top to bottom: NS, EW and Z component.

SPECTRAL COMPUTATIONS

As a main engineering output of our simulation we consider the response velocity spectrum, PRV, with a damping coefficient 0.05, and we further discuss only the frequencies below the uppermost accurately simulated frequency of 3 Hz, or periods longer than 0.3 s. As a reference we use the published smooth PRV spectrum proposed by [9] as an upper reasonable limit for ground motion near Messina straits bridge due to a 1908-like event, from now on called SM-PRV.

Our results (see Figure 4, 5 and 6) suggest that ground motions under Torre Sicilia dominate over those under Torre Calabria; thus only the former will be discussed in detail. For any individual sample function of our model ground motion, we see that at some frequencies (periods), there are horizontal spectral ordinates that are above the reference spectrum. If we consider the median (average log or geometric average) spectrum over many sample functions, we observe (see Figure 6) that the median PRV is slightly above the SM-PRV spectrum at all frequencies below 0.25 Hz (periods above 4 s). Between 0.25 and 1 Hz (T=1-4 s), the levels of N and E components of ground motion differ. Whereas the median E component is close to the SM-PRV spectrum, the N component (one oriented roughly along fault slip) is significantly larger, up to 2 times at T=2 s. The r.m.s. deviation of individual spectra corresponds, roughly, to the variation of +65% / -40% with respect to the median. Therefore, if a 84% upper quantile (median+1 r.m.s. dev.) motion is selected as a safe upper limit, it must be positioned at 1.65 times above our median spectrum, or from 1.8 to 2.8 times above the SM-PRV spectrum. The considerable variability of the accelerograms in the individual simulations can be noticed in Figure 7.

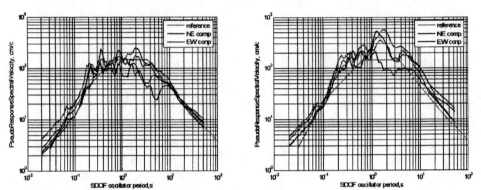

FIGURE 4. PSV Response spectra for horizontal components only, both piers on one plot. variants 222 (left) and 224(right)

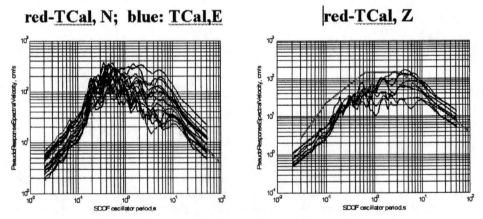

FIGURE 5. PSV Response spectra Torre Calabria only, 8 variants.

Torre Sicilia: **N**   **E**

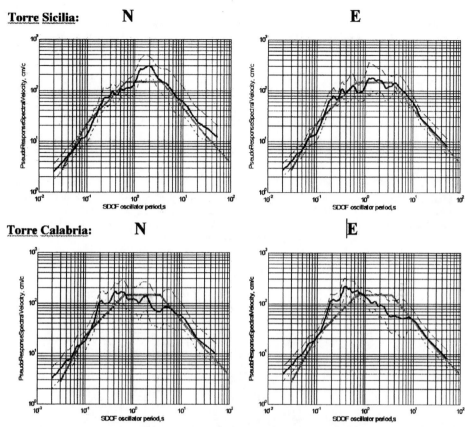

Torre Calabria: **N**   **E**

FIGURE 6 .Median, and +/- 1 r.m.s. deviation corridor, for PRV for two horizontal components and the two piers.

We investigate the uniqueness of the results as follows. We test the variant with the use, instead of Gusev [3] spectral model, of the more traditional one of Brune (as modified by [10]), with stress drop of 30 bar. This resulted in increase of about 30% in the amplitudes of LP ground motion, as we could expect, because the Brune's scaling law is known to overestimate the LP spectra. Therefore we consider the lower predictions based on Gusev's model as more realistic. We verified the natural result that the amplitudes of LP motion can be significantly reduced by selecting another nucleation point. The selected mean value of the rupture velocity, of about 80% of ambient S-wave velocity, has been perturbed, resulting in a systematic variation of the strong motion amplitudes. The selected value is however quite representative for several studied earthquakes, and its use in simulation seems reasonable. From Figure 8 one can appreciate the importance of taking into account the realistic stratigraphy.

A well-known cause of uncertainty in the predicted ground motion is the variation of the stress drop. However, with a fixed fault dimensions and moment magnitude, stress drop variations are essentially suppressed. Therefore, we can consider our estimates as relatively reliable.

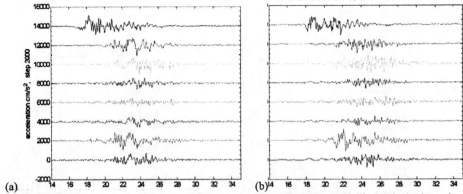

(a) (b)

FIGURE 7. Eight sample functions of the ground acceleration at Torre Sicilia for the horizontal components: (a) NS and (b) EW. Vertical interval between zero-lines of traces is 2000 cm/s^2 . The first trace is for the less usual source sample function, when a large asperity happened to coincide with the spot with the highest permitted propagation velocity.

blue: halfspace; red: layered

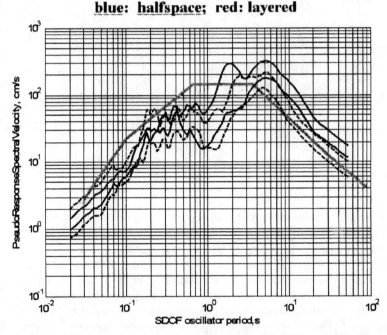

FIGURE 8. Effect of a layered crustal model against the case of a half-space with average-crust properties. Case of Torre Sicilia pier, solid line component NS , dashes - component EW, blue: halfspace; red: layered. The strongest resonance is at 1.5-2 s period, it makes amplification as high as 4-5 times with respect to the average crust. With respect to the "engineering hard rock", such an effect will be significantly reduced.

The reference spectrum SM-PRV in fact has done a quite reasonable job in the determination of the median horizontal PRV spectral shape at T> 3-4 s, and especially in the range T=20-50 s, where accelerogram database provides practically no support.

Our main contribution is the accounting for the statistical variation in individual spectra. In our view, such variations must be taken into account in deterministic engineering estimates of future ground motion. As an example, one can use a 84% upper quantile of the distribution of the spectral ordinates generated from a sufficiently large set of simulated accelerograms.

CONCLUSIONS

We compare a set of response horizontal velocity spectra (PRV) obtained from our calculations with a reference PRV that is considered as a reasonable upper bound for the possible ground motion near the piers. Our results suggest that the seismic ground motion under Torre Sicilia dominates over this under Torre Calabria and that the median (average log) PRV is generally above the reference one, about 1.1-1.3 times for $T>4$ s, and up to 2 times for $1<T\leq4$ s. The use of advanced fault and medium models, accounting also for the natural scatter of individual PRV due to events with the same gross source parameters, provides a sound basis for the deterministic engineering estimates of future seismic ground motion.

ACKNOWLEDGMENTS

The study was supported by the SAND Group of the Abdus Salam International Centre for Theoretical Physics ESP section and in part by the Russian Foundation for Basic Research (Grant 07-05-00775)

REFERENCES

1. L. Valensise and D. Pantosti, "A 125 kyr-long geological record of seismic source repeatability: the Messina Straits (Southern Italy) and the 1908 earthquake (Ms=7.5)", *Terra Nova*, 4, 472-483 (1992).
2. G. F. Panza, A.,Peccerillo, A. Aoudia, and B. Farina, "Geophysical and petrological modeling of the structure and composition of the crust and upper mantle in complex geodynamic settings: The Tyrrhenian Sea and surroundings", *Earth-Science Reviews*, 80, 1-46 (2007).
3. A.A. Gusev, V.M. Pavlov "Broadband Simulation of Earthquake Ground Motion by a Spectrum-Matching, Multiple-Pulse Technique". *Earthquake Spectra*, in press, (2008).
4. A. A. Gusev, "Descriptive statistical model of earthquake source radiation and its application to an estimation of short period strong motion", *Geophys. J. R. Astron. Soc.*, 74, 787-800 (1983).
5. A. Bottari, F. Broccio, B. Federico and E. Lo Giudice, "Preliminary crustal model from seismological observations at the Messina Straits Network", *Annali di Geofisica*, 32, 91-111 (1979).
6. A. Bottari, E. Lo Giudice and D. Schiavone, "Geophysical study of a crustal section across the Straits of Messina", *Annali di Geofisica*, 32, 241-261 (1979).
7. E. Faccioli, "Seismic ground amplification, stability analyses and 3-dimensional SSI studies for the 3300m one-span suspension bridge across the Messina Straits" in *10th European Conf. on Earthquake Eng.*, edited by G. Duma, Balkema, Rotterdam, 1994.
8. G. Neri, G. Barberi, B. Orecchio and M. Aloisi, "Seismotomography of the crust in the transition zone between Thyrrenian and Sicilian tectonic domains", *Geophys. Res. Let.*, 29, n. 23, 2135 (2002).
9. Stretto di Messina, "Approfondimenti relativi al terremoto di progetto per l'opera di attraversamento", Ref. DT/ISP/S/E/R1/001 (2004).
10. W.B.Joyner, "A scaling law for the spectra of large earthquakes", *Bull. Seism. Soc. Am.*, 74, 1167-1188 (1984)

Evaluation of liquefaction potential for building code

C. Nunziata[a], G. De Nisco[a], G. F. Panza[b,c]

[a]Dipartimento di Scienze della Terra, Univ. Napoli Federico II, Italy. E-Mail: conunzia@unina.it
[b]Dipartimento di Scienze della Terra, Univ. Trieste, Italy.
[c]The Abdus Salam International Center for Theoretical Physics, ESP-SAND Group, Trieste, Italy. E-Mail: panza@units.it

Abstract. The standard approach for the evaluation of the liquefaction susceptibility is based on the estimation of a safety factor between the cyclic shear resistance to liquefaction and the earthquake induced shear stress. Recently, an updated procedure based on shear-wave velocities (Vs) has been proposed which could be more easily applied.
These methods have been applied at La Plaja beach of Catania, that experienced liquefaction because of the 1693 earthquake. The detailed geotechnical and Vs information and the realistic ground motion computed for the 1693 event let us compare the two approaches. The successful application of the Vs procedure, slightly modified to fit historical and safety factor information, even if additional field performances are needed, encourages the development of a guide for liquefaction potential analysis, based on well defined Vs profiles to be included in the italian seismic code.

Keywords: Rayleigh waves, Shear-wave velocities, ground motion simulation, liquefaction evaluation.

INTRODUCTION

Semi-empirical procedures for evaluating liquefaction potential of cohesionless soils during earthquakes basically consist of analytical approaches to explain experimental findings of past case histories, and the development of a suitable in-situ index to represent soil liquefaction characteristics. The original simplified procedure for predicting liquefaction resistance of soils was developed by using the Standard Penetration Test (SPT) blow counts correlated with a parameter representing the seismic loading on the soil, called *cyclic stress ratio* (CSR) [1]. Since 1971, procedures also based on the Cone Penetration Test (CPT), Becker Penetration Test (BPT), and small-strain shear-wave velocity (Vs) measurements have been developed. An exhaustive update on the semi-empirical field-based procedures for evaluating liquefaction potential can be found in Idriss and Boulanger [2].

SPTs and CPTs are generally preferred because of the more extensive databases and past experience, but they cannot be applied at sites underlain by gravelly sediments or where access by large equipment is limited. Instead, detailed Vs measurements can be obtained at whichever site, provided that the right methods are used. In the last few years, the Vs-based procedures for evaluating the liquefaction

CP1020, *2008 Seismic Engineering Conference Commemorating the 1908 Messina and Reggio Calabria Earthquake,*
edited by A. Santini and N. Moraci
© 2008 American Institute of Physics 978-0-7354-0542-4/08/$23.00

potential have been revised and updated. A detailed discussion of advantages and disadvantages of in situ V_S test methods can be found in Andrus et al. [3].

Aim of this paper is to illustrate the application of the procedure recently proposed by Andrus et al. [3] in a case study for which detailed geotechnical, V_S models, beside realistic ground motion evaluation, have allowed to validate scientifically the liquefaction of saturated sands occurred for the 1693 earthquake.

SHEAR WAVE VELOCITY MODELS

The key point of V_S-based procedures for evaluating the liquefaction potential is of course the definition of accurate V_S models, capable to handle thin layers too. Detailed V_S profiles with depth can be measured with standard borehole logging and measurements in hole, like down-hole and cross-hole. Such measurements are expensive, very local (point measurements) and may be not representative of large areas. Powerful methods for Vs measurements, that do not need drillings, are all based on the dispersion properties of Rayleigh wave phase and group velocities. Among the appropriate (for liquefaction evaluation) methods for phase velocity measurement of surface waves, we remind the methods SASW [4], and MASW [5]. All these methods need recordings along dense arrays, with small geophone spacing, to avoid spatial aliasing, or, in case of 2 receivers (SASW method), there is the problem of getting the right number of cycles and, hence, the analysis may lead to wrong values (e. g. [6]).

The FTAN (Frequency Time Analysis) method, successfully employed in seismological field, is an innovative method in engineering field. It represents a significant improvement [7] of the multiple filter analysis [8] and can be applied to a single channel to measure group velocity, and, if the source is known and the minimum frequency sufficiently low (< 0.1Hz), phase velocity, even when there is higher modes contamination. FTAN is appropriate to process surface waves data both for the identification and the separation of signals and for the measurement of signal characteristics other than phase and group velocities, like attenuation, polarization, amplitude and phase spectra.

At very local scale (about 100-200m), the analysis consists in extracting the group velocity of the fundamental mode of Rayleigh waves from seismic signals artificially generated by a vertical impact of a weight on the ground and recorded by 1 or more vertical geophones. Very recently, it has been experimented also on signals of ambient noise cross correlations [9]. Detailed V_S profiles with depth are then obtained from the non-linear inversion with the hedgehog method [10, 11, 12] of the average dispersion curve of the fundamental mode of Rayleigh group velocities. The hedgehog method is an optimized Monte Carlo non-linear search of velocity-depth distributions. In the inversion, the unknown Earth model is modeled as a stack of N homogeneous isotropic layers, each one defined by four parameters: V_P, V_S, density and thickness. In the inversion problem of V_S modeling, the parameter function is the dispersion curve of group velocities of Rayleigh fundamental mode. Given the error of the experimental group velocity data, it is possible to compute the resolution of the parameters, computing partial derivatives of the dispersion curve with respect to the parameters to be inverted. The estimation of the resolution of a parameter gives the 'weight' of that parameter in the resulting earth model, and all parameters with low resolution are held fixed. The theoretical velocities computed during the inversion are compared with the corresponding experimental ones. If the root mean square error of the entire data set is less than a value defined a priori on the basis of the quality of the data and if, at a given frequency, no individual computed velocity differs from its experimental counterpart by more than an assigned error, depending upon the accuracy of the measurements, the model is accepted as a solution. From the set of solutions, we accept as representative solution either the one with rms error closest to the average rms error of the solution set (to reduce the projection of possible systematic errors into the structural model [12]) or the one closer to the, when available, reliable geological information.

FTAN measurements have been performed, at engineering scale, in italian urban areas with different soil and rock environments and high seismic risk [13, 14]. Moreover, comparisons between FTAN measurements and down-hole tests have been extensively done and they have indicated a good internal consistency (e.g. [14]).

LIQUEFACTION EVALUATION PROCEDURE

The V_S-based liquefaction potential evaluation procedure [3] requires the calculation of three parameters: (1) the level of cyclic loading on the soil caused by the earthquake, expressed as a cyclic stress ratio; (2) the stiffness of the soil, expressed as a stress-corrected shear-wave velocity; and (3) the resistance of the soil to liquefaction, expressed as a cyclic resistance ratio. Guidelines for calculating each parameter are presented below.

Cyclic Stress Ratio

The cyclic stress ratio, CSR, at a particular depth in a level soil deposit can be calculated [1]:

$$CSR = \frac{\tau_{avg}}{\sigma_v^{'}} = 0.65 \left(\frac{a\max}{g} \right) \left(\frac{\sigma_v}{\sigma_v^{'}} \right) r_d$$

where τ_{avg} is the average equivalent uniform cyclic shear stress caused by the earthquake, $a_{\max}$ is defined as the peak value in a horizontal ground acceleration record that would occur at the site, σ'_v is the initial effective vertical stress at the depth in question, σ_v is the total overburden stress at the same depth, and r_d is a shear stress-reduction coefficient. The variation of r_d with depth z, from the ground surface, may be calculated analytically using site-specific layer thicknesses and stiffnesses, or, alternatively, using the following equations [15]:

$r_d= 1.0-0.00765\ z$ for $z \leq 9.15$m; $r_d=1.174-0.0267\ z$ for 9.15m$< z \leq 23$m

Stress-Corrected Shear-Wave Velocity

Following the traditional procedures for correcting penetration resistance, V_S should be corrected to a reference overburden stress by [3]:

$$V_{S1} = V_S C_{VS} = V_S \left(\frac{Pa}{\sigma_v'} \right)^{0.25}$$

where V_{S1} is stress-corrected shear-wave velocity, C_{VS} is a factor to correct measured V_S for overburden pressure, Pa is a reference stress of 100 kPa. For soil deposits where the coefficient of effective earth pressures at rest, K_0, is significantly different from 0.5, the suggested correction is:

$$V_{S1} = V_S C_{VS} = V_S \left(\frac{Pa}{\sigma_v'} \right)^{0.25} \left(\frac{0.5}{K_0} \right)^{0.125}$$

Cyclic Resistance Ratio

The cyclic resistance ratio, CRR, can be thought as the value of CSR separating liquefaction and nonliquefaction occurrences for a given V_{S1}. The case history data are limited to sites with Holocene uncemented soils (10,000 years), average depths less than 10 m, groundwater table depths between 0.5 m and 6 m, and V_S measurements performed below the water table. The CRR-V_{S1} curves are defined by [3]:

$$CRR = MSF \left\{ 0.022 \left(\frac{K_{a1} V_{S1}}{100} \right)^2 + 2.8 \left(\frac{1}{V_{S1}^* - (K_{a1} V_{S1})} - \frac{1}{V_{S1}^*} \right) \right\} K_{a2}$$

where MSF is the magnitude scaling factor, V_{S1}^* is the limiting upper value of V_{S1} for liquefaction occurrence, K_{a1} and K_{a2} are factors to correct for influence of age on V_{S1} and CRR, respectively. The assumption of a maximum upper value of V_{S1} for liquefaction occurrence is equivalent to the assumption commonly made in the penetration-based procedures dealing with clean sands, where liquefaction is considered not possible above a corrected SPT blow count of about 30 [16] and a corrected cone tip resistance of about 160 [17]. A V_{S1} value of 210 m/s, yielding a CRR value of about 0.6, is considered equivalent to a corrected SPT blow count of 30 in clean sands. Based on penetration-V_S equations and on the case histories, V_{S1}^* is defined depending on the average fines content (FC) of soils as:

$$V_{S1}^* = 215 m/s \qquad \text{for} \qquad FC \leq 5\%$$

$$V_{S1}^* = 215 - 0.5(FC - 0.5) m/s \quad \text{for} \quad 5\% < FC \leq 35\%$$

$$V_{S1}^* = 200 m/s \qquad \text{for} \qquad FC \geq 35\%$$

The magnitude scaling factor is equal to 1 for earthquakes with a moment magnitude of 7.5, and calculated as $MSF = (M_w/7.5)^{-2.56}$ for magnitudes other than 7.5. The factors K_{a1} and K_{a2} are 1.0 for uncemented soils of Holocene age, and, for older soils, they can be estimated with methods based on local SPT-V_{S1} equations and measurements, as illustrated in Andrus and Stokoe [3], Figure 2.

APPLICATION OF THE RECOMMENDED PROCEDURE

In the framework of the Catania Project supported in 1997 by CNR (Consiglio Nazionale delle Ricerche) through the GNDT (Gruppo Nazionale per la Difesa dai

Terremoti), geological, geotechnical and seismic studies were performed in order to define a risk scenario in the town of Catania for a destructive earthquake like that happened in 1693, which caused extensive structural damages, thousands of victims and liquefaction at the La Plaja beach. The 1693, intensity above X, event represents one of the most destructive earthquakes in Sicily and its magnitude should be larger than 7 [18]. The evaluation of liquefaction susceptibility was carried out at the La Plaja beach by the estimation of safety factor for scenario earthquake simulations and the 1990 S. Lucia earthquake [13]. In the following we present the application of the Andrus et al. procedure [3].

The investigated site is located in the northern corner of the Catania plain where recent sediments of fluvial and marine origin are present (Fig. 1). Drillings have been made up to a depth of 20m, both for the lithostratigraphies and for the mechanical CPT (Cone Penetration Test) and SPT (Standard Penetration Test) tests. The typical stratigraphy consists of fine and coarse sands, with natural unit weight of 18.1-18.7 kN/m^3, friction angle of 36-43° and mean grain size diameter of 0.13-0.23mm. The water table is 1.3-2.0m below ground surface, and the clay horizon is underneath the sand layer, as reconstructed by close deep drillings [18].

Seismic measurements with forward and backward spreading of 32 and 64m have been carried out, the source being a vertical impact of a 20 kg weight on the ground surface and the receivers being 4.5 Hz Sensor SM-6 vertical geophones. The recorded signals have been analyzed with FTAN method and an average group velocity dispersion curve of the Rayleigh wave fundamental mode has been obtained. Vs models have been then obtained with the non linear hedgehog method [10, 11, 12] gives an S-wave velocity profile ranging from 120m/s to 350m/s at a depth of 20m. For these models synthetic seismograms have been computed to estimate the gross anelastic properties. A qualitative satisfactory match of the signal shapes has been reached between synthetic and field data for Q values of 70 [13].

Geotechnical data put in evidence a sharp decrease of the cone penetration resistance values q_{c1}, corrected of the overburden pressure and fine content [19], corresponding to a changing of relative density Dr from 75% to about 40%. This trending seems to be correlated with Vs models showing an increase of velocities at 2-3 m of depth, then a decrease of them at 5 m of depth, and again an increase at 11 m of depth (Fig. 2). All models present Vs values of 220 m/s at 5-11 m of depth and 280 m/s up to 20 m of depth [13].

The liquefaction susceptibility has been evaluated both for the recorded S. Lucia earthquake in 1990 and for the scenario earthquake modelling the 1693 (M=7) event. The S. Lucia earthquake (M_L=5.4) was generated by a rupture along the transverse segment of the Iblean-Malta escarpment and recorded at Catania on alluvial deposits 61 m thick. Such accelerogram has been reduced to the seismic basement (Vs=800 m/s) represented by the blue clay horizon through the known soil column with SHAKE program [20]. It has also been simulated an earthquake with M=7-7.5 following the attenuation law after Sabetta and Pugliese [21].

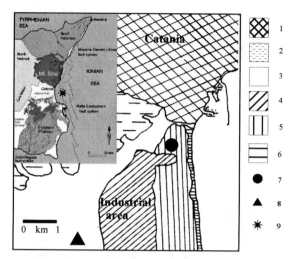

FIGURE 1. Geological sketch map of the investigated site at Catania. Legend: 1) Etna volcanic products; 2) Fluvial deposits of conglomerates and sands; 3) Present and recent alluvial deposits; 4) Marshy deposits; 5) Shore dunes; 6) Present beach deposits; 7) Location of the investigated area; 8) Catania accelerometer station; 9) Epicentre of the December 13, 1990 earthquake.

The scenario earthquake has been simulated starting from the normal Iblean-Malta escarpment (Fig. 1). Two synthetic seismograms have been taken into account and both computed for 2D structural models. That computed by Romanelli and Vaccari [22] is relative to 1D reference model (a=0.44g) and has been attributed to the outcropping seismic basement. Instead, the accelerogram computed by Priolo [23] is relative to a receiver on the top of 75 m soil column. Hence, it has been reduced to 75 m of depth to suppress the amplification effect of the soil column and later assumed as input to our seismo-stratigraphic model (Fig. 2). Following the updated procedures for estimating liquefaction [2, 3], safety factor F is greater than 1 (here a value of 1.2 is assumed after [1]) if we consider as seismic input the S. Lucia accelerogram, both recorded and scaled to magnitude 7-7.5 by assuming the attenuation law [21]. Shore sands become susceptible to liquefaction if scenario earthquakes are assumed. The results are practically the same for the two examined synthetic accelerograms [22, 23] down to 10 m of depth (Fig. 2). At greater depths, the F factor computed with the accelerograms [22] is higher than 1.2, and it is in agreement with the improvement of the geotechnical and geophysical properties at depths higher than 12 m (Fig. 2).

The investigated site fits the requested conditions by the Andrus et al. [3] procedure being Holocene incoherent soils, with groundwater level at 2 m, average depths less than 12 m, and earthquake magnitude of 7-7.5. All the computed cyclic stress values fall in the no liquefaction area, but the CSR values computed for the scenario earthquake, and relative to 5-11 m depths, are very close to the bounding curves.

Taking into account the historical chronicles reporting liquefaction, if we increase slightly (<10%) the V_{s1}^* limiting upper value of V_{s1} for liquefaction occurrence, we

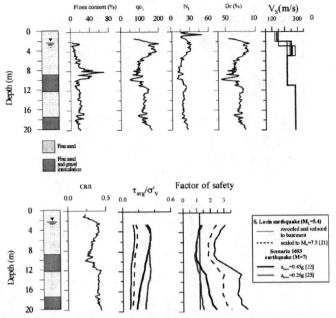

FIGURE 2. Geotechnical and shear seismic properties of shore sands of Catania at La Plaja beach. Top: the fines content, the normalized cone penetration resistance corrected for overburden stress (q_{c1}), the hammer blow count corrected for the effective vertical stress (N_1), and the relative density (Dr) are shown together with the hedgehog solutions and the chosen solution (bold line). Bottom: the cyclic resistance ratio (CRR) [13], the cyclic stress ratio τ_{avg}/σ_v' induced by earthquake [1], and the factor of safety, computed for the Catania accelerogram reduced to the bedrock and for the scenario earthquake.

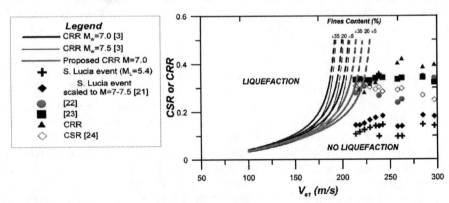

FIGURE 3. Liquefaction resistance curves for magnitudes 7 and 7.5 [3] and the proposed curves in this study. Symbols indicate CSR values computed for the examined earthquakes (see text).

can draw curves that can fit our data (Fig. 3). This implies to assume a V_{S1} of 220-230 m/s for which CRR is about 0.6 and no liquefaction can occurr. The procedure can be expeditiously applied even by assuming the a_{max} from the seismic building code for Catania [24], that is 0.31g for soil type C (180< V_{S30}<360m/s), based on the measured

V_{S30}. The closeness to the Andrus et al. [3] curves can help to make decision for further simulations of ground motion.

CONCLUSIONS

Shear-wave velocity measurements provide a promising approach to liquefaction potential evaluation. The important thing is to use methods like FTAN and hedgehog that are appropriate both for screening large areas and for site-specific evaluations. For evaluating liquefaction potential, of course they must be integrated with geotechnical data and realistic ground motion simulations. Further applications are needed, anyway it could be suggested in the seismic code as a method of screening if the liquefaction hazard is high or not based on the closeness to the Vs- cyclic stress or resistance curves.

REFERENCES

1. H.B. Seed and I.M. Idriss, *J. Geotech. Eng. Div.*, ASCE **97** (3), 458-482 (1971).
2. I. M. Idriss and R. W. Boulanger, "Semi-empirical procedures for evaluating liquefaction potential during earthquakes", 11th ICSDEE & 3rd ICEGE Conferences Proceedings, 2004, pp 32 – 56.
3. R. D. Andrus, K. H. Stokoe II and C. Hsein Juang, *Earthquake Spectra* **20** (2), 285–308 (2004).
4. K. H. Stokoe II and S. Nazarian, "Use of Rayleigh waves in liquefaction studies, Measurements and Use of Shear Wave Velocity for Evaluating Dynamic Soil Properties", ASCE, 1985, pp. 1–17.
5. C. B. Park, R. D. Miller and J. Xia, *Geophysics* **64**, 800–808 (1999).
6. C. Nunziata, *J. of Technical and Env. Geol.* **3**, 25-43 (2005).
7. A. Levshin, V.F. Pisarenko and G.A. Pogrebinsky, *Ann. Geophys.* **28** (2), 211-218 (1972).
8. A. Dziewonski, S. Bloch and M. Landisman, *Bull. Seism. Soc. Am.* **59**, 427-444 (1969).
9. G. De Nisco, C. Nunziata, F. Vaccari, M. Guidarelli, G.F. Panza "Shear wave velocities from noise correlation at local scale", (this iusse).
10. V.P. Valyus, V.I Keilis-Borok,. and A.L. Levshin, "Determination of the upper-mantle velocity cross-section for Europe", Proc. Acad. Sci. USSR, 185, 1969, pp. 3.
11. L. Knopoff and G.F. Panza, *Ann. Geophys.* **30**, 491–505 (1977).
12. G.F. Panza, "The resolving power of seismic surface wave with respect to crust and upper mantle structural models," in *The solution of the inverse problem in Geophysical Interpretation* edited by R. Cassinis, New York: Plenum Press, 1981, pp. 39-77.
13. C. Nunziata, G. Costa, M. Natale and G.F. Panza, *J. Seismology* **3** (3), 253-264 (1999).
14. C. Nunziata, M. Natale, and G.F. Panza, *Pageoph.* **161** (5/6), 1285-1300 (2004)
15. S. S. C. Liao and R. V. Whitman, *J.Geotech. Eng,* ASCE, **112** (3), 373–377 (1986).
16. H. B. Seed, K. Tokimatsu, L. F. Harder and R. M. Chung, *J. Geotech. Eng.,* ASCE **111** (12), 1425–1445 (1985).
17. P. K. Robertson and C. E. Wride, *Can. Geotech. J.* **35** (3), 442–459 (1998).
18. "The Catania project: studies for an earthquake damage scenario", *J. Seismology* (special issue) **3** (3), 211-350 (1999).
19. P.K. Robertson *Can. Geot. Journal* **27** (1), 151-158 (1990).
20. B. Schnabel, J. Lysmer and H. Seed, "Shake: a computer program for earthquake response analysis of horizontally layered sites", Rep. E.E.R.C. 70-10, Univ. California, Berkeley, 1972.26.
21. F. Sabetta, and A. Pugliese, *Bull. Seism. Soc. Am.* **5**, 1491-1513 (1987).
22. F. Romanelli and F. Vaccari, *Journal of Seismology* **3** (3), 311-326 (1999).
23. E. Priolo, *Journal of Seismology* **3** (3), 289-309 (1999).
24. O.P.C.M. 3274/03 "Primi elementi in materia di criteri generali per la classificazione sismica del territorio nazionale e di normative tecniche per le costruzioni in zona sismica".

Realistic Ground Motion Scenarios: Methodological Approach

C. Nunziata[a], A. Peresan[b], F. Romanelli[b], F. Vaccari[b], E. Zuccolo[b], G. F. Panza[b,c]

[a]*Dipartimento di Scienze della Terra, Univ. Napoli Federico II, Italy. E-Mail: conunzia@unina.it*
[b]*Dipartimento di Scienze della Terra, Univ. Trieste, Italy.*
[c]*The Abdus Salam International Centre for Theoretical Physics, ESP-SAND Group, Trieste, Italy. E-Mail: panza@units.it*

Abstract. The definition of realistic seismic input can be obtained from the computation of a wide set of time histories, corresponding to possible seismotectonic scenarios. The propagation of the waves in the bedrock from the source to the local laterally varying structure is computed with the modal summation technique, while in the laterally heterogeneous structure the finite difference method is used. The definition of shear wave velocities within the soil cover is obtained from the non-linear inversion of the dispersion curve of group velocities of Rayleigh waves, artificially or naturally generated. Information about the possible focal mechanisms of the sources can be obtained from historical seismicity, based on earthquake catalogues and inversion of isoseismal maps. In addition, morphostructural zonation and pattern recognition of seismogenic nodes is useful to identify areas prone to strong earthquakes, based on the combined analysis of topographic, tectonic, geological maps and satellite photos. We show that the quantitative knowledge of regional geological structures and the computation of realistic ground motion can be a powerful tool for a preventive definition of the seismic hazard in Italy. Then, the formulation of reliable building codes, based on the evaluation of the main potential earthquakes, will have a great impact on the effective reduction of the seismic vulnerability of Italian urban areas, validating or improving the national building code.

Keywords: ground motion simulation, microzonation

INTRODUCTION

The prediction of the intensity of shaking due to an earthquake before it occurs can prevent damage. Doing this rapidly after an earthquake can be useful for emergency rescue. Understanding and predicting the ground motion reduces the seismic risk.

The seismic response (in terms of both peak ground acceleration and spectral amplification) depends upon the mechanical characteristics of the local soil conditions and the source characteristics of the expected earthquake. The definition of realistic seismic input can be obtained from the computation of a wide set of time histories, corresponding to possible scenarios associated with different seismic sources and structural models. Modeling allows us to account for the large set of potential sources, yet unobserved, which might affect a given site. Moreover, the availability of realistic numerical simulations enables us to estimate the local effects in complex structures exploiting the available geotechnical, lithological, geophysical parameters, topography of the medium, tectonic, historical, paleoseismological data, and seismotectonic

CP1020, *2008 Seismic Engineering Conference Commemorating the 1908 Messina and Reggio Calabria Earthquake,*
edited by A. Santini and N. Moraci
© 2008 American Institute of Physics 978-0-7354-0542-4/08/$23.00

models. The realistic modeling of the ground motion is a very important base of knowledge for the preparation of ground shaking scenarios that represent a valid and economic tool for the seismic microzonation.

In the framework of the UNESCO-IUGS-IGCP project 414 "Seismic Ground Motion in Large Urban Areas", an innovative modeling technique has been proposed that takes into account source, propagation and local site effects. This is done using first principles of physics about wave generation and propagation in complex media, and does not require to resort to convolutive approaches, that have been proven to be quite unreliable, mainly when dealing with complex geological structures, the most interesting from the practical point of view. The general plan [1] included a group of Large Urban Areas and Megacities in the world representative of a broad spectrum of seismic hazard severity, that require different efforts to reach a satisfactory level of preparedness. Objects have been deliberately chosen not situated very close to known seismogenic zones. In fact the condition of being some tens of kilometers from the epicenter allowed us an optimum exploitation of the results of microzoning, and filled in a gap in preparedness, since, usually, most of the attention is focused on very near seismogenic zones.

Aim of this paper is to present an overview of the methodological approach used for the realistic estimation of the seismic ground motion, and its application in Italy, both at regional scale and local scale, with the microzoning of Napoli.

METHOD

Strong earthquakes are very rare phenomena and it is therefore very difficult to prepare a representative database of recorded strong motion signals that could be analyzed to define generally valid ground parameters, to be used in seismic hazard estimations. While waiting for the improvement of the strong motion data set, a very useful approach to perform immediate microzonation is the development and use of modeling tools based, on one hand, on the theoretical knowledge of the physics of the seismic source and of wave propagation and, on the other hand, on the rich database of geological, tectonic, historical information already available.

The flow chart of the algorithm is given in Figure 1. In the hybrid method ([2] and references therein), modal summation is applied along the bedrock (1D) model that represents the average path between the assumed source and the local, laterally heterogeneous (2D) structure beneath the area of interest. Double-couple point source of seismic waves is assumed, described by the strike, dip of the causative fault, its rake and the depth of focus. The Earth's model is defined as a stack of horizontal layers, each of which is characterized by the longitudinal and transversal wave velocity within the layer, the density, thickness and the Q-factor controlling the anelastic attenuation. The seismograms are computed for frequencies up to 5–10 Hz. These signals are numerically propagated through the laterally varying local structure by the finite differences method. Synthetic seismograms of the vertical, transversal and radial components of ground motion are computed at a predefined set of points at the surface. After scaling the signal's spectra to the assumed seismic moment by using the curves proposed by [3] as reported in [4], the ratios of peak ground acceleration, PGA(2D)/PGA(1D), and the response spectra ratio, RSR, i.e. the response spectra computed from the signals synthesized along the laterally varying section normalized by the response spectra computed from the corresponding signals, synthesized for the bedrock, are routinely extracted from the computed seismograms.

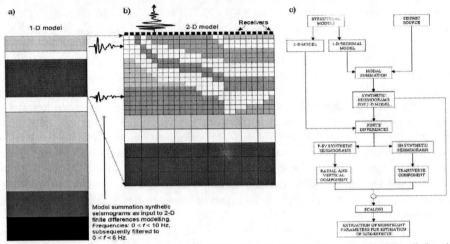

FIGURE 1. Schematic representation of the hybrid technique. The modal summation is applied to the bedrock model (a) to compute the input signals for the laterally varying part (b) where signals are propagated by finite differences to a set of sites on the surface. (c) Flow chart of the algorithm.

GROUND MOTION SCENARIOS AT BEDROCK

The realistic modelling of ground motion allows us the computation of complete seismograms, corresponding to a wide set of possible scenarios associated with different structural models and seismic sources, yet unobserved, which might affect a given site. According to the proposed methodology, the potential seismic source models are defined accounting both for the seismogenic zones defined by GNDT [5] and for the information obtained from historical seismicity, based on earthquake catalogues and inversion of isoseismal maps. In addition, morphostructural zonation and pattern recognition of seismogenic nodes are used to identify areas prone to strong earthquakes [6,7], based on the combined analysis of topographic, tectonic, geological maps and satellite photos. Further constraints about the space and time of occurrence of the impending strong earthquakes can be included, as provided by formally defined and globally tested algorithms for intermediate-term middle-range earthquake predictions (e.g. CN and M8S algorithms), following the approach proposed by [8,9].

For a first-order seismic zonation (Fig. 2a) of the Italian territory, a database of synthetic seismograms is generated, by means of full waveforms modeling at bedrock [2]. In the first-order zoning, seismic sources are defined considering the earthquake catalogue and the seismogenic zones, with their representative focal mechanisms [5]. Synthetic seismograms are then computed for sources and receivers placed at the nodes of a grid with step 0.2°x0.2°, covering the Italian territory.

The lateral heterogeneity of the medium is taken into account by making use of polygons with different structural models [2]. The synthetic signals (radial and transverse component) are computed for an upper frequency limit of 1 Hz, and are properly scaled using the moment-magnitude relation given by [10] and the spectral scaling law proposed by [3] as reported in [4]. The deterministic results are extended to higher frequencies by using design response spectra [11], for instance Eurocode 8 [12], which defines the normalized elastic acceleration response spectrum of the ground motion, for 5% critical damping. Since the used regional structural models for Italy [1] are all of type A, we can immediately determine the design ground acceleration (DGA) using the EC8 parameters for soil A (Fig. 2a).

In a complementary step, the areas prone to strong earthquakes are identified based on the Morphostructural Zonation (MZS) Method [13] that delineates a hierarchical block structure of the studied region, using tectonic and geological data. Among the defined nodes, represented as circles of radius R=25 km surrounding each point of intersection of the morphostructural lineaments, those prone to strong earthquakes are then identified by the pattern recognition on the basis of the parameters characterising indirectly the intensity of neo-tectonic movements and fragmentation of the crust at the nodes. The procedure has been applied in Italy by [6,7] for two magnitude thresholds, M≥6.0 and M≥6.5.

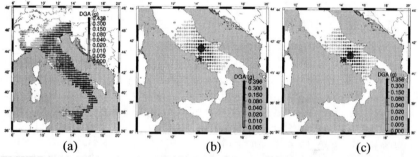

FIGURE 2. (a) Acceleration wave-field computed for the first order zoning of Italy extended to frequencies > 1Hz by using Eurocode 8 design response spectra [11]. Maps of DGA generated for M=6.5 earthquakes at the nodes (full circle) (b) I073 and (c) I084 [6], that are responsible of the highest hazard for the city of Napoli (star).

In the second-order zoning, seismogenic nodes are used instead of seismogenic zones in the definition of sources for the computation of synthetic seismograms. The same procedure followed for the first-order zoning is applied, treating each seismogenic node separately. If the magnitude assigned to the cells that belong to a node, based on historical data, is lower than the magnitude threshold of the node indicated by the morphostructural analysis, the used magnitude is the magnitude threshold of the node. This choice allows us to consider strong earthquakes for areas where they are not yet observed but which are recognized prone to strong earthquakes. In fact, a number of nodes where strong events have not been recorded to date, have been recognized to be prone to large earthquakes and they naturally indicate that the seismic potential is generally higher than the one consistent with historical and instrumental records. According to the neo-deterministic procedure [1,9,14], each node can be associated with maps describing the associated potential seismic ground motion.

We supply examples of two scenarios corresponding to nodes prone to magnitude M≥6.5 earthquakes, close to the city of Napoli (Fig. 2b,c). The two nodes (I073 and I084 in [6]), are located in the Sannio area, nearby or including the 1456 and 1688 earthquake epicentres, which caused major damage in Napoli. Figure 2b,c shows the possible scenarios of the maximum design ground acceleration (DGA). However, each site is associated to synthetic seismograms of many different sources and any parameter of interest can be extracted from such complete time series, therefore different maps can be produced. To apply the innovative engineering techniques for seismic isolation, for example, it is particularly relevant to estimate the maximum ground displacement (PGD), which is provided in Table I.

TABLE 1. Peak ground displacement (PGD), peak ground velocity (PGV) and maximum estimated intensity (I_{max} computed), calculated for each of the two nodes considered for the city of Napoli. Intensities are computed using the conversion tables proposed by [15], based on the observed intensities from ING [16] and ISG [17] data sets.

Node	PGD (cm)	PGV (cm/s)	I_{max} computed		I_{max} observed
			ING	*ISG*	
I073	3.5 – 7.0	4.0 – 8.0	IX	VIII	VIII
I084	7.0 – 15.0	8.0 – 15.0	X	IX	VIII

SEISMIC MICROZONING: THE EXAMPLE OF NAPOLI

Shear wave velocity (V_S) plays a fundamental role in the seismic ground motion amplification and it is, in fact, an essential ingredient of the seismic building codes. Standard borehole logging and measurements in hole, like down-hole and cross-hole, are expensive and not suitable to urbanized settings. Such measurements are point measurements and may be not representative of the properties along paths effectively crossed by seismic waves. Detailed V_S profiles with depth are then obtained from the non-linear inversion ([18,19] and references therein) of the average dispersion curve of the fundamental mode of Rayleigh group velocities. FTAN measurements have been performed at the urban area of Napoli [20], and comparisons with down-hole tests have indicated a good internal consistency.

Microzoning for the 1980 Earthquake

Based on the historical record, the largest experienced intensity at Napoli is VIII on the MCS scale. Therefore Napoli is in the particular situation where an appropriate action of retrofitting can drastically reduce structural damage. In fact intensity VIII corresponds to large structural damage, while intensity VII corresponds to minimal damage. The 1980, Irpinia earthquake is a good example of a strong shaking at Napoli since it caused intensity about VIII (MCS). The 1980 earthquake (M_S=6.9, M_L=6.5) is the only strong event recorded close to Napoli, at the seismic station Torre del Greco. The recorded seismograms at Torre del Greco have been used to validate the numerical simulations of ground motion at Napoli, and they have been modeled by [21]. The seismic ground motion in the urban area of Napoli for the 1980 earthquake, distant about 90-100 km, has been computed within each of the 6 recognized geological zones (Fig. 3). A parametric study has been performed considering different shear wave velocity profiles, based on down- and cross-hole tests and inversion of Rayleigh group velocities [20]. The historical centre (zone5) is noticeably important because of its concentration of monuments and it is characterized by the presence of several man made cavities, whose effect on seismic ground motion is negligible [22]. For several V_S models, parameters of the ground motion have been computed and average and maximum values have been plotted (Fig. 3). From the correlation, obtained for the Italian territory, between synthetic PGA and MCS intensity [15], the computed accelerations correspond to intensity around VIII, in agreement with observed data [22].

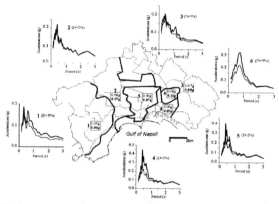

FIGURE 3. Average (thin line) and maximum (thick line) response spectra of the radial component of the ground motion computed at each geological zone of Napoli for the 1980 quake. Damping has been assumed according to the building material characteristics.

Scenario Earthquake

Major damage at Napoli was caused by earthquakes from Molise and Benevento areas. In addition, the pattern recognition of seismogenic nodes (RSN) in the Italian region [6,7] identifies two nodes (nodes I073 and I084) in the Sannio area, prone to magnitude M≥6.5 earthquakes, containing or in the proximity of the 1456 and 1688 earthquake epicentres. Taking into account the long seismic quiescence of the Sannio seismogenic area from 1805, only broken by the 1962 earthquake, the area can be realistically considered prone to strong earthquakes. Both the 1456 and 1688 earthquakes can be considered representative of the Sannio seismicity, but the former is believed to have been composed by at least 3 sub-events. The uncertainty on 1456 earhquake complex source, and the highest (in the central-southern Apennines) magnitude of 7.3 attributed to 1688 event, have suggested to consider the 1688 seismic event as the strongest earthquake scenario to simulate the ground motion at Napoli ([23] and references therein). Based on the available information, for the simulation of the ground motion at Napoli for the 1688 earthquake, strike (135.85°) and dip (84.77°) have been assumed and the rake has been varied in order to simulate transcurrent (rake=356.61°) and a normal (rake=90°) faulting, respectively. Hypocentre depths of 7 and 10 km have been assumed. Acceleration time series for P-SV and SH-waves have been computed along a NE-SW geological section oriented towards the epicentre and crossing the area occupied by the most damaged buildings, that is present-day historical centre (Fig. 4). Seismograms have been scaled for magnitudes 7.3 (NT4.1 catalogue), 6.7 (CPTI04 catalogue), and 6.1 (CCI1996 catalogue). The analysis of computed seismograms shows that the peak ground acceleration and the response spectra are similar for strike-slip and normal faults, being different only the component of the ground motion where the maximum is recorded, transverse and radial respectively. With respect to the bedrock structure, the pyroclastic soil cover is responsible for the doubling of PGA that reaches maximum values of 0.1 g and 0.3 g, for 6.7 and 7.3 magnitudes, respectively. A similar trend of spectral amplifications is seen along the cross section, but at receivers S21-S24, where no building existed (Fig. 4) at the time of the quake. From the synthetic PGA-intensity correlation valid for the Italian territory [15], one obtains that computed and observed intensities are comparable if seismograms are scaled for M=6.4. Anyway, a scenario earthquake like the 1688 earthquake with M=7.3 must be taken into consideration when reinforcement of ancient buildings is performed, to take into account their degradation in the last three centuries.

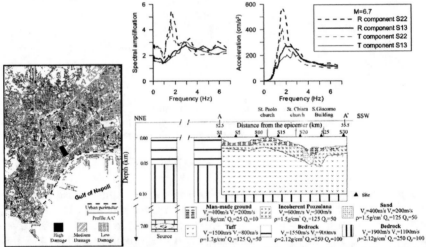

FIGURE 4. Geological cross section at the historical centre of Napoli crossing the damaged area by the 1688 earthquake (on the left modified from [23]). A schematic representation of the hybrid method is also shown. In the top, the spectral amplifications and response spectra (5% damping) computed for the radial and transverse components, scaled for a source magnitude of 6.7, representative of the majority of receivers (S13) and the receiver S21-S24 (S22) are also shown.

CONCLUSIONS

The approach described is capable to provide, in a reasonable amount of time, large sets of realistic seismic signals and related quantities of earthquake engineering interest. Thus it makes it possible to obtain the definition of the seismic input at low cost and exploiting large quantities of existing data (e.g. geotechnical, geological, seismological). Various earthquake scenarios, i.e. earthquakes with different focal mechanisms, epicentral distances, azimuths and magnitudes, can be assumed compatible with the active faults and seismogenic nodes in the neighboring areas. A key point for detailed microzonation studies is a robust definition of Vs profiles with depth, which can be achieved from the non linear inversion of Rayleigh group velocities. The method is particularly expeditious and suitable in highly noisy and urbanized centers. Napoli is a good example of application of the hybrid method as it is a large town close enough to a seismogenic area to suffer serious damage both because of the degraded conditions of the historical built environment and because severe local site amplification occurs. The high density of population and the kind of built environment increase the vulnerability of some parts of Napoli. A preventive definition of the seismic hazard can be obtained immediately, without having to wait for another strong event to occur, from the systematic computation of time histories corresponding to possible seismotectonic scenarios for different sources and structural models.

REFERENCES

1. G. F. Panza, L. Alvarez, A. Aoudia, A. Ayadi, H. Benhallou, D. Benouar, Chen Yun-Tai, C. Cioflan, Ding Zhifeng, A. El-Sayed, J. Garcia, B. Garofalo, A. Gorshkov, K. Gribovszki, A. Harbi, P. Hatzidimitriou, M. Herak, M. Kouteva, I. Kuznetzov, I. Lokmer, S. Maouche, G. Marmureanu, M. Matova, M. Natale, C. Nunziata, I. Parvez, I. Paskaleva, R. Pico, M. Radulian, F. Romanelli, A. Soloviev, P. Suhadolc, P. Triantafyllidis, and F. Vaccari, *Episodes* 25, No. 3, 160-184 (2002).
2. G. F. Panza, F. Romanelli and F. Vaccari, *Advances in Geophysics* 43, 1-95 (2001).
3. A. A. Gusev, *Geophys. J.R. Astron. Soc.* 74, 787-800 (1983).
4. K. Aki, *Strong motion seismology: Strong Ground Motion Seismology*, NATO ASI Series C., D. Reidel Publishing Company, M.O. Erdik and M.N. Toksoz (eds.), Dordrecht, 1987, 204, pp. 3-39.
5. C. Meletti and G. Valensise, Zonazione sismogenetica ZS9 – App.2 al Rapporto Conclusivo per il Dipartimento della Protezione Civile, INGV, Milano-Roma, aprile 2004.
6. A. I. Gorshkov, G. F. Panza, A. A. Soloviev and A. Aoudia, *J. Seism. Earthq. Eng.* 4, 1-24 (2002).
7. A. I. Gorshkov, G. F. Panza, A. A. Soloviev and A. Aoudia, *Boll. Soc. Geol. It.* 123, 3-18 (2004).
8. A. Peresan, G. F. Panza, A. Gorshkov and A. Aoudia, *Boll. Soc. Geol. It.* 1 (parte 1), 37-46 (2002).
9. A. Peresan, E. Zuccolo, F. Vaccari and G. F. Panza, *Boll. Soc. Geol. It.*, In press (2008).
10. H. Kanamori, *J. Geophys. Res.* 82, 2981-2987 (1977).
11. G. F. Panza, F. Vaccari, G. Costa, P. Suhadolc and D. Fäh, *Earthquake Spectra* 12, 529-566 (1996).
12. EC8, Eurocode 8 structures in seismic regions, Doc TC250/SC8/N57A, 1993.
13. M. A. Alekseevskaya, A. M. Gabrielov, A. D. Gvishiani, I. M. Gelfand and E. Ya. Ranzman, *J. Geophys.* 43, 227-233 (1977).
14. E. Zuccolo, F. Vaccari, A. Peresan, A. Dusi, A. Martelli, G. F. Panza, *Eng. Geol.*, In press (2008).
15. G. F. Panza, F. Vaccari and R. Cazzaro, *Annali di geofisica*, 15, 1371-1382 (1997).
16. E. Boschi, P. Favalli, F. Frugoni, G. Scalera and G. Smriglio, *Mappa Massima Intensità Macrosismica risentita in Italia*, Istituto Nazionale di Geofisica, Roma, 1995.
17. D. Molin, M. Stucchi, G. Valensise, *Massime intensità macrosismiche osservate nei comuni italiani*, elaborato per il Dipartimento della Protezione Civile, GNDT, ING, SSN, Roma, 1996.
18. G. F. Panza, "The resolving power of seismic surface wave with respect to crust and upper mantle structural models", In *The solution of the inverse problem in Geophysical Interpretation.*, R. Cassinis (ed.), Plenum press, 1981, pp. 39-77.
19. G. F. Panza, A. Peccerillo, A. Aoudia and B. Farina, *Earth-Science Reviews* 80, 1-46 (2007).
20. C. Nunziata, M. Natale, G. F. Panza, *Pure and Applied Geophysics* 161, 1285-1300 (2004).
21. C. Nunziata, G. Costa, F. Marrara and G. F. Panza, *Earthquake Spectra*, 16, 643-660 (2000).
22. C. Nunziata, *Pure and Applied Geophysics*, 161, 1239-1264 (2004).
23. C. Nunziata, M. Natale, F. Imperio, G. F. Panza, "Realistic ground motion scenarios for Napoli" in Proc. 8th U.S. Conference on Earthquake Engineering, San Francisco, 2006, Paper n. 819.

Seismic Vulnerability Assessment for Massive Structure: Case Study for Sofia City

Ivanka Paskaleva[a], Gergana Koleva[b], Franco Vaccari[c]
and Giuliano F. Panza[d]

[a] CLSMEE-BAS, 3 Acad. G. Bonchev str, 1113 Sofia, Bulgaria, E-mail:paskalev_2002@yahoo.com
Tel: +359-2-9793341, Fax: +359-2-9712407.

[b] CLSMEE-BAS, 3 Acad. G. Bonchev str, 1113 Sofia, Bulgaria, E-mail: gvkoleva@gmail.com
Tel: +359-2-9793319, Fax: +359-2-9712407.

[c] DST - University of Trieste, E. Weiss 4, 34127 Trieste, Italy, E-mail: vaccari@dst.units.it
ICTP, Tel: +39-040-5582119, Fax: +39-040-5582111, Trieste, Italy.

[d] DST - University of Trieste, E. Weiss 4, 34127 Trieste, Italy, E-mail: panza@units.it,
Tel: +39-040-5582117, Fax: +39-040-22407334, ICTP-SAND group, Trieste, Italy.

Abstract. An advanced modeling technique, which allows us to compute realistic synthetic seismograms, is used to create a database of synthetic accelerograms in a set of selected sites located within Sofia urban area. The accelerograms can be used for the assessment of the local site response, represented in terms of Response Spectra Ratio (RSR). The result of this study, i.e. time histories, response spectra and other ground motion parameters, can be used for different earthquake engineering analyses. Finally, with the help of 3D finite elements modeling, the building structural performance is assessed.

Keywords: seismic hazard, synthetic seismograms, damage estimations.

INTRODUCTION

Seismic hazard and seismic risk are high in Sofia city and the surrounding area. Tall buildings, big dams and social structures can be subject to considerable social and environmental risk, and generate casualties and huge economic losses if hit by strong seismic events. Seismic loading is one of the approaches to seismic risk reduction in engineering and building practice. The retrofitting of structures can be properly designed by means of seismic loading computation [11], based on models derived from accurate geological and geo-technical site studies.

SITE-DEPENDENT SEISMIC MOTION COMPUTATION

According to the Bulgarian seismic code'87, Sofia has been included in a seismic category characterized by intensity IX (MSK), which corresponds to a horizontal acceleration of 0.27g for the anchoring of the elastic response spectrum. A detailed ground motion modeling has been done for some scenario earthquakes, chosen according to the seismotectonic regime of the area. Broadband synthetic seismograms

CP1020, *2008 Seismic Engineering Conference Commemorating the 1908 Messina and Reggio Calabria Earthquake,*
edited by A. Santini and N. Moraci
© 2008 American Institute of Physics 978-0-7354-0542-4/08/$23.00

have been computed in a laterally heterogeneous anelastic medium with the hybrid approach [9], [22]. The modal part [23], [10] is used to model wave propagation from the source to the beginning of the local profile of interest. It is based on the computation of eigenvalues (phase velocity) and eigenfunctions associated with Rayleigh (P-SV motion) and Love (SH motion) waves for a laterally homogeneous medium (layered anelastic halfspace). The phase velocity-frequency space is efficiently explored [22] also at high frequencies, where nearby modes get extremely close to one another. Anelasticity is taken into account considering, for each layer, a frequency-independent Q value, which expresses the attenuation term in space and time. In the hybrid approach, the seismograms obtained at the bedrock with the modal summation are the input for the propagation of the wavefield along the selected local profiles, where the finite difference technique allows us to deal with complicated lateral heterogeneities, and therefore to adequately estimate site effects.

Here, using the specific knowledge about geology and geotechnical properties described in the cartographic material available for the Sofia area, three profiles (local 2D sections) have been considered to compute several ground shaking scenarios, with magnitude equal to 7 and different source positions. One bedrock model has been taken into account for propagating the wavefield from the source to the beginning of the 2D sections, [26]. This model has been used for the computation of the reference signal (seismograms at the bedrock), to be compared with the results of the detailed modeling for the definition of the site effects, described by the ratios of the response spectra (2D/bedrock). The three-component synthetic seismograms, computed in the domain of displacement, velocity and acceleration, have been processed to extract some parameters significant from the engineering point of view.

BACKGROUND

In the computations carried out in this study, based on the earthquake history at Sofia and on the available seismic hazard assessments provided in the literature, one shallow earthquake scenario is considered. A computation is made with respect to a local seismic source that can strike just beneath the city. The earthquake epicenter corresponds to a real seismic event that struck Sofia in 1858. A generalized scheme of the model adopted for the numerical experiments is shown in Fig.1. The seismic waves propagation path consists of the traveled path between the source and the target site ("bedrock structure") and the target local cross sections. The data used to build-up the local structural models, up to 1000 m below the surface, are obtained from a large set of boreholes and geological cross sections [17], [18], [19], [28]. The deep part of the local model, describing the structure below 1 km, coincides with the bedrock velocity model available in the literature and is assumed to be the same for the whole Sofia Kettle [32].

To model such a large event at relatively small distances – 10 km – the simple scaling by Gusev [15], as reported by Aki [1], is not capable to model the influence of the rupture process. Therefore, the algorithm for the simulation of the source radiation from a fault of finite dimensions, named PULSYN (PULse-based wide band SYNthesis) is applied [16]. The source is modeled with a cluster of point sources, distributed along linear segments, each characterized by its own time function.

FIGURE 1. Sofia area with the three profiles superimposed ("Sofia 1", "Sofia 2", "Sofia 3"), [20].

For producing the synthetic seismogram at a specific site, the time function of each sub-force is multiplied by the proper Green function, and seismic moment tensor of the corresponding point source. This approach produces broadband signals which reproduce directivity effects and the rupture process of the source.

Realistic synthetic seismograms are computed for all sites of interest along the profiles shown in Fig. 1 (~100 sites per profile). By "realistic" we mean that the modeling takes into account simultaneously the geotechnical properties of the site, the position and geometry of the seismic source and the mechanical properties of the propagation medium, following basic laws of physics and avoiding standard convolutive approaches that, as described in [25], have a quite questionable validity. Acceleration, velocity and displacement time histories are obtained for transverse (TRA), radial (RAD) and vertical (VERT) components considering three different azimuths (0°, 90°, 180°) with respect to the rupture propagation.

This work is focused on profile "Sofia 2"(2-2 see Figure 1).

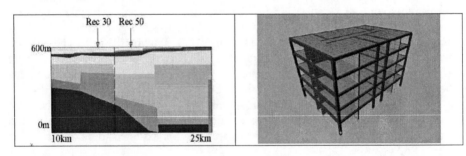

FIGURE 2. Laterally varying 2D model "Sofia 2" cross the town in WE - SN direction.

FIGURE 3. Toggle view of the 3D building model.

The building calculations have been carried out for two characteristic sites (at epicentral distances 13km – Rec. 30, 15km – Rec. 50). The acceleration time histories are scaled according to the BG code' 87.

REGIONAL STRUCTURAL MODEL

The input data, necessary for the ground motion simulation with the hybrid approach, consist of the regional bedrock model, the laterally heterogeneous local model, and the earthquake source model.

Following the investigations carried out on the recent geodynamic features of the Sofia complex, a geological and geophysical 3D model of the Earth's crust for this region has been derived [29]. The analysis of the seismotectonic setting and of the structure of the basement of the Sofia depression is used to specify the reference structural model. The P-wave velocities of the 1D regional structural model are taken from [29], and to be conservative for the S-wave velocities we assume $V_P=2V_S$. This leads to slightly lower S-wave velocities than the ones valid for Poissonian solids and therefore higher ground motion are expected at the free surface.

No specific information exists about the quality factor for P-waves (Qp) for the region, so the values adopted have been taken from [8] and the widely applied rule Qp=2.2*Qs, [33] has been used to derive the quality factor for S-waves (Qs).

GROUND SHAKING SCENARIO FOR THE PROFILES

The maximum macroseismic intensity at Sofia, I=IX (MSK), observed in 1858 [3], can be expected to occur within a period of 150 years [5], i.e. it could correspond to the strong scenario earthquake. Recently seismic hazard maps of the Circum-Pannonian Region [13], [24], show that Sofia is placed in a node having potential for the occurrence of an earthquake with M>6.5 and that it could suffer macroseismic intensity up to X. The intensity expected in the Sofia region is associated with earthquakes occurring in the upper 17-20 km of the lithosphere. Macroseismic intensity I about IX can be expected in Sofia [12], if an earthquake with magnitude M=7 located at 10 km-15 km distance from the center of the city in the West-South direction [2], [5], [30], [21], [31] occurs. In the modeling, the source parameters are chosen to approximate the seismic event that hit Sofia in 1858. The parameters of the source mechanism adopted are: strike angle 340°, dip angle 78° and rake angle (with respect to strike) 285° [25].

BUILDING MODEL

The building considered is a five storey reinforced concrete building (H=15m), with fundamental period T=0.40 s. The construction is regular both in plan and in elevation (L=22m; B=15m). The modeling and calculations have been performed with the help of SAP2000 [6].

For material was chosen concrete B20 with material properties: Young's modulus $E=2.75\times10^7 kN/m^2$ and Poisson's ratio – 0.2.

The analysis case was defined as linear modal time history analysis. The load is applied as acceleration time histories in both horizontal directions (RAD Figure 4 and TRA Figure 8). The time histories are first normalized to the maximum of the accelerograms and then normalized to the Bulgarian code'87 with maximum horizontal acceleration for the area of Sofia city – 2.7 m/s². The respective

389

acceleration response spectra of the input accelerograms are given in Figure 4 and Figure 7. The output extracted from the building response represents the influence of the geological site conditions along the Model 2, on one almost regular building representative for the Sofia building stock. The output response spectrum curves with damping value of 5% (spectral acceleration versus period), for each of two sites are shown in Figures 6(a), 9(a) (Rec. 30) and 6(b), 9(b) (Rec. 50). It can be seen that the peak value is approximately equal to 0.4s, which matches with the fundamental period of the construction.

To find the combined effect due to the simultaneous loading with the two horizontal components (RAD and TRA) their vectorial combination [7] has been used. Spectral acceleration responses versus period for Rec. 30 and Rec 50. (RAD, TRA components) and the vectorial combination curve (COMB) are shown in Figure 10.

The maximum values for the spectral accelerations generated at the base (SA$_b$) of the structure (column 4) and at the roof (column 7) (SA$_r$) are given in Table 1. In columns 5 and 8, the maximum period in which the spectral maximums, is given. The dynamic factor (amplification) (SA$_r$/(SA$_b$) is shown in column 9.

RADIAL COMPONENT

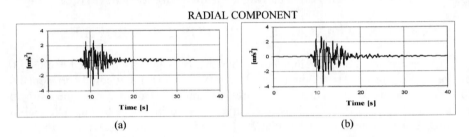

(a)   (b)

FIGURE 4. Input time history acceleration (PULSYN approach) at base at Rec 30 (a) and Rec 50 (b).

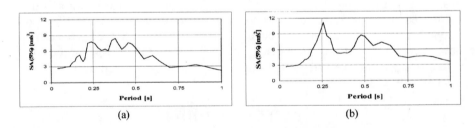

(a)   (b)

FIGURE 5. Input spectral acceleration response (PULSYN approach) at the base at Rec30 (a) and Rec 50 (b).

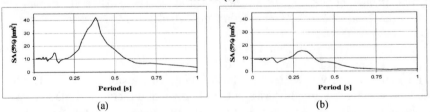

(a)   (b)

FIGURE 6. Output roof spectral acceleration response from 3D model at Rec 30 (a) and Rec 50 (b).

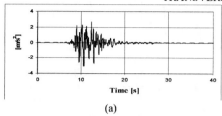
(a)

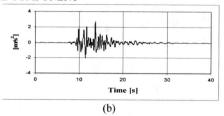
(b)

FIGURE 7. Input time history acceleration (PULSYN approach) at base at Rec 30 (a) and Rec 50 (b).

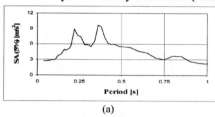
(a)

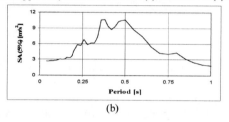
(b)

FIGURE 8. Input spectral acceleration response (PULSYN approach) at the base at Rec30 (a) and Rec 50 (b).

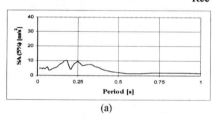
(a)

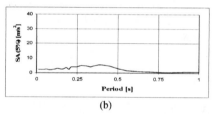
(b)

FIGURE 9. Output roof spectral acceleration response from 3D model at Rec 30 (a) and Rec 50 (b).

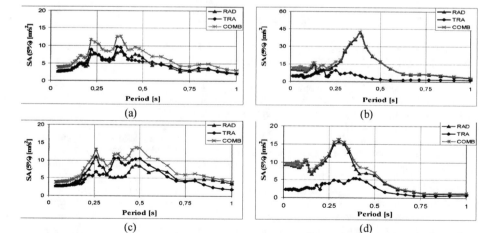
(a)

(b)

(c)

(d)

FIGURE 10. Spectral acceleration responses for RAD, TRA and vectorial comb. curve COMB components, at the base (a), (c) and at the roof level of 3D model (b), (d) for Rec 30 (a), (b), and for Rec. 50 (c), (d).

TABLE 1. Maximum values for the spectral accelerations in both cases base from generation (SA$_b$) and from SAP-model calculation on the roof of the structure (SA$_r$).

Load TH	Receiver №	Generated			3D model-roof			Dynamic factor
		T = 0s [m/s^2]	max SA [m/s^2]	T [s]	T= 0s [m/s^2]	max SA [m/s^2]	T [s]	
1	2	3	4	5	6	7	8	9
RAD	r 30	2.57	8.39	0.38	10.59	42.09	0.38	5.017
	r 50	1.82	11.21	0.26	9.17	15.66	0.3	1.397
TRA	r 30	1.60	9.66	0.36	4.69	10.15	0.18	1.051
	r 50	0.90	10.64	0.38	2.27	5.45	0.4	0.512
COMB	r 30	3.03	12.79	0.38	11.58	43.30	0.38	3.384
	r 50	2.03	15.46	0.48	9.45	16.58	0.3	1.073

CONCLUSION

Synthetic accelerations, generated in a laterally heterogeneous anelastic model representative of the local site conditions in Sofia city has been used as seismic input to a typical building of the city.

We show that the data base constructed from generated time histories and seismic response spectra can assist local authorities in minimizing future losses. For building use it is necessary to define (to map) seismic hazard zones to regulate the new construction seismic safety. The local governments must require site specific seismic hazards evaluations to validate the hazard level at the site which allow making appropriate recommendations for mitigation.

ACKNOWLEDGEMENT

The support of the CEI-SAND fellowship ICTP, Trieste and NATO project N.980468 is acknowledged.

REFERENCES

1. Aki, K., *Strong motion seismology*, in M. Erdik and M. Toksoz (ed) Strong ground motion seismology, NATO ASI Series, Series C: Mathematical and Physical Sciences, D. Reidel Publishing Company, Dordrecht, Vol.204, 1987, pp.3-39.
2. Alexiev G, Georgiev Tz., J. Problems of geography, 1-2, 1997, pp.60-69.
3. Bonchev, E, Bune V., Christoskov L., Karagyuleva J., Kostadinov V., Reisner G., Rizikova S., Shebalin N., Sholpo V., Sokerova D., Geol. Balkanica, 12, 1982, pp 3-48.
4. Bulgarian seismic code'87.
5. Christoskov L, Georgieva Tzv., Deneva D., Babachkova B., In: Proc. Of the 4th Int. Symposium on the Analysis of Seismicity and seismic risk, Bechyne castle, CSSR, 1989,pp. 448-454.
6. Computers and structures, *SAP Manual Integrated Finite Element analysis and design of structures*, Inc, Berkeley, USA, 2000.
7. Douglas, J. *Earthquake ground motion estimation using strong-motion records:a review of equations for the estimation of peak ground acceleration and response spectral ordinates*, Earth-Science Reviews 61 43–104 53 (2003).
8. Dziewonski A., Anderson D.: Phys Earth Planet Int., 25, 1981, pp 297-356.

9. Fäh, D., Suhadolc, P. and Panza, G.F, *Variability of seismic ground motion in complex media: the Friuli area (Italy)*. In Geophysical Exploration in Areas of Complex Geology, II (eds Cassinis, R., Heibig, K. and Panza, G.F.), J. Appl. Geophys., 3, 1993, pp. 131-148.

10. Florsch, N., Fäh, D., Suhadolc, P. and Panza, G. F., *Complete synthetic seismograms for high–frequency multimode SH-waves*, PAGEOPH, 136, 1991, pp. 529-560.

11. Foutch, D. A., *State-of- the-Art Report on Performance Prediction and Evaluation of Moment-Resisting Steel Frame Structures, FEMA 355*, Federal Emergency Management Agency, Washington, D.C (2000).

12. Glavcheva R.,: Bulg. Geoph. Jurn., 16, 1990, pp. 38-44.

13. Gorshkov A., Kuznetzov I., Panza G.F, Soloviev A., 2000. *Identification of Future Earthquake Sources in the Carpatho-Balkan Orogenic Belt using Morphostructural Criteria. Seismic Hazard of the Circum-pannonian Region*, Pageoph Topical Vols, Birkhauser Verlag, pp. 79-85.

14. Gusev A.A.: *A program PULSYN02 for wide-band simulation of source radiation from a finite earthquake source/fault.*, Abdus Salam ICTP, Trieste, Italy (2003).

15. Gusev, A. A., *Descriptive statistical model of earthquake source radiation and its application to an estimation of short period strong motion*, Geophys. J. R. Astron. Soc. 74, 1983, pp. 787-800.

16. Gusev, A. A., Pavlov. V., *Wideband simulation of earthquake ground motion by a spectrum-matching, multiple-pulse technique.* First European Conference on Earthquake Engineering and Seismology, Geneva, Switzerland, 3-8 September 2006.Paper Number: 408.

17. Ivanov Pl., In: Proc. Intern. IAEG Conference, Athens, Balkema, Rotterdam, 1997 pp. 1265-1270.

18. Ivanov, P., Frangov, G., Yaneva, M. *Distribution, composition and properties of Quaternary deposits in Sofia kettle.* In: - XVI Congress Carpathian-Balkan Geological Association, August 30-September 2, Vienna, Austria, Abstracts, 1998, pp. 328.

19. Kamenov B., Kojumdjieva N, 1983. Paleontology, Stratigraphy and Lithology, 18, pp 69-85.

20. Koleva G., Vaccari F., Paskaleva Iv. Zuccolo E., Panza G., *An approach of microzonation of Sofia city*, Acta. Geod. Geoph. Hung. (2008)(in print).

21. Matova, M.,:In Proc.Final Conf. Of UNESCO-BAS Proc. on land subsidence.2001, pp.93-98.

22. Panza, G. F., Romanelli F., Vaccari F., *Seismic wave propagation in laterally heterogeneous media: theory and applications to seismic zonation*, Advances in geophysics, 2001, pp 43, 1-95.

23. Panza, G.F. *Synthetic seismograms: The Rayleigh waves model summation.* Journal of Geophysics, 58, 1985; pp 125-145.

24. Panza, G.F., Vaccari F., *Introduction in seismic hazard of the Circum-pannonian region* (Editors: G. F. Panza, M. Radulian and C. Trifu), Pageoph Topical Volumes, Birkhauser Verlag, 2000; pp 5-10.

25. Paskaleva I., Dimova S., Panza G.F., Vaccari F., *An earthquake scenario for the microzonation of Sofia and the vulnerability of structures designed by use of the Eurocodes, Journal* Soil Dynamics and Earthquake Engineering Vol. 27, Issue 11, 2007, pp. 1028-1041.

26. Paskaleva, I., Panza, G. F., Vaccari, F., Ivanov, P., *Deterministic modeling for microzonation of Sofia – an expected earthquake scenario*, Acta. Geod. Geoph. Hung., Vol.39(2-3), 2004, pp. 275-295.

27. Paskaleva,I., Matova, M., Frangov, G., *Expert assessment of the displacement provoked by seismic events: case study for the Sofia metropolitan area*, PAGEOPH, 2004, Vol. 161.

28. Petkov P, Iliev I., In: Proc. Of the 3rd Eur. Symposium on EE, Sofia, 1970, pp 79-86.

29. Shanov, S., Tzankov Tz., Nikolov, G., Bojkova, A., Kurtev, K.,: Review of the Bulgarian Geological Society, Vol. 59, part I, 1998, pp 3-12.

30. Slavov, Sl., Paskaleva, I., Kouteva, M., Vaccari, F., Panza, G.F., *Deterministic earthquake scenarios for the city of Sofia*, PAGEOPH, Vol. 161 Birhauser Verlag, Basel, 161, 2004, pp.1221-1239.

31. Solakov D, Simeonova S, Christoskov L, :Annali di geofisica,44, 2001, pp 541-556.

32. Stanishkova I, Slejko D., *Seismic Hazard of Bulgarian Cities*, Atti Del 10 Convegno Annuale del Gruppo Nazionale di geofisica della solisda, Roma (1991).

33. Stein, S. and Wysession, M., An introduction to seismology, earthquakes, and earth structure; Blackwell Publishing, Oxford, UK (2003).

Application of the Neo-Deterministic Seismic Microzonation Procedure in Bulgaria and Validation of the Seismic Input Against Eurocode 8

Paskaleva Ivanka[a], Kouteva Mihaela[a,c], Vaccari Franco[b] and Panza Giuliano F.[b,c]

[a] CLSMEE - BAS, 3 Acad. G. Bonchev str, 1113 Sofia, Bulgaria,
[b] DST - University of Trieste, Via E. Weiss 4, 34127 Trieste, Italy
[c] ESP - SAND, ICTP, Trieste, Italy

Abstract. The earthquake record and the Code for design and construction in seismic regions in Bulgaria have shown that the territory of the Republic of Bulgaria is exposed to a high seismic risk due to local shallow and regional strong intermediate-depth seismic sources. The available strong motion database is quite limited, and therefore not representative at all of the real hazard. The application of the neo-deterministic seismic hazard assessment procedure for two main Bulgarian cities has been capable to supply a significant database of synthetic strong motions for the target sites, applicable for earthquake engineering purposes. The main advantage of the applied deterministic procedure is the possibility to take simultaneously and correctly into consideration the contribution to the earthquake ground motion at the target sites of the seismic source and of the seismic wave propagation in the crossed media. We discuss in this study the result of some recent applications of the neo-deterministic seismic microzonation procedure to the cities of Sofia and Russe. The validation of the theoretically modeled seismic input against Eurocode 8 and the few available records at these sites is discussed.

Keywords: Strong Motion, Seismic Wave Propagation, Seismic Input, Seismic Microzonation, Eurocode 8.

INTRODUCTION

Bulgaria, situated in the Balkan Region as a part of the Alpine-Himalayan seismic belt, characterized by high seismicity, is exposed to a high seismic risk. Over the centuries, Bulgaria has experienced strong earthquakes. In historical time we mention the 1818 (about IX MSK) and the 1858 (IX MSK, $M_S > 6.3$) earthquakes near the capital Sofia. Some of Europe's strongest earthquakes in the 20th century occurred on the territory of Bulgaria. At the beginning of the last century (from 1901 to 1928) five earthquakes with magnitude $M_S >= 7.0$ occurred there. In central Bulgaria a sequence of three destructive earthquakes occurred in 1928. The 1986 earthquake of magnitude $M_S = 5.7$ in central northern Bulgaria (near the town Strazhitza) is the strongest quake after 1928. The seismicity of the neighboring areas, from Greece, Turkey, former Yugoslavia and Romania (especially the strong Vrancea intermediate depth earthquakes), contribute significantly to the seismic hazard in Bulgaria. Typical

CP1020, *2008 Seismic Engineering Conference Commemorating the 1908 Messina and Reggio Calabria Earthquake,*
edited by A. Santini and N. Moraci

examples of long period, i.e. far-reaching seismic effects, are the Vrancea intermediate-depth earthquakes. The Bulgarian territory is regularly suffering these quakes and during the last century the earthquakes in 1908, 1940, 1977, 1986 and 1990 caused significant damage. The wave field radiated by the intermediate-depth (70 to 170 km) Vrancea earthquakes, mainly at long periods (T > 1s), attenuates less with distance, compared to the wave field generated by the earthquakes located in other seismically active zones in Bulgaria. The quake of March 4, 1977, (M_w = 7.5) caused significant damage in Bulgaria and was felt up to Central Europe [1]. Therefore the Vrancea intermediate-depth sources represent a regional danger, since large industrial areas can be seriously affected by the strong events originating in this seismogenic area. In fact, even if the Vrancea 1977 event motivated some changes in the Bulgarian Code for Design and Construction in Seismic Regions'87, the seismic excitation from the 1986 and 1990 events exceeded the prescribed seismic loading in the BG Code'87.

The recent analysis of the seismicity depth distribution in Bulgaria and the neighboring territories [2] recognized that the earthquakes in all zones occurred in the earth's crust (h < 60 km) with the exception of the events in the Vrancea (Romania) intermediate depth zone. The depths distribution shows that the earthquakes in Bulgaria are mainly located in the upper crust, and only a few events are related to the lower crust. The maximum depth is about 50 km in SW Bulgaria; outside, the foci affect only the upper 30–35 km. The maximum density of seismicity is found in the layers between 5 and 25 km. The seismicity within the Vrancea (Romania) region consists of two depth horizons: normal depth (less than 60 km) and intermediate depth (60 – 180 km) earthquakes. The extreme irregularities of the isoseismals of intermediate depth earthquakes and the available damage record in Bulgaria [1] have shown that the regional seismic hazard in NE Bulgaria is controlled mainly by the Vrancea intermediate-depth events. Urban areas located at rather large distances from earthquake sources may thus be prone to severe earthquake hazard as well as the near field sites.

Worldwide, many large urban areas are settled in regions where different natural disasters have been observed. Two of the most vulnerable cities in Bulgaria, which are exposed to significant earthquake hazard, are Sofia and Russe. The available strong ground motion database [3] is too limited to reliably quantify the magnitude scaling and the attenuation characteristics of large magnitude earthquakes. The purpose of this paper is to illustrate the computed seismic input at Sofia and Russe due to local and remote strong intermediate-depth Vrancea earthquakes correspondingly.

CASE STUDY: SOFIA CITY

The city of Sofia is the main administrative centre in Bulgaria, with the densest population. Large industrial zones are located in its vicinity. If a strong earthquake should occur in the Sofia area it could produce disastrous damages in a large region, followed by numerous heavy consequences for the whole country (communications, lifelines). Strong earthquakes with M up to 7 did shake Sofia in the past centuries [1, 2, 3]. The maximum macroseismic intensity in Sofia, I = IX (MSK), observed in 1858 [4, 5], is expected to occur with a return period of 150 years [6], i.e. it could correspond to the strong earthquake scenario. Recently seismic hazard maps of the Circum - Pannonian Region [7, 8], show that Sofia is placed in a node having potential for the occurrence of an earthquake with M > 6.5 and that it could suffer macroseismic intensity up to X. The seismicity of the Sofia region is limited to the upper 20 - 30 km of the lithosphere. A maximum macroseismic intensity I = VIII can be expected at Sofia [9], if an earthquake with maximum magnitude $M_{max} = 7$ [5] occurs at a depth of about 20 km, and a maximum macroseismic intensity IX (and higher) can be provoked by an event with $M_{max} = 7$ and focal depth around 10 km.

Recently, a synthetic database containing more than 2700 accelerograms, velocigrams and seismograms has been constructed by Paskaleva et al. [10] considering four earthquake scenarios with magnitudes $M_w = 3.7$, $M_w = 6.3$ and $M_w = 7$. Synthetic ground motions along three selected geological cross sections [10] have been generated applying the neo-deterministic hybrid technique [11, 12, 13]. It combines the modal summation technique [14, 15, 16, 17] used to describe the seismic wave propagation in the anelastic bedrock structure with the finite difference method [18, 19, 20] used for the computation of wave propagation in the anelastic, laterally inhomogeneous sedimentary media [21]. For the considered earthquake scenarios with magnitude Mw = 7.0 the extended source with bilateral rupture propagation is considered, and the observation point is on a line at 90° from the rupture propagation direction [22]. The earthquake scenarios considered in this study are chosen to correspond to a seismic source, located at 10 km distance west or southwards from the city center, respectively [10 and references therein]. The assumed source parameters are chosen to approximate the seismic event which hit Sofia in 1858, Sce1_all in Table 1. To estimate the effect of the change of the seismic source mechanism on the site response, one more set of seismic source parameters have been used - Sce1_3a in Table 1. Both focal mechanisms, Sce1_all and Sce1_3a in Table 1, are consistent with the available geological studies performed within the epicentral area.

Table 1. Earthquake scenarios used for the computations.

Name of the scenario	Name of the geological profile	Magni tude M	Strike angle (°)	Dip angle (°)	Rake angle (°)	Focal depth (km)	Epicentral distance to the nearest profile (km)
Sce1_all	M1, M2, M3	7	340	77	285	10	10
Sce1_3a	M3	7	0	44	309	10	10

The obtained results are regionalized in three groups by epicentral distances: 10 - 12 km, 12 - 16 km and 16 - 24 km. Elastic acceleration and displacement response spectra have been extracted and validated against the seismic input recommended in Eurocode 8 (EC8). The Dynamic Coefficients are computed as the ratio of the elastic acceleration response spectra amplitudes for 5% (SA 5%) damping and the Peak Ground Acceleration of the seismic signal. The comparison between the computed quantities (PGA) and the corresponding values, recommended by the EC8, are shown in Figures 1 and 2. The synthetic seismic signals are computed for the period range 0 - 10s. The period interval 0 – 2.5s has been chosen as the most interesting for engineering practice, and the comparisons of the computed displacements response spectra with the Eurocode 8 (EC8) ones have been performed over this period.

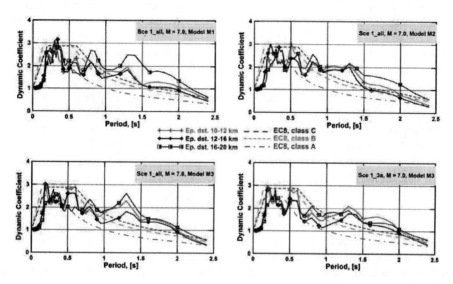

FIGURE 1. Comparison of the seismic input (dynamic coefficient) computed at Sofia with the recommended Eurocode 8 values. Each graph corresponds to a separate geological profile [10].

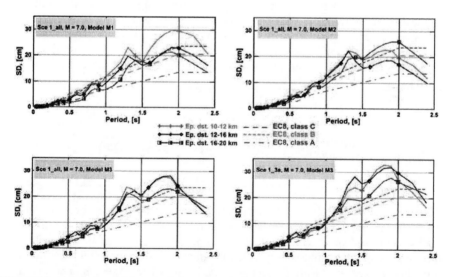

FIGURE 2. Comparison of the seismic input (elastic displacement response spectrum, 5% damping) computed at Sofia with the recommended Eurocode 8 values. Each graph corresponds to a separate geological profile [10].

The comparisons of the computed dynamic coefficients shown in Figure 1 indicate that for the low period range, 0.04 - 0.7s, the theoretically computed values for all models are slightly lower than the EC 8. For the periods larger than 0.7-0.8s, i.e. for periods at which observations are very scarce, the computed dynamic coefficients overestimate the EC 8 ones. The comparisons of the computed displacement design spectra with the recommended EC8 design spectra (Figure 2) show that the synthetic spectral values for all models follow the EC 8 amplitudes for the period range T = 0.05 - 1s. The synthetic amplitudes overestimate the EC 8 [23] ones for periods T = 1 - 2s, at which very few reliable data are available. The comparison among the four graphs in Figures 1 and 2 shows the significant contribution of the local geology along the investigated profiles on the seismic input concerning both, signal amplitudes and frequency content. The change of the seismic source mechanism Sce1_all and Sce1_3a (bottom of Figures 1 and 2) shows that the seismic source mechanism visibly influence the spectral amplitudes, but not the frequency content of the seismic input.

CASE STUDY: RUSSE

Russe is one of the main Bulgarian cities. Russe is the biggest Bulgarian port on the Danube River and it is the most important industrial, administrative and cultural centre in NE Bulgaria, where the earthquake hazard is controlled by the strong intermediate - depth Vrancea earthquakes. Theoretical modelling of the seismic loading at Russe, due to the Vrancea intermediate depth events, has been performed by Kouteva et al. [24, 25] applying the neo-deterministic analytical procedure [17]. The procedure models the wavefield, generated by a realistic earthquake source, as it propagates through a laterally varying anelastic medium. Details on the computation model and on the geological information used were published by Paskaleva et al. [26]. In this study the target site of Russe is represented by three generalized local geological models corresponding to the soil classes A, B and C according to the EC 8 soil classification. The earthquake scenario considered is described in Table 2.

Table 2. Earthquake scenarios used for the computations [32]

Name of the scenario	Name of the geological profile	Magnitude Mw	Strike angle (°)	Dip angle (°)	Rake angle (°)	Focal depth (km)	Ep. dist. (km)
VR901	Russe	6.9	240	63	101	74	236.64

The comparison of the normalized elastic response spectra for 5% damping, extracted from the computed signals with the available record and the recommended in the EC 8 spectral values, is shown in Figure 3. There is one three-component strong motion accelerogram available at this site, the quake of May 30, 1990 [3]. This signal, low passed filtered at cut-off frequency 3Hz, is shown in Figure 3. The parametric analyses performed by varying the parameters describing the geometry and the movement at the seismic source [25, 27], quantifies the contribution of the seismic source to the seismic signal amplitude. The spectral amplitudes of the generalized horizontal earthquake excitation H = (TRA²+RAD²)^{1/2} remain in the strip, limited by +/- one standard deviation of the theoretically computed signal spectra.

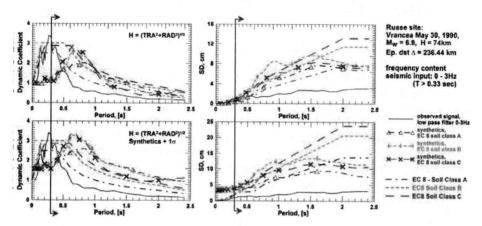

FIGURE 3. Comparison of the seismic input at Russe (dynamic coefficients and elastic displacement, SD, response spectra, 5% damping) with the Eurocode 8 recommendations. Upper left graph: Dynamic Coefficient; Lower left graph: theoretical Dynamic Coefficient+1σ; Upper right graph: SD; Lower right graph: synthetics SD+1σ

The results, plotted in Figure 3, show that the computed Dynamic Coefficients are consistent with the recommended EC 8 values, particularly in the period range T > 1s, which correspond to that part of the seismic signal, which is practically not influenced by the local site geology. The computed Dynamic Coefficients + one standard deviation (SA + 1σ) follow the EC 8 and the observation trend with higher amplitudes. The peak amplitudes of the synthetic signal occur at larger periods, compared to the observation and the EC8 values. The amplitudes of the computed elastic displacement spectra are significantly higher than the observed ones. The synthetics follow the EC 8 graphs for periods 0.3 < T < ~ 1.5s and for periods T > 1.5s the synthetics fall below EC 8. Considering the recent experience on the Vrancea earthquakes observation and modeling [27, 28], the difference between synthetics, observations and the EC values can be due to the relatively shallow local geological models considered.

FINAL REMARKS

The task and results discussed in this work have been provoked by the need of reliable procedures, capable of predicting realistic seismic input. A synthetic ground motion database, containing more than 3000 site and source dependent accelerograms, velocigrams and seismograms is available for the cities of Sofia and Russe. The theoretical results are validated against the corresponding quantities, recommended in Eurocode 8. The results show that the seismic source and the local geological conditions contribute significantly to the characteristics of acceleration and displacement response spectra.

Both case studies have shown that the neo-deterministic seismic hazard assessment procedure is a capable tool for the construction of realistic synthetic strong motion databases, particularly for regions which are characterized by high seismicity and lack of instrumental earthquake record.

ACKNOWLEDGMENTS

This research investigation has been carried out in the framework of the bilateral cooperation between DST - UNITS, Trieste, Italy and CLSMEE - BAS, Sofia, Bulgaria. The financial support from Project INTAS-Moldova 200505-104-7584 "Numerical Analysis of 3D seismic wave propagation using Modal Summation, Finite Elements and Finite Differences Method", NATO SfP Project N980468, CEI Projects "Deterministic seismic hazard analysis and zonation of the territory of Romania, Bulgaria and Serbia" and "Geodynamical Model of Central Europe For Safe Development Of Ground Transportation Systems", and the CEI university network are gratefully acknowledged.

REFERENCES

1. "Vrancea Earthquake in 1977. Its After-Effects in the People's Pepublic of Bulgaria", 1983, Editor: Brankov G. VV AA, BAS, Sofia, 300 (in Bulgarian).
2. Simeonova S., Solakov D., Leydecker G. Busche H., Schmitt T., Kaiser D. (2006) Probabilistic Ceismic Hazard Map for Bulgaria as a Basis for a New Building Code, Nat. Hazards Earth Syst. Sci., 6, 881–887, www.nat-hazards-earth-syst-sci.net/6/881/2006.
3. Nenov D, Georgiev G, Paskaleva I, Lee V W, Trifunac M. 1990 "Strong Ground Motion Data in EQINFOS: Accelerograms Recorded in Bulgaria between 1981-1987", Bulg. Acad. of Sci., Centr. Lab. for Seism. Mech. And Eq. Eng., & Dept. of Civ. Eng. Dpt. No 90-02, Univ. of S. California, L. A., California.
4. Watzov, Sp. (1902) Tremblements de terre en Bulgaria au XIX siecle, IMPR. DE L'ETAT, Sofia, Bulg., 95 (in Bulgarian).
5. Bonchev, E., Bune, V., Christoskov, L., Karagyuleva, J., Kostadinov, V., Reisner, G., Rizikova, S., Shebalin, N., Sholpo, V., Sokerova, D. (1982) A Method for Compilation of Seismic Zoning Prognostic Maps for the Territory of Bulgaria, Geologica Balkanica, 12; 2: 3-48.
6. Christoskov, L., Georgiev, Tzv., Deneva, D., Babachkova, B. (1982) On the Seismicity and Seismic Hazard of Sofia Valley, Proc. Of the 4th Int. Symposium on the Analysis of seismicity and seismic risk, Bechyne castle, CSSR, IX: 448-454.
7. Panza, G.F., Vaccari F. (2000) Introduction in Seismic Hazard of the Circum-Pannonian Region (Editors: Panza, GF., Radulian, M., Trifu. C.), Pageoph Topical Volumes, Birkhauser Verlag: 5-10.
8. Gorshkov, A., Kuznetzov, I., Panza, G.F., Soloviev, A. (2000) Identification of Future Earthquake Sources in the Carpatho-Balkan Orogenic Belt Using Morphostructural Criteria. Seismic Hazard of the Circum-pannonian Region, PAGEOPH Topical Volumes, Birkhauser Verlag: 79-85.
9. Glavcheva R. and Dimova S. (2003) Seismic Risk of the Wide-spread Residential Buildings in Bulgaria. Proceedings of the First International Conference "Natural Risks: Developments, Tools and Technologies in the CEI Area", Sofia, 4-5 of November (on CD).
10. Paskaleva, I., Kouteva, M., Vaccari, F., Panza, G.F. (2008) Characterization of the Elastic Displacement Demand: Case Study - Sofia City, AGGH, (in print).
11. Fäh, D., Iodice, C., Suhadolc, P., Panza, G.F. (1993) A New Method for the Realistic Estimation of Seismic Ground Motion in Megacities: The Case of Rome, Earthquake Spectra 9: 643-668.
12. Fäh, D., Iodice, C., Suhadolc, P., Panza, G.F. (1995a) Application of Numerical Simulations for a Tentative Seismic Microzonation of the City of Rome, Annali di geofisica, Vol XXXVIII, N.5-6, Nov. Dec 1995: 607-615.

13. Fäh, D., Suhadolc, P., Mueller, St., Panza, G.F. (1995b) A Hybrid Method for the Estimation of Ground Motion in Sedimentary Basins; Quantitative Modelling for Mexico City, BSSA,84:383-399.
14. Panza GF. (1985) Synthetic Seismograms: the Rayleigh Waves Modal Summation. J Geophys;58:125–45.
15. Panza, G.F., Suhadolc, P. (1987) Complete Strong Motion Synthetics, in Seismic Strong Motion Synthetics, B. A. Bolt (Editor), Academic Press, Orlando, Computational Techniques 4: 153-204.
16. Panza GF., Romanelli F., Vaccari F. (2000) Realistic Modelling of Waveforms in Laterally Heterogeneous Anelastic Media by Modal Summation. Geophys J Int;143:1–20.
17. Panza, G.F., Romanelli, F., Vaccari, F. (2001) "Seismic Wave Propagation in Laterally Heterogeneous Anelastic Media: Theory and Applications to the Seismic Zonation", Advances in Geophysics, Academic press; 43: 1-95.
18. Virieux, J. (1984) SH - Velocity-Stress Finite-Difference Method: Velocity-Stress Finite-Difference Method, Geophysics; 49: 1933-1957.
19. Virieux, J. (1986) P-SV Wave Propagation in Heterogeneous Media: Velocity-Stress Finite-Difference Method", Geophysics; 51: 889-901.
20. Levander, A.R. (1988) Fourth-Order Finite-Difference P-SV Seismograms. Geophysics, 53: 1425-1436.
21. Stein, S. and Wysession, M. (2003) An Introduction to Seismology, Earthquakes, and Earth Structure, Blackwell Publishing, Oxford, UK.
22. Gusev, A. A., Pavlov. V. (2006) Wideband Simulation of Earthquake Ground Motion by a Spectrum-Matching, Multiple-Pulse Technique. First European Conference on Earthquake Engineering and Seismology (a joint event of the 13th ECEE & 30th General Assembly of the ESC). Geneva, Switzerland, 3-8 September 2006. Paper Number: 408.
23. Eurocode 8 - Design of Structures for Earthquake Resistance. Part 1: General Rules, Seismic Actions and Rules for Buildings. prEN 1998-1, Brussels, 2002.
24. Kouteva, M., Panza, G.F., Romanelli, F., Paskaleva, I. (2004) Modelling of the Ground Motion at Russe site (NE Bulgaria) Due to the Vrancea Earthquakes. JEE, 8, 2, 209-229.
25. Kouteva, M., Panza, G.F., Paskaleva, I. and Romanelli, F., (2004). A View to the Intermediate-Depth Vrancea Earthquake of May 30, 1990 - Case Study in NE Bulgaria. AGGH., 39 (2-3), 223-231.
26. Paskaleva, I., Kouteva, M., Panza, G.F., Evlogiev, J., Koleva, N. and Ranguelov, B., (2001). Deterministic Approach of Seismic Hazard Assessment in Bulgaria; Case Study Northeast Bulgaria - The Town of Russe. The Albanian Journal of Natural & Technical Sciences, 10, 51-71.
27. Kouteva M., Panza, G.F., Paskaleva, I., (2000) An Example for Strong Ground Motion Modelling in Connection with Vrancea Earthquakes (Case Study in NE Bulgaria, Russe site). 12WCEE 2000, Wellington, NZ.
28. Aldea, A., Kashima, T., Poiata, N., Kajiwara, T. (2006) A New Digital Seismic Network in Romania with Dense Instrumentation in Bucharest, 1st ECEES, Geneva, Switzerland, 3-8 September 2006, Paper Number: 515.

Recent Achievements of the Neo-Deterministic Seismic Hazard Assessment in the CEI Region

Panza G.F.[a,b], Kouteva M.[c,b], Vaccari F.[a], Peresan A.[a], Cioflan C.O.[d], Romanelli F.[a], Paskaleva I.[c], Radulian M.[d], Gribovszki K.[e], Herak M.[f], Zaichenco A.[g], Marmureanu G.[d], Varga P.[e], Zivcic M.[i]

[a] DST - University of Trieste, Via E. Weiss 4, 34127 Trieste, Italy
[b] ESP - SAND, ICTP, Trieste, Italy
[c] CLSMEE - BAS, 3 Acad. G. Bonchev str, 1113 Sofia, Bulgaria,
[d] NIEP - Magurele-Bucharest, 12 Calugareni str., Ilfov, Romania
[e] Geodetic and Geophysical Research, Institute of HAS, Sopron, Hungary
[f] Department of Geophysics, Faculty of Science, University of Zagreb,
Horvatovac bb, 10000 Zagreb, Croatia
[g] IGG, Chisinau, Moldova
[i] ARSO - Seismology and Geology Office, Ljubljana, Slovenia

Abstract. A review of the recent achievements of the innovative neo-deterministic approach for seismic hazard assessment through realistic earthquake scenarios has been performed. The procedure provides strong ground motion parameters for the purpose of earthquake engineering, based on the deterministic seismic wave propagation modelling at different scales - regional, national and metropolitan. The main advantage of this neo-deterministic procedure is the simultaneous treatment of the contribution of the earthquake source and seismic wave propagation media to the strong motion at the target site/region, as required by basic physical principles. The neo-deterministic seismic microzonation procedure has been successfully applied to numerous metropolitan areas all over the world in the framework of several international projects. In this study some examples focused on CEI region concerning both regional seismic hazard assessment and seismic microzonation of the selected metropolitan areas are shown.

Keywords: Seismic hazard assessment, deterministic approach, neo-deterministic approach, probabilistic approach, seismic microzonation.

INTRODUCTION

Deterministic versus probabilistic approaches have differences, advantages, and disadvantages in assessing earthquake hazards and risks that often make use of one advantage over the other. The factors that influence the choice of seismic hazard assessment procedure including the decision to be made (i.e. the purpose of the hazard or risk assessment), the seismic environment (whether the location is in a high, moderate, or low seismic risk region), the available input data and the scope of the assessment (whether one is assessing a site risk, a multi-site risk, or risk to a region) have been recently discussed in the literature.

CP1020, 2008 Seismic Engineering Conference Commemorating the 1908 Messina and Reggio Calabria Earthquake,
edited by A. Santini and N. Moraci
© 2008 American Institute of Physics 978-0-7354-0542-4/08/$23.00

THE NEO-DETERMINISTIC SEISMIC HAZARD ASSESSMENT (NDSHA) PROCEDURE - MAIN ADVANTAGES COMPARED TO THE PROBABILISTIC PROCEDURE

McGuire [1] proposed a general rule stating the more qualitative the decision to be made, the more appropriate deterministic hazard assessment is. The classification of "highly qualitative" related to the deterministic assessments is debatable and in this paper, realistic and highly quantitative deterministic seismic hazard assessments at regional and local urban levels in the CEI region, are shown. The seismic environment plays also a strong role in the appropriateness of the deterministic assessments. For high seismic regions, where the largest earthquakes may occur every 100-300 years, the 475-year shaking could be considered as design ground motion. This may correspond to the largest magnitude on the closest fault to the site, which is particularly relevant to a site located next to an active fault. A deterministic scenario for this event would allow to examine such details as ground motion effects caused by rupture propagation, leading to insights on the risk for a particular lifeline or city that might not be available from more encompassing probabilistic analyses. Regional assessments often benefit most from deterministic models, where the probability of occurrence of the scenario in, for example, one city is small, but is large for the region. The concept of multiple deterministic scenarios allows rational preparation, even if the details of the forecast earthquake may be wrong.

The seismic hazard evaluation, which is based on the traditional Probabilistic Seismic Hazard Analysis (PSHA) relies on the probabilistic analysis of earthquake catalogues and of ground motion, macroseismic observations and instrumental recordings. Recently PSHA showed its limitation in providing a reliable seismic hazard assessment, possibly due to insufficient information about historical seismicity, which can introduce relevant errors in the purely statistical approach mainly based on the seismic history. The comparison between the observed peak ground accelerations (PGA) and the PGA predicted by PSHA (the GSHAP Project) for the recent examples of the Kobe, 17.1.1995; Bhuj, 26.1.2001; Boumerdes, 21.5.2003 and Bam, 26.12.2003 has shown significant disagreement [2]. The probabilistic map, for the 475 years return period, gives a maximum PGA in the range 0.6-0.8 g, centered on the Vrancea region (GSHAP). Observations clearly indicate that Vrancea sits on a relative minimum.

The major uncertainties and sources of errors of the PSHA have recently been analyzed by Klügel et al., [3, 4, 5]. The standard error of empirical attenuation equations (or the difference between observation and estimates of a physical predictive model) is one of the main sources of errors in the traditional PSHA, since it is interpreted as aleatory uncertainty and not understood as an (limited by our knowledge) estimate of total variability of the stochastic process attenuation equation. In view of the limited seismological data, it seems more appropriate to resort to a scenario-based deterministic approach, as it allows us to realistically define hazard in scenario-like format accompanied by the determination of advanced hazard indicators as, for instance, damaging potential in terms of energy. In 2006 Klügel et al. [3] proposed a scenario-based procedure for seismic risk analysis. This scenario-based approach allows us to incorporate all available information collected in a geological, seismotectonic and geotechnical database of the site of interest as well as advanced physical modelling techniques providing a reliable and robust deterministic design basis for civil infrastructures. At the same time a scenario-based seismic hazard analysis allows to develop the required input for probabilistic risk assessment (PRA), as required by safety analysts and insurance companies. The scenario-based approach removes the ambiguity in the results of probabilistic seismic hazard analysis (PSHA) which relies on the projections of the Gutenberg–Richter (G–R) equation. The problems in the validity of G–R projections, because of the incomplete to total absence of data for making the projections, are still unresolved. The scenario-based methodology is strictly based on observable facts and data and complemented by physical modelling techniques, which can be submitted to a formalized validation process. By means of sensitivity analysis, knowledge gaps related to lack of data can be dealt with easily, due to the limited amount of scenarios to be investigated [3]. The comparative analyses of the recently published results on regional seismic hazard assessments, obtained via PSHA and NDSHA [2; 6] have shown, as appropriate, the suggestion to limit the probabilistic analysis to the definition, for a given area, of the magnitude of the different scenario earthquakes: (a) disastrous (return period about 500 years); (b) very strong (return period about 250 years); (c) strong (return period about 125 years); (d) frequent (return period about 60 years) and to use them for deterministic computations [6].

TABLE 1. Summary of the discussed seismic hazard assessment procedures, PSHA, DSHA and NDSHA.

Procedure description	PSHA	DSHA	NDSHA
Step 1	Seismic sources		
	Identification of Seismogenic Zones and Capable Faults;		
	Epicenters; Geometry and Focal mechanism;		
Step 2	Recurrence rate can be represented by a linear relation only if the size of the study area is large with respect to linear dimensions of sources.	**Fixed magnitude Fixed distance** Choice of the Controlling Earthquake	Scenario Earthquakes – fixed magnitudes, distances and specific seismic source properties. Choice of the Controlling Earthquake
Step 3	Attenuation relations - they represent the functional dependency of the random spectral acceleration on the random variables, magnitude, distance and measurement error [3] and thus are source of systematic error in the seismic hazard assessment		Synthetic ground motions. NO NEED OF ATTENUATION RELATIONS.
Step 4	Seismic hazard assessment in terms of *Probability of exceedance of a given ground motion measure*	Seismic hazard assessment in terms of *Fixed Ground Motion Measure*	Seismic hazard assessment *Envelopes of PGA or other Ground Motion Measure*

The main advantage of the proposed neo-deterministic procedure is the simultaneous treatment of the contribution of the seismic source and seismic wave propagation media to the strong motion at the target site/region, as required by basic physical principles [7]. A brief comparative analysis of the traditionally used seismic hazard approaches, probabilistic (PSHA) and deterministic (DSHA), and the neo-deterministic seismic hazard assessment (NDSHA) is provided in Table 1.

NDSHA – Applications for the CEI Region

In the framework of the UNESCO-IUGS-IGCP Project 414 and taking the benefits of the existing CEI Network, neo-deterministic hazard computation for some CEI countries have been performed at national and regional scales. The peak values of the bedrock acceleration (frequency content 0-1 Hz), computed in Slovenia for the case of the Mt. Sneznik events were simulated and exhibited considerable similarity with the response spectra of the 1995 Mt. Sneznik [8]. Computation of realistic synthetic seismograms for the Croatian territory has yielded the also meaningful results, thus providing a powerful and economically valid scientific tool for seismic zonation and hazard assessment [9]. The results obtained for Hungary have shown that considerable seismic hazard can be expected in three major Hungarian cities, but for the significant part of the Great Hungarian Plain the seismic hazard can be practically neglected [10]. The deterministic seismic zoning of Romania [11] leads to effective peak acceleration (EPA) estimates in the range 0.6 - 1.0 g. These values are shifted with respect to the hypocentral zone, in agreement with observations. Using numerically simulated ground motion, a first-order deterministic evaluation of the seismic hazard of Romania has been proposed [11]. Design Ground Accelerations (DGA) values greater than 0.3g have been obtained over an extended area. The distribution of the peak values numerically determined correlates with the values recorded in the areas situated eastwards and southward of the Carpathians arc. The results of the innovative integrated neo-deterministic seismic hazard assessment for the Italian territory have been recently discussed by Peresan et al. [6]. The obtained impending macroseismic intensities (ISG) [12], estimated in terms of peak ground velocities and displacements, and DGA as well, are consistent with the observed ones, showing intensities in the range of VII - X over the Italian territory.

Maps of neo-determinstic seismic hazard (DGA, Peak Ground Velocities and Peak Ground Displacements) for some European countries have been published [13] and an example is shown in Figure 1. Shallow seismicity has been considered, limiting the computations to epicentral distances shorter than 90 km. The hypocentral depth considered is 10 km for events with magnitude Mw < 7 and 15 km for larger events. In the case of the Vrancea intermediate depth events spectral properties especially determined for the Romanian intermediate-depth earthquakes has been considered and the computations have been performed over the Romanian, Northeastern Croatian and Hungarian territory, within a circle of 350 km of radius centered on Vrancea. The hypocentral depths considered are 90 km for magnitude less than 7.4 and 150 km for larger quakes. According to this map DGA, up to 0.675g, can be expected in the investigated region.

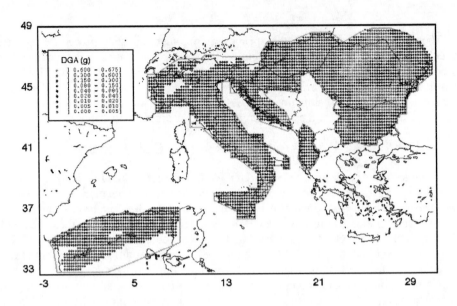

FIGURE 1. Design Ground Acceleration computed for the CEI Region by Panza and Vaccari [2000].

NDSHA – Applications for Seismic Microzonation Purposes

The neo-deterministic approach has been successfully applied to numerous metropolitan areas e.g. Delphi (India), Beijin (China), Naples (Italy), Algiers (Algeria), Cairo (Egypt), Santiago de Cuba (Cuba), Thessaloniki (Greece) [14].

Complete synthetic seismograms in terms of displacement, velocity and acceleration are computed separately for the SH and P-SV waves in the frequency range of interest. For urban areas where the numerical modelling of ground motion has been successfully compared with strong motion records, the computations of synthetic seismograms permit a detailed microzoning based upon the set of possible scenario earthquakes. The results obtained for all discussed case studies have shown that for areas where very limited or no recordings are available the synthetic time series can be used to estimate the expected ground motion, thus leading to a pre-disaster microzonation without having to wait for an earthquake to occur. The use of synthetic computations is also necessary to overcome the fact that the local site response can be strongly dependent upon the properties of the seismic source generating the seismic input. The constructed data set of synthetic seismograms can be fruitfully used and analyzed by civil engineers for design and reinforcement actions, and therefore supply a particularly powerful and economical tool for the prevention aspects of Civil Defense.

The well-documented distribution of damage in Rome, caused by the Fucino earthquake, is, in fact, successfully compared by Fäh et al. [15] with the results of a series of different numerical simulations, using PGA and Arias intensity. The good correlation between PGA and the damage statistics made it possible to extend the zoning to the entire city of Rome, thus providing a basis for the prediction of the expected damage from future strong events.

The hybrid approach was successfully applied to compute P-SV (radial and vertical components) and SH (transversal component) synthetic displacement time series, velocity time series and accelerograms along the constructed eleven local models at Debrecen, Hungary, considering theoretically the macroseismic effects of intensity VI - VII (MSK) [16]. The obtained results are in agreement with the observed macroseismic intensities due to the devastating 1834 earthquake, which occurred in the Érmellék seismoactive region.

The numerical synthesis of ground motion, simultaneously accounting for source, wave propagation path and local site geology, has been particularly useful for the microzonation of Bucharest, the capital of Romania. The synthetic seismic signals (frequency range up to 1 Hz) have been computed for three recent strong intermediate-depth Vrancea earthquakes (1986 and 1990) along three representative profiles crossing the Bucharest area. The computed acceleration response spectra at the Magurele recording station have been successfully compared with the available registrations. The analysis of the characteristics of the computed response spectra, in correlation with the geological features of the shallow structure, results in a preliminary seismic zonation of the city [17].

Using the hybrid method, complete synthetic accelerograms (up to frequencies of 6 Hz) have been computed along a selected known geological profile crossing Zagreb, the capital and largest city of Croatia, considering the Kasina 1880 earthquake [18]. The computed PGA amplification along the investigated profile has been consistent with the reported distribution of intensities for this earthquake, which exhibited a uniform band of intensity I = VIII (MSK) stretching from the epicenter southwestwards along the investigated profile. Using a set of realistically chosen scaled sources [19] has shown that even variations of the order of commonly observed uncertainties of only dip and rake angles of the seismogenic fault cause the amplification to vary at some sites by more than a factor of two. They concluded that, especially for strongly laterally heterogeneous structures, local effects must be determined for each of the relevant sources considering all associated uncertainties as completely as possible.

The city of Sofia, which is exposed to a high seismic risk, is a good example when a reliable definition of the seismic input for earthquake engineering purposes in the region are necessary in the lack of both instrumental data and the possibility to accurately quantify the magnitude scaling and the attenuation characteristics of the large magnitude earthquakes. The obtained results, applying the hybrid approach, are consistent with the few macroseismic data available for the city. The elastic displacement demand computed for Sofia has been consistent with the available near-field observations worldwide [20]. Synthetic time histories and site response along a selected geological profile crossing Russe, NE Bulgaria, for five recent strong intermediate-depth Vrancea earthquakes (1940, 1986, 1977 and 1990) have been computed applying the mode summation combined with the mode coupling technique [21]. The obtained results have been successfully validated against the available accelerograms and the derived acceleration response spectra. The analyses of the seismic record and the computed site response along the investigated profile at Russe have confirmed that urban areas located at large epicentral distances with respect to the seismic source may be prone to severe earthquake hazard and that all ground motion components give a significant contribution to the design seismic input.

CONCLUSIONS

The necessity of preventive action arises from the increasing vulnerability of the society in the processes of running globalization and urbanization due to the high and underestimated earthquake risk. The recently completed UNESCO - IUGS - IGCP Project 414, and the still running CEI Projects, the Central Europe Regional Geodynamics Project (CEGROP) and the Unification of Gravity Systems in Central and Eastern Europe (UNIGRACE) brought about, through their scientific conferences and publications, a new view on the dynamics of the Central European area. For areas where very limited or no recordings are available the synthetic time series can be used to estimate the expected ground motion, thus leading to a pre-disaster microzonation without having to wait for an earthquake to occur. The use of synthetic computations is also necessary to overcome the fact that the local site response can be strongly dependent upon the properties of the seismic source generating the seismic input.

The scenario-based methodology is strictly based on observable facts and data and is complemented by physical modelling techniques, which can be submitted to a formalized validation process. By means of sensitivity analysis, knowledge gaps related to lack of data can be dealt with easily, due to the limited amount of scenarios to be investigated. The applied neo-deterministic procedure provides realistic and reliable seismic input, which could be directly used in earthquake engineering practice, urban planning, land using, the insurance industry against natural disasters - e.g. design of new constructions, estimates of the earthquake resistant capacity of the existing built stock and metropolitan seismic microzonation.

ACKNOWLEDGMENTS

The Environment and Sustainable Development Programme, Contract number: EVK2-CT-2000-57002,the UNESCO - IGCP Project 414, the INTAS-Moldova 200505-104-7584 Project "Numerical Analysis of 3D seismic wave propagation using Modal Summation, Finite Elements and Finite Differences Method", the CEI university network and the CEI Project "Geodynamical Model of Central Europe For Safe Development Of Ground Transportation Systems" are gratefully acknowledged.

REFERENCES

1. McGuire R., Deterministic vs Probabilistic Earthquake Hazards and Risks, Soil Dynamics and Earthquake Engineering 21 (2001), 377 - 384.
2. Panza G.F., Cioflan C.O., Kouteva-Guentcheva M., Paskaleva I., Marmureanu G., Radulian M. (2007) Preventive Actions to the Impactof Large Vrancea Events Carried on in the Framework of CEI; 3rd CEI Conference Science and Technology for Safe Development of Life Line Systems, October 24th – 26th, 2007, Bucharest, Romania, http://www.infp.ro/workshops/CEI/news.html

3. Klügel J.-U. Mualchin L., Panza G.F. (2006) A scenario-based procedure for seismic risk analysis, Engineering Geology 88 (2006) 1–22, available online at www. sciencedirect.com

4. Klügel, J. -U. (2007 a) Error inflation in Probabilistic Seismic Hazard Analysis, Engineering Geology 90 (2007) 186-192, available online at www. sciencedirect.com

5. Klügel, J. -U. (2007 b) Comment on "Why Do Modern Probabilistic Seismic-Hazard Analyses Often Lead to Increased Hazard Estimates" by Julian J. Bommer and Norman A. Abrahamson, BSSA, Vol. 97, No.6, pp.-, comment, doi: 10.1785/0120070018.

6. Peresan A., Panza G.F., Romanelli F., Vaccari F. (2008) Deterministic seismic hazard assessment and earthquake prediction; http://www.geophysik.uni-kiel.de/~geo43/downloads/aperesan3.ppt#2

7. Panza, G.F., Romanelli, F., Vaccari, F. (2001) "Seismic wave propagation in laterally heterogeneous anelastic media: theory and applications to the seismic zonation", Advances in Geophysics, Academic press; 43: 1-95.

8. Zivcic Ml., Suuhadolc P., Vaccari F. (2000) Seismic Zoning of Slovenia Based on Deterministic Hazard Computations. In "Seismic Hazard of the Circum-Pannonian Region (Editors: Panza, GF., Radulian, M., Trifu. C.), Pageoph Topical Volumes, Birkhauser Verlag: 171-184.

9. Markusic S., Suhadolc P., Herak M., Vaccari F. (2000) A Contribution to Seismic Hazard Assessment in Croatia from Deterministic Modeling. In "Seismic Hazard of the Circum-Pannonian Region (Editors: Panza, GF., Radulian, M., Trifu. C.), Pageoph Topical Volumes, Birkhauser Verlag: 185-204.

10. Bus Z., Szeidovitz G., Vaccari F. (2000) Synthetic Seismogram Based Deterministic Seismic ZOning for the Hungarian Part of the Pannonian Basin. In "Seismic Hazard of the Circum-Pannonian Region (Editors: Panza, GF., Radulian, M., Trifu. C.), Pageoph Topical Volumes, Birkhauser Verlag: 205-220.

11. Radulian M., Vaccari F., Manderscu N., Panza G.F., Moldoveanu C.L. (2000) Seismic Hazard of Romania: Deterministic Approach. In "Seismic Hazard of the Circum-Pannonian Region (Editors: Panza, GF., Radulian, M., Trifu. C.), Pageoph Topical Volumes, Birkhauser Verlag: 205-220.

12. Panza, G.F., Vaccari F., Cazzaro R. (1999) Deterministic seismic hazard assessment. In F. Wenzel et al. (Eds), Vrancea Earthquakes: Tectonics, Hazard and Risk Mitigation, 269-286. Kluwer Academic Publishers, The Netherlands.

13. Panza, G.F., Vaccari F. (2000) Introduction in seismic hazard of the Circum-pannonian region (Editors: Panza, GF., Radulian, M., Trifu. C.), Pageoph Topical Volumes, Birkhauser Verlag: 5-10.

14. Panza, G.F., Alvarez, L., Aoudia, A., Ayadi, A., Benhallou, H., Benouar, D., Bus, Z., Chen, Y.-T., Cioflan, C., Ding, Z., El-Sayed, A., Garcia, J., Garofalo, B., Gorshkov, A., Gribovszki, K., Harbi, A., Hatzidimitriou, P., Herak, M., Kouteva, M., Kuznetzov, I., Lokmer, I., Maouche, S., Marmureanu, G., Matova, M., Natale, M., Nunziata, C., Imtiyaz A. Parvez, Paskaleva, I., Pico, R., Radulian, M., Romanelli, F., Soloviev, A., Suhadolc, P., Szeidovitz, G., Triantafyllidis, P., Vaccari F., (2002). Realistic Modeling of Seismic Input for Megacities and Large Urban Areas (the UNESCO/IUGS/IGCP project 414). Episodes, 25(3), 160-184.

15. Fäh, D., Iodice, C., Suhadolc, P., Panza, G.F. (1993) A new method for the realistic estimation of seismic ground motion in megacities: The case of Rome, Earthquake Spectra 9: 643-668

16. Gribovszki K., Panza G.F. (2004): Seismic microzonation with the use of GIS (Case study for Debrecen, Hungary). *Acta Geod. Geoph. Hung.,* **39(2-3).** pp. 177-190.

17. Cioflan, C.O., Apostol, B.F., Moldoveanu, C.L., Panza, G.F., Marmureanu, Gh. (2004) Deterministic Approach for the Seismic Microzonation of Bucharest, PAGEOPH 161, 1149-1164.

18. Lokmer I, Herak M, Panza GF, Vaccari F. (2002): Amplification of strong-ground motion in the city of Zagreb, Croatia, estimated by computation of synthetic seismograms, Soil Dynamics and Earthquake Engineering, 22, 105-113.

19. Herak M, Lokmer I, Vaccari F, Panza GF (2004): Linear amplification of horizontal strong ground motion in Zagreb (Croatia) for a realistic range of scaled point sources, PAGEOPH, 161, 1021-1040.

20. Paskaleva, I., Kouteva, M., Vaccari, F., Panza, G.F. (2008) Characterization of the Elastic displacement demand: cASE STUDY - Sofia City, AGGH, (in print).

21. Kouteva, M., Panza, G.F., Romanelli, F., Paskaleva, I. (2004) Modelling of the ground motion at Russe site (NE Bulgaria) due to the Vrancea earthquakes. Journal of Earthquake Engineering, 8, 2, 209-229.

SOIL LIQUEFACTION AND
LIQUEFACTION COUNTERMEASURES

Prediction of Liquefaction Potential of Dredge Fill Sand by DCP and Dynamic Probing

Md. Jahangir Alam[a], Abul Kalam Azad[b] and Ziaur Rahman[b]

[a]Assistant Professor, Department of Civil Engineering, Bangladesh University of Engineering and Technology, Dhaka-1000, Bangladesh. E-mail: mjahangiralam@ce.buet.ac.bd
[b]Graduate Students, Department of Civil Engineering, Bangladesh University of Engineering and Technology, Dhaka-1000, Bangladesh

Abstract. From many research it is proved that liquefaction potential of sand is function of mainly relative density and confining pressure. During routine site investigations, high-quality sampling and laboratory testing of sands are not feasible because of inevitable sample disturbance effects and budgetary constraints. On the other hand quality control of sand fill can be done by determining in situ density of sand in layer by layer which is expensive and time consuming. In this paper TRL DCP (Transportation Research Laboratory Dynamic Cone Penetration) and DPL (Dynamic Probing Light) are calibrated to predict the relative density of sand deposit. For this purpose sand of known relative density is prepared in a calibration chamber which is a mild steel cylinder with diameter 0.5 m and height 1.0 m. Relative density of sand is varied by controlling height of fall and diameter of hole of sand discharge bowl. After filling, every time DPL and DCP tests are performed and for every blow the penetration of cone is recorded. N10 is then calculated from penetration records. Thus a database is compiled where N10 and relative densities are known. A correlation is made between N_{10} and relative density for two types of sand. A good correlation of N_{10} and relative density is found.

Keywords: Relative density and liquefaction

INTRODUCTION

From many research it is proved that liquefaction potential of sand is function of mainly relative density and confining pressure. In general higher the relative density of sand lesser is the possibility of liquefaction of sand deposit. However confining pressure also play an important role in defining the dilatancy behavior of sand during shearing or earthquake motion. In higher confining pressure of sand positive dilatancy changed to negative dilatancy. Negative dilatancy increases the liquefaction potential of sand. In an alluvial sand deposit or dredge fill sand, confining pressure can be easily calculated from the over burden pressure. If the relative density of sand can be determined in an easy and cheapear method, liquefaction potential of sand deposit can be predicted. During routine site investigations, high-quality sampling and laboratory testing of sands are not feasible because of inevitable sample disturbance effects and budgetary constraints. On the other hand quality control of sand fill can be done by determining in situ density of sand in layer by layer which is expensive and time consuming.

CP1020, *2008 Seismic Engineering Conference Commemorating the 1908 Messina and Reggio Calabria Earthquake*, edited by A. Santini and N. Moraci

TABLE 1. Calibration of sand discharge bowl to prepare a deposit of known relative density.

Type of Sand	Diameter of discharge hole (mm)	Height of fall (m)	Relative Density (%)
Dredge fill sand (Fineness Modulus = 0.8)	5	0.15	22.3
	5	0.30	24.9
	5	0.46	32.1
	5	0.61	35.9
	5	0.91	44.1
	5	1.07	48.5
	5	1.22	57.1
	3	0.15	58.8
	3	0.30	83.5
	3	0.46	88.7
	3	0.69	95.9
	3	0.91	95.6
	3	0.91	95.5
Jamuna Sand (Fineness Modulus = 0.8)	5	0.76	69.2
	5	0.91	74.6
	5	1.07	79.6
	3	0.46	89.0
	3	0.30	85.8
	3	0.61	93.5
	7	0.76	46.9
	7	0.91	54.2
	7	1.22	57.8
	5	0.61	53.0
Sylhet Sand (Fineness Modulus = 2.2)	7	0.15	21.6
	7	0.30	47.2
	7	0.46	57.6
	7	0.61	68.9
	7	0.76	76.5
	7	0.91	88.0
	7	1.07	90.5
	7	1.22	91.2

Now a day land reclamation using dredge fill sand became popular in Bangladesh. Sand fill is used in construction of road embankment, large industrial floor area and development of model towns etc. To ensure the nonliquefiable sand fill, widely used and accepted method in Bangladesh is determination of field density of sand by sand cone method. Then relative density and percent compaction is calculated using maximum and minimum dry density determined in the laboratory. Sand cone method is time consuming, expensive, inconvenient and limited to certain depth of fill. In this paper TRL DCP (Transportation Research Laboratory Dynamic Cone Penetration) and

DPL (Dynamic Probing Light) are calibrated to predict the relative density of sand deposit.

From experience it is observed that minimum relative density of dredge fill sand is about 30%. If the dredge fill sand is tried to compact, it is very easy to achieve upto 70% relative density. However, relative density below 30% and above 70% is very difficult to achieve. Calculation of relative density and percent compaction shows that percent compaction become above 75-80% while the sand is very loose (Dr = 30%). So, Percent compaction should never be used for sandy deposit which can mislead to engineers about quality and liquefaction potential of sand deposit.

TESTING PROGRAM

For the purpose calibration of DCP and DPL, sand of known relative density is prepared in a calibration chamber which is a mild steel cylinder with diameter 0.5 m and height 1.0 m. Relative density of sand is varied by controlling height of fall and diameter of hole of sand discharge bowl. Before filling calibration chamber hole diameter of discharge bowl and height of fall were calibrated for various relative densities. Table 1 summarizes the calibration of sand discharge bowl to prepare a sand deposit of known relative density three types of sand; Dredge fill sand from a reclamation site, Jamuna sand and Sylhet sand. Figure 1 shows one of the sand discharge bowls used in the sand preparation. Grain size distribution of three types of sand (Dredge fill sand, Jamuna Sand and Sylhet sand) are shown in Figure 2. Dredge fill sand and Jamuna sand is more or less similar in particle size distribution. However, Jamuna sand contains significant amount of mica.

Then calibration chamber is filled at different relative densities of sand as shown in Figure 3. After filling, every time DPL and DCP tests are performed and for every blow the penetration of cone is recorded. Components of DCP are shown in Figure 4. DPL is also has similar arrangements except differences in cone size, weight of anvil, weight of drop hammer and height of fall. Table 2 shows the differences in DCP and DPL. In both case N_{10} is the number of blows required for 10 cm penetration of the cone. Figure 5 shows the top view of the calibration after completion of two DCP and two DPL tests. After completion of tests, sands are taken out from the calibration chamber and weighed to check the density and relative density of sand.

N_{10} is then calculated from penetration records. A typical N_{10} plot with depth is shown in Figure 6. Representative N_{10} is taken from where N_{10} values are constant. Thus a database is compiled where N_{10} and relative densities are known. A correlation is made between average N_{10} and relative density for two types of sand. A good correlation of N_{10} and relative density is found. Figure 7 and Figure 8 show the N10 and relative density plot with exponential curve fitting. It is found that the correlation is dependent on the particle size of sand. Cone size of DPL is larger than that of DCP. That is why DPL is more sensitive in determining relative density of loose sand.

After calibration of DCP and DPL, the proposed method is applied in a reclamation site where Jamuna sand was dredged to fill the low land near Jamuna Bridge. At four locations DCP and DPL tests were performed and insitu density of fill were determined using sand cone method. N_{10} is plotted against relative density for the

TABLE 2. Comparision between
DCP and DPL

	DCP	DPL
Hammer (kg)	8	10
Height of fall (m)	0.66	0.5
Mass of anvil (kg)	--	6
Cone diameter (mm)	22.5	35.7

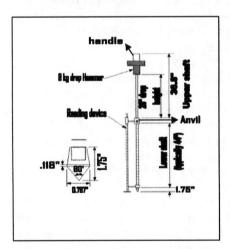

FIGURE 1. Photograph of a typical
discharge bowl with 5 mm diameter hole

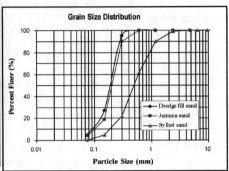

FIGURE 2. Grain size distribution of sands used
in the study.

FIGURE 4. Components of a TRL DCP

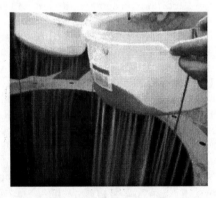

FIGURE 3. Preparation of sand deposit in
calibration chamber by discharging
sand from discharge bowl maintaining a
constant height of fall

FIGURE 5. Top view of calibration chamber after tests

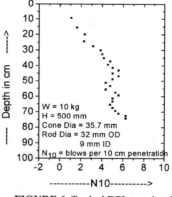

FIGURE 6. Typical DPL test data in calibration chamber.

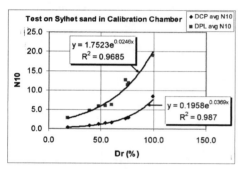

FIGURE 7. Calibration of DCP and DPL for Sylhet Sand

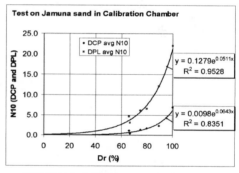

FIGURE 8. Calibration of DCP and DPL for Jamuna Sand

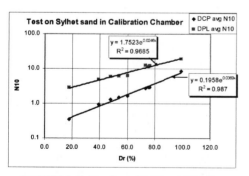

FIGURE 9. Calibration of DCP and DPL for Sylhet sand (N10 in logarithmic scale)

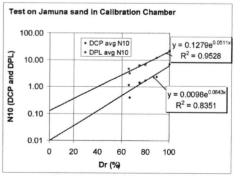

FIGURE 10. Calibration of DCP and DPL for Jamuna sand (N10 in logarithmic scale)

reclamation site. Similar trend of N_{10} vs. relative density was found but not the same which may be due to error in sand cone method. Logarithm plot of N_{10} against relative density are shown in Figure 9 and Figure 10 which shows a better correlation between N_{10} and relative density.

CONCLUSIONN

DCP and DPL are found to be useful in determining the relative density of sand fill and thereby assess the quality of compaction. As liquefaction potential is a function of relative density. This method can be used to assess the liquefaction potential.

REFERENCES

1. D. J. V. Vuuren, "Rapid determination of CBR with the portable Dynamic Cone Penetrometer", *The Rhodesian Engineer*
2. C. R. Jones and J. Rolt, "Operating instructions for the TRL dynamic cone penetrometer (2nd edition)" *Information note,* Crowthorne: Transportation research laboratory.

Some Important Aspects of Physical Modelling of Liquefaction in 1-g Shaking Table

Md. Jahangir Alam[a] and Ikuo Towhata[b]

[a]Assistant Professor, Department of Civil Engineering, Bangladesh University of Engineering and Technology, Dhaka-1000, Bangladesh. E-mail: mjahangiralam@ce.buet.ac.bd
[b]Professor, Department of Civil Engineering, University of Tokyo, Japan

Abstract. Physical modeling of liquefaction in 1-g shaking table and dynamic centrifuge test become very popular to simulate the ground behavior during earthquake motion. 1-g shaking table tests require scaled down model ground which can be prepared in three methods; water sedimentation, moist tamping and dry deposition method. Moist tamping and dry deposition method need saturation of model ground which is expensive and very difficult to achieve. Some model tests were performed in 1-g shaking table to see the influence of preparation method of model ground. Wet tamping and water sedimentation method of ground preparation were compared in these tests. Behavior of level ground and slope were also examined. Slope and level ground model test increased the understanding of excess pore pressure generation in both cases. Wet tamping method has a possibility of not being fully saturated. Pore pressure transducers should be fixed vertically so that it can not settle down during shaking but can move with ground. There was insignificant difference in acceleration and excess pore pressure responses between wet tamping and water sedimentation method in case of level ground. Spiky accelerations were prominent in slope prepared by water sedimentation method. Spiky accelerations were the result of lateral displacement induced dilatancy of soil.

Keywords: Liquefaction, Physical modeling, Shaking table and Preparation method.

INTRODUCTION

Seismic liquefaction induced damage of geotechnical structures like embankment, retaining walls and slopes [1-6] posed challenges to civil engineers. Before revising the design guidelines and codes it is necessary to understand the behavior of those structures during earthquake. Numerical modeling of those structures simulating earthquake load has got many limitations. That is why, physical modeling of liquefaction in 1-g shaking table and dynamic centrifuge test become very popular to simulate the ground behavior during earthquake motion. 1-g shaking table tests require scaled down model ground which can be prepared in three methods; water sedimentation, moist tamping and dry deposition method. Moist tamping and dry deposition method need saturation of model ground which is expensive and very difficult to achieve. Some model tests were performed in 1-g shaking table to see the influence of preparation method of model ground. Wet tamping and water sedimentation method of ground preparation were compared in these tests. Behavior of level ground and slope were also examined.

CP1020, *2008 Seismic Engineering Conference Commemorating the 1908 Messina and Reggio Calabria Earthquake,*
edited by A. Santini and N. Moraci
© 2008 American Institute of Physics 978-0-7354-0542-4/08/$23.00

MODEL PREPARTION

Model test number TRIAL33 consisted 3 sections; one is prepared by water sedimentation method (Dr = 30%), another is middle section prepared by wet tamping method (Dr = 20%) and another section is prepared by wet tamping method (Dr = 40%). Figure 1 and Figure 2 shows the cross section and transducer locations of model TRIAL33 respectively. At first middle and right section were prepared by wet tamping method using Toyoura sand, pore airs were expelled by circulating CO_2 from the bottom of container and then saturated by water supply. At this stage, the left section was prepared by water sedimentation method using Toyoura sand. Accelerometers were installed on top of ground surface and at 10 cm depth from surface. Pore pressure transducers were installed only at 10 cm depth from ground surface. Input motion applied in this test consists of 23 cycles at 10 Hz having maximum acceleration amplitude of 350 Gal (Figure 5).

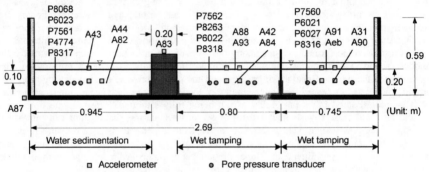

FIGURE 2. Accelerometers and pore pressure transducer locations in TRIAL33.

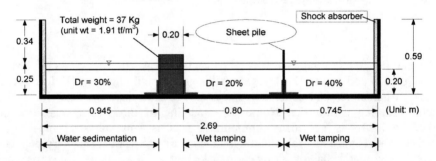

FIGURE 1. Section of trial test TRIAL33 of level ground prepared by two methods

Density of model ground

In a 1-g shaking table test of reduced scale model, the similitude rule in terms of stress and strain against the prototype can not be fully satisfied because of the stress dependency of the stress-strain behavior of soil. Verdugo [7, 8] showed the effect of confining pressure and void ratio by undrained triaxial tests on loose Toyoura sand. When the initial confining pressure was 1960 kPa and initial void ratio was 0.908, strain softening occurred after the peak strength suggesting contracting behavior of specimen. When the confining pressure was reduced to 98 kPa and the initial void ratio was maintained same at 0.908, the softening behavior never appeared, suggesting dilative behavior. A third test was run under 98 kPa and increased initial void ratio 0.949. Strain softening behavior appeared again in this third specimen, suggesting that to reproduce strain softening and contractive behavior of soil in scaled down models, density of sand should be less than prototype density. However, till now, there is no established similitude rule regarding the density of sand in scaled down model. Considering that confining pressure is an important parameter (Verdugo and Ishihara, 1996), the model ground was prepared at Dr = 20% and 40% by wet tamping method and Dr = 30% (minimum possible relative density of Toyoura sand by water sedimentation method) to keep the contractive behavior of sand at very low confining pressure of 1-g model test.

Shock absorber

At the ends of model container shock absorbers were attached to minimize the compression wave (P-wave) generation from two ends of container. To minimize the P-wave generation from the both end of rigid model container, a dry mattress made of polymer wires encapsulated with impermeable polythene sheet was used as interface between sand and end wall of container. As this mattress material was highly permeable, the excess pore pressure generated in model ground could be easily dissipated through these shock absorbers. That was why the shock absorbers were wrapped around by impermeable sheets.

Accelerometers and Pore pressure transducers

Accelerometers and pore pressure transducers were used on top and into model ground to monitor the responses during dynamic loading. Maximum capacity of pore pressure transducers was 20 kPa. These transducers were placed inside the soil and fixed vertically by using narrow aluminum U-channels supported on bottom of container. This system prevents the subsidence of transducers without preventing any flow of soil. Some pore pressure transducers were kept free into the ground to see the differences in response records.

The soil acceleration during shaking was measured with accelerometers with capacity of 2g. These accelerometers were stabilized against horizontal rotation and vertical subsidence during liquefaction of ground. The stabilizer was tiny square platform made of 2-3 mm thick acrylic sheets. The platform size was almost double the width of accelerometers. The square platform was supported on four acrylic walls of size 20 mm by 15 mm. A stabilized accelerometer is shown in Figure 3. The important point is that accelerometers with stabilizer can follow the flow of ground with tilting in the direction of flow which represents the soil movement in two dimensions of vertical plane. On the other hand it helps to prevent horizontal rotation against vertical axis. The accelerometer reading affected by the forward inclination of transducer caused by liquefied soil flow can be easily corrected by applying the principle explained in Figure 4.

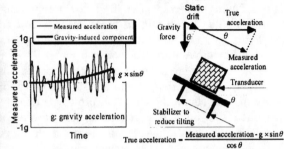

True acceleration $= \dfrac{\text{Measured acceleration} - g \times \sin\theta}{\cos\theta}$

FIGURE 3. A stabilized accelerometer.

FIGURE 4. Stabilization of accelerometers and tilting correction principle.

RESULT AND DISCUSSION ON LEVEL GROUND MODEL

Acceleration and pore pressure response in three sections of TRIAL33 were almost same. Typical responses are shown in Figure 5. After the first cycle, excess pore pressure reached 100% of the initial overburden pressure. Acceleration amplitude disappeared at the ground surface after the first cycle and at depth 10 cm during the second cycle. At level ground condition, there was no dilatant behavior of Toyoura sand in both water sedimentation and wet tamping portion.

Excess pore pressures in three sections are shown in Figure 6, Figure 7, and Figure 8. Except P8316, other transducers were found to have no problem in records. However, pore pressure transducer P8316 was relatively new and had shown no problem in any other tests. So the problem of P8316 pore pressure record can be attributed to partial or no saturation of the transducer. P7560 recorded less excess pore pressure than others in the same section. This can be due to less saturation level of soil around that transducer.

Vertically fixed pore pressure transducers (P8317 and P8318) were more reliable than free transducers in determining level of excess pore pressure ratio. Because free transducers settled downward during shaking which increase hydrostatic pore pressure. All good transducers were found to give reliable and consistent pore pressure records. Water sedimentation method did not exhibit any problem in pore pressure records. Wet tamping with 20% relative density also did not exhibit any problem in pore pressure records, while wet tamping with 40% relative density showed less excess pore pressure generation in P7560 and inconsistent pore pressure in P8316. Wet tamping has possibility of saturation problem of soil and/or pore pressure transducers.

RESULT AND DISCUSSION ON SLOPING GROUND MODEL

A slope model test SEDIM1 was performed by making a slope of 1:2. SEDIM1 was prepared by water sedimentation method. Model configuration is shown in Figure 9. The model was shaken by shake1 and shake2. Shake1 consists of 23 cycles at 10 Hz having maximum acceleration amplitude 110 Gal. Shake2 consists of 23 cycles at 10 Hz having maximum acceleration amplitude 330 Gal.

Acceleration and excess pore pressure time histories of test SEDIM1 during shake1 and shake2 are displayed in Figure 10 and Figure 11. Excess pore pressure could not reach 100% of initial overburden pressure in shake1 due to small initial static shear stress caused by slope. However, maximum excess pore ratio reached 80% and excess pore pressure ratio had cyclic portion throughout the shaking which varied from 45% to 80% in all cycles. Lateral movement of sand observed during shaking. Lateral deformation induced dilatancy of sand caused cyclic portion of pore pressure variation. Due to this dilatancy, acceleration did not de-amplify during shaking (see Figure 10). However dilatancy induced spiky acceleration directed to upslope was not seen during shake1. Maximum acceleration amplitude of shake1 was 100 Gal which might not produce the required dilatancy of soil.

During shake2 (maximum acceleration 330 Gal), maximum excess pore pressure ratio reached about 80% of initial overburden pressure which was same as in shake1. However, unlike shake1, this time cyclic variation of excess pore pressure reached as low

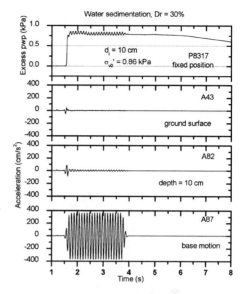

FIGURE 5. Typical acceleration and excess pressure time histories in TRIAL33 taken from water sedimentation portion.

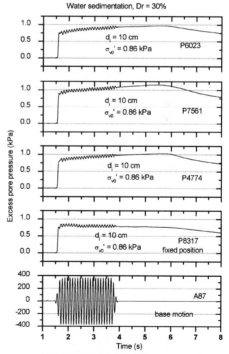

FIGURE 6. Excess pore pressure time histories in water sedimentation portion of TRIAL33

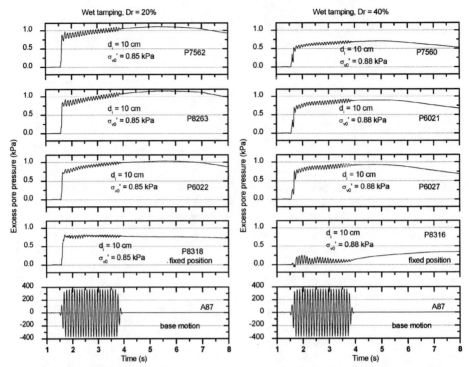

FIGURE 7. Excess pore pressure time histories in wet tamping portion (Dr=20%) of TRIAL33

FIGURE 8. Excess pore pressure time histories in wet tamping portion (Dr=40%) of TRIAL33

as -200% of initial overburden pressure. This negative excess pore pressure increased at every cycle of shake2. Acceleration also exhibited negative spiky response during shake2 while negative acceleration is directed to upslope, the deformation of soil is directed to down slope. Figure 12 shows the coincidence of dilation spikes and seaward displacement peaks of soil.

The negative spiky acceleration is explained in Figure 13. In these nearly sinusoidal shaking acceleration and cyclic displacement of soil and accelerometer is always out of phase having phase difference 180°. When soil moves downslope (Figure 14), acceleration is negative means directed to upslope. This is exemplified by a car which is going left direction having acceleration in right direction, if it start break to stop somewhere. If the car is stopped by a hidden bar before its target stop, sudden change of velocity would cause spiky acceleration which is explained in Figure 13. The hidden bar is the dilation spike of excess pore pressure response.

CONCLUSION

Slope and level ground model test increased the understanding of excess pore pressure generation in both cases. Wet tamping method has a possibility of not being fully saturated. Pore pressure transducers should be fixed vertically so that it can not settle down during shaking but can move with ground. There was insignificant difference in acceleration and excess pore pressure responses between wet tamping and water sedimentation method in case of level ground. Spiky accelerations were prominent in slope prepared by water sedimentation method. Spiky accelerations were the result of lateral displacement induced dilatancy of soil.

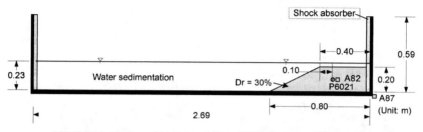

FIGURE 9. Section of trial test of slope SEDIM1 prepared by water sedimentation method.

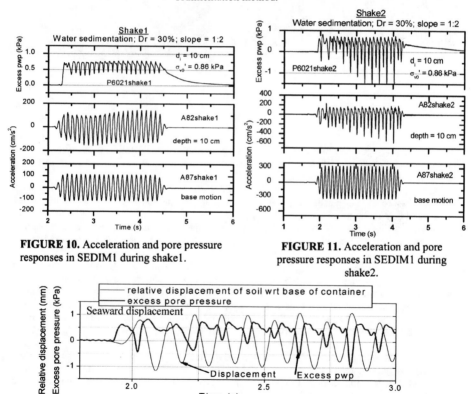

FIGURE 10. Acceleration and pore pressure responses in SEDIM1 during shake1.

FIGURE 11. Acceleration and pore pressure responses in SEDIM1 during shake2.

FIGURE 12. Relative cyclic displacement of soil wrt base and excess pore pressure time history.

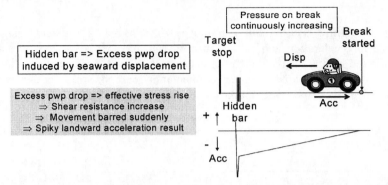

FIGURE 13. Explanation of spiky acceleration during down slope movement liquefied sand.

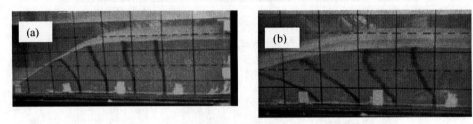

FIGURE 14. Photograph of slope model after shake1 (a) and shake (b).

REFERENCES

1. Adalier K, Aydingun O (2003). "Numerical analysis of seismically induced liquefaction in earth embankment foundations. Part II. Application of remedial measures", Canadian Geotechnical Journal, Vol. 40, No. 4, pp. 766-779.
2. Adalier, K., Elgamal, A. W., Meneses, J. and Baez, J. I. (2003): Stone columns as liquefaction countermeasure in non-plastic silty soils, Soil Dynamics and Earthquake Engineering, Elsevier Science Ltd., Vol. 23, pp 571-584.
3. Adalier, K., Elgamal, A.W., Martin, G.R. (1998): Foundation Liquefaction Countermeasures for Earth Embankments, Journal of Geotech. and Geoenvironmental Engg., ASCE, vol. 124, no. 6, pp. 500–517.
4. Alam, M. J., Fukui, S., Towhata, I., Honda, T., Tamate, S., Tanaka, T., Uchiyama, J., and Yasuda, A. (2004c). "Centrifuge model tests on mitigation effects of underground walls on liquefaction-induced subsidence of embankment", Proc. 11th Int. Conf on Soil Dynamics and Earthquake Engineering and the 3rd Int. Conf. Earthquake Geotechnical Engineering, Berkeley, Vol.2, 537-544.
5. Matsuo, O. (1996): Damage to river dikes, Soils and Foundations Journal., Special Issue on Geotech. Aspects of the January 17, 1995 Hyogoken-Nambu Earthquake. Japanese Geotechnical Society, Tokyo, Japan. vol. 36, no. spec issue, pp. 235–240.
6. Miura, K., Kohama, E., Inoue, E., Ohtsuka, N., Sasajima, T., Hayashi, T. and Yoshida, N. (2000). "Behavior of gravity quay walls during earthquake regarding dynamic interaction between caisson and backfill during liquefaction" Proc of 12th World Conference on Earthquake Engineering, paper no. 1737.
7. Verdugo, R., (1992): "Characterization of Sandy Soil Behaviour under Large Deformation", Ph. D. Thesis, The University of Tokyo.
8. Verdugo, R., and Ishihara, K. (1996): "The steady state of sandy soils", Soils and Foundations, Vol. 36, No. 2, pp. 81-91.

A Potential Cost Effective Liquefaction Mitigation Countermeasure: Induced Partial Saturation

Hanbing Bian, Yun Jia, and Isam Shahrour

Laboratoire de Mécanique de Lille (UMR 8107), Université des Sciences et Technologies de Lille 59655 Villeneuve d'Ascq, France

Abstract. This work is devoted to illustrate the potential liquefaction mitigation countermeasure: Induced Partial Saturation. Firstly the potential liquefaction mitigation method is briefly introduced. Then the numerical model for partially saturated sandy soil is presented. At last the dynamic responses of liquefiable free filed with different water saturation is given. It shows that the induced partial saturation is efficiency for preventing the liquefaction.

Keywords: Liquefaction Mitigation, Induced Partial Saturation, Numerical Model

INTRODUCTION

Soil liquefaction constitutes a major cause of damage induced by earthquakes. During the past decades, intensive efforts have been made by the geotechnical research community to understand the mechanism of liquefaction, and to develop methods for evaluating liquefaction potential at a site for a given seismic event. Many kinds of remediation methods against liquefaction have been developed and applied to structures since the Niigata earthquake in 1964. However, only a few remediation methods which can be applied to existing structures have been developed [1]. A potential cost effective liquefaction mitigation countermeasure: Induced Partial Saturation is very attractive; it could be applied for both new construction and existing structures.

The main idea of this potential method is to reduce the initial water saturation in liquefaction potential area. The idea is coming firstly from the effects of system compliance on liquefaction tests [2]. It shows that a small reduction in the degree of water saturation of sand soil can result in a significant increase in shear strength against liquefaction. Later, several other researchers also indicated that the water saturation has great influence on the liquefaction resistances. Even the quantitative conclusions differ from author to author, but the qualitative results are the same: The decrease in water saturation will result in higher liquefaction resistances.

The in-situ observation and the laboratory experiments also conform and encourage the researches on the potential liquefaction mitigation countermeasure. Shiraishi reported that 312 examples of structures having pneumatic caisson foundation were prevented from fatal damage by liquefaction in Kobe earthquake, because the ground surrounding the foundation were unsaturated by air entrapped during the construction

CP1020, *2008 Seismic Engineering Conference Commemorating the 1908 Messina and Reggio Calabria Earthquake*, edited by A. Santini and N. Moraci
© 2008 American Institute of Physics 978-0-7354-0542-4/08/$23.00

[3]. Singh et al. [4] examined the damage of 6 earth dams in Bhuj earthquake. The 6 reservoirs were nearly empty prior to earthquake, and the liquefiable soil layers were in a state of partially saturated sate. The in-situ observations indicated that the downstream area of the dam has less damage than the upstream. The observation shows that the partially saturation reduces the liquefaction potential. Based on the laboratory experimental results, Yegian et al. [5] indicates that the induced partial saturation may prevent the liquefaction in loose sand.

Clearly, the partial saturation in liquefiable soil layer will reduce the risk. This encourages the researchers to develop a new liquefaction mitigation method: the induced partial saturation. One of the key factors of this countermeasure is: how to generate the partial saturation. Yegian et al. [5] suggested that the Electrolysis may be a good method to induce partial saturation in ground. He also suggested a drained-recharge method for inducing partial saturation during the laboratory research; this may be used to construction site too. Nagao et al. [3] however advised a micro-bubble injection method to induce partial saturation in ground, he argued that as the micro-bubble is an independent small bubble of 10-100μm in diameter, it can easily permeate voids among sand particles.

The other key factor is the stability of partial saturation. Because the underground water is whole system, it is important to insure the sustainability of Induced Partial Saturation. Yegian et al. [5] have conducted a test referred as "long term air diffusion test" to investigate stability of the partial saturation. This test shows that under hydrostatic conditions, small well-distributed air bubbles could remain trapped for long time. The site observation reported by Okamura et al. [6] indicates that the partial saturation in sandy soil may be maintained for 26 years and further.

All these insure that the Induced Partial Saturation could be applied as liquefaction mitigation method. When the new remediation method is applied, the first question needs to answer is which water saturation should be used. After the remediation, it needs to evaluate the liquefaction risk with certain water saturation. The followed work will answer these problems. It focuses on the influence of water saturation of liquefiable soil layer on the liquefaction potential. Firstly, the numerical model for partially saturated sands is presented. And then dynamic response of partially saturated free filed is analyzed.

THE NUMERCIAL MODEL FOR UNSATURATED SANDY SOILS

Most numerical researches on liquefaction are supposed that the soil are fully saturated and suppose that the pore-water with constant compressibility, for example the work of Zienkiewicz et al [7], in the U-P model, the constant Biot modulus is used. However, as indicated in the first section, the soil layer where liquefaction is potential is not as usual assumed full saturated, but in a partially saturated sate. The influence of water saturation should be taken into consideration, especially for evaluating the liquefaction risk when Induced partial saturation is used. Bian and Shahrour [8] based on the theory of Biot and Coussy developed a formulation which can take the partially saturated condition into consideration for studying the liquefaction problem. The numerical model for partially saturated sandy soil is described in the followed.

The partially saturated soils can be viewed as the supposition of mechanically interacted media: two saturating fluids plus one deformable skeleton. It is considered partially saturated for a referenced fluid, the pore-water. Under isothermal condition, the interaction of water/vapour and dry air/water-vapour is so weak that it could be neglected in the liquefaction problem. For the sake of simplicity, the water-vapour and the dry air will not be distinguished. So based on the thermodynamic theory of Coussy [9], the fluid components of unsaturated soils include pore-water (index w) and pore-air (index a). Since the pore-air and pore-water occupy the total porous space, the porosity can be divided into two parts, as follows:

$$n = n_w + n_a \tag{1}$$

The saturation of each fluid can then be defined as:

$$S_w = \frac{n_w}{n}; S_a = \frac{n_a}{n} \tag{2}$$

The state equations for partially saturated soils then can be arranged in the incremental form as:

$$\delta\sigma_{ij} = \lambda\delta\varepsilon_{kk}\delta_{ij} + 2\mu\delta\varepsilon_{ij} - \left\{\frac{\delta m_w}{\rho_w^0} \quad \frac{\delta m_a}{\rho_a^0}\right\}\begin{bmatrix} M_{ww} & M_{wa} \\ M_{aw} & M_{aa} \end{bmatrix}\begin{Bmatrix} b_w S_w \\ b_a S_a \end{Bmatrix}\delta_{ij} \tag{3}$$

$$\begin{Bmatrix} \delta p_w \\ \delta p_a \end{Bmatrix} = \begin{bmatrix} M_{ww} & M_{wa} \\ M_{aw} & M_{aa} \end{bmatrix}\left(-\begin{Bmatrix} b_w S_w \\ b_a S_a \end{Bmatrix}\delta\varepsilon_{kk} + \begin{Bmatrix} \delta m_w/\rho_w^0 \\ \delta m_a/\rho_a^0 \end{Bmatrix}\right) \tag{4}$$

δ_{ij} is the Kronecker's delta. λ and μ are undrained elastic parameters. m is the flux of pore-fluid. b is the Biot coefficients for pore-fluid. ρ^0 is the initial density of fluid. $M_{ww}, M_{wa}, M_{aw}, M_{aa}$ depend essentially on the porosity, the compressibility of the soil grains, the pore-water and the pore-air, the fluid saturations and the soil water characteristic curve.

Equations (3) and (4) are general state equations for partially saturated soils. However, in case of sandy soils, some assumptions can be made for the simplicity of the problem. They are listed as follows: 1) The solid matrix is supposed as incompressible; 2) Under unsaturated condition, the pore-water is supposed as pure saturated water; 3) The pore-air is supposed as ideal gas; 4) The soil suction is neglected; 5) The interaction between pore-water and pore-air is neglected; 6) Null air flux is assumed. The last three assumptions are based on the laboratory and in-situ observations.

Combination of the assumptions listed above and the stress partition concept allows us to derive the constitutive equations for partially saturated sandy soils, where the parameters for Biot modulus can be expressed as:

$$M_{aa} = M_{aw} = M_{wa} = M_{ww} = M' \tag{5}$$

With the Biot modulus for partially saturated sandy soil M' expressed as:

$$\frac{1}{M'} = \frac{n_a}{K_a} + \frac{n_w}{K_w} = \frac{n(1-S_w)}{p_w + p_{a0}} + \frac{nS_w}{K_w} \tag{6}$$

And the governing equations can be arranged as [8]:

$$\delta\sigma = C : \delta\varepsilon - M' \frac{\delta m_w}{\rho_w} I$$

$$\delta p_w = \delta p_a = M'\delta\varepsilon_v - M' \frac{\delta m_w}{\rho_w}$$

(7)

Which are similar to that used for saturated soil (U-P model of Zienkiwicz), but with a Biot modulus depending on the porosity, pore pressure and water saturation. It is of interest to indicate that, in Equation (6), if the water saturation is unit, that means the soil is saturated, the formulation for saturated soil (U-P model) is recovered. So this formulation can be used for both saturated and unsaturated problems. The water saturation in the constitutive equation should be determined. Generally, for unsaturated soils, the water saturation is determined by the water retention curve, however for sandy soils, the ideal gas law could be used in stead of the water retention curve. The assumption of zero air flux used for the formulation of the model will be used to determine the water saturation too.

THE DYNAMIC RESPONSE OF LIQUFIABLE FREE FILED

The proposed numerical model for partially saturated sandy soils in section 2 was programmed into a finite element calculation code, where the Newmark schema is used to solve the dynamic problem in time domain. It has been used to evaluate the liquefaction risk of a liquefiable soil layer, with a thickness of 30 meters. The water table is supposed at the top of the soil layer. The initial water saturation is supposed uniform. The boundary conditions are defined as following: at the base of soil layer, the vertical displacement is blocked, and it is impermeable; at the lateral boundaries, the equivalent displacement and pore pressure are imposed; at the top of the soil layer, null water pressure is applied, and it is permeable. The soil layer is composed of loose Nevada sand with the relative density as Id=0.4. The physical properties of this sand are extracted from the VELCAS experiment report [10] and listed in Table 1.

TABLE 1. Material Properties of Nevada sand [10].				
Relative density	Void ratio	Porosity	Permeability	Density
0.4	0.736	0.42	6.6E-5m/s	1540kg/m3

The robust constitutive equations for granular material MODSOL [11] which is base on the effective stress conception is used to describe the mechanical behaviour of sand. The model parameters of the Nevada sand for MODSOL were determined by Khoshravan [12] to simulate the centrifuge tests of Nevada sand in the framework of VELCAS; the simulation results show that this constitutive model can well describe the mechanical behaviour of Nevada sand under dynamic loading.

For the partially saturated soil layer, the water saturation varies as Sw =100%, 99%, 95% and 90%. The loads frequency is equal to 2Hz and its amplitude is 0.25g. It is applied at the base of the soil layer during 5s. The problem considered is nonlinear; the initial stresses may have a great influence on the numerical results. The initial stress used is due to the gravity action.

Figure 1 presents the distribution of the excess pore pressure at the end of excitation for different initial water saturations. The pore-pressure is normalized by the initial effective stresses; while the depth is normalized by the total thickness of the soil layer. It can be observed that with the decrease in water saturation, the liquefaction zone will decrease. It confirms that Induced Partial Saturation is efficient in mitigating the liquefaction. For example, for fully saturated case, the liquefaction zone concerns about 20% of the layer. In case of water saturation Sw =99%, the liquefaction zone is reduced to 10%. The liquefaction resistance of partially saturated soil is greater than that of saturated soil. For instance, at the depth Z=0.6H, the ratio of excess pore-water pressure to effective stresses is about 0.42, while this ratio decreases to 0.25, 0.1 and 0.05 for the initial water saturation Sw =99%, 95% and 90% respectively. It is of interest to remark that: in Figure 1, at the top the soil layer the excess pore pressure, sometimes is greater than the initial confining pressure. This could be due to the very low value of the effective stresses.

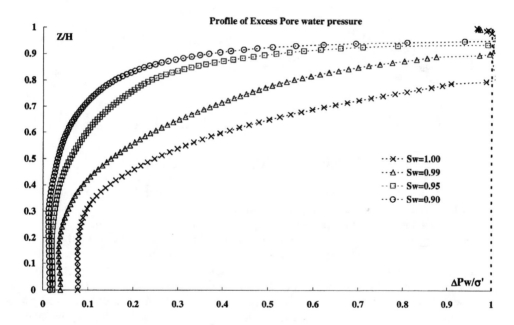

FIGURE 1. Influence of water saturation on the distribution of excess pore-pressure

Figure 2 shows the process of excess pore-pressure generation during the excitation at the depth of 5meter. For fully saturated case, at time about 2.66s, the excess pore-pressure is equal to the initial effective stresses (liquefaction occurs), while for the partially saturated cases, there is no liquefaction. The slope decreases with the decrease in the initial water saturation. It can be seen that the presence of gas in the pore-water reduces the excess pore-water generation. The increase in the amount of gas in soil voids leads to a reduction in the excess pore pressure generation rate.

The calculation results agree well with the in-situ observation, and correspond well with the laboratory results. It confirms that the Induced Partial Saturation could be used as liquefaction mitigation. It also serves as a tool to evaluate the partial saturation need, and asses the efficient of the remediation method.

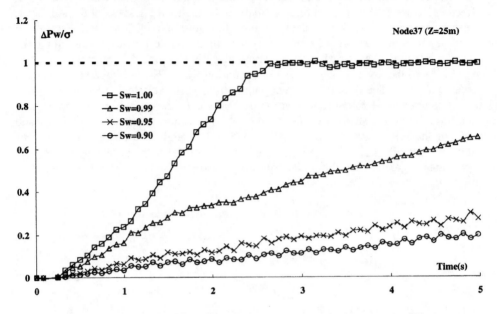

FIGURE 2. History of the excess pore-water pressure at depth of 5 meters for different

DISCUSSION AND CONCLUSION

The initial water saturation has significant influence on the dynamic response of free filed. With the decrease in water saturation, the potential liquefaction zone reduced. And the excess pore-pressure generation rate is decrease with the decrease in the initial water saturation. In the example described in section above, when the water saturation lower than 0.9, the liquefaction could not occurs. That suggests that lowering the water saturation is efficient in liquefaction mitigation. The induced partial saturation could be a potential liquefaction mitigation countermeasure.

From the technical point view, the Induced Partial Saturation could be used as liquefaction mitigation countermeasure. However, when applied to engineering project, there are still so many questions to be answered. The most important are: which partial saturation should be used? What is the efficiency of this method? Larger scale dynamic experiment in laboratory is possible, but it cost too much. The in-situ experiment will be more complicate and costly, and it is nearly impossible. The numerical predication and simulation would be a good choice. The proposed numerical model together with in-situ investigation and several necessary classical laboratory experiments will reinforce the potential liquefaction mitigation countermeasure.

ACKNOWLEDGMENTS

This researcher is conduced in Laboratory of Mechanics of Lille. The authors thank the laboratory for all the help and finance.

REFERENCES

1. S. Yasuda, "Remediation methods against liquefaction which can be applied to existing structures", *Earthquake geotechnical engineering*, Edited by Kyriazis D. P., Springer. 2007.
2. G. R. Martin, W. D. L. Finn and H. B. Seed, "Effects of system compliance on liquefaction testes", *Journal of the geotechnical engineering division*, Vol. 104, No. GT4, 1978.
3. K. Nagao, Y. Azegami, S. Yamada, N. Suemasa and T. Katada, "A Micro-bubble injection method for a countermeasure against liquefaction", *4th International Conference on Earthquake Geotechnical Engineering*, June 25-28, 2007.
4. R. Singh, D. Roy, S. K. Jain, "Analysis of earth dams affected by the 2001 Bhuj Earthquake", *Engineering Geology*, Vol. 80, pp. 282-291, 2005.
5. M. K. Yegian, E. Eseller-Bayat, A. Alshawabkeh and S. Ali, "Induced-Partial saturation for liquefaction mitigation: Experimental investigation", *Journal of geotechnical and geoenvironmental engineering*, ASCE, Vol. 133, No. 4, pp. 372-380, 2007.
6. M. Okamura, M. Ishihara, K. Tamura, "Degree of saturation and liquefaction resistances of sand improved with sand compaction pile", *Journal of geotechnical and Geoenvironmental engineering*, ASCE, Vol. 132, No. 2, pp. 258-264, 2006.
7. O. C. Zienkiewicz, A. H. C. Chan., M. Pastor, B. A. Schrefle and T. Shiomi, "Computational Geomechanics with special reference to earthquake engineering", John-Wiley, 1999.
8. H. B. Bian and I. Shahrour, "A numerical model for dynamic response of unsaturated sands", *4th International Conference on Earthquake Geotechnical Engineering*, June 25-28, 2007.
9. O. Coussy, Poro mechanics, John-Wiley, 2004.
10. K. Arulmoli, K. K. Muraleetharan, M. M. Hossain and L. S. Fruth, "VELACS Verification of liquefaction analyses by centrifuge studies laboratory testing program soil data report", 1992.
11. I. Shahrour, W. Chehade, "Development of a constitutive elastoplastic model for soils", *11th international congress on rheology*, Bruxelles, Elsevier, 1992.
12. A. A. Khoshravan, "Problèmes de sols satures sous chargement dynamique : Modèle cyclique pour les sols et validation sur des essais en centrifugeuse", *Thèse de doctorat*, Université des Sciences et Technologies de Lille, 1995.

433

Liquefaction Potential Assessment Of Silty And Silty-Sand Deposits: A Case Study

Diego C. F. Lo Presti and Nunziante Squeglia

Department of Civil Engineering, University of Pisa, via Diotisalvi 2 – 56126 Pisa

Abstract. The paper shows a case study concerning the liquefaction potential assessment of deposits which mainly consist of non plastic silts and sands (FC > 35 %, Ip < 10 %, CF negligible). The site under study has been characterized by means of in situ tests (CPTU, SPT and DPSH), boreholes and laboratory tests on undisturbed and remolded samples. More specifically, classification tests, cyclic undrained stress-controlled triaxial tests and resonant column tests have been performed. Liquefaction susceptibility has been evaluated by means of several procedures prescribed by codes or available in technical literature. The evaluation of liquefaction potential has been carried out by means of three different procedure based on in situ and laboratory tests.

Keywords: Liquefaction, cyclic triaxial test.

INTRODUCTION

Since 2003, the Tuscany Seismic Survey has started an investigation plan for retrofitting and repair of existing Public Buildings (Schools, Hospitals etc.) and for the design of the new ones, in the most seismic areas of Tuscany. Investigations for the existing buildings concerned the structure, the structural materials, the geology of the site and the geotechnical characterization of the soil deposits. Obviously for the design of new buildings only geological and geotechnical investigations have been done. Different levels of investigation have been undertaken. The first level consisted in geological surveys and seismic refraction tests in P and SH waves, in order to have geological maps (1:2000) and geological sections. The second level usually consisted in a borehole (at least) with SPT and down-hole measurements. The borehole extended down to the seismic bedrock or at least down to 30 m depth. In some cases, undisturbed samples have been retrieved. In the framework of these activities, the geotechnical investigations undertaken for the construction of a new Primary School located in Fornaci di Barga, in the northern part of Tuscany, indicated that the subsoil was susceptible to liquefaction. Therefore additional investigations have been carried out in order to have a better evaluation of liquefaction susceptibility and liquefaction hazards. The designed building is a one-storey construction with a reinforced concrete cast-in-situ structure. The paper shows the results of the investigations and analyses. Moreover the paper comments on the prescriptions of [1] and [2] in the light of the case study.

CP1020, *2008 Seismic Engineering Conference Commemorating the 1908 Messina and Reggio Calabria Earthquake*,
edited by A. Santini and N. Moraci

GROUND INVESTIGATION

Figure 1 shows the location in plan of preliminary and integrative investigations. The ground investigations consist in: 3 boreholes up to 52 m (S4) or 15 m (S15, S16); 9 Standard Penetration Tests; a down-hole test in borehole S4; a seismic refraction test (ST4); a super heavy dynamic probing (DPSH4) up to 19 m; 3 cone penetration tests, CPTU1, CPTU2 and CPTU3, carried out by means of a piezocone up to 10, 11.3 and 16.4 m, respectively. Nine undisturbed samples have been retrieved from boreholes. Several laboratory tests have been carried out, including resonant column test (CR), torsional shear test (TTC) and triaxial cyclic test (TXC). Table 1 lists the laboratory tests performed on undisturbed samples. In addition to listed tests, several determination of grain size distribution (Fig. 2a) and plastic index (Fig. 2b) have been carried out on remolded specimens.

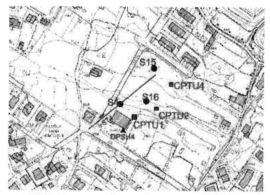

FIGURE 1. Location in plan of ground investigation.

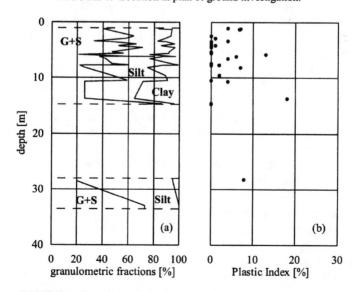

FIGURE 2. Granulometric fractions (a) and Plastic Index (b) profiles.

TABLE 1. List of laboratory tests performed on undisturbed samples.

Sample	Depth [m]	Tests
S4 C1	1.30	ED-IL, direct shear (TD), CR, TTC
S4 C2	7.80	CR, TTC
S4 C4	33.30	TD
S15 C1	1.20	ED-IL, TD, CR
S15 C2	4.30	TXC
S16 C1	1.15	ED-IL, TD
S16 C2	4.30	TXC
S16 C3	8.30	CR, TXC

EVALUATION OF LIQUEFACTION SUSCEPTIBILTY

Liquefaction susceptibility has been evaluated by means of procedures prescribed by [1], which is very similar to Italian Code [2], Chinese Code [3] [4] [5] and the criterion suggested by [6].

[1] and [2] consider a simplified and very conservative approach to exclude the occurrence of liquefaction. This approach is based on the expected peak ground acceleration (PGA), on soil composition and on soil state. As for the expected PGA, [1] and [2] assume that liquefaction hazard analysis can be omitted if PGA < 0.15g. In addition the considered sandy soils should met, at least one of the following conditions:

 − clay fraction greater than 20% and Ip greater than 10%;
 − fine content greater than 35% and $(N_1)_{60}$ greater than 20;
 − $(N_1)_{60}$ greater than 25.

The compositional criterion reported in the [1] and [2] leads to the following considerations:

 − [6] and [7] stated that cyclic strength of fine-grained soils with Ip > 10% is greater than that of non-plastic fine-grained soils. As a consequence it seems that the plasticity of fine content is more important of its quantity in defining cyclic strength. In fact the greater is the plastic index, the lower is liquefaction susceptibility;
 − The limit of 20% for clay fraction seems too much conservative;
 − The parameter $(N_1)_{60}$ is strongly affected by grain size distribution [8]. As a consequence the same value for $(N_1)_{60}$ refers to a high relative density for a fine sand, whereas refers to a very low relative density for a medium coarse sand. Also in this case the criterion reported above seems too much conservative.

In the Chinese Code, soils are susceptible of liquefaction if all the condition listed in the following simultaneously occur:

 − fine content (d < 0.005 mm) minor than 15%
 − liquid limit (LL) minor than 35%
 − water content (w_n) greater than 0.9 LL.

An equivalent way to stress the condition about fine content and liquid limit are [3]:

 − clay fraction (d < 0.002 mm) minor than 10%
 − liquid limit (LL) minor than 32 %.

In this criterion there is no reference to plastic index, although liquid limit is strongly correlated to it.

Lastly, in [6] a criterion based on plastic properties of soils has been suggested, as reported in Fig. 3.

The evaluation of liquefaction susceptibility carried out by means of Eurocode criterion is positive in 85 % of considered cases (the expected PGA has not been considered). A similar result has been obtained with the criterion reported in Fig. 3, in which almost all (92 %) the analyzed specimens are susceptible to liquefaction.

A different result has been obtained by means of Chinese code, in which only 8 of 26 analyzed specimens resulted susceptible to liquefaction. Since the susceptibility of soil to liquefaction does not means that liquefaction will occur, these results impose to evaluate the liquefaction potential.

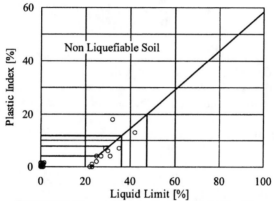

FIGURE 3. Compositional criterion proposed by [6].

ANALYSIS OF LIQUEFACTION POTENTIAL

In a stress approach the first step consists in determination of seismic action. Following the simplified procedure proposed by Italian Code, the seismic action in terms of PGA is expressed as:

$$a_{max} = \frac{\gamma_I Sa_g}{g} = 0.375 \tag{1}$$

in which a_{max} is the PGA, γ_I is a factor which takes into account the "importance" of the building ($\gamma_I = 1.2$), S is the soil factor (S = 1.25 for type C soil) and a_g/g is the PGA at the rock outcrop prescribed by Code according to macrozonation rules. In addition, determination of number of cycles due to earthquake is essential and subordinate to definition of earthquake Magnitude.

An alternative procedure consists in:
  - definition of PGA at rock outcrop by means of a Probabilistic Seismic Hazard Approach (PSHA) for a return period of 975 years [9];

437

- definition of site effect by means of 1-D seismic response analysis using EERA [10], in which ground profile characterization has been based on in situ and laboratory tests and a group of seven natural free-field accelerograms on rock has been selected after deaggregation of PSHA [11]. The selected accelerograms match, on average, the prescribed spectrum on rock;
- deaggregation of PSHA in order to obtain the most likely earthquake in terms of Magnitude and the most likely distance [11].

Following this procedure a value of a_{max} equal to 0.257 has been estimated, with a reduction greater than 30 % with respect to value calculated by means of Eq. 1. The deaggregation leads to couples Magnitude/Distance [km] equal to 5.4/13 or 5.8/20. Shear stress induced by earthquake can be estimated with the following relationship:

$$CSR = \frac{\tau_{av}}{\sigma_{v0}'} = 0.65 a_{max} \frac{\sigma_{v0}}{\sigma_{v0}'} r_d \qquad (2)$$

in which τ_{av} is the average shear stress and r_d is a stress reduction factor which takes into account the reduction of shear stress with depth.

The normalized cyclic shear stress that causes liquefaction (CRR) has been estimated by means of three different procedures based on in situ tests or laboratory tests.

With reference to a Magnitude equal to 7.5 the CRR can be evaluated using dynamic probing (SPT and DPSH) by [12]:

$$CRR_{M=7.5} = \frac{1}{34 - (N_1)_{60cs}} + \frac{(N_1)_{60cs}}{135} + \frac{50}{[10(N_1)_{60cs} + 45]^2} - \frac{1}{200} \qquad (3)$$

where $(N_1)_{60cs}$ is the value of number of blows per foot corrected in order to take into account both the confining stress and the fine content. DPSH tests have been processed by means of the same expression after converting N_{20} into N_{SPT} by a factor equal to 1.83, which takes into account the difference in penetration length and efficiency of equipments.

Cone penetration tests can be processed in order to obtain CRR by the expressions [12] [13]:

$$CRR_{M=7.5} = 0.833 \frac{(q_{c1N})_{cs}}{1000} + 0.05 \quad \text{if } (q_{c1N})_{cs} < 50 \qquad (4)$$

$$CRR_{M=7.5} = 93 \left[\frac{(q_{c1N})_{cs}}{1000} \right]^3 + 0.08 \quad \text{if } (q_{c1N})_{cs} < 160 \qquad (5)$$

where $(q_{c1N})_{cs}$ is the cone penetration resistance corrected in order to take into account both the confining stress and the fine content. It is worthwhile to stress that, according

438

to the suggested procedure [12], the soil could be classified as clay (Ic > 2.6) which contrasts the laboratory soil classification.

Lastly, CRR can be obtained by means of laboratory tests. In particular eight undrained triaxial cyclic tests have been carried out on samples retrieved at depth showing the lowest penetration resistance. Figure 4 shows the results of such tests. The number of cycles reported in abscissa has been determined for each test applying the condition $\varepsilon_{a,DA}$ = 5%. Since a Magnitude equal to 5.8 corresponds to a number of equivalent uniform stress cycles equal to 4, the value of CRR deduced by laboratory tests is 0.280.

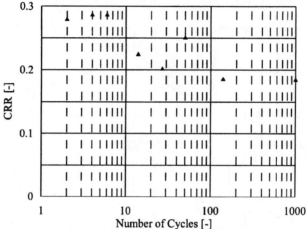

FIGURE 4. CRR as a function of number of cycles in triaxial tests.

Starting from the considerations reported above, an estimation of factor of safety against liquefaction has been computed by means of the following relation

$$FS_L = \frac{CRR_{M=7.5}}{CSR} MSF \qquad (6)$$

where MSF = $10^{2.24} \cdot M^{-2.56}$ [12] which takes into account the differences in Magnitude. This factor have not been applied at the case of CRR deduced by laboratory tests.

Figure 5 shows the profiles of FS_L deduced by in situ and laboratory tests. The results reported in the above figures lead to the following considerations:

- the simplified definition of seismic actions, in terms of CSR, lead to conservative estimation of FS_L; this is often due to the hidden introduction of margin of safety both in definition of PGA at outcrop and site effect;
- dynamic penetration tests lead to locate a liquefiable stratum between 3 and 4 m depth;
- cone penetration tests lead to locate liquefiable strata at different depth (4 ÷ 9 m CPTU1, 4.5 ÷ 10 m CPTU2, 8.5 ÷ 14 m CPTU3);

- the use of laboratory tests to determine the liquefaction resistance together with the definition of seismic action through a ground response analysis lead to values of FS_L always greater than 1.
- thickness of liquefiable soil is always lower than thickness of above non-liquefiable soil.

Historically, the site under consideration has experienced a number of earthquakes with Magnitude and distance equal or greater than those obtained from deaggregation of PSHA as shown in Table 2 [14]. Nonetheless, liquefaction phenomena have never been observed in the study area. Therefore it is possible to conclude that a true liquefaction can be excluded. On the other hand it is not possible to exclude the occurrence of localized phenomena (e.g. sand boils, water spouts).

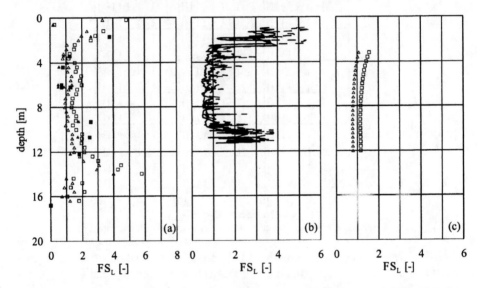

FIGURE 5. Profiles of FS_L deduced by: a) SPT and DPSH tests; b) CPTU tests and c) laboratory tests. Lower series of FS_L are related to simplified evaluation of PGA.

TABLE 2. List of earthquakes with an epicentral distance minor than 20 km.

Date	Location	I_0 (MCS)	Mw	Distance [m]
6 March 1740	Garfagnana	VII	5.18	9975
23 July 1746	Garfagnana	VI	4.83	3758
5 March 1902	Garfagnana	VII	5.17	2432
27 July 1916	Fosciandora	VI	4.83	3131
25 September 1919	Fosciandora	V – VI	4.63	7230
7 September 1920	Garfagnana	IX – X	6.48	19904
15 October 1939	Garfagnana	VI – VII	5.20	18475
12 August 1951	Barga	V – VI	4.74	10245
30 June 1934	Abetone	IV – V	4.38	17641
7 June 1980	Bagni		4.70	9841
23 January 1985	Garfagnana	VI	4.69	10192

CLOSING REMARKS

The paper presents an analysis of liquefaction hazard in a site devoted to construction of a school. The analysis has been carried out by means of different approaches based on in situ and laboratory test, for the aspects concerning the resistance, and on simplified coded procedures and PSHA with 1D-GRA, for the aspects concerning the seismic action. Simpler approaches often introduces hidden margin of safety both in definition of resistances and actions.

A qualitative estimation of possible damages to shallow structures can be carried out on the basis of indication contained in [15]. The presence of a unliquefiable and resistant stratum from ground surface to 3 m depth, in addition to the condition that the thickness of liquefiable soil is always lower than thickness of above non-liquefiable soil, reduces the vulnerability of structures. This last aspect is crucial in managing the problem of liquefaction. In fact, an estimation of earthquake consequences on a construction in that area reported in [16] shows that the possible damages are negligible.

REFERENCES

1. Eurocode 8, *Design Provisions for Earthquake Resistance of Structures - Part 1-1:General Rules for the Representation of Seismic Actions." Part 5: Foundations, Retaining Structures and Geotechnical Aspects*, 1998.
2. OPCM 3274 "Primi elementi in materia di criteri generali per la classificazione sismica del territorio nazionale e di normative tecniche per le costruzioni in zona sismica" in *Gazzetta Ufficiale della Repubblica Italiana*, 8 maggio 2003, n. 108, 2003.
3. W. Wang, "Some findings in soil liquefaction", Research Report of Water Conservancy and Hydroelectric Power Scientific Research Institute, Beijing, 1979.
4. H. B. Seed and I. M. Idriss, *Ground motion and soil liquefaction during earthquakes*, Oakland, EERI, 1982.
5. D. C. Andrews and G. R. Martin, "Criteria for Liquefaction of Silty Soils" in 12th WCEE Proceedings, Auckland, New Zealand, 2000.
6. R. B. Seed, O. Cetin, R. E. S. Moss, A. M. Kammarer, J. Wu, J. M. Pestana, M. F. Riemer, R. B. Sancio, J. D. Bray, R. E. Kayen and A. Faris, "Recent Advances in Soil Liquefaction Engineering: a Unified and Consistent Framework", 26th Annual ASCE Los Angeles Geotechnical Spring Seminar, 2003.
7. K. Ishihara "Liquefaction and flow failure during earthquakes", *Géotechnique* 43(3), 351-415, 1993.
8. M. Cubrinovski and K. Ishihara, "Empirical Correlation between SPT N-value and Relative Density of Sandy Soils", *Soils and Foundations*, 5, 61-71, 1999.
9. INGV, http://esse1-gis.mi.ingv.it/ , 2006.
10. J. P. Bardet, K. Ichii and C. H. Lin, "EERA – A Computer Program for Equivalent-Linear Earthquake Site Response Analyses of Layered Soil Deposits", Department of Civil Engineering, University of Southern California, http://geoinfo.usc.edu/gees, (2000).
11. C. Lai, C. Strobbia and Dall'Ara, "Convenzione tra Regione Toscana e Eucentre. Parte 1. Definizione dell'Input Sismico per i Territori della Lunigiana e della Garfagnana", Departement of Structural and Geotechnical Engineering, Technical University of Turin, 2005.
12. T. L. Youd and I. M. Idriss, "Liquefaction Resistance of Soils: Summary Report from the 1996 NCEER and 1998 NCEER/NSF Workshops on Evaluation of Liquefaction resistance of Soils", *Journal of Geotechnical and Geoenvironmental Engineering*, 127(4), 297-313, 2001.
13. P. K. Robertson and C. E. Wride, "Evaluating Cyclic Liquefaction Potential Using the Cone Penetration Test", *Canadian Geotechnical Journal*, 35, 442 – 459, 1998.
14. G. Fialdini, "Indagini per la valutazione del rischio di liquefazione e dei suoi effetti: un caso reale", Thesis, University of Pisa, 2008.
15. Technical Committee for Earthquake Geotechnical Engineering, TC4, ISSMGE, "Manual for Zonation on Seismic Geotechnical Hazards (Revised Version)", Tokyo: Japanese Geotechnical Society, 1999.
16. F. Benelli, "Valutazione del rischio di liquefazione e dei suoi effetti: un caso reale", Thesis, University of Pisa, 2008.

Efficiency of Micro-fine Cement Grouting in Liquefiable Sand

Mojtaba Mirjalili[a], Alireza Mirdamadi[b] and Alireza Ahmadi[c]

[a] *Doctoral student, Dept. of Civil & Earth Resources Eng., Graduate School of Eng., Kyoto University, Kyoto 615-8540, JAPAN. Email: m.mirjalili@at4.ecs.kyoto-u.ac.jp*
[b, c] *M.Sc. in Geotechnical Eng, Faculty of Civil Eng, University College of Eng, University of Tehran Tehran, IRAN.*

Abstract. In the presence of strong ground motion, liquefaction hazards are likely to occur in saturated cohesion-less soils. The risk of liquefaction and subsequent deformation can be reduced by various ground improvement methods including the cement grouting technique. The grouting method was proposed for non-disruptive mitigation of liquefaction risk at developed sites susceptible to liquefaction. In this research, a large-scale experiment was developed for assessment of micro-fine cement grouting effect on strength behavior and liquefaction potential of loose sand. Loose sand samples treated with micro-fine grout in multidirectional experimental model, were tested under cyclic and monotonic triaxial loading to investigate the influence of micro-fine grout on the deformation properties and pore pressure response. The behavior of pure sand was compared with the behavior of sand grouted with a micro-fine cement grout. The test results were shown that cement grouting with low concentrations significantly decreased the liquefaction potential of loose sand and related ground deformation.

Keywords: Liquefaction mitigation, Grouting, Large-scale experiment, Loose sand.

INTRODUCTION

Recent major seismic events such as 1964 Niigata, 1989 loma Prieta, and the 1995 Kobe earthquakes, continue to demonstrate the damaging effect of liquefaction-induced loss of soil strength and associated lateral spreading [1]. Densification methods, modifications leading to improving the cohesive properties of the soil (hardening or mixing), removal and replacement, or permanent dewatering can reduce or eliminate liquefaction potential. Often a mitigation measure may involve the implementation of a combination of techniques or concepts such as solidification, densification, reinforcement, and mixing. A combination of techniques may provide the most cost-effective ground improvement solution for preventing damages. However, there is a need to investigate the effect of each technique on liquefaction mitigation and enhancing the resistance.

In this paper the experimental investigation of micro-fine cement-based grout on cyclic and monotonic behavior of saturated loose sand is presented. Cyclic and monotonic triaxial tests were carried out to assessment of deformability and strength properties of treated sand sample in comparison with pure sand specimens. Although several experimental studies have dealt with effect of grouting on characteristics and liquefaction potential of loose sand in the laboratory [2, 3], but in most of them, a one-

CP1020, *2008 Seismic Engineering Conference Commemorating the 1908 Messina and Reggio Calabria Earthquake*,
edited by A. Santini and N. Moraci

dimensional grouting set-up was used for modeling of grouting in the laboratory which cannot represent the real conditions similar to in-situ grouting procedure. Hence, in the present study, a large-scale experiment was developed to examine the nature pattern of cement grouting into saturated loose sand close to real operation.

EXPERIMENTAL SET-UP

The RHEOCEM 650 micro-fine cement was used as grouting material in this research program. According to RHEOCEM product information, RHEOCEM 650 is superfine Portland cement, an average particle size of 6 μm and a specific surface area of 6940 cm^2/g. the initial setting time is between 60-120 min, and final setting is achieved after 120-150 min. To disperse the cement particles in the grout, a superplasticizer (RHEOBUILD® 2000 SF) which is an admixture base on water soluble sulphonated polymers of different molecular weight, was utilized. The dosage of superplasticizer was about 1.5 – 3 mass percent of cement material.

The sand used in the testing program was Firouzkooh No. 131 sand, commercially available sand from Iran. Firouzkooh No. 131 sand is poorly graded, medium to fine sand, with angular grains. Over 98% passes the No. 20 sieve (0.84 mm) and is retained on the No. 100 sieve (0.15 mm). Figure 1 shows the gradation curve of Firouzkooh No.131 sand, RHEOCEM® 650 and the gradation range of most liquefiable sand [4].

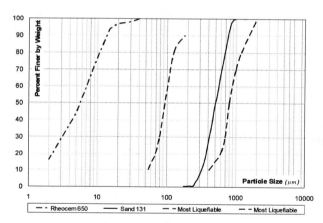

FIGURE 1. Grain size distribution for Firouzkooh No. 131 sand, RHEOCEM® 650, and most liquefiable sands

The samples of grouted sand were obtained from the three-dimensional large-scale injection model. A cylindrical box with 1.20 m inside diameter and approximately 1.00 m length, was filled with Firouzkooh sand by moist tamping method to obtain reproducible models at a relative density equal to 25%, corresponding to loose to very loose soils category, and strongly susceptible to liquefaction. Figure 2 presents the cross-section view of large-scale experimental set-up. After water saturation of the model, the constant mechanical loading equal 5.0 ton/m^2 was uniformly distributed over the top of the models by pneumatics jack to simulate a ground depth of about 3.5 m during the tests.

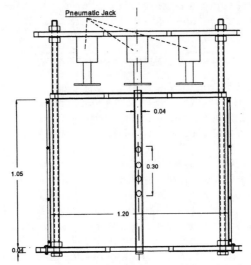

FIGURE 2. Cross-section view of experimental set-up

A fine cement grout was then injected by a perforated pipe which located at the center of cylindrical box. The water/cement mass ratio, denoted by W/C, was controlled during injection. In the present study, the values W/C = 3, 6, and 9 have been considered. After injection of grout and setting time of grout, the sampling of sand in grouted area, was carried out using thin-walled sampler at various depths and distances from injection pipe. The samples was kept under wet condition during 28 days, then were levelled to meet the dimensional of the experimental apparatus (50 mm in diameter and 100 mm in length).

The samples were then water saturated and subjected to a confining pressure in a triaxial apparatus. Cyclic and monotonic triaxial compression tests under different confining pressures were performed. In addition, triaxial tests have been performed on pure Firouzkooh No.131 sand samples prepared at the same relative density.

TEST RESULTS

The purpose of the experimental study was to evaluate the influence of cement grouting on strength and liquefaction potential of saturated loose sand, especially in large-scale experiment under conditions close to in-situ grouting procedure. Identical triaxial tests have been performed for each model with different water-cement ratios.

The cyclic tests were run in general, accordance with ASTM D5311 standard test method for load controlled cyclic triaxial strength of soil. Certain modifications to the test procedure were required because of the special properties of the treated sand. The pore pressure response and the axial strain during cyclic loading were measured to quantify the results of the treated soils. The confining stress used for all cyclic testing was 100 kPa. The equipment used for the cyclic triaxial testing was an automated triaxial testing system. The loading is controlled using closed-loop feedback systems capable of stress or strain controlled testing. The success of the micro-fine cement grout treatment was evaluated by comparing the cyclic deformation resistance of treated and untreated samples. In grouted sand because of filling the void spaces of sand medium by cement grout, the response of pore water pressure variation could not be compared very well; hence the axial deformation results were used for grout effect assessment and evaluation of liquefaction resistance.

The deformation resistance of the treated sand was measured in terms of DA axial strain, which is the largest difference in strain that develops during an entire cycle of compression and extension. If the stabilized sands accumulate less strain during cyclic loading than untreated sands at a given CSR, then the treatment would be considered successful. The CSR is defined as the ratio of the maximum cyclic shear stress to the initial effective confining stress.

In general, samples stabilized with higher concentrations of micro-fine cement experienced less strain during cyclic loading than those stabilized with lower concentrations. When tested at a CSR of 0.50, samples with W/C=9 showed very little deformation during more than 500 cycles of loading. However, all samples remained intact during loading and were able to be tested for other tests like unconfined compressive strength after cyclic loading. Figure 3 presents the axial deformation of pure sand and treated sand sample. As shown in these graphs, the axial deformation of pure sand under cyclic loading with CSR of 0.22, exceeded the DA of 5% after about 7 cycles while the treated sand still remained in DA of 0.7% after 200 cycles with CSR equal to 0.8.

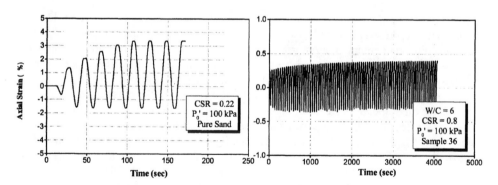

FIGURE 3. Axial deformation of pure sand (left) and grouted sand with W/C=6 (right) under cyclic loading

On the other hand, the triaxial monotonic tests were done according to ASTM D4767 for evaluation of strength parameters of grouted sand samples and better comparison with pure sand. The results are presented in terms of deviator stress q versus axial strain. The axial strain was calculated from the sample shortening measured on an external micrometer, which induces an experimental uncertainty due to bedding and compliance errors. This uncertainty is more pronounced in the case of stiff grouted sands. An external load cell was used to determine the axial stress.

As shown in figure 4, the results of these tests illustrate the beneficial effect that the grout injection has on the strength and on the stiffness of the soil. They also confirm the following general trends for the cement-treated soils: Stiffness and strength increase as the binder content increases, the volumetric behavior is typically contractive–dilatant and the post-peak behavior is more brittle for high cement contents. Failure occurs with visible vertical cracks for high cement-to-water ratios and low confining pressures, whereas strain localization with inclined shear bands occurs for low cement-to-water ratios and high confining pressures

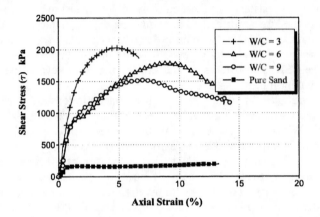

FIGURE 4. Shear stress vs. axial strain for pure sand and grouted sand samples in confining pressure equal to 150 kPa

The results confirmed that the strength of grouted sand under compression loading is correctly described by a Mohr-Coulomb criterion (friction angle and cohesion). The cohesion due to cemented inter granular bonds and the friction angle due to inter particle contacts were determined by plotting the failure envelope in the stress path. For un-cemented granular soils, a straight line failure envelope with zero cohesion was apparent. For grouted sands, a straight-line failure envelope that was almost parallel to that for the un-cemented granular soil was apparent. However, the value of the friction angle was a little more, since permeation grouting with low injection pressures causes some disturbance of particle assembly especially for samples near injection line. The values obtained for cohesion and friction angle of treated and pure sand are shown in table 1. Similar results have been found by Dupas and Pecker (1979) on mortars, and by Clough and Sitar (1981) on natural weakly cemented sands [5, 6].

TABALE 1. Cohesion and friction angle value of pure and grouted sand samples

	φ	C (kPa)
Pure Sand	34.5	0
W/C = 9	41.8	134
W/C = 6	41.8	188
W/C = 3	41.8	235

CONCLUSION

In the present study, the influence of micro-fine cement grouting on liquefaction potential and strength of saturated loose sand by using a multidirectional experimental model for injection in soil medium, was presented. It was considered that the grouting set-up had most similarity to in-situ conditions. Cyclic and monotonic triaxial tests were performed to investigate the influence of micro-fine cement grout on the deformation properties and liquefaction resistance of loose sand. The grouting process can improve the soil features by strengthening the touching surface of grains and making a stronger and stiffer skeleton than that of the primary soil. It can be observed by increasing the cohesion of cemented sand in comparison with pure sand. Overall, treatment with micro-fine cement grout with low concentrations significantly increases the deformation resistance of loose sand to cyclic loading.

ACKNOWLEDGMENT

The writers would like to acknowledge Dr. Moradi and Dr. Ghalandarzadeh, assistant professors of faculty of civil engineering, university of Tehran, for their help and guidance in this research.

REFERENCES

1. Seed H.B., Idriss I.M., "Ground motions and soil liquefaction during earthquakes", Berkeley, CA: *Earthquake Engineering Research Institute* (1982).
2. Dano C., et al, "Engineering properties of grouted Sands", *J. Geotech. and Geoenviron. Eng.*, Vol. 130, No 3, (2004).
3. Delfosse R. E. et al, "Shear modulus and damping ratio of grouted sand", *Soil Dynamics and Earthquake Engineering*, Vol. 24, 461-471 (2004).
4. Tsuchida H. "Prediction and countermeasure against the liquefaction in sand deposits", *Abstract of the Seminar in the Port and Harbor Research Institute*, pp 3.1-3.3 (1970).
5. Clough, G. W., and Sitar, B. "Cemented sands under static loading". *J. Geotech. Eng. Div., Am. Soc. Civ. Eng.*, 107(6), 799–817(1981).
6. Dupas, J. M., and Pecker, A. "Static and dynamic properties of sand-cement", *J. Geotech. Eng. Div., Am. Soc. Civ. Eng.*, 105(3), 419–436 (1979).
7. Mirjalili M., "Experimental investigation on influence of cement grouting on strength, deformability parameters and liquefaction potential of sand", M.Sc. Thesis, Faculty of Civil Eng, University College of Eng., University of Tehran, IRAN, 157p (2007). (In Persian)

Effect of Initial Fabric on Cyclic and Monotonic Undrained Shear Strength of Gioia Tauro Sand

Daniela Porcino[a] and Vincenzo Marcianò[b]

[a]Researcher, Department of Mechanics and Materials, University "Mediterranea" of Reggio Calabria, Via Graziella (Feo di Vito)-89060 Reggio Calabria (Italy).
[b]PhD Student, Department of Mechanics and Materials, University "Mediterranea" of Reggio Calabria, Via Graziella (Feo di Vito)-89060 Reggio Calabria (Italy).

Abstract. An experimental study is presented aimed at a direct comparison of the undrained behaviour of a natural coarse sand using specimens reconstituted by different techniques. Undrained monotonic and cyclic triaxial compression tests were carried out on reconstituted specimens of Gioia Tauro sand, as well as on truly undisturbed specimens retrieved by in-situ ground freezing.
It is worth noting that Gioia Tauro Plain, on the Calabrian side of the Messina Strait, manifested various types of geotechnical hazards related to soil liquefaction during several catastrofic earthquakes.
Two different preparation methods were employed, namely air pluviation and water sedimentation. Soil fabrics resulting from the above methods appear to exhibit different undrained response during monotonic straining.
The behaviour of truly undisturbed specimens (and hence with their natural fabric) appears to be similar to that exhibited by water sedimentation reconstituted specimens in both undrained monotonic and cyclic triaxial tests. Finally, in view of assessing equivalent simple shear or in-situ response of the natural deposit, undrained cyclic shear strength characteristics of the tested sand from triaxial tests were compared with those gathered from the simple shear device.

Keywords: sand fabric, undrained shear strength, undisturbed sampling, triaxial and simple shear.

INTRODUCTION

Fundamental studies of sand behaviour require laboratory tests on homogeneous reconstituted specimens which are deemed to replicate depositional conditions in the field. Profound differences may be noted in the undrained shear behaviour of sand depending on the method of reconstitution which controls the ensuing fabric of in-situ sands (Tatsuoka et al., 1986; Vaid & Sivathayalan, 1999).

Air pluviation (*AP*), moist tamping (*MT*) and water sedimentation (*WS*) have been used by many researchers for sand sample reconstitution in laboratory studies by using triaxial, simple shear and torsional shear apparatus.

Such studies demonstrated that modelling water deposited in-situ sands, such as natural alluvial and hydraulic fill sands, by their WS equivalent may be appropriate if the sands are uncemented and unaged. However, for high risk projects, for which the consequences of liquefaction can be severe, consideration should be given to carrying out laboratory tests on high-quality undisturbed samples. The ground freezing technique has been used successfully to recover undisturbed samples of sandy and gravelly soils deposits (Ghionna et al., 2001; Ghionna & Porcino, 2006).

The paper is aimed at investigating the influence of sample preparation methods on the monotonic and cyclic undrained stress-strain-strength behaviour of a coarse sand by means of triaxial compression tests. The behaviour of truly undisturbed sands obtained from in-situ ground freezing is also presented in view of evidencing the similarity between the behaviour of intact samples and that exhibited by the corresponding water sedimentation reconstituted specimens. Finally, in order to explore the influence of different deformation mode (e.g. triaxial and simple shear) on in-situ cyclic liquefaction characteristics of tested sand, water-sedimented reconstituted counterparts were also tested in an NGI simple shear apparatus.

EXPERIMENTATION

The sand used in the experimentation is Gioia Tauro sand, a coarse silica sand with an average particle size $D_{50}= 2mm$, $G_s= 2.69$ and $C_u=2.10$ (Ghionna & Porcino, 2006). The maximum and minimum void ratios in accordance with ASTM Test Methods are 0.69 and 0.45, respectively. The investigated sand was derived from a relatively recent natural deposit, which is considered to have been formed by the marine water environment.

Monotonic and cyclic undrained triaxial tests on 70-mm-diameter by 140-mm high specimens were carried out by adopting two different methods, namely air pluviation (AP) and sedimentation in water (WS). All reconstituted specimens were prepared at the same density index (after isotropic consolidation) of the undisturbed ones ($I_r \cong 42\%$); test procedure of the frozen and reconstituted specimens are described in detail in Ghionna & Porcino (2006).

In addition, cyclic undrained simple shear (SS) tests on 80-mm-diameter by approximately 20-mm high specimens were carried out in a modified NGI type device. The specimen is laterally confined by a reinforced rubber membrane in the simple shear test (K_0 consolidation) and constant volume conditions during shear are enforced by arresting the vertical displacement. In this test the changes in vertical stress to maintain constant volume are equivalent to the changes in pore water pressure in the corresponding undrained test (Finn, 1985; Dyvik et al., 1987).

In cyclic SS tests the reconstituted specimens were prepared at a density index I_r equal to 42% on average by using water sedimentation method, which was shown to closely replicate the fabric of the natural sand deposit. Additional information on the SS device and test procedure are provided by Porcino et al. (2006) and Porcino & Ghionna (2008). It is noteworthy that all SS tests were interpreted assuming that the horizontal plane is the plane of maximum shear stress (Roscoe, 1970).

Both cyclic triaxial and simple shear tests were performed under an initial vertical effective consolidation stress (σ'_{v0}) equal to 40 kPa, corresponding to the estimated in-situ effective overburden stress.

INFLUENCE OF SAMPLE RECONSTITUTION METHOD

The basic requirement for any of the adopted methods is firstly to obtain homogeneous samples with a uniform distribution of void ratio. It was preliminary ascertained (Ghionna & Porcino, 2006) that the two adopted reconstitution methods allow uniformity features to be achieved, as reported also by other authors (Ishihara, 1996; Vaid et al., 1999).

As shown in fig. 1, undrained triaxial tests performed at about the same void ratio on Gioia Tauro sand may give very different behaviours, depending on the preparation method of the specimens. Figure 1 evidences that the specimen prepared by water sedimentation behaves in a more strain-hardening (dilative) manner when it is compared with the air pluviated specimen. However, at large strains (approaching critical state) the stress-paths move towards its critical state without showing significant differences.

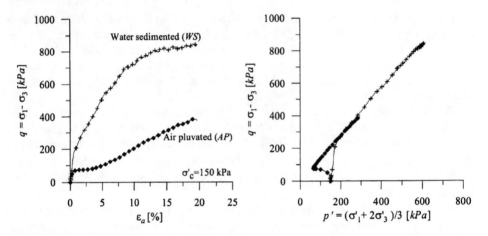

FIGURE 1. Effect of specimen reconstitution method on undrained monotonic triaxial response.

The comparative cyclic undrained triaxial response of tested sand is shown in fig. 2 for the two adopted preparation methods (i.e. *AP* and *WS*). Figure 2 shows trends with numbers of cycles (N_c) of cyclically-induced pore pressures (Δu), normalized to σ'_c ($R_u = \Delta u / \sigma'_c$) for specimens tested at the same cyclic stress ratio $CSR = q / 2 \cdot \sigma'_c = \sigma_d / 2 \cdot \sigma'_c = 0.22$.

As regards AP specimen (fig. 2a), the increase in pore water pressure rapidly reduces the effective mean stress (p') to zero. On the other hand, the water sedimented specimen illustrated in fig. 2b has a relatively long and stable life before triggering initial liquefaction condition ($p' \cong 0$ and $\varepsilon_{DA} = 5\%$). In view of the diversity of undrained shear strength of sand samples reconstituted by different methods of preparation, the authors underline the importance of running laboratory tests on reconstituted specimens which should simulate as closely as possible the depositional conditions of the in-situ natural deposits.

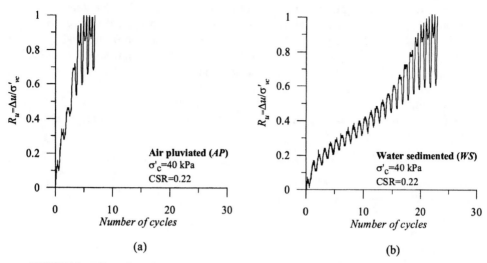

FIGURE 2. Effect of specimen reconstitution method on undrained cyclic triaxial response.

BEHAVIOUR OF UNDISTURBED SOIL SAMPLES

Undrained monotonic and cyclic triaxial response was assessed for the undisturbed sand specimens recovered by in-situ freezing technique. Figure 3 compares monotonic undrained response of the undisturbed sand specimens and the corresponding reconstituted ones prepared by *AP* and *WS* methods at the same void ratio and effective stress state.

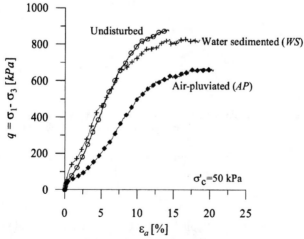

FIGURE 3. Comparative undrained monotonic triaxial response of undisturbed and reconstituted specimens of Gioia Tauro sand.

Looking at the stress-strain plot it is possible to observe that the monotonic undrained triaxial behaviour of the undisturbed frozen sample is very close to that of the specimen reconstituted by water sedimentation.

The effective stress failure envelope at ultimate/critical (US/CS) state in triaxial loading conditions are illustrated in figure 4a for both undisturbed and reconstituted specimens. It is noteworthy that all data points lie along an approximate straight line passing through the origin. This implies that the internal friction angle mobilized at US state (ϕ'_{US}) is essentially invariant with respect to initial soil fabric (natural and reconstituted fabric) and was calculated as approximately 33°.

The initial state ratio (r_c) defined by Ishihara (1993) is a parameter of prime importance for characterization of the undrained behaviour of sand. It is defined as:

$$r_c = p'_c / p'_{PT} \tag{1}$$

where p'_c and p'_{PT} are the effective mean stress after isotropic consolidation and at the states of phase transformation, respectively. It is apparent that when r_c decreases, the sample becomes less and less contractive.

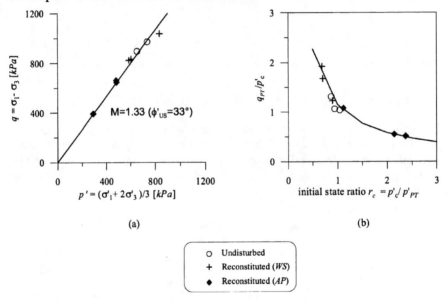

(a)   (b)

o	Undisturbed
+	Reconstituted (*WS*)
◆	Reconstituted (*AP*)

FIGURE 4. Undrained monotonic triaxial tests on Gioia Tauro sand: a) Ultimate state strength envelope; b) normalized shear strength at phase transformation state.

A threshold value of r_c approximately equal to 2, separating flow-type from non flow type response for several sands, was suggested by several authors (Sladen et al., 1985; Ishihara, 1993). In figure 4b the normalized strength corresponding to the phase transformation, q_{PT} / p'_c, is plotted against initial state ratio. It is apparent that the intact specimens, as well as the reconstituted ones, show an initial state ratio less than 2 to locate them in a zone of non-flow. This is comparable with the stress-strain curves shown in fig. 3. Furthermore, figure 4b evidences that all data under different initial fabrics are correlated uniquely with the r_c parameter.

Results from undrained cyclic triaxial tests on undisturbed frozen and reconstituted specimens of Gioia Tauro sand are shown in figure 5a, where cyclic stress ratio required to cause 5% double-amplitude (D.A.) axial strain is plotted against number of cycles (N_c). It can be seen that there is a large range in which the cyclic resistance of sand can vary depending upon the fabric structure.

In a manner similar to undrained monotonic response, the undrained cyclic shear strength of Gioia Tauro sand reconstituted by *WS* may also be seen to be very close to that of the undisturbed samples. On the other hand, reconstituted specimens by *AP* were more susceptible to liquefaction under cyclic loading than their water-sedimented counterparts. The findings reported above are consistent with those obtained by other authors from similar experiences for fluvial and hydraulic fill sands (Oda et al., 1978; Vaid et al., 1999).

COMPARISON BETWEEN CYCLIC UNDRAINED TRIAXIAL AND SIMPLE SHEAR STRENGTHS

Even though triaxial apparatus is the most commonly used laboratory device to study cyclic liquefaction characteristics of sands, it is widely accepted that in-situ stress conditions for level sand deposits and sand slopes under earthquakes are more correctly simulated in laboratory by adopting cyclic simple shear apparatus.

Accordingly, figure 5b shows the influence of different deformation modes (e.g. triaxial or simple shear) on the measured liquefaction resistance of tested sand. The cyclic stress ratios $\sigma_d / 2 \cdot \sigma_c'$ causing liquefaction in triaxial compression tests (ε_{DA}=5%) are clearly higher than the cyclic stress conditions $\tau_{cyc} / \sigma_{v0}'$ causing failure in laboratory SS tests (γ_{SA}=3.75%).

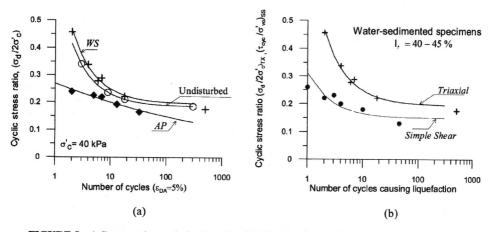

FIGURE 5. a) Comparative undrained cyclic triaxial response of undisturbed and reconstituted specimens of Gioia Tauro sand; b) . Cyclic liquefaction resistance curves in simple shear and triaxial tests (WS specimens).

In order to verify the applicablity of laboratory test results for predicting field performance under cyclic loading, figure 6 shows the results reported by Seed & Peacock (1970) together with those gathered from the present study. Data refer to cyclic stress conditions causing failure in a number of cycles equal to 10 under both triaxial and simple shear tests for different loose to medium dense sands. The selected number of loading cycles is approximately equivalent to a design earthquake magnitude M=7.3 at Gioia Tauro site, drawn from a site specific seismicity analysis. Figure 6 evidences a good correspondence between simple shear and field performance even for coarser sands, such as that tested in the present study.

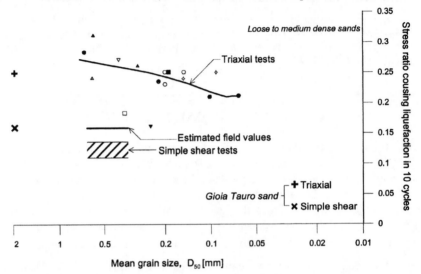

FIGURE 6. Comparison of stress conditions causing liquefaction in cyclic loading under simple shear, triaxial and field conditions (modified from Seed and Peacock, 1970).

The simple shear strength is often estimated from the triaxial strength basing on correction factors proposed in literature (Seed & Peacock, 1970; Ishihara, 1996). However, as outlined by Tatsuoka et al. (1986), the ratio of cyclic strength between the triaxial and the simple shear tests for a given sand element depends on several factors, such as: the kind of sand, the sample preparation method, the relative density, the strain value for which failure is defined, the number of loading cycles. Accordingly, it can be concluded that the accurate estimation of the simple shear strength from the triaxial strength is in general rather difficult.

CONCLUSIONS

An experimental study was undertaken with the aim of investigating:
1. The influence of sample-reconstituting technique and, hence, the ensuing fabric on the monotonic and cyclic undrained behaviour of a coarse sand through triaxial compression tests. A direct comparison of the undrained behaviour of undisturbed specimens retrieved by in-situ ground freezing was also presented.
2. The influence of loading modes (triaxial or simple shear) for the assessment of in-situ cyclic shear strength of tested sand.

The main conclusions which can be drawn from the investigation may be summarised as follows:

- Undrained monotonic and cyclic triaxial response is greatly influenced by reconstitution method. Reconstituted sand specimens by water sedimentation (*WS*) exhibit a pattern of behaviour which was found to be very close to that of the undisturbed frozen specimens. On the other hand, sand specimens prepared by air pluviation (*AP*) evidence a weaker and more compressible undrained response.
- The effective stress conditions at ultimate/critical (*US/CS*) state in triaxial loading conditions are essentially independent of initial soil fabric (natural and reconstituted fabric) of tested sand.
- When the comparative response of sand specimens prepared by *WS* method under different loading modes (e.g. triaxial or simple shear) is analysed, it is found that the expected response of in-situ sand deposit under earthquakes can be predicted more properly through simple shear tests.

REFERENCES

1. R. Dyvik, T. Berre, S. Lacasse and B. Raadim, "Comparison of truly undrained and constant volume direct simple shear tests", *Geotechnique* 37 (1), pp. 3-10 (1987).
2. W.D.L. Finn, "Aspects of constant volume cyclic simple shear, Advances in the art of testing soils under cyclic conditions", *ASCE Convention*, (ed. by Khosla), Detroit, 1985, pp. 74-98.
3. V. N. Ghionna, S. Pedroni, and D. Porcino, "Undisturbed sampling by ground freezing at Gioia Tauro site for seismic liquefaction analyses" in *XVth ICSMGE- Earthquake Geotechnical Engineering Satellite Conference*, edited by Atilla M. Ansal Istanbul, Turkey, 2001, pp. 249-254.
4. V.N. Ghionna and D. Porcino, "Liquefaction Resistance of undisturbed and reconstituted samples of a natural coarse sand from undrained cyclic triaxial tests", *Journal of Geotechnical and Geoenvironmental Engineering, ASCE* 132 (2), pp. 194-202 (2006).
5. K. Ishihara, "Liquefaction and Flow Failure during Earthquakes". *Geotechnique* 43 (3), pp. 351-415.
6. K. Ishihara, *Soil Behaviour in Earthquake Geotechnics*, Oxford Engineering Science Series, 46 Oxford University Press. Inc., New York. 1996.
7. M. Oda, I. Koishikawa and T. Higuehi. "Experimental study of anisotropic shear strength of sand by plane strain test". *Soils and Foundations* 18 (1), pp. 25-39 (1978).
8. D. Porcino, G. Caridi, M. Malara and E. Morabito, "An automated control system for undrained monotonic and cyclic simple shear tests" in *GeoCongress06,* Atlanta (GA), 26 February-1 March 2006.
9. D. Porcino, G. Caridi and V.N. Ghionna, "Undrained monotonic and cyclic simple shear behaviour of carbonate sand. *Accepted for publication in Geotechnique* (accepted on 25 [th] Sept. 2007) (2008).
10. K.H. Roscoe, "The influence of strains in soil mechanics. Tenth Rankine Lecture", *Geotechnique* 20 (2), pp. 129-170.
11. H. B. Seed and W.H. Peacock, "Applicability of laboratory test procedures for measuring soil liquefaction characteristics under cyclic loading". *Earthquake Engineering Research Center Report EERC 70-8,* Nov. 1970.
12. J.A. Sladen, R.D. D'Hollander and J. Krahn, "The liquefaction of sands, a collapse surface approach". *Can. Geotech. J.,* **22,** pp. 564-578 (1985).
13. F. Tatsuoka, K. Ochi, S. Fujii and M. Okamoto, "Cyclic undrained triaxial and torsional shear strength of sands for different sample preparation methods", *Soils and Foundations* 26 (3), pp. 23-41, (1986).
14. Y.P. Vaid and S. Sivathayalan, "Fundamental factors affecting liquefaction susceptibility of sands". in *Physics and Mechanics of Soil Liquefaction,* edited by P.V. Lade and J.A. Yamamuro, Int. Workshop Proceedings, Baltimore, Maryland (USA), 1998, pp. 105-120.
15. Y.P. Vaid, Sivathayalan S. and D. Stedman, "Influence of specimen-reconstituting method on the undrained response of sand", *Geotechnical Testing Journal, GTJODJ* **22** (3), pp. 187-195. (1999).

Estimation of the Uplift Displacement of a Sewage Manhole in Liquefied Ground

Tetsuo Tobita[a], Susumu Iai[a], Gi-Cheon Kang[b], Shoji Hazazono[c], and Yasuhiko Konishi[c]

[a]*Disaster Prevention Research Institute, Kyoto University, Gokaho, Uji, 611-0011,Kyoto, Japan.*
[b]*Graduate School of Engineering, Department of Earth Resources Engineering, Kyoto University, Nishi-Kyo-Ku, 615-8530, Kyoto, Japan*
[c]*Nihon Suido Consultants Co., Ltd., 22-1 Nishi-Shinjuku 6, Shijuku-ku, Tokyo 163-1122, Japan*

Abstract. A simplified method to estimate the maximum uplift displacement of a manhole and settlements of backfill soil under liquefaction is proposed. The method is derived based on the mechanism of uplifting of a manhole under undrained condition of backfill soil. It also has a capability of estimating effectiveness of countermeasures against uplifting by considering excess pore water pressure ratio and/or unit weight of backfill soil. In the present study, however, estimation for the case of no countermeasure is verified through comparison with experimental results. Results show that measured uplift displacements and settlements are confined within the boundary predicted by the proposed method.

Keywords: Manhole, Liquefaction, Centrifuge modeling, Earthquake

INTRODUCTION

Uplifting of sewage manholes (Fig. 1) is one of the typical and striking damage pattern observed in the area being hit by large earthquakes (e.g., Yasuda and Kiku, 2006; Yasuda et al., 1993). It has been typically observed in the areas where soft soils, such as clayey or organic soils, are dominant and ground water level is relatively high. When manholes and buried pipes are installed, sandy soils are normally used as backfill material because of its ease of handling. With recent investigation and research, liquefaction of these materials appears to be a major cause of the uplifting (Yasuda and Kiku, 2006). Mechanism of the uplift is explained as follows; firstly uplifting force is initiated by the increase of excess pore water pressure due to liquefaction of backfill soil during large earthquakes. Once a manhole is slightly uplifted, then, a void of negative pressure underneath a manhole is replaced by liquefied soils. If duration of shaking is sufficiently long, uplifting continues until the equilibrium of uplifting force and weight of a manhole is achieved.

So far, many experimental and analytical studies are conducted to study the mechanism of uplift and to develop its countermeasures (e.g., Kiku et al., 2007a; Kiku et al., 2007b). For countermeasures against the uplift on newly installed manholes, a number of methods have been proposed; compaction and improvement of backfill soils, backfill with stones, or discharging pressured water into a manhole to reduce excess pore water pressure. Some of the methods have been already put into practice.

CP1020, *2008 Seismic Engineering Conference Commemorating the 1908 Messina and Reggio Calabria Earthquake,*
edited by A. Santini and N. Moraci

However, for existing manholes, economical and effective countermeasures are still under development.

Supposing situation just after large earthquakes, Honda et al. (2002) made questionnaire survey to fire fighters asking allowable uplift displacement of a manhole for them to drive emergency vehicles. Results showed the allowable amounts to be 13 cm for narrow streets, and 23 cm for wide streets. Estimation of uplift displacement is important for designing manholes and buried pipes under the framework of the performance based design.

However, reasonable method for the estimation of an uplift displacement of a manhole has not been established yet. In this study, a method to estimate the maximum uplift displacement of a manhole and settlement of backfill soil is derived based on the mechanism of uplifts. The proposed method has a capability of estimating effectiveness of countermeasures against uplifting by considering excess pore water pressure ratio and/or unit weight of backfill soil. Estimation by the method is verified through comparison with experimental results.

FIGURE 1. Observed uplifted manhole after 2003 Tokachi-oki, Japan, earthquake.

ESTIMATION OF THE MAXIMUM UPLIFT OF A MANHOLE

Here we propose a simple method to estimate the maximum uplift of a manhole. A manhole is uplifted when excess pore water pressure at the bottom of it exceeds stresses due to a self weight of the manhole and side friction. By considering this condition, factor of safety against uplift is obtained. This factor has been used in practice for design and construction of a manhole and buried pipes.

Another condition to be considered here is the equivalence between uplifted volume of a manhole and settled volume of backfill soil. On derivation, following assumptions are made: (1) undrained condition during liquefaction of backfill soil, i.e., no volume change is allowed in backfill soil, (2) settlement of backfill soil is uniform, (3) no tilting of a manhole, (4) ground water depth is kept constant, and (5) pipes attached to a manhole is neglected for simplicity.

Figure 12 illustrates a manhole and an excavated area in which liquefiable backfill soils are filled. A manhole is simplified as a cylinder solid whose height and diameter are, respectively, h and c. It is installed in an excavated area whose area is set as $a \times a$ and depth is taken arbitrary but larger than h. Ground water depth which is equivalent to a thickness of non-liquefiable layer is specified by d.

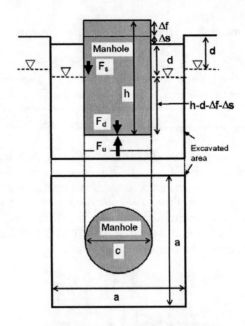

Figure 2. Simplified illustration of a manhole and excavated area.

When a manhole is uplifted, a void underneath the manhole is replaced by liquefied soils. By the assumption given above, volume of the void may be equal to the amount of settled volume of the backfill soil. Let Δf and Δs be, respectively, the displacement of uplift and settlement of the backfill soil, then equating these volumes,

$$\pi\left(\frac{c}{2}\right)^2 \Delta f = \left\{a^2 - \pi\left(\frac{c}{2}\right)^2\right\} \Delta s \tag{1}$$

Solving Eq. (1) for Δf gives,

$$\Delta f = \left\{\frac{1}{\pi}\left(\frac{2a}{c}\right)^2 - 1\right\} \Delta s \tag{2}$$

Next, let us derive the equilibrium of force at the bottom of a manhole. Forces applied to a manhole are, as shown in Fig. 2, self-weight of a manhole F_d and the frictional force F_s, and uplift force F_u due to the excess pore water pressure. Hence the following equilibrium is established,

$$F_d + F_s = F_u \tag{3}$$

The force due to a self-weight of a manhole is given as,

$$F_d = \pi \left(\frac{c}{2}\right)^2 \{\gamma_m h - \gamma_w (h - d - \Delta f - \Delta s)\} \tag{4}$$

where, γ_m is a unit weight of a manhole, and γ_w is a unit weight of water.

If we assume that the frictional force on the side of a manhole acts only above the ground water level, then it is written,

$$F_s = 2\pi \frac{c}{2} dK\sigma_v' \tan\delta = \pi c d K \sigma_v' \tan\delta \tag{5}$$

where, K is the coefficient of lateral earth pressure, σ_v' is the effective vertical stress, and δ is the angle of friction between manhole and soils.

The uplift force due to liquefaction is,

$$F_u = \pi \left(\frac{c}{2}\right)^2 (\gamma_t d + \beta \sigma_v') = \pi \left(\frac{c}{2}\right)^2 \{\gamma_t d + \beta\gamma'(h - d - \Delta f - \Delta s)\} \tag{6}$$

where γ_t is a unit weight of backfill soil above the ground water table, $\gamma' = \gamma_{sat} - \gamma_w$ is a submerged unit weight of backfill soil, and β is the effective stress ratio. By substituting Eq. (4) to (6) into Eq. (3), we obtain,

$$\Delta s = \frac{1}{\beta\gamma' + \gamma_w} \left\{ -\gamma_m h - F_s \frac{1}{\pi}\left(\frac{2}{c}\right)^2 \right\} + \frac{\gamma_t}{\beta\gamma' + \gamma_w} d + h - d - \Delta f \tag{7}$$

Then, from Eq. (7) and Eq. (2), the maximum uplift displacement of a manhole is given as,

$$\Delta f = \left\{ 1 - \pi\left(\frac{c}{2a}\right)^2 \right\} \left\{ \left(1 - \frac{\gamma_m}{\beta\gamma' + \gamma_w}\right)h - \left(1 - \frac{\gamma_t}{\beta\gamma' + \gamma_w}\right)d - \frac{F_s}{\beta\gamma' + \gamma_w}\frac{1}{\pi}\left(\frac{2}{c}\right)^2 \right\} \tag{8}$$

Substituting Eq. (8) into Eq. (7), then, settlement of backfill soil can be written as,

$$\Delta s = \pi\left(\frac{c}{2a}\right)^2 \left[\left\{1 - \frac{\gamma_m}{\beta\gamma' + \gamma_w}\right\}h - \left\{1 - \frac{\gamma_t}{\beta\gamma' + \gamma_w}\right\}d \right] - \frac{F_s}{a^2(\beta\gamma' + \gamma_w)} \tag{9}$$

Now let us study Eq. (8) under the following conditions,
(1) Backfill soil is totally liquefied ($\beta = 1.0$)
(2) Ground water level is at the ground surface ($d = 0$)
(3) Area of excavation is large ($c/2a = \infty$)
Then Eq. (8) can be simplified as follows,

TABLE 1. Parameters for soils

Max. void ratio	e_{max}	1.19
Min. void ratio	e_{min}	0.710
Relative density	Dr	50 %
Unit weight of water	γ_w	9.8 kN/m³
Void ratio	e=	0.95
Density	G_s	2.66
Degree of saturation	S_r	30 %
Unit weigh of wet sand	γ_w	14.8 kN/m³
Unit weight of saturated sand	γ_{sat}	18.1 kN/m³
Unit weight of submerged sand	γ'	8.3 kN/m³

TABLE 2. Parameters for manholes (Prototype scale)

Aluminum	Unit weight	γ_c	26.5 kN/m³
Manhole	Length	h	3 m
	Diameter	c	1.1 m
	Wall thickness	t	0.1 m
	Mass of sensors	m	68 kg
	Mass of base slab		1.7 kN
	Total weight		27.3 kN
	Volume		2.85 m³
	Unit weight	γ_m	9.57 kN/m³
Excavation width		a	2 m
Factor of safety against liquefaction		F_L	1.0
Excess pore water pressure ratio		β	1.0
Friction angle b/w concrete and soil		δ	10 deg
Coefficient of lateral earth pressure		K	0.5

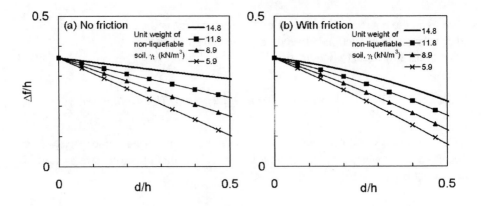

Figure 3. Normalized uplift amount, $\Delta f/h$, versus normalized ground water depth, d/h: (a) No frictional force at the side of a manhole is assumed; (b) Frictional force is assumed.

$$\Delta f = \left(1 - \frac{\gamma_m}{\gamma_{sat}}\right) h \qquad (10)$$

From Eq. (10), the maximum uplift amount is expressed as a function of a ratio of a unit weight of a manhole and saturated unit weight of backfill soil.

Figure 3 shows the normalized maximum uplift displacement versus the normalized ground water depth. Length of a manhole, h, is employed for the normalization. Figure 3(a) is the case with no frictional force, i.e., $\tan\delta=0$, while Fig. 3(b) is the case with frictional force. Values of parameters to be input in Eq. (8) are listed in Table 1 and 2. These parameters are taken form the centrifuge experiment that will be explained later. In Fig. 3, curves relative to various unit weights of back fill soil are also plotted for comparison. With the increase of ground water depth, d, the uplift displacement is decreasing. While the increase of a unit weight of backfill soil increases the uplift displacement. With the frictional force at the side of a manhole, the uplift displacement slightly reduced compared with the case of no frictional force.

VERIFICATION OF THE METHOD BY MODEL TESTING

Predicted uplift displacements are compared with those obtained from centrifuge model tests. The geotechnical centrifuge at the Disaster Prevention Research Institute, Kyoto University was employed. Model tests were conducted under 20 G (gravitational force) with the size of a manhole 1/20 scale. Material parameters of the model ground and the prototype of a manhole are specified in Table 1 and 2. For comparison purposes, a model manhole with no countermeasure was shaken simultaneously with a model manhole having some countermeasures (Fig. 4). However, in this study, only the former results are used. Detail of model tests including the effects of countermeasures will be found in upcoming reports.

Model ground was prepared in a rigid box (450 × 150 × 300 mm) by compacting moist silica sand up to 260 mm lift (model scale). To install model manhole, ground was excavated with a volume of 100 × 100 × 160 mm (model scale), then it was placed on gravels at the bottom of the excavated ground. Silica sand as backfill soil were pluviated into viscous water to make loose deposit (Dr=50%). Viscous water with a viscosity 20 times larger than that of water was used to precisely simulate behavior of liquefied sands. Sinusoidal wave with frequency of 1.25 Hz and acceleration amplitude of 6.0 m/s^2 (prototype scale) was employed as an input (see Fig. 5: A0). As shown in Fig. 4, accelerometers (A0-A3), laser displacement transducers (D1), and pore water pressure transducers (P1-P3) were installed at the specified position. Total 7 tests were conducted to study the mechanism of uplift and examine the effects of countermeasures. As shown in Table 3, ground water depth was kept 1 m for all the cases except for Case 1 in which it was set at the ground surface. Uplift displacements and settlements were directly measured by a ruler before and after each experiment. Relatively large variation of the measured uplift displacement, from 0.28 m to 0.82 m as shown in Table 3, might be due to some errors on model construction, and/or mechanical instability of input motions.

As shown with a vertical broken line in Fig. 5 (a), manhole started uplifting (D2) when the excess pore water pressure exceeded the initial vertical effective stress level at 7 s [P1: Fig. 5(b)]. However, pore water pressure at the bottom of the manhole did not reach to the initial vertical effective stress [approx. 30 kPa shown with horizontal line in Fig. 5 (c)]. Primary cause of this low pore water pressure may be the negative pressure by suction due to uplift of a manhole.

Measured uplift displacements of a manhole and settlements of backfill soils are compared with the prediction in normalized form in Fig. 6. In the present study, the ratio of d/h is 0 for Case 1 and 0.33 for others. As shown in Fig. 6(a), although there are some variations in measured data, all the data are plotted within the boundary of predicted uplifts and settlements. For Case 1, measured uplift displacement ($\Delta f/h$) is plotted far below the predicted line because excess pore pressure was dissipated from the ground surface, while in derivation of the method, dissipation of pore water pressure is not allowed.

461

TABLE 3. Measured uplift, settlement, and total vertical displacements of the model manhole with no countermeasures (Prototype scale).

Case	Uplift displacement Δf (m)	Surface settlements Δs (m)	Total vertical displacements $\Delta F + \Delta s$ (m)	Ground water depth d (m)
1	0.34	0.10	0.44	0.0
2	0.78	0.16	0.94	1.0
3	0.52	0.16	0.68	1.0
4	0.28	0.14	0.42	1.0
5	0.82	0.06	0.88	1.0
6	0.63	0.12	0.75	1.0
7	0.65	0.07	0.72	1.0

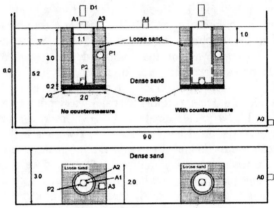

Figure 4. Centrifuge model setup

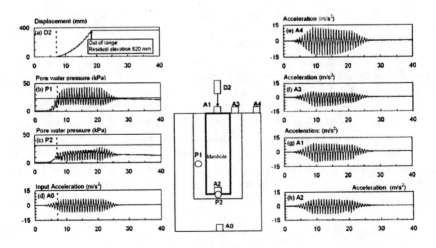

Figure 5. Results of centrifuge model tests: Case 5.

462

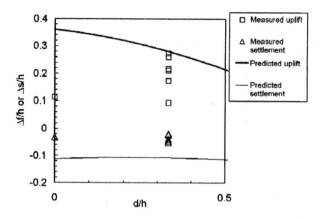

Figure 6. Predicted and measured normalized uplift displacements of a manhole and settlements of backfill soils versus normalized ground water depth.

CONCLUSIONS

A method to estimate the maximum uplift displacement of a manhole and settlements of backfill soils under liquefaction was proposed. The method was derived based on the mechanism of uplifting of a manhole with undrained condition of backfill soils. It also has a capability of estimating effectiveness of countermeasures against uplifting by considering excess pore water pressure ratio and/or unit weight of backfill soil. In the present study, however, estimation for the case of no countermeasure was verified through comparison with experimental results. Results showed that measured uplift displacements and settlements were confined within the boundary predicted by the proposed formula.

REFERENCES

1. A. Honda, H. Nakase, S. Yasuda, and T. Suehiro, "Study on permissible uplift of buried structures," *Proceedings of the 57th Annual Conference of the Japan Society of Civil Engineers.* Sapporo: JSCE, pp 1439-1440 (2002) (in Japanese).
2. H. Kiku, D. Fukunaga, M. Matsumoto, and M. Takahashi, "Shaking table tests on countermeasures against uplift of manhole due to liquefaction," *The 42nd National Conference on Geotechnical Engineering.* Nagoya, Japan, pp 1889-1890 (2007a) (in Japanese).
3. H. Kiku, R. Fukunaga, M. Kimura, M. Takahashi, and M. Matsumoto, "Shaking table test on mechanism of uplift of manhole due to liquefaction," *The 42nd Japan National Conference on Geotechnical Engineering.* Nagoya, pp 1887-188 (2007b) (in Japanese).
4. S. Yasuda, and H. Kiku, "Uplift of sewage manholes and pipes during the 2004 Niigataken-Chuetsu earthquake," *Soils and Foundations, Japanese Geotechnical Society,* Vol. **46**, No. 6, pp 885-894 (2006).
5. S. Yasuda, Y. Sakamoto, and M. Miyajima"10. Damage to lifeline facilities," *The 1993 Kushiro-oki and Notohanto-oki, Japan, Earthquake*: The Japanese Society of Soil Mechanics and Foundations, pp 277-315 (1993) (in Japanese).

Relationship between recurrent liquefaction-induced damage and subsurface conditions in Midorigaoka, Japan

Kazue Wakamatsu[a] and Nozomu Yoshida[b]

[a] Department of Civil and Environmental Engineering, Kanto-Gakuin University Yokohama, Japan
[b] Department of Civil and Environmental Engineering, Tohoku-Gakuin University, Tagajyo, Japan

Abstract. Midorigaoka, Kushiro City, northeast Japan, suffered liquefaction-induced ground failures during four successive earthquakes in the past thirty years. This paper presents the ground failures and their effects to structures observed in Midorigaoka during the earthquakes, and examines the relationships between recurrent liquefaction-induced damage and subsurface conditions. As a result, thick liquefiable fill, slope of the ground surface, and subsurface water conditions, which resulted primarily from filling a marshy valley, are found to be responsible on the damage.

Keywords: Liquefaction, recurrence of liquefaction, liquefaction-induced damage.

INTRODUCTION

Midorigaoka is a residential area situated in Kushiro City, Hokkaido, Japan, approximately 900 km northeast of Tokyo. It was developed for housing in the 1960s by cutting and filling a Pleisto-cene terrace with a maximum elevation of approximately 30 meters above sea level. The district has been frequently affected by large earthquakes and suffered various kinds of ground deformation and damage to structures in every earthquake. The earthquakes include the 1973 Nemuro-hanto-oki M7.4 quake, the 1993 Kushiro-oki M7.5 quake, the 1994 Hokkaido-toho-oki M8.2 quake, and the 2003 Tokachi-oki M8.0 quake. We conducted field survey in these earthquakes and found that the damage was repeatedly concentrated in particular areas in Midorigaoka. In order to investigate the causes of concentration of the damage on such local areas, a history of land reclamation and construction in Midorigaoka was investigated and geotechnical data were collected to determine the subsurface conditions in the area. Old topographic maps were collected and photogrammetric analyses were conducted to establish the profiles of the land forms before and after the residential development.

GEOLOGY, HISTORIC DEVELOPMENT OF MIDORIGAOKA

Midorigaoka is located on a Pleistocene terrace in the eastern part of Kushiro City that faces the Pacific Ocean on the southern end, Hokkaido, Japan (Figure 1). The terrace is compose of, from upper to lower, volcanic ash known as the Kussharo

CP1020, *2008 Seismic Engineering Conference Commemorating the 1908 Messina and Reggio Calabria Earthquake*,
edited by A. Santini and N. Moraci
© 2008 American Institute of Physics 978-0-7354-0542-4/08/$23.00

Pumice Flow Deposit, volcanic sandy soil known as the Otanoshike Formation, and sand, gravel, silt or clay known as the Kushiro Group on a base of Paleogene rock. Figure 1 is an old topographic map surveyed in 1958. It should be noted that the terrace is dissected by a deep valley in the eastern part of Midorigaoka. The bottom of this valley is marsh and extends towards Kushiro Wetland.

The development of the terrace started from the northwestern part of the area around the estuary of Old Kushiro River in the early 1900s and gradually extended toward the northeast. Midorigaoka was completed in 1972-73. The area was developed for construction by cutting and leveling the terrace and then filling the valleys with soils from the terrace. Therefore there are two general types of subsurface conditions in Midorigaoka: 1) natural soils associated with the volcanic pumice flow deposit overlying the original or excavated surface of the terrace and 2) fills consisting mainly with the volcanic pumice flow deposit placed on the valleys.

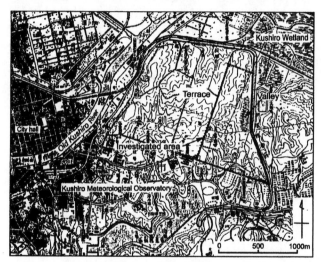 

FIGURE 1. Topographic map surveyed in 1953 before development

GROUND FAILURES AND THEIR EFFECTS ON STRUCTURES DURING PAST EARTHQUAKES

The 1993 Kushirro-oki and the 1973 Nemuro-hanto-oki earthquakes

The Kushiro-oki earthquake occurred at 8:06 PM local time, on January 15, 1993, and registered 7.5 on the Japan Meteorological Agency (JMA) magnitude scale. The focus was located at, longitude 42°55.2' N and latitude 144°21.2' E, with depth of 101 km. Kushiro city is located only 10 km from the epicenter. Strong motions record with peak acceleration 711 cm/s^2 and 919 cm/s^2 were recorded at ground surface and at ground floor of the Kushiro Meteorological Observatory, respectively [2], which are located on the Pleistocene terrace about 2.5 km southwest of Midorigaoka.

Figure 2 shows locations of major ground failures and damage to structures in Midorigaoka caused by the 1993 Kushiro-oki earthquake. The ground failures include slope failures, collapses of retaining walls, ground cracks, ground settlements and sand boiling. The locations of damaged structures coincided with these of ground failures, which imply that most of the damage was caused by the ground failure.

Manholes uplifted 5 to 20 cm at many sites. Sand boils were observed at fewer locations compared with those during the 1994 Hokkaido-toho-oki earthquake, as described later, although ground shaking is stronger in this earthquake. The ground surface was frozen with 50-100 cm thickness at the time of the earthquake, which was supposed to prevent sand boils to eject the ground surface.

Underground pipelines and foundations of wooden houses settled differentially and were displaced laterally where the ground cracks were abundant. This implies that a large amount of permanent ground deformation occurred in these areas. The magnitudes of the vertical and horizontal ground displacements were estimated to be in the range of approximately 30 to 50 cm on the basis of widths of ground cracks and displacements of pipelines and foundations of the houses. According to the residents, sand boils and ground cracks and associated damage to houses were also observed at the time of the 1973 Nemuro-hanto-oki earthquake of magnitude 7.4 at the same locations as shown in Figure 2.

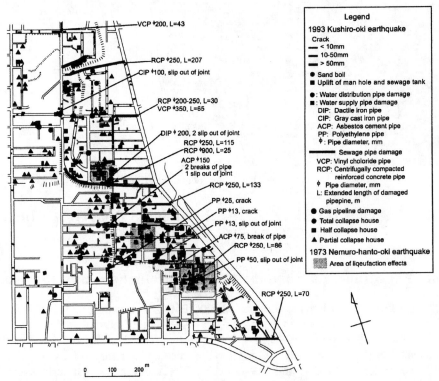

FIGURE 2. Map of Midorigaoka showing locations of ground failures and damaged lifelines and buildings during the 1993 Kushiro-oki earthquake and area of liquefaction during the 1973 Nemuro-hanto-oki earthquake.

466

The 1994 Hokkaido-toho-oki earthquake

The Hokkaido-toho-oki earthquake occurred at 10:23 PM local time on October 4, 1994, and registered 8.2 both in the JMA Magnitude Scale and in the Surface Wave Magnitude Scale. The focus was located at latitude 43°22' N and longitude 147°40' E, with depth of 30 km. The epicenter of the earthquake was located approximately 270 km east-northeast from Kushiro city. A peak horizontal acceleration of 473 cm/s^2 was recorded at Kushiro Meteorological Observatory and 269 cm/s^2 was recorded at the reclaimed land of Kushiro Port, 5 km west-northwest of Midorigaoka [3].

Figure 3 shows locations of ground failures, and damaged houses and pipelines in Midorigaoka. Although the epicenter was much far distance and the peak acceleration was down by half as compared with that in the 1993 Kushiro-oki earthquake, the similar ground failures were also observed; sand boiling occurred at the sites where liquefaction-induced damage was serious in the 1993 earthquake. Despite less damage to structures than that due to the previous 1993 quake, sand boiling was observed at more locations. This implies that the frozen layer of ground at the surface was prevented the ejection of sand boils at the time of the 1993 Kushiro-oki earthquake that occurred in midwinter.

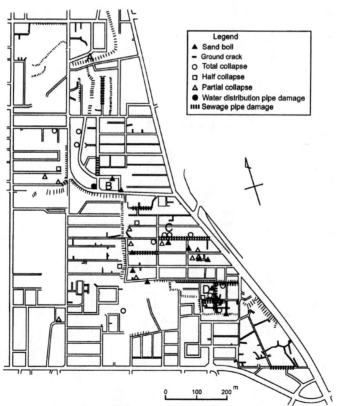

FIGURE 3. Map of Midorigaoka showing locations of ground failures, damaged houses and lifelines during the 1994 Hokkaido-toho-oki earthquake.

The 2003 Tokachi-oki earthquake

The primary JMA Magnitude 8.0 earthquake occurred at 4:50 AM local time on September 26, 2003, and was followed by a M7.1 and 6.2 tremors at 6.08 AM and 3:27 PM, respectively. The focus of the primary shock was located at longitude 41°46.7' N and latitude 144°04.7' E, with depth of 45 km. A peak horizontal acceleration of 355 cm/s^2 was recorded at Kushiro Meteorological Observatory and 288 cm/s^2 and 407 cm/s^2 were also recorded at Kushiro Joint Government Building and K-NET Kushiro both about 3km west of Midorigaoka, respectively [4].

Despite of far distance from the epicenter and smaller acceleration compared with that in the 1993 and 1994 earthquakes, the same pattern of the liquefaction effects as in the previous two earthquakes occurred. Figure 4 shows locations of ground failures, uplift of manholes and collapsed houses in Midorigaoka. A number of houses were affected by the earthquake. Damage was primarily in the form of cracks of walls and foundation, and differential settlement and horizontal displacement of the foundation. The most houses that were damaged due to the 1993 and 1994 earthquakes were significantly affected by the present earthquake.

FIGURE 4. Map of Midorigaoka showing locations of ground failures, damaged houses and water distribution pipes during the 2003 Tokachi-oki earthquake.

468

RELATION BETWEEN SUBSURFACE SOIL CONDITIONS AND DAMAGE

Figure 5 shows a cross section oriented in the north-south direction. Loose fill with SPT N-values less than 5 extends along the section with a thickness ranging from 1 to 9 m. Underlying the fills is very soft peat, which was originally at the bottom of the marshy valley. The depth of the water table is several meters beneath the ground surface in the terrace, whereas it is shallow ranging from 0.3 to 2 m deep in the valley, which indicates that the water table has risen up into the fill. This water has come from neighboring area because the valley forms water catchment.

Sand boils were observed in the gentle slope of the valley wall where 1 to 2 m thick fill overlies the water table. Major ground cracks and settlements with sand boils occurred in the upper part of the slope. This pattern of ground deformation implies that lateral movement toward the center of the valley occurred due to the liquefaction of the loose fill.

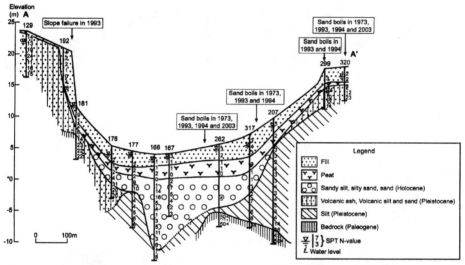

FIGURE 5. Soil profile along cross-section A-A' in Midorigaoka and location of sand boils (modified from [1]).

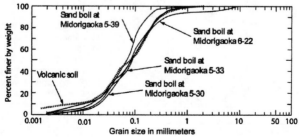

FIGURE 6. Grain size distribution curves for volcanic ash used as fill material, and sands collected from sand boils due to the 1993 Kushiro-oki earthquake in Midorigaoka [1].

Figure 6 is a comparison of the grain size distribution curves for sands from sand boil deposits during the 1993 Kushiro-oki earthquake at different locations in Midorigaoka and natural volcanic ash which is generally used as fill material in this area. Both soils have similar grain size characteristics, which supports that the fill liquefied during the earthquakes.

To examine further the relationships between ground deformation and subsurface conditions, land form changes due to development are investigated by analyzing a pair of aerial photographs taken before and after the development. A total of 13 sections were selected for analysis; 12 sections through Midorigaoka and 1 outside Midorigaoka where few effects of the earthquakes were observed. The accuracy of the measurements is about ±1 m in the vertical direction.

The typical examples of the photogrammetric analysis were shown in Figure 7, in which the locations of streams (terrace runoff) before the development and the ground failures and damage to structures during the 1993 Kushiro-oki earthquake are plotted. It can be seen from the figure that the heaviest concentration of the damage to

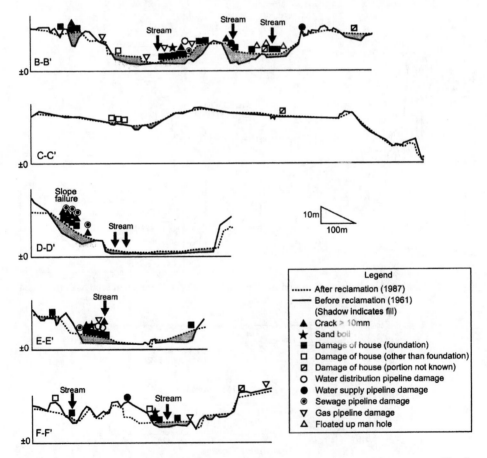

FIGURE 7. Cross-sections showing land form changes due to development and damage caused by the 1993 Kushiro-oki earthquake [1]

470

structures was observed in the filled area where ground failures such as slope failure, ground cracks and sand boils were concentrated. The slope failure occurred at the boundary between the terrace and the valley where fill is thick and the slope of ground surface is steep as shown in section D-D', whereas liquefaction effects such as sand boils and uplift manholes occurred in the valley where fill is thick, slope is gentle and streams flowed down before the development. Ground cracks also occurred on both steep and gentle slopes in the filled area. Damage to pipelines and foundation of houses occurred in the locations of all ground failures, whereas structural damage was absent in the valley bottom where the fills are not so thick and the ground surface is flat, as shown in section D-D'. The data imply that permanent ground deformation and/or movement associated with ground failures was the principal cause of the damage to structures. In contrast, damage other than that in foundation of house such as collapse of chimney and cracks of exterior walls, a few in numbers, did occur in all part of sections, in both filled and cut areas, which implies that primary cause of the damage was strong ground shaking.

CONCLUDING REMARKS

There was substantial damage to underground pipelines and buildings in Midorigaoka, Kushiro City during the successive earthquakes of the 1973 Nemuro-hanto-ok, the 1993 Kushiro-oki, the 1994 Hokaido-toho-oki and the 2003 Tokachi-oki. The damage was caused primarily by ground deformations and/or ground movements associated with soil liquefaction and liquefaction-induced flow, and partially by ground shaking and slope failure.

In summary, Midorigaoka had several adverse conditions such as thickness and density of fills, slope of the ground surface, and subsurface water conditions. The conditions resulted primarily from the filling on the marshy valley, which lead to induce various types of ground failures. They were strongly affected to not only locations and pattern of the ground failures but also resulting damage modes of structures. Although the level of damage was different for intensity of ground motion and seismic capacity of each structure, it is notable that locations and patterns of ground failures were very similar in the all earthquakes.

REFERENCES

1. K. Wakamatsu, and N. Yoshida, "Ground deformations and their effects on structures in Midorigaoka Districts, Kushiro City, during the Kushiro-oki earthquake of January 15, 1993," *Proceedings of the 5th U.S.-Japan Workshop on Earthquake resistant Design of Lifeline Facilities and Countermeasures for Soil Liquefaction*, Snowbird, U.S.A. Technical Report NCEER, 1995.
2. Japan Gas Association. *The 1994 Kushiro-oki and Hokkaido Nansei-oki Earthquakes and Urban Gas Facilities*,(in Japanese), 1994.
3. The National Research Institute for Earth Science Disaster Prevention Science and Technology Agency. *Prompt Report on Strong-Motion Accelerograms* No.41, 1993.
4. The National Research Institute for Earth Science Disaster Prevention Science and Technology Agency. *Prompt Report on Strong-Motion Accelerograms* No.44, 1994.
5. The National Research Institute for Earth Science Disaster Prevention Science and Technology Agency. *Prompt Report on Strong-Motion Accelerograms* No.112, 2003.

SLOPE EMBANKMENTS, DAMS AND WASTE FILLS

Prediction of Seismic Slope Displacements by Dynamic Stick-Slip Analyses

Ernesto Ausilio[a], Antonio Costanzo[b], Francesco Silvestri[c], Giuseppe Tropeano[d]

[a] Assistant Professor, DDS, University of Calabria, Italy, Email: ausilio@dds.unical.it
[b] Ph.D., DDS, University of Calabria, Italy, Email: acostanzo@dds.unical.it
[c] Professor, DIGA, University of Naples "Federico II", Italy, Email: francesco.silvestri@unina.it
[d] Ph.D. Student, DDS, University of Calabria, Italy, Email: tropeano@dds.unical.it

A good-working balance between simplicity and reliability in assessing seismic slope stability is represented by displacement-based methods, in which the effects of deformability and ductility can be either decoupled or coupled in the dynamic analyses. In this paper, a 1D lumped mass "stick-slip" model is developed, accounting for soil heterogeneity and non-linear behaviour, with a base sliding mechanism at a potential rupture surface. The results of the preliminary calibration show a good agreement with frequency-domain site response analysis in no-slip conditions. The comparison with rigid sliding block analyses and with the decoupled approach proves that the stick-slip procedure can result increasingly unconservative for soft soils and deep sliding depths.

Keywords: seismic slope stability, displacements, numerical analysis

BACKGROUND

The usual design procedures to analyse the seismic slope stability typically refer to two classes of approaches:
- pseudo-static method, in which the seismic action is represented by an "equivalent acceleration", used in a conventional limit equilibrium analysis;
- displacement-based analysis, in which the permanent displacements induced by earthquake acceleration-time history are predicted.

The Newmark's model [1] is still the basis of most engineering approaches used to compute earthquake-induced sliding displacements. The basic procedure considers the sliding mass as a rigid block, accounting for the slope ductility, i.e. the capacity to sustain permanent displacements. In the application to earth structures, Makdisi and Seed [2] suggested to account for the deformable response of the slope soil, which can induce asynchronous motion and significant reduction of inertia forces for softer soils.

By following a "decoupled approach", the effects of deformability and ductility are separated, since a dynamic site response analysis is carried out prior to the sliding block analysis. Bray *et al.* [3] extensively developed a displacement-based decoupled procedure for natural and artificial slopes; such methods were recently calibrated and extended to the Italian seismicity by the Authors [4].

In a "coupled approach", the wave propagation and sliding block analyses are performed simultaneously. Lin and Whitman [5] used a lumped-mass shear beam

CP1020, *2008 Seismic Engineering Conference Commemorating the 1908 Messina and Reggio Calabria Earthquake*,
edited by A. Santini and N. Moraci

475

model with linear elastic soil properties, finding out that the decoupled approximation overestimates the coupled permanent displacements by an average of 20%. However, it is not generally possible to assess which analysis is more conservative.

Chopra and Zhang [6] used a generalized single degree of freedom (SDOF) procedure, with mass and elasticity distributed along the height of a dam. The displacements predicted by the decoupled analysis resulted more conservative for low values of the ratio, a_y/a_{max}, between the yield and peak acceleration; the opposite trend was found at larger values of the same ratio.

Gazetas and Uddin [7] implemented the coupled sliding into a 2D linear-equivalent finite element program, and again found coupled displacements lower than those computed through the decoupled approach.

Kramer and Smith [8] used a discrete mass model with two masses, connected by a spring and a dashpot, with the lower mass that can slide along an underlying inclined plane. Their study revealed that the coupled displacement analysis could be unconservative when the natural period of the system is large.

Rathje and Bray [9] proposed a layered "stick-slip" model to compute coupled sliding displacements of geotechnical structures, with linear or equivalent-linear material properties. The same authors updated the model to a multi-degree-of-freedom (MDOF) system (Fig. 1a), accounting for the fully non-linear material response [10]. Their conclusion was that, overall, the decoupled predictions can be even strongly conservative, unless when the period of the earth structure, T_S, is significantly greater than the median period, T_m, of the input ground motion.

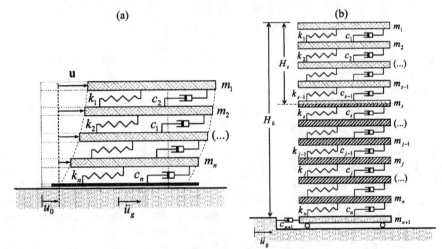

FIGURE 1. The MDOF system by Rathje & Bray [10] (a) and that used in this study (b).

THE STICK-SLIP MODEL

In this paper a lumped-mass stick-slip model with more generalised assumptions is introduced. In the layered subsoil, the depth of the sliding surface, H_s, does not necessarily coincide with that of the bedrock, H_b, and can be located in a generic layer

s (Fig. 1b). Moreover, the bedrock can be considered deformable; a n-layers subsoil can be therefore idealized as an $(n+1)$-DOF model.

In no-slip ("stick") conditions, the seismic response in terms of absolute displacements, $\mathbf{u_a}$, to a base ground motion, u_g, can be computed by integrating the following equation:

$$\mathbf{M} \cdot \ddot{\mathbf{u}}_a + \mathbf{C} \cdot \dot{\mathbf{u}}_a + \mathbf{K} \cdot \mathbf{u}_a = \mathbf{i} \cdot c_{n+1} \cdot \ddot{u}_g \tag{1}$$

where $\mathbf{M}$, $\mathbf{C}$ and $\mathbf{K}$ are the mass, damping and stiffness matrices, and $c_{n+1} = \rho_r V_{Sr}$ is the seismic impedance of the bedrock; $\mathbf{i}$ is a vector with each element equal to zero, except for the $(n+1)$-th that is equal to unity.

The elements of the matrices $\mathbf{M}$ and $\mathbf{K}$ are defined from the mass, m_j, and the spring stiffness, k_j, for a generic layer j, as follows:

$$m_1 = \frac{\rho_1 h_1}{2} ; \; m_j = \frac{\rho_j h_j + \rho_{j-1} h_{j-1}}{2} ; \; m_{n+1} = \frac{\rho_n h_n}{2} \tag{2}$$

$$k_j = \frac{G_j}{h_j} \tag{3}$$

where ρ_j, h_j, and G_j are the density, thickness and shear stiffness of the j-th layer.

The viscous damping matrix, $\mathbf{C}$, is defined according to the full Rayleigh damping formulation [11]:

$$\mathbf{C} = \alpha_R \mathbf{M} + \beta_R \mathbf{K} \tag{4}$$

The constants α_R and β_R are functions of the soil damping ratio, D, the first natural frequency of the subsoil profile, and the predominant frequency of input motion, defined as the inverse of the modal value of the response spectrum.

The frictional strength at the sliding surface can be written as $m_T \cdot a_y$, where m_T is the total mass above the sliding surface, and a_y is the yield acceleration of the slope. The forces at the sliding interface (Fig. 2) are:

$$-m_T \cdot \ddot{u}_s - \mathbf{1}^T \mathbf{M_s} \cdot \ddot{\mathbf{u}} \tag{5}$$

The first term is the inertial force induced by the absolute acceleration at s-th layer, $\ddot{u}_s$, while the second is that resulting from the non-uniform relative acceleration profile, $\ddot{\mathbf{u}}$, within the sliding mass, referred to $\ddot{u}_s$ (i.e.: $\ddot{\mathbf{u}} = [\ddot{\mathbf{u}}_a]_s - \mathbf{1} \ddot{u}_s$). $\mathbf{1}$ is the unity vector, while the subscript s indicates sub-matrices and sub-vectors with index from 1 to $(s-1)$ for the sliding part of the dynamic system.

Permanent displacements, u_0, will be triggered when the forces indicate in Eq. 5 will equal the resisting force $m_T \cdot a_y$; in such a case ("slip" conditions), the equation of motion for the deformable mass above the sliding surface is given by the expression:

$$\mathbf{M_s} \ddot{\mathbf{u}} + \mathbf{C_s} \dot{\mathbf{u}} + \mathbf{K_s} \mathbf{u} = -\mathbf{M_s} \cdot \mathbf{1} \cdot (\ddot{u}_s + \ddot{u}_0) \tag{6}$$

477

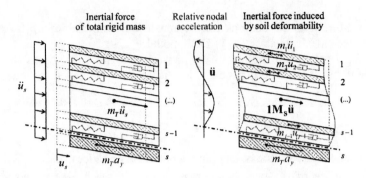

FIGURE 2. Forces acting on the sliding deformable mass.

During the sliding, the equilibrium is governed by the condition:

$$- m_T \cdot \left(\ddot{u}_s + \ddot{u}_o \right) - \mathbf{1}^T \mathbf{M_S} \cdot \ddot{\mathbf{u}} = m_T \cdot a_y \qquad (7)$$

Substituting Eq. 7 in Eq. 6, the equation for the "slip" conditions is obtained as:

$$\mathbf{M}^* \ddot{\mathbf{u}} + \mathbf{C_S} \dot{\mathbf{u}} + \mathbf{K_S} \mathbf{u} = \mathbf{M_S} \cdot \mathbf{1} \cdot a_y \qquad (8)$$

where $\mathbf{M}^*$ is defined as:

$$\mathbf{M}^* = \mathbf{M_S} - \frac{1}{m_T} \mathbf{M_S} \cdot \mathbf{1} \cdot \mathbf{1}^T \mathbf{M_S} \qquad (9)$$

Solving Eq. 8 in terms of nodal relative acceleration, $\ddot{\mathbf{u}}$, the sliding acceleration time histories are obtained from Eq. 7 as:

$$\ddot{u}_o = -a_y - \frac{1}{m_T} \cdot \mathbf{1}^T \mathbf{M_S} \cdot \ddot{\mathbf{u}} - \ddot{u}_s \qquad (10)$$

The sliding displacement, u_0, is computed by integrating twice Eq. 10 until sliding velocity, $\dot{u}_0$, becomes equal to zero.

NUMERICAL SOLUTION

Many step-by-step integration methods can be used to solve Eq. 1 and Eq. 8. In this study, an explicit method proposed by Chan [12] for MDOF systems has been used. The integration method can be expressed as:

$$\mathbf{M} \cdot \ddot{\mathbf{u}}_{t+1} + \mathbf{C} \cdot \dot{\mathbf{u}}_{t+1} + \mathbf{K} \cdot \mathbf{u}_{t+1} = \mathbf{q}_{t+1} \qquad (11)$$

$$\mathbf{u}_{t+1} = \mathbf{u}_t + (\Delta t) \cdot \boldsymbol{\beta}_1 \cdot \dot{\mathbf{u}}_t + (\Delta t)^2 \cdot \boldsymbol{\beta}_2 \cdot \ddot{\mathbf{u}}_t \qquad (12)$$

$$\dot{\mathbf{u}}_{t+1} = \dot{\mathbf{u}}_t + \frac{1}{2}(\Delta t) \cdot (\ddot{\mathbf{u}}_t + \ddot{\mathbf{u}}_{t+1}) \qquad (13)$$

where t is the time step index and $\mathbf{q}$ is the external load vector. For "stick" conditions (i.e. Eq. 1), $\mathbf{q} = \mathbf{i}\, c_{n+1} \ddot{u}_g$, while $\mathbf{M}$, $\mathbf{C}$ and $\mathbf{K}$ are the full matrices of the lumped-mass system; instead, for "slip" conditions (i.e. Eq. 8) $\mathbf{q} = \mathbf{M}_S\, \mathbf{1}\, a_y$, while $\mathbf{M} = \mathbf{M^*}$, $\mathbf{C} = \mathbf{C}_S$ and $\mathbf{K} = \mathbf{K}_S$.

The integration coefficients β_1 and β_2 are defined as:

$$\boldsymbol{\beta}_1 = \left[\mathbf{I} + \frac{1}{2}(\Delta t) \cdot \mathbf{M}^{-1}\mathbf{C} + \frac{1}{4}(\Delta t)^2 \cdot \mathbf{M}^{-1}\mathbf{K}_0\right]^{-1} \left[\mathbf{I} + \frac{1}{2}(\Delta t) \cdot \mathbf{M}^{-1}\mathbf{C}\right] \qquad (14)$$

$$\boldsymbol{\beta}_2 = \left(\frac{1}{2}\right) \cdot \left[\mathbf{I} + \frac{1}{2}(\Delta t) \cdot \mathbf{M}^{-1}\mathbf{C} + \frac{1}{4}(\Delta t)^2 \cdot \mathbf{M}^{-1}\mathbf{K}_0\right]^{-1} \qquad (15)$$

with $\mathbf{K}_0$ = initial stiffness matrix, while $\mathbf{K}$ and $\mathbf{C}$ are updated in each load-unload cycle, accounting for soil non-linear behaviour.

The velocity vector can be explicitated by substituting Eq. 11 in Eq. 13, obtaining the following expression:

$$\dot{\mathbf{u}}_{t+1} = \left[\mathbf{M} + \frac{1}{2}(\Delta t) \cdot \mathbf{C}\right]^{-1} \cdot \left[\mathbf{M} \cdot \dot{\mathbf{u}}_t + \frac{1}{2}(\Delta t) \cdot \mathbf{M} \cdot \ddot{\mathbf{u}}_t + \frac{1}{2}(\Delta t) \cdot (\mathbf{q}_{t+1} - \mathbf{K} \cdot \mathbf{u}_{t+1})\right] \qquad (16)$$

The acceleration vector can be computed by solving Eq. 11 as:

$$\ddot{\mathbf{u}}_{t+1} = \mathbf{M}^{-1} \cdot [\mathbf{q}_{t+1} - \mathbf{C} \cdot \dot{\mathbf{u}}_{t+1} - \mathbf{K} \cdot \mathbf{u}_{t+1}] \qquad (17)$$

and introducing Eq. (12) for $\mathbf{u}_{t+1}$ and Eq. (17) for $\dot{\mathbf{u}}_{t+1}$.

This procedure has the same characteristics of stability, numerical damping and period distortion of Newmark's average acceleration method for linear systems.

APPLICATION AND RESULTS

After the preliminary calibration, the above method was applied to an infinite slope model with an angle of 20°, constituted by a medium-plasticity clay. Two subsoil profiles were considered, the first normally consolidated (NC) and second heavily over-consolidated (OC). The basic soil parameters are summarized in Table 1.

TABLE 1. Parameters for the NC and OC clay.

	NC	OC
Unit weight, γ (kN/m³)	19.00	
Plasticity Index, I_P (%)	25	
Friction angle, φ' (°)	21	
Cohesion ratio, $c'/\gamma H_S$	0	0.025

The constitutive parameters were assumed with values typical for a clay soil with I_P = 25%. The same friction angle, φ', and different effective cohesion ratio, $c'/\gamma H_S$, were assumed for the NC and OC clay subsoils. Thus, the yield acceleration values, computed for an infinite slope model, resulted a_y = 0.017g and 0.042g, respectively.

A non-homogenous variation of the small strain stiffness, G_0, with depth (Fig. 3a) was assumed as dependent on the stress state and history. The stiffness parameters were again correlated to I_P, using the relationships proposed by Rampello *et al.* [13]. The discretization of the soil profiles was optimized in order to get the best reproduction of earthquake frequencies as high as 25 Hz [14]. The strain-dependent shear modulus and damping curves, representing the non-linear soil behaviour, were based on those suggested by Vucetic and Dobry [15] for clays with I_P = 25% (Fig. 3b).

(a)   (b)

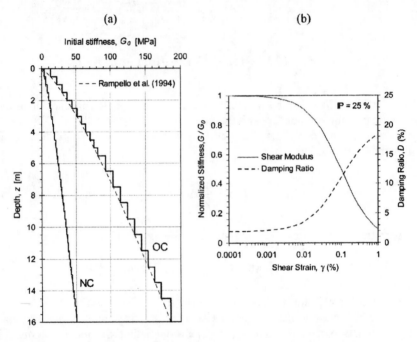

FIGURE 3. Small strain stiffness profile (a) and variation of normalised stiffness and damping with shear strain (b) assumed for the clay subsoils.

In these calibration analyses, the bedrock was assumed at a constant depth (H_b = 16 m), with a unit weight of 22 kN/m^3 and a shear wave velocity equal to 1000 m/s. The depth of the sliding surface, H_s, was taken equal to 5, 10 and 15 m. The bedrock input motion was the NS component of the Borgo Cerreto-Torre record (a_{max} = 0.185g, T_m = 0.154 s) of the 1997 Umbria-Marche earthquake (M_w = 5.7).

The seismic site response, in no-slip conditions, predicted using the numerical code implemented in this study (ACST) was compared to the results of linear equivalent analysis in the frequency domain obtained by EERA [16]. Figs. 4 and 5 show the time histories around the critical phase of the motion (a, b) and the peak values of acceleration (c) and shear strain (d) for NC and OC subsoils, respectively.

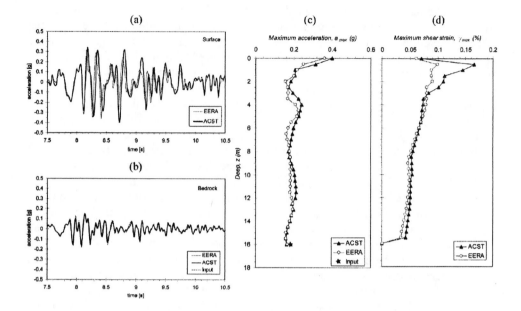

FIGURE 4. 1D seismic site response computed by EERA and ACST for NC clay subsoil: accelerograms at surface (a) and at bedrock (b); profiles of peak acceleration (c) and shear strain (d).

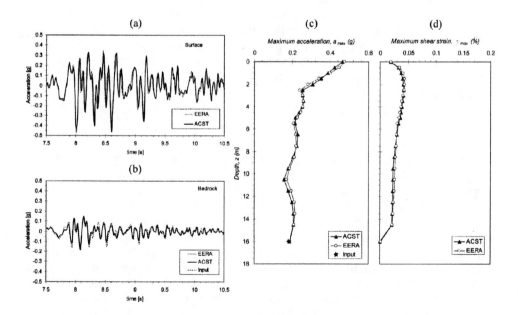

FIGURE 5. 1D seismic site response computed by EERA and ACST for OC clay subsoil: accelerograms at surface (a) and at bedrock (b); profiles of peak acceleration (c) and shear strain (d).

The agreement between the two solutions is satisfactory for both soil profiles. For the softer NC soil, both methods induce a base acceleration slightly lower than the reference motion amplitude, due to the high impedance contrast (Fig. 4b). In the uppermost layers, higher motion amplitudes are predicted by ACST, due to high frequency amplification and more pronounced non-linear effects, being the shear strain even higher than 0.1%. On the other hand, the behaviour of the OC clay predicted by the two codes is about the same, due to its stiffer and more linear behaviour ($\gamma < 0.05\%$).

The permanent sliding displacements obtained using the stick-slip model are compared in Fig. 6 with those resulting from the rigid block model and the decoupled approach, for shallow (a, d), intermediate (b, e) and deep (c, f) sliding surfaces.

The Newmark displacements predicted using the reference input motion (black dashed lines) are higher for the NC clay, because of the lower yield acceleration, but independent of the sliding depth; those calculated with the accelerogram computed at the depth of the sliding surface (blue lines) result systematically higher, due to the propagation up to $z=H_S$ and the related amplification, which is higher for the NC soil.

When the soil deformability above the sliding depth is accounted for, the slope displacements depend on the combination of surface amplification and asynchronous motion. The green lines represent the cumulated displacements computed by the equivalent accelerogram, i.e. that corresponding to the resultant inertia force on the unstable slope mass. The displacements of the OC clay are further increased by the amplification above the sliding surface, which increasingly affects the slope deformation as the ratio between the natural period of the unstable soil and the median period of the accelerogram (T_S/T_m) increases with the sliding depth (Fig. 6 d,e,f). In the case of soft NC clay, conversely, the asynchronous motion contributes to reduce the predicted displacements with respect to those calculated by the rigid block model (Fig. 6 a,b,c). By comparing the decoupled and coupled approach (red thick lines), it appears that the "stick-slip" model confirms to be conservative for the softer soil, while it results increasingly unconservative for the stiffer OC soil, as the sliding depth and T_S/T_m increase.

CONCLUSIONS

A lumped mass model accounting for non-linear soil behaviour was implemented and calibrated to compute the dynamic ground motion and the sliding displacements of a one-dimensional soil mass. The application can be addressed to seismic response analysis and instability of different geotechnical structures, such as natural and artificial soil slopes and solid-waste landfills, which can also include simplified structural elements. The formulation was inspired by previous literature models [8 - 10] and updated accounting for bedrock deformability and variability of the sliding depth; accurate methods of step-by-step integration with strain-dependent equivalent parameters were introduced in the numerical solution of the dynamic equilibrium.

The results in no-slip conditions were successfully compared to 1D site response analysis in the frequency domain; the displacements predicted for a sliding mass showed to be significantly different from those obtained by the direct application of the conventional rigid block model.

482

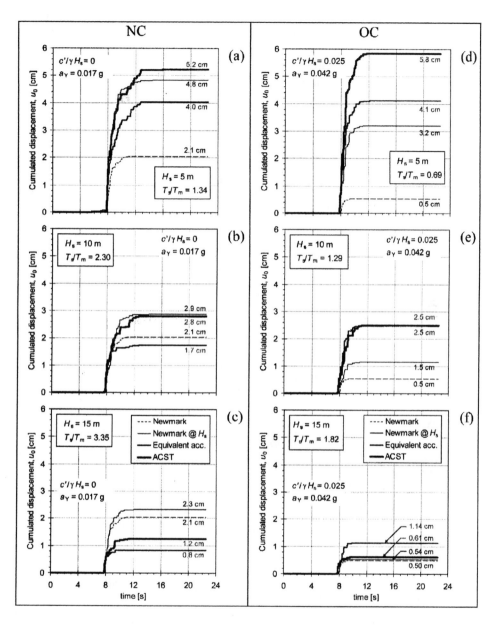

FIGURE 6. Cumulated displacements computed using the rigid block model with the input accelerogram at the bedrock (black line), and at the *s*-th layer (blue line), versus the prediction of the decoupled approach (green line) and stick-slip model (red line) at variable depths of the sliding surface.

The predictions of such "coupled model" were also compared to those obtained by uncoupling the seismic response analysis from that of the sliding block model: the preliminary results appear to confirm the conclusions by Rathje and Bray [10], i.e. that this method predicts conservative displacements for soft soils and results increasingly unconservative for stiff soils, as the sliding depth increases. Future developments with more extended selections of seismic input motions and subsoil profiles should be specifically addressed to this issue.

ACKNOWLEDGMENTS

This work is a part of a Research Project funded by ReLUIS (Italian University Network of Seismic Engineering Laboratories) Consortium.

REFERENCES

1. N.W. Newmark, "Effects of earthquakes on dams and embankments", in The V Rankine Lecture of the British Geotechnical Society, *Géotechnique*, 15 (2), 139-160 (1965).
2. F.I. Makdisi, H.B. Seed, "Simplified procedure for estimating dam and embankment earthquake-induced deformations" *Journal of Geotechnical Engineering*, ASCE, 104 (7), 849-867 (1978).
3. J.D. Bray, E.M. Rathje, A.J. Augello, S.M. Merry, "Simplified seismic design procedure for geosynthetic-lined, solid-waste landfills", *Geosynthetics International*, 5(1-2), 203-235 (1998).
4. E. Ausilio, F. Silvestri, A. Troncone, G. Tropeano, "Seismic displacement analysis of homogeneous slopes: a review of existing simplified methods with reference to italian seismicity", *Proc. IV International Conference on Earthquake Geotechnical Engineering*, Thessaloniki, 2007. Springer, NY. ID 1614.
5. J.S. Lin, R. V. Whitman, "Decoupling approximation to the evaluation of earthquake-induced plastic slip in earth dams", *Earthquake Engineering and Structural Dynamics*, 11, 667–678 (1983).
6. A. K. Chopra, L. Zhang, "Earthquake-induced base sliding of concrete gravity dams", *Journal of Structural Engineering*, ASCE, 117(12), 3698–3719 (1991).
7. G. Gazetas, N. Uddin, "Permanent deformation on preexisting sliding surfaces in dams", *Journal of Geotechnical Engineering*, ASCE, 120(11), 2041–2061 (1994).
8. S. L. Kramer, M. W. Smith, "Modified Newmark model for seismic displacements of compliant slopes", *J. Geotech. Engrg.*, ASCE, 123(7), 635–644 (1997).
9. E. M. Rathje, J. D. Bray, "An examination of simplified earthquake-induced displacement procedures for earth structures", *Canadian Geotechnical Journal*, Ottawa, 36, 72–87 (1999).
10. E. M. Rathje, J. D. Bray, "Nonlinear coupled seismic sliding analysis of earth structures", *Journal of Geotechnical and Geoenvironmental Engineering*, 126(11), 1002–1014 (2000).
11. Y.M.A. Hashash, D. Park, "Viscous damping formulation and high frequency motion propagation in nonlinear site response analysis", *Soil Dynamics and Earthquake Engineering*, 22(7), 611–624 (2002).
12. S. Y. Chang, "Improved explicit method for structural dynamics", *Journal of Engineering Mechanics*, 133 (7), 748-760 (2007).
13. S. Rampello, F. Silvestri, G. Viggiani, "The dependence of small strain stiffness on stress and history for fined grained soils: the example of Vallericca clay", *Proc. 1st Int. Symp. on Pre-Failure Deformation Characteristics of Geomaterials*, IS-Hokkaido, Sapporo, 1994. Balkema, Rotterdam, 1: 273-278.
14. R. L. Kuhlemeyer, J. Lysmer, "Finite Element Method accuracy for wave propagation problems", *Journal of Soil Mechanics & Foundations Division*, ASCE, 99(SM5), 421-427 (1973).
15. M. Vucetic, R. Dobry, "Effect of soil plasticity on cyclic response", *Journal of Geotechnical Engineering*, ASCE, 117(1), 89-107 (1991).
16. J.P. Bardet, K. Ichii, C.H. Lin, "EERA a computer program for equivalent-linear earthquake site response analyses of layered soil deposits", *University of Southern California, Dept. of Civil Engineering* (2000).

A *GLE* multi-block model for the evaluation of seismic displacements of slopes

V. Bandini[a], E. Cascone[a], G. Biondi[b]

[a]Università di Messina, Dipartimento di Ingegneria Civile, Contrada di Dio 98166 Messina, Italy.
[b]Università di Catania, Dip. di Ingegneria Civile e Ambientale, Viale A. Doria 6, 95125 Catania, Italy.

Abstract. The paper describes a multi-block displacement model for the evaluation of seismic permanent displacements of natural slopes with slip surface of general shape. A rigorous limit equilibrium method of stability analysis is considered and an application to an ideal clay slope is presented including the effect of excess pore pressure build-up on the displacement response.

Keywords: Multi-block displacement model, GLE method, clay slopes, excess pore pressure.

INTRODUCTION

In the analysis of seismic stability conditions and post-seismic serviceability of natural slopes and earth structures the sliding block approach is considered the better compromise between computational effort and result accuracy. The effectiveness in simulating inertial and/or weakening instabilities through a single sliding or rotating block model is documented in a number of studies [1, 2] and this approach is suggested in many codes and guidelines.

In many real cases, at failure, combined rotational and translational movements of the slope may occur along a slip surface formed by a combination of curved and planar segments. In this case rigid body movements are not admissible unless contact between soil mass and slip surface is lost or internal deformation of the rigid block is considered to ensure kinematic compatibility of displacements.

Different slope conditions, such as stratigraphic irregularities and concentration of strain in weak soils, may produce deformation of distinct soil blocks rather than a global slope failure. This may happen even in a homogeneous slope as a consequence of earthquake induced cyclic straining and related effects of shear strength reduction and stiffness degradation. In all these cases the use of the single-block model can not be reliably applied and more blocks moving along different straight-line segments, approximating the actual slip surface, must be considered. The multi-block discretization of the slope requires introducing further hypothesis on shear strength mobilization along the slip surface and the inter-block boundaries.

Recently Sarma & Chlimintzas [3] proposed a limit equilibrium multi-block model for the evaluation of seismic and post-seismic slope displacements occurring along slip surfaces of general shape. The procedure is based on an approximated method of slices with kinematically admissible failure mechanism. No excess pore water pressure is accounted for in the analysis which is carried out assuming that soil shear strength is

CP1020, *2008 Seismic Engineering Conference Commemorating the 1908 Messina and Reggio Calabria Earthquake*,
edited by A. Santini and N. Moraci

constant during the motion.

In this paper a multi-block displacement model based on a rigorous limit equilibrium method is derived for the evaluation of seismic permanent displacements of slopes. The proposed solution algorithm was checked against a commercial stability analysis code and application to cohesive slopes is presented including the effect of soil shear strength reduction due to seismic-induced excess pore pressures.

GLE-BASED MULTI-BLOCK DISPLACEMENT MODEL

In the proposed multi-block displacement model the rigorous General Limit Equilibrium [4] (*GLE*) method of stability analysis is adopted. In the *GLE* method both global horizontal force and moment equilibrium conditions as well as local horizontal and vertical force equilibrium conditions in any slice of the slope are satisfied, assuming that the inclination of the inter-slice forces is described by the product of a scalar unknown coefficient λ times an arbitrary function $f(x)$ that takes values in the interval 0-1 with distance x in the slope. For a given $f(x)$ the slope safety factor F corresponds to the value of λ for which the safety factors F_f and F_m, computed through the force and moment equilibrium condition respectively, coincide.

In the present work the *GLE* method was used to detect the slope critical acceleration coefficient k_c. Imposing the condition $F_f=F_m=1$, the corresponding values $k_{c,m}$ and $k_{c,f}$ may be computed through the force and moment equilibrium equation, respectively. Expressions of $k_{c,m}$ and $k_{c,f}$ may be found in Cascone & Bandini [5].

Following Sarma & Chlimintzas [3], the displacement analysis is based on the following assumptions: *i*) the shear strength along the slip surface segments and the inter-slice boundaries is fully mobilized during the motion; *ii*) the slip surface and internal shear surfaces do not change position and direction during sliding; *iii*) each block slides along the underlying segment of the slip surface without any change of the block mass; *iv*) adjacent blocks are rigidly connected to ensure that they do not separate or overlap between each other; *v*) displacements are small with respect to the slope length so that overlapping of the blocks with the slip surface is negligible and the mass of each block can be considered constant.

Under these assumptions, referring to the scheme of Figure 1 the displacements component h of two adjacent blocks normal to the inter-block boundary coincide. Then, the displacement u_j of the jth block can be related to the displacement u_{j+1} of the adjacent block through the geometric parameters β_j, δ_j, β_{j+1} and δ_{j+1}. For a multi-block system the recursive use of the relationship between the displacements of adjacent blocks allows describing the displacement u_i of the ith block as a function of the displacement u_r of an arbitrary reference block through a displacement conversion factor $q_{i,r}$:

$$u_i = u_r \cdot q_{i,r} \quad \text{where} \quad q_{i,r} = \begin{cases} \displaystyle\prod_{j=r}^{i-1} \frac{\cos\left(\beta_j + \delta_j\right)}{\cos\left(\beta_{j+1} + \delta_j\right)} & \text{for } i > r \\ 1 & \text{for } i = r \\ \displaystyle\prod_{j=i}^{r-1} \frac{\cos\left(\beta_{j+1} + \delta_j\right)}{\cos\left(\beta_j + \delta_j\right)} & \text{for } i < r \end{cases} \quad (1)$$

486

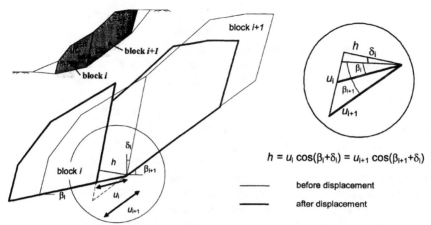

FIGURE 1. Relationship between displacements of adjacent blocks.

The equation of motion of the generic block of the slope can be written equating the resultant of the unbalanced forces parallel to the block base to the inertia force due to the acceleration $\ddot{u}_i$ of the block relative to firm soil. Using Eqn. (1) it is possible to combine the equations of motion of all blocks of the system and to express the equation of motion of the reference block in the form:

$$\ddot{u}_r = g \cdot [k(t) - k_c] \cdot S \tag{2}$$

where $k(t)$ is the ground acceleration expressed as a fraction of gravity acceleration g and S is a shape factor depending on the system geometry and soil shear strength parameters at the base of the blocks [5]. Except for the shape factor S, Eqn.(2) represents the equation of motion of a single block sliding on a horizontal plane, implying that under the imposed constraints the multi-block system reduces to an equivalent single block model.

EFFECT OF EXCESS PORE PRESSURE

Saturated clay soils subjected to uniform or irregular cyclic loading exhibit excess pore water pressures whose magnitude changes during each cycle depending on the amplitude γ of the cyclic shear-strain. If cyclic shear strains greater than the volumetric threshold γ_v are attained in the soil, excess pore pressures build up and do not dissipate after each strain cycle. Residual excess pore pressures induce a reduction of the soil shear strength and, consequently, of the critical acceleration coefficient, therefore affecting the seismic and post-seismic stability condition of the slope and its displacement response. To evaluate the occurrence of these phenomena an excess pore pressure generation model must be introduced in the analysis. The excess pore pressure build-up in the multi-block displacement analysis is evaluated through a simplified procedure that is synthetically outlined in the following.

For given slope, failure surface and input motion, the time-history $\gamma_i(t)$ of the shear strain induced at the base of each slice of the slope is computed solving iteratively the following equation:

$$\gamma_i(t) = \frac{\Delta\tau_i(t)/G_{0,i}}{G(\gamma)/G_0} \qquad (3)$$

where $\Delta\tau_i(t)$ and $G_{0,i}$ represent the pseudo-static increment of the shear stress with respect to the static case and the small strain shear modulus at the base of the ith slice, respectively, while $G(\gamma)/G_0$ represents the modulus reduction curve. In the analysis the modulus reduction curves proposed by Vucetic & Dobry [6] for different values of the soil plasticity index PI are adopted; G_0 is computed as a function of PI and mean effective pressure at the base of each slice using the relationship by Rampello et al. [7] and assuming that the soil is normally consolidated.

For each of the computed shear strain time-histories $\gamma_i(t)$ 'significant straining cycles' are selected detecting the cycles for which the mean cyclic amplitude $\gamma_{c,i}$ (obtained averaging the positive and negative peak values of the strain cycle) is greater than γ_v. The residual pore-water pressure ratio $\Delta u_{i,j}{}^*$ (i.e. the ratio of the excess pore pressure and the static effective stress $(\sigma'_{n,o})_i$ normal to the base of the slice) at the base of the ith slice after the jth significant straining cycle is then evaluated through Eqn.(4), obtained combining the maximum value of excess pore pressure induced by the input motion, evaluated according to Matsui et al. [8] with a weighting factor $\varepsilon_{i,j}$, the latter derived assuming that the pore pressure build-up can be distributed over the number $N_{i,v}$ of significant cycles using Miner's law of cumulative damage:

$$\Delta u_{i,j}^* = \beta \cdot \log\frac{(\gamma_{c,i})_{max}}{\gamma_v} \cdot \varepsilon_{i,j} \qquad \text{with} \qquad \varepsilon_{i,j} = \frac{\log(\gamma_{c,i}/\gamma_v)_j}{\sum\limits_{s=1}^{N_{i,v}}\log(\gamma_{c,i}/\gamma_v)_s} \qquad (4)$$

In Eqn.(4), the material constant β is equal to 0.45 and γ_v is computed as a function of plasticity index PI and over-consolidation ratio OCR [8]. Due to the generality of the proposed procedure any other pore pressure generation model available in the literature can be considered in the analysis.

Once the time-history of the excess pore pressure is known for each slice of the slope, the current value $k_c(t)$ of the slope critical acceleration coefficient can be computed applying the GLE method for each time step of the input accelerogram assuming the pore water pressure at the base of each slice as the sum of the static and the seismic-induced pore pressure. Thus, $k_c(t)$ reduces from the initial value $k_{c,o}$, computed before the excess pore water pressures take place in the slope, to the minimum value $k_{c,min}$ attained when the maximum excess pore-water pressure is achieved in all the slices. The values of $k_{c,o}$ and $k_{c,min}$ do not depend on the shape of the considered accelerogram, however, the shape of the accelerogram affects the reduction path of the critical acceleration coefficient.

VALIDATION OF THE STABILTY ANALYSIS ALGORITHM

The proposed GLE multi-block model was implemented in the code *Mathematica v.4*. To check the accuracy of the algorithm different ideal slope schemes were solved and results were compared with those provided by the code *SLOPE/W*. The comparison of results was performed in terms of critical acceleration coefficient $k_{c,o}$ and corresponding values of λ and in terms of distributions of forces and stresses along the slip surface and at the inter-slice boundaries.

Figure 2 shows a 90 m long and 20 m high slope together with a potential failure surface consisting of three segments with different inclinations. The soil strength parameters assumed in the analysis together with the shape of the piezometric surface are shown in the figure. The results of the analysis performed discretizing the slope into 90 slices and neglecting excess pore pressures are reported in Table 1 in terms of $k_{c,o}$ and λ and in Figure 3 in terms of distributions of inter-slice forces and stresses acting along the failure surface. It is apparent that the results obtained with the two codes are in excellent agreement: differences lower than 4% and 2.5% on the values of $k_{c,o}$ and λ, respectively, were observed; similarly, a good agreement in the values and in the distributions of normal (E) and shear (X) forces at each inter-slice boundary and total normal (σ) and shear (τ) stresses at the base of each slice was obtained.

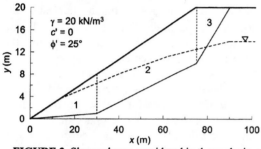

FIGURE 2. Slope scheme considered in the analysis.

TABLE 1. Comparison between the results obtained with the implemented algorithm and *SLOPE/W*.

Code	k_c	λ
Mathematica v.4	0.121	0.549
SLOPE/W	0.123	0.555

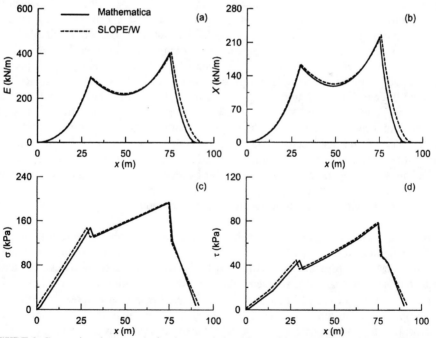

FIGURE 3. Comparison between the implemented algorithm and *SLOPE/W* in terms of: a) normal and b) shear forces at inter-slice boundaries; c) total normal and b) shear stresses at the base of slices.

DISCUSSION OF RESULTS

The results of multi-block displacement analyses performed accounting for the occurrence of excess pore pressure build-up are presented herein including, for comparison, also results of displacement analyses carried out neglecting the soil shear strength reduction. Two acceleration records were used as input motions: the Hollywood Storage record of the 1994 Northridge (M_w=6.7) earthquake (accelerogram #1) and the El Centro#6 record of the 1979 Imperial Valley (M_w=6.5) earthquake (accelerogram #2). The two accelerograms have about the same peak value $k_{max} \cong 0.36$ but different strong motion duration.

Figure 4 shows the results obtained applying accelerogram #1 to the slope in Figure 2 ($k_{c,o}$=0.121) for which PI=15% (γ_v=0.04%) was assumed. In particular, for three slices of the slope (slices A, B and C) the time histories of the mean shear strain γ_c exceeding the volumetric threshold shear strain γ_v (Fig. 4a) and the corresponding time-histories of the pore pressure ratio Δu^* at the base of the slices (Fig. 4b) are plotted. In Figure 4c the time histories of the critical acceleration coefficient together with its initial value $k_{c,o}$ are superimposed to the input accelerogram. The computed displacement responses of the three blocks are shown in Figure 4d. Dashed lines represent the results of a multi-block displacement analysis performed neglecting excess pore pressures and therefore assuming a constant value of the critical acceleration coefficient. Solid lines represent the results of a displacement analysis carried out including the reduction of the slope critical acceleration coefficient due to pore pressure build-up. If no excess pore pressure is taken into account the computed displacement responses show the stepwise trend typical of inertial instabilities. In this case for block n.2, assumed as reference block in the analysis ($q_{2,2}$=1), a permanent displacement of 13 cm is predicted; for blocks n.1 and 3 the permanent displacement reaches 13 cm and 15.5 cm, respectively, being $q_{1,2}$=0.98 and $q_{3,2}$=1.17 the displacement conversion factors computed using Eqn.(1). If the excess pore pressure is accounted for in the analysis, larger permanent displacements of the three blocks are predicted: blocks n.1 and 2 undergo permanent displacements of about 150 cm while the displacement computed for block n.3 is about 175 cm. Also in this case the displacement time-histories follow a stepwise pattern; however, during the strong motion phase, larger displacement increments occur at each pulse of the accelerogram due to the significant contemporary reduction of the slope critical acceleration coefficient k_c. Comparing the values of block displacements obtained in the two analyses, it can be observed that displacements computed accounting for pore pressure build-up are about 11 times larger than those computed neglecting excess pore pressure.

For the same slope, Figure 5 shows the results obtained using accelerogram #2 (Fig. 5c). The response of the slope is described in terms of *significant straining cycles* (Fig. 5a) and excess pore pressure ratio (Fig. 5b) evaluated at the base of slices A, B and C and in terms of permanent displacements computed neglecting (dashed lines in Fig. 5d) or accounting for (solid lines in Fig. 5d) the excess pore pressures. In both cases the computed displacements are significantly smaller than those obtained using accelerogram #1 and the effect of seismic induced pore pressure build-up in the slope response is almost negligible.

490

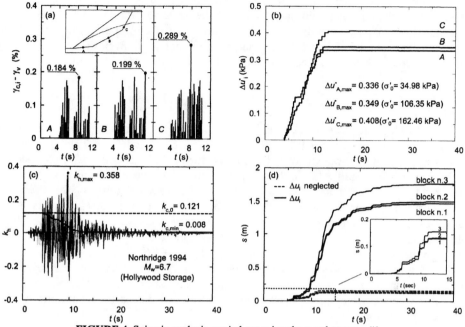

FIGURE 4. Seismic analysis carried out using the accelerogram #1.

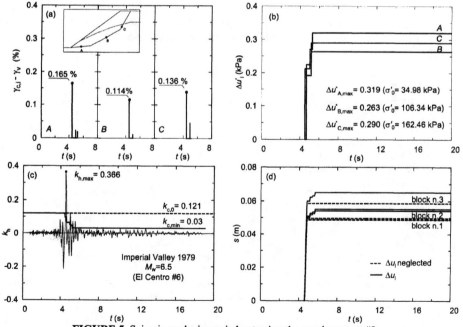

FIGURE 5. Seismic analysis carried out using the accelerogram #2.

491

It should be observed that due to the simplified pore pressure generation model assumed in the analysis, similar values of the excess pore pressure ratios were computed for the two records (Figs. 4b and 5b) even if the input acceleration time-histories (Figs. 4c and 5c) and the computed significant strain cycles (Figs. 4a and 5a) are considerably different in terms of amplitude distribution and duration.

CONCLUDING REMARKS

A multi-block displacement model based on a rigorous limit equilibrium method of stability analysis is proposed for the evaluation of earthquake-induced permanent displacements of natural slopes with sliding surface of general shape. Numerical validation of the solution algorithm is presented. Using an empirical pore pressure generation model for clay soils, a simplified procedure accounting for the effects of soil shear strength on the slope displacement response is presented. The results of the analyses showed that pore pressure build-up may significantly affect the slope displacements depending on the strong motion duration of the input accelerogram. Though the pore pressure generation model implemented in the proposed procedure is rather simplified, it allows capturing general patterns of slope behavior. Moreover, the procedure for the evaluation of seismic induced shear strains, though simplified, is a useful tool and can embody more sophisticated models for earthquake-induced pore pressure prediction.

ACKNOWLEDGMENTS

The Authors wish to thank the ReLUIS (University Network of Seismic Engineering Laboratories) Consortium which funded the Research Project "Linea 6.3 Stabilità dei pendii".

REFERENCES

1 Sarma S.K. (1981). *Seismic displacement analysis of earth dams.* Journal of Geotechnical Engineering, ASCE, Vol. 107, No. 12, pp. 1735-1739.
2 Biondi G., Cascone E., Maugeri M. (2002). *Flow and deformation failure of sandy slopes.* Soil Dynamics and Earthquake Engineering, 22, pp. 1103-1114.
3 Sarma S.K., Chlimintzas G.O. (2001). *Co-seismic and post seismic displacement of slopes.* XV ICSMEG TC4, Satellite Conference on "Lesson learned from recent strong earthquakes", Istanbul, Turkey, pp.183-188.
4 Fredlund D.G., Krahn J. (1977). *Comparison of slope stability methods of analysis.* Canadian Geotechnical Journal, Vol. 14, pp. 429-439.
5 Cascone E., Bandini V. (2007). *Analisi agli spostamenti per pendii con superfici di scorrimento mistilinee.* Programma Quadro Dipartimento di Protezione Civile – Reluis, Progetto di Ricerca n.6. Linea di Ricerca 6.3. Rapporto di Ricerca 2° Anno.
6 Vucetic M., Dobry R. (1991). *Effects of the soil plasticity on cyclic response.* Journal of Geotechnical Engineering, ASCE, Vol. 117, No.1, pp.89-107.
7 Rampello S., Silvestri F., Viggiani G. (1994). *The dependence of G_0 on stress state and history in cohesive soils.* Int. Symp. on Pre-failure deformation characteristics of geomaterials, Sapporo, pp.1155-1160.
8 Matsui T., Ohara H., Ito T. (1980). *Cyclic stress-strain history and shear characteristics of clay.* Journal of Geotechnical Engineering, ASCE, Vol. 103, No. 7, pp.757-768.

Hydromechanics behavior of dam with core by taking into account the effect of contact

A. Bekkouche[a], Z. Benadla[a], Y. Houmadi[b], M.Ghefir[a]

(a) Department of Civil Engineering FSI University Aboubekr Belkaid Tlemcen Algeria
BP 230 (13000) Tlemcen ALGERIA
(b) Department of Civil Engineering ISNV University Center of Saïda Algeria

Abstract. Forces acting on the thin cores of earth dams could be reduced by the effect of contact with the refills. Thus the effective stress could be reduced and in turn will induce cracks at the base of the dams. This phenomenon is called hydraulic fracturing.

The modeling of this phenomenon, using ANSYS program, by taking into account the effect of contact will make possible the prediction of global behavior of the dam and in the meantime will allow the assessment of the thickness of the core under which the effect of contact will have an influence.

A parametric study has been performed to understand the relationship between the effect of contact and the variation of the effective stress.

Keywords: hydraulic fracturing, thermal, dam, core, Ansys.

INTRODUCTION

In geotechnical, applications involving interactions soil-structures are many (soil-foundation, soil-retaining wall, soil-pile, soil-core...). In recent years, considerable efforts have been made to improve understanding of these phenomena, because of their importance and complexity, which has resulted in a high number of contributions on the subject .

In the small ones and average dams earth, the contact between the clay core and the refills represent the principal cause of reduction in the effective constraints, in this case we speak about hydraulic fracturing. Indeed, several ruptures of dams have been attributed to this phenomenon. Hence it is of great importance to design the earth dams, in particular those which have a core, to avoid the hydraulic fracturing.

The objective of this study firstly, is to define the dimensions of cores affected and the rate of reduction in these constraints. For that, two approaches are to be analyzed. The first consists in studying the model by supposing a discontinuity with the interface core-filter by introducing elements of contact. The second consists to remove the discontinuity which can exist with the interface core-filter and to suppose it non-existent. Indeed, the two interfaces are supposed are assumed pasted.

CP1020, 2008 Seismic Engineering Conference Commemorating the 1908 Messina and Reggio Calabria Earthquake,
edited by A. Santini and N. Moraci
© 2008 American Institute of Physics 978-0-7354-0542-4/08/$23.00

Then, an analogy thermo-hydraulics is considered, which makes it possible to determine the hydraulic loads, the leak-flows as well as the water pressures pore. In this case, the flow in the dike of dam is governed by the law of Darcy.

MODELING

This study concerns the earth dam El Izdihar, located in Tlemcen city, in north-west of Algeria. The capacity of the dam is 110 HM^3 and its depth is of 60m (Ras ,2003). The length of the dam is much more important than its width; the problem is studied in distortions previsions. The figure.1 presents the structure of the dam, consisting of four materials. The contact is simulated at the interfaces core-refills. The dam is assumed stable without the possibility of sliding.

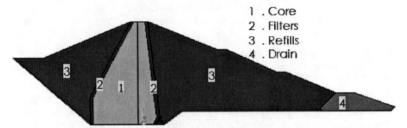

FIGURE 1. Modeling of the contact core-refills.

An automatic mesh was chosen for the general structure with elements quadratic iso-parametric of 4 nodes. The latter presents a good compromise between the time and the quality of results (Figure 3.). The interfaces core-filters are modeled by elements of two-dimensional contact that simulate the contact between two areas in 2D. CONTA172 is defined by three nodes (Figure.2). The behaviour law used to model this dam is of elastoplastic type by using the Drucker Prager model. (Chen et al. 1988).

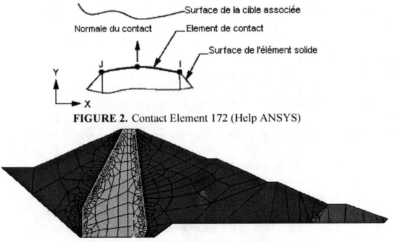

FIGURE 2. Contact Element 172 (Help ANSYS)

FIGURE 3. Adopted grid (564 elements).

MECHANICAL ANALYZES OF EL IZDIHAR DAM

The two deformed corresponding to two approaches are represented on figures (4.a and 4.b). The dam is subject only to its own weight and the refills. The core is deformed involving with him the adjacent embankments according to the two instability modes. The first case represents the cases where no element of contact is considered. In this case the settlement is valued at 1,79 m on the ridge. This method to mark by the absence of sliding refills, a certain continuity of settlement is noticed on the whole of the structure (core-filters-refills).

In the second method, where elements of contact are considered, the soil has deformed to 2.22m. This is mainly with the friction and the rigidity of contact considered with weakest of rigidities constituting the interface at knowing that of clay. Remain that the slip appears always at the top.

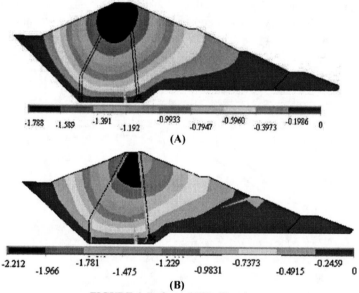

FIGURE 4. Deformed El Izdihar dam.
(A). without the contact effect.
(B). with the contact effect.

Vertical displacements in the points registers in figure 5, are given in the table 1, by taking account of the contact effect and without contact.

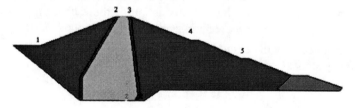

FIGURE 5. The points register in EL IZDIHAR dam.

TABLE 1. Vertical displacements in the points registers in the figure 5

Points in the dam	Vertical displacement without the contact effect (figure 4a)	Vertical displacement without the contact effect (figure 4b)
1	-0.1764	-0.2157
2	-1.4732	-1.7179 (refills) -2.3708 (core)
3	-1.4749	-1.0318 (refills) -2.3708 (core)
4	-0.4870	-0.4852
5	-0.1798	-0.2361

A Parametric study was carried out on the width, where it has been reduced to (2/3), a half and then a (1/3). Then it was expanded to twice, three and half to three times the width of the base, in keeping the same dimensions of dam.

After obtaining results of vertical constraints in the different approaches and with the calculation of the vertical analytical constraint, we can do assessment of the average rate of differences between constraints calculated by the two approaches, with and without contact. It has plotted a curve representative of the rate of reduction in the constraints according to the width ratio/height of the dam in a point of the base, without and by considering of contact effect (Figure 5, 6 and Table 1).

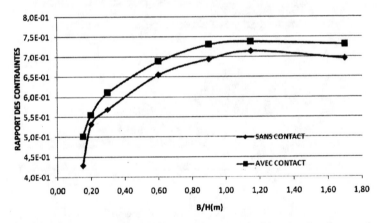

FIGURE 6. The report of the constraints (without contact and with contact) and constraints analytical according to the relationship between the base B core and its height H

It appears clear in this representation which the width of the core is the major element in the process of weakening of the constraints under the dam apart from the phenomenon of contact itself. Since the reduction of these constraints passes from a value of $0.5\sigma_{vAnalitique}$ for a thin core and stabilizes itself around $0.7\sigma_{vAnalitique}$ for large cores. It concluded that the contact does not influence the large cores. The analysis without taking account of the effect of contact remains in the safety interval, because the corresponding constraints remain always lower than the case of contact.

THERMAL ANALYSES

In this part, an analogy thermo-hydraulics is presented. This step has enables to us to pass from a thermal problem to a hydraulic problem by analogy. Indeed, the software used cannot treat hydraulic cases.

The infiltration in the porous environments is generated by the law of Darcy (Balazova et al, 2002).

$$v = \frac{Qs}{A} = -Ks\frac{\partial h}{\partial L} = Ks\,J \tag{1}$$

$$Qs = AKs\frac{\Delta h}{L} \tag{2}$$

Qs : Seepage   L : Percolation path length
A : Sectional Area   v : Average velocity
h : Piezometric head   J : osmotic gradient
Ks : Permeability coefficient

In thermals, the conduction is generated by the equation of Fourier.

$$Q = \frac{Qv}{A} = -i_y\frac{dT}{\partial x} \tag{3}$$

$$Qv = AK_v\frac{dT}{\partial x} \tag{4}$$

Q: Heat (flow): flux   Q: Thermal condition heat flux intensity
A : Sectional Area   Kv : Heat transfer coefficient

$\frac{dT}{\partial x}$: Temperature gradient

By analogy thermo-hydraulics, the equations of Darcy and Fourier are Similar where one can make the following analogy:

TABLE 2. thermo-hydraulics Analogy (Balazova et al, 2002)

behavior Law	Transferred quantity	Potential	Constant of proportionality
Fourier	Heat by Conduction	Temperature	Thermal conductivity
Darcy	Fluid in a porous environment	Hydraulic load	Hydraulic conductivity

Two types of laws of physics are used to solve a problem hydraulic and thermal:

1) A law of behavior (Darcy or Fourier)
2) A law of conservation.

Expressed mathematically and combined, these laws allow for drift equations to partial derivatives which will represent the physical problem to solve (He Xiaoming, 2006; Henryk Z. 1993):

Differential equation of the flow without presence of source and the storage

$$\frac{\partial}{\partial x}\left(K_{xx}\frac{\partial h}{\partial x}\right)+\frac{\partial}{\partial y}\left(K_{yy}\frac{\partial h}{\partial y}\right)+\frac{\partial}{\partial z}\left(K_{zz}\frac{\partial h}{\partial z}\right)=0 \tag{5}$$

The presence of storage term, the equation of the flow becomes

$$\frac{\partial}{\partial x}\left(K_{xx}\frac{\partial h}{\partial x}\right)+\frac{\partial}{\partial y}\left(K_{yy}\frac{\partial h}{\partial y}\right)+\frac{\partial}{\partial z}\left(K_{zz}\frac{\partial h}{\partial z}\right)=S_s\frac{\partial h}{\partial t} \tag{6}$$

In the thermal case, the differential equation of thermal conduction

$$\frac{\partial}{\partial x}\left(K_{xx}\frac{\partial t}{\partial x}\right)+\frac{\partial}{\partial y}\left(K_{yy}\frac{\partial t}{\partial y}\right)+\frac{\partial}{\partial z}\left(K_{zz}\frac{\partial t}{\partial z}\right)=C\frac{\partial t}{\partial t} \tag{7}$$

In the same equation and with the absence of storage

$$\frac{\partial}{\partial x}\left(K_{xx}\frac{\partial t}{\partial x}\right)+\frac{\partial}{\partial y}\left(K_{yy}\frac{\partial t}{\partial y}\right)+\frac{\partial}{\partial z}\left(K_{zz}\frac{\partial t}{\partial z}\right)=0 \tag{8}$$

Initial Conditions and boundary conditions

The principal boundary conditions of the flows are four (figure 7):

1. Equipotential surface (AB), on which the hydraulic load is constant;
2. The impermeable surface (AF), through which the flow is null;

$$\frac{\partial h}{\partial n}=0 \tag{9}$$

3. Free surface flow (BE), which to check two conditions simultaneously: it is tangent with the flow velocity vector and the pore water pressure is equal to the atmospheric pressure.

$$\frac{\partial h}{\partial n}=0 \text{ et } u=0 \text{ ou } h=z \tag{10}$$

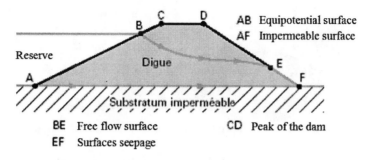

AB   Equipotential surface
AF   Impermeable surface

BE   Free flow surface   CD   Peak of the dam
EF   Surfaces seepage

FIGURE 7. Boundary conditions in dam (Magnan, 2004)

4.   The seepage surface (EF), on which the water pressure is null but the vector velocity of flow is directed towards the outside of the solid mass.

$$\frac{\partial h}{\partial n} > 0 \text{ et } u = 0 \text{ ou } h = z \tag{11}$$

The same geometric modeling is made in mechanical analysis with an automatically mesh. The element type chosen is "solid plane55". Figures 8 and 9 represent the variation of the hydraulic load in the dam body, the thermal flux and the hydraulic gradient.

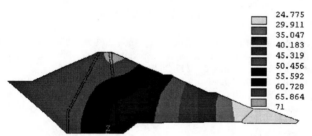

| 24.775 |
| 29.911 |
| 35.047 |
| 40.183 |
| 45.319 |
| 50.456 |
| 55.592 |
| 60.728 |
| 65.864 |
| 71 |

FIGURE 8. Variation of the hydraulic radial force in the dam body

| -.756E-04 |
| -.627E-04 |
| -.499E-04 |
| -.371E-04 |
| -.242E-04 |
| -.114E-04 |
| .146E-05 |
| .143E-04 |
| .271E-04 |
| .400E-04 |

FIGURE 9. Variation of the hydraulic flux in the dam body (Y)

FIGURE 10. Variation of the hydraulic gradient in the dam body (Y)

For the maximum value of the hydraulic gradient is 0.755 (figure.10); and the computed value of the critical gradient is of 0.85.

CONCLUSION

In conclusion after mechanical analysis, the contact has an influence on the thin cores, and does not on the broad kernels. The analysis without taking account the contact effect remains within the interval of security because the corresponding constraints remain below the contact case. For analogy thermo-hydraulic, thermal analysis has given the variation of the hydraulic load, hydraulic gradient. This last will be useful for a thermo-mechanical coupling and to determine the effective constraints in the dam.

REFERENCES

1. Balazova A., Barokova D, Mikula K, Pfender D., Anda. Soltesz,2002 "Numerical modelling of the groundwater flow in the left floodplain area danube river", Proceedings of Algoritmy 2002 Conference on Scientic Computing, pp. 237-244
2. Chen W.F. and Han D.J.,1988,"plasticity for structural engineers" Springer-Verlag, New Work.
3. He Xiaoming Li Zhi Wang Shuxian recorded , 2006 "Research of the seep-field problem by using the ANSYS' thermal analysis module" (Xi'an University of Technology School of hydropower Xi'an 710048)
4. Help ANSYS Version 8 (UNIX)
5. Henryk zaradny 1993 "Groundwater flow in saturated and unsaturated soil" Editet by R.B. Zeidler, A.A. Balkema/Roterdam/Brookfield/ 1993.
6. Magnan J-P 2004, "L'eau dans le sol" Technique de l'ingénieur, traité de construction, Ref C212
7. Ras A.,2003 "Contribution à l'étude de l'effet de contact dans les problèmes de géotechnique par le logiciel ANSYS", Master Thesis, Civil Engineering department, Tlemcen University, 2003.

Evaluation of Pseudo-Static Coefficients According to Performance-Based Criteria

Giovanni Biondi[a], Ernesto Cascone[b], Michele Maugeri[a]

[a]*Università di Catania, Dip. di Ingegneria Civile e Ambientale, Viale A. Doria 6, 95125 Catania, Italy.*
[b]*Università di Messina, Dipartimento di Ingegneria Civile, Contrada di Dio 98166 Messina, Italy.*

Abstract. A rational procedure is presented for the selection of the equivalent seismic coefficient to be introduced in the pseudo-static analysis of geotechnical systems which, at failure, behave as a 1-degree of freedom system. It is shown that although pseudo-static and displacement analyses may be regarded as alternative methods of analysis, the seismic coefficient may be related to earthquake-induced permanent displacements and, then, to the expected level of damage. Following the proposed procedure a pseudo-static analysis in accordance with performance based design can be carried out.

Keywords: Equivalent pseudo-static coefficient, performance-based design.

INTRODUCTION

In current practice seismic stability analyses of retaining walls and earth slopes are performed through the limit equilibrium or limit analysis methods and the earthquake effects are accounted for according to the pseudo-static methods introducing an equivalent static force that should reproduce the overall earthquake effects.

In the case of rigid retaining walls, equivalent static forces acting on the wall and on the retained soil are computed multiplying the weight of the wall and the weight of the soil wedge involved in the potential failure mechanism by an equivalent seismic coefficient k_{eq}. Usually, both horizontal ($k_{h,eq}$) and vertical ($k_{v,eq}$) components of k_{eq} are introduced. Once the seismic problem is converted into an equivalent static one, classical earth pressure theories can be utilized to estimate the active earth pressure behind the wall. The seismic stability conditions of the wall are quantified through a pseudo-static safety factor computed using a force-balance approach with reference to the possible failure mode.

Also in the seismic stability analysis of slopes the equivalent static force is computed multiplying the weight of the sliding soil mass by the equivalent seismic coefficient k_{eq}. Again both horizontal $k_{h,eq}$ and vertical $k_{v,eq}$ components of k_{eq} are considered acting on the soil mass potentially involved in the failure mechanism. Once the equivalent static problem is defined, for a given potential failure surface, the seismic stability condition of the slope is measured by a pseudo-static factor of safety computed as the ratio of the soil resistance to the earthquake-imposed demand. For both walls and slopes, the result of the pseudo-static analysis critically depends on the adopted value of k_{eq}. In the paper a rational procedure is presented for the selection of

CP1020, *2008 Seismic Engineering Conference Commemorating the 1908 Messina and Reggio Calabria Earthquake*,
edited by A. Santini and N. Moraci

the equivalent seismic coefficient to be adopted for geotechnical system that at failure can be assimilated to a 1-degree of freedom system. Firstly, the proposed procedure is outlined with reference to an ideal geotechnical system consisting in a rigid block sliding on a horizontal plane. Then, application to an infinite slope and a gravity retaining wall is described using recently published solutions [1, 2].

FACTORS AFFECTING THE SELECTION OF k_{eq}

In principle, the equivalent seismic coefficient should represent the amplitude of the earthquake-induced inertia forces. Since these forces are time- and space-dependent the selection of $k_{h,eq}$ and $k_{v,eq}$ should require an appropriate weighting procedure of acceleration time histories computed in a seismic response analysis and, thus, should depend on the main factors typically affecting the dynamic response of geotechnical systems: input motion and bedrock depth, system geometry, soil profile, stiffness and damping, site hydraulic conditions. Moreover, different values of k_{eq} may be obtained depending on the analysis conditions, assumptions for soil behaviour, and criteria adopted for evaluating the equivalent accelerogram and selecting an appropriate value for k_{eq}. This kind of procedure, however, is complex and is seldom applied and the selection of k_{eq} is, to a certain extent, arbitrary and lacks a clear rationale.

In current practice, values of k_{eq} are given in seismic codes and, usually, depend on the horizontal peak ground acceleration coefficient $k_{h,max}$ at a given site.

Generally, the values of $k_{h,eq}$ to be adopted for the seismic stability analysis of slopes is assumed to be a portion η_h of $k_{h,max}$ while $k_{v,eq}$ is assumed to be a portion Ω of $k_{h,eq}$. Similar expressions are adopted in the pseudo-static analysis of rigid retaining walls; in this case, according to Eurocode 8 (EC8) [3], the selection of η_h is also related to the permanent displacements d_{lim} that the soil-wall system can undergo. A value of $\eta_h = 0.5$ is often considered as appropriate for most slope and wall schemes and is assumed as a reference in many seismic codes; values of η_h ranging from 1 to 0.5 are suggested in EC8 depending on the values of d_{lim}. Few indications are available in the literature concerning $k_{v,eq}$ and values of Ω in the range ± 0.33 to ± 0.5 are usually suggested. In this way, the only parameter describing the earthquake effect is the peak value of the horizontal acceleration coefficient $k_{h,max}$, the coefficient $k_{h,eq}$ being defined independently from the characteristics of the considered geotechnical system and from the effects induced by the ground shaking.

The tendency to adopt equivalent seismic coefficients related to tolerable displacement levels has emerged, as an attempt to link the traditional pseudo-static approach to a performance-based analysis of both slopes and walls.

According to performance-based design criteria, Biondi et al. [1, 2] showed that a proper selection of the equivalent seismic coefficient $k_{h,eq}$ to be adopted in a pseudo-static analysis of slopes and rigid retaining walls should take into account the earthquake characteristics (frequency and amplitude content and strong motion duration), threshold or limit values of permanent displacements d_{lim} which may be suffered without reaching an ultimate or a serviceability limit state, the horizontal component of the critical acceleration coefficient $k_{h,c}$, and finally the ratio $k_{h,c}/k_{h,max}$ which represents a start out parameter for evaluating the occurrence of permanent

502

displacement in a sliding block analysis. It was also shown that although pseudo-static and displacement analyses may be regarded as alternative methods of analysis, the seismic coefficient to be used in a pseudo-static analysis may be related to earthquake-induced displacements and, then, to the expected level of damage.

In the following the proposed procedure was proposed in a generalized form which can be applied to any geotechnical system characterized by a 1-degree of freedom failure mechanism under seismic loading condition.

PROCEDURE OUTLINE

To outline the proposed procedure a rigid block sliding on a horizontal plane under both horizontal and vertical component of seismic acceleration is assumed as a reference scheme (Figure 1). Four steps are involved in the procedure: 1)Estimate of the earthquake-induced permanent displacement; 2) Definition of a safety factor in terms of acceleration level; 3) Definition of an equivalent seismic coefficient.

1) In Figure 1, the equation of motion of the reference scheme together with the expressions giving the pseudo-static safety factor F_{psd} and the horizontal component of the critical acceleration coefficient $k_{h,c}$ are shown. In eq.(1) S_f is a shape factor and g is the gravity acceleration. If k_v points upwards the sign (-) in eqs. (2) and (3) and the sign (+) in eq.(4) must be used. Several geotechnical systems, such as slopes, dams and embankment, landfill cover systems, gravity retaining walls, earth-reinforced walls, quay walls and shallow foundations, under earthquake loading condition can be studied using the sliding block analogy. In many of these cases the equation of motion is similar to those of the rigid block sliding on a horizontal plane (eq. 1) differences being embodied only in the shape factor S_f. Thus, the proposed procedure, outlined herein with reference to the simple scheme of Figure 1, can be easily extended to the analysis of several geotechnical systems.

For a given earthquake record the maximum value d_{max} of the cumulated permanent displacement of the block can be obtained through a one-way Newmark-type analysis. Many empirical relationships relating d_{max} to some seismic parameters are available deriving from best-fit regression analyses of permanent displacements computed using given sets of accelerograms. The confidence level assumed in the best-fit analysis, the characteristics of the considered accelerogram database and, finally, the seismic parameters selected for representing the seismic shaking are the main factors affecting the reliability of the prediction of d_{max}.

2) Once d_{max} is estimated it can be compare with a threshold or limit value d_{lim} of permanent displacement and a safety factor $F_d = d_{lim} / d_{max}$ may be introduced. Values of $F_d \geq 1$ mean that earthquake-induced displacements are smaller than the limit value and the considered limit state is not achieved. However, despite this definition of safety factor seems to be appropriate for a displacement-based analysis, two limit cases in which it fails can be distinguished: i) if no displacement of the system is allowed ($d_{lim} = 0$), whatever the induced displacement d_{max}, F_d is zero; ii) if the design earthquake does not induce any permanent displacement ($d_{max} = 0$), F_d diverges. An alternative measure of the system performance must then be detected to check the results of a displacement analysis.

Instead of computing d_{max} for a given $k_{h,max}$, any of the available displacement regression models can be adopted to evaluate a limit value $k_{h,lim}$ of the horizontal seismic coefficient associated to a given limit value of permanent displacement d_{lim}. Consistently with the meaning of the regression model, $k_{h,lim}$ represents the peak horizontal seismic acceleration coefficient required to induce a permanent displacement equal to the limit value d_{lim}. Thus, values of $k_{h,max}$ lower than $k_{h,lim}$ yield earthquake-induced displacements smaller than d_{lim}, the assumed limit state not being achieved. If the system cannot sustain any permanent displacement ($d_{lim} = 0$), the tolerable acceleration coefficient $k_{h,lim}$ equals $k_{h,c}$. Then, for values of the limit displacement d_{lim} greater than zero, $k_{h,lim}$ can be thought of as a generalised value of the critical acceleration coefficient for the assumed limit state being satisfied.

A factor of safety F_k actually representing a measure of the system performance can be defined as the ratio between the tolerable acceleration coefficient $k_{h,lim}$ and the maximum acceleration coefficient $k_{h,max}$ to which the system is subjected: $F_k = k_{h,lim} / k_{h,max}$. Since $k_{h,lim}$ and $k_{h,max}$ are related to the limit and to the earthquake-induced displacements, d_{lim} and d_{max}, respectively, F_k evaluates safety against a limit state using an acceleration ratio rather than a displacement ratio. Values of $F_k > 1$ indicate that the limit state related to d_{lim} is not achieved since if $k_{h,lim} > k_{h,max}$ the condition $d_{lim} > d_{max}$ occurs.

Differently from F_d, the safety factor F_k always provides a finite measure of the system performance: i) if no permanent displacement is allowed ($d_{lim} = 0$), it is $k_{h,lim} = k_{h,c}$ and F_k reduces to $k_{h,c}/k_{h,max}$, yielding a finite measure of the displacement response; ii) if the selected earthquake does not induce any permanent displacement ($k_{h,c} > k_{h,max}$ and then $d_{max} = 0$), F_k is always greater than the ratio $k_{h,lim}/k_{h,c}$ (because $k_{h,c} > k_{h,max}$) thus providing a finite measure of safety against a limit state.

3) To achieve a matching between the results of the pseudo-static and the displacement-based analysis an equivalence between the two approaches must be introduced. A rational criterion to define this equivalence consists in the evaluation of the equivalent seismic coefficient $k_{h,eq}$ for which the two approaches provide the same factor of safety ($F_k=F_{psd}$). In this way, although no displacement analysis is carried out, even using the pseudo-static approach a measure of the safety condition congruent with a more reliable performance-based analysis can be obtained.

Using eq.(3) the expression for $k_{h,eq}$ can be derived imposing the condition $F_{psd} = F_k$ and solving for k_h. The obtained expression of $k_{h,eq}$ is shown in Figure 1 as eq.(5).

Reference scheme:	Equation of motion and shape factor:	Critical acceleration coefficient:
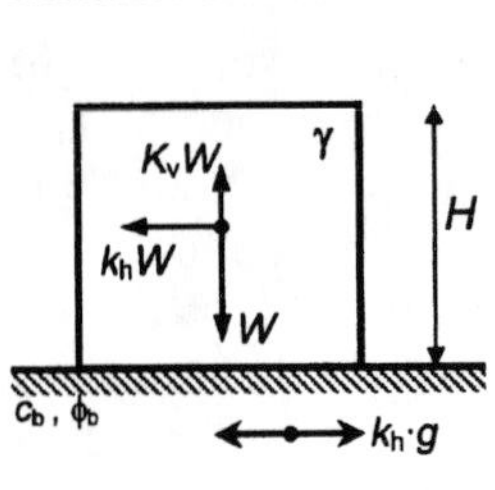	$\ddot{d}_o = (k_h - k_{h,c}) \cdot S_f$ (1) $S_f = g \cdot (1 \pm \Omega \cdot \tan\phi_b)$ (2)	$k_{h,c} = \dfrac{\dfrac{c_b}{\gamma \cdot H} + \tan\phi_b}{1 \pm \Omega \cdot \tan\phi_b}$ (4)
	Pseudo-static safety factor:	**Equivalent seismic coefficient:**
	$F_{psd} = \dfrac{\dfrac{c_b}{\gamma \cdot H} + \tan\phi_b \cdot (1 \pm \Omega \cdot k_h)}{k_h}$ (3)	$k_{h,eq} = \dfrac{\dfrac{c_b}{\gamma \cdot H} + \tan\phi_b}{\dfrac{k_{h,lim}}{k_{h,max}} \pm \Omega \cdot \tan\phi_b}$ (5)

FIGURE 1. Reference scheme and expressions giving the relevant parameters.

Due to its definition, $k_{h,eq}$ satisfies the following conditions:

- $k_{h,eq} = k_{h,c}$ if $F_k = 1$ (due to the definition of $k_{h,c}$ and the condition $F_k = F_{psd} = 1$);
- $k_{h,eq} < k_{h,c}$ if $F_k > 1$ ($d_{lim} > d_{max}$);
- $k_{h,eq} > k_{h,c}$ if $F_k < 1$ ($d_{lim} < d_{max}$).

Eq.(5) shows that the equivalent seismic coefficient depends on the parameters affecting the static stability condition of the system (c_b, ϕ_b, γ, H), on the effects induced by the earthquake ($k_{h,max}$, Ω) and on the adopted limit displacement d_{lim} which is implicitly taken into account through $k_{h,lim}$. It is worth noting that the expression for $k_{h,eq}$ (eq.5) does not depend on the adopted displacement regression model which is only involved in the computation of $k_{h,lim}$. If a displacement regression model involving frequency, amplitude and strong motion parameters is considered, the effects of these factors is accounted for in the evaluation of $k_{h,lim}$ and, then, being implicitly considered in the computation of F_k and $k_{h,eq}$. Finally, once the expression for $k_{h,eq}$ is known, the reducing factor η_h can be easily computed.

SOLUTIONS FOR SLOPES AND RIGID RETAINING WALLS

A closed form solution for the evaluation of $k_{h,eq}$ and η_h for the infinite slope scheme was recently derived by Biondi et al. [1] applying the proposed procedure to the slope scheme shown in Figure 2. The equation of motion of the slope subjected to both horizontal and vertical seismic inertia forces and can be written in the same form of eq.(1). Eq.(2) in Figure 2 shows the corresponding expressions of the shape factor S_f and the expressions derived for the other relevant parameters involved in the procedure: the pseudo-static safety factor F_{psd}, the horizontal critical and equivalent seismic coefficient, $k_{h,c}$ and $k_{h,eq}$ respectively, the reducing factor η_h. The expressions obtained for $k_{h,eq}$ and η_h clearly show their dependence on the parameters affecting the static slope stability (c', ϕ', γ, D, β, r_u), on the effects induced by the earthquake ($k_{h,max}$, Ω, Δu^*) and on the adopted limit displacement d_{lim}. Using the displacement regression model derived by Rampello et al. [4] a parametric analysis was carried out [1]. In Figure 3, for an example slope scheme, η_h is plotted against $k_{h,max}$ together with the ratio $k_{h,c}/k_{h,max}$ (thick line) and the value of $\eta_h = 0.5$ (thin dashed line). For a given $k_{h,max}$ the reducing factor η_h clearly depends on the adopted limit displacement d_{lim}: the greater is the value of d_{lim} that can be undergone by the slope, the lower is the factor η_h to be adopted in the pseudo-static analysis. As a consequence a single value of $\eta_h = 0.5$, as suggested in EC8, can lead to erroneous evaluations of slope stability. The influence of the vertical component of the seismic acceleration on η_h is shown in Figure 4 for the same slope of Figure 3. In the figure η_h and the ratio $k_{h,c}/k_{h,max}$ are plotted against $k_{h,max}$ for $\Omega = 0$ and $\Omega = \pm\frac{1}{2}$. From the plots it is evident that a vertical acceleration pointing upwards (positive values of Ω) leads to lower values of $k_{h,c}$ and to higher values of η_h representing the most conservative condition irrespective of d_{lim}. Biondi et al. [1] showed that: i) the expression of $k_{h,eq}$ can reflect the effect of as many seismic parameters as the displacement regression model accounts for; ii) the proposed solution can be used to plot displacement-based stability charts.

The seismic displacement response of rigid retaining walls is generally evaluated referring to the analogy with the rigid block sliding on an horizontal plane. However, such analogy does not hold since the equations of motion of the block and the wall are different [2]. When permanent displacements of the wall occur, the backfill slides outward and the soil-wall system actually consists of two bodies: the wall sliding along its base and the active wedge sliding along the failure surface.

Notation

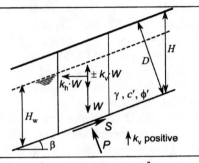

β: slope angle
H, D: failure surface depth, thickness of unstable soil
c', ϕ': effective cohesion and angle of shear strength
γ, γ_w: soil and water unit weight
r_u: pore pressure ratio $= (\gamma_w \cdot H_w)/(\gamma \cdot H)$
Δu^*: earthquake-induced pore pressure ratio
k: seismic coefficient
k_h, k_v: horizontal and vertical component of k
$S_f = g \cdot \cos(\phi - \beta + \omega)/\cos\phi$: shape factor (with tg $\omega = \Omega$)

Pseudo-static safety factor:	$F_{psd} = \dfrac{\dfrac{c'}{\gamma \cdot D} + \left[\cos\beta \cdot (1 - r_u) \cdot (1 - \Delta u^*) - k_h \cdot (\sin\beta \mp \Omega \cdot \cos\beta)\right] \cdot \tan\phi'}{\sin\beta + k_h \cdot (\cos\beta \pm \Omega \cdot \sin\beta)}$
Horizontal critical acceleration coefficient:	$k_{h,c} = \dfrac{\dfrac{c'}{\gamma \cdot D} + \cos\beta \cdot (1 - r_u) \cdot (1 - \Delta u^*) \cdot \tan\phi' - \sin\beta}{(\sin\beta \mp \Omega \cdot \cos\beta) \cdot \tan\phi' + (\cos\beta \pm \Omega \cdot \sin\beta)}$
Horizontal equivalent acceleration coefficient:	$k_{h,eq} = \dfrac{\dfrac{c'}{\gamma \cdot D} + \cos\beta \cdot (1 - r_u) \cdot (1 - \Delta u^*) \cdot \tan\phi' - \dfrac{k_{h,lim}}{k_{h,max}} \cdot \sin\beta}{(\sin\beta \pm \Omega \cdot \cos\beta) \cdot \tan\phi' + \dfrac{k_{h,lim}}{k_{h,max}} \cdot (\cos\beta \mp \Omega \cdot \sin\beta)}$
Equivalent acceleration reducing factor:	$\eta_h = \dfrac{k_{h,eq}}{k_{h,max}} = \dfrac{\dfrac{c'}{\gamma \cdot D} + \cos\beta \cdot (1 - r_u) \cdot (1 - \Delta u^*) \cdot \tan\phi' - \dfrac{k_{h,lim}}{k_{h,max}} \cdot \sin\beta}{k_{h,max} \cdot (\sin\beta \pm \Omega \cdot \cos\beta) \cdot \tan\phi' + k_{h,lim} \cdot (\cos\beta \mp \Omega \cdot \sin\beta)}$

FIGURE 2. Infinite slope scheme considered by Biondi et al. [1]

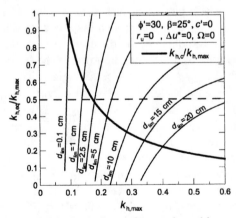

FIGURE 3. Dependence of η_H on d_{lim} and $k_{h,max}$

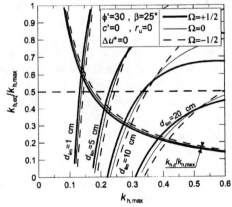

FIGURE 4. Influence of Ω on η_h

The analysis of the dynamic equilibrium condition of this two-wedge system shows that a time-dependent shape factor must be introduced to account for the kinematic compatibility of the displacements of the two wedges [2]. Neglecting the effect of the change in the system geometry, or accounting for it in a simplified way, a constant value of the shape factor can be introduced and, thus, displacement regression models derived for a 1-way sliding block analysis can be adopted [2].

Under these assumptions solutions for the evaluation of $k_{h,eq}$ and η_h for gravity retaining walls were recently derived [2] for various loading and boundary conditions. Figure 5 shows one of the considered schemes together with the expressions giving the more relevant parameters of the procedure. The expressions of $k_{h,eq}$ and η_h depend on the parameters affecting static stability of the soil-wall system (c_b, ϕ_b, γ, H, δ and β), on the active earth-pressure coefficient K_{ae} (i.e. on the values of ϕ, δ, β, i and Ω) and on the adopted limit displacement d_{lim} of the wall through $k_{h,lim}$. In Figure 6 η_h and the ratio $k_{h,c}/k_{h,max}$ (thick dashed line) are plotted against Γ for $\phi=35°$, $\delta=\phi_b=2\phi/3$, $\beta=i=0$ and $\Omega=0$. For a given value of Γ the reducing factor η_h depends on d_{lim} and the greater is the limit displacement that can be undergone by the wall, the lower is the factor η_h to be adopted in the pseudo-static analysis; as a consequence a single value of η_h as suggested in EC8 leads to erroneous evaluations of wall stability.

For a given soil-wall system it is possible to introduce a maximum value $\eta_{h,max}$ of η_h which allows to perform a conservative stability analysis characterised by $F_k=1$ and $d_{max}=d_{lim}$ [2]. Figure 7 shows the plot of $\eta_{h,max}$ versus d_{lim} computed using the displacement regression models derived by Rampello et al. [4] for different soil types.

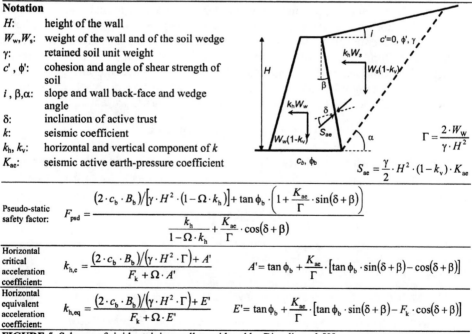

Notation

H:	height of the wall
W_w, W_s:	weight of the wall and of the soil wedge
γ:	retained soil unit weight
c', ϕ':	cohesion and angle of shear strength of soil
i, β, α:	slope and wall back-face and wedge angle
δ:	inclination of active trust
k:	seismic coefficient
k_h, k_v:	horizontal and vertical component of k
K_{ae}:	seismic active earth-pressure coefficient

$$\Gamma = \frac{2 \cdot W_w}{\gamma \cdot H^2}$$

$$S_{ae} = \frac{\gamma}{2} \cdot H^2 \cdot (1-k_v) \cdot K_{ae}$$

Pseudo-static safety factor:	$F_{psd} = \dfrac{(2 \cdot c_b \cdot B_b)/\left[\gamma \cdot H^2 \cdot (1-\Omega \cdot k_h)\right] + \tan\phi_b \cdot \left(1 + \dfrac{K_{ae}}{\Gamma} \cdot \sin(\delta+\beta)\right)}{\dfrac{k_h}{1-\Omega \cdot k_h} + \dfrac{K_{ae}}{\Gamma} \cdot \cos(\delta+\beta)}$
Horizontal critical acceleration coefficient:	$k_{h,c} = \dfrac{(2 \cdot c_b \cdot B_b)/(\gamma \cdot H^2 \cdot \Gamma) + A'}{F_k + \Omega \cdot A'}$ $A' = \tan\phi_b + \dfrac{K_{ae}}{\Gamma} \cdot \left[\tan\phi_b \cdot \sin(\delta+\beta) - \cos(\delta+\beta)\right]$
Horizontal equivalent acceleration coefficient:	$k_{h,eq} = \dfrac{(2 \cdot c_b \cdot B_b)/(\gamma \cdot H^2 \cdot \Gamma) + E'}{F_k + \Omega \cdot E'}$ $E' = \tan\phi_b + \dfrac{K_{ae}}{\Gamma} \cdot \left[\tan\phi_b \cdot \sin(\delta+\beta) - F_k \cdot \cos(\delta+\beta)\right]$

FIGURE 5. Scheme of rigid retaining wall considered by Biondi et al. [2].

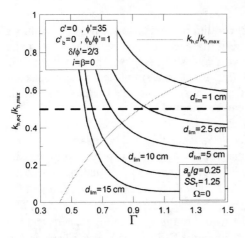
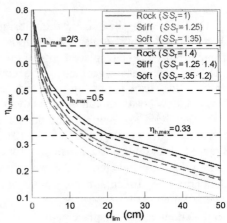

FIGURE 6. Dependence of η_H on d_{lim} and Γ **FIGURE 7.** Influence of d_{lim} on η_H

The calculation was carried out including (green curves) and neglecting (red curves) the effect of the change in the system geometry. It is apparent that the amplitude of the tolerable wall displacement d_{lim} significantly affects $\eta_{h,max}$; conversely, the change in the system geometry has a small influence on $\eta_{h,max}$.

CONCLUDING REMARKS

A procedure is presented for the evaluation of the equivalent seismic coefficient to be adopted in a pseudo-static analysis which leads to a safety measure in accordance with performance-based criteria. The procedure requires the use of a suitable relationship for simplified evaluation of earthquake-induced displacement and the definition of limit values of tolerable displacements related to ultimate and/or serviceability limit states. The proposed solution can be applied to any geotechnical system which at failure behaves as a 1-degree of freedom system. Applications to the infinite slope scheme and to sliding gravity retaining walls are presented.

ACKNOWLEDGMENTS

The Authors wish to thank ReLUIS (University Network of Seismic Engineering Laboratories) Consortium which funded the Research Project "Linea 6.3 Stabilità dei pendii".

REFERENCES

[1] Biondi G, Cascone E., Rampello S. (2007a). *Performance-based pseudo-static analysis of slopes.* Proc. 4th Int. Conf. Earthquake Geotechnical Engineering, 4ICEGE, Thessaloniki, 25-28 June.
[2] Biondi G., Maugeri M., Cascone E. (2007b). *Displacement-based seismic analysis of rigid retaining walls.* Proc. 14th ECSMGE, ERTC12 Workshop "Geotechnical aspect of EC8", Madrid.
[3] EN 1998-1 (2003). *Eurocode 8: Design of structures for earthquake resistance.* CEN European Committee for Standardisation, Bruxelles, Belgium.
[4] Rampello S., Callisto L., Fargnoli P. (2006). *Valutazione del coefficiente sismico equivalente.* Report of Task 6.3 'Slope stability', ReLUIS Consortium, 2006, pp. 23-38 (in Italian).

One-dimensional Seismic Analysis of a Solid-Waste Landfill

Francesco Castelli[a], Valentina Lentini[b] and Michele Maugeri[c]

[a]Department of Civil and Environmental Engineering, University of Catania, Viale Andrea Doria no.6, 95125, Catania, Italy
[b]Department of Civil and Environmental Engineering, University of Catania, Viale Andrea Doria no.6, 95125, Catania, Italy
[c]Department of Civil and Environmental Engineering, University of Catania, Viale Andrea Doria no.6, 95125, Catania, Italy

Abstract. Analysis of the seismic performance of solid waste landfill follows generally the same procedures for the design of embankment dams, even if the methods and safety requirements should be different. The characterization of waste properties for seismic design is difficult due the heterogeneity of the material, requiring the procurement of large samples. The dynamic characteristics of solid waste materials play an important role on the seismic response of landfill, and it also is important to assess the dynamic shear strengths of liner materials due the effect of inertial forces in the refuse mass. In the paper the numerical results of a dynamic analysis are reported and analysed to determine the reliability of the common practice of using *1D* analysis to evaluate the seismic response of a municipal solid-waste landfill. Numerical results indicate that the seismic response of a landfill can vary significantly due to reasonable variations of waste properties, fill heights, site conditions, and design rock motions.

Keywords: Solid-waste landfill, dynamic response, seismic loading, damping.

INTRODUCTION

Geotechnical engineering practice has expanded far beyond the traditional areas of Soil Mechanics and Foundation Engineering to include effective protection of environment against man-made pollution caused by all kinds of waste fills and waste heaps, with contaminated ground and underground water.

Stability analysis during and after seismic loading is a key point for design of solid waste landfills in seismic areas. From the lessons learned from past earthquakes, such as Loma Prieta earthquake [1],[2],[3] and Northridge earthquake [4],[5],[6] modern solid waste landfills have generally shown a good ability to withstand strong earthquakes without damages to human health and environment.

Experience has shown that well-built waste landfills can withstand moderate peak accelerations up to at least 0.2g with no harmful effects. Nonetheless, the integrity of solid waste landfills during strong earthquakes to achieve environmental and public health objectives deserves more consideration. The most dangerous manifestation concerning landfill stability and integrity is the surface fault breaking, and intersecting the landfill site.

CP1020, *2008 Seismic Engineering Conference Commemorating the 1908 Messina and Reggio Calabria Earthquake*,
edited by A. Santini and N. Moraci
© 2008 American Institute of Physics 978-0-7354-0542-4/08/$23.00

In the paper the numerical results of a dynamic analysis are reported and analysed to determine the reliability of the common practice of using *1D* analysis to evaluate the seismic response of a municipal solid-waste landfill.

The results of the seismic response analyses (for different levels of base motion) are examined to evaluate the impact of stiffness and height of the refuse fill on the overall response. Numerical results indicate that the seismic response of a landfill, particularly the seismic loading for the cover, can vary significantly due to reasonable variations of waste properties, fill heights, site conditions, and design rock motions.

MODEL DEFINITION

Analysis of the seismic performance of solid waste landfill follows generally the same procedures for the design of embankment dams, but the methods and safety requirements should be different. The characterization of waste material properties for seismic design is difficult due the heterogeneity of the material, requiring the procurement of large samples. It is important to stress that the dynamic characteristics of solid waste materials play an important role on the seismic response of landfill, and this area deserves more consideration [7]. It also is important to assess the dynamic shear strengths of liner materials due the effect of inertial forces in the refuse mass.

Several finite element computer programs assuming an equivalent linear model in total stress have been developed for *1D* [8],[9] or *2D* [10],[11] and pseudo *3D* [12] analysis.

The seismic responses obtained by computer finite element *1D* programs are considered reasonable. Since the slopes of landfills are usually flatter than slopes of earth dams, and landfill decks are larger than dam crests, two-dimensional response effects in landfills should be less significant than in earth dams.

For the soil behaviour the equivalent linear method is used and the shear modulus and damping ratio are adjusted in each iteration until convergence has occurred. In this case the variation of shear modulus G and damping ratio D with shear strain γ (Figure 1) can be derived by the following relations proposed by Yokota et al. [13]:

$$\frac{G(\gamma)}{G_0} = \frac{1}{1+\alpha\gamma(\%)} \tag{1}$$

$$D(\gamma)(\%) = \eta \cdot \exp\left[-\lambda\frac{G(\gamma)}{G_0}\right] \tag{2}$$

in which $\alpha = 15$, $\beta = 1.28$ and $\lambda = 1.95$.

Similarly, the solid-waste modulus reduction and damping curves used in the numerical dynamic analysis should be the "recommended" modulus reduction and damping curves developed, for example, by Matasovic and Kavazanjian [14] based upon laboratory testing and back-analysis of recorded accelerograms (Figure 1). When solid-waste landfills incorporate construction demolition debris the curves proposed for rockfill and gravel materials can be used.

CASE STUDY

The paper presents the numerical results for the seismic response of the behaviour of a solid-waste landfills under vertical propagation of seismic waves. Using the computer codes SHAKE91 [8], a one-dimensional (*1D*) dynamic analysis was carried out. The solid-waste municipal landfill (Figure 2) is located approximately 8 km northwest of Gela (Catania, Italy).

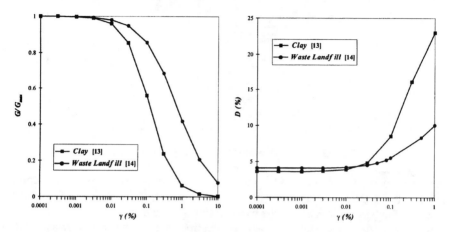

FIGURE 1. Shear modulus degradation and damping for soil and waste materials.

The study of the behaviour of municipal solid-waste landfills subjected to a *free-field* peck horizontal ground acceleration provides the most reliable means of validating and calibrating seismic performance analysis for landfill design.

The waste containment regulations in most, if not all, European countries do not give specific requirements about the design of landfill in seismic areas, because the seismicity is moderate in most European countries, especially in the northern and in the central Europe.

In Italy, where seismicity is moderate in the north and relatively high in the south, no restrictions are stipulated for MSW landfills, whereas the noxious and toxic (hazardous) waste landfills are forbidden in seismic areas with maximum horizontal acceleration equal to or greater than 0.1g.

To predict landfill response, knowledge of the bedrock motion and duration and the dynamic property of waste materials is required. In the case of landfills resting on soft clay or alluvial soil, the site amplification effect must be considered to select a design earthquake. In regard to the waste material characterisation, knowledge of the unit weight, shear wave velocity, shear modulus and damping ratio, dynamic shear strength of waste material and dynamic geosynthetic interfaces behaviour is required.

The shear strength properties of waste landfills are not easily determined since the physical composition of the mixture makes it unsuitable for the conventional laboratory strength testing. To overcome this situation, the solid-waste properties are generally established based on the type of waste, the waste processing and the placement procedures.

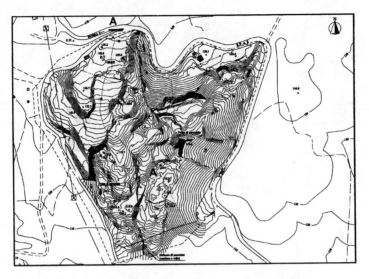

FIGURE 1. Plan view of the municipal solid-waste landfill.

Some properties are measured directly, such as dry density and water contents, whereas other properties, due the difficulties related with sampling, are obtained from indirect methods combining with the existent knowledge of waste properties.

Selecting a design earthquake and evaluating the dynamic soil waste landfill properties, the seismic response of a solid waste landfills can be performed by mathematical methods.

The peck ground horizontal accelerations used in the numerical analyses are on the order of 0.32g, 0.11g and 0.16g recorded during the Sturno earthquake on 23th November 1980, Loma Prieta earthquake on 18th October 1989, and finally, Catania earthquake on 13th Dicember 1990, respectively (Figure 3). The section of the solid-waste landfills reported in Figure 4, corresponding in a zone in which there is the soil below the solid waste, was taken into consideration.

Results of the seismic response analyses (for different levels of base motion) are examined to evaluate the impact of stiffness and height of the refuse fill on the overall response. Numerical results indicate that the seismic response of a landfill, particularly the seismic loading for the cover, can vary significantly due to reasonable variations of waste properties, fill heights, site conditions, and design rock motions.

PREDICTION OF THE LANDFILL RESPONSE

The one-dimensional seismic site response analysis was carried out by the SHAKE91 computer program [8],[9] to evaluate the amplification of earthquake motions at the top of the landfill in terms of peak ground acceleration a_{max}. A representative 45 m high column of landfill solid-waste and soil material was employed in the numerical analysis to model landfill response.

The SHAKE91 site response analysis was carried out employing the average shear wave velocity (V_s) profile shown in Figure 5. The measurement of the shear wave velocity by cross-hole and down-hole techniques provides values of V_s ranging between 100 and 200 m/sec with depth, in reasonable agreement with the landfill stratigraphy, being the landfill height about 17 m.

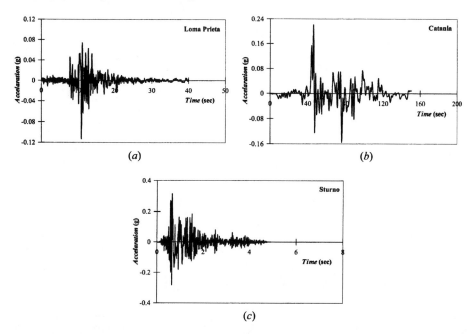

(a)

(b)

(c)

FIGURE 3. Acceleration *vs* time for the Loma Prieta (*a*), Catania (*b*), Sturno (*c*) earthquakes.

The unit weight profile was developed from the average shear wave velocity profile using the equation proposed by Kavazanjian et al. [15], relating the unit weight to shear wave velocity of municipal solid waste.

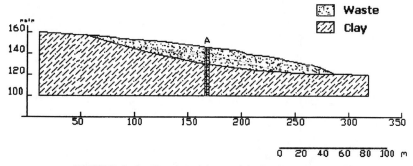

FIGURE 4. Section *A-A* of the municipal solid-waste landfill.

In the seismic response analyses, both 1989 Loma Prieta accelerogram, 1980 Sturno (Italy) accelerogram, and 1990 Catania (Italy) accelerogram were applied as bedrock motions.

The acceleration response spectra at the top of the landfill were compared to the response spectra for the *free-field* input motions to evaluate the amplification of earthquake motions at the top of the landfill.

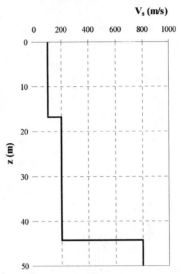

FIGURE 5. Shear waves velocity profile.

In Figures 6 (*a*), (*b*) and (*c*) the acceleration profiles versus depth obtained for the three earthquake motions are reported. An increasing of the peak ground acceleration a_{max} at the top of the landfill can be observed. Figures 7 (*a*), (*b*) and (*c*) shows the amplification of the spectral acceleration at zero period of the solid-waste mass from 0.16 to 0.54 g, and attenuation of the spectral acceleration in the vicinity of 1 sec.

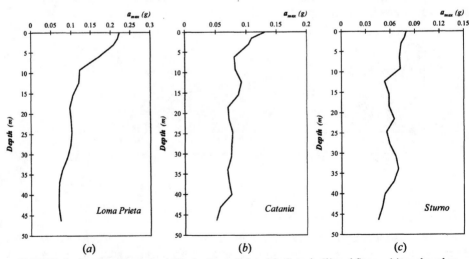

FIGURE 6. Acceleration *vs* depth for the Loma Prieta (*a*), Catania (*b*) and Sturno (*c*) earthquakes.

CONCLUDING REMARKS

Geotechnical Engineering has become a interdisciplinary applied science dealing with several problems of practical significance and importance.

Municipal solid-waste landfills, in particular, are complex engineered systems with a multitude of components. Solid-waste fill is extremely heterogeneous and it's properties involve significant uncertainties.

Municipality waste landfills owners, regulatory authorities and consultants typically are interested in carrying out a seismic risk analysis. The purpose of this analysis is to identify the risks associated with each type and height of landfill.

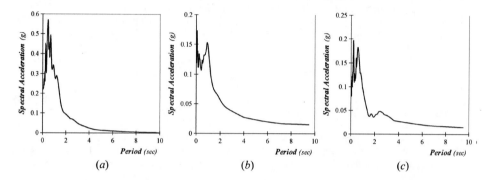

FIGURE 7. Spectral acceleration *vs* period for Loma Prieta (*a*), Catania (*b*) and Sturno (*c*) earthquakes.

In the paper the results of a one-dimensional dynamic analysis are reported to determine the reliability of the common practice of using *1D* analysis to evaluate the seismic response of a municipal solid-waste landfill. Numerical results indicate that the seismic response of a landfill can vary significantly due to reasonable variations of waste properties, fill heights, site conditions, and design rock motions.

REFERENCES

1. M. Johnson, M. Lew, J. Lundy, and M.E. Ray, "Investigation of sanitary slope performance during strong ground motion from the Loma Prieta Earthquake of October 17, 1989", 2nd ICRAGEESD, St. Louis, 1991, **2**, pp. 1701-1708.
2. D. Buranek and S. Prasad, "Sanitary landfill performance during the Loma Prieta earthquake", 2nd ICRAGEESD, St. Louis, 1991, **2**, pp. 1655- 1660.
3. H.D. Sharma, and H.K. Goyal, "Performance of a hazardous waste and sanitary landfill subjected to Loma Prieta earthquake", 2nd ICRAGEESD, St. Louis, 1991, **2**, pp. 1717- 1725.
4. N. Matasovic, E.J. Kavazanjian, A.J. Augello, J.D. Bray, and R.B. Seed, "Solid Waste Landfill Damage Caused by 17 January 1994 Northridge Earthquake", in Woods, Mary C. and Seiple, Ray W. Eds., *The Northridge, California, Earthquake of 17 January 1994*: California Department of Conservation, Division of Mines and Geology Special Publication 116, Sacramento, California, 1995, pp. 43-51.
5. J. P. Stewart, J. D. Bray, R.B. Seed, and N. Sitar, "Preliminary report on the principal geotechnical aspects of the January 17, 1994 Northridge Earthquake", *Report N°. UCB/EERC-94/08*, College of Engineering, University of California at Berkeley, Berkeley, California, 1994, 238 p.

6. J.D. Bray, A.J. Augello, G.A. Leonards, P.C. Repetto, and R. J. Byrne, "Seismic stability procedures for solid waste landfills". *JGE*, **121**, (2), 1995, pp 139-151.

7. Sêco e Pinto, P.S. Lopes, L. Agostinho, and A. Vieira, "Seismic analysis of solid waste landfills". 3rd International Congress on Environmental Geotechnics, Lisbona, 1998, edited by Pedro S. Sêco e Pinto, published by A. Balkema.

8. P.B. Schnabel, J. Lysmer, and H.B. Seed, "Shake: A computer program for earthquake response analysis of horizontally layered sites", Report n° UCB/EERC 72-12. University of California, Berkeley, 1972.

9. I. M. Idriss and J. I. Sun, "User's manual for Shake91", Centre for Geotechnical Modelling, Department of Civil and environmental engineering, University of California, Davis, California, USA, 1992, 12 p.

10. I.M. Idriss, J. Lysmer, R. Hwang, and H.B. Seed, "QUAD-4: A computer program for evaluating the seismic response of soil-structures by variable damping finite element procedures". Earthquake Engineering Research Centre, University of California, Berkeley, Report No. EERC 73-16, 1973

11. J. Lysmer, T. Udaka, H.B. Seed, and R. Hwang, "FLUSH 2-A computer program for complex response analysis of soil-structure systems". Report n° UCB/EERC 75-30. University of California, Berkeley, 1974.

12. J. Lysmer, T. Udaka, C. Tsai and H.B. Seed, "FLUSH-A computer program for approximate 3D analysis of soil-structure interaction problems". Report n.UCB/EERC 74-4. University of California, Berkeley, 1975.

13. K. Yokota, T. Imai, and T. Kanemori, "Dynamic deformation characteristics of soils determined by laboratory tests. *OYO* Technical Report n.3, 1981, pp.13-37.

14. N. Matasovic and E. Kavazanjian, "Cyclic characterization of OII landfill solid waste", *Journal of Geotechnical and Geoenvironmental Engineering, ASCE*, **124**, (3), 1998, pp.197-210.

15. N. Matasovic and E. Kavazanjian, "Dynamic Properties of Solid Waste from Field Observations", First International Conference on Earthquake Geotechnical Engineering, **1**, 1996, pp.549-554.

Use of Reinforced Lightweight Clay Aggregates for Landslide Stabilisation

Vitezslav Herle

SG GEOTECHNIKA, Geologicka 4, CZ 152 00 Praha 5, Czech Republic

Abstract. In spring 2006 a large landslide combined with rock fall closed a highway tunnel near Svitavy in NE part of Czech Republic and cut the main highway connecting Bohemia with Moravia regions. Stabilisation work was complicated by steep mountainous terrain and large inflow of surface and underground water. The solution was based on formation of a stabilisation fill made of reinforced free draining aggregates at the toe of the slope with overlying lightweight fill up to 10 m high reinforced with PET geogrid and steel mesh protecting soft easily degrading sandstone against weathering. Extensive monitoring made possible to compare the FEM analysis with real values. The finished work fits very well in the environment and was awarded a special prize in the 2007 transport structures contest.

Keywords: Landslide, stabilization, reinforced soil, lightweight clay aggregates, monitoring.

LANDSLIDE CHARACTERISATION

Description of the Landslide

In early hours of April 1st 2006 an extensive landslide of slope debris and colluvial soil buried the eastern portal of the Hrebec highway tunnel near Svitavy town in E Bohemia, Czech Republic and interrupted the traffic on an important highway (Fig. 1). The landslide pulled down a four meter high gabion wall in the upper part of the slope collecting the falling rocks from the surroundings cliffs. The landslide was classified as translational with the shear zone at the contact between slope debris and intact clay bedrock. The lower part of the landslide shifted the storage of salt (for winter treatment of pavement) to the other side of the highway. The soil slope movement was accompanied by the collapse of a part of the rock cliff on left side of the tunnel portal.

The landslide was triggered by a sudden rise of temperature and accompanied by heavy rain that melted the snow cover. Rain water that filled the system of open joints in the sandstone rock exerted the hydrostatic pressure on colluvial deposits as well as produced uplift on a potential failure plane. These factors, together with low shear strength on the potential failure plane led to the landslide.

Geology and Morphology of the Site

The landslide movement developed at the pronounced terrain chine composed of cretaceous sediments. Sub horizontal rock layers forming the face of the landslide have a slight dip to the W. Turonian marlstone and cenomanian glauconitic sandstone

CP1020, *2008 Seismic Engineering Conference Commemorating the 1908 Messina and Reggio Calabria Earthquake,*
edited by A. Santini and N. Moraci

are exposed at the rock face. Cretaceous sediments are underlain by perm sandstone and siltstone.

FIGURE 1. Landslide at the E portal of the Hrebec tunnel. Situation on April 2, 2006

The height difference between the upper edge of the rock face and the toe of the landslide was around 33 m, length of the slide in the plane projection was 65 m, thickness of the sliding mass was between 3 and 6 m and volume of the slid mass was estimated on 10.000 m^3. The central and lower part of the slide mass was formed by sandy clay soil of soft to very soft consistency. Upper part of the landslide was composed of fallen sandstone rock blocks mixed with sandy clay and clay sand, products of weathering of the rock face. Very important springs of underground water in the amount of 5 to 10 l/s were detected at the upper part of the landslide on the contact between overlying highly fractured turonian marlstone and underlying soft glauconitic sandstone.

Initial Phase of the Landslide Study

Within 2 days following the landslide SG Geotechnika made a 3-D topographic model of the landslide by means of the laser scanning device (Fig. 2) and proposed a system of fixed points all over the area that was immediately realized. The fixed points were continuously topographically monitored and measured results entered in the BARAB® data base that was accessible on line via internet for the Consultant as well as for the Client. The need for continuing surveillance was particularly due to presence of the existing houses near the upper landslide edge that were at imminent danger. In this way the behavior of the landslide and its vicinity was under control.

Practically at the same time the geological study of the site was undertaken with sinking test pits, boreholes and penetrometer probes and taking samples from the land slid mass as well as from the underlying intact clay subsoil and exposed rock outcrops.

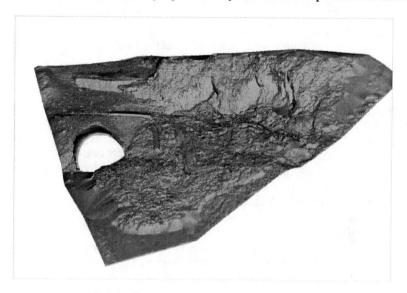

FIGURE 2. 3-D model of the landslide area made by the laser scanner

The clay bedrock under the landslide mass had a high plasticity ($w_L = 50$, $I_P = 22$) and effective shear strength measured in the direct shear test varied from peak value of ($\varphi' = 26^0$), to critical or constant volume ($\varphi_{cv} = 12^0$) to residual strength ($\varphi_{res} = 8^0$). These shear strength values were used in the stability analysis by classical limit equilibrium methods. The computation model is presented at Fig. 3.

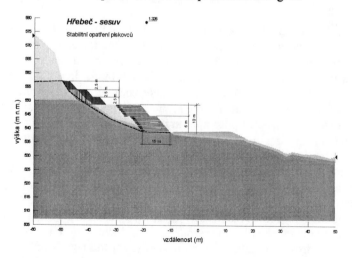

FIGURE 3. Slope stability model with proposed stabilization measures

In the same time the work on temporary stabilisation of the landslide were started. The aim was to stop further movement and spreading of the landslide that would have complicated stabilisation measures. The following steps were taken:
- realisation of open drainage ditches in the toe of the slide,
- removal of loose rock blocks and boulders from the rock cliff by mountaineers,
- cutting trees from the rock edge,
- excavation and taking away soft soil from the road pavement in order to enable partial traffic on the main highway.

The open ditches were later enlarged and extended through the slipped soil mass into the clay bedrock and filled with aggregates in order to fulfil the drainage as well as stabilisation function. A short mining gallery was uncovered during excavation of the slipped soil mass. The gallery was filled with coarse aggregates and used as one of the drainage elements for evacuation of underground water outside the unstable area.

After analysis of the preliminary data the following criteria were considered for preparation of alternatives of stabilisation measures, i.e.:
- high plasticity and low shear strength of the clay mass,
- existence of old shear zones within the clay mass,
- high inflow of underground water from the system of joints into the landslide zone, particularly from the contact between underlying compact glauconitic sandstone and overlying jointed marlstone,
- high rate of weathering of glauconitic sandstone,
- intensive jointing of the marlstone cliff forming unstable blocks and outcrops,
- environmental aspects of the stabilisation measures.

LANDSLIDE STABILISATION

Selected Solution

From comparison of the alternative solutions (stabilisation fill, concrete retaining wall, gabions, piles, anchors) the stabilisation fill at the slope toe was chosen as economic and environmental friendly alternative. In order to get maximum weight at the toe it was necessary to use reinforced fill which made possible to increase the fill slope angle in the lower part to 80^0. Coarse rock aggregates were placed in the stabilisation fill structure to secure free drainage of underground water from the rock mass and prevent build-up of pore pressure in the potential shear zone. PET geogrids 80/30 in 0,6 m spacing were used for the fill reinforcement. Total height of the stabilisation fill was 10 to 12 m.

One of the main problems that had to be solved was protection of the glauconitic sandstone layer outcropping in the rock cliff. Continuing weathering of this layer would lead to undercutting of the marlstone cliff and subsequent fall of rock blocks. Covering of the glauconitic sandstone layer by shotcrete was not accepted due to problem of drainage of this layer (need of horizontal drains drilled through the shotcrete), low frost protection and unfriendly appearance. Application of gabion protective wall would have brought a relatively high weight at the crest of the slip plane and increased the driving forces. At the end an application of expanded clay aggregates in the form of a lightweight fill was considered as a suitable solution. Expanded clay aggregates that sell in Czech Republic under the name "Liapor" is free draining frost resistant granular material that is at least five times lighter than common soil or rock aggregates. It also provides very good thermal insulation. Due to its low weight it reduced substantially the driving forces at the upper part of the slope and guaranteed sufficient stability. The fraction 8 – 16 mm was used for this purpose. As the expanded clay aggregates are predominantly of spherical shape it is necessary to provide a certain outer protection to the expanded clay aggregate body. In our case we used the steel hexagonal mesh (Terramesh) with non woven geotextile on the inner side to prevent expanded clay aggregates from falling out. The expanded clay aggregates were transported and deposited into the mesh cages by pneumatic way (Fig. 4).

FIGURE 4. Depositing the expanded clay aggregates in the Terramesh cages

Before installation of the first layer of expanded clay aggregates two pneumatic cells for measurement of vertical stress have been installed. Width of the lightweight fill with steel mesh facing varied between 4 and 6 m, depending on the shape of the glauconitic sandstone rock face. Maximum total height of the lightweight fill reached 10 m. The fill was internally reinforced with PET geogrids 55/55 kN/m at 0.7 m vertical spacing. The PET geogrids were fixed by nails to the rock face at 1 m horizontal distance.

The upper part of the fractured marlstone rock face that was almost vertical on 15 m height was cleaned from vegetation and loose boulders and rock fragments by mountaineers. The clean rock face was covered with hexagonal wire mesh commonly used for gabions and fixed by high strength HEA panels with nails and bolts to the rock face. This system is sufficiently stiff to hold in place big boulders. A general view of the whole stabilized area is on the Fig. 5 clearly showing wire mesh protected rock face at the top, stepped lightweight fill of expanded clay aggregates in the middle and the gravel toe embankment reinforced with PET geogrids and covered by antierosion grid with topsoil and hydroseed.

FIGURE 5. General view of the stabilized landslide area

Geotechnical Monitoring

Due to highly exposed project on the main highway the Client agreed with installation of geotechnical monitoring system. This system was based on the following measurements:

- pore pressure measurements in the highly plastic clay subsoil during building of the stabilizing fill reinforced by PET geogrids,
- extensometers on the geogrids reinforcing the stabilizing drainage fill at the slope toe were installed at two elevations for measurement of mobilization of tensile strength in geogrids,
- contact pressure of the light expanded clay aggregate fill on the upper part of the stabilizing fill,
- inclinometric measurements within the sliding area as well as outside (below the highway),
- topographic measurements of fourty fixed points and bench marks within and outside the sliding area.

Evaluation of one and a half year monitoring of the sliding area can be summarized as follows:

Three pore pressure cells recorded maximum pore pressure built-up during construction of the stabilizing granular fill and reached 20 kPa. In one year the pore pressure dropped and is between 8 and 15 kPa. This value confirms the stable character of the subsoil.

Extension of the geogrids reinforcements was measured by two four-stage extensometers with fixed points at 1 m, 3 m, 5 m and 7 m distance from the slope surface. The extension linearly increased during construction of the stabilizing fill and max. extension reached 3 to 5 mm (relative extension $\Delta l/l = 0.2$ %). After one and a half year from completion of the construction the maximum extension is between 0.1 and 0.3 % i.e. less than 3 % of the max. tensile strength or ten times less than calculated.

Pressure of the lightweight fill is measured by two contact pressure cells installed below the max. thickness of the expanded clay aggregates fill. In one place the measurement is influenced by high water inflow from the rock face into the lightweight fill, the other cell is in dry conditions. Max. measured value of the vertical stress under 8 m thick wet lightweight fill was 65 kPa or approx. 8 kN/m^3 which is about double the value of compacted dry fill. The other cell recorded max. 43 kPa under 9.6 m of lightweight fill i.e. 4.5 kN/m^3 which is rather in good agreement with the estimated compacted weight plus steel mesh and PET geogrid.

Inclinometers installed in the stabilizing fill within the former sliding area confirm stable character of the slope. Max. movement within the subsoil has been 6 to 9 mm. However, outside the stabilized area below the road one inclinometer has showed slight continuous increase of movement at 2 to 4 m depth. The movement has been influenced by the road fill the toe of which is in contact with the inclinometer. Max. recorded movement has been over 20 mm. Drainage trenches with coarse aggregates will be constructed if the movement continues.

A net of fixed points or bench marks has been distributed over the rock face, stabilizing fill, tunnel portal as well as outside the sliding area. The fixed points are measured topographically (vertical and horizontal component of movement) every 3 months. The rock face has been stable all the time since the end of the stabilizing work. Vertical and horizontal movements have not exceeded 3 mm. Max. settlement of the stabilizing fill was 16 mm, however majority of the measured points were below 10 mm since the end of construction. Horizontal movement of the fill was in average between 10 and 15 mm with max. value of 19 mm on one extensometer head.

CONCLUSION

Stabilisation of the Hrebec landslide area near Svitavy town in Czech Republic was a complex task. The landslide cut the traffic on the main highway connecting NW and NE part of the country so there was a great pressure to stabilise the area as soon as possible. Due to complicated broken terrain it was necessary to use special mountaineering technique for stabilisation of the rock face. A pioneer solution is a high lightweight fill from expanded clay aggregates that protects the soft glauconitic sandstone against weathering and in the same time provides drainage of outflowing water from the rock face.

The solution adopted for stabilisation of the landslide matches very well in the surrounding environment and received a special prize last year from Czech Highway Society for an outstanding project in the road infrastructure.

ACKNOWLEDGMENTS

I would like to express high appreciation to Geoprojekt Brno, Ltd., particularly to Mr. S. Stabl for preparation of a perfect architectural design of the landslide stabilization measures and to courageous Mrs. Bohatkova of SG Geotechnika for management of the stabilization works which needed her presence even on rope hanging from the rock cliff during checking of the wire mesh installation.

REFERENCES

1. L. Bohatkova, V. Herle, P. Mitrenga and J. Pospisil "Highway I/35. Landslide in front of the Hrebec tunnel. Preliminary technical report" (in Czech) (4/2006). Archive of SG Geotechnika.
2. P. Kucera and Z. Sekyra "Numerical modeling of the landslide on I/35 near Hrebec tunnel" (in Czech) (5/2006). Archive of SG Geotechnika.
3. S. Stabl "Design of stabilization measures for landslide on I/35 near Hrebec tunnel" (in Czech) (6/2006) Archive of SG Geoprojekt.
4. L. Bohatkova, V. Herle, and P. Mitrenga "Highway I/35. Stabilization of the landslide near the Hrebec tunnel. Concise report on monitoring" (in Czech) (11/2006). Archive of SG Geotechnika.
5. L. Bohatkova, V. Herle, P. Mitrenga and J. Pospisil "Highway I/35. Landslide in front of the Hrebec tunnel. Periodical report on monitoring" (in Czech) (11/2007). Archive of SG Geotechnika.
6. BARAB® data base system. On line data on geotechnical monitoring (in Czech) (4/2004 till today)

Slope Stability Analysis In Seismic Areas Of The Northern Apennines (Italy)

D. Lo Presti[a] T. Fontana[b] D. Marchetti[c]

[a]Department of Civil Engineering, University of Pisa, via Diotisalvi 2, 56126 Pisa Italy.
[b]Consultant Cittàfutura Ltd ,V. S. Chiara n°9 Lucca Italy.
[c]Department of Earth Sciences, University of Pisa, via S. Maria 53, 56126 Pisa Italy

Abstract. Several research works have been published on the slope stability in the northern Tuscany (central Italy) and particularly in the seismic areas of Garfagnana and Lunigiana (Lucca and Massa-Carrara districts), aimed at analysing the slope stability under static and dynamic conditions and mapping the landslide hazard ([16] , [5], [6], [9]). In addition, in situ and laboratory investigations are available for the study area, thanks to the activities undertaken by the Tuscany Seismic Survey. Based on such a huge information the co-seismic stability of few ideal slope profiles have been analysed by means of Limit equilibrium method LEM - (pseudo - static) and Newmark sliding block analysis (pseudo-dynamic). The analysis - results gave indications about the most appropriate seismic coefficient to be used in pseudo-static analysis after establishing allowable permanent displacement. Such indications are commented in the light of the Italian and European prescriptions for seismic stability analysis with pseudo-static approach. The stability conditions, obtained from the previous analyses, could be used to define microzonation criteria for the study area.

Keywords: Slope Stability, Debris, Seismicity, Pseudo-static analysis, Block analysis

INTRODUCTION

The present paper mainly deals with the stability of debris deposits which are very diffused in the Northern Apennines which is an area of relatively high seismicity. The purpose is to define microzonation criteria for the study areas and verify the prescriptions of European and Italian Technical Codes.

From geological evidences the debris deposits have limited thickness (few meters) and can be regarded as infinite slopes. Even though debris landslides are frequently triggered by heavy rainfall, normally the deposits are in dry conditions. The paper is aimed at defining the stability conditions in the presence of an earthquake. Therefore, the co-seismic stability of an infinite dry homogeneous slope with a non –degradable strength has been considered. In doing that the following steps have been done:

- definition of a limited number of ideal geological profiles;
- assessment of the peak ground acceleration (PGA) for given return periods (475 & 975 years) in the study areas on a probabilistic basis [11];
- selection of appropriate natural free-field accelerometers according to the criteria of European [8] and Italian [18] technical codes . This activity has been defined and carried out by Lai et al (2005) [12] and Paolucci & Lai (2007) [19];

CP1020, *2008 Seismic Engineering Conference Commemorating the 1908 Messina and Reggio Calabria Earthquake*,
edited by A. Santini and N. Moraci

- transfer of the selected accelerograms from rock outcrop to the top of the soil deposit by means or EERA [1] ;
- parametric study to define the yield acceleration (ky) for a homogeneous deposit by the Limit Equilibrium Method (LEM). Analyses have been carried out by means of the commercial code SLIDE (RocScience) assuming a curvilinear failure surface, corrected Janbu method of analysis, Coulomb-type material (c' − φ'), different inclination (β) and height (H) of the slope and different level of water in the slope (Hw). Nonetheless for the study area the hypotheses of dry infinite slope seem reasonable, it was decide to consider more general conditions to make the approach applicable in different situations;
- parametric study to define the accumulated average displacement as a function of the ky/kmax ratio, by means of a Newmark − type approach [17] The permanent displacement has been obtained by double integration of the accelerograms at the top of the soil deposits;
- use of the parametric studies to define microzonation criteria for Garfagnana & Lunigiana debris deposits;
- after fixing allowable permanent displacements, use of the sliding block analysis results to define the seismic coefficient to be used in pseudo-static analysis.

IDEAL PROFILES

The study areas are shown in **FIGURE 1**. The already mentioned works together with geological evidences and surveys clearly show the existence (in the study areas) of three different types of debris having the following main geological features:
- debris of Macigno Toscano (Sandstone) which mainly consists of silty clayey sands with rare blocks of sandstone. Observed thicknesses of debris involved in landslides are from 0.5 to 2.0 m;
- debris of Scaglia Toscana (Claystone) which mainly consists of clay or clay and silt and sand. Observed thicknesses of debris involved in landslides are less than 3.0 m;
- debris of Argille & Calcari (Stratified Claystone & Calcarenite) which mainly consists of blocks of Calcarenite in a matrix of silty clay or clayey −sandy silt. Observed thicknesses of debris involved in landslides are from 0.2 to 5.0 m

Observed inclinations of the debris slopes is mainly between 30° and 40° (60 % of cases), while inclinations larger than 40° have been observed in another 25% of cases. The remaining 15% exhibits inclinations lower than 30°.

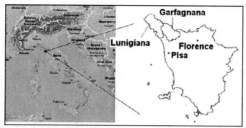

FIGURE 1. Seismic areas of Garfagnana and Lunigiana in the northern Tuscany (central Italy).

SEISMICITY OF THE STUDY AREA

Since 2003 big efforts have been done in Italy to improve the macrozonation of the national territory. Actually the PGA for different return periods are available at the apexes of a square net of 0.05° of side [11] which really represents an extremely advanced tool in the a - seismic design. For the whole study area the following values have been obtained:

- $PGA_{475} = 0.20g$ (i.e. the PGA corresponding to a return period of 475 years or a 90% of probability of non exceedance in 50 years);
- $PGA_{975} = 0.26g$ (i.e. the PGA corresponding to a return period of 975 years or a 95% of probability of non exceedance in 50 years);

According to European [8] and Italian [18] technical codes 475 years is the return period to be considered for an ordinary construction, while for schools, theatres, etc. the considered return period should be of 975 years. A greater return period should be considered for Hospitals and strategic buildings.

De-aggregation of the seismic hazard in the study areas has been done by Lai et al. (2005) [12]. A group of seven free-field natural accelerograms compatible with the obtained Magnitude and distance has been selected from existing database [12]. The capability of the selected accelerograms of reproducing on average the prescribed spectrum on rock European [8] and Italian [18] technical codes has also been verified. Paolucci and Lai (2007) [11], following an approach similar to that above described, selected two groups of accelerograms capable of reproducing the prescribed spectrum on rock. These other groups of accelerograms have been used in the present study. **TABLES** 1 a, b, c summarize the two groups of accelerograms that have been used and the criteria adopted by Paolucci and Lai (2007) [19] for their selection.

TABLE 1. (a) Accelerograms Group 3 [19].

Group 3 Name	Acc. Number	d	Date	Ml	Ms	Mw
Friuli (after shock)	1	16	11/09/1976	5.7	5.52	5.6
Montenegro	2	16	15/04/1979	-	7.04	-
Kalamata (Southern Greece)	3	10	13/09/1986	5.5	5.75	-
Erzincan (Turkey)	4	13	13/03/1992	-	6.75	-
Ionian (Greece)	5	18	23/03/1983	5.5	6.16	-
Parkfield	6	11.6	28/09/2004	-	-	6.0
Parkfield	7	14	28/09/2004	-	-	6.0

TABLE 1. (b) Accelerograms Group 4 [19].

Group 4 Name	Acc. Number	d	Date	Ml	Ms	Mw
Kalamata (Southern Greece)	1	10	13/09/1986	5.5	-	-
Erzincan (Turkey)	2	13	13/03/1982	-	-	-
Chalfant Valley	3	18	21/07/1986	6.4	-	-
North Palm Spring	4	11	08/07/1986	-	-	6.2
Whittier Narrows	5	14	01/10/1987	-	-	6.1
Parkfield	6	20	28/09/2004	-	-	6.0
Parkfield	7	14	28/09/2004	-	-	6.0

TABLE 1. (c) Selection criteria [19].

Group	Tr return period (years)	PGA (g)	M	d(Km)	Suggested reale factor
3	475	0.2-0.3	5.5-6.0	<20	0.7-1.4
4	975	0.28-0.42	6.0-6.5	<20	0.7-1.4

GEOTECHNICAL CHARACTERIZATION

Geotechnical characterization of the soil of interest took advantage of the existing database of the Tuscany Seismic Survey [20]. The database has been developed by the Tuscany Seismic Survey in order to accomplish the seismic retrofitting of existing buildings (mainly schools). As for the geotechnical characterization, it consists of the following data:

- boreholes with SPT measurements, extending down to the bedrock or at least down to 30 m in the case of deeper bedrock;
- down hole tests performed in the boreholes;
- seismic refraction tests in SH and P waves;
- some laboratory Resonant Column tests and cyclic triaxial tests on undisturbed samples.

For this specific study, data from 18 boreholes at 14 different sites have been used. Range of thickness and of shear wave velocity for the three ideal profiles are shown in TABLE 2.

According to European [8] and Italian [18] technical codes the three ideal profiles are classified as type B soil.

TABLE 2. Ideal profiles

IDEAL PROFILE A (Macigno Toscano debris)	Hmin / Hmax[m]	Vs[m/s]	γ[KN/mc]
Debris	2 / 7	VsD=160-450	18.5
Transition zone (Fractured bedrock)	7 / 17	VsT=450-800	21
Bedrock	-	VsB=800-1500	21
Vs30=732 m/s (Type B soil – EC8 2003)			
IDEAL PROFILE B (Scaglia Toscana debris)	Hmin / Hmax[m]	Vs[m/s]	γ[KN/mc]
Debris	1 / 10	VsD=140-400	19.1
Transition zone (Fractured bedrock)	7 / 35	VsT=400-1000	21
Bedrock	-	VsB=1000-1780	23
Vs30=748 m/s (Type B soil – EC8 2003)			

PROFILE C (Argille e Calcari debris)	Hmin / Hmax[m]	Vs[m/s]	γ[KN/mc]
Debris	2.5 / 8	VsD=180-400	18.4
Transition zone (Fractured bedrock)	10 / 28	VsT=400-810	21
Bedrock	-	VsB=810-1250	23
Vs30=611 m/s (Type B soil – EC8 2003)			

NEWMARK TYPE PARAMETRIC STUDY

The procedure followed to determine the permanent displacement is essentially similar to that adopted by other researchers ([15], [10], [3]). Anyway, some differences exist among the above mentioned procedures.

That adopted for the present study is outlined below. Each group of accelerograms was scaled to the corresponding PGA. After that, the selected accelerograms were transferred from rock outcrop to the top of the soil deposit by means of EERA [1]. In order to consider the range of thicknesses and shear wave velocity, for each ideal profile 12 different cases have been considered:

- the thickness of debris (H_D) and that of debris plus the underlying fractured rock (H_T), as indicated in **TABLE 2** assumed in one case their minimum values and in another case the maximum;
- three different combination of shear wave velocities of debris, fractured rock and intact bedrock were considered **TABLE 3**;
- as for the shear modulus and damping ratio curves, an existing database for the study area was used [13]. More specifically the lower and upper envelopes for stiffness and damping ratio of each geological formation were used.

TABLE 3. Three different combinations of shear wave velocities.		I	II	III
Debris	VSD	min	max	min
Transition zone (Fractured bedrock)	VST	min	max	min
Bedrock	VSB	max	max	min

For each ideal profile and for each group of input accelerograms on rock, a total of 84 accelerograms were obtained at the top of the soil deposit. Double integration of the 84 accelerograms was done by considering first the positive and after the negative values, assuming different yield accelerations. **FIGURE 2** shows the permanent displacement (s) for different values of the ky/kmax ratio for each input accelerogram (group 3) in the case of the deal profile A. Similar results have been obtained for the other profiles and for group 4 accelerograms.

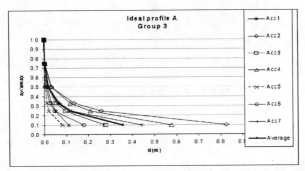

FIGURE 2. Permanent displacement (s) for different values of the ky/kmax ratio Group 3 – Ideal Profile A.

PARAMETRIC STUDY FOR K_Y

Available simplified approaches to compute ky assume [4] curvilinear failure surface for a homogeneous dry slope with c' and φ', or [3] an infinite dry slope with c' and φ', or [2] an infinite slope with c' and φ' and different height of the water table parallel to the slope. In the present study the following assumptions have been done:
- curvilinear failure surface;
- Mohr-Coulomb failure criterion with c' = 0 – 50 kPa and φ' = 20 – 40°;
- homogeneous slope with inclination β= 20 – 40° and height (H) from 5 to 300 m;
- dry conditions plus four different water table levels as shown in **FIGURE 3**.

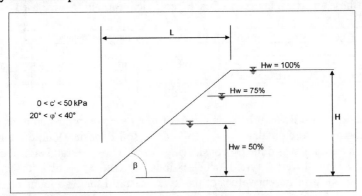

FIGURE 3. Considered parameters.

As an example **FIGURE 4** shows the values of ky which give a unit safety factor as a function of β, for friction angle of 30°, water table (Hw = 50) and the c'/γH ratio (0.0 to 0.5). As expected, for a zero cohesion the height of the slope resulted to be non influent. The complete set of results is reported by Morelli (2007) [14]

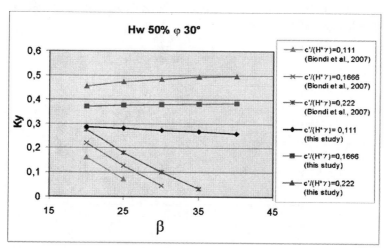

FIGURE 4. Values of ky which give a unit safety factor (LEM) as a function of β, for different values the c'/γH ratio; (φ' = 30°, Hw = 50%).

The obtained values are always higher than those predicted for infinite slope and agree quite well with those proposed by Chang et al. 1984 [4]. Anyway, the proposed solution (unlike the Chang et al. 1984 [4] results) accounts for the presence of water table.

PRACTICAL IMPLICATIONS

Generally, it is suggested to perform pseudo-static analysis, using the static strength parameters and a seismic coefficient reduced with respect to the PGA [7]. Obviously, soils which loose more than 20% of their strength during the earthquake are excluded from the above suggested procedure. Anyway, indications about the reduction coefficient to be adopted are very different and mainly based on the behaviour of earth dams [7]. Usually, the reduction coefficient can be inferred from a sliding block analysis after establishing allowable displacements and.

Obviously, very different allowable displacements can be defined depending on the type of construction or infrastructure resting on the slope. Anyway, a displacement of less than 0.1 m is generally accepted. On the other hand the occurrence of a generalised failure is considered possible for displacements larger than 1.0 m.

Going back to the **FIGURE 2**, which shows the permanent displacement (s) vs. the ky/kmax ratio, it is possible to see that s = 0.1 m corresponds (on average) for the whole set of cases to ky/kmax = 0.3.

A coefficient of 0.5 is prescribed by European [8] and Italian [18] technical codes for pseudo-static analysis. Such a prescription appears very severe for the case under study and probably is very conservative with the only exception of soils exhibiting a great strength decay.

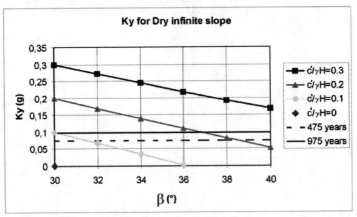

FIGURE 5. (a) Microzonation criteria for the study area φ' = 30°.

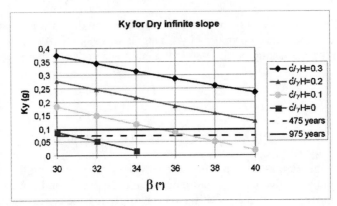

FIGURE 5. (b) Microzonation criteria for the study area φ' = 35°.

FIGURES 5 a, b show the proposed microzonation criteria. According to European [8] and Italian [18] technical codes the following range of values has been assumed for the design ground acceleration:

- $0.3*S*PGA_{475} = 0.3*1.25*0.20g = 0.075g$ (i.e. the PGA corresponding to a return period of 475 years or a 90% of probability of non exceedance in 50 years for type B soil – S = 1.25, considering allowable displacement of 0.1 m);
- $0.3*S*PGA_{975} = 0.3*1.25*0.26g = 0.098g$ (i.e. the PGA corresponding to a return period of 975 years or a 95 % of probability of non exceedance in 50 years for type B soil – S = 1.25, considering allowable displacement of 0.1 m);

The above range for design ground acceleration has been reported in **FIGURE 5**. As for the strength parameters, **FIGURE 5** a considers φ' = 30° and **FIGURE 5** b φ' = 35°, while the c'/γH parameter is considered to range from 0 to 0.3. An infinite dry slope has been considered when computing the critical acceleration Ky. The Figures show the strength parameters required to guarantee stability, for typical slope inclination (β = 30-40°). The Figures indicate that assuming an angle of shear resistance of 30-35° a cohesion of about 15 kPa is necessary to have stability even for the steepest slopes.

CONCLUDING REMARKS

The co-seismic stability of debris slopes, well diffused in the Northern Apennines (Central Italy), has been analysed by the (pseudo-static) Limit Equilibrium Method and (pseudo-dynamic) Newmark type method. The study area is affected by a relatively high seismicity. The seismic hazard has been defined on the basis of a conventional probabilistic approach leading to the selection of a number of free-field natural accelerograms. The parametric study took advantage of an existing geological and geotechnical characterization of such debris deposits. Conventional pseudo-dynamic method has been applied (i.e. the block sliding analysis has been accomplished after computation of the seismic response). The following conclusions can be drawn:

- assuming an allowable permanent displacement of 0.1 m, the parametric study suggests that, for the study area, it is possible to consider a seismic coefficient, for pseudo-static analysis, equal to 0.3*PGA;
- microzonation criteria were defined, for the study area, assuming an infinite dry slope. The strength parameters ranged in the following intervals: $\varphi' = 30 - 35°$, $c' = 0 - 25$ kPa;
- the study gave a number of charts for the determination of the critical acceleration coefficient Ky (i.e. the acceleration which leads to a safety factor equal to 1.0 in pseudo-static analyses). Such a coefficient is for homogeneous slopes and depends on slope inclination (β), angle of shear resistance (φ'), cohesion parameter ($c'/\gamma H$) and water level Hw. The obtained values refer to a curvilinear failure envelope and obviously are greater than that obtained in the case of infinite slopes. The values well agree with those obtained from other analogous solutions available in literature but which consider only dry slopes.

ACKONWLEDGEMENTS

The research activities have been supported by PRIN 2005 (Geological and Geotechnical Characterization of slopes and stability analysis in seismic areas of the Northern Apennines) of the Italian Ministry of Education. The help given by the undergraduate students Clara Ciurli & Elena Morelli is highly appreciated. The authors would like to thank dr. Ferrini, head of the Tuscany Seismic Survey, for the data base concerning the geotechnical characterization of the debris.

REFERENCES

1. Bardet, J.P., Ichii, K. & Lin C.H. (2000). *"EERA – A Computer Program for Equivalent-Linear Earthquake Site Response Analyses of Layered Soil Deposits."*, Department of Civil Engineering, University of Southern California, http://geoinfo.usc.edu/gees.
2. Biondi G., Cascone E. and Rampello S. 2007 *Performance based pseudo-static analysis of slope*. Proceedings of 4th ICEGE, paper ID: 1645. Thessaloniki Greece 25-28 June 2007.
3. Bray J.D., Rathje E.M., Augello A.J. and Merry S.M. 1998 *Simplified seismic design procedure for geosynthetic - lined solid waste landfills*. Geosynthetics International, 5((1-2): 203-235.
4. Chang C.J., Chen W.F. and Yan J.T.P. 1984 *Seismic displacements in slopes by limit analysis*. Journal of Geotechnical Engineering, Vol. 110(7): 850-874.
5. D'Amato Avanzi, G., Giannecchini, R., Puccinelli, A., 2004. *The influence of the geological and geomorphological settings on shallow landslides*. An example in a temperate climate environment: the June 19, 1996 event in northwestern Tuscany (Italy). Engineering Geology 73, 215–228.
6. D'Amato Avanzi, G., Puccinelli, A., 1997. *Deep-seated gravitational slope deformations in north-western Tuscany (Italy): remarks on typology, distribution and tectonic connections*. Geografia Fisica Dinamica Quaternaria 19, 325–334.
7. Duncan J. M. and Wrigth S.G. 2005 *Soil strength and slope stability*. John Wiley and Sons, Inc. Hoboken New Jersey.
8. Eurocode 8 (2003) *Design Provisions for Earthquake Resistance of Structures* - Part 1-1:General Rules for the Representation of Seismic Actions." Part 5: Foundations, Retaining Structures and Geotechnical Aspects.
9. Giannecchini R. & Pochini A. (2003) - *Geotechnical influence on soil slips in the Apuan Alps (Tuscany):* first results in the Cardoso area. Proc. Int. Conf. on fast slope movements-prediction and prevention for risk mitigation (IC-FSM 2003), Napoli, 11-13 maggio 2003, 241-245.
10. Hynes-Griffin M.E. and Franklin A.G. 1984 *Rationalising the seismic coefficient method*. Paper GL-84-13. US Army Corps of Engineers. Waterways Experiment Station. Vicksburg.
11. INGV (2005) (http://essel-gis.mi.ingv.it/.)
12. Lai C., Strobbia C. & Dall'Ara (2005) Convenzione tra Regione Toscana e Eucentre. Parte 1. *Definizione dell'Input Sismico per i Territori della Lunigiana e della Garfagnana*. L.R. 56/97 programma VEL
13. Lo Presti D., Squeglia N., Mensi E., Pallara O. (2007) *Assessment of EC8 and Italian Code Prescriptions for Seismic Actions*. 4th International Conferenze on Earthquake Geotechnical Engineering, Thessaloniki (Greece) 25-28 June 2007
14. Morelli E. (2007) Analisi di stabilità sismica dei pendii: formule speditive per l'accelerazione critica. M. Sc. Thesis, University of Pisa, Department of Civil Engineering.
15. Makdisi F.I. and Seed H.B. 1978 *A simplified procedure for estimating dam and embankment earthquake induced deformations*. Journal of Geotechnical Engineering Division , Vol. 104(7): 849-867.
16. Nardi, R., Puccinelli, A., D'Amato Avanzi, G., 2000. *Carta della franosita` del bacino del Fiume Serchio*. Autorita` di Bacino del Fiume Serchio, Lucca, http://www.serchio-autoritadibacino.it/carto/index.htmlS.
17. Newmark N., [1965]. *Effects of earthquakes on dams and embankments* Geotechnique, 15, 2, 139-160.
18. NTC (2005) *Norme Tecniche per le Costruzioni*, Ministero delle Infrastrutture e dei Trasporti. Decreto 14 Settembre 2005.
19. Paolucci R and Lai C. 2007 *Limiti di Applicabilità dei Metodi Pseudo-Statici nelle Analisi di Stabilità delle Opere di Sostegno dei Terreni in Zona Sismica:* Confronto tra Analisi Rigorosa e Metodi Semplificati alla Luce dell'Eurocodice 8 e della Recente Normativa Sismica Italiana – PRIN 2005 – Meeting 30th October 2007. Politecnico di Milano. Department of Structural and Geotechnical Engineering.
20. VEL (2007) http://www.rete.toscana.it/sett/pta/sis

Assessment of Slope Stability and Interference of Structures Considering Seismity in Complex Engineering-Geological Conditions Using the Method of Finite Elements

Papuna Menabdishvili, Nelly Eremadze

Kiriak Zavriev Institute of Structural mechanics and Earthquake Engineering
C 8,M.Alexidze street, 0193, Tbilisi, Georgia.
Email: smee@.acnet.ge, Phone/Fax: 995 32 335613

Abstract. There is elaborated the calculation model of slope deformation mode stability and the methodic of calculation considering the interference of structures to be built on it using the method of finite elements. There is examined the task of slope stability using the soil physically nonlinear finite element considering the seismicity 8. The deformation mode and field of coefficients of stability are obtained and slope supposed sliding curve is determined. The elaborated calculation methodic allows to determine the slope deformation mode, stability and select the optimum version of structure foundation at any slant and composition of slope layers.

Keywords: slope, stability, seismisity, geology, finite elements.

STATEMENT

Ensuring of building and structure reliability is particularly important at their reconstruction (building on of additional storey, increasing of equipment weight and so on), strengthening and liquidation of wrecking. For instance, at construction of new buildings and structures the near located buildings often suffer inadmissible deformation. The primary reasons of additional deformation development are following: soil compacting due to effect of loadings passed by new building or structures, replacement of piles, development of negative friction on piles, dynamical effect on untied soils at pile arranging and so on.

In order to obtain rational solutions for foundation strengthening and reconstruction the thoroughly investigation of foundations and soils is carried out. After the analysis of received results the recommendations of soil and foundation strengthening are elaborated.

As a rule, for specific engineering-geological conditions the different versions of soils and foundations are examined considering their structural properties. The preference is given to technically and economically expedient version.

All complex of works of foundation and soil investigations are divided to following stages:

CP1020, *2008 Seismic Engineering Conference Commemorating the 1908 Messina and Reggio Calabria Earthquake,*
edited by A. Santini and N. Moraci
© 2008 American Institute of Physics 978-0-7354-0542-4/08/$23.00

- Gathering of building and structure construction and operation data and their generalization;
- Investigation of the site and overhead building and structure technical conditions (examination of existing deformation and observation of dynamical processes).
- Investigation of building and structure soils and foundations;

When loadings by foundations are passed to basis, the complex mechanical processes have place in it. At local loading the detached lot of soil suffers the effect of normal and tangent (shear) stresses. These tangent stresses, after reaching a certain value, cause the irreversible process of local sliding – shear. Beside this, at local loading effect (with the definite value) in foundation soil there have place the extinguished deformation of soil compacting , as well as unextinguished shear deformation that in conformable conditions turns to plastic flow, outflanking and other deformations of foundation soil. This can be explained by the fact that, when a bearing capacity is exhausted, under foundation basis the sphenoid compact core is formed that replaces the soil basis and causes the foundation slump in small period of the time.

At present, due to lack of parcels of land in central parts of cities and development of infrastructure, the active process of existing building reconstruction have place: a functional area is changed, building on and different extensions are carried out.

Any reconstruction, particularly extensions lead to change of existing deformation mode of building or structure. The closer buildings are located, and the more types of loading pass to soil , the more their interference becomes important and dangerous. At this time the set of problems arise: whether at reconstruction the inadmissible deformations can be developed, or the bearing capacity of existing buildings can be lost. Difference between foundation lumps can cause the significant increasing of their stresses and lead to destruction. In result all building will be destructed.

Ensuring of reliability of buildings and structures, to be built, is noticeable complicated in disadvantageous engineering-geological and seismic conditions. At this time the assessment of slope stability considering above mentioned problems becomes necessary.

SURVEY OF EXISTING METHODS

The assessment of slope stability by most spread methods is connected to significant schematization of all the process, that often causes the deflection of calculation scheme from real conditions that finally leads to incorrect assessment of slope stability. This basically is caused by more complex configuration of slipping surface, than it is considered in calculation, or real displacement occurs not on one surface, but in confinements of the zone, where soils have low strength, or tangent stresses have increased values.

In the most of calculations the deformation mode in sliding zone, redistribution of stresses and its influence on soil forming are considered not enough fully. Despite the comparative simplicity of such calculations, received assessment of slope stability in most cases does not include the necessary reliability. Most true results are obtained in simplest cases: at homogenous composition of rocks in solid, and at comparatively simply configuration of slope surface. Due to necessity of consideration of subsurface geology nonuniformity and lack of replacement mechanism and condition of sliding surface, above mentioned calculations are complicated and often become fruitless. All these calculations can be used only in cases, when the simplifications and assumptions, on which it is based, are strictly met.

By our opinion, the methods those are based on matching of rock strength to field of stresses and in result obtaining the field of stability, are more prospective. So, the task is brought to determination of integral values of tangent stresses τ_1 and τ_2 in each point of examinable solid, to their conditionality, and obtaining of the special values and complete field of K_{md} (stability coefficient $K_{md} = \tau_1/\tau_2$) [1].

For solution of deformation mode of solids there are most prospective the methods of simulation based on solution of differential equations those describe this conditions using the physical modeling and computer. At present as a basis of this research there is used the method of finite elements that represents an universal powerful modern numerical method for solution of scientific-technical problems [2].

PROPOSED METHODIC OF SLOPE STABILITY CALCULATION BY THE METHOD OF FINITE ELEMENTS AND CALCULATION OF REAL OBJECT

In represented work the methodic of slope calculation using the method of finite elements is elaborated. The calculation of structures, to be built on it, considering their interference is carried out by space-iteration method using the physically nonlinear finite elements. Consideration of soil properties is achieved using the More-Coulomb dependence for tangent stresses [3].

$$\sigma_1 - \sigma_2 \le -\sin\varphi(\sigma_1 + \sigma_2) + 2R_s\cos(\varphi) \qquad (1)$$

When the main N_1 and N_2 meet the equations:

$$N_{1 \le}R_S, \quad \text{and} \quad N_1 \le R_S, \qquad (2)$$

than the linear calculation is carried out, if not, calculation is conducted by iteration process [4,5].

As an initial data in soil finite elements following values are used: E –module of elasticity, v - Poison coefficient, R_0 – relative density, R_C – shear coherence, R_S – limit tension stress, φ - angle of inner friction.

The task is examined considering the shear by the plane deformation scheme.
The object of investigation is slope containing three zones: very weathered rocks, weathered rocks and slightly weathered rocks.

Here two many-storied buildings are constructed, on Southern and North mountainsides Distance between them consists 10,5 m.

Before primary model elaboration a few testing tasks were realized with different boundary conditions, where as sand-stone, as mudstone layers were modeled. In these models sizes of models for sand-stones and mudstones consisted 0,7 m lengthwise and 0,5m widthway. In mentioned testing tasks for assessment of model work many versions of calculation were carried out. Namely, as boundary conditions, as initial data of sand-stone and mudstone layers were varied.

Received results show apparently that mudstone layers represent the possible sliding surfaces.

In result of thoroughly analysis there was elaborated the mountainside primary calculation model that is created on the basis of data obtained by geological investigations.

The preference of finite element method against other methods was established. Namely, the calculation model maximally was advanced to real condition: The model precisely describes the profile of mountainside, inclination of layers that consists 40°, dislocation of pales on it and their geometry. In full profile the interchange of sand-stone and mudstone layers is modeled as well.

According to deformation mode the main stresses in centre of gravity of element middle surface are calculated:

$$N_{1,2} = \frac{N_a + N_z}{2} \pm ((\frac{N_a + N_z}{2})^2 + T^2{}_{xz})^{1/2} \tag{3}$$

The inclination angle of main stress N_1 with maximum value to axis X_1:

$$\varphi = arct(\frac{N_1 - N_a}{T_{XZ}}). \tag{4}$$

On the basis of obtained main stresses profile stability is assessed.

Boundary conditions are following: the bottom of cross-section is embedded in Z direction, and lateral sides – in X direction.

The calculation is carried out for two cases of loading. In first case the slope suffers loading passed from building considering the seismicity 8, and in second case –the gravity load only.

For slope stability examination there were carried out 12 versions of calculation: there were examined the slope without building loading, with buildings located on different levels, with building loadings, second and third EGE –s containing mudstone only, second and third EGE-s with 30% sand-stone containing, 15 % containing, partially or full cutting of the territory between buildings, first building with, and without piles, and second building with slab foundation and with piles.

As testing, as different versions of calculation have shown that the preciseness and reliability of calculation results are defined by initial data, namely by values of $E, v, R_0, R_C, R_S, \varphi$, hence the trustworthiness of calculation results fully depends on data obtained in result of engineering-geological research and on design loads passed from building to soil.

The analysis of obtained results makes clear that at loadings passed from building to soil considering seismisity 8 the coefficient of slope stability consists 1,8 that meets the admissible value of slope stability defined by building codes, if:
- the weathered layer of elevated road narrow stripe, existing between the buildings will be cut out;
- first building will be founded on slightly weathered layer;
-second building will be founded on the slab.

538

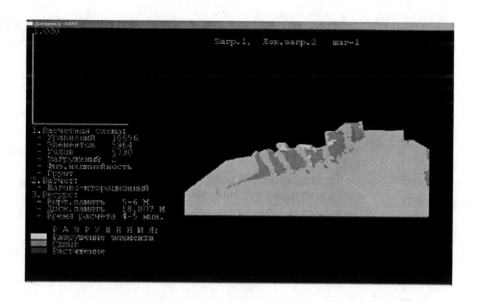

FIGURE 1. Shear of finite elements (green colour), when foundation is implemented by piles, I and II layers contain only mudstones (without sand stones)

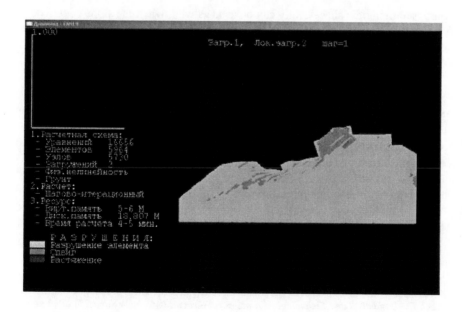

FIGURE 2. Shear of finite elements (green colour), when foundation is implemented without piles.

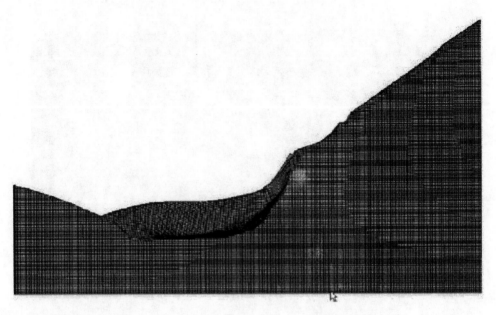

FIGURE 3. Deformed state of mountainside, when its bottom part consists of filled soils

The proposed methodic successfully can be used as well, when the slope consists of cliffy rocks (loam, rebblestone and so on). On fig. 3 there is shown the deformed state of mountainside, when its bottom part consists of filled soils.

INFERENCES

As testing, as different versions of calculation have shown that the preciseness and reliability of calculation results are defined by initial data, namely by values of $E, v, R_0, R_C, R_S, \varphi$, hence the trustworthiness of calculation results fully depends on data obtained in result of engineering-geological research and on design loads passed from building to soil.

The elaborated methodic gives possibility of determination of the deformation mode and stability of slope at any inclination and composition of layers and selection of the optimum version of foundations.

REFERENCES

1. P.Z. Medabdishvili, "Investigation of Deformation Mode and Assessment of Stability of Slopes and Banks". *The Dynamical and Technological problems of Structure Mechanics and Solid Bodies,* X[th] International Symposium Proceedings, Moscow, 2004. (in Russian)
2. P.Z. Menabdishvili, *Using of Method of Finite Elements for Calculation of Thin-Walled Ribbed Shell Structures,* Tbilisi, edited by "Technical University, 2002. (in Russian)

3. PC LIRA, version 9. Program Complex for Analysis and Design of Structures. Edited by Academician of Ukrainian AN   A.S. Gorodetski, Moscow-Kiev, Publishing House "Fact", 2003, 464 p.
4. A.A. Iliushin, *Mechanics of Continuum,* edited by Moscow University, 1990. (in Russian)
5. S.B. Ukhov, V.V. Semionov et al., *Mechanics of Soils, Bases and Foundations.* Moscow, ACB 1994 , p. 524. (in Russian).

Dynamic Slope Stability Analysis of Mine Tailing Deposits: the Case of Raibl Mine

Meriggi Roberto, Del Fabbro Marco, Blasone Erica and Zilli Erica

Department of Georesources and Territory, University of Udine
Via Cotonificio, 114, 33100 Udine, Italy.

Abstract. Over the last few years, many embankments and levees have collapsed during strong earthquakes or floods. In the Friuli Venezia Giulia Region (North-Eastern Italy), the main source of this type of risk is a slag deposit of about $2 \times 10^6 m^3$ deriving from galena and lead mining activity until 1991 in the village of Raibl. For the final remedial action plan, several in situ tests were performed: five boreholes equipped with piezometers, four CPTE and some geophysical tests with different approaches (refraction, ReMi and HVSR). Laboratory tests were conducted on the collected samples: geotechnical classification, triaxial compression tests and constant head permeability tests in triaxial cell. Pressure plate tests were also done on unsaturated slag to evaluate the characteristic soil-water curve useful for transient seepage analysis. A seepage analysis was performed in order to obtain the maximum pore water pressures during the intense rainfall event which hit the area on 29th August 2003. The results highlight that the slag low permeability prevents the infiltration of rainwater, which instead seeps easily through the boundary levees built with coarse materials. For this reason pore water pressures inside the deposits are not particularly influenced by rainfall intensity and frequency. Seismic stability analysis was performed with both the pseudo-static method, coupled with Newmark's method, and dynamic methods, using as design earthquake the one registered in Tolmezzo (Udine) on 6th May 1976. The low reduction of safety factors and the development of very small cumulative displacements show that the stability of embankments is assured even if an earthquake of magnitude 6.4 and a daily rainfall of 141.6mm occur at the same time.

Keywords: mine tailings, dynamic stability analysis, seepage, safety factor, partial saturation.

INTRODUCTION

The mining village of Raibl stands along the "Rio del Lago" valley, 9 km south of Tarvisio (Udine), in a seismic and rainy mountainous region, on the border between Italy, Austria and Slovenia. From the 1980s up to 1991, when the mine closed, the waste of the flotation enrichment process of Pb, Zn sulphide and iron minerals was discharged, in slurry form, into four impoundment surface areas confined by levees, built using the coarse material removed when excavating mine galleries, while the base consists of the alluvial deposits of the torrent.

The deposits are situated along the course of the torrent and extend for about 130000m^2, with a longitudinal length of 1.2km and a height of the dikes ranging between 22m and 15m. The volume involved is almost 2 million m^3. The territory of Raibl, now called Cave del Predil, is classified as a second category seismic zone, with a maximum acceleration of 0.25g.

This paper describes the results of a dynamic stability analysis of the mine tailing deposits performed using the accelerogram of the Friuli earthquake that occurred on

CP1020, *2008 Seismic Engineering Conference Commemorating the 1908 Messina and Reggio Calabria Earthquake*,
edited by A. Santini and N. Moraci

6th May 1976 as seismic input. The pore water pressure state was defined by means of a previous seepage analysis carried out using the rainfall data of the local weather station during the event on 29th August 2003 as input. All the FEM analyses were carried out on 4 cross sections, which are considered more representative of the real domain.

IN SITU INVESTIGATIONS

To integrate the information deriving from a previous site investigation campaign [1] conducted on the levees in 1989-90, further geognostic investigations were done in the summer of 2007, for the geotechnical characterization of the fine mine tailings confined in the basins. The in situ investigations consisted of 5 boreholes equipped with piezometers to measure the water table levels, and 5 CPTE tests. In the boreholes, undisturbed samples were collected at different depths, and the SPT and Lefranc permeability tests performed (Fig.1). The main dynamic properties of tailings were identified on the basis of both SPT tests and geophysical field measurements (ReMi and refraction methods), with 60-120m long seismic lines.

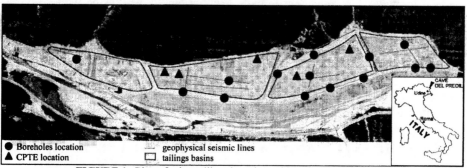

FIGURE 1. Plan of mine tailing deposits and location of in situ tests.

GEOTECHNICAL PROPERTIES OF LEVEES AND SLAG

The fine material in the basins is mainly composed of sequences of sandy and silty layers with very low plasticity. The unit weights of solid particles are generally high due to the presence of heavy metals inside the tailings confined in the basins.

The shear strength and conductivity parameters were obtained in a triaxial apparatus, with CIU triaxial and constant head permeability tests respectively.

The shear strength properties show that the fine sediments have a low cohesion c' and peak shear friction angle ϕ' ranging between 33° and 37°; these values are confirmed by the CPT data processing. The range of permeability values obtained from the constant head triaxial test ($10^{-6} \div 10^{-7}$m/s) point out that the increasing of the finer fraction of sediments induces a reduction in hydraulic conductivity up to 2 orders of magnitude. The main geotechnical parameters evaluated on undisturbed samples are summarized in Tab.1.

TABLE 1. Main geotechnical properties of tailings.

Grain size parameters			Atterberg Limits		
sand	(%)	6 - 55	w_P	(%)	13.9 - 17.3
silt	(%)	45 - 79	w_L	(%)	17.2 - 25.2
clay	(%)	0 -22	I_P	(-)	2.1 - 8.0
Index properties			Oedometer tests		
w_n	(%)	17.2 - 36.6	C_c	(-)	0.058 - 0.225
γ	(kN/m^3)	20.8 - 22.9	C_s	(-)	0.012 - 0.024
G_S	(-)	2.97 - 3.08	C_v	(cm^2/s)	$3.2 \cdot 10^{-2}$ - $6.9 \cdot 10^{-3}$
γ_d	(kN/m^3)	16.9 - 18.4	C_α	(%)	0.041 - 0.079
n	(-)	0.38 - 0.62	Triaxial tests		
e	(-)	0.61 - 1.09	c'	(kPa)	5 - 28
S	(%)	86 - 100	φ'	(°)	33 - 37

At the end of the main laboratory tests, measurements of the water retention properties of the sediments were taken with a Richards apparatus for a range of negative pore pressures between 0kPa and 1518kPa. These evaluations are useful to obtain the characteristic soil-water curve of the materials involved in the seepage analysis, to model their behaviour in partial saturation regime. Lastly, the CPTE tests pointed out cone resistance of between 0.8MPa and 7MPa, typical of loose and medium-dense granular soils respectively. The geotechnical characterization of coarse dike materials was based on the results of the 1989-90 in situ and laboratory campaign. The USCS classification defines this coarse material as GM, instead other finer sandy-silt layers are classified as SM-ML. The levee material shows a sandy fraction ranging between 12% and 59%, a finer fraction between 1% and 37%, a null cohesion and an effective friction angle of about 40°.

SEISMIC INPUT AND DYNAMIC CHARACTERIZATION

To analyze the dynamic behaviour of the basins the accelerogram was used recorded in Tolmezzo on the occasion of the 6th May 1976 earthquake, which had a magnitude of 6.4 on the Richter scale. The seismic wave propagated in a N-S direction, the seismic signal lasted 36.53s and a peak ground surface acceleration of 0.357g after 4s. Only the first 10 seconds of accelerogram data were used for the dynamic analysis simulation, as after this the seismic action produces negligible stresses. The main parameters used to synthetically represent the seismic action are the Arias intensity I_A = 0.75 m/s [2] and destructive potential P_D = 0.062 m·s. [3].

For the dynamic characterization of the materials involved, the representative values of maximum shear moduli G_0 were defined on the basis of geophysical test results as well as the SPT interpretation and processing data.

TABLE 2. Main dynamic properties of materials.

	G_0 (MPa)	E_0 (MPa)	ν
levees coarse material	150	400	0.334
mine tailings	60	160	0.334
bank protection work	31220	83295	0.334

The G_0 reduction functions and damping ratio functions due to the response to cyclic shear deformations were obtained from data in the literature for both the levees material [4] and the fine tailings [5] and are represented in Fig.2.

The pore water function was introduced to estimate the excess pore water pressures due to the shaking. It defines the trend of the pore water pressures coefficient r_u as a function of the N/N_L ratio (Fig.3a). The N parameter was chosen on the basis of the earthquake magnitude (6.4) [6], instead the N_L parameter, evaluated in post processing of the dynamic analysis, depends on the cyclic shear ratio (CSR) (Fig.3b), proportional to the cyclic deviatoric stress and the initial static effective minor principal stress ratio.

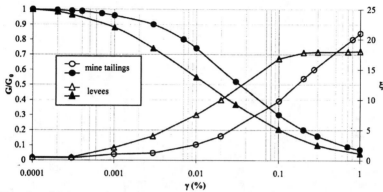

FIGURE 2. G_0 reduction functions and dumping ratio functions for mine tailings and levee materials.

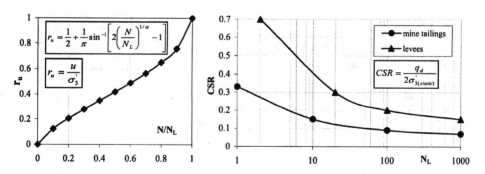

FIGURE 3. Pore pressure ratio and cyclic stress ratio functions for mine tailings and levee materials.

PORE WATER PRESSURES DURING INTENSE RAINFALL

The main purpose of transient seepage analysis was to compute the deposits response to the intense rainfall on 29th August 2003 (92.2 mm in one hour), evaluate the maximum values of pore water pressures during the event and their influence on the basins' stability.

Combinations of analyses with different hydraulic conductivity and anisotropy characteristics of materials were conducted for this; a summary is given in Fig.4.

For fine mine tailings, the choice of modelling both the isotropic and anisotropic behaviour ($k_x/k_y=5$) is justified by the discharge method of the slurry into the basins during the mining activity, which favoured a sedimentation of the materials in thin layers of different grain size.

Regarding the levee materials, the decision to vary the saturated horizontal conductivity coefficient from a maximum value of 10^{-4} m/s to a minimum of 10^{-6} m/s is justified by the fact that the levees were affected, during the years of tailings and slurry production, by an intense seepage phenomenon that could have caused the migration and transport of the finer fraction towards and into the levees.

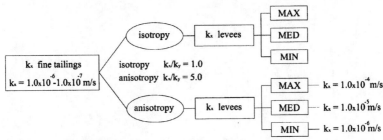

FIGURE 4. Summary of the analysis combinations.

The seepage phenomenon inside the basins is ruled by the interaction of two very different materials: that of the levees, coarse and more pervious, which rapidly drains the flows that pass through it, and the one confined in the basins, which can remain in conditions of partial saturation (Fig.5). This material, even many hours after the rain has stopped, tends to hold and slow the flows that move to the more pervious lateral sides. Water tends mainly to move from the core of the basins towards the lateral levees in order to reach the torrent, instead of moving in the direction of the bottom of the basins.

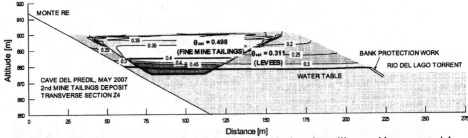

FIGURE 5. Distribution of volumetric water content inside the mine tailings and levees materials.

The pore water pressures are not particularly influenced by the intensity and frequency of rainfalls, as is also confirmed by the groundwater level variations measured by piezometers over six months. In the end the analysis results showed that the anisotropic case introduces a certain delay in the basins response, but not as important as expected; this could be due to the natural tendency of the seepage flow to have a prevalently horizontal component.

DYNAMIC STABILITY ANALYSIS

The maximum values of the pore water pressures in the tailings deposits, during the modelled rainfall, were computed and used as initial condition for the subsequent stability analysis. This analysis was developed in three phases: pre-seismic, dynamic and post-seismic, using the main geotechnical properties of the materials, which are summarized in Tab.3.

TABLE 3. Main geotechnical properties of material involved in stability analysis.

	γ (kN/m^3)	ϕ' (°)		γ (kN/m^3)	ϕ' (°)
levees	22.5	37	alluvial deposits	22.5	45
mine tailings	21 - 22	33 - 38	bank protection work	25.5	45

The results of the pre-seismic static analysis, performed on the most representative sections with the Morgenstern & Price method, show the remarkable stability of the embankments (Fs>1.51 without the suction contribution and Fs>1.66 with this resistance contribution), even varying the permeability and anisotropy characteristics of the materials.

In order to evaluate the dynamic behaviour and stability due to the concomitance of intense rainfall and seismic action (Friuli, 6th May 1976), the most critical of the modelled sections was chosen. This section, for each levee's permeability condition, was modelled with a commercial software [7], defining both the seismic input and the dynamic properties of the materials involved (Tab.2).

The dynamic analysis identified the time steps that show the maximum shear stresses (Fig.6), the greatest horizontal displacements (Fig.7) and the distribution of the excess pore water pressures induced by the shaking in the whole analyzed domain. Fig.6 also shows the time history acceleration in two representative points situated at the top and toe of the slope.

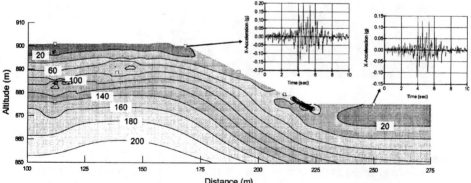

FIGURE 6. Distribution of maximum shear stress at critical time step on dynamic analysis.

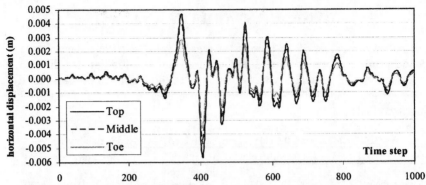
FIGURE 7. Trend of instantaneous horizontal displacements during the seismic action.

The calculated acceleration values at the toe, middle section and top of the slope never reach the peak ones of the accelerogram and there is then a general phenomenon of deamplification of the seismic motion.

The maximum deamplification values are registered at the toe and instead decrease moving towards the top of the slope.

These results are surely due to the dynamic behaviour of the modelled materials characterized by high stiffness and regular geometries.

In the post-seismic phase the safety of dikes for the most critical section was evaluated with Newmark's method; the downstream levee is never affected by the formation of critical sliding surfaces, not even in some instants of the modelled accelerogram (Fig.8), so there are no permanent displacements because the maximum acceleration is less than the critical one.

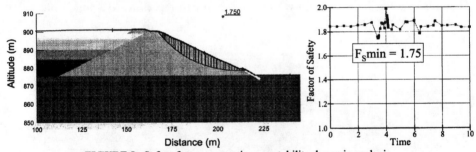
FIGURE 8. Safety factor versus time on stability dynamic analysis.

Confirmation of the remarkable dynamic stability of the dikes and confined sediments is provided by the safety factors values, which are always greater than 1.75; this value was reached at the time step in which, in the advanced dynamic analysis, the maximum horizontal displacement was computed.

CONCLUSIONS

The stability analyses performed on the levees and mine tailings basins of the Raibl Mine permit some conclusions to be drawn:

1. Because of the low permeability of fine sediments, the basins are affected by water flows that permit even intense rainfall to drain slowly, without rapid variations in water table levels. The pore water pressures inside the basins are not particularly influenced by the intensity and frequency of rainfall.
2. Velocity vectors show the tendency of the external flows to move, in a sub-horizontal way, from the core of the basins towards the lateral levees in order to reach the torrent, instead of moving in the direction of the bottom of the basins.
3. The static stability analyses performed for the maximum pore pressure values show safety factors greater than 1.5, for sliding surfaces passing both inside and at the base of the levees.
4. The dynamic analyses, performed combining the rainfall event and the seismic action point out that there are no sensible reductions in the basins' stability, because of the high stiffness of materials, regularity of geometries and low excess pore water pressures generated by the cyclic action.
5. The displacement analysis performed using Newmark's method confirms the significant safety of the downstream levee even at the time step in which, in the advanced dynamic analysis, maximum horizontal displacements were computed.
6. The calculated accelerations at the toe, middle, and top of the slope show a general deamplification phenomenon of the seismic motion.

REFERENCES

1. Ismes & Aquater S.p.A. Studio sulla verifica della sicurezza delle dighe, degli invasi minori, delle vasche d'accumulo e dei canali pensili. Studio di dettaglio dei bacini di filtrazione della miniera di Raibl. Committente Direzione Regionale della Protezione Civile - Friuli Venezia Giulia, 1990.
2. A. Arias, "A Measure of Earthquake Intensity", in *Seismic Design of Nuclear Power Plants*, MIT Press, Cambridge, Massachusetts, 1970, pp.438-468.
3. G.R. Saragoni, "Response Spectra and Earthquake Destructiveness", Proceedings of IV U.S. National Conference on *Earthquake Engineering*, Palm Springs, Berkeley, California, Earthquake Engineering Research Institute (EERI), 1990, pp.35-43.
4. H.B. Seed, R.T. Wong, I.M. Idriss, K. Tokimatsu. Moduli and damping factors for dynamic analyses of cohesionless soils. Report n.UCB/EERC-84/14, University of California, Berkeley, 1984.
5. H.B. Seed, I.M. Idriss. Soil moduli and damping factors for dynamic response analyses. Report n. EERC 70-10, University of California, Berkeley, 1970.
6. H.B. Seed, K. Mori, C.K. Chan. Influence of seismic history on the liquefaction characteristics of sands. Report EERC 75-25, Earthquake Engineering Research Center, University of California, Berkeley, 1975.
7 Geostudio. Dynamic Modeling with QUAKE/W. An Engineering Methodology. GEO-SLOPE International Ltd., 2007.

Seismic Analysis of a Rockfill Dam by FLAC Finite Difference Code

Livia Miglio, Alessandro Pagliaroli, Giuseppe Lanzo,
and Salvatore Miliziano

Dept. of Structural and Geotechnical Eng., Sapienza Università di Roma, Italy.

Abstract. The paper presents the results of numerical analyses carried out with FLAC finite difference code aiming at investigating the seismic response of rockfill dams. In particular the hysteretic damping model, recently incorporated within the code, coupled with a perfectly plastic yield criterion, was employed. As first step, 1D and 2D calibration analyses were performed and comparisons with the results supplied by well known linear equivalent and fully non linear codes were carried out. Then the seismic response of El Infiernillo rockfill dam was investigated during two weak and strong seismic events. Benefits and shortcomings of using the hysteretic damping model are discussed in the light of the results obtained from calibration studies and field-scale analyses.

Keywords: rockfill dam, numerical analyses, hysteretic damping model, calibration study, El Infiernillo dam.

INTRODUCTION

A new hazard model for the Italian territory has been recently proposed [1] leading to updated PGA hazard maps for different return periods. This re-classification has led to the necessity of evaluating the seismic safety of existing dams. As matter of fact, many Italian earth and rockfill dams were designed without adequately considering the seismic actions, since sites were classified as non-seismic or low seismicity areas at the construction time.

Traditionally the seismic safety assessment of earth dams has been evaluated through simple pseudostatic and sliding block methods. However, in the past fifteen years the numerical procedures have substantially evolved and the use of more sophisticated approaches, such as fully coupled effective stress analyses sometimes based on very complex constitutive models, has gradually increased. Several numerical codes are nowadays available for the simulation of the dynamic behavior of earth dams [2]. In particular, in the finite difference code FLAC has been recently implemented a new dynamic module that incorporates the hysteretic damping model [3,4,5] which directly accounts for nonlinear and dissipative soil behavior. This model was so far applied only in few literature studies for analyzing the seismic response of earth dams [6,7] and therefore it needs to be further validated through calibration studies.

This paper presents the results of numerical analyses carried out with FLAC for investigating the dynamic response of El Infiernillo Dam (Mexico) for which

CP1020, *2008 Seismic Engineering Conference Commemorating the 1908 Messina and Reggio Calabria Earthquake*,
edited by A. Santini and N. Moraci

extensive static and dynamic monitoring data are available [2]. The hysteretic damping model, coupled with a simple Mohr-Coulomb failure criterion, was employed for the study. Preliminary calibration studies were undertaken with increasing levels of complexity with the purpose of comparing the FLAC results with those obtained by well known 1D [8,9] and 2D [10] computer codes.

FLAC MODELING OF HYSTERETIC BEHAVIOR

Two approaches are conventionally used to simulate the non linear and dissipative behaviour of soils when subjected to cyclic loading: equivalent linear methods and nonlinear methods. The first approach represents the material as visco-elastic; nonlinearity is simulated by an iterative process in which the shear modulus and damping ratio are adjusted to be compatible with the level of strain. Nonlinear methods use nonlinear constitutive models to represent the hysteretic behaviour; these models can be quite complex and require many material parameters. If a simple elastic or elastic-plastic model is employed, an additional damping must be added to represent the hysteretic characteristics, usually through the Rayleigh formulation. This causes a significant reduction in the time step in a explicit solution scheme, employed by codes like FLAC, and thus higher computational times.

The hysteretic damping formulation, recently available in the numerical code FLAC [3,4], can be added to simple constitutive models as an alternative to Rayleigh damping. It allows to adjust the tangent shear modulus for each zone in the model thus reproducing the hysteretic cyclic response. It is well known that the relationship between shear stress τ and the corresponding shear strain γ may be expressed as:

$$\frac{\tau}{G_0} = \frac{G_s(\gamma)}{G_0}\gamma = M_s(\gamma)\,\gamma \tag{1}$$

where G_S is the secant shear modulus, G_0 the small strain shear modulus and M_S the normalized shear modulus. If the relationship $M_S(\gamma)$ is known from an appropriate modulus decay curve, the tangent normalized shear modulus M_t can be evaluated as:

$$M_t = \frac{d\tau}{d\gamma} = M_s + \gamma\frac{dM_s}{d\gamma} \tag{2}$$

and then the tangent shear modulus for each zone of the model can be computed as $G_t = M_t\,G_0$. Several built-in $M_S(\gamma)$ functions are available in FLAC; in the present study the *sigmoidal3* expression was employed:

$$M_s = \frac{a}{1 + \exp[-(L - x_0)/b]} \tag{3}$$

where $L = \log_{10}\gamma$ while a, b and x_0 are parameters to be determined by fitting the reference G_S/G_0-γ and damping D-γ curves. In addition to the backbone curve provided by applying equation (2), the two standard Masing rules are used to specify the behavior at reversal points thus modeling the unloading-reloading loops. The formulation above presented, well defined for 1D straining, is therefore extended to 2D-3D conditions by generalizing the notions of strain amplitude γ to update G_t and reversal points [3,4]. This hysteretic mechanism is stationary, i.e. stress depends only on strain and not on the number of cycles. It has been found that analyses performed employing hysteretic damping run up to 7-8 times faster than the comparable simulations with Rayleigh damping [4].

1D CALIBRATION ANALYSES

As first step of the calibration study, 1D site response analyses were carried out. A soil profile characterized by a shear wave velocity V_S increasing with depth from about 90 to 300 m/s, overlying a bedrock having $V_S=914$ m/s [11], was considered (Fig. 1). The nonlinear soil behavior was expressed by the standard G_S/G_0-γ and D-γ Vucetic and Dobry curves [12] corresponding to a plasticity index PI=50 (Fig. 1a). The WE component of the accelerogram recorded at Yerba Buena station during the 1989 Loma Prieta earthquake, scaled to 0.2g, was applied as input outcropping motion. The results obtained with FLAC, employing the *sigmoidal3* hysteretic damping model, were compared with those supplied by Proshake and D-MOD_2. The first one performs equivalent linear analyses in the frequency domain; the full nonlinear D-MOD_2 employs the modified Kondner-Zelasko hyperbolic model to describe the backbone curve coupled with standard Masing rules [9,11].

The results of the calibration study are illustrated in terms of response spectra at surface (Fig. 1b), maximum acceleration a_{max} (Fig. 1c) and maximum shear strain γ_{max} profiles (Fig. 1d). A very satisfactory agreement among the codes can be observed both in term of peak values and response spectra at surface. D-MOD_2 computes slightly lower surface PGA with respect to the other codes because of higher damping ratio at the maximum shear strain amplitude (0.1%) reached in the analyses (Fig. 1a).

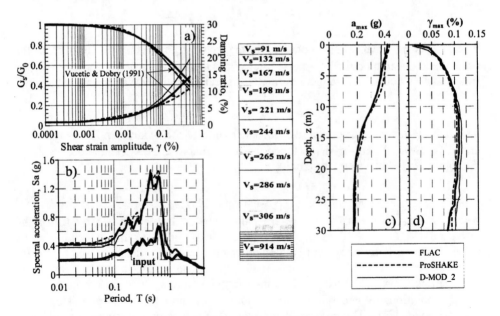

FIGURE 1. 1D calibration analyses: G_S/G_0-γ and D-γ curves (a), acceleration response spectra at surface (b), maximum acceleration (c) and maximum shear strain (d) profiles.

2D CALIBRATION ANALYSES

The 2D calibration analyses were carried out on an homogeneous triangular-shaped ridge characterized by height and width at the base equal to 30 m and 60 m, respectively. The ridge has V_s=300 m/s and the nonlinear behavior is expressed by the average curves proposed by Seed et al. [13] for cohesionless soils. The ridge overlies a V_S=900 m/s bedrock located 30 m below the ridge toe (Fig. 2). The results obtained with FLAC *sigmoidal3* model were compared with those by 2D QUAD4M finite element code, in which soils are modeled as equivalent linear visco-elastic materials being the viscous damping of the full Rayleigh type [10].

The NS component of the accelerogram recorded at Gilroy #1 station during the 1989 Loma Prieta earthquake, scaled to 0.1g, was applied as input outcropping motion. The results are shown in terms of response spectra at the crest (Fig. 2a), PGA profiles at the surface (Fig. 2b) and along the vertical across the ridge crest (Fig. 2c). Linear analyses with damping ratio D=2% are also shown for comparison. It can be noted that for the linear case FLAC almost perfectly matches the QUAD4M results with the exception of the peak of spectral acceleration at the crest (Fig. 2a). This minor discrepancy can be ascribed to the differences in the boundary conditions, input motion application at the base of the mesh and Rayleigh formulations employed by the two codes [14]. For the nonlinear case significant differences can be noted especially in the PGA values in the upper 10 m of the ridge; in particular, PGA at the crest computed by FLAC is about 25% higher than the QUAD4M corresponding value (Fig. 2b). Conversely, the agreement in terms of response spectra, with exception at high frequencies, is quite satisfactory (Fig. 2a). As the level of nonlinearity is the same of that experienced in the 1D simulations ($\gamma = 0.1\%$), for which FLAC perfectly matches the other codes, the discrepancies observed in the 2D simulation can be presumably ascribed to the procedures adopted to compute the reversal points of the loops and the cyclic strain amplitude employed to update the stiffness modulus [4].

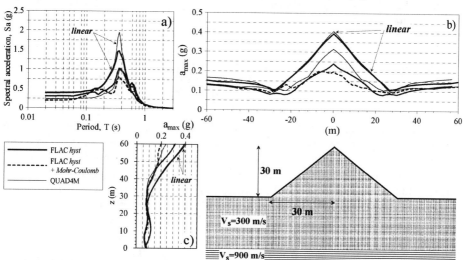

FIGURE 2. 2D calibration analyses: acceleration response spectra at the crest (a), maximum acceleration at the surface (b) and along the vertical across the crest (c).

An additional analysis was executed with FLAC by coupling the hysteretic damping formulation with a Mohr-Coulomb failure criterion. A cohesion c'=10 kPa and a friction angle φ'=35° were assigned to the material. During the shaking irreversible strain accumulation took place in the upper part of the ridge with permanent displacement of about 10 cm at crest. It can be observed that this plastic deformations produce an additional energy dissipation with about 35% PGA reduction close to the crest with respect to elastic non linear analysis (Fig. 2b-c). A significant decrement can be also observed in terms of spectral accelerations (Fig. 2a).

ANALYSIS OF SEISMIC RESPONSE OF EL INFIERNILLO DAM

The El Infiernillo dam is a rockfill dam located in Southern Mexico. The core is a compacted sandy silt of medium plasticity (PI=26), the filters are made of washed and screened sand from alluvial deposits while the shells consist of compacted and dumped rockfill, a dioritic rock mixed to silicified conglomerate [2]. The bedrock is constituted by silicified conglomerates. Monitoring data are available for both static and dynamic loading conditions. Displacement profiles during the construction and the reservoir impoundment were measured. Moreover, during its life the dam experienced several earthquakes, recorded by the seismic monitoring system. In particular, the strong motion event of March 14, 1979 (M=7.6) and the weak motion event (M=5.5) of May 30, 1990 were considered. These earthquakes produced a recorded PGA at the outcropping bedrock of about 0.1g and 0.02g, respectively. The corresponding transversal acceleration values recorded at crest were 0.36g and 0.096g. During the 1979 event a crest settlement of 11 cm was measured while the weak motion event induced no significant displacements [2].

The dynamic analyses were carried out on the main cross section of the dam having height and width at the base equal to about 140 m and 600 m, respectively (Fig. 3). It was discretized in about 34,000 elements, whose height ranges from 1 m in the core and rockfill to 4.5 m in the bedrock. The aspect ratio (i.e. the ratio between width and height of the elements) is comprised between 0.67 and 3. The size of elements was chosen in order to achieve a satisfactory solution accuracy for frequencies up to 10 Hz. Shells and filters were considered as a unique material in the numerical simulations.

The initial state of effective stress, before the application of dynamic excitation, was obtained from static 2D decoupled analyses in which multistage dam construction, the reservoir impoundment at the maximum level, and the unconfined seepage through the dam were modeled [15]. The soils were assumed to behave as elasto-plastic material with a Mohr-Coulomb failure criterion. The shear modulus (G) an the bulk modulus (K) were varied to back-calculate the displacement observed during dam construction. In particular, constant values were assumed for the rockfill (G=14 MPa, K=43 MPa) and for the core (G=8 MPa, K=24 MPa) which allowed to perfectly reproduce settlement profiles measured during the construction [15].

For dynamic simulations the Mohr-Coulomb elasto-plastic model was coupled with the *sigmoidal3* hysteretic damping formulation. A small amount (0.5%) of Rayleigh damping was also added to provide a non-zero damping at very small strains. The stiffness of soil was updated from "static" values to small strain ones. In particular, different V_S profiles were assumed for the materials. In the following the results are only shown for two hypotheses: a "step profile" and a "continuous profile" this latter computed according to Duncan and Chang relation [16] as a function of initial state of effective stress obtained from the static analyses (Fig. 3). A V_S=1200 m/s was assigned to the bedrock. Regarding the G_S/G_0-γ and D-γ curves, reference was made to literature data relative to materials of similar characteristics. In particular, the Vucetic and Dobry [12] curves for PI=30 was adopted for the core while those proposed by Kokusho and Esashi [17] for crushed rock were assigned to rockfill material. It should be noted that the adopted constitutive model does not take into account for possible pore pressure increment development in the dam. Anyway, in lack of experimental data, the analyses carried out by Sica [2] showed that quite low increment could be generated during the strong motion event, reaching a maximum value of 50 kPa in the upper part of the upstream filter. The parameters adopted in the dynamic analyses, resulting from laboratory tests [2] and the selected V_S profiles are listed in Table 1.

The transversal and vertical acceleration time histories recorded at the outcropping bedrock during both the seismic events were applied as vertically propagating SV and P waves, respectively.

TABLE 1. Physical and mechanical properties of the dam construction materials and bedrock.

Material	γ (kN/m^3)	G (MPa)	K (MPa)	c' (kPa)	φ' (°)
Core	20	5-180	15-550	10	28
Shells-filters (Rockfill)	18	135-535	400-1600	10	43
Bedrock	23	3380	5630	linear elastic	

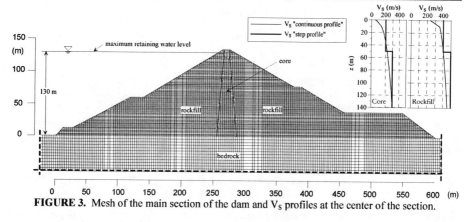

FIGURE 3. Mesh of the main section of the dam and V_S profiles at the center of the section.

A comparison between computed and measured horizontal and vertical displacements and response spectra at crest for the 1979 earthquake is reported in Fig. 4a-b. It can be noted that the calculated displacements are fairly close to the field-measured ones. The recorded a_{max} is also satisfactorily matched by numerical analyses while a severe underestimation of the spectral acceleration values can be observed, especially in the range 0.5-1.0 s (Fig. 4c). In this respect, it is worth noting that the signals recorded at crest was longer than 60 seconds while only the higher amplitude 10s-lasting motion was recorded at the rock outcrop and used as input motion. Therefore, the ground motion recorded at crest cannot be completely reproduced.

Moreover, the numerical results showed that the dam deformation is due to plastic strain development at the contact between the upper core and the upstream rockfill [15] where the shear strain amplitude reached peak values between 0.1% and 0.6%.

The comparison between recorded and predicted data for the 1990 event is shown in Fig. 5. As for the 1979 event, the a_{max} at crest is captured while appreciable underestimation of spectral acceleration values are observed. Elastic horizontal and vertical displacements take place at crest with no significant permanent deformations according to monitoring data. The numerical analyses therefore confirm that during the 1990 event the dam substantially remained in the elastic domain.

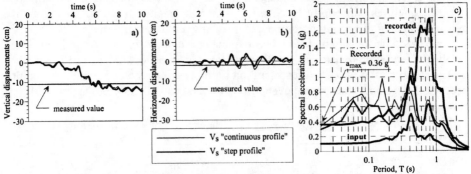

FIGURE 4. Vertical (a) and horizontal (b) displacement time-histories, acceleration response spectra (c) computed at crest for the 1979 seismic event.

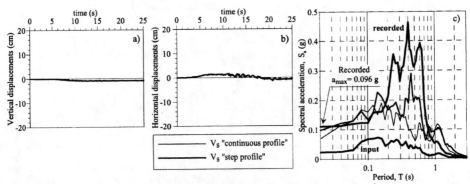

FIGURE 5. Vertical (a) and horizontal (b) displacement time-histories, acceleration response spectra (c) computed at crest for the 1990 seismic event.

CONCLUSIONS

Calibration analyses have shown that the recently implemented FLAC hysteretic damping model can successfully reproduce the seismic response of typical 1D and 2D configurations. In the dynamic analysis of El Infernillo rockfill dam the hysteretic damping model was coupled with the Mohr-Coulomb failure criterion. The results show that the code allows to capture the real behaviour of the dam in terms of deformation mechanism, permanent displacements and maximum acceleration at crest. Anyway, severe underestimation of recorded spectral acceleration at crest was found.

This discrepancy could be related to possible 3D effects, uncertainties on mechanical properties of the involved materials and to the limitations of the constitutive models. In this respect, the procedure adopted to compute the shear strain amplitude employed to update the stiffness modulus should be further validated. Moreover, it should be investigated if the hysteretic damping model, coupled with an elasto-plastic model, could lead to an overestimation of the material damping when appreciable plastic deformations take place. It can be concluded that the employed constitutive model represents an useful tool for the routine seismic safety assessment of earth dams, being characterized by few input parameters and a relatively high computation speed.

REFERENCES

1. Ordinanza PCM n. 3519 del 28 aprile 2006 "Criteri generali per l'individuazione delle zone sismiche e per la formazione e l'aggiornamento degli elenchi delle medesime zone" (G.U. n.108 del 11/05/06)
2. S. Sica, "Analisi del comportamento dinamico di dighe in terra", *PhD Thesis*, Consorzio Università di Roma "La Sapienza" e Università di Napoli "Federico II" (2001). (in italian)
3. ITASCA, "FLAC – Fast Lagrangian Analysis of Continua – Version 5.0". User's Guide, Itasca Consulting Group, Minneapolis, USA (2005).
4. P. A. Cundall, "A simple hysteretic damping formulation for dynamic continuum simulations", *4th Int. FLAC Symp. on Numerical Modeling in Geomechanics*, Madrid, May 29-31, (2006).
5. Y. Han and R. Hart, "Application of a simple hysteretic damping formulation for dynamic continuum simulations", *4th Int. FLAC Symp. on Numerical Modeling in Geomechanics*, Madrid, May 29-31, (2006).
6. I. Escuder, L. Altarejos and M. G. Membrillera, "FLAC numerical models applied to safety assessment of dams", *4th Int. FLAC Symp. on Numerical Modeling in Geomechanics*, Madrid, May 29-31, (2006).
7. Z. Y. Feng, Y. H. Chang, P. H. Tsai and J. N. Li, "Dynamic response of Li-yu-tan earthdam subjected to the 1999 Chi-Chi earthquake in Taiwan", *4th Int. FLAC Symp. on Numerical Modeling in Geomechanics*, Madrid, May 29-31, (2006).
8. EduPro Civil System, Inc. "ProShake – Ground Response Analysis Program", EduPro Civil System, Inc., Redmond, Washington (1998).
9. N. Matasovic, "D-MOD_2 A Computer Program fos Seismic Response Analyses of Horizontally Layered Soil Deposits, Earthfill Dams and Solid Waste Landfills", GeoSyntec Consultants, Huntington Beach, California (1995).
10. M. Hudson, I. M. Idriss and M. Beikae, "QUAD4M: a computer program to evaluate the seismic response of soil structures using finite element procedures and incorporating a compliant base", Dep. of Civil and Environmental Eng., University of California Davis, Davis California (1994).
11. N. Matasovic, "Seismic response of composite horizontally-layered soil deposits", *PhD Thesis*, Department of Civil Engineering, University of California, Los Angeles (1993).
12. M. Vucetic and R. Dobry, Effect of soil plasticity on cyclic response. ASCE, *Journal of Geotech. Eng.* **117**, 89-107 (1991).
13. H.B. Seed, R.T. Wong, I.M. Idriss and K. Tokimatsu, "Moduli and damping factors for dynamic analyses of cohesionless soils", *J. of Soil Mech. and Found. Divis. ASCE* **112**, 1016-1032 (1986).
14. A. Pagliaroli, "Studio numerico e sperimentale dei fenomeni di amplificazione sismica locale di rilievi isolati", *PhD Thesis*, Sapienza Università di Roma, Roma, (2006). (in italian)
15. L. Miglio, "Analisi sismica di una diga in terra mediante un codice di calcolo alla differenze finite", *Master Thesis*, Sapienza Università di Roma "", (2007). (in italian)
16. J. M Duncan and C. Y. Chang, "Nonlinear analysis of stress and strain in soils", *Proc. ASCE, Journal of Geotechnical Engineering Division* **96**, N. SM5. (1970).
17. T. Kokusho and Y. Esashi, "Cyclic triaxial tests on sands and coarse materials", *X ICSMFE*, Vol. 1, 673-676, (1981).

Effect of Sediment on Dynamic Pressure of Gravity Dams Using an Analytical Solution

Majid Pasbani Khiavi[a], Ahmad R. M. Gharabaghi[a], Karim Abedi[a]

[a]Department of Civil Engineering, Sahand University of Technology, Tabriz, Iran

Abstract. This paper presents an analytical solution to get a reliable estimation of the earthquake-induced hydrodynamic pressure on gravity dams by proposing closed-form formulas for the eigenvalues involved when solving the fluid and dam interaction problem. A new analytical technique is presented for calculation of earthquake-induced hydrodynamic pressure on rigid gravity dams allowing for water compressibility and wave absorption at the reservoir bottom. This new analytical solution can take into account the effect of bottom material on seismic response of gravity dams. The obtained results are in good agreement with other classical solutions. The main capability of proposed analytical solution is direct calculation of eigenvalues, without any need for numerical solution. In addition the method can be easily incorporated in dynamic analysis of a dam.

Keywords: Dam, Reservoir, Analytical solution, Bottom absorption, Earthquake.

INTRODUCTION

There are a large number of concrete gravity dams all over the world. Some of these dams are built in seismically active areas. The analysis of dams is a complex problem due to the dam-water-foundation interaction. An important factor in the design of dams in seismic regions is the effect of hydrodynamic pressure exerted on the face of dam as a result of earthquake ground motions. For a rational analysis of a dam-reservoir system, it is essential that the hydrodynamic effects and interaction between the dam and reservoir are properly considered. The dynamic behavior of a concrete gravity dam is affected by the adjacent reservoir and the flexible strata consisting of porous sediments on which it rests. The developed hydrodynamic pressure on dam is dependent on the physical characteristic of the boundaries surrounding the reservoir including reservoir bottom. In various methods proposed by different researchers for simplification of the analytical procedures, the reservoir bottom is generally considered to be rigid. This assumption does not represent the actual behavior of the system. The hydrodynamic pressure in the reservoir is usually affected by radiation of waves towards infinity and wave absorption at the reservoir bottom. When the reservoir bottom is considered to be rigid, the pressure waves are reflected from the reservoir bed and consequently the hydrodynamic pressure is over-estimated. Due to the absorption at the reservoir bottom, the magnitude of hydrodynamic pressure due to ground motion will be reduced. Therefore, the hydrodynamic pressure exerted on the upstream face of dams will be less and the displacement and stress field in the dam will be affected. Thus, an accurate evaluation

CP1020, *2008 Seismic Engineering Conference Commemorating the 1908 Messina and Reggio Calabria Earthquake*,
edited by A. Santini and N. Moraci

of hydrodynamic pressure on the dam must consider the effect of sediments of the reservoir bottom.

This paper presents an analytical solution to get a reliable estimation of the earthquake-induced hydrodynamic pressure on gravity dams by proposing closed-form formulas for the eigenvalues involved when solving the fluid and dam interaction problem. In this paper, a formulation based on a new analytical method is used to simulate the two dimensional domain of the reservoir including an absorbing bottom.

THEORETICAL FORMULATION

Let us consider a gravity dam with a vertical upstream face, impounding a reservoir of constant depth and extending to infinity in the upstream direction. It is assumed that the dam and reservoir are resting on a flexible foundation which is modeled as a viscoelastic half-plane. Assuming the water in the reservoir to be inviscid, compressible, irrotational and its motion to be of small amplitude, the hydrodynamic pressure equation will be:

$$\Delta P = \frac{1}{C^2}\ddot{P} \quad \text{in} \quad \Gamma \tag{1}$$

where Δ, P, C and Γ are the Laplacian operator, hydrodynamic pressure in the reservoir, velocity of sound in water and reservoir domain, respectively. Two-dimensional form of Eq. 1 can be written as:

$$\frac{\partial^2 P}{\partial x^2} + \frac{\partial^2 P}{\partial y^2} = \frac{1}{C^2}\frac{\partial^2 P}{\partial t^2} \tag{2}$$

where x and y are the cartesian coordinates and t is the time variable. Eq. 2 together with the appropriate boundary conditions, defines completely the hydrodynamic aspects of the problem. Figure 1 shows reservoir domain and boundaries.

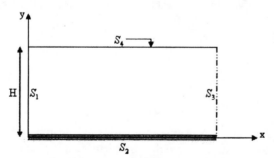

FIGURE 1. Reservoir domain and boundaries

The boundary conditions to be satisfied are as follows:
(a) At the dam-reservoir interface:

$$\frac{\partial P}{\partial n} = -\rho_w a_n \quad \text{on} \quad S_1 \tag{3}$$

559

in which n denotes the inward normal direction to a boundary, a_n the normal component of boundary acceleration and ρ_w the mass density of water.

(b) At the reservoir-bed interface assuming the absorbing boundary for reservoir bottom, the condition will be:

$$\frac{\partial P}{\partial n} + q\,\dot{P} = \rho a_y(x,t) \quad \text{on} \quad S_2 \tag{4}$$

where
q is the damping coefficient of the reservoir bottom

$$q = \frac{\rho_w}{C_s \rho_s} \tag{5}$$

ρ_s and C_s are mass density and sound wave velocity in sediment. The portion of the wave amplitude reflected back to the reservoir can be represented by the wave reflection coefficient α defined by

$$\alpha = \frac{1 - Cq}{1 + Cq} \tag{6}$$

where α may vary from 0 for full wave absorption to 1 for full wave reflection.
$a_y(x,t)$ is the vertical acceleration of ground motion at foundation. For a horizontal upstream excitation, the vertical acceleration is zero. In this case, the bottom boundary condition leads to an eigenvalue problem.

(c) At the reservoir farfield, perfect damping is assumed:

$$\lim_{x \to \infty} P = 0 \quad \text{on} \quad S_3 \tag{7}$$

(d) At the free surface, neglecting the effects of water surface waves, the boundary condition is:

$$P = 0 \quad \text{on} \quad S_4 \tag{8}$$

When an excitation is caused by a harmonic horizontal acceleration, the pressure field in the rectangular reservoir is governed by the following boundary conditions:

$$\frac{\partial P(0, y, \omega)}{\partial x} = -\rho_w a_x(0, y, \omega) \tag{9}$$

$$-\frac{\partial P(x, 0, \omega)}{\partial y} = i\omega q P \tag{10}$$

$$P(x, H, \omega) = 0 \tag{11}$$

$$\lim_{x \to \infty} P(x, y, \omega) = 0 \tag{12}$$

Using the standard method of separation of variables, it can be shown that the solution of Eq. 2 leads to a Sturm Liouville problem, with complex-valued frequency dependent eigenvalues of the impounded water.

For harmonic ground motion, the pressure in the reservoir can be expressed in the frequency domain as $P(x, y, t) = \overline{P}(y)e^{i(\omega t - kx)}$, where ω is the excitation frequency, k the wave number and $\overline{P}(y)$ the complex-valued frequency response function for hydrodynamic pressure. Substitution of this expression into Eq. 2 yields to the classical Helmholtz equation:

$$\frac{\partial^2 \overline{P}}{\partial y^2} + \lambda^2 \overline{P} = 0 \tag{13}$$

where

$$\lambda^2 = \frac{\omega^2}{C^2} - k^2 \tag{14}$$

The general solution of Eq. 13 is obtained as:

$$\overline{P}(y) = A\cos \lambda y + B\sin \lambda y \tag{15}$$

In the above equation, A and B are constant and are determined using boundary conditions. Using the mentioned boundary conditions for free surface and reservoir bottom, one can obtain:

$$B = -A\cot \lambda H \tag{16}$$

and

$$\cot \lambda H = -\frac{i\omega q}{\lambda} = -i\frac{\omega q H}{\lambda H} \tag{17}$$

The roots of this equation yield to frequency-dependent complex eigenvalues. According to recent equation, it is clear that the problem only involves the dimensionless quantities of $\omega q H$ and λH. Note that, if foundation absorption is considered, the eigenvalues and eigenvectors will be frequency-dependent. With an arbitrary scaling ($A=1$) the corresponding eigenvectors will be

$$\overline{P}_n(y) = \cos \lambda_n y - \cot \lambda_n H Sin\lambda_n y \tag{18}$$

In this equation λ_n is the frequency-dependent eigenvalues for each mode.

The differential Eq. 3 with the assumed boundary conditions constitutes a Sturm-Liouville problem and hence the eigenvectors are orthogonal. By a linear combination of the eigenvectors the complete solution for the hydrodynamic pressure may be written as following:

$$P = e^{i\omega t} \sum_{n=1}^{\infty} A_n e^{-ik_n x} \overline{P}(y) \tag{19}$$

which is determined up to the coefficient A_n. These coefficients can be calculated from the acceleration boundary condition at the dam. For a horizontal time-harmonic acceleration in the form of

$$a_x(y,t) = \hat{a}_x(y)e^{i\omega t} \tag{20}$$

using dam-reservoir interface boundary condition, the amplitude A_n for each mode is given by

$$A_n = -\frac{\hat{a}_m}{ik_n} \tag{21}$$

where

$$\hat{a}_m = \frac{\int_0^H \hat{a}_x \overline{P}_n(y)dy}{\int_0^H \overline{P}_n^2(y)dy} \tag{22}$$

and

$$ik_n = \sqrt{\lambda_n^2 - \frac{\omega^2}{C^2}} \qquad (23)$$

The pressure distribution in the reservoir is obtained in non-dimensional form as:

$$\frac{P(x,y,t)}{H\rho_w \hat{a}_x} = \left[\sum_{n=1}^{\infty} \frac{1}{ik_n H} \frac{E_1}{E_2} \overline{P}_n(y) e^{-ik_n x} \right] e^{i\omega t} \qquad (24)$$

in which

$$E_1 = \int_0^H \overline{P}_n(y) dy = \int_0^H (\cos \lambda_n y + \frac{i\omega q}{\lambda_n} \sin \lambda_n y) dy = H(1 - \frac{\omega^2 q^2}{\lambda_n^2}) \frac{\sin \lambda_n H}{\lambda_n H} + \frac{i\omega q}{\lambda_n^2} \qquad (25)$$

and

$$E_2 = \int_0^H \overline{P}_n^2(y) dy = \int_0^H (\cos \lambda_n y + \frac{i\omega q}{\lambda_n} \sin \lambda_n y)^2 dy =$$

$$\frac{H}{2}(1 - \frac{\omega^2 q^2}{\lambda_n^2})(1 + \frac{\sin^2 \lambda_n H}{\lambda_n H} \frac{i\omega q}{\lambda_n}) \qquad (26)$$

Solution of Eigenvalues

In this section Eq. 17 is solved directly for each excitation frequency ω by using proposed solution. For this purpose we can consider two cases. In the first case we assume that $\omega q H \leq n\pi$ and the solution of Eq. 17 for this case is:

$$\lambda_n H = (n - \frac{1}{2})\pi + \text{arctanh} \left[\frac{\omega q H}{(n - \frac{1}{2})\pi} \right] i \qquad (27)$$

In the second case, assuming $\omega q H \geq n\pi$ gives the solution of Eq. 17 as follows:

$$\lambda_n H = -\text{arctanh}(\frac{n\pi}{\omega q H})i + n\pi \qquad (28)$$

Model Verification

To assess the effectiveness and ability of the proposed analytical solution, two examples are considered in this section. By observing the solution leading to Eq. 24 one can follow that the accuracy of the proposed simplified formulation depends essentially on the wave reflection coefficient that represents the reservoir bottom absorption. To have a good understanding of the effect of this parameter, several situations were studied where the wave reflection coefficient is varied over a wide range. Finally, the distribution of hydrodynamic pressures at vertical upstream face of the gravity dam are obtained and compared with those obtained by the numerical formulations of Humar et al. (1988) and Maity et al. (1999) to check the validity and feasibility of the present analytical solution.

Example 1: Maity et al.(1999) assumed the dam to be rigid and subjected to harmonic excitation in the horizontal direction. The full depth of the reservoir was considered to be $70\,m$. The mass density of water was assumed to be $1000\,kg/m^3$ and the acoustic velocity of water $1440 \, m/s$. The amplitude of the external sinusoidal excitation was assumed to be equal to the gravitational acceleration. Similar assumptions are used in this study. The hydrodynamic pressure at the upstream face of dam is determined for two excitation frequencies that equal to $\omega = 0.04\Omega$ and $\omega = \Omega$, where Ω is the natural frequency of the reservoir given by $\Omega = \pi C / 2H$ and C is the acoustic velocity. In this example, the effect of reservoir bottom absorption is included by considering different values of reflection coefficients as 0.75 and 0.5. The normalized hydrodynamic pressure $P / \rho g H$ was selected as variable to assess the accuracy and effectiveness of the proposed analytical method. Figure 2 shows the obtained results from proposed analytical solution for different conditions of bottom absorption and comparison is made with the numerical model developed by Maity et al. (1999).

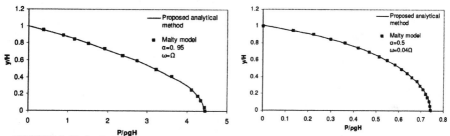

FIGURE 2. Hydrodynamic pressure distribution on dam height for different values of α and ω

Example 2: Humar et al. (1988) used the following data in their computation of hydrodynamic pressure:
Height of reservoir was considered as $100 \, m$, the mass density of water as 1000 kg / m^3 and the acoustic velocity of water as $1440 \, m/s$. The amplitude of the external horizontal acceleration was assumed to be equal to unit. Excitation frequency was assumed to be equal to natural frequency of the reservoir. The same situation is considered in this case. The results are obtained for two different conditions of partial damping due to wave absorption in the alluvial deposit at the bottom of the reservoir. These condition are presented by $\alpha = 0.5$ and $\alpha = 0.75$.

The results shown in Fig. 3 have been obtained from proposed analytical solution for different condition of bottom absorption and compared with the numerical results developed by Humar et al. (1988).

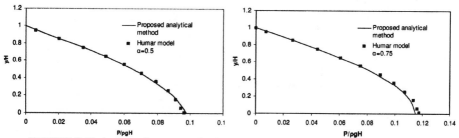

FIGURE 3. Hydrodynamic pressure distribution on dam height for different values of α

According to Fig. 2 and 3, the results obtained by new proposed analytical method are in very in good agreement with the results obtained by Maity (1999) and Humar (1988).

Results

Finally, the reservoir of example 2 is selected to illustrate the influence of the frequency ratio ω/Ω and reflection coefficient α on the variation of maximum hydrodynamic pressure exerted on dam. Figures 4-6 show the effect of ω/Ω and α on maximum hydrodynamic pressure for different cases.

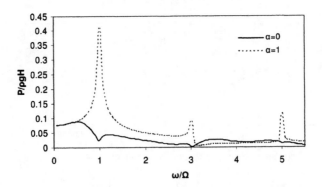

FIGURE 4. Maximum hydrodynamic pressure variation due to ω/Ω for full absorption and reflection

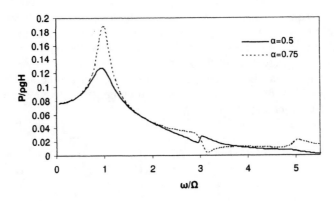

FIGURE 5. Maximum hydrodynamic pressure variation due to ω/Ω for moderate values of α

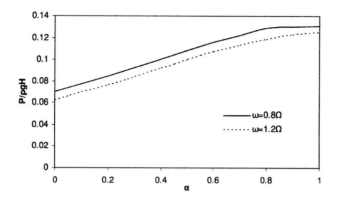

FIGURE 6. Maximum hydrodynamic pressure variation due to α for different excitation frequencies

It is obvious from Fig. 4 and 5 that the response becomes so much complicated when excitation frequencies are close to natural frequency of reservoir, where resonant is observed. With increasing wave absorption at the reservoir bottom and decreasing α, the fundamental resonant peak due to horizontal ground motion decreases in amplitude. Figure (6) represents similarity in behavior of maximum pressure variation due to α for both case of excitation frequencies (less and more than the natural frequency of the reservoir). In addition the magnitude of hydrodynamic pressure exceeds when $\omega < \Omega$ compared with the case of $\omega > \Omega$.

Conclusion

The paper has presented a new analytical technique to estimate the earthquake-induced hydrodynamic pressure on gravity dams allowing for water compressibility and wave absorption at the reservoir bottom. This new analytical solution can take into account the effect of bottom material on seismic response of gravity dams. The obtained results are in good agreement with other classical solutions. The main capability of proposed analytical solution is direct calculation of eigenvalues for different values of reflection coefficient without any need for numerical solution. In addition the method can be easily incorporated in dynamic analysis of dam.

REFERENCES

1. H. M. Westergaard, "Water pressure on dams during earthquake", Transactions, ASCE, Vol. **98,** 418-433 (1933).
2. G. Fenves and A. K.Chopra, "Effect of reservoir bottom absorption and dam-water-foundation rock interaction on frequency response functions for concrete gravity dams", Earthquake engineering and structural dynamics, Vol. **13,** 13-31 (1985).
3. J. L. Humar and A. M. Jablonski' "Boundary element reservoir model for seismic analysis of gravity dams", Earthquake engineering and structural dynamics, Vol. **16,** 129-1156 (1988).
4. D. Maity and S. K. Bhattacharyya, "Time-domain analysis of infinite reservoir by finite element method using a novel far-boundary condition", Finite element in analysis and design, Vol. **32,** 85-96 (1999).

FOUNDATIONS AND
SOIL-STRUCTURE INTERACTION

Finite element modeling of a shaking table test to evaluate the dynamic behaviour of a soil-foundation system

G. Abate[a], M.R. Massimino[b] and M. Maugeri[c]

[a] Ph.Dr. in Geotechnical Engineering, Department of Civil and Environmental Engineering, University of Catania, Viale A. Doria No. 6, 95125 Catania, Italy; glenda.abate@dica.unict.it
[b] Researcher in Geotechnical Engineering, Department of Civil and Environmental Engineering, University of Catania, Viale A. Doria No. 6, 95125 Catania, Italy; mmassimi@dica.unict.it
[c] Full Professor in Geotechnical Engineering, Department of Civil and Environmental Engineering, University of Catania, Viale A. Doria No. 6, 95125 Catania, Italy; mmaugeri@dica.unict.it

Abstract. The deep investigation of soil-foundation interaction behaviour during earthquakes represent one of the key-point for a right seismic design of structures, which can really behave well during earthquake, avoiding dangerous boundary conditions, such as weak foundations supporting the superstructures. The paper presents the results of the FEM modeling of a shaking table test involving a concrete shallow foundation resting on a Leighton Buzzard sand deposit. The numerical simulation is performed using a cap-hardening elasto-plastic constitutive model for the soil and specific soil-foundation contacts to allow slipping and up-lifting phenomena. Thanks to the comparison between experimental and numerical results, the power and the limits of the proposed numerical model are focused. Some aspects of the dynamic soil-foundation interaction are also pointed out.

Keywords: soil-foundation interaction, dynamic behaviour, shaking table test, FEM modeling, constitutive model, foundation rocking.

INTRODUCTION

Different recent earthquakes (1985 Mexico City, 1995 Kobe, 1999 Turkey) clearly demonstrate the fundamental role of Geotechnical Engineering for a good behaviour of all the kinds of structures during earthquakes. In particular, during the 1999 Turkey earthquake many buildings considerable tilted, losing their functionality but maintaining their structural integrity. Only sometimes the reasons of this geotechnical damage can be related to the very often quoted liquefaction phenomena, in many cases this damage is due to neglected soil-foundation interaction phenomena. The deep investigation of soil-foundation interaction behaviour during earthquakes represents one of the key-point for a right seismic design of structures, avoiding this kind of structural behaviour, or in general dangerous weak foundations supporting the superstructures.

CP1020, *2008 Seismic Engineering Conference Commemorating the 1908 Messina and Reggio Calabria Earthquake*,
edited by A. Santini and N. Moraci

The aim of this paper is to present the results of FEM modeling of a shaking table test involving a concrete shallow foundation resting on a Leighton Buzzard sand deposit. The test was performed using a six-degree-of-freedom shaking table, available at the University of Bristol and a flexible soil container for the sand deposit. The foundation was subjected to an eccentric vertical load and the whole foundation-soil system was subjected to an unidirectional horizontal excitation, characterized by constant frequency and variable amplitude, applied at the base of the soil container through the shaking table. The dynamic response of the whole system was deeply investigated by means of many accelerometers and displacements transducers located in the soil and on the concrete block.

The numerical simulation of the test is performed using a cap-hardening Drucker-Prager elasto-plastic constitutive model for the soil and specific soil-foundation contacts to allow slipping and up-lifting phenomena. Thanks to the comparison between experimental and numerical results, the power and the limits of the proposed numerical model are focused. Finally, some aspects of the dynamic soil-foundation interaction are pointed out.

PHYSICAL MODEL

The tested physical model consists of a concrete block resting on a dry Leighton Buzzard sand deposit. The concrete block, 0.4m wide x 0.4 m high with a length of 0.95 m, was embedded in the sand for a depth of 0.1 m. The sand was pluviated into a flexible soil container (named shear stack) 1.0 m wide x 5.0 m long x 1.2 m deep, for a total high of 1.0 m, in order to obtain an average void ratio $e = 0.6$ and a relative density D_r of about 52%, which leads to a shear strength angle $\varphi = 40°$, according to the following expression [1]:

$$\varphi(°) = 0.238 \cdot D_r\,(\%) + 28.4 \tag{1}$$

The utilized sand was an uncemented sand with sub-rounded particles, characterized by an average particle size $D_{50} = 0.94$ mm and a uniformity coefficient $C = D_{60}/D_{10} = 2.128$ corresponding to a narrow grading. It has been carefully investigated by many researchers in both static [2, 3] and dynamic fields [4, 5]. Index properties reported by [5] include: $G_s = 2.679$, $\gamma_{max} = 17.94$ kN/m^3, $\gamma_{min} = 15.06$ kN/m^3. As regards the sand dynamic characteristics, G and ξ, they are here fixed according to [6] concerning previous numerical simulation of similar shaking table tests. In particular, the shear modulus G is fixed equal to 2.9 MPa, according to that suggested by [6] for the shallower sand layers, and the damping ratio ξ is fixed equal to 20%.

The concrete block was embedded in the sand so that its short side is along the long side of the shear stack. The shear stack consists of a rectangular, laminar box made from aluminum rectangular hollow section rings separated by rubber layers. The dimensions and the materials of the shear stack were established in order to give it low elastic stiffness and natural frequency for reducing as much as possible boundary effects [7]. The shear stack was put on 3.0 x 3.0 m aluminum shaking table, available at the EERC Laboratory of the University of Bristol. It is characterized by 6-degree-of-freedom controlled using 8 servo-hydraulic actuators, and is capable of carrying a maximum load of about 15 t with an operative frequency range of 0-200 Hz.

As regards the load conditions, the concrete block was subjected to a vertical load given by two steel plates (10 kN each). The lower one is symmetrically placed on the concrete block, the second one is placed on the previous one with an eccentricity of 0.05 m; furthermore, the whole system was subjected to dynamic excitation applied through the shaking table only along the long side of the shear stack. The dynamic excitation consisted of a 5Hz sine-dwell having an amplitude, which increased from 0 to the maximum value during the 5 five cycles, was constant for the following 10 cycles, then decreased down to 0 in the last 5 cycles. Six shakes (named runs), characterized by different peak amplitude, were applied to the system (Table 1). Due to the lack of space, in the present paper only some results of run II and run V are discussed. In particular, for run II the dynamic response of the physical model is analyzed referring to the displacements, while for run V it is analyzed referring to the accelerations. Fig. 1 shows the concrete block supporting the two steel plates significantly tilted at the end of run V.

TABLE 1. Peak amplitude of the input acceleration at the shaking table for the different runs.

Run	Peak acceleration (g)
I	± 0.10
II	± 0.15
III	± 0.20
IV	± 0.25
V	± 0.30
VI	± 0.35

The whole physical model was widely monitored using: 10 Setra accelerometers on the shaking table, the shear stack and the concrete block; 8 Dytran accelerometers in the sand; 4 Celesco displacement transducers for the horizontal displacements of the shaking table, the shear stack and the concrete block; 4 LVDT displacement transducers for the vertical displacements of the concrete block (Fig. 2).

More detail on the experiment can be found in [8].

FIGURE 1. Concrete block, supporting the two steel plates, at the end of run V [8].

571

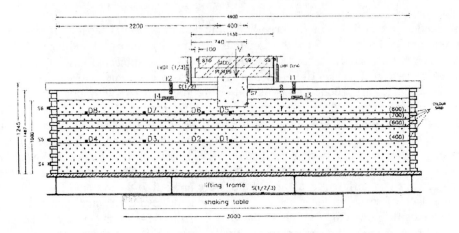

FIGURE 2. Instrumentation layout: S = Setra accelerometers; D = Dytran accelerometers; I = Indikon displacement transducers; LVDT = LVDT displacement transducers [8].

FEM MODELING

In order to simulate the shaking table experiment described in the previous paragraph, a 2-D FEM model, according to plane strain conditions, is created (Fig. 3), using ADINA code [9]. The whole model consists of 640 8-noded 2-D solid elements and three groups of elements: the soil, the concrete block, the steel plates. Totally there are 2102 nodes.

The mesh reproducing the soil is 4.8 m wide and 1.0 m high, the mesh reproducing the concrete block is 0.4 m wide and 0.4 m high. The concrete block mesh is embedded 0.1 m into the soil mesh. The steel plate mesh is 1.1 m wide and 0.1 m high and is drawn so that it has an eccentricity in respect to the concrete block of 0.05 m on the right side (this is an approximation: only the second plate actually presented the eccentricity). The shear stack is not modeled, but specific conditions are applied to the vertical boundaries of the soil mesh. In particular, the nodes of the two vertical boundaries of the soil mesh are linked by "constrain equations" that impose the same horizontal y-translation at the same z-depth, while they are completely free in the vertical direction to allow sand settlements. All the nodes at the soil mesh base, which represents the shear-stack base, are blocked along the z-direction.

Moreover, two load conditions are applied: i) a "mass proportional load", which is applied to the whole system, in order to take into account the weight of all the involved materials; ii) a sinusoidal input motion, applied at the horizontal bottom boundary of the finite element model, perfectly equal to that applied during the experiment at the shaking table. Thus, the incremental analysis is performed according to a number of steps equal to 4000, with a constant magnitude equal to 0.004 s, in order to compare directly the numerical results with the experimental ones, characterized by the same steps.

It is, also, important to underline the definition of special contacts between the soil and the foundation block, in order to guarantee soil-foundation relative displacements when the friction action is exceeded, as well as foundation block uplifting when tensile stresses should be applied to the sand. Special contacts are also applied between the concrete block and the steel plate meshes. Fig. 3 shows the mesh and the imposed boundary conditions.

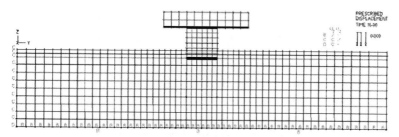

FIGURE 3. FEM model.

As regards the adopted constitutive models, the steel plates are modeled using the elastic linear isotropic constitutive model ($E = 2 \times 10^8$ kPa, $v = 0.2$, $\rho = 9.3$ kNs^2/m^4); the concrete block is also modeled by the elastic linear isotropic constitutive model ($E = 2.85 \times 10^7$ kPa, $v = 0.15$, $\rho = 2.65$ kNs^2/m^4); while for the soil an elastic isotropic work-hardening plastic cap constitutive model, available in the ADINA code, is used ($E = 5712$ kPa, $v = 0.3$, $\rho = 1.7$ kNs^2/m^4, a $= 0.31$ ($\varphi = 40°$), k $= 0$ (c $= 0$), W $= -0.015$, D $= -0.1$; T $= 0$, Initial cap position $= 0$, Cap ratio $= 0$). In particular, as regards this last constitutive model, its isotropic behaviour in the elastic range is governed by Young's modulus and Poisson's ratio; while, for the plastic behaviour, the loading function is assumed to be isotropic and consists of two parts: the ultimate, linear Drucker-Prager failure envelope and an elliptical strain-hardening cap, which produces plastic volumetric and shear strains. An associated flow rule is also considered. The chose of a work-hardening plastic constitutive model is necessary to catch the great accumulation of block displacements, caused by significant vertical and horizontal forces due to the steel plates and concrete block weights, showed in the following.

Furthermore, the Rayleigh damping factors α and β, necessary for a "Transient Dynamics" analysis, are determined according the relations $\alpha = \xi_{min} \cdot \omega$ and $\beta = \xi_{min} / \omega$ for $\xi_{min} = 20\%$ and $\omega = 2\pi \cdot 5$ r/s (see paragraph "Physical model").

EXPERIMENTAL AND NUMERICAL RESULTS

Due to the lack of space, for run II the comparison between experimental and numerical results is presented in terms of displacements regarding the sand surface around the concrete block (Fig. 4) as well as the concrete block (Fig. 5). While for run V the comparison between experimental and numerical results is presented in terms of accelerations recorded in the sand (Fig. 6) as well as on the concrete block (Fig. 7). For the sign convention adopted during the recording the displacements downwards are negative [8]; this convention is maintained also in this paper to a clearer comparison between experimental and numerical results.

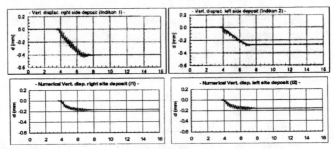

FIGURE 4. Comparison between experimental (first row) and numerical (second row) vertical displacements of the sand surface around the concrete block.

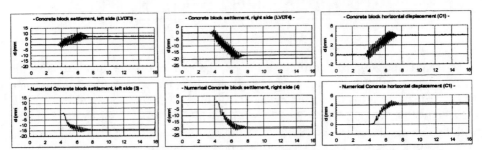

FIGURE 5. Comparison between experimental (first row) and numerical (second row) displacements of the concrete block.

As it is possible to observe from Fig. 4, the significant weight of the concrete block, its dimension in comparison with its embedment and the sand relative density caused settlements of the sand surface around it. This settlement was more significant on the right side, probably due to the steel plates, which slipped down (Fig. 1). The sand surface settlement is well captured by the numerical simulation on the left side of the concrete block and underestimated on the right side, because the FEM simulation does not predict the effects of the steel plate down slipping.

As regards the concrete block displacements (Fig. 5), the LVDT3 and LVDT4 transducers recorded essentially a rocking phenomenon, which caused the up-lifting of the block left side (LVDT3); while the FEM analysis shows a less evident rocking phenomenon and a predominant downwards movement of the block. In any case the settlement time-history recorded by LVDT4 transducer is perfectly captured by the FEM analysis. A very good agreement between experimental and numerical results exist also for the block horizontal displacement.

As regards the dynamic response of the sand deposit in terms of accelerations, Fig. 6 reports the experimental and numerical acceleration time-histories at 40 cm (D1, D3 and D4) and 80 cm (D5, D7 and D8) from the shear stack base. Due to the lack of space the accelerometers D2 and D6 are not considered in the present paper; the experimental results recorded by these accelerometers are reported in [8]. The alignment D1-D5 is taken into account because it allows us to investigate on the influence of the block in the sand dynamic response; the alignment D3-D7 because it shows the sand free-field response; the alignment D4-D8 because it allows us to observe possible boundary effects due to the shear stack walls. Comparing the acceleration time-histories recorded by D5 with those recorded by the other Dytran accelerometers it is possible to observe a significant influence of the block on the dynamic response of the sand. No amplification/ deamplification phenomena in free-field condition as well as near the shear stack walls, moving from the sand base to the sand surface can be observed. Finally, none influence of the shear stack walls can be observed.

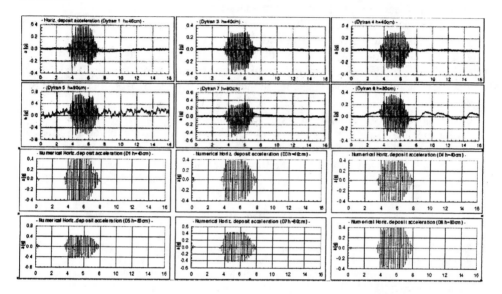

FIGURE 6. Dynamic sand response in terms of acceleration: experimental results (first and second rows) and numerical results (third and fourth rows).

The FEM results are globally very close to the experimental ones, even if the FEM analysis does not allow us to catch the increasing of sand acceleration near the block. Finally, in Fig. 7 the experimental response of the concrete block in terms of acceleration is compared with the numerical prediction. From this figure it is possible to observe, once more, a very good agreement between experimental and numerical results.

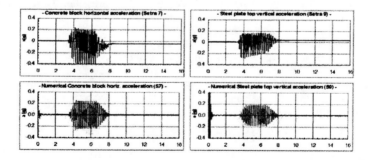

FIGURE 7. Dynamic concrete block response in terms of acceleration: experimental results (first row) and numerical results (second row).

575

Results similar to those reported in Figs. 4-7 are achieved for the other runs not discussed in the present paper.

CONCLUSIONS

The presented FEM simulation of a shaking table test on a concrete block/sand physical model leads to the following considerations:
- Transient dynamic FEM simulation of shaking table tests involving soil-foundation interaction can be successfully performed;
- Well known elasto-plastic soil constitutive model, such as the considered elastic, isotropic, work-hardening cap constitutive model can be appropriately used dealing with a heavy loaded massive structure for which the effects of isotropic hardening is predominant;
- For the considered sinusoidal wave (f = 5Hz, A_{max}=0.10-0.35g) moving through a 1.0 m thick deposit of Leighton Buzzard sand (D_r=50%) no amplification/deamplification phenomena are observed. An important increasing of the dynamic response in terms of acceleration is observed only below the foundation block.

REFERENCES

1. D.C.F. Lo Presti, S. Pedroni and V. Crippa, "Maximum dry density of cohesionless soils by pluviation and by ASTM-D4253-83: a comparative study. GTJ, XV(2), pp. 180-189 (1992).
2. M. A. Stroud, "The behaviour of sand at low stress levels in the simple shear apparatus", *PhD Thesis* ambridge University, UK (1971).
3. D. M. Wood and M. Budhu, "The behaviour of Leighton Buzzard sand in cyclic simple shear tests," *Soils under cyclic and transient loading,* edited by G.N. Pande and O.C. Zienkiewicz, Publisher Rotterdam: Publisher Balkema, 1980, pp. 9-21.
4. A. Cavallaro, M. Maugeri. and R. Mozzarella, "Static and dynamic properties of Leighton Buzzard sand from laboratory tests". *Proc. 4th Int. Conf. on Recent Advances in Geotechn. Earth. Eng. and Soil Dynamic and Symp. in honour of Prof. W.D. Liam Finn,* San Diego, California, march 26-31, 2001, Paper No. 1.13 (2001).
5. M. Dietz and D. Muir Wood, "Shaking table evaluation of dynamic soil properties". *Proc. 4th Int. Conf. on Earthquake Geotechnical Engineering,* June 25-28, 2007, Paper No. 1196 (2007).
6. A. Gajo and D. Muir Wood, "Numerical analysis of behaviour of shear stacks under dynamic loading". *Report on ECOEST Project, EERC Laboratory,* Bristol University (1997).
7. A. W. Crewe, M. L. Lings, C. A. Taylor, A. K. Yeung and R. Andrighetto, "Development of a large shear-stack for resting dry sand and simple direct foundations on a shaking table". *Proc. 5th SECED Conf. on European Seismic Design Practice,* Chester, Balkema, (1995).
8. M. Maugeri, G. Musumeci, D. Novità, C.A. Taylor, "Shaking table test of failure of a shallow foundation subjected to an eccentric load". *Soil Dynamics and Earthquake Engineering,* **20** (2000), pp. 435-444.
9. K.J. Bathe, "Finite Element Procedures". Prentice Hall, Englewood Cliffs, NJ, (1996).

Vibration Based Wind Turbine Tower Foundation Design Utilizing Soil-Foundation-Structure Interaction

Mohamed Al Satari, Ph.D., P.E.[a] and Saif Hussain, S.E.[b]

[a] Structural Engineer
[b] Managing Principal
Coffman Engineers, Inc., 16133 Ventura Blvd., Suite 1010, Encino, California, USA.

Abstract. Wind turbines have been used to generate electricity as an alternative energy source to conventional fossil fuels. This case study is for multiple wind towers located at different villages in Alaska where severe arctic weather conditions exist. The towers are supported by two different types of foundations; large mat or deep piles foundations. Initially, a Reinforced Concrete (RC) mat foundation was utilized to provide the system with vertical and lateral support. Where soil conditions required it, a pile foundation solution was devised utilizing a 30" thick RC mat containing an embedded steel grillage of W18 beams supported by 20"-24" grouted or un-grouted piles. The mixing and casting of concrete in-situ has become the major source of cost and difficulty of construction at these remote Alaska sites. An all-steel foundation was proposed for faster installation and lower cost, but was found to impact the natural frequencies of the structural system by significantly softening the foundation system. The tower-foundation support structure thus became near-resonant with the operational frequencies of the wind turbine leading to a likelihood of structural instability or even collapse. A detailed 3D Finite-Element model of the original tower-foundation-pile system with RC foundation was created using SAP2000. Soil springs were included in the model based on soil properties obtained from the geotechnical consultant. The natural frequency from the model was verified against the tower manufacturer analytical and the experimental values. Where piles were used, numerous iterations were carried out to eliminate the need for the RC and optimize the design. An optimized design was achieved with enough separation between the natural and operational frequencies to prevent damage to the structural system eliminating the need for any RC encasement to the steel foundation or grouting to the piles.

Keywords: Frequency, Soil-Foundation-Structure, Interaction, Steel.

INTRODUCTION

Wind towers have to sustain continuous vibration-induced forces throughout their operational life. The operating frequency of the three-blade turbine could potentially cause dynamic amplification of these forces significantly posing a threat to the overall structural integrity. Sufficient separation of the structural system natural frequency from the turbine operational frequencies is key to avoiding potentially catastrophic failures. The turbine operating frequency is typically lower than the structural system natural frequency, but could approach it as higher turbine output is obtained. Idealized assumptions of fixity at the base of the tower are un-conservative; a more realistic analysis accounting for foundation flexibility yields lower estimates of the natural

CP1020, *2008 Seismic Engineering Conference Commemorating the 1908 Messina and Reggio Calabria Earthquake*,
edited by A. Santini and N. Moraci
© 2008 American Institute of Physics 978-0-7354-0542-4/08/$23.00

frequency for the system. In such cases, soil-foundation-structure interaction needs to be considered.

Case Study: Wind Towers, Multiple Village Locations, Alaska

Structural Description

The utilized towers are out-of-commission prefabricated models that were donated to and/or purchased by the state of Alaska to generate electricity in the following remote villages: *Hooper Bay, Chevak, Gambell, Savoonga, Mekoryuk, Kasigluk Akula Bay, Toksook Bay, Danwin and Nordtank.* The tower models differ in height (23-46m), weight and wall thickness. Furthermore, the three-bladed turbines vary in weight (7812-7909km), blade diameter (19-27m) and power output (100-225kW). All towers were supplied by the same manufacturer; Distributed Energy Systems (DES), formerly Northern Power Systems (NPS), while the turbines were supplied by NPS and Vestas. Figure 1 below shows some of the installed towers.

FIGURE 1. Operational wind towers.

Operational Issues (Hazards)

As the wind turbine blades start to rotate from rest, their circular speed increases and the induced vibration frequency increases. Depending on its power output capacity, the turbine blades rotate at maximum rotational (circular) speeds that rage from 45 to 60 rpm corresponding to 0.75 to 1.00 Hz. These operational frequencies are very close to the range of natural frequencies of the entire Soil-Foundation-Tower-Turbine system.

If more output power is desired, higher rotational speeds have to be accommodated. A poor design decision would involve a maximum rotational speed that is very close to the natural frequency of the structural system resulting in a high likelihood of resonant amplification causing structural instibility. Another poor design would have a rotational speed not very close to yet higher than the natural frequency of the structural system. In such cases, the structure would have to endure violent near-resonance vibrations as the operational frequency approaches the natural frequency while speeding up to and down from the maximum speed. This situation would result in very high dynamic forces which could cause immediate damage to the structure. Even if these dynamic forces do not exceed the structure's strength capacity, fatigue-induced failures could also be encountered.

A sound design would avoid allowing the operational frequency to approach the vicinity of the natural frequency by a certain safety factor. A safety factor of 15% of the natural frequency was recommended by the turbine vendor and adopted by the authors for this project.

Design Objective

In order to develop a sound overall structural system that meets the structural performance requirements of the wind towers, the dynamic interaction of the supporting soil, foundation and super-structure needs to be considered. Since the tower and turbine are prefabricated and manufactured, once selected for a certain installation location, only the foundation can be designed and fine-tuned in accordance with the site soil conditions and desired system frequency.

Depending on the soil conditions, the optimum foundation system needs to be selected (spread footing, deep piles, micro-piles, etc.). Additionally, the foundation must have adequate stiffness in order to maximize the system natural frequency within practical limits. A suitably stiff soil-foundation-structure system will allow for higher power output generated by the turbines.

Foundation Design

Based on the geotechnical conditions at the different sites, two types of foundations were selected; large spread foundation and deep piles. A 5' deep, 12'x12' Reinforced Concrete (RC) spread footing was utilized to provide the system with vertical and lateral support as well as damping and stiffness. Where soil conditions necessitated it, a pile foundation solution was devised utilizing a 30" thick mat of RC foundation embedded with a steel grillage of W18 beams founded on 20" grouted piles.

After some installations were made, it was determined that the mixing and casting of concrete in-situ is the major source of cost and difficulty of construction. An all-steel foundation was proposed for faster installation and lower cost, but such a foundation system impacted the natural frequency and significantly softened the system. Consequently, the foundation design was driven by the system natural frequency. Multiple solutions combining different pile sizes, grouted and un-grouted, and different beam sizes were devised. The optimum design was selected for each location based on the highest practically obtainable natural frequency and cost effectiveness of the design.

Modeling and Analysis

A detailed 3D Finite Element Analysis (FEA) model of the tower-foundation-pile system was created using SAP2000. The tower was modeled using a fine mesh of thin shell elements, while the steel grillage and piles were assigned the appropriate cross-sectional properties. Thick plate elements were utilized to model the RC foundation.

In order to capture the soil-foundation-structure interaction, compression-only springs were devised to mimic the soil around the piles. Soil damping properties were conservatively neglected and the turbine mass was lumped at the hub height above the top of the tower. The natural frequency from the model was verified against the tower manufacturer analytical and the experimental values.

Discretization of FEA elements into sub-elements is a not as straight forward a task as some may believe. Unfavorable Discretization can give rise to subsequent numerical difficulties. In vibration analysis, for example, abrupt changes in element size should be avoided, as such changes tend to produce spurious wave reflections and numerical noise [4]. Consequently, a simplified tapered frame element was devised to model the tower instead of the thin-walled shell elements. Negligible deviations of the results from the two modeling techniques were observed. Thus, all design optimization runs utilized the tapered frame element. Figures 2 below shows the two different modeling techniques.

The piles were modeled using frame elements, meshed into 1' segments. The large spread footing, on the other hand, was modeled using a 3D solid element with RC properties (Figure 3). The solid element was meshed into sub-elements using an intelligent algorithm consistent with the object-based FEA modeling of the SAP2000 program. Figures 4 through 6 show the discretization of the piles and footing with the application of the soil springs to the meshed surfaces.

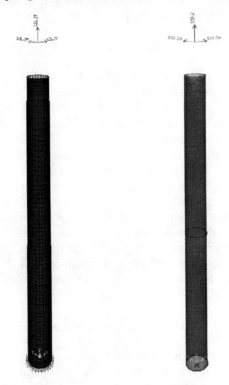

FIGURE 2. a) Meshed shell element model with lumped turbine mass b) Tapered frame element model with lumped turbine mass.

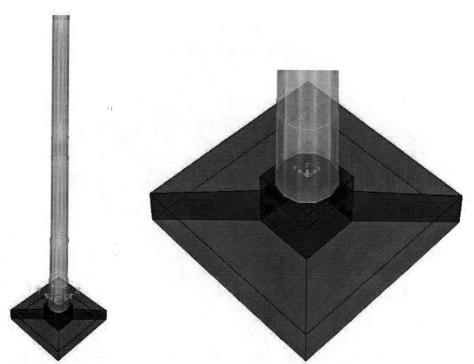

FIGURE 3. a) Tapered frame model on RC spread footing   b) Close-up of the 3D footing element.

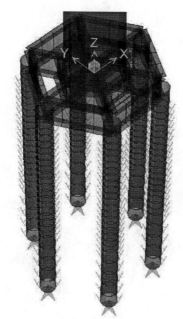

FIGURE 4. Steel tower support on top of piles which are laterally constrained by soil springs.

FIGURE 5. Meshed 3D solid element with vertical and horizontal compression-only soil springs.

FIGURE 6. Underside view of foundation.

Results

Where piles are used, numerous iterations were carried out to eliminate the need for the RC and optimize the design once a comfort level with the modeling technique was reached. The foundation system design was optimized through a parametric sensitivity-based approach in which the radius of the pile group, grillage beams and piles sizes were varied to produce comparable alternatives. It was found that he radius of the pile group had the most impact on the system frequency. A favorable radius was selected using a set of typical grillage beam and pile sizes. A series of further variations to the beam/pile sizes and different combinations yielded an optimized foundation design for each site. The optimized designs were achieved with enough separation (15%) between the natural and operational frequencies to prevent damage to the structural system. The optimization eliminated the need for any RC encasement to the steel foundation or grouting to the piles in many cases.

In most cases, an optimized foundation system design for a particular site was also found to be satisfactory for other locations. Thus, a small library of universally applicable *standard* designs we compiled in an effort to keep the fabrication cost low. Table 1 summarizes the final design for two of the tower locations and demonstrates how one optimized design is adequate in two locations with different geotechnical conditions. Figure 7 shows one of the optimized all-steel tower support foundation.

TABLE 1. Design summary for Savoonga and Mekoryuk villages.

Savoonga	
Tower Height	29m
Turbine C.G.	1.28 m (50") above top of tower
Combined Turbine and Blades Mass	7812 kg-mass (535.292 slugs)
Foundation Beams	W36X170
Pile Section	24" steel pipe, 3/4" thickness
Number of Piles	6
Point of Fixity	Varies; 18-11 ft below soil surface
Modulus of Horizontal Subgrade Reaction	Varies with depth; 19-319 kip/in
System Natural Frequency	1.128 Hz
Recommended Maximum rpm*	57 rpm
Mekoryuk	
Tower Height	29m
Turbine C.G.	1.28 m (50") above top of tower
Combined Turbine and Blades Mass	7812 kg-mass (535.292 slugs)
Foundation Beams	W36X170
Pile Section	24" steel pipe, 3/4" thickness
Number of Piles	6
Point of Fixity	30 ft below soil surface
Modulus of Horizontal Subgrade Reaction	Varies with depth; 19-1070 kip/in
System Natural Frequency	1.148 Hz
Recommended Maximum rpm*	58 rpm

*The recommended rpm incorporates a 15% safety factor.

FIGURE 7. Steel tower support.

CONCLUSIONS

The foundation system design was controlled by the natural frequency of the Soil-Foundation-Structure system rather than by strength or serviceability considerations. Taking into account the Soil-Foundation-Structure interaction yielded a more realistic estimate of the natural frequency. Had a fixed-base-tower assumption been adopted, significantly under-designed systems would have been incorporated.

ACKNOWLEDGMENTS

The authors would like to thank the following engineers at Coffman Engineers, Inc. for their contributions to this project: Steve Cegelka, Logan Haines, Ben Momblow, Scott Thompson, Paul Van Benschoten and Will Veelman.

REFERENCES

1. ACI 351-04, 2004, *Foundations for Dynamic Equipment,* American Concrete Institute, Farmington Hills, Michigan.
2. Vestas V27-225kW Specifications and Technical Data, *J.P. Saylor & Associates, Consultants Ltd.,* Des Moines, Iowa.
3. Geotechnical Reports, *Golder Associates, Inc.,* Anchorage, Alaska.
4. R. D. Cook, D. S. Malkus and M. E. Plesha, *Concepts and Applications of Finite Element Analysis*, New York: John Wiley & Sons, 1989, pp. 553-582.

The Impact of Dam-Reservoir-Foundation Interaction on Nonlinear Response of Concrete Gravity Dams

AliReza Amini[a], Mohammad Hossein Motamedi[b], Mohsen Ghaemian[c]

[a] Graduate Student, Department of Civil Engineering, Sharif University of Technology, Tehran, Iran
[b] Graduate Student, Department of Civil Engineering, Sharif University of Technology, Tehran, Iran
[c] Associate Professor, Department of Civil Engineering, Sharif University of Technology, Tehran, Iran

Abstract. To study the impact of dam-reservoir-foundation interaction on nonlinear response of concrete gravity dams, a two-dimensional finite element model of a concrete gravity dam including the dam body, a part of its foundation and a part of the reservoir was made. In addition, the proper boundary conditions were used in both reservoir and foundation in order to absorb the energy of outgoing waves at the far end boundaries. Using the finite element method and smeared crack approach, some different seismic nonlinear analyses were done and finally, we came to a conclusion that the consideration of dam-reservoir-foundation interaction in nonlinear analysis of concrete dams is of great importance, because from the performance point of view, this interaction significantly improves the nonlinear response of concrete dams.

Keywords: Concrete gravity dam, Dam-reservoir-foundation interaction, Finite element method, Nonlinear analysis

INTRODUCTION

In the past years, numerous researches have been conducted in order to determine how dams behave nonlinearly against the seismic loads. Although many achievements were obtained in the process of analysis and design of concrete dams, there are still many important questions unsolved. One of these crucial questions is the dam-reservoir-foundation interaction during an earthquake.

When subjected to earthquake, the analysis of dam-reservoir interaction effects is a complex problem. Westergaard [1] introduced an approach to determine approximately the linear and nonlinear response of the dam-reservoir system by a number of masses that are added to the dam body. Ghaemian and Ghobarah [2] showed that the added mass approximation may not be a suitable approach for nonlinear analysis of dam-reservoir systems.

The dam-reservoir system can be categorized as a coupled field system in a way that these two physical domains interact only at their interface. The staggered solution is a partitioned solution procedure that can be organized in terms of sequential execution of a single field analyzer. Ghaemian and Ghobarah [3] proposed two unconditionally stable methods of staggered solution procedure for the dam-reservoir interaction problem.

To simplify and economize the finite element modeling of an infinite reservoir, the far-end boundary of the reservoir has to be truncated. As a rule for the truncated boundary, there is no reflection for the outgoing wave. Sommerfeld boundary

CP1020, *2008 Seismic Engineering Conference Commemorating the 1908 Messina and Reggio Calabria Earthquake*,
edited by A. Santini and N. Moraci
© 2008 American Institute of Physics 978-0-7354-0542-4/08/$23.00

condition [4] is the most common one that is based on the assumption that at long distance from the dam, the water wave can be considered as a plane wave. Therefore, a radiation damping is introduced in the system. In the mentioned condition, the fluid is assumed to be incompressible. Sharan proposed a damper radiation boundary condition for the time domain analysis of a compressible fluid with small amplitude [5]. This Boundary condition is found to be very effective and efficient for a wide range of excitation frequencies.

The consideration of dam-foundation interaction and modeling the foundation makes the problem more challenging. In the past two decades, some solutions have been introduced for this question. Fenves and Chopra [6] studied the dam-water-foundation rock interaction in a frequency domain linear analysis. Then, Leger and Bhattacharjee [7] presented a methodology which is based on frequency-independent models to approximate the representation of dam-reservoir-foundation interaction problem. In the work presented by Gaun, Moore and Lin [8], an efficient numerical procedure has been described to study the dynamic response of a reservoir-dam-foundation system directly in the time domain. Later, Ghaemian, Noorzad and Moghaddam [9] showed that the effects of foundation's shape and mass on the linear response of arch dams are considerable.

Just like the reservoir, it is wise and economical to truncate the far-end boundaries of the foundation. Probably the most widely used model for soil radiation damping is the one of Lysmer and Kuhlemeyer [10]. In this model the foundation is wrapped by dashpots tuned to absorb the S and P waves. Saouma [11] proposed another boundary condition called Boulder Recommendation using some springs and dampers at the vertical truncated boundaries. Although Boulder Recommendation benefits the same Lysmer formulation for dampers, it introduces an efficient formulation for utilizing springs. Finally, it is necessary to mention the work done by Wilson [12] that presents a method called Soil Structure Interaction Method (SSI) for estimation of free field earthquake motions at the site of dams. This method neglects the presence of structure (dam) during the earthquake and assumes that the relative (added) displacement at the truncated boundary is zero, and shows that under these circumstances, the foundation just bears the inertia force and does not bear the earthquake force.

In the present article, a two dimensional dam-reservoir-foundation system is analyzed nonlinearly using finite element method and smeared crack approach. The dam-reservoir interaction is solved by staggered solution procedure while the Sharan Boundary condition is applied at the reservoir's far-end truncated boundary. The foundation is defined as a part of the structure and some different boundary conditions are applied at its truncated boundaries. Moreover, in order to estimate the free field motion at the site of dam, we use the SSI method.

DESCRIPTION OF THE STUDIED DAM-RESERVOIR-FOUNDATION SYSTEM AND THE INPUT GROUND MOTION

A two dimensional model of the highest monolith of Pine Flat dam, a part of its foundation and a part of its reservoir was established. This monolith is 122 meter high and 96.8 meter wide at the base. The reservoir and foundation length are ten and three

586

times as long as the monolith's width respectively and also the foundation's depth is equal to the monolith's width. Figure 1 shows the meshed model of the dam-foundation system and Table 1 introduces the assumed material properties of the dam and its flexible massed foundation.

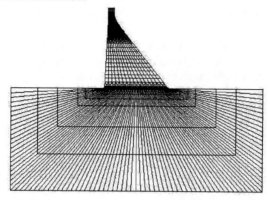

Figure 1. The meshed model of the dam-foundation system

Table 1. The assumed material properties of the dam and its foundation

	Modulous of Elasticity (E)	Poisson Ratio (v)	Unit Weigth (γ)	Tensile Strength (f_t)	Fracture Energy (G_f)
	MPa		KN/m^2	MPa	N/m
Dam Concrete	27580	0.2	23.52	2.7	250
Foundation Rock	14000	0.2	22.54	1.4	150

As shown in Fig. 1, the elements located in the potential crack zones (at the dam's base and neck) were defined so small so as to attain favorable aspect ratios.

In this study, two levels of earthquake motion were used. The horizontal component of Taft-Lincoln record of Kern County earthquake (PGA= 0.18g) was scaled up by the factors of 1.5 and 3 to act as moderate (PGA= 0.27g) and extreme (PGA= 0.54g) ground motions respectively.

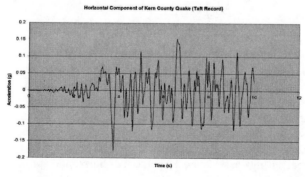

Figure 2. The horizontal component of the Taft-Lincoln record of Kern County earthquake

587

INTRODUCTION OF SOME CONDITIONS FOR FOUNDATION FAR-END TRUNCATED BOUNDARIES

We have studied and compared four different foundation boundary conditions in order to introduce the one which can absorb the outgoing waves as much as possible.

Lysmer Boundary Condition

This boundary condition is defined using Lysmer's theory about radiation damping. According to this theory, the surrounding boundaries could be modeled using normal and tangential dampers. In a finite element model, the damping factor of these dampers can be calculated as follows:

$$C_{11}^i = V_P \rho \int_{l_e} N_i dl \tag{1}$$

$$C_{22}^i = V_S \rho \int_{l_e} N_i dl \tag{2}$$

C_{11}^i and C_{22}^i are the damping factors in the normal and tangential directions respectively.

Figure 3-a introduces the defined boundary condition based on Lysmer's theory. The few supports which are placed at the base and both sides of the foundation prevent the instability of the dam-foundation system.

Boulder Recommendation

As shown in Fig. 3-b, this boundary condition has two specifications:
a) Setting horizontal springs at one of the foundation sides and horizontal dampers at the base and both sides of the foundation
b) Setting roller supports at the foundation base

The dampers' damping factor can be calculated from equation 1 and the springs' stiffness is computed by equation 3:

$$K_m = \frac{EA}{h} \tag{3}$$

where E is the foundation's modulus of elasticity, A the tributary area of the node connected to the spring, and h is a representative equivalent depth of the foundation.

Boulder Recommendation has two influential problems: firstly it does not constrain the vertical movement of the nodes at the foundation sides, and secondly it does not consider the foundation's radiation damping in the vertical direction.

To solve these problems, the Boulder Recommendation was modified as shown in Fig. 3-c. The damping factor of vertical dampers is calculated by equation 2 and the stiffness of vertical springs is set to be 10 percent of horizontal springs' stiffness.

In the finite element method, the properties of the constrained nodes are deleted from all matrices. Therefore, placing the vertical dampers at the nodes of foundation base is of no use. We placed these symbolic dampers just to mention that we should find a way for absorbing the vertical outgoing waves.

To solve the problems of Boulder Recommendation, we decided to define new boundary conditions in such a way that it can cover the mentioned weak points. Two of them are introduced in the following items.

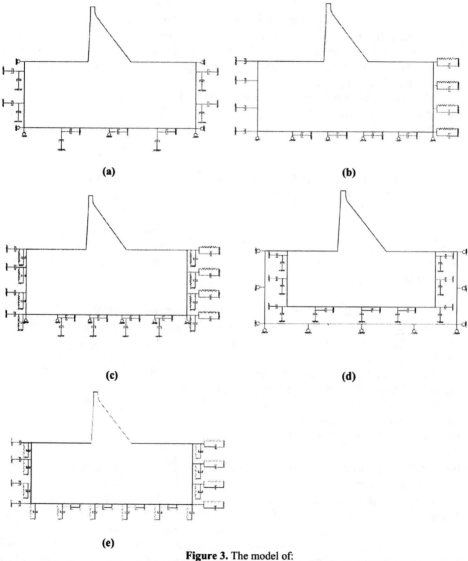

(a)

(b)

(c)

(d)

(e)

Figure 3. The model of:
a)Lysmer boundary condition, b) Boulder Recommendation, c) Modified Boulder boundary condition, d) Twofold boundary condition, e) Harmonic boundary condition

Twofold Boundary Condition

In order to maintain the stability of the dam-foundation system, some roller supports are put around the truncated boundary, and for the purpose of providing radiation damping, some horizontal and vertical dampers are set at an inner layer. The damping factors can be computed using equations 1 and 2. Figure 3-d shows that how this boundary is defined.

The appellation "Twofold" originates in this reality that two different conditions are defined in two different layers.

Harmonic Boundary Condition

In another try to solve the Boulder Recommendation problems, we decided to substitute the roller supports of Modified Boulder boundary condition (at the base of the foundation) with vertical springs (Fig. 3-e). Equation 3 is suitable for calculating the stiffness of these springs.

In this way, a proper harmony is made in the bilateral performance of springs and dampers. They will cooperate in the process of energy absorption and the springs will limit the truncated boundary displacements.

RESPONSE OF THE DAM-RESERVOIR-FOUNDATION SYSTEM TO A MODERATE EARTHQUAKE

In this section we study the response of the dam-foundation-reservoir system (defined in section 2) to a moderate earthquake with PGA= 0.27g (also introduced in section 2). The four mentioned foundation boundary conditions will be tested in order to report the best one.

When Lysmer Boundary Condition Is Applied

Figure 4-a shows the dam's condition at the end of the seismic excitation (As indicated in Fig. 2, the assumed excitation is 10 Sec long) when the Lysmer boundary condition is applied.

The cracks occurred inside the foundation near the heel of the dam, they are not that much long, and totally, the system response is good.

Inasmuch as the foundation rock's modulus of elasticity, tensile strength and fracture energy are less than the dam concrete's same parameters, the cracks predictably form inside the foundation.

When Modified Boulder Boundary Condition Is Applied

As shown in Fig. 4-b, there is no evident difference of the system response between this case and the previous one, hence the system response when using Modified Boulder boundary condition is favorable. On the other hand, this claim could be put forward that the Modified Boulder boundary condition is more real than Lysmer, because in this case all nodes around the truncated boundary bear stresses, but in Lysmer boundary condition there are many unconstrained nodes in the truncated layer bearing no stress.

When Twofold Boundary Condition Is Applied

Despite the fact that Twofold boundary condition was defined in order to cover the weak points of Modified Boulder, the results show that it does no act properly. For the first time, the energy balance error of the analysis overpasses the admissible quantity (15%), in such a case the analysis is supposed to become unstable (although our solution method- HHT-α method- is unconditionally stable, we do not admit the errors over 15%) .

Figure 4-c shows the dam's condition at the time of analysis instability. Compared to previous cases, both the length and depth of cracks are more.

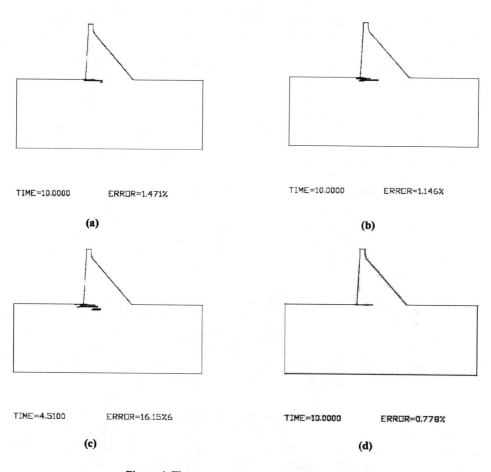

TIME=10.0000   ERROR=1.471%

(a)

TIME=10.0000   ERROR=1.146%

(b)

TIME=4.5100   ERROR=16.15%6

(c)

TIME=10.0000   ERROR=0.778%

(d)

Figure 4. The system response to a moderate earthquake:
a) when Lysmer boundary condition is applied
b) when Modified Boulder boundary condition is applied
c) when Twofold boundary condition is applied
d) when Harmonic boundary condition is applied

When Harmonic Boundary Condition Is Applied

Unlike Twofold boundary condition, the Harmonic boundary condition's performance is desirable. The analysis does not become unstable and the cracks are not that much long. It seems that this Boundary condition can compete with Lysmer and Modified Boulder boundary conditions. The analysis result is shown in Fig. 4-d.

RESPONSE OF THE DAM-RESERVOIR-FOUNDATION SYSTEM TO AN EXTREME EARTHQUAKE

This time we study the response of the dam-foundation-reservoir system to an extreme earthquake with PGA= 0.54g (introduced in section 2).

When Lysmer Boundary Condition Is Applied

The analysis becomes unstable after 4 seconds. Almost the whole of dam's base has cracked, and also an interesting event happened: the dam's base cracked from its toe. Of course considering the intensity of the ground motion, the system response is not weak. Figure 5-a shows the results at the time of analysis instability. As you see in Fig. 5-a, after 4 seconds of analysis, the error is less than 15%, but it reaches 25% at the time of 4.01 Sec, hence we decided to report the results of Time= 4 Sec.

When Modified Boulder Boundary Condition Is Applied

The analysis becomes unstable at the tome of 3.87 Sec, sooner than the case of Lysmer boundary condition's application. Therefore, it seems that Lysmer boundary condition acts a little better than Modified Boulder (Fig. 5-b). On the other hand, as mentioned before, it should not be neglected that Modified Boulder models the boundary condition in a more real way.

When Twofold Boundary Condition Is Applied

This boundary condition does not act well again. The instability time is 3.6 Sec which is sooner than the last two cases, in addition, the dam's heel cracks severely (Fig. 5-c). The problem is that the viscous dampers are set in the immediate vicinity of the roller supports, in the area where the nodes' velocities are not high. The lower the nodes' velocities are, the less the viscous dampers absorb energy.

When Harmonic Boundary Condition Is Applied

This boundary condition acts the best. The time of instability (4.12 Sec) is later than the other cases and also the cracks' configurations are more desirable, because the cracks are not as long as the previous ones but they are deeper (Fig. 5-d).

Therefore, Harmonic boundary condition absorbs the outgoing waves' energy very well and performs better than Lysmer and Modified Boulder boundary conditions.

Moreover, the model of Harmonic boundary condition is so similar to the reality, because in addition to energy absorption, it takes the both stresses and displacements of the truncated boundary nodes into consideration.

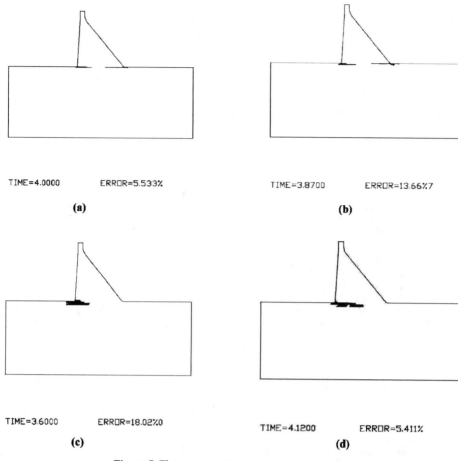

TIME=4.0000   ERROR=5.533%

(a)

TIME=3.8700   ERROR=13.66%7

(b)

TIME=3.6000   ERROR=18.02%0

(c)

TIME=4.1200   ERROR=5.411%

(d)

Figure 5. The system response to an extreme earthquake:
a) when Lysmer boundary condition is applied
b) when Modified Boulder boundary condition is applied
c) when Twofold boundary condition is applied
d) when Harmonic boundary condition is applied

CONCLUSIONS

The former researches have illustrated that if the impacts of dam-reservoir and dam-foundation interactions are neglected, even a moderate earthquake can cause considerable damage to the both parts of a well-designed gravity dam's base and neck This seems to be an overestimation.

When the nonlinear analysis includes the dam-reservoir-foundation interaction, in other words, when the reservoir's radiation damping and incompressibility and the foundation's mass, flexibility and radiation damping are cared with, the gravity dam's response will be so much better. The results of the present study show that moderate earthquakes can not badly damage a well-designed concrete gravity dam (like Pine

Flat dam). Thus the consideration of dam-reservoir-foundation interaction in numerical nonlinear solutions will outstandingly reduce the amount of damage.

To accurately model the reservoirs and foundations, special attention should be paid to the boundary conditions. There are some famous reservoir boundary conditions that perform favorably but the debates about the proper foundation boundary conditions have not finished yet. The analyses results indicated that the presented "Harmonic boundary condition" can act properly at the foundation's truncated boundaries. Therefore, the application of this boundary condition is advisable.

REFERENCES

1. Westergaard H.M., "Water Pressure on Dams During Earthquakes", Transactions, American Society of Civil Engineers, 1931, 98: 1303-1318
2. Ghaemian M.& Ghobarah A., "Nonlinear Seismic Response of Concrete Gravity Dams with Dam-Reservoir Interaction", Engineering Structures, 1999, 21: 306-315
3. Ghaemian M., Ghobarah A., "Staggered Solution Schemes for Dam-Reservoir Interaction", Journal of Fluids and Structures, 1998, 12: 933-948
4. Sommerfeld A., "Partial Differential Equations in Physics", Academic Press, New York, 1949
5. Sharan S.K., "Time-Domain Analysis of Infinite Fluid Vibration", International Journal of Numerical Methods in Engineering, 1987, 24: 945-958
6. Fenves G. & Chopra A.K., "Earthquake Analysis of Concrete Gravity Dams Including Reservoir Bottom Absorption and Dam-Water-Foundation Rock Interaction", Earthquake Engineering and Structural Dynamics, 1984, 12(5): 663-680
7. Leger P. & Bhattacharjee S.S., "Reduced Frequency-Independent Models for Seismic Analysis of Concrete Gravity Dam", Computer and Structures, 1992, 44(6): 1381-1387
8. Gaun F., Moore I.D. & Lin G., "Seismic Analysis of Reservoir-Dam-Soil Systems in the Time Domain", Computer Methods and Advances in Geomechanics, Siriwardane & Zaman (Eds), 1994, 917-922
9. Ghaemian M., Noorzad A. & Moghaddam R.M., "Foundation Effect on Seismic Response of Arch Dams Including Dam-Reservoir Interaction", Europe Earthquake Engineering, 2005, Ⅲ: 49-57
10. Lysmer J. & Kuhlemeyer R.L., "Finite Dynamic Model for Infinite Media", Journal of Engineering Mechanics Division, ASCE, 1969, 95: 859-877
11. Saouma V., "Course Note, Chapter 8: Nonlinear Dynamic Analysis of Dams"
12. Wilson E.L., "Three Dimensional Static and Dynamic Analysis of Structures, A Physical Approach with Emphasis on Earthquake Engineering", 4th Ed., Computers and Structures Inc., 2000

Influence of the Soil-Structure Interaction on the Design of Steel-Braced Building Foundation

Alireza Azarbakht[a] and Mohsen Ghafory Ashtiany[b]

[a] Assistant Professor, Department of Civil Engineering, Arak University, Arak, Iran 38156-879.
Email: a-azarbakht@araku.ac.ir
[b] Professor, International Institute of Earthquake Engineering and Seismology, (IIEES), Tehran, Iran.
Email: ashtiany@iiees.ac.ir

Abstract. The modeling and analysis of the superstructure and the foundation for the seismic lateral loads are traditionally done separately. This assumption is an important issue in the design/rehabilitate procedures especially for the short period structures, i.e. steel braced or shear wall systems, which may result to a conservative design. By using more advance procedures, i.e. nonlinear static method, and the incorporation of the soil–structure interaction (SSI), the seismic demand in the lateral resisting system decreases and the design will become more economic. This paper includes an investigation about the influence of the SSI effect on the design of the steel-braced building foundation. The presented example is a three-bay three-storey steel braced frame. Three design methods based on the FEMA 356 guideline and the UBC 97 code are taken in to consideration. The three methods are: (1) linear static analysis based on the UBC 97 code assuming the fixed based condition; (2) linear static analysis based on the FEMA 356 guideline assuming the fixed based condition; and (3) nonlinear static analysis assuming both fixed and flexible based assumptions. The results show that the influence of the SSI on the input demand of the short period building foundations is significant and the foundation design based on the linear static method with the fixed base assumption is so conservative. A simple method is proposed to take the SSI effect in to consideration in the linear static procedure with the fixed base assumption, which is a common method for the engineers. The advantage of this proposed method is the simplicity and the applicability for the engineering purposes.

Keywords: soil-structure interaction, FEMA 356, UBC 97, foundation, fixed base, flexible base.

INTRODUCTION

The actions for the design of foundation are defined to be force-controlled based on the FEMA 356 guideline [4]. It means that the total seismic lateral force shall be reduced by the factor of $C_1 C_2 C_3 J$ and should be combined with the gravity load effects (Equation 3-21 of [4]). The J is the force-delivery reduction factor. The resultant force shall be compared with the element resistance considering the lower bound material properties and the partial safety factors equal to the unity. In other words, the ultimate force is compared with the lower bound of the resistance.

It worth to compare the seismic base shears based on the FEMA 356 guideline and a well known design code such as UBC 97 code [6]. The comparison result is shown in Equation (1). The Equation (1) is derived for the structures with the fundamental period greater than T_0, where T_0 is the period at which the constant acceleration region of the design response spectrum begins. The structures, with shorter natural periods

CP1020, *2008 Seismic Engineering Conference Commemorating the 1908 Messina and Reggio Calabria Earthquake*,
edited by A. Santini and N. Moraci
© 2008 American Institute of Physics 978-0-7354-0542-4/08/$23.00

less than T_0, are not in this paper's context. For clarify of exposition, we can consider an ordinary steel braced frame ($R = 5.6$), which is a school building ($I = 1.25$), located in a high seismic zone (Tehran, Iran, $Z = 0.4$) and the soil type of S_c. The total ultimate base shear for the structure based on the UBC 97 code [6] and the force-controlled base shear based on the FEMA 356 guideline [4] are equalized to derive the Equation (1). The only difference will remain in the resistance side, where the partial safety factors are equal to the unity in FEMA 356 guideline [4] but it is less than one in the case of design codes, i.e. UBC 97 code [6]. This difference in the resistance side does not mostly affect on the final results, because the final conclusion strongly depends on the foundation uplift rather than the resistance partial safety factors.

$$\frac{C_v I}{RT} W = \frac{C_1 C_2 C_3 C_m S_a}{C_1 C_2 C_3 J} W \Rightarrow J = \begin{cases} (\dfrac{T}{T_s}) \dfrac{C_m R}{I} & for \quad T_0 \le T \le T_s \\ \dfrac{C_m R}{I} & for \quad T > T_s \end{cases} \tag{1}$$

where C_m is the effective mass factor, I is the importance factor, C_v is the seismic coefficient, R is the numerical coefficient representative of the inherent overstrength and global ductility capacity of lateral force-resisting system, T is the elastic fundamental period of vibration, T_s is the period at which the constant acceleration region of the design response spectrum ends and S_a is the spectral response acceleration at the fundamental period of structure.

For the illustrated example in the paper, the J factor can be written as Equation (2).

$$J = \begin{cases} (\dfrac{0.36}{0.56}) \dfrac{0.9 \times 5.6}{1.25} = 2.59 & for \quad T_0 \le T \le T_s \\ \dfrac{0.9 \times 5.6}{1.25} = 4.03 & for \quad T > T_s \end{cases} \tag{2}$$

In means, the above values for the J factor are implicitly used in the design code. On the other hand the FEMA 356 guideline [4] recommends $J = 2$ for the high seismic zones which is obviously less, at least with the prescribed assumptions, than the values that is implicitly used in the design procedures (as derived in Equation (2)). Thus, the foundation based on the design codes (e.g., UBC 97 code [6]) may be vulnerable based on the rehabilitation guidelines (e.g., FEMA 356 guideline [4]). In other words, the assessment of the foundation based on the FEMA 356guideline [4], at least for the linear static procedure with the fixed base assumption, is so conservative. This conclusion is not correct, if the movement of the base (soil-structure interaction) is taken in to account. The fixed-base modeling assumption is inappropriate for many structures though [2]. Structural systems that incorporate stiff vertical elements for lateral resistance (e.g., shear walls, braced frames) can be particularly sensitive to even small base rotations and translations that are neglected with a fixed base assumption [2]. This problem is reported also in FEMA 357, "*overturning calculations at pseudo lateral force levels appear to be overly conservative and can predict overturning stability problems that are not well correlated with observed behaviour*" [3].

METHODOLOGY AND TEST STRUCTURE

A reference structure is selected which is an existing school located in Tehran, Iran. For the purpose of simplicity, one of the three braced frames, that is shown in Figure (1), in the shorter direction of the school plan is chosen for the two-dimensional analysis. It is assumed that one third of the seismic lateral load is carrying by this braced frame. The storey height is equal to 340 centimeter and the outer and the inner bays length are, respectively, equal to 600 centimeter and 360 centimeter. The diaphragms are assumed to be rigid. The allowable soil stress under foundation is 1.7 kg/cm^2. The frame view is shown in Figure (1).

The braced frame is designed/checked using the below conditions:

1- The existing frame is checked for the combination of gravity and seismic load based on the UBC 97 code [6]. It is assumed that the base is fixed and the linear static procedure is used.
2- The frame is rehabilitated using UBC 97 code [6]. The fixed base assumption as well as the linear static procedure is used.
3- The frame is rehabilitated using FEMA 356 guideline [4]. Both of the fixed base and the flexible based assumptions are used in combination with the linear static and non-linear static procedures.
4- A simple method is proposed for the foundation design to avoid from the prescribed conservation.

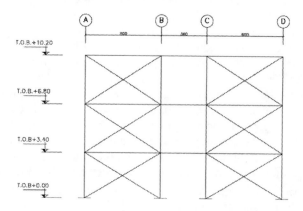

FIGURE 1. The elevation of the frame under investigation.

PROPOSED METHOD

The use of R_{OT} is proposed as an alternative method for the assessment of the overturning effect as well as the soil pressure checking in FEMA 356 guideline [4]. The alternative procedure is intended to provide a method that is consistent with prevailing practice specified in current codes for new buildings [4]. On the other hand, the allowable foundation and lateral pressures, in the design code, shall be checked using the allowable stress load combinations (Section 1805 of [6]). It means that the use of R_{OT} in FEMA 356 guideline [4] can be interpreted as the allowable stress design force combinations in the design codes (e.g., reference [6]). The seismic lateral force for the ultimate strength design is 1.4 times of the allowable stress design force (compare sections 16.12.2 and 16.12.3 of reference [6]). Hence, 1.4 times of the combinations corresponding to the force-controlled seismic forces, by substituting the J factor by R_{OT} factor, are reasonable choices for the design of the foundation. The seismic lateral force can be simply combined with the gravity load effects as shown in Equation (3).

$$Q_{UF} = Q_G \pm 1.4 \times \frac{Q_E}{C_1 C_2 C_3 R_{OT}} \tag{3}$$

where Q_G is the gravity force and $1.4 \times \dfrac{Q_E}{C_1 C_2 C_3 R_{OT}}$ is the seismic lateral force corresponding to the ultimate strength design. The results, which are explained in the next section, confirm that the use of the proposed force combinations results to the much more reasonable aspects for the design of foundation.

RESULTS USING DIFFERENT METHODS

The frame is designed using four different procedures. The results, as shown in Figure (2), are categorized for the analysis, substructure and superstructure parts. In the analysis part, the natural period of the structure, the design base shear, the overturning moment and the roof displacement are presented. The natural period of the vibration is equal to 0.57 second in the case of the flexible base condition which is 58% greater than the empirical value corresponding to the design code. This means the J factor can be taken, at least, equal to 4 based on the Equation (2). The base uplift is a function of the gravity loads as well as the foundation overturning moment. Thus, the influence of the foundation overturning moment on the design procedure is more important than the base shear. For example, the foundation overturning moment is a little bit less than the value corresponding to the nonlinear static case with the flexible based assumption; however, the design results are in the same order.

In the substructure part, as explained in Figure (2), the overturning safety factor is calculated with two different methods. The method 1 and 2 are, respectively, based on the J and R_{OT} factors. The soil pressure safety factor, the maximum design compressive reactions in the A and B axis, see Figure (1), and the rehabilitated foundation results are shown in this part. The designed foundation based on the design code is vulnerable based on the linear static procedure with the fixed based assumption. However, the foundation based on the nonlinear static procedure is more economic.

In the superstructure part, the bracing sections as well as the average columns area in each story is presented. The design elements are, obviously, more economic by implementing more advanced procedures, i.e., nonlinear static procedure with flexible based assumption. The designed frame is the same as the frame in the linear static procedure with the fixed base assumption, but, the foundation is approximately similar to the foundation of the nonlinear static case with the flexible based assumption.

		Analysis Results				SubStructure								SuperStructure					
	Analysis Period, (s)	For Foundation		Roof Disp. (cm)	Overturning Safety Factor		Soil Pressure Safety Factor	Max Design Comp. Reaction, (ton)		Rehabilitated Foundation			Bracing			Average Columns Area, (cm^2)		
		Design Base Shear (ton)	Design Overturning Moment (t.m)		Method 1	Method 2		A	B	Width (m)	Height (m)	Rebar	1st Storey	2nd Storey	3rd Storey	1st Storey	2nd Storey	3rd Storey
Existing Structure / UBC97 / Fl.B	1.08	250	3209	8.97	1.1		0.20	196	206	—	—	—	L60X6	L60X6	L40X4	44	44	44
UBC97 / Fl.B.	Anal.=0.42, Emp.=0.36	250	3216	1.34	1.75		1.20	193	209	3.9	1	31Φ20top 62Φ20bot.	2U200	2U180	2U140	128	62	44
Linear Static: FEMA356 / Fl.B.	0.39	420	4503	2.83	1.1	4.0	4.25	304	313	5.7	1.5	33Φ20top 61Φ20bot.	2U220	2U200	2U160	128	62	44
Nonlinear Static: FEMA356 / Fl.B.	0.46	275	3397	7.54	-		3.22	209	223	3.6	1	37Φ20top 40Φ20bot.	2U160	2U140	2U140	106	59	44
Nonlinear Static: FEMA356 / FL.B.	0.57	182	2675	11.5	-		—	201	113	1.8	0.7	26Φ20top 32Φ20bot.	2U160	2U140	2U120	72	44	44
Proposed Method	0.39	147	2422	1	0.5	2.0	1.86	130	139	1.8	1	19Φ20top 32Φ20bot.	2U220	2U200	2U160	128	62	44

"Fl.B.": Fixed Base Assumption
"FL.B.": Flexible Base Assumption

FIGURE 2. The results of the existing and the rehabilitated structure using different methods.

THE NONLINEAR STATIC ANALYSIS RESULTS

The frame is rehabilitated, in the last section, using the nonlinear static procedure considering the fixed base and the flexible base assumptions. The comparison of the pushover curves and the corresponding target displacements are shown in Figure (3). In the case of flexible based assumption, the target displacement is greater and the input base shear is less than the case of the fixed base assumption. This effect comes from the base flexibility which elongates the structural natural period of vibration and obviously decreases the seismic demand.

The frames (superstructure) which are designed by the four different methods, as explained in Table (1), are analyzed using the nonlinear static procedure once considering the fixed based assumption and second using the flexible based condition. The input base shears for the four frames, as shown in the left graph in Figure (4), are different. But, by using the flexible base assumption, as shown in the right graph in Figure (4), the difference in the base shears becomes practically negligible for all of the structures.

Another interesting fact is that, the frames which are designed based on FEMA 356 guideline [4] using the linear static procedure with the fixed based assumption as well as the nonlinear static procedure with the flexible based assumption cannot satisfy the acceptance criteria, if the nonlinear static procedure with the fixed base assumption is utilized. This fact makes an inconsistency between the different rehabilitation procedures in the FEMA 356 guideline [4]. However, the frames which are designed by any procedure with the fixed based assumption will satisfy the acceptance criteria, if the same procedure with the flexible based assumption is used, as seen in Figure (4).

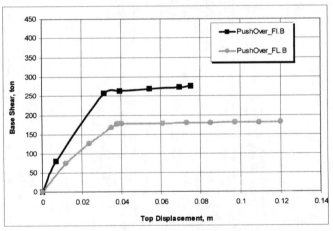

FIGURE 3. The comparison of the nonlinear static curves (and the corresponding target displacements) for the structures designed by the nonlinear static analysis assuming both the fixed and the flexible based assumptions.

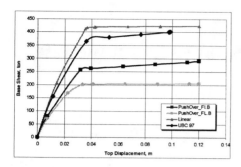 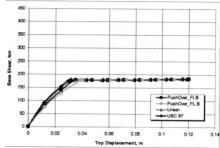

FIGURE 4. The comparison of the nonlinear static curves for the structures which are designed by the four different methods, (Left): fixed based assumption and (Right): flexible based assumption.

CONCLUSION

The design or rehabilitation of foundation, for the structures with stiff lateral resisting systems using the linear static procedure with the fixed based assumption, results to a relatively conservative design. By taking the soil-structure-interaction in to account, the seismic demand decreases and the foundation design will be more economic.

A simple procedure is proposed to modify the design of foundation in the linear static procedure with the fixed based assumption within the FEMA 356 guideline [4]. The proposed method is based on the philosophy of the alternative method for the overturning control which is available in the guideline. It is shown that the foundation design based on the proposed procedure is economic without explicitly taking the base flexibility in to consideration.

REFERENCES

1. AISC. Allowable Stress Design Manual of Steel Construction. American Institute of Steel Construction, Inc. Chicago, IL (1989a).
2. FEMA. Improvement of Nonlinear Static Seismic Analysis Procedures. Report No. FEMA-440, Federal Emergency Management Agency, Washington, DC, (2005).
3. FEMA. Global Topics Report on the Prestandard and Commentary for the Seismic Rehabilitation of Buildings. Report No. FEMA-357, Federal Emergency Management Agency, Washington, DC, (2000).
4. FEMA. Prestandard and commentary for the seismic rehabilitation of buildings. Report No. FEMA-356, Federal Emergency Management Agency, Washington, DC, (2000).
5. FEMA. NEHRP Commentary on the Guidelines for the Seismic Rehabilitation of Buildings. Report No. FEMA-274, Federal Emergency Management Agency, Washington, DC, (1997).
6. UBC 97. Uniform Building Code. International Conference of Building Officials, 5360 Workman Mill Road Whittier, California, 90601-2298, (1997).

Nonlinear Seismic Response Of Single Piles

R. Cairo[a] , E. Conte[a] and G. Dente[a]

[a]University of Calabria, Dipartimento di Difesa del Suolo, Rende (CS), ITALY

Abstract. In this paper, a method is proposed to analyse the seismic response of single piles under nonlinear soil condition. It is based on the Winkler foundation model formulated in the time domain, which makes use of p-y curves described by the Ramberg-Osgood relationship. The analyses are performed referring to a pile embedded in two-layer soil profiles with different sharp stiffness contrast. Italian seismic records are used as input motion. The calculated bending moments in the pile are compared to those obtained using other theoretical solutions.

Keywords: soil-pile interaction, bending moment, nonlinear behaviour.

INTRODUCTION

The analysis of the seismic behaviour of piles is strongly influenced by a complicated problem of interaction among piles, soil and supported structure. In particular, the pile response to earthquake loading includes important aspects of wave propagation in soil, involving reflection, refraction and radiation damping phenomena. In other words, the particular stratigraphic conditions of the subsoil govern the characteristic of the seismic motion, and affect the behaviour of the piles embedded. The loading pattern is a three-dimensional problem and seismic waves may propagate with different direction and various angles of incidence. Besides, under severe earthquake shaking, piles behave in a nonlinear way, the soil undergoes plastic deformations and pile separation (gapping), slippage and friction phenomena may occur.

Under these circumstances, a comprehensive rigorous solution which could incorporate all aspects of the soil-structure interaction problem is extremely hard to achieve. The finite element method [1-5] provides the most powerful and versatile technique to analyse soil-structure interaction, taking into account nonlinearity and heterogeneity of the soil, although it is very expensive and requires sophisticated boundaries conditions to simulate the radiation damping.

From a computational point of view, a more efficient procedure is represented by the boundary element method [6-8]. It needs far less discretization and the condition of wave propagation towards infinity is automatically satisfied. This techinique is generally formulated in the frequency domain and, in principle, is valid under the assumption of linear behaviour of the soil.

Among approximated approaches, the methods based on the Winkler foundation model [9-11] reveal quite accurate despite the modest computational effort. Formulated in the time domain, they permit the nonlinear behaviour of the soil to be

CP1020, *2008 Seismic Engineering Conference Commemorating the 1908 Messina and Reggio Calabria Earthquake*,
edited by A. Santini and N. Moraci

easily incorporated [12-15] by using the approach of the unit load transfer curves (also known as *p-y* curves), often confining soil nonlinearity in an inner field around the pile [16-18].

In this paper, the Winkler formulation originally developed by Conte and Dente [19, 20] is reviewed, and the seismic response of single piles is investigated. The soil is assumed to behave as either a linear viscoelastic medium or a nonlinear material. In this regard, *p-y* curves based on the Ramberg-Osgood model are employed. The analyses focus on the influence of soil layers with sharp stiffness contrast on the response of single piles in terms of maximum bending moments. The scheme adopted and the seismic motions used as input are part of the RELUIS research project.

METHOD OF ANALYSIS

The used method (Fig. 1) models the pile as a linearly elastic beam with length L and diameter d, discretized into segments connected to the surrounding soil by springs and dashpots, which provide the interaction forces in the lateral direction. The stiffness k of the springs is related to the soil modulus; the dashpots, with coefficient c, are considered to account for both material and radiation dampings. The involved soil properties are the shear modulus G_s, Poisson's ratio v_s, and mass density ρ_s. They can vary with depth.

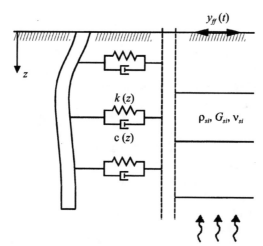

FIGURE 1. The Winkler foundation model used in this study.

The dynamic equilibrium of an infinitesimal element of pile is expressed as

$$EI \frac{\partial^4 y}{\partial z^4} + m \frac{\partial^2 y_t}{\partial t^2} + p = 0 \tag{1}$$

where t is the time, E is the Young's modulus of the pile, I the moment of inertia of the pile, m the pile mass per unit length, y_t is the total displacement and y is the relative displacement of the pile, p is the soil reaction on a unit length of pile, which is provided by the equation

$$p = p_1 + p_2 = k(y_t - y_{ff}) + c\frac{\partial(y_t - y_{ff})}{\partial t} \qquad (2)$$

where y_{ff} is the free-field displacement of the soil at prescribed depth and time. The stiffness k can be expressed as [13]

$$k = 2(1 + v_s)G_s\delta_1 \qquad (3)$$

with G_s depending on the strain level induced by wave propagation, so that the nonlinearity effects are directly accounted for. In the hypothesis of linear soil behaviour, G_s coincides with the small-strain shear modulus G_0. The coefficient δ_1 [13] is a function of the pile flexibility factor EI/E_sL^4, and pile slenderness ratio L/d, being E_s the Young's modulus of the soil.

The nonlinear and hysteretic behaviour of the soil is modelled using the Ramberg-Osgood constitutive law [19], the skeleton curve of which is provided by

$$\frac{\gamma}{\gamma_y} = \frac{\tau}{\tau_y} + \alpha\left(\frac{\tau}{\tau_y}\right)^R \qquad (4)$$

where γ and τ are the shear strain and shear stress, respectively; γ_y and τ_y are their reference values, so that τ_y/γ_y is equal to G_0. This ratio approximately indicates when significant departure from linearity begins to occur. The parameters α and R govern the shape of the τ-γ curve and can be evaluated to achieve a best fit to experimental data. The Masing rules are invoked to describe the unloading and reloading branches of the hysteresis loops.

The soil reaction p_2 accounts for the radiation damping, which can be expressed using approximate relationships developed in literature [21-23]. The details of the influence of the different formulation of radiation damping on the pile response are not given here as they can be found elsewhere [19].

In order to account for material damping under elastic soil condition, a Voigt-type viscous damping is introduced as

$$c = 2\frac{\beta_s k}{\omega_n} \qquad (5)$$

where ω_n denotes the natural frequency of the soil deposit, and β_s is the damping ratio of the soil.

The free-field displacement at any time and depth is first calculated using the characteristic method [24]. Then, Eq. 1 is solved numerically to obtain the response of the pile in terms of its lateral displacements.

CASE STUDIES AND RESULTS

The analyses performed in this study refers to a single fixed-head pile with length L=20 m, diameter d=0.6 m, Young's modulus E=2.5·10^7 kN/m², and mass density ρ=2.5 Mg/m³ (Fig. 2). The pile is embedded in a two-layer soil deposit resting on a stiffer bedrock. The thickness of the layers is assumed to be 15 m; V_{s1} and V_{s2} are the shear wave velocities of the upper and the lower layers, respectively; Poisson's ratio, mass density and damping ratio of the soil are: v_s=0.4, ρ_s=1.9 Mg/m³, and β_s=0.10. The shear wave velocity of the rock is 1200 m/s. The shear wave velocity V_{s2} of the lower layer is kept constant and equal to 400 m/s, whereas two different values of V_{s1} are considered, i.e. 100 and 150 m/s. On the basis of the average shear wave velocity V_{s30}, which results 160 and 218 m/s, the two soil profiles can be classified as ground type D and C, respectively, according to Eurocode 8 [25]. Italian seismic records [26], scaled to 0.35g, are used as excitation at the rock level and assumed to consist solely of vertically propagating shear waves.

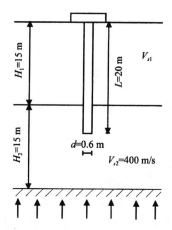

FIGURE 2. The soil-pile system studied.

The linear elastic behaviour of the soil is first examined. The pile response is presented in terms of the envelopes of the maximum bending moment along the pile. Figure 3 shows the results obtained considering three different accelerograms recorded during the 1976 Friuli (A-TMZ000), 1980 Irpinia (A-STU270), and 1997 Umbria-Marche (A-AAL018) earthquakes.

The results calculated in this study are compared to those obtained using the boundary element approach developed by Cairo and Dente [6]. Despite the different methods used, the agreement is very satisfactory. As expected, the bending moment diagrams exhibit a pronounced peak at the interface between the two layers. The corresponding values are greater than the bending moment at the pile head, depending on the stiffness contrast (V_{s2}/V_{s1}=4 for D soil profile; V_{s2}/V_{s1}=2.67 for C soil profile), and the seismic excitation considered.

Nonlinear analyses are also performed employing the soil data provided by Maiorano et al. [27], which are shown in Fig. 4 in terms of G_s/G_0-γ curves. On the basis of these data, the parameters α and R of the Ramberg-Osgood model are obtained under the assumption that the reference strain γ_y for the upper layer of soft clay and the lower layer of gravel is 0.5% and 0.067%, respectively. Specifically, values of α=19.89 and R=2.33 are determined for the clay, α=17.11 and R=2.09 for the gravel.

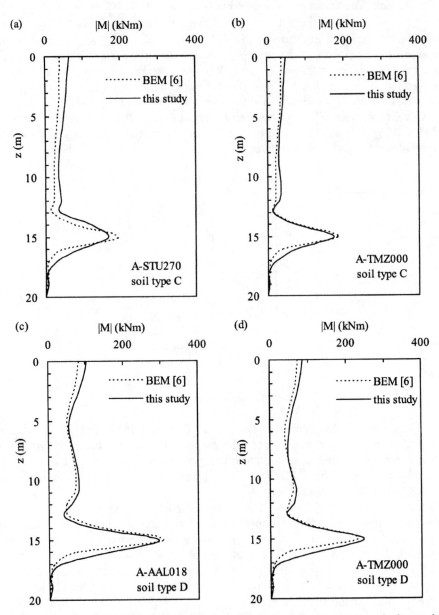

FIGURE 3. Envelopes of the bending moment along the pile for different seismic excitations and two soil profiles.

The seismic response of the pile, under nonlinear soil condition, is illustrated in Fig. 5. For comparison, the moments at the layer interface and pile head calculated by Maiorano et al. [27] using a finite element method are also shown. Like the linear elastic solution, the bending moment diagram presents a peak at the interface between the two layers. Nevertheless, the portion of pile embedded in the upper softer layer exhibits more severe bending moments, in general.

Finally, it is worth noting that the linear elastic analysis leads to underestimate the maximum bending moment in the pile, except for the case documented in Fig. 5c, where the nonlinear solution reveals slightly less onerous.

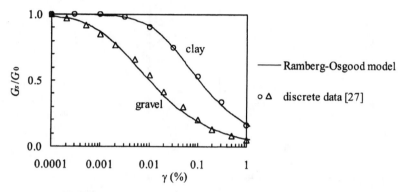

FIGURE 4. Modulus reduction curves used in the analyses.

CONCLUDING REMARKS

A method has been proposed to analyse the seismic response of single piles in layered soils. It is formulated as a dynamic-Winkler type approach in the time domain, and makes use of nonlinear p-y curves based on the Ramberg-Osgood model. Italian seismic records have been used as input motion in the analyses. Comparisons with other existing procedures have been performed and satisfactory agreement between the results has been found. In addition, the analyses have highlighted the importance of including the nonlinear and hysteretic behaviour of the soil in the soil-structure interaction under seismic loading.

ACKNOWLEDGMENTS

The present work is part of the RELUIS research project "Innovative methods for the design of geotechnical systems", funded by the Italian Department of Civil Protection (DCP) and coordinated by the Italian Geotechnical Association (AGI). The authors wish to thank the DPC and the project team coordinators.

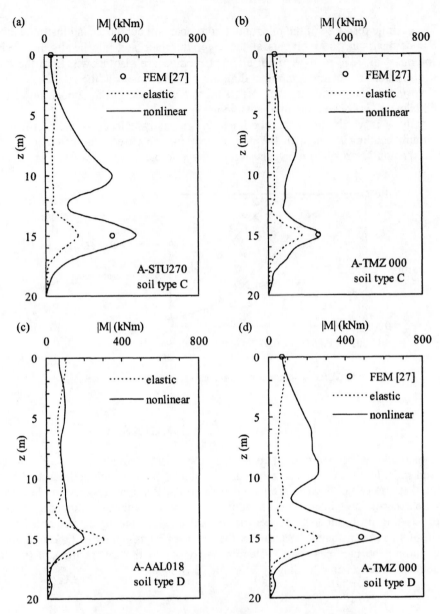

FIGURE 5. Envelopes of the bending moment along the pile.

REFERENCES

1. G. W. Blaney, E. Kausel and J. M. Roësset, "Dynamic stiffness of piles", Proc. 2nd Int. Conf. Num. Methods in Geomech., Blacksburg, 1976, pp. 1001-1012.
2. Y. X. Cai, P. L. Gould and C. S. Desai, "Nonlinear analysis of 3-D seismic interaction of soil-pile-structure systems and application", *Eng. Struct.* **22**, 191-199 (2000).

3. M. Kimura and F. Zhang, "Seismic evaluation of pile foundations with three different methods based on 3D elasto-plastic finite element analysis", *Soils and Foundations* **40**, 113-132 (2000).
4. B. K. Maheshwari, K. Z. Truman, M. H. El Naggar and P. L. Gould, "Three-dimensional finite element nonlinear dynamic analysis of pile groups for lateral transient and seismic excitations", *Can. Geotech. J.* **41**, 118-133 (2004).
5. G. Wu and W. D. L. Finn, "Dynamic nonlinear analysis of pile foundations using finite element method in the time domain", *Can. Geotech. J.* **34**, 44-52 (1997).
6. R. Cairo and G. Dente, "Kinematic interaction analysis of piles in layered soils", ISSMGE-ERTC 12 Workshop "Geotechnical Aspects of EC8", Madrid, 2007, paper No. 13.
7. A. M. Kaynia and E. Kausel, "Dynamic behavior of pile groups", Proc. 2nd Int. Conf. Num. Meth. in Offshore Piling, Austin, Texas, 1982, pp. 509-532.
8. S. M. Mamoon and P. K. Banerjee, "Response of piles and pile groups to travelling SH waves", *Earthq. Eng. and Struct. Dyn.* **19**, 597-610 (1990).
9. M. Kavvadas and G. Gazetas, "Kinematic seismic response and bending of free-head piles in layered soil", *Géotechnique* **43**, 207-222 (1993).
10. N. Makris and G. Gazetas, "Dynamic pile-soil-pile interaction. Part II: lateral and seismic response", *Earthq. Eng. and Struct. Dyn.* **21**, 145-162 (1992).
11. G. Mylonakis, A. Nikolaou and G. Gazetas, "Soil-pile-bridge seismic interaction: kinematic and inertial effects. Part 1: soft soil", *Earthq. Eng. and Struct. Dyn.* **26**, 337-359 (1997).
12. M. H. El Naggar, M. A. Shayanfar, M. Kimiaei and A. A. Aghakouchak, "Simplified BNWF model for nonlinear seismic response analysis of offshore piles with nonlinear input ground motion analysis", *Can. Geotech. J.* **42**, 365-380 (2005).
13. T. Kagawa and L. M. Kraft, "Lateral load-deflection relationships of piles subjected to dynamic loadings", *Soils and Foundations* **20**, 19-36 (1980).
14. H. Matlock, S. H. C. Foo and L. M. Bryant, "Simulation of lateral pile behavior under earthquake motion", Spec. Conf. on Earthq. Eng. and Soil Dyn., Pasadena, 1978, pp. 600-619.
15. H. Tahghighi and K. Konagai, "Numerical analysis of nonlinear soil-pile group interaction under lateral loads", *Soil Dyn. Earthq. Eng.* **27**, 463-474 (2007).
16. R. W. Boulanger, C. J. Curras, B. L. Kutter, D. W. Wilson and A. Abghari, "Seismic soil-pile-structure interaction experiments and analyses", *J. of Geotech. Eng.* **125**, ASCE, 750-759 (1999).
17. M. H. El Naggar and K. J. Bentley, "Dynamic analysis for laterally loaded piles and dynamic *p-y* curves", *Can. Geotech. J.* **37**, 1166-1183 (2000).
18. B. K. Maheshwari and H. Watanabe, "Nonlinear dynamic behavior of pile foundations: effects of separation at the soil-pile interface", *Soils and Foundations* **46**, 437-448 (2006).
19. E. Conte and G. Dente, "Effetti dissipativi nella risposta sismica del palo singolo", Deformazioni dei terreni ed interazione terreno-struttura in condizioni di esercizio, Monselice, 1988, vol. 2, pp. 19-38 (in Italian).
20. E. Conte and G. Dente, "Il comportamento sismico del palo di fondazione in terreni eterogenei", Atti XVII Convegno Nazionale di Geotecnica, Taormina, 1989, vol. 1, pp. 137-145 (in Italian).
21. E. Berger, S. A. Mahin and R. Pyke, "Simplified method for evaluating soil-pile-structure interaction effects", Proc. 9th Offshore Tech. Conf., Houston, 1977, pp. 589-598.
22. R. Dobry, E. Vincente, M. J. O'Rourke and J. M. Roësset, "Horizontal stiffness and damping of single piles", *J. of Geotech. Eng.* **108**, ASCE, 439-458 (1982).
23. G. Gazetas and R. Dobry, "Horizontal response of piles in layered soils", *J. of Geotech. Eng.* **110**, ASCE, 20-40 (1984).
24. G. Dente, "La risposta dinamica dello strato eterogeneo attraversato da onde di taglio", Atti XV Convegno Nazionale di Geotecnica, Spoleto, 1983, pp. 153-161 (in Italian).
25. EN1998-1, *Eurocode 8: Design of structures for earthquake resistance – Part 1: General rules, seismic actions and rules for buildings*, CEN TC 250, Brussel, Belgium, 2003.
26. G. Scassera, G. Lanzo, F. Mollaioli, J. P. Stewart, P. Bazzurro and L. D. Decanini, "Preliminary comparison of ground motions from earthquakes in Italy with ground motion prediction equations for active tectonic regions", Proc. 8th US Conf. Earthq. Eng., San Francisco, 2006, paper No. 1824.
27. R. M. S. Maiorano, S. Aversa and G. Wu, "Effects of soil non-linearity on bending moments in piles due to seismic kinematic interaction", Proc. 4th Inter. Conf. Earthq. Geotech. Eng., Greece, 2007, paper No. 1574.

Soil-structure Interaction in the Seismic Response of Coupled Wall-frame Structures on Pile Foundations

S.Carbonari[a] F.Dezi[b] and G.Leoni[c]

[a]*Dept of Architecture, Constructions, Structures, Università Politecnica delle Marche, Italy*
[b]*Dept of Materials, Environment Engineering, Physics, Università Politecnica delle Marche, Italy*
[c]*Dept ProCAm, Università di Camerino, Ascoli Piceno, Italy*

Abstract. This paper presents a study on the seismic response of coupled wall-frame structures founded on piles. A complete soil-structure interaction analysis is carried out with reference to a case study. Three different soils and seven real accelerograms are considered. Local site response analyses are performed in order to evaluate the incoming free-field motion at different depths and the ground motion amplifications. A numerical model, accounting for the pile-soil-pile interaction and for material and radiation damping, is used to evaluate the impedance matrix and the foundation input motion. The domain decomposition technique is adopted to perform time-domain seismic analyses introducing Lumped Parameter Models to take into account the impedance of the soil-structure system. Applications show that the rocking phenomena affect the behaviour of the structure by changing the base shear distribution within the wall and the frame and by increasing the structural displacements.

Keywords: local response analyses, lumped parameter models, pile foundations, soil-structure interaction, wall-frame systems.

INTRODUCTION

In the conventional building design, structures are generally assumed to be fixed at their bases and the input motion is defined at the outcropping soil according to suitable design spectra. In reality, the deformability of the soil-foundation system makes the problem more complex. Firstly, the input motion of the structure is filtered by the embedded pile foundation subjected to ground motions variable with depth; secondly, the foundation deformability reduces the overall stiffness of the structure and increases the damping due to the energy radiation into the soil.

For wall and coupled wall-frame structures it is important to guarantee a suitable degree of restraint at the base. Commonly, pile foundations are considered to be rigid enough against rocking motions and a fixed base model is supposed to be conservative because it overestimates the actual seismic action. In reality the wall rocking may affect the behaviour of the structure by changing the distribution of the ductility demand and of the shear resisted by the wall and the frame (Ultimate Limit State), and by increasing the overall structural displacements responsible of damage (Damageability Limit State). A reliable analysis on the seismic response of coupled-

CP1020, *2008 Seismic Engineering Conference Commemorating the 1908 Messina and Reggio Calabria Earthquake,*
edited by A. Santini and N. Moraci
© 2008 American Institute of Physics 978-0-7354-0542-4/08/$23.00

wall frame systems should account for the main features previously described by carrying out a complete soil-structure interaction study (Fig.1).

The first step consists in performing a local site response analysis of the soil profile (for usual purposes a one-dimensional analysis is satisfactory) by deconvoluting the accelerograms defined at the outcropping bedrock and studying the polarized shear wave propagation [1]. This permits to evaluate the effects of the free-field motion variable with depth. The second step consists in the structural analysis including the soil-foundation system. The problem is governed, in the frequency domain, by a system of complex linear equations

$$
\begin{bmatrix}
K_{SS} & K_{SF} & 0 \\
K_{FS} & K_{FF}^{[S]} + K_{FF}^{[F]} & K_{FE} \\
0 & K_{EF} & K_{EE}
\end{bmatrix}
\begin{bmatrix}
d_S \\
d_F \\
d_E
\end{bmatrix}
=
\begin{bmatrix}
0 \\
f_F \\
f_E
\end{bmatrix}
\tag{1}
$$

where the partition descends from the separation of the displacements in the structure (d_S), foundation (d_F) and embedded piles (d_E) components [2-3]. Foundation impedances have to account for pile-soil-pile interaction and radiation damping and should be able to describe also the foundation rocking response. The known term collects the forces descending from the local site response analysis previously carried out. In the sequel the influence of soil-structure interaction on the seismic response of coupled wall-frame structures is investigated according to the methodology previously introduced. The soil-pile-structure interaction analysis is performed by means of the substructure method which consists in studying separately the soil-foundation system (kinematic interaction analysis) and the superstructure on deformable restraints (inertial interaction analysis) [4].

SEISMIC RESPONSE ANALYIS OF A COUPLED WALL-FRAME SYSTEM

The problem of the seismic response of coupled wall-frame systems is investigated with reference to a 6-storey 4-bays building. The building is supposed to have a square

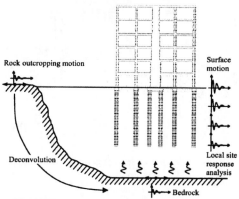

FIGURE 1. Soil-structure interaction problem

611

plan with a side length of 18.00 m. Four coupled wall-frame systems, placed at the boundaries in order to achieve a torsionally restrained system, constitute the seismic-resistant structure.

The total height is 19.20 m and the storey height is 3.20 m. The structure was designed by considering current geotechnical and structural design practice with fixed restraints at the base and assuring interstorey drifts less than 5‰ (DLS).

Thanks to the plan symmetry, the structural analysis is carried out by considering only one seismic-resistant wall-frame system. The masses associated to each floor of the system, shown in Fig. 2a, refer to half storey and are not comprehensive of the structural masses that are automatically accounted for in the model.

The wall-frame system is founded on piles: a single pile foundation is considered for each column of the frame, while a 2x2 pile group foundation is assumed for the wall. The piles are floating and characterized by the same diameter ($\phi = 0.8$ m). Fig. 2b shows the geometry of the wall foundation used in the numerical application.

The soil profile consists of a deformable homogeneous soil layer of 30 m thickness overlying a bedrock which is supposed to be elastic. Three kinds of soils, characterized by different shear wave velocities and densities, are considered for the deformable layer. A constant hysteretic material damping ratio $\xi = 10\%$ and a constant Poisson's ratio $\nu = 0.4$ are assumed for all the soils. Table in Fig. 2b shows the soil properties chosen for each soil profile.

The pile lengths, varying with soil properties, are 14 m for soil profile A, 16 m for soil profile B and 22 m for soil profile C.

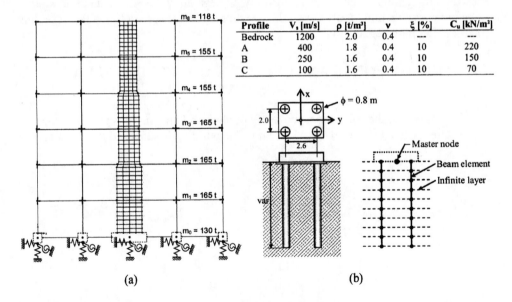

Profile	V_s [m/s]	ρ [t/m³]	ν	ξ [%]	C_u [kN/m²]
Bedrock	1200	2.0	0.4	---	---
A	400	1.8	0.4	10	220
B	250	1.6	0.4	10	150
C	100	1.6	0.4	10	70

(a)   (b)

FIGURE 2. (a) Coupled wall-frame system model; (b) geometry and model of the wall foundation and soil properties

Structural Modelling

Structural models are separately developed for the foundation and the superstructure according to the substructure method.

The foundation is modelled by considering the piles as beam elements embedded in a Winkler-type layered soil [5]. The numerical procedure proposed by the authors [2, 6] for the kinematic analysis of pile groups is used to effectively study the soil-foundation and the pile-to-pile interaction. The procedure is used to compute the foundation input motion and the frequency-dependent impedances of the soil-foundation systems. Each pile is modelled by 1 m long finite elements to provide a suitable level of accuracy. Concerning the wall foundation, a rigid cap is considered and a master node is introduced at the centroid of the pile group at the level of pile heads (Fig. 2b).

Two structural models were considered for the superstructure: one is fully restrained at the base (Fixed-Base Model) while the other is supported by deformable restraints (Soil-Structure Interaction Model). The finite element models were developed in SAP2000® [7]. Frame members were modelled by beam elements while shell elements were used to simulate the concrete wall (Fig. 2a). Rigid links, automatically introduced by the program, were used to simulate the beam-to-column joints; in particular, the actual length of the rigid link was assumed to be 50% of the joint region to take into account the strain penetration effect. To incorporate the in-plane rigidity of the floor, a stiff diaphragm constraint was introduced at each storey. Mass of the foundations and tie-beams at the pile cap level were also incorporated in the flexible base model.

The concrete (C30/37) was considered to be linearly elastic, with a Young's modulus $E_c = 3.5 \times 10^4$ N/mm². The effects of members cracking was taken into account by considering 75% of the concrete Young's modulus for the wall and the columns and 50% for beams. A structural damping is introduced in terms of Rayleigh damping: stiffness and mass proportional terms were evaluated to provide a 5% effective damping for the first and second modes.

Kinematic Interaction Analysis

The incoming seismic action at the outcropping bedrock was represented trough seven real accelerograms, obtained from the European Strong Motion Database [8], matching the Type 1 elastic response spectrum of the Eurocode 8 [9] for soil type A and PGA 0.25g.

A one-dimensional ground response analysis was performed by using EERA [1]. The three soil profiles and the seven accelerograms were considered. The time histories of the free-field motion at the ground surface and at different depths corresponding to the finite element nodes of the foundation model were calculated. The time histories obtained from the site response analyses were used as input, at different depths, for the kinematic interaction analysis, whereas the motions obtained at the ground surface were used for FBMs. Fig. 3 shows the mean response spectra of the motion at the ground surface where amplifications in correspondence of the fundamental periods of the soil deposits are evident.

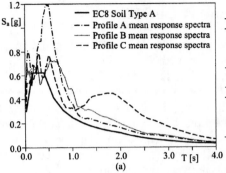

Fundamental structural periods		
Model	T_1 [s]	T_2 [s]
FB	1.098	0.316
SSI on soil profile A	1.125	0.327
SSI on soil profile B	1.140	0.334
SSI on soil profile C	1.198	0.356

Modal participating mass		
Model	M_1 [%]	M_2 [%]
FB	64.6	19.0
SSI on soil profile A	65.3	19.2
SSI on soil profile B	65.9	19.3
SSI on soil profile C	68.7	19.6

(a)   (b)

FIGURE 3. (a) Mean response spectra at the outcropping soil for the three profiles, (b) fundamental structural periods and modal participating mass

Inertial Interaction Analysis

Suitable LPMs are used to catch the frequency dependent behaviour of foundations in the time domain analysis. The simplest of these models consists of one spring, one damper and one mass for each degree of freedom of the rigid cap. The relevant constants were calibrated so that both the real part and the imaginary part of the foundation impedances were well approximated in the frequency range of interest.

The dynamic stiffness relevant to the j-th dof of the foundation is thus approximated by

$$Z_j(\omega) = \left[\left(K_j - \omega^2 M_j \right) + i\omega C_j \right]$$

(2)

where the real part varies with frequency as a second order parabola and the imaginary part varies linearly with frequency.

The frequency-independent coefficients of springs, viscous dampers and masses are calibrated to better reproduce, in the low and medium frequency range (0 ÷ 10 Hz), the trend of the frequency dependent impedances obtained from the soil-pile-foundation interaction analysis previously described.

In SAP2000® the lumped parameter models were introduced as visco-elastic grounded springs and additional masses applied to the master node of each foundation.

MAIN RESULTS

The main response quantities describing the overall behaviour of the structure, namely the base shears and the foundation rocking, are shown and discussed with reference to ULSs seismic actions. Concerning the DLSs the interstorey drifts are presented; in this case the accelerograms are divided by a factor 2.5 in order to reduce the action to the damage limitation level hazard.

Figure 4a shows the time histories of the rocking of the wall foundation obtained for the different soils for one of the accelerograms used in the analyses.

In Fig. 5a the rocking of the wall foundation and of the column foundation closest to the wall are presented for each soil type. Dots are representative of all the analyses results while the relevant mean values are plotted with curves. The rocking of the wall

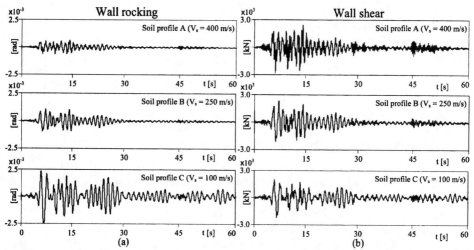

FIGURE 4. (a) Time histories of the foundation rocking for the different soils, (b) time histories of the wall base shear for the different soils

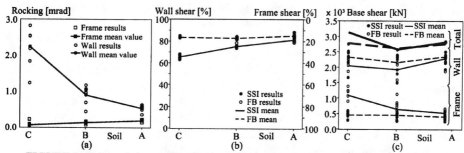

FIGURE 5. (a) Foundation rocking for the different soils; (b) distribution of the base shears, (c) absolute base shear values

foundation, directly related to the vertical dynamic impedance of the single pile, dramatically increases as the dynamic properties of the foundation soil decrease. The maximum mean value of 2.24 mrad is obtained for soil profile C. On the contrary the foundation of the column undergoes minor rotations for soft soils. This appears to be a contradiction but it is justified by the presence of stiff tie-beams that not only restrain the column foundation but also exert recall forces increasing with the wall foundation rocking.

Fig. 5b shows the percentage distribution of the seismic base shear between the wall and the frame, comparing results obtained from the FBMs and the SSIMs. When the model is fully restrained at the base the wall absorbs about 85% of the maximum seismic base shear; considering the soil-foundation flexibility in the analysis produces a migration of the shear stresses from the wall to the frame. In the case of soil profile C the wall base shear reduces to 65% of the total shear and thus results in an increasing frame base shear. Figure 4b shows the time histories of the wall base shear obtained for the different foundation soils.

Figure 5c shows the absolute values of the seismic base shear absorbed by the wall

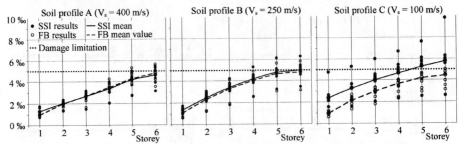

FIGURE 6. Interstorey drifts obtained from all the accelerograms used in the analyses

and the frame; the analyses results (dots) and the relevant mean values (curves) are presented. The total seismic base shear is also plotted. As already stated a migration of the base shear from the wall to the frame can be observed in the SSIMs with respect to the FBMs, particularly for soil profile C. The total base shear obtained from SSIMs and FBMs is almost the same for soil profiles A and B while for soil profile C the total base shear obtained from the SSIM is higher than that relevant to the FBM. An attempt to understand this result can be made starting from the observation of the fundamental periods of the structures (Fig. 3b) and the response spectra presented in Fig. 3a deriving from the local site response analyses.

As expected, the SSIMs are characterized by the increase of the fundamental periods with respect to the FBMs. For soil profile A the increase of the structural period leads to a lower pseudo-acceleration in correspondence of the first period and to a higher pseudo-acceleration associated to the second period that is very close to the first peak of the spectrum. This results in the increase of the total base shear in the SSIM as a consequence of the contribution of the second structural mode that is characterized by a substantial participating mass (Fig. 3b).

For soil profile B a decrease of the pseudo-accelerations associated to the first and second periods of the SSIM can be observed with respect to the FBM and consequently a reduction of the total base shear is expected.

Finally for soil profile C the increase of the first and second structural periods lead to higher pseudo-accelerations (first and third peak of the spectrum, respectively) and consequently a higher total base shear is observed when considering SSI effects.

In Fig. 6 the interstorey drifts obtained from the 14 analyses performed for each soil profile at the DLS are presented (dots) together with the relevant mean values (curves). It can be observed that for the fixed base analyses the mean curves (dashed) are all below the limiting line of 5‰ for every soil profile.

For the soil profile A and B (V_s = 400 m/s and 250 m/s) slightly differences are observed between the FBM and that including the SSI and the mean values are still lower than the limit of 5‰.

In the case of soil profile C, the mean interstorey drifts obtained with the two models are very different. The mean value obtained at the sixth storey with the SSI model is more than 30% higher than that relevant to the FBM and reaches more than 100% at the first storey. Differences are due to the foundation rocking previously described that adds more than 1‰ storey drift corresponding to the maximum rocking of about 1 mrad.

CONCLUSIONS

In this paper the seismic response of wall-frame systems founded on piles was investigated. In order to understand the significance of the foundation deformability, a complete SSI analysis was performed. The domain decomposition technique was adopted to perform time-domain seismic analyses introducing LPMs to take into account the frequency dependent impedance of the soil-foundation system.

Local site response analyses were performed to compute the free-field input motion by considering seven real accelerograms at the outcropping bedrock. The foundation kinematic interaction analysis was performed by means of a numerical model proposed by the authors accounting for pile-soil-pile interaction.

The following conclusions may be drawn:

– the local site response analysis is necessary to catch the free-field motion variable with depth and the spectral amplification of the input signals;
– the rocking of the wall foundation is directly related to the vertical dynamic impedance of the single pile and thus increases as the dynamic properties of the foundation soil decrease;
– the rocking of the wall induces a redistribution of the seismic base shear among the structural elements;
– SSI phenomena induce an additional drifts that are almost constant for all the storeys. In the case of soft soils these effects cannot be neglected in the verification of the DLSs.

Although the previous remarks are derived from the linear analysis of a case study they support evidence on the significance of considering SSI effects on seismic response analysis of coupled wall-frame structures.

REFERENCES

1. J. P. Bardet, K. Ichii and C. H. Lin, EERA: Equivalent-linear Earthquake site Response Analyses of Layered Soil Deposit. University of Southern California.
2. F. Dezi, S. Carbonari, A. Dall'Asta and G. Leoni, "Dynamic spatial response of structures considering soil-foundation-structure interaction: analytical model", *A.N.I.D.I.S. 2007 - 12° Convegno Nazionale – L'ingegneria sismica in Italia*, Pisa, Italy, paper n. 310 (2007).
3. F. Dezi, A. Dall'Asta, G. Leoni and G. Scarpelli, "Influence of the soil-structure interaction in the seismic response of a railway bridge", *ICEGE 2007 - 4th International Conference on Earthquake Geotechnical Engineering*, Thessaloniki, Greece, paper n. 1711 (2007).
4. J. P. Wolf, *Soil-structure Interaction Analysis in Time Domain*, Englewood Cliffs, Prentice-Hall, N, 1988.
5. N. Makris, G. Gazetas, "Dynamic Pile-Soil-Pile Interaction - Part II: Lateral and Seismic Response", *Earthquake Engineering Structural Dynamics*, 21(2), pp.145-162 (1992).
6. G. Leoni, F. Dezi, and S. Carbonari, "A model for 3D kinematic interaction analysis of pile groups in layered soils", *Submitted to Earthquake Engineering and Structural Dynamics*.
7. SAP2000 (October 2005). "CSI analysis reference manual". Computer and Structures, Inc., Berkeley, California, U.S.A.
8. European Strong Motion Database - www.isesd.cv.ic.ac.uk.
9. prEN 1998-1 EUROCODE8: Design of structure for earthquake resistance – Part 1: General rules, seismic actions and rules for buildings, (2003).

Numerical analysis of kinematic soil - pile interaction

Francesco Castelli[a], Michele Maugeri[b], and George Mylonakis[c]

[a]Department of Civil and Environmental Engineering, University of Catania, Viale Andrea Doria no.6, 95125, Catania, Italy
[b]Department of Civil and Environmental Engineering, University of Catania, Viale Andrea Doria no.6, 95125, Catania, Italy
[c]Department of Civil Engineering; University of Patras, Rio GR-26500, Patras, Greece

Abstract. In the present study, the response of singles pile to kinematic seismic loading is investigated using the computer program SAP2000[@]. The objectives of the study are: (1) to develop a numerical model that can realistically simulate kinematic soil-structure interaction for piles accounting for discontinuity conditions at the pile-soil interface, energy dissipation and wave propagation; (2) to use the model for evaluating kinematic interaction effects on pile response as function of input ground motion; and (3) to present a case study in which theoretical predictions are compared with results obtained from other formulations. To evaluate the effects of kinematic loading, the responses of both the *free-field* soil (with no piles) and the pile were compared. Time history and static pushover analyses were conducted to estimate the displacement and kinematic pile bending under seismic loadings.

Keywords: piles, lateral, dynamic, kinematic loading, numerical modeling.

INTRODUCTION

Recent destructive earthquakes have highlighted the need for increased research into the revamping of design codes and building regulations to prevent catastrophic losses in terms of human life and economic assets.

European seismic code (Eurocode-8) and Italian seismic normative states that piles shall be designed for the following two loading conditions: *a)* inertia forces in the superstructure transmitted on the heads of the piles in the form of axial and horizontal forces and bending moments; *b)* soil deformations arising from the passage of seismic waves which impose curvatures and, thereby, lateral strains on the piles along their whole length. Accepting these lines, kinematic effects should be taken into account in pile design.

While there is an ample experience in carrying out dynamic or equivalent static analyses for inertial loading (generally the inertial forces acting on the pile head are obtained from the product of mass and spectral acceleration), no universally accepted methods or procedures are available to predict deformations and bending moments from kinematic loading.

To estimate the maximum internal forces on piles subjected to lateral seismic excitation, some Authors [1] [2] propose an analysis in which the pile is subjected to the *free-field* soil displacements at each node along its length. These displacements are

CP1020, *2008 Seismic Engineering Conference Commemorating the 1908 Messina and Reggio Calabria Earthquake*,
edited by A. Santini and N. Moraci

obtained from a separate *free-field* site response analysis carried out, say, by the well known SHAKE [3] program or similar computer codes. Then a *"p-y method"* is adopted to simulate the response of the pile to the lateral soil movements.

According to this approach, a pseudo static pushover analysis can be adopted to simulate the behavior of piles subjected to kinematic loading.

In this paper, numerical analyses using the computer code SAP2000$^{@}$ are presented, aiming at investigating practical analysis approaches for kinematic pile bending considering typical subsoil conditions.

To this end, kinematic pile interaction is studied by means of a numerical model in which the pile is connected to the soil through continuously-distributed springs and dashpots, specifically developed for the seismic response of piles in layered soil.

The numerical results obtained in terms of displacement profile and bending moment distribution along the pile length, and are compared with those derived from a pseudo static pushover analysis. It is shown that the proposed numerical analyses can be successfully utilized to simulate kinematic interaction of piles.

MODEL DEFINITION

The passage of seismic waves through soft soil during strong earthquakes may cause significant strains develop in the soil. In the presence of embedded piles, curvatures will be imposed to the piles by the vibrating soil which, in turn, will generate bending moments. These moments will develop even in the absence of a superstructure and are referred to as *"kinematic"* moments, to be distinguished from moments generated by lateral loads imposed at the pile head (so-called *"inertial"* moments [4]).

Kinematic pile bending tends to be amplified in the vicinity of interfaces between soft and stiff soil layers (Figure 1). The reason is, soil shear strain is discontinuous across interfaces because of the different shear moduli between the soil layers and, thereby, the associated soil curvature (the derivative of strain) is infinite. In fact most of the pile damage observed deep below the soil surface is concentrated close to such discontinuities [4].

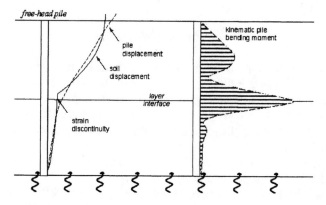

FIGURE 1. Kinematic bending of a free-head pile in a two layer soil profile.

The most widely used model to analyze pile response to lateral loads treats the pile as a series of elements, and pile-soil interaction by concentrated springs oriented perpendicular to the pile axis (Discrete Winkler Model). In order to consider non-linear soil behavior, the springs should have varying stiffness specified through a non-linear load-deflection relationship that depends on the type of soil and pile, known as "*p-y curves*".

To adequately address pile response under earthquake action, it is often required to perform a dynamic analysis of the pile for lateral excitation. The literature review shows that soil stiffness and damping properties can be included in a dynamic analysis through a *BDWF* (Beam on Dynamic Winkler Foundation) model [5] [6], in which the combinations of springs and dashpots represent the soil-pile stiffness and damping at each particular layer (Figure 2).

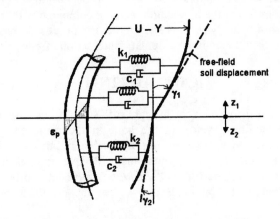

FIGURE 2. *BDWF* representation of the soil-pile interaction.

The first step in a kinematic pile-soil interaction analysis consists of a seismic soil response analysis for the soil profile specific to each pile location in the absence of the foundation structure (analysis of the *free-field* conditions). Then, to evaluate the kinematic effects, the analysis of the seismic interaction of the pile with the adjacent soil is performed.

Numerical Analysis

The series of numerical analyses were carried out by using the computer package SAP2000@. Although SAP2000@ is not a geotechnical software packages for pile-soil interaction problems, it was selected due to its capability for carrying out the present study. SAP2000@ software allows for strength and stiffness degradation in the components by providing the force-deformation criteria for hinges used in the numerical analyses.

For a dynamic analysis it is critical to have an adequate representation of the system stiffness and adequate representation of the system mass involved in the vibration phenomena. Then, the first objective of the investigation was to perform an analytical and numerical study of the seismic response of the pile-soil system, to assess the importance of the soil mass in the system response, and develop a rational method that includes the soil contribution to the system inertial properties through a series of lumped masses, consistent with the Discrete Winkler Model.

Free - Field Soil Response

To analyze the effects of the kinematic loading, the responses of both the *free-field* soil (with no piles) and the soil-pile system were studied. To evaluate the seismic response in *free-field* conditions, the soil is modeled through a series of lumped masses (Figure 3). Soil layers are assumed to be elastic, characterized by a value of the Young modulus E_i, of the shear modulus at small strain G_i, of the shear waves velocity V_i, of the soil density ρ_i and, finally, of the Poisson's coefficient v_i. The soil, considered as a column with sectional area A_{soil}, was discretized in a series of elements having thick Δ_i, and mass M_i (Figure 3).

Frame/cable elements with shear stiffness were adopted to connect each to other the masses representing the soil discretization.

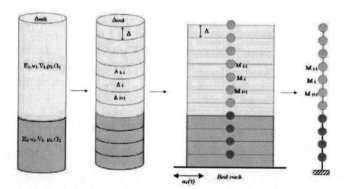

FIGURE 3. *Representation of the soil by lumped masses.*

Dynamic Soil - Pile Interaction

To perform dynamic analysis of the pile for earthquake excitation, soil stiffness and damping properties were included in the model, in which the soil masses are attached to by linear springs of stiffness K and linear dashpots of constant C (Figure 2). The pile is modeled as a linearly elastic beam discretized in elements having length Δ_i, corresponding to the soil discretization.

In the SAP2000$^@$ numerical simulations, two types of link elements were used to model the soil-pile interaction in dynamic analysis: *damper link* (Figure 4a) and *linear link* (Figure 4b), for the response of the soil-pile system in the frequency domain (designated as *Steady State Analysis - SSA*) or in the time domain (designated as *Time History - THA*) respectively. In the first case (*SSA*), a horizontal harmonic load with a constant vibration frequency can be used as input motion in the analyses, in the second case (*THA*), the input acceleration time history can be selected from a database of recorded seismic events.

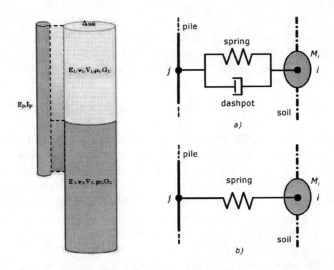

FIGURE 4. *Soil-pile interaction: damper link (a) and linear link (b).*

ANALYSIS RESULTS

To verify the model implementation in SAP2000[@] (i.e. definition and assignment of the link elements required to model the spring and dashpot soil elements), the results obtained with this program were compared with those computed from other existing analytical and numerical methods.

The numerical simulation regards the results presented and discussed during the recent workshop on *"Recent advances in Codes"* held in Thessaloniki during the *4th International Conference on Earthquake Geotechnical Engineering.* The aim of the workshop was to investigate on the state of the art in geotechnical earthquake engineering codes. One of the benchmark problems was the evaluation of the kinematically induced bending moments in a pile, drilled through a very soft clay layer, and subjected to a strong and long duration shaking.

Numerical analyses were performed for a free-head pile having a length of 28.5 m. It is a hollow-cylinder of 40 cm external diameter and 30 cm internal and it is made of pre-stressed high strength concrete. The soil deposit consists of 20 meters very soft clay and peat, underlain by a 10 m thick, much stiffer sandy silt layer. Dense gravel exists between the depths of 30 to 132 m, where rock is encountered. The gravel is interrupted by a thick layer of mudstone between the depths of 40 to 52 m. The relevant soil properties used in the numerical analysis are given in the Table 1. More information's are reported in the proceedings of the conference and/or in *Summary of Workshop 3* available on the web site of the *4th ICEGE.*

An earthquake ground motion recorded at a depth of 153 m (in the sand stone bedrock) in Atsuma by the Kik-net array, was used as input motion. It was shaken by the 2003 Tokachi-oki Earthquake. Figure 5 shows the acceleration time history.

In the following, the results of the numerical simulations carried out by Fugro West Inc. & AMC [7] and Murono & Hatanak [8] are taken into consideration. These results are obtained by two approaches in which a separate free field site response analysis is carried out (in the first case using the code SHAKE), and the free field displacement profile is applied to model the kinematic loading induced by the propagation of the seismic waves.

Pseudo Static Pushover Analysis

The soil-pile modeling has been developed with the aim to simulate dynamic and pseudo static loading conditions. In particular, the possibility to evaluate the kinematic pile bending by a pseudo static pushover analysis has been cheeked.

The static pushover analysis is a technique for estimating the strength capacity of building in the post-elastic range. The procedure involves to apply along the structure a distributed and predefined lateral load pattern.

According to the soil model reported in Figure 3 and to the soil-pile interaction model reported in Figure 4, the model represented in Figure 6 has been adopted for the pushover analysis.

A pseudo static pushover load distribution is applied to the soil strata to evaluate a static soil displacement profile. To model the kinematic loading, this displacement profile is applied to the pile by a bed of static spring.

TABLE 1. Soils Properties used in the Numerical Analysis.

Depth (m)	Soil	V_s (m/s)	Density (kN/m³)
0-6	Peat	60	13
6-20	Clay	90	15
20-30	Sandy Silt	190	18
30-40	Gravel	320	20
40-52	Mudstone	210	22
52-76	Gravel	310	20
76-132	Gravel	430	20
132-153	Sandstone	520	23

Considering the soil properties in Table 1 and using as input motion the acceleration time history of Figure 5, two numerical simulations have been carried out: a dynamic analysis in the time domain (*THA*) and a pseudo static pushover analysis. In this last case, the distributed lateral loading (Figure 6) has been computed according to the variation of the soil density with depth and to the peak acceleration profile.

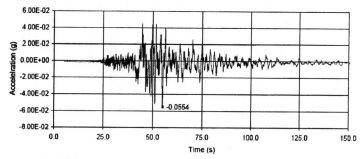

FIGURE 5. *Input ground motion: acceleration time history.*

The numerical simulations have been carried out in elastic conditions and the results obtained are reported in Figure 7 in terms of *free-field* soil displacement profile (*a*), normalized respect to u_{max}, and bending moment distribution (*b*) along the pile length. In the dynamic analysis (*THA*), the profile reported in Figure 7(*a*) represent the envelope of peak *free-field* soil displacements.

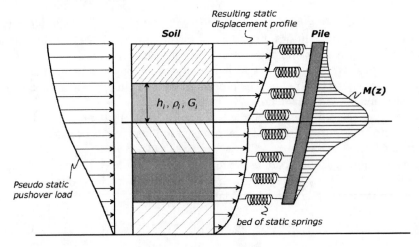

FIGURE 6. *Model adopted for the pseudo static pushover analysis.*

The site stratigraphy, soil properties and shear wave velocity profile were used as provided in the problem statement (Table 1). Taking into account the reduction of the shear modulus of soil with the shear strain level, the values of the shear modulus at small strain G_i were degraded to obtain numerical results close to those computed by

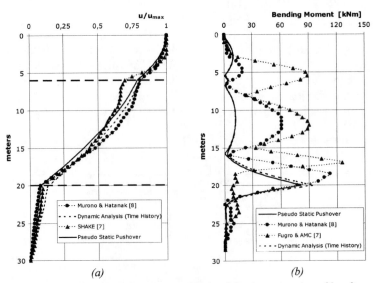

FIGURE 7. *Numerical results: profiles of normalized soil displacement a) and bending moment b).*

the other authors (Figure 7). The profiles reported in Figure 7 are obtained for a reduction of about 50 % respect to the initial value.

The agreement between the present numerical analyses and the results provided by the other formulations is found to be satisfactory, including those provided by the pseudo static approach.

CONCLUSIONS

In the present study, the possibility to develop a numerical model that can simulate the kinematic response of piles to has been investigated.

In particular, a pseudo static pushover approach was outlined for the kinematic bending of a single pile under the passage of vertically-propagating seismic waves. This approach could be attractive for practicing engineers because it is simpler than dynamic analyses, and permits to evaluate a soil peak displacement and bending distribution which is independent of the time history.

A further improvement in the approach is the possibility to take into account the nonlinear characteristics of the beam elements representing the pile, through the properties of a bending moment-curvature curve.

ACKNOWLEDGMENTS

This work is part of the RELUIS Research Project *"Innovative methods for the design of geotechnical systems"*, promoted by the Department of Civil Protection of the Italian Government.

REFERENCES

1. A. Abghari J. and Chai, "Modeling of soil-pile super-structure interaction for bridge foundations", in *Performance of deep foundations under seismic loading*, J.P. Turner ed., ASCE, New York, 1995, pp. 45-59.
2. A. Tabesh and H.G. Poulos, "Pseudo-static approach for seismic analysis of single piles", *Journal of Geotechnical and Geoenvironmental Engineering*, ASCE, **127**, (9), 2001, pp.757-765.
3. B. Schnabel, J. Lysmer and H. B. Seed, "SHAKE a computer program for earthquake response analysis of horizontally layered sites", Report No. EERC 72-12, Earthquake Engineering Res. Ctr., National Information System for Earthquake Engineering, 1972.
4. G.E. Mylonakis, "Strain transfer as fundamental concept for seismic design of pile foundations", XIV European Conference on Soil Mechanics and Geotechnical Engineering Proceedings, ERCT 12 Workshop on *"Geotechnical Aspects of EC8"*, Madrid, 2007, 6 p.
5. N. Makris and G. Gazetas, "Dynamic pile-soil-pile interaction. Part II: Lateral and seismic response", *Earthquake Engineering and Structural Dynamics*, **21**, 1992, pp.145-162.
6. M. Kavvadas and G. Gazetas, "Kinematic seismic response and bending of free-head piles in layered soil", *Géotechnique*, **43**, (2), 1993, pp. 207-222.
7. Fugro West, Inc. and AMC Consulting Engineers, "Pile in soft clay subjected to long duration seismic shaking", 4th International Conf. on Earthquake Geotechnical Engineering Proceedings, Workshop on *"Earthquake Geotechnical Codes"*, Thessaloniki, Greece, 2007.
8. Y. Murono and H. Hatanak, "Evaluation of kinematically induced moment and shear force in pile by Seismic Deformation Method", 4th International Conf. on Earthquake Geotechnical Engineering Proceedings, Workshop on *"Earthquake Geotechnical Codes"*, Thessaloniki, Greece, 2007.

Kinematic Interaction and Rocking Effects on the Seismic Response of Viaducts on Pile Foundations

F. Dezi[a] S. Carbonari[b] and G. Leoni[c]

[a]Dept of Materials, Environment Engineering and Physics, Università Politecnica delle Marche, Italy
[b]Dept of Architecture, Constructions, Structures, Università Politecnica delle Marche, Italy
[c]Dept ProCAm, Università di Camerino, Ascoli Piceno, Italy

Abstract. This paper is aimed at providing a contribution for a more accurate and effective design of bridges founded on piles. A numerical model is employed herein to determine the stresses and displacements in the piles taking into account soil-foundation-structure interaction. A 3D finite element approach is developed for piles and superstructure whereas the soil is assumed to be a Winkler-type medium. The method is applied to single piers representative for a class of bridges. Varying the soil layers characteristics and the pile spacing (from 3 to 5 diameters), bending and axial stresses along piles as well as the pier base shear are computed. A comparison with respect to a fixed-base model is provided. Special issues such as the contribution of the soil profile, of the local amplification and of the rocking at the foundation level are discussed. Soil-structure interaction is found to be essential for effective design of bridges especially for squat piers and soft soil.

Keywords: bridges, foundation rocking, frequency domain, local response analyses, pile foundations, soil-structure interaction.

INTRODUCTION

The importance of Soil-Structure Interaction (SSI) under dynamic loads was recognized in the 1960s in the design of shallow foundations for machines and mechanical equipments. During the last twenty five years, the major thrust of SSI research has been targeted to the understanding of pile-supported structures subjected to seismic waves [1, 2, 3, 4, 5]. In the case of bridges, where the substructures are often constituted by squat abutments and piers, the problem may be extremely significant. Modern seismic codes [6] have acknowledged these features and suggest accounting for soil-structure interaction effects in the foundation and superstructure design. These are conventionally divided into kinematic and inertial effects: the firsts derive from soil-foundation interaction due to the seismic wave propagation through the soil, the seconds are due to the presence of the vibrating superstructure. Two main approaches for analyzing soil-structure interaction are currently available, namely the direct method and the substructure method where kinematic and inertial interactions are separately studied [7].

The aim of this study is to capture the real dynamic behaviour of multispan bridges and of their foundation on pile groups by using the direct method. To this purpose the

CP1020, *2008 Seismic Engineering Conference Commemorating the 1908 Messina and Reggio Calabria Earthquake*,
edited by A. Santini and N. Moraci

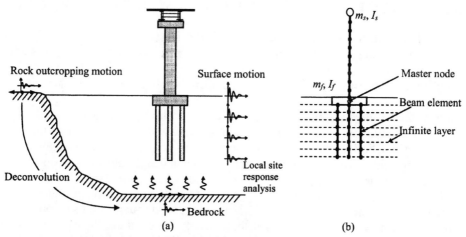

FIGURE 1. (a) Soil-structure interaction problem; (b) discrete model

effects of the local site configuration on the free-field motion starting from the earthquake motion at the bedrock level are considered (Fig. 1). A parametric analysis is performed to investigate the importance of the geometry of the pile group, of the soil conditions and of the structural rigidity.

ANALYSIS PROCEDURE

In this paper, the analysis of soil-structure interaction is performed by considering the whole structure-foundation system assuming linear behaviours for the structure, the piles and the soil. The analysis can be divided into three steps:

- local site response analysis providing the three translational displacement components of the free-field motion at the location of the embedded piles;
- computation of the seismic response of the whole soil-foundation-structure system subjected to the free-field motion;
- calculation of the stress resultants in the piles due to both the kinematic and the inertial interactions.

In the first step, the analysis should be performed with the suitable refinement according to the site configuration. In the simpler case of horizontal layer deposits a one-dimensional analysis is satisfactory.

For the second step the problem is approached in the frequency domain by considering for the pile-soil system a model proposed by the authors [8] where the piles are assumed to be beams and the soil to be a Winkler-type medium constituted by independent infinite layers with uncoupled in-plane and out-plane mechanics (Fig. 1b). Simplified elastodynamic Green's functions [9] are introduced for the soil layers, accounting for pile-soil-pile interaction and for material and radiation damping, and the finite element method is considered to integrate to problem on the piles and the superstructure.

SOIL-STRUCTURE INTERACTION OF MULTISPAN BRIDGES

The effects of soil-structure-interaction on the seismic response of bridges are studied considering different heights of piers, soil conditions and foundation geometries. The compliance-base model was developed according to the previous formulation and the results obtained were compared with those given by a fixed-base model.

Single bridge piers are analyzed. This case is realistic for relatively short bridges, with the deck laying on sliding supports at abutments and fixed supports at the piers, and also for long bridges with a high number of spans and piers of equal height.

The bridge pier is constituted by a single column and is founded on four piles. The values m_s and I_s are the lumped masses of the deck and are representative of a steel-concrete composite deck with span of approximately 40 m; similarly, m_f and I_f are the lumped masses of the foundation cap. The pier has a solid square cross section of edge 2.00 m and its mass is consistently distributed along the column. Three piers' heights equal to 5, 10 and 15 m (labelled by H05, H10 and H15, respectively) are considered in the analyses. The concrete is of grade C30/37 and is considered to be linearly elastic with Young's modulus $E_c = 3.5 \times 10^4$ N/mm². Cracking effects are accounted for by considering an effective modulus of elasticity $E_{ceff} = 0.75E_c$.

The pier is discretized by 1 m long beam finite elements. A structural damping is introduced in terms of Rayleigh damping: stiffness and mass proportional terms were evaluated to provide a 5% effective damping for the first and second mode of the fixed-base structures.

The soil profile consists of a 40 m thick deposit overlying a bedrock which is supposed to be elastic. Two different soil profiles, labelled by S1 and S2, having stiffness increasing with depth are considered in the analyses (Fig. 2).

The foundation is on a 2x2 floating pile group and is placed at depth of 2.5 m. The piles have diameter 1 m and length 30 m. Three pile spacings of 3, 4 and 5 diameters (labelled by D3, D4 and D5, respectively) were adopted in the analyses. Each pile is modelled by 1 m long finite elements to provide a suitable level of accuracy. The cap at the top of the pile group is considered to be rigid with a master node placed in correspondence of its centroid.

The seismic action is defined at the bedrock outcropping with three artificial accelerograms matching the Type 1 elastic response spectrum for soil type A and PGA 0.25g of Eurocode 8 [6].

To predict the free field motion and capturing the effects of the soil profiles, one-dimensional ground response analyses for the two soil profiles were performed by using the computer program EERA [10]. The obtained motion, variable with depth, was input at the level of each pile node. Furthermore, the motion obtained at the ground surface was used as input motion for the fixed-base model. Fig. 2 shows the elastic response spectra for the free field motion obtained at the outcropping soil for profiles S1 and S2. These are the maximum of the three spectral values obtained for each profile by processing the relevant artificial accelerograms with the local site response analysis. They are superimposed to the Type 1 elastic response spectrum for soil type D and PGA 0.25g [6] that should be considered in the case examined. It is worth noting that the spectra do not match the one proposed by Eurocode 8 as

	Profile S1				Profile S2			
	V_s	ρ	ν	ξ	V_s	ρ	ν	ξ
Layer	m/s	t/m³		%	m/s	t/m³		%
L1	100	1.8	0.4	10	200	1.8	0.4	10
L2	200	1.9	0.4	10	350	1.9	0.4	10
B	800	2.0	0.4	---	800	2.0	0.4	---

	Period	Period
Mode	[sec]	[sec]
1	1.01	0.50
2	0.40	0.21

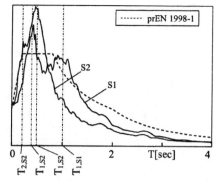

FIGURE 2. Soil properties and elastic response spectra at ground surface obtained for the two soil profiles by processing the accelerograms with a local site response analysis

amplifications are observed for systems with natural period close to the first two fundamental natural periods of the soil deposits (see table in Fig. 2).

SEISMIC RESPONSE: MAIN RESULTS

In Fig. 3 the time histories of the foundation rocking are reported. As expected, the major rotations are obtained in the case of soft soil, piles spaced of three diameters and slender pier (S1H15D3). This is due to the major deformability of the base restraints against rotation and to the major rotational inertia of the pier-deck system. Also, the frequency content of the responses is strongly affected by the soil profile, the pile spacing and the structure.

Figure 4 shows the maximum values of the pier base shear obtained with the three accelerograms in the eighteen studied cases. They are compared with those obtained by considering fixed-base models. In the case of soft soil S1 it is evident that the compliance-base models provide base shears lower than those obtained with the fixed-base models in the case of pier with an intermediate height (H10); in the case of a slender pier (H15) the shears are almost the same whereas, for squat piers (H05), shears obtained accounting for soil structure interaction are higher. The pile spacing seems to have not a significant influence on the results. In the case of profile S2, characterised by a higher stiffness, two different behaviours are evident for the squat-medium piers and the slender pier. The results obtained with pile spacing D4 and D5 demonstrate that the foundation can be considered to be rigid while in the case of spacing D3 the soil-structure interaction induces higher shears than those obtained with the fixed-base model. For slender piers the behaviour is strongly affected by the pile spacing and in all the cases the fixed-base model provides higher base shear.

The increase of the fundamental period is commonly expected leading to an increase of the base shear for squat structures and to a reduction of the base shear for slender ones. These trends are not evident in the applications because of the importance of the local site response analysis that, as already stated, leads to multi-peaks response spectra and results can be understand just comparing the fundamental periods of the structures (Table 1) with the response spectra of Fig. 2. As expected, the compliance-base models are characterised by the increase of the fundamental

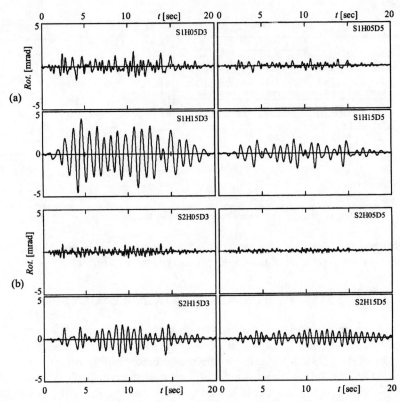

FIGURE 3. Rocking of the pier foundation: (a) soil profile S1; (b) soil profile S2 (response obtained with the first accelerogram)

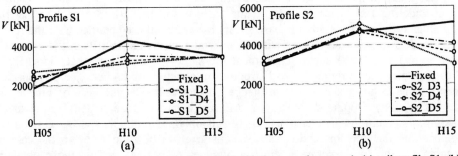

FIGURE 4. Base shear (maximum values obtained with three accelerograms): (a) soil profile S1; (b) soil profile S2

TABLE 1. Fundamental periods of the structure

		Profile S1			Profile S2		
	fixed	D3	D4	D5	D3	D4	D5
	sec	sec	sec	sec	sec	sec	sec
H05	0.12	0.32	0.29	0.28	0.22	0.20	0.19
H10	0.32	0.56	0.50	0.47	0.45	0.41	0.39
H15	0.60	0.86	0.78	0.74	0.74	0.70	0.67

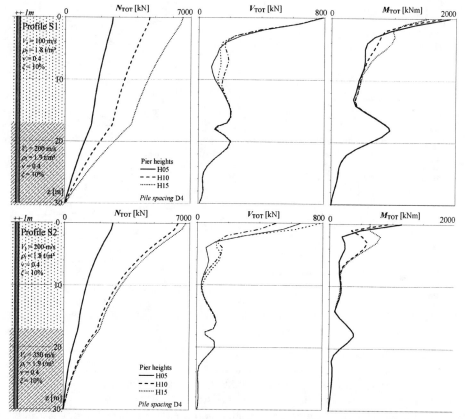

FIGURE 5. Base shear (maximum values obtained with three accelerograms): (a) soil profile S1; (b) soil profile S2

period with respect to the fixed-base models. Due to the peaks of the response spectra, period variation may lead to an increase or a decrease of the spectral acceleration that has to be evaluated case by case. This confirms the importance of performing local response analysis to predict more realistically the structure seismic response.

Figure 5 shows the stress resultants in one pile of the group with spacing of 4 diameters. Diagrams for the three pier heights are the envelopes of the results obtained with the three accelerograms. With reference to the bending moments and the shear forces, it is worth noticing that the diagrams are superimposed for depth greater than 14 m, for soil profile S1, and 10 m for profile S2. Furthermore, at the interface between the two soil layers peaks of the stress resultants are registered. These are due only to the kinematic interaction whereas differences in diagrams at minor depths are due to the superstructure (inertial interaction). It is also interesting to observe the presence of a second peak at few meters from the pile head due to the foundation rocking arising in the cases of slender piers. The effect of rocking is also evident in the axial force in the pile that significantly increases for slender piers.

In order to provide a more efficient comprehension of the response at the foundation level, the decomposition of the SSI results into kinematic and inertial

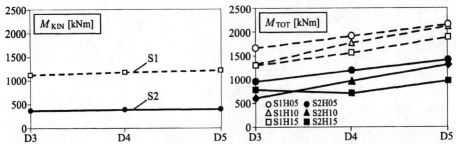

FIGURE 6. Kinematic and total bending moment at piles head (maximum values obtained with three accelerograms) for soil profile S1 and S2.

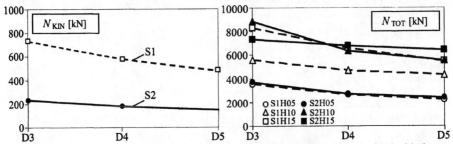

FIGURE 7. Kinematic and total axial force at piles head (maximum values obtained with three accelerograms) for soil profile S1 and S2.

components was operated. The first one was calculated setting the mass of the superstructure and the pile cap equal to zero and considering only the effects induced by the free-field ground motion. The second component is the difference between the total value and the kinematic value of the stress resultants. In fig. 6 and 7 the maximum values of the kinematic component and the total value of bending moments and axial forces at the pile head are reported. They are the maximum values obtained with the three accelerograms. Three different diagrams of the total response quantities are obtained considering the three pier heights. It is worth noticing that total bending moments are strongly affected by the kinematic interaction effects. For soil profile S1 kinematic bending moment component reaches values between 60% and 80% of the total bending moment and for soil profile S2 values between 30% and 60%. The effects of kinematic interaction are less evident in the axial forces where the kinematic component is about 10% of the total axial force.

CONCLUSIONS

A study, comprehensive of local site response behaviour and of a coupled kinematic and inertial soil-structure interaction analysis, was presented with reference to a class of bridges founded on piles. The main conclusions of the study are reported hereafter:

– the seismic response of bridges on pile foundation is very sensitive to the soil-structure dynamic interaction and to the site response that amplifies frequency contents close to the first two free vibration frequencies of the deposit;

- for squat piers the effects of soil structure interaction generally leads to an increase of the base shears; for piers with medium height it is not possible to find trends that could be considered of general validity and a site response analysis should always be coupled with a soil-structure interaction one. For slender piers and deposit with medium stiffness, the fixed-base model are conservative and a strong reduction of the base shear can be obtained by considering compliance-base models;
- soil-structure interaction induces, in the foundation system, bending moments with maximum values at the head of the capped pile and at the interface between layers characterized by different stiffness: the first are mainly related to the inertial interaction, the second are due only to the kinematic interaction; secondary peaks at few meters from the pile head may also arise due to the foundation rocking in the case of slender piers;
- rocking is an important component of the foundation motion; compliance-base model should account for it in the case of soft soil and in the case of slender piers even with medium stiffness soils; for squat piers, rocking becomes relatively significant only with very soft soil end piles closely spaced.

Finally fixed-base models should be carefully considered since they may be not conservative. Furthermore a preliminary local site response analysis should always be performed in order to catch the effects of local amplifications that sometimes are related even to second vibration modes of the deposit.

ACKNOWLEDGEMENTS

The authors are grateful to Mr Diego Baldarelli, for his assistance in performing the computer simulations.

REFERENCES

1. F. Dezi, "Soil-Structure Interaction Modelling", Ph.D. Thesis, Università Politecnica delle Marche, 2006.
2. F. Dezi, A. Dall'Asta, G. Leoni and G. Scarpelli, "Influence of the soil-structure interaction in the seismic response of a railway bridge", *ICEGE 2007 - 4th International Conference on Earthquake Geotechnical Engineering*, Thessaloniki, Greece, paper n. 1711 (2007).
3. K. Fan, G. Gazetas, A. Kaynia, E. Kausel and S. Ahmad, "Kinematic seismic response of single piles and pile groups", *J. of Geotechnical Engineering*, 117(12), pp.1860-79 (1991).
4. A. Kaynia and E. Kausel, "Dynamic stiffness and seismic response of sleeved piles", *Rep. R80-12*, MIT, Cambridge, Mass (1982).
5. G. Mylonakis, "Contributions to static and seismic analysis of piles and pile-supported bridge piers", Ph.D. Thesis, Faculty of the Graduate School of State University of New York, 1995.
6. prEN 1998-1 (2003), "EUROCODE 8: Design of structures for earthquake resistance - Part 1: General rules, seismic actions and rules for buildings".
7. J. P. Wolf, *Soil-structure Interaction Analysis in Time Domain*, Englewood Cliffs, Prentice-Hall.
8. G. Leoni, F. Dezi, and S. Carbonari, "A model for 3D kinematic interaction analysis of pile groups in layered soils", *Submitted to Earthquake Engineering and Structural Dynamics*.
9. N. Makris, G. Gazetas, "Dynamic Pile-Soil-Pile Interaction - Part II: Lateral and Seismic Response", *Earthquake Engineering Structural Dynamics*, 21(2), pp.145-162 (1992).
10. J. P. Bardet, K. Ichii and C. H. Lin, EERA: Equivalent-linear Earthquake site Response Analyses of Layered Soil Deposit. University of Southern California.

The Frequency and Damping of Soil-Structure Systems with Embedded Foundation

M. Ali Ghannad[a], Mohammad.T. Rahmani[b] and Hossein Jahankhah[c]

[a] Associate Professor, Department of Civil Engineering, Sharif University of Technology, P.O.Box 11155-9313, Tehran, Iran.
[b] M.Sc. Student, Department of Civil Engineering, Sharif University of Technology, Tehran, Iran.
[c] Ph.D. Student, Department of Civil Engineering, Sharif University of Technology, Tehran, Iran.

Abstract: The effect of foundation embedment on fundamental period and damping of buildings has been the title of several researches in three past decades. A review of the literature reveals some discrepancies between proposed formulations for dynamic characteristics of soil-embedded foundation-structure systems that raise the necessity of more investigation on this issue. Here, first a set of approximate polynomial equations for soil impedances, based on numerical data calculated from well known cone models, are presented. Then a simplified approach is suggested to calculate period and damping of the whole system considering soil medium as a viscoelastic half space. The procedure includes both material and radiation damping while frequency dependency of soil impedance functions is not ignored. Results show that soil-structure interaction can highly affect dynamic properties of system. Finally the results are compared with one of the commonly referred researches.

Keywords: soil-structure interaction, embedded foundation, impedance functions, equivalent damping

INTRODUCTION

Soil-structure interaction has significant effect on dynamic behavior of the super structure. This effect is usually investigated in two separate sections of kinematic Interaction (KI) and Inertial Interaction (II). To account for II, it is common to calculate effective period and effective damping of the system as the dynamic properties of replacement oscillator [1-5]. Yet, many researches have been done to make rational estimations of effective dynamic parameters. Some of the applied methods for this purpose were, eigen value analysis[5], numerical calculation of transfer function[4] and equivalent single degree of freedom approach[2,3]. It is note worthy to mention that in every method for effective parameter calculation, soil impedance functions play important role. These functions may be calculated from analytical[6] or numerical[7] methods or some times combination of both[8]. The estimated effective period and damping of soil-structure system may take different values with various soil impedance functions and different methods of dynamic parameter extraction. In this study, dynamic parameters of soil-structure systems, with embedded foundation, are investigated. The impedance functions were calculated through approximate analytical cone model solutions. This method has an attractive scene of being used in future works on embedded foundation in layered stratum[9]. Implying the above impedances, equations of motion are formulated and target dynamic parameters are calculated in an eigen value extraction procedure. Then a comparison with a commonly referred previous works is presented.

CP1020, 2008 Seismic Engineering Conference Commemorating the 1908 Messina and Reggio Calabria Earthquake, edited by A. Santini and N. Moraci

The interacting system investigated here constituted of SDOF structure mounted on cylindrical rigid foundation embedded in viscoelastic half space.

BASIC SOIL-STRUCTURE MODEL

Fig1. shows the model used in this research. The super structure is considered as a single mass, m_{str}, with moment of inertia, I_{str}, mounted on a spring having stiffness, K_{str}, at an elevation of, $\bar{h}$. These parameters can be representatives for first modal mass and inertia, effective stiffness and effective height of a MDOF structure.

The foundation is represented by a rigid cylindrical mass, m_f, with moment of inertia, I_f, radius r and depth of embedment e and is in full contact with surrounding soil. The flexibility of the surrounding soil is modeled by coupled sway-rocking springs. S_{hh}, S_{rr}, S_{hr} and S_{rh}, in Fig2., are representatives of soil stiffness matrix components.

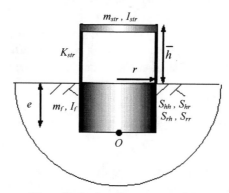

Figure(1): Basic soil-structure model

The equilibrium equation can then be written as follows:
$$M\ddot{X} + KX = F \tag{1}$$
Substituting M, K and F leads to:

$$\begin{bmatrix} m_{str} & 0 & m_{str}(\bar{h}+e) \\ 0 & m_f & m_f(e/2) \\ m_{str}(\bar{h}+e) & m_f(e/2) & I_{str}+I_f+m_{str}(\bar{h}+e)^2 \end{bmatrix}\begin{Bmatrix} \ddot{x}_s \\ \ddot{x}_f \\ \ddot{\phi}_f \end{Bmatrix} + \begin{bmatrix} K_{str} & -K_{str} & 0 \\ -K_{str} & K_{str}+S_{hh} & S_{hr} \\ 0 & S_{rh} & S_{rr} \end{bmatrix}\begin{Bmatrix} x_s \\ x_f \\ \phi_f \end{Bmatrix} =$$

$$-\begin{bmatrix} m_{str} & & m_{str}(\bar{h}+e) \\ m_{str}+m_f & & m_{str}(\bar{h}+e)+m_f(e/2) \\ m_{str}(\bar{h}+e)+m_f(e/2) & & m_{str}(\bar{h}+e)^2+I_{str}+I_f \end{bmatrix}\begin{Bmatrix} \ddot{x}_g^e \\ \ddot{\phi}_g^e \end{Bmatrix} \tag{2}$$

Where x_s, x_f and ϕ_f are deformation of structure plus foundation translation, foundation translation with respect to ground, foundation rotation with respect to ground respectively. x_g^e and ϕ_g^e are effective translation input motion and effective rotational input motion, respectively. By introducing five dimensionless parameters, $a_0 = \omega r/V_s$, $\bar{h}/r$, e/r, $\alpha = m_f/m_{str}$ and $\gamma = m_{str}/(\pi\rho r^2\bar{h})$, Eq.(2) can be interpreted in new form that provides deeper insight into formulation and improves its generality. Parameters ω and V_s

are vibration frequency and soil shear wave velocity respectively. It should be added that the material damping in the soil and structure may be included in the formulations as hysteretic form of damping by using the correspondence principle, i.e. just by multiplying the stiffness of the structure and soil by $(1+2i\xi_{str})$ and $(1+2i\xi_{soil})$ respectively. Where, ξ_{str} and ξ_{soil} are the material damping ratios in the structure and soil.

IMPEDANCES OF SOIL-FOUNDATION SYSTEM

In this section, the impedances of rigid cylindrical foundation embedded in a homogeneous half space are investigated. For this case soil stiffness matrix includes S_{hh}, S_{rr}, S_{hr} and S_{rh} as horizontal, rocking and coupling constituents. Definition of each term can be stated as follows:

$$S_{hh} = K_{HH}\left[k_{hh} + ia_0 c_{hh}\right]$$
$$S_{rr} = K_{RR}\left[k_{rr} + ia_0 c_{rr}\right] \qquad (3)$$
$$S_{rh} = S_{hr} = K_{HR}\left[k_{hr} + ia_0 c_{hr}\right]$$

Where K_H, K_R and K_{HR} are horizontal, rocking and coupling terms of static stiffness matrix and k_{hh}, k_{rr}, k_{hr}, c_{hh}, c_{rr} and c_{hr} are impedance coefficients of soil-foundation system. Different components of static stiffness matrix can be expressed as follows [8]:

$$K_{HH} = \frac{8Gr}{2-\upsilon}\left(1 + e/r\right)$$

$$K_{RR} = \frac{8Gr^3}{3(1-\upsilon)}\left(1 + 2.3\,e/r + 0.58(e/r)^3\right) \qquad (4)$$

$$K_{HR} = \frac{e}{3}K_{HH}$$

In which G is soil shear modulus. The equivalent radius may be used in the case of cubic foundation by matching the area or moment of inertia of the foundation with a replacement cylindrical foundation for the sway and rocking DOF's respectively. The method to derive soil impedance coefficients is approximate analytical formulation using cone models. Such type of formulation can be applied to more complex problems including layered soil texture and foundation irregularities in depth [9]. In this method the foundation is divided into several disks. Calculating related stiffness functions and implying geometric constraint leads to final form of soil-foundation stiffness matrix. The impedance coefficients, calculated by this method, are all frequency dependent and differ with variation of embedment ratio.

Here for wide ranges of embedment ratio and vibration frequency, the impedance coefficients are numerically calculated. The results are used to complete stiffness matrix in Eq.(2). As an alternative, using two dimensional regressions for same data, a set of approximate formulations for impedance coefficients is introduced. It can be used as substitutive for the above multi-stage procedure. The general form of represented formulation is described in (3).

$$\left(k_{ij} \text{ or } c_{ij}\right) = C_1 a_0^2 + C_2 a_0\left(\frac{e}{r}\right) + C_3\left(\frac{e}{r}\right)^2 + C_4 a_0 + C_5\left(\frac{e}{r}\right) + C_6 \quad i,j = (r \text{ or } h) \qquad (5)$$

Where C_1 to C_6 are the coefficients calculated from regression. Fig(2) shows fitted surfaces and original data for different impedance coefficients. Typical results are sketched for Poison's ratio of 0.4 as functions of a_0 and e/r.

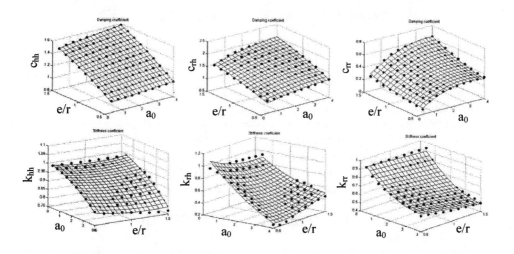

Figure(2): The accuracy of regression surfaces to match numerical results for impedance coefficients for Poison's ratio, 0.4 as functions of non-dimensional frequency and embedment ratio.

Similarly for Poison's ratios of 0.25 and 0.33, the coefficients C_1 to C_6 are estimated and final results for different terms are tabulated in table (1). The last column in this table shows an index of how represented curves match numerical data. As it is seen the agreement is good enough.

Table(1): Dynamic stiffness coefficients of embedded foundation

		C_1	C_2	C_3	C_4	C_5	C_6	R^2
$\nu = 0.25$	k_{hh}	-0.0086	-0.0116	0.0657	-0.0108	-0.1384	1.0621	0.985
	k_{rr}	0.0282	-0.0383	0.0119	-0.1721	-0.0607	1.0480	0.984
	k_{hr}	0.0190	0.1168	0.3022	-0.3876	-0.8710	1.5955	0.973
	c_{hh}	-0.0016	0.0151	-0.1117	0.0237	0.7273	0.6728	0.999
	c_{rr}	-0.0336	0.0401	0.1422	0.1878	-0.1306	0.0676	0.986
	c_{hr}	-0.0213	0.0551	-0.2070	0.1563	1.0491	0.4096	0.990
$\nu = 0.33$	k_{hh}	-0.0079	-0.0079	0.0641	-0.0131	-0.1355	1.0606	0.983
	k_{rr}	0.0285	-0.0394	0.0133	-0.1729	-0.0608	1.0476	0.984
	k_{hr}	0.0198	0.1194	0.2912	-0.3831	-0.8422	1.5771	0.971
	c_{hh}	-0.0015	0.0132	-0.1119	0.0240	0.7214	0.6402	0.999
	c_{rr}	-0.0336	0.0401	0.1492	0.1879	-0.1483	0.0786	0.986
	c_{hr}	-0.0209	0.0508	-0.2055	0.1552	1.0369	0.3818	0.990
$\nu = 0.45$	k_{hh}	-0.0070	-0.0025	0.0616	-0.0162	-0.1307	1.0581	0.979
	k_{rr}	0.0231	-0.0006	-0.0101	-0.2162	-0.0449	1.0701	0.989
	k_{hr}	0.0209	0.1231	0.2751	-0.3761	-0.8003	1.5503	0.969
	c_{hh}	-0.0014	0.0103	-0.1122	0.0245	0.7125	0.5945	0.999
	c_{rr}	-0.0313	0.0370	0.1292	0.1786	-0.1031	0.0263	0.990
	c_{hr}	-0.0204	0.0445	-0.2031	0.1535	1.0188	0.3432	0.990

DYNAMIC PARAMETERS OF THE SYSTEM

The method that is chosen here to extract dynamic parameters of system is eigen value analysis using matrices M and K in an iterative process. Components of dynamic stiffness matrix, S_{hh}, S_{rr}, S_{hr} and S_{rh}, are updated in primary frequency of soil-structure system in each loop. Using this procedure the damped frequency and damping ratio of the system are evaluated from real and imaginary parts of the principal eigen value respectively. The method is efficient and just a few iterations are required for convergence into the results. Among the dimensionless parameters, a_0, $\bar{h}/r$ and e/r have been selected as the key parameters here. Parameter α is evaluated as a function of the previous three. The next parameter, γ, is set to a typical value of (0.15) for ordinary buildings[3,4]. Material dampings are set to common values of (0.0) and (0.05).

REPRESENTATIVE RESULTS

In this section, the effect foundation embedment on dynamic properties of soil-structure systems is investigated. First the graphs of effective period and damping of soil-structure systems for extensive variation in non-dimensional parameters are presented. Then the results are compared with one of the commonly used procedures of dynamic parameter extraction.

Fig.(3) shows effective period ratios for three embedment ratios e/r=0, 0.5, 1 and three aspect ratios h/r=1, 2, 4 as a function of fixed base non-dimensional frequency $(a_0)_{fix}$.

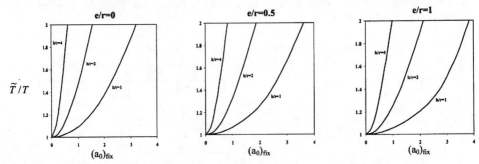

Figure(3): Effective system-to-structure period ratio with respect to system, aspect ratio, εμβεδμεντ ρατιο ν=0.4, ξ_{soil}=0.

As it can be seen, supposing a specific non-dimensional frequency, increasing in aspect ratio lead to substantial increase in effective period ratios. Also increase in embedment ratio slightly decreases effective period ratios.

Focusing on damping of the system, β_0, Fig.(4) shows the results for a wide range of $\tilde{T}/T$ covering systems with no SSI, $\tilde{T}/T$ =1, to systems with severe SSI effect, $\tilde{T}/T$ =2, where $\tilde{T}$ is the main period of the total system and T is the period of the super structure. Also a range of short squat buildings to slender ones are considered by varying h/r from 1 to 4 while the embedment ratios, *e/r*, have taken values of 0, 0.5 and 1. Soil material dampings of 0 and 0.05 are tried.

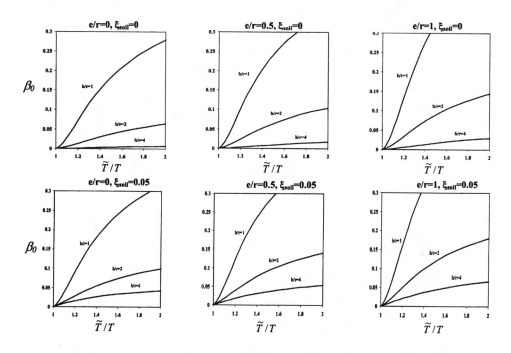

Figure(4): Effective damping of soil-structure systems with respect to system-to-structure period ratio, aspect ratio, embedment ratio, material damping of soil and ν=0.4

It can be seen the higher the aspect ratio, the lower the system damping. A reverse trend is observed for embedment ratio, increasing in which lead to higher system dampings. Though possessing increasing effect, material dampings don't affect total damping severely.

As last a comparison of results of this paper with one of the commonly referred researches, Veletsos[4], has been made. Formulations of that reference were originally derived for surface foundation; But has been recently implemented in the case of embedded foundations [10] which will be sited as "Modified Veletsos" here. Fig.(5) shows system dampings derived in this study besides "Modified Veletsos" related data. As it can be seen for high aspect ratios, i.e. h/r=2, 4, both methods propose essentially same results. But in low aspect ratios, i.e. h/r=1, results of this study are up to 20% higher than the other method.

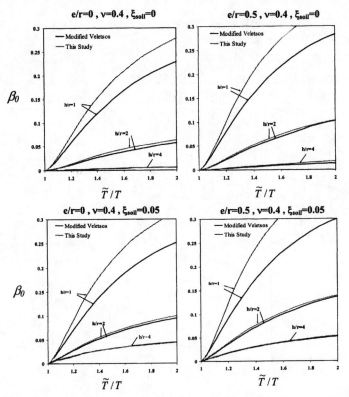

Figure(5): Effective damping of soil-structure systems calculated from two different methods with respect to system-to-structure period, aspect ratio, embedment ratio and material damping of soil.

CONCLUSION

The issues discussed in this paper include two major parts. At first a set of approximate formulations for impedance coefficients for soil-embedded foundation systems are presented. The equations were derived based on numerical data, calculated from well known cone models. Next, using the same impedances, a general equilibrium equation for soil-structure system is presented. Then using the method of complex eigen value analysis, main dynamic parameters of the system are calculated for different values of system-to-structure period ratio, aspect ratio, embedment ratio and material damping of the soil. Results show sharp increase in effective period ratio as the aspect ratio increases and slight decrease in effective period ratio when embedment ratio rises. On the other hand, any increase in aspect ratio has decreasing effects on damping while increase in embedment ratio results in higher damping. Finally, the results are compared with one of the commonly referred researches. For high aspect ratios both methods show good agreement. However, for low aspect ratios, the results of the proposed method in this study are up to 20% higher than that of the other method.

REFERENCES

1. A.S. Veletsos, and J.W. Meek, "Dynamic behavior of building-foundation systems," *Journal of Earthquake Engineering and Structural Dynamics*, Vol. 3,No. 2, 1974, pp. 121-138.
2. A.S. Veletsos and V.V. Nair, "Seismic interaction of structures on hysteretic foundations," *Journal of Structural Engineering, ASCE*, Vol. 101, No. 1, 1975, pp 109-129.
3. J. Bielak, "Dynamic Behavior of Structures with Embedded Foundations", *Journal of Earthquake Engineering and Structural Dynamics*, Vol. 3, 1975, pp. 259-274.
4. J. Aviles, M. Suarez, "Effective Period and Damping of Building Foundation Systems Including Seismic Wave Effects", *Journal of Engineering Structures*, No. 15, 2001, pp 553-562.
5. M.A. Ghannad, N. Fukuwa, R. Nishizaka "A study on the Frequency and Damping of Soil-Structure Systems Using a Simplified Model", *Journal of Structural Engineering, AIJ*,1998, , pp. 85-93.
6. J. W. Meek, and J. P. Wolf, "Cone Models for Embedded Foundation", *Journal of Geotechnical Engineering, ASCE*, V. 120, No. 1, 1994, pp. 60-80.
7. Apsel, R. J., and Luco, J. E. "Impedance Functions For Foundations Embedded In A Layered Medium: An Integral Equation Approach." *Journal of Earthquake Engineering and Structural Dynamics*, Vol. 2,1987, pp. 213-231.
8. A. Pais, E. Kausel, "approximate formulas for dynamic stiffnesses of rigid foundations", *Journal of soil dynamics and Earthquake Engineering* , Vol. 3, No. 4, 1988, pp 213-227.
9. J. P. Wolf, "Foundation Vibration Analysis: A Strength-of-Materials Approach", *Elsevier*, 2004.
10. J. P. Stewart, G. L. Fenves,R. B. Seed, "Seismic Soil-Structure Interaction in Buildings. I: Analytical Methods", *Journal of Structural Engineering, ASCE*, Vol. 125, No. 1, 1999, pp 26-37.

Effects of Soil-Structure Interaction on Response of Structures Subjected to Near-Fault Earthquake Records

M. Ali Ghannad[a], Asghar Amiri[b] and S. Farid Ghahari[c]

[a] Associate Professor, Department of Civil Engineering, Sharif University of Technology,
P.O.Box 11155-9313, Tehran, Iran.
[b] M.Sc. Student, Department of Civil Engineering, Sharif University of Technology, Tehran, Iran.
[c] Ph.D. Student, Department of Civil Engineering, Sharif University of Technology, Tehran, Iran.

Abstract. Near-fault ground motions have notable characteristics such as velocity time histories containing large-amplitude and long-period pulses caused by forward directivity effects and acceleration time histories with high frequency content. These specifications of near-fault earthquake records make structural responses to be different from those expected in far-fault earthquakes. In this paper, using moving average filtering, a set of near-fault earthquake records containing forward directivity pulses are decomposed into two parts having different frequency content: a Pulse-Type Record (PTR) that possesses long period pulses, and a relatively high-frequency Background Record (BGR). Studying the structural response to near-fault records reveals that elastic response spectra for fixed-base systems, in contrast to their response to ordinary earthquakes, show two distinct local peaks related to BGR and PTR parts. Also, the effect of Soil-Structure Interaction (SSI) on response of structures subjected to this type of excitations is investigated. Generally, the SSI effect on the response of structures is studied through introducing a replacement single-degree-of-freedom system with longer period and usually higher damping. Since this period elongation for the PTR-dominated period range is greater than that of the BGR-dominated one, the spectral peaks become closer in the case of soil-structure systems in comparison to the corresponding fixed-base systems.

Keywords: Near-fault earthquake records; Directivity effects; Moving average filter; Elastic spectra; Soil-structure interaction; Cone models.

INTRODUCTION

Ground shaking near a fault rupture may be characterized by a short-duration impulsive motion. This pulse type motion is particular to the forward direction, where the fault rupture propagates towards the site at a velocity close to shear wave velocity. The radiation pattern of the shear dislocation on the fault causes this large pulse of motion to be oriented in the direction perpendicular to the fault, i.e., the fault-normal component of the motion is more severe than the fault-parallel component [1]. In addition, near-fault earthquake records are rich in high frequencies because the short travel distance of the seismic waves would not allow enough time for the high-frequency content to be damped as is normally observed in the far-fault records [2].

CP1020, *2008 Seismic Engineering Conference Commemorating the 1908 Messina and Reggio Calabria Earthquake*,
edited by A. Santini and N. Moraci
© 2008 American Institute of Physics 978-0-7354-0542-4/08/$23.00

Elastic and inelastic response of fixed-base structures to near-fault pulse-type records has been the subject of many studies up to now [3]. However, the role of high frequency part of near-fault records has not been studied in detail. Moreover, the effect of Soil-Structure Interaction (SSI) when the structure is subjected to near-fault records has attracted much less attention. In this paper, the response of soil-structure systems to long-period pulse type and high frequency parts of the record is investigated simultaneously. Hence, using a moving average filter, original near-fault ground motions are decomposed into two components having different frequency contents, i.e., the Pulse-Type Record (PTR) that possesses long period pulses, and the relatively high-frequency Background Record (BGR) [4]. Then the contribution of each part on elastic response of soil-structure systems is investigated.

Since distance to the site can not be a suitable criterion, near-fault records collection with forward directivity effects requires a simple procedure to determine pulse-containing records. Quantitative methods used to date can be summarized by two general requirements. First, the site must fill the geometric and seismologic requirements of forward directivity, and second, visual inspection of the velocity time history reveals a pulse-like shape [5]. Recently, a quantitative algorithm to identify ground motions containing velocity pulses, such as those caused by near-fault directivity effect, is proposed by Baker [6]. In this paper, a combination of all mentioned criteria has been applied to collect near-fault earthquake records containing directivity pulses.

RECORD DECOMPOSITION

By applying a middle-assigned moving average filter on collected near-fault database, all original records are decomposed into PTRs and BGRs using the method proposed in [7]. As is mentioned there, the PTR is considered as the low-frequency smoothed part of each record which may have the same duration as the original record. The number of points for moving average filtering, which is related to the cut-off frequency of the filter, is directly dependent on pulse period and inversely on the length of the time intervals of input record, i.e.:

$$m = \alpha \frac{T_P}{dt} \tag{1}$$

where m, T_P and dt are the number of points for moving average filter, period of the dominant pulse, and the length of the time intervals of input record, respectively, and α is a coefficient which is determined empirically and set to 0.25. T_P is calculated using Short-Time Fourier Transform (STFT) [7]. BGR would be considered as the difference between original record and PTR. Figure 1 presents the results of this decomposition for two records. This figure reveals the appropriate performance of moving average filtering for records with only one main pulse (left) or multiple pulses (right).

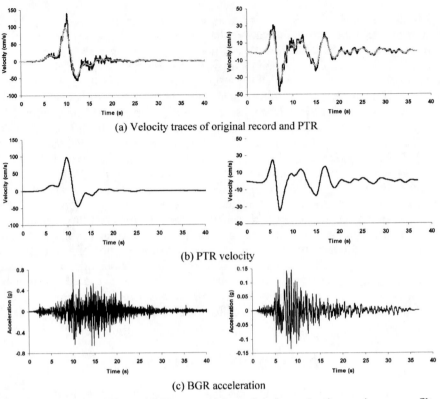

(a) Velocity traces of original record and PTR

(b) PTR velocity

(c) BGR acceleration

FIGURE 1. Samples of PTR and BGR extracted from original records using moving average filter.

STRUCTURAL RESPONSE

Fixed-Base Structures

Elastic response spectra computed for all records containing their PTR and BGR components confirm the idea that their elastic response spectra, in contrast to ordinary earthquakes, have two distinct local maximum regions. In one region, usually short periods, response of structures is dominated by BGR, and in the second part, typically related to long periods, by PTR. Figure 2 represents pseudo acceleration response spectra of four near-fault earthquake records and their BGR-PTR components, in which two mentioned regions are apparent. Therefore, considering directivity pulses alone as a representative of near-fault earthquake records for computing response of structures may be unsafe in some situations, especially for short buildings. Following the discussion above, we can say that there is usually a frequency gap in elastic response spectrum of structures against near-fault earthquake records. In other words, these spectra are a combination of two conventional design response spectrums, with two constant-acceleration regions corresponding to BGR and PTR components. Note that in some cases, these regions may overlap, or even cover each other.

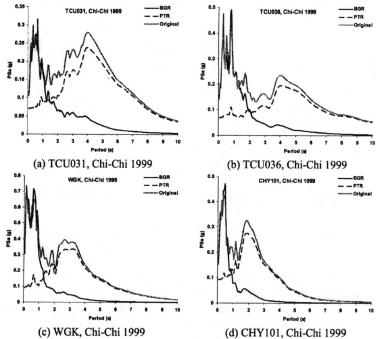

(a) TCU031, Chi-Chi 1999   (b) TCU036, Chi-Chi 1999

(c) WGK, Chi-Chi 1999   (d) CHY101, Chi-Chi 1999

FIGURE 2. Examples of pseudo acceleration response spectra for near-fault earthquake records.

Soil-Structure Systems

In this section, the effects of SSI on response of linear structures subjected to near-fault earthquake records are investigated. For this purpose, a simplified discrete model shown in Fig. 3 is used to represent the real soil-structure system. This model is based on the following assumptions:

1. The structure is replaced by an equivalent elastic SDOF system.
2. The foundation is replaced by a circular rigid disk of mass m_f, and mass moment of inertia I_f.
3. The soil beneath the foundation is considered as a homogeneous half-space and replaced by a simplified 3DOF system based on the concept of Cone Models [8].

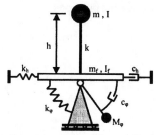

FIGURE 3. Soil-structure model [9].

The non-dimensional parameters for studying SSI effects are summarized in Table1.

TABLE 1. SSI non-dimensional parameters.

Parameter	Definition	Assigned Values
$a_0 = \omega_s h / V_s$	non-dimensional frequency	0, 1, 2
ω_s	natural frequency of fixed-base structure	
h	height of the structure	
V_s	shear wave velocity of the soil	
h/r	aspect ratio of the building	1, 3
$\overline{m}$	structure-to-soil mass ratio index	0.5
m_f/m	mass ratio of the foundation to structure	0.1
υ	Poisson's ratio of the soil	0.25
ξ_0, ξ_s	material damping ratios of the soil and the structure	0.05

More details of parameters are presented in reference [9]

The first two items, i.e. a_0 and h/r, are the key parameters that define the principal SSI effects. The other parameters, however, are those with less importance and are set to some typical values for ordinary buildings [10]. As mentioned in literature, the effects of SSI on elastic response of a structure can be summarized in two general aspects: first, it increases the period of the system because of the soil softness which is ignored in fixed-base models; second, it usually increases the damping ratio of the system via introducing new sources of damping in the soil, including hysteretic and radiation damping [11].

Elastic response spectra for soil-structure systems having three different values of non-dimensional frequency (a_0=0, 1, 2) and two values of aspect ratio (h/r=1, 3) subjected to all records containing PTR and BGR components are calculated. Note that a system with a_0=0 represents a fixed-base structure such as those presented in the previous section.

As an illustration, Fig. 4 shows pseudo acceleration response spectra of different soil-structure systems subjected to the four near-fault accelerograms recorded during the Chi-Chi, Taiwan earthquake. As it can be seen, increasing in structure-to-soil stiffness, i.e. a_0, results in movement of response spectrum to lower periods and reduction in spectral ordinates. Because this period shifting for PTR-dominated region is greater than BGR-dominated region, frequency gap between these two spectral peaks decreases in comparison to the fixed-base systems. Moreover, the greater is the value of h/r, the lower is the damping effect of SSI. Therefore, as the value of a_0 and h/r get larger, the structures which are located in gap region and have low fixed-base response may be located in peak spectral regions.

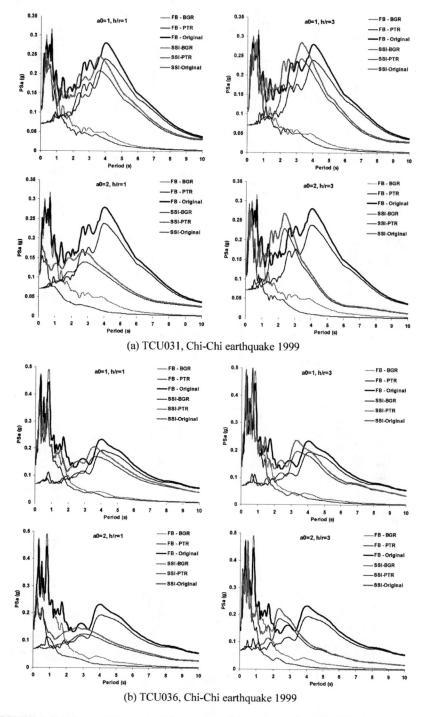

(a) TCU031, Chi-Chi earthquake 1999

(b) TCU036, Chi-Chi earthquake 1999

FIGURE 4. Pseudo acceleration response spectra of SSI systems subjected to near-fault records.

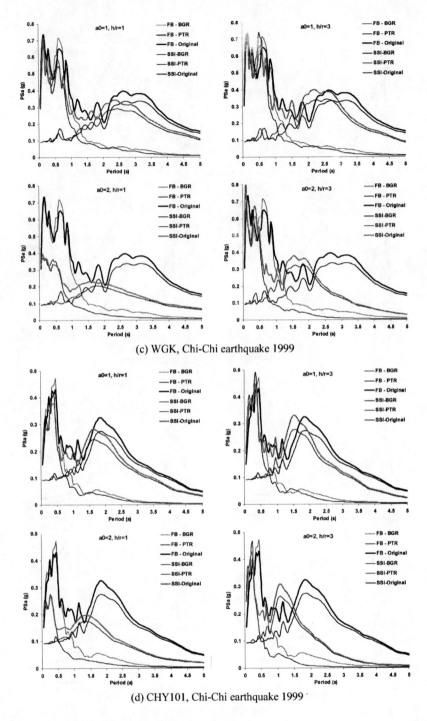

(c) WGK, Chi-Chi earthquake 1999

(d) CHY101, Chi-Chi earthquake 1999

FIGURE 4. (cont.) Pseudo acceleration response spectra of SSI systems subjected to near-fault records.

CONCLUSION

In this paper, using moving average filtering, the near-fault velocity records were decomposed into two components; one, PTR with large-amplitude long-period pulses, and the other, pulse-free BGR with higher frequency content. The elastic response spectra for the original near-fault records and their decomposed parts showed that in contrast to ordinary earthquakes, response spectra of near-fault records typically have two distinct local peak regions, each of them is representative of periods most affected by high and low frequency components, i.e., BGR and PTR, respectively. Then, using a simplified soil-structure model for surface foundations, based on the concept of Cone Models, elastic response spectra for different soil-structure systems subjected to near-fault records were calculated and compared with those of fixed-base structures. Results show that the spectral peaks become closer in the case of soil-structure systems in comparison to the corresponding fixed-base systems. This phenomenon is more pronounced for systems with lower soil-to-structure stiffness ratios. As a result, the elastic response of soil-structure systems in which the period of structure, in the fixed-base state, falls between the two mentioned peaks, may be greater than expected based on the fixed-base assumption. This effect is a more significant consideration in the case of slender buildings where the radiation damping is not so high.

REFERENCES

1. Somerville P, Smith N, Graves R, Abrahamson N. Modification of empirical strong ground motion attenuation relations to include the amplitude and duration effects of rupture directivity. *Seismological Society Letters* 1997; Vol. 68, No. 1, 180-203.
2. Ghobarah A. Response of structures to near-fault ground motion. *13th World Conference on Earthquake Engineering*, Vancouver, B.C., Canada, Aug. 1-6, 2004, Paper No. 1031.
3. Alavi B, Krawinkler H. Effects of near-fault ground motions on frame structures. *BLUME*, John A. Blume Earthquake Engineering Center Stanford, California, No. 138, 2001.
4. Ghahari SF, Jahankhah H, Ghannad MA. The effect of background record on response of structures subjected to near-fault ground motions. *First European Conference on Earthquake Engineering and Seismology*, Geneva, Switzerland, 3-8 September, Paper Number: 1512, 2006.
5. Cox KE, Ashford SA. Characterization of large velocity pulses for laboratory testing. *Pacific Earthquake Engineering Research Center* (PEER), Report No. 22, April 2002.
6. Baker JW. Quantitative classification of near-fault ground motions using wavelet analysis. *Bulletin of the Seismological Society of America*, Vol. 97, No. 5, pp. 1486–1501, October 2007.
7. Ghahari SF, Ghannad MA, Jahankhah H. Decomposition of near-fault earthquake records and their elastic response spectra. *Earthquake Eng. and Structural Dynamics*, Submitted, 2008.
8. Wolf JP. Dynamic soil-structure interaction. *Prentice-Hall*, Englewood cliff, New Jersey, 1985.
9. Ghannad MA, Jahankhah H. Site dependent strength reduction factors for soil-structure systems. *Soil Dynamics and Earthquake Engineering*, Vol. 27, No.2, pp.99-110, 2007.
10. Ghannad MA, Fukuwa N, Nishizaka R. A study on the frequency and damping of soil-structure systems using a simplified model. *Journal of Structural Engineering*, Architectural Institute of Japan (AIJ), 1998; 44B: 85–93.
11. Veletsos AS. Dynamic of structure-foundation systems Structural and geotechnical mechanics. In: *Hall WJ, editor. A volume honoring N.M. Newmark*. Englewood Cliffs, NJ: Prentic-Hall, Inc.; 1977. p. 333–61.

Comparing of Normal Stress Distribution in Static and Dynamic Soil-Structure Interaction Analyses

Alireza Kholdebarin[a], Ali Massumi[b], Mohammad Davoodi[c], Hamid Reza Tabatabaiefar[d]

[a] M.Sc. Student of Geotechnical Engineering, Graduate School of Engineering, Tarbiat Moallem University of Tehran (Kharazmi),Tehran, Iran.

[b] Assistant Professor of Structural Engineering, Graduate School of Engineering, Tarbiat Moallem University of Tehran (Kharazmi), Tehran, Iran, e-mail: massumi@tmu.ac.ir

[c] Assistant Professor, Dept. of Geotechnical Earthquake Engineering, International Institute of Earthquake Engineering and Seismology, IIEES, Tehran, Iran.

[d] M.Sc., Geotechnical Engineer, Tehran, Iran.

Abstract. It is important to consider the vertical component of earthquake loading and inertia force in soil-structure interaction analyses. In most circumstances, design engineers are primarily concerned about the analysis of behavior of foundations subjected to earthquake-induced forces transmitted from the bedrock. In this research, a single rigid foundation with designated geometrical parameters located on sandy-clay soil has been modeled in FLAC software with Finite Different Method and subjected to three different vertical components of earthquake records. In these cases, it is important to evaluate effect of footing on underlying soil and to consider normal stress in soil with and without footing. The distribution of normal stress under the footing in static and dynamic states has been studied and compared. This Comparison indicated that, increasing in normal stress under the footing caused by vertical component of ground excitations, has decreased dynamic vertical settlement in comparison with static state.

Keywords: Dynamic vertical settlement, Shallow footing, FLAC software, Time history analyses.

INTRODUCTION

Deliberation history of foundation bearing capacity and settlements goes back decades ago and development of theoretical evaluation and relevant analysis from Terzaghi's epoch (1943) until now have led to publishing tens of articles and technical reports. In addition, several contemporary researchers like Caquot & Kerisel (1953), Meyerhof (1963), Vesic (1973) and Chen (1975) worked on this specific subject [1]. However, in mentioned researches, evaluation of earthquake motion and related phenomena in geotechnical ambit took into consideration; a large portion of researches, analysis, experimental results and reports has been available in soil static states, while seismic geotechnical subjects hasn't been analyzed and evaluated as it deserved.

Dynamic loads applied to foundation might be caused by miscellanies reasons like earthquake, explosion, equipments vibration and water wave effects. When a

CP1020, *2008 Seismic Engineering Conference Commemorating the 1908 Messina and Reggio Calabria Earthquake,*
edited by A. Santini and N. Moraci
© 2008 American Institute of Physics 978-0-7354-0542-4/08/$23.00

structural system has been influenced by seismic waves of earthquake, applied loads during earthquake are included cyclic horizontal and vertical loads as well as cyclic moments around one or more main axes. In addition inertial forces in foundation caused by earthquake have influenced on footing settlements.

In conducted researches and studies about subsoil normal stresses of foundation, Richards (1993) proclaimed that under the influence of horizontal component, bearing capacity of foundation has been decreased because of subsoil inertial forces and shear transferring at interface of Soil-Structure [2]. In 1995, Dormieux and Pecker by considering failure surface as a Prandle curves under foundation and soil model boundaries reached the conclusion which decreasing in bearing capacity generally has been happened by inclined loads and subsoil inertial forces has had minor effect on foundation bearing capacity [3]. In this basis, other researchers like Budehu & Al-Kerni (1993) [4], Kumar & Mohan Rao (2002)[5], Soubra (1997)[6] and Sarma & Isossifelis (1990) by considering failure surface under foundation and designating ultimate load, found decreasing proportions for coefficients of bearing capacity formula and represented modified diagrams for Nc, Nq and Nγ in conjunction with soil friction angle. In addition Okamoto in 1973 modeled reduction of friction angle about $i = \tan^{-1} k_h$, caused by earthquake, consequently it is decrease bearing capacity of foundation [8].

In recent years, considering earthquake vertical acceleration in dynamic analysis and experimental observations has been taking into account intensely. For instant experimental observations conducted by Varadi & Saxena (1980) pointed out this trend. In these experiments, a rectangular foundation located on a sandy soil was used and transferring of vertical load under foundation was studied. In addition, vertical soil stresses in relation with depth in static and dynamic states with and without foundation was studied and compared which results is shown in Figure 1 [4].

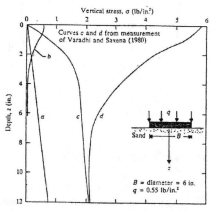

FIGURE 1. Normal stress under circular footing on sandy subsoil caused vertical component of earthquake

In accordance with mentioned explanations and in order to study and develop this subject, a numerical model of foundation has been modeled in FLAC software and effects of different factors on foundation settlements in different states has been studied.

CHARACTERISTICS OF NUMERICAL MODEL

Geometrical characteristics of numerical model have been selected as if boundary conditions haven't had any effect on analysis results. Subsoil horizontal distances between soil boundaries and centre of foundation have been assumed 15m from each side and vertical boundary extended to the bed rock. The bed rock depth has been assumed to be 30m and earthquake acceleration applied to the bed rock.

Subsoil dimensions have been postulated 30x30 meters with 0.5x0.5 meters rectangular mesh grids near the footing as well as 0.5x1.0 and 1.0x1.0 meters far from footing so as to study accuracy of model functioning. Footing lay in the middle of upper boundary of clayey sand soil with 2 meters width and 50cm depth (Figure 2). In addition, soil model with Mohr-coulomb failure criterion has been exploited in order to model soil stress-strain behavior because of simplicity of the criterion. Furthermore, main soil parameters like C and φ could have been obtained via ordinary soil mechanic laboratory tests.

In FLAC software, soil parameters like ρ (Soil density), K (Bulk module), G (Shear module), C (Cohesion), φ (soil friction angle) and ψ (Soil dilation angle) have to be introduced for soil model. Bulk and Shear module which are essential inputs for software was extracted from following formulas [10]:

$$K = \frac{E}{3(1-2v)} \tag{1}$$

$$G = \frac{E}{2(1+v)} \tag{2}$$

In above mentioned formulas, E and υ are Elastic module and Poisson ratio.

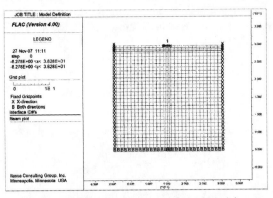

FIGURE 2. Schematic of Analyzed Model

Foundation has been modeled as a bending elastic element (beam element) which elastic module, density, area and inertial moment assigned to it. For modeling discontinues boundaries between soil and foundation, boundary elements or interface with zero depth has been used, which connected to beam element from one side and rectangular mesh grids of soil from another side (Figure 3). Normal and shear stiffness of this element is about 10 times greater than equal stiffness of smallest appressed mesh of soil which could be calculated with following formula [10]:

$$k_n = k_s = 10 \max\left[\frac{K + \frac{4}{3}G}{\Delta z_{\min}}\right] \qquad (3)$$

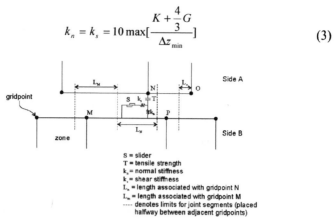

S = slider
T = tensile strength
k_n = normal stiffness
k_s = shear stiffness
L_N = length associated with gridpoint N
L_M = length associated with gridpoint M
---- denotes limits for joint segments (placed halfway between adjacent gridpoints)

FIGURE 3. Interface of soil-structure system connected with shear and normal spring

Specified model for Interface, postulated as prefect easto-pastic Mohr-Coulomb model which this model has been included Cohesion, soil friction angle, Soil dilation angle, tensional strength, normal and shear stiffness.

One of the most common methods for seismic analysis in FLAC software is Time History Analysis which has been exploited in this research so as to evaluate and analysis the model under vertical acceleration of earthquake. In this regard three distinct earthquake accelerogram included Kojoor vertical earthquake records (M = 6.1 Richter and PGA= 0.289g), Bam vertical earthquake records (M = 6.5 Richter and PGA= 0.231g) and Zarand vertical earthquake records (M = 6.3 Richter and PGA= 0.125g), recorded on the bedrock, has been applied to the models which related accelerograms and Fourier amplitude spectrum is shown in Figures 4 to 6.

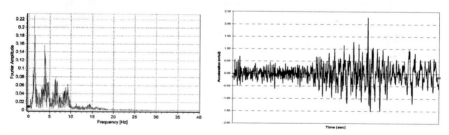

FIGURE 4. Time history of acceleration (Vertical component) and Fourier Spectrum for kojoor earthquake

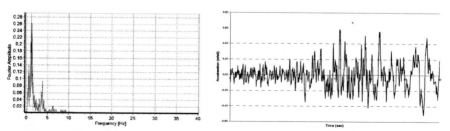

FIGURE 5. Time history of acceleration (Vertical component) and Fourier Spectrum for Bam earthquake

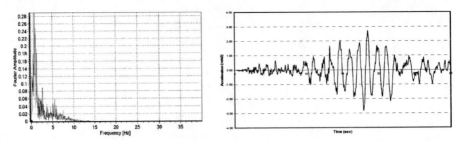

FIGURE 6. Time history of acceleration (Vertical component) and Fourier Spectrum for Zarand earthquake

Footing Elastic parameters:

$$A = 1m^2 \quad ; \quad \rho = 2500 kg / m^3 \quad ; \quad E = 2.5e10 pa \quad ; \quad I = 0.0208 m^4$$

Where: A: Area; ρ: Density; E: Elastic modulus; and I: Inertial moment
Plastic parameters of subsoil:

$$K = 1.36e8 pa \quad ; \quad \rho = 1715 kg / m^3 \quad ; \quad \varphi = 30°$$

$$G = 6.3e7 pa \quad ; \quad C = 2000 pa \quad ; \quad T \& di = 0$$

Where: C: Cohesion; φ: soil friction angle; di: soil dilation angle; and T: tensional strength
In order to dynamic modeling of soil, Rayleigh damping factor assumed D = 0.05 and central frequency assumed f = 1.6Hz in addition Beam elements has been used for foundation modeling which its Nodes has been connected to the subsoil mesh.

ANALYSIS RESULTS

Studying of Normal Stress in Static and Dynamic States

Depicting of normal stresses diagram in conjunction with depth in dynamic state has pointed out; existing of the footing would increase normal stresses in subsoil foundation up to 8 times more than static state in accordance with bousinesq equation (Figure 7).

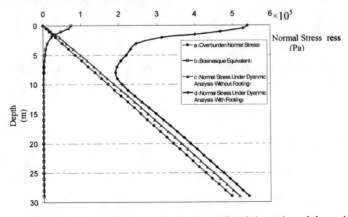

FIGURE 7. Comparing to Normal stress under footing vs Depth in static and dynamic state

For evaluating of this trend in this research a 350kpa static load has been used and applied to the 2 meters width strip footing which average vertical stress of subsoil in dynamic state has been obtained 5.4e5Pa under the influence of three mentioned earthquakes. In deeper depths of subsoil media, existence of footing has had fewer effects on the normal stresses in so far as, in the depth of 4B (B = width of foundation), existence of footing hasn't had any effect on this specific factor whatsoever.

Effects of Normal Stress Increasing on Dynamic Settlements

In this part, changeless load effect has been used in order to compare settlement counters. Figure 8 is shown settlements counter in static state with 450kpa load. Maximum settlement happened under the foundation about 61cm. The more far from foundation, the less settlement has been shown in so far as in 5B depth reached to 10mm.

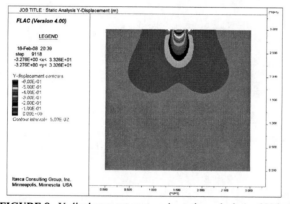

FIGURE 8. Y-displacement contour in static analysis (q=450 kpa)

In dynamic analysis, as it shown in Figures 9, 10 and 11 for 450kpa changeless static load under influence of three distinct earthquake records, increasing of stress led to some changes in Load-Settlement curves in comparison with static state. Maximum settlement in dynamic settlements history has been recorded 6mm for Kojoor earthquake, 14mm for Bam earthquake and 30mm for Zarand earthquake which comparatively have been reduced 10 times in Kojoor earthquake, 4.4 times in Bam earthquake and 2 times in Zarand earthquake (Figure 8).

According to maximum acceleration of each earthquake records in the bed rock, increasing of settlements with decreasing of maximum earthquake acceleration have been predictable.

Figure 12 is shown a comparison between Load-Settlement curves in static and dynamic states under influence of three mentioned earthquake. Zarand and Bam earthquake for changeless static loads from 100 to 450kpa have been showing more settlements in comparison with Kojoor earthquake. In this settlement rang, static settlements with increasing static load would be increased in comparison with dynamic settlements and lastly in 500kpa curves would be converged.

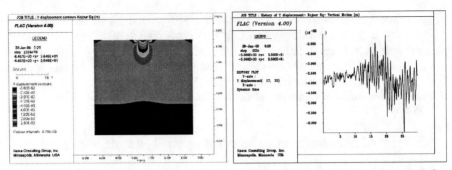

FIGURE 9. Y-displacement contour and Time history of Y-displacement in dynamic analysis for Kojoor earthquake (q=450 kpa)

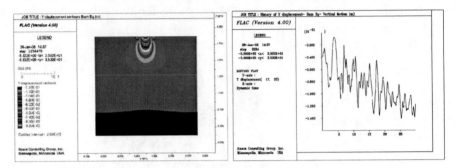

FIGURE 10. Y-displacement contour and Time history of Y-displacement in dynamic analysis for Bam earthquake (q=450 kpa)

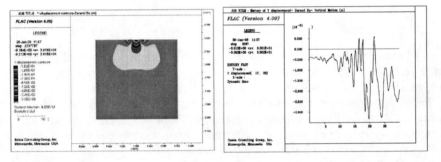

FIGURE 11. Y-displacement contour and Time history of Y-displacement in dynamic analysis for Zarand earthquake (q=450 kpa)

Depicting of Load-Settlement curves in different static loads under influence of earthquake records, indicate the role of increasing stress on dynamic settlements (Figure 12).In this basis, for static loads less than fracture load, because of increasing normal stresses, maximum settlement in dynamic state would be decreased in comparison with static state.

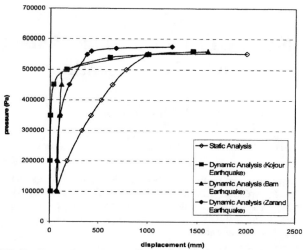

FIGURE 12. Comparison of load-settlement curves in dynamic and static states

CONCLUSIONS

The following conclusions may be drawn from the analytical investigation reported in this paper on theoretical evaluation of different soil parameters effects on dynamic settlement of shallow foundations:

- Maximum normal stress under foundation in dynamic state has been about 8 times greater than static state.
- In about 4B (B: width of foundation) depth, effect of footing on the normal stress distribution has been negligence.
- With increasing normal stress under the foundation caused by seismic loads, settlements have been decreased in comparison with static state.
- Amount of dynamic settlements decreasing have been connected to maximum earthquake acceleration. The more acceleration would lead to fewer settlements.

REFERENCES

1. S. Abdul-Hamid, "Upper-Bound Solution for Bearing Capacity of Foundations", *Journal of Geotechnical Eng, ASCE,* **125** (No.1), p. 59-68 (1999).
2. R.J. Richards, D.G. Elms and M. Budhu, "Seismic Bearing Capacity and Settlements of Foundations", *Journal of Geotechnical Eng, ASCE,* **119** (No.4), p. 662-674 (1993).
3. L. Dormieux and A. Pecker, "Seismic Bearing Capacity of Foundation on Cohesion Less Soil", *Journal of Geotechnical Eng, ASCE,* **121**, pp. 300-303 (1995).
4. M. Budhu and A. Al-karni, "Seismic Bearing Capacity of Soils", *Journal of Geotechnical Eng, ASCE,* **43**, pp. 181-187 (1993).
5. J. Kumar and V.B.K. Mohan Rao, "Seismic Bearing Capacity Factors for Spread Foundations", *Journal of Geotechnical Eng, ASCE,* **25**, pp. 79-88 (2002).
6. A. H. Soubra, "Discussion on Seismic Bearing Capacity and Settlement of Foundations", *Journal of Geotechnical Eng, ASCE,* **119**, pp. 1634-1640 (1993).
7. S. K. Sarma and I. S. Iossifelis, "Seismic Bearing Capacity Factors of Shallow Strip Footing", *Journal of Geotechnical Engineering, ASCE,* **40**, pp. 265-273 (1990).
8. R. Paolucci and A. Pecker, "Seismic Bearing Capacity of Shallow Strip Footing on Dry Soil", *Journal of Soil and Foundation Eng,* **37**, pp. 68-79 (1997).
9. B. M. Das, *Principles of soil Dynamics,* PWS-KENT Publishing Company, 1992, Boston.
10. Flac, *Fast Lagrangian Analysis of Continua,* User Manual, 2000.

Analysis Of The Interface Behavior
Under Cyclic Loading

Giuseppe Mortara

Department of Mechanics and Materials, University of Reggio Calabria, Italy

Abstract. This paper analyses the frictional behavior between soil and structures under cyclic loading conditions. In particular, the attention is focused on the stress degradation occurring in sand-metal interface tests and on the relevant parameters playing a role in such kind of tests. Also, the paper reports the analysis of the experimental data from the constitutive point of view with a two-surface elastoplastic model.

Keywords: interface, sand, cyclic loading, constitutive modeling.

INTRODUCTION

The behavior of many geotechnical applications is based on the frictional interaction that develops through the interface between the soil and the structural surface. The volumetric behavior of the interface soil is recognized to have a key role in such interaction [1, 2, 3, 4], which is particularly important if the interface is subjected to cyclic loading [5, 6, 7, 8, 9, 10]. For sand-structure interfaces, repeated loading causes progressive shear stress degradation due to the gradual densification of the sand layer close to the structural surface. Therefore, it is important to quantify such degradation in order to predict the loss of the structure skin friction capacity. This paper analyzes certain sand-metal direct shear tests discussing some peculiarities of the interface behavior under cyclic loading. Then, a constitutive model [11, 12] is used to predict the behavior observed in the laboratory.

SAND-STRUCTURE INTERFACE BEHAVIOR

The behavior of sand-structure interfaces depends on several factors. First of all, such behavior does not depend only on the sand properties, but also on the characteristics of the structural surface. This is clear in figure 1 where the graphical interpretation of roughness is shown. In figure 1a the size of sand particle is comparable to that of the pile surface asperities. Such a condition ensures the maximum mobilization of the sand-structure friction and this interface is referred to as "rough". In figure 1b, instead, the surface is completely smooth so that particles slide on the surface. Consequently, the minimum sand-structure friction coefficient is obtained and this interface is referred to as "smooth". An evaluation of the relative roughness between structure and soil particles is provided by the relationship proposed by Kishida and Uesugi [13].

CP1020, *2008 Seismic Engineering Conference Commemorating the 1908 Messina and Reggio Calabria Earthquake*,
edited by A. Santini and N. Moraci

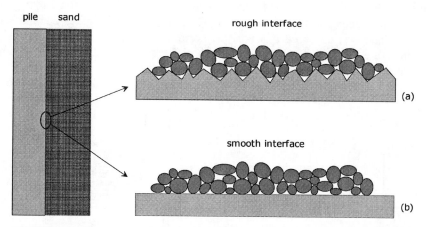

FIGURE 1. Graphical interpretation of roughness. (a) rough interface; (b) smooth interface

The coupling of surface asperities and sand particle size is not the only factor affecting interface behavior. Another key aspect is that under frictional interaction an intense shear localization of the sand takes place at the interface. This implies that the behavior of the sand layer close to the structural surface is different from that of the sand mass confining this layer.

The confinement exerted by the sand mass on the interface plays an important role because it interacts with the volumetric behavior of the interface layer. The result of this interaction is that the stress normal to the structural surface increases when the interface sand dilates and decreases when the interface sand contracts. If K is the normal stiffness imposed by the confining medium, the normal stress variation is given by the following relation

$$d\sigma_n = -Kdu \tag{1}$$

where du is the variation of normal displacement (positive for contraction). A continuous increase in normal displacement is the reason for the cyclic shear stress degradation when the stiffness K is positive.

Sand-structure interfaces are tested in the laboratory using a modified direct shear apparatus which is able to impose a constant normal stiffness reaction on the interface [3, 4]. With this apparatus all the relevant factors influencing the sand-structure interface behavior can be investigated and the parameters for the prediction of true scale application can be derived.

The interface tests shown in this paper deal with interfaces between Toyoura sand and aluminum plates having different roughness [14]. The most important effect caused by cyclic loading on sand-structure interfaces is the continuous accumulation of volumetric strains even if sand is in dense initial conditions. Figure 2 shows this characteristic effect in zero stiffness tests on rough interfaces. In both tests it can be observed that the gradual accumulation of normal displacement corresponds to the increase in shear stress (figures 2b and 2d). This effect (opposite to degradation) is consistent with the increase in the friction coefficient due to the increase in relative density caused by the sand densification. After the cyclic stage a consistent part of normal displacement accumulated during cycling is recovered (figures 2a and 2c). A different behavior is observed if a positive normal stiffness condition is applied (as shown in figure 3). The effect of the normal stiffness is evident in figure 3b which shows the gradual decrease in the shear stress as a result of the normal displacement accumulation. The post-cyclic dilation observed in figure 3a enables a partial recovery of shear stress.

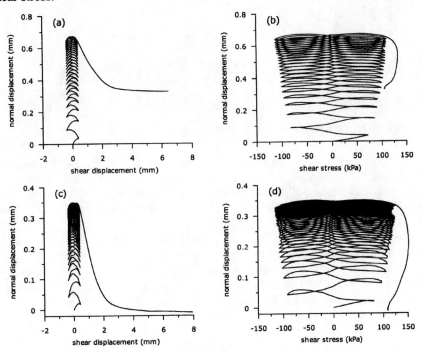

FIGURE 2. Comparison between zero stiffness interface tests (data from [14]).
(a), (b) loose Toyoura sand (*DR*=35%); (c), (d) dense Toyoura sand (*DR*=85%)

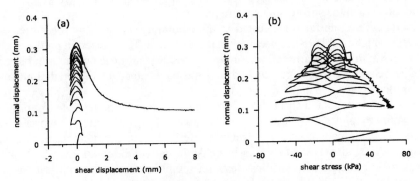

FIGURE 3. Effect of stiffness (*K*=500 *kPa/mm*) on a rough interface (data from [14]).
(a), (b) loose Toyoura sand (*DR*=35%).

The test results shown in figure 4 refer to a smooth interface while the test conditions are the same as those considered in figure 3. The comparison between figure 3 and 4 highlights the following considerations: (i) the post-cyclic dilation is not observed for the smooth interface (figure 4a); (ii) low values of shear stress are mobilized for the smooth interface and no recovery of shear stress is observed in the post-cyclic stage (figure 4b).

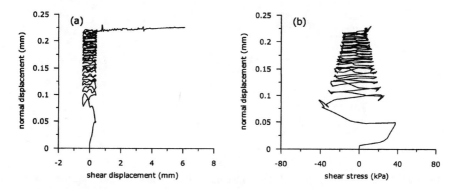

FIGURE 4. Effect of stiffness (K=500 kPa/mm) on a smooth interface (data from [14]). (a), (b) loose Toyoura sand (DR=35%).

INTERFACE CONSTITUTIVE MODELING

In this section the elements of an interface constitutive model [11, 12] will be analyzed and discussed. This 2D model is based on multi-surface elastoplasticity, it is formulated in terms of interface stresses and displacements and the interface is represented as a continuum surface subjected to kinematic discontinuities [15]. The main characteristics of this model are:
- linear elastic behavior;
- conical isotropic plastic surface;
- hardening/softening rule for the isotropic surface;
- distinct flow rules for hardening and softening;
- cyclic conical surface subjected to rotational hardening;
- bounding hardening modulus for the cyclic conical surface;
- continuous variation of the flow rule with densification.

Figure 5 shows the plastic surfaces and the plastic potentials of the model. The equations of f, f_0 and g are the following:

$$f = \sqrt{\tau^2} - \alpha\sigma_n = 0 \qquad (2)$$

$$f_0 = \sqrt{\overline{\tau}^2} - \alpha_0\overline{\sigma}_n = 0 \qquad (3)$$

$$g = \sqrt{\tau^2} - \frac{b}{1+a}\sigma_n\left[1 + a\left(\frac{\sigma_n}{\sigma_c}\right)^{-\frac{1+a}{a}}\right] = 0 \qquad (4)$$

Plastic potential (4) is obtained by integrating a linear stress-dilatancy function [15]. The cyclic plastic potential g_0 has the same mathematical expression of g but its parameters vary according to densification.

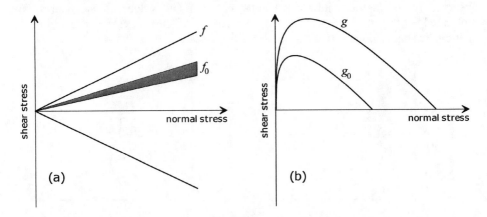

FIGURE 5. Plastic surfaces (a) and plastic potentials (b) for monotonic and cyclic conditions.

In previous papers [9, 12, 16] the model was used to model the behavior of rough interfaces while in this paper it is used to predict the behavior of smooth interfaces obtained by testing Toyoura sand (having mean diameter of 0.22 mm) and an aluminum plate having a roughness of 3 microns [14].

As shown in figures 3 and 4, the main difference between tests on rough and smooth interfaces is that in the latter, almost complete lack of dilation is observed. To model this aspect the softening part of the hardening function originally formulated [11, 12] has been ignored, given that non-softening behavior is observed in static tests. Consequently, the flow rule gives only non-negative values of the dilatancy rate (compression). The following hyperbolic expression is used here for the hardening function

$$\alpha(w^p) = \alpha_c - \alpha_c \exp\{-L\,w^p\} \qquad (5)$$

In Eq. 5 w^p is the plastic shear displacement, α_c is the maximum value of the stress ratio and L is a constitutive parameter. Figure 6 shows the influence of parameter L on the hardening function. For increasing values of L a stiffer behavior is observed in the shear stress – shear displacement curve. With respect to the original model [11, 12] the present model requires a smaller number of parameters.

Figure 7 shows the comparison between a cyclic test on an interface between loose Toyoura sand ($DR=35\%$) and a smooth aluminum plate, subjected to a normal stiffness $K = 500$ kPa/mm. The progressive degradation of the shear stress is observed in the experimental test (figure 7a). It is worth noting that the ultimate value of the shear stress is about the 20% of that of the first cycle. Figure 7b shows that the experimental data are well interpreted by the model. Similar comments follow for the comparison in terms of normal displacement (figures 7c and 7d).

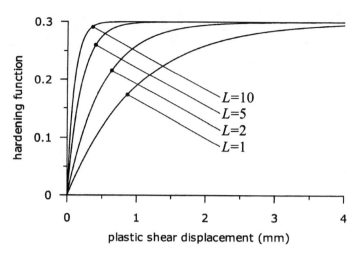

FIGURE 6. Effect of parameter L on the hardening function.

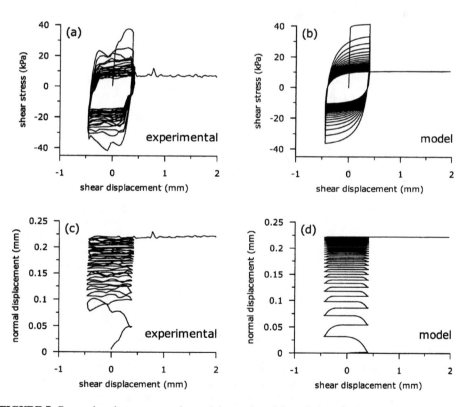

FIGURE 7. Comparison between experimental data and model predictions for a cyclic interface test on a smooth interface. Loose Toyoura sand (DR=35%), K=500 kPa/mm.

Figure 8 shows the comparison between an experimental test carried out on the same interface shown in figure 7, but subjected to a stiffness $K = 1000$ kPa/mm. As in figure 7, the normal displacement accumulation is well interpreted by the model that predicts a final displacement of 0.12 mm (figures 8a and 8b). As shown in figures 8c and 8d, where the experimental and predicted stress-path are reported, such compression causes a consistent reduction in normal stress from 150 to 30 kPa.

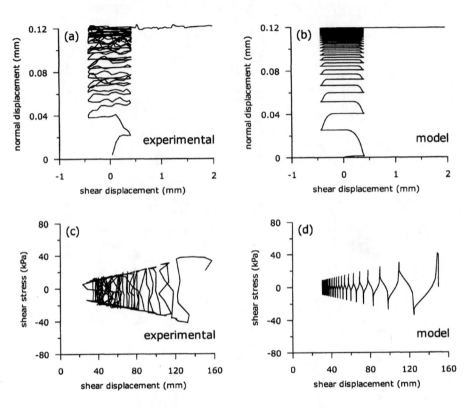

FIGURE 8. Comparison between experimental data and model predictions for a cyclic interface test on a smooth interface. Loose Toyoura sand (DR=35%), K=1000 kPa/mm.

CONCLUSIONS

This paper has shown some experimental results dealing with the behavior of interfaces between sand and structures. The experimental tests have evidenced that the normal stiffness acting on the interface is responsible for the degradation of shear stress under cyclic loading. The key aspect in these tests is the progressive accumulation of contractive volumetric displacement. The experimental results of a smooth interface have been interpreted with an existing elastoplastic constitutive model [11, 12]. Some changes have been made in the original model in order to capture the non-dilating behavior evidenced in these tests.

REFERENCES

1. E. Wernick, "Skin friction of cylindrical anchors in non-cohesive soils" in *Symposium on Soil Reinforcing and Stabilising Techniques in Engineering Practice*, Sydney, Australia, 1978, pp. 201-219.
2. F. Schlosser and A. Guilloux, "Le frottement dans le renforcement des sols", *Revue Française de Géotechnique* **16**, 65-79 (1981).
3. M. Boulon and P. Foray, "Physical and numerical simulation of lateral shaft friction along offshore piles in sand" in *Third International Conference on Numerical Methods in Offshore Piling*, Nantes, France, 1986, pp. 127-147.
4. I. W. Johnston, T. S. K. Lam and A. F. Williams, "Constant normal stiffness direct shear testing for socketed pile design in weak rock". *Géotechnique* **37**, 83-89 (1987),
5. J. T. Tabucanon, D. W. Airey and H. G. Poulos, "Pile skin friction in sands from constant normal stiffness tests", *Geotechnical Testing Journal* **18**, 350-364 (1995).
6. K. Fakharian and E. Evgin, "Cyclic simple-shear behavior of sand-steel interfaces under constant normal stiffness condition", *Journal of Geotechnical and Geoenvironmental Engineering* **123**, 1096-1105 (1997).
7. V. N. Ghionna, G. Mortara, G. P. Vita, "Sand–structure interface behaviour under cyclic loading from constant normal stiffness direct shear tests" in *Third International Conference on Deformation Characteristics of Geomaterials*, Lyon, France, 2003, pp. 231-237.
8. J. T. DeJong, M. F. Randolph and D. J. White, "Interface load transfer degradation during cyclic loading", *Soils and Foundations* **43**, 81-93 (2003).
9. G. Mortara, A. Mangiola and V. N. Ghionna, "Cyclic behaviour of sand-structure interfaces", in *International Conference on Cyclic Behaviour of Soils and Liquefaction Phenomena*, Bochum, Germany, 2004, pp. 173-178.
10. G. Mortara, A. Mangiola and V. N. Ghionna, "Cyclic shear stress degradation and post-cyclic behaviour from sand-steel interface direct shear tests", *Canadian Geotechnical Journal* **44**, 739-752 (2007).
11. G. Mortara, "An elastoplastic model for sand-structure interface behaviour under monotonic and cyclic loading", Ph.D. Thesis, Politecnico di Torino, 2001.
12. G. Mortara, M. Boulon and V. N. Ghionna, "A 2-D constitutive model for cyclic interface behaviour", *International Journal for Numerical and Analytical Methods in Geomechanics* **26**, 1071-1096 (2002).
13. K. Kishida and M. Uesugi, "Tests of the interface between sand and steel in the simple shear apparatus", *Géotechnique* **37**, 45-52 (1987).
14. G. P. Vita, "Comportamento delle interfacce tra terreni sabbiosi ed inclusioni solide in campo dinamico", Degree Thesis, University of Reggio Calabria, 1998.
15 M. Boulon and R. Nova, "Modelling of soil-structure interface behaviour, a comparison between elastoplastic and rate type laws", *Computers and Geotechnics* **9**, 21-46 (1990).
16 G. Mortara, A. Mangiola, V. N. Ghionna, "Experimental and theoretical study of sand-structure frictional interaction" in *11th International Conference of IACMAG*, Torino, Italy, 2005, Vol. 1, pp. 425-432.

Design Of Bridges For Non Synchronous Seismic Motion

Camillo Nuti[a] and Ivo Vanzi[b]

[a] *Dipartimento di Strutture, Dis, Università di Roma 3, Via Segre 4-6, 00146, Roma,*
c.nuti@uniroma3.it
[b] *Dipartimento di Progettazione, Pricos, Università di Chieti, Viale Pindaro 42, 65127, Pescara,*
i.vanzi@unich.it

Abstract. this paper aims to develop and validate structural design criteria which account for the effects of earthquakes spatial variability. In past works [1, 2] the two simplest forms of this problem were dealt with: differential displacements between two points belonging to the soil or to two single degree of freedom structures. Seismic action was defined according to EC8 [3]; the structures were assumed linear elastic sdof oscillators. Despite this problem may seem trivial, existing codes models appeared improvable on this aspect. For the differential displacements of two points on the ground, these results are now validated and generalized using the newly developed response spectra contained in the new seismic Italian code [4]; the resulting code formulation is presented. Next, the problem of statistically defining the differential displacement among any number of points on the ground (which is needed for continuos deck bridges) is approached, and some preliminary results shown. It is also shown that the current codes (e.g. EC8) rules may be improved on this aspect

Keywords: Bridge design, Earthquake, Non synchronous motion, Support design, Random field, Probability

INTRODUCTION

Models defining the spatial variability of earthquakes have been developed in the last twenty years, departing from experimental observations of simultaneous recordings of earthquakes [5, 6]. From the classical work of Luco and Wong [7], different statistical descriptions have been proposed and fit to the experimental data [8, 9], with varying degree of complexity and accuracy. The effects on structures have been also investigated, either in the linear field, with random vibration tools [10, 11], or in the non linear one, via numerical simulations or equivalent linearization procedures [12, 13, 14, 15]. The most important outcome of the studies has been unambiguous: apart from a few cases, non synchronous action decreases the structural stresses with respect to the case with synchronous actions. There are however situations in which non-synchronism negatively influences structural behavior, e.g. deck unseating; and some of the current design rules provided by the codes appear improvable on this aspect. Departing from these observations, and from the results of previous studies by the authors, this paper aims to validate structural design rules which account for the effects of earthquakes spatial variability. In past works [1, 2] the two simplest forms of this problem were dealt with: differential displacements

CP1020, *2008 Seismic Engineering Conference Commemorating the 1908 Messina and Reggio Calabria Earthquake,*
edited by A. Santini and N. Moraci

between two points belonging to the soil or to two single degree of freedom structures. Sesimic action was defined according to EC8 [3]; the structures assumed linear elastic sdof oscillators. For the differential displacements of two points, these results are now validated and generalized using the newly developed response spectra contained in the new seismic Italian code [4]; the resulting code formulation is presented. Next, the problem of statistically defining the differential displacement among any number of points (which is needed for continuos deck bridges) is approached, and some preliminary results shown. It is also shown that the current codes (e.g. EC8) rules may be improved on this aspect.

MODEL AND DEFINITION OF EARTHQUAKES THAT VARY IN SPACE

For the sake of completeness, a short summary of the model used in this work is presented. Readers are however referred to the works of Nuti and Vanzi [1, 2] for a more detailed presentation of the mathematical aspects. An earthquake acceleration recording at point P in space can be represented via its Fourier expansion as a sum of sinusoids [9]:

$$A_P(t) = \sum_k \left[B_{PK} \cdot \cos(\omega_K \cdot t) + C_{PK} \cdot \sin(\omega_K \cdot t) \right]$$

(0.1)

In equation (0.1), A is the measured acceleration in point P at time t, k is an index varying from 1 to the number of circular frequencies ω_k considered, B_{Pk} and C_{Pk} are the amplitudes of the k-th cosine and sine functions. Assume the acceleration $A_P(t)$ is produced by a wave moving with velocity V towards a different point in space, say Q, at distance X_{PQ} from P. At point Q, and at time t, one would have:

$$A_Q(t) = \sum_k \left[B_{QK} \cdot \cos\left(\omega_K \cdot (t - \tau_{PQ})\right) + C_{QK} \cdot \sin\left(\omega_K \cdot (t - \tau_{PQ})\right) \right]$$

$$\tau_{PQ} = \frac{X_{PQ}}{V} = X_{PQ} \cdot \left(\frac{\cos(\psi)}{v_{app}} \right)$$

(0.2)

In equation (0.2) ψ is the angle between the vector of surface wave propagation and the vector that goes from P to Q, τ_{PQ} is the time delay of the signal and v_{app} is the surface wave velocity. The amplitudes B_{Qk} and C_{Qk} would be respectively equal to B_{Pk} and C_{Pk} if the medium through which the waves travel did not distort them. In a real medium, B_{Pk} is correlated with B_{Qk} and C_{Pk} is correlated with C_{Qk} and the B's and C's are independent. The amplitudes B_{PK} and C_{QK} are statistically independent, for any points P and Q, and any circular frequency ω_K, with the only exception of B_{PK} and B_{QK} i.e. same circular frequency but different points in space. The same holds for C_{PK} and C_{QK}. The amplitudes are assumed normally distributed with zero mean and this assumption is experimentally verified. In order to quantify the acceleration time histories in different points in space, equations (0.1) and (0.2), all is needed is definition of the correlation between amplitudes and of their dispersion, as measured by the variance or, equivalently, of the covariance matrix of the amplitudes. The covariance matrix Σ of the amplitudes B and C is assembled via independent definition, at each circular frequency ω, of its diagonal terms, the variances in each space point and frequency, and of the correlation coefficients. The diagonal terms Σ_{PP} are quantified via a power spectrum. A traditional choice is the Kanai-Tajimi power spectrum, as modified by Clough and Penzien [16]:

$$G_{PP}(\omega) = G_0 \cdot \frac{\omega_f^4 + 4 \cdot \beta_f^2 \cdot \omega_f^4 \cdot \omega^2}{\left(\omega_f^2 - \omega^2\right)^2 + 4 \cdot \beta_f^2 \cdot \omega_f^4 \cdot \omega^2} \cdot \frac{\omega^4}{\left(\omega_g^2 - \omega^2\right)^2 + 4 \cdot \beta_g^2 \cdot \omega_g^4 \cdot \omega^2} \qquad (0.3)$$

$$\Sigma_{PP} = G_{PP}(\omega) \cdot d\omega$$

The Kanai-Tajimi power spectrum is herein adopted. The correlation coefficient between the amplitudes is expressed via the coherency function: the form originally proposed by Uscinski [17] on theoretical grounds and Luco [18] is retained:

$$\rho = \exp\left(-\omega^2 \cdot X^2 \cdot \left(\frac{\alpha}{v}\right)^2\right) \qquad (0.4)$$

The correlation decreases with increasing distance X and circular frequency ω and increases with increasing soil mechanical and geometric properties as measured by v/α. α is the incoherence parameter, v the shear wave velocity. The incoherence parameter α is the most difficult aspect in the coherency function assessment. For a more detailed discussion the reader is referred to [1]; however, values in a range as wide as 0.02-0.5 are reported in past experimental studies. Departing from the above earthquake spatial model, using random vibration concepts, it may be shown [2] that the distribution of the maximum differential displacement can be found with the peak factor formulation [19], by setting :

$$Z_{s,p}^* = \sigma_{Z^*} \cdot r_{s,p} \qquad (0.5)$$

where $Z_{s,p}^*$ is the displacement value which is not exceeded with probability p during an earthquake of duration s, and σ is the standard deviation of Z^*. Typical values of the peak factor $r_{s,p}$ lie within 1.20-3.5; $r_{s,p}$ is computed as set out in [19], in which proper account is taken for the non-stationarity of the response via the use of the equivalent damping.

DIFFERENTIAL DISPLACEMENTS BETWEEN TWO POINTS ON THE SOIL: CODE PROVISIONS VS. PREVIOUS AND CURRENT FINDINGS

In this section a comparison is made between some of the code provisions and the findings of past [1, 2] and new analyses by the authors. Only the case of differential displacement between two points on the ground is considered. In more detail, the codes considered are:

- The European seismic code Ec8 [3], partially adopted by the Italian seismic code of 2003 [20]. This code will be referred to with EC8/ICPC, meaning EuroCode 8 / Italian Civil Protection Code.
- The new Italian seismic code [4]. This code will be referred to as ICB, meaning Italian Code for Bridges. This code, for non synchronism, has been drafted following also the results in [2]

The analyses presented are:
- A summary of the results obtained in [1, 2] using EC8/ICPC response spectra with soils type A, B, D (respectively rock, stiff soil, loose soil)
- Some of the results obtained using the DICB for soil types A and B (corresponding to soil types A and B of EC8).

For both codes, reference is made to the ultimate limit state. For this one, the codes state the ground differential displacements be computed as:

$$\left\{ \begin{array}{l} u_{PQ}^I = X_{PQ} \cdot pga \cdot \dfrac{T_C}{v_{app}} \cdot \left(\varepsilon \cdot \dfrac{1}{2 \cdot \pi} \right) \le u_{PQ}^{I\_MAX} \\[3mm] u_{PQ}^{I\_MAX} = 0.025 \cdot pga \cdot \sqrt{\left(\varepsilon_P \cdot T_{PC} \cdot T_{PD} \right)^2 + \left(\varepsilon_Q \cdot T_{QC} \cdot T_{QD} \right)^2} \end{array} \right\} \text{EC8/ICPC} \qquad (0.6)$$

$$\left\{ \begin{array}{l} u_{PQ}^{II} = u_{PQ}^{II\_MIN} + \left(u_{PQ}^{II\_MAX} - u_{PQ}^{II\_MIN} \right) \cdot \left\{ 1 - \exp\left[-1.25 \cdot \left(X_{PQ}/v \right)^{0.7} \right] \right\} \\[3mm] u_{PQ}^{II\_MIN} = 1.25 \cdot 0.025 \cdot pga \cdot \left| \left(\varepsilon_P \cdot T_{PC} \cdot T_{PD} \right) - \left(\varepsilon_Q \cdot T_{QC} \cdot T_{QD} \right) \right| \\[3mm] u_{PQ}^{II\_MAX} = 1.25 \cdot 0.025 \cdot pga \cdot \sqrt{\left(\varepsilon_P \cdot T_{PC} \cdot T_{PD} \right)^2 + \left(\varepsilon_Q \cdot T_{QC} \cdot T_{QD} \right)^2} \end{array} \right\} \text{ICB}$$

with:

$$X_{PQ} = \text{distance between points P and Q} \; ; \; \varepsilon_P = \text{soil coefficient in point P}$$
$$pga = \text{peak ground acceleration} \; ; \; T_{PC}, T_{PD} = \text{periods defining the response spectra in point P} \qquad (0.7)$$
$$v_{app} = \text{surface wave velocity} \; ; \; v = \text{shear wave velocity}$$

In all the analyses, the most severe condition for non synchronism, i.e. highest uncorrelation, has been studied; therefore, α in equation (0.4) has been taken equal to 0.5. The reader is referred to [1, 2] for further discussion. In figure 1 the response spectra of EC8/ICPC and the ICB are shown. A few words, compatibly with the sake of brevity and space, about the ICB spectra are convenient. The spectra are defined by nearly the same relationships as the EC8, with three important exceptions: the maximum spectral acceleration amplification is soil and site dependent; the periods defining each interval (Tb, Tc, Td) depend on the soil type and on the maximum site spectral velocity; topographic effects are explicitly accounted for. In order to make a comparison between the model results obtained with the EC8/ICPC spectra, and those of the ICB, the above dependancies have been drastically simplified: the minimum value of the topographic effect (i.e. multiplicative parameter for topography = 1) has been adopted. Further, the maximum spectral velocity and maximum spectral acceleration amplification have been assumed constant and equal to the median values computed by Newmark and Hall [21] for rock soil. These values are: PGV/PGA=0.91 [m/sec/g]; maximum spectral acceleration amplification equal to 2.12; maximum

669

spectral velocity equal to 1.65 PGV. In this way, the ICB spectra (pictured in fig. 1, right) depend only on the ground type and the peak ground acceleration.

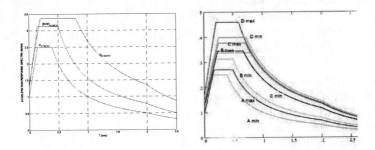

FIGURE 1. Acceleration response spectra of EC8 (left) and the DICB (right) ; pga=0.1g. The min and max suffixes in the DICB spectra are relative to the (minimum and maximum) topographic effects

The first result shown is in figure 2 (left). The figure, excerpted from [2], shows the comparison between the soil differential displacements of EC8, versus those computed using the model shortly described in section 2 [1, 2]. Notice that the results coming from the analysis shortly described in section 2 have been cast in the form expressed by equations (0.6), lower part, for inclusion in the DICB. Examining figure 2 (left), one can see that the maxima differential displacements computed with EC8 and this model differ by about 1.25; further, the trend is very different. EC8/ ICPC increases linearly up to the maximum, the analyses results (and the DICB prescriptions) grow in a parabolic fashion. In the range of distances where most civil engineering structures are, between 5 and 100 m, from buildings columns to long bridges piers, the differences are large: at 20 m distance, EC8/ICPC gives 2 mm or less while ICB forecasts differential displacements from 2 mm to about 40 mm, depending on the soil coupling.

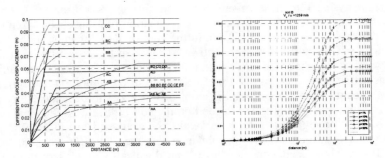

FIGURE 2. Left:: soil differential displacements; thicker lines are for EC8/ICPC; remaining lines are the results of the theoretical model of section 2. Right: differential displacement on soil type B. ICB spectra with 0.1g pga, and adoption of the Newmark and Hall relationship (described before) between PGA, PGV, maximum spectral acceleration amplification

The relative displacements computed with the ICB spectra for soil B, and with the Newmark and Hall simplification described before, are next shown in figure 2, right.From figure 2, one can notice that the increase of differential displacement with the distance is the same (the abscissae of figure 2, left, are in natural scale; those of figure 2, right, in logarithmic scale). The maximum values of the differential displacements appear to indicate dependence on the spectral shape: with the B soil type, the maximum (at high distance) differential displacement is equal to 72 mm, 58 mm and 83 mm respectively for EC8 and ICB. These results indicate that there is indeed a dependence of the differential displacements on the spectral shape, although it must be investigated which part of the spectra this is due to.

DIFFERENTIAL DISPLACEMENTS BETWEEN MULTIPLE POINTS ON THE SOIL: PRELIMINARY FINDINGS

Bridges on multiple supports must be checked also for the case of ground displacements occurring in opposite directions at adjacent piers. For this case, simple verification rules are given by the EC8. The bridge must be verified for the displacement set, occurring at the base of the piers, pictured in fig. 3.

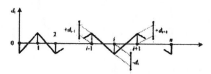

Figure 3.2 : Displacement Set B

FIGURE 3. displacement set for verification of multiple support bridges under ground displacements occurring in opposite directions (taken from EC8)

The displacement set consists in opposite direction displacements; the relative displacement between two adjacent piers equals the maximum differential displacement u_{PQI_max} (see equation (0.6)) times the ratio between the piers distance and Lg, the distance beyond which ground motion may be considered uncorrelated, ranging from 600 m (soil A) to 300 m (soil D). For instance, on soil type D, with piers distance equal to 30 m, and pga of 0.1g, the relative displacement equals (see fig. 2, left) 78 mm x 30 / 300 = 7.8 mm. This rule appears unconservative on one side (i.e. 7.8 mm appears too small a value) and far too conservative on the other (the probability that all the piers are displaced in opposite directions by the same amount is zero, from an engineering view – point).

A preliminary analysis has then been carried out via Montecarlo sampling of the earthquakes generated with the model in [1, 2], shortly described in section 2. Two soil types, A and D, as defined by EC8, have been assumed; the peak ground acceleration has been taken equal to 0.1g. The results of this analysis are shown in fig. 4. The statistics of soil curvatures, sampled at the base of the piers, show negative correlation (equal or higher than -0.5) between adjacent piers and no significant correlation thereafter (fig. 4, top). The statistics of curvatures may be therefore easily computed since those for two adjacent piers suffice to define the entire curvature field (see fig. 4, middle and bottom figures for the statistics of curvatures).

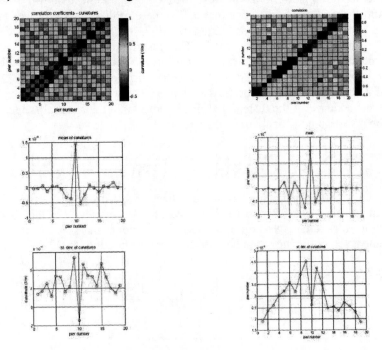

FIGURE 4. Statistics of soil curvatures and displacements for 21 piers at 20 m distance. 30 earthquake samples; soil D (left column) and A (right column) of EC8; pga=0.1g. Top: correlation; middle: mean value; bottom: standard deviation.

CONCLUSIONS

Based on well known expressions for spatial variability of seismic motion, a theoretical model founded on basic random vibration theory, has been developed in [1, 2]. The model is here used to compute the differential displacements of points on the grounds, both for two and multiple points cases, and with different response spectra shapes. Preliminary results indicate that the design codes can be strongly improved on this topic, both for the two points (e.g. simply supported decks) and the multiple points (e.g. continuous decks on multiple piers) cases. The results, in terms of differential displacements, have further shown sensitivity to the spectral shape, an aspect which must be carefully investigated. As a final remark, it is highlighted that earthquake spatial variability does appear to be a significant problem for failure modes governed by differential displacements, also for structures of minor importance like small bridges. Since its inclusion in the design phase brings about small or no extra cost for most situations, it is worth to stress the importance of a rapid code update on this subject.

REFERENCES

1. Nuti, C., Vanzi, I., Influence of earthquake spatial variability on the differential displacements of soil and single degree of freedom structures, *Rapporto del Dipartimento di Strutture*, DIS 1/2004, Università di Roma Tre, Rome, Italy, 2004

2. Nuti, C., Vanzi, I., Influence of earthquake spatial variability on differential soil displacements and sdf system response, Earthquake Engineering and Structural Dynamics, John Wiley and Sons, Volume 34, Issue 11, September 2005, 1353-1374

3. Comité Européen de Normalisation, CEN, *Eurocode 8: design of structures for earthquake resistance*, Draft n. 2, doc cen/tc250/sc8/n320, May 2002

4. Norme Tecniche per le Costruzioni, DM 14 gennaio 2008, Gazzetta Ufficiale n. 29 del 4 febbraio 2008 - Suppl. Ordinario n. 30

5. Abrahamson, N. A., Schneider, J. F., Stepp, J. C., Empirical spatial coherency functions for application to soil-structure interaction analyses, *Earthquake Spectra*, 7, 1, Feb. 1991, pages 1-27

6. Oliveira, C. S., Hao, H., Penzien, J., Ground motion modeling for multiple-input structural analysis, *Structural Safety*, 10, 1-3, May 1991, pages 79-93

7. Luco, J. E., Wong, H. L., Response of a rigid foundation to a spatially random ground motion, *Earthquake Engineering & Structural Dynamics*, 14, 6, Nov.-Dec. 1986, pages 891-908

8. Santa-Cruz, S., Heredia-Zavoni, E., Harichandran, R. S., Low-frequency behavior of coherency for strong ground motions in Mexico City and Japan, *12th World Conference on Earthquake Engineering*, New Zealand Society for Earthquake Engineering, Upper Hutt, New Zealand, 2000, Paper No. 0076

9. Vanmarcke, E. H., Fenton, G. A., Conditioned simulation of local fields of earthquake ground motion, *Structural Safety*, 10, 1-3, May 1991, pages 247-264

10. Der Kiureghian, A., Neuenhofer, A., A response spectrum method for multiple-support seismic excitations, UCB/EERC-91/08, Berkeley: *Earthquake Engineering Research Center*, University of California, Aug. 1991, 66 pages

11. Der Kiureghian, A., Neuenhofer, A., Response spectrum method for multi-support seismic excitations, *Earthquake Engineering & Structural Dynamics*, 21, 8, Aug. 1992, pages 713-740

12. Monti, G., Nuti, C., Pinto, P.E., Vanzi, I., Effects of non Synchronous Seismic Input on the Inelastic Response of Bridges, *II international workshop on seismic design of bridges*, Queenstown, New Zeland, 1994

13. Monti, G., Nuti, C., Pinto, P. E., Nonlinear response of bridges under multisupport excitation, *Journal of Structural Engineering*, 122, 10, Oct. 1996, pages 1147-1159

14. Hao, H., A parametric study of the required seating length for bridge decks during earthquake, *Earthquake Engineering & Structural Dynamics*, 27, 1, Jan. 1998, pages 91-103

15. Sextos, A.G., Pitilakis K.D., Kappos A. J., Inelastic dynamic analysis of RC bridges accounting for spatial variability of ground motion, site effects and soil-structure interaction phenomena. Part 1: Methodology and analytical tools. Part 2: Parametric study, *Earthquake Engineering & Structural Dynamics*, 32, 4, Apr. 2003, pages 607-652

16. Clough, R. W., Penzien, J., Dynamics of structures, *McGraw-Hill*, Inc., New York, 1975, 634 pages

17. Uscinski, B. J., The elements of wave propagation in random media, Mc Graw-Hill, New York, 1977

18. Luco, J. E., Mita, A., Response of circular foundation to spatially random ground motion, *Journal of Engineering Mechanics*, 113, 1, Jan. 1987, pages 1-15

19. Vanmarcke, E.H., Fenton, G.A., Heredia-Zavoni, E., SIMQKE-II, conditioned earthquake ground motion simulator : user's manual, version 2.1, *Princeton University*, [Princeton, N.J.], 1999, 25 pages

20. Presidenza del Consiglio dei Ministri, Primi elementi in materia di criteri generali per la classificazione sismica del territorio nazionale e di normative tecniche per le costruzioni in zona sismica, Ordinanza 3274, March 2003, (in Italian, updated after 2003)

21. Newmark, N. M., Hall, W. J., Earthquake spectra and design, *Earthquake Engineering Research Center Institute*, Berkeley, California, 1982

Analytical solutions for the seismic response of underground structures under SH wave propagation

C. Smerzini[a], J. Avilés[b], F. J. Sánchez-Sesma[c] and R. Paolucci[d]

a ROSE School, EUCENTRE, Via Ferrata 1, Pavia, Italy, csmerzini@roseschool.it
b Instituto Mexicano de Tecnología del Agua, javiles@tlaloc.imta.mx
c Universidad Nacional Autónoma de México, Instituto de Ingenería, Cd. Universitaria, Circuito Escolar s/n, Coyoacán 04510, Mexico D. F., Mexico, sesma@servidor.unam.m
d Department of Structural Engineering, Politecnico di Milano
P.za Leonardo da Vinci 32, 20133, Milano, ITALY, paolucci@stru.polimi.it

Abstract A theoretical approach is presented to study the antiplane seismic response of underground structures subjected to the incidence of plane waves. The structure is assumed to be a circular inclusion embedded in a homogenous, isotropic and linear visco-elastic halfspace and its mathematical formulation is approached through the theory of multiple scattering and diffraction. The inclusion may consist either of a cavity, with or without a ring-shaped boundary, or it may be filled in with a linear-elastic material, without loss of generality. The seismic response of the inclusion and its influence on surface ground motions are analyzed in both frequency and time domains. The dependence of the transfer function amplitudes on several parameters, such as the angle of incident SH waves, the frequency content of the excitation, the impedance contrast between the inclusion and the surrounding medium and the position along the ground surface, is underlined. Considering the lack of analytical solutions for quantifying the modification of ground motions induced by subterranean inhomogeneities, the results of this study can be used, on one side, as benchmark for both geophysical investigations and numerical dynamic soil-structure interaction studies, and, on the other side, to support the formulation of simplified approaches and/or formulas for the seismic design and assessment of underground structures.

Keywords: Underground structures, elastic inclusion, images technique, Graf's theorem, soil-structure interaction.

INTRODUCTION

The assessment of the seismic response of underground structures and subsurface irregularities is of emerging interest for engineering practice, especially when soil-structure interaction effects are likely to affect significantly the surface ground motion. Analogously, the identification and the characterization of subsurface obstacles, which may be either natural or artificial anomalies, such as cavities or petroleum reservoirs, constitute a challenging issue for geophysical subsurface investigations. From the engineering point of view, there is a lack of analytically-derived closed-form solutions suitable for quantifying the seismic response of subterranean inclusion. Though in the last thirty years the analysis of the scattered field produced by a cavity within the soil has been investigated in depth, limited effort has been dedicated to the development of

CP1020, *2008 Seismic Engineering Conference Commemorating the 1908 Messina and Reggio Calabria Earthquake*, edited by A. Santini and N. Moraci

practical tools to account for the influence of subsurface inhomogeneities on seismic ground motions (e.g. site correction factors for design response spectra).

Following the pioneering works of [4, 5, 6, 7, 10], in this paper we illustrate a theoretical approach to study the response of underground structures and/or subsurface irregularities of simple shape when subject to plane SH waves. The problem is formulated by modelling the cross-section of the subsurface irregularities as an ideal circular inclusion embedded in a homogenous, isotropic and linear visco-elastic halfspace, and it is approached through the theory of multiple scattering and diffraction. The method is versatile in dealing with various kinds of subsurface obstacles and type of excitation. The choice of a suitable set of cylindrical coordinates allows us to derive the exact analytical solution. In each reference system, the diffracted or refracted wavefield is represented with the aid of the expansion of wave functions in terms of Bessel and Hankel functions. The boundary conditions are satisfied with the aid of the Graf' addition theorem [1], analogously to what employed in [9, 3, 14] as well. The results are presented both in time and frequency domains, giving particular emphasis to their dependence on the type of inclusion (cavity/tunnel or elastic inclusion), the frequency content of the input wavefield, the angle of incidence and the relative position of the observation point.

MATHEMATICAL MODEL

The physical model under consideration is depicted in Figure 1. It consists of a circular cylindrical strip inclusion of radius a_C embedded in a homogenous, isotropic and linear visco-elastic halfspace. The inclusion may be either vacuous, i.e. the simplest case of a cavity, or filled in with a linear visco-elastic material, and its cross-section may be modelled with or without a ring-shaped boundary of variable thickness without loss of generality in the mathematical model. In the sequel, accordingly to the notation used in Figure 1, the halfspace will be denoted by index S, whilst the interior region of the subterranean anomaly with C. Superimposed in Figure 1 are two reference systems, the former located on the free surface and at distance L from the centre of the inclusion and the latter centered in the inclusion itself at depth H. The total displacement wavefield in the halfspace is a scalar quantity and has to satisfy the reduced wave equation, i.e. Helmholtz equation [2] as follows:

$$\nabla^2 w_S + k_S^2 w_S = 0 \qquad \text{with} \qquad \nabla^2 = \frac{\partial^2}{\partial^2 x} + \frac{\partial^2}{\partial^2 y} \tag{1}$$

where $\beta_S = \sqrt{\mu_S^{(0)}/\rho_S^{(0)}} \cdot \left(1 + \frac{i}{2Q}\right)$ is the complex shear wave velocity of the halfspace with density, shear modulus and quality factor given by μ_C, ρ_C and Q, respectively, and $k_S = \omega/\beta_S$ is the wavenumber associated to shear waves. Note that throughout the work it has been assumed that $Q_S = Q_C = Q$.

Relying on the theory of diffraction [10] the wavefield w_S within the halfspace results from the free-field $w^{(0)} = w^{(i)} + w^{(r)}$ given by the incident and reflected wavefields in the absence of the anomaly C, the wavefield $w_f^{(d)}$ diffracted by the surface of the obstacle and the wavefield $w_i^{(d)}$ diffracted by the surface of the image of the inclusion:

$$w_S = w^{(i)} + w^{(r)} + w_f^{(d)} + w_i^{(d)} \qquad (2)$$

A sketch of the rationale behind Eq. (2) is provided by Figure 2. The method of images, applicable to the scalar case, is used in this framework. If the incident wavefield is represented by a train of SH plane waves, polarized in the z direction, the free-field $w^{(0)}$ is given by the following expression:

$$w^{(0)} = w^{(i)} + w^{(r)} = 2e^{-ik_S x_1 \cos\psi} \cos(k_S y_1 \sin\psi) \qquad \text{with } i = \sqrt{-1} \qquad (3)$$

where ψ is the incidence angle. From here on, the time harmonic factor $e^{i\omega t}$ is understood and the wavefields are reported normalized with respect to the displacement amplitude w_0. The diffracted fields induced by the inclusion and its image are obtained by solving Eq. (1) with the method of separation of variables, yielding:

$$w_f^{(d)} = \sum_{m=0}^{\infty} \left\{ H_m^{(2)}(k_S r_2) \left[A_m \cos m\theta_2 + \hat{\delta}_{m0} B_m \sin m\theta_2 \right] \right\} \qquad (4)$$

and

$$w_i^{(d)} = \sum_{m=0}^{\infty} \left\{ H_m^{(2)}(k_S r_3) \left[A_m \cos m\theta_3 + \hat{\delta}_{m0} B_m \sin m\theta_3 \right] \right\} \qquad (5)$$

where $H_m^{(2)}(\circ)$ is the Hankel's function of second kind and order m and the symbol $\hat{\delta}_{m0}$ is equal to 1 for $m \neq 0$ and does not exist for $m=0$. The expansions (4) and (5) satisfy the stress-free condition at $y_1 = 0$ and the Sommerfeld radiation condition at infinity [11]. On the other hand, the displacement wavefield refracted and trapped within the obstruction can be expressed as follows:

$$w_C = \sum_{m=0}^{\infty} \left\{ J_m(k_C r) \left[C_m \cos m\theta_2 + \hat{\delta}_{m0} D_m \sin m\theta_2 \right] \right\} \qquad (6)$$

with $k_C = \omega / \beta_C$ and $J_m(\circ)$ being the Bessel's function of first kind and order m.

The coefficients A_m and B_m as well as C_m and D_m are determined by enforcing the boundary conditions regarding the continuity of displacements and stresses around the soil-inclusion interface:

$$w_S \big|_{r_2 = a_C} = w_C \big|_{r_2 = a_C} \quad \text{and} \quad \mu_S \frac{\partial w_S}{\partial r_2} \bigg|_{r_2 = a_C} = \mu_C \frac{\partial w_C}{\partial r_2} \bigg|_{r_2 = a_C} \quad \forall \theta_2 \in [0 \quad 2\pi] \quad (7)$$

Note that C_m and D_m are not significant as we are interested predominantly in reproducing the surface ground motion. Behind Eq. (7) there is the assumption of a perfect bounding between the inclusion and the surrounding medium.

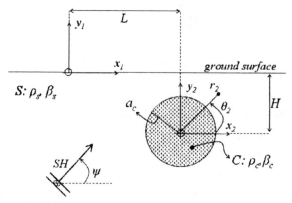

FIGURE 1 Sketch of the physical model of this work. Notice that the subsurface cylindrical irregularity, denoted by the index C, can be either vacuous, with or without a ring-shaped boundary (i.e. a tunnel), or filled in with a linear visco-elastic material of density ρ_C and shear-wave velocity β_C.

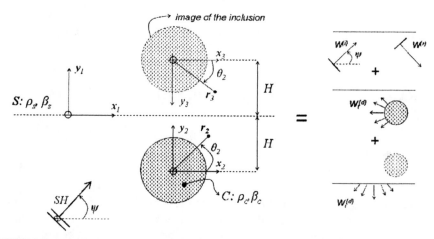

FIGURE 2 Superposition technique used to solve the problem depicted in Figure 1.

SOLUTION TECHNIQUE

The exact solution of the model depicted in Figure 2 is obtained with a boundary method which involves series expansion of incident and reflected SH waves in terms of cylindrical wave functions, as well as coordinate transformations between any two reference systems. In this contribution, it is convenient to express the total wavefield given by Eq. (2) in terms of the polar coordinates (r_2, θ_2). Firstly, the incident and reflected fields are expressed with the aid of the Neumann expansion [1] as series of cylindrical waves:

677

$$w^{(i)(r)} = e^{\varphi^{(i)(r)}} e^{-ik_S r_2 \cos(\theta_2 \mp \psi)} \qquad \text{with} \qquad \varphi^{(i)(r)} = -ik_S\left(L\cos\psi \mp H\sin\psi\right) \qquad (8)$$

such that the free-field takes the form

$$w^{(0)} = 2e^{-ik_S L\cos\psi}\left\{\cos(k_S H\sin\psi)\sum_{m=0}^{\infty}\left[(-1)^m \varepsilon_m J_m(k_S r_2)\cos m\psi\cos m\theta_2\right]+\right.$$

$$\left. 2i\sin(k_S H\sin\psi)\sum_{m=1}^{\infty}\left[(-1)^m J_m(k_S r_2)\sin m\psi\sin m\theta_2\right]\right\}$$

$$(9)$$

where ε_m is the Neumann factor ($\varepsilon_m = 1$ if $m=0$, $\varepsilon_m = 2$ elsewhere).

Finally, the diffracted wavefield $w_i^{(d)}$ can be expressed in terms of (r_2, θ_2) with the aid of Graf's addition theorem [1], leading to:

$$w_i^{(d)} = \sum_{m=0}^{\infty}\left\{J_m(k_S r_2)\left[\frac{\varepsilon_m}{2}\Lambda_{nm}^+ \cos m\theta_2 + \hat{\delta}_{m0}\Lambda_{nm}^- \sin m\theta_2\right]\right\}$$

$$\text{with} \quad \Lambda_{nm}^{\pm} = \sum_{n=0}^{\infty}(-1)^n\left[K_{nm}^{\pm} A_n \mp \hat{\delta}_{n0} L_{nm}^{\pm} B_n\right] \qquad (10)$$

The transport factors $K_{nm}^{\pm}$ and $L_{nm}^{\pm}$, which directly arise from transferring wave solution from system (r_3, θ_3) to (r_2, θ_2), are given by the following expressions:

$$K_{nm}^{\pm} = \cos\left((n+m)\frac{\pi}{2}\right)H_{n+m}^{(2)}(2k_S H)\pm(-1)^m \cos\left((n-m)\frac{\pi}{2}\right)H_{n-m}^{(2)}(2k_S H) \qquad (11a)$$

$$L_{nm}^{\pm} = \sin\left((n+m)\frac{\pi}{2}\right)H_{n+m}^{(2)}(2k_S H)\pm(-1)^m \sin\left((n-m)\frac{\pi}{2}\right)H_{n-m}^{(2)}(2k_S H) \qquad (11b)$$

Substituting eqs. (4), (9) and (10) into Eq. (2), the total displacement wavefield is referred to the reference system (r_2, θ_2). Enforcing the boundary conditions (7) and taking into account the orthogonal properties of the trigonometric functions, we obtain four infinite linear systems of equations for the unknowns A_n, B_n, C_n and D_n. The field within the inclusion is of no interest here and therefore C_n and D_n can be eliminated by substitution. The resulting system can be written in matrix compact notation as follows:

$$\begin{bmatrix} G_{mn}^{11} & G_{mn}^{12} \\ G_{mn}^{21} & G_{mn}^{22} \end{bmatrix}\begin{Bmatrix} A_n \\ B_n \end{Bmatrix} = \begin{Bmatrix} I_m^1 \\ I_m^2 \end{Bmatrix}, \qquad (12)$$

where the independent terms $I_m^{1,2}$ are defined as:

$$I_m^1 = -4\cos(k_S H \sin\psi)e^{-ik_S L\cos\psi}(-i)^m \cos m\psi, \tag{13a}$$

$$I_m^2 = -4i\sin(k_S H \sin\psi)e^{-ik_S L\cos\psi}(-i)^m \sin m\psi. \tag{13b}$$

The boundary conditions are those of continuity of displacements and tractions. It comes out, after solving the inclusion problem, that the coefficients for the diffracted field admit simple limiting forms if the inclusion is a void or a lined tunnel. After some straightforward, but lengthy algebra, the sub-matrices G^{ij} can be expressed as:

$$G_{mn}^{11} = \left[(-1)^n K_{nm}^+ + \frac{2}{\varepsilon_m}\delta_{mn}\Delta_m\right] \tag{14a}$$

$$G_{mn}^{12} = \left[-(-1)^n L_{nm}^+\right] \tag{14b}$$

$$G_{mn}^{21} = \left[(-1)^n L_{nm}^-\right] \tag{14c}$$

$$G_{mn}^{22} = \left[(-1)^n K_{nm}^- + \delta_{mn}\Delta_m\right] \tag{14d}$$

where δ_{mn} is the Kronecker delta ($= 1$ if $m = n$; $= 0$ if $m \neq n$) while the factor Δ_m, which only affects the terms along the principal diagonal, is given by:

$$\Delta_m = \frac{H_m'^{(2)}(k_S a_C) - F_m H_m^{(2)}(k_S a_C)}{J'_m(k_S a_C) - F_m J_m(k_S a_C)} \tag{15}$$

Δ_m depends on the inclusion factor F_m defined as:

$$F_m = C\frac{J'_m(k_C a_C)}{J_m(k_C a_C)} \quad \text{with } C = \sqrt{\frac{\mu_C \rho_C}{\mu_S \rho_S}} \qquad \text{for an elastic inclusion} \tag{16a}$$

$$F_m = C\frac{J'_m(k_C a_C) - \alpha_m Y'_m(k_C a_C)}{J_m(k_C a_C) - \alpha_m Y_m(k_C a_C)}, \quad \alpha_m = \frac{J'_m(k_C(1-t)a_C)}{Y'_m(k_C(1-t)a_C)} \text{ for a lined tunnel} \tag{16b}$$

$$F_m = 0 \qquad \text{for a cavity} \tag{16c}$$

We come up with a well-structured system, straightforward to generalize not only to different kinds of buried anomalies, but also to various incident fields (see following section for further details). Note that in Eq. (16b) $t = (a_C - a_i)/a_C$ is the dimensionless thickness of the wall of the tunnel, a_i and a_C being the inner and external radius, respectively, while $Y_m(\circ)$ is the Bessel's function of second kind and order m. The prime in $H_m'^{(2)}(\circ)$ and $J'_m(\circ)$ denotes differentiation with respect to the argument.

Infinite systems like that in Eq. (12) can not in general be solved. Truncation to finite size is therefore necessary to obtain the sought solution at any point of the model considered. In other words, the order of the expansions is set equal to MN, such that the unknown coefficients are $2MN+1$. Sensitivity analyses with respect to MN, as it will be illustrated in detail in the sequel, show that the solution is rather robust especially at low frequencies.

NUMERICAL RESULTS

From the engineering and seismological perspectives, it is of interest to analyze the variations in response amplitudes on the ground surface induced by the nearby underground structure. The results are shown both in time and frequency domain. The wavefields are represented in terms of the dimensionless frequency

$$\eta = \frac{\omega}{\pi} \frac{a_C}{\beta_S} = \frac{2a_C}{\lambda} \tag{17}$$

which is merely the inclusion-diameter to wavelength ratio, in analogy with the terminology commonly used in soil-structure interaction studies. In the absence of the underground structure, the modulus of ground displacement is assumed to be 1 (i.e. free-field solution), neglecting the free surface amplification factor. The presence of the irregularity induces displacement amplitudes which may depart relevantly from the free-field solution and may vary significantly over short distances. Throughout the work the dynamic properties of the inclusion have been normalized with respect to those of the uniform halfspace. Since the exact solution is obtained by truncating to finite size the infinite system of equations (12), a sensitivity analysis with respect to the parameter MN has been carried out. As reasonably expected, the minimum order of the expansions, MN_{min}, is frequency dependent providing the following rule of thumb: $MN_{min} \sim 8$ for $\eta_{max} = 4.8$; $MN_{min} \sim 12$ for $\eta_{max} = 9.6$ and $MN_{min} \sim 18$ for $\eta_{max} = 19.2$. Figure 3 illustrates an example of this convergence study for $\eta_{max} = 4.8$ in terms of synthetic seismograms at the position $x_1 = 0$ affected by a cylindrical cavity of radius $a_C = 1.5$ m and embedded at a depth $H = 2.5$ m, under the incidence of plane waves.

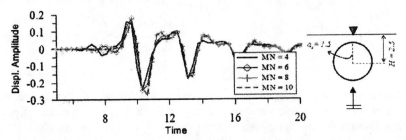

FIGURE 3 Sensitivity analysis with respect to the order of truncation of the wave function expansions for $\eta_{max} = 4.8$. Comparison of the synthetic seismograms calculated at the highlighted receiver for different values of the parameter MN.

In the sequel, some examples of synthetic seismograms are shown for the cases of a vacuous and rigid circular inclusion. The incident wavefield is represented by a Ricker pulse with characteristic period $t_p = 1.25$ sec, offset $t_s = 5.0$ sec, frequency step $\Delta\eta = 0.075$ and total duration of the signals $T = 40$ sec. Figure 4 depicts the synthetic seismograms calculated at 80 equally spaced receiver along the ground surface for vertical plane waves impinging on a cylindrical cavity of radius $a_C = 1.5$ m (see sketch on the right-hand side of Figure 4). Superimposed on the uppermost graph are the most relevant phases deriving from a physical interpretation of the time-space sections: the direct arrival (denoted with D), the reflected phase (R) and the multiples M_1, M_2, corresponding most probably to waves traveling along one quarter or the entire surface of the cavity, respectively. Note that the direct phase is strongly attenuated at the stations located on the projection of the center of the cavity along the free surface, creating a shadow zone. The two-dimensional transfer function corresponding to the time-space traces of Figure 4 is depicted in Figure 5 as function of the dimensionless distance x/a_C and the normalized frequency η. It is apparent that phenomena of constructive and deconstructive interference contribute to generate complex amplification/de-amplification patterns along the surface of the halfspace. The surface displacement peaks may occur at different locations on the ground surface, depending on the frequency of interest. Furthermore, the surface displacement field depends significantly on the location of the observation point, such that the amplification level may vary by a factor of approximately 10 solely considering different relative distances. The transfer function amplitude shows several peaks of amplitude either larger or smaller than 1.0 due to focusing and de-focusing effects of the elastic waves, probably induced by complex interaction effects between elastic waves scattered by the free surface and the embedded obstacle.

Figure 6 compares the transfer functions calculated at $x_1 = 0$ under vertical plane incidence when different types of inhomogeneity are taken into account, namely: a circular cavity, a tunnel (modeled as a cavity with a ring-shaped boundary of rigid properties) and a rigid/soft inclusion. While the surface amplification factor shows several peaks larger than 1 for the cases of a cavity and soft inclusion ($\beta_C / \beta_S = 0.5$ and $\rho_C / \rho_S = 0.8$), the presence of either a tunnel of normalized thickness $t = 1 - a_i/a_c$ $=0.2$ or a rigid inclusion ($\beta_C / \beta_S = 2.5$ and $\rho_C / \rho_S = 1.5$) induces a nearly constant attenuation of the surface ground displacement with respect to the free-field solution with an average de-amplification factor of about 0.5, at least under the incidence of vertical plane waves.

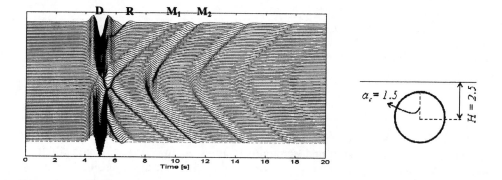

FIGURE 4 Synthetic seismograms obtained at 80 equally-spaced receivers located on the ground surface under vertical plane waves. Superimposed is the input Ricker pulse (dashed line). The geometrical model referred to is depicted on the right-hand side.

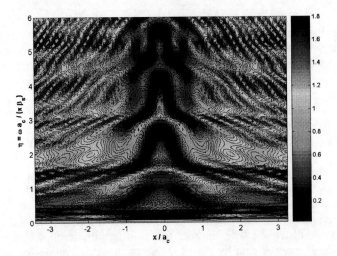

FIGURE 5 Transfer function amplitude as function of frequency η and distance x/a_C for the case study illustrated in Figure 5. Note that both these quantities are normalized with respect to the radius of the cavity.

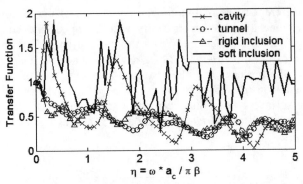

FIGURE 6 Transfer function calculated at the receiver located at the position corresponding to the projection of the centre of the irregularity on the ground surface under vertical plane wave incidence for different type of irregularities (cavity, tunnel, rigid or soft inclusion)

CONCLUSIONS

A number of analytical solutions have been presented in order to study the antiplane seismic response of underground structures of various type, from cylindrical cavities, to tunnels up to inclusions filled in with material of arbitrary rigidity, subjected to plane waves. The method of solution is based on the expansion of wave functions in terms of Bessel's and Hankel's functions, and their transport within a suitable set of cylindrical coordinate systems is obtained thanks to the application of Graf's addition theorem.

Strong ground motion results indicate that transfer function amplitudes may be consistently influenced by the presence of the subsurface irregularity, departing significantly from the free-field solution and presenting pronounced variations over short distances. The surface ground response depends upon several parameters: i) the frequency content of the input motion; ii) the incident angle ψ; iii) the type of inclusion; iv) the properties of the inhomogeneity and v) the location of the observation point along the surface of the halfspace. The surface amplification level turns out to vary by a factor of approximately 10 depending on the relative position between the anomaly and the observation point. Although these exact solutions are obviously limited by the simplified hypotheses and geometrical properties of the rationale model under consideration, they can be useful as benchmark solutions for either the validation of numerical models or geophysical investigations. Beside this, they can be used to develop approximated formulae for studying soil-structure interaction phenomena and empirical factors for modulating the response spectra when subsurface topographies might be of relevant interest from the engineering point of view.

ACKNOWLEDGMENTS

This work was done while the senior author was on leave from the ROSE School, of Pavia, Italy, at the Institute of Engineering of UNAM, Mexico. G. Sánchez N. and her team of Unidad de Servicios de Información (USI) helped us with useful references. Partial support from DGAPA-UNAM, Mexico, under Project IN114706 is gratefully acknowledged.

REFERENCES

1. Abramowitz M., Stegun I.A., Handbook of mathematical functions, National Bureau of Standards, Washington, DC, 1964.
2. Achenbach J.D., Wave propagation in elastic solids, North-Holland Publishing Co., Amsterdam, 1973.
3. Avilés J. and Sánchez-Sesma F.J. (1983), Piles as barriers for elastic waves, J. of Geotechnical Eng., 109 (9), 1133-1146.
4. Avilés J. and Mora-Orozco L.S. (1990), Modificación del movimiento sísmico por obstrucciones subterráneas, Sismodinámica, 1 (3), 147-170.
5. Datta S.K. and El-Akily N. (1978), Diffraction of elastic waves by cylindrical cavity in a half-space, J. Acoust. Soc. Am., 64 (6), 1692-1978.
6. Dravinski M. (1982), Scattering of SH waves by subsurface topography, J. Eng. Mech. Div. - ASCE, 108, 1-17.
7. Lee, V. W. (1977), On deformations near circular underground cavity subjected to incident plane Sh waves, Proc., Symp. on Applications of Comp. Methods in Engrg., 122–129.
8. Lee V.W. and M. D. Trifunac (1979), Response of tunnels to incident SH-waves, J. Eng. Mech. Div. - ASCE, 105, 643-659.
9. Lee V.W., Chen S. and Hsu I. R. (1999), Antiplane diffraction from canyon above subsurface unlined tunnel, J. of Geotechnical Eng., 125 (6), 668-675.
10. Pao H.Y. and Mow C.C. (1973), The Diffraction of Elastic Waves and Dynamic Stress Concentrations, Crane-Russak, New York.
11. Sommerfeld A., Partial differential equations in physics, Academic Press, Inc., New York, N.Y., 1949.

EARTH RETAINING STRUCTURES AND GEOSYNTHETICS

The Influence of Seismic Amplification and Distanced Surcharge on the Active Thrust on Earth-Reinforced Walls

Giovani Biondi[a], Francesco Grassi[a], Michele Maugeri[a]

[a] *Università di Catania, Dip. di Ingegneria Civile e Ambientale, Viale A. Doria 6, 95125 Catania, Italy.*

Abstract. The paper describes a closed form pseudo-static solution for the estimation of the active earth-pressure coefficient for an earth-reinforced wall assuming a non-uniform profile of the seismic coefficients along the wall height and a distanced uniformly-distributed surcharge on the backfill surface. The static and seismic hydraulic conditions of the backfill are also accounted for. A parametric analysis is carried out and the obtained results are discussed.

Keywords: earth-reinforced wall, distanced surcharge, soil amplification, seismic earth pressure.

INTRODUCTION

Recent experience with large earthquakes has shown that earth-reinforced walls can be considered a suitable alternative to concrete retaining structures. In fact, even for earth-reinforced walls not specifically designed to withstand earthquakes, complete or catastrophic failures seldom occur and post-earthquake reports generally describe minor damage with respect to concrete retaining structures [1, 2].

For both internal and external stability analyses, the seismic design of earth-reinforced walls is traditionally based on the pseudo-static approach [3]. In this framework the limit equilibrium or limit analysis approaches are usually adopted to detect the most critical failure mechanism and all the uncertainties involved in the analyses are accounted for by means of partial or global safety factors [4]. Hence, the attention of designers is mainly focused on the satisfaction of force-balance conditions neglecting the effects of the earthquake-induced strain in both the soil and reinforcements.

In the last decade, performance-based criteria have been increasingly used in many fields of geotechnical earthquake engineering. Accordingly, when designing earth-reinforced walls against large earthquakes we may need to accept that some residual displacement can occur which should, however, be within a certain allowable limit. To rationally estimate the residual permanent displacement both sophisticated numerical procedures or simplified Newmark-type analysis can be carried out [5]. Since numerical approaches are still considered with skepticism, at present the displacement approach represents the usual means for performance-based design. However, designers are still not very familiar with this approach and usually prefer the pseudo-static one.

CP1020, *2008 Seismic Engineering Conference Commemorating the 1908 Messina and Reggio Calabria Earthquake,*
edited by A. Santini and N. Moraci
© 2008 American Institute of Physics 978-0-7354-0542-4/08/$23.00

Despite the fact that displacement and pseudo-static analyses are usually considered as alternative methods, they can lead to the same safety measurement through an appropriate selection of the pseudo-static coefficients based on performance criteria [6]. Thus, despite the large number of shortcomings involved in a pseudo-static analysis, this approach is still of interest in geotechnical earthquake engineering.

The paper describes a closed form pseudo-static limit-equilibrium solution to estimate the active earth pressure coefficient on an earth-reinforced wall including the presence of distanced uniformly-distributed surcharge on the backfill surface and a non-uniform profile of the seismic coefficients along the wall height. The effects of the hydraulic conditions are also accounted for even if they are not discussed.

PROPOSED SOLUTIONS

Figure 1 shows the scheme of the earth-reinforced wall assumed in the paper with the notation adopted for the parameters involved in a tie-back wedge analysis. Except for the distanced uniformly-distributed surcharge q, the scheme is similar to those considered by Biondi et al.[7] to analyze the influences of the static pore pressure and of seismic-induced excess pore pressure on the earth-pressure coefficient and on the corresponding critical wedge angle at active limit state.

In the analysis, the inertial effect arising in the soil mass is accounted for through a couple of horizontal F_h and vertical F_v pseudo-static forces computed as the product of the weight W of the soil wedge potentially involved in the failure mechanism and of horizontal k_h^* and vertical k_v^* equivalent pseudo-static coefficients, respectively:

$$k_h^* = k_h \cdot (1 + 2 \cdot \delta)/3 \qquad k_v^* = k_v \cdot (1 + 2 \cdot \delta)/3 \qquad (1)$$

where k_h and k_v are the horizontal and vertical seismic coefficients. Eq.1 were derived assuming a linear variation of the seismic coefficients along the wall height (see Figure 1), with minimum values k_h and k_v at the base of the wall and maximum values $\delta \cdot k_h$ and $\delta \cdot k_v$ at the top of the wall [7]. If a unity value of the amplification ratio δ is considered the traditional pseudo-static scheme is obtained.

Limit equilibrium conditions of the wall are analyzed according to the hypothesis described by Biondi et al.[7] concerning the length and the axial behavior of reinforcements and the distribution of the static pore pressure ratio r_u and of the excess pore pressure ratio Δu^* along the base of the potential failure wedge.

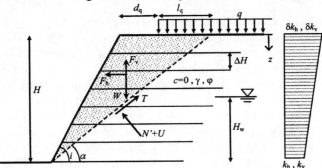

FIGURE 1. Scheme adopted in the analysis including the distanced surcharge.

The equilibrium conditions of the potential failure wedge together with the Mohr-Coulomb failure criterion lead to the following system of equilibrium equations:

$$\begin{cases} S_{ae,q}(\alpha) = F_h - (N-U)\cdot \tan\varphi \cdot \cos\alpha + N\cdot\sin\alpha \\ \overline{W} = F_v + (N-U)\cdot\tan\varphi\cdot\sin\alpha + N\cdot\cos\alpha \end{cases} \tag{2}$$

where T and N are the frictional tangential force and the normal force at the base of the wedge, respectively, $S_{ae,q}(\alpha)$ is the result of the reinforcement forces required for the equilibrium of the wedge, U is the result of the pore pressures at the base of the wedge:

$$U = \frac{\gamma_w \cdot H_w^2}{2\cdot\sin\alpha}\cdot\left[1 + 2\cdot\Delta u^* \cdot(r_u - 1)\right] \quad \text{with} \quad r_u = \frac{\gamma_w \cdot H_w}{\gamma \cdot H} \tag{3}$$

In Eq.(3) γ_w is the water unit weight, Δu^* is the ratio of the earthquake-induced excess pore pressure and the static effective stress normal to the base of the wedge and:

$$\overline{W} = W + q\cdot l_q = \frac{\gamma\cdot H^2}{2}\cdot\left[(\cot\alpha - \cot i)\cdot(1 + n_q) - \lambda\cdot n_q\right] \quad \text{with} \quad \lambda = \frac{d_q}{H},\ n_q = \frac{2\cdot q}{\gamma\cdot H} \tag{4}$$

Eqs.(2) solved with respect to $S_{ae,q}(\alpha)$ lead to:

$$S_{ae,q}(\alpha) = \frac{\gamma\cdot H^2}{2}\cdot(1 - k_v)\cdot K_{ae,q}(\alpha) \tag{5}$$

where:

$$K_{ae,q}(\alpha) = (\cot\alpha - \cot i)\cdot\left(\frac{1 - \cot\alpha\cdot\tan\varphi}{\cot\alpha + \tan\varphi} + \frac{k_h^*}{1 - k_v^*}\right) + \ldots$$

$$\ldots + \frac{H_w}{H}\cdot\left[r_u + 2\cdot\Delta u^*\cdot(1 - r_u)\right]\cdot\frac{\tan\varphi}{1 - k_v^*}\cdot\left(\frac{1 - \cot\alpha\cdot\tan\varphi}{\cot\alpha + \tan\varphi} + \cot\alpha\right) + \ldots \tag{6}$$

$$\ldots + n_q\cdot(\cot\alpha - \cot i - \lambda)\cdot\left[\frac{1 - \delta\cdot k_v^*}{1 - k_v^*}\cdot\frac{1 - \cot\alpha\cdot\tan\varphi}{\cot\alpha + \tan\varphi} + \frac{\delta\cdot k_h}{1 - k_v^*}\right]$$

According to the limit equilibrium approach the critical value α_c for which $K_{ae,q}(\alpha)$ attains its maximum must be detected. Through Eq.(6) the conditions for $K_{ae,q}(\alpha)$ to attain the maximum ($\partial k_{ae,q}/\partial\alpha = 0$ and $\partial^2 k_{ae,q}/\partial^2\alpha < 0$) lead to:

$$\cot\alpha_c = \left(-b^* + \sqrt{b^{*2} - 4\cdot a^*\cdot c^*}\right)/(2\cdot a^*) \tag{7}$$

where:

$$a^* = -\sin\phi\cdot\cos\varphi + \cos^2\varphi\cdot\left[k_h^* + n_q\cdot\delta\cdot k_h + R\cdot\tan\varphi\right]/\left[1 - k_v^* + n_q\cdot(1 - \delta\cdot k_v)\right]$$

$$b^* = 2\cdot\sin\varphi\cdot\cos\varphi\cdot\left[k_h^* + n_q\cdot\delta\cdot k_h + R\cdot\tan\varphi\right]/\left[1 - k_v^* + n_q\cdot(1 - \delta\cdot k_v)\right] - 2\cdot\sin^2\varphi \tag{8}$$

$$c^* = \cot i - a^* + \left[k_H^* + n_q\cdot\delta\cdot k_h + \lambda\cdot n_q\cdot(1 - \delta\cdot k_v)\right]/\left[1 - k_v^* + n_q\cdot(1 - \delta\cdot k_v)\right]$$

The active earth-pressure coefficient $K_{ae,q}$, including the earthquake effects in a pseudo-static way and the influence of the distanced surcharge q through n_q and λ, can be obtained from Eq.(6) for $\alpha = \alpha_c$.

The obtained expression of $K_{ae,q}$ is valid until the following condition is satisfied:

$$\cot\alpha_c \geq \cot i + \lambda \tag{9}$$

689

Otherwise the computed critical wedge does not intersect the distanced surcharge and the actual solution is not affected by λ and n_q. In this case $K_{ae,q}$ and α_c reduces to the active earth pressure coefficient K_{ae} and to the corresponding critical wedge angle computed for the a wall without surcharge [7], respectively.

In the case of surcharge close to the wall ($\lambda=0$), q does not affect the geometry of the critical wedge, Eqs.(6) and (8) coincide with those for the wall without surcharge [7] and, finally, $K_{ae,q}$ reduces to:

$$K_{ae,q} = K_{ae} \cdot (1 + n_q) \tag{10}$$

Finally, for the static case ($k_h = k_v = 0$) with $i = 90°$, $\lambda = 0$ and dry backfill ($H_w = 0$) Eqs. (6) and (9) reduce to the well-known Rankine's solution.

PARAMETRIC ANALYSIS AND DISCUSSION OF RESULTS

To analyze the influence of the parameters involved in the proposed solution, a parametric analysis was performed. In this paper attention is focused on the influence of the distanced surcharge and of the seismic amplification. In detail, the influence of the surcharge factor n_q and of the normalized surcharge distance λ on the values of $K_{ae,q}$ and α_c are investigated including, separately, the effects of the vertical component k_v of seismic coefficient and of the seismic amplification along the wall. The influence of the static and seismic hydraulic condition of the backfill is not discussed in the paper and all the analyses are carried out assuming $r_u = \Delta u^* = 0$. A parametric study on the influence of both r_u and Δu^* can be found in Biondi et al.[7].

Figure 2 shows the plots of $K_{ae,q}$ and α_c versus the angle of shear strength φ of the backfill soil, computed for two different values of the face angle ($i = 60°$ and $i = 90°$) and for several values of the surcharge factor n_q ranging from 0 to 0.75. In all the cases the surcharge is located close to the wall ($\lambda = 0$) and the horizontal and vertical component of the seismic coefficient are $k_h = 0.20$ and $k_v = 0$, respectively. The influence of the surcharge can be quantified through the difference between the values of $K_{ae,q}$ computed for $n_q = 0$ and those computed for $n_q > 0$.

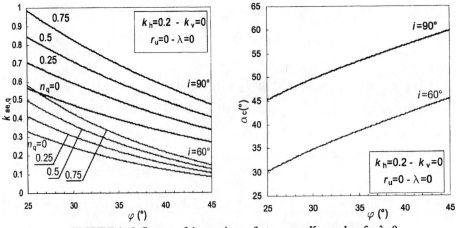

FIGURE 2. Influence of the surcharge factor n_q on $K_{ae,q}$ and α_c for $\lambda = 0$.

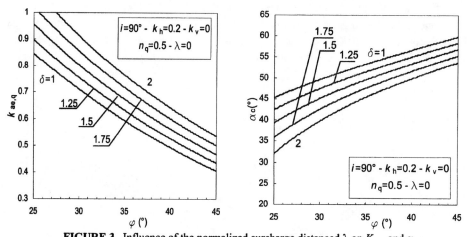

FIGURE 3. Influence of the normalized surcharge distanced λ on $K_{ae,q}$ and α_c.

In this case since it is $\lambda=0$, $K_{ae,q}$ reduces to K_{ae} and the increase in $K_{ae,q}$ due to the surcharge q depends linearly on n_q being not affected by the other parameters involved in the solution . Contemporarily, for $\lambda=0$ the critical wedge angle does not depend on the magnitude of the surcharge; thus, in Figure 2 α_c is plotted regardless of n_q.

For the previous wall scheme with $i=90°$, Figure 3 shows the influence of the amplification factor δ on both $K_{ae,q}$ and α_c for the case $n_q=0.50$. In the figure, $K_{ae,q}$ and α_c are plotted for δ varying in the range 1-2 according to the experimental evidence of shaking table and centrifuge test results available in published literature [8]. Then the curves plotted for $\delta=1$ represent the solutions for the case of uniform values of seismic coefficients and can be assumed as a reference to estimate the effect of the soil amplification on $K_{ae,q}$ and α_c. As shown in the figure, regardless of φ, values of δ greater than unity lead to higher values of the active earth pressure coefficient and of the corresponding critical wedge angle. As an example, for $\varphi=35°$ increases of $K_{ae,q}$ of about 13% and 29%, with respect to the reference values ($\delta=1$) are predicted for $\delta=1.5$ and $\delta=2.0$ respectively. Accordingly, α_c reduces from about 53° (for $\delta=1$) to 50° and 45° for $\delta=1.5$ and $\delta=2.0$, respectively.

The influence of the vertical component k_v of the seismic coefficient is described in Figure 4, where $K_{ae,q}$ and α_c are plotted versus φ for the same set of parameters as in Figure 3 with $n_q=0.50$ and $\delta=1.5$; values of the ratio k_v/k_h ranging from $-\frac{1}{2}$ to $+\frac{1}{2}$ are considered. As shown in the figure, the influence of k_v is less remarkable with respect to those observed for δ and n_q and the condition with the vertical inertia force directed upward (positive values of k_v) is the most critical. For the considered case, regardless of φ, the increment in $K_{ae,q}$ with respect to the values computed for $k_v=0$, is always less than 6%; numerical analyses, not described herein, showed increments of about 26% for $k_h=0.3$. Conversely in the case of $\delta=1$ increments of about at least 12% are estimated for k_h varying in the range 0.1-0.3.

In Figure 5 $K_{ae,q}$ and α_c are plotted versus φ for a vertical wall ($i=90°$) assuming $k_h=0.20$, $k_v=0$, $\delta=1$, $n_q=0.50$ and several values of λ. As shown in the figure the higher

λ is, the larger the critical wedge is, being characterized by a smaller α_c; correspondingly, the higher λ is, the lower $K_{ae,q}$ is. As an example, assuming as a reference the values of $K_{ae,q}$ and α_c computed for $\lambda=0$, for $\varphi=35°$ $K_{ae,q}$ reduces from about 0.6 (for $\lambda=0$) to 0.53, 0.47 and 0.42 for $\lambda=0.25$, 0.50 and 0.75 respectively; correspondingly, α_c reduces from 53° (for $\lambda=0$) to 50°, 48° and 45° for $\lambda=0.25$, 0.50 and 0.75 respectively.

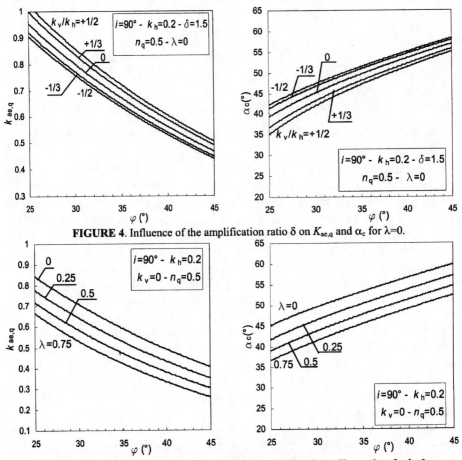

FIGURE 4. Influence of the amplification ratio δ on $K_{ae,q}$ and α_c for $\lambda=0$.

FIGURE 5. Influence of the vertical seismic coefficient k_v on $K_{ae,q}$ and α_c for $\lambda=0$.

Figure 6 shows the influence of the amplification ratio δ on $K_{ae,q}$ and α_c for the same set of parameters adopted in Figure 5 with $n_q=0.50$ and $\lambda=0.5$. As in the case of $\lambda=0$ (Fig. 3), the plots of $K_{ae,q}$ and α_c versus φ of Figure 6 clearly show that in the case of non-uniform profiles of the seismic coefficients ($\delta>1$) the values of $K_{ae,q}$ and α_c are significantly affected by δ. As an example, assuming, again, the condition $\delta=1$ as a reference, for $\varphi=35°$, increases in $K_{ae,q}$ of about 17% and 36%, with respect to the reference values ($\delta=1$) are predicted for $\delta=1.5$ and $\delta=2.0$ respectively. Accordingly, α_c reduces from 48°(for $\delta=1$) to 44° and 40° for $\delta=1.5$ and $\delta=2.0$, respectively.

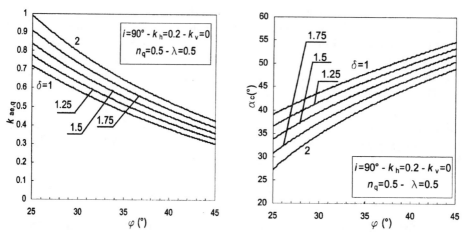

FIGURE 6. Influence of the amplification ratio δ on $K_{ae,q}$ and α_c for distance surcharge.

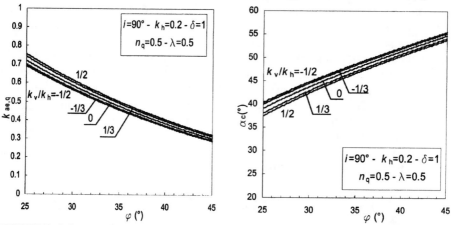

FIGURE 7. Influence of the vertical seismic coefficient k_v on $K_{ae,q}$ and α_c for distanced surcharge.

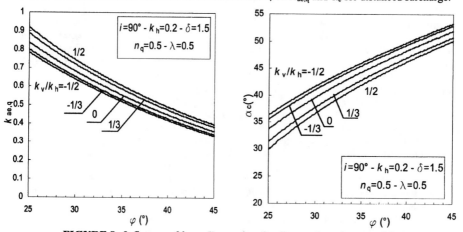

FIGURE 8. Influence of k_v on $K_{ae,q}$ and α_c for distanced surcharge and δ=1.5

The effect of k_v on $K_{ae,q}$ and α_c in the case of distanced surcharge ($\lambda>0$) is shown in Figures 7 and 8 assuming the same set of parameters of Figure 6 and $\delta=1$ and $\delta=1.5$, respectively. Similarly to the case of $\lambda=0$ (Fig. 4), the influence of k_v can be analyzed comparing the values of $K_{ae,q}$ and α_c computed for $k_v=0$ with those computed for k_v/k_h equal to $\pm1/2$ and $\pm1/3$. In the case of a uniform profile of the seismic coefficients (Fig. 7) the influence of k_v on the values of $K_{ae,q}$ and α_c is negligible; in the case of $\delta=1.5$ (Fig. 8) the effect of k_v is more remarkable. However, similarly to the case of $\lambda=0$ (Fig. 4), regardless of φ the differences between the values computed for $k_v=0$ and for $k_v\neq0$ are generally lower than 9% for $K_{ae,q}$ and 12% for α_c. Again, in all the cases ($\delta=1$ and $\delta=1.5$) the condition with k_v direct upward is the most critical.

CONCLUDING REMARKS

A closed form solution for the evaluation of the active earth-pressure coefficient and of the corresponding critical wedge angle is proposed for a pseudo-static tie-back wedge analysis of an earth-reinforced wall. A non-uniform profile of the seismic coefficients and a distanced uniformly-distributed surcharge is considered. The results of a parametric analysis point out that the amplification of the seismic actions arising in the soil mass produces a significant increase in the active earth pressure coefficient with respect to those computed neglecting the possibly occurring seismic amplification. Accordingly, the extension of the corresponding critical wedge can be underestimated. These effects were observed for both distanced surcharge and surcharge close to the wall. In the case of non-uniform profile of seismic coefficients the influence of the vertical seismic coefficient is more remarkable and, as in the case of a uniform profile of seismic coefficients, the inertia force directed upwards represents the most critical condition.

REFERENCES

1 F. Tatsuoka, M. Tateyama, J. Koseki, (1995), *Behavior of geogrid reinforced soil retaining walls during the great Hanshin-Awaji earthquake*, Proc. 1st Int. Symposium on Earthquake Geotechnical Engineering, Tokyo, pp.55–60.
2 D. Sandri (1998), *A performance summary of reinforced soil structures in the Greater Los Angeles Area after the Northridge earthquake*, Geotextiles and Geomembranes, 15(4), pp.235-253.
3 R.J. Bathrust and Z. Cai (1995), *Pseudo-static seismic analysis of geosynthetic-reinforced segmental retaining wall*, Geosynthetic International, Vol.2(5), pp.787-830.
4 J.G. Zomberg, N. Sitar, J.K. Mithchell (1998), *Limit equilibrium as basis for design of geosynthetic reinforced slopes*, Journal of Geotech. Geoenviron. Eng., 124(8), 684-698.
5 Z. Cai and Bathurst R.J., *Seismic-induced permanent displacement of geosynthetic-reinforced segmental retaining walls*, Canadian Geotechnical Journal, 33, pp. 935-955 (1996).
6 G. Biondi, M. Maugeri, E. Cascone (2007), *Displacement-based seismic analysis of rigid retaining walls*. Proc. 14th European Conference on Soil Mechanics and Geotechnical Engineering, ISSMGE-ERTC12 Workshop "Geotechnical aspect of EC8", Madrid, Spain, 24-27 September 2007.
7 G. Biondi, F. Grassi, M. Maugeri (2007), *Earthquake effect on the earth pressure coefficient at active limit state for a geogrid reinforced soil wall*, Proc. 1st Pan American Geosynthetic Conference & Exhibition, 2-5 March, Cancun, Mexico.
8 L. Nova-Roessig and N. Sitar (2006), *Centrifuge studies of the seismic response of reinforced soil slopes*, Journal of Geotechnical and Geoenvironmental Engineering, ASCE, Vol. 132, No. 3

The DDBD Method In The A-Seismic Design of Anchored Diaphragm Walls

Cecconi Manuela, Pane Vincenzo and Vecchietti Sara

Department of Civil and Environmental Engineering, University of Perugia, Italy

Abstract. The development of displacement based approaches for earthquake engineering design appears to be very useful and capable to provide improved reliability by directly comparing computed response and expected structural performance. In particular, the design procedure known as the Direct Displacement Based Design (DDBD) method, which has been developed in structural engineering over the past ten years in the attempt to mitigate some of the deficiencies in current force-based design methods, has been shown to be very effective and promising ([1], [2]). The first attempts of application of the procedure to geotechnical engineering and, in particular, earth retaining structures are discussed in [3], [4] and [5]. However in this field, the outcomes of the research need to be further investigated in many aspects. The paper focuses on the application of the DDBD method to anchored diaphragm walls. The results of the DDBD method are discussed in detail in the paper, and compared to those obtained from conventional pseudo-static analyses.

Keywords: displacement-based design, diaphragm walls, dynamic active and passive thrust, equivalent damping.

1. MOTIVATION

The design procedure known as Direct Displacement Based Design (DDBD) method has been developed over the past ten years ([1], [2]) in the structural engineering field in the attempt to mitigate some of the deficiencies in current force-based methods for earthquake engineering design, with the aim of improving the reliability of the system at hand, by more directly comparing computed response and expected structural performance.

As a starting point, in the DDBD method the structure is modelled as an *equivalent*, single-degree-of-freedom (SDOF) elastic system. The design process is developed by assuming that the dynamic performance of the equivalent SDOF system is characterized by its maximum displacement response, rather than its initial tangent elastic properties (see the *Substitute Structure Approach* in [6], [7]). The design steps of the DDBD procedure are deeply described in [8] for a frame-wall building. Nonetheless, to the Authors knowledge, the fundamentals may be appropriate for many different structural types (see [2], [9]).

The application of DDBD to geotechnical engineering and, in particular, *free* earth retaining structures is pursued in [3], [4] and [5]. In this paper we propose the same methodology for the geotechnical design of *anchored* diaphragm walls. The main objective of the study is the evaluation of the dynamic earth pressure, based on a more reliable seismic performance of the soil-structure system.

CP1020, *2008 Seismic Engineering Conference Commemorating the 1908 Messina and Reggio Calabria Earthquake*,
edited by A. Santini and N. Moraci

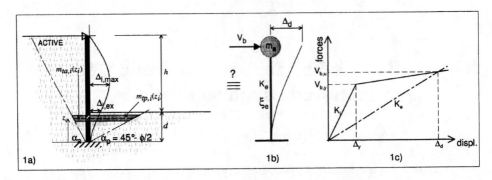

FIGURE 1. Design problem and SDOF system with equivalent mass, m_e, damping, ξ_e, and stiffness, K_e

Figure 1 represents the position of the problem. An embedded anchored diaphragm wall retains a vertical excavation in coarse-grained soils. The main – and, perhaps, ambitious - goal of the proposed method is to substitute the complex soil/structure system with a simpler SDOF system, defined by equivalent values of mass, m_e, damping, ξ_e, and global secant stiffness, K_e (Figures 1b) and c)). The maximum displacement $\Delta_{i,max}$ is set by the designer to ensure an acceptable level of displacement for a given risk event. Obviously, the choice of an acceptable deformation value results from due consideration of the damage/collapse of both structural (wall, anchors, struts, etc...) or non-structural items. The definition of the soil masses involved in the seismic event and the evaluation of the damping ratio of the whole system have been carefully investigated in [3] and [5], since they affect sensibly the numerical results.

The application of DDBD procedure to anchored diaphragm walls is described step-by-step in the following sections.

2. GEOMETRY, GROUND MOTION AND MATERIALS

The geometry of the problem is schematically represented in Figure 1a), while the physical and mechanical properties of the soil and of the wall are summarized in Table 1. A diaphragm wall made of r.c. piles retains a 5 m high vertical excavation in cohesionless sand by means of an embedment length d = 2 m. An anchor/strut is located at the top of the wall. For simplicity, it is assumed that ground water is absent.

In choosing an appropriate displacement shape of the structure, the wall is supposed to behave according to the *fixed earth support* scheme, with no rotation or horizontal/vertical displacements being permitted at the toe. The anchor is supposed to be rigid and totally efficient.

The sand deposit is assumed to fall within Ground Type C of the Eurocode EC8–EN1, i.e., characterized by a shear wave velocity $V_{S,30} = 200$ m/s and by a Soil Factor S = 1.25. It is further assumed that the area belongs to the Italian seismic zone 1. For such a zone, the peak ground acceleration on Type A ground is rather high ($a_g = 0.35g$), leading to the elastic acceleration response spectrum depicted in Figure 2a) and to a value of the horizontal seismic coefficient, $k_h = 0.44$.

TABLE 1. Soil properties and wall geometry .

Soil: medium dense sand	symbol	numerical value
unit volume weight	$\gamma \, [kN/m^3]$	17
friction angle	$\phi \, [°]$	35
friction angle at the contact soil/wall		
active	δ_a	$2/3 \, \phi$
passive	δ_p	0
Wall: concrete anchored diaphragm wall	**symbol**	**numerical value**
excavation height	h [m]	5.0
embedment length	d [m]	2.0
total height of the wall	H [m]	7.0
unit volume weigth	$\gamma_c \, [kN/m^3]$	25.0
pile diameter	D [m]	0.5
spacing	s [m]	0.6

Also, the elastic displacement spectrum (Figure 2b) is expressed as a function of the vibration period, according to the analytical equations reported in EC8. The numerical values of representative periods that identify different portions of the response spectra are given in the Italian OPCM 2003 (see also EC8-EN-1).

Before describing the main steps of the calculation procedure, it is useful to mention the results obtained by means of a conventional pseudo-static analysis based on Mononobe-Okabe (MO) limit equilibrium method. In this case, for d = 2m, by assuming that the seismic increments of active thrust (ΔS_{ae}) and passive resistance (ΔR_{pe}) act respectively at H/2 and d/2– as prescribed by EC8–EN5 – a factor of safety against rotation around the anchor FS = $M_{stab}/M_{overturning}$ = 0.6 is achieved. In order to provide a reasonable value of the global safety factor (FS ≥ 1.5), an embedded length d = 6 m is needed, corresponding to a ratio d/h = 1.2.

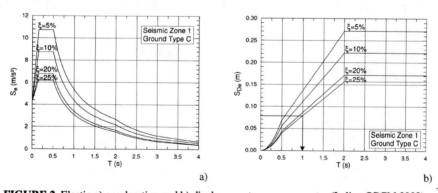

FIGURE 2. Elastic a) acceleration and b) displacement response spectra (Italian OPCM 2003)

3. THE DDBD PROCEDURE

The fundamentals of the DDBD method consist in choosing a suitable single-degree-of-freedom system, characterized by a secant effective stiffness K_e at a maximum displacement $\Delta_{i,max}$ (Figure 1c). The latter value represents a design value which must be set by the designer to ensure an acceptable level of displacement for a given risk event. By choosing an appropriate displaced shape, this is scaled until a critical section in the structure reaches its limit displacement and attains its yield curvature. At this section, a plastic hinge is formed and the ultimate strength of the section is mobilized. If the retaining wall is designed with the proper steel reinforcement, the system can develop a safe displacement capacity by relying on its capacity of dissipating energy both in the structural members (plastic hinge) and in the soil (soil damping).

3.1 Evaluation of the Participating Masses

Figure 1a) shows also the assumed discretised equivalent system. The soil masses participating in the seismic event are conventionally represented by the active soil wedge behind the wall and the passive wedge in front of it. Based on the outcomes discussed in [3], [4] their inclination – evaluated in static condition - is [1]$\alpha_a = f(\phi, \delta_a)$ and [2] $\alpha_p = 45° - \phi/2$ ($\delta_p = 0$, see EC8). The weight of the structure is considered too, even if its contribution is small with respect to the soil mass. The ith-soil mass at a depth (H - z_i) from the ground table is expressed by:

$$m_{ta,i}(z_i) = \frac{\gamma/g \cdot z_i d_i}{\tan \alpha_a} \quad \text{if} \;\; 0 < z_i < h+d$$

$$m_{tp,i}(z_i) = \frac{\gamma/g \cdot z_i d_i}{\tan \alpha_p} \quad \text{if} \;\; 0 < z_i < d$$

(1)

and for the pile-wall mass

$$m_{p,i}(z_i) = \gamma_c/g \cdot (\pi D^2/4s) \cdot d_i = \text{constant}$$

(2)

where z_i is the height of the centre of mass of the ith soil layer of thickness d_i, while D and s represent the piles diameter and pile spacing, respectively.

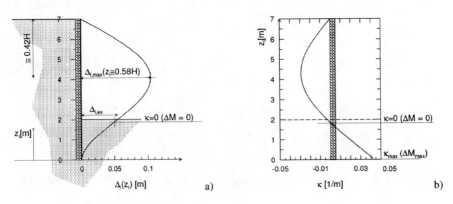

FIGURE 3. a) Displacement pattern and b) section curvature *vs* height ($\Delta_{i,max}$ = 100 mm)

---

[1] According to Mayniel's solution of the limit equilibrium method
[2] Rankine solution for the passive limit state.

3.2   Definition of the Displacement Shape

The shape of the displacement pattern of the system has been assumed to match with the elastic 1st mode of vibration of a supported-cantilever beam (Figure 3a), given by the following equation:

$$\Delta_i(z_i) = C_1 \left\{ \sin(k_1 z_i) - \sinh(k_1 z_i) + \frac{\sin(k_1 H) - \sinh(k_1 H)}{\cos(k_1 H) - \cosh(k_1 H)} [\cos(k_1 z_i) - \cosh(k_1 z_i)] \right\} \quad (3)$$

where Δ_i is the horizontal displacement for the ith layer, $k_1 H = 3.927$, C_1 is the scaling constant and $k_1 H = 3.927$. The maximum displacement, $\Delta_{i,max}$, occurs roughly at 0.58 H from the toe of the wall (Figure 3a). The displacement $\Delta_i(z_i)$ attains a null value at the base where the section curvature, κ, is maximum (see Figure 3b). Thus, for the given displaced shape of the structure at maximum response, a design displacement, Δ_d and an equivalent mass, m_e, are defined as [2]:

$$\Delta_d = \frac{\sum_{i=1}^{n}(m_i \Delta_i^2)}{\sum_{i=1}^{n}(m_i \Delta_i)} \quad (4a) \qquad m_e = \frac{\sum_{i=1}^{n}(m_i \Delta_i)}{\Delta_d} \quad (4b) \qquad (4)$$

3.3   Choice of Equivalent Damping

An important point of the design process lies in the evaluation of an *equivalent damping* of the system, ξ_e, which should correspond to an expected level of ductility demand. The equivalent damping ξ_e of the soil-structure system has been defined as:

$$\xi_e = \frac{\xi_{ta}\sum_{i=1}^{n}m_{ta,i} + \xi_{tp}\sum_{i=1}^{n}m_{tp,i} + \xi_p\sum_{i=1}^{n}m_{p,i}}{\sum_{i=1}^{n}(m_{ta,i} + m_{tp,i} + m_{p,i})} \quad (5)$$

i.e., by "weighting" the soil damping in the active and passive wedges (ξ_{ta}, ξ_{tp}) and the wall damping (ξ_p) over the relative masses. As a first approximation, the relationship proposed in Figure 4 was used to estimate the soil damping in both active and passive states.

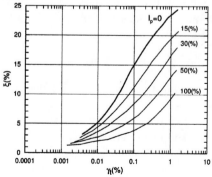
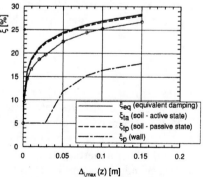

FIGURE 4. Soil damping vs. shear strain γ_t [10]   **FIGURE 5.** Damping coeff. *vs* max. displacement

To this aim an average shear strain (γ_t) within the two wedges was calculated, according to:

$$\gamma_{ta} = \frac{\Delta_{i,max}}{0.42H} \quad \text{and} \quad \gamma_{tp} = \frac{\Delta_{i,ex}}{d} \tag{6}$$

where $\Delta_{i,ex}$ represents the horizontal displacement at the excavation level (See Figure 3a). In order to evaluate the wall damping coefficient, ξ_p, the simplified equation proposed by [11] has been adopted:

$$\xi_p \cong 0.05 + 0.5\left(\frac{\mu-1}{\mu\pi}\right) \tag{7}$$

where μ is the ductility coefficient defined as the ratio between the design wall displacement and the yield displacement (see [3] for details).

By following this procedure the equivalent damping of the system - calculated through Eq. 5 – increases with the maximum expected displacement ($\Delta_{i,max}$), as shown in Figure 5. In the plot, a value of $\xi = 5\%$ was imposed at zero displacement in accordance with EC8 - EN 1998-1. It is noted that for $\Delta_{i,max} > 30$ mm the equivalent damping of the system reaches values larger than 20%, while the pile-wall damping coefficient ξ_p overcomes 5%.

3.4 Calculation of Equivalent Stiffness and Dynamic Forces

Given the design displacement Δ_d and the corresponding value of damping ξ_e, the effective period T_e of the soil-structure system can be directly read from the elastic displacement spectra as shown in Figure 2b). Since the soil-structure system is assimilated to an equivalent SDOF oscillator (see Figure 1b), the equivalent secant stiffness is related to the effective period, T_e, and mass, m_e, as follows:

$$T_{SDOF-oscillator} = 2\pi\sqrt{\frac{M}{K}} \quad \Rightarrow \quad K_e = \frac{4\pi^2 m_e}{T_e^2} \tag{8}$$

Finally, the design dynamic increment, V_b, acting as a lateral force on the anchored diaphragm wall, is calculated as:

$$V_b = K_e \Delta_d \tag{10}$$

Such horizontal force V_b represents the sum of the active dynamic increment ΔS_{ae}, the passive dynamic decrement ΔR_{pe} and the diaphragm wall inertia $k_h m_p g$. The force V_b can then be distributed to the various discretized masses of the structure according to

$$V_i = V_b \frac{m_i \Delta_i}{\sum_{i=1}^{n} m_i \Delta_i} \tag{11}$$

where V_i is the design force at i^{th} mass. In particular, for the problem at hand, the force V_b may be usefully distributed into three parts and precisely:

- V_{ta}: the seismic increment pertaining to the soil mass in the active state;
- V_{tp}: the seismic decrement pertaining to the soil mass in the passive state;
- V_p: the lateral inertia force of the wall.

The active seismic increment is formulated as:

$$V_{ta} = \sum_{i=1}^{n} V_{ta,i} \qquad V_{ta,i} = V_b \dfrac{m_{ta,i}\Delta_i}{\sum_{i=1}^{n}(m_{ta,i} + m_{tp,i} + m_{p,i})\Delta_i} \qquad (12)$$

Equivalent equations hold for the horizontal forces V_{tp} and V_p, which can be calculated by simply substituting $m_{ta,i}$ at the numerator with $m_{tp,i}$ and $m_{p,i}$ respectively.

As already observed in [5], Eq. 12 is very useful because it makes it possible to obtain the entire distribution of the active/passive seismic forces along the wall or, alternatively, the position of their application point. This represents one of the advantages of the method with respect to conventional pseudo-static approaches, in which such application point is unknown.

4. RESULTS

In Figure 6 the results of the DDBD method are plotted in terms of dynamic forces V_b, V_{ta} and V_{tp} as a function of the maximum displacement $\Delta_{i,max}$, which occurs approximately at the mid-height of the structure. It is noted that the assumption of a displacement dependent damping (Eq.s 5, 6, 7) mostly affects the gentle decrease of V_b in the range $0.01 < \Delta_{i,max} < 0.05$ m.

In the same figure, the results of the DDBD calculations are compared with those obtained from Mononobe-Okabe (MO) pseudo-static analyses (horizontal dashed lines). For rather small displacements ($\Delta_{i,max} < 0.05$ m), an excellent agreement is found between the dynamic increment V_{ta} applied to the active soil wedge, and the corresponding pseudo-static MO increment, $\Delta S_{ae}\cos\delta_a$. For larger displacements, the DBD procedure, by taking into account the ductility capacity of the soil-structure system, leads to lower dynamic forces. The reduction is rather significant; in fact for $\Delta_{i,max} = 0.1$m, the active seismic increment is approximately 40% of the pseudostatic MO force ($\Delta S_{ae}\cos\delta$).

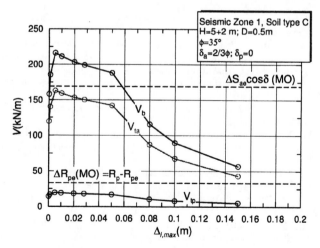

FIGURE 6. Dynamic forces, V, as a function of maximum displacement $\Delta_{i,max}$

Figure 6 also shows the dynamic decrement V_{tp} pertaining to the passive soil wedge and the corresponding pseudo-static MO decrement, ΔR_{pe}. In the whole range of investigated displacements, the values of V_{tp} are systematically lower than those obtained through a MO pseudo-static analysis. This implies a larger total passive resistance exerted by the soil during the seismic event.

Finally, the values of safety factor against rotation around the anchor are shown in Table 2 as a function of maximum displacement and compared to the one obtained from the MO pseudo-static analysis. It is noted that the chosen value of the embedment length (d = 2m) leads to values of FS slightly above unity (FS=1.0-1.1), provided that maximum displacements of the order of 8 – 10 cm can be tolerated. Smaller displacements lead to larger dynamic thrusts and values of FS below unity, thus requiring larger values of embedment length.

TABLE 2. Factors of safety against rotation around the anchor
(d = 2m)

FS_{-MO}	$\Delta_{i,max}$ [m]	FS_{-DBD}
	0.01	0.7
	0.05	0.8
0.6	0.08	1.0
	0.10	1.1

REFERENCES

1. M.J.N. Priestley, "Myths and fallacies in earthquake engineering: conflicts between design and reality", *Bulletin of NZ National Society for Earthquake Engineering*, New Zealand **26**(3), 329-341. (1993).
2. Priestley, M.J.N., G.M. Calvi and M.J. Kowalsky, *"Displacement based seismic design of structures"*. IUSS Press, Pavia, Italy (2007).
3. V. Pane, M. Cecconi, S. Vecchietti, "Metodo DDBD per il progetto agli spostamenti di strutture di sostegno" in *XII Convegno Nazionale Anidis - L'Ingegneria Sismica in Italia*, Pisa, on CD rom (2007).
4. S. Vecchietti, M. Cecconi, V. Pane, "Displacement-methods for the design of earth retaining structures" in *4th International Conference on Earthquake Geotechnical Engineering*, Thessaloniki, Greece, on CD rom (2007).
5. M. Cecconi, S. Vecchietti, V. Pane, "The DDBD method in the design of cantilever diaphragm walls" in *60th Canadian Geotechnical Conf. & 8th Joint CGS/IAH-CNC Groundwater Conf. –* Ottawa 2007, The Diamond Jubilee, Ottawa, Ontario, Canada, pp. 912-919 (2007).
6. P. Gulkan and M. Sozen, "Inelastic response of reinforced concrete structures to earthquake motions", *ACI Journal* 71/12, 604-610 (1974).
7. A. Shibata and M. Sozen, "Substitute structure method for seismic design in reinforced concrete", *Journal Structural Division, ASCE* **102**(12), 3548-3566 (1976).
8. T.J. Sullivan, M.J.N. Priestley and G.M. Calvi, "Development of an innovative seismic design procedure for frame-wall structures", *Journal of Earthquake Engineering*, **9**, Special Issue 2, 279-307 (2005).
9. Priestley, M.J.N. and J.D. Petting, "Dynamic behaviour of reinforced concrete frames designed with direct displacement-based design", *Journal of Earthquake Engineering*, **9**, Special Issue (2005).
10. M. Vucétic and R. Dobry, "Effects of the soil plasticity on cyclic response". *Journal of Geotechnical Engineering Division, ASCE*, **117**, n. 1 (1991).
11. D. N. Grant, C. Blandon and M.J.N. Priestley, *"Modelling Inelastic Response in Direct Displacement-Based Seismic Design"*, IUSS Press, Pavia, Italy (2005).

Parametric study of cantilever walls subjected to seismic loading

Cesare Comina[a], Mirko Corigliano[b], Sebastiano Foti[a], Carlo G. Lai[b], Renato Lancellotta[a], Francesco Leuzzi[a], Giovanni Li Destri Nicosia[b], Roberto Paolucci[c], Alberto Pettiti[a], Prodromos N. Psarropoulos[b] and Omar Zanoli[c]

[a] *Dept. of Structural and Geotechnical Eng., Politecnico di Torino, c.so Duca degli Abruzzi 24, IT – 10129 Torino*

[b] *European Centre for Training and Research in Earthquake Engineering (EUCENTRE), via Ferrata 1, IT – 27100 Pavia*

[c] *Dept. of Structural Eng., Politecnico di Milano, Piazza Leonardo da Vinci 32, IT – 20133 Milano*

Abstract. The design of flexible earth retaining structures under seismic loading is a challenging geotechnical problem, the dynamic soil-structure interaction being of paramount importance for this kind of structures. Pseudo-static approaches are often adopted but do not allow a realistic assessment of the performance of the structure subjected to the seismic motions. The present paper illustrates a numerical parametric study aimed at estimating the influence of the dynamic soil-structure interaction in the design. A series of flexible earth retaining walls have been preliminary designed according to the requirements of Eurocode 7 and Eurocode 8 - Part 5; their dynamic behaviour has been then evaluated by means of dynamic numerical simulations in terms of bending moments, accelerations and stress state. The results obtained from dynamic analyses have then been compared with those determined using the pseudo-static approach.

Keywords: dynamic soil-structure interaction, pseudo-static analyses, cantilever walls, seismic design.

1. INTRODUCTION

The seismic design of earth retaining structures is currently based on the theory of Mononobe-Okabe (Okabe, 1926; Mononobe and Matsuo, 1929), the so-called pseudo-static (P-S) approach. This approach was originally developed to allow the application of limit equilibrium analysis to rigid earth retaining structures, but it is far from capturing the actual conditions experienced by a flexible earth retaining wall during an earthquake. Indeed the flexible nature of the wall may, for instance, inhibit the development of active and passive wedges and, as shown by Steedman and Zeng (1990, 1993), the phase difference plays an important role. Furthermore, this approach does not provide an estimate of the displacements experienced by the structure during an earthquake.

Therefore, validation of existing pseudo-static approaches, or the introduction of novel simplified methods for design, requires the preliminary parametric investigation of the problem by fully dynamic soil-structure interaction analyses, involving the

CP1020, *2008 Seismic Engineering Conference Commemorating the 1908 Messina and Reggio Calabria Earthquake*,
edited by A. Santini and N. Moraci
© 2008 American Institute of Physics 978-0-7354-0542-4/08/$23.00

numerical determination of stress state, displacements, and spatial and temporal variability of accelerations. However, dynamic analyses are time consuming and require experienced users.

As an introductory example of numerical analyses needed to validate the current approaches for design, this work is intended to evaluate numerically the behavior of cantilever retaining structures subjected to seismic loading and to compare the results with those determined using a pseudo-static approach. In particular, a sensitivity analysis has been carried out to highlight the influence of peak ground acceleration (PGA), soil stiffness profile and height of excavation on the behavior of the walls.

2. MODEL DESCRIPTION

The earth retaining structure considered in this study is a cantilever wall, consisting of a reinforced-concrete diaphragm wall, retaining a dry coarse-grained soil. Various walls have been preliminary designed by using limit equilibrium methods varying excavation depth, soil classes and seismic zones in order to have a wide frame of reference for the interpretation of the results; after the preliminary design both dynamic and P-S analyses have been carried out.

2.1 Preliminary Design of the Walls

The preliminary design is based on a limit equilibrium analysis associated to a P-S approach, according to the rules of Eurocode 7 and Eurocode 8 - Part 5, and soil factors according to Eurocode 1. The wall has been analyzed with the classical scheme of fixed earth support. Eurocode 8 suggests the application of M-O equations both on active and passive side, neglecting for the latter the influence of soil-wall friction. In this study, the M-O equation was retained only for the active thrust, whereas for passive resistance an alternative formulation (Lancellotta, 2007), based on the lower bound theorem of plasticity, was applied. The reason for this assumption is that M-O formula for passive state fails to provide a consistent estimate when friction between wall and soil (δ) is to be taken into account, in order to avoid uneconomical design (Comina et al., 2007). Both active and passive formulae require the assessment of seismic coefficients (EC8-5):

$$k_h = \frac{S \cdot a_g / g}{r} \tag{1}$$

$$k_v = \pm 0.5 \cdot k_h \tag{2}$$

where S is a soil factor, a_g is the PGA on rock, and r is a factor ranging from 1 to 2, depending on the type of structure and movement allowed.

TABLE 1. Soil factor S

Soil B	Soil C	Soil D
1.2	1.15	1.35

For the preliminary design, the values $r = 2$ and $k_v = 0$ were assumed. Reference was made to the two zones of Italian Seismic Classification: zone 1 ($a_g / g = 0.35$) and zone 2 ($a_g / g = 0.25$). Moreover, three different soil classes have been taken into account: B, C, and D; the values for S suggested by EC8-5 are shown in Table 1.

Three different excavation heights have been assumed ($h = 3, 5, 7$ m). The results obtained in the different conditions have been compared in terms of maxima of embedment (d) and bending moment (M), for every soil and zone, showing that $d \cong 1.5 \cdot h$ can be considered reasonable for all seismic zones and soil factors considered. The thicknesses of the walls have been selected in order to keep constant the ratio of flexural stiffness to total height, i.e. $EJ / h \cong \text{const}$.

2.2 Numerical Model

The numerical model for dynamic and pseudo-static analyses, implemented in the finite difference code FLAC v.5 (Itasca, 2005), is a symmetric plane strain model with respect to the axis of excavation. Therefore, it includes a pair of cantilever walls retaining a 30 m excavation front (Fig. 1).

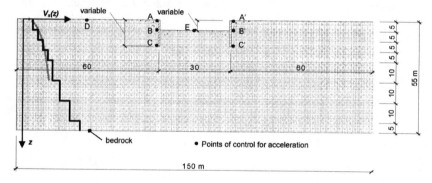

FIGURE 1. Numerical model for dynamic and pseudo-static analyses.

An elastic-perfectly plastic constitutive model with Mohr-Coulomb failure criterion and non associated flow rule (dilatancy angle $\psi = 0$) has been adopted for the soil, the characteristic shearing resistant angle ($\varphi'_k = 34°$) has been used in the simulations. The behaviour of the retaining walls has been modelled as linear elastic Bernoulli beam and no relative movements have been allowed at the contact surface between the soil and the walls ("glued" interface). For the three different soil classes the soil profile is characterized by a smooth increase of shear-wave velocity with depth, up to the bedrock ($V_s = 800$m/s) located at the base of the model. To account for dissipative behaviour of soils at small to moderate strains, a Rayleigh damping ($\xi = 0.5\%$ at 1 Hz) was applied to the model. The initial excavation phases (1 m depth each) were carried out under static conditions and, to account for soil non linearity, a reduced soil stiffness was assumed: $G = G_0 / 5$.

Dynamic analyses have been performed at the end of excavation updating the stiffness moduli to the observed level of deformation. Two groups of seven spectrum-compatible real accelerograms (one for each zone) have been used as seismic input for the models according to Eurocode 8 prescriptions. Indeed, if at least seven accelerograms are selected, the average of the response in terms of envelope of maxima could be selected. The accelerograms that were applied at the bedrock have been scaled to the PGA of the relative zone, divided by a factor of two.

Standard dynamic boundaries available in FLAC were used: (a) on the lateral sides the grid was connected to "free-field" boundaries, (b) at the base "quiet" boundaries were applied, in order to ensure the absence of reflections due to the boundary itself. Seismic loading was introduced in the model as a stress history at the lower boundary.

In order to have a consistent comparison with P-S approaches, the P-S load was applied rotating the gravity inside the whole model, by introducing a series of calculation steps in which the horizontal seismic coefficient was slowly increased (Δk_h = 0.001) up to the value suggested by Eq. 2, considering both the case of $r = 1$ and $r = 2$. Although this approach is time consuming, it is necessary to achieve the convergence of the calculation.

3. ANALYSIS OF RESULTS

Some results of the sensitivity analysis are summarized in the sequel. In particular, the attention is focused on soil class C and for the wall $h = 5$ m, $d = 8$ m. The results are discussed in terms of acceleration time histories (as obtained in the reference points of Fig. 1), bending moments and earth pressures distributions.

3.1 Acceleration

Fig. 2 shows the amplification ratios (AR) at point D (presumed to be representative of far field conditions, Fig. 1) and point A (top of diaphragm wall) with respect to the bedrock acceleration.

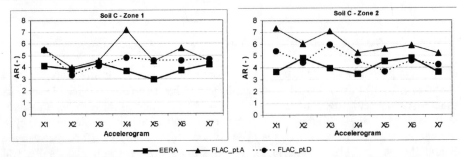

FIGURE 2. Amplification ratio over the model and comparison with simplified analyses.

The average trend obtained at point D is in reasonable agreement with that resulting from simplified linear equivalent analyses performed with the code EERA (Bardet et al., 2000) on the same soil profile, assuming standard decaying curves for sands. These results justify the choice of the value assumed for Rayleigh damping. Results for Point A in most situations show higher amplification, which can be justified by the presence of the excavation.

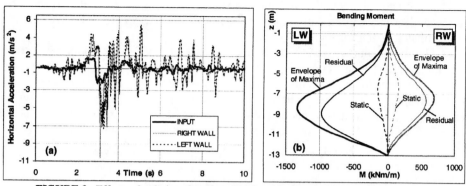

FIGURE 3. Effects of polarity of motion: (a) Time histories of horizontal accelerations
(b) Bending moments on left (LW) and right wall (RW).

Moreover, the choice of a symmetric model allows the comparison between the response of both the walls and their behaviour to the seismic input, which is strongly affected by the polarity of motion. Fig. 3a shows the time histories of horizontal acceleration of point A and A' (Fig. 1), on the top of left wall (LW) and right wall (RW), respectively. The accelerogram considered was recorded during 1982 Erzincan (Turkey) earthquake. As it is clearly noticeable, the change of sign of the input accelerogram (which is strongly polarized, i.e. maximum positive acceleration is sensibly different from maximum negative acceleration) induces remarkably different effects for right and left walls.

3.2 Bending Moments

Fig. 4 shows a comparison between dynamic and P-S bending moments obtained on the same model for the wall $h = 5$m in soil class C.

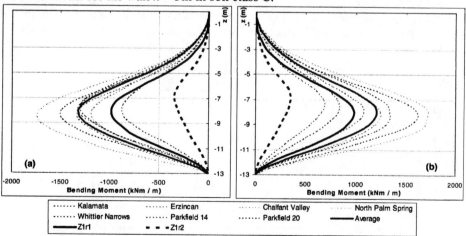

FIGURE 4. Dynamic vs. pseudo-static bending moments for soil type C, $h = 5$ m, seismic zone 1
(a) Left wall; (b) Right wall.

The average of dynamic bending moments is substantially higher than the correspondent P-S values, particularly for the case of $r = 2$, whereas the dynamic and P-S results approach for $r = 1$.

In Fig. 5 a comparison of maxima of dynamic and P-S bending moments is presented for all the cases considered in soil C as a function of excavation height. The shadow area envelopes the values of dynamic bending moments as a function of wall height. Results are consistent with the expectation that the excavation height induces not only increasing values of bending moment along the wall, but a stronger influence as well of the type of input motion, with a wider dispersion of results.

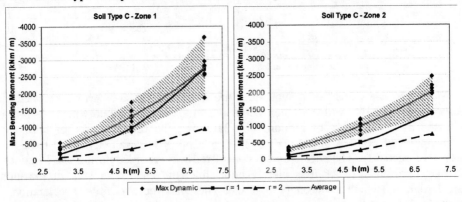

FIGURE 5. Dynamic vs. pseudo-static bending moments: influence of height of excavation.

The influence of the soil class is investigated in Fig. 6, where dynamic and P-S bending moments are normalized to the correspondent static values. There is an evident increase in dynamic bending moments with the compliance of soil (i.e. from soil B to D). On the contrary, P-S bending moments do not increase monotonically with compliance, since they are rigidly related to the site amplification factors S prescribed by EC8 (see Table 1), with a dependence on the soil profile that is relatively poor and is not sufficient to properly account for the dynamic amplification effects.

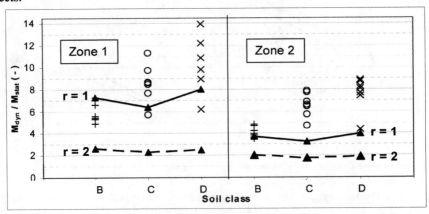

FIGURE 6. Dynamic vs. pseudo-static bending moments, $h = 5$ m: influence of soil class.

Recent researches based on the statistical analysis of a wide digital records dataset (Cauzzi and Faccioli, 2008), mostly based on the Japanese K-Net records and with a careful quantitative soil characterization, suggest larger site amplification factors than prescribed by current seismic norms, especially for soil classes C and D, and in much better agreement with the findings of this study.

Probably the most important issue that arises from the analysis of results is the high ratio between dynamic (and also P-S) and static bending moments. The adoption in seismic design of the large bending moments predicted by dynamic analyses would satisfy the requirement that the embedded structure would remain elastic under seismic loading. Such prescription would be in agreement with general requirements for piles which are often used as retaining structures. However, an alternative paradigm could also be adopted tolerating a temporary and localized yielding of the structure, as long as the permanent displacement at the end of the earthquake remains acceptable, in order to eventually take into account some ductility of the retaining structure.

3.3 Earth pressure distribution

Fig. 7 shows an example of the horizontal pressures on left wall after the last excavation phase (7a), and after the attainment of maximum acceleration during seismic shaking (7b) for the same 1982 Erzincan (Turkey) earthquake.

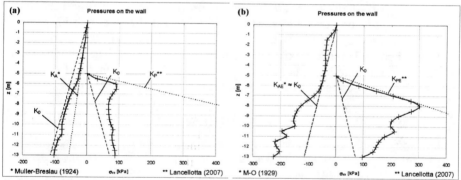

FIGURE 7. Example of distribution of horizontal pressure on LW (soil C, $h = 5$ m) - 1982 Erzincan (Turkey) earthquake (a) static distribution: (b) seismic distribution after a_{max}.

In both cases, active and passive states are fully reached in the zones of soil where the movement of structure is higher. The assumptions made in the phase of preliminary design on the active and passive state are hence substantially corroborated by the execution of fully dynamic analyses.

5. CONCLUSIONS

This paper presented some results of a comprehensive numerical study on the behaviour of flexible earth retaining structure subjected to seismic loading. The validation of the numerical results against a physical model on a shaking table is planned in the near future. The execution of dynamic analyses has pointed out some of the limits of simplified pseudo-static analyses in the design of such structures. The dynamic analyses confirmed the requirements for the definition of the pseudo-static action proposed by EC8, although in some situations an underestimation of the maximum bending moment has been observed. In particular in some cases this underestimation is apparently associated to stratigraphic amplification factors which appear not to be adequate for softer soils.

The high levels of stress inside the structure (e.g. bending moments) observed in both dynamical and pseudo-static analyses requires the adoption of very large amount of reinforcement in order to comply with the requirements of avoiding plastic hinges in soldier piles and diaphragm walls. Such problems would make the structural scheme of cantilever walls not suitable from an economic point of view for high seismicity zones, unless the structure is temporary and reduced peak ground acceleration can be assumed in the design.

The adoption of temporary yielding of the structure during the seismic event could be an alternative paradigm for the design. Clearly, this assumption requires some additional considerations on the amount of permanent displacement that the structure can accommodate in order to retain its functions or to prevent collapse. However, for the practical application of the latter concepts, it must be remarked that reliable approaches for simplified calculations of permanent seismic displacements of flexible retaining structures are still missing.

ACKNOWLEDGMENTS

This work is part of the *PRIN 2005* research program on the seismic behaviour of earth retaining structures and slopes, which has been funded by the Italian Ministry for Research and University (MIUR). Comments and suggestions by Dr. Alberto Callerio and Prof. Michele Jamiolkowski are gratefully acknowledged. The authors are also grateful to the Eucentre Laboratory for the support in the development of the physical model that will be used for the validation of the results.

REFERENCES

1. Bardet, J. P., Ichii, K. and C. H. Lin (2000) *A Computer Program for Equivalentlinear Earthquake site Response Analyses of Layered Soil Deposits* Department of Civil Engineering, University of Southern California.
2. Comina C., Foti S., Lancellotta R., Leuzzi F. and Pettiti A. (2007) *On the seismic design of diaphragm walls according to EC8-5* XIV European Conference on Soil mechanics and Geotechnical Engineering, September 24th-27th, 2007 Madrid, Spain, Patron Editore.
3. Cauzzi C. and E. Faccioli (2008). *Broadband (0.05 s to 20 s) prediction of displacement response spectra calibrated on worldwide digital records.* Accepted for publication in Journal of Seismology.
4. Itasca (2005) "FLAC Fast Lagrangian Analysis of Continua" v. 5.0. User's Manual.
5. Lancellotta R. (2007) *Lower-bound approach for seismic passive earth resistance* Géotechnique, vol. 57, n.3, pp 319-321.
6. Mononobe N. and Matsuo H. (1929) *On the determination of earth pressures during earthquakes* Proceedings, World Engineering Congress, vol. 9, p 275.
7. Okabe S. (1926) *General theory of earth pressures* J. Japan Soc. Civil Eng., Tokyo, vol.12 (1)
8. Steedman R. S. and Zeng X. (1990) *The influence of phase on the calculation of pseudo-static earth pressure on a retaining wall* Géotechnique vol. 40, n. 1, 103–112.
9. Steedman R. S. and Zeng X. (1993) *On the behaviour of quay walls in earthquakes* Géotechnique, vol.43, n. 3, 417–431.

SEISMIC BEHAVIOR OF GEOGRID REINFORCED SLAG WALL

Ayşe Edinçliler[a], Gökhan Baykal[b] and Altuğ Saygılı[c]

[a]Accoc.Prof., Boğaziçi University, Kandilli Observatory and Earthquake Research Institute,
Department of Earthquake Engineering, Çengelköy – Istanbul-Turkey.
[b]Prof.Dr. Boğaziçi University, Department of Civil Engineering, Bebek-Istanbul-Turkey
[c]Research Assistant, Boğaziçi University, Department of Civil Engineering, Bebek-Istanbul-Turkey

Abstract. Flexible retaining structures are known with their high performance under earthquake loads. In geogrid reinforced walls the performance of the fill material and the interface of the fill and geogrid controls the performance. Geosynthetic reinforced walls in seismic regions must be safe against not only static forces but also seismic forces. The objective of this study is to determine the behavior of a geogrid reinforced slag wall during earthquake by using shaking table experiments. This study is composed of three stages. In the first stage the physical properties of the material to be used were determined. In the second part, a case history involving the use of slag from steel industry in the construction of geogrid reinforced wall is presented. In the third stage, the results of shaking table tests conducted using model geogrid wall with slag are given. From the results, it is seen that slag can be used as fill material for geogrid reinforced walls subjected to earthquake loads.

Keywords: Reinforced retaining wall, geogrid, slag, shaking table.

INRODUCTION

Reinforced soil walls offer significant technical advantage over conventional reinforced concrete retaining structures at the sites with poor foundation conditions when compared to conventional reinforced concrete retaining walls, they can undergo larger displacements due to their flexible behavior.

Post earthquakes have provided numerous case studies of reinforced soil wall performance under dynamic loading. These cases have increased the confidence in the reinforced retaining systems. In general reinforced soil structures have performed well in earthquakes. Numerous cases have been reported where reinforced soil structure performance in major earthquakes have been documented [1,2,3,4,5,6 and 7].

In geogrid reinforced walls, the performance of the fill material and the interface of the fill and geogrid controls the performance. Available industrial wastes can be used as a fill materials for reinforced retaining walls. Using of these waste materials prevents depletion of the natural resources, avoids the disposal costs of these wastes. Aim of this study is to analyze the behavior of the geogrid reinforced slag wall under seismic loadings. The objective of performing the shaking table tests is to investigate any potential interlocking problem between the slag and the geogrid during shaking. In the model tests, the same material properties are used as the slag wall constructed in a valley in the Karabuk city of Black Sea region of Turkey.

CP1020, 2008 Seismic Engineering Conference Commemorating the 1908 Messina and Reggio Calabria Earthquake,
edited by A. Santini and N. Moraci

BACKGROUND

Sakaguchi (1996) performed shaking table tests on a 1.5 m high model test of a geogrid wall [8]. The tests were constructed with lightweight blocks and five layers of geogrid reinforcement. It is found that the reinforced zone was observed to cut as a monolithic body with no evidence of a yield surface propagating across the reinforcement layers even after large wall displacement developed.

Matsuo et al (1998) carried out shaking table tests on six geosynthetic-reinforced soil retaining wall models. In their models, the geogrid reinforcement length, wall height, wall facing type, wall slope, and input acceleration waveform were varied in order to understand the behavior and reinforcement mechanisms [9].

Large scale shaking table tests were reported by Ling et al. (2005) [10]. Three large scale 2.8 m high modular-block geosynthetic-reinforced soil walls were subjected to Kobe earthquake motions. Each wall was excited with one-dimensional horizontal maximum acceleration of 0.4g followed by 0.86g. The reinforced soil retaining walls showed negligible deformation under simulated earthquake (peak acceleration of 0.4g). Under very strong shaking of 0.86g the walls performed well.

Laboratory and mathematical models are used to analyze the seismic behavior of reinforced retaining walls. Common laboratory models used for this purpose are shaking table tests and centrifuge tests. The shaking tables which were primarily designed for modeling of structures, are adopted to perform geotechnical modeling [11 and 12]. A common problem to both centrifuge and shaking table testing is the design of the specimen container. A simple, rigid walled box will not allow the specimen to deform uniformly. To overcome this problem, several researchers have used variations of the "shear stack" concept. The boxes that contain the specimens are categorized in two groups; rigid sided shear boxes and flexible, layered shear stacks. The flexible shear stack is composed of alternating rubber layers and metal frames stacked over each other. Taylor et al (1994) have developed a flexible shear stack of 1.2 m long, 0.6 m wide and 0.9 m high [13]. They have conducted several experiments using dry sand under seismic loading. They have measured consistent acceleration values in the horizontal plane. The vertical profile of the measured accelerations was sinusoidal as desired. With the experience gained by Taylor et al (1994), the results of this study revealed that the shear stack tests provided representative values for field conditions.

MATERIALS AND METHODS

A geogrid-slag wall was designed and constructed in one of the development sites in the Karabuk city of Black Sea region of Turkey. The high rise apartment blocks are located in a valley and the slope of the valley is unstable creating landslide hazard. A typical cross-section of the geogrid-slag wall is presented in Figure1. At this location the maximum elevation difference is 25 meters. The horizontal length of slope is 25 meters. Two geogrid walls of 8.40 m. height x 15.40 m. length at the bottom and 8.00 m. height x 10.00 m. length at the top is designed and constructed. Geogrids of tensile strengths (45, 60, 90 kN/m - Tenax) are used at 0.4 m spacing. A total of 3000 square meters of geogrid wall with maximum height of 16 meters is completed. The construction site is located at first degree earthquake zone. An evaluation of the wall under earthquake loads has been completed by limit equilibrium analysis.

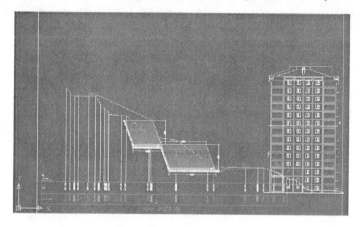

FIGURE 1. Cross-Section View of the Constructed Geogrid Wall

Material Properties

The blast furnace slag is supplied from Kardemir Iron and Steel Ltd., Co. located in the Black Sea region of Turkey. Blast furnace slag, the nonmetalic core product produced in the molten iron production consists of silicates, aluminosilicates and calcium-alumina-silicates. The air cooled blast furnace slag used in this study is presented in Figure 2. The slag disposal site was only 3km to the wall constructed site.

Samples were prepared at optimum moisture content for the large size direct shear tests with the dimensions of 30 x 30x30 cm. The samples are compacted into the shear box with modified effort to apply modified compaction energy (2700kN m/m^3). From the results of large size direct shear test of compacted blast furnace slag, the internal friction angle corresponding to peak values is 49 degrees. The optimum water content and maximum dry density of the blast furnace slag are 10% and 23 kN/m^3, respectively.

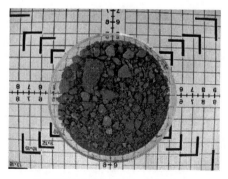

FIGURE 2. Blast Furnace Slag

Shaking Table Tests

A shear stack was constructed by using the experience from the studies performed at the Bristol University by Taylor et al (1994) (Figure 3). The length of the stack is 1.45 m, the width is 0.75 m and the height is 0.78 m. Eight rubber layers were specially molded to target specifications. Eight aluminum frames were manufactured using tubular aluminum profiles. The rubber frames allow the soil specimen to deform freely while the aluminum frames provide the horizontal boundaries for the specimen. The size of the frames and material properties were determined after performing finite element analysis to achieve the criteria set by Gazetas (1982) [14]. This essential criteria that the shear stack had to meet were: (i) lateral motions should be uniform on any horizontal plane through the soil and the shear stack itself, and (ii) lateral motions over the depth of the stack should follow a near sinusoidal profile.

Flexible shear stack for shaking table testing of geogrid reinforced slag specimens is described. The shear stack develops boundary conditions closely resembling those of a classical one dimensional soil deposit for up to moderately high strain levels allowing the soil specimen to deform freely under simulated seismic loads.

FIGURE 3. Experimental Setup

Set Up System

Compaction of the soil and slag was achieved by applying an energy comparable to standard compaction proctor energy. The lower geogrid layer is placed 0.3 m above the base. The upper layer is placed 0.52m above the base. A 0.10 m thick layer of slag is placed above the upper geogrid layer. The vertical pressure of 40 kPa is applied by airbags above this top layer. The instrumentation consisted of eleven accelerometers (A1,A2,A3,A4,A6,A7,A8,A9,A10,A11 and A12), eight displacement transducers (1 on the table, seven on the shear stack- L1,L2,L3,L4,L5,L6,L7 and L8) and four force transducers (F1,F2,F3 and F4) (Figure 4). One accelerometer is fixed on the shaking table (A1). At the front face of the shear stack three accelerometers are placed at the first, fourth and eight aluminum frames from the bottom (A2,A3 and A4, respectively). At the level of lower geogrid layer three accelerometers are installed 0.24m, 0.72 and 1.20m from the front face (A9,A10 and A11). The same accelerometer configuration is applied at the upper geogrid elevation (A6, A7 and A8).

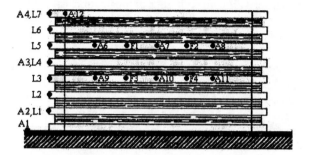

FIGURE 4. Instrumentation Layout

The air bags are tied down with ropes on a circular rollers sliding on steel rods. The sliding steel rod is fixed on the shaking table platform with steel screws. The rollers provided minimal friction during the shaking table motion (Figure 3). The loading mechanism has been used for previous tests and performed successfully [15]. Four force transducers were placed at two geogrid elevations to measure the applied vertical load. The displacement of eight aluminum frames forming the shear stack were measured using eight displacement transducers fixed on the reaction frame. The displacement transducers are calibrated before the tests.

EXPERIMENTAL STUDY

The large uniaxial shaking table facility at the Shaking Table Laboratory of Kandilli Observatory and Earthquake Research Institute (KOERI) at Bogazici University was used. The uniaxial shaking table (3mx3m) is driven by a servo-hydraulic actuator. Test objects up to 10 tons force can be accumulated over a frequency range of 0-50 Hz.

In order to gain better insight into dynamic behavior of a geogrid reinforced slag wall under earthquake loads, 1/4 reduced–scale shaking table tests were carried out using the shaking table facility at the KOERI. According to the scaling laws the original El Centro earthquake record was scaled by √4. El Centro Earthquake motions were used to excite the models at a maximum acceleration of 0.3 g followed by 0.6g.

A total of nine models were tested: one test with uniform sand (Model 1), only slag (Model 2), 2 layers of geogrid fold up at the back wall of the stack (Model 3a), the same configuration with Model 3a with no vertical pressure (Model 3b), two layers of geogrid with geogrid trimmed near the back wall instead of folding up (Model 4). For the Models 1 through 4, ElCentro earthqauke record with 0.3g were applied as input motion. For models 5a, 5b, 6a, 6b, and 6c, Model 5 refers to folded geogrid, Model 6 refers to trimmed geogrid. The letters refers to number of repetitions. The maximum accelerations measured and corresponding time (t) values are tabulated in Tables 1 through Table 6.

Results of the Tests

At the end of every shaking, sandy soil was compacted to a higher relative density level and thus the stiffer soil caused higher amplification values in the structures than the previous observed values. Stiff soils are strong and tend to amplify the accelerations, soft soils are ideal dampers and they decrease the acceleration amplification [10].

The maximum amplification factor (the ratio of the maximum acceleration in the structure to the peak input base acceleration) was evaluated for all of the El Centro Earthquake motions. Since it is the peak input base acceleration that it is needed to focus on primarily to be able to measure the amplification factors, the peak values of the base acceleration at certain times instances were taken into consideration (Tables 1 through Table 6).

TABLE 1. Maximum Acceleration Values at Various Locations on the Shear Stack (0.3g)

| | Base Acc. | | Accelerations on Shear Stack | | | | | | | | |
| | A1 | | A2 (Botton) | | | A3 (Middle) | | | A4 (Top) | | |
Model No.	Time (Sec)	Acc. (g)	Time (Sec)	Acc. (g)	Amp	Time (Sec)	Acc. (g)	Amp	Time (Sec)	Acc. (g)	Amp
1	139	0.32	1.29	0.35	1.07	1.12	0.39	1.18	2.95	0.35	1.06
2	2.31	0.31	2.31	0.31	1.00	1.61	0.34	1.10	1.64	0.42	1.35
3a	1.64	0.30	1.64	0.31	1.03	1.64	0.34	1.13	1.64	0.38	1.27
3b	2.32	0.26	1.46	0.26	1.00	2.32	0.29	1.12	2.32	0.33	1.27
4	2.31	0.27	2.32	0.28	1.04	2.32	0.31	1.15	2.33	0.35	1.30

TABLE 2. Maximum Acceleration Values at Upper Level Geogrids (0.3g)

| | Base Acc. | | Accelerations on Upper Level | | | | | | | | |
| | A1 | | A6 | | | A7 | | | A8 | | |
Model No.	Time (Sec)	Acc. (g)	Time (Sec)	Acc. (g)	Amp	Time (Sec)	Acc. (g)	Amp	Time (Sec)	Acc. (g)	Amp
1	139	0.32	2.94	0.34	1.06	2.94	0.35	1.09	2.94	0.33	1.03
2	2.31	0.31	1.64	0.33	1.07	1.64	0.34	1.10	1.64	0.33	1.06
3a	1.64	0.30	1.64	0.34	1.13	1.64	0.34	1.13	1.64	0.34	1.13
3b	2.32	0.26	2.32	0.29	1.12	2.32	0.30	1.15	2.32	0.28	1.08
4	2.31	0.27	2.32	0.31	1.15	2.32	0.32	1.19	2.32	0.30	1.11

TABLE 3. Maximum Acceleration Values at Lower Level Geogrids (0.3g)

| | Base Acc. | | Accelerations on Lower Level | | | | | | | | |
| | A1 | | A6 | | | A7 | | | A8 | | |
Model No.	Time (Sec)	Acc. (g)	Time (Sec)	Acc. (g)	Amp	Time (Sec)	Acc. (g)	Amp	Time (Sec)	Acc. (g)	Amp
1	139	0.32	1.31	0.32	1.00	1.31	0.33	1.03	1.31	0.34	1.06
2	2.31	0.31	2.28	0.31	1.00	2.28	0.33	1.06	2.28	0.33	1.06
3a	1.64	0.30	1.64	0.32	1.07	1.04	0.32	1.07	1.64	0.32	1.07
3b	2.32	0.26	2.31	0.27	1.04	2.32	0.27	1.04	2.32	0.29	1.12
4	2.31	0.27	2.32	0.29	1.07	2.32	0.31	1.15	2.32	0.31	1.15

TABLE 4. Maximum Acceleration Values at Various Locations on the Shear Stack (0.6g)

Model No.	Base Acc. A1 Time (Sec)	Acc. (g)	A2 (Bottom) Time (Sec)	Acc. (g)	Amp	A3 (Middle) Time (Sec)	Acc. (g)	Amp	A4 (Top) Time (Sec)	Acc. (g)	Amp
5a	2.23	0.64	2.29	0.66	1.03	2.29	0.76	1.19	2.29	0.84	1.39
5b	1.64	0.59	1.66	0.63	1.07	1.66	0.72	1.22	1.66	0.84	1.42
6a	1.10	0.61	1.04	0.61	1.00	1.10	0.63	1.03	1.11	0.75	1.23
6b	2.32	0.61	2.31	0.62	1.02	2.31	0.69	1.13	2.33	0.78	1.28
6c	2.32	0.61	2.32	0.65	1.07	2.46	0.73	1.20	2.46	0.90	1.48

TABLE 5. Maximum Acceleration Values at Upper Level Geogrids (0.6g)

Model No.	Base Acc. A1 Time (Sec)	Acc. (g)	A6 Time (Sec)	Acc. (g)	Amp	A7 Time (Sec)	Acc. (g)	Amp	A8 Time (Sec)	Acc. (g)	Amp
5a	2.23	0.64	2.29	0.75	1.17	2.29	0.78	1.22	2.29	0.77	1.20
5b	1.64	0.59	1.66	0.74	1.25	1.66	0.76	1.29	1.66	0.75	1.27
6a	1.10	0.61	1.11	0.62	1.02	1.11	0.63	1.03	1.11	0.61	0.98
6b	2.32	0.61	2.33	0.68	1.10	2.31	0.71	1.16	2.33	0.66	1.08
6c	2.32	0.61	2.46	0.74	1.21	2.46	0.77	1.27	2.46	0.73	1.20

TABLE 6. Maximum Acceleration Values at Lower Level Geogrids (0.6 g)

Model No.	Base Acc. A1 Time (Sec)	Acc. (g)	A9 Time (Sec)	Acc. (g)	Amp	A10 Time (Sec)	Acc. (g)	Amp	A11 Time (Sec)	Acc. (g)	Amp
5a	2.23	0.64	2.29	0.69	1.08	1.29	0.71	1.11	2.29	0.72	1.13
5b	1.64	0.59	1.66	0.71	1.20	1.66	0.72	1.22	1.66	0.72	1.22
6a	1.10	0.61	1.11	0.61	0.98	1.10	0.62	1.01	1.10	0.64	1.05
6b	2.32	0.61	2.33	0.66	1.08	2.32	0.67	1.10	2.32	0.68	1.11
6c	2.32	0.61	2.32	0.68	1.11	2.32	0.71	1.16	2.32	0.71	1.11

The maximum accelerations were observed at the time instances of t=1.04 sec and t=2.94 for 0.3g and t=1.02sec. and t=2.29sec. for 0.6g. The amplification ratio of the models was less than 1.35 under a peak acceleration of 0.3g and 1.48 under the peak acceleration of 0.6g. Amplification factors range between 1.00 and 1.35 for 0.3g and 0.98 and 1.48 for 0.6g. At the top of the frame higher amplification values were observed for slag when compared to that of sand. This is due to higher stiffness of slag which results higher amplification. The results of the tests showed the interlocking of slag and geogrid was not affected during shaking.

CONCLUSIONS

When potential use of waste materials or by-products are investigated for geosynthetic reinforcement, it may be a good practice to test the integrity of the composite during shaking. From the test results, no interlocking problem and no segragation of slag were observed after shaking. Slag can be successively used as fill material for geogrid reinforced walls subjected to earthquake loads.

REFERENCES

1. J.G. Collin, V.E. Chouery-Curtis and R.R. Berg, "Filed Observations of Reinforced Soil Structures Under Seismic Loading", Earth Reinforcement practice, *International Symposium on Earth Reinforcement Practice*, IS Kyusku'92, Balkema, Vol.1, pp.223-228, 1992.
2. U.Eliahu and S.Watt, "Geogrid-Reinforced Wall Withstands Earthquake" *Geotechnical Fabric Report*, Vol.9, No.2, pp.8-13, 1991.
3. P.C. Franlenberger, R.A., R.A. Bloomfield and P.L. Anderson, "Reinforced Earth Walls Withstand Northridge Earthquake" Earth Reinforcement, *Third International Symposium on Earth Reinforcement*, IS Kyushu'96, Balkema, Vol.1, pp.345-350, 1996.
4. B.L. Kutter, J.A, Casey and K.M. Romstad, "Centrifuge Modelling and Field Observations of Dynamic Bahaivor of Reinforced Soil and Concrete Cantilever Retaining Walls", *Fourth U.S. National Conference on Earthquake Engineering*, EERI, May 20-24, Vol.3, pp.663-672, 1990.
5. H.I. Ling, D.Lechchinsky and NNS. Chou, "Post-earthquake Investigation on Several Geosythetic-reinforced Soil Retaining Walls and Slopes During the Ji-Ji Earthquake of Taiwan, *Soil Dynamics and Earthquake Engineering*, Vol.21, No.4, pp.29313, 2001.
6. F. Tatsuoka, J. Koseki and M. Tateyama,"Performance of Geogrid-reinforced Soil Retaining Walls During the Great Hanshin-Awaji Earthquake, January 17, 1995", *First International Conference on Earthquake Geotechnical Engineering*, IS Tokyo'95, Balkema, Vol.1, pp.55-62, 1995.
7. D.M. White and RD.Holtz, "Performance of Geosynthetic-reinforced Slopes and Walls During the Northridge, California Earthquake of January 17, 1997", *Earth Reinforcement, Third International Symposium on Earth Reinforcement*, IS Kyushu'96, Balkeme, Vol.2, pp.965-972, 1996.
8. M., Sakaguchi, "A Study of the Seismic Behavior of Geosynthetics Reinforced Walls in Japan", *Geosynthetics Int.*, Vol.3, No.1, pp.13-30, 1996.
9. O.T, Matsuo, T. Tsutsumi, K. Yokoyama and Y. Saito, "Shaking Table Tests and Analysis of Geosynthetic-Reinforced Soil Retaining Walls, *Geosynthetics International*, Vol.5, No.1-2, pp.97-126, 1998.
10. HIY, Ling, Y.Mohri, D.,Leshchinsky, C.Burke, K.Matsushima and H.Liu,"Large Scale Shaking Table Tests on Modular – Block Reinforced Soil Retaining Walls",*Journal of Geotechnical and Geoenvironmental Engineering*, ASCE Vol.131, No.4, pp.465-476, 2005.
11. K. Fukutake, A. Ohtsuki and Y., Shamoto Y, "Analysis of Saturated Dense Sand – Structure System and Comprarison with Results from Shaking Table Test", *Earthquake and Structural Dynamics*, Vol. 19, 977-992 , 1990.
12. R. Richards, DG, Elms and M. Budhu"Dynamic Fluidization of soils", *Journal of Geotechnical Engineering Division*, ASCE, Vol. 116, No.5, 740-759, 1991.
13. CA, Taylor, AR. Dar and AJ. Crewe, "Shaking Table Modelling of Seismic Geotechnical Problems", *Proc. 10 th European Conference on Earthqauke Engineering*, Vianne, Austria, Rottersa:AA, Balkema, 441-446, 1994.
14. G. Gazetas, "Vibrational Characteristics of Soil Deposits With Variable Wave Velocity", Int. *Journal of Numerical and Anal. Math. in Geomechanics*, pp:1-6, 1982.
15. A. Edinçliler, G. Baykal and M. Erdik, "Shaking Table Modeling of Solid Waste Landfills", Project Report, TÜBİTAK-MAG, Project No. 03I010, 2007.

Effective Parameters on Seismic Design of Rectangular Underground Structures

G. Ghodrati Amiri[a], N. Maddah[b] and B. Mohebi[c]

[a]Professor, Center of Excellence for Fundamental Studies in Structural Engineering, College of Civil Engineering, Iran University of Science & Technology, Narmak, Tehran 16846, Iran
E-mail: ghodrati@iust.ac.ir
[b]MSc Student, College of Civil Engineering, Iran University of Science & Technology, Tehran, Iran
[c]PhD Student, College of Civil Engineering, Iran University of Science & Technology, Tehran, Iran

Abstract. Underground structures are a significant part of the transportation in the modern society and in the seismic zones should withstand against both seismic and static loadings. Embedded structures should conform to ground deformations during the earthquake but almost exact evaluation of structure to ground distortion is critical. Several two-dimensional finite difference models are used to find effective parameters on racking ratio (structure to ground distortion) including flexibility ratio, various cross sections, embedment depth, and Poisson's ratio of soil. Results show that influence of different cross sections, by themselves is negligible but embedment depth in addition to flexibility ratio and Poisson's ratio is known as a consequential parameter. A comparison with pseudo-static method (simplified frame analysis) is also performed. The results show that for a stiffer structure than soil, racking ratio decreases as the depth of burial decreases; on the other hand, shallow and flexible structures can suffer greater distortion than deeper ones up to 30 percents.

Keywords: Underground Structures, Racking ratio, Embedment depth, Flexibility ratio

INTRODUCTION

The increased construction of underground facilities and transportation in the seismic zones requires safety provision against motions caused by earthquakes. Generally, underground structures have lower earthquake risk as compared to surface structures [1]. Nevertheless, significant damages are known to have been caused to underground structures like Daikai subway station in Kobe, Japan which was collapsed in the 1995 Hyogoken-Nambu earthquake [2]. Therefore, seismic analysis should not be overlooked. Most of rectangular structures are built with the cut and cover method in which an open excavation is made, the structure is constructed and fill is placed over the finished structure. So these structures commonly are placed in shallow depth relatively. Subway stations, portal structures, and highway tunnels are examples of these structures.

CP1020, *2008 Seismic Engineering Conference Commemorating the 1908 Messina and Reggio Calabria Earthquake*,
edited by A. Santini and N. Moraci
© 2008 American Institute of Physics 978-0-7354-0542-4/08/$23.00

CURRENT STATE OF KNOWLEDGE

The problem of the underground structure under seismic loading has so far been approached in three ways: (a) by using a criterion of resisting a specified dynamic earth pressure as in the Mononobe-Okabe method, as suggested by Seed and Whitman [3], and the Japanese Society of Civil Engineers [4], applied to buried structures or the Wood method [5], (b) to carry out dynamic, non-linear soil–structure interaction analysis using finite element or finite difference methods and (c) by specifying the loading in terms of deformation and ensure that it can be absorbed by the structure. The latter, known as the free-field deformation method for the estimation of structure racking was presented by Kuesel [6] and has been widely used by engineers. St. John and Zahrah [7] developed this method and conducted a series of two and three-dimensional analyses for underground box type structures applying the loading statically. His results predicted racking deformations of 50-65 percent of the free-field deformation, depending on soil properties and structure flexibility. Analytical solutions exist to provide some relationships to evaluate the magnitude of seismic-induced displacements or strains in underground structures. Wang [8] and Penzien [9] investigated and proposed this method for estimating the racking deformation of rectangular structures subjected to ground shaking. Both researchers are oriented towards a pseudo-static approach. Their difference lies in the calculation of said force and, to be more exact, in the calculation of the normalized structure racking ratio R = structure horizontal deformation / free-field horizontal shear deformation, a parameter used for the soil-structure interaction effect.

Wang [8] conducted a series of dynamic analyses using the computer program FLUSH depicting the effect of various parameters to the structure racking but mainly focusing on the flexibility ratio, i.e. the soil-structure relative stiffness. His conclusions were that the effects of rectangular structure geometry, ground motion characteristics and stiffer foundation are negligible, while the effect of relative stiffness is the most significant followed by the effect of embankment depth only for small depths.

The flexibility ratio is expressed as [8]:

$$F = \frac{G \times W}{S_1 H} \tag{1}$$

Where W and H are the height and wide of the structure respectively, G is the shear modulus of the ground, and $S_1 = 1/D_1$ with D_1 equal to the displacement produced on the structure by a unit lateral concentrated force applied to the top of the structure. Having determined the normalized racking distortion for various flexibility ratios, Wang proposed the use of diagrams containing all his results as a means to determine the normalized structure racking R and from then on the actual distortion of the structure $\Delta_s = R\Delta_{\text{free-field}}$. The free-field ground deformation should be determined by use of the Newmark method [10] by use of one-dimensional site response analysis. Hashash [1] has extensively presented Wang's proposed approach for rectangular structures.

Penzien [9] proposed an approximate method to evaluate the racking deformation of deep rectangular tunnels subjected to a far field shear stress. Penzien showed that the deformations of the structure depend on the relative stiffness, or the flexibility ratio, between the ground and the structure. The relative stiffness is defined with the parameter k_{str}/k_{soil}, which is the ratio between k_{str}, the stiffness of the structure and k_{soil} the stiffness of the soil. k_{str} is equal to the magnitude of a uniform shear stress applied to the perimeter of the structure that produces a unit displacement of the structure; and $k_{soil}=G/H$, where G is the shear modulus of the soil and h is the height of the structure. The relative stiffness which is used in the Penzien [9] equation is the inverse of Wang's flexibility ratio. The normalized deformation of the structure, or the ratio between the structure deformation (Δ_{str}) and the free-field ground deformation (Δ_{ff}), $\Delta_{stru}/\Delta_{ff}$, can be obtained by:

$$R = \frac{4(1-v_s)}{1+\alpha_s} \tag{2}$$

In which α_s is:

$$\alpha_s = (3-4v_s) \times \frac{k_{str}}{k_{soil}} \tag{3}$$

PARAMETRIC STUDY

Modeling Assumptions

Rectangular sections of underground structures under vertically propagating shear seismic excitation are considered.

The soil layer is a homogeneous isotropic elastic medium on rigid bedrock. In order to examine the peak sectional racking due only to the excitation, gravitation is excluded. However, According to the principle of superposition, the total solution can be obtained by adding the complementary to the particular solution that satisfies gravitation. The lining properties are those of reinforced concrete and no-slip condition along the soil and structure is assumed. Fig. 1 shows an example of the finite difference soil-structure model.

Earthquake Accelerograms

For dynamic analysis of the model, two rock outcrop ground motion accelerograms employed. These accelerograms was modified to abide by the design response spectra of the Iranian earthquake code [11] (see Fig. 2). Horizontal earthquake accelerograms were converted to the velocity and were input at the rigid base to simulate the vertically propagating shear waves.

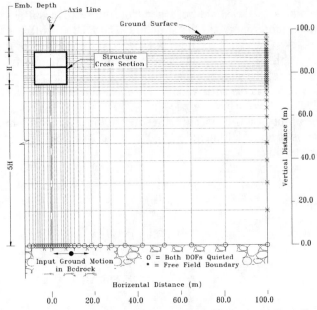

FIGURE 1. Typical soil-structure finite difference model (Structure Type-3)

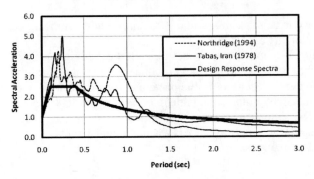

FIGURE 2. Spectral Acceleration

Cases Analyzed

The main parameters that might influence the response of a square underground structure are the cross section of the structure, the relative stiffness between soil and structure, Poisson's ratio and thickness of the soil layer above the structure.

Three different cross sections are considered to investigate the influence of structure geometry. Fig. 3 illustrates the three types of structures configurations that were analyzed.

It is anticipated that the relative stiffness between soil and structure is the dominating factor that governing the soil/structure interaction. Thus, in each series of analysis shear modulus altered by changing in the shear wave propagation in the range of the 50 to 500 m/s. This can be a reasonable range of the shear wave propagating in the soil or rock. Poisson's ratio also might influence the results. For comparison, three series of analysis by varying the Poisson's ratio between the 0.25 and 0.45 are applied, albeit use of these values with the whole range of the shear modulus may not applicable. However, a wide spectrum of shear modulus and diverse Poisson's ratios are considered to complete the diagrams in unrealistic way and compare the results.

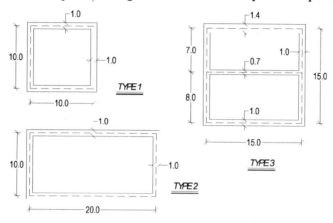

FIGURE 3. Three types of structure geometry are used in study

As it mentioned, some rectangular structures are built at shallow depths. To study the effect of the depth factor, some sets of analyses were performed with varying soil cover thickness. Here, the burial depth of the structure is varied. Embedment depth ratio D=h/H, as shown in Fig. 4, was varied between 0.5 and 3.5. It is evident, when embedment depth ratio is equal to 0.5, soil cover thickness will be equal to zero.

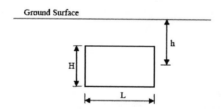

FIGURE 4. Defination if embedment depth parameter (h/H)

NUMERICAL RESULTS AND COMPARISONS

24 numerical soil structure models by changing the soil properties were analyzed. Both flexibility ratio and Poisson's ratio were investigated and racking of the structure for each analysis calculated. In this set of analysis embedment depth ratio was remained constant equal to 3. The results are obtained for type-1 structure under the Tabas earthquake time history. Fig. 5 provides a comparison between analytical solution with application of Eq.2 and numerical results. This figure shows a good consistency in racking response between the numerical and analytical solution.

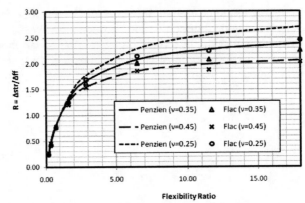

FIGURE 5. Comparison between analytical and numerical solutions, for discrete values of Poisson's ratio

The effect of structure geometry was studied by using three different types of box structure configurations (Fig. 3) and 22 cases were analyzed. Here constant parameters are embedment depth ratio equal to 3 and Poisson's ratio equal to 0.35. The results were obtained for Tabas earthquake time history. Fig. 6 provides a comparison between analytical solution and numerical results. Fig. 6 clearly demonstrates that the normalized racking deformations are insensitive to the structure geometry.

To determine the effect of embedment depth on the normalized racking response, finite difference analysis was conducted by use of Type-1 structure under both Tabas and Northridge ground motion loadings. Here, the burial depths of the structure were varied. For each case analysis, the average of the racking ratio was obtained from two responses of seismic excitation. Fig. 7 illustrates average racking ratio for different structure flexibility ratio in various depths. Fig. 7 clarifies that for a flexibility ratio, for example 2.39, racking ratio will be varied in deferent embedment depths.

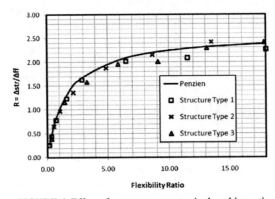

FIGURE 6. Effect of structure geometry in the raking ratio

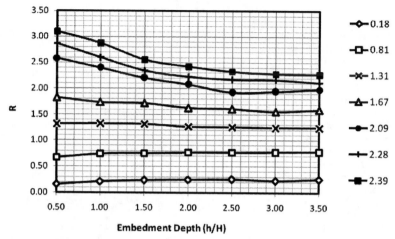

Based on the results, it is evident that:

- The normalized racking distortion, R, is relatively independent of the depth of burial for h/H>2. At this burial depth, the structure can be considered to respond as a deeply buried structure.
- For structures were stiffer than the surrounded soil, while embedment depth is less than 2, the racking distortion decreases as the depth of burial decreases, implying that design based on Eq. 2 is on the safe side for structures with little to no soil cover.
- For flexible structures with embedment depth of lower than 2, the racking distortion increases as the depth of burial decreases. Thus design of structure by application of the Eq. 2 is underestimation.

SUMMERY AND CONCLUSIONS

Numerous rectangular underground structures under seismic shear waves were analyzed. Effective parameters in estimation of the racking deformation ratio for underground structures were investigated. For this purpose, a profound parametric study using dynamic finite differences soil-structure interaction analysis was conducted. Diverse models demonstrate that not only flexibility ratio and Poisson's ratio are dominant factors (as in previous studies have been shown), but also embedment depth may have a strong effect in the soil-structure interaction. While the structure is stiffer than the surrounding soil, effect of embedment depth is negligible; on the contrary, if the structure is more flexible than soil, the influence of embedment depth is significant. In fact, deep and flexible structures experiment lower degree of racking, nonetheless, shallow and flexible underground structures suffer more lateral drifts in compare with previous studies.

REFERENCES

1. Y. M. A. Hashash, J. J. Hook, B. Schmidt and J. I. Yao, "Seismic design and analysis of underground structures", *Tunneling and Underground Space Technology* **16**, 247-293 (2001).
2. S. Nakamura, N. Yoshida and T. Iwatate, *Damage to Daikai Subway Station During The 1995 Hyogoken-Nambu Earthquake and its Investigation*, Japan Society of Civil Engineering, Committee of Earthquake Engineering, 1996, pp. 287-295.
3. H. B. Seed and R. V. Whitman, "Design of earth retaining for dynamic structures for dynamic loads", *in Proceedings of the ASCE Specialty Conference on Lateral Stresses in the Ground and Design of Earth Retaining Structures*, Japan Society of Civil Engineering, Committee of Earthquake Engineering, 1970.
4. Japanese Society of Civil Engineering, *Specifications for Earthquake Resistant Design of Submerged Tunnels*, 1975.
5. J. Wood, *Earthquake-Induced Soil Pressures on Structures*, Report EERL 73-05, California Institute of Technology, California, 1973.
6. T. R. Kussel, "Earthquake design criteria for subway", *Journal of Structural Division – ASCE*, **95 (ST6)**, 1213–1231 (1969).
7. St. C. M. John and T.F. Zahrah, "Aseismic design of underground structures", *Tunneling and Underground Space Technology* **2 (2)**, 165-197 (1987).
8. J. N. Wang, *Seismic Design of Tunnels: A State-of-the-Art Approach*, Parsons Brinckerhoff Quade & Douglas, Inc., New York, NY, Monograph 7, 1993.
9. J. Penzien, "Seismically induced racking of tunnel linings", *Int. J. Earthquake Engineering Structure Dynamics* **29**, 2000, 683–691 (2000).
10. N. M. Newmark, "Problems in wave propagation in soil and rock", *in Proceedings of the International Symposium on Wave Propagation and Dynamic Properties of Earth Materials*, 1968.
11. Building and Housing Research Center, *Iranian Code of Practice for Seismic Resistant Design of Buildings*, Standard No. 2800-05, Third Edition, BHRC Publication No. S-253, 2005.

Calcium Stabilized And Geogrid Reinforced Soil Structures In Seismic Areas

Pietro Rimoldi[a] and Edoardo Intra[a]

[a] Professional Engineer. World Tech Engineering Srl, Milano, Italy.

Abstract. In many areas of Italy, and particularly in high seismic areas, there is no or very little availability of granular soils: hence embankments and retaining structures are often built using the locally available fine soil. For improving the geotechnical characteristics of such soils and/or for building steep faced structures, there are three possible techniques: calcium stabilization, geogrid reinforcement, and the combination of both ones, that is calcium stabilized and reinforced soil. The present paper aims to evaluate these three techniques in terms of performance, design and construction, by carrying out FEM modeling and stability analyses of the same reference embankments, made up of soil improved with each one of the three techniques, both in static and dynamic conditions. Finally two case histories are illustrated, showing the practical application of the above outlined techniques.

Keywords: Geogrid, Calcium, Stabilization, Reinforcement, FEM modelling

INTRODUCTION

In many areas of Italy, and particularly in high seismic areas, there is no or very little availability of granular soils: hence, for avoiding the cost and environmental impact of sourcing sand and gravel from very long distance, embankments and retaining structures are often built using the locally available fine soil.

For improving the geotechnical characteristics of such soils and/or for building steep faced structures, there are three possible techniques: calcium stabilization, geogrid reinforcement, and the combination of both ones, that is calcium stabilized and reinforced soil.

The present paper aims to evaluate these three techniques in terms of performance, design and construction.

Performances are evaluated using FEM modeling and stability analyses of the same reference embankments, made up of soil improved with each one of the three techniques, both in static and seismic conditions. Preliminary data show interesting results, which need to be confirmed by physical modeling and in situ testing.

Then case histories are illustrated, showing the stability analyses, the design layouts and pictures taken during construction, thus affording a complete picture of the design and building activities associated with calcium stabilized and reinforced soil structures in seismic areas.

CP1020, *2008 Seismic Engineering Conference Commemorating the 1908 Messina and Reggio Calabria Earthquake*,
edited by A. Santini and N. Moraci
© 2008 American Institute of Physics 978-0-7354-0542-4/08/$23.00

NUMERICAL ANALYSIS

The evaluation of the three techniques above outlined is carried out by analyzing the behavior of a simple reference structure, made up of a road embankment with trapezoidal cross section, 10 m high, placed on soft soil. Three typical configuration are considered, namely embankment with 2V:3H, 1V:1H, 2V:1H side slopes (that is 34°, 45°, 63° side slopes), with 20 kPa uniform surcharge on top. Fig. 1 shows the reference structures. Let's suppose that the in situ soil is a normally consolidated silt and clay, and that the same soil shall be used for building the embankments. The three reference embankment will be stabilized with three methods: 3 % calcium addition; geogrid reinforcement; 3 % calcium addition plus geogrid reinforcement. Moreover the unreinforced unstabilized 1V:2H (that is 34° side slopes) embankment will be considered.

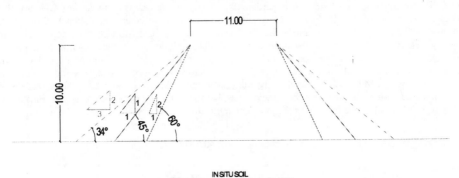

FIGURE 1. The reference embankments for the analytical study

Regarding the geogrid reinforced embankment, we will consider the use of woven polyester geogrids; internal stability analyses indicate that the 34° side slopes embankment can be reinforced with geogrids having 35 kN/m tensile strength, equally spaced at 60 cm vertical centers; the 45° side slopes embankment can be reinforced with geogrids having 55 kN/m tensile strength, equally spaced at 60 cm vertical centers; the 63° side slopes embankment can be reinforced with geogrids having 80 kN/m tensile strength, equally spaced at 60 cm vertical centers.

The performances of the reference embankments will be evaluated with both global stability analyses and FEM modeling.

The global stability analyses allows to verify that no failure mechanism may occur, involving the embankment soil mass and the foundation soil. The analyses, performed using the well known ReSSA software by Adama Engineering Inc. [1], were:

- rotational analyses (circular failure surfaces): the Bishop method is used, where the soil mass is divided into many wedges limited at bottom by a circular surface;

- translational analyses (horizontal sliding surfaces): two-part wedge mechanism, commonly known as Spencer's analysis, is used, where the reinforced soil mass and a portion of the retained soil may slide as a block along the base or along one of the reinforcing layers; hence many bilinear surfaces are investigated.

All the reference structures have been checked also in seismic conditions with $a_g = 0,15$ g, by adding a pseudo-static horizontal force $F_{PS} = a_g \times W_i$ to each wedge, applied to the wedge center of gravity, where W_i is the weight of the i-th wedge.
In the ReSSA software the translational analysis can be performed only if geogrid layers are specified; hence, for performing such analysis on the calcium stabilized embankment, dummy geogrid layers were introduced, with 0.10 kN/m tensile strength. Table 1 summarizes all the soil parameters used for the global stability and FEM analyses, while Table 2 shows the geogrids parameters, where:
- RFid   = reduction factor for installation damage;
- RFd   = reduction factor for chemical/biological damage;
- RFc   = reduction factor for creep;
- Cds-phi   = coefficient of soil – geogrid interaction applied to friction angle
- Cds-c   = coefficient of soil – geogrid interaction applied to cohesion

TABLE 1. Soil parameters used for the global stability analyses

SOIL TYPE	soil density γ (kN/m³)	friction angle ϕ'(deg)	cohesion c' (kPa)	E (MPa)	ν (–)	ψ (deg)
in situ soil	18.0	22.0	10.0	15.0	0.28	0
calcium stabilized embankment	19.0	22.0	20.0	50.0	0.28	0
geogrid reinforced embankment	19.0	22.0	5.0	30.0	0.25	0
calcium stab. + geogrid reinf. emb.	19.0	22.0	20.0	50.0	0.25	0

TABLE 2. Geogrids parameters used for the global stability analyses

GEOGRID	Tensile strength (kN/m)	RFid	RFd	RFc	Available strength (kN/m)	Cds-phi	Cds-c
GG 35	35,00	1.20	1.10	1.67	15.88	1.00	0.50
GG 55	55,00	1.20	1.10	1.67	24.95	1.00	0.50
GG 80	80,00	1.20	1.10	1.67	36.29	1.00	0.50
DUMMY	0,10	1.00	1.00	1.00	1.00	1.00	1.00

TABLE 3. Results of global stability analyses on reference embankments

SIDE SLOPE	TYPE	SEISMIC/ STATIC	GG	FS ROTATION	FS TRANSLATION
34°	CALCIUM	STATIC	0	1.44	1.54
34°	CALCIUM	SEISMIC	0	1.12	1.15
34°	GG	STATIC	35	1.38	1.31
34°	GG	SEISMIC	35	1.03	1.04
45°	CALCIUM	STATIC	0	1.24	1.29
45°	CALCIUM	SEISMIC	0	1.02	1.00
45°	GG	STATIC	55	1.52	1.32
45°	GG	SEISMIC	55	1.05	1.03
60°	CALCIUM	STATIC	0	0.95	1.03
60°	CALCIUM	SEISMIC	0	0.82	0.87
60°	GG	STATIC	80	1.59	1.30
60°	GG	SEISMIC	80	1.23	1.03
60°	CALCIUM + GG	STATIC	35	1.50	1.33
60°	CALCIUM + GG	SEISMIC	35	1.24	1.08

Results of global stability analyses in static and seismic conditions are summarized in Table 3, while Figures 2, 3, 4 show the critical failure surfaces for the reference embankments in seismic conditions. It can be noted that the calcium stabilized embankment shows adequate Factors of Safeties (FS) only with 34° side slopes, while with 45° side slopes in both static and seismic conditions it results just on the verge of instability; with 60° slopes it is unstable; the geogrid reinforced embankment always shows adequate FS. It is also interesting to note that the calcium stabilized and geogrid reinforced embankment with 60° side slopes, reinforced with 35 kN/m geogrids, shows almost the same FS of the 60° side slopes embankment reinforced with 80 kN/m geogrids.

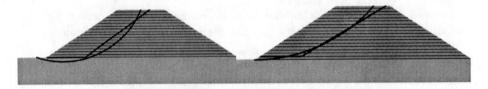

FIG. 2 – Critical failure surfaces of 2V:3H embankment in seismic conditions: left the calcium stabilized one; right the geogrid reinforced one

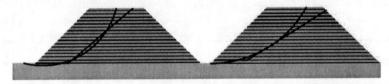

FIG. 3 – Critical failure surfaces of 1V:1H embankment in seismic conditions: left the calcium stabilized one; right the geogrid reinforced one

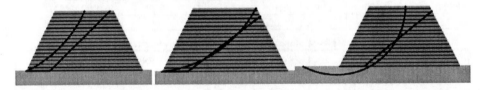

FIG. 4 – Critical failure surfaces of 2V:1H embankment in seismic conditions: left the calcium stabilized one; center the geogrid reinforced one; right the calcium stabilized and geogrid reinforced one

The reference embankments have been analyzed also by finite elements modeling, using the well known Plaxis software [2]. Soil parameters are listed in Table 2. All soils have been modeled with Mohr – Coulomb failure criterion. Geogrids have been characterized by the secant modulus at 2 % strain: 500 kN/m, 750 kN/m, 1,000 kN/m modulus have been applied to 35 kN/m, 55 kN/m, 80 kN/m geogrids. For seismic conditions a harmonic wave has been applied to the bottom boundary, with frequency f = 2 Hz and duration equal to 30 s, which applies the acceleration $a_g = 0.15$ g.

Results of FEM analyses are summarized in Table 4, in terms of maximum total displacement. The displacement in seismic condition is the result only of the harmonic wave, hence such seismic displacement shall be added to the static one to get the total displacement in the modeled seismic conditions.

TABLE 4. Results of FEM analyses on reference embankments

SIDE SLOPE	TYPE	Geogrid tensile strength (kN/m)	total displacement at end of construction (mm)	total displacement at surcharge application (mm)	Seismic displacement (mm)
34°	unstabilized unreinforced	-	FAILURE	FAILURE	-
34°	calcium stabilized - static	-	194,39	214,57	103,11
34°	geogrid reinforced	35 kN/m	218,03	245,36	259,24
45°	calcium stabilized	-	181,72	FAILURE	-
45°	geogrid reinforced	55 kN/m	241,95	287,94	166,30
45°	calcium + geogrids	55 kN/m	181,44	203,97	183,58
60°	calcium stabilized	-	181,53	FAILURE	-
60°	geogrid reinforced	80 kN/m	269,55	324,63	153,96
60°	calcium + geogrids	35 kN/m	178,03	207,52	176,79

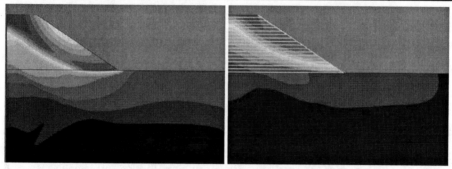

FIG. 5 – FEM displacements of the calcium stabilized (left) and the 35 kN/m geogrid reinforced (right) 2V:3H embankment in seismic conditions

FIG. 6 – FEM displacements of the 80 kN/m geogrid reinforced (left) and the calcium stabilized + 35 kN/m geogrid reinforced (right) 2V:1H embankment in seismic conditions

It can be noted that the 34° slopes unstabilized embankment undergoes failure even at end of construction, while the 45° and 63° slopes calcium stabilized embankments

731

undergo failure when the surcharge is applied. Such results are in very good agreement with the global stability analyses results, listed in Table 3. Figure 5 shows the total displacements pattern for the 2V:3H (34°) embankment in seismic condition: the critical surfaces are in excellent agreement with the ones from global stability analyses shown in Fig. 3. Fig. 6 shows the total displacements pattern for of the 80 kN/m geogrid reinforced and the calcium stabilized + 35 kN/m geogrid reinforced 2V:1H (63°) embankment in seismic conditions: again the critical surfaces are in excellent agreement with the ones from global stability analyses shown in Fig. 4.

It is interesting to note that in general the total displacement is lower for the calcium stabilized embankments than for the geogrid reinforced ones. Taking into account also the results of the global stability analyses, it seems that the 60° slopes calcium stabilized and geogrid reinforced embankment (35 kN/m geogrids) is a better technical option than the 60° slopes with only geogrid reinforcement (80 kN/m geogrids). Anyway, considering also the environmental problems associated with the dust and potential pollutants migration produced by calcium, the geogrid reinforcement alone shall be the preferred solution in many cases.

CASE HISTORIES

The project of the Provincial Road Ex SS 277 "Trasversale Alta Basentana – Bradanica", close to the town of Grassano (Matera Province) in Southern Italy, included tall geogrid reinforced embankments, using the wrap-around technique with sacrificial steel mesh formworks and vegetated facing. Geogrid reinforced embankments have a total length of 840 m, height between 2.10 m and 9.30 m, for a total of almost 8.000 m^2 face in vertical projection. The cross- section includes the 65° geogrid reinforced slopes, on both sides of embankments, a 2.0 m wide horizontal berm at crest, and on top a 5.0 m high unreinforced embankment with 2V:3H (34°) side slopes, which carries the road structure, providing 20 kPa uniform surcharge. All embankments had to be built with the locally available soil, that is silt and clay with variable sand content. The project is located in a highly seismic area: for the tallest embankments the design acceleration was $a_g = 0,156$ g. As shown in Fig. 7, finally the embankments were designed with a calcium stabilized and geogrid reinforced lower body, while the top unreinforced embankment is made up of compacted silty sand. Table 5 shows the soil parameters used for design. Fig. 8 shows the results of the global stability analyses in seismic conditions for the tallest embankment. Figure 9 shows pictures of the Bradanica highway embankments during construction. It is easy to note that the soil parameters, embankment geometry, and seismic accelerations are closely related with the values used in the analyses of reference structures. For the Bradanica highway project the calcium stabilized and geogrid reinforced embankments proved to be the best solution both in technical and economical terms.

TABLE 5. Soil parameters used for the global stability analyses of the Bradanica highway project

SOIL TYPE	γ (kN/m³)	φ'(deg)	c' (kPa)
in situ soil (sandy silt and clay)	20.0	25.0	15.0
calcium stabilized embankment	20.0	25.0	30.0
top unreinforced embankment	20.0	25.0	5.0

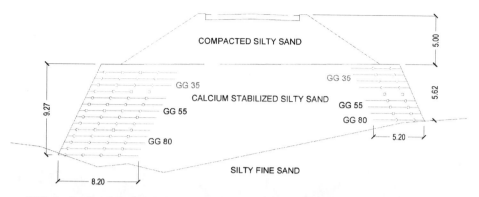

COMPACTED SILTY SAND

5.00

GG 35

GG 35

CALCIUM STABILIZED SILTY SAND

GG 55

GG 55

GG 80

5.62

GG 80

5.20

9.27

SILTY FINE SAND

8.20

FIG. 7 – Designed cross-section of the tallest embankment of the Bradanica highway project

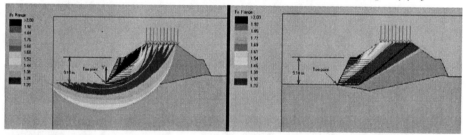

FIG. 8 – Results of the global stability analyses in seismic conditions for the embankment in Fig. 7.

FIG. 9 – Pictures of the Bradanica highway embankments during construction

The Bologna – Firenze highway, in Central Italy, is under reconstruction and many tunnels have to be excavated in the Apennines mountains, producing huge amount of debris. In the "Fienile" location outside the town of Barberino del Mugello, a 55.000 m^2 depression in hilly area was selected for dumping part of the debris, thus forming a 46 m high hill. The in.situ soil (silt and clay) is sloping downward and the debris show very low friction angle ($\phi = 20°$). Hence a dike, made up of the same debris, had to be designed to stabilize the toe. Barberino is in seismic area with design acceleration $a_g = 0,156$ g. Both calcium stabilization and geogrid reinforcement were considered for the dike: in this case environmental consideration of the effects of calcium powder forced

733

to select only geogrid reinforcement. Fig. 10 shows the cross section, the global stability analyses in seismic conditions and pictures taken during construction of the dike.

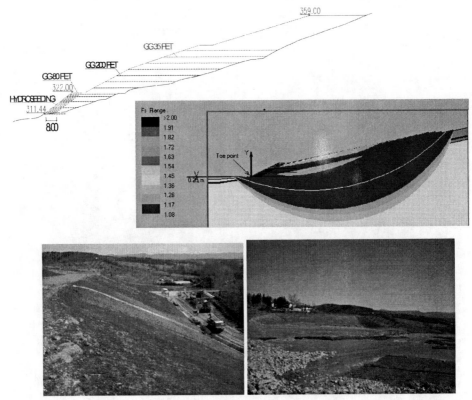

FIG. 10 – Design and construction of the dike for debris stabilization in Barberino del Mugello

CONCLUSIONS

From numerical analyses it seems that calcium stabilized and geogrid reinforced embankments are a good technical option when only silt and clay soils are available. Anyway, considering also the environmental problems associated with the dust and potential pollutants migration produced by calcium, geogrid reinforcement embankments shall be the preferred solution in many cases. The suitability of both techniques in seismic areas is demonstrated by complex projects already built in Italy.

REFERENCES

1. ReSSA - Reinforced Slope Stability Analysis. Version 2.0. Adama Engineering Inc. Newark, Delaware, USA.
2. Plaxis – Finite Element Code for Soil and Rock Analyses. 2D Version 8.2. Plaxis bv. Delft, Netherlands.

CODES AND GUIDELINES

Seismic Safety Of Simple Masonry Buildings

Mariateresa Guadagnuolo and Giuseppe Faella

Dipartimento di Cultura del Progetto, Seconda Università di Napoli
Abbazia di S. Lorenzo ad Septimum, 81031, Aversa (CE)
Tel: 081.8120156 – Fax: 081.8149266 - E-mail: m.guadagnuolo@unina2.it

Abstract. Several masonry buildings comply with the rules for simple buildings provided by seismic codes. For these buildings explicit safety verifications are not compulsory if specific code rules are fulfilled. In fact it is assumed that their fulfilment ensures a suitable seismic behaviour of buildings and thus adequate safety under earthquakes. Italian and European seismic codes differ in the requirements for simple masonry buildings, mostly concerning the building typology, the building geometry and the acceleration at site. Obviously, a wide percentage of buildings assumed simple by codes should satisfy the numerical safety verification, so that no confusion and uncertainty have to be given rise to designers who must use the codes. This paper aims at evaluating the seismic response of some simple unreinforced masonry buildings that comply with the provisions of the new Italian seismic code. Two-story buildings, having different geometry, are analysed and results from nonlinear static analyses performed by varying the acceleration at site are presented and discussed. Indications on the congruence between code rules and results of numerical analyses performed according to the code itself are supplied and, in this context, the obtained result can provide a contribution for improving the seismic code requirements.

Keywords: Simple masonry buildings, Seismic response, Nonlinear static analysis, Code provisions.

INTRODUCTION

Several seismic codes devote a section to simple masonry buildings, for which the safety against collapse is deemed to be verified without explicit safety verification if buildings comply with some code provisions and rules. In fact, it is assumed that the respect of these requirements assures a suitable seismic behaviour, and thus an adequate safety. Such an exemption is usually applicable both to the new buildings and to the existing ones, if after possible retrofit they respect the code provisions.

Eurocode 8 (EC8) [1] limits the exemption of the explicit safety verification to the only ordinary buildings belonging to lower importance categories, located in zones of low and medium seismicity and satisfying the provisions and the rules specified in detail in the Section 9.7. Therefore, the code subordinates the exception to the building use, excluding both the buildings whose integrity during earthquakes is of vital importance (e.g. hospitals and fire stations) and the buildings whose seismic resistance is important in consideration of the consequences associated with collapses (e.g. schools and assembly halls). Furthermore, the acceleration at site must be lower than $0.15 \cdot k \cdot g$, where k is a corrective factor depending on the shear walls length.

The recent Italian National Building Technical Code (NTC_2008) [2] considers "simple" the buildings that comply both with the general criteria for regularity in plan

CP1020, *2008 Seismic Engineering Conference Commemorating the 1908 Messina and Reggio Calabria Earthquake,*
edited by A. Santini and N. Moraci
© 2008 American Institute of Physics 978-0-7354-0542-4/08/$23.00

and in elevation and with some specific provisions in terms of building geometry, masonry stress and constructive rules, detailed in the Section 7.8.1.9 for the new buildings, and integrated in the Section C8.7.1.7 of the related Instructions for the existing ones. No limitation dependent on the building importance category is introduced, whereas the building must not be located in zone 1, even though the acceleration at site is not limited.

The specific provisions provided by the two above seismic codes are rather similar, but EC8 is more demanding for what concerns the shear walls length. In fact, a minimum of two walls in two orthogonal directions, each having length greater than 30% of the building length in the direction under consideration, is required in EC8 (except in cases of low seismicity), whereas in NTC_2008 the required wall length (greater than 50% of the building dimension) may be provided by the cumulative length of the piers separated by openings.

With regard to the exemption of the seismic verification it can be noticed that the code requirements that define the simple buildings must be absolutely suitable for structures of smaller importance and complexity also. Specifically, it needs a substantial certainty that the respect of code provisions and rules ensures a suitable building safety against earthquakes, without yielding confusion in the designers that employ the code. This means, for instance, that a large percentage of the buildings held simple from the seismic code, if subjected to numerical seismic verification, must provide a result congruent with the assumption. The above considerations assume special importance by considering that a great share of the Italian one-story and two-story masonry buildings are "simple" (more than 50% according to recent ratings).

Moreover, several code provisions drafted in Italy after the main earthquakes of 20th century advised numerous limitations to the unreinforced masonry buildings to be retrofitted or built in seismic zones. As an example, the Royal Decree [3] issued after the Reggio and Messina earthquake of december 1908 restricted the building total height, the number of storeys, the storey height, the masonry typology, the wall thickness and distance, the floor type, the distance of openings from the building corners, the dead load of roofs.

This paper focuses the attention on the seismic response of some unreinforced masonry structures that comply with the requirements provided by the NTC_2008 for the simple buildings. The buildings have two stories and the same plan configuration, but are different in wall distance, pier width and opening geometry. Nonlinear static analyses are carried out through masonry-type frame models and safety verifications according to the code provisions are performed.

The simplified modelling adopted doesn't enable an in-depth evaluation of the building seismic response and thus of the code provisions reliability, which need analyses based on more sophisticated modelling techniques. Nevertheless, the masonry-type frame model being the most used modelling in the everyday practice, the achieved results provide a clear indication on the congruence between code rules and numerical results of analyses performed according to the procedure advised by code itself. Besides, as far as the procedure is approximate and not quantitatively correct, broadly fulfilled safety verifications as well as significantly not satisfied verifications are inevitably representative of the actual seismic response of the buildings examined, and, within these limits, can provide a contribution to evolution and improvement of code provisions.

BUILDINGS ANALYSED

The analyses have been carried out with reference to two-story existing unreinforced tuff masonry buildings, frequently found in seismic areas of Southern Italy. The buildings have the same plan configuration, illustrated in Figure 1.

The floor height is assumed equal to 3.5 m, while the distance between walls, the pier width and the opening geometry are varied in order to analyze buildings differently complying with the code provisions on simple buildings. Namely, the distance L_2 between two longitudinal walls is assumed equal to 6 m and to 7 m, while L_1 is supposed equal to 4.5 m. The distances L_3 and L_4 are equal to 4.5 m and 4.8 m respectively. The wall thickness t is constant for both the stories and equal to 0.6 m.

The opening width d_o is varied between 1.2 m and 2.4 m, so that the pier width is significantly varied. Such a variation implies that the provisions regarding the cumulative length of shear walls in both the x and y directions, the percentage of the shear wall gross section on the total floor area and the mean stress in piers are differently satisfied.

Table 1 shows the initials used to identify the buildings analysed as the parameters L_2 and d_o vary. Table 1 also contains the percentage p_A of the total cross-sectional area of walls in x and y direction to the total floor area. The values of p_A range from 7.44 % to 12.16 %, showing that all the buildings widely comply with the limit values provided by NTC_2008 and EC8, independently of the acceleration at site $S \cdot a_g$.

The analyses are carried out taking into account both buildings having openings 2.5 m high (h_o) and buildings with windows 1.5 m high above parapets 1 m high. This difference leads to different pier squatness and, thus, to different seismic building response.

On the basis of experimental tests and reference values of tuff masonry, a masonry having compression strength $f_{mc} = 2$ MPa, pier masonry shear strength $f_{pv} = 0.035$ MPa,

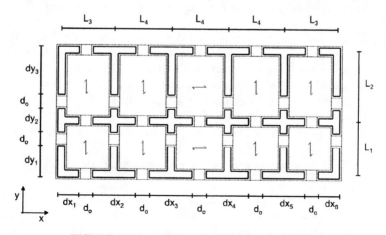

FIGURE 1. Plan configuration of buildings analysed.

739

TABLE 1. Main parameters changed in the plan configuration of buildings

Building	L_2 [m]	D_0 [m]	$p_{A,x}$ [%]	$p_{A,y}$ [%]	σ_{max} [MPa]
S2-A/120	6	1.2	12.16	11.76	0.19
S2-A/180	6	1.8	10.14	10.14	0.20
S2-A/240	6	2.4	8.04	8.51	0.22
S2-B/120	7	1.2	11.16	12.02	0.19
S2-B/180	7	1.8	9.29	10.54	0.20
S2-B/240	7	2.4	7.44	9.04	0.22

spandrel masonry shear strength $f_{sv} = 0.025$ MPa, Young modulus $E_m = 1000$ MPa and tangent modulus $G_m = 350$ MPa has been taken into account (the masonry is identified by M1). In order to evaluate the influence of masonry characteristics, a masonry having strengths equal to $f_{mc} = 3$ MPa, $f_{pv} = 0.050$ MPa, $f_{sv} = 0.037$ MPa, Young modulus $E_m = 1500$ MPa and tangent modulus $G_m = 450$ MPa (identified by M2) has also been considered. Therefore, masonries characterized by low strengths and moduli are taken into account, since they are frequently found in tuff masonry buildings of Southern Italy. According to the NTC_2008 provisions, the above strength values have not been reduced in performing the nonlinear static analyses ($\gamma_m = 1$ and FC = 1). The weight per unit volume has been assumed equal to 17 kN/m^3.

In addition to the self-weight of masonry, extra loads are considered acting at the floor levels of the buildings: specifically, dead loads equal to 5 kN/m^2 and live load equal to 2 kN/m^2 are supposed. Tributary areas at each floor are assumed to contribute to the vertical loading of each wall according to the floor beams direction illustrated in Figure 1. In the analyses, the distributed masses are lumped at the story levels and summed to the masses associated to the floors weights and loads.

The last column of Table 1 contains the maximum value of the average stress σ computed according to the NTC_2008 provisions at the first floor as the ratio of the total vertical load to the total cross-sectional area of walls: the values show that the buildings S2-A/240 and S2-B/240 hardly comply with the code limit value ($\sigma_{max} = 0.25$ MPa) for the masonry M1, whereas the margin is larger for the masonry M2 ($\sigma_{max} = 0.37$ MPa).

The building safety verification is carried out in terms of displacements by the capacity spectrum method as required by NTC_2008. The displacement demand is evaluated assuming the response spectrum for soil B/C (coefficient S = 1.25) and a peak ground acceleration a_g of 0.15 g, 0.20 g and 0.25 g. Therefore, the acceleration at site S·a_g is equal to 0.19 g, 0.25 g and 0.31 g respectively. The first value fulfils the requirements both of NTC_2008 and of EC8 (in the cases examined the corrective factor k is equal to 1) whereas the other ones are allowed by NTC_2008 only.

MODELING AND ANALYSES PERFORMED

The analyses have been performed using a masonry-type frame model. Therefore, the structure is considered as an assemblage of column-beam elements [4÷8], according to the customary routine in practical applications and to the requirements of

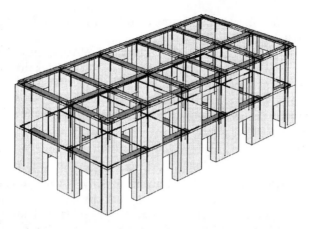

FIGURE 2. Masonry-type frame model of the building S2-A/180.

seismic codes. The masonry portions connecting piers and spandrel beams are considered greatly stiff and resistant by introducing rigid off-sets at the ends of columns and beams [4,9,10]. The beams reproducing spandrels are located at the floor level and not at the spandrel barycentric axis since it was demonstrated that this assumption does not notably influence the nonlinear response of buildings [11]. Under the hypothesis that during building seismic response no disconnection takes place among orthogonal walls, appropriate rigid offsets are located at the floor level in order to reproduce the good arrangement in walls connection and to impose the continuity of displacements. In addition, the connection between the floors and walls is adequately provided by reinforced concrete ring beams. The floors are assumed to be stiff in its own plane, and thus to provide an effective diaphragm action. Figure 2 shows the masonry-type frame model of the building S2-A/180.

Piers have an elastic-perfectly-plastic behaviour and their response is based on the plastic hinge concept. Hinges are located at both ends of each column and have strength threshold values deriving from both flexural and shear failure mechanisms. The failure criteria taken into account in the new Italian seismic code are considered. Accordingly, values of drift equal to 0.6% of the deformable pier height h_p and to 0.4% of h_p are assumed for reproducing flexural and shear pier collapse respectively.

Likewise to the pier elements, an elastic-plastic behaviour is assumed for the spandrel beams, but a lower shear strength is assumed, as previously reported. Furthermore, the spandrel collapse is defined through a pre-fixed ductility (assumed equal to 2) due to the lack of experimental researches on their nonlinear behaviour.

The loads are imposed onto the structure in a two-step sequence. Firstly, the vertical loads are applied and subsequently the lateral loads are monotonically increased. An invariant height-wise lateral load distribution proportional to the floor tributary masses is used to perform the pushover analysis. The mass centre of the top story is assumed as control point. No additional accidental eccentricity is taken into account in performing the pushover, that is the lateral forces are applied at the mass centre of each floor. The horizontal components of the seismic action are not considered as acting simultaneously since the buildings satisfy the regularity criteria in plan and the masonry walls are the only horizontal load resisting components.

Therefore, the seismic forces are assumed to act separately along the two main orthogonal horizontal axes of the structure.

Each analysis is performed until the building lateral capacity is reduced of 20% with respect to the maximum strength, due to the loss of contribution of the piers that achieve the limit failure displacement, or until any further increment in lateral load is impossible. Quite obviously, possible softening branches of the structure capacity curve are not captured due to the adopted force-controlled procedure.

RESULTS

Figure 3 shows the ratio η of displacement capacity to demand computed for the buildings S2-A in both the x and the y direction as the opening width d_o varies. The buildings have openings 2.5 m high and the results obtained for both the masonry M1 and M2 are plotted. The values concern the safety verification performed assuming the structure in seismic zones with acceleration at site $S \cdot a_g$ equal to 0.25 g.

Assuming the masonry M1, the values of η show that in the buildings analysed the safety verification in terms of displacement should not be assumed satisfied independently of the opening width. Obviously, the better masonry leads to larger η values and wider openings to be allowed: specifically, in the cases examined the masonry M2 leads to fulfilled safety verifications for all the buildings, except the building S2-A/240 loaded in the x-direction only. Small differences are found by varying the seismic input direction.

Figure 4 contains the η ratios computed for the buildings S2-A assuming the masonry M2 and varying the acceleration at site $S \cdot a_g$ from 0.19 g to 0.31 g. The figure shows that for the lowest considered value of $S \cdot a_g$ (0.19 g), all the buildings are verified independently of the opening width, while for $S \cdot a_g$ equal to 0.31 g the safety verification is regularly not satisfied.

The results of Figures 3 and 4 clearly show that if the acceleration at site $S \cdot a_g$ is equal or larger than 0.25 g the safety verification in terms of displacement is not verified. Therefore, it seems more appropriate, and safe, the EC8 assumption that in

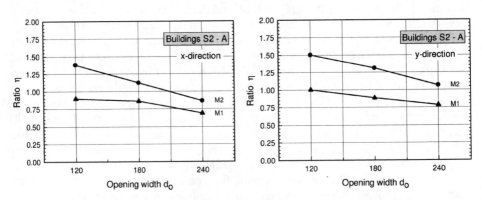

FIGURE 3. Ratio η of displacement capacity to demand for buildings S2-A
[Opening height h_o = 2.5 m - Acceleration at site $S \cdot a_g$ = 0.25 g]

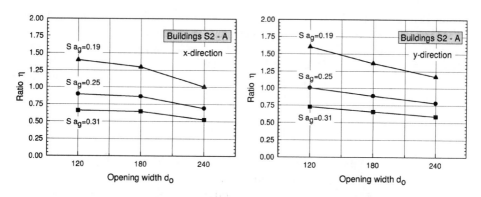

FIGURE 4. Ratio η of displacement capacity to demand for buildings S2-A
[Opening height h_o = 2.5 m – Masonry M1]

these cases does not allow the buildings to be included within the simple ones, requiring an explicit safety verification. Besides, the percentage p_A of the minimum wall cross-sectional area to the total floor area (depending on d_o in the cases examined) seems to be less important, being the increment in p_A of about 50% not sufficient to compensate for the increase in acceleration at site.

It can also be highlighted that the building lateral strength computed is about the double of the design base shear if the masonry has the characteristic M2 and the acceleration at site is lesser than 0.20 g. On the contrary, for masonry M1 and $S·a_g$ larger than 0.30 g the building lateral strength and the design base shear are comparable and thus the building seems not to have suitable overstrength. Therefore, these results confirm the previous ones and specifically the more suitability of the EC8 code provisions, at least for buildings built using masonry of low mechanical characteristics, such as the ones examined.

Figure 5 shows the η ratios evaluated for the buildings S2-B in both the x and the y direction as the opening width d_o and the masonry characteristics vary. These buildings also have openings 2.5 m high and the values concern the safety verification performed assuming $S·a_g$ equal to 0.25 g. Comparing the curves of Figure 5 with the

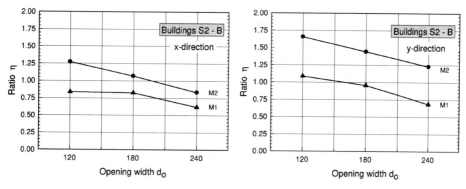

FIGURE 5. Ratio η of displacement capacity to demand for buildings S2-B
[Opening height h_o = 2.5 m - Acceleration at site $S·a_g$ = 0.25 g]

743

ones of Figure 3 it can be stated that, in the cases examined, the distance L_2 between the walls (ranging from 6 m to 7 m) does not assume a noteworthy role.

Finally, the opening height h_o also does not significantly influence the previous remarks, since smaller heights equal to 1.5 m lead to slight increment in lateral capacity counterbalanced by smaller building displacement capacity and thus safety verification even more not satisfied.

CONCLUSIONS

The seismic response of some unreinforced tuff masonry buildings that comply with the requirements provided by the new Italian seismic code for the simple buildings is presented in the paper. The analyses are mainly aimed at evaluating the congruence between code rules and numerical results of analyses performed according to the procedure advised by the code itself, as well as at comparing the Italian code provisions with the ones provided by Eurocode 8. The buildings analysed have two stories and the same plan configuration, but are different in wall distance, pier width and opening geometry. Nonlinear static analyses are carried out by varying the masonry characteristics and the acceleration at site also. Safety verifications in terms of displacements according to the code provisions are performed. With reference to the buildings examined, the results show some inconsistencies in the Italian seismic code framework whereas the Eurocode 8 assumptions seem to be more appropriate and safe.

REFERENCES

1. CEN European Committee for Standardization, *Eurocode 8: Design of structures for earthquake resistance* - prEN 1998-1. FINAL DRAFT, December (2003).
2. Ministero delle Infrastrutture e dei Trasporti, *Norme Tecniche per le Costruzioni*, D.M. 14.01.2008, Official Bulletin n. 29, February 4 (2008). (in Italian)
3. Regio Decreto 18 aprile 1909 n. 193, *Norme tecniche ed igieniche obbligatorie per le riparazioni ricostruzioni e nuove costruzioni degli edifici pubblici e privati nei luoghi colpiti dal terremoto del 28 dicembre 1908 e da altri precedenti elencati nel R.D. 15 aprile 1909*, Official Bulletin n. 95, Aprile 22 (1909). (in Italian)
4. P.B. Lourenço, "Computations on historic masonry structures" in *Progress in Structural Engineering and Materials*, 4, pp: 301-319 (2002).
5. G. Magenes, "Masonry building design in seismic areas: recent experiences and prospects from a european standpoint" in Proc. *First European Conference on Earthquake Engineering and Seismology* (13th ECEE and 30th General Assembly of ESC), Geneva, Switzerland, Sept. 3-8 (2006).
6. A.J. Kappos, G.G. Penelis and C.G, Drakopoulos, "Evaluation of simplified models for lateral load analysis of unreinforced masonry buildings" in *J. of Struct. Eng.*, 128 (7), pp: 890-897 (2002).
7. T. Salonikios, C. Karakostas, V. Lekidis and A. Anthoine , "Comparative inelastic pushover analysis of masonry frames" in *Engineering Structures*, 25, pp:1515-1523 (2003).
8. A. Penna, S. Resemini, A. Falasco and S. Lagomarsino, "Non linear seismic analysis of masonry structures" in *13th World Conference on Earthquake Engineering*, Vancouver, Canada, Aug. (2004).
9. G. Magenes, "A Method for Pushover Analysis in Seismic assessment of Masonry Buildings" in *12th World Conference on Earthquake Engineering*, Auckland, New Zealand, (2000).
10. G. Magenes, D. Bolognini and C. Braggio, *Metodi semplificati per l'analisi sismica non lineare di edifici in muratura*, CNR- Gruppo Nazionale per la Difesa dai Terremoti, Roma, (2000). (in Italian)
11. M. Guadagnuolo, A. Giordano and G. Faella, "Comparative seismic response of low-rise unreinforced tuff masonry buildings", submitted to *Engineering Structures*, Elsevier, (2007).

Experimental Data and Guidelines for Stone Masonry Structures: a Comparative Review

Alessandra Romano

Department of Structural Engineering, University of Naples Federico II, Italy

Abstract. Indications about the mechanical properties of masonry structures contained in many Italian guidelines are based on different aspects both concerning the constituents material (units and mortar) and their assemblage. Indeed, the documents define different classes (depending on the type, the arrangement and the unit properties) and suggest the use of amplification coefficients for taking into account the influence of different factors on the mechanical properties of masonry. In this paper, a critical discussion about the indications proposed by some Italian guidelines for stone masonry structures is presented. Particular attention is addressed to the classification criteria of the masonry type and to the choice of the amplification factors. Finally, a detailed analytical comparison among the suggested values and some inherent experimental data recently published is performed.

Keywords: Stone Masonry, Experimental Data, Italian Guidelines, Mechanical Properties.

INTRODUCTION

Structural analysis on the great historical and architectural heritage of existing masonry structures requires essentially the knowledge of the mechanical properties of the "masonry material". Generally, for this type of constructions, it is rather difficult to deduce these parameters through in-situ or laboratory tests. For these reasons, several Italian guidelines and standard codes furnish indications about the main mechanical and elastic properties of different kinds of masonry and some studies have been performed in the last years in order to improve the proposed data.

Starting from the guideline "Circolare n. 21745" issued in 1981 [1], resumed in the "OPCM 3431" of 2005 [2] until the last version of the code "NTC" of 2008 [3] currently enforced, the parameters of compression strength, shear strength, normal modulus and tangential modulus are listed taking into account the masonry type, the arrangement and the unit. Naturally, the above values changes times to times. In this paper, a critical examination of the parameters reported in different codes and guidelines has been carried out also comparing the examined values with some experimental ones deduced from literature.

MASONRY CHARACTERISTICS IN THE GUIDELINES

A table in [1] lists some kinds of masonry indicating the compressive strength σ_k, the shear strength τ_k (both expressed in t/m^2) and defining the shear modulus G = 1100

CP1020, *2008 Seismic Engineering Conference Commemorating the 1908 Messina and Reggio Calabria Earthquake,*
edited by A. Santini and N. Moraci

τ_k and the normal modulus E = 6G. The following types of un-reinforced and un-cracked masonry are contemplated: 1) Rubble masonry; 2) Faced masonry; 3) Rough organized masonry; 4) Tuff masonry. In the same document, no amplifying coefficients are mentioned in case of good quality of masonry with the exception of stripes allowing a 30% increase of the four parameters and masonry retrofitted through mortar injections or reinforced plaster.

Table 11.D.1 in the Annex 11.D [2] specifies the same parameters (indicated as f_m and τ_0 expressed in N/cm^2; E and G expressed in MPa). The numerical values refer to the condition of: i) mortar with low strength characteristics; ii) lack of stripes; iii) lack of connection between the walls; iv) un-reinforced masonry. Masonry types are the same of [1] adding a fifth class of squared masonry. In this case, contrarily to [1], a vast list of details (a. good mortar, b. stripes, c. transversal connection; d. mortar injections; e. reinforced plaster) is reported in order to increase the proposed values through the use of some amplification factors.

Table 8.B.1 [3] recalls the same classes, symbols and unit of measure of [2] modifying, however, some numerical values (especially the compressive strength and the shear modulus). In case of good quality of masonry, in addition to the details reported in [2], other two options are contemplated: f) thin joints; g) bad or wide core in faced masonry.

The values of the compressive strength (indicated as f_m), the shear strength (uniformed into f_{to}), the elastic modulus (E) and the shear modulus (G) suggested by [1], [2] and [3] are reported in the next Tables 1 and 2 for stone masonry choosing MPa as unit of measure.

TABLE 1. Comparison of f_m and f_{to} for NTC – OPCM – Circ.

Masonry type	f_m [MPa]			f_{to} [MPa]		
	NTC	OPCM	Circ.	NTC	OPCM	Circ.
1. Rubble Masonry	1.0÷1.8	0.6÷0.9	0.5	0.020÷0.032	0.020÷0.032	0.020
2. Faced Masonry	2.0÷3.0	1.1÷1.6	1.5	0.035÷0.051	0.035÷0.051	0.039
3. Rough Organized Masonry	2.6÷3.8	1.5÷2.0	2.0	0.056÷0.074	0.056÷0.074	0.069
4. Tuff Masonry	1.4÷2.4	0.8÷1.2	2.5	0.028÷0.042	0.028÷0.042	0.098
5. Squared Masonry	6.0÷8.0	3.0÷4.0	--	0.090÷0.120	0.078÷0.098	--

TABLE 2. Comparison of E and G for NTC – OPCM – Circ.

Masonry type	E [MPa]			G [MPa]		
	NTC	OPCM	Circ.	NTC	OPCM	Circ.
1. Rubble Masonry	690÷1050	690÷1050	132	230÷350	115÷175	22
2. Faced Masonry	1020÷1440	1020÷1440	257	340÷480	170÷240	43
3. Rough organized Masonry	1500÷1980	1500÷1980	455	500÷660	250÷300	76
4. Tuff Masonry	900÷1260	900÷1260	647	300÷420	150÷210	108
5. Squared Masonry	2400÷3200	2340÷2820	--	780÷940	390÷470	--

Nevertheless, some differences characterize the indications contained in the examined codes, i): the compressive strength of [2] is similar to the one indicated in [1] with the exception of tuff masonry which values are one half of [1]. [3] reviewed completely the values assuming for all the categories nearly twice the values of [2], coming back for tuff units to the values of [1]. The pure shear strength is substantially the same for the three guidelines, although in [1] the tuff type is rather emphasized.

With reference to E, [3] is equivalent to [2], whilst [1] has values notably low. Finally, for G, [3] increases the values compared to [2] and [1] indicates values too low.

ii): Class 4, dedicated to tender stones as tuff is almost scarce since many kinds of workmanship exist into reality, from scabbled to rough to squared masonry but they are all included in a unique category, contrarily to the other types.

About the amplifying factors in [2] and [3], the following comments can be assigned:

i): Class 4 lacks of the amplifying coefficient in presence of stripes. Figure 1 shows some examples of existing tuff walls with striped bricks in three or four rows not considered in the examined guidelines. Since the tuff units are rather squared, the adopted solution seems to improve the overall compressive strength instead of the classical function of smoothing the surface.

ii): for Class 5, it is opinion of the author that the intervention with mortar injections is not suited for squared masonry due to the lack of voids among the stones.

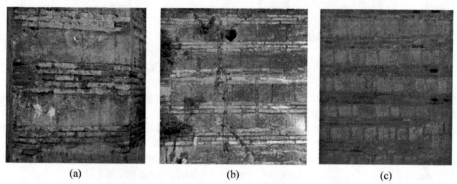

(a)   (b)   (c)

FIGURE 1. Striped Tuff Masonry: some examples.

EXPERIMENTAL DATA VS GUIDELINE SUGGESTIONS

Some experimental data on stone masonry published recently were examined in order to check the reliability of the numerical values proposed by the last enforced Italian guideline [3]. The comparison was displayed in bar charts so composed:

- several bars standing for values deduced by the consulted papers; different hatches vary the scatter from the guideline values: solid black in case of values internal or within 30% of the range; double skew striped for values within 50% of the range; simple skew striped for values within 100% of the range; solid white for values beyond 100% of the range;

- two horizontal lines, representing the minimum and the maximum value in [3];

In the following, the plots of compressive strength f_m, shear strength f_{to}, normal modulus E and tangential modulus G are reported vary the masonry class.

Type 1: Rubble Masonry

In this type the experimental tests performed on pebbles and generally irregular stones were considered. In Figure 2.a the values of f_m according to [4] within the range for generic stones and lime mortar and [5] with a scatter of about 30% for sandstones and stones in two faces, are reported. The values of the modulus of elasticity E are plot in Figure 2.b and contemplate the experience of [6] besides the previous authors. Also in this case, [5] measured values lower than the reference quantities in two cases, whilst [6] and [4] are near them. Finally, in Figure 2.c the tangential modulus G calculated by [4], [5] and [7,8] is reported. The values are almost inside the range with the exception, again, of [5]'s values probably due to the poor quality of the specimens.

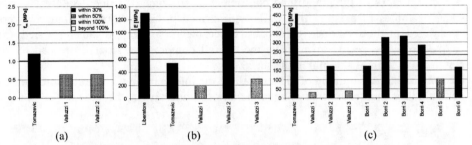

(a)   (b)   (c)

FIGURE 2. Rubble Masonry - a): f_m; b): E; c): G.

Type 2: Faced Masonry

For this type of masonry the faced travertine and limestone walls with hydraulic and lime mortar were considered. The reference authors were [9], [5] and [10]. In Figure 3 only f_m and E are represented. In both the cases, [9] and [5] almost agree with the reference values, and [10] is rather far from them due to the better quality of the specimens and the hydraulic mortar. For f_{to} no experimental data were found with the exception of [10] who determines, anyway, values much higher of [3].

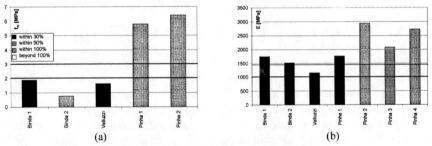

(a)   (b)

FIGURE 3. Faced Masonry - a): f_m; b): E.

Type 3: Rough Organized Masonry

Limestones and scabbled stones were grouped in this masonry type. In the determination of f_m (Fig. 4.a) the experience of [10] with Noto and Serena limestone shows values slightly greater the maximum value. Figure 4.b for E, shows [11] and [10] (with Serena stone) inside the range, whilst [10] with Noto stone is extremely beyond the range.

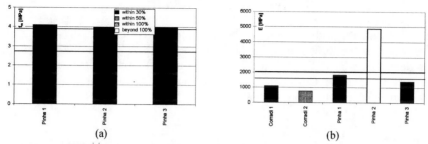

FIGURE 4. Rough Organized Masonry, Limestone - a): f_m; b): E.

In the case of scabbled masonry, no results for f_m and E were found. On the contrary, pure shear tests were conducted by [11] and [7,8]. About f_{to} (Fig. 5.a), their experience agree completely with [3]. The same can not be asserted for G (Fig. 5.b), which values are almost underneath the minimum threshold.

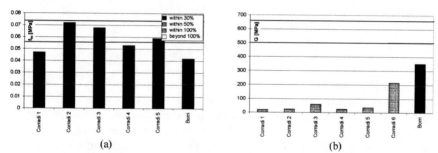

FIGURE 5. Rough Organized Masonry - Stones - a): f_m; b): E.

Type 3 + Stripes

According to [3], an increasing of 10% of the basic values may be estimated for this masonry type. The experimentation carried out by [11] and [7,8] in order to determine f_{to} and E (Figs. 6.a, 6.b) on masonry made by different kinds of units (limestones, stones and sandstones) and mortar (hydraulic and lime) coupled to bricks agree with [3]. Furthermore, the great range of variability allows to assert that the check is duly satisfied. Finally G (Fig. 6.c) presents a great range of variability and the values are considerably underneath the minimum of [3].

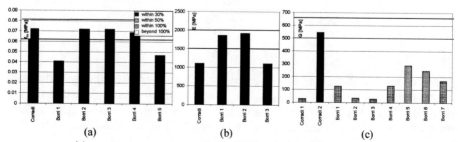

FIGURE 6. Rough Organized Masonry + Stripes - a): f_{to}; b): E; c): G.

Type 4: Tuff Masonry

In this class, all the natural easily to be cut stones were considered (tuff, lava and calcarenite). For the three quantities f_m (Fig. 7.a), E (Fig. 7.b) and G (Fig. 7.c), the values determined by [12], [13] and [14] adopting tuff stones and pozzolana mortar agree perfectly with [3]. Considering the lava stone [6], f_m is within the range whilst E is widely greater than the proposed values.

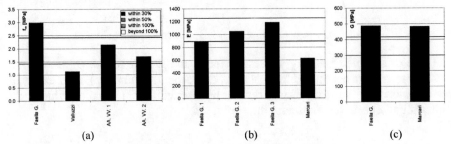

(a)   (b)   (c)

FIGURE 7. Tuff Masonry - a): f_m; b): E; c): G.

Type 4 + Good Mortar

According to [3] it is possible to increase the basic values of 50% if the binder of tender stones is cementicious mortar. Many experimental tests were performed on tuff and good mortar. In Fig. 8.a, the compression strength deduced by [15] is on the minimum threshold (vary the presence of vertical mortar joints) whilst [16], [17] and [18] are slightly upon the maximum value (for regular workmanship and one face arrangement). Also for the pure shear strength (Fig. 8.b), although in a great variability, the values are close to [3]. Finally E values (Fig. 8.c) generally agree with some exceptions of good workmanship.

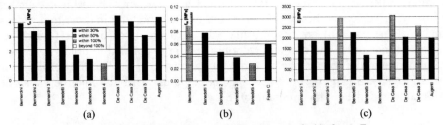

(a)   (b)   (c)

FIGURE 8. Tuff Masonry + good mortar - a): f_m; b): f_{to}; c): E.

Type 4 + inner core

The presence of an inner core for tender masonry walls decreases the overall values of 10% according to [3]. In Figure 9, f_m and E are reported. It is evident that for f_m the values of [15] are slightly underneath the minimum reference threshold due to the bad manufacturing of the specimens whilst [12] and [13] agree completely. Slightly the opposite occurs for E, although the average values generally agree.

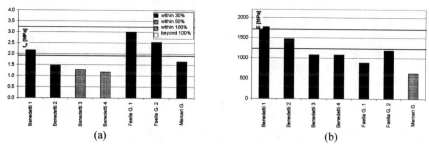

FIGURE 9. Tuff Masonry + good mortar + inner core - a): f_m; b): E.

Type 5: Squared masonry

Inside this section all the limestone masonry characterized by regular and squared manufactory were included. In Fig. 10.a, f_m is represented: [19] (white Siracusa limestone and bastard mortar) and [10] (Noto limestone) are close to the [3] range. On the contrary, f_{to} (Fig. 10.b) deduced by experiments is much higher than the ones proposed by [3]. About E, although with a great heterogeneity (Fig. 10.c), the values are close to the range in [3] with the exception of [19]. Finally, G (Fig. 10.d) shows a great disagreement between experimental data and [3].

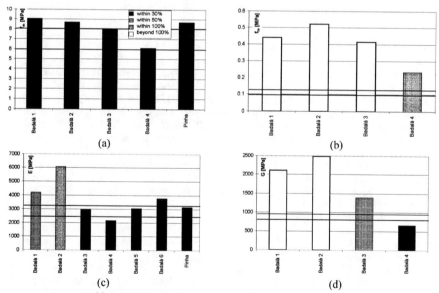

FIGURE 10. Squared Masonry - a): f_m; b): f_{to}; c): E; d): G.

CONCLUSIONS

In this paper a critical analysis of the foremost mechanical and elastic characteristics necessary to define stone masonry behaviour was performed. Numerical values indicated in the last three Italian guidelines were considered and compared. The experimental data related to the selected categories and recently published were assessed. Scope of the paper was the comparison among the last enforced guideline and the values obtained on field. It can be asserted that many numerical values fit with the proposed references except some cases. Furthermore, some categories related to tender stones and some amplifying coefficients should be introduced or revised.

ACKNOWLEDGMENTS

The author would like to thank her friend Ernesto for his support and the helpful comments to this paper.

REFERENCES

1. Circolare M. LL. PP. n. 21745, *Istruzioni per l'applicazione della Normativa Tecnica per la riparazione ed il rafforzamento degli edifici danneggiati da sisma*, Roma, 1981 (in Italian).
2. Presidenza del Consiglio dei Ministri, O.P.C.M. n. 3431 del 3 Maggio 2005, *Ulteriori modifiche ed integrazioni all'O.P.C.M. n. 3274 del 20 marzo 2003*, Roma, 2005, (in Italian).
3. D.M. 14.01.2008, *Norme Tecniche per le Costruzioni*, Roma, 2008 (in Italian).
4. M. Tomazevic, A. Znidaric et Al., "The influence of traffic induced vibrations on seismic resistance of historic stone masonry buildings" in *Proceedings of the XII ECEE*, London, 2002 paper n. 631.
5. M.R. Valluzzi, *Consolidamento e recupero delle murature*, Faenza, Gruppo Editoriale Faenza Editrice, 2003 (in Italian).
6. D. Liberatore (a cura di), *Progetto Catania. Indagine sulla risposta sismica di due edifici in muratura* Roma, GNDT, 1998 (in Italian).
7. A. Borri, M. Corradi et Al., "Analisi sperimentali e numeriche per la valutazione della resistenza a taglio delle murature" in *Ingegneria Sismica*, Bologna: Pàtron Editore, 2004, **3** pp. 50-68 (in Italian).
8. A. Borri, M. Corradi et Al., "Nuove sperimentazioni per la valutazione della resistenza a taglio delle murature prima e dopo rinforzo" in *Proceedings of XI ANIDIS*, Genova, 2004 (in Italian).
9. L. Binda, C. Tiraboschi et Al., "On site investigation on the remains of the Cathedral of Noto" in *Construction and Building Materials*, Elsevier Science, 2003, **17** pp. 543-555.
10. J. Pinha-henriques, P.B. Lourenço et Al., "Testing and modelling of multiple-leaf masonry walls under shear and compression" in *Proceedings of the IV SAHC*, Padova, 2004, pp. 299-310.
11. M. Corradi, A. Borri et Al., "Experimental study on the determination of strength of masonry walls" in *Construction and Building Materials*, Elsevier Science, 2003, **17** pp. 325-337.
12. G. Faella, G. Manfredi at Al., "Experimental evaluation of mechanical properties of old tuff masonry subjected to axial loading" in *Proceedings of the IX IB²MaC*, Berlin, 1991, pp. 172-179.
13. G. Marcari, "Rinforzo sismico di murature di tufo con materiali fibrorinforzati", Ph.D. Thesis, University of Naples Federico II, 2005 (in Italian).
14. AA.VV., "Indagini propedeutiche alla progettazione definitiva ed esecutiva dei lavori di adeguamento di alcuni spazi del Complesso di S. Patrizia", Private communication, 2005 (in Italian).
15. D. Benedetti and G. Benzoni, "Esperienze a taglio su pannelli in tufo" in *Ingegneria Sismica*, Bologna: Pàtron Editore, 1985, **3** pp. 15-24 (in Italian).
16. A. Bernardini, R. Mattone, et Al., "Determinazione delle capacità portanti per carichi verticali e laterali di pannelli murari in tufo" in *Proceedings of the II ASS.I.R.CO.*, Ferrara, 1984, pp. 345-360 (in Italian).
17. G. De Casa and G. Giglio, "Contributo alla conoscenza del comportamento delle murature in blocchi di tufo vulcanico" in *Proceedings of the IX IB²MaC*, Berlin, 1991, pp. 141-148 (in Italian).
18. N. Augenti and A. Romano, "Preliminary experimental results for advanced modelling of tuff masonry structures", in *Proceedings of the VI SAHC*, Bath, 2008.
19. A. Badalà, M. Cuomo, "Determinazione delle proprietà meccaniche della muratura come solido composto. Risultati sperimentali su muretti in scala 1:4 di quattro diverse tipologie" *in Proceedings of La meccanica delle murature tra teoria e progetto*, Messina, 1996, pp. 85-94 (in Italian).

Comparative Study on Code-based Linear Evaluation of an Existing RC Building Damaged during 1998 Adana-Ceyhan Earthquake

A. Emre Toprak[a], F. Gülten Gülay[a] and Peter Ruge[b]

[a]Istanbul Technical University, Department of Civil Eng., Maslak, Istanbul, 34469, Turkey
[b]Technische Universitaet Dresden, Institut für Statik und Dynamik der Tragwerke,
Dresden, 01062, Germany

Abstract. Determination of seismic performance of existing buildings has become one of the key concepts in structural analysis topics after recent earthquakes (i.e. Izmit and Duzce Earthquakes in 1999, Kobe Earthquake in 1995 and Northridge Earthquake in 1994). Considering the need for precise assessment tools to determine seismic performance level, most of earthquake hazardous countries try to include performance based assessment in their seismic codes. Recently, Turkish Earthquake Code 2007 (TEC'07), which was put into effect in March 2007, also introduced linear and non-linear assessment procedures to be applied prior to building retrofitting. In this paper, a comparative study is performed on the code-based seismic assessment of RC buildings with linear static methods of analysis, selecting an existing RC building. The basic principles dealing the procedure of seismic performance evaluations for existing RC buildings according to Eurocode 8 and TEC'07 will be outlined and compared. Then the procedure is applied to a real case study building is selected which is exposed to 1998 Adana- Ceyhan Earthquake in Turkey, the seismic action of Ms =6.3 with a maximum ground acceleration of 0.28g It is a six- storey RC residential building with a total of 14.65 m height, composed of orthogonal frames, symmetrical in y direction and it does not have any significant structural irregularities. The rectangular shaped planar dimensions are 16.40m x 7.80m= 127.90 m^2 with five spans in x and two spans in y directions. It was reported that the building had been moderately damaged during the 1998 earthquake and retrofitting process was suggested by the authorities with adding shear-walls to the system. The computations show that the performing methods of analysis with linear approaches using either Eurocode 8 or TEC'07 independently produce similar performance levels of collapse for the critical storey of the structure. The computed base shear value according to Eurocode is much higher than the requirements of the Turkish Earthquake Code while the selected ground conditions represent the same characteristics. The main reason is that the ordinate of the horizontal elastic response spectrum for Eurocode 8 is increased by the soil factor. In TEC'07 force-based linear assessment, the seismic demands at cross-sections are to be checked with residual moment capacities; however, the chord rotations of primary ductile elements must be checked for Eurocode safety verifications. On the other hand, the demand curvatures from linear methods of analysis of Eurocode 8 together with TEC'07 are almost similar.

Keywords: Eurocode 8; Turkish Earthquake Code; Seismic Assessment, Linear Static Analysis.

CP1020, *2008 Seismic Engineering Conference Commemorating the 1908 Messina and Reggio Calabria Earthquake*,
edited by A. Santini and N. Moraci
© 2008 American Institute of Physics 978-0-7354-0542-4/08/$23.00

INTRODUCTION

Performance based design and assessment in structural engineering is becoming more important in the past several years. The decision of the analysis method for performance-based assessment is being a new topic and linear elastic methods of analysis have been used for a long time.

Structural assessment and design concept with the principle of performance criteria based on the displacement and strain are especially put forward and developed for the realistic safety and rehabilitation of structures in the United States' earthquake regions.

The damage caused by the 1989 Loma Prieta and 1994 Northridge, in California – USA, made it possible to reconsider not only the current performance criteria regarding the strength of materials but also add more realistic criteria based on displacement and strain. With this concept, Guidelines and Commentary for Seismic Rehabilitation of Buildings – the ATC 40 [1] Project by the Applied Technology Council (ATC), and NEHRP Guidelines for the Seismic Rehabilitation of Buildings – FEMA 273 [2] and FEMA 356 [3] by the Federal Emergency Management Agency (FEMA) have been developed. Later on, in order to examine the results further on, the ATC 55 and FEMA 440 [4] have been developed. Besides these organizations, different projects like Building Seismic Safety Council (BSSC), American Society of Civil Engineers (ASCE) and Earthquake Engineering Research Center of University of California at Berkeley (EERC-UCB) contributed them. With the aid of these projects and papers, the assessment of the performance the existing structures at the quake zones and the redesigning of buildings according to their earthquake performances could be possible.

On the other hand, there exist also some researches regarding the performances of structures according to Eurocode 8.3 [5] which is among the standards of the European Union. Eurocode 8 (EC8) proposes displacement-based approaches for the seismic assessment and retrofit of existing buildings.

Recent earthquakes which occurred in Turkey made it compulsory to assess the safety of structures. Thus, in addition to Turkish Earthquake Code of 1998, the new version of Turkish Earthquake Code (TEC'07) was issued in March 2007 [6] in which the assessment and rehabilitation of structures have been added. The researches state that both linear and non-linear static analysis methods under the scope of TEC'07 generally result with same performance levels. However, it is noted that linear analysis method is relatively more conservative on the basis of component performance damage level [7, 8, 9]. Numerical studies comparing FEMA 356 and TEC'07 using non-linear static analysis method shows that both codes result in almost similar damage levels on the basis of structural elements [10].

The aim of this study is to investigate the code-based procedure of seismic performance assessments of existing buildings and to determine the seismic performance levels of a case study reinforced concrete building, which represents typical existing building stock in Turkey, as well as comparing the consequences of linear static analysis procedures. according (TEC'07) and EC8.

CODE-BASED PERFORMANCE ASSESSMENT PROCEDURES

Performance Requirements

Building performance levels or limit states are chosen discrete levels of building damage under earthquake excitation.

Eurocode 8 Part 3 defines three limit states, related to structural damage:

Damage Limitation (DL): The structure is only slightly damaged with insignificant plastic deformations. Repair of structural components is not required, because their resistance capacity and stiffness are not compromised. Cracks may present on non-structural elements, but they can be economically repaired. The residual deformations are unnecessary.

Significant Damage (SD): The structure is significantly damaged and it has undergone resistance reduction. The non-structural elements are damaged, yet the partition walls are not failed. The structure consists of permanent significant drifts and generally it is not economic to repair.

Near Collapse (NC): The structure is heavily damaged; on the other hand, vertical elements are still able to carry gravity loads. Most non-structural elements are failed, and remained ones will not survive under next seismic actions, even for slight horizontal loads.

Turkish Earthquake Code 2007 defines the seismic performance as the expected structural damage under considered seismic actions. Seismic performance of a building is determined by obtaining storey-based structural member damage ratios under a linear or non-linear analysis. Member damage levels are classified as shown in Figure 1. The building performances are as in the following:

Immediate Occupancy (IO): For each main direction that seismic loads affect, at any storey at most 10% of beams can be at moderate damage level; however, the rest of the structural elements should be at slight damage level. With the condition of brittle elements to be retrofitted (strengthened), the buildings at this state are assumed to be at *Immediate Occupancy Performance Level*.

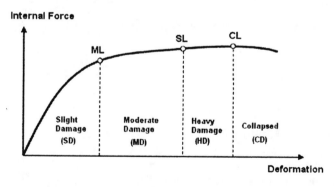

FIGURE 1. Cross-sectional Member Damage Limits (TEC'07)

Life Safety (LS): For each main direction that seismic loads affect, at any storey at most 30% of beams and some of columns can be at heavy damage level; however, shear contributions of overall columns at heavy damage must be lower than 20%. The rest of the structural elements should be at slight or moderate damage levels. With the condition of brittle elements to be retrofitted, buildings at this state are assumed to be at Life Safety Performance Level. For the validity of this performance level, the ratio between the shear force contribution of a column with moderate or higher damage level from both ends and the total shear force of the corresponding storey must be at most 30%. This ratio can be permitted up to 40% at the top storey.

Collapse Prevention (CP): For each main direction that seismic loads affect, at any storey at most 20% of beams can collapse. Rest of the structural elements should be at slight damage, moderate damage, or heavy damage levels. With the condition of brittle elements to be retrofitted, the buildings at this state are assumed to be at Collapse Prevention Performance Level. For the validity of this performance level, the ratio between the shear force contribution of a column with moderate or higher damage level from both ends and the total shear force of the corresponding storey must be at most 30%. Functionality of a building at this performance level has risks for life safety and it should be strengthened. Cost-effective analysis is also recommended for such seismic rehabilitation.

A target performance assessment objective for a given building consists of one or more performance level for given earthquake hazard level. European countries check the return periods due to the various limit states and define it in its National Annex. Recommended return periods to corresponding limit states are given in Table 1. Required performance levels according to TEC'07 to corresponding existing building types are given in Table 2.

TABLE 1. Eurocode 8 Recommended Return Periods

Limit States	Return Period	Probability of Exceedance
LS of Damage Limitation	225 years	20% / 50 years
LS of Significant Damage	475 years	10% / 50 years
LS of Near Collapse	2457 years	2% / 50 years

TABLE 2. TEC'07 Required Seismic Performance Levels

| Purpose of Occupancy | Probability of Exceedance | | |
	50% / 50 years	20% / 50 years	2% / 50 years
Operational After Earthquake	-	IO	LS
Crowded for Long-term	-	IO	LS
Crowded for Short-term	IO	LS	-
Contains Hazardous Material	-	IO	CP
Other	-	LS	-

Linear Static Analysis Procedures

There are four types of displacement-based analysis procedures described in EC8. Depending on the structural characteristics of the building, *Lateral Force Method Of Analysis or Modal Response Spectrum Analysis* may be used as linear-elastic methods. Static procedure may be used whenever participation of higher modes is negligible.

756

TABLE 3. Linear-Static Methods of Analysis Acceptance Criteria

EC8 Lateral Force Method	TEC'07 Equivalent Seismic Load Method
Structural systems must be continuous to the top	Height of the building < 25 m
Storey stiffness and mass must be constant or gradually decreasing	Number of storey < 8
Individual floor setbacks on each side < 10% of underlying storey	At each storey torsional irregularity factor in plan must be < 1.40
Unsymmetrical setbacks < 30% of base in total	
Single setbacks at lower 15% of building < 50% of base	
$T_1 < \min (4\ T_C\ ;\ 2\ \text{sec})$	
The ratio of max. to min. value of DCR over all ductile members that go inelastic < 2.50	

The load patterns, used for static analyses, are not able to represent deformed shape of the structure when higher modes are put into effect. The participation of higher modes depends generally on regularity of mass and stiffness and on the distribution of natural frequencies of the building with respect to seismic fundamental frequencies. Linear procedures (lateral force method of analysis and modal response spectrum) are applicable when the structure remains almost elastic or when expected plastic deformations are uniformly distributed all over the structure. The ratio of maximum to minimum value of demand-capacity ratio over all ductile members in a storey that go inelastic must not exceed the value of 2.50 according to EC8.

The *Equivalent Seismic Load Method* and *Mode Superposition Method* are suggested in TEC'07. The main objective of these methods is to compare demands by using unreduced elastic response spectrum with the existing capacity of elements, then to evaluate damage levels on the basis of elements with obtained demand-capacity ratios, and to determine the seismic performance level of the overall building. The conditions of using the equivalent seismic load method according to EC8 and TEC'08 is summarized in Table 3. In determination of base shear force, unreduced (elastic) response spectrum is utilized.

Distribution of the horizontal seismic forces according to EC8 *Lateral Force Method* depends on modal shape of the structure at the fundamental period. On the other hand, in *Equivalent Seismic Load Method* lateral force distribution is related to storey masses and their elevation:

$$F_{i,EC8} = F_b \cdot \frac{s_i \cdot m_i}{\sum s_i \cdot m_j} \qquad , \qquad F_{i,TEC} = (Fb - \Delta F) \cdot \frac{h_i \cdot m_i}{\sum h_j \cdot m_j} \qquad (1)$$

According to EC8, chord rotation capacity limits of ductile components are checked for safety. For *limit state of near collapse*, the demand rotation must be lower than the value of the total chord rotation capacity (elastic plus inelastic part) at ultimate θ_u. The cord rotation corresponding to significant damage θ_{SD} may be assumed to be ¾ of the ultimate chord rotation capacity. The capacity for *limit state of damage limitation* used in the verifications corresponds to the yielding bending moment under the design value of the axial load.

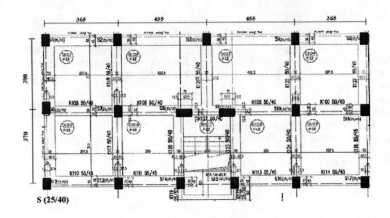

FIGURE 2. Typical Storey Plan

In TEC'07, denoting by (r), the ratio of the demand obtained from the analysis under the seismic loads, over the capacity of the same ductile element is used in order to determine the damage level of the corresponding element. Demand – capacity ratio (DCR) is obtained by dividing moments from unreduced seismic actions at element end cross-sections to residual moment capacities. Residual moment capacity is the difference between cross-sectional total bending moment capacity minus the demand moments under vertical loads. Due to the verifications for horizontal reinforcement configuration acceptance criteria, element ends are classified as "confined" and "unconfined". The calculated (r) values are to be compared with damage level limit values (r_s) to determine the damage levels of each structural member.

CASE STUDY ON AN EXISTING RC BUILDING

The considered building was exposed to seismic action of Adana Ceyhan Earthquake in 1998 (Ms = 6.3 with PGA= 0.28 g) and reported as moderately damaged under that seismic action. The case study building has six storeys with a total of 14.65 m height and it is composed of orthogonal frames, symmetrical in y direction and does not have any structural irregularities. The planar dimensions are 16.4 x 7.8 m = 127.9 m² with five spans in x and two spans in y directions, (Figure 3). It was initially designed and constructed according to the 1975 Turkish Seismic Code.

Storey heights are 2.15 m for the first storey and 2.50 m for the other storey. Slabs are having a thickness of 12 cm and they are modeled as rigid diaphragm at each storey level. The column dimensions are 25/45 cm, 25/50 cm, 25/70 cm at each storey. The in-situ tests for material properties reports that the characteristic compression capacity of the concrete is 10 MPa and the characteristic yielding capacity of the reinforcement is 220 MPa which are lower values than the ones given in the original project.

The computed base shear value according to EC8 is much higher than the TEC'07 while the selected ground conditions represent the same characteristics (Figure 3). The main reason is that the ordinate of the horizontal elastic response spectrum for Eurocode 8 is increased by the soil factor as shown in Figure 3.

Demand curvatures obtained from linear methods of analysis of EC8 together with TEC'07 is given in Figure 4. Curvatures from linear methods of analysis are determined on the basis of equal displacement rule.

The top storey displacements obtained from linear methods of analysis of Eurocode 8 together with TEC'07 is given in Figure 5.

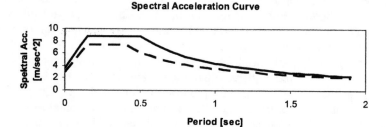

FIGURE 3. Horizontal Elastic Response Spectrum Curves

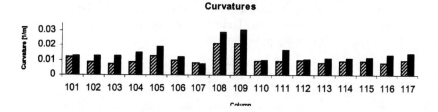

FIGURE 4. Curvatures at First Storey Columns

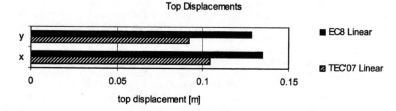

FIGURE 5. Top Displacements

CONCLUSIONS

In this study, performance based assessment methods and basic principles given in TEC'07 and Eurocode 8 are investigated. After the linear elastic approach is outlined as given in two codes, the procedures of seismic performance evaluations for existing RC buildings according to EC8 and TEC'07 are applied on a real three dimensional case study building and the results are compared.

The computations show that the performing methods of analysis with linear approaches using either EC8 or TEC'07 independently produce a very similar performance levels for the critical storey of the structure. The case study building is found to be as in collapse performance level. The computed base shear value according to Eurocode is much higher than the Turkish Earthquake Code while the selected ground conditions represent the same characteristics. The main reason is that the ordinate of the horizontal elastic response spectrum for EC8 is increased by the soil factor. In TEC'07 force-based linear assessment, the seismic demands at cross-sections are to be checked with residual moment capacities; however, the chord rotations of primary ductile elements must be checked for Eurocode safety verifications. It is also observed that the demand curvatures obtained from linear methods of analysis for both codes are almost similar. Higher curvatures obtained from EC8 procedure is the consequence of having higher ordinate for the horizontal elastic response spectrum.

ACKNOWLEDGEMENTS

The results presented in this paper are based on research work carried out and funded by The Scientific and Technological Research Council of Turkey (TUBITAK). The authors are also grateful to Istanbul Technical University Structural and Earthquake Research Center for providing the existing building datum.

REFERENCES

1. ATC-40, 1996. *Seismic Evaluation and Retrofit of Concrete Buildings*, ATC, California.
2. FEMA-273, 1997. *NEHRP Guidelines for Seismic Rehabilitation of Buildings*, Washington.
3. FEMA-356, 2000. *Prestandard and Commentary for Seismic Rehabilitation of Buildings*, Washington.
4. FEMA-440, 2005. *Improvement of Nonlinear Static Seismic Analysis Procedures*, Washington.
5. European Committee for Standardization, 2004. *Design of Structures for Earthquake Resistance-Assessment and Retrofitting of Buildings*, Eurocode 8-3.
6. *Turkish Seismic Design Code*, Ministry of Public Works, Official Gazette, March 2007.
7. Toprak, A.E., 2008. *Code-based Evaluation of Seismic Performance Levels of Reinforced Concrete Buildings with Linear and Non-linear Approaches*, Top Industrial Manager Europe (TIME) Double Degree Program MS Thesis, ITU Institute of Science and Technology, Istanbul, Technische Universitaet Dresden, Rehabilitation Engineering, Dresden.
8. Kuran, F., Demir, C., Koroglu, O., Kocaman, C., İlki, A., 2007. *Seismic Safety Analysis of an Existing 1502 Type Disaster Building Using New Version of Turkish Seismic Design Code*, ECCOMAS Thematic Conference on COMPDYN 2007, Rethymno, Crete, Greece, June 13-15, 2007
9. Gulay, F.G. Bal, I.E., Gokçe, T. 2008 '*Correlation Between Detailed And Preliminary Assessment Techniques In The Light Of Real Damage States*'(accepted to be published in Special Issue of Journal of Earthquake Engineering)
10. Yılmaz, H.E., 2006. *A Comparative Numerical Study on Seismic Performance Evaluation of Existing Reinforced Concrete Buildings on FEMA 356 and 2006 Turkish Seismic Code Non-linear Analysis Approaches*, MS Thesis, ITU Institute of Science and Technology, Istanbul.

Calculation Method of Lateral Strengths and Ductility Factors of Constructions with Shear Walls of Different Ductility

Nobuyoshi Yamaguchi[a], Masato Nakao[b], Masahide Murakami[c] and Kenji Miyazawa[d]

[a] Senior research engineer, Building Research Institute, Tachihara 1 Tsukuba-City Ibaraki Pref. Japan 305-0802, yamaguch@kenken.go.jp, Phone :++81-29-879-0653, Fax:++81-29-864-6772

[b] Research Associate, Faculty of Engineering, Yokohama National University, M. Eng

[c] Professor, Department of Architecture, Graduate School of Science & Engineering, Kinki University, Dr. Eng.

[d] Professor, Department of Architecture, Kogakuin Unibersity, Dr. Eng.

Abstract. For seismic design, ductility-related force modification factors are named R factor in Uniform Building Code of U.S, q factor in Euro Code 8 and Ds (inverse of R) factor in Japanese Building Code. These ductility-related force modification factors for each type of shear elements are appeared in those codes. Some constructions use various types of shear walls that have different ductility, especially for their retrofit or re-strengthening. In these cases, engineers puzzle the decision of force modification factors of the constructions. Solving this problem, new method to calculate lateral strengths of stories for simple shear wall systems is proposed and named 'Stiffness - Potential Energy Addition Method' in this paper. This method uses two design lateral strengths for each type of shear walls in damage limit state and safety limit state. Two lateral strengths of stories in both limit states are calculated from these two design lateral strengths for each type of shear walls in both limit states. Calculated strengths have the same quality as values obtained by strength addition method using many steps of load-deformation data of shear walls. The new method to calculate ductility factors is also proposed in this paper. This method is based on the new method to calculate lateral strengths of stories. This method can solve the problem to obtain ductility factors of stories with shear walls of different ductility.

Keywords: Ductility factors, R factor, q factor, Ds factor, Stiffness, Potential energy.

INTRODUCTION

Ultimate strength design method for the seismic design of constructions utilizing simple shear wall systems requires engineers to decide ductility factors from the design tables in building code books. When engineers select ductility factors from the tables for the combined structural systems that have different ductility, they will be puzzled in Japan. Engineers need advanced method to calculate ductility factors for those combined systems. This paper proposes a new method to calculate ductility

CP1020, *2008 Seismic Engineering Conference Commemorating the 1908 Messina and Reggio Calabria Earthquake,* edited by A. Santini and N. Moraci
© 2008 American Institute of Physics 978-0-7354-0542-4/08/$23.00

factors theoretically for the structural system combined shear wall systems that have different ductility.

Design Loads in Building Codes

Countries in seismic regions have seismic design codes in their building code. These seismic design codes include seismic design loads for the ultimate strength design method using ductility related factors. Japanese building code[1], Uniform building code(UBC)[2] and Euro code 8[3] utilize Ds factor, R factor and q factor for those ductility related factors respectively.

Seismic design method for severe earthquakes in Japanese Building Code utilizes ultimate strength design method and design shear forces for i-th story defined by Eq.1. Ds values in Eq.1 range from 0.25 to 0.5 for steel or wood structures, from 0.3 to 0.55 for reinforced concrete structures or others. Inverses of these Ds values are from 2 to 4, and from 1.8 to 3.3 respectively.

$$Qui = \frac{Fes \cdot Z \cdot Rt \cdot Ai \cdot C_0 \cdot \sum Wi}{\dfrac{1}{Ds}}. \tag{1}$$

Qui: Design lateral seismic shear of i-th Story, Fes: Shape factor (from 1.0 to 3.0), Ds: Structural coefficient (Inverse of ductility factor, from 0.25 to 0.55), ΣWi: Weight of building of upper than i-th stories, Z: Seismic hazard zoning coefficient (from 0.7 to 1.0), Rt: Design spectrum factor normalized (less than 1.0), Ai: Lateral shear distribution factor along stories normalized (less than 1.0), C_0: Seismic shear coefficient (more than 1.0)

Ductility Related Factors in Building Codes

There are several concepts of ductility related factors in some building codes. Ds factor defined in Eq.2 is utilized for ductility related factors in Japanese Building Code from 1981. Eq.2 indicates inverse of Ds factor is adjusted by D_h. Adjusted factor D_h is defined in Eq.3. Eq.3 indicates, when h is 0.05, D_h will be 1. Newmark and Hall defined R factor as Eq.4 in ATC-19 [4]. These relations induce inverse of Ds factor in Japanese Building Code is comparable to the R factor in Uniform Building Code.

$$\frac{1}{Ds} = \frac{\sqrt{2\mu-1}}{D_h} \tag{2}$$

$$D_h = \frac{1.5}{1+10 \cdot h} \tag{3}$$

$$R = \sqrt{2\mu-1} \tag{4}$$

μ: Ductility ratio, D_h: Coefficient for damping factor, h: Damping factor, R: R Factor

ATC-19(1995) note R factor is composed of strength factor, ductility factor and damping factor[4]. R factor in UBC(1997) is defined as numerical coefficient representative of the inherent overstrength and global ductility capacity[2]. 'q factor' in Euro Code 8 is composed of behavior factor, factor reflecting ductility class, factor reflecting the structure and factor reflecting the prevailing failure[3]. These factors are not same, but it is clarified that all of them is based on ductility of their lateral-force-resisting systems.

UBC has the tables to indicate R factors for each structural system. Where combination of different structural systems is used along the same axis, it descries the value of R shall not be greater than the minimum value for any of the systems utilized in that same direction (UBC-1630.4.4) [2]. These procedures clarified in UBC are simple and convenient, but conservative a little.

METHOD

New Method to Calculate Lateral-Strength of Stories

New method to calculate lateral strength of shear resistant systems was proposed by M. Murakami, and completed by M. Nakao, M. Murakami, N. Yamaguchi and K. Miyazawa[5] and named 'Stiffness - Potential Energy Addition Method'. This method was developed to decide lateral strengths of structures in the damage limit state and safety limit state. The procedure of this new method is introduced as follows.

Limit State Deformation of Damage and Safety

Two limit states of damage and safety for seismic design methods are defined utilizing relative story deformations of structures. Damage limit state means that structures make elastic responses but have no damage by the relative story deformations. Safety limit state means that structures make inelastic responses and have damage by the relative story deformations but not collapse and not harm people in and around the structures. The allowable strengths of each type of wooden shear walls have been defined by special procedures based on strengths in Japan, but this new method proposes to define the limit states using relative story deformations. The damage limit state is defined by the relative story deformation R_d of 1/200(0.5%) of story heights, and also the safety limit state is defined by the relative story deformation R_s of 1/30(3.3%) in case of wood constructions. Because most of shear walls for wood constructions make their maximum strengths around 1/100(1%) to 1/40(2.5%), these constructions will not make story collapses under the relative deformation of 1/30(3.3%). These proposals are obtained from lateral cyclic loading tests and shaking table tests of various shear walls.

Unit Elastic Stiffness and Unit Potential Energy

Load-deformation curves of various shear walls are obtained from lateral cyclic loading test of them. Unit load (kN/m) normalized by wall length is used instead of loads. The relative story deformation (radian) normalized by story heights is also used instead of displacements. Then, skeleton curve of the unit load-deformation relationships are obtained. The skeleton curves are transformed to idealized bi-linear model using the equal-energy method. Figure 1 illustrates idealized bi-linear model. This method assumes that the area enclosed by the curve above the bilinear

763

approximation is equal to the area enclosed by the curve below the bilinear approximation.(1997-UBC) In this transformation, the first slope of the bilinear model pass through the damage limit state points(R_d, P_d) and reached the yield points(R_v, P_s) of them. The first slope indicates the unit elastic stiffness K (kN/rad./m) of the shear walls. Unit energy (Area below the bilinear model) from the origin to the deformation of safety limit state R_s is defined as unit potential energy E (kN rad./m).

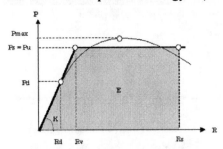

FIGURE 1. Idealized Bilinear Model of Load-Deformation Curves for Shear Elements

Lateral Strengths of Each Shear Walls

In case that the length of shear wall-i is Li, damage limit state strength P_{di} of shear wall-i is given by Eq.5. In Eq.5, K_i is a unit elastic stiffness of shear wall-i. R_{vi} defined in Eq.6 is a relative story deformation at the yield points of bilinear model. E_i is a unit potential energy of shear wall-i. Safety limit state strength P_{si} of shear wall-i is given by Eq. 7. R_d, R_s in Eq.5 and Eq.7 are relative story deformations of the story normalized by the story heights.

$$Pdi = Ki \cdot Li \times Rd \qquad (5)$$

$$Rvi = \frac{Psi}{Ki} \qquad (6)$$

$$Psi = \left(Rs \cdot Ki - \sqrt{Rs^2 \cdot Ki^2 - 2Ei \cdot Ki}\right) \cdot Li \qquad (7)$$

Lateral Strengths of Stories

It is assumed that simple shear wall systems have lateral strength of stories proportional to the lengths of shear walls in stories. Elastic stiffness and potential energy of story-j are calculated by Eq.8 and Eq.9. K_j is a elastic stiffness of j-th story. E_j is a potential energy of j-th story. Lateral strengths of j-th story in the both states are calculated by Eq.10 and Eq.12.

$$Kj = \sum_i Ki = \frac{\sum_i (Pdi)}{Rd} \qquad (8)$$

$$Ej = \sum_i Ei = \sum_i \left(Psi \cdot \left(Rs - \frac{Psi \cdot Rd}{2Pdi} \right) \right) \qquad (9)$$

$$Pdj = Kj \cdot Rd \tag{10}$$

$$Rvj = \frac{Psj}{Kj} \tag{11}$$

$$Psj = Rs \cdot Kj - \sqrt{Rs^2 \cdot Kj^2 - 2Ej \cdot Kj} \tag{12}$$

Method to Calculate Ductility Factors of Stories

Ductility ratio μ_j of j-th story is indicated in Eq.13. Ductility factor ($1/Ds$) of j-th story with shear walls of various ductility is proposed in Eq.14 using Eq.15. Eq.15 indicates ratio of potential energy E_j to elastic stiffness K_j of j-th story. Figure 2 illustrates relationship between ductility factor $1/Ds$ and γ_{ek}.

$$\mu_j = \frac{Rsj}{Rvj} \tag{13}$$

$$\frac{1}{Ds} = \frac{1}{D_h} \cdot \sqrt{\frac{1 + \sqrt{1 - \gamma_{ek}}}{1 - \sqrt{1 - \gamma_{ek}}}} \tag{14}$$

$$\gamma_{ek} = \frac{2\,Ej}{Kj \cdot Rs^2} \tag{15}$$

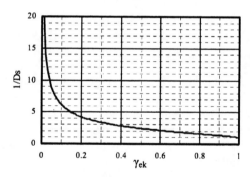

FIGURE 2. Relationship between ductility factor $1/Ds$ and γ_{ek}

ANALYSIS USING EXAMPLE CALCULATION

Model for Calculation

New Method to calculate lateral strength of stories is applied for a conventional post & beam wood construction in Japan. Model plan is shown in figure 3. Unit strengths in the both states, unit elastic stiffness and unit potential energy of shear walls used in the model are summarized in table 1.

TABLE 1. Strengths and Ductility Factors (*1/Ds*) of Shear Walls used in the Model

WALL TYPE	K (kN/rad./m)	E (kN rad/m)	Pd (kN/m)	Ps (kN/m)	Rv (rad.)	μ	Ds	1/Ds
Brace	485	0.201	2.425	8.022	0.0165	2.015	0.574	1.741
Plywood	1000	0.219	5.000	7.400	0.0074	4.504	0.353	2.830
Gypsum board	276	0.040	1.380	1.306	0.0047	7.045	0.276	3.618

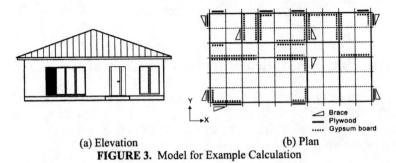

(a) Elevation   (b) Plan
FIGURE 3. Model for Example Calculation

Lateral Strengths of Stories by Old and New Method

The old method to obtain lateral strength of a story uses procedure to add loads of load-deformation curves of the shear walls. Therefore this method needs multi steps of load-deformation data of skeleton curves of shear walls. After adding load-deformation curves of shear walls in a story, added curve is possible to be transferred to idealized bilinear model. Lateral strengths P_d and P_s of a story in the both states are calculated by Eq.5 and Eq.7 like as shear walls.

The new method to obtain lateral strength of stories uses aforementioned procedures. P_{di} and P_{si} of shear wall-i are calculated by Eq.5 and Eq.7. After K_j and E_j of j-th story are calculated by the Eq. 8 and Eq. 9, P_{dj} and P_{sj} of j-th story are calculated by Eq.10 and Eq.12.

Figure 4 indicates the difference of process to obtain the idealized bilinear models by these two methods. In case that each length of shear walls in a story is 1 meter for simplicity, the processes to obtain the idealized bilinear models by the old method and new method are illustrated in figure 4(a) and figure 4(b) respectively. The bold lines in figure 4(a) and figure 4(b) indicate idealized bilinear models obtained by these two methods are same naturally.

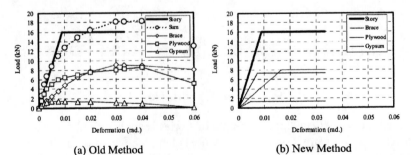

(a) Old Method   (b) New Method
Figure 4. Processes to Obtain Lateral Strengths of Stories by Two Methods

RESULTS AND DISCUSSION

Strengths of Stories

Lateral strengths of the story in the both states obtained by old and new methods are summarized in table 2 and table 3. Results by old and new methods in table 2 and table 3 indicate these old method and new method give same values. Although old method needs many steps of loads-deformations data for a shear wall, but new method requires only two values for each shear wall. Researcher is possible to use old method, but engineers are hard to use this old method in actual design works. These results clarifies new method is superior to the old method and valuable for engineers.

TABLE 2. Lateral Strengths and Ductility Factors of Stories by Old Method

Direction	item	Pd (kN/m)	Ps (kN/m)	Rv (rad.)	μ	Ds	1/Ds
X	story	71.925	91.058	0.00633	5.266	0.324	3.087
Y	story	61.815	103.330	0.00836	3.988	0.379	2.641

TABLE 3. Lateral Strengths and Ductility Factors of Stories by New Method

Direction	item	length (m)	K (kN/rad./m)	E (kN rad/m)	Pd (kN/m)	Ps (kN/m)	Rv (rad.)	μ	Ds	1/Ds
X	Brace	1	485	0.201						
	Plywood	7	7000	1.535						
	Gypsum board	25	6900	1.011						
	story		14385	2.747	71.925	91.058	0.0063	5.266	0.324	3.087
Y	Brace	7	3395	1.407						
	Plywood	4	4000	0.877						
	Gypsum board	18	4968	0.728						
	story		12363	3.013	61.815	103.330	0.0084	3.988	0.379	2.641

Ductility Factors of Stories

The old and new method can calculate ductility factors ($1/Ds$) of stories with combined shear walls of different ductility. Calculated ductility factors of the story are summarized in table 2 and table 3. Same ductility factors of X and Y directions of the story are given by these two methods. This result indicates these old method and new method provide same values. Although the old method is difficult to use in actual design works by engineers, the new method requires only two values P_{di} and P_{si} for each type of shear walls. It is clarified the new method is valuable for actual design works. If the method of UBC1997 were applied for obtaining ductility factors of this model, the minimum value of ductility factors of shear walls in table 1 would be used.

Required Performance for Shear Walls

The new method is developed for low-rise constructions using simple shear wall systems. It is assumed that all of shear walls in stories responses in a same manner during earthquakes. Damage limit state and safety limit state of stories are also assumed to be defined using relative story deformations. This method uses fixed damage limit state and safety limit state for shear walls of different ductility. Satisfying these assumptions, shear walls are required to provide stable load-deformation performance not only in damage limit state and also in safety limit state.

Application for Structures Other than Wood Structures

These methods would be possible to apply not only for these low-rise wood constructions but also the other constructions using the addition of their lateral strengths of each shear walls. These methods would contribute to the progress of ultimate strength design method used in many seismic countries.

CONCLUSIONS

New method to calculate lateral strengths of stories for simple shear wall systems is proposed and named 'Stiffness - Potential Energy Addition Method'. This method needs only two design lateral strengths for each type of shear walls in damage limit state and safety limit state. Two lateral strengths of stories in both limit states are calculated using two design strengths of each type of shear walls in both limit states. Calculated strengths have the same quality as values obtained by the old strength addition method using many steps of load-deformation data of shear walls.

The new method to calculate ductility factors is also proposed. This method is based on the new method to calculate lateral strengths of stories. This method is possible to calculate ductility factors of stories with shear walls of different ductility.

These methods would contribute to the progress of ultimate strength design method not only for wood constructions but also for the other constructions using simple shear wall systems.

ACKNOWLEDGEMENTS

This research was studied for the development of 'Simplified Seismic Calculation Method of Timber Constructions' in Architectural Institute of Japan (AIJ). We would like to express our gratitude for the members of sub-committee and work group for this development.

REFERENCES

1. Building Standard Law of Japan, Notification No.1792 of 1980, Regulations for Seismic Design A world List-1996, International Association for Earthquake Engineering, June,1996
2. 1997 Uniform Building Code, Volume 2, Division 4 Earthquake Design, Section 1630-Minimum Design Lateral Forces and Related Effects, 1630.2 Static Force Procedure, International Conference of Building Officials, April 1997, pp.2-14
3. Euro Code 8, Regulations for Seismic Design A world List-1996 Supplement-2000, International Association for Earthquake Engineering, January 2000
4. Structural response modification factors, ATC-19, Applied Technology Council, California, 1995
5. Masato NAKAO, Masahide MURAKAMI, Nobuyoshi YAMAGUCHI, Kenji MIYAZAWA, Horizontal Resistance and Ds Values of Stories in Timber Construction with Shear Walls, -Evaluation method of seismic performance of shear resisting elements using horizontal stiffness and potential energy-, Journal of Structural and Construction Engineering, AIJ, Vol.73, No.624, pp.291-298, Feb., 2008

STRUCTURAL ENGINEERING

The Damaging Effects of Earthquake Excitation on Concrete Cooling Towers

Farhad Abedi-Nik[a], Saeid Sabouri-Ghomi[b]

[a]*Asst. Professor of Civil Engineering, SADRA Institute of Higher Education, Tehran, Iran*
Farnikan@aut.ac.ir
[b]*Associate Professor of Civil Engineering, K.N.T University of Technology, Tehran, Iran*
Sabouri@kntu.ac.ir

Abstract. Reinforced concrete cooling towers of hyperbolic shell configuration find widespread application in utilities engaged in the production of electric power. In design of critical civil infrastructure of this type, it is imperative to consider all the possible loading conditions that the cooling tower may experience. an important loading condition in many countries is that of the earthquake excitation, whose influence on the integrity and stability of cooling towers is profound. Previous researches have shown that the columns supporting a cooling tower are sensitive to earthquake forces, as they are heavily loaded elements that do not possess high ductility, and understanding the behavior of columns under earthquake excitation is vital in structural design because they provide the load path for the self weight of the tower shell. This paper presents the results of a finite element investigation of a representative "dry" cooling tower, using realistic horizontal and vertical acceleration data obtained from the recent and widely-reported Tabas, Naghan and Bam earthquakes in Iran. The results of both linear and nonlinear analyses are reported in the paper, the locations of plastic hinges within the supporting columns are identified and the ramifications of the plastic hinges on the stability of the cooling tower are assessed. It is concluded that for the (typical) cooling tower configuration analyzed, the columns that are instrumental in providing a load path are influenced greatly by earthquake loading, and for the earthquake data used in this study the representative cooling tower would be rendered unstable and would collapse under the earthquake forces considered.

Keywords: Cooling Towers, Earthquakes, Nonlinear Behavior.

INTRODUCTION

Reinforced Concrete Cooling Towers are kind of special structures that could be seen in most of power-plants and substations. To design these structures, a variety of loading conditions should be taken into account. It is important to check all of loading conditions. All these loading conditions will be used in designing an element of the structure. Earthquake does have a crucial effect on the stability of these structures. Previous studies of Gran and Young [1], Castiau [2], Kratzig [3] and Sabouri-Ghomi [4] made sign towards damage of columns of RC cooling towers during earthquake. Due to the importance of columns in study of seismic behavior of cooling towers, current research concentrates on studying the behavior of columns of RC cooling towers during earthquake. Also in finite element modeling of the whole structure a particular attention was focused on the columns of structure.

CP1020, *2008 Seismic Engineering Conference Commemorating the 1908 Messina and Reggio Calabria Earthquake,*
edited by A. Santini and N. Moraci
© 2008 American Institute of Physics 978-0-7354-0542-4/08/$23.00

In order to decrease computation time, critical sections of accelerometers of Tabas, Naghan and Bam earthquakes were selected and used. Maximums of horizontal and vertical accelerations in these accelerometers are listed in Table 1.

TABLE 1 - Maximum horizontal and vertical acceleration of Tabas, Naghan and Bam earthquakes

EARTHQUAKES	Horizontal PGA (%g)	Vertical PGA (%g)
TABAS	0.82	0.57
NAGHAN	0.74	0.5
BAM	0.79	1.01

CASE STUDIES

In order to study the nonlinear behavior of RC cooling towers,The RC cooling tower of Shazand thermal power plant has been selected. It is one of the tallest reinforced concrete cooling towers in Iran with the height of 134 meters. An important characteristic of this structure is its long columns in comparison with the columns in similar structures. The structure consists of a 106 m height hyperbolical shell above 36 pair of X-type RC columns. The general specifications of this structure is shown in Figure 1.

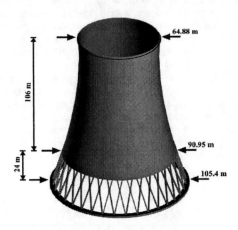

FIGURE 1. The general specification of the case study concrete cooling tower

Dynamic analysis has been performed in two separate sections. First the finite element model of structure with elastic elements was set up and dynamic analysis accomplished. Then, the finite element model with nonlinear elements established and the results of dynamic analysis obtained. A comparison is made based on the force and displacement responses of model in both cases. By this way we will have a better understanding of the structural behavior.

After a wide study on nodal displacements, finally four key nodes A, B, A', and B' were selected. Displacement responses of the model found at these nodes (A and B)

were compared to displacements of nodes A' and B'. This will make the distortion of top and bottom parts available. (See Figure 2)

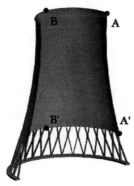

FIGURE 2. The chosen key points for displacement control

FINITE ELEMENT MODELING

In FE modeling with ANSYS [5], foundation is modeled by selecting linear three-dimensional SOLID45 elements. For nonlinear modeling of the columns, SOLID65 three dimensional concrete elements are invoked. These types of elements are capable of cracking in tension and crushing in compression. Shell of the structure with the height of 106m is modeled by using three different types of elements, SOLID65, SOLID45, and SHELL63. SOLID65 is used to model one meter of height of the shell above the columns where the risk of stress concentration and local non-linearity is quite likely. Five meters above this level is modeled by SOLID45, a linear three-dimensional element. Finally, the rest of the shell with 100meters height is modeled by thin elastic-shell element SHELL63. All the columns and one meter the height of the shell above the columns (SOLID65) are reinforced by LINK8 three-dimensional spar, considering nonlinear material properties for them.

STUDYING THE NONLINEAR BEHAVIOR OF THE MODEL IN EARTHQUAKE

The selected accelerometers were used to apply acceleration into the model. It was seen that concrete of the column was cracked. Moreover, in some cases the reinforcing bars of columns have passed proportional limit of elastic behavior and experienced nonlinearity. Cracking of concrete and yielding of reinforcing bars took place in lower parts of the shell, near columns. Figure 3 demonstrates the sequence of plastic hinge occurrence and their position in the analyzed model. In this figure, plastic hinges happen where the reinforcing bars in cross-section yield and enter into nonlinear phase. As it can be seen, plastic hinges almost happen in columns, in contact region of columns and foundation, in the interface of columns and shell, and in intersection of two columns. Plastic hinges in shell always happen after formation of other hinges in structure. This almost ensue after serious damage to environs columns.

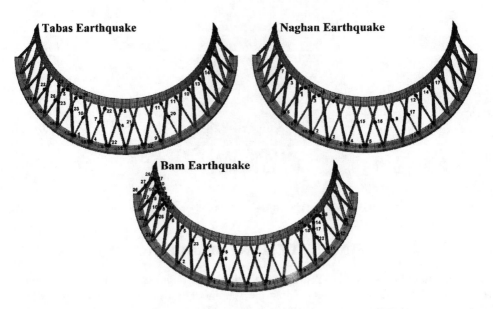

FIGURE 3. Position and sequence of plastic hinges in the model subject to Tabas, Naghan, and Bam earthquakes

Figure 4 highlights the comparison that has been made between linear and nonlinear time history of base shear. Moreover, chronological position of plastic hinges is depicted with filled circles in Figure4. As it can be seen linear and nonlinear responses are coinciding in each other. However, there is an apparent decrease in nonlinear response after formation of plastic hinges. It can be described as the stiffness of the structure decreases, income force to the structure and consequently the base shear decreases. In current nonlinear analysis, based on Tabas and Bam records, the analysis in some time steps after 2.4 and 3 seconds was diverged. As well, there was a sever decrease in base shear nonlinear response of the model after aforesaid stage. This can be explicated due to instability of the structure at this time.

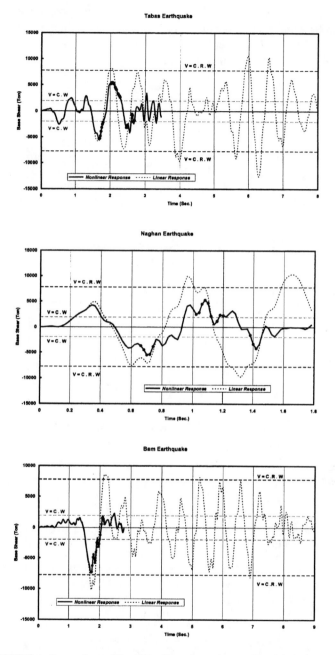

FIGURE 4. Comparison of linear nonlinear base shears of modeled structure
By Tabas, Naghan and Bam records

Owing to the behavior of bottom stiffener ring before incidence of instability in structure, importance of columns in structure, and drift of the top of the columns, a comparison between linear and nonlinear displacements has been made. This is almost helpful in understanding the behavior of structure. For this, elastic and inelastic displacements of point B' under aforementioned earthquake records were found. Figure 5 shows the comparison of responses.

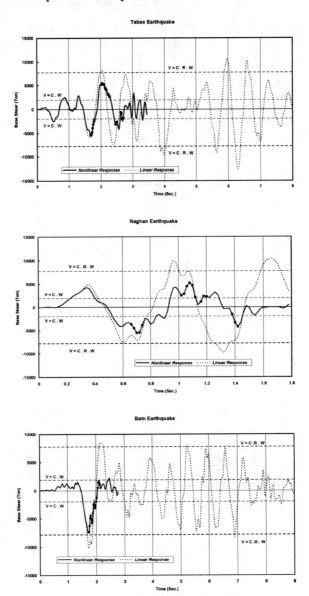

FIGURE 5. Comparison of horizontal elastic and inelastic displacement of point B' under Tabas, Naghan and Bam earthquake records

Figure 6 shows the base shear versus displacement of columns of this structure in both linear and nonlinear states. Due to the effect of the bottom stiffener ring, the displacements of the columns at the top assumed to be equal to each other. Based on the results, the location of force and deformation covers particular zone in elastic state. On the other hand, hysteretic diagram of columns in nonlinear state shows an increase in displacement and a decrease in correspondent force.

Inner area of hysteretic loops renders the dissipated energy. This dissipated energy will be higher for wider hysteretic loops. As it can be seen, absorbed energy by the system is not considerable. There is a reduction in stiffness of the model in large deformations. Hence, we can claim that the structure does not show a stable and proper hysteretic behavior.

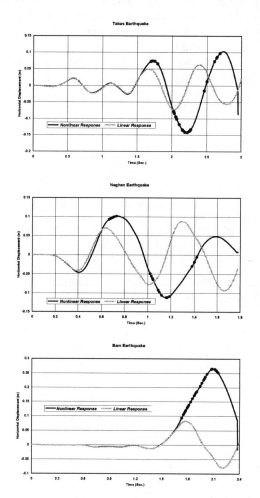

FIGURE 6. Force-displacement graph of columns, Tabas, Naghan and Bam records.

CONCLUSIONS

1. Positions and sequences of plastic hinges formation on columns depend on the nature of earthquake record. However, plastic hinges mainly form on columns at the junction of columns, columns and foundation, and columns and shell. Plastic hinges of shell were almost the very last plastic hinges formed in the model. They come about after serious damages to adjacent columns.
2. Inelastic behavior of structure due to formation of plastic hinges reduces stiffness of the model and consequently decreases applied base-shear and increases displacements compared to the elastic state.
3. Bottom stiffener ring prior to incident of instability in structure due to earthquake, exhibited a rigid behavior. Then, the displacement of the columns at the top could be considered equal to each other. Upper parts of the shell experience a minor deterioration. This may affect longitudinal and transverse forces at the top of the shell.
4. Hysteretic loops of the modeled structure do not represent apposite energy absorption. Stiffness and rigidity of the model decreases when the displacement starts to increase.

REFERENCES

1. Gran C. S., Yang T. Y. "Nastran and Sap IV applications on the seismic response of column supported cooling towers". Computers & Structures, Vol 8. pp761- 768. Pergamon press Ltd. 1978, Printed in Great Britain.
2. Castiau Th, Gaurois R. "The design of cooling towers in extremely serve earthquake condition". Eng. Structure. 1991, Vol. 13, January.
3. Kratzig W. B. '7oward safe and economic seismic design of cooling tower of extreme height, 1979.
4. Sabouri-Ghomi.S. "The effect of earthquake on stability of concrete cooling towers", K.N.T University of technology, Tehran, Iran, 1999.
5. Ansys (Ver. 8. 00). Users manual. Swanson Analysis System. Inc.

Non Linear Step By Step Seismic Response and the Push Over Analysis Comparison of a Reinforced Concrete of Ductile Frames 25 Level Building

Jorge A. Avila[a] and Eduardo Martinez[a]

[a] *Institute of Engineering, National University of Mexico (UNAM)*
Ciudad Universitaria, Coyoacan 04510, Mexico, D.F.; javr@pumas.iingen.unam.mx

Abstract. Based on a ductile frames 25 level building, a non-linear analysis with increased monotonically lateral loads (Push-Over) was made in order to determine its collapse and its principal responses were compared against the time-history seismic responses determined with the SCT-EW-85 record. The seismic-resistance design and faced to gravitational loads was made according to the Complementary Technical Norms of Concrete Structures Design (NTC-Concrete) and the NTC-Seismic of the Mexico City Code (RDF-04), satisfying the limit service states (relative lateral displacement between story height maximum relations, story drifts $\leq$ 0.012) and failure (seismic behavior factor, Q= 3). The compressible (soft) seismic zone III_b and the office use type (group B) were considered. The non-lineal responses were determined with nominal and over-resistance effects. The comparison were made with base shear force–roof lateral displacement relations, global distribution of plastic hinges, failure mechanics tendency, lateral displacements and story drift and its distribution along the height of the building, local and global ductility demands, etc. For the non-linear static analysis with increased monotonically lateral loads, was important to select the type of lateral forces distribution.

INTRODUCTION

México City suffers in a great way earthquake effects because it is located mostly in a highly compressible soil that provokes the amplification of the seismic waves that arrive from the Pacific coasts, causing higher intensity telluric movements. During the 1985 earthquakes in Mexico City, there were many structures between 5 and 17 levels that did not behave in a good way in the excessive lateral flexibility direction. The design criteria used in Mexico City is ruled by the Mexico City Construction Code, used since 2004 (RDF-04) [1], and its Complementary Technical Norms (NTC).

Objectives

In this work the elastic and inelastic behavior (time-history and static with lateral loads monotonically increased Push-Over procedure) of a reinforced concrete 25 level building, with office use type (group B) and located in Mexico City's III_b compressible seismic zone, is compared. The spectral modal dynamic analysis was used for the design, checking that, in the service condition, the maximum relations of

CP1020, *2008 Seismic Engineering Conference Commemorating the 1908 Messina and Reggio Calabria Earthquake,*
edited by A. Santini and N. Moraci
© 2008 American Institute of Physics 978-0-7354-0542-4/08/$23.00

the relative horizontal displacement between story height (drifts) did not exceed the permissible limit of γ_p = 0.012 (design condition in which the non-structural elements are not linked to the main structure), and also that the failure limit state (resistances) were accomplished, assuming that the elastic seismic forces were reduced by the seismic behavior factor, Q=3; the designs satisfy the special requirements of ductile frames, in addition to those general requirements, as the NTC-Seismic [2] and NTC-Concrete [3] specify when Q = 3 or 4 is used. When the designs were known, inelastic step-by-step dynamic analysis were performed with the SCT accelerogram, EW component, recorded during the September 19th of 1985 earthquake in Mexico City. For the Push-over analysis, different options for the seismic forces to be used were checked: a) according to the spectral modal dynamic analysis results; b) according to the accepted static method in the NTC-Seismic, RDF-04; c) according to the results of the initial elastic stage of the step-by-step dynamic analysis; and, d) according to the most intense stage in the inelastic range of the time-history dynamic analysis, without and with over-resistance effects. During the performance of the Push-over analysis, and without the necessity of achieving the failure mechanism, the instant in which the supposed maximum local ductility values could be reached (30 for beams and/or 20 for columns) was monitored.

STRUCTURE DESCRIPTION

Reinforced concrete building has 25 levels. Its foundation is made of a two level rigid box with a parking basement, foundation beams grid, and under the foundation slab, a group of point piles. The plant is rectangular with three 6 meters bays in Y direction (short direction), and five 9 meters bays in X direction (longitudinal) (see Fig. 1). The story height is 3.5 meters, from the roof level to the first level; the ground level story height is 4.5 meters (PB-N1), and 3.5 meters for the basement story height (PB-basement). Figure 2a compares the design spectra for the compressible seismic zone type III_b (Q= 1 a 4), specified in the NTC-Seismic.

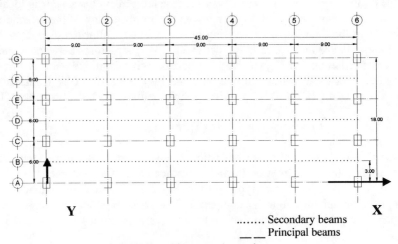

FIGURE 1. Plant-type (meters)

780

ELASTIC RESPONSES

Vibration Periods

Table 1 presents the periods for the first three modes vibration for the directions: X (long), Y (short) and torsion of the 25 levels building, according to the NTC-Seismic criteria. For both directions the fundamental periods of vibration result very alike.

Ratios of Relative Lateral Displacement Between the Story Height

Figure 2b shows the maximum story distortion of the 25 level building, with the earthquake acting in both directions X and Y, respectively; they are compared against the permissible limit of 0.012.

TABLE 1. Vibration periods, 25 levels building.

Direction	Period, T_1 (s)		
	Mode 1	Mode 2	Mode 3
X	1.96 (63.33)	0.66 (9.57)	0.37 (3.65)
Y	1.98 (60.02)	0.63 (12.92)	0.33 (4.47)
θ	1.61 (60.44)	0.54 (11.13)	0.29 (4.28)

( ) Effective modal mass, %

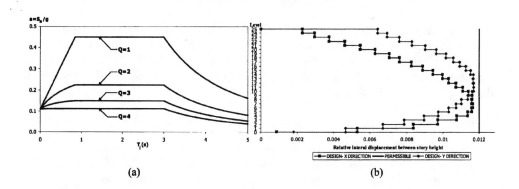

(a)   (b)

FIGURE 2. (a) Design spectra for the compressible seismic zone type III_b (Q= 1 to 4), group B (offices), specified in the NTC-Seismic, RDF-04. (b) Maximum story distortion of the 25 level building, earthquake acting in both directions X and Y, respectively, spectral modal dynamic analysis.

INELASTIC RESPONSES, STEP-BY-STEP ANALYSIS

The inelastic responses were calculated for the structural 2 (transversal direction) and C (longitudinal direction) axes; in this paper are shown the C axe responses. The step-by-step inelastic dynamic responses are determined with the SCT-EW record (see Fig. 3a). Each axe was calibrated with the mass and stiffness dynamic properties that were representative for each direction of the structure. The fundamental periods of vibration of both directions are near of the critical zone of the SCT-EW-85 record elastic spectrum (see Fig. 3b).

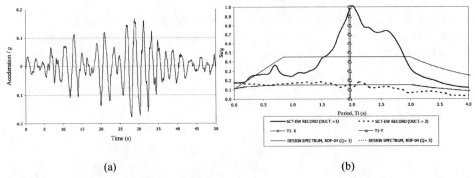

(a)   (b)

FIGURE 3. (a) SCT-EW record, September 19th of 1985 earthquake in Mexico City. (b) Location of the fundamental periods of vibration of the 25 level building, seismic zone III$_b$, design spectra RDF-04 and response spectra SCT-EW (ξ= 5%).

Maximum Horizontal Displacements and Maximum Global Ductility Demands

The inelastic responses are no quite different from the elastic ones. The maximum lateral displacements of the C axe (longitudinal direction) tend to have a similar behavior in elevation with slightly minor values (see Fig. 4a). Table 2 compare the maximum global ductility demands of both step-by-step analyzed axes, versus with the design level (Q=3); they are slightly minor with the over-resistance effects.

TABLE 2. Maximum global ductility demands of C and 2 axes (long and short directions), without and with over-resistance effects, 25 levels building.

Case	Nominal-resistance			Over-resistance		
	Δ_Y (cm)	$\Delta_{máx\ inel}$ (cm)	μ_G	Δ_Y (cm)	$\Delta_{máx\ inel}$ (cm)	μ_G
C axe	25.28	55.05	2.2	38.10	61.63	1.6
2 axe	11.11	57.44	5.2	16.82	62.33	3.7

Maximum Relations of Relative Lateral Displacement Between Story Height and Ratios of Base Shear Force-Roof Lateral Displacement

Figure 4b shows the maximum relations of the relative lateral displacement between story height of the same axe in the longitudinal direction (C axe), elastic and inelastic behavior with nominal resistance and over-resistance effects. The behavior pattern of this kind of response is quite similar to those of the maximum lateral displacements; the step-by-step responses tend to be away from the permissible level of 0.012. When the inelastic comparisons are done, it is observed a behavior in the non-lineal range, with higher hysteretic area with nominal resistances; the help that over-resistances offer makes the response to tend to be elastic.

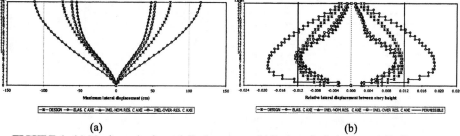

(a)   (b)

FIGURE 4. (a) Maximum horizontal displacements. (b) Ratios of relative lateral displacement between story height, spectral modal dynamic and time-history analysis (without and with over-resistance effects).

Plastic Hinges Global Distribution and Local Ductility Demands

The plastic hinges get diminished in number and plastic rotation amplitude when we take notice of the possible over-resistance sources (see Figs. 5 and 6). When we compare the maximum local ductility demands developed in the beams and columns of C axe, without and with over-resistance effects, it is corroborated that in both directions we have a similar inelastic behavior.

PUSH-OVER ANALYSIS

Figure 7 presents the different lateral forces distributions (F_i) used to analyze C axe (longitudinal direction), with the Push-over procedure. After checking the form of each forces distribution F_i and comparing versus the step-by-step analysis results, only the case a) results are showed. In C axe the local ductility maximum limit, established for the columns of PB-N1 story, predominated; the beams developed, practically in a generalized way in all the levels, local ductility under 20, inferior to the assumed permissible limit (see Fig. 8). In general, the results are congruent with the observed behavior in the responses of the inelastic analysis in time-history, versus the SCT-EW record effects.

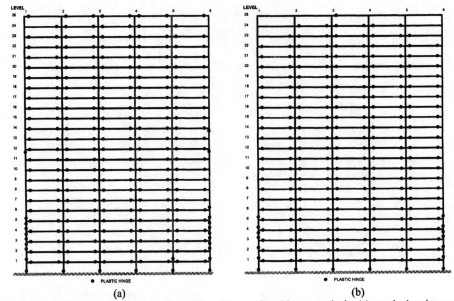

(a)   (b)

FIGURE 5. Plastic hinges global distribution of C axe, time-history analysis: (a) nominal resistance, (b) over-resistance effects.

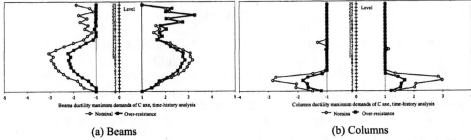

(a) Beams   (b) Columns

FIGURE 6. Beam and columns ductility maximum demands of C axe, time-history analysis (without and with over-resistance effects).

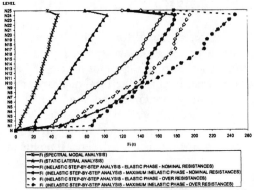

FIGURE 7. Different lateral forces distributions (F_i) used to analyze C axe (longitudinal direction), with the Push-over procedure, without and with over-resistances effects.

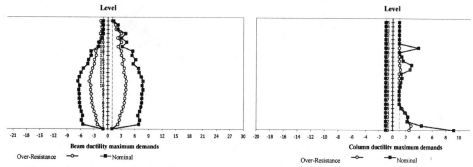

FIGURE 8. Beam and columns ductility maximum demands of C axe, Push-over analysis (without and with over-resistance effects).

Figures 9 and 10 show the comparisons of the hysteretic step-by-step inelastic analysis curves, in function of the ratios of base shear force–roof lateral displacement, versus to those determined in the Push-over analysis. The failure mechanism was never reached, because the established yield maximum limit dominated in beams or columns; the roof maximum lateral displacements reported are slightly under the maximum lateral displacement associated to the failure mechanism. It is to be noticed in these graphics that certainly there is a correspondence between resistance and lateral stiffness (equivalent slope) available determined with both type of non-linear analysis (step-by-step and static Push-over type) and that, in the other hand, with the time-history analysis, it is not possible to determine the real deformation capacity of each structure before the collapse mechanism is formed, limited in this work if the local ductility maximum demands showed in beams and columns, over the acceptable values in a practical design point of view.

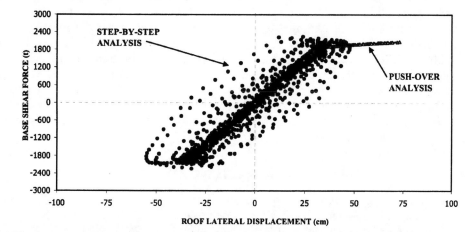

FIGURE 9. Comparisons of the hysteretic step-by-step inelastic analysis curves, in function of the ratios of base shear force–roof lateral displacement, versus to those determined in the Push-over analysis (nominal resistance effects).

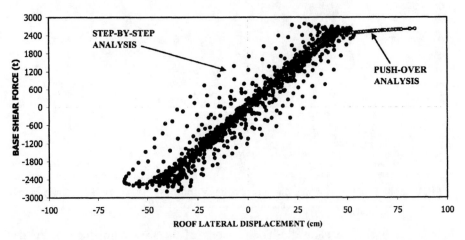

FIGURE 10. Comparisons of the hysteretic step-by-step inelastic analysis curves, in function of the ratios of base shear force–roof lateral displacement, versus to those determined in the Push-over analysis (over-resistance effects).

CONCLUSIONS

The 25 levels structure, with frames in both directions, results with fundamental periods of vibration underneath of the maximum acceleration critic zone of the elastic spectra of SCT-EW record. The structure presents inelastic behavior, mainly in the beam ends; the general tendency of the failure mechanism of both directions accomplish with the philosophy of "strong column–weak beam", as it is recommended in the modern Codes. In general, the acting shear forces in the different structural members do not overpass the proportioned resistance according to the designs developed with the NTC-Concrete; this gives us the certitude that any fragile failure won't be present at all. With the step-by-step dynamic analysis we are able to estimate in a better way the structural behavior closest to reality in case of a severe earthquake. With the Push-over analysis the failure mechanism was never reached, because the established maximum limit of yield dominated in beams or columns. The effects of the possible over-resistance sources that structures have, not included in a conventional design, give us an additional security margin in the response of the structures in front of intense solicitations, allowing us to estimate in a better way the seismic-resistant behavior of the structures.

REFERENCES

1. Mexico City Construction Code, 2004 (RDF-04).
2. Complementary Technical Norms (NTC), NTC-Seismic, RDF-04.
3. Complementary Technical Norms (NTC), NTC-Concrete, RDF-04.

Seismic Response Of Masonry Plane Walls: A Numerical Study On Spandrel Strength

Michele Betti[a], Luciano Galano[a] and Andrea Vignoli[a]

[a]*Department of Civil and Environmental Engineering (DICeA)*
University of Florence, Via di S. Marta 3, I-50139, Florence (ITALY)

Abstract. The paper reports the results of a numerical investigation on masonry walls subjected to in-plane seismic loads. This research aims to verify the formulae of shear and flexural strength of masonry spandrels which are given in the recent Italian Standards [1]. Seismic pushover analyses have been carried out using finite element models of unreinforced walls and strengthened walls introducing reinforced concrete (RC) beams at the floor levels. Two typologies of walls have been considered distinguished for the height to length ratio h/l of the spandrels: a) short beams (h/l=1.33) and b) slender beams (h/l=0.5). Results obtained for the unreinforced and the strengthened walls are compared with equations for shear and flexural strength provided in Standards [1]. The numerical analyses show that the reliability of these equations is at least questionable especially for the prediction of the flexural strength. In the cases in which the axial force has not been determined by the structural analysis, Standards [1] seems to overestimate the flexural strength of short spandrels both for the unreinforced and the strengthened wall.

Keywords: Masonry plane walls, Nonlinear analysis, Pushover analysis, Spandrel strength.

INTRODUCTION

A large amount of masonry Italian buildings are located in earthquake prone areas and most of them have been built before the provision of specific rules for aseismic design. These buildings are composed by load bearing masonry walls arranged along two orthogonal directions and connected to flexible floor diaphragms. Past earthquakes show that the collapse of these buildings is due to out-of-plane failures, originated by large deflections of the diaphragms and poor connection between adjacent walls [2]. If the out-of-plane failure modes are prevented, failure modes in the plane of the walls will be observed. Shear walls with regular geometry are composed of piers (masonry columns) interconnected by spandrels (masonry beams) at the floor levels so, their behavior under in-plane seismic actions is similar to the behavior of the RC framed structures.

The seismic capacity of a building activated by the in-plane failure modes is quite difficult to evaluate because it depends on many geometrical and mechanical factors, among which: a) the pier and the spandrel geometry; b) the opening sizes; c) the presence of architraves; d) the actual stresses due to vertical loads; e) the mechanical characteristics of mortar and blocks, and f) the constructive techniques.

In this paper the attention is focused on the spandrel strength in regular brick masonry walls under in-plane seismic loads. During major earthquakes these elements

CP1020, *2008 Seismic Engineering Conference Commemorating the 1908 Messina and Reggio Calabria Earthquake,*
edited by A. Santini and N. Moraci
© 2008 American Institute of Physics 978-0-7354-0542-4/08/\$23.00

are subjected to high shear forces and bending moments. Depending on the height to length ratio and the amount of axial force three failure modes have been observed: a) flexural failure (initial cracking of the masonry on the tensile side and subsequent crushing of the compressed corner); b) shear failure for diagonal cracking and c) shear failure for sliding. Several approaches have been proposed to predict the ultimate load capacity of the masonry beams (finite element models, macro-elements modeling, equivalent frame methods, etc.). The recent Italian Standards for seismic analysis of existing masonry buildings suggest simplified equations to this purpose [1].

Presented in this paper are the results of a numerical investigation on brick masonry walls subjected to in-plane seismic loads, to verify the reliability of the equations given in [1]. Nonlinear finite element models of the walls and the pushover analysis have been used. Two masonry plane walls are analyzed varying the spandrel height to length ratio, namely: a) short beams (h/l=1.33, h=spandrel height, l=spandrel length) and b) slender beams (h/l=0.5). Furthermore, two typologies of walls have been considered, i.e. unreinforced walls with architraves in the upper zone of the openings, and reinforced walls by the introduction of RC beams at each floor levels.

The results of the pushover analyses are presented here in terms of maximum shear forces and bending moments in masonry beams and are compared with the ultimate strengths predicted by Standards [1].

STRENGTH OF MASONRY SPANDRELS

Consider a masonry wall loaded in its plane by horizontal forces. Standards [1] make a distinction between the case in which the axial force N in a spandrel section has been determined by the structural analysis and the case in which N is unknown. In the first hypothesis the shear strength is expressed by:

$$V_s = ht \frac{f_t}{b} \sqrt{1 - \frac{\sigma_0}{f_t}} \qquad (1)$$

Symbols h and t refer to the height and the thickness of the spandrel section; f_t is the masonry tensile strength; b is a shape coefficient variable from 1.0 up to 1.5 and σ_0 is the average normal stress negative in sign when N is a compressive force [$\sigma_0 = N/(ht)$]. Equation (1) was originally proposed by Turnšek and Cacovic [3] and corresponds to the limit state in which the higher principal stress at the center of the panel equals the masonry tensile strength f_t. If the compression is high this leads to:

$$V_s = ht \frac{f_m}{b} \sqrt{1 + \frac{\sigma_0}{f_m}} \qquad (2)$$

Here f_m is the masonry compressive strength. Equation (2) corresponds to the limit state in which the lower principal stress at the center of the panel equals the masonry compressive strength f_m. Standards [1] do not consider Eqn. (2) because during earthquakes the stress σ_0 in masonry beams is rather low and even positive.

The ultimate bending moment M_u is given by [1]:

$$M_u = -\frac{h^2 t \sigma_0}{2}\left(1 + \frac{\sigma_0}{\chi f_h}\right) \qquad (3)$$

where f_h is the masonry compressive strength in the horizontal direction and χ a shape coefficient equal to 0.85. Eqn. (3) is valid if $\sigma_0 \leq 0$ while M_u is set equal to zero when $\sigma_0 > 0$. Other simplified formulae have been proposed to predict the flexural strength of a masonry panel. If the compression is high the next expression can be used:

$$M_u = \frac{h^2 t}{6}\left(\chi f_h + \sigma_0\right) \qquad (4)$$

If the compression is low or if $\sigma_0 > 0$ it is possible to use:

$$M_u = \frac{h^2 t}{6}\left(f_t - \sigma_0\right) \qquad (5)$$

where the tensile strength of the masonry is taken into account. Equations (4) and (5) are not considered in [1].

When the axial force N is unknown the shear strength of a spandrel with an architrave or a tie on it (RC beam or chain) is evaluated by [1]:

$$V_s = min \{V_t, V_p\} \qquad (6)$$

where V_t is the ultimate shear related to the diagonal cracking failure mode and V_p is the ultimate shear related to the flexural failure mode. V_t and V_p are evaluated by:

$$V_t = ht \, \tau_0; \quad V_p = \frac{2M_u}{l} \qquad (7)$$

$$M_u = \frac{H_p h}{2}\left(1 - \frac{H_p}{\chi f_h ht}\right) \qquad (8)$$

$$H_p = min \{0.4 f_h \, ht, T_c\} \qquad (9)$$

In Eqns. (7), (8) and (9), τ_0 is the masonry shear strength without compression, T_c the ultimate tensile force of the tie and, M_u is the ultimate bending moment of the spandrel.

In Fig. 1 the shear strength $\tau_u = V_s/(ht)$ and the flexural strength $m_u = M_u/(h^2 t)$ are plotted versus the normal stress σ_0 for the parameters $f_m = f_h = 0.757$ MPa, $f_t = 0.165$ MPa, $\tau_0 = f_t/b$, $b = 1.5$, $\chi = 0.85$ and assuming $H_p = 0.4 f_h ht$. In this case the ultimate shear of Eqn. (6) is $V_s = V_t$. Furthermore, the flexural strength given by Eqn. (8) is almost equal to the maximum value provided by Eqn. (3).

789

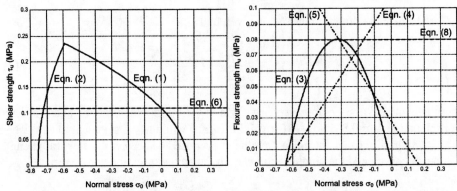

FIGURE 1. Shear and flexural strengths of a masonry spandrel versus the normal stress σ_0.

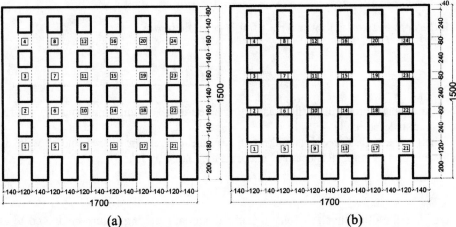

(a)   (b)

FIGURE 2. Geometry of the two walls: a) masonry wall with short spandrels; b) masonry wall with slender spandrels (measures are in cm).

NUMERICAL INVESTIGATION

Figure 2 shows the geometry of the two plane walls considered in this study. They have total length of 17.0 m, total height of 15.0 m and thickness equal to 0.45 m. These models aim to reproduce two traditional 5-stories masonry façades with pier width equal to 1.4 m and opening width equal to 1.2 m. The height of the windows has been modified to reproduce the cases of slender and short beams.

Pushover analyses based on finite element models have been carried out to predict the seismic response of the walls. The general purpose *ANSYS* code has been used [4]. The models have been created by means of 8 nodes three-dimensional isoparametric elements (*Solid65*), whose dimensions are about 0.2×0.2×0.2 m. The mechanical behavior of the masonry material has been represented by the elastic perfectly-plastic formulation of Drucker-Prager (DP). The Willam-Warnke (WW) failure surface simulates the crushing and cracking phenomena [5]. The material parameters required to define the DP model are the cohesion c and the internal friction angle φ. In case of plane stress, the DP yield criterion is:

$$\sqrt{\frac{1}{3}\left(\sigma_x^2+\sigma_y^2-\sigma_x\sigma_y\right)+\tau_{xy}^2}+\alpha\left(\sigma_x+\sigma_y\right)-k=0 \qquad (10)$$

where the parameters α and k are related to c and φ by the following equations:

$$\alpha=\frac{2\sin\varphi}{\sqrt{3}\,(3-\sin\varphi)}; \qquad k=\frac{6c\cos\varphi}{\sqrt{3}\,(3-\sin\varphi)} \qquad (11)$$

Parameters α and k allow to evaluate the yield stresses in case of uniaxial tension and compression (f_{tDP} and f_{cDP}).

The elements *Solid65* have been used also to model the RC beams at the floor levels in the reinforced walls. The values assumed for the definition of the constitutive models of masonry material (PLA1) and reinforced concrete material (RC) are reported in Table 1. F_t of the WW criterion provides the tension cut-off of the DP yield surface, while F_c has been selected greater than f_{cDP} to reproduce the plastic behavior of the masonry in compression. The seismic horizontal forces have been distributed on the whole masonry façade and their intensities have been chosen proportional to the masses.

Figure 3 shows the ultimate cracking patterns for the four cases analyzed. As general remark it is possible to observe that the failure in the unreinforced wall with slender beams is due to a poor whole performance. Here each pier behaves as a single masonry columns. The presence of short spandrels allows collaboration between the piers; so, the resistance against the seismic loads is improved. The RC beams at the floor levels increase this effect allowing a good collaboration between the piers and the spandrels. The ratios between the ultimate base shears (V_b) and the corresponding weights of the walls (P_t) are (see Fig. 3): (a) V_b/P_t= 0.325; (b) V_b/P_t= 0.170; (c) V_b/P_t= 0.382 and (d) V_b/P_t= 0.229.

TABLE 1. Elastic parameters, DP yield criterion and WW failure surface parameters.

Material parameters	Material PLA1	Material RC
E_m (modulus of elasticity)	1962 MPa	32373 MPa
υ (Poisson coefficient)	0.25	0.20
γ_m (masonry specific weight)	17.66 kN/m³	24.53 kN/m³
DP yield criterion		
c (cohesion)	0.177 MPa	2.75 MPa
η (flow angle)	40°	37°
φ (friction angle)	40°	32°
f_{cDP} (uniaxial compressive strength)	0.757 MPa	-
f_{tDP} (uniaxial tensile strength)	0.223 MPa	-
WW failure criterion		
F_c (uniaxial compressive strength)	3.92 MPa	37.28 MPa
F_t (uniaxial tensile strength)	0.165 MPa	2.84 MPa
β_c (shear transfer coefficient for close cracks)	0.75	0.75
β_t (shear transfer coefficient for open cracks)	0.25	0.25

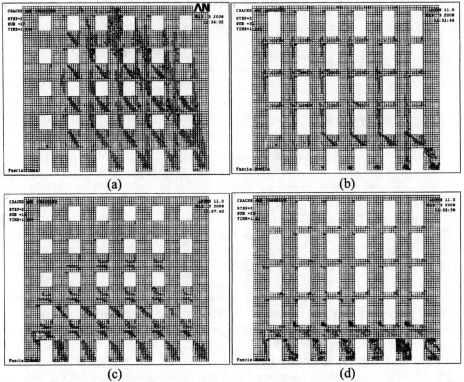

FIGURE 3. Ultimate cracking patterns of the four walls; (a) unreinforced wall with short beams, (b) unreinforced wall with slender beams; (c) reinforced wall with short beams and (d) reinforced wall with slender beams.

DISCUSSION OF THE RESULTS

Pushover analyses provide the maximum shear force and bending moment in the two sections delimiting each masonry spandrel. We distinguish the spandrels in two groups on the basis of their cracking patterns. The spandrels which belong to the first group are fully elastic up to the collapse of the wall (o); therefore, these elements are subjected to low forces and are stressed far away from their ultimate capacity. The spandrels which belong to the second group are characterized by medium-to-wide damage in terms of diagonal or flexural cracking patterns (×). Figure 4 shows the specific maximum shear τ_u and bending moment m_u versus the corresponding average normal stress σ_0 for all the spandrel sections. They are represented together with the regions delimited by Eqns. (1) to (9) for the different cases [it has been fixed $f_t=F_t$, $f_m=f_h=f_{cDP}$, while other parameters are the same as in Fig. 1)].

For the unreinforced wall with short spandrels the flexural strength predicted by Eqn. (8) is largely overestimated (Fig. 4b) because the bending moment is greater than the limit furnished by Eqn. (8) in only one spandrel section. Small increases of spandrel bending moments have been obtained in the reinforced wall (Fig. 4d).

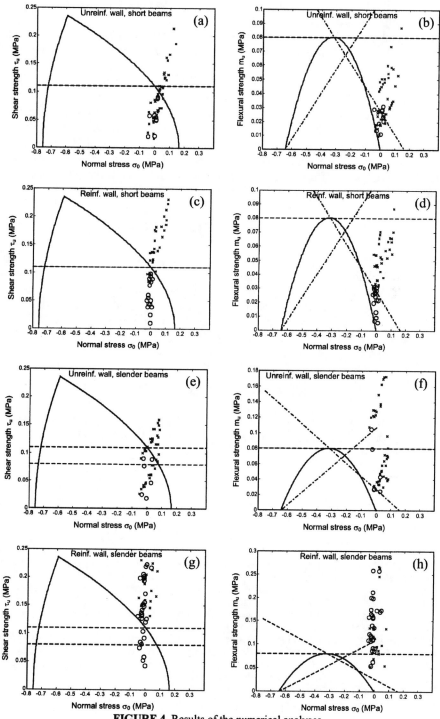

FIGURE 4. Results of the numerical analyses.

This result is due to the position of the RC beams that cross the spandrels approximately at midheight.

On the contrary Eqn. (3) underestimates the flexural spandrel strength because it neglects the tensile masonry strength f_t. More reliable appears Eqn. (5) especially for the reinforced wall. In the case of the reinforced wall and slender beams (Fig. 4h), all Eqns. predict flexural strengths which are lower than the values obtained by the pushover analysis. A similar trend is true in the case of the unreinforced wall (Fig. 4f) in which the application of Eqn. (5) leads again to results from the safe side. Nevertheless, the application of Eqn. (8) is questionable because in many damaged spandrels the bending moments are lower than those predicted by this formula.

Shear strengths are well predicted by Eqns. (1) and (6) for short and slender beams in the reinforced walls (Figs. 4c and 4g); more difficult is the interpretation of results in the case of the unreinforced walls (Figs. 4a and 4e).

CONCLUSIVE REMARKS

The results of the numerical analyses herein presented offer useful indications on the subject. Reliability of formulae provided by Standards [1] for flexural strength seems questionable for two main reasons. When the axial force is known, Standards [1] neglect the tensile strength of the masonry [(Eqn. (3)], so the flexural strength of a spandrel is largely underestimated. If the axial force is unknown, the application of Eqn. (8) is unsafe for short masonry beams. The application of formulae given in [1] to evaluate the ultimate shear of the spandrels is acceptable.

ACKNOWLEDGMENTS

Founding from Research Projet ReLuis (Research Line 1: "Valutazione e riduzione della vulnerabilità di edifici in muratura") is gratefully acknowledged.

REFERENCES

1. O.P.C.M. 3274 del 20/3/2003. Primi elementi in materia di criteri generali per la classificazione sismica del territorio nazionale e di normative tecniche per le costruzioni in zona sismica. G.U. 8/5/2003, n. 105, S.O. n. 72 and O.P.C.M. 3431 del 3/5/2005. "Ulteriori modifiche ed integrazioni all'O.P.C.M. 3274 del 20/3/2003" (in Italian).
2. G. Magenes and G. M. Calvi, "In-plane seismic response of brick masonry walls", *Earthq. Engng. and Struct. Dyn.* **26**, 1091-1112 (1997).
3. V. Turnšek and F. Cacovic, "Some Experimental Results on the Strength of Brick Masonry Walls" in *Proceedings of the 2nd International Brick Masonry Conference*, Stock on Trent, 1971, pp. 149-156.
4. Ansys Inc. *Users's manual*, Swanson analysis systems, 1992.
5. K. J. Willam, and E. D. Warnke, "Constitutive model for the triaxial behaviour of concrete" in *Proc. Int. Ass. for Bridge and Struct. Engng.*, ISMES, Bergamo, 1975, pp. 174-186.

Investigations On Historic Centers In Seismic Areas: Guidelines For The Diagnosis

Luigia Binda[a], Giuliana Cardani[a], Antonella Saisi[a], Claudio Modena[b], Maria Rosa Valluzzi[b]

[a]Dept. of Structural Engineering, Politecnico di Milano, P.zza L. Da Vinci 32 ,20133 Milano, Italy.
[b]Dept. of Construction and Transportation Engineering, University of Padua, v. Marzolo 9, 35131 Padova, Italy

Abstract. After the earthquake that hit central Italy in 1979, many small historic centers were restored. A subsequent seismic event occurred in 1997 in Umbria-Marche regions revealed that some techniques used in the previous interventions were not successful due to low durability of new materials and/or incompatibility between the new and the existing materials and structures. An extensive investigation on four small typical historic centers in Umbria was carried out. The objectives of the research were: (i) to define a methodology for the vulnerability analysis of historic buildings at the level of the historic centre, (ii) to collect information on the effectiveness of the repair techniques both traditional and new, (iii) to set up Databases storing the information useful to prepare rescue plans, (iv) to use the collected knowledge for the implementation of reliable models for the vulnerability analysis, (v) to prepare guidelines for investigation and vulnerability analysis.

Keywords: historic centers, masonry, vulnerability, diagnostic investigation, methodology.

INTRODUCTION

The 1997 earthquake which hit the Umbria and Marche Regions in the center of Italy, gave the occasion to learn about the effectiveness of the repair and retrofitting techniques applied on historic masonry buildings after the previous seismic event of 1979. Most retrofitting mainly performed with upgrading interventions (substitutions of timber floors and roofs with r.c., jacketing, etc.), have caused unforeseen and serious out-of-plane effects (large collapses, local expulsions), due to the "hybrid" behavior activated between the new and the old structures [1], not predictable, at the time, by the existing assessment methods suggested by the Italian standards.

It appeared immediately clear that the main cause of inappropriate choice for the intervention techniques was due to the lack of knowledge on the material and structural behavior of the peculiar type of construction techniques used in the past for the historic buildings. The aim of the research supported by the Italian Civil Protection Department was to set up, for historic centers, systematic Data-bases storing the information useful to prepare rescue plans and to design interventions for the preservation of the cultural heritage. The object of the research was not the single building, but the whole historic centre (even if small), frequently with a large number of complex buildings. These buildings are the results of the addition over centuries of volumes differently connected one to the other, forming irregular geometries and

CP1020, *2008 Seismic Engineering Conference Commemorating the 1908 Messina and Reggio Calabria Earthquake*,
edited by A. Santini and N. Moraci

structures. Their complexity, as they were considered "minor architecture", even if with meaningful testimonies of cultural heritage, was totally ignored by the previous codes and by the applied repairs. For all the above mentioned reasons there was the necessity of defining a "minimal" investigation program, in order to support the designers in choosing the right analytical models for the safety definition and the appropriate intervention techniques for their projects.

The research was focused on four sample areas: Montesanto di Sellano, Roccanolfi di Preci, Campi Alto di Norcia and Castelluccio di Norcia, all located in the Perugia province. They are characterized by different typologies of buildings (simply isolated buildings in Montesanto, row buildings in Campi and complex buildings in Roccanolfi and Castelluccio) and by different levels of damage [2]. The extensive survey of the buildings was useful to produce an abacus of the typical damages occurring in constructive typologies, as already previously done for churches [3]. The better knowledge of damages led to the consequent systematization of the mechanical models able to describe their specific behavior by kinematics models, both for in-plane and out-of-plane mechanisms. Procedures for the evaluation of the seismic vulnerability, based on the application of single or combined kinematics models involving the equilibrium of macro-elements [4], have been developed by direct comparison with the real damage occurred [5,6].

A MULTILEVEL APPROACH TO THE ANALYSIS OF THE HISTORICAL BUILDINGS

The seismic vulnerability assessment of historical buildings should consist of an articulated procedure which takes advantage of two sources of information: (*i*) indirect (e.g. archives and bibliographic information in order to reconstruct the evolution of the building, its load history and the list of the past earthquakes) and (*ii*) direct, as: a) geometrical and photographic survey; b) typological analysis of the building, aimed at understanding their formation and growing; c) chronological information through direct survey of layering of masonry, renderings and details (stratigraphic survey) ; d) survey of the masonry section and surface texture; e) survey of the crack pattern; f) analysis of the main structural elements including load-bearing walls, roofs, floors and vaults, staircases, and of their connections, g) damages, h) effectiveness of past repair; k) laboratory characterization of material samples; i) on site tests).

When the investigation phase is completed, proper procedures for the evaluation of the seismic vulnerability have to be considered and calibrated on the basis of the observations and of the experimental results detected on site. The identification of the most typical failure mechanisms activated by the earthquake add more information to choose appropriate analytical models, which can be used both for prediction analyses and for design of proper interventions.

The villages of Montesanto and Roccanolfi, seriously damaged by two earthquakes in the last twenty years, show an interesting scenario for the study of the behavior of historic masonry buildings repaired and "upgraded" in the '80s following the Italian Code. Some buildings presented retrofitting interventions made before the 1979 seismic event: timber rods with steel connections to the walls, steel rods, buttresses and buttressed walls. These techniques can be defined "traditional" and were commonly and efficiently applied in the past before the use of reinforced concrete. Furthermore, buildings repaired according to the recent regional plan of intervention were identified and the following repair techniques were applied: stiffening or substitution of timber floors with r.c. beams or plates, tie concrete beam insertion in the walls, cementitious grout injections, reinforced injections, roof substitution, jacketing, local replacement in the walls, mortar joints re-pointing, etc. In some buildings traditional and modern interventions were mixed. The survey allowed also to detect many failures of the more recent repairs due to incompatibility of the new materials or techniques with the existing ones.

Building Typology

When dealing with historic centers the detection of the building typologies is the first step to be carried out by an accurate geometrical survey. Effort should also be made to find through historic documents, and also by the on site observation, the modifications the building was subjected to over time: born as an isolated building, it could have become a row building or a complex one, after the addition of several volumes. The more complex the building is the more difficult is the detection of its vulnerability.

All data (direct and indirect information) were systematically collected in a survey form called "Template for the description of the building typology and the seismic damages" realized by DIS-Politecnico di Milano, considering different level of analysis and useful to be recorded in a Data-base for consultation. Representative typologies of a centre can be easily recognizable through similar features (number of stores, exposure, type of facade, architecture details as doors and windows, material and structural elements, etc.). Blocks or parts of buildings may be identified and surveyed, also adopting axonometric representations which can better show the different levels of the ground. The stratigraphical method [7] was adopted to subdivide the complex buildings into homogeneous blocks characterized by relative chronological relationships (Fig. 1).

FIGURE 1. Structural evolution based on volumes of a complex building

Every block corresponds to a unique building phase, recognized from constructive details and different masonry typologies.; its relationship with the other blocks may be "preceding" or "subsequent", even if there is no possibility of an absolute dating. Critical connections between blocks have been investigated, so to clarify the phases of expansion and transformation of the complex. The most diffused use of the buildings observed in the studied area is a single family unit inserted in a row or in a more complex aggregate, with two or more storey, built with a simple technique in stonework and with timber roof and floors. Sometimes, due to the ground slope, buildings are generally with three storey: the first one with an entrance at the lower street (for stables or deposits), one in the middle and the last with the entrance at the upper street (for living places). The lowest storey is partially excavated in the natural rock (Fig. 2). The ground floor rooms are covered by barrel vaults, that, despite the several seismic events, are still well preserved, even in the partially collapsed buildings.

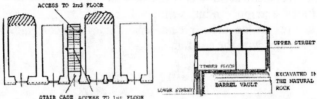

FIGURE 2. The row building typology of Campi Alto di Norcia (PG)

Materials and construction techniques

The structural performance of a masonry wall structure can be understood provided the following factors are known: i) its geometry; ii) the characteristics of its masonry texture (single or multiple leaf walls, connections between the leaves, joints empty or filled with mortar, physical, chemical and mechanical characteristics of the components (mortar, brick, stone); iii) the characteristics of masonry as a composite material. In the case of multiple leaf masonry, the masonry texture, which strongly influences the bearing capacity of the wall, often can not be easily identified. Furthermore the characteristic strength and stiffness of a highly non-homogeneous material is difficult to be experimentally determined being the strength and the elastic parameters of the components not representative of the global strength and deformability of the masonry (f.i. made by small pebbles or by two external leaves even well ordered but not mutually connected and containing a rubble infill).

In order to evaluate the characteristic of the masonry, a classification of the different cross sections locally recognizable should be carried out, especially of the multiple leaf ones [4,8]. The presence or not of some characteristics, as the leaves transversal connection elements called "diatoni", can be a discriminating parameter for the evaluation of the wall mechanical behavior. In the proposed methodology, the masonry is surveyed by pictures, obtained as parallel as possible to the masonry surface, and by placing a graduated stick close to the texture for the wall dimension. The schematic draw should represent also the presence of voids, useful for possible strategies of intervention (e.g. grout injection). Figure 3a shows an example of masonry texture with the corresponding cross section, which was possible to survey on a damaged stable after the seismic event [9]. It is worth to remark that textures regular on façade often do not correspond to regular morphology in the section and, therefore, the analysis can not disregard the proper investigation of the arrangement of materials in the thickness of the wall. In case the cross section is not completely accessible, it is suggested to carefully and slowly disassemble a small portion of a wall in dry conditions, over two thirds of the wall section and survey the section morphology (Fig.3b).

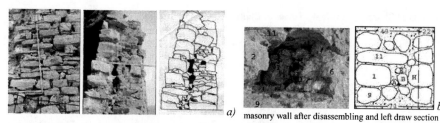

a)   masonry wall after disassembling and left draw section   *b)*

FIGURE 3. Survey of the wall constructive typology: a) survey of a cross section on a damaged building; b) reconstruction of the wall section through a small disassembling

When working at a urban scale, a "minimal" investigation program is suggested to collect significant information, as for the building typologies by sampling materials from representative buildings. The aim is to identify the different materials used for the masonry walls and their mechanical and physical properties and behavior. This investigation is also useful to detect compatible materials and techniques for representative prevention and repair measures. On the basis of the geometric and stratigraphical surveys and of the surveys of the crack patterns, the in situ tests should be carried out on chosen strategic points. The masonry investigations can be performed through: i) flat jacks tests; ii) sonic tests; iii) sampling of materials for their chemical-physical-mechanical characterization [10]. A diagnostic investigation was possible on some private (houses and stables) and religious buildings of the four centers. In Figure 4a,b the results of the sonic tests and simple and double flat jack tests performed on some sampled buildings of Campi di Norcia are reported. Results pointed out the differences in stress-strain behavior of masonry belonging to important buildings or monuments (churches or the bell towers) in comparison with private poor buildings. In particular, it is possible to see that both sonic velocity and stress-strain curves obtained by flat jack tests are in agreement, assuming e.g. the elastic modulus higher for more important constructions.

Structural damage survey, abacus of collapse mechanisms

The assessment of seismic vulnerability of masonry buildings requires the identification of the damage and collapse mechanisms activated by the earthquake. The current practice in Italy is to take into account only a limited number of modes of failure; some of them are neglected implicitly, assuming in advance a strength capacity of certain structural typologies after appropriate retrofitting measures. On the contrary, the possibility of damage prediction is related to the knowledge of the highest number of possible mechanisms of progressive deterioration or sudden failure. The survey and drawing of the crack patterns is with no doubts important. The interpretation of the crack pattern can be of great help in understanding the state of damage of the structure, its possible causes. Damages, which are frequently attributed to the earthquake, can have a different nature and can be caused by excessive dead loads or soil settlements, or simply by lack of maintenance. A complete survey of the structural damages, reported on the front walls and not only in the plans, can help in understanding the vulnerable points of the structure and also the possible future mechanisms.

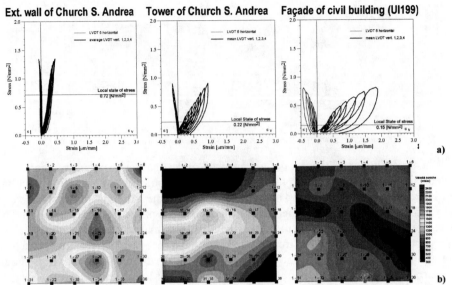

FIGURE 4. Results obtained on three different masonry specimens: a) stress-strain behaviour with flat-jack tests and b) sonic pulse velocity. The masonry specimens refer to the external wall of (left)a church, (middle) of a bell tower and (right) of a civil building.

The interpretation of failure or of the damage mechanisms in the case of complex buildings is difficult. Blocks or parts of buildings should be identified and surveyed, better adopting axonometric representations. It was observed that blocks placed at the free extremities of a complex building are less constrained and therefore more severely damaged, with local collapses and large cracks [4]. When collapses occur in the internal part of a complex they frequently affect non repaired building blocks, adjacent to the repaired ones. In the central part of row buildings with presence of decayed floors and roofs, large continuous deformations and tilting of the walls are generally detected. Once the damage mechanisms have been singled out and defined, appropriate calculations should be adopted for modeling the observed behavior, which is one of the most difficult tasks, due to the complexity of the structure. The extensive survey, carried out by the authors in Umbria after the 1997 earthquake, allowed to set up an abacus of failure mechanisms referred to different building typologies, and taking into account also the effects of previous repairs. In Figure 5 some examples of the different mechanisms are given distinguishing between: a) the presence of traditional and b) modern intervention techniques [10]. The adopted diagnostic approach is based on the recognition of local and global collapse mechanisms traceable to in-plane or out-of-plane seismic action.

SEISMIC VULNERABILITY ANALYSIS

The methodology proposed for assessment of seismic vulnerability of existing buildings in historical centers concerns the application of simple kinematics models able to describe the mechanical behavior of structural components and assemblages (macro-models) [3,4]. The analysis is performed at global level by using the program "Vulnus", set-up at the University of Padova; at local level the most significant elements in the buildings are selected and describing the single mechanisms ("C-sisma" program), both for in-plane and out-of-plane collapses (Figure 6).

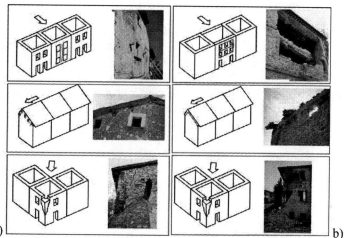

a)   b)

FIGURE 5 - Some examples from the collapse mechanisms abacus in presence of: a) traditional, b) modern intervention techniques

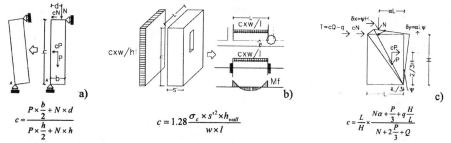

$$c = \frac{P \times \frac{b}{2} + N \times d}{P \times \frac{h}{2} + N \times h}$$

$$c = 1.28 \frac{\sigma_c \times s'^2 \times h_{wall}}{w \times l}$$

$$c = \frac{L}{H} \times \frac{N\alpha + \frac{P}{3} + q\frac{H}{L}}{N + 2\frac{P}{3} + Q}$$

FIGURE 6 - Examples of simple kinematics models for out-of-plane (a: overturning of a solid wall) (b: crushing of the masonry) and in-plane mechanisms (c: in-plane overturning).

The Vulnus methodology is able to define two indexes, I1 and I2, concerning the in-plane shear resistance and out-of-plane collapse mechanisms, respectively. It is able to combine different mechanisms for global vulnerability analyses of buildings with sufficient regularity (both in plane and in elevation) and limited height (three storey or less), and take into account the type of connection among the structural elements. The significant parameter describing the kinematics models is the collapse coefficient $c=a/g$, which corresponds to the seismic masses multiplier characterizing the limit of the equilibrium conditions for the considered element. Preliminarily, if the seismic degree of the zone is given, it is possible to execute safety assessments of the buildings in seismic conditions according to the current standards prescriptions (e.g. $c_{min}=0.28$ for the Umbria region). The proposed procedures, applied to different typologies of buildings in this region (isolated, rows and complexes), showed their reliability, in comparison with systematic adoption of typical assessment methods based on the "box behavior" of the structure, which take into account only the in-plane shear strength of the masonry panels composing the walls, as also suggested by Italian standards until now [11].

CONCLUSIONS

The investigation program applied on the center of Italy after the '97 seismic event has allowed to gather a great amount of data useful to calibrate analytical procedures able to assess the seismic vulnerability of historical centers. These procedures can also be used for prediction analyses in non-damaged sites in order to design suitable interventions. In collaboration with the Italian Civil Protection Agency, a Data-base, built on the basis of filled templates for each building and with many different research keys, was realized and will be available on its website. This data-base could be linked to other existing Data-base. A tool like the Data-basecan be used and implemented by the Civil Protection Agency also with data coming from other historic centers, by municipalities to plan and control interventions and by local designers as a data source. The methodology applied to the four centers is now well calibrated and can be used for other similar cases.

REFERENCES

1. D. Penazzi, M.R. Valluzzi, G. Cardani, L. Binda, G. Baronio, C. Modena, "Behaviour of Historic Masonry Buildings in Seismic Areas: Lessons Learned from the Umbria-Marche Earthquake", Proc. *12th IBMaC*, Madrid, Spain, 2000, pp. 217-235.
2. L. Binda, G. Cardani, A. Saisi, C. Modena, M.R. Valluzzi, "Multilevel Approach to the Analysis of the Historical Buildings: Application to Four Centers in Seismic Area finalized to the Evaluation of the Repair and Strengthening Techniques", Proc. *13th IBMaC*, RAI Amsterdam, the Netherlands, on CD-ROM, 2004.
3. F. Doglioni, A. Moretti, V. Petrini, *Le chiese e il terremoto. Dalla vulnerabilità constatata nel terremoto del Friuli al miglioramento antisismico nel restauro, verso la politica di prevenzione*, ed. LINT, Trieste, Italy, 1994.
4. A. Giuffrè, *Sicurezza e conservazione dei centri storici: il caso di Ortigia*, Ed. Laterza, Bari, 1993
5. M.R. Valluzzi, E. Michelon, L. Binda, C. Modena, C., "Modellazione del comportamento di edifici in muratura sotto azioni sismiche: l'esperienza Umbria–Marche", 10th Conf. ANIDIS *L'ingegneria sismica in Italia*, Potenza, 2001.
6. A. Avorio, A. Borri, M. Corradi, *Ricerche per la ricostruzione. Iniziative di carattere tecnico e scientifico a supporto della ricostruzione*, Regione Umbria, DEI, Rome, Italy, 2002.
7. T. Mannoni, *Caratteri costruttivi dell'edilizia storica*, Ed. Sage, Genova, Italy, 1994.
8. L. Binda, D. Penazzi, A. Saisi, "Historic Masonry Buildings: necessity of a classifi¬cation of structures and masonries for the adequate choice of analytical models", 6th Int. Symp. *Computer Methods in Structural Masonry*, Rome, Italy, 2003, 168-173.
9. L. Binda, G. Cardani, A. Saisi, Application of a multidisciplinary investigation to study the vulnerability of Castelluccio (Umbria), *9th Int. Conf. STREMAH*, Eds. C.A. Brebbia, A. Torpiano, Malta, Section 4, ISBN 1743-3509, 2005, 311-322.
10. L. Binda, G. Cardani, A. Saisi, C. Modena, M.R. Valluzzi, L. Marchetti, Guidelines for Restoration and Improvement of Historical Centers in Seismic Regions: the Umbria Experience, *IV Int. Seminar, SAHC*, Padova, Vol. 2, 2004, 1061-1068.
11. M.R. Valluzzi, G. Cardani, L. Binda, C. Modena, Seismic vulnerability methods for masonry buildings in historical centres: validation and application for prediction analyses and intervention proposals, *13th WCEE*, Vancouver, B.C., Canada , 2004.

Basis of Design & Seismic Action for Long Suspension Bridges: the case of the Messina Strait Bridge.

Franco Bontempi[a]

[a]University of Rome "La Sapienza", School of Engineering
Via Eudossiana 18- 00184 Roma ITALY
franco.bontempi@uniroma1.it

Abstract. The basis of design for complex structures like suspension bridges is reviewed. Specific attention is devoted to seismic action and to the performance required and to the connected structural analysis. Uncertainty is specially addressed by probabilistic and soft-computing techniques. The paper makes punctual reference to the work end the experience developed during the last years for the re-design of the Messina Strait Bridge.

Keywords: Long Suspension Bridges, Seismic Action, Structural Analysis, Bridge Engineering, Structural Design, Performance-based Design.
PACS: 89.20.Kk

INTRODUCTION

The general framework for the design of an extraordinary structure like the Messina Strait Bridge can be set with reference to the scheme of Fig.1. Here are collected the phases necessary to find in a constructive approach the solution for the design problem:

a)   definition of the <u>structural domain,</u> as bridge geometrical and material characteristics;

b)   definition of the <u>design environment</u> where the structure is immersed with specific attention to the specifications of the
   i.   <u>natural actions</u> (wind & temperature and soil & earthquake);
   ii.   <u>antropic actions</u> (related to highway & train traffic);

c)   <u>assessment of the performances</u> obtainable by the current structural design configuration, resulting from accurate and extensive structural analysis developed on models, both analytically or experimentally based;

d)   alignment of expert judgments and emergence of decision about the <u>soundness of the design,</u> first in qualitative terms then in quantitative terms;

e)   <u>negotiation and reframing of the expected performances,</u> in comparison with what has been obtained by the analysis and by the knowledge acquired working on the problem.

CP1020, *2008 Seismic Engineering Conference Commemorating the 1908 Messina and Reggio Calabria Earthquake,*
edited by A. Santini and N. Moraci
© 2008 American Institute of Physics 978-0-7354-0542-4/08/$23.00

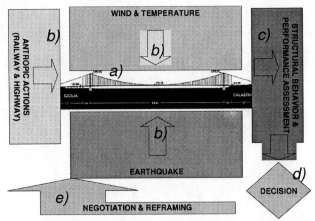

FIGURE 1. Performance-based Design framework for the Messina Strait Bridge

This scheme is recognized as a *Performance-based Design* approach. It is worth to note two features:

1) the strongly affection by heuristics and experience of the problem formulation and of the recognition of the solution; particularly, the engineering deontology is the only capable to correctly address the interest of all the stakeholders;

2) the central role of the numerical modeling, as the unique knowledge engine able to connect all the details of the theory and of the experimentation in a truly comprehensive representation of the problem and of its solution.

STRUCTURAL DECOMPOSITION

The whole structure of the bridge is organized hierarchically as shown in Fig.2, where one has considered structural parts categorized in three levels:

I. MACROSCOPIC, related to geometric dimensions comparable with the whole construction or with general role in the structural behavior; the parts so considered are called structural systems: one has essentially three systems,
   - principal, connected with the main resistant mechanism,
   - secondary, connected with the structural part loaded directly by highway and railway traffic,
   - auxiliary, related to specific operations that the bridge can normally or exceptionally face during its design life: serviceability, maintainability and emergency.

II. MESOSCOPIC, related to geometric dimensions still relevant if compared to the whole construction but connected with specialized role in the structural system; the parts so considered are called structures or substructures;

III. MICROSCOPIC, related to smaller geometric dimensions and specialized structural role: these are components or elements.

The meaning of this subdivision is manifold:

a) <u>the organization of the structure is first of all naturally connected with the load paths that must be developed by the structure itself</u>; in this way, this subdivision can clear the vision of the design team about the duties of each part of the structure; this identification is essential in the Conceptual Design, and it is implicitly a precondition for the accomplishment of the Performance-based Design, where the importance of form is strongly emphasized, leading, for example, to concepts like integral bridges;

b) <u>parts belonging to different levels of this organization require different reliability properties</u>; with regard to structural failure conditions, this decomposition allows single critical mechanisms to be ranked in order of risk and consequences of the failure mechanism: for example, in Fig.2 there are, indicated by yellow, orange and red frames, different level (decreasing) of permissible damage; these qualitatively assumed requirements can be quantitatively translated defining different levels of stress in the different bridge parts; all these considerations lead to the so-called crisis canalization;

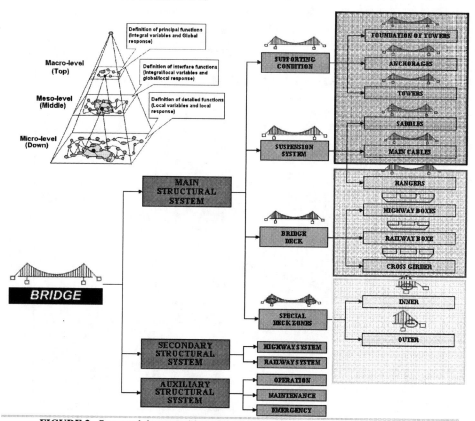

FIGURE 2. Structural decomposition of the bridge: left, macro-level, right meso-level

c) <u>there are strong relationships between life cycle and maintenance of the different parts</u>: with reference to their structural function, the safety required levels and their reparability, structures and sub-structures are distinguished in primary components (critical, non-repairable or which require the bridge to be placed out of service for a consistent period in order for them to be repaired), and secondary components (repairable with minor restrictions on the operation of the bridge). As specific case, one can consider the whole hanger system, which can be classified as a main structural component in relation to the global structural safety of the bridge, whereas a single hanger group can be considered a secondary component due to its reparability and/or replacement ability.

d) for the operative aspects, Fig.3 is the manifestation of the strategy introduced with Fig.2: in fact, the whole structural analysis can be subdivided in coordinated phases as shown. There, one has the connections among different performance levels and different design variables, while the link is established by efficient modeling, at different structural scale but globally related, being possible that the results from model at one level are the input for another model at another scale. To fix the concept, think for example to the fatigue checks that requires fine description of the hot spots at a very micro-level scale coupled with the description of wind global structural response.

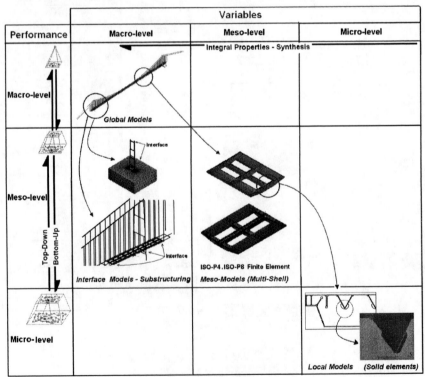

FIGURE 3. Relationship between performance and variables as stemming from the structural decomposition of the bridge

DEPENDABILITY

The quality of such a large bridge is multifaceted. The holistic and comprehensive measure of the quality of this complex structure is called <u>dependability</u>. This concept can be synthetically defined as the grade of confidence on the safety and on the performance of a structural system. It is an integrative concept that, for a construction, encompasses the following attributes:

1) <u>availability</u>: readiness for correct serviceability;
2) <u>reliability</u>: continuity of correct serviceability;
3) <u>safety</u>: absence of catastrophic consequences on the users and the environment;
4) <u>security</u>: absence of catastrophic consequences for illegitimate antropic actions;
5) <u>integrity</u>: absence of improper system state alterations;
6) <u>maintainability</u>: ability to undergo repairs and modifications.

The means to attain dependability can be summarized as:
- <u>fault prevention</u>: how to prevent the occurrence or introduction of faults;
- <u>fault tolerance</u>: how to deliver correct service in the presence of faults;
- <u>fault removal</u>: how to reduce the number or severity of faults;
- <u>fault forecasting</u>: how to estimate the present number, the future incidence, and the likely consequences of faults.

Absence of catastrophic consequences and fault tolerance are guaranteed by the <u>structural robustness</u>. This is the capacity of the construction to undergo only limited reductions in its performance level in the event of departures from the original design configuration as a result of a) local damage due to accidental loads, b) secondary structural elements, c) being out of service for maintenance purpose, d) degradation of their mechanical properties.

In general terms, the following recommendations apply:
a) appropriate contingency scenarios shall be identified, i.e. scenarios of possible damage together with suitable load scenarios, able to characterize the structural robustness in the various conditions of service;
b) analyses shall be conducted in order to explore and to bound structural safety and performance levels of the structure in these conditions; this analysis can be correctly developed in a fuzzy based framework.

Specifically, it was requested that:
c) for <u>Ultimate Limit State</u> (ULS), in addition to the accidental loads specifically defined, it was necessary to consider the contingency scenarios that envisage the failure of the support of one extremity of a cross beam, at the most unfavorable location along the structure; the analysis had to be done in the dynamic field, assuming the instantaneous rupture of the support;
d) for <u>Structural Integrity Limit State</u> (SILS), in addition to the accidental loads specifically defined, it was necessary to consider the contingency scenario of the failure of one crossbeam and the section of main longitudinal deck girders connected to it; the analysis had to be done in the dynamic range, considering the sudden detachment of a section of the main deck 60 m long, at the most unfavorable location along the structure.

The spreading out of the reliability basis of the bridge requires:
a) the definition of the design life: Ld = 200 years;
b) the identification of the return periods for the actions shown in Tab.1 and the connections with the Limit States: specifically, Level 1 regards the Serviceability Limit State (globally denoted by SLS) with further distinction in two grades (SLS1 and SLS2) of increasing loss of functionality; Level 2 is the Ultimate Limit State (ULS), which refers to the attainment of the ultimate strength of a structural component; Level 3 is the Structural Integrity Limit State (SILS), which refers to the survival of the primary structure even if significant damage may have occurred.

TABLE 1. Loading level and return period

Loading Level	Limit States	Acronym	Return Period
1	Serviceability	*SLS1*	50 years
		SLS2	200 years
2	Ultimate	*ULS*	2000 years
3	Structural Integrity	*SILS*	Accordingly to the contingency scenarios considered

The organization of the safety of such a complex structure cannot be defined only in quantitative terms. It is typical of real composite situations to express judgments by fuzzy terms instead of crisp ones. In this sense, the safety requirements are arranged by the following phases:

I)   Definition of the increasing grades of damage levels of Tab.2.

II) Association of the structural parts defined by the decomposition of Fig.3 to different damage grades at the reaching of the different limit states as shown in Tab.3. In this way, for example, for the towers, there will be the following sequence of states on the increase of the actions intensity:
   i.   before the reaching of the SLS, there is no damage (ND);
   ii.   passing SLS, until ULS, towers suffer minor damage (MD) and, later on, repairable damage (RD);
   iii. over the SILS, towers reach significant damage (SD).
   Specific attention is devoted to the components that can be substituted by maintenance process, without serviceability interruptions.

III) Indications of the compatible stress levels, as in Tab. 4 for the crucial main suspension systems and for the hangers.

IV) Remarks of the role of the structural robustness. In fact, the structural configuration of the bridge must prevent the progressive propagation of failure mechanisms, by means of a suitable definition, both at the local and at the global levels, of structural details and the provision of appropriate lines of defense. A suitable structural compartmentalization must therefore be sought, if necessary by means of an appropriate arrangement of connections. In particular, the local collapse of a section of the deck structure as a consequence of the failure of the corresponding hangers and cross beams shall not propagate along the whole deck.

TABLE 2. Definition of damage levels

Damage grades	Acronym	Description
1 NO DAMAGE	*ND*	All structural elements and restraint systems retain their nominal performance capacity remaining in the elastic field and do not present any significant degradation due to fatigue.
2 DEGRADATION DAMAGE	*DD*	Degradation of mechanical properties of materials after an appropriate period of service due to environmental actions (corrosion) or cyclical actions (fatigue). These effects shall be allowed for the over sizing of structural sections and shall be eliminated or minimized through scheduled maintenance activities.
3 MINIMAL DAMAGE	*MD*	Occurrence of localized slight inelastic behavior which does not alter the overall performance capacities of the bridge. This damage can be repaired by means of ordinary maintenance operations, guaranteeing the road and rail traffic.
4 REPAIRABLE DAMAGE	*RD*	Occurrence of localized inelastic behavior which alters the overall performance capacities of the bridge. This damage can be repaired by extraordinary maintenance operations, involving partial and temporary closures of the bridge.
5 SIGNIFICANT DAMAGE	*SD*	Occurrence of inelastic behavior which significantly alters the overall performance capacities of the bridge. It corresponds to a serious damage of the structure which may require the reconstruction of entire structural components. The damage can be repaired by significant extraordinary maintenance operations, which may involve prolonged closures of the bridge.

TABLE 3. Association between damage grades and limit states conditions for different structural parts of the bridge

Macro-Level Structural Systems	Structures	Meso-Level Sub-structures	SLS	ULS	SILS	
Main	Restraint / support system	Foundations of towers	ND	MD	RD	SD
		Anchor blocks	ND	MD	RD	SD
		Towers	ND	MD	RD	SD
	Main suspension system	Main cables	ND	MD	RD	SD
		Saddles	ND	MD	RD	SD
	Secondary suspension system	Hangers system	ND	MD	RD	SD
		Single hanger	DD	RD	SD	SD
	Main standard deck	Cross girders	ND	MD	SD	SD
		Rail box girders	ND	MD	SD	SD
		Road box girders	ND	MD	SD	SD
	Special deck regions	End restraint regions and expansion joints	DD	MD	SD	SD
		Internal restraint regions and Restraint devices	DD	MD	SD	SD

TABLE 4. Definition of compatible stress levels for the main structural systems

Compatible stress levels	Serviceability Limit States (*SLS*)	Ultimate Limit States (*ULS*)
Main cables	Ultimate stress / 2.10	Ultimate stress / 1.67
Hangers	Ultimate stress / 1.67	Ultimate stress / 1.40

SEISMIC ACTION AND SAFETY ASESSMENTS

As said, long suspension and cable stayed bridges are complex structures because its design involves problems regarding nonlinearity, uncertainties and interactions among loads and environmental [Sgambi & Bontempi 2004]. For these reasons, the seismic analysis of a long bridge is a classical ill-conditioned problem. It is important to observe that some (but not all) of these parameters are referred to the composition of the soil. However, the complete knowledge of the soil does not make the problem well-structured because a great amount of uncertainties persists in the problem. In fact, it is impossible to define with precision the real position of the epicenter of the seismic event, or the mass of the structure (with trains and cars) during the earthquake.

In the following just one main aspect of the structural response of the Messina Strait Bridge under seismic action is shown: one refers to the behavior of a so long bridge connected to potentially partially uncoupled excitations between the Sicily side and the Calabria side. With the schemes of Fig.4, actually one considers:

a) asynchronous seismic excitations; 50 different simulations of earthquake events artificially generated were developed for different peak ground acceleration (PGA) and applied to the model of Fig.5; results are summarized represented in Fig.6 and Fig.7;

b) the possibility to have crustal displacements; one considers 4 kind of imposed displacement, between the two legs of Sicily and the two legs of Calabria bridge towers: Ced S – full symmetric; Ced U – only Sicily; Ced D partially symmetric; Ced C partially unsymmetrical; the size of the single displacement is 1.00 m; in this case, the hierarchical sub-structured model of Fig.8 was used, leading to the stress status reported in Tab.5.

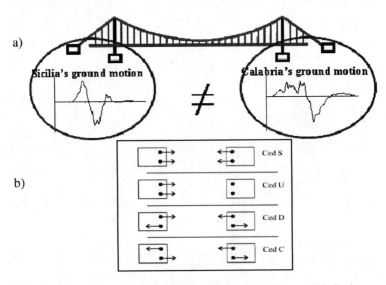

FIGURE 4. Two main aspects for seismic action: a) asynchronous explicit displacement time-histories, b) crustal displacements development.

FIGURE 5. Global model of the Messina Strait Bridge for the time-domain seismic analysis.

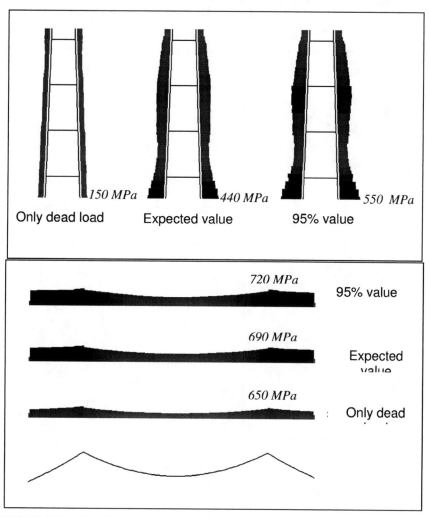

150 MPa   *440 MPa*   *550 MPa*

Only dead load   Expected value   95% value

720 MPa   95% value

690 MPa   Expected
                   value

650 MPa   Only dead

FIGURE 6. Statistics of the stress status on towers and main cables for seismic action PGA=5.70.

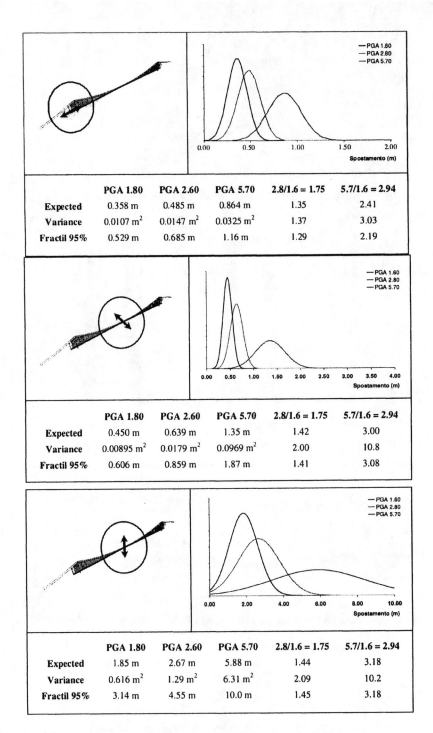

	PGA 1.80	PGA 2.60	PGA 5.70	2.8/1.6 = 1.75	5.7/1.6 = 2.94
Expected	0.358 m	0.485 m	0.864 m	1.35	2.41
Variance	0.0107 m^2	0.0147 m^2	0.0325 m^2	1.37	3.03
Fractil 95%	0.529 m	0.685 m	1.16 m	1.29	2.19

	PGA 1.80	PGA 2.60	PGA 5.70	2.8/1.6 = 1.75	5.7/1.6 = 2.94
Expected	0.450 m	0.639 m	1.35 m	1.42	3.00
Variance	0.00895 m^2	0.0179 m^2	0.0969 m^2	2.00	10.8
Fractil 95%	0.606 m	0.859 m	1.87 m	1.41	3.08

	PGA 1.80	PGA 2.60	PGA 5.70	2.8/1.6 = 1.75	5.7/1.6 = 2.94
Expected	1.85 m	2.67 m	5.88 m	1.44	3.18
Variance	0.616 m^2	1.29 m^2	6.31 m^2	2.09	10.2
Fractil 95%	3.14 m	4.55 m	10.0 m	1.45	3.18

FIGURE 7. Statistical synthesis of the results for the main displacements response.

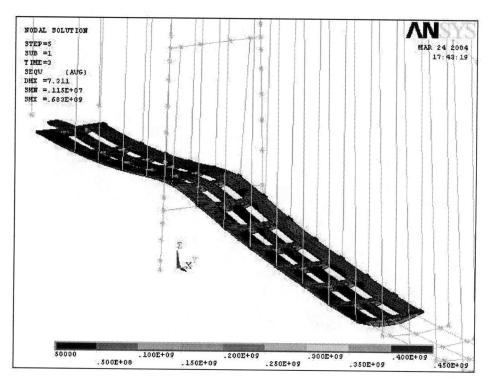

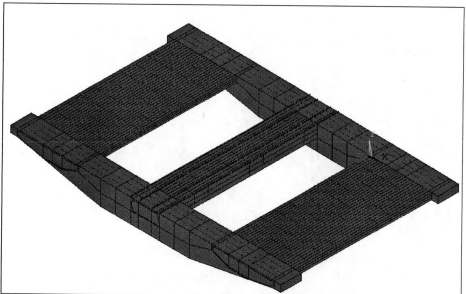

FIGURE 8. Global frame model with local shell based refined modeling for the stress analysis connected with crustal displacements: top) global arrangement, bottom) details of the generic deck span showing highway, railway and transverse girders.

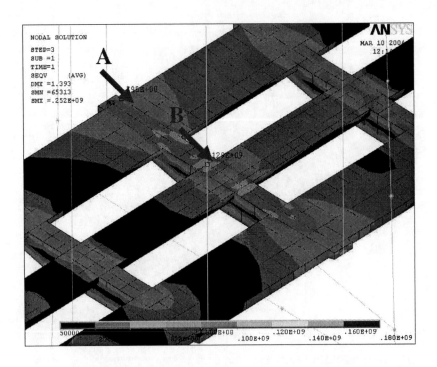

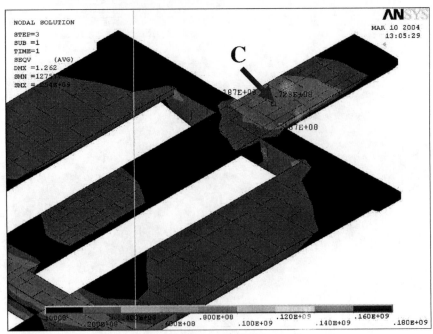

FIGURE 9. Local stress results at the beginning of the bridge main deck.

TABLE 5. Stress levels for the main bridge deck under crustal displacements (MPa).

	Mean state	Ced U	Ced S	Ced C	Ced D
Point A	79	94	107	131	90
Point B	130	146	156	113	108
Point C	72	86	91	81	82

CONCLUSIONS

Behind the Messina Strait Bridge project there is a huge amount of work developed by an enormous number of persons in several years. Some aspects of the crucial formulation of the basis of design and of the expected performances of this suspension bridge have been briefly considered, from a very personal point of view, also in relation with seismic actions. The main ideas appear:

1. the Performance-based Design approach for the overall definition of the bridge qualities;
2. the necessity to deal with the complexity of the structural system and to recognize the strong interactions among different parts of the design and among different structural parts, with scale size effects;
3. the systemic approach to correctly deal with all the aspects of the design;
4. the structural decomposition as the main tool to assure the governance of the whole design process, specifically in order to impose coherence among different levels of modeling (multilevel analysis) and to canalize the structural behavior, first of all in relation to structural crisis developments;
5. the description of safety and performance requirements in a format that can be described in mathematical terms as fuzzy;
6. the development of the loading systems, both from natural and from antropic origins, which take into account the size of the structure, which are iteratively tuned evaluating the structural response obtained by the structural analysis;
7. pervasiveness of structural robustness in a general dependability oriented design.

ACKNOWLEDGMENTS

The author wish to express his gratitude to Professors R. Calzona, F. Casciati, R. Casciaro, P.G. Malerba, G. Muscolino of the Scientific Committee of the Messina Strait Bridge for fundamental considerations.

REFERENCES

1. R. Calzona, "Epistemological Aspects of Safety concerning the challenge of Future Construction: the Messina Strait Bridge", Proc. 10th Int. Conf. on Civil, Structural and Environmental Engineering Computing, 1-3 September, 2005 Rome (Italy).
2. F. Bontempi, "Basis of Design and expected Performances for the Messina Strait Bridge". Proc. of the Int. Conf. on Bridge Engineering – Challenges in the 21st Century, Hong Kong, 1-3 Nov. 2006.
3. L. Sgambi, F. Bontempi, "A fuzzy approach in the seismic analysis of long span suspension bridge", Proc. of the 13th World Conference on Earthquake Engineering, Vancouver, B.C., Canada, 2004.

Numerical Investigations On The Seismic Behaviour Of Confined Masonry Walls

Chiara Calderini[a], Serena Cattari[a] and Sergio Lagomarsino[a]

aUniversity of Genoa, Department of Civil, Environmental and Architectural Engineering

Abstract. In the last century, severe earthquakes highlighted the seismic vulnerability of unreinforced masonry buildings. Many technological innovations have been introduced in time in order to improve resistance, ductility, and dissipation properties of this type of constructions. The most widely diffused are reinforced masonry and confined masonry. Damage observation of recent earthquakes demonstrated the effectiveness of the response of confined masonry structures to seismic actions. In general, in this type of structures, reinforced concrete beams and columns are not main structural elements, however, they have the following functions: to confine masonry in order to increase its ductility; to bear tensile stresses derived from bending; to contrast the out-of-plane overturning of masonry panels. It is well evident that these functions are as much effectively performed as the connection between masonry and reinforced concrete elements is good (for example by mean of local interlocking or reinforcements). Confined masonry structures have been extensively studied in the last decades both from a theoretical point of view and by experimental tests Aims of this paper is to give a contribution to the understanding of the seismic behaviour of confined masonry walls by means of numerical parametrical analyses. There latter are performed by mean of the finite element method; a non-linear anisotropic constitutive law recently developed for masonry is adopted. Comparison with available experimental results are carried out in order to validate the results. A comparison between the resistance obtained from the numerical analyses and the prevision provided by simplified resistance criteria proposed in literature and in codes is finally provided.

Keywords: confined masonry, resistance criteria, seismic performance, numerical analyses.

INTRODUCTION

In the last century, severe earthquakes highlighted the seismic vulnerability of unreinforced masonry buildings. Many technological innovations have been introduced in time in order to improve resistance, ductility, and dissipation properties of this type of constructions. The most widely diffused are *Reinforced Masonry* (especially in the USA, Japan, New Zealand) and *Confined Masonry*. This latter building technique has been widely employed in many countries starting from the beginning of XX century. In Italy, a strong impulse was given by the first national seismic regulations [1], published immediately after Messina and Reggio Calabria earthquake, which explicitly suggested it for both the design of new buildings and the repair of existing ones. Today, *Confined Masonry* buildings are particularly diffused in the countries of South America, North Africa and in China.

In *Confined Masonry*, the performance is improved by the introduction of reinforced concrete (r.c.) "frames" built within the masonry walls. This latter building technique can be easily confused with the most common technique of infilled r.c.

CP1020, *2008 Seismic Engineering Conference Commemorating the 1908 Messina and Reggio Calabria Earthquake*,
edited by A. Santini and N. Moraci
© 2008 American Institute of Physics 978-0-7354-0542-4/08/$23.00

frames. In *Confined Masonry*, masonry is the main resistant component of the structure, while r.c. frames provide only a contribution to its seismic horizontal load response. In infilled r.c. frames, frames are the main resistant elements of the structure, while masonry infills may act as bracing (depending on their stiffness and strength). Due to the different role played, r.c. frames may strongly differ in stiffness in the two cases. Moreover, in *Confined Masonry* the r.c. frames are "grouted within" masonry walls (thus, they are built after the masonry), while, in infilled r.c. frames, the r.c. frames are "filled with" masonry (thus, they are built before the masonry) [2]. In the first case, adherence between masonry and concrete is guaranteed (leading to a cooperation of the two materials). In the second case, scarce adherence may develop.

In this paper, the influence of the confinement in the seismic response of *Confined Masonry* walls is analyzed by mean of numerical parametrical analyses. These latter are performed by mean of the finite element method, by adopting a non-linear anisotropic constitutive law recently developed for masonry [3]. In particular, the following issues will be faced: the role of steel reinforcements and concrete; the role of adherence between masonry and concrete. Comparisons with available experimental tests [4] are carried out in order to validate the obtained results. Furthermore, the main verification criteria proposed in literature and codes are discussed and their strength previsions are compared with the experimental and numerical results obtained.

INTERPRETATION OF THE SEISMIC BEHAVIOUR THROUGH AVAILABLE MODELS

Many national codes provide simple rules for the design and construction of *Confined Masonry* structures. However, in most of the cases, such rules are of technical and practical nature (e.g. maximum admissible distance between vertical confinement elements), rather than of analytical one. The evaluation of the internal forces is usually performed by idealizing the combined masonry-r.c. system as a frame with a diagonal bracing simulating the masonry panel as an equivalent strut.

Although it seems reasonable to suppose that the confinement supplied by the r.c. frame may influence both the flexural and shear behaviour of the masonry panel, codes provide contrasting requirements. As an example, the code adopted by Mexico City [5] explicitly allows to consider the contribution of vertical r.c. elements to the flexural resistance of the panel, while EC6 [6] and EC8 [7] allows to consider their contributions only to the shear resistance. The contribution to the flexural resistance, if kept into account, can be computed analogously to the *Reinforced Masonry*: r.c. column in tension produces an enlargement of the in-plane failure domain analogous to that produced by an equivalent tensile strength of masonry. Such enlargement is meaningful mainly in the range of low axial loads, for which the flexural failure tends to become prevailing. Moreover, the confinement due to the r.c. frame leads to a increasing of the compressive stresses acting on the masonry, thus inducing a transition from flexural to shear failure mechanisms. The prevalence of shear failure is also confirmed by the observation of seismic damages on complex masonry structures, as well as the results from laboratory experimental tests.

817

In the following, some resistance criteria presented in the literature and codes for the prediction of the shear capacity of confined masonry panel are discussed. Table 1 summarize them. Each expression is adimensionalised with respect to τ_m which represents the shear strength under zero compressive stress obtained from a diagonal compression test.

TABLE 1. Failure criteria proposed in the literature and in codes.

Eq.	Reference	Failure Model [*]
(1)	INPRES-CIRSOC 103 [8]	$\tau/\tau_m = \beta\left(0.6 + 0.3(\sigma_0/\tau_m)\right)$
		where $\beta = A_T/A_m$
(2)	Decanini and D'Amore [9]	$\tau/\tau_m = \beta K_T\left(0.6 + 0.3(\sigma_0/\tau_m)\right)$
		where: $K_T = f\left(H/L; M/VL\right)$
(3)	Moroni et al. [10]	$\tau/\tau_m = 0.37 + 0.43(\sigma_0/\tau_m)$
(4)	Tomazevic and Klemenc [11]	$\tau/\tau_m = \left(\dfrac{f_t/\tau_m}{b}\right)\left(C + \sqrt{C^2 + 1 + \dfrac{\sigma_0}{\tau_m}\dfrac{\tau_m}{f_t}}\right)$
		where: $C = H/2bL\alpha'$; $\alpha' = 5/4$; $b = f(H/L)$
(5)	Lafuente et al. [12]	*Sliding failure:*
		$\tau = \dfrac{a/L}{1 - \mu tg\alpha}(c + \mu\sigma_0)$
		Diagonal failure:
		$\tau/\tau_m = \dfrac{f_t}{\tau_m}\dfrac{a}{L}\left(tg\alpha + \sqrt{tg\alpha + 4 + 4\dfrac{\sigma_0}{\tau_m}\dfrac{\tau_m}{f_t}}\right)$
		where: $tg\alpha = H/L$; a= width of the masonry equivalent strut.

[*] Legend of symbols: masonry panel of height H, width L and thickness t; A_m transversal cross section of masonry (Lt); A_T transversal cross section inclusive of confining r.c. elements; V and M overall horizontal force and bending moment, respectively; σ_0 mean vertical stress ($=P/A_m$, P being the vertical force applied); c and μ local cohesion and friction coefficient of mortar bed joints, respectively; f_t reference tensile strength of masonry.

Some of these criteria is based on an empirical approach [8,9,10], considering a Mohr-Coulomb-type criterion in which the constant coefficients have been calibrated on the basis of experimental results. Other criteria are based on an analytical approach [11,12]; among these, the most interesting appears that formulated by Tomazevic and Klemenc [11]. They proposed a tri-linear relationship in which the maximum resistance is obtained as a sum of two contributions: the first one supplied by the masonry panel (Equ. (4)), the second one by the r.c. confining elements through the dowel action of reinforcing steel. The evaluation of masonry contribution is derived by the formulation of Turnšek and Čačovič [13] for unreinforced masonry, by modifying the compressive stress in order to keep into account the interaction forces developed between the confining elements and masonry at the contact zones.

In general, the following observations could be made on the different approaches proposed:

- In the criteria based on an empirical approach, the constants adopted are strictly dependent on the experimental databases on which they have been calibrated. Such databases are clearly connected to the technical practice of the countries where they have been developed.

818

- Analytical criteria, even if make an effort to provide a mechanical interpretation of the response of the confined panels, are anyway dependent on some empirical parameters. For example, the coefficient α', introduced in Eq. (4) in order to modify the compressive stresses acting in masonry, let be considered. The value 5/4 has been proposed by Tomazevic and Klemenc on the basis of experimental results carried out on confined panels of slenderness 1.5 [11]. However, it reasonable to argue that such coefficient is influenced by many factors (such as the slenderness of the panel or the effectiveness of the adherence between r.c. elements and masonry).

The reliability of the criteria here presented will be further on discussed in the following paragraph on the basis of the parametrical analyses performed.

PARAMETRICAL ANALYSES

In order to calibrate the numerical model and to validate results, the experimental tests carried out by Yanez et al. [4] are assumed as reference ones. In particular, tests on walls without openings, 3600 mm width, 2200 height and 140 mm thick, made up of hollow clay units arranged in running bond pattern, are considered. The longitudinal reinforcement of the confinement columns and beam was 4 bars of 10 mm diameter, while the transverse reinforcement was a hoops of 6 mm diameter at 150 mm spacing. A horizontal cyclic load was applied along the axis of the top r.c. beam and controlled by displacement. The top beam was free to rotate (cantilever test). There was no vertical load applied. Two specimens were tested. By considering the highest peak of the envelop of the cyclic force-displacement (V-dx) diagram obtained, the following maximum horizontal loads were defined: 172 kN (Specimen 1) and 199 kN (Specimen 2).

The numerical analyses have been performed by the finite element method, adopting non linear constitutive laws for both masonry and concrete. In particular, for masonry, a constitutive law recently developed by Calderini and Lagomarsino have been employed [3]. It considers both friction and cohesive resistant mechanisms of masonry, on the basis of a micromechanical analysis of the composite continuum. Masonry walls and concrete beam and columns have been modeled by means of 3-nodes plane elements (plane stress). Steel reinforcements have been modeled by means of linear truss elements. The adherence between masonry and concrete has been modeled by introducing a joint of limited thickness between the elements of these two different material. Table 2 shows the parameters employed; in bold are marked those parameters which have been directly defined on the basis of the information given by the authors of the experimental tests [4]. The cohesion of mortar joints has been defined on the basis of the shear strength of masonry provided by the diagonal compression tests. The numerical simulations were carried out by applying a monotonically increasing horizontal displacement on the r.c. beam at the top of the wall.

In a first time, the model was calibrated by comparing the results obtained with those of the two experimental tests available. In particular, the properties of the joint simulating the interface between masonry and concrete were calibrated.

TABLE 2. Principal parameters employed in the numerical analyses.

Mass densities	Value
Masonry	1600 kg/m^3
Concrete	2200 kg/m^3
Steel	7800 kg/m^3
Elastic Parameter	
Elastic modulus of masonry	**4 450 MPa**
Shear modulus of masonry	**528 MPa**
Normal elastic modulus of concrete	30 000 MPa
Normal elastic modulus of steel	210 000 MPa
Inelastic parameters of masonry	
Friction coefficient	0.6
Tensile strength of mortar joints	0.164 MPa
Cohesion of mortar joints	**0.327 MPa**
Compressive strength of masonry	**6.89 MPa**
Interlocking ratio (width to height ratio of units)	**2**
Inelastic parameters of concrete	
Friction coefficient	1
Tensile strength	2.4 MPa

After the calibration of the base model, other models were investigated. In particular, the following variations were introduced in the confined masonry panels (CF): the percentage of reinforcement in the r.c. elements, the properties of the interface joint. Moreover, these further cases were considered: unconfined panel (NC); unconfined panels only reinforced with vertical steel bars (RF). They are summarized in Table 3, while Figure 1 illustrate the geometry adopted in the different cases.

TABLE 3. Description of the models analysed.

Distinctive features of models	NC	RF		CF			
	NC	RF(R1)	RF(R2)	CF(JA-R1)	CF(JB-R1)	CF(JC-R1)	CF(JA-R2)
Concrete in columns	-	-	-	X	X	X	X
Steel reinforcement in r.c. elements	-	4ϕ10	8ϕ10	4ϕ10	4ϕ10	4ϕ10	8ϕ10
Presence of interface joint	-	-	-	X	X	-	X
Cohesion of the interface joint [MPa]	-	-	-	0.327	0.164	-	0.327
Friction coeff. of the interface joint	-	-	-	0.6	0.6	-	0.6

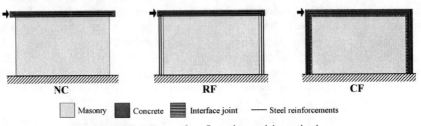

      NC   RF   CF

☐ Masonry   ■ Concrete   ▤ Interface joint   — Steel reinforcements

FIGURE 1. Types of configuration model examined.

The analysis of force-displacement (V-dx) curve (Figure 2) allows to highlight as: just the insertion of the vertical steel bars (RF) leads to an increase of 162% in the maximum overall resistance (V_u) with respect to the NC case; the complete confinement (CF) produces a further increase of 23÷43% in V_u (depending on the properties of the interface joint) and a significant stiffening; the resistance increase related to reinforcement percentage variation in the r.c. elements is strongly less meaningful than the influence of the configuration typology (RF or CF).

The damage pattern (Figure 3) attests the progressive transition from a failure mode classifiable as *Rocking* (NC case) to *Diagonal Cracking* (both for RF and CF cases, although with different spread); this phenomenon is also evidenced by the progressive reduction of the effective cracked section length on the base section, as showed in Figure 4.

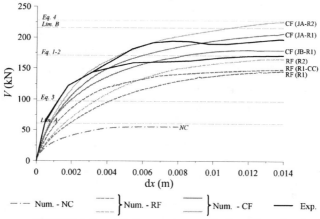

FIGURE 2. Force-displacement curves for the walls analysed.

In Figure 2, the maximum overall resistances obtained in the numerical analyses are compared to these provided by the criteria summarized in Table 1. It is worth pointing out that the limit values A and B are reference theoretical values. They are calculated considering a *Rocking* failure; an equivalent rectangular stress block distribution at the masonry compressed toe is assumed, keeping into account the contribute of the tensile strength of mortar joints in the NC case, and that of the tense reinforcement (assumed at yielding) in the RF case, respectively. It can be observed that: among the criteria, Eqs. (1)-(2) provide the better estimate of resistance of the experimental panels (actually designed according with standards of INPRES-CIRSOC 103), while Eq.(4) significantly overcomes it; the limit B overcomes the actual resistance occurred in the RF(R1) case (in fact the panel shows a shear failure). However, the analysis of the stress distribution evolution along the top horizontal and vertical left sections (Figure 4), allowed to verify that, in case CF(JA-R1), the value of the parameter α' introduced in the criterion of Tomazevic and Klemenc (Eq. (4)) is 1.93; this value would lead to an much more accurate estimate of the resistance than 5/4.

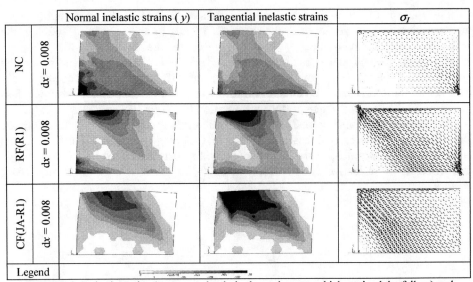

FIGURE 3. Inelastic strains (represented only in those elements which attained the failure) and compressive principal stresses (σ_I) acting on the walls.

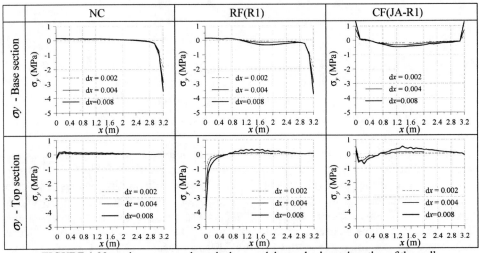

FIGURE 4. Normal stresses σ_y along the base and the top horizontal section of the walls

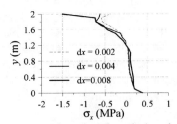

FIGURE 5. Normal stresses σ_x along the left vertical section in CF(JA-R1) case.

CONCLUSIONS

Confined masonry structures are mainly designed according to rules of technical and practical nature. Also the criteria presented in literature and codes are mainly based on empirical parameters, calibrated on experimental tests. Since such tests are affected by local building practices, their generalized use may lead to unreliable previsions.

The parametrical analyses performed showed in particular that: the steel reinforcements has a fundamental role in the resistance of masonry panels, increasing significantly the flexural resistance, and inducing thus a transition towards a shear type failure mode (for low levels of axial loads); moreover, the presence of concrete in adherence with steel and masonry produces a further contribution of strength and stiffness; the amount of such contribution is strictly related to the properties of the interface joint between the confining elements and masonry.

ACKNOWLEDGMENTS

The authors acknowledge the partial financial contribution by Italian Network of Seismic Laboratories (RELUIS), in the frame of the 2005-2008 Project "Evaluation and reduction of seismic vulnerability of existing masonry buildings" – Line 1.

REFERENCES

1. Royal Decree no.193, 18 April 1909 (in Italian).
2. Liberatore L., Decanini L.D., Benedetti S., "Le strutture miste muratura-cemento armato: uno stato dell'arte", XII Convegno Nazionale "L'ingegneria sismica in Italia, 2007(in Italian).
3. Calderini, C., Lagomarsino, S., "A continuum model for in-plane anisotropic inelastic behaviour of masonry." *Journal of Structural Engineering - ASCE*, **134** (**2**), 2008, 209-220.
4. Yanez F., Astroza M., Holmberg A., Ogaz O., "Behaviour of confined masonry shear walls with large openings", *Proc. of the 13th WCEE*, Vancouver, Canada, 2004.
5. Gaceta Oficial del Distrito Federal de Mexico City, 6 de Octubre de 2004. Tomo II, No. 103-BIS. Normas Técnicas complementarias para diseño por sismo.
6. EC6, prEN 1996-1-1, 2002. Design of Masonry Structures, Part 1-1: Common rules for reinforced and unreinforced masonry structures.
7. EC8, prEN 1998-1, 2003. Design of structures for earthquake resistance. Part 1: General rules, seismic actions and rules for buildings.
8. INPRES-CIRSOC 103, 1983. Normas Argentinas para construcciones sismorresistentes Parte III Construcciones de mamposteria.
9. D'Amore, E., Decanini, L., "Considerazioni sul comportamento sismico delle costruzioni in muratura intelaiata". VI Convegno Nazionale "L'ingegneria sismica in Italia", **3**, 1993, 1255-1264.
10. Moroni, M.O., Astroza, M., Tavonatti, S., "Nonlinear models for shear failure in confined masonry walls", Masonry Society Journal, **12**, 1994, 72-78.
11. Tomaževic, M. and Klemenc, I., "Seismic behaviour of confined masonry walls", *Earthquake Engineering and Structural Dynamics*, **26**, 1997,1059-1071.
12. Lafuente,M. Genatios. C., Lorrain, M., "Analytical studies of masonry walls subjected to monotonic lateral loads". *Proc. of the 10th ECEE*, **3**, 1995, 1751-1756. Vienna.
13. Turnšek V., Čačovič, F. "Some experimental results on the strength of brick masonry walls." Proc. of the Second International Brick Masonry Conference, Stoke-on-Trent, 149-1,1970.

Experimental Analyses of Yellow Tuff Spandrels of Post-medieval Buildings in the Naples Area

B. Calderoni[a], E.A. Cordasco[a], L. Guerriero[b], P. Lenza[a]

[a]Dept. of structural engineering, University Federico II, P.zzale Tecchio, 80, Naples, Italy.
[b]Dept. of Restoration and Construction of Architecture and Environment, 2nd University of Naples, Italy

Experimental analyses have been carried out on tuff masonry specimens in order to investigate the structural behaviour of historical buildings in the Naples area (Southern Italy). Spandrels of post-medieval buildings (late XVI to early XX century) have been analysed, with emphasis on morphological characteristics according to chronological indicators. Results of the experimentation on scaled models (1:10) are discussed and the better behaviour of historical masonry typologies on respect to the modern one is highlighted. Comparison with theoretical formulations of ultimate shear resistance are provided too.

Keywords: spandrel, shear strength, masonry wall, seismic behaviour

INTRODUCTION

During the last century, historical masonry buildings were frequently modified, particularly with reference to the floor, which is known to have an important role in the overall seismic behaviour of the building. When floors connect effectively the walls and, if rigid enough, distribute seismic loads, walls are stressed by shear and bending in their own plane, leading to reduced seismic vulnerability. In this case the structural characteristics of piers and spandrels, the constitutive elements of the masonry walls, take on a particular importance.

More specifically the spandrels play a fundamental role in the resistance and deformability of the wall itself. As shape and/or structural typologies of the spandrels vary, different degrees of coupling between the piers occur, so that the structural behaviour of the walls, characterized by the same geometry and loads applied, may be very different. Current seismic codes for URM buildings do not seem sufficiently detailed in relation to the resistance assessment of masonry spandrels, due to inadequate theoretical and experimental analysis. In particular, EC8 allows for the coupling capacity of masonry spandrels considering only whether they have lower and upper reinforcement (i.e. beam behaviour). The American code (FEMA 375) does not specify any estimate of the spandrel capacity of connection, though it does consider shear-type behaviour of the masonry walls possible. Finally, Italian codes (OPCM 3274 and 3431) provide the means to estimate the bending and shear resistance of the spandrels, provided at least one horizontal tensile-resistant element (tie, beam floor, lintel effectively connected to adjoining walls, etc) is present, leaving the case of spandrels with beam behaviour to reinforced masonry.

CP1020, 2008 Seismic Engineering Conference Commemorating the 1908 Messina and Reggio Calabria Earthquake,
edited by A. Santini and N. Moraci

FIGURE 1. a) Eighteenth century spandrel, b) Nineteenth century spandrel

MORPHOLOGICAL CHARACTERISTICS

In the last decade the metrological characters of modern and contemporary tuff masonry in the Naples area have been defined. They can be dated with precision by means of morphological, chronological and dimensional indicators [1]: *cantieri* masonry; dating back to the XVI and XVII centuries, *bozzette* masonry, common in the XVIII century, and *sacco* masonry, used in the XIX century and in the first half of the XX century. Further studies have concurred to verify experimentally the mechanical behaviour of the different masonry types identified [2].

In historical building in the Naples area two basic types of masonry spandrel were generally adopted. In the first, a wooden beam, as wide as the entire thickness of the wall, directly supports the masonry spandrel, built according to one of the chronotypes described above. In the second, above the opening a small arch with horizontal lower surface is prepared, built with stones purposely worked, often occupying a significant part of the masonry spandrel depth. Sometimes, the lower surface of the arch structure is slightly raised and is regulated by wedge-shaped stones, supported by thin wooden board, similar (in appearance only) to the previous typology. In some cases, the wooden lintels were replaced by steel or r.c. beams during restoration work.

In synthesis, during the reign of the Spanish viceroys (XVI-XVII century) masonry spandrels were constructed using *cantieri* typology, while lintels were often made up of simple wooden boards (first type). Sometimes, however, arch lintels have been found with small yellow tuff, roughly worked, with considerably thick mortar joints.

Eighteenth century spandrels were built with using *bozzette* typology, with arch lintels made up of wedge-shaped stones, obtained from *spaccatelle,* roughly the same size as the stones used for vertical walls, with fairly large mortar joints. Wooden lintels were also widespread at the time (Figure 1a).

In the XIX and the XX century lintels were made with carefully worked blocks and thin mortar joints, alternating short and very long stones (up to 50 cm). The lower part of the spandrel was almost always made with a wooden board. (Figure 1b)

STRUCTURAL TYPOLOGIES

In general terms three classes of structural behaviour can be defined for masonry spandrels:

a) *weak-spandrel* are poorly anchored to the adjoining vertical wall element. These are typical of older buildings with vaults or floors with parallel wooden and steel beams, lacking practically any tensile-resistant element. This "weakness" means that the floor is unable to provide rigid diaphragm behaviour and that the wall is unable to provide resistance to shear and bending, and thus to structurally connect two adjoining wall elements;

b) *truss-spandrels* have elements equipped with at least one effectively anchored horizontal tensile-resistant element. Thus, compressive horizontal stress (of unknown value) may arise within the spandrel, and so the spandrel behaviour becomes similar to a vertical wall element, with equivalent truss behaviour. The main difference is that within the spandrel the value of normal stress which arises internally and the shear value that the spandrel itself transmits are directly related, while in the vertical wall elements normal stress is known and does not depend on acting shear;

c) the spandrels with lower and upper adherent reinforcement (such as floor tie beam, well fixed lintels, adherent ties) have beam-behaviour (*beam-spandrel*). It is then possible that flexural beam behaviour may be activated, with shear and bending resistance to be evaluated as for a reinforced masonry beam. In the absence of any specific reinforcement, shear resistance of the spandrel is limited by the shear resistance of masonry, which constitutes the central part of the beam.

EXPERIMENTAL ACTIVITY

The experimental tests have been carried out on scaled-down masonry spandrels (1:10). The reduced scale is known to limit research to some degree as it is impossible to apply rigorous geometrical and mechanical similitude, particularly as regards the thickness of the mortar joints. However, qualitative appraisal of the phenomena is surely correct, while, in comparative terms between the tested specimens, quantitative evaluation is possible too.

The experimental program included the preparation of some specimens for two chronological masonry types: XVIII century masonry (EM) and XIX/XX (NM) century masonry. In addition, samples with positioning of blocks on horizontal rows (ordinary masonry - OM) were prepared (Fig.2). In particular, historical samples were worked by hand in order to simulate ancient roughly-squared stones. Mortar was made up of lime (one part in volume) and volcanic sand (three parts in volume) (Fig.3).

Three different slenderness ratios have been considered: slender spandrels ($H/L \cong 0.5$), intermediate spandrels ($H/L \cong 0.7$) and stocky spandrels ($H/L \cong 1.1$).

A horizontal steel frame with a 25 kN cyclical electrical jack (Figure 3) was the equipment used in order to simulate the behaviour of a masonry spandrel during a seismic event. Rotating together the two arms of the equipment stress the specimen, housed between the two opposite extremities of the same arms, to bending and shear. The mobile arms reproduce well the behaviour of the (rigid) node panels of the masonry wall, while the sample represents the (deformable) masonry spandrel, when the masonry wall of which it is a part is subjected to horizontal forces. The extremities of the spandrel cannot be moved between them as the two points around which the steel arms rotate are fixed. In this way the real situation of a masonry spandrel with a tensile-resistant element placed at half height, is simulated.

Preliminarily it should be noted that previous and appropriate experimentation on materials were carried out in order to evaluate the compressive strength of tuff, mortar and ordinary masonry [3]. The corresponding mean compressive strength values are

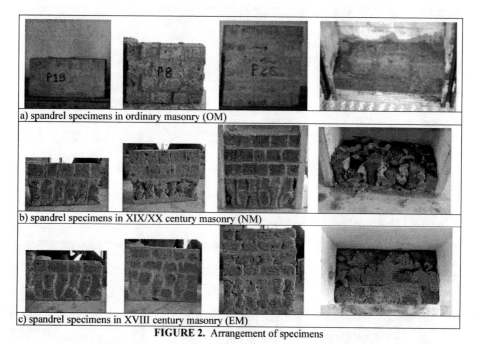

a) spandrel specimens in ordinary masonry (OM)

b) spandrel specimens in XIX/XX century masonry (NM)

c) spandrel specimens in XVIII century masonry (EM)

FIGURE 2. Arrangement of specimens

respectively 4.7 N/mm^2, 2.1 N/mm^2 and 2.5 N/mm^2. While the mean value of compressive strength of historical masonries is equal to 1.9 N/mm^2.

For each geometrical typology two cyclical tests and one monotonic test were carried out under displacement control. Specifically, rotation was applied (equal for the two extremities) obtained by means of horizontal displacement imposed from the jack to the steel arms. In the cyclical tests nine steps of loading of three cycles each were carried out, with maximum rotation varying from a minimum of 0.002 rad to a maximum of 0.018 rad. In the monotonic tests the maximum rotation of 0.042 rad was reached, corresponding to the maximum displacement capacity of the test equipment.

In this paper, the test results have been especially analysed in terms of maximum strength. Then in case of cyclic tests the shear resistance of the specimen was obtained indirectly from the corresponding envelope curve M-φ.

During tests all samples showed cracks typical of shear failure, though the shape and evolution of the cracks varied in relation to the slenderness ratio of the panel. Fig.4 shows cracking during testing for each slenderness ratio and masonry typology.

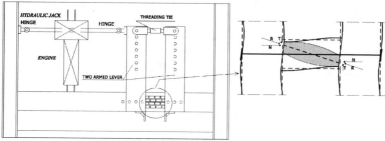

FIGURE 3. Scheme of testing equipment

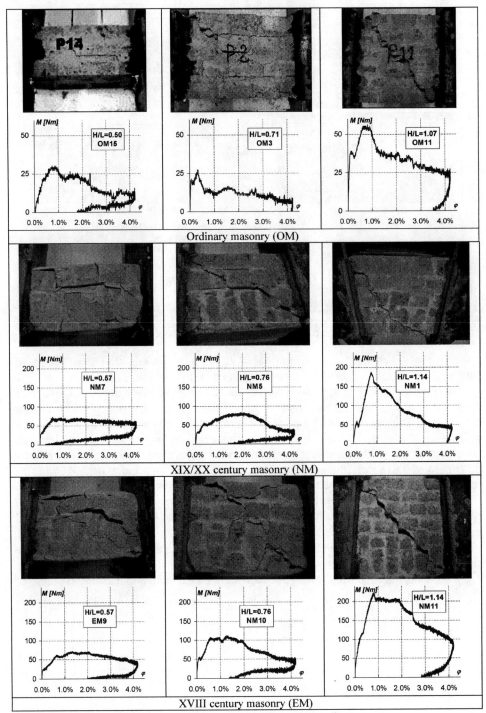

FIGURE 4. Collapse mechanisms and M-φ diagrams

TABLE 1. Experimental results

Ordinary spandrels				Nineteenth masonry				Eighteenth masonry		
	H/L	V_{max} [N]	M_{max} [Nm]		H/L	V_{max} [N]	M_{max} [Nm]		H/L	V_{max} [N]	M_{max} [Nm]
OM13	0.50	416	29	NM7	0.57	1041	73	EM9	0.57	1027	72
OM14	0.50	376	26	NM8	0.57	1053	74	EM13	0.57	1318	92
OM15	0.50	430	30	NM12	0.57	847	59	EM14	0.57	915	64
OM1	0.71	430	30	NM4	0.77	1194	83	EM10	0.77	1596	112
OM2	0.71	430	30	NM5	0.77	-	-	EM15	0.77	1446	101
OM3	0.71	388	27	NM6	0.77	973	68	EM16	0.77	1374	96
OM9	1.07	958	67	NM1	1.14	2680	187	EM11	1.14	3194	224
OM10	1.07	804	56	NM2	1.14	2344	164	EM18	1.14	2218	155
OM11	1.07	804	56	NM3	1.14	2150	150	EM17	1.14	2370	166

In general terms, however, cracks appeared essentially at the interface between the mortar joints and tuff stones; they almost always formed initially in the central zone of the sample, spreading outwards under increasing load towards the corners. Only for the slender samples made of historical masonry, compression cracks at the corner of the panel have been observed (Fig.4).

Each test gave an F-δ diagram, from which the M-φ diagram of the spandrel was gained analytically. The diagrams are drawn in Fig.4, for the monotonic tests only.

It can be noted that in all cases masonry panels behave substantially linearly during the load phase until maximum strength is reached. As rotation increases a fairly large descending branch occurs, showing a significant, though gradual, decline in strength, together with the capacity to maintain a certain residual strength even with quite high rotations. Only the stocky specimens showed a more accentuated degrading branch, while the slender historical samples exhibited a very ductile behaviour.

Maximum shear and bending moment reached during tests are reported in Table 1.

CHRITICAL REMARKS ON SHEAR RESISTANCE

The experimental results showed that shear strength and collapse mechanisms depend on both slenderness ratio and arrangement of stones in the spandrel.

Three different collapse mechanisms, all correlated to the formation of a diagonal compressed truss, have been observed: a) compressive cracking at panel edges; b) slipping along horizontal mortar beds c) diagonal cracking. Vertical sliding (d) have been never observed, even if it can not be in principle excluded.

For each collapse mechanism (Fig.5) the corresponding shear strength (as reported more in detail in [4]) is given by the following equations:

$$V_{max} = \left(f_{vo} \cdot t \cdot H \right) / \left(1 - H/L \cdot tg\varphi \right) \qquad \text{(horizontal slipping)} \quad (1)$$

$$V_{max} = \left(f_{vo} \cdot t \cdot H \right) / \left(1 - L/H \cdot tg\varphi \right) \qquad \text{(vertical slipping)} \quad (2)$$

$$V_{max} = \left(f_{vo} \cdot t \cdot H \right) \cdot \frac{L}{2 \cdot H \cdot k^2} \cdot \left(1 + \sqrt{1 + \frac{4 \cdot H^2 \cdot k^2}{L^2}} \right) \qquad \text{(diagonal crack)} \quad (3)$$

with: $\qquad \sigma_o = N_{max}/H \cdot t$ $\qquad\qquad\qquad$ $k=L/H$ $\quad$ and $\quad$ $1< k < 1.5$

In these equations $tg \ \varphi$ is the frictional coefficient and f_{vo} is the pure shear resistance, considered equal to tensile strength. If the frictional effect is neglected ($tg \ \varphi=0$) Eqs.1-2 give back the shear strength laid down by OPCM 3431:

$$V_{OPCM} = f_{vo} \cdot t \cdot H \qquad\qquad\qquad\qquad\qquad (4)$$

The shear resistance related to collapse mechanism (a), can be evaluated by means of formula provided by OPCM 3431, in which f_c is the compressive strength of masonry:

$$V_p = N_p \cdot \frac{H}{L} \cdot \left(1 - \frac{N_p}{0,85 f_c \cdot H \cdot t} \right) \qquad \text{with} \qquad N_p =0.4 \cdot f_c \cdot H \cdot t \qquad (9)$$

This formula is obtained considering the ultimate compressive and bending behaviour of the masonry spandrel analogously to masonry piers.

The shear resistance (V) obtained by testing are represented versus the slenderness ratio (H/L) in Fig.5 in not-dimensional format ($V^*=V/V_{OPCM}$). In the same diagrams the curves of shear resistance corresponding to the various failure mechanisms above considered are drawn. The horizontal line indicates V_{OPCM}, which is independent of slenderness ratio.

Note that in the left diagram V_{OPCM} has been evaluated considering the pure shear resistance of the material (f_{vo}) equal to $f_c/27$, while in the right one $f_{vo}=f_c/13$ has been considered.

In the left diagram the triangular shaped dots indicate the experimental results of OM specimens and the circular and squared dots refers to NM and EM specimens respectively. It is worth to note that the OM results quite well correspond to the theoretical shear values of horizontal slipping and diagonal cracking failure mechanisms, depending on H/L in accordance to the experimental evidences. On the contrary the NM and EM results are very far to the curves corresponding to the observed collapse mechanisms.

This means that the adopted pure shear resistance ($f_{vo}=f_c/27$) is too low for the spandrel made with historical arrangements of masonry. In fact if we double it

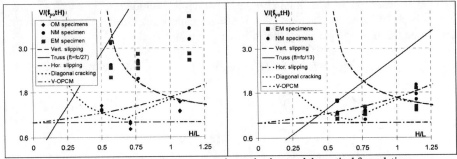

FIGURE 5. Comparison between experimental values and theoretical formulations

($f_{vo}=f_c/13$) the dots representing NM and EM results move very close to the theoretical curves, as showed in the right diagram, particularly to the one corresponding to diagonal cracking, which is the kind of failure mainly observed during testing for intermediate and stock spandrels. Instead for slender specimens the observed failure mechanism was compressive cracking at panel edges and the experimental values of maximum shear are correctly on the corresponding curve (solid line in the figure).

The possibility to adopt an increased shear resistance for historical typologies is related to the different arrangements of tuff stones: if they are positioned in order to form a small arch, which occupies a significant part of the spandrel depth, the horizontal or diagonal crack does not meet the weaker mortar bed but the stronger stones and than they can form when the acting shear has reached an higher value.

CONCLUSIONS

Experiments have been performed on tuff masonry samples of spandrels aimed to evaluate their shear resistance. Both modern and historical arrangements of masonry have been considered.

The observed failure mechanisms were diagonal cracking mainly, horizontal slipping or compressive cracking at edges of panels in some cases only, depending on the slenderness ratio and masonry typology.

Theoretical formulations give ultimate shear resistance of the panels in good accordance with the experimental results, provided that timely value of pure shear resistance (f_{vo}) of masonry is considered. For ordinary masonry f_{vo} equal to $1/25 \div 1/30$ of compressive strength (f_c) (as commonly indicated in literature and Italian code) seems to be adequate. The shear resistance of historical masonry has been used two times longer than the one of ordinary masonry; this assumption is related to the better behaviour of ancient typologies. This fact is essentially due to the arch arrangement of the tuff stones in the spandrel, which makes more difficult the development of diagonal crack and/or horizontal slipping.

Finally, the obtained results confirm once again that the ultimate shear resistance of spandrels provided by OPCM 3431 is underestimated.

REFERENCES

1. G. Fiengo and L. Guerriero, (Eds) *Murature tradizionali napoletane. Cronologia dei paramenti dal XVI al XIX secolo*, Napoli: Arti tipografiche, 1988.
2. B. Calderoni, E.A. Cordasco, L. Guerriero, P. Lenza, "Experimental tests on post-medieval and modern tuff masonry wall", *Proceedings of Tenth North American Masonry Conference*, St.Louis, Missouri, USA:, The Masonry Society, 3-6 June 2007, pp.921-932.
3. B. Calderoni, E. A. Cordasco, P. Lenza, "Analisi teorico sperimentale del comportamento della fascia di piano delle pareti murarie per azioni sismiche", *Proceedings of XII convegno ANIDIS*, Pisa, Italy, June 2007 (on CD-Rom).
4. B. Calderoni, E. A. Cordasco, P. Lenza, "La capacità di accoppiamento delle fasce di piano nelle pareti murarie", *Procedings of Workshop on design for rehabilitation of masonry structures - WONDERmasonry2*, Ischia, Italy, October 2007.
5. V. Turnsek and F. Cacovic, "Some Experimental Results on the Strength of Brick Masonry Walls", *Proceedings of the 2nd International Brick Masonry Conference*, Stoke-on-Trent, 1970.
6. U. Andreaus, "Failure Criteria for Masonry Panels under In-Plane Loading", *Journal of Structural Engineering*, n°1-1996, pp.37-46.

A new discrete-element approach for the assessment of the seismic resistance of composite reinforced concrete-masonry buildings

I. Caliò[a], F. Cannizzaro[a], E. D'Amore[b], M. Marletta[a], B. Pantò[a]

[a]Department of Civil and Environmental Engineering, University of Catania
Viale Andrea Doria, 6 – 95125 Catania (Italy)
[b]Department of Mechanics of Materials, University of Reggio Calabria
Via Graziella – Feo di Vito – 89124 Reggio Calabria (Italy)

Abstract. In the present study a new discrete-element approach for the evaluation of the seismic resistance of composite reinforced concrete-masonry structures is presented. In the proposed model, unreinforced masonry panels are modelled by means of two-dimensional discrete-elements, conceived by the authors for modelling masonry structures, whereas the reinforced concrete elements are modelled by lumped plasticity elements interacting with the masonry panels through nonlinear interface elements. The proposed procedure was adopted for the assessment of the seismic response of a case study confined-masonry building which was conceived to be a typical representative of a wide class of residential buildings designed to the requirements of the 1909 issue of the Italian seismic code and widely adopted in the aftermath of the 1908 earthquake for the reconstruction of the cities of Messina and Reggio Calabria.

Keywords: Macro-element, composite reinforced concrete-masonry building, framed-masonry, Messina 1908 earthquake.

INTRODUCTION

Composite reinforced concrete-masonry building structures, so defining structural systems where both reinforced concrete elements and unreinforced masonry panels participate to structural strength and stiffness, represent a relevant percentage of the existing building particularly in Italy. These structures are often difficult to be classified because in many cases they represent non engineered structures built before the emanation of seismic codes and/or they are the by-product of partial modification of pre-existing masonry buildings. Engineered composite reinforced concrete-masonry was introduced in Italy after the 1908 Messina and Reggio Calabria earthquake by the 1909 seismic code. The structural system introduced by this code, named *framed-masonry* became the most widespread seismic-resistant structural system for residential buildings during the first forty years of the last century [1]. The analytical modelling of the nonlinear behaviour of composite reinforced concrete-masonry structures can be conducted by detailed nonlinear finite element analyses or by simplified approaches. Detailed finite element analyses generally require constitutive laws for reinforced concrete elements, usually modelled with diffused or lumped

CP1020, *2008 Seismic Engineering Conference Commemorating the 1908 Messina and Reggio Calabria Earthquake,*
edited by A. Santini and N. Moraci

plasticity elements, and for the masonry elements taking into account for the limited tensile strength of simple masonry. Two- or three-dimensional inelastic elements are usually adopted for the modelling of unreinforced masonry. A detailed finite element approach, even though capable at giving a deep insight on the nonlinear behaviour of the component materials, on their interaction and on the local and global collapse mechanisms, is extremely time consuming during both the modelling and the results interpretation phases. Moreover, its complexity and some convergence issues, usually make the described approach not suitable for nonlinear dynamic analysis of real three-dimensional buildings. To partially overcome the complexity of detailed finite element analysis, some simplified analytical models where proposed especially for simple masonry buildings. Generally these simplified approaches adopt equivalent nonlinear beam elements, or more complex mechanical sub-assemblages, for the analysis of unreinforced masonry panels. With reference mixed structures the simplified models usually consider reinforced concrete and masonry elements arranged in series or in parallel, without taking into account for the confining effects, which, at least in case of framed and confined masonry systems, play a relevant role to the seismic response of the structure.

In the present study a new analytical approach for the evaluation of the seismic resistance of composite reinforced concrete-masonry structures is presented. In the proposed model, unreinforced masonry panels are modelled by two-dimensional macro-elements [2, 3, 4], whereas the reinforced concrete elements are modelled by lumped plasticity elements interacting with the masonry through nonlinear interface elements. The results of the analysis conducted on the case study of a confined masonry structure proposed by D'Amore [5] representative of a wide class of residential structures built in Italy after the 1908 Messina and Reggio Calabria earthquake are presented.

THE PROPOSED MODEL

The proposed simplified approach is based on a discrete-element originally conceived for the simulation of nonlinear response of masonry buildings, [2, 3, 4]. The basic element, i.e. the *panel*, exhibits a simple and understandable mechanical scheme, represented in figure 1. It is an articulated quadrilateral with four rigid edges connected by hinges whose deformability is controlled by two nonlinear diagonal springs. Each edge of the panel can interact with other elements or external restraints by means of a discrete distribution of nonlinear springs (*interface*). Each interface consists of *n* springs perpendicular to the edges of the elements which connects (*transversal springs*) and a single nonlinear spring which controls the motion in the direction parallel to the edges which connects (*sliding spring*).

The proposed mechanical model, in spite of its simplicity, is able to effectively model the main failure mechanisms of a portion of masonry subjected to in-plane, vertical and horizontal, action.

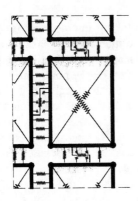 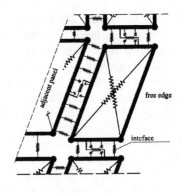

(a)   (b)

FIGURE 1. A portion of masonry schematised with the proposed macro-element:
(a) initial configuration; (b) an example of deformed configuration.

The mechanical model of an entire plane masonry wall can be regarded as an assemblage of such panels connected by interfaces. The accuracy of the model can be improved by adopting a refined mesh in order to better describe the kinematics and the failure mechanisms of the masonry wall.

The proposed macro-element is based on a mechanical equivalent scheme where deformability is lumped in the nonlinear springs or *NLinks* (Nonlinear Links). This choice leads to great simplifications, both conceptual and computational, since it is based on uniaxial constitutive laws. The transversal springs have the role of simulating the flexural behaviour of the masonry portion represented by the interface, in which each spring is calibrated according to the corresponding influence volume similarly to the well known fibre models adopted for the simulation of the nonlinear behaviour of reinforced concrete structures. The nonlinear force-displacement relationship of the transversal springs depends on the constitutive law chosen for the axial-flexural behaviour of masonry. Namely in the case of compressive rupture, a *crushing* behaviour has been considered, i.e. the material loses the capacity to resist to further compressive and tensile loads; in the case of tensile failure a *cracking* behaviour is adopted, i.e. the tensile resistance drops to zero when the rupture tensile tension is reached, however the material can still resist to compressive loads.

The simple procedure which permits to transfer the masonry properties to the interface springs is based on the geometry of the portion of masonry which is intended to represent and on the deformability and resistance parameters of the masonry along the main directions. The procedure enforces an equivalence between a single transversal spring and its masonry influence volume. In the considered case an elastic-plastic law with different behaviour in traction and compression has been adopted according to the procedures reported in [3].

The sliding spring has the role of simulating the failure mechanisms associated to the sliding of the masonry portion. This mechanism is ruled by a rigid-plastic Mohr-Coulomb law.

The diagonal springs simulate the in-plane shear deformability and the diagonal cracking failure mechanism. The ultimate load relative to this mechanism can be evaluated according to a specific Mohr-Coulomb law or according to the Turnsek-Cacovic criterion [6], or with reference to other criteria proposed in the literature.

In the analyses presented in the following an elastic-plastic law with limited deformability according to the Mohr-Coulomb criterion has been considered.

For mixed masonry-reinforced concrete structures, beam/column elements are included in the model as uniaxial finite elements with lumped plasticity. Each of these frame elements interacts with the adjacent masonry for its whole length by means of the interfaces. The proposed modelling approach allows the simulation of the confinement effect in framed masonry.

THE CASE-STUDY

The case-study building, described in full detail by D'Amore [5] was conceived on the basis of detailed investigation on eleven two-storeys framed-masonry residential buildings, built in Reggio Calabria (Italy) after the 1908 Messina Earthquake.

The typical floor plan of the case-study building, shown in figure 2, is rectangular with dimensions of 22.84 m and 11.78 m alongside the longitudinal and transversal directions respectively. The case-study structure represents a wide class of confined masonry residential buildings widely adopted in the cities of Messina and Reggio Calabria after the 1908 earthquake.

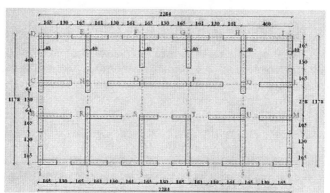

FIGURE 2. Plan of the case-study building.

THE MODEL OF THE BUILDING

The three-dimensional model of the building has been realised with the computer software *3DMacro* [7] developed by a research group of the University of Catania. This computer code allows the implementation of the approach previously described for composite reinforced concrete-masonry buildings. In order to investigate different classes of structures, two different models have been considered: (1) the *SH model*, in which solid brick masonry have been considered at the first level of the building and hollow brick masonry has been considered at the second level; (2) the *SS model*, where solid brick masonry has been considered at both levels.

The mechanical characteristics of the masonry considered in the two models are reported in table 1; the friction angle φ has been set to 0.3.

TABLE 1. Mechanical characteristics of the masonry considered in the numerical analyses.

	Young's modulus, E MPa	Transv. elastic mod., G MPa	Compr. strgth., σ_s, MPa	Tens. strgth., σ_t, MPa	Shear strgth., τ_o, MPa	Ultim. shear strain, γ_u	Density Kg/m^3
Solid bricks	2000	350	3	0.2	0.1	0.006	1800
Hollow bricks	900	150	1.5	0.1	0.05	0.003	810

In the following table 2 the mechanical characteristics of concrete and steel rebars are summarised.

TABLE 2. Mechanical characteristics of concrete and steel considered in the numerical analyses.

Young's modulus, E MPa	Cubic resistance of concrete, R_{ck} MPa	Ultim. strain of concrete, ε_{cu}, MPa	Yield. strgth of steel, f_y kN/cm^2	Ultimate strain of steel, ε_{yu}
20000	1.2	0.003	24	0.01

RESULTS OF NUMERICAL ANALYSES

For the estimation of the seismic resistance of the considered models, nonlinear static (pushover) analyses have been performed; two load distributions have been considered: (1) a *mass-proportional* distribution and (2) an *inverse-triangular* distribution.

In order to investigate the resistance capacity as a function of the earthquake direction, many loading directions and verses have been considered. Namely the principal directions of the building (X and Y) and the directions which form a 45° angle with respect to the principal directions in both verses have been taken into account.

In the following, the results obtained for the two considered models loaded along the principal directions are reported together with some images of the collapse mechanisms. The results are expressed as storey shears as a function of the relative inter-storey drifts.

The SH Model

In figures 3 and 4 the results of the analyses obtained for the SH model loaded in its two principal directions according to the two load distributions considered are reported. Each plot reports the storey shear, normalized respect to the building weight W=6662 kN, as a function of the corresponding relative inter-storey drift $\Delta u_i / h_i$.

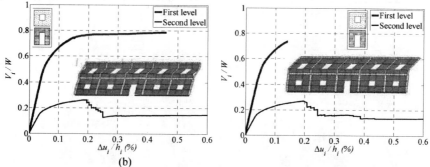

(b)

FIGURE 3. Pushover analyses in the X direction for the SH model.
(a) mass-proportional load distribution; (b) inverse-triangular force distribution.

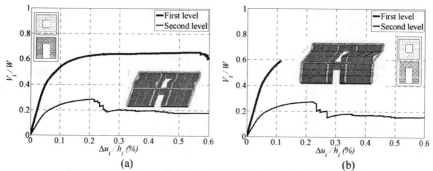

(a)   (b)

FIGURE 4. Pushover analyses in the Y direction for the SH model.
(a) mass-proportional load distribution; (b) inverse-triangular force distribution.

The obtained results highlight that the weaker direction is Y and that the collapse mechanisms is localized at the second level.

The SS Model

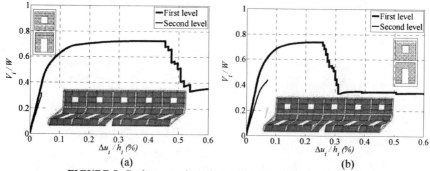

(a)   (b)

FIGURE 5. Pushover analyses in the X direction for the SS model.
(a) mass-proportional load distribution; (b) inverse-triangular force distribution.

In figures 5 and 6 the results relative to the SS model are presented. The total weight of the SS model has been estimated equal to W=7363 kN.

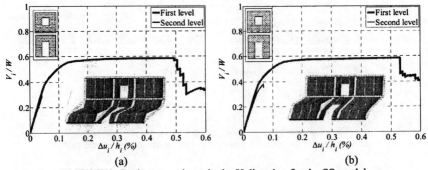

FIGURE 6. Pushover analyses in the Y direction for the SS model.
(a) mass-proportional load distribution; (b) inverse-triangular force distribution.

From the observation of the results, it is apparent that in this case the collapse mechanism is concentrated at the first level for both the load distributions considered.

Three-Dimensional Dominia

The pushover curves obtained for different loading directions and verses can be synthetically represented by means of the three-dimensional *capacity dominia* [8].

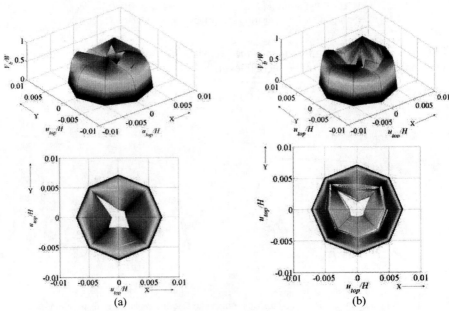

FIGURE 7. Capacity dominia for the mass-proportional load distribution.
(a) SH model; (b) SS model.

In such a 3D plot (figure 7) each push-over curve is represented along the input direction, in a plane perpendicular to the XY plane, so that the base shear coefficient is reported in the Z axis. Such a representation allows to identify the stronger and weaker directions of the building. Furthermore the resistance levels is represented in a colour scale. The ductility of the structure in all the investigated directions is also readable in the graph. The directions that has not been investigated has been linearly interpolated.

ACKNOWLEDGEMENTS

This work has been financially supported by the Executive Project 2005-2008 of the *ReLUIS* consortium (*Rete dei Laboratori Universitari di Ingegneria Sismica*), research line 1: *"Evaluation and reduction of the vulnerability of masonry buildings"*, coordinated by Professors S. Lagomarsino and G. Magenes.

REFERENCES

1. D'Amore E. (1989). "Considerazioni sulla vulnerabilità sismica delle costruzioni di Messina," Atti del IV Convegno Nazionale "L'Ingegneria sismica in Italia", Vol I, pp. 154÷163, Milano, Ottobre 1989 (in Italian).
2. I. Caliò, M. Marletta, B. Pantò, "Un semplice macro-elemento per la valutazione della vulnerabilità sismica di edifici in muratura" in *XI Convegno ANIDIS L'Ingegneria Sismica in Italia - Genoa (Italy), 25-29 Gennaio 2004*, Conference Proceedings, 2004 (in Italian).
3. I. Caliò, M. Marletta, B. Pantò, "A simplified model for the evaluation of the seismic behaviour of masonry buildings" in *10th International Conference on Civil, Structural and Environmental Engineering Computing, Rome (Italy), 30 August - 2 September 2005*, edited by B. J. Topping, Conference Proceedings, Paper no. 195, 2005.
4. I. Caliò, M. Marletta, B. Pantò, "Un macro-elemento in grado di cogliere il comportamento nel piano e fuori piano di pareti murarie" in *XII Convegno ANIDIS L'Ingegneria Sismica in Italia, Pisa (Italy), 10-14 Giugno 2007*, Conference Proceedings, 2007 (in Italian).
5. E. D'Amore, "Gli edifici in muratura intelaiata realizzati in Italia dopo il terremoto di Messina del 1908. Definizione di un caso di studio per la valutazione della vulnerabilità sismica" in *XII Convegno ANIDIS L'Ingegneria Sismica in Italia, Pisa (Italy), 10-14 Giugno 2007*, Conference Proceedings, 2007 (in Italian).
6. V. Turnsek, F. Cacovic F., "Some experimental results on strength of brick masonry walls" in *2nd International Brick Masonry Conference, Stoke-on-Trent (UK)*, Conference Proceedings, 1971, pp. 149-156.
7. 3DMacro: a 3D computer program for the seismic assessment of masonry building. Web: www.3dmacro.it.
8. I. Caliò, F. Cannizzaro, D. Grasso D., M. Marletta, B. Pantò, D. Rapicavoli, "Progetto TREMA: Schema modello fisso alla base", in Report del Progetto TREMA – Linea 1 Progetto ReLUIS. Web: www.unibas.it/trerem/TREREMDW/Blindtests/UNICT/4.4_UR14relazione%20base%20fissa.pdf, 2006 (in Italian).

Some comments on the experimental behavior of FRC beams in flexure

Giuseppe Campione, Lidia La Mendola, Maria Letizia Mangiavillano and
Maurizio Papia

Dipartimento di Ingegneria Strutturale e Geotecnica, Università di Palermo, Viale delle Scienze,
90128 PALERMO

Abstract. In the present paper the experimental results, recently obtained by the authors, regarding the monotonic and the cyclic flexural response of normal and high-strength concrete beams reinforced with steel bars and discontinuous fibers, are shown. From the experimental results, all referred to low values of shear-to-depth ratios, it emerges clearly that the shear failure is brittle especially under cyclic actions highlighting the role of the fibers in the flexural behavior of the beams. The cyclic action produces a significant decay in the stiffness and in the strength capacity of the beams, and the addition of fibers reduces these negative effects. Form theoretical point of view good agreement can be found utilizing the recent analytical model proposed by the authors.

Keywords: beams, steel fibers, shear, cyclic actions, shear and flexural strength, design.

INTRODUCTION

Ductile design of reinforced concrete beams is generally related to flexural failure in bending, but very often the presence of high shear forces, especially in deep members, reduces their flexural capacity. In the presence of cyclic actions, or when high strength concrete is utilized, further penalization occurs in achievement of the complete flexural capacity and ductility.

Many reports published over the past 25 years have considered the possibility of using fibrous concrete in order to improve shear resistance [1-4] and they have stressed the importance in using fibers coupled with traditional steel reinforcement to improve to flexural behavior of R.C. members affected by shear failure in the presence of cyclic actions.

The current paper refers to the above mentioned problems and it consists of two parts: - the first one in which experimental flexural tests recently carried out by the authors on plain and fibrous reinforced concrete beams under monotonic and cyclic actions are presented and discussed; - and a second part in which an analytical expression proposed in [6] to predict the bearing capacity of beams both in shear and flexure is discussed and verified against available data.

CP1020, *2008 Seismic Engineering Conference Commemorating the 1908 Messina and Reggio Calabria Earthquake,*
edited by A. Santini and N. Moraci
© 2008 American Institute of Physics 978-0-7354-0542-4/08/$23.00

REFERENCE EXPERIMENTAL INVESTIGATIONS

The experimental researches here mentioned are those presented in [1-4]. In Figure 1 the reinforcement arrangements and the load scheme adopted for the beams are shown.

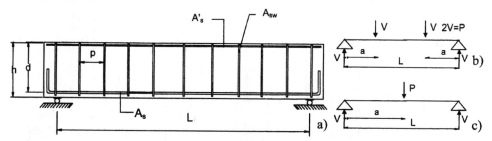

FIGURE 1. Experimental test: a) reinforcement arrangement; b)four-point bending test; and c) three-point bending test.

These experimental tests refer to plain and fibrous concrete reinforced beams tested in four-point (Fig. 1 b) or three-point (Fig. 1 c) bending test, under monotonic and cyclic actions. The beams, have rectangular cross-sections of base b, height h and length between two supports L. All the tests were carried out by using a universal testing machine operating in a controlled displacement mode. It must be noted that the cyclic tests were carried out with and without changes of sign.

The main longitudinal steel bars have area A_s, while the transversal bars, constituted by close stirrups, have area A_{sw} and they were placed at pitch p. Additionally steel bars were utilized in the upper zone of the beam. In the case of cyclic actions double and symmetric reinforcing bars were utilized.

In Table 1 the geometry of the beams, the percentage of main and secondary steel adopted and the mechanical properties of constituent materials are given. In this table the steel percentage is furnished in terms of geometrical ratios: longitudinal steel $\rho_s=A_s/(b\cdot d)$, being d the effective depth of the cross-section; transverse steel $\rho_{sw}=A_{sw}/(b\cdot p)$ in the pitch p of the stirrups.

Normal or high strength concretes (NSC, HSC) were characterized by the compressive cylindrical strength f_c and splitting tensile strength f_t. Steel reinforcement had yielding stress of longitudinal bars f_y and of transverse stirrups f_{yw}. Different types of fibers were utilized (hooked steel, crimped steel, carbon, polyolefin) with volume percentage v_f. For all the beams Table 1 gives the values of ultimate shear strength v_u, defined as the ratio between the maximum shear force V_u and the shear area $(b\cdot d)$. The ratios adopted between the shear span a and the effective depth d of the beam were also given in Table 1. From the experimental researches mentioned it emerges clearly that the addition of fibers increases significantly the shear strength, and in some cases allows one to change the failure mode from shear to flexure. In Fig. 2 the load-deflection curves of high-strength beams, tested in [2], under monotonic and cyclic actions, without changes of sign and for two different values of pitch between the stirrups are reported. It must be noted that P is the load applied by the testing machine and δ is the displacement in the middle.

TABLE 1. Details of tested beams

Reference	b (mm)	h (mm)	ρ_s (%)	a/d	f_y (MPa)	ρ_sw (%)	f_yw (MPa)	f_c − f_t (MPa)	Fiber type	v_f (%)	V_u,exp (MPa)	V_u,cal (MPa) Eq. (6) or (7)
Campione and Mindess (1999)	100	125	3.20	2.5	300	0.35	550	70.0 − 4.7	/	/	4.85	4.13
						0.70					6.05	5.11
						0.35			hooked steel		8.09	5.23
						0.70					9.25	5.77
						0.35			crimped steel		8.10	5.76
						0.70				2	8.80	6.21
						0.35			carbon		7.60	5.25
						0.70					8.24	5.77
						0.35			polyolefin		7.64	5.17
						0.70					7.74	5.72
Campione et al. (2003)	150	250	1.82	2.0	610	0.00	510	41.2 − 2.0	/	/	1.09	1.25
						0.00					1.45	1.25
						0.19					1.81	2.20
						0.19					1.90	2.20
						0.63					4.06	4.10
						0.63					4.78	4.10
						0.19		40.9 − 4.8	hooked steel	1	3.19	2.60
						0.19		42.2 − 5.5		2	3.48	3.32
						0.19		40.9 − 4.8		1	3.77	2.60
						0.19		42.2 − 5.5		2	3.91	3.32
Campione et al. (2004)	150	150	1.13	1.95	467	0.75	450	30.0 − 2.9	/	/	3.42	2.37
								32.0 − 3.5	hooked steel	1	3.66	2.56
			1.22	2.12				30.0 − 2.9	/	/	3.55	2.34
								32.0 − 3.5	hooked steel	1	3.80	2.54

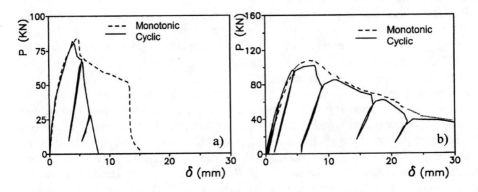

FIGURE 2. Load-deflection curves of beams with pitch: a) p=180 mm; b) p=98 mm.

It is evident that for beams with pitch between the stirrups of 198 mm brittle shear failure occurs, while for beams with pitch 98 mm flexural rupture occurs, with brittle concrete crushing failure due to the high percentage of steel reinforcement adopted.

Figure 3 shows load-deflection curves (for beams with pitch 180 and 98 mm) referred to fibrous concrete beams with v_f = 2 %. It emerges clearly that the addition of fibers improves the behavior with respect to that of ordinary concrete members. In particular, the addition of steel fibers in the beams produces ductile flexural failure.

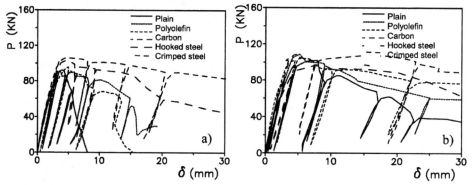

FIGURE 3. Load-deflection curves of fibrous concrete beams with: a) p=180 mm; b) p=98 mm.

Figure 4 shows the load-deflection curves of normal strength beams, tested in [4], under monotonic actions. Fig. 4 a) shows the experimental results of beams reinforced with stirrups at pitch 200 mm and 60 mm, while Fig. 4 b) shows the analogous results for fibrous reinforced concrete beams (without stirrups) at volume percentage of 1 and 2 %.

In Fig. 4 a) the continuous lines indicate the load-deflection curves for beams with lower pitch of stirrups, while the dashed lines are relative to the beams with higher pitch. In Fig. 4 b) the continuous lines indicate the load-deflection curves for beams with lower percentage of steel fibers ($v_f = 1\%$), while the dashed lines are relative to the beams with higher percentage of steel fibers ($v_f = 2\%$).

In the same diagrams the load-deflection curves for beams reinforced with only main steel bars (thin continuous lines) are reported.

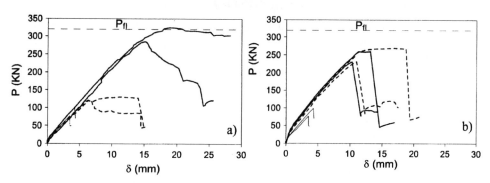

FIGURE 4. Load-deflection curves for beams reinforced with main steel bars: a) in absence and presence of stirrups; b) in absence and presence of steel fibers.

Higher values of strength are shown in beams with high percentage of fibers though the shear failure always occurs due to the low a/d ratio chosen and to the high percentage of main steel bars adopted.

843

It must be noted that the addition of fibers increases the ultimate strength of beams ensuring to obtain similar behavior to observed for beams reinforced with stirrups at close space.

The straight line in Fig. 4 at $P = P_{fl}$ is corresponding to the complete flexural capacity calculated by the flexural model proposed in ACI 318-02 [5].

Figure 5 shows the load-deflection curves of beams under cyclic actions with a cover of 5 or 15 mm [3]. Under cyclic actions a significant reduction of strength, stiffness and energy dissipated is observed, but in the case of fibrous reinforced concrete better behavior is observed.

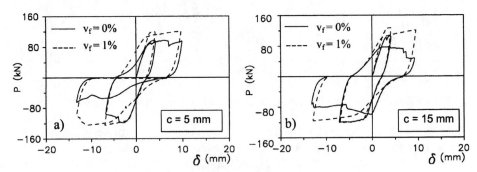

FIGURE 5. Load-deflection curves of beams under cyclic actions.

ULTIMATE STRENGTH

In the following section the expressions recently obtained by the authors [5] to calculate the shear and the flexural strength of the beams are utilized for a comparison with experimental data.

Flexural strength

To calculate the ultimate bending moment M_{fl} the translation and the rotational equilibrium conditions, under the plane section hypothesis are utilized.

Figure 6 shows the possible failure modes for a rectangular cross-section in the presence of single reinforcement and when ductile behavior is expected.

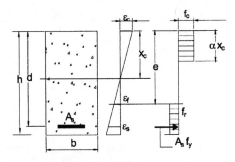

FIGURE 6. Hypothesis of failure of the section in flexure.

It must be observed that the maximum strain of the steel bars is assumed equal to 1 %, as suggested in the most common code also for fibrous concrete beams, and the post-cracking tensile strength of the material f_r, is also considered.

For this value, in absence of experimental data, some researchers, such as in [6] or in [7], have proposed analytical expressions validated on the basis of existing experimental data.

In [7] it was shown that the maximum tensile strength f_t of FRC material with strain softening behavior can be assumed the one of plain concrete, while the post-cracking tensile strength can be assumed as:

$$f_r = 0.3475 \cdot v_f \cdot \frac{L_f}{D_f} \cdot (f_c)^{0.66} \qquad \text{in MPa} \qquad (1)$$

being L_f and D_f the length and the diameter of fibers adopted, respectively.

As suggested in [6] the value of f_r can be assumed as:

$$f_r = 0.2 \cdot v_f \cdot \frac{L_f}{D_f} \cdot (f_c)^{0.5} \qquad \text{in MPa} \qquad (2)$$

In the following sections, Eq. (2), which is more conservative than Eq. (1), will be adopted. With reference to the symbols in Fig. 6, the distance e between the extreme fiber in compression and the fiber at which the strain ε_f, corresponding to the post-cracking tensile strength of concrete is:

$$\frac{e}{x_c} = \frac{\dfrac{f_t}{E_{ct}} + \varepsilon_c}{\varepsilon_c} \qquad (3)$$

in which $\varepsilon_f = f_t / E_{ct}$ because of the tensile response assumed for the fibrous concrete as proposed in [8], in which a bilinear curve constituted by a linear branch up to the maximum tensile strength f_t and a second horizontal branch having constant post-cracking strength f_r after the vertical dropt of the maximum strength is adopted.

By using the translation equilibrium condition, if the yielding of the steel is assumed and the maximum strain of concrete ε_c is settled equal to 0.003, the position of the neutral axis results from equilibrium condition in the form:

$$\frac{x_c}{d} = \frac{1}{\alpha} \cdot \frac{\rho_s \cdot f_y + f_r \cdot \dfrac{h}{d}}{f_c + f_r \cdot \dfrac{f_t / E_{ct} + 0.003}{\alpha \cdot 0.003}} \qquad (4)$$

In Eq. (4) α is a parameter that defines the depth of the rectangular stress-diagram of the compressive concrete and it is assumed, as suggested in [9], equal to 0.85 e 0.65 for normal (NSC) and high (HSC) strength, respectively.

By using the rotational equilibrium condition across the compressive centroid it is possible to obtain the ultimate bending moment in the form:

845

$$\frac{M_{fl}}{b \cdot d^2} = \rho_s \cdot f_y \cdot \left(1 - 0.5 \cdot \frac{\alpha \cdot x_c}{d}\right) + f_r \cdot \left(\frac{h}{d} - \frac{e}{d}\right) \cdot \left(\frac{h}{d} - \frac{h - e}{2 \cdot d} - 0.5 \cdot \frac{\alpha \cdot x_c}{d}\right) \quad (5)$$

By using the equilibrium conditions of the beams in the shear span a, known the ultimate bending moment M_{fl} (see Eq. 5) it is possible to obtain the shear strength v_u, defined as $v_u = V_u/(b \cdot d)$, in the following form:

$$\frac{M_{fl}}{b \cdot d^2} = \frac{a}{d} \cdot v_u \quad (6)$$

The strength of the beam can be assumed as the minimum value among the flexural strength expressed as in Eq. (6) and the shear strength, the latter determined as shown in the following section.

Shear strength

Several studies, focused on the shear strength prevision of FRC beams, propose semi-empirical and analytical expressions calibrated and validated on the basis of the available experimental data. Among them the focus here is on the expression recently proposed by the authors in [6], calibrated on the basis of experimental data available in literature, for fibrous reinforced beams with hooked steel fibers, in the form:

$$v_u = \frac{1}{\sqrt{1 + \frac{d}{25 \cdot d_a}}} \cdot \left\{ j_0 \cdot \left[A \cdot \rho_s^{1/2} \cdot \sqrt{f_c} + B \cdot \varepsilon \cdot f_y \rho_t \cdot \left(\frac{d}{a}\right)^{1.8} \right] + 0.27 \cdot \frac{L_f \cdot v_f}{D_f} \cdot \sqrt{f_c} \right\} + \Phi_f \cdot \rho_{sw} \cdot f_{yw} \quad (7)$$

It must be noted that Eq. (7) is expressed in MPa, d_a is the maximum aggregate size of the concrete, $\rho_t = \rho_s + \rho_f$ is the fictitious equivalent geometrical ratio of main steel with expression:

$$\rho_f = \frac{f_r}{0.3 \cdot f_y} \cdot \frac{h}{d} \cdot \left(1 - \frac{e}{h}\right) \quad (8)$$

taking into account the presence of the fibers. In Eq. (7) A and B are two empirical coefficient assumed 1.3 and 0.3 for NSC members and 1.75 and 0.65 for HSC members.

In Eq. (7) ε is a corrective coefficient, originally proposed in [10], equal to 2.5 d/a when a/d $\leq$ 2.5 and 1 when a/d > 2.5, j_0 is the dimensionless arm of internal forces given in [6] and Φ_f is an effectiveness function ((see [6]) giving the share of the yielding stress reached at beam failure. This function shows the share of beam action strength contribution on the global strength of the beam calculated including the effect of fibers.

In Table 1 the ultimate strength values of tested beams calculated by utilizing Eq. (6) or (7), respectively for flexural or shear failure, are also indicated. In particular the ultimate strength, reported in Table 1 is the minimum value between the shear and the flexural strength.

846

It has to be observed that if the crimped steel fibers or to carbon and polyolefin fibers are utilized, the strength contribution due to fibers is calculated referring to the post-cracking tensile stress reported as in [11]. The comparison among the experimental data and the predicted ones shows good agreement.

CONCLUSIONS

The experimental results here mentioned and referred to normal and high strength concrete beams, reinforced with steel bars and reinforcing fibers, show that shear failure if occurs is very brittle and the addition of fibers in some cases can change the mode of failure from shear to flexure ensuring more ductile behavior. Under cyclic actions a rapid decay of the strength and stiffness occurs with more brittleness if compared with the monotonic response. With the addition of fibers the flexural behavior of the beams is improved. Finally from the analytical point of view, and only in terms of ultimate strength, the semi-empirical analytical expression recently proposed by the authors give accurate predictions of the experimental data.

REFERENCES

1 D. Dupont and L. Vandewalle, "Shear capacity of concrete beams containing longitudinal reinforcement and steel fibers", ACI SP-216 Innovations in fiber-reinforced concrete for value, Editors N. Banthia, M. Criswell, P. Tatnall and K. Folliard, 78-94 (2003).
2 G. Campione and S. Mindess, "Fibers as shear reinforcements for high strength reinforced concrete beams containing stirrups", Proceeding of the Third International HSPFRC - RILEM Workshop, Mainz, Germany, 16-19 May, 519-529 (1999).
3 G. Campione, N. Miraglia and N. Scibila, "Indagine sperimentale su travi in calcestruzzo fibroso sottoposte a carichi monotonici e ciclici", Convegno CTE Novembre 2004. Bari, 1, 83-192 (2004) (only avaialabe in Italian).
4 G. Campione, C. Cucchiara and L. La Mendola, " Role of fibres and stirrups on the experimental behaviour of reinforced concrete beams and flexure and shear", CD-Rom of Int. Conf. on Composites in Construction, september 2003, Rende, Italy.
5 ACI Committee 318, " Building code requirements for structural concrete (ACI 318-02) and Commentary (318R-02)", American Concrete Institute, Farmington Hills, Michigan, 443 pp. (2002).
6 G. Campione, L. La Mendola and M. Papia, "Ultimate shear strength of fibrous reinforced concrete beams with stirrups", Structural Engineering and Mechanics, 24 (1), 107-136 (2006).
7 S. J. Foster and M. M. Attard, "Strength and ductility of fiber-reinforced high-strength concrete columns", Journal of Structural Engineering, ASCE, 127 (1), 28-34 (2001).
8 T. Y. Lim, P. Paravasivam and S.L. Lee, "An analytical model for tensile bahaviour of steel concrete", ACI Material Journal, 84 (4), 286-298 (1987).
9 G. Campione, L. La Mendola and G. Zingone, "Flexural shear-moment interaction in high strength fibre reinforced concrete beams", Proceeding of Fifth RILEM symposium on Fibre-reinforced concretes (FRC), BEFIB 2000, 13-15 September, Lyon, France, 451-460 (2000).
10 T.C. Zsutty, "Beam shear strength prediction by analysis of separate categories of simple beam tests", ACI Structural Journal, 68 (11), 943-951 (1971).
11 A.E. Naaman, "Strain hardening and deflection hardening fiber reinforced cement composites", Proc. of IV Intern. RILEM Workshop "High performance fiber reinforced cement composites (HPFRCC4), Ann Arbor, MI, US, 95-113 (2003).

Seismic Monitoring To Assess Performance Of Structures In Near-Real Time: Recent Progress

Mehmet Çelebi

USGS (MS977), 345 Middlefield Rd., Menlo Park, CA. 94025, USA

Abstract. Earlier papers have described how observed data from classical accelerometers deployed in structures or from differential GPS with high sampling ratios deployed at roofs of tall buildings can be configured to establish seismic health monitoring of structures. In these configurations, drift ratios[1] are the main parametric indicator of damage condition of a structure or component of a structure.

Real-time measurement of displacements are acquired either by double integration of accelerometer time-series data, or by directly using GPS. Recorded sensor data is then related to the performance level of a building. Performance-based design method stipulates that for a building the amplitude of relative displacement of the roof of a building (with respect to its base) indicates its performance.

Usually, drift ratio is computed using relative displacement between two consecutive floors. When accelerometers are used, a specific software is used to compute displacements and drift ratios in real-time by double integration of accelerometer data from several floors. However, GPS-measured relative displacements are limited to being acquired only at the roof with respect to its reference base. Thus, computed drift ratio is the average drift ratio for the whole building. Until recently, the validity of measurements using GPS was limited to long-period structures (T>1 s) because GPS systems readily available were limited to 10-20 samples per seconds (sps) capability. However, presently, up to 50 sps differential GPS systems are available on the market and have been successfully used to monitor drift ratios [1,2] – thus enabling future usefulness of GPS to all types of structures. Several levels of threshold drift ratios can be postulated in order to make decisions for inspections and/or occupancy.

Experience with data acquired from both accelerometers and GPS deployments indicates that they are reliable and provide pragmatic alternatives to alert the owners and other authorized parties to make informed decisions and select choices for pre-defined actions following significant events.

Keywords: Monitoring, structures, performance, seismic, drift, GPS.

INTRODUCTION

Following an earthquake, rapid and accurate assessment of the damage condition or performance of a building is of paramount importance to stakeholders, including owners, leasers, permanent and/or temporary occupants, and city officials and rescue teams that are concerned with safety of those in the building and those that may be affected in nearby buildings and infrastructures. These stakeholders will require answers to key questions such as: (a) is there visible or hidden damage?, (b) if damage occurred, what is the extent?, (c) does the damage threaten other neighboring structures?, (d) can the structure be occupied immediately without compromising life safety? Economic loss due to non-occupancy of a building may be significant.

---

[1] Drift ratio is defined as relative displacement between any two floors divided by the difference in elevation of the two floors. Usually, this ratio is computed for two consecutive floors.

CP1020, *2008 Seismic Engineering Conference Commemorating the 1908 Messina and Reggio Calabria Earthquake,*
edited by A. Santini and N. Moraci
2008 American Institute of Physics 978-0-7354-0542-4/08/$23.00

Until recently, assessments of damage to buildings following an earthquake were essentially carried out through inspections by city-designated engineers following procedures similar to ATC-20 tagging requirements [3]. Tagging usually involves visual inspection only and is implemented by colored tags indicative of potential hazard to occupants[2]. However, with visual inspection, some serious damage may not be noticed due to the presence of building finishes and fireproofing. Thus, most steel or reinforced concrete moment-frame buildings may be tagged based on visual indications of building deformation, such as damage to partitions or glazing. Lack of certainty regarding the actual deformations that the building experienced may typically lead an inspector toward a relatively conservative tag leading to expensive and time-consuming intrusive inspections (e.g., it is reported that, after the [M_w=6.7] 1994 Northridge, CA earthquake, approximately 300 buildings ranging in height from 1 to 26 stories were subjected to costly intrusive inspection of connections [4]).

This paper describes an alternative to tagging that is now available to owners and their designated engineers by configuring real-time response of a structure instrumented as a health monitoring tool. As Porter and others [5] state, most new methods do not utilize real-time response measurements of a building for assessments of a building's performance during an event with the exception outlined by Çelebi and Sanli [6] and Çelebi and others [7]. In these applications, differential GPS [6]with high sampling ratios and classical accelerometer deployed structures [7] are configured to obtain data in real-time and compute drift ratios as the main parametric indicator of damage condition of a structure or one or more components of a structure. The rationale here is that a building owner and designated engineers are expected to use the response data acquired by a real-time health monitoring system to justify a reduced inspection program as compared to that which would otherwise be required by a city government for a similar non-instrumented building in the same area[3]. It is possible, depending on the deformation pattern and associated damage indicators observed in a building, to direct the initial inspections toward specific locations in the building that experienced large and potentially damage-inducing drifts during an earthquake.

Examples of and data from either type of sensor deployment (GPS or accelerometers) indicate that these methods are reliable and provide requisite information for owners and other parties to make informed decisions and select pre-defined actions following significant events. Furthermore, recent additional adoptions of such methods by financial and industrial enterprises validate its usefulness.

REQUISITES

The most relevant parameter to assess performance is the measurement or computation of actual or average story drift ratios. Specifically, the drift ratios can be related to the performance-based force-deformation curve hypothetically represented in Figure 1 (modified from Figure C2-3 of *FEMA-274* [9]). When drift ratios, as

---

[2] Green tag indicating the building can be occupied - that is, the building does not pose a threat to life safety; yellow indicates limited occupation - that is, hazardous to life safety but not to prevent limited entrance to retrieve possessions; and red indicating entrance prohibited - that is, hazardous to life.

[3] The City of San Francisco, California, has developed a "Building Occupancy Resumption Program" [8]) whereby a pre-qualified Occupancy decision-making process, as described in this paper, may be proposed to the City as a reduced inspection program and in lieu of detailed inspections by city engineers following a serious earthquake.

computed from relative displacements between consecutive floors, are determined, the performance (and "damage state" of the building can be estimated as in Figure 1.

Except for tests conducted in a laboratory using displacement transducers, measuring displacements of real-life structures very difficult. For structures with long-period responses, such as tall buildings, displacement measurements using GPS are measured directly at the roof only; hence, only average drift ratio is computed. Thus, recorded sensor data is related to performance level of a building and therefore to performance-based design, which stipulates that for a building the amplitude of relative displacement of the roof of a building with respect to its base indicates its performance. For accelerometer-based systems, the accelerometers must be strategically deployed at specific locations on several floors of a building to facilitate real-time measurement of the actual structural response, which in turn is used to compute displacements and drift ratios as the indicators of damage.

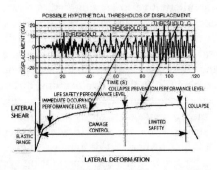

FIGURE 1. Hypothetical displacement time-history as related to performance [modified from Figure C2-3 of *FEMA-274(also known as ATC1997 [9])*].

Table 1 shows typical ranges of drift ratios that define threshold stages for steel moment resisting framed buildings as developed from FEMA 352 [4]. For reinforced concrete framed buildings, the lower figures may be more appropriate to adopt.

TABLE 1. Summary of Suggested Typical Threshold Stages and Ranges of Drift Ratios

Threshold Stage	1	2	3
Suggested Typical Drift Ratios	0.2-0.3%	0.6-0.8%	1.4-2.2%

GPS FOR DIRECTLY MEASURING DISPLACEMENTS

Until recently, use of GPS was limited to long-period structures (T>1 s) because differential GPS systems readily available were limited to 10-20 sps capability[4]. Currently, the accuracy of 10-20 Hz GPS measurements is ±1 cm horizontal and ±2 cm vertical. Furthermore, with GPS deployed on buildings, measurement of displacement is possible only at the roof.

---

[4] Recently, 50 sps differential GPS systems are available on the market and have been successfully used [2,10].

A schematic and photos of a pioneering application using GPS to directly measure displacements is shown in Figure 2. Two GPS units are used in order to capture both the translational and torsional response of the 34-story building in San Francisco, CA [6]. Tri-axial accelerometers are co-located to compare the displacements measured by GPS with those computed by double-integration of accelerations. Both acceleration and displacement data stream into the monitoring system as shown also in Figure 2.

To date, strong shaking data from the deployed system has not been recorded. However, ambient data obtained from both accelerometers and GPS units (Figures 3a-d) have been analyzed. Sample cross-spectra (Sxy) and coherency and phase angle plots of pairs of parallel records (N-S component of north deployment [N_N] vs. N-S component of south deployment [S_N], from accelerometers are shown in Figures 3e-f. The same is repeated for the differential displacement records from GPS units (Figures 3g-h). The dominant peak in frequency at 0.24-.25 Hz seen in cross-spectra (Sxy) plots from both acceleration and displacement data are compatible with expected fundamental frequency for a 34-story building. A second peak in frequency at 0.31 Hz in the acceleration data belongs to the torsional mode.

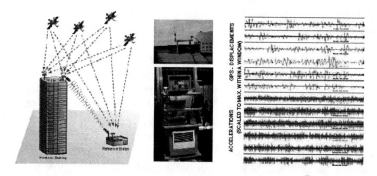

Figure 2. (Left)- Schematic of the overall system using GPS and accelerometers (San Francisco, CA.): (Center)- GPS and radio modem antenna and the recorders connected to PC, (Right)- streaming acceleration and displacement data in real-time.

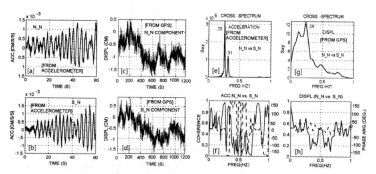

Figure 3. [a,b] 60-second windowed accelerations and [c,d] 1200 second windowed GPS displacement data in the north-south orientation and at N (North) and S (South) locations (acceleration data sampled at 200 sps and GPS at 10 sps). Cross-spectra (Sxy) and associated coherency and phase angle plots of horizontal, and parallel accelerations [e,f] and GPS displacements [g,h]. [Note: In the coherency-phase angle plots, solid lines are coherency and dashed lines are phase-angle].

At the fundamental frequency at 0.24 Hz, the displacement data exhibits a 0° phase angle; however, the coherencies are low (~0.6-0.7). The fact that the fundamental frequency (0.24 Hz) can be identified from the GPS displacement data (amplitudes of which are within the manufacturer specified error range) and that it can be confirmed by the acceleration data, is an indication of promise of better results when larger displacements can be recorded during strong shaking.

DISPLACEMENT VIA REAL-TIME DOUBLE INTEGRATION

A general flowchart for an alternative strategy based on computing displacements and drift ratios in real-time from signals of accelerometers strategically deployed throughout a building is depicted in Figure 4 (left) [7]. Although ideal, deploying multiple accelerometers in every direction on every floor level is not a feasible approach, not only because of the installation cost, but also from the point of view of being able to robustly, and in near real-time, (a) stream accelerations, (b) compute and stream displacements and drift ratios after double-integration of accelerations, and (c) visually display threshold exceedences, thus fulfilling the objective of timely assessment of performance level and damage conditions.

A flowchart of a health monitoring system utilizing these principles is shown in Figure 4 (right) [7]. The distribution of accelerometers provides data from several pairs of neighboring floors to facilitate drift computations. The system server at the site (a) digitizes continuous analog data, (b) pre-processes the 1000 sps digitized data with low-pass, anti-alias filters (c) decimates the data to 200 sps and streams it locally, (d) monitors and applies server triggering threshold criteria and locally records (with a pre-event memory) when prescribed thresholds are exceeded, and (e) broadcasts the data continuously to remote users by high-speed internet. Data can also be recorded locally on demand also. A "Client Software" remotely acquires acceleration data that can then be used to compute velocity, displacement and drift ratios. Figure 5 shows two PC screen snapshots of the client software display configured to stream acceleration or velocity or displacement or drift ratio time series. The amplitude spectrum for one of the selected channels is periodically recomputed and clearly displays several identifiable and distinct frequency peaks. In the lower left, time series of drift ratios are shown.

Corresponding to each drift ratio, there are four stages of colored indicators. When only the "green" color indicator is activated, it indicates that the computed drift ratio is below the first of three specific thresholds. The thresholds of drift ratios for selected pairs of data must also be manually entered in the boxes. As drift ratios exceed the designated three thresholds, additional indicators are activated, each with a different color (see Figure 4 left). The drift ratios are calculated using data from any pair of accelerometer channels oriented in the same direction. The threshold drift ratios for alarming and recording are computed and determined by structural engineers using structural information and are compatible with the performance-based theme, as previously illustrated in Figure 1.

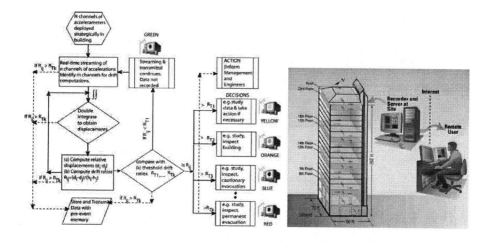

Figure 4. (left)Flowchart for observation of damage levels based on threshold drift ratios, and (right) schematic of real-time seismic monitoring of the building.

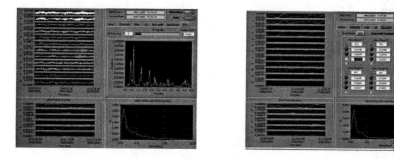

Figure 5. Screen snapshots of sample client software displays: (left) acceleration streams and computed amplitude and response spectra, and (right) displacement and corresponding drift ratios and alarm systems corresponding to thresholds.

A set of low-amplitude accelerations (largest peak acceleration ~ 1 % g) recorded in the building during the December 22, 2003 San Simeon, CA. earthquake (Mw=6.4, epicentral distance 258 km) are exhibited in Figure 6 for one side of the building. Figure 6 (center) also shows accelerations at the roof and corresponding amplitude spectra for the (a) two parallel channels (Ch12 and Ch21), (b) their differences (Ch12-Ch21), and (c) orthogonal channel (Ch30). The amplitude spectra depicts the first mode translational and torsional frequencies as 0.38 Hz and 0.60Hz respectively. The frequency at 1.08 Hz belongs to the second translational mode. At the right of Figure 6, a 20 s window of computed displacements starting 20 s into the record reveals the propagation of waves from the ground floor to the roof. The travel time is about 0.5 seconds. Since the height of the building is known (262.5 ft [80m]), travel velocity is computed as 160 m/s. One of the possible approaches for detection of possible damage to structures is by keeping track of significant changes in the travel time, since such travel of waves will be delayed if there are cracks in the structural system [11].

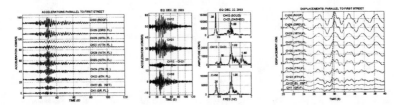

Figure 6. Accelerations (left) at each instrumented floor on one side of the building for the San Simeon, CA earthquake of 12/22/2003, (center) from parallel channels (CH12, CH21) at the roof, their difference (CH12-CH21), and orthogonal CH30, and corresponding amplitude spectra indicate fundamental frequency at 0.38 Hz. (right) A 20-s window starting 20-s into the record of computed displacements shows propagating waves [travel time of ~ 0.5 s - indicated by dashed line] from the ground floor to the roof.

MONITORING MULTIPLE (CAMPUS) STRUCTURES

Rather than having only one building monitored, there may be situations where some owners desire to monitor several buildings simultaneously, such as on industrial campus. Figure 7 schematically shows a campus-oriented monitoring configuration. Depending on the choice of the owner and consultants, a campus system may have building-specific or central-monitoring systems and as such is highly flexible in configuration. As can be stipulated, potential variations and combinations of alternatives for a campus-wide monitoring system are tremendous. There can be central-controlled or building-specific monitoring (or both). A wide variety of data communication methods can be configured to meet the needs (Figure 7).

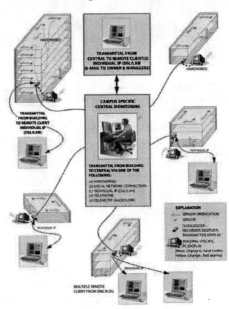

Figure 7. A schematic of campus-oriented monitoring system. Each building within a campus may have its own monitoring system or there may a central monitoring unit.

CONCLUSIONS

Capitalizing on advances in global positioning systems (GPS), in computational capabilities and methods, and in data transmission technology, it is now possible to configure and implement a seismic monitoring system for a specific building with the objective of rapidly obtaining and evaluating response data during a strong shaking event in order to help make informed decisions regarding the health and occupancy of that specific building. Displacements, and in turn, drift ratios, can be obtained in real-time or near real-time through use of GPS technology and/or double-integrated acceleration. Drift ratios can be related to damage condition of the structural system by using relevant parameters of the type of connections and story structural characteristics including its geometry. Thus, once observed drift ratios are computed in near real-time, technical assessment of the damage condition of a building can be made by comparing the observed with pre-determined threshold stages. Both GPS and double-integrated acceleration applications can be used for performance evaluation of structures and can be considered as building health-monitoring applications. Although, to date, these systems were not tested during strong shaking events, analyses of data recorded during smaller events or low-amplitude shaking are promising.

REFERENCES

1. M. Panagitou, J. I. Restrepo, J.I., J. P. Conte and R. E. Englekirk, "Seismic Response of Reinforced Concrete Wall Buildings", 8NCEE (paper no. 1494), San Francisco, Ca. April 18-22, 2006.
2. J.I. Restrepo, (private communication), 2007.
3. Applied Technology Council (ATC), "Procedures for Post-Earthquake Safety Evaluation of Buildings",ATC-20, Redwood City, CA., 1989.
4. Federal Emergency Management Agency, "FEMA-352: Recommended Post-earthquake Evaluation and Repair Criteria for Welded Steel Moment-Frame Buildings (also SAC 2000 prepared by SAC Joint Venture), 2000.
5. K. Porter, J. Mitrani-Reiser, J., J. L. Beck and J. Ching, "Smarter Structures: Real- time loss estimation for instrumented buildings," 8NCEE (no.1236), San Francisco, Ca. Apr.18-22, 2006.
6. M. Çelebi and A. Sanli, "GPS in Pioneering Dynamic Monitoring of Long-Period Structures, " Earthquake Spectra, Journal of EERI,. Volume 18, No. 1, pp. 47–61, February 2002.
7. M. Çelebi, A. Sanli, M. Sinclair, S.Gallant, and D. Radulescu, D., "Real-Time Seismic Monitoring Needs of a Building Owner and the solution – A Cooperative Effort," Journal of EERI, Earthquake Spectra, v.19, Issue 1, pp.1-23, 2004.
8. City and County of San Francisco, Department of Building Inspection, "Building Occupancy Resumption Program (BORP), , Emergency Operation Plan, 2001. [www.seaonc.org/member/committees/des_build.html).
9. Applied Technology Council (ATC), "FEMA 274: NEHRP Commentary on the Guidelines for the Seismic Rehabilitation of Buildings, prepared for the Building Seismic Safety Council", published by the Federal Emergency Management Agency, , Washington, D.C., 1997.
10. M. Restrepo, (private communication), 2007.
11. E. Safak, Wave-propagation formulation of seismic response of multistory buildings, ASCE, Journal of Structural Engineering, vol. 125, no. 4, April 1999, pp. 426-437.

Seismic Energy Demand of Buckling-Restrained Braced Frames

Hyunhoon Choi[a] and Jinkoo Kim[b]

[a]*Post-doctoral researcher, Sungkyunkwan University, Suwon, Korea 440-746*
[b]*Associate Professor, Sungkyunkwan University, Suwon, Korea 440-746*

Abstract. In this study seismic analyses of steel structures were carried out to examine the effect of ground motion characteristics and structural properties on energy demands using 60 earthquake ground motions recorded in different soil conditions, and the results were compared with those of previous works. Analysis results show that ductility ratios and the site conditions have significant influence on input energy. The ratio of hysteretic to input energy is considerably influenced by the ductility ratio and the strong motion duration. It is also observed that as the predominant periods of the input energy spectra are significantly larger than those of acceleration response spectra used in the strength design, the strength demand on a structure designed based on energy should be checked especially in short period structures. For that reason framed structures with buckling-restrained-braces (BRBs) were designed in such a way that all the input energy was dissipated by the hysteretic energy of the BRBs, and the results were compared with those designed by conventional strength-based design procedure.

Keywords: Input energy, hysteretic energy, energy-based seismic design

INTRODUCTION

The energy-based design approach started from the work of Housner [1]. After publication of his work, a lot of research on energy method has followed; in this paper only a few works are mentioned in the following. Zahrah and Hall [2] carried out seismic analyses of structures using 8 earthquake records, and reported that the effect of ductility, damping, and post-yield stiffness on the input and hysteretic energy is not significant. Fajfar and Vidic [3] computed seismic input and hysteretic energy using 40 earthquake records, and found that the input energy and the ratio of the hysteretic to input energy decrease as ductility ratio increases. For energy-based design of structures, Akbas et al. [4] proposed a procedure in which the seismic input energy demand is dissipated by the accumulated plastic deformation at beam ends.

Although a lot of research have been carried out in the field of energy-based seismic design, there has yet been discrepancies on some issues, probably due to the differences in the number and characteristics of earthquake records used in the analysis. Also the validity of the energy-based design compared to the conventional strength-based design procedure still needs to be investigated.

In this study seismic analyses of steel structures were carried out to examine the effect of ground motion characteristics such as site conditions and strong motion duration of earthquakes, and of the structural characteristics such as natural period, ductility ratio. The results were compared with the results of previous research. Then

CP1020, *2008 Seismic Engineering Conference Commemorating the 1908 Messina and Reggio Calabria Earthquake*,
edited by A. Santini and N. Moraci
© 2008 American Institute of Physics 978-0-7354-0542-4/08/$23.00

framed structures with buckling-restrained-braces (BRBs) were designed in such a way that all the input energy is dissipated by the hysteretic energy of the BRBs, and the performance of the structures were compared with those of structures designed based on the conventional strength-based procedure.

VARIATION OF INPUT AND HYSTERETIC ENERGY

The difference between the relative and the absolute energy is contributed from the difference in input and kinetic energy. However, the total amount of energy becomes the same at the end of the vibration. In the energy-based seismic design the hysteretic energy contributed from the plastic deformation of structural members computed based on the relative displacement is one the most important design parameters. Therefore in this study the relative energy is used in the formulation of energy equation.

Earthquake ground motions used in the analysis

The 60 earthquake records, originally developed in SAC steel project [5] for three different soil conditions; stiff soil, soft soil, and near-fault conditions in Los Angeles, are used in this study. These ground motions for LA have 10 % probability of exceedance in 50 years, which are for design level ground motions. For each soil condition, 20 earthquake records (10 sets) are used.

In the evaluation of the damage accumulated in a structure as a result of an earthquake, the magnitude of strong motion duration of the earthquake is an important factor, which is obtained as follows [6]:

$$t_{sd} = t_{0.95} - t_{0.05} \tag{1}$$

where $t_{0.05}$ and $t_{0.95}$ denote the time that the Arias intensity (I_A), obtained as follows [6], reaches 5% and 95%, respectively:

$$I_A = \frac{\pi}{2g} \int_0^{t_{td}} \ddot{x}_g^{\,2}(t)dt \tag{2}$$

where t_{td} is the total duration of earthquake, g and $\ddot{x}_g$ are the gravity acceleration and the ground acceleration, respectively. The earthquake records were classified into two sets of records with long and short strong motion duration.

Input and hysteretic energy

From analyses using 8 earthquake records, Zahrah and Hall [2] concluded that ductility demand (μ) does not have a significant influence on the earthquake input energy. However, according to the analysis results using mean response of 40 earthquake records by Fajfar and Vidic [3], the input energy decreases as ductility increases for period range larger than about 0.4 second. On the other hand, though the input energy was not much affected by the change of ductility for short periods less than 0.4 second, it was observed that the input energy increases as ductility increases.

It can be observed in the review of the previous research that there is inconsistency whether ductility has a negligible effect on input energy or not. Much part of the discrepancy is considered to be originated from how the results were obtained; e.g. results from individual record or from averaging many results for individual records. In the view point of the authors, considering the non-stationary nature of earthquakes, it seems to be more reasonable to use mean values using ensemble of earthquake records in the investigation of the effect of various parameters on the seismic input and hysteretic energy. The analysis model structures used in this study have bi-linear force-displacement relationship with zero post-yield stiffness. Time-history analyses were carried out and the energy responses for the 20 earthquakes in each site condition were averaged.

Figure 1 presents the variation of earthquake input energy per unit mass for various ductility ratios. The input energy spectra increase up to period of 1.5 second and decrease as period increases larger than 1.5 second. For period range less than approximately 0.5 second, input energy is not affected by ductility; however the input energy spectra decrease with the increase of ductility in period range longer than 0.5 second. Figure 2 shows that the shapes and variation of hysteretic energy spectra are similar regardless of the site conditions. According to the figure, the hysteretic energy demand increases as the ductility ratio increases in short period region. On the other hand, this trend is reversed in the long period region of the energy spectra. As the target ductility ratio increases, the period at which this reversal occurs becomes shorter.

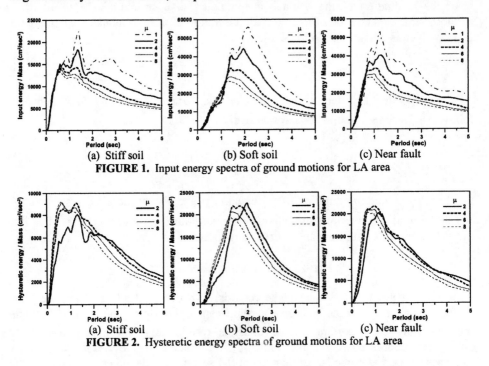

FIGURE 1. Input energy spectra of ground motions for LA area

(a) Stiff soil   (b) Soft soil   (c) Near fault

(a) Stiff soil   (b) Soft soil   (c) Near fault

FIGURE 2. Hysteretic energy spectra of ground motions for LA area

In order to investigate the effect of the post yield stiffness ratio on input and hysteretic energy, the energies of structures with =0, 0.05 and 0.1 were estimated for the constant ductility ratio of μ =4.0. Analysis results indicate that has negligible influence on the input and hysteretic energy. This result is similar to the result reported by Zahrah and Hall [2]. They have concluded that post yield stiffness ratio do not affect the input and hysteretic energy.

Ratio of hysteretic to input energy

Figure 3 shows that the ratio of the hysteretic to input energy, E_h / E_i, generally increases as the ductility ratio increases, but decreases as the natural period of the structure increases. It can be observed that the ratio reaches upper bound up to the natural period of about 2.5 second. It also can be noticed that in the long-period region of the spectrum the effect of ductility is insignificant because the ductility demand is small. Therefore it would be unreasonable or uneconomical to design a medium to high-rise structure to dissipate earthquake energy by plastic deformation. The ratios in all site conditions are similar to one another regardless of the ductility demands for structures with natural period less than about 3 seconds, whereas the ratios of records developed for near fault condition become smaller than those of records corresponding to the other site conditions when the natural periods are larger than 3 seconds.

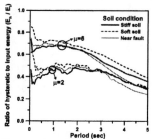

FIGURE 3. Ratio of hysteretic to input energy for various soil conditions

Figure 4 shows the ratios for earthquake records with long and short strong motion durations. It can be observed that the energy ratios remain stable for the records with long strong motion duration time, whereas the energy ratios decrease significantly in

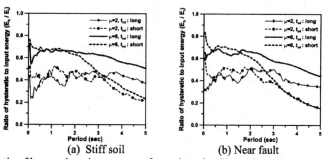

(a) Stiff soil   (b) Near fault

FIGURE 4. Ratio of hysteretic to input energy for various ductility ratios and strong motion durations

long period region when the earthquake records with relatively short strong motion duration are used in the analysis. This implies that earthquakes with long strong motion duration inflict plastic deformation (damage) on structures with wide spectrum of natural periods, whereas those with short strong motion duration damage structures with only short natural periods.

STRENGTH-BASED AND ENERGY-BASED SEISMIC DESIGN

Pseudo-acceleration and energy spectra

Figure 5 shows that in the pseudo-acceleration spectrum for the LA10 earthquake the maximum value corresponds to the natural period of 0.25 second, but that the maximum input energy occurs at the period of 1.4 second at which the relative velocity is maximized. Figure 6 plots the natural periods at which the maximum pseudo-acceleration and input energy spectra occur. It can be observed that the maximum pseudo-accelerations for the 20 earthquake records occur at quite similar natural periods. However in the case of input energy the maximum values occur at longer natural periods. The dominant period of hysteretic energy spectrum is close to that of input energy spectrum.

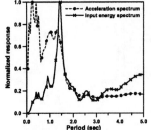

FIGURE 5. Normalized pseudo-acceleration and input energy spectra

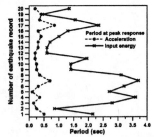

FIGURE 6. Natural periods at maximum pseudo-acceleration and input energy spectra of stiff soil

Design of model structures

A buckling-restrained brace (BRB) exhibits a stable hysteretic behavior with excellent energy dissipation capacity both in tension and compression, and consequently the hysteretic behavior of BRB can easily be modeled and the amount of dissipated energy can be computed using simple equations [7]. To compare the seismic performance of structures designed based on both strength and energy, BRB frames are prepared. As shown in Fig. 7 the girders are pin-connected to columns, and only the BRBs dissipate seismic energy through plastic deformation.

For strength based design of BRB frame, the seismic load was computed using the mean response spectrum of stiff soil. The importance factor of 1.25 was used to obtain design base shear. The BRBs were designed per Chapter 8 of FEMA-450 [8], in which the response modification factor of buckling-restrained braced frames is specified as 7. The yield stress of the structural steel is 240 MPa. In the strength-based design the ratio of stress demand to capacity is kept above 0.9 for beams and columns and especially above 0.95 for BRBs.

For energy based design of BRB frame, the procedure proposed previously by the authors [7] is followed: The target displacement (1.5% of story height) and the ductility ratio at the target displacement are determined first. Then the hysteretic energy spectrum and the accumulated ductility spectrum corresponding to the ductility ratio are constructed, and the hysteretic energy and the accumulated ductility ratio corresponding to the natural period of the model structures are obtained. By assuming that all the seismic input energy is dissipated by BRBs, the cross-sectional area of BRBs can be computed.

In this study it is assumed that the hysteretic energy is distributed in each story proportional to the story-wise distribution ratio of hysteretic energy obtained from nonlinear dynamic analysis [7]. Figure 7 show the size of structural members selected using the code-based and the energy-based procedures, respectively.

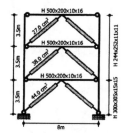
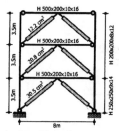

(a) Structure designed based on strength   (b) Structure designed based on energy balance
FIGURE 7. Geometry and sectional properties of 3-story model structure

Seismic performance of model structures

Figure 8 shows the pseudo accelerations and the hysteretic energy demands of model structures averaged over the results for the 20-earthquake records used previously. For the given model structures, the acceleration demand, which is used for strength-based design of model structures, decreases as the number of story increases, whereas the energy demand increases as the number of story increases.

Figure 9 presents the pushover curves of the model structures. The nonlinear analysis code DRAIN-2D+ [9] was used in the analysis. It is assumed that yield stress of BRB is equal in both tension and compression. In the curves it can be observed that the yield strength of structures designed based on energy is much smaller than that of the structures designed based on strength. The over-strength factors, the ratio of the yield to design stress, are 3.20 (1-story), 2.24 (3-story), and 1.97 (6-story) for strength-designed structures, which are larger than or equal to the over-strength factor of 2.0 specified in the FEMA-450 [8]. However, the over-strength factors of the structures designed based on the energy are 2.10 (1-story), 1.16 (3-story), and 0.69 (6-story), which are significantly smaller. The results imply that when the model structures designed based on energy, especially the six-story one, are subjected to design earthquake, the structures may undergo significant plastic deformation or damage.

Figure 10 shows the maximum inter-story drifts of the model structures obtained from time-history analysis. The mean values for the 20 earthquake records are plotted in the figure. It can be observed that the energy-designed structures generally meet the target displacements as the structures were designed so that the input energy and the hysteretic energy are balanced at the target displacement. The inter-story drifts of structures designed based on strength, however, are much less than the target displacements. The inter-story drifts are largest in the first story, which implies that large damage is concentrated there, whereas in energy-based design lesser damage is distributed throughout the stories.

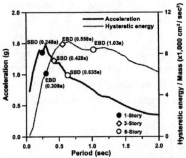

FIGURE 8. Pseudo-accelerations and hysteretic energy demands of model structures

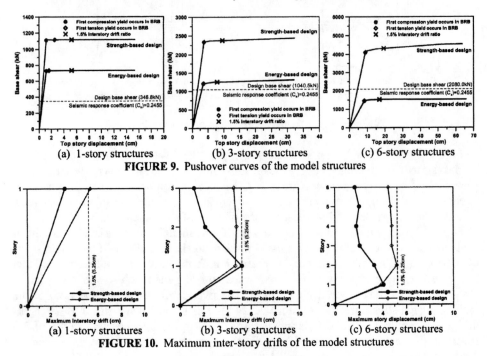

(a) 1-story structures   (b) 3-story structures   (c) 6-story structures

FIGURE 9. Pushover curves of the model structures

(a) 1-story structures   (b) 3-story structures   (c) 6-story structures

FIGURE 10. Maximum inter-story drifts of the model structures

5. CONCLUSIONS

In this study the influences of ground motion characteristics and structural properties on energy demands were evaluated using 60 earthquake ground motions recorded in different soil conditions, and the results were compared with those of previous works. Then framed structures with buckling-restrained-braces (BRBs) were designed in such a way that all the input energy is dissipated by the hysteretic energy of the BRBs, and the results were compared with those by conventional strength-based design procedure.

The analytical results showed that ductility ratios and the site conditions had significant influence on the input energy. The ratio of hysteretic to input energy was considerably affected by the ductility ratio and the strong motion duration. In the comparison of pseudo-acceleration and input energy spectra, it was found that the natural periods corresponding to the maximum input energy were about three times longer than those at the maximum pseudo-acceleration. This implies that the structural design based on the input energy may not satisfy the strength demanded by earthquake loads. The static and dynamic nonlinear analysis of structures designed based on both strength and energy showed that the energy-based design, which is basically performance-based, resulted in smaller member size and strength, and more damage compared to structures designed based on strength. Therefore if a structure, designed based on energy, is to meet various performance criteria, i.e. both strength and displacement criteria, it is necessary to check whether enough strength and stiffness are secured. A good way for this is to design the structure in such a way that the story-wise inelastic damage is nearly uniform.

ACKNOWLEDGMENTS

This work was supported by grant No. R0A-2006-000-10234-0 from the Basic Research Program of the Korea Science & Engineering Foundation.

REFERENCES

1. Housner, G., "Limit design of structures to resist earthquakes," Proceedings of the First World Conference on Earthquake Engineering, Berkeley, California, 1956.
2. Zahrah, T. and Hall, J., "Earthquake energy absorption in SDOF structures," Journal of Structural Engineering, Vol. 110, No. 8, 1984, pp. 1757-1772.
3. Fajfar, P. and Vidic, T., "Consistent inelastic design spectra: Hysteretic and input energy," Earthquake Engineering and Structural Dynamics, Vol. 23, No. 5, 1994, pp. 523-537.
4. Akbas, B., Shen, J., and Hao, H., "Energy approach in performance-based seismic design of steel moment resisting frames for basic safety objective," The structural design of tall buildings, Vol. 10, 2001, pp. 193-217.
5. Somerville, P., Smith, H., Puriyamurthala, S., and Sun, J., "Development of Ground Motion Time Histories for Phase 2 of the FEMA/SAC Steel Project," SAC Joint Venture, SAC/BD 97/04, 1997.
6. Trifunac, M.D. and Brady, A.G., "A study on the duration of strong earthquake ground motion," Bulletin of the Seismological Society of America, Vol. 65, No. 3, 1975, pp. 581-626.
7. Hyunhoon Choi and Jinkoo Kim, "Energy-based seismic design of buckling-restrained braced frames using hysteretic energy spectrum," Engineering Structures, Vol. 28, No. 2, 2006, pp. 304-311.
8. Building Seismic Safety Council, "NEHRP Recommended provisions for seismic regulations for new buildings and other structures, 2003 Edition, Part 1: Provisions," Report No. FEMA-450, Federal Emergency Management Agency, Washington, D.C., 2004.
9. Tsai, K.C., and Li, J.W., "DRAIN2D+, A general purpose computer program for static and dynamic analyses of inelastic 2D structures supplemented with a graphic processor," Report No. CEER/R86-07, National Taiwan University, Taipei, Taiwan, 1997.

A Methodology for Assessing the Seismic Vulnerability of Highway Systems

Francis Cirianni[a], Giovanni Leonardi[a] and Francesco Scopelliti[a]

[a]University of Reggio Calabria - Faculty of Engineering
Department of Computer Science, Mathematics, Electronics and Transportation - Italy

Abstract. Modern society is totally dependent on a complex and articulated infrastructure network of vital importance for the existence of the urban settlements scattered on the territory. On these infrastructure systems, usually indicated with the term *lifelines*, are entrusted numerous services and indispensable functions of the normal urban and human activity.

The systems of the lifelines represent an essential element in all the urbanised areas which are subject to seismic risk. It is important that, in these zones, they are planned according to opportune criteria based on two fundamental assumptions: a) determination of the best territorial localization, avoiding, within limits, the places of higher dangerousness; b) application of constructive technologies finalized to the reduction of the vulnerability.

Therefore it is indispensable that in any modern process of seismic risk assessment the study of the networks is taken in the rightful consideration, to be integrated with the traditional analyses of the buildings.

The present paper moves in this direction, dedicating particular attention to one kind of lifeline: the highway system, proposing a methodology of analysis finalized to the assessment of the seismic vulnerability of the system.

Keywords: Lifelines, Highway, Seismic risk assessment, Vulnerability, AHP.

INTRODUCTION

Catastrophic events as earthquakes cause unavoidably consistent socio-economic losses: direct losses due to damage to the physical infrastructure, indirect losses due to business interruption and social losses due to casualties, injuries and homelessness.

Italy is a country with elevated seismic activity, in terms of frequency of the events that have interested its territory and of the intensity with which some of them have struck, causing significant effects on the social and economic structure.

Furtehrmore, in Italy the relationship between the damages produced from earthquakes and the energy released in the course of the events is higher than in other countries with elevated seismic grade, as California or Japan. As an example, the earthquake which took place in Umbria and Marche in 1997 has produced a situation of damage (roofless buildings: 32000, economic damage: 5 billions of Euro actuated to 2002) comparable with that one of California of 1989 (14.5 billions of USD), in spite it was characterized from an energy approximately 30 times inferior.

That is due mainly to the fact that Italy's buildings are characterized from a remarkable fragility, caused above all by age and typological and constructive characteristics, as for the poor state of maintenance.

CP1020, *2008 Seismic Engineering Conference Commemorating the 1908 Messina and Reggio Calabria Earthquake*,
edited by A. Santini and N. Moraci

In order to develop appropriate risk management and loss mitigation strategies it is necessary to have a reliable method for loss & risk assessment.

Risk assessment procedures essentially consist of the following two steps:
- *Hazard Assessment*: assessing the probability that a natural disaster occurs.
- *Vulnerability Analysis*: evaluating the correlation of loss and casualties with the hazard's intensity.

Then, the *societal risk* R can be defined as:

$$R = P \cdot V \cdot N \tag{1}$$

where:
- P is the probability that an emergency event occurs (*dangerousness*);
- V is the vulnerability;
- N is he exposure.

The seismic *dangerousness* P of an area is the probability that, in a given time interval, it is interested by strong earthquakes which can induce damage.

The vulnerability of a structure is its tendency to endure a damage as a result of an earthquake.

The exposure expresses the value of the losses caused from the earthquake: economic, artistic, cultural, and human in terms of injured, casualties and homeless.

In literature there is no clear distinction between *vulnerability* and *exposure*, especially in transportation systems.

The vulnerability of a transportation system can be defined as the resistance of the infrastructures when the emergency occurs.

The exposure of the system can be defined as the equivalent homogeneous weighted value of people, goods and infrastructures affected during and after the event [1]. It is, therefore, in some way connected to the value of how much it can be destroyed from the earthquake. Such factor, therefore, in Italy is extremely relevant, in consideration of the high population density, and the presence of buildings of historical and artistic value.

The transport system physically is made of the road network and infrastructures covering the territory and its relations.

It is therefore necessary, in order to formulate a judgment of total vulnerability of a transport network, to know beforehand the vulnerability of the system, where the system is considered as a complex body constituted from various components each with its own characteristics of vulnerability.

The analysis of seismic vulnerability of each component becomes, therefore, a fundamental and indispensable phase in the vulnerability assessment of the system itself.

The analysis of vulnerability of a highway system can be simplified, limiting the assessment only to the structural components, as *bridges*, *galleries*, *trenches*, and *retaining walls* [1].

VULNERABILITY ANALYSIS

Vulnerability, represents the capacity of the buildings, of transport infrastructures and, more in general, of the civil constructions to resist to the eventual catastrophic events [2].

In this paragraph a methodology for the analysis of the ability of resistance of a road system is proposed, which holds in account also the functional criticalities connected to the vulnerability of transport infrastructures.

In fact, in presence of catastrophic events it is necessary to optimize the transport system in order to allow two principal operations: the evacuation of the population towards assistance, medical and reunion centers; and the access of vehicles, means of aid and civil forces.Such a process of optimization can be faced in advance, regarding the event, if the happening of the event itself does not modify the functional state of the infrastructure or if vulnerability scenarios after the event can be forecasted.

Vulnerability Assessment

The proposed methodology in assessing the highway system vulnerability is based on the techniques of MCA (Multi-Criteria Analysis). The referred data (input) are combined and transferred into a resultant vulnerability score VS (output).

In particular, the process can be divided into five steps [3]:

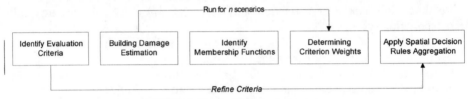

FIGURE 1. Vulnerability assessment steps.

FIGURE 2. Example of highway component collapse.

- The first step is *identifying the measures* that determine the scope of the analysis.
- The second stage is *estimating system components collapse*.
- The third step is *standardization of the evaluation criteria* by appropriate membership functions.

- In the fourth, the *criteria are compared pair wise* using the analytical hierarchy process (AHP) developed by Saaty [4].
- In the fifth stage, *components vulnerability values* are aggregated into *infrastructure.*

Identifying evaluation criteria

The starting step of the process is identifying evaluation criteria. It is the most important part of the process. Because, criteria that are selected will be used in spatial vulnerability analysis and also different criteria give different results leading different spatial vulnerability distribution. They must be suitable for vulnerability definition.
Malczewski [5] recommends that a criterion is considered good if it is:
- *comprehensive* (i.e. clearly indicates the achievement of the associated objectives);
- *measurable* (i.e. lends itself to a quantification/measurement).

Beside, this Rashed and Weeks [6] state that a set of criteria is good if it is:
- *complete* (i.e. covers all aspects of decision problem);
- *operational* (i.e. is meaningful for a decision definition);
- *decomposable* (i.e. is amenable to partitioning into subsets of criteria, which may be necessary to facilitate a hierarchical approach to decision analysis);
- *non-redundant* (i.e. avoids the double counting of decision consequences);
- *minimal* (i.e.) has the property of the smallest set of complete set if criteria characterizing the consequences of decisions).

We considered four main criteria (Figure 3).
Criteria for *social risks*, these include:
1. Flow (population) Distribution (short-term social losses);
2. Component Collapse (long-term social losses).
Criteria for *systematic vulnerability*, which may influence the emergency response and management activities following the earthquake:
3. Health Center Distance;
4. Accesibility.

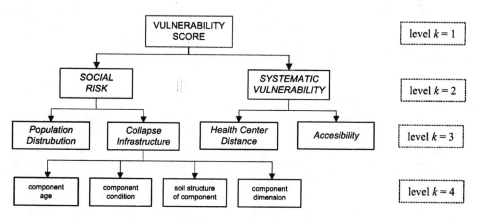

FIGURE 3. Hierarchical structure of the considered criteria.

The component damage is estimated using four criteria which are compared pair wise using AHP [4]. The four criteria are:

1. Component age;
2. Component condition;
3. Soil structure of component;
4. Component dimension.

Standardizing the Criteria

Evaluation criteria can be in different measurement scales. Therefore they can be standardized into a common scale. Identifying membership functions $\eta(x)$ for each criteria give such standardization. Some of those membership functions can be defined with expert-knowledge.

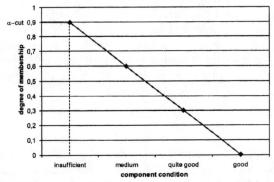

FIGURE 4. Membership degrees of component condition.

Moreover, some of these criteria are of qualitative nature and often can be expressed only through judgements of value for their subjective nature and therefore, are vague and with uncertain confinements: the explicit recognition of the vague and inaccurate nature of same criteria, can be faced by using a methodological approach based on the polyvalent logic, and in particular on the fuzzy set theory.

In fact, vulnerability can be considered as an indicator in the interval [0, 1], that indicates the vulnerability of the component when the event happens: zero stands for infinite resistance; one stands for zero resistance.

Nevertheless, considering the zero resistance condition as a purely theoretical condition, as it has been shown that even if strongly damaged, each component retains a minimum value of resistance to which corresponds a vulnerability factor of 0.9, which represents the maximum value that can be reached in reality. Referring to the theory of fuzzy sets it is possible to consider the real conditions of maximum vulnerability introducing a new fuzzy set obtained with an α-cut equal in value to 0.9.

As an example, in figure 4 is shown the fuzzy function relative to the criterion "component condition". It is shown that the functional condition expressed from the linguistic variable "insufficient" corresponds to a state of greater criticality for what the vulnerability is concerned.

Determining Criteria's Weights

Multi Criteria Decision Analysis-Pairwise Comparison Method

Multi-criteria decision analysis (MCDA) is a quantitative approach in evaluating decision problems that involve multiple variables.

Multi-criteria decision problem has six elements: a goal or a set of goals the decision maker attempts to achieve, the decision maker or makers, a set of evaluation criteria (objectives and/or attributes), a set of decision alternatives, set of uncontrollable variables, set of outcomes.

Multi-criteria decision making problems typically involve criteria of verifying importance to decision makers. This is usually achieved by assigning a weight to each criterion.

A weight can be defined as a value assigned to an evaluation criterion that indicates its importance relative to other criteria consideration. The larger the weight, the more important is the criteria in the overall utility. The weights are usually normalized to sum to 1 [7].

The pairwise comparison method was developed by Saaty [4] in the context of the analytical hierarchy process.

The AHP approach allows assessing the relative weight of multiple criteria in an initiative manner. The fundamental input to the AHP is the decision maker's or expert's answers to a series of questions of general form:"How important is criteria A relative to Criteria B?". These are termed pairwise comparison.

Importance Value	Definition
1	Equal importance
3	Weak importance of an objective over the another
5	Strong importance of an objective over the another
7	Demonstrated importance of an objective over the another
9	Absolute domination of an objective over the another
2, 4, 6, 8	Intermediate values between the two adjacent (values of compromise)

FIGURE 5. Saaty semantic scale.

Responses are gathered in verbal form and subsequently codified on nine-point intensity scale (figure 5). Saaty's basic method to identify the value of the weights depends on matrix algebra and calculates the weights as the elements in the eigenvector associated with the maximum eigienvalue of the matrix. Final results will include the weight of each criteria in addition the measure of inconsistency which informs if or not the preferences assignment needs to be revised.

TABLE 1. Pairwise comparison matrix.

Criteria	Age	Condition	Soil structure	Dimension
Age	0.083	0.112	0.055	0.059
Condition	0.417	0.561	0.655	0.471
Soil structure	0.333	0.187	0.218	0.353
Dimension	0.167	0.140	0.073	0.118
TOTAL	1.000	1.000	1.000	1.000

TABLE 2. Critera weights (level 3).

Criteria	W
Age	0.077
Condition	0.526
Soil structure	0.273
Dimension	0.124
TOTAL	1.000

Aggregation Criteria

Once determined all the vectors of the weights $\vec{W}$, it is possible to aggregate the vulnerability weighted values of each criterion for each level k.
For the aggregation we use the Ordered Weighted Averaging Operators (OWA) originally introduced by Yager [8].

$$OWA_{w_1,...,w_m}(a_1,...,a_m) = \sum_{i=1}^{m} w_i \cdot a_{\sigma(i)} \qquad (2)$$

where $\sum_{i=1}^{m} w_i = 1$, and σ is a permutation that orders the elements; $a_{\sigma(1)} \le a_{\sigma(2)} \le ... \le a_{\sigma(m)}$. The Ordered Weighted Averaging operators are commutative, monotone, idempotent, they are stable for positive linear transformations and they have a compensatory behaviour. This last property translates the fact that the aggregation done by an OWA operator always is between the maximum and the minimum. The OWA operators provide a parameterized family of aggregation operators, which include many of the well-known operators such as the maximum, the minimum, the k-order statistics, the median and the arithmetic mean. In order to obtain these particular operators we should simply choose particular weights. So we have:

$$\mu_j^k(x) = OWA_{w_1,...,w_m}\left(\mu_1^{k+1}(x),...,\mu_i^{k+1}(x),...,\mu_m^{k+1}(x)\right) \qquad (2)$$

where: $\mu_j^k(x)$ $(0 \le \mu_j^k(x) \le 1)$ is the vulnerability value of the j-th element of the hierarchy at level k, that is linked with the m element at level $k+1$, and w_1, w_2, ..., w_m are the normalized values of the m components of weights vector $\vec{W}^{level \; k+1}$.
Proceeding from the lower levels to the higher ones it is possible to determine the global vulnerability score $(0 \le VS \le 1)$ of the considered highway component.
Final step is aggregation of vulnerability scores from components to system. Aim of this operation is to identify earthquake sensitive highway system in the considered area. Also these sensitive systems can be of operative importance for authorities to create earthquake resistant transportation systems [9] [10].
For this aggregation, many methods can be used. Moreover, in this study, the highest vulnerability value of component in the highway system will be accepted as system's vulnerability value.

This acceptance will give the most correct idea about the system. Because, if there will be only one component which has a higher vulnerability value than other components in other systems, it will create a sensitive condition for itself and its environment [3].

CONCLUSIONS

The analysis methodology proposed places the objective to estimate the seismic vulnerability of a transport system, considering the highway system, as a complex system constituted from various components each with its own characteristics of vulnerability. The analysis of the seismic vulnerability of each component becomes, therefore, a fundamental and indispensable phase in the assessment of vulnerability of the infrastructure. The illustrated approach allows estimating the total vulnerability of the system starting from the vulnerability of each component.

The method for the unavoidable approximations which it involves, does not allow an absolute allocation of the probability of collapse of the structure but rather a relative hierarchical scale of values. In such sense it supplies useful information regarding the intervention priorities and above all of deepening of inspection surveys and analyses of structural nature which can find the effective level of vulnerability and risk of damage of the infrastructure, and the most opportune mitigation actions for the extensive damage. In this context the scale of values assumed from the vulnerability index has to be read as an indicator of the intervention priorities.

REFERENCES

1. D'Andrea A. and A. Condorelli (2006), "Metodologie di valutazione del rischio sismico sulle infrastrutture viarie", *AIPCR*.
2. Russo F. and A. Vitetta (2005), "Risk Evaluation in a Transportation System" in *Modelli e Metodi per l'Analisi delle Reti di Trasporto*, Edited by Ermes, 2005, pp. 09-38.
3. Servi M. (2004), "Assessment of Vulnerability to Earthquake Hazards Using Spatial Multicriteria Analysis: Odunpazari, Eskisehir Case Study" in *Thesis Submitted to the Graduate School of Natural and Applied Sciences of Middle East Technical University*.
4. Saaty T. L. (1980), "The Analytic Hiearchy Process", New York: McGraw-Hill.
5. Malczewski J. (1999), "GIS and Multicriteria Decision Analysis". New York J. Wiley &Sons.
6. Rashed T. and J. Weeks, "Assessing Vulnerability to Earthquake Hazards Through Spatial Multicriteria Analysis of Urban Analysis" in *Geographical Information Science*, Vol:17, 2003, pp:547-56.
7. Leonardi G. (2001). "Using the multi-dimensional fuzzy analysis for the project optimization of the infrastructural interventions on transport network". *Supplementi dei Rendiconti del Circolo Matematico di Palermo*, vol. 70, pp. 95-108.
8. Yager, R. R. and Kacprzyk J. (Eds.), The Ordered Weighted Averaging Operators, *Theory and Applications*, Kluwer Academic Publisher : Boston, Dordrecht, London, 1997
9. Werner, S.D., C.E. Taylor, J.E. II Moore and J.S. Walton (2000), "A Risk-Based Methodology for Assessing the Seismic Performance of Highway Systems", Report MCEER-00-0014, Multidisciplinary Center for Earthquake Engineering Research, University at Buffalo, (Dec. 31).
10. Carausu A. and A. Vtdpe (2001), "Updating Fuzzy Models for Seismic Risk Assessment", *Transactions*, SMiRT 16, Washington D. C.

Experimental Tests and FEM Model for SFRC Beams under Flexural and Shear Loads

Piero Colajanni[a], Lidia La Mendola[b], Salvatore Priolo[b] and Nino Spinella[a]

[a]*Dipartimento di Ingegneria Civile, Università di Messina,*
C.da di Dio 1 – 98166, Vill. S. Agata, Messina – Italy
[b]*Dipartimento di Ingegneria Strutturale e Geotecnica, Università di Palermo,*
Viale delle Scienze – 90128, Palermo - Italy

Abstract. The complete load-vs-displacement curves obtained by four-point-bending tests on Steel Fiber Reinforced Concrete (SFRC) beams are predicted by using a nonlinear finite element code based on the Modified Compression Field Theory (MCFT) and the Disturbed Stress Field Model (DSFM) suitably adapted for SFRC elements. The effect of fibers on the shear-flexure response is taken into account, mainly incorporating tensile stress-strain analytical relationship for SFRC. The numerical results show the effectiveness of the model for prediction of the behavior of the tested specimens reinforced with light amount of stirrups or with fibers only.

Keywords: Experimental tests, fiber-reinforced concrete, shear and flexure, FEM analysis.

INTRODUCTION

Many reports published during the last few decades have considered the possibility to add fibers in the concrete mixture to help the post-cracking constitutive behavior of the final composite. Fibers influence the mechanical properties of concrete in all failure modes, showing a consistent ductile branch of constitutive curves under uniaxial compression, direct shear and, especially, direct tension.

In fiber reinforced concrete structures, fibers furnish a noticeable contribution to transfer stress between different parts of matrix, specially in presence of cracks. Stress is shared by the fiber and matrix in tension until the matrix cracks, and then the total stress is progressively transferred to the fibers [1].

A popular valuable and effective application of fibrous concrete is in shear reinforcement of members, allowing a total or partial substitution of transversal reinforcements. Fibers increase the first cracking load and improve the ability of concrete to sustain higher stress level corresponding to high strain values, especially in tension.

First, this paper presents the results of experimental tests carried out on rectangular simply supported beams made of hooked steel reinforced high strength concrete with and without stirrups, subjected to two-point symmetrically placed loads. The tests show that the inclusion of fibers in adequate volume percentage can change the brittle mode of failure, typical of shear collapse, into a ductile flexural mechanism.

Secondly, the behavior of the beams is analyzed herein using a nonlinear finite element code based on the modified compression field theory (MCFT) [2] and the

CP1020, *2008 Seismic Engineering Conference Commemorating the 1908 Messina and Reggio Calabria Earthquake,*
edited by A. Santini and N. Moraci
© 2008 American Institute of Physics 978-0-7354-0542-4/08/$23.00

disturbed stress field model (DSFM) [3]. The analytical model is suitable adapted for steel fiber reinforced concrete (SFRC), introducing several modifications to take into account the different constitutive behavior of fibrous concrete and his complex interaction with concrete matrix.

EXPERIMENTAL PROGRAMME

Specimen geometry

Among the 18 couples of flexure and shear tests carried out at University of Palermo [4], the following discussion will focus on those experiments conducted to investigate the influence of only a limited amount of fibers (volume percentage V_f = 1%) and the interaction mechanism between fibers and stirrups, with spacing s = 200 mm, employed as shear reinforcement.

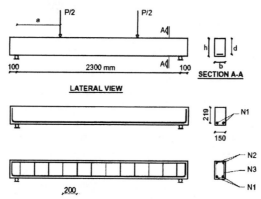

FIGURE 1. Geometry and reinforcement details: N1=2 Φ20 mm; N2=2 Φ10 mm; N3= Φ6 mm.

Figure 1 shows the details of the beams. All the beams had a rectangular nominally identical cross-section of dimensions b= 150 mm, h= 250 mm, effective depth d= 219 mm and length of the span L = 2300 mm. The flexural reinforcements (two longitudinal deformed bars with diameter 20 mm) were designed to obtain a shear failure for members without transversal reinforcement, and bars were hooked upwards beyond the supports to preclude the possibility of anchorage failure, which can be important in practice. Finally, when a lightly amount of stirrups were used, two longitudinal bars with diameter 10 mm were inserted in the top zone to ensure an adequate operation of stirrups.

To evaluate the influence of shear span-effective depth ratio (a/d), beams have divided in two groups: A series with a/d = 2.8 and B series with a/d = 2.0. The former shear span-effective depth ratio was chosen to obtain a shear failure for diagonal tension, where the beam mechanism and the tensile strength of material govern the collapse mode. The latter was chosen to obtain a shear-compression failure, governed by arch mechanism and principally depending by compression strength of concrete.

873

Six couples of beams with the same setup, three for each series, are herein analyzed. Each beam is marked with the letter that represents the shear span chosen (A or B); the numerical indexes that follow refer to the presence of fibers (0 in absence of fibers and 1 for $V_f = 1\%$) and to the stirrups (0 in absence of stirrups and 1 for s = 200 mm, corresponding to a geometrical transverse reinforcement ratio $\rho_w = A_w /(bs) = 0.188\%$) respectively.

Material properties

The concrete used to cast the beams is made by Portland cement type 42.5 (450 kg/m^3), sand (850 kg/m^3), aggregate with a maximum size of 10 mm (1050 kg/m^3) and water (160 kg/m^3). The hooked-end steel fibers have the following characteristics: length $l_f = 30$ mm; diameter $d_f = 0.55$ mm (aspect ratio l_f/d_f=54.5), and nominal tensile strength 1100 MPa. The yield stress (f_y) of longitudinal and transverse reinforcement is 599 MPa and 473 MPa respectively.

TABLE 1. Properties of concrete mixtures.

V_f (%)	f_c' (MPa)	ε_0 (mm/m)	E_c (MPa)	f_{sp} (MPa)
0	75.53	2.55	43385	4.19
1	75.41	2.83	37343	8.34

The measured values of compressive strength f_c', the corresponding strain value ε_0 and the initial tangent modulus E_c for concrete with and without fibers are presented in Table 1. In particular they are obtained as the mean value of results of three compressive tests carried out for each type of concrete. As shown in Table 1, the addition of 1% of fibers didn't change the compressive strength respect the value obtained for plain concrete, but it allows a more gradual micro-cracking process with a consequent increasing of the ε_0 value and of the ductility in the post-peak phase. These constitutive characteristics of fibrous concrete play an important role in the global response of structural members under shear loading.

In addition some splitting tension-tests were performed to determine increasing in the tensile strength due to the fibers. The mean values of maximum stress (f_{sp}) obtained from the three tests are reported in Table 1 for plain and fibrous concrete showing the considerable increasing induced by fibers in the mixture.

Test results

Two equal loads were applied to the specimen using a steel spreader beam and 100 mm-wide loading plates with spherical joints between them. Each test was controlled by gradually increasing the beam deflection. Total load P and the beam deflection at midspan (δ), were recorded continuously until failure.

The experimental force-deflection curves are shown in Figure 2 for the six beams with a/d = 2.8 and a/d = 2.0 respectively. It can be observed that in all cases the presence of fibers allows to increase the load capacity and the ultimate deflection with respect to the values recorded for the plain concrete members. In particular the SFRC beams A10 and B10 reach ultimate loads higher than those of the corresponding beams A00 and B00 by 158% and 96% respectively.

However, as shown in Figure 3, the amount of fibers employed for the beams A10 and B10 was not able to change the collapse mechanism and failure was characterized by a diagonal crack extended from the load point to the support.

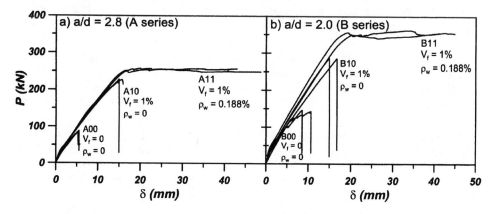

FIGURE 2. Load-deflection curves for beams with a) a/d = 2.8 and b) a/d = 2.0.

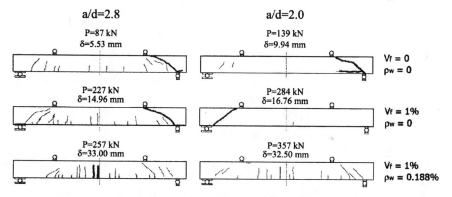

FIGURE 3. Crack patterns at ultimate conditions.

Nevertheless, for beams A00 and A10 the beam effect is predominant in the rupture mechanism and failure was due to the overcoming of the tensile strength of the material, with a more spread cracking in the members with fibers; for the beams B00 and B10 the arc effect governs the failure and slippage along the predominant crack was observed.

As shown in Figure 2, when beams are reinforced with both fibers and stirrups (beams A11 and B11), the ultimate deflection increases more than the shear resistance of specimens. For example, the addition of stirrups with geometrical ratio equal to 0.188% increased the load capacity of beam B11 by only 17%, but increased the ultimate deflection by 130%. Further, beams A11 and B11 were able to reach full flexural capacity and crack patterns were characterized by vertical cracks close to the midspan.

NUMERICAL ANALYSES

This section is focused on numerical analyses carried out by a finite element model (FEM), namely VecTor, based on the MCFT and DSFM [5]. These represent two well known models for prediction of nonlinear behavior of reinforced plain concrete members, in particular subjected to shear and torsion loads. They are smeared, rotating crack models where the cracked concrete and the steel reinforcements are treated as an unique orthotropic material with stress-strain relationships that are dependent on the amount of reinforcements eventually placed. Several constitutive laws can be implemented in the FEM code, developed on the basis of results obtained by series of experimental tests on reinforced plain concrete panels subjected to different load conditions.

To extend the above models at SFRC members is needed to update the constitutive relationships, aiming to better representing the real constitutive behavior of SFRC.

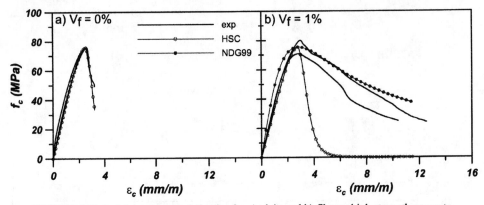

FIGURE 4. Uniaxial compressive behavior for a) plain and b) fibrous high strength concrete.

In Figures 4a) and 4b) the entire curves obtained from the compression tests performed on the concrete employed for the beams described in the previous section are shown for plain and SFR concrete respectively. The fibrous concrete behavior in compression is similar to that of plain concrete until the peak stress value (f_c') is attained, and it can be easily modeled using the same constitutive laws available in literature [3] for plain concrete (Figure 4a). By contrast, the post-peak branch of the stress-strain curve in compression of SFRC (Figure 4b) is more ductile than plain concrete one. This branch of the f_c-ε_c curve is modeled by using the following relationship proposed by Nataraja et al. [6] (NDG99) for fibrous concrete:

$$\frac{f_c}{f_c'} = \frac{\beta\left(\varepsilon_c/\varepsilon_0\right)}{\beta-1+\left(\varepsilon_c/\varepsilon_0\right)^\beta} \tag{1}$$

where the parameter $\beta = 0.5811+1.93RI^{-0.7406}$ is evaluated on the basis of the index by weight of hooked steel fibers $RI = W_f l_f/d_f$, W_f being the weight of fiber for weight unit of the mixture.

Concrete stress-strain curve in direct tension is assumed to be linear up to the tensile strength. Unfortunately direct tension tests couldn't be performed on the concrete. Therefore, an analytical formulation proposed by Bentz [7] for high strength plain concrete is used to link tensile strength to compressive strength: $f_t' = 0.45f_c'^{0.4}$ (in MPa), while for fibrous concrete the tensile strength is calculated by the simple mixture rule as proposed by Lim et al. [8]. The post-peak behavior is modeled by the Variable Engagement Model (VEM06) suggested by Foster et al. [9]. The constitutive tensile law, expressed in terms of tensile tension and crack opening displacements (w), is the simple summation of stress contributions by matrix and fibers: $\sigma_{cf}(w) = \sigma_c(w) + \sigma_f(w)$. The VEM06 considers that slippage between the fibers and the concrete matrix occurs before the full bond stress is developed, and the fibers can be broken before being pulled out across a crack. According to VEM06 the fibers are mechanically anchored to the matrix and some slips, between fiber and matrix, must occur before the anchorage is engaged. The crack opening w for which the fiber becomes effectively engaged in the tension carrying mechanism is termed the engagement length $w_e = \alpha \tan\theta$, where $\alpha = d_f/3.5$ is a material parameter and θ is the fiber inclination angle, evaluated respect to the crack plane. The bridging tension stress across the crack provided by fibers is:

$$\sigma_f(w) = f_t'[K_f(w)F_\tau]$$

(2)

where $F_\tau = \beta_\tau V_f (l_f/d_f)$ with $\beta_\tau = \tau_f/f_t'$, being τ_f the mean value of the bond fiber-matrix interface strength. The function $K_f(w)$ in Eq. (2) is the following global orientation factor which depends by w:

$$K_f(w) = \frac{\tan^{-1}(w/\alpha)}{\pi}\left(1 - \frac{2w}{l_f}\right)^2$$

(3)

To predict the value of residual tensile strength of fibrous concrete at shear failure of the beam, the contribute given by the matrix, $\sigma_c(w)$, is computed by a simple linear law [3].

Because the FEM used is a smeared cracking model, the stress constitutive law (VEM06) expressed in terms of crack opening w is reduced in terms of tensile strain ε_t by an appropriate characteristic length [10]: $\varepsilon_t = w/S_\theta$, where S_θ is the average crack spacing that is function of the crack angle value at the generic load step and of the reinforcement pattern. When transverse reinforcement is not present, the average crack spacing becomes $S_\theta = S_{xef}/\sin\theta$. S_{xef} for fibrous concrete is equal to [10]:

$$S_{xef} = 0.9d\left(\frac{35}{a_g+16}\right)\underbrace{\left(\frac{50}{l_f/d_f}\right)}_{\leq 1}$$

(4)

with a_g the maximum dimension of aggregate (that have to be assumed equal to zero for concrete with strength up to 70 MPa).

A two dimensional plane stress model is developed for all specimens (Fig. 5). The numerical analyses are performed with a displacement controlled procedure, by imposing an increasing displacement in the node located at the middle of the load transfer steel plate.

Figure 6 shows the comparison between numerical and experimental load-displacement curves for all the beams considered.

The slope of numerical curves obtained for plain concrete beams (A00 and B00) is lightly greater than the experimental one, probably due to the effect of tension stiffening, which plays an important role at the onset of cracking.

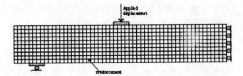

FIGURE 5. FE mesh adopted for all specimens.

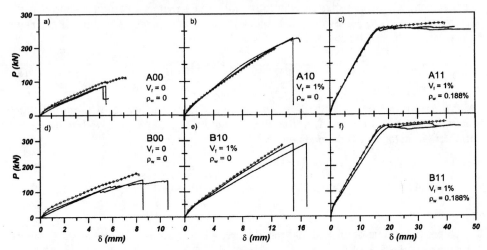

FIGURE 6. Numerical and experimental load-deflection curves for: a) A00; b) A10; c) A11; d) B00; e) B10; and f) B11 beams.

For fibrous concrete beams (A10 and B10) the DSFM is able to capture the stiffness of the specimens and to estimate load capacity and ultimate deflection.

The tension softening law, used with the adequate characteristic length, is appropriate to reproduce the mechanical role played by fibers in the shear capacity of the SFRC beams.

The numerical model is very effective in representing the behavior of beams with stirrups, as shown in Figures 6c) and 6f). A11 and B11 beams showed a flexural failure, with a considerable post-peak branch that is captured with great accuracy by the analytical model. In this case the main role is played by the compression constitutive law adopted. During the ductile phase of the beam response, characterized by high level of compressive strains, fibers help the concrete to sustain considerable levels of compressive stresses.

In Figure 7 the numerical crack patterns for fibrous beams are depicted. An appreciable agreement between them and the experimental ones (Fig. 3) is found. A10 and B10 specimens exhibited a shear failure with a predominant crack and some secondary cracks that are well reproduced by the analytical model. Further, the FEM is capable to reproduce the flexural failure mode of fibrous beams with stirrups (A11 and B11), characterized by the vertical cracks at the midspan.

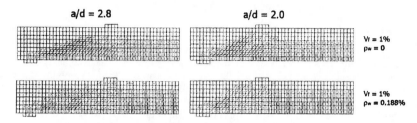

FIGURE 7. Numerical crack pattern for fibrous beams (half length of the beam).

CONCLUSIONS

The complete load-vs-displacement curves obtained by four-point-bending tests on Steel Fiber Reinforced Concrete (SFRC) beams has been predicted by using a nonlinear finite element code based on the Modified Compression Field Theory (MCFT) and the Disturbed Stress Field Model (DSFM) suitably adapted for SFRC elements. The adopted constitutive relationships and the modifications introduced in the analytical model allow an accurate prediction of the beam responses in terms of strength, stiffness, crack patterns and failure modes.

REFERENCES

1. ACI 544, Technical Committee, "Design Consideration for Steel Fiber Reinforced Concrete" (ACI544-88). Technical report, American Concrete Institute - Detroit - Michigan (USA) 1988.
2. F. J. Vecchio and M. P.Collins, "The modified compression field theory for reinforced concrete elements subjected to shear". *ACI Struct. J.* **83** (2), 219–231 (1986).
3. F. J. Vecchio, "Disturbed stress field model for reinforced concrete: Formulation". *ASCE J. of Struct. Eng.* **9**, 1070–1077 (2000).
4. S.Priolo, "Influenza delle fibre di rinforzo sul comportamento di travi in c. a. sottoposte a taglio e flessione" (in Italian). Ph. D. Thesis, University of Palermo - Italy (2007).
5. P. S. Wong and F. J. Vecchio, "VecTor2 and Form Works Users Manual". *Technical report.* Department of Civil Engineering, University of Toronto – Canada (2002).
6. M. C. Nataraja, N. Dhang and A. P. Gupta, "Stress-strain curve for steel-fiber reinforced concrete under compression". *Cement and Concrete Composites* **21**, 383-390 (1999).
7. E. C. Bentz, "Sectional Analysis of Reinforced Concrete Membrane". Ph. D. Thesis, University of Toronto - Canada (2000).
8. T. Y. Lim, P. Paramasivam and S. L. Lee, "Analytical Model for Tensile Behaviour of Steel-Fiber-Concrete". *ACI Mat. J.* **84** (4), 286-298 (1987).
9. S. J. Foster, Y. L. Voo, and K. T. Chong, "FE analysis of steel fiber reinforced concrete beams failing in shear: Variable Engagement Model", *ACI SP-237*, 55–70 (2006).
10. N. Spinella, "Modelli per la risposta a taglio e flessione di travi in calcestruzzo rinforzato con fibre d'acciaio" (in Italian). Ph. D. Thesis, University of Messina - Italy (2008).

Influence Of Lateral Load Distributions On Pushover Analysis Effectiveness

P. Colajanni[a] and B. Potenzone[a]

[a]Dipartimento di Ingegneria Civile, Università di Messina, Contrada Di Dio, S. Agata, 98166 Messina

Abstract. The effectiveness of two simple load distributions for pushover analysis recently proposed by the authors is investigated through a comparative study, involving static and dynamic analyses of seismic response of eccentrically braced frames. It is shown that in the upper floors only multimodal pushover procedures provide results close to the dynamic profile, while the proposed load patterns are always conservative in the lower floors. They over-estimate the seismic response less than the uniform distribution, representing a reliable alternative to the uniform or more sophisticated adaptive procedures proposed by seismic codes.

Keywords: pushover curve, load distributions, non liner static analysis, seismic codes.

INTRODUCTION

Current civil engineering practice now tends to use non-linear static procedures or pushover analysis (POA) to estimate seismic demands, as opposed to non-linear response history analysis (NRHA). In the past few years, several researchers have discussed the underlying assumptions and limitations of pushover analysis [1]. It has been found that, if a unique invariant force distribution proportional to the fundamental mode of vibration is assumed, satisfactory predictions of seismic demands are mostly restricted to regular plane, low and medium-rise elevation structures, for which inelastic demand is distributed through the height of the structure, and higher mode effects are likely to be minimal. Invariant force distributions are not able to take into account the redistribution of inertia forces due to yielding, and the associate change in the mode shape. Moreover, force distribution and displacement pattern, related to the fundamental period of vibration, do not account for the contribution of higher modes.

To overcome these limitations, and with the aim of bounding the likely distribution of interstorey drifts and local ductility demands along the height of the structure, seismic codes [1,2,3] require that analysis is performed enveloping the results obtained by using two different seismic force patterns: - a load pattern aimed at reproducing the distribution of seismic forces acting on the structure in the elastic state; - a uniform or adaptive load pattern aimed at bounding or reproducing the change in distribution of seismic forces due to progressive yielding of the structure.

However, numerical analyses, performed in the last decade, have shown that the uniform load pattern is too conservative for the estimation of the response parameters for the lower floors of buildings, while all the adaptive load patterns proposed in literature, sometimes improve the effectiveness of pushover procedure, but cannot

CP1020, *2008 Seismic Engineering Conference Commemorating the 1908 Messina and Reggio Calabria Earthquake,*
edited by A. Santini and N. Moraci

provide a better estimation of seismic response for all structures, as they do not provide suitable solutions for conservative bounding of seismic response.

Recently [4], two very simple load distributions have been proposed. When used in conjunction with a load pattern aimed at reproducing the elastic load distribution, both were found to be effective in bounding seismic response of the structure without introducing the large overestimation at the lower floors that characterizes uniform load distribution.

Most of the numerical analyses in literature aimed at investigating the influence of load distribution on the assessment of seismic response by pushover analysis are performed on moment resisting frames (MRF). MRF are characterized by high redundancy and show a smooth variation of structure stiffness when a plastic hinge is activated. Therefore, load distributions, derived on the basis of dynamic properties of the elastic system, are effective for assessment of seismic demand by POA. By contrast, in steel braced frames, often only a few braced frames are designed to withstand seismic load. When yielding of the dissipative zone occurs, seismic behaviour is affected by a sharp variation of the dynamic properties, namely mode shapes and vibration periods associated to the tangent stiffness matrix. Therefore, POA load distribution must reflect these behaviour.

Here, a comparison of the effectiveness of several load distributions for pushover analysis in assessment of seismic demand of eccentric braced steel frames (EBF) is performed, assuming non-linear response history analysis as the benchmark. It will be shown that the load distributions proposed in [4] are effective in bounding the seismic response, when employed in conjunction with the results provided by a load pattern derived from elastic properties of the system or by the Modal Pushover Analysis (MPA) method proposed by Chopra and Goel [5].

LOAD PATTERNS IN PUSHOVER ANALYSES

Recent seismic codes [1,2,3] suggest the use of two different groups of load distributions to assess the response of systems with weakly and strongly non-linear behaviour, respectively.

In particular, for the former group, the load pattern must be selected among the following load distributions: a) a triangular distribution, proportional to the product of the seismic weight W_i at storey i and the height at the same storey; b) a load distribution proportional to the shape of the fundamental mode (indicated in the following numerical applications by 1M); c) a distribution, corresponding to the Modal storey Shear (MS) of the building, evaluated by a modal response spectrum analysis. Thus, the force at the i level is given by the following equations:

$$ F_i = \alpha (Q_i - Q_{i+1}) \ (i = 1, 2...., n-1); \qquad F_n = Q_n; \qquad Q_i = \sqrt{\sum_{j=1}^{N \, mod \, i} Q_{ij}^2} \qquad (1a,b,c) $$

where Q_{ij} is the seismic shear at the i level due to the j-mode and α the load multiplier.

The second group of patterns, useful for providing the response of strongly non-linear behaviour systems, consists of the following load patterns:

d) a uniform distribution (U) for which the force at the i level is proportional to the total mass W_i at the same storey; e) adaptive load distributions in which the load shape changes during analysis according to progressive stiffness degradation.

A well-known multi-modal force-based adaptive procedure has been proposed by Gupta and Kunnath [6] and Elnashai [7], indicated in [8] as FAPM, which provides an incremental update of the load distribution. According to this, the storey force at a given analysis step "k", is obtained by adding a new load increment ΔF_i to the load $F_{i,k}$ of the previous step "k-1", as follows:

$$F_{i,k} = F_{i,k-1} + \alpha \, \Delta F_i; \quad \Delta F_{i,j} = \Gamma_j M_i S_{pa}(T_{j,k}); \quad \Delta F_i = \sqrt{\sum_{j=1}^{n \, mod \, i} \Delta F_{ij}} \qquad (2a,b,c)$$

where Γ_j is the modal participation factor for the jth mode, $S_{pa}(T_{j,k})$ the acceleration response spectrum ordinate corresponding to the period of the jth mode at step k, and M_i is the mass of the ith storey.

Notwithstanding the superiority of adaptive procedures, compared to standard procedures, sometimes they do not yield conservative results, underestimating seismic demand. Instead, a conservative demand can be obtained with results from the envelope of the response assessed by uniform distribution and by a distribution belonging to the first group of patterns described above. However, in most cases the use of the uniform distribution strongly overestimates seismic demands at the lower storeys.

According to these indications, two very simple load distributions, belonging to the second group of patterns, have been proposed in a previous work [4]. They can be seen as an alternative to uniform distribution, or to the more complex adaptive procedures.

Extensive numerical investigations [9] has shown that, during an adaptive POA; the MS load pattern is the most accurate in providing maxima member forces of structures in the elastic phase. A criterion of theoretical derivation to define the load distribution, when structural elements overcome the yielding strength, is not yet available. By contrast, when the results are used according to the enveloping procedure, the uniform distribution is found to be conservative. Therefore, a simplified conservative adaptive load pattern (indicated with PropA) was proposed, as follows:

$$F_i = \alpha_1 \left(Q_i - Q_{i+1} \right) \qquad 0 \le \alpha_1 \le \alpha_y \qquad (3a)$$

$$\Delta F_i = \alpha_2 W_i \qquad F_i = \alpha_y \left(Q_i - Q_{i-1} \right) + \Delta F_i \qquad (3b,c)$$

where α_1 is the load factor of the MS distribution, α_y its value at the first yielding, and α_2 the load factor of the uniform distribution. Let us stress that in Eqs. 3 any other load distribution belonging to the first group can be used.

Alternatively, an invariant load distribution of the second group has been proposed (PropI), by combining MS and uniform distributions, proportionally to load factors of the MS distribution at the first yielding α_y and at ultimate state α_u, as follows:

$$F_i = \alpha \left[\alpha_y \left(Q_i - Q_{i+1} \right) + \beta \left(\alpha_u - \alpha_y \right) \left(Q_1 \middle/ \sum_{i=1}^{n} W_i \right) W_i \right] \qquad (4)$$

where Q_1 is the base shear, and β a coefficient that amplifies the uniform load distribution counterpart. The greater the coefficient β, the more conservative the seismic demand in the lower storeys, thus approaching the results provided by the uniform load pattern. In particular, by means of numerical investigations carried out on many structural typologies, it has been shown [9] that, for $\beta = 2$, a sufficient conservative demand in all the analyzed cases can be obtained.

The effectiveness of the proposed procedures will be verified, according to seismic code guidelines, by comparing seismic demand obtained by the envelope between the results from a couple of distributions of the first and second groups, including the proposed ones, with the results of the non-linear dynamic analysis, which is the benchmark. The Modal Pushover Analysis (MPA) method, proposed in [5] and mentioned in [1], has been also considered, along with the load patterns and procedures described above, because it has been found to be highly accurate for the assessment of weakly non-linear structure seismic response.

COMPARISON OF LOAD DISTRIBUTION EFFECTIVENESS

The effectiveness of the load distributions described in the previous sections, in reproducing the seismic demands evaluated by NRHA is compared for three EBFs, with 4, 8, and 12 storeys. Each structure is made up of pinned steel frames having four sides, bay length 8m and storey height 3.2 m (Fig.1). Seismic action is withstood by K eccentrically braced frames only located in the central bays of the external frames, with short links, length $e=0.1 \, l_t$, pinned beam to column joints and column pinned at the base. The structures have been designed to carry dead and live storey loads of $G_k=4.4$ kN/m^2 and $Q_k=2.0$ kN/m^2 respectively, and seismic design action evaluated according to EC8 for soil type B and peak ground acceleration (PGA) of 0.35g, assuming a behavior factor $q=6$. In Table 1, designed element cross sections are reported. Modeling the seismic action with spettrocompatible accelerograms [10], for each structure, the PGA that induces collapse rotation of the links ($\gamma_u=0.09$ rad) is evaluated by NRHA. Mean values of maximum response parameters for 50 samples of seismic excitation are compared with the results of pushover procedures, performed by imposing a top storey displacement equal to the mean value found by NRHA. Thus, the distribution of response parameters along structure height can be compared.

In Figure 2, the storey displacements U, the storey drifts ΔU and the plastic rotations of the links γ_{pl}, evaluated by NRHA and by POA with the six load distributions and by MPA procedure for the four storey EBF are shown. The structure exhibits a global collapse mechanism, as can be recognized by the values of the storey drifts, which are almost constant along the height. This deformed shape is similar to that exhibited in the elastic phase, and only the uniform distribution fails in the assessment of seismic demand, except at the top storey, where none of the distributions is able to predict the storey drift.

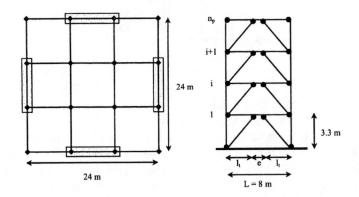

FIGURE 1. Plan and frame structural scheme of the Eccentric Braced Frames

TABLE 1: Structural sections

	STOREY	LINK	COLOUMN	BRACES	STOREY	LINK	COLOUMN	BRACES
4 ST. FRAME	4	HEA 180	HEA 160	HEM 140	12	HEA 160	HEA 180	HEM 120
	3	HEA 220	HEA 160	HEM 140	11	HEB 180	HEA 180	HEM 140
	2	HEA 260	HEA 260	HEM 160	10	HEB 220	HEB 240	HEM 140
	1	HEA 280	HEA 260	HEM 160	9	HEB 240	HEB 240	HEM 160
8 STOREY FRAME	8	HEA 180	HEA 180	HEM 120	8	HEB 260	HEB 300[1]	HEM 160
	7	HEB 200	HEA 180	HEM 140	7	HEB 280	HEB 300[1]	HEM 180
	6	HEB 240	HEB 240	HEM 140	6	HEB 300	HEB 340[1]	HEM 180
	5	HEB 280	HEB 240	HEM 160	5	HEB 300	HEM 340[1]	HEM 180
	4	HEB 300	HEB 320	HEM 160	4	HEB 320	HEM 340[1]	HEM 180
	3	HEB 320	HEB 320	HEM 180	3	HEB 340	HEM 340[1]	HEM 180
	2	HEB 320	HEM 320	HEM 180	2	HEB 340	HEM 340[1]	HEM 200
	1	HEB 340	HEM 320	HEM 180	1	HEB 340	HEM 340[1]	HEM 200
Section steel: FE 360, except sections marked by superscript [1] that have steel FE 430.								

In Figures 3, storey percentage errors in the estimation of storey drifts $s_{i,\Delta U}$ and link plastic rotations $s_{i,\gamma pl}$ given by the 7 procedures (Figs. 3a and 3b), and by the envelope prescribed by [3], are depicted (Fig.3c). The results clearly show that the proposed load distributions give an accurate and conservative estimation of the seismic demands at the lower storeys, strongly reducing overestimation provided by uniform distribution. The best performance is obtained if the envelope of the results provided by the proposed adaptive load distribution and by the MPA is retained.

In Figures 4 and 5, the corresponding results for the eight storey structure are shown. In this case, a concentration of the storey drift at the top storey is obtained in the NRHA analysis. Only MS distribution and MPA procedures are able to capture this phenomenon. In the lower storey, the proposed load distributions are the most accurate, providing errors for both the interstorey drifts and plastic rotations, more than three times smaller than those provided by uniform or FAPM load distributions.

Load distributions or procedures derived on the basis of elastic behaviour, namely 1M, MS and MPA, fail at the lower storeys, giving an underestimation of the seismic demand. In this case, the lowest errors are obtained by enveloping the results provided by the proposed adaptive and MS distributions; but also the envelope with MPA procedure provides satisfactory prediction of the response.

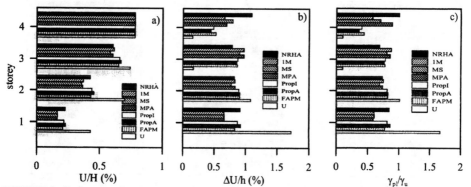

FIGURE 2. Four storey frame: a) storeys displacement; b) storeys drifts; c) plastic link rotations

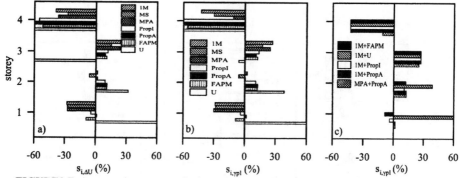

FIGURE3 Four storey frame: errors in the assessment of: a) storeys drifts; b and c) plastic link rotations

Lastly, in Figures 6 and 7, the results for the twelve storey frame are shown. In this case also, a global collapse mechanism is obtained, and the ultimate link rotation is attained at the first and tenth storey almost simultaneously. At the top floor, only MPA is able to provide an accurate prediction of interstorey drifts and plastic link rotation. However, the latter is small (less than 15% of the ultimate value), and its prediction is not particularly useful for design purpose. MS distribution is effective in prediction of response parameters between 8th and 11th storey. In the lower storey, the proposed procedures lead to a very small underestimation of storey drifts, and are very accurate in the assessment of link plastic rotations. The envelope of the results obtained by the proposed adaptive load and MS distributions gives the better prediction of link plastic rotations along structure height, but also the envelope with MPA procedure provides satisfactory prediction of the response.

CONCLUSION

The effectiveness of two new load patterns in the assessment of the seismic demand by pushover analysis has been shown by investigation of seismic response of eccentric braced frames. The proposed load patterns are able to provide an accurate and conservative estimation of the seismic demand when the results are enveloped with those provided by modal shear load pattern. The results shown here encourages further

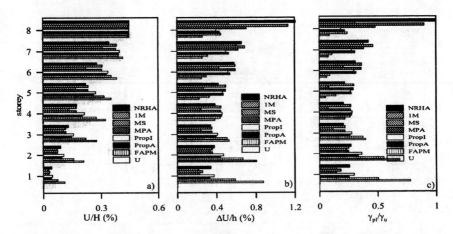

FIGURE 4 Eight storey frame: a) storey displacement; b) storey drifts; c) plastic link rotations

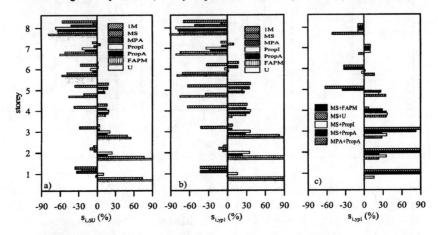

FIGURE 5 Eight storey frame: errors for: a) storey drifts; b and c) plastic link rotations

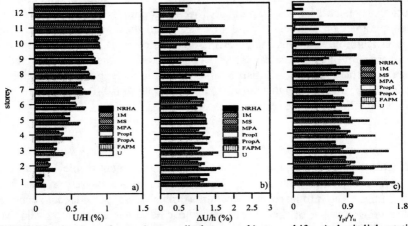

FIGURE 6 Twelve storey frame: a) storey displacement; b) storey drifts; c) plastic link rotations

886

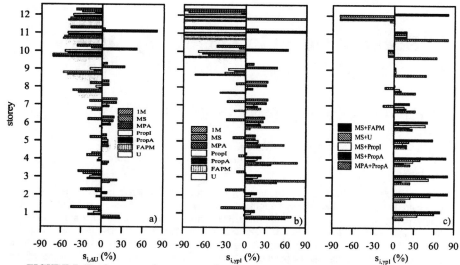

FIGURE 7: Twelve storey frame: errors for: a) storey drifts; b and c) plastic link rotations

investigation to prove that the proposed load distributions give safe and accurate results for all structural systems, and are eligible to be included in seismic codes.

REFERENCES

1. Fema 440 (2005): "Evaluation and improvement of inelastic seismic analysis procedure", Federal Emergency Management Agency, Washington (USA).
2. Fema 356 (2000): "Prestandard and commentary for the seismic rehabilitation of buildings", *Federal Emergency Management Agency*, Washington (USA).
3. Ministero delle Infrastrutture, dell'Interno e Dip. Protezione Civile (2008)."Norme tecniche per le costruzioni"
4. Colajanni P., Potenzone B.(2008): "Due proposte per i profili di carico nell'analisi pushover", Proc. Valutazione e riduzione della vulnerabilità sismica di edifici esistenti in c.a. Reluis, Roma, Maggio.
5. Chopra A. K. and Goel R.K. (2002): "A modal pushover analysis procedure for estimating seismic demands for buildings", *Earthquake Engineering and Structural Dynamics*, 31, pp: 561-582.
6. Gupta B. And Kunnath S.K. (2000): "Adaptive spectra-based pushover procedure for seismic evaluation of structures", Earthquake Spectra, 16 (2), pp: 367-391.
7. Elnashai A. S. (2000): "Advanced inelastic static (pushover) analysis for earthquake application", *G.Penelis Int. Symp. on concrete masonry structures*, Aristotle University of Thessaloniki, Greece.
8. Antoniou S., and Pinho R. (2004): "Advantages and limitations of adaptive and non-adaptive force-based pushover procedures", *Journal of Earthquake Engineering*, Vol 8 - No 4, 497-522 Imperial College Press
9. Potenzone B. (2008): "Analisi statica non lineare per la valutazione della risposta sismica di strutture intelaiate", *Ph.D. Thesis*, Dipartimento di Ingegneria Civile, Università di Messina.
10. Cacciola P., Colajanni P. and Muscolino G. (2004): "Combination of modal responses consistent with seismic input representation", *Journal of Structural Engineering*, ASCE, 130(1), pp: 47-55.

Shear Strength Prediction By Modified Plasticity Theory For SFRC Beams

Piero Colajanni[a] Antonino Recupero[a] and Nino Spinella[a]

[a]*Dipartimento di Ingegneria Civile, Università di Messina,*
C.da di Dio 1 – 98166, Vill. S. Agata, Messina - Italy.

Abstract. the plastic Crack Sliding Model (CSM) is extended for derivation of a physical model for the prediction of ultimate shear strength of SFRC beams, by assuming that the critical cracks is modeled by a yield lines. To this aim, the CSM is improved in order to take into account the strength increases due to the arch effect for deep beam. Then, the effectiveness factors for the concrete under biaxial stress are calibrated for fibrous concrete. The proposed model, able to provide the shear strength and the position of the critical cracks, is validate by a large set of test results collected in literature.

Keywords: shear strength, SFRC beams, Crack Sliding Model, plastic model, yield lines.

INTRODUCTION

In the last twenty years many models have been proposed in literature for evaluation of the ultimate shear capacity of SFRC beams without stirrups. Several of them are obtained on the basis of well-known relations for plain concrete beams by providing an additional contribute that depends on the fibers amount and the concrete matrix and fibers mechanical properties.

An attractive alternative approach is the plasticity theory that has long been applied with good success to reinforced concrete members [1]. Usual plastic approach assumes that stress fields transfer the loads to the bearing supports by satisfying the yield material criteria. Recent works [2, 3] on plain concrete shear beams illustrated as slips along the crack can delay or prevent the development of direct strut action spanning between the loads and the bearing supports. This failure mechanism is typical of slender beams and it is taken into account by plasticity theory in the CSM, proposed by Zhang [2] to determine the ultimate shear strength of plain concrete beams without stirrups.

In present work, by taking into account the "arch action", the classic formulation of CSM is firstly improved to evaluate the shear capacity of short beams. Then the model is extended to fibrous concrete members by a calibration of the effectiveness factors for fibrous concrete under biaxial stresses.

The proposed formulation (CSMf) is validated by favorable comparison against a large database of experimental test results collected in literature, and a comparison with several known relationships for shear capacity prediction of normal and high strength fibrous concrete element is presented.

CP1020, *2008 Seismic Engineering Conference Commemorating the 1908 Messina and Reggio Calabria Earthquake,*
edited by A. Santini and N. Moraci

MODELS FOR SHEAR CAPACITY IN SFRC BEAMS

In literature numerous empirical or semi empirical relationship, founded on wide experimental researches, have been suggested (Narayanan e Darwish (ND87) [4], Ashour et alt. (AWH92) [5], Kwak et alt. (KEKK02) [6], Khuntia et alt. (KS99) [7], Sharma (Sh86) [8] and Campione et alt. (CLP06) [9].) Some of them are derived by increasing the shear capacity of plain concrete members by an additional contribution, which depends on the geometric and the mechanic characteristics of fibers, generally expressed by the fiber factor $F= \beta V_f (l_f/d_f)$, where β is the fiber bond factor, V_f the fiber volume percentage, and l_f/d_f the fiber aspect ratio (ratio between fiber length and diameter). Thus, on the basis of the simple equilibrium equations of free body, the shear capacity is expressed as $V_{uf} = V_{cc} + V_b = \tau_{uf} b \, d$, where V_{cc} is the contribute of a plane concrete, V_b the contribute of fibers, b and d base and depth of cross section, τ_{uf} the ultimate average shear stress. More details on the analytical expressions of above mentioned models can be found in the papers listed in References.

Here, a model is proposed, derived by applying the theory of plasticity to SFRC elements. The prediction of the shear capacity of plain concrete beams is obtained assuming that the reinforcements react in the axial direction only with yield stress f_y, and the concrete behave as a rigid, perfectly plastic material, obeying the modified Coulomb failure criterion.

At failure the cracked concrete in compression is simultaneously subjected to compression and tensile strains, the latter in the direction normal to the compression. Therefore, it exhibits a reduced strength compared to the uncracked mono-axial compressed concrete. This phenomenon is referred to as "compression softening" and it can be recognized in the plasticity theory by the concrete effectiveness factor. In the usual plastic solution the effective compressive strength is $f_{c,ef} = v_c f_c$, where the effectiveness factor is given by:

$$v_c = \left(0.35/\sqrt{f_c}\right)\left[0.27\left(1+1/\sqrt{h}\right)\right]\left(0.15r+0.58\right)\left[1.0+0.17\left(a/h-2.6\right)^2\right] \quad (1)$$

with f_c = compressive cylinder concrete strength; h= height of beam's cross section; $r = 100 \, A_s / b \, h$; and a = shear span. Equation (1) shows that v_c is function of shear span-depth ratio and this dependency was explored and solved by Zhang [2] in the CSM, on the whole range of a/h values. The low values of v_c for a/h around 2.5 are due to sliding in initial cracks.

Due to the dramatically reduced sliding resistance in a crack, sliding along a crack originated in a generic section of the shear span may be more dangerous than sliding along the theoretical yield line between support and load point, like in the usual plastic solution. The crack pattern at the failure state is schematically shown in the Figure 1. The first cracks are usually formed in the region with maximum moment and are vertical. Progressively, additional diagonal cracks appear in the shear span closer to the support, along a line that approximately intersects the top face at the loading point.

The load required for developing these cracks is the higher, the less the distance x to the support (see the marked cracking load curve in Fig. 1); by contrast, the load needed to develop a sliding failure through a crack is lower, the less the distance is

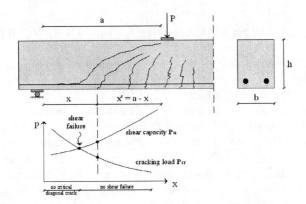

FIGURE 1. Typical crack pattern in a beam without stirrups under shear load.

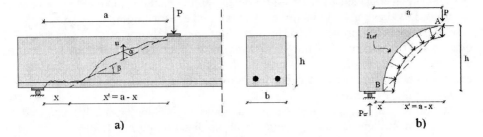

a)   b)

FIGURE 2. a) Ideal crack pattern in a beam under shear load; b) Stress field at the formation of a crack.

from the support, like in the usual plastic solution. The shear capacity curve in Figure 1 shows that higher the shear span, the lower the load capacity. According to the CSM, when the two curves intersect the crack may develop, in terms of the plastic theory, into a yield line and a shear failure takes place. The last diagonal crack is referred to as the critical diagonal crack.

In the CSM it is assumed that diagonal cracks develops following the straight lines from the bottom face to the loading point. Thus the starting crack sections may be denoted by their horizontal projection x. Further assumptions are that the beam is assumed to be over reinforced in the longitudinal direction, and the relative displacement u along the critical diagonal crack is vertically oriented (Fig. 2a).

Using the upper bound approach of plastic theory, and on the basis of the beam's failure mechanism in Figure 2a, the average shear stress at failure is:

$$\tau_u = \frac{P_u}{bh} = \frac{1}{2} f_{c,ef} \left[\sqrt{1 + \left(\frac{a-x}{h} \right)^2} - \frac{a-x}{h} \right] \tag{2}$$

where $\alpha = (90°-\beta)$, and $\cot\alpha = (a-x)/h$.

890

In order to evaluate the cracking load curve, the moment equation about point A for the beam with a semicircular crack (Fig. 2b) is considered. Thus, the average cracking stress τ_{cr}, assuming a statically equivalent constant tensile stress $f_{t,ef}$, is:

$$\tau_{cr} = \frac{P_{cr}}{bh} = \frac{1}{2} f_{t,ef} \frac{1+\left[(a-x)/h\right]^2}{a/h} \tag{3}$$

where $f_{t,ef} = 0.156 f_c^{2/3} (h/0.01)^{-0.3}$ is the effective tensile strength.

Introducing this new concept for the identification of critical diagonal crack, Zhang eliminated the dependence of v_c by the shear span-depth ratio, and proposed the evaluation of the effectiveness factor for concrete in compression by the product of two terms:

$$v_c = v_s v_0 = v_s \frac{0.56}{\sqrt{f_c}}\left[0.27\left(1+1/\sqrt{h}\right)\right](0.15r+0.58) \tag{4}$$

where $v_s = 0.50$ is the sliding reduction factor due to the reduced cohesion of cracked concrete when the yield line follows the diagonal crack path or crosses many cracks; and v_0 is partly adhere to the empirical formula obtained in the original plastic solution in Eq. 1.

The CSM is a mechanical model to determine the ultimate shear load of plain concrete beams without stirrups. It has been validate by Zhang on a large database of data collected in literature [2]. The tests considered by the author for the model corroboration are characterized by values of a/h higher than 2; thus the most of specimens collapse for diagonal tension and the beam action is the principal shear resistance mechanism.

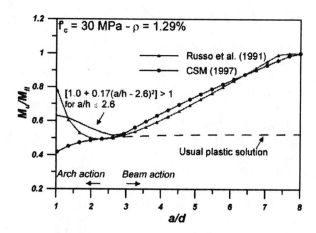

FIGURE 3. Relative flexural capacity evaluated with CSM and Russo et al.'s model.

The Figure 3 shows how the relative flexural capacity (M_u/M_{fl}) depends on the shear span-effective depth (a/d) for a plain concrete beam, evaluating the nominal flexural capacity as suggested by ACI [10]: $M_{fl} = bd^2 \rho f_y [1-(\rho f_y)/(1.7 f_c)]$, where f_y=yield steel strength and ρ= geometrical ratio of longitudinal reinforcements. The ultimate moment M_u is calculated with CSM and by a formulation known in literature [11] for plain concrete beams. The Russo et al.'s model provides the contribution of both beam and arch resistance mechanisms in the whole range of a/d values. The CSM is in good agreement with numerical results only for a/h values higher than 2 and fails for a/h values lower than 2 because it's not able to provide a good estimation of the arch action. This is due to the Zhang's choice of eliminating the dependency from a/h of the effectiveness factor for concrete in compression. This assumption provides numerical results far from the experimental values observed for beams with $a/h < 2$.

In order to eliminate this drawback, the CSM is modified retaining the correlation of the efficiency factor by the a/h ratio for a/h lower than 2, i.e. assuming an additional term $[1.0+0.17(a/h-2.6)^2]$ in (5) for $a/h \leq 2.6$ The accuracy increment obtained by the modified version of the CSM is shown by the solid line in Figure 3, where the assessment of the noticeable increment in the relative flexural capacity for the deep beams is shown.

THE CSM FOR FIBROUS CONCRETE BEAMS

Flatten stress-strain relationship in the post peak range of fibrous concrete in compression and tension make the SFRC more suitable than plain concrete for the application of the plasticity theory. Moreover, the presence of fibers in the matrix induces the reduction of the slips along cracks. To extend the CSM formulation to fibrous concrete beams, the most important issue is the use of reliable values of effective compressive and tensile strengths for SFRC. Values of effectiveness factors for fibrous concrete higher than plain concrete are expected.

The residual tensile stress of SFRC plays an important role in the shear failure mechanism of the beam. The analytical relationship proposed by Foster et al. [12] for fibrous concrete in direct tension, namely the Variable Engagement Model (VEM06), is proposed for evaluation of the effective tensile strength ($f_{t,ef}= v_{tf}f_{ct}$). The VEM06 considers that slippage between the fibers and the concrete matrix occurs before the full bond stress is developed and the fibers can be broken before being pulled out across a crack. The constitutive tensile law, expressed in terms of tensile tension and crack opening displacements (w), is the simple summation of stress contribution by matrix and fibers: $\sigma_{cf}(w)= \sigma_c(w)+\sigma_f(w)$.

According VEM06 the fibers are mechanically anchored to the matrix and some slips, between fiber and matrix, must occur before the anchorage is engaged. The crack opening w for which the fiber becomes effectively engaged in the tension carrying mechanism is termed the engagement length $w_e= \alpha tan\theta$, where $\alpha= d_f/3.5$ is a material parameter and θ is the fiber inclination angle, evaluated respect to the crack plane. When w is equal or higher than w_e, the force in a single fiber is P_f. The bridging tension stress across the crack provided by fibers is obtained by a simple integration of P_f over a plane of unit area, as follows:

$$\sigma_f(w) = f_{ct}\left[K_f(w)F_r\right] \tag{5}$$

where $F_\tau = \beta_\tau V_f (l_f/d_f)$ is a parameter analogous to the fiber factor, $\beta_\tau = \tau_f/f_{ct}$, and $K_f(w)$ is the following global orientation factor which depends by w:

$$K_f(w) = \frac{\tan^{-1}(w/\alpha)}{\pi}\left(1 - \frac{2w}{l_f}\right)^2 \tag{6}$$

To predict the value of residual tensile strength of fibrous concrete at shear failure of the beam, the contribute given by the matrix, $\sigma_c(w)$, is computed by a simple linear law [13], where the energy fracture of plain concrete is evaluated as suggested in [14]. The crack opening at shear collapse of the beam (w_m) was evaluated by Casanova and Rossi [15] like $h/100$, where h is the height of the beam.

Finally the analytical expression of the effectiveness tensile factor is $v_{tf} = \sigma_{cf}(w_m)/f_{ct}$.

COMPARISON BETWEEN PROPOSED AND CURRENT MODELS

A large database (109 data) of experimental tests results on SFRC beams without stirrups was compiled from literature to validate the proposed model (CSMf) for prediction of shear strength of rectangular fiber reinforced concrete beams. Beam specimens failing in shear, or with a crack patterns indicating that shear failure mode is predominated, are considered only in the database [16].

Firstly, for the validation of the CSMf, data have been splitted in two groups, depending on the concrete compression strength. In the Figure 4 the values of the ratio between experimental results given in literature and the values predicted by using the presented analytical model, are reported together with the mean value and Coefficient Of Variation (COV). After an accurate procedure of best fitting, the values of crack sliding factor (v_{sf}) equal to 0.77 has been used, in substitution of the original value 0.50 proposed by Zhang for plain concrete beams. The latter is too conservative for normal and high strength concrete, providing also high value of COV for both the two concrete compression strengths. By contrast, the former value provides an accurate prediction of experimental results with a mean value of 1.01 and COV value of 0.32 for normal strength concrete and of 0.18 for high strength concrete. The choice of a v_{sf} value for fibrous concrete higher than plain concrete is explained by the capacity of fiber to limit the crack slips.

In Figures 5-6, a comparison of the predicted shear strength using the empirical and semi empirical formulations known in literature [4, 5, 6, 7, 8, 9] and the experimental measured failure shear stress has been performed. The results show that Kwak et alt. model [6] and Campione et alt. [9] model (CLP06) give a good estimation of shear capacity of high strength SFRC beams, with mean values close to 1.00 and the COV greater than 0.20. For normal SFRC beams Narayanan e Darwish [4] model (ND87) provides a sufficient accurate prediction of the shear capacity, but it is less conservative than CSMf.

Furthermore the comparison shows that CSMf provides the best prediction for normal and high strength SFRC beams.

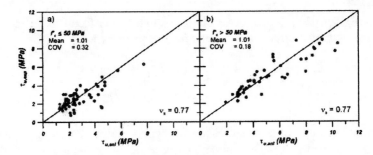

FIGURE 4. Comparison between experimental and analytical results with proposed model (CSMf).

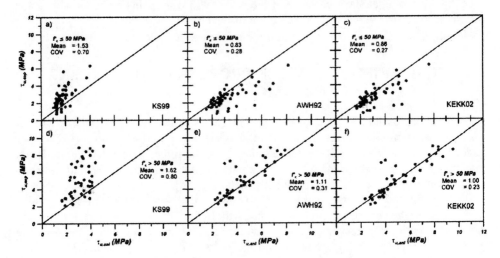

FIGURE 5. Comparison between experimental and analytical results with different formulas.

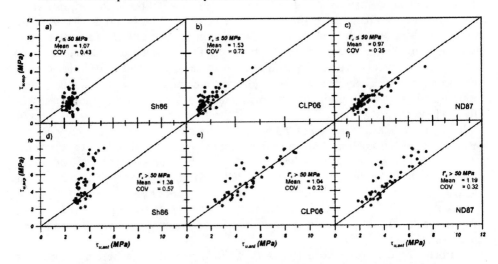

FIGURE 6. Comparison between experimental and analytical results with different formulas.

CONCLUSIONS

For providing the shear strength capacity of fibrous concrete beams without stirrups under transversal loads, in the present paper a mechanical model is proposed that is based on plastic theory and limit analysis. It takes into account the fiber concrete contribute to shear strength including the high residual post cracking tensile strength of SFRC and at this aim the VEM06 [12] constitutive law was used.

In the proposed model, the effectiveness factor of fiber concrete in compression was modified for deep beams, by introducing an additional term depending on the shear span-depth ratio. The reduction slide factor for fiber concrete, v_{sf}, was increased to 0.77, in order to take into account the ability of fibers in reducing slips along shear cracks. Further study might be necessary to evaluate more accurately the contribute of fiber onto the shear resistance mechanism of short beams (arch action).

A comparison of the predicted shear strength using the empirical and semi empirical formulations known in literature [4, 5, 6, 7, 8, 9] and the experimental measured failure shear stress has been performed and it shows that the proposed CSMf provides the best prediction for normal and high strength SFRC beams.

REFERENCES

1. Nielsen, M. P. "Limit analysis and concrete plasticity" (2nd ed.). Boca Raton - Florida: CRC, 1999
2. Zhang, J.-P. "Diagonal cracking and shear strength of reinforced concrete beams". *Magazine of Concrete Research* **178**, 55–65 (1997).
3. Vecchio, F. J. "Analysis of shear critical reinforced concrete beams". *ACI Struct. J.* **1**, 102–110 (2000).
4. Narayanan, R. and I. Y. S. Darwish "Use of steel fibers as shear reinforcement". *ACI Struct. J.* **3**, 2066–2079 (1987).
5. Ashour, S. A., G. S. Hasanain, and F. F. Wafa "Shear behaviour of high strength fiber reinforced concrete". *ACI Struct. J.* **2**, 176–184 (1992).
6. Kwak, Y. K., M. O. Eberhard, W. S. Kim, and J. Kim "Shear strength of steel fiber-reinforced concrete beams without stirrups". *ACI Struct. J.* **4**, 530–538 (2002).
7. Khuntia, M., B. Stojadinovic, and C. G. Subhash "Shear strength of normal and high-strength fiber reinforced concrete beams without stirrups". *ACI Struct. J.* **2**, 282–289 (1999).
8. Sharma, A. K. "Shear strength of steel fiber reinforced concrete beams". *ACI J.* **4**, 624–628 (1986).
9. Campione, G., L. La Mendola, and M. Papia "Shear strength of fiber reinforced beams with stirrups". *Structural Engineering and Mechanics* **1**, 107–136 (2006).
10. ACI 318 Technical Committee "Building code requirements for reinforced concrete" (ACI318-83). Technical report, American Concrete Institute - Detroit - Michigan (USA), 1983.
11. Russo, G., G. Zingone, and G. Puleri "Flexure-shear interaction model for longitudinally reinforced beams". *ACI Struct. J.* **1**, 66–68 (1991).
12. Foster, S. J., Y. L. Voo, and K. T. Chong "FE analysis of steel fiber reinforced concrete beams failing in shear: Variable Engagement Model", *ACI SP-***237**, 55–70 (2006).
13. Vecchio, F. J. "Disturbed stress field model for reinforced concrete: Formulation". *ASCE J. of Struct. Eng.* **9**, 1070–1077 (2000).
14. Marti, P., T. Pfyl, V. Sigrist, and T. Ulaga "Harmonized test procedures for steel fiber-reinforced concrete". *ACI Mat. J.* **6**, 676–685 (1999).
15. Casanova, P. and P. Rossi "Analysis and design of steel fiber reinforced concrete beams". *ACI Mat. J.* **5**, 595–602 (1997).
16. Spinella N. "Modelli per la risposta a taglio e flessione di travi in calcestruzzo rinforzato con fibre d'acciaio" (in Italian). Ph. D. Thesis, University of Messina, 2008

Shaking Table Tests Validating Two Strengthening Interventions on Masonry Buildings

Gerardo De Canio[a], Giuseppe Muscolino[b], Alessandro Palmeri[c],
Massimo Poggi[a] and Paolo Clemente[a]

[a] Research Centre "Casaccia," ENEA, Roma, Italy
[b] Department of Civil Engineering, University of Messina, Italy
[c] School of Engineering, Design & Technology, University of Bradford, UK

Abstract. Masonry buildings constitute quite often a precious cultural heritage for our cities. In order to future generations can enjoy this heritage, thence, effective projects of protection should be developed against all the anthropical and natural actions which may irreparably damage old masonry buildings. However, the strengthening interventions on these constructions have to respect their authenticity, without altering the original conception, not only functionally and aesthetically of course, but also statically. These issues are of central interests in the Messina area, where the seismic protection of new and existing constructions is a primary demand. It is well known, in fact, that the city of Messina lies in a highly seismic zone, and has been subjected to two destructive earthquakes in slightly more than one century, the 1783 Calabria earthquake and the more famous 1908 Messina-Reggio Calabria earthquake. It follows that the retrofitting projects on buildings which survived these two events should be designed with the aim to save the life of occupants operating with "light" techniques, i.e. respecting the original structural scheme. On the other hand, recent earthquakes, and in particular the 1997 Umbria-Marche sequence, unequivocally demonstrated that some of the most popular retrofitting interventions adopted in the second half the last century are absolutely ineffective, or even unsafe. Over these years, in fact, a number of "heavy" techniques proliferated, and therefore old masonry buildings suffered, among others, the substitution of existing timber slabs with more ponderous concrete slabs and/or the insertion of RC and steel members coupled with the original masonry elements (walls, arches, vaults). As a result, these buildings have been transformed by unwise engineers into hybrid structures, having a mixed behaviour (which frequently proved to be also unpredictable) between those of historic masonry and new members. Starting from these considerations, a numerical and experimental research has been carried out, aimed at validating two different strengthening interventions on masonry buildings: (i) the substitution of the existing roof with timber-concrete composite slabs, which are able to improve the dynamic behaviour of the structure without excessively increase the mass, and (ii) the reinforcement of masonry walls with FRP materials, which allow increasing both stiffness and strength of the construction. The experimental tests have been performed on a 1:2 scale model of a masonry building resembling a special type, the so-called "tipo misto messinese", which is proper to the reconstruction of the city of Messina after the 1783 Calabria earthquake. The model, incorporating a novel timber-concrete composite slab, has been tested on the main shaking table available at the ENEA Research Centre "Casaccia," both before and after the reinforcement with FRP materials. Some aspects related to the definition of the model and to the selection of an appropriate seismic input will be discussed, and numerical results confirming the effectiveness of the interventions mentioned above will be presented.

Keywords: Timber-Concrete Composite, Light Slabs, Shaking Table test, FRP Reinforcement.

CP1020, *2008 Seismic Engineering Conference Commemorating the 1908 Messina and Reggio Calabria Earthquake*,
edited by A. Santini and N. Moraci
© 2008 American Institute of Physics 978-0-7354-0542-4/08/$23.00

THE "TIPO MISTO MESSINESE" ARCHITECTURE

The Messina's masonry building named "Tipo Misto Messinese" (Messina's Mixed Type) is a characteristic nineteenth century architecture developed during the reconstruction of the city after the 1783 catastrophic earthquake. It is a combination of modular aligned cell houses and masonry palace elements, like the high arch with large front door of each cell house and the uniform façade of entire aligned body structure. The "Tipo Misto Messinese" house is hierarchically structured, starting from the ground level used as warehouse or shop, the intermediate gallery for the servants and the last floor as family house. The condominium named "Case Cicala" (Fig. 1) is a well preserved example of "Tipo Misto Messinese" masonry houses brick architecture. Each house unit (7 m length, 5 m width, 14 m high) is independent, with intermediate vaults.

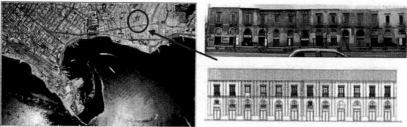

FIGURE 1. Location of the condominium "Case Cicala" in Via Garibaldi, Messina, Italy.

The shaking table experiments on the 1:2 scale model of a two level cell house have been performed at the ENEA Research Centre "Casaccia." The model was provided with a novel timber-concrete composite roof and Carbon Fiber Reinforced Polimeric (CFRP) reinforcement. The experimental campaign consisted of two series of tests before and after the reinforcement with the CFRP. In the following will be illustrated the main aspects related to the model definition, the test design, the seismic input and the validation of the numerical model with the experimental results.

TEST CAMPAIGN ON THE 1:2 SCALE MODEL

Different kind of repair or strengthening for many houses of the "Case Cicala" condominium have made in the last century. For instance, in Fig.2 three examples of past interventions on the vaults and slabs of three different houses are shown.

FIGURE 2. Case Cicala interventions: a) replacing the vault with horizontal steel-brick slab; b) FRP reinforcement of the vault extrados; c) wood bars for the horizontal strengthening.

In this series of tests a particular aspect of the seismic improvement regarded the roof. It is composed by a concrete-timber roof and the connection between the GluLam timber beam and the reinforced concrete slab is guaranteed by a particular groove in which a shrinkage compensated mortar is first cast and the lower bars of reticular reinforcing steel are inserted. The advantage of this floor are its light self-weight, which is half than that of a usual concrete floor, and the presence of the concrete slab, usually missing in wooden floors, which guarantees a suitable diaphragm between the vertical walls.

The scope of the Shaking Table tests was to evaluate the combined efficacy of this new "light" timber-concrete composite roof slabs plus the CFRP wrapping to increase ductility at the vault contrast of the thrust. Starting from these considerations, the design of the 1:2 model has been made to preserve as more as possible the main aesthetic and structural characteristics of the actual "Case Cicala" cells.

Due to the shaking table [1] characteristics (4 m x 4 m large, 300 KNm overturning moment payload) the model dimensions was 3.5 m x 2.6 m, 3.30 m high and 195 KN weight. The payload limitation of the crane (200 KN) handling the model forced to eliminate the second intermediate level illustrated in the previous paragraph, in so achieving the 195 KN weight (see the Fig. 3). The Case Cicala walls are 0.4 to 0.6 m thick, thus the walls in the 1:2 scale model were 0.25 m thick, made by 0.25 x 0.12 x 0.055 bricks (double head wall), while the intermediate vault was 0.12 m thick (one head wall). Tendons have been used to reproduce the strengthening effects of the adjacent house cells. Arches in the model, finally, are geometrically similar to the originals.

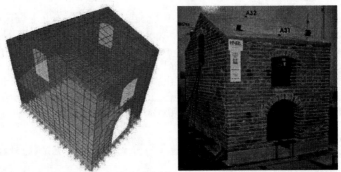

FIGURE 3. a) FEM numerical model, b) realisation of the 1:2 scaled mock-up.

Great attention has been paid in the test design configuration, specially for the base table seismic forcing. Due to the experimental limits in terms of available base displacements, a series of natural acceleration time histories was selected to activate the limit states of the model. Two types of time histories have been considered: (i) spectral compatible synthetic time histories for the Italian First category seismic Zone and B-C-E ground type [2], generated according to the procedure described in Refs. [3] and [4]; (ii) near-source accelerograms recorded in sites close to seismogenetic faults like the city of Messina. The choice of these time histories was due to the crescent attention [5, 6] to the effects of near-source earthquakes on structures [6, 7].

Generally those acceleration time histories show impulsive components, which can be very dangerous for civil engineering structures. These components are evident analyzing the ground motion components of near-source seismic waves, having pseudo velocity values of 50-200 cm/s within the range of 0.5 to 2.0 s period. In Fig. 4a the displacement and velocity time histories recorded by a near field station during the 1978 TABAS earthquake are shown, while in Fig. 4b comparison of the correspondent pseudo acceleration with the first seismic zone, BCE Ground Type Italian Spectrum, is depicted.

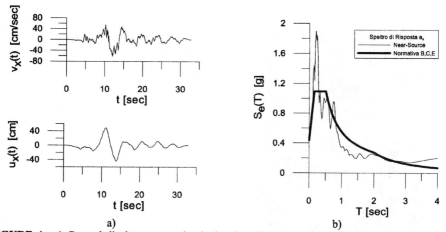

a)   b)

FIGURE 4. a) Ground displacement and velocity time histories recorded at the station N° 9101 of TABAS, Iran (September 16th 1978, I= 7.4) at 3.0 km distance from the epicentre; b) comparison between the correspondent spectral accelerations and the Italian spectrum for BCE ground type.

As shown in the sketches of Fig. 5, the model was monitored by 6 displacement sensors and 11 accelerometers (5 in the longitudinal direction, 5 in the lateral direction and 1 in the Vertical direction on the vault extrados named A11z)

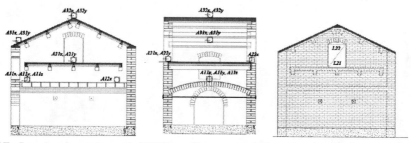

FIGURE 5. Accelerometer and LVDT positions: e.g. **A11x**= Accel._ level N°1_ Position N°1_DirectionX, **A21y**= Accelerometer_Level N°2_Position N°1_DirectionY; Displacement sensors: e.g. **L21**= Linear Variable Displacement Transducer_Level N°2_Position N°1

The model was subjected to 43 mono and bi-directional seismic tests, with increasing intensity. Three time histories were used: (i) Random tests (single directional, lateral and longitudinal) for the dynamic identification after each seismic test; (ii) Synthetic time histories compatible with the Italian spectrum for the BCE ground type; (iii) The lat-lon horizontal components of ground motion recorded at the station N° 9101 of TABAS. The time histories have been opportunely reduced due to the 1:2 scale of the model.

The July 8th 2005 a first phase of seismic tests were performed using incremental acceleration steps of 0.01g for BCE and TABAS time histories up to 0.5g Peak Ground Accelerations (PGA), the first damages occurs at 0.35g as shown in Fig. 6. At this stage the first phase of the experimental campaign was suspended to allow the CFRP wrapping strengthening

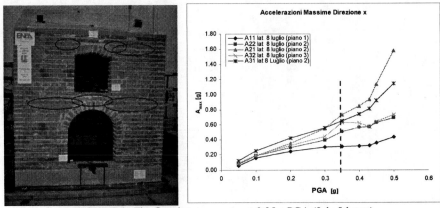

FIGURE 6. The first damages occur at 0.35 g PGA (July 8th test)

FRP wrapping reinforcement

The FRP wrapping was made using Carbon fibres composite (CFRP) as shown in Fig. 7. To test the efficacy of the reinforcement the application was intentionally waste (no preparation of the surfaces, direct application of the primer, etc...).

FIGURE 7. Steps of the CFRP reinforcement: the white painting was applied to enhance the cracks pattern during the test

The July 14th 2005, other 33 mono-and bi-directional tests on the reinforced model continued until the collapse of the structure (Fig. 8). Fig. 9 shows the peak accelerations recorded by the accelerometers A11X (Level_1-Position_1-Longitudinal) and A31X (Level_3-Position 1_Longitudinal) in the first (July 8th) and second (July 14th) phase of the test campaign.

FIGURE 8. Sequence of the collapse: note the increased ductility due to the CFRP wrapping.

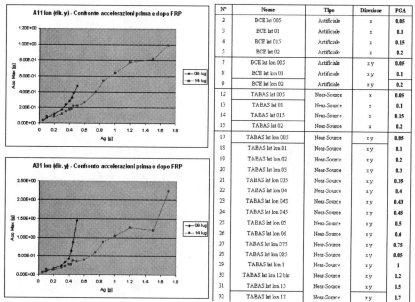

N°	Nome	Tipo	Direzione	PGA
2	BCE lat 005	Artificiale	x	0.05
3	BCE lat 01	Artificiale	x	0.1
4	BCE lat 015	Artificiale	x	0.15
5	BCE lat 02	Artificiale	x	0.2
7	BCE lat lon 005	Artificiale	x y	0.05
8	BCE lat lon 01	Artificiale	x y	0.1
9	BCE lat lon 02	Artificiale	x y	0.2
12	TABAS lat 005	Near-Source	x	0.05
13	TABAS lat 01	Near-Source	x	0.1
14	TABAS lat 015	Near-Source	x	0.15
15	TABAS lat 02	Near-Source	x	0.2
17	TABAS lat lon 005	Near-Source	x y	0.05
18	TABAS lat lon 01	Near-Source	x y	0.1
19	TABAS lat lon 02	Near-Source	x y	0.2
20	TABAS lat lon 03	Near-Source	x y	0.3
21	TABAS lat lon 035	Near-Source	x y	0.35
22	TABAS lat lon 04	Near-Source	x y	0.4
23	TABAS lat lon 043	Near-Source	x y	0.43
24	TABAS lat lon 045	Near-Source	x y	0.45
25	TABAS lat lon 05	Near-Source	x y	0.5
26	TABAS lat lon 06	Near-Source	x y	0.6
27	TABAS lat lon 075	Near-Source	x y	0.75
28	TABAS lat lon 085	Near-Source	x y	0.85
29	TABAS lat lon 1	Near-Source	x y	1
30	TABAS lat lon 12 bis	Near-Source	x y	1.2
31	TABAS lat lon 15	Near-Source	x y	1.5
32	TABAS lat lon 17	Near-Source	x y	1.7

FIGURE 9. Peaks of accelerations recorded on the vault extrados and on the top roof by the sensors A11Y (longitudinal) and A31Y (Longitudinal) during the first and second test campaigns. The left table shows the test sequence before and after the FRP reinforcement.

The presence of the light timber-concrete composite slab reduced the roof accelerations. Moreover, the model damages increases with the base table PGA as shown by the non-linear shape of the A11 and A31 peaks VS the peak base accelerations in abscissa. The different shape of blue and magenta curves demonstrates that the CFRP wrapping increases the ductility of the model. This is also evident looking at the collapse sequence of Fig. 8, where the CFRP wrapping increased the ultimate strength of the structure. Table I compares the peak accelerations with and without CFRP wrapping during the test N° 25 (TABAS, PGA=0.5g) and during the 8th July test, respectively.

TABLE I. Comparison between the peak accelerations during the test N° 25 (TABAS, PGA=0.5g).

TABAS bidirectional PGA=0.5g

	Acceleration max [g]	
	No CFRP	With CFRP
A32 lat	0.74	0.65
A32 lon	0.63	0.44
A31 lat	1.14	0.74
A31 lon	1.45	0.36
A21 lat	1.58	0.83
A21 lon	0.93	0.27
A11 lat	0.43	0.45
A11 lon	0.47	0.22

Fig. 10 shows the maximum displacements recorded by L01 (front door deformation) during first and second test phases, i.e. before and after the FRP reinforcement. The numerical values are reported in Table II, along with those recorde by L11 (window).

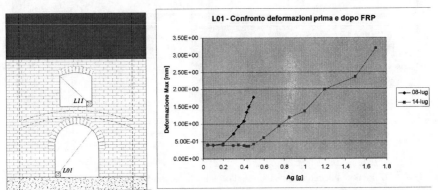

FIGURE 10. Maximum displacements recorded by L01 (front door deformation) during first and second test phases.

TABLE II. Maximum displacements recorded by L01 and L11 during first and second test phases.

	displacements [mm]			
	L01		L11	
Ag	July 8	July 14	July 8	July 14
0.1	0.39	0.39	0.28	0.08
0.3	0.72	0.37	0.94	0.06
0.4	1.1	0.38	1.59	0.06
0.5	1.8	0.42	2.42	0.23
0.6	-	0.60	-	0.11
0.75	-	0.94	-	0.27
0.85	-	1.18	-	0.48
1	-	1.36	-	0.52
1.2	-	1.99	-	0.63
1.5	-	2.36	-	1.10
1.7	-	3.21	-	1.60

CONCLUSIONS

The performances of two alternative strengthening interventions for masonry structures, namely the introduction of a new timber-concrete composite slab and the reinforcement of damaged walls with CFRP wrappings, have been investigated both numerically and experimentally. The results presented in previous sections demonstrate that these interventions are very effective in improving the dynamic behaviour of the model under investigation. More precisely, the timber-concrete composite slab induces a useful diaphragmatic behaviour, while the CFRP wrappings increase both stiffness and strength of the structure. Moreover, being intrinsically light, both techniques lend themselves to be applied to preserve old masonry buildings in highly-seismic zones.

ACKNOWLEDGMENTS

The Authors wish to acknowledge the ENEA researchers Nicola Ranieri, Alessandro Colucci, Massimiliano Baldini, Stefano Bonifazi, Francesco Di Biagio, Gianni Fabrizi, Alessandro Picca, Santino Spadoni for the precious contributions to the Shaking Table experiment set up and realization; Angelo Tatì and Nicola Labia for the thermograph and ultrasound analysis of the FRP support interface and Salvatore Iraci Sareri and Carmelo Sturiale for the hard work thesis.

The contribution by Coperlegno s.r.l. (Rome, Italy) is also gratefully acknowledged

REFERENCES

1. G. De Canio, "Large Scale experimental facilities at ENEA for seismic tests on structural elements of the historical/monumental cultural Heritage". *9th Int. Congress on Deterioration and Conservation of Stone*, Venice 19-24 June 2000
2. Presidenza del Consiglio dei Ministri, "Ordinanza n. 3274, Primi elementi in materia di criteri per la classificazione sismica del territorio nazionale e di normative tecniche per le costruzioni in zona sismica", 2003.
3. P. Cacciola, P. Colajanni, G. Muscolino, "Combinazione dei massimi modali coerente con lo spettro di potenza spettro-compatibile" , *IX Convegno Nazionale ANIDIS "L'Ingegneria Sismica in Italia"*, 1999 (atti pubblicati su CD Rom).
4. G. Muscolino, "Dinamica delle Strutture" McGraw-Hill, Milano, 2001.
5. N. Makris, Y. Roussos, "Rocking response of rigid blocks under near source ground motions" *Geotechnique*, 2000, Vol. 50, pp. 243-262.
6. G.P. Mavroeidis, A.S. Papageorgiou. "A mathematical representation of near-fault ground Motions", *Bulletin of the Seismological Society of America*, 2003, Vol. 93, pp. 1099-1131.
7. T. Psychogios, N. Makris "Dimensional response analysis of yielding structures" *Università di Patrasso (Grecia), Report Series in Earthquake Engineering and Applied Mechanics*, 2005/01.
8. A. Palmeri, N. Makris "Response analysis of rigid structures rocking on viscoelastic foundation" *Università di Patrasso (Grecia), Report Series in Earthquake Engineering and Applied Mechanics*, 2005/02.
9. Coperlegno SRL: Solaio Compound®. <http://www.coperlegno.com/>.
10. Computers & Structures, Inc.: SAP 2000 V9. <http://csiberkeley.com/>.

Earthquake Induced Damage Mechanism of Long Period Structures Using Energy Response

Yongfeng Du [a, b] and Hui Li [a, b]

[a] Western Engineering Research Center for Disaster Mitigation in Civil Engineering of Ministry of Education, Lanzhou, 730050, PR China
[b] Institute of Earthquake Protection and Disaster Mitigation, Lanzhou University of Technology, Lanzhou, 730050, PR China

Abstract. This paper presents a method of expounding the damage of RC long period frame structure using energy analysis method. Since the damage of structures usually occurs under major earthquakes, the structure is assumed to be in elasto-plastic state, and degraded Bouc-Wen model is used to describe the hysteretic component of the restoring force. A double index damage criterion defined by the maximum drift and energy absorption is used as the damage criterion. The energy transferring relation in a structure is derived, and both momentary and cumulative energy response is used to reflect the delay of the collapse of a long period structure. The mechanism of collapse delay of the long period structure is suggested through a numerical example combing the energy response and time history response.

Keywords: collapse mechanism; energy analysis; RC frame structure; response of long period structure; damage analysis

INTRODUCTION

Energy analysis method has found wide application in earthquake engineering since the middle of the previous century. Whether controlled or uncontrolled, the seismic resistant ability of a structure may be judged by its energy response. Energy analysis method may be used to track the transmission, absorption and transformation of the energy flow in the structure induced by earthquake action.

Energy response, being an important supplementary criterion for the seismic design in addition to the traditional strength and displacement criteria, is chosen as a quantitative index to measure the level of seismic response by scholars worldwide and has a long history. Housner proposed the concept of energy analysis in the 1956, and derived an energy equilibrium relation for the design of a simple structure. A rapid development of energy analysis method took place in the past half century, resulting in numerous of literatures. The cumulative energy was used as a quantitative indicator to measure the level of the structural damage by Akiyama [1] and a momentary energy input indicator was established by Inoue et al.[2] to identify the peak response of the structure effectively. Tso et al. [3] discussed the seismic energy demands on the RC frames. Xiong and Shi [4] studied the collapse mechanism of RC frame structures using energy time history based on Q model. Zhu and Shen [5] investigated the collapse of RC frame structures using hysteretic energy and push-over method considering multi-mode energy response. The author's research group developed an energy response

CP1020, *2008 Seismic Engineering Conference Commemorating the 1908 Messina and Reggio Calabria Earthquake,*
edited by A. Santini and N. Moraci
© 2008 American Institute of Physics 978-0-7354-0542-4/08/$23.00

method to judge the vibration mitigation effect in passive isolated structures[6], and compared the control effect of different control algorithms in smart isolated structures[7].

This paper presents an investigation into the energy response of the structure with long period. The structure is assumed to be in elasto-plastic state under major earthquakes, and energy transferring relation is derived for an ordinary frame structure. The energy response is employed through numerical simulation of time history of energy response to explain the phenomenon of collapse delay in long period structures which was revealed in some practical engineering project[8].

EQUATION OF MOTION UNDER MAJOR EARTHQUAKES

The structure is assumed to be in elasto-plastic state under major earthquakes, and degraded Bouc-Wen model is used to describe the hysteretic component of the restoring force. To meet the requirement of Bouc-Wen hysteretic model, the inter story drift of every story is used to describe the vibration of the whole structure. With the positive direction of the restoring forces assumed to be as shown in Fig. 1a and 1b, the equation of motion (EOM) of the structure can be derived as

$$[M_{x0}][T_{L0}]\{\ddot{x}_0\}+[C_{x0}]\{\dot{x}_0\}+[K_{x0}]\{x_0\}+[K_{z0}]\{z_0\}=-\ddot{u}_g[M_{x0}]\{\delta_0\} \qquad (1)$$

$$\dot{z}_j=(1/\upsilon_1)\left[A_{\upsilon j}\dot{x}_j-\upsilon_2(\beta_j\,|\,\dot{x}_j\,|\cdot|\,z_j\,|^{\eta-1}\,z_j+\gamma_j\dot{x}_j\,|\,z_j\,|^{\eta})\right], \qquad j=1,2,\cdots,n \qquad (2)$$

$$A_{\upsilon j}=A_j+\varsigma_A E_{hj}(t),\quad \upsilon_1=1+\varsigma_1 E_{hj}(t),\quad \upsilon_2=1+\varsigma_2 E_{hj}(t) \qquad (3)$$

$$E_{hj}(t)=(1-\alpha_j)k_j\int_0^{t_A} z_j\dot{x}_j dt \qquad (4)$$

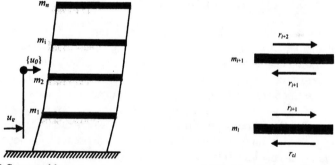

(a) Structural layout   (b) Positive direction of restoring forces
FIGURE 1. MDOF model for dynamic analysis of hysteretic structures

In Eq. (1), $\{x_0\}$ and $\{z_0\}$ are the inter-story drift and hysteretic displacement component vector, respectively, and A_j、 β_j、 γ_j、 α_j in Eq. (2) are the parameters of Bouc-Wen model for the j th story of superstructure, η is the parameter representing the degree of smoothness of the hysteretic curve. $[M_{x0}]$、 $[C_{x0}]$、 $[K_{x0}]$ and $[K_{z0}]$ are the mass, damping, stiffness and hysteretic stiffness matrices, respectively, of superstructure. υ_1, υ_2, ς_A, ς_1, ς_2 are the degrading parameters. k_j

and c_j are the stiffness ratio and the base line damping ratios of the jth story, respectively. $[M_{x0}]$ is expressed in diagonal matrix, other important variables are given by:

$$[C_{x0}] = \begin{bmatrix} c_1 & -c_2 & \cdots & 0 \\ 0 & c_2 & \ddots & \vdots \\ \vdots & & \ddots & -c_n \\ 0 & 0 & \cdots & c_n \end{bmatrix}, \quad [Q_{f0}] = \begin{bmatrix} 1 & -1 & \cdots & 0 \\ 0 & 1 & \ddots & \vdots \\ \vdots & & \ddots & -1 \\ 0 & 0 & \cdots & 1 \end{bmatrix} \tag{5}$$

$$[L_T] = \begin{bmatrix} 1 & 0 & \cdots & 0 \\ 1 & 1 & \ddots & \vdots \\ \vdots & & \ddots & 0 \\ 1 & 1 & \cdots & 1 \end{bmatrix}, \quad [K_{x0}] = \begin{bmatrix} \alpha_1 k_1 & -\alpha_2 k_2 & \cdots & 0 \\ 0 & \alpha_2 k_2 & \ddots & \vdots \\ \vdots & & \ddots & -\alpha_n k_n \\ 0 & 0 & \cdots & \alpha_n k_n \end{bmatrix} \tag{6}$$

$$\{\delta_0\} = \begin{Bmatrix} 1 \\ 1 \\ \vdots \\ 1 \end{Bmatrix}, \quad [K_{z0}] = \begin{bmatrix} (1-\alpha_1)k_1 & -(1-\alpha_2)k_2 & \cdots & 0 \\ 0 & (1-\alpha_2)k_2 & \ddots & \vdots \\ \vdots & & \ddots & -(1-\alpha_n)k_n \\ 0 & 0 & \cdots & (1-\alpha_n)k_n \end{bmatrix} \tag{7}$$

DAMAGE CRITERION OF RC FRAME STRUCTURES

Damage index of RC Frame Structures

Damage in a RC structure may have a broad sense ranging from the crazing of certain structural element to the collapse of the whole structures. Since this paper mainly investigates the energy response associated with collapse of the structures, the nonlinear flexural resistant damage would be of major interest. While the crazing of a structure is usually defined by the first passage damage criterion, the damage criterion based on hysteretic energy would be more suitable for representing the flexural resistant damage and collapse damage. This paper uses a damage criterion proposed by Park and Ang[9], in which the damage index is expressed in terms of the combination of maximum drift and cumulative hysteretic energy

$$D = \frac{\delta_m}{\delta_u} + \frac{\beta}{Q_y \delta_u} \int dE \tag{8}$$

Where, a larger value of D represents higher degree of damage. δ_m is the earthquake induced maximum deformation, and δ_u is the ultimate deformation under monotonic loading. Q_y is the yield strength of the structure, and dE is the increment of hysteretic energy dissipated by the structure. β is a constant given by

$$\beta = (-0.447 + 0.073 m_Q + 0.24 n_0 + 0.314 \rho) \times 0.7^{\rho_\omega} \tag{9}$$

Where, m_Q is shear span ratio, n_0 is the normalized axial compression ratio, and ρ_ω is the volume reinforcement ratio. The value of β is usually set to be 0.15.

Energy Transferring Relation and Energy Response

In the relative coordinate system, the cumulative energy under major earthquake can be expressed by different energy components as
1) The kinetic energy

$$E_K(t) = \int_0^t \left\{\dot{u}_g\{\delta_0\} + [T_{L0}]\{\dot{x}_0\}\right\}^T [M_{x0}]\left\{\ddot{u}_g\{\delta_0\} + [T_{L0}]\{\ddot{x}_0\}\right\}dt \tag{10}$$

2) The energy dissipated by base line damping

$$E_D(t) = \int_0^t \left([T_{L0}]\{\dot{x}_0\}\right)^T [C_{x0}]\{\dot{x}_0\}dt \tag{11}$$

3) The strain energy

$$E_S(t) = \int_0^t \left([T_{L0}]\{\dot{x}_0\}\right)^T [K_{x0}]\{x_0\}dt \tag{12}$$

4) The energy dissipated by hysteretic damping

$$E_H(t) = \int_0^t \left([T_{L0}]\{\dot{x}_0\}\right)^T [K_{z0}]\{z_0\}dt \tag{13}$$

6) The total energy input

$$E_I(t) = \int_0^t \dot{u}_g\{\delta_0\}^T [M_{x0}]\left\{\ddot{u}_g\{\delta_0\} + [T_{L0}]\{\ddot{x}_0\}\right\}dt \tag{14}$$

From Eqs. (10)~(13), equation of energy equilibrium of hysteretic structure under major earthquake can be obtained

$$E_K(t) + E_D(t) + E_S(t) + E_H(t) = E_I(t) \tag{15}$$

Substituting Eqs. (10)~(13) into Eq. (9), and referring to EOM (1), together with the following relationship

$$\{\delta_0\}^T [M_{x0}]\left\{\ddot{u}_g\{\delta_0\} + [T_{L0}]\{\ddot{x}_0\}\right\} = -c_1\dot{x}_1 - \alpha_1 k_1 x_1 - (1-\alpha_1)k_1 z_1 \tag{16}$$

one can verify that the relation shown in Eq. (15) holds true.

Main Energy Index Reflecting the Seismic Resistant Ability

Substituting EOM (1) into Eq. (16), yields another format of expression of total amount of energy input to the whole structure

$$dE_I(t) = du_g[-c_1\dot{x}_1 - \alpha_1 k_1 x_1 - (1-\alpha_1)k_1 z_1] \tag{17}$$

Generally speaking, the story drift, the relative velocity and hysteretic displacement in Eq. (17) are not in-phase with each other, therefore the damping force proportional to the base relative velocity can not be eliminated, which infers that a long period (soft) structure may have less energy input to the whole structure. The vibration energy relationship shown in Eq. (15) can be expressed as:

1) The total energy input to the system is partly transformed into kinetic energy of the masses, and partly transformed into strain energy stored in the equivalent springs represented by Eq. (12).

2) The part of the input energy will be dissipated by hysteretic damping and base line damping. For a structure closed to collapsing stage, the energy dissipated by hysteretic damping plays a dominant role.
3) The kinetic energy and strain energy will be feed back to the system during the subsequent vibration. The hysteretic energy will be closely related to the collapsing of the structure, while at the same time closely related to the vibration mitigation of the system.

Combining Eq. (8) and Eqs. (10)~(13), the hysteretic energy and total input energy may be of major concern in this study.

NUMERICAL EXAMPLE AND DISCUSSION

A 2-story building which was conceived according to some of the typical buildings collected from a strong earthquake event[10] is chosen as numerical example, the main structural parameters are shown in table 1, and the parameters of Bouc-Wen model are shown in table 2. The Kobe (1995) earthquake signal is adopted as the input excitation, and the PGA of the natural earthquake is adjusted to 6.20ms^{-2}, representing major earthquake at the construction site with fortification intensity of 9 degree according to the requirement specified by the Seismic Design Code of China[11].

TABLE 1. Structural parameters

Story No. j	Vibration parameters		Geometric parameter /m		
	m_j /kg	κ_j /kNm^{-1}	H_j	b_j	h_j
1	803772	379430	3.9	0.45	0.60
2	477166	284260	3.9	0.45	0.60

TABLE 2. Parameters of Bouc-Wen model

Story No. j	basic parameters				Degrading parameter		
	Q_{yj} /MN	α_j	β_j	γ_j	ζ_A	ζ_1	ζ_2
1	3.94607	0.02	13319.2	-4439.7	0.0	0.0000001	0.0000001
2	2.95630	0.02	13319.2	-4439.7	0.0	0.0000001	0.0000001

The hysteretic loop of the degraded Bouc-Wen model, associated with the parameters which is listed in the above tables is shown in Fig. 2.

In a structure similar to the above listed tables, a special phenomenon of collapse delay for structures with long periods have been observed from several real earthquakes. The collapse delay may be referred to as the situation when the collapse of the structure with long period occurred not at the moment

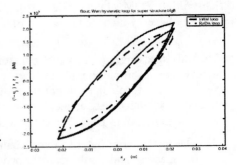

FIGURE 2. Hysteretic loop of the degraded Bouc-Wen model

with peak energy response, but occurred at a delayed time much later than the peak energy response moment, in some cases even occurred after the earthquake had ended. The main task of this paper is to provide numerical simulation of the time domain response and energy response using the given example, and some of the response indices related to the collapse of the structure will be presented to explain the collapse delay phenomenon.

Substituting the parameters into Eq. (10)~(14), one obtains the time history of the momentary and cumulative input energy to the whole structure, as shown in Fig. 3. One may easily observe from the figure that the total energy input to a structure with a long period is less than the corresponding value in a structure with a short period. This phenomenon can be explained using the similar mechanism of an isolated structure, where a soft structure would usually isolate large part of the earthquake energy.

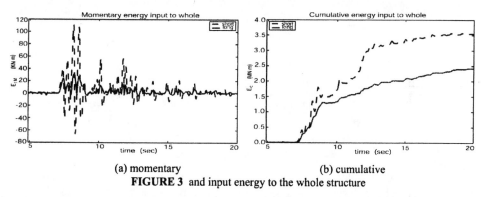

(a) momentary   (b) cumulative

FIGURE 3 and input energy to the whole structure

Correspondingly, the momentary and cumulative hysteretic energy dissipated by the structure in a structure with a long period is less than the corresponding value in a structure with a short period, as shown in Fig. 4. The dissipated hysteretic energy is among the most important indices reflecting the collapse resistant ability during strong earthquakes, this figure shows that the structures with long periods might be less likely to collapse.

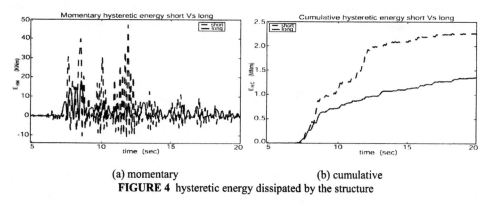

(a) momentary   (b) cumulative

FIGURE 4 hysteretic energy dissipated by the structure

Similar to the many ordinary structures, the momentary kinetic and strain energy fluctuate with time, the values of these two types of energy component in a structure with a long period are less than the corresponding values in a structure with a short period, as shown in Fig. 5. The lower part of this figure is the normalized values of the time history of kinetic and strain energy, defined by dividing the direct values of these two types of energy component by their corresponding maximum values. Though the normalized values may not act as an index to make judgment on the magnitude of the vibration energy response, they do reflect the phase of the energy response. From the normalized values, one observes a delay both in the peak response of kinetic and strain energy, which has been revealed by several other researchers.

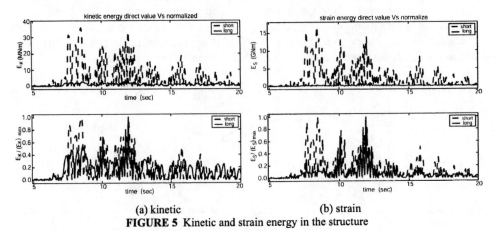

(a) kinetic   (b) strain

FIGURE 5 Kinetic and strain energy in the structure

One of the other important things concerning the property of a structure with a long period is that it has a larger inter-story drift in the time domain response, as shown in Fig. 6. One may also observe, on the other hand, that a less number of cycles of vibration in the time domain over the whole length of excitation duration, this might indicate a more favorable fatigue resistant ability because the load bearing ability of the RC structure would decay swiftly with the increase of number of cycles of the loading[12]. Combing Fig. 5a and Fig. 6a, one may notice that the vibration of the structure with long periods would last longer which is inferred by a larger value of the remaining kinetic energy and a slow vanishing of the inter-story drift after the excitation ended. This might be another reason for the delay of the collapse.

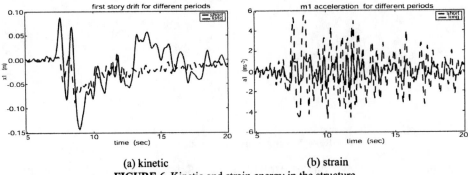

(a) kinetic   (b) strain

FIGURE 6 Kinetic and strain energy in the structure

910

CONCLUSIONS

This paper presents a discussion on the damage of RC frame structure with long period using numerical method combing the energy response with the time history response. The writer suggested the following three things might be able to constitute the reason for the collapse delay of the reinforced concrete structures with long periods during strong earthquakes:

1) the longer lasting vibration;
2) the delayed accumulation of the number of vibration cycles over the duration of the excitation acting.
3) the delay of peak value of momentary kinetic and strain energy response reflecting the delay of cumulative energy input to the structural system.

ACKNOWLEDGMENTS

This research is supported by Natural Science Foundation of China under Project No. 50778087 (Project director: Ming Lei and Jiping Ru). This support is gratefully acknowledged.

REFERENCES

1. Akiyama, Hiroshi. *Earthquake-Resistant Limit-State Design for Buildings.* Tokyo: University of Tokyo Press, 1985.
2. Norio Hori,Norio Inoue. Damaging properties of ground motions and prediction of maximum response of structures based on momentary energy response. *Earthquake Engineering and Structural Dynamics,* 2002, 31, pp. 1657-1679.
3. Tso, W.K., Zhu T.J. and Heidebrecht, A.C. Seismic energy demands on reinforced concrete moment- resisting frames. Earthquake Engineering and Structural Dynamics, 1993, 22, pp. 533-545.
4. Xiong, Z. and Shi, Q. Study on evaluation theory for collapse mechanism of RC frame structures with energy method. *Journal of Vibration and Shock,* 2003, 22(4), pp. 212-213.
5. Zhu, JH. and, and Shen, PS. Collapse analysis of reinforced concrete frames based on energy method. *Science Technology and Engineering,* 6(8), 2006, pp. 1146-1149
6. Dang, Y, Du, YF, and Li, H. Energy dissipation analysis of base isolated structures. *World Seismic Engineering,* 2005, 21(3), pp. 100-104. (in Chinese)
7. Du, YF, and Li, H. An investigation to energy response of smart isolated structures and control efficiency evaluation. *Proc. 1st European Conference on Earthquake Engineering and Seismology,* Geneva, Sept, 2006, No. 1232.
8. Xu, ZX, and Hu, ZL. Analysis of seismic response of structures. *Higher Education Press, 1993.* (in Chinese)
9. Park, Y.J. Ang, A.H.-S. Mechanistic seismic damage model for reinforced concrete. *Journal of Structural Engineering,* ASCE, 111(4), 1985, pp. 740-757
10. Du, Yongfeng, Li, Hui and Spencer, Billie F, Jr. Effect of Non-Proportional Damping on Seismic Isolation. *Journal of Structural Control,* 2002, 9(3):205~236.
11. Du, YF, and Liu, KY. An investigation to energy response of smart isolated structures under major earthquake energy. *Earthquake Resistant Engineering and Retrofitting, 2008, 30(1),* pp. 31-34. (in Chinese)
12. Du, YF, Li, H, Li, WR, and Fan, PP. A damage criterion of RC structure considering fatigue induced degrading. To appear in *Proc. 14th National Conference on Fatigue and Fracture Mechanics, 2008.* (in Chinese)

Influence of shear in the non-linear analysis of RC members

Pier Paolo Diotallevi[1], Luca Landi[1], Filippo Cardinetti[1]

1 DISTART, Department of Civil Engineering, University of Bologna, Italy

Abstract. The purpose of this study is to develop an analytical model characterized by a beam-column finite element which is able to reproduce the non-linear flexural-shear behavior of RC structures. The paper shows a brief description of the finite element formulation, the theory used for modeling the constitutive relationship and the scheme of the algorithm, transformed in a computer program, which was developed for implementing the theoretical model. Finally it illustrates a comparison with available experimental results for the calibration and validation of the model and a study on the influence of the non-linear shear response.

Keywords: RC members, non-linear analysis, fiber beam-column model, shear-flexure interaction

INTRODUCTION

The behavior of Reinforced Concrete (RC) squat elements was investigated in the past according to different approaches, based in most cases on bi-dimensional finite elements. The main objective of this study is to develop a beam-column element which is able to account for the shear-flexure interaction in the non-linear response. In the category of fiber elements, few models have been developed to date [1] able to predict the behavior of squat structures including non-linear shear deformations. An important aspect is that most of proposed models consider flexure and shear uncoupled. A way to avoid this assumption is to use biaxial constitutive relationships characterized by fix or variable smeared cracks, but most models that implement these relationships are based on bi-dimensional finite elements, which can be not useful to study complex frame structures [1].

To reach the goal a fiber beam-column element that uses "Modified Compression Field Theory" (MCFT) [2,3] as constitutive relationship has been developed. Just the combination between the fiber element (for modeling the structure) and the biaxial constitutive law (for modeling the behavior of each fiber) allows to describe the flexure-shear interaction in the non-linear response [4,5]. The element was implemented in a special purpose FEM program which is able to model generic frames. This computer program was also tested to check its working and its field of applicability.

CP1020, *2008 Seismic Engineering Conference Commemorating the 1908 Messina and Reggio Calabria Earthquake,*
edited by A. Santini and N. Moraci

MODEL FORMULATION

Finite element

The beam-column finite element, schematically shown Figure 1, is based on the flexibility method. The formulation of the finite element is based on the fiber beam-column model described in Filippou et al. [5]. Both deformation and force fields along the element are evaluated using shape functions and interpolation matrices. Deformation shape functions $\mathbf{a}(x)$ are determined as follows:

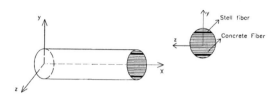

FIGURE 1. Beam-column fiber model.

$$\mathbf{a}(x) = \mathbf{f}^{i-1} \cdot \mathbf{b}(x) \cdot \left[\mathbf{F}^{i-1} \right]^{-1} \qquad (1)$$

where $\mathbf{b}(x)$ are the force interpolation functions, $\mathbf{f}^{i-1}$ is the section flexibility matrix, $\mathbf{F}^{i-1}$ is the element flexibility matrix and superscript i indicates the iteration of structure state determination. These interpolation functions allows to correlate increment of nodal displacements $\Delta\mathbf{q}^i$ with increment of section deformations $\Delta\mathbf{d}^i$:

$$\Delta\mathbf{d}^i(x) = \mathbf{f}^{i-1} \cdot \mathbf{b}(x) \cdot \left[\mathbf{F}^{i-1} \right]^{-1} \cdot \Delta\mathbf{q}^i \qquad (2)$$

With these interpolation functions the element equilibrium equation becomes:

$$\left[\mathbf{F}^{i-1} \right] \cdot \Delta\mathbf{q}^i = \mathbf{P}^i - \mathbf{Q}^{i-1} \qquad (3)$$

where $\mathbf{P}\text{-}\mathbf{Q}^{i-1}$ are the applied unbalanced forces.

The Modified Compression Field and its Development.

The Modified Compression Field Theory (MCFT) [2] and its next development "Disturbed Stress Field Model" (DSFM) [6, 7] allow consistent and accurate simulation of RC behavior. Average and local stress conditions are treated separately.

In the original MCFT theory crack shear slips associated with shear stresses are not explicitly calculated nor accounted for in the element deformation. Just for this reason the DSFM was formulated, attempting to provide a better representation of concrete behavior by explicitly including crack shear slip according to two mechanisms.

Equilibrium Condition

Let us consider a RC element (Fig. 2) subjected to uniform stresses $\boldsymbol{\sigma} = [\sigma_x \, \sigma_y \, \tau_{xy}]$.

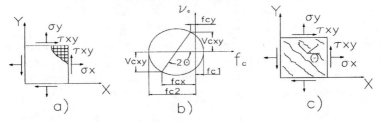

a)   b)   c)

FIGURE 2. a), b) Element stresses, c) principal stress directions.

Equilibrium is examined on two levels: in terms of average stress smeared over the element area and of local condition along the crack surfaces. The principal stresses f_{c1} and f_{c2} are parallel and perpendicular to the crack direction defined by angle θ in Figure 2. The equilibrium equation for the reinforcements are:

$$\sigma_x = f_{cx} + \rho_x \cdot f_{sx}; \qquad \sigma_y = f_{cy} + \rho_y \cdot f_{sy}; \qquad \tau_{xy} = \nu_{cxy} \tag{4}$$

On crack surfaces the equilibrium condition is:

$$\sum_{i=1}^{n} \rho_i \cdot \left(f_{scri} - f_{si} \right) \cdot \cos^2 \theta_{ni} = f_{c1} \tag{5}$$

where ρ_i is the reinforcement ratio, f_{si} is the average stress in steel, f_{scri} is the local stress of the i-th reinforcement relating to ε_{scri} and the angle $\theta_{ni} = \theta - \alpha_i$, where α_i is the angle of reinforcement. The local increase in reinforcement stress, at crack location, leads to the development of shear stress along the crack surfaces ν_{ci}:

$$\nu_{ci} = \sum_{i=1}^{n} \rho_i \cdot \left(f_{scri} - f_{si} \right) \cdot \cos \theta_{ni} \cdot \sin \theta_{ni} \tag{6}$$

Compatibility Relations

The continuum strain results from smearing of crack over a finite area, while the slip component results from rigid body movement along a crack interface. Total or "apparent" strain are denoted as $\boldsymbol{\varepsilon} = [\varepsilon_x \, \varepsilon_y \, \gamma_{xy}]$. The apparent inclination and principal strains can be calculated using Mohr circle. The shear slip and the associated deformation components are calculated as follows:

$$\gamma_s = \frac{\delta_s}{s} \tag{7}$$

$$\varepsilon_x^s = -\frac{\gamma_s}{2} \cdot \sin(2\theta), \quad \varepsilon_y^s = \frac{\gamma_s}{2} \cdot \sin(2\theta), \quad \gamma_{xy}^s = \gamma_s \cdot \cos(2\theta)$$

being δ_s the slip along the crack surface and s the crack spacing.

Constitutive Relations

The principal compressive stress f_{c2} in the concrete is a function not only of the principal compressive strain, but also of coexisting principal tensile strain. The influence is captured by the reduction factor β_d, which is calculated using expressions derived from experimental results. This factor is used to define both peak stress f_p and strain at peak stress ε_p:

$$f_p = -\beta_d \cdot f_c', \quad \varepsilon_p = -\beta_d \cdot \varepsilon_0 \qquad (8)$$

The compression response curve of concrete is represented in Figure 3 (a) while tension behavior of concrete is illustrated in Figure 3 (b) and (c). Reinforcement constitutive relationship is illustrated in Figure 3 (d). Slip δ_s is calculated according to two approaches [6]: as a function of shear stress v_{ci} through a formulation which depends on crack width and compressive strength and as a function of angle of principal stress. The maximum of two obtained values is considered.

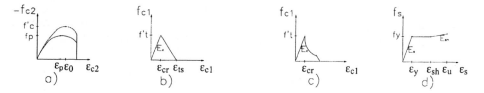

FIGURE 3. Constitutive relationships.

IMPLEMENTATION OF THE MODEL

The model described so far was implemented in a special purpose program organized on three iteration levels. A Newton-Rapson (NR) iteration loop at structural level, an element state determination at element level (necessary for flexibility formulation) and another iteration loop at each fiber for application of Disturbed Stress Field Model (DSFM). The program is organized as follows:

1. Creation of initial stiffness structural matrix.

2. Application of load increment and NR iteration. Each NR iteration is indicated by a superscript i.

3. Calculation of nodal element displacements trough a condensation and a rotation matrix.

4. Beginning of element state determination procedure: calculation of nodal forces. Each iteration of element state determination is indicated by a superscript j.

5. Calculation of section forces in control sections:

6. Calculation of section deformations.

7. Calculation of fiber deformations.

8.Beginning DSFM at the fiber level. Each fiber is characterized by a deformation:

$$\varepsilon = \begin{bmatrix} \varepsilon_x & \varepsilon_y = 0 & \gamma_{xy} \end{bmatrix} \tag{9}$$

Initially it is assumed $\varepsilon = \varepsilon_c$. With application of Mohr circle principal strains ε_1 and ε_2 for concrete are obtained. The average strains ε_{sx} and ε_{sy} for steel are set equal to those of concrete along x and y axis. After calculating average stresses in concrete and steel through constitutive relations, local deformations in reinforcements ε_{sxcr} and ε_{sycr} on crack location are calculated through an iterative procedure:

$$\varepsilon_{sxcr} = \varepsilon_{sx} + \Delta\varepsilon_{1cr} \cdot \cos^2(\theta_\sigma); \quad \varepsilon_{sycr} = \varepsilon_{sy} + \Delta\varepsilon_{1cr} \cdot \cos^2\left(\theta_\sigma - \frac{\pi}{2}\right) \tag{10}$$

At beginning of procedure $\Delta\varepsilon_{1cr} = 0$, then $\Delta\varepsilon_{1cr}$ is increased at each iteration until subsequent equilibrium equation is satisfied:

$$\rho_x \cdot (f_{sxcr} - f_{sx}) \cdot \cos^2(\theta_\sigma) + \rho_y \cdot (f_{sycr} - f_{sy}) \cdot \cos^2\left(\theta_\sigma - \frac{\pi}{2}\right) = f_{c1} \tag{11}$$

where stresses f_{sxcr} and f_{sycr} are functions of ε_{sxcr} and ε_{sycr} through constitutive relationship of steel. Then shear stress along crack surfaces are calculated:

$$v_{ci} = \rho_x \cdot (f_{sxcr} - f_{sx}) \cdot \cos(\theta_\sigma) \cdot \sin(\theta_\sigma) + \rho_y \cdot (f_{sycr} - f_{sy}) \cdot \cos\left(\theta_\sigma - \frac{\pi}{2}\right) \cdot \sin\left(\theta_\sigma - \frac{\pi}{2}\right) \tag{12}$$

where θ_σ is the angle of principal stresses. Being s_x and s_y crack spacings in x and y directions it is possible to determine the crack spacing s and the crack width w:

$$s = \frac{1}{\dfrac{\sin\theta_\sigma}{s_x} + \dfrac{\cos\theta_\sigma}{s_y}}; \quad w = \varepsilon_{c1} \cdot s \tag{13}$$

Once the shear slip is calculated, it is possible to know the value of γ_s and the strain components due to shear slip ε_s (Eq. 7). Then strain components $\varepsilon_c = \varepsilon - \varepsilon_s$ are obtained.

Considering the new ε_c, an iterative procedure begins, that stops when the difference between subsequent values of ε_c are smaller than a fixed tolerance.

Once the convergence is reached, values of tangent modulus for the two principal directions are calculated form equations of constitutive laws. These values are introduced into diagonal matrices referred to principal directions. Trough rotation matrix it is possible to pass from system of principal axes to original system. The stiffness matrices of each fiber are assembled in order to obtain the modulus matrices $\mathbf{E}_s$ and $\mathbf{E}_c$ of all fibers of the section, which become bandwidth when flexure-shear coupling begins. From these matrices the stiffness matrix of section is obtained.

916

9. Calculation of section resisting forces $\mathbf{D}^j{}_R(x)$.
10. Calculation of unbalanced section forces $\mathbf{D}^j{}_u(x) = \mathbf{D}^j(x) - \mathbf{D}^j{}_R(x)$.
11. Determination of section residual deformations.
12. Determination of residual nodal displacements and then check of the convergence by energy criterion. If convergence is reached $\mathbf{Q}^i$ is set equal to $\mathbf{Q}^j$ and $\mathbf{K}^i{}_{ele}$ to $\mathbf{K}^j{}_{ele}$. Then another element is examined. When convergence is reached for all elements the procedure continues from step 13. If convergence is not achieved j is incremented to $j+1$ and a new iteration begins.
13. Caculation of resisting nodal forces $\mathbf{F}^i{}_R$ and of stiffness matrix of structure.
14. Calculation of unbalanced nodal forces $\mathbf{F}^i{}_u = \mathbf{P} - \mathbf{F}^i{}_R$ then check of the convergence at structural level. If convergence is achieved the NR procedure is stopped and a new load increment is applied, otherwise i is set equal to $i+1$ and another NR iteration is performed.

APPLICATIONS OF THE MODEL

Comparison between numerical and experimental results

The comparison is carried out on a shear wall loaded by a force at the top. Considered experimental data are those of Vulcano and Bertero [8]. The comparison is performed between numerical results obtained with proposed model, experimental results and numerical results obtained by the three vertical line element (TVLE) proposed by Vulcano and Bertero. Geometric data are illustrated in Figure 4. With regard to mechanical characteristics, for wall reinforcements f_y =507 Mpa and ε_u =0.12 while for flange reinforcements f_y =444 Mpa and ε_u =0.15. The cylinder compressive strength of concrete is f_c =34.8 Mpa and the elastic modulus is E_c =22500 Mpa. The comparison between numerical and test results, illustrated in Figure 5 in terms of base shear-top displacement curve, allows calibration and validation of model.

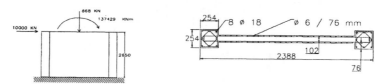

FIGURE 4. Geometric data of test specimen.

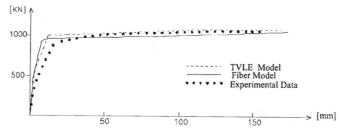

FIGURE 5. Comparison between numerical and experimental results.

Investigation on the influence of flexure-shear interaction

The flexure-shear coupling in non-linear analysis is investigated through a parametric study. The study regards a cantilever shear wall loaded by a force at its free end (Figure 6). The analyses are repeated changing the height of the wall L and keeping the section unaltered. The results referred to the subsequent cases are illustrated: $L/d = 1$ and $L/d = 10$, where L is the wall length and d the height of section.

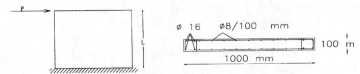

FIGURE 6. Geometry of the wall in parametric study.

Analysis of slender wall (L/d =10)

The wall is modeled with 20 finite elements. The length of the elements is fixed as a function of the length of the plastic hinge expected at the base. The results of the analyses are shown in Figure 7 in terms of shear-shear deformation (V-γ) curve. The two curves in Figure 7 are both related to the model, but one has been obtained accounting for the flexure-shear interaction in non-linear range, the other one considering only flexure non-linearity and keeping shear response in linear range. Since shear forces are constant along the height and V-γ diagram does change, the variation of bending moment along the height have affected shear deformations.

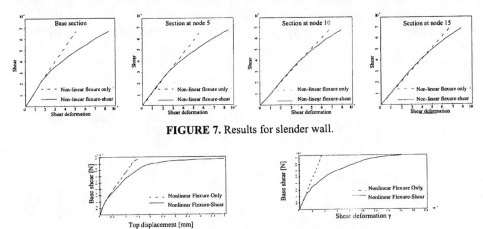

FIGURE 7. Results for slender wall.

FIGURE 8. Results for short wall.

Analysis of short wall (L/d =1)

In this second example the wall was modeled with 2 elements. For this wall the shear non-linear deformations are decisive. Before cracking, the model based on nonlinear shear deformations and the model based on elastic shear response produce the same results. On the contrary the post-cracking structural behavior is strongly influenced by non-linear shear response (Fig. 8). In Figure 9 diagrams of stress along the section are illustrated for the base section.

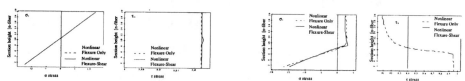

FIGURE 9. Stress along the base section for short wall.

CONCLUSION

A new fiber beam-column element was proposed and implemented in order to simulate non-linear flexure-shear behavior. The main characteristics of this model are substantially the flexibility formulation and constitutive relationship characterized by rotating smeared-crack model. The model was calibrated and validated through comparison between numerical and experimental results. Analyses underlined the capability of the finite element to reproduce non-linear flexure-shear interaction, and the importance of non-linear shear deformations for squat structures. The use of DSFM as constitutive relationship allowed to obtain the flexure-shear interaction in non-linear range.

REFERENCES

1. CERESA P., PETRINI L., PINHO R. - (2007), "Flexure-shear fiber beam-column elements for modelling frame structures under seismic loading-state of the art", *Journal of Earthquake Engineering*, Vol. 11, Special Issue 1.
2. VECCHIO, F.J., COLLINS, M.P. - (1986), "The Modified Compression Field Theory for Reinforced Concrete Element Subjected to Shear", *Journal of American Concrete Institute*, 83(2), 219-231.
3. VECCHIO, F.J., COLLINS, M.P. - (1988), "Predicting the Response of Reinforced Concrete Beams Subjected to Shear Using the Modified Compression Field Theory", *ACI Structural Journal*, 85, 258-268.
4. CERESA P., PETRINI L., PINHO R., AURICCHIO F. - (2006), "Development of a Flexure-Shear Fibre Beam-Column Element for Modelling of Frame Structures Under Seismic Loading." *First European Conference on Earthquake Engineering and Seismology*, Geneva, Switzerland 2006, paper number 1425.
5. FILIPPOU, F.C., SPACONE, E., TAUCER, F.F. - (1996), "A Fiber Beam-Column Model for Nonlinear Analysis of R/C Frames: Part.1 Formulation" *Earthquake Engineering and Structural Dynamics*, 25, 711-725.
6. VECCHIO, F.J. - (2000), "Disturbed Stress Field Model of Reinforced Concrete: Formulation", *Journal of Structural Engineering*, ASCE, 126(9), 1070-1077.
7. VECCHIO, F.J. - (2001), "Disturbed Stress Field Model of Reinforced Concrete: Implementation", *Journal of Structural Engineering*, ASCE, 127(1), 12-20.
8. VULCANO, A., BERTERO, V.V. - (1987), "Analytical Models for Predicting the Lateral Response of RC Shear Walls: Evaluation of their Reliability". Report EERC No. 87/19, University of California, Berkeley.

Numerical Investigation of Reinforced Concrete Frames Considering Nonlinear Geometric, Material and Time Dependent Effects with Incremental Construction Loadings

Omid Esmaili[a] Siamak Epackachi[a] Rasoul Mirghaderi[a,b]
and Ali A. Taheri Behbahani[b]

[a] School of Civil Engineering, College of Engineering, University of Tehran, Enghelab Avenue, Tehran, Iran
[b] The Committee for Revising the Iranian Code of Practice for Seismic Resistant Design of Buildings, Tehran, Iran

Abstract. Reliable prediction of forces, stresses, and deflections in reinforced concrete structures has long been the goal of structural engineers. In fact analytical determination of the displacements, internal forces, stresses and deformations of reinforced concrete structures throughout their load histories has been complicated by a number of factors. Non homogeneity of the material, continuously changing topology of the structural system due to cracking of concrete under increasing loads, nonlinear stress-strain relationship of concrete and reinforcing steel, variation of concrete deformations due to creep, shrinkage, and applied load history are classified as complicating factors. Due to mentioned difficulties, engineers in the past have been relying heavily on empirical formulas derived from numerous experiments for the design of concrete structures. In this study, it has been tried to determine the real behavior and vulnerability of reinforced concrete frames in the basis of developed numerical nonlinear analytical procedure, considering nonlinear response of the structure. In fact, in order to obtain more accurate results, material and geometric nonlinearities plus time dependent effects of creep and shrinkage in addition to incremental construction loadings are included in the analyses. More over the structural response of the structure is traced through its elastic, inelastic and ultimate load ranges. The analyses are adopted with a finite element displacement formulation coupled with a time step integration solution. Also an incremental and iterative scheme based upon a constant imposed displacement is used so that local instabilities or strain softening can also be analyzed accurately.

Keywords: RC Frames, Geometric Nonlinearity, Material Nonlinearity, Concrete Time Dependency, Construction Sequence Loading.

INTRODUCTION

The study of the influence of the construction process on the short and long term structural response is generally reserved to special or long spans structures such as bridges built by the free cantilever method. However, today's means of lifting and transportation, combined with the use in a same structure of different materials and techniques, such as pre-tensioning and post-tensioning, prefabrication and cast in place construction, allow building of efficient structural systems in which the

CP1020, *2008 Seismic Engineering Conference Commemorating the 1908 Messina and Reggio Calabria Earthquake,*
edited by A. Santini and N. Moraci
© 2008 American Institute of Physics 978-0-7354-0542-4/08/$23.00

evolutionary construction largely influences their structural behavior along their service life.

Examples of the above mentioned structures are bridges composed of long span pre-cast prestressed girders erected simply on supports and subsequently made continuous by casting in place top slabs and closures at the supports. In these and similar cases especially in reinforced concrete structures, the interaction between cracking, creep and shrinkage of concrete and relaxation of the prestressing steel can lead to important instantaneous and time dependent redistributions of internal forces and stresses, making it difficult to accurately estimate the deflections, stresses and cracks width, as required to satisfy the serviceability limit states. Many analytical models have been developed to predict the nonlinear and time dependent structural response of concrete frame structures, such as those proposed by Kang and Scordelis [1], Buckle and Jackson [2], Chan [3], Marí [4], and Ulm [5], based on layer or filament beam element approaches. The effects of segmental construction in the longitudinal structural scheme were included in the models developed by Ghali and Elbadry [6], Kang [7], Abbas and Scordelis [8] and Ketchum [9].

Recently, Murcia and Herkenhoff [10], and Cruz et al. [11], included the effects of segmental construction in both longitudinal and transverse directions for planar frames. General reviews of the mathematical modeling and structural effects of creep, shrinkage and segmental structures have been presented by Bazant [12, 13] and Scordelis [14].

In this paper a numerical model for the nonlinear and time dependent analysis of a three dimensional reinforced and composite concrete frame is presented, which takes into account the effects of longitudinal scheme variations and possible changes on the cross section along the construction process. An illustrative example shows the capabilities of the model to simulate the structural effects of a complex construction process, along with the nonlinear and time dependent materials behavior. The quality of the numerical results is assessed by comparison with those obtained from an experimental model test and current analytical software.

DESCRIPTION OF THE ANALYTICAL MODEL

Material Properties

In order to incorporate the varied material properties within a frame in evaluating element properties, the element is divided into a discrete number of concrete and reinforcing steel filaments, as shown in Fig. 1, each of which is assumed to be in a uniaxial stress state. It is assumed that plane sections remain plane and the deformations due to shearing strains are neglected. The total strain at a given time and point in the structure $\varepsilon(t)$, is taken as the direct sum of mechanical strain $\varepsilon^m(t)$, and non-mechanical strain $\varepsilon^{nm}(t)$, consisting of creep strain $\varepsilon^c(t)$, shrinkage strain $\varepsilon^{sh}(t)$, ageing strain $\varepsilon^a(t)$, and thermal strain $\varepsilon^t(t)$.

$$\varepsilon(t) = \varepsilon^m(t) + \varepsilon^{nm}(t) \tag{1}$$
$$\varepsilon^{nm}(t) = \varepsilon^c(t) + \varepsilon^{sh}(t) + \varepsilon^a(t) + \varepsilon^t(t) \tag{2}$$

921

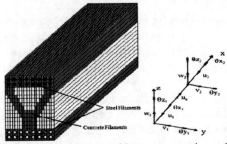

FIGURE 1. A filament beam element with an arbitrary cross section and 13 degrees of freedom

The short time uniaxial stress-strain relationship governing the response of the individual concrete filaments is based on the one suggested by Hognestad [15], including cracking at tensile strength and load reversals. When the mechanical strain of a concrete filament reaches the tensile strain corresponding to the tensile strength, cracking takes place. Then, the tensile stress drop to zero, as shown in Fig. 2.a. Such procedure is implemented taking into account reloading and unloading processes. For the reinforcing steel, a bilinear stress-strain relationship is assumed with load reversals (Fig. 2.b).

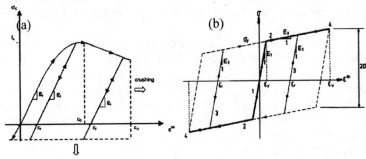

FIGURE 2. Uniaxial stress-strain curve of concrete (a), of reinforcing steel (b).

Age Dependent Integral Formulation for Creep and Shrinkage

Creep strain $\varepsilon^c(t)$ of concrete is evaluated by an age dependent integral formulation based on the principle of superposition. Thus,

$$\varepsilon^c(t) = \int_0^t C(\tau, t - \tau) \frac{\delta\sigma(\tau)}{\delta\tau} d\tau \qquad (3)$$

Where $(\tau, t - \tau)$, is the specific creep function, dependent on the age at loading τ, and $\delta\sigma(\tau)$ is the stress applied at instant τ. Numerical creep analysis may be performed by subdividing the total time interval of interest into time intervals Δt, separated by time steps. The integral Eq. (3) can then be approximated by a finite sum involving incremental stress change over the time steps. The adopted form for the specific creep function $C(\tau, t - \tau)$ is a Dirichlet series:

$$C(x) = \sum_{i=1}^m a_i \times \left[1 - e^{-\lambda_i x}\right] \qquad (4)$$

In which the values of m, λ_i and a_i are coefficients to be determined from experimental or empirical creep formulas, as recommended by international codes, by a least square fit. In this work it is considered that sufficient accuracy is obtained using three terms of the series $(m=3)$, and adopting $\lambda_i = 10^{-i}$.

The value of the increment of creep strain $\Delta\varepsilon^c{}_n$ in a time interval Δt_n is obtained by a recursive relation, as follows;

$$\Delta\varepsilon^c{}_n = \sum_{i=1}^{m} A_{i,n}\left[1 - e^{-\lambda_i[\phi(T_{n-1})\Delta t_n]}\right] \tag{5}$$

$$A_{i,n} =$$
$$\Delta\sigma_1 a_i(t_1)e^{-\lambda_i[\phi(T_1)\Delta t_2+\phi(T_2)\Delta t_3+\cdots+\phi(T_{n-2})\Delta t_{n-1}]} +$$
$$\Delta\sigma_2 a_i(t_2)e^{-\lambda_i[\phi(T_2)\Delta t_3+\phi(T_3)\Delta t_4+\cdots+\phi(T_{n-2})\Delta t_{n-1}]} + \cdots +$$
$$\Delta\sigma_{n-2} a_i(t_{n-2})e^{-\lambda_i[\phi(T_{n-2})\Delta t_{n-1}]} + \Delta\sigma_{n-1} a_i(t_{n-1})$$

$$A_{i,n} = A_{i,n-1} \times e^{-\lambda_i[\phi(T_{n-2})\Delta t_{n-1}]} + \Delta\sigma_{n-1} \times a_i(t_{n-1}) \tag{6}$$

$$[A]_{16\times3}[a]_{3\times1} = [C]_{16\times1}, [A]^T{}_{3\times16}[A]_{16\times3}[a]_{3\times1} = [A]^T{}_{3\times16}[C]_{16\times1}$$

$$\left[[A]^T[A]\right]_{3\times3}[a]_{3\times1} = \left[[A]^T[C]\right]_{3\times1}, [a]_{3\times1} = \left[[A]^T[A]\right]^{-1}_{3\times3}\left[[A]^T[C]\right]_{3\times1} \tag{7}$$

The value of the increment of shrinkage strain δS in a short time interval δt is obtained as follows for a $(2a \times 2b)$ rectangular cross section (Fig. 3);

$$K \times \left[\frac{\delta^2 S}{\delta X^2} + \frac{\delta^2 S}{\delta Y^2}\right] = \frac{\delta S}{\delta t}, \frac{\delta S}{\delta X} = \pm\frac{f}{K}(S_\infty - S) @ X = \pm a, \frac{\delta S}{\delta Y} = \pm\frac{f}{K}(S_\infty - S) @ Y = \pm b$$

$$\frac{S}{S_\infty} = \phi_x + \phi_y - \phi_x \times \phi_y,$$

$$\phi_x = 1 - \sum_{n=1}^{\infty} F_n \times \frac{\cos(\beta_n\frac{X}{a})}{\cos(\beta_n)} \times e^{-T\beta_n^2}, \phi_y = 1 - \sum_{n=1}^{\infty} F_n \times \frac{\cos(\beta_n\frac{Y}{b})}{\cos(\beta_n)} \times e^{-T\beta_n^2} \tag{8}$$

Where $T = \frac{Kt}{a\times b}$, $B_x = \frac{f\times a}{K}$, $B_y = \frac{f\times b}{K}$ and $\beta_{nx} = n^{th}$ root of $\beta \times \tan(\beta) = B_x$, $\beta_{ny} = n^{th}$ root of $\beta \times \tan(\beta) = B_y$, $F_{nx} = \frac{2\times B_x}{B_x^2+B_x+\beta_{nx}^2}$, $F_{ny} = \frac{2\times B_y}{B_y^2+B_y+\beta_{ny}^2}$.

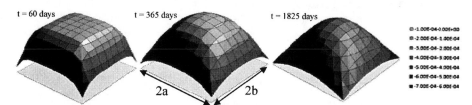

FIGURE 3. Shrinkage strain cross sectional distribution after different elapsed times

To adequately characterize the time dependent concrete properties, the required parameters to fit the curves of specific creep and shrinkage functions, CEB-FIP 1970, CEB-FIP 1978, ACI-209, BAZANT-PANULA (B3) and AS3600 were applied in the analytical model.

A typical three dimensional frame element has a length and a prismatic cross section of arbitrary shape which is assumed to be composed by a number of concrete and steel filaments, as shown in Fig. 1. Each filament is geometrically defined by their area and position with respect to the sectional local axes. An efficient algorithm has been developed to define the filament dimensions, for a given geometry of the cross section. Each filament can be made from a different concrete or steel, making it possible to study the effects of arbitrary distributions of shrinkage or thermal strains (Fig. 3). The elastic stiffness of a beam element is obtained by the following known expression:

$$K_e = \iiint_v B^T.E.B.dv \tag{9}$$

Where E represents the material tangent modulus associated either with flexural or torsional degrees of freedom, and B is the matrix relating nodal displacements and strains at the Gauss points. The beam geometric stiffness is obtained by considering the second order terms in the strain-displacement relationship and following the standard procedure. The geometric stiffness matrix is added to the elastic one, to get a better approach of the element stiffness. The element internal resisting load vector due to stress in the concrete and steel filaments, can be evaluated by:

$$R^i = \iiint_v B^T.\sigma.dv + \int_L B_g.T_x.dx \tag{10}$$

In which T_x is the torque in the beam and B_g is the component of the B matrix associated with the torsional degree of freedom. The equivalent load vector due to non-mechanical strain ε_{nm} is calculated by the equation:

$$R^{nm} = \iiint_v B^T.E.\varepsilon^{nm}dv \tag{11}$$

The material properties of the concrete and steel at any time and load level depend on the nonlinear stress-strain relationship, cracking, yielding and crushing of the concrete and yielding of the steel.

To account for geometric nonlinearity, an updated Lagrangian formulation is used, in which the direction of the local reference system is continuously updated as the structure deforms and gradually constructed. The approach used for the large displacement analysis, is restricted to small strains and small incremental rotations, which is considered adequate for concrete frame structures. Internal forces and stiffness are calculated in local coordinate system and transformed to fixed global coordinate system, where the equilibrium equations for the entire structure are assembled by the direct stiffness method and solved. Thus, the continuously changing displacement transformation matrix for each element takes into account the effect of geometric nonlinearity along with the nonlinear form of the strain displacement relationship.

Solution Strategy for Nonlinear Problems

In order to incorporate the nonlinear and time dependent behavior of a concrete structure, the time domain is divided into time intervals and a step forward integration is performed in which increments of displacements and strains are successively added to the previous total as we march forward in the time domain. A number of construction steps are defined along the time domain. A construction step is a situation of the structure in which any variation on its geometry, loading or boundary conditions takes place. For each construction step, external joint loads, temperature distributions, short time stress-strain curves, time dependent materials properties, internal and external boundary conditions and also structure scheme are given.

The time elapsed from one construction step to the next is subdivided into time intervals, separated by time steps. At each time step the material properties, the stiffness matrix and the load vector are updated. Increments of non-mechanical strains $\Delta\varepsilon^{nm}$ are evaluated, due to creep and shrinkage of the concrete, and temperature changes occurred during the time interval from t_{n-1} to t_n. The equivalent joint load increments ΔR^{nm} at time t_n, summed for the concrete, reinforcing steel, are then calculated from their respective non-mechanical strain increments $\Delta\varepsilon^{nm}$. Thus, at a time t_n, the load increment ΔR_n to be applied to the structure is obtained by adding the external joint load increment ΔR_n^i and the unbalanced load R_{n-1}^u left over from time t_{n-1} to the equivalent joint load increments ΔR^{nm} due to non mechanical strains

$$\Delta R_n = \Delta R_n^i + \Delta R_n^{nm} + R_{n-1}^u \tag{12}$$

The total load obtained at each time step is divided into load increments, so that the load-displacement curve can be traced along the elastic, cracking, inelastic and ultimate ranges. At each load step, the load vector is updated according to specified factors, which can be different for external loads, imposed deformations and imposed displacements. At each load step, a direct stiffness finite element method based on the displacement formulation is used, in which the resulting equilibrium equations will be nonlinear for the current state of material properties and geometry. An iterative procedure is used in which tangential equilibrium equations are solved for global displacement increments at each iteration. At the beginning of each iteration, all the joint displacements, total strains, total non-mechanical strains and stresses at every point in the structure are known. The increments of strain at any concrete or steel filament are obtained by first transforming the increment of global displacements to element local coordinates and then, by means of the incremental form of the strain displacements equations. Total strains are obtained by adding the increment of strain to the previous total. Mechanical strain is calculated by subtracting non-mechanical strain from total strain, and stress is calculated by the nonlinear stress-strain curve. The internal resisting load vector is obtained from Eq. (10), and the unbalanced load vector is obtained by subtracting it from the current total external load vector.

The present model has been designed to simulate most of the structural changes that take place during the construction process and along the entire service life of structures. At each construction step, changes in the longitudinal or transverse geometry of the structure can be accounted for by adding or removing elements or filaments of the cross section.

External supports and connections between elements can be modified at any time. Each filament of a given element cross section can be made of a different concrete type. For each concrete type the instant of casting and the instant of demolition are specified at the beginning of the process. Similarly, for each steel filament, the instant of placement in the structure and the instant of removal are specified. When a concrete or steel filament is placed in the element, its stiffness is included in the element stiffness; when it is removed, its stiffness and its contribution to the internal force vector is not taken into account any more in the step by step analysis. In this case, an unbalanced load vector will appear in the next time step that is automatically introduced in the iterative procedure until equilibrium is obtained. Variations in the external boundary conditions are also recognized in the model. Due to the incremental procedure used, no special problems are associated when a new support is introduced.

The model has been implemented into a computer program called CONSCREEP, written in FORTRAN language, which can be run on personal computers. The general primary input data include geometry of the structure, finite element mesh, support conditions, materials properties, reinforcement layout, construction steps, environmental conditions, convergence criteria and output control information. At each construction step secondary input data includes variations on the geometry, boundary conditions, loading, as well as the number of time intervals between construction steps and load factors for load steps.

NUMERICAL EXAMPLES

Washa-Fluck beam is presented to show the capabilities of the analytical model to reproduce the effects of the structural behavior of a 2 bay simple reinforced concrete beam under sustained uniformly distributed load in which cracking, creep and shrinkage of concrete take place (Fig. 4). The analytical results are compared with those obtained from an experimental full scale model tested by Washa and Fluck, 1956 [16].

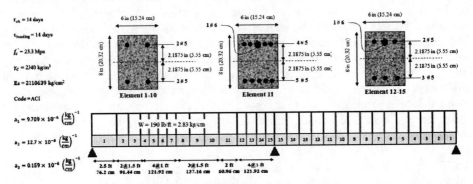

FIGURE 4. Washa-Fluck beam properties

Table 1 shows the typical results of the beam. Generally, good agreement is obtained from the results [17].

TABLE 1. Washa-Fluck beam results

Displacement at Node 7 (cm)				Strains at Maximume Positive Moment (x 0.0001)				
					Compressive Steel Level		Tensile Steel Level	
	at Loading	after 6 months	after 2.5 years		at Loading	after 2.5 years	at Loading	after 2.5 years
Exp.	1.422	2.692	2.896	Exp.	-2.20	-6.70	4.10	5.10
Kang (Method)	1.499	2.896	2.997	Kang (Method)	-1.40	-6.70	5.65	6.50
CONSCREEP	1.482	2.907	3.06	CONSCREEP	-2.83	-6.14	5.97	6.36

Vertical Reaction at Middle Support (kg)				Strains at Maximume Negative Moment (x 0.0001)				
					Compressive Steel Level		Tensile Steel Level	
	at Loading	after 6 months	after 2.5 years		at Loading	after 2.5 years	at Loading	after 2.5 years
Exp.	2214	2263	2250	Exp.	-3.90	-9.30	7.00	8.00
Kang (Method)	2190	2210	2210	Kang (Method)	-2.47	-8.59	6.78	8.00
CONSCREEP	2198	2214	2216	CONSCREEP	-4.29	-7.26	7.26	7.90

926

A 10 Story 2D Reinforced Concrete Frame

Another numerical example is presented to show the capabilities of the analytical model to reproduce the effects of a complex construction process on the structural behavior of a 10 story 2D reinforced concrete structure in which cracking, creep and shrinkage of concrete and take place. The analytical results are compared with those obtained from traditional analytical model (Fig. 5) [18]. Figure 5 also shows the applied load history and material properties due to time dependency of concrete.

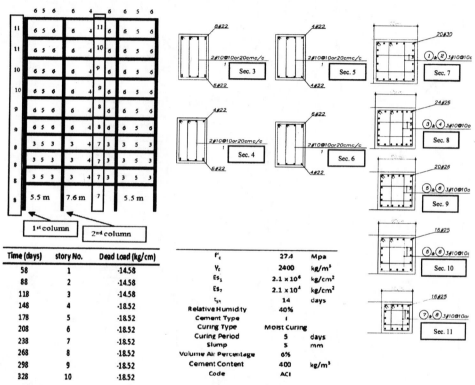

Time (days)	story No.	Dead Load (kg/cm)
58	1	-14.58
88	2	-14.58
118	3	-14.58
148	4	-18.52
178	5	-18.52
208	6	-18.52
238	7	-18.52
268	8	-18.52
298	9	-18.52
328	10	-18.52

f'c	27.4	Mpa
γc	2400	kg/m³
Es₁	2.1 x 10⁶	kg/cm²
Es₇	2.1 x 10⁴	kg/cm²
t₀ₕ	14	days
Relative Humidity	40%	
Cement Type	I	
Curing Type	Moist Curing	
Curing Period	5	days
Slump	5	mm
Volume Air Percentage	6%	
Cement Content	400	kg/m³
Code	ACI	

FIGURE 5. A 10 story 2D reinforced concrete frame, applied load history and material properties

CONCLUSION

A numerical model for the nonlinear and time dependent analysis of three-dimensional reinforced and prestressed concrete frames segmentally constructed has been presented. The structural effects of the load and temperature histories, materials nonlinear behavior, creep, shrinkage, ageing of concrete are taken into account, as well as the nonlinear geometric effects. Possible changes on the structural geometry, boundary conditions and loading at any time are also recognized, making it possible to simulate most of the current construction methods of buildings.

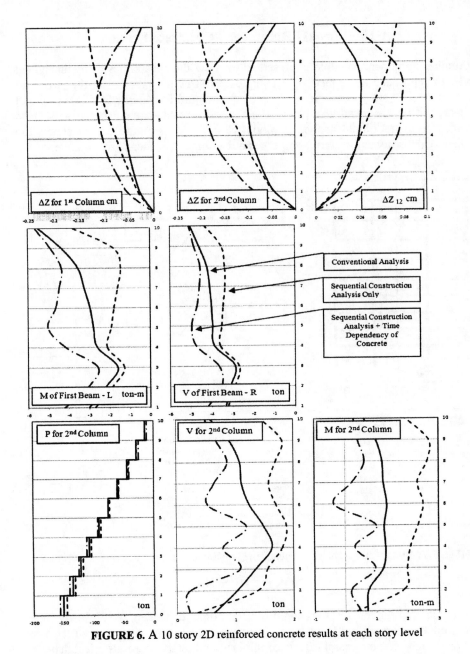

FIGURE 6. A 10 story 2D reinforced concrete results at each story level

A realistic formulation of concrete creep, especially suited for structures subjected to relaxation of stresses, such as those segmentally erected, is utilized in the time dependent analysis. A numerical example has been presented in which the analytical model has been used to predict the behavior during construction and under permanent loads after 2.5 years of an experimental full scale model test, in which the evolutionary construction largely influences the structural response. Another model has shown how to constitute a powerful tool to be used either at the design or construction stages or for the assessment, evaluation, retrofit or demolition of concrete structures.

REFERENCES

1. Kang YJ, Scordelis AC. Nonlinear analysis of prestressed concrete frames. Journal of Structural Division, ASCE 1980; 106:445–62.
2. Buckle IG, Jackson AT. A filamented beam element for the nonlinear analysis of reinforced concrete shells with edge beams. Department of Civil Engineering, University of Auckland, New Zealand, 1981.
3. Chan EC. Nonlinear geometric, material and time dependent analysis of reinforced concrete shells with edge beams, Report UCB-SESM 82/8. University of California at Berkeley, December 1982.
4. Marı́ AR. Nonlinear geometric, material and time dependent analysis of three dimensional reinforced and prestressed concrete frames, Report UCB-SESM-84/12. Berkeley (USA): University of California, June 1984.
5. Ulm FJ, Clement JL, Guggen berger J. Recent advances in 3-D nonlinear FE-analysis of R/C and P/C beam structures. Proc. ASCE Structures Congress XII, Atlanta (GA), New York, 1994, pp. 427–1433.
6. Ghali A, Elbadry MM. User's manual and computer program CPF: cracked plane frames in prestressed concrete. Department of Civil Engineering, The University of Calgary, Research Report CE 85-2, Calgary (Alberta, Canada), 1985.
7. Kang YJ, Scordelis AC. Non-linear segmental analysis of reinforced and prestressed concrete bridges. 3rd International Conference on Short and Medium Span Bridges. Toronto (Ontario, Canada), August 6 1990:229–40.
8. Abbas S, Scordelis AC. Nonlinear geometric, material and time dependent analysis of segmentally erected three-dimensional cable stayed bridges, Report UCB/SEMM-93/09. University of California at Berkeley: 1993.
9. Ketchum MA. Redistribution of stresses in segmentally erected prestressed concrete bridges, Report UCB-SESM 86/07. University of California at Berkeley: 1986.
10. Murcia J, Herkenhoff L. Time dependent analysis of continuous bridges composed by precast elements. Hormigo'n y Acero, 1994; 192:55–71.
11. Cruz PJS, Marı́ AR, Roca P. Nonlinear time-dependent analysis of segmentally constructed structures. J Structural Engineering ASCE 1998;124(3):278–87.
12. Bazant ZP, Wittmann FH, editors. Creep and shrinkage in concrete structures. New York: John Wiley and Sons, 1982.
13. Bazant ZP, editor. Mathematical modeling of creep and shrinkage of concrete. John Wiley and Sons: 1988.
14. Scordelis AC. Recent developments at Berkeley in nonlinear analysis of prestressed concrete structures. Proceedings of the International Federation for Prestressing (FIP) Symposium, Jerusalem (Israel). September 1988:369–76.
15. Hognestad E. A study of combined bending and axial load in reinforced concrete members. University of Illinois Eng. Experimental Station, Bulletin Series No. 399, Bulletin No. 1, November 1951.
16. Washa, G.W., Fluck, P.G., Plastic Flow (Creep) of Reinforced Concrete Continuous Beams, ACI Journal, Proc. Vol. 52, January 1956.
17. Epackachi, S., Consideration of the Effects of Time - Dependent Parameters Such as Creep and Shrinkage on the Behavior of Reinforced Concrete Structures, University of Tehran, College of Engineering, MSc. Thesis Project, September 2007.
18. Esmaili, O., Considering the Sequential Construction Methods of Reinforced Concrete High-rise Buildings with Material Time-Dependency Effects, University of Tehran, College of Engineering, MSc. Thesis Project, February 2008.

Seismic Assessment of R/C Building Structure through Nonlinear Probabilistic Analysis with High-performance Computing

M. Faggella[a,b], A. Barbosa[a], J. P. Conte[a], E. Spacone[b], J. I. Restrepo[a]

[a]Department of Structural Engineering University of California, San Diego
9500 Gilman Drive, MC 0085 La Jolla, CA, USA, 92093-0085- Fax: 001 858 822-2260.
[b]Department PRICOS. University of Chieti-Pescara, Pescara, Italy

Abstract. This paper presents a probabilistic seismic demand analysis of a three dimensional R/C building model subjected to tri-axial earthquake excitation. Realistic probability distributions are assumed for the main structural and material properties and for the ground motion Intensity Measure (IM) Sa(T$_1$). Natural ground motions are used in the analyses to represent the inherent randomness in the earthquake ground motion time histories. Monte Carlo simulations are performed to account for the record-to-record variability and Tornado diagrams are used to represent the uncertainty induced in the response by the basic uncertainties in the structural properties. In order to perform a probabilistic study on three-dimensional engineering demand parameters (EDPs), a large number of ensemble time history analyses were carried out using the TeraGrid high-performance computing resources available at the San Diego Supercomputer Center. Early results show that for the testbed building used in this study, uncertainty in the structural parameters contribute little to the uncertainty of the EDPs, while large variations in the EDPs are due to the variability of the ground motion intensity measure and the record-to-record variability.

Keywords: Nonlinear analysis, seismic response, probability analysis, fiber beam-column elements, Monte Carlo simulation, Tornado sensitivity analysis, Parallel computing, TeraGrid.

INTRODUCTION

Performance based engineering approaches are based on the probabilistic treatment of demand uncertainties accounting for the propagation of inherent uncertainties to the response parameters through the structural model. Recently, a Performance Based Earthquake Engineering methodology (PBEE) has been developed by the Pacific Earthquake Engineering Research Center (PEER) (Porter et. al 2002). In some cases, the methodology has been used to investigate the structural demand sensitivity to input parameters and ground motions (Lee and Mosalam 2006).

The present work focuses on the seismic performance assessment and demand sensitivity analysis of a three-dimensional reinforced concrete (R/C) framed building structure. A suite of natural ground motion recordings is used in the probabilistic analysis. Structural properties, ground motion time history and ground motion intensity measure are treated as random variables. Uncertainties in the input variables propagate through nonlinear time history analysis based on the finite element frame model into uncertainty in the response or engineering demand parameters. A

CP1020, *2008 Seismic Engineering Conference Commemorating the 1908 Messina and Reggio Calabria Earthquake*,
edited by A. Santini and N. Moraci

sensitivity analysis method, referred to as tornado analysis, is used to investigate the dispersion in the engineering demand parameters (EDPs) using a three-dimensional nonlinear finite element frame model under three directional seismic excitation.

BASIC RANDOM VARIABLES FOR 4 STORY BUILDING

The testbed structure used in this study is an existing four-story R/C building in Bonefro, Italy, whose skeletal model is shown in Figure 1. A nonlinear 3D frame model of the building structure was created using the structural analysis software framework OpenSees (McKenna 1997). The frame is modelled using force-based nonlinear fiber-section frame elements for beams and columns (Spacone et al. 1996), while the presence of the infills is neglected at this stage.

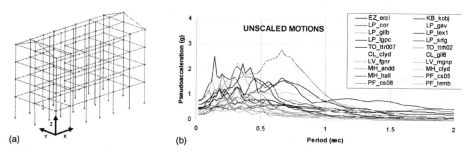

FIGURE 1. (a) Building model and (b) elastic response spectra of selected unscaled natural ground motions.

In this study, both the seismic input and the structural parameters are treated as random variables. Ground motion uncertainties include both the ground motion Intensity Measure (IM) and the details of the earthquake ground motion time history. In the PBEE methodology, different levels of seismic intensity are used assuming as IM the spectral ordinate $S_a(T_1)$ of the elastic spectrum at the fundamental period (T_1) of the structure. The ground motions used here are those selected within the PEER testbed program for the building site at the University of California, Berkeley (Somerville 2001). The fundamental period T_1 of the building in the bare frame configuration is 1.07sec (it corresponds to the first sway mode in the transversal/short direction). Therefore, the 20 ground motions were scaled so as to yield the same spectral acceleration at T_1. According to this methodology, the scaling is made only in one direction, which in this case corresponds to the building transversal direction. The seismic hazard analysis for the Berkeley site provides uniform hazard spectra for three hazard levels with return periods of 72 years, 475 years, and 2475 years, respectively. These uniform hazard spectra are provided for a discrete number of periods, as shown in Figure 2. Intermediate values are obtained through log-linear interpolation. These return periods correspond to a probability of exceedance of 50% in 50 years, 10% in 50 years, and 2% in 50 years, respectively. These three levels of $S_a(T_1)$ are then least-square fitted with a lognormal distribution. Based on this probability distribution for IM = $S_a(T_1)$, the intensity measure can be perturbed for the purpose of sensitivity analysis in order to scale the selected ground motions at different intensity measures. The ground motions used in the analysis presented here are scaled for an IM with a probability of exceedance of 50% in 50 years.

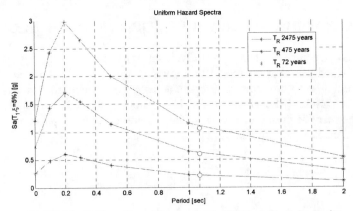

FIGURE 2. Uniform hazard spectra for the UC Berkeley Science site.

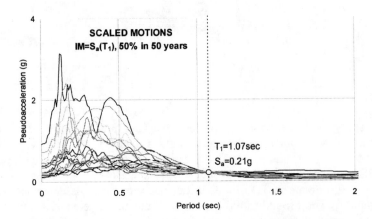

FIGURE 3. Elastic response spectra of scaled natural ground motions.

Uncertainty in the ground motion time history is difficult to represent, unlike uncertainties in the intensity measure and in structural properties which can be modeled as single random variables. Therefore, the record-to-record variability is taken into account by performing Monte Carlo simulation based on an ensemble of 20 tri-axial records. The approach used here for sensitivity analysis was first used by Lee and Mosalam (2006) for a 2D shear wall frame structure.

Realistic probability distributions are assumed for structural properties such as building mass, structural damping, concrete and steel strength, concrete and steel stiffness. Measures of concrete strength from 36 concrete cube samples are used in this study and fitted with a normal probability distribution. For the steel rebars, only 4 strength measures are available, therefore a lognormal distribution is used with mean value equal to the average steel strength of the four samples, and a 10% coefficient of variation (consistent with the literature).

A normal probability distribution is used for the concrete elastic modulus, with mean value corresponding to the concrete mean strength (through an empirical relation between elastic modulus and compressive strength), and a coefficient of variation of 8% (consistent with the literature). A normal distribution is assumed for the steel Young's modulus, with a mean elastic modulus of 210 GPa and a coefficient of variation of 3.3%. Table 1 summarizes the probability distributions and characteristics of the input random variables.

TABLE 1. Distributions adopted for the basic random variables of seismic action and structural properties, and mean value (X_m) and coefficient of variation (COV).

	GM	$S_a(g)$	Damping	Mass (ton/m²)	F_y (MPa)	F_c (MPa)	E_s (GPa)	E_c (GPa)
Distrib.	MCS	Logn.	Norm.	Norm.	Logn.	Norm.	Norm.	Norm.
X_m	On EDP	0.2931	0.03	0.87	451	25.8077	210	28.956
COV %		84	40	10	10	6.4	3.3	8

MONTE CARLO SIMULATION AND "TORNADO" DIAGRAMS

INPUT
Probability distribution of Variabile Xi

NL FE ANALYSIS

OUTPUT
Probability distribution of EDP j

X_{LO} X_M X_{HI}

$EDP(X_{LO})$ $EDP(X_M)$ $EDP(X_{HI})$

FIGURE 4. Scheme of the propagation of uncertainties from input random variables to EDPs through deterministic nonlinear FE analysis.

Figure 4 schematically represents the procedure for determining the sensitivity of a given EDP to a selected input random variable. In this study, the deterministic function/mapping between input parameters and EDPs is represented by the nonlinear finite element model analysis procedure. Uncertainty in the basic random variables is propagated through nonlinear FE analysis to the uncertainty of the EDPs. The sensitivity of an EDP to an input variable is computed based on the deterministic function evaluated at different fractile values of this input variable. In the Tornado analysis, distances between two given fractiles (low and high) of each random variable are transformed into "swings" of EDPs. In this study, the swings in the EDPs are computed based on the response obtained using the 10% and 90% fractiles of the input random variable of interest. The deterministic nonlinear FE analysis is performed by perturbing one random variable at a time about their median value.

The number of FE runs strictly necessary for the Tornado analysis depends on the number of both the number of input parameters and EDPs of interest. The initial 20 FE runs for the Monte Carlo simulation with all random variables set at their median values are performed regardless of the number of EDPs of interest. The median ground motion for each EDP is then determined after sorting the results of the Monte Carlo simulation. In this study, it is assumed that the median ground motion for a given EDP is the one that produces the 11th absolute value of that EDP as in (Lee and Mosalam 2006). This sensitivity analysis is based on a mixed use of Monte Carlo simulation and Tornado sensitivity analysis. In this particular study, in order to study the probabilistic response using multiple global and local "three-dimensional" EDPs, a large number of nonlinear response history analyses must be carried out. Therefore a nonlinear dynamic analysis is run for each ground motion and for each of the three fractiles (10%, 50% and 90%) of each single random variable. A total of 300 analyses are assigned one per cpu of the TeraGrid cluster at the San Diego SuperComputer Center, making use of the Multiple Parallel OpenSees Interpreter (McKenna et al. 2007). The use of parallel supercomputers leads to a considerable reduction in computation time.

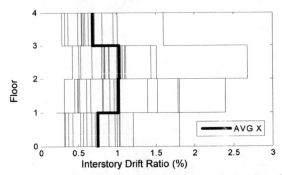

FIGURE 5. Monte Carlo simulation using 20 tri-axial ground motions. All random variables are set at their median values, including IM. Envelopes of the X (longitudinal) component of the Interstory drift ratios from each time history.

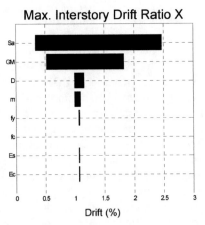

FIGURE 6. Tornado diagram of the EDP defined as the maximum overall interstory drift ratio IDR in the X (longitudinal) direction.

934

The envelopes of the interstory drift ratio (IDR) in the X-direction for each of the 20 ground motions of the Monte Carlo Simulation show a significant scatter of the IDR (see Figure 5). The results presented here correspond to ground motions scaled to the median value of IM, which has a probability of exceedance of 50% in 50 years. It is important to point out that the IM considered here corresponds to the horizontal ground motion component in the Y- (transversal) direction and that for each tri-axial ground motion, the same scaling factor is applied to each of the three components.

Finally, the Tornado diagrams can be represented for each of the selected EDPs. In Figure 6, the Tornado diagram for the maximum (over all stories) IDR in the X- (longitudinal) direction shows that the ground motion Intensity Measure $S_a(T_1)$ and record-to-record variability (GM) are the predominant sources of uncertainty. The damping and the mass produce small swings in the maximum IDR, while the remaining structural properties induce very small swings in the maximum IDR.

CONCLUSIONS

The seismic performance of an existing RC building structure is assessed through nonlinear ensemble time history analyses on a 3D model based on state-of-the-art nonlinear force-based fiber-section frame elements. The probabilistic response analysis enables to study the effects and the relative importance of the basic sources of uncertainty. In order to perform the large number of analyses required to follow the probabilistic approach on a three-dimensional model with multiple "three-dimensional" global and local EDPs, the parametric analyses were performed using the Multiple Parallel version of the OpenSees computational platform recently developed within NEES and the TeraGrid cluster at the University of California at San Diego.

Initial results of the tornado analyses show that, for the testbed building used in this study, structural properties contribute little to the response uncertainty in terms of interstory drift. The ground motin intensity measure and record-to-record variability are the predominant sources of uncertainty and therefore deserve the most attention.

ACKNOWLEDGMENTS

The authors would like to acknowledge the support of NEESit and of Dr. Frank McKenna for the use of the TeraGrid supercomputing resource and of the Opensees Multiple Parallel Interpreter. Partial financial support from the ReLUIS program of the Italian Civil Protection Agency is also acknowledged.

REFERENCES

1. L. D. Decanini, A. De Sortis, A. Goretti, L. Liberatore, F. Mollaioli, and P. Bazzurro, 2004, "Performance of Reinforced Concrete Buildings During the 2002 Molise, Italy, Earthquake". *Earthquake Spectra*, Volume 20, Issue S1 pp. S221-S255..

2. Lee, T.-H. and K.M. Mosalam, 2006, "Seismic Demand Sensitivity of Reinforced Concrete Shear-Wall Building Using FOSM Method," Earthquake Engineering and Structural Dynam-ics, Vol. 34, No. 14 , pp. 1719-1736.
3. McKenna F., 1997, Object-oriented finite element analysis: frameworks for analysis, algo-rithms and parallel computing. Ph.D. dissertation, University of California, Berkeley..
4. McKenna F, Fenves GL., 1997, Using the OpenSees Interpreter on Parallel Computers. NEESit. TN-2007-16.
5. Porter, K.A., 2003. "An overview of PEER's performance-based earthquake engineering methodology." Proc. Ninth International Conference on Applications of Statistics and Prob-ability in Civil Engineering (ICASP9), Volume 2, pp. 973-980, San Francisco, CA, USA, July 6-9.
6. Somerville, P.G., 2001, Ground Motion Time Histories for the UC Lab Building, URS Corporation, Pasadena, CA.
7. Spacone, E., Filippou, F.C. and Taucer, F., 1996, Fiber Beam-Column Model for Nonlinear Analysis of R/C Frames: I. Formulation, International Journal of Earthquake Engineering and Structural Dynamics, Vol. 25, No. 7, pp. 711-725.

Estimation of Characteristic Period for Energy Based Seismic Design

Baykal Hancıoğlu, Zekeriya Polat and Murat Serdar Kırçıl

Yıldız Technical University, Department of Civil Engineering, Beşiktaş 34349 Istanbul-Turkey

Abstract. Estimation of input energy using approximate methods has been always a considerable research topic of energy based seismic design. Therefore several approaches have been proposed by many researchers to estimate the energy input to SDOF systems in the last decades. The characteristic period is the key parameter of most of these approaches and it is defined as the period at which the peak value of the input energy occurs. In this study an equation is proposed for estimating the characteristic period considering an extensive earthquake ground motion database which includes a total of 268 far-field records, two horizontal components from 134 recording stations located on both soft and firm soil sites. For this purpose statistical regression analyses are performed to develop an equation in terms of a number of structural parameters, and it is found that the developed equation yields satisfactory results comparing the characteristic periods calculated from time history analyses of SDOF systems.

Keywords: predominant period, characteristic period, seismic energy

INTRODUCTION

Estimation of input energy using approximate methods has been always a considerable research topic of energy based seismic design. For SDOF systems, input energy spectrum can be investigated by dividing the spectrum into two characteristic parts (Fig.1). For the systems which have shorter periods than the period at which the peak value of spectral input energy occurs, an ascending-linear spectral shape can be assumed, while a descending-curved spectral shape can be assumed for the systems which have longer periods. Thus, the period at which the peak value of spectral input energy occurs – which is called *characteristic period* in the remaining part of the paper - is the key parameter in such method of approaches.

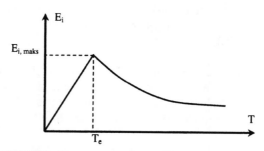

FIGURE 1. Input energy spectrum (drawn schematically).

CP1020, *2008 Seismic Engineering Conference Commemorating the 1908 Messina and Reggio Calabria Earthquake*,
edited by A. Santini and N. Moraci

It is worth to note that the characteristic period can be assumed to coincide with the predominant period of ground motion. Such an approach has also an analytical meaning since it is known that for undamped systems, equivalent input energy velocity spectrum equals to the Fourier amplitude spectrum of the ground acceleration (Eq. 1)[1,2].

$$V_e = |FS(\omega)| = \sqrt{\frac{2E_i}{m}} \qquad (1)$$

However, for a given ground motion, there is no unique characteristic period as it depends on the lateral strength of the system and, to a lesser extent, on the damping of the system [3]. But it is found that the change in lateral strength has not significant affect on characteristic period, thus for the approach proposed in this paper it is neglected and characteristic period is assumed as the period at which the peak value of %5 damped elastic spectral input energy occurs.

Statistical Regression

To obtain a reliable statistical evaluation of characteristic period T_e, regression analyses are carried out considering 268 far-field earthquake ground acceleration records (two horizontal components from 134 recording stations located on both soft and firm soil sites) given in detail in Table 1. The best representative equation which allows estimating T_e is obtained with performing following steps:

- An exponential type formulation is adopted:

$$T_e = a \cdot T_s \cdot e^{b \cdot (T_s / T_1)} \qquad (2)$$

- A statistical regression is performed comparing the values of T_e obtained by time history analyses of %5 damped SDOF systems with those given by the proposed formulation.
- The best coefficients minimizing the standard error are selected.
- The most statistical significant coefficients are tried to be obtained. (p-level<0.05, t-value>2.0)

Where T_1 is the transition period between the acceleration-controlled and velocity-controlled response spectrum and T_s is the period at which the peak value of %5 damped spectral pseudo velocity occurs.

The transition period T_1 can be estimated by considering relation between idealized pseudo-velocity and pseudo-acceleration response spectra (Fig. 2) [4]. It is clear that the peak values of spectral responses do not always occur at the same period. However, the period obtained from Eq. 3 could take close values with characteristic period.

$$T_1 = 2\pi \frac{PSV_{maks}}{PSA_{maks}} \qquad (3)$$

938

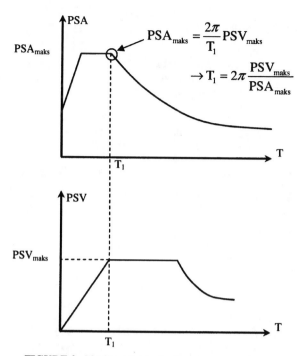

FIGURE 2. Idealized linear elastic response spectra.

TABLE 1. Records used in the regression analysis.

Event		Station name	ID[*1]	M[*2]	R_{epc}[*3]	SC[*4]	Vs30[*5]	Owner
Big Bear-01	1992	Rancho Cucamonga - Deer Can	23598	6.5	69	B	822	CDMG
Chi-Chi, Taiwan	1999	TAP065	99999	7.6	173	B	1023	CWB
		TAP077	99999	7.6	170	B	1023	CWB
		TCU085	99999	7.6	107	B	1000	CWB
		TTN042	99999	7.6	105	B	845	CWB
Chi-Chi, Taiwan-05	1999	TTN042	99999	6.2	92	B	845	CWB
Denali, Alaska	2002	Carlo (temp)	Carl	7.9	68	B	964	ANSS/UA
Irpinia, Italy-01	1980	Arienzo	99999	6.9	77	B	1000	ENEL
Loma Prieta	1989	Piedmont Jr High	58338	6.9	92	B	895	CDMG
		Point Bonita	58043	6.9	104	B	1316	CDMG
		SF - Pacific Heights	58131	6.9	96	B	1250	CDMG
		SF - Rincon Hill	58151	6.9	94	B	873	CDMG
		So. San Francisco, Sierra	58539	6.9	84	B	1021	CDMG
Morgan Hill	1984	Gilroy Array #1	47379	6.2	39	B	1428	CDMG
Norcia, Italy	1979	Bevagna	99999	5.9	36	B	1000	ENEL
Northridge-01	1994	Anacapa Island	25169	6.7	77	B	822	CDMG
		Antelope Buttes	24310	6.7	64	B	822	CDMG

Continued.

TABLE 1. *Continued.*

Event		Station name	ID[*1]	M[*2]	R$_{epc}$[*3]	SC[*4]	Vs30[*5]	Owner
Northridge-01	1994	Lake Hughes #4 - Camp Mend	24469	6.7	50	B	822	CDMG
		Littlerock - Brainard Can	23595	6.7	61	B	822	CDMG
		Mt Wilson - CIT Seis Sta	24399	6.7	46	B	822	CDMG
		Rancho Cucamonga - Deer Can	23598	6.7	90	B	822	CDMG
		Sandberg - Bald Mtn	24644	6.7	62	B	822	CDMG
		Vasquez Rocks Park	24047	6.7	38	B	996	CDMG
		Wrightwood - Jackson Flat	23590	6.7	78	B	822	CDMG
San Fernando	1971	Pasadena - Old Seismo Lab	266	6.6	39	B	969	USGS
Sierra Madre	1991	Vasquez Rocks Park	24047	5.6	40	B	996	CDMG
Whittier Narrows-01	1987	LA - Wonderland Ave	90017	6.0	28	B	1223	USC
		Vasquez Rocks Park	24047	6.0	54	B	996	CDMG
Big Bear-01	1992	Newport Bch	13160	6.5	118	C	405	CDMG
Chi-Chi, Taiwan	1999	HWA029	99999	7.6	77	C	614	CWB
		HWA038	99999	7.6	69	C	643	CWB
		HWA046	99999	7.6	88	C	618	CWB
		ILA031	99999	7.6	132	C	649	CWB
		KAU012	99999	7.6	117	C	474	CWB
		TAP035	99999	7.6	140	C	438	CWB
		TAP052	99999	7.6	148	C	474	CWB
		TAP075	99999	7.6	160	C	553	CWB
		TTN025	99999	7.6	108	C	705	CWB
		TTN032	99999	7.6	90	C	474	CWB
		TTN044	99999	7.6	100	C	474	CWB
		TTN046	99999	7.6	107	C	474	CWB
Drama, Greece	1985	Kavala	99999	5.2	47	C	660	ITSAK
Irpinia, Italy-01	1980	Torre Del Greco	99999	6.9	80	C	660	ENEL
		Tricarico	99999	6.9	72	C	460	ENEL
Kern County	1952	Pasadena - CIT Athenaeum	80053	7.4	126	C	415	CIT
		Santa Barbara Courthouse	283	7.4	88	C	515	USGS
Landers	1992	Arcadia - Campus Dr	90093	7.3	148	C	368	USC
		Glendale-Las Palmas	90063	7.3	165	C	446	USC
		Glendora-N Oakbank	90065	7.3	133	C	446	USC
		LA - Fletcher Dr	90034	7.3	167	C	446	USC
		La Habra - Briarcliff	90074	7.3	145	C	361	USC
		Puerta La Cruz	12168	7.3	100	C	371	CDMG
Landers	1992	Puerta La Cruz	12168	7.3	100	C	371	CDMG
Loma Prieta	1989	Berkeley LBL	58471	6.9	98	C	597	CDMG
		Hayward - BART Sta	58498	6.9	72	C	371	CDMG
		SF - Cliff House	58132	6.9	99	C	713	CDMG

Continued.

940

TABLE 1. *Continued.*

Event		Station name	ID[*1]	M[*2]	R$_{epc}$[*3]	SC[*4]	Vs30[*5]	Owner
Loma Prieta	1989	SF - Diamond Heights	58130	6.9	92	C	583	CDMG
		SF - Presidio	58222	6.9	98	C	594	CDMG
		SF - Telegraph Hill	58133	6.9	97	C	713	CDMG
		Sunol - Forest Fire Station	1688	6.9	62	C	401	USGS
N. Palm Springs	1986	Anza - Tule Canyon	5231	6.1	60	C	685	USGS
		Murrieta Hot Springs	13198	6.1	66	C	685	CDMG
		Puerta La Cruz	12168	6.1	76	C	371	CDMG
		Temecula - 6th & Mercedes	13172	6.1	75	C	371	CDMG
Northridge-01	1994	Glendora - N Oakbank	90065	6.7	62	C	446	USC
		Huntington Beach - Lake St	13197	6.7	79	C	371	CDMG
		Newport Bch - Irvine Ave. F.S	13160	6.7	88	C	405	CDMG
		Newport Bch - Newp & Coast	13610	6.7	87	C	371	CDMG
		Palmdale - Hwy 14 & Palmdale	24521	6.7	57	C	552	CDMG
		Rancho Palos Verdes - Hawth	14404	6.7	53	C	478	CDMG
		Rancho Palos Verdes - Luconia	90044	6.7	56	C	509	USC
		Riverside Airport	13123	6.7	106	C	371	CDMG
		Seal Beach - Office Bldg	14578	6.7	66	C	371	CDMG
San Fernando	1971	Upland - San Antonio Dam	287	6.6	75	C	446	ACOE
		Wrightwood - 6074 Park Dr	290	6.6	72	C	486	USGS
Whittier Narrows-01	1987	Castaic - Old Ridge Route	24278	6.0	77	C	450	CDMG
		Huntington Beach - Lake St	13197	6.0	44	C	371	CDMG
		Leona Valley #5 - Ritter	24055	6.0	63	C	446	CDMG
		Malibu - Las Flores Canyon	90050	6.0	51	C	623	USC
		Moorpark - Fire Sta	24283	6.0	78	C	405	CDMG
		Pacific Palisades - Sunset	90049	6.0	44	C	446	USC
Chi-Chi, Taiwan	1999	CHY065	99999	7.6	116	D	273	CWB
		KAU085	99999	7.6	119	D	261	CWB
		TAP026	99999	7.6	147	D	215	CWB
		TAP090	99999	7.6	156	D	324	CWB
		TAP095	99999	7.6	158	D	215	CWB
Dinar, Turkey	1995	Cardak	99999	6.4	50	D	339	ERD
Friuli, Italy-01	1976	Conegliano	8005	6.5	90	D	275	
Imp. Valley-06	1979	Coachella Canal #4	5066	6.5	84	D	345	USGS
Irpinia, Italy-01	1980	Bovino	99999	6.9	52	D	275	ENEL

Continued.

TABLE 1. *Continued.*

Event		Station name	ID[*1]	M[*2]	R$_{epc}$[*3]	SC[*4]	Vs30[*5]	Owner
Irpinia, Italy-02	1980	Mercato San Severino	99999	6.2	48	D	350	ENEL
Kern County	1952	LA-Hollywood S.FF	24303	7.4	118	D	316	CDMG
Kobe, Japan	1995	HIK	99999	6.9	136	D	256	
Kocaeli, Turkey	1999	Atakoy	99999	7.5	100	D	275	ITU
		Botas	99999	7.5	171	D	275	KOERI
		Cekmece	99999	7.5	108	D	346	KOERI
		Fatih	99999	7.5	94	D	339	KOERI
		Zeytinburnu	99999	7.5	95	D	275	ITU
Landers	1992	Amboy	21081	7.3	75	D	271	CDMG
		Boron Fire Station	33083	7.3	143	D	345	CDMG
		Burbank - N Buena Vista	90012	7.3	174	D	271	USC
		Compton - Castlegate St	90078	7.3	166	D	309	USC
		Fort Irwin	24577	7.3	121	D	345	CDMG
		Fountain Valley - Euclid	90002	7.3	149	D	270	USC
		LA - Obregon Park	24400	7.3	162	D	349	CDMG
		LB - Orange Ave	90080	7.3	164	D	270	USC
		Lakewood - Del Amo Blvd	90084	7.3	158	D	235	USC
		Pomona - 4th & Locust FF	23525	7.3	122	D	230	CDMG
		San Bernardino - E & Hospitality	23542	7.3	80	D	271	CDMG
		Tarzana - Cedar Hill	24436	7.3	193	D	257	CDMG
Lazio-Abruzzo, Italy	1984	Garigliano-Centrale Nucleare	99999	5.8	51	D	200	ENEL
Loma Prieta	1989	Oakland - Outer Harbor Wharf	58472	6.9	94	D	249	CDMG
		Oakland - Title & Trust	58224	6.9	92	D	306	CDMG
		Olema - Point Reyes Station	68003	6.9	138	D	339	CDMG
		Richmond City Hall	58505	6.9	107	D	260	CDMG
Manjil, Iran	1990	Rudsar	99999	7.4	87	D	275	BHRC
Morgan Hill	1984	Los Banos	56012	6.2	80	D	271	CDMG
		SF Intern. Airport	58223	6.2	71	D	190	CDMG
N. Palm Springs	1986	Anza Fire Station	5160	6.1	50	D	339	USGS
		Colton Interchange - Vault	754	6.1	64	D	275	CDOT
		Indio - Coachella Canal	12026	6.1	53	D	345	CDMG
Northridge-01	1994	Anaheim - W Ball Rd	90088	6.7	70	D	235	USC
		Baldwin Park - N Holly	90069	6.7	55	D	309	USC
		Brea - S Flower Av	90087	6.7	69	D	309	USC
		Featherly Park - Maint	13122	6.7	86	D	309	CDMG
		Hemet - Ryan Airfield	13660	6.7	151	D	339	CDMG
		Huntington Bch - Waikiki	90083	6.7	71	D	235	USC
		San Bernardino	23542	6.7	117	D	271	CDMG
		San Jacinto-CDF Fire Sta	12673	6.7	154	D	271	CDMG

Continued.

TABLE 1. *Continued.*

Event		Station name	ID[*1]	M[*2]	R$_{epc}$[*3]	SC[*4]	Vs30[*5]	Owner
Northridge-01	1994	Tustin - E Sycamore	90089	6.7	86	D	235	USC
		Loma Linda; VA Hospital, North Freefield	5229	6.7	121	D	275	USGS
		Loma Linda; VA Hospital, South Freefield	5229	6.7	121	D	275	USGS
San Fernando	1971	Gormon - Oso Pump Plant	994	6.6	50	D	308	CDWR
Whittier Narrows-01	1987	Lancaster - Med Off FF	24526	6.0	71	D	271	CDMG
		Rosamond - Goode Ranch	24274	6.0	88	D	271	CDMG

[*1] ID: Station ID number
[*2] M: Moment magnitude of earthquake
[*3] R$_{epc}$: Distance from the recording site to epicenter
[*4] SC: NEHRP Site Classification
[*5] Vs30: Average shear wave velocity down to 30m depth (m/s)

Examples of the computation of the periods T_1 and T_s for a ground acceleration record (Gilroy Array #1 ground acceleration record – G01230 horizontal component, Morgan Hill Earthquake, 1984) are shown in Fig. 6. Characteristic period is computed T_e=0.24s by performing time history analysis to SDOF systems while the transition period estimated by Eq. 3 is T_1=0.16s and the period at which the peak value of spectral pseudo velocity occurs is T_s=0.32s.

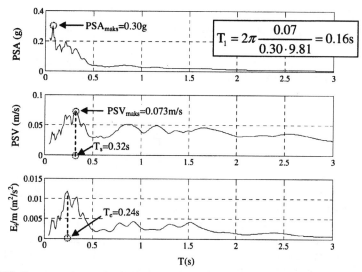

FIGURE 3. Response spectra for Gilroy Array #1 ground acceleration record (G01230 horizontal component, Morgan Hill Earthquake, 1984), ξ=0.05.

As a result of regression analysis, Eq. 4 which estimates T_e is obtained. The proportion of variance accounted for proposed equation is R^2=0.82 and the standard error is SE=0.23. The coefficients are given in Table 2 with their standard errors and the predicted and observed values are drawn in Fig. 4.

$$T_e = 1.23 \cdot T_s \cdot e^{-0.18 \cdot (T_s / T_1)} \qquad (4)$$

TABLE 2. Statistical parameters.

Coefficient	Predicted value	Standard Error	t-value
a	1.23	0.039	31
b	-0.18	0.016	-11

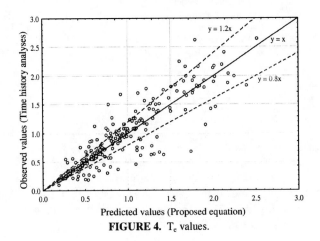

FIGURE 4. T_e values.

Comparison with Different Approaches

Chai et al. [3] assumed that the characteristic period corresponds to the transition period T_1, and they estimated the transition period by Eq. 5 proposed by Vidic et al. [5].

$$T_1 = 2\pi \frac{c_v}{c_a} \frac{PGV}{PGA} \qquad (5)$$

Where c_v corresponds to the ratio of the spectral elastic response velocity to peak ground velocity in the velocity-controlled (medium) period range, and c_a corresponds to the ratio of the spectral elastic response acceleration to peak ground acceleration in the acceleration-controlled (short) period range. Chai et al. [3] assumed c_a and c_v as 2.0 and 2.5, respectively, proposed by Chai et al. [6]. Furthermore, many researchers [3,6,7,8] have estimated seismic energy by assuming that the transition period proposed by Vidic et al [5] can be considered as the characteristic period at which the peak value of input energy occurs. Thus, the proposed equation in this paper is needed to compare with the Eq. 5 proposed by Vidic et al. [5].

Fajfar et al [9] estimated the transition period T_1 by Eq. 6 proposed by Heidebrecht:

$$T_1 = 4.3 \frac{PGV}{PGA} \tag{6}$$

Miranda and Garcia [10] estimated the predominant period of ground motion using the approach proposed by Miranda [11] in which the predominant period of the ground motion is defined as the period at which the peak value of spectral velocity occurs.

In Fig. 5, Fig. 6 and Fig. 7, the value of the transition periods calculated by Eq. 5, Eq. 6 and the T_s periods proposed by Miranda are drawn for all records given in Table.1, respectively, in comparison with the values of characteristic periods obtained from time history analyses. Standard errors for each approaches and for proposed equation in this paper is given in Table. 3.

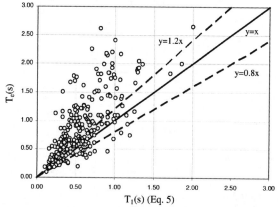

FIGURE 5. T_1 (obtained by Eq. 5) and T_e.

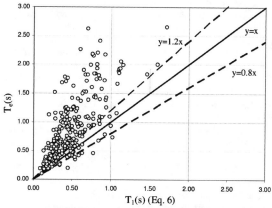

FIGURE 6. T_1 (obtained by Eq. 6) and T_e.

945

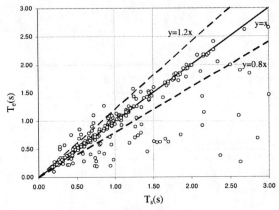

FIGURE 7. T_e vs T_s.

TABLE 3. Standard errors.

	Proposed Eq. 4	Eq. 5	Eq. 6	T_s
Standard Error	0.23	0.38	0.40	0.35

REFERENCES

1. Kuwamura H, Kirino Y, Akiyama W., *Prediction of earthquake energy input from smoothed Fourier amplitude spectrum*, Earthquake Engineering and Structural Dynamics, 1994, 23:1125-1137.
2. Ordaz, M., Huerta, B., Reinoso, E., *Exact computation of input-energy spectra from Fourier amplitude spectra*, Earthquake Engineering and Structural Dynamics, 2003, 32:597-605.
3. Chai, Y.H., Fajfar, P., *Procedure for estimating input energy spectra for seismic design*, Journal of Earthquake Engineering, 2000, 4(4):539-561.
4. Lam, N., Wilson, J., Chandler, A., Hutchinson, G., *Response spectrum modeling for rock sites in low and moderate seismicity regions combining velocity, displacement and acceleration predictions*, Earthquake Engineering and Structural Dynamics, 2000, 29:1491-1525.
5. Vidic, T., Fajfar, P., Fischinger, M., *Consistent inelastic design spectra: Strength and displacement*, Earthquake Engineering and Structural Dynamics, 1994, 23:507-521.
6. Chai, Y.H., Fajfar, P., Romstad, K.M., *Formulation of duration-dependent inelastic seismic design spectrum*, J. Struct. Div., ASCE, 1998, 124(8):913-921.
7. Manfredi, G., *Evaluation of seismic energy demand*, Earthquake Engineering and Structural Dynamics, 2001, 30:485-499.
8. Kunnath, S.K., Chai, Y.H., *Cumulative damage-based inelastic cyclic demand spectrum*, Earthquake Engineering and Structural Dynamics, 2004, 33:499-520.
9. Fajfar, P., Vidic, T., Fischinger, M., *Seismic demand in medium- and long-period structures*, Earthquake Engineering and Structural Dynamics, 1989, 18:1133-1144.
10. Miranda, E., Garcia, J.R., *Influence of stiffness degradation on strength demands of structures built on soft soil sites*, Engineering Structures, 2002, 24:1271-1281.
11. Miranda, E., *Site-dependent strength reduction factors*, J of Structural Engineering, ASCE, 1993, 119(12):3505-19.
12. StatSoft Inc., STATISTICA V.6.0 for Windows. Tulsa, OK, USA; 1995.

Static Nonlinear Analysis In Concrete Structures

Ali Hemmati

Lecturer, Department of Civil Engineering, Islamic Azad University, Semnan Branch, Iran,
ahemmati2000@yahoo.com

Abstract. Push-over analysis is a simple and applied approach which can be used for estimation of demand responses influenced by earthquake stimulations. The analysis is non-linear static analysis of the structure affected under increasing lateral loads and specifying the displacement – load diagram or structure capacity curve, draw the curve the base shear values and lateral deflection on the roof level of the building will be used. However, for estimation of the real behavior of the structure against earthquake, the non-linear dynamic analysis approaches and various accelerographs should be applied. Of course it should be noted that this approach especially in relation with tall buildings is complex and time consuming. In the article, the different patterns of lateral loading in push-over analysis have been compared with non-linear dynamic analysis approach so that the results represented accordingly. The researches indicated the uniformly – distributed loading is closer to real status.
Keywords: Push over analysis, Dynamic nonlinear analysis, Concrete, Buildings.

INTRODUCTION

Push-over analysis means static analysis of the structure under increasing lateral loads. The output of this analysis is displacement – load diagram or in other words the structure capacity curve. To draw the curve the base shear values and the lateral deflection of a reference point (the point at the level of the structure roof) will be used [1]. Designing of the structures to be resistant against earthquake and preserve the residents involving with the natural disaster have been as main targets of engineers and researchers. However in recent years many efficient and rapid paces have been accomplished towards designing the resistant structures against the earthquake. Thus, the general viewpoints of researchers and designers, in providing safety and reliable structures against ruinous loads, have been completely changed. In other word the older and previous approaches upon designing based on force changed to new approaches on the basis of behavior. The new approach is so called the designing based on the performance or performance based design [2].

The reinforced concrete buildings are more concerned because of their most appropriate features such as resistant against fire joint continuity, more economic (cost effective) compare to steel structures. Among the structures, those buildings in which the resistant systems to be medium reinforced concrete flexural frame are more highly approved by designers and engineers, since in addition to open spaces and fulfillment of architecture demands, they have actually high plasticity and the relevant operation are fairly simple. In this paper, the buildings having flexural frame of medium reinforced concrete will be monitored. These buildings have different stories and analyzed upon both analysis approaches i.e push-over and non-linear dynamic. These

CP1020, *2008 Seismic Engineering Conference Commemorating the 1908 Messina and Reggio Calabria Earthquake,*
edited by A. Santini and N. Moraci
© 2008 American Institute of Physics 978-0-7354-0542-4/08/$23.00

models consisting of 2, 5 and 10 story frames included short, medium and tall buildings. These kinds of the buildings, which have several stories, are popular in the country so the derived results can be applied in operations and computations [3].

CHARACTERISTICS OF THE MODELS

In this paper the reinforced concrete buildings upon the system of the medium reinforced flexural frame are investigated. The investigations are focused on a concrete frame which load bearing width is 5m. Figure 1 shows the plan of investigated models and the specified frame.

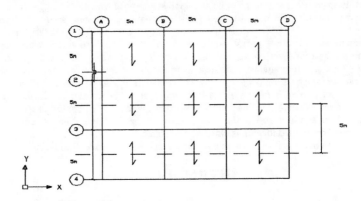

FIGURE 1. Plan of Investigated Models and the Specified Frame

Some of the features of the investigated models are as follows:
- Reinforced concrete building upon medium flexural frame
- Site location: Semnan city, Iran
- Site soil: type 2
- Behavior factor: 7
- Usage: Residential
- Ceiling system: Block Joist
- Dead load of the stories and roof: 550 kg/m^2
- Live load of the stories: 200 kg/m^2
- Live loads of the roof: 150 kg/m^2
- Partition load: 50 kg/m^2
- Concrete compressive strength: 250 kg/m^2
- Shear yield strength of longitudinal steel: 4000 kg/m^2
- Shear yield strength of transverse steel: 240 kg/m^2

NAME OF THE MODELS

Model 1: Loading pattern in push-over analysis is triangular pattern (similar to static approach)

Model 2: Loading pattern in push-over analysis is the point at the roof level pattern

Model 3: Loading pattern in push-over analysis is rectangular pattern

Model 4: Loading pattern in push-over analysis is mode 1 pattern

Model 5: Loading pattern in push-over analysis is exerting of acceleration on the base

The whole loading patterns are applied on 2, 5 and 10 story structures and the derived results represented accordingly. In the columns, plastic hinges type PMM, M3 & P and in the beams, type M3 have been defined [4,5].

RESULTS AND DIAGRAMS

Figures 2, 3, 4 and 5 indicate the capacity curves of 2-storey structure in the different patterns of the loading as triangular, the point at the roof level, rectangular and base acceleration. As the figures indicated the base acceleration loading pattern and rectangular pattern are highly close to each other while the difference is approx. 3%. The patterns triangular and point are fairly close to each other while the difference is approx 10%. Mean while, in 2-storey structure, the loading patterns as base acceleration, rectangular, point and triangular will compute more capacity for the structure respectively.

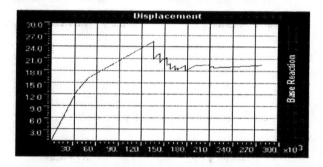

FIGURE 2. Capacity Curve of the 2-Storey Structure (Triangular Pattern)

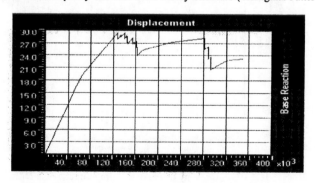

FIGURE 3. Capacity Curve of the 2-Storey Structure (Point Loading Pattern)

949

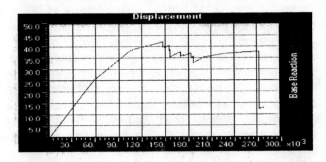

FIGURE 4. Capacity Curve of the 2-Storey Structure (Rectangular Pattern)

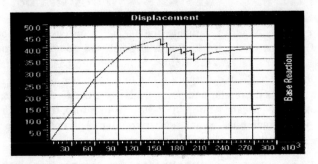

FIGURE 5. Capacity Curve of the 2-Storey Structure (Base Acceleration Pattern)

Figure 6 represents the comparisons of the capacity curves of 2-storey buildings with different loading patterns. As indicated in the figure, the lateral loading patterns of the point and triangular loads are fairly close to each other (in linear range are completely fitted on each other). But the loading patterns, rectangular and base acceleration are also close to each other and indicate more capacity in compare to previous aspects.

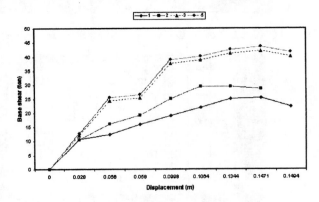

FIGURE 6. Capacity Curve of 2-Storey Structures in Different Loading Patterns

950

In figure 7 the capacity curves of 5-storey structures under various loading patterns have been compared to each other. As indicated in this figure, the lateral loading pattern of point load is completely different from other loading patterns. The triangular loading pattern is far from point load state and is closing to base acceleration and rectangular loading slowly.

Two states of rectangular and base acceleration close extremely each other and indicate more capacity in compare to previous aspects.

In figure 8 the capacity curves of 10-storey structures with various loading patterns have been compared to each other. As indicated in fig.8, the lateral loading pattern of point load is completely different from other loading patterns. The triangular loading pattern is also far from point load state and closes more to rectangular and base acceleration loading patterns.

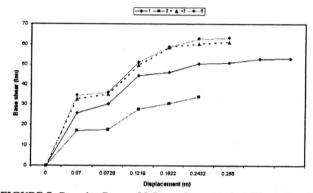

FIGURE 7. Capacity Curve of 5-Storey Structures in Different Loading Patterns

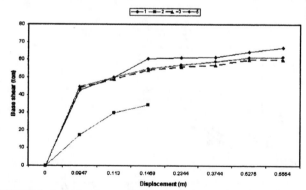

FIGURE 8. Capacity Curve of 10-Storey Structures in Different Loading Patterns

For better evaluation of derived results the procedures of non-linear static, will be used by three earthquake accelerographs. The whole accelerographs regulated coequally on the basis of maximum gravity equal to 0.3g. In the figures 9, 10 and 11 the distribution of maximum drift ratio indicated in 2, 5 and 10 storey buildings.

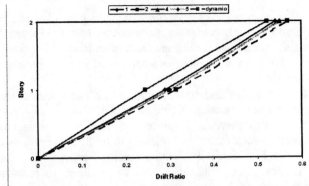

FIGURE 9. Distribution of Maximum Drift Ratio in 2 Storey Building

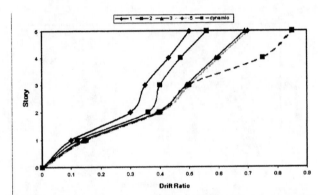

FIGURE 10. Distribution of Maximum Drift Ratio in 5 Storey Building

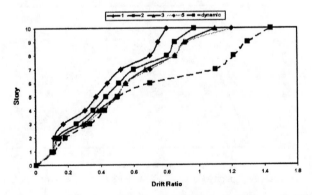

FIGURE 11. Distribution of Maximum Drift Ratio in 10 Storey Building

As indicated in above figures, maximum drift ratio of the structures will be increased proportionally with number of the stories. By increasing the number of the stories, the difference between derived results upon non-linear static and dynamic procedures will also increased. As in model 2-storey the push-over analysis represented approved results and maximum difference on floor level is approx. 5%. In 5-storey structure the derived results of push-over procedure up to the third storey are compatible with those of non-linear dynamic procedure. But from the third storey to higher ones the difference equal to approx. 15% will be produced. In 10-storey structure, the derived results of push-over procedure in one-third of top story's of the structure is low estimate while at roof level, the difference reaches to approx. 20%.

Furthermore, the uniform distribution (rectangular) pattern among other various loading patterns represents accurate responses.

CONCLUSIONS

In 2-storey structure, the loading patterns, base acceleration and rectangular are mostly close to each other and the difference is approx. 3%. The loading patterns, triangular and point are fairly close to each other while the difference is approx. 10%. Furthermore, in 2-storey structure the loading patterns as base acceleration, rectangular, point and triangular will compute higher capacity for the structure respectively. In 5-storey structure, the loading patterns, base acceleration and rectangular are close to each other while the difference is approx. 3%. The triangular loading patterns are fairly close to these patterns and the difference is approx 15%. But loading pattern of point at roof level differs with rectangular state approx. 40%.

Meanwhile in 5-storey structure, loading patterns as base acceleration, rectangular, triangular and point respectively compute more capacity for the structure. In 5-storey structure the amount or rate of difference between those procedures in compare to 2-storey structure increases while the point pattern or procedure accomplish the largest difference.

In 10-storey structure, the patterns, base acceleration and rectangular are mostly close to each other and the difference is approx. 4%. The loading patterns, triangular, are fairly close to these patterns and the difference is approx. 15%. But the loading pattern, point, has difference; at roof level; rectangular state approx. 45% while upon triangular is approx. 35%.

Meanwhile, in 10-storey structure, the loading patterns, base acceleration, rectangular and point, respectively compute more capacities for the structure. In 10-storey structure the difference rate among these procedures in compare to 2-storey one increases while the point procedure represents the highest rate. But its rate is less than 5-storey structure.

By increasing the number of the story's the difference among the derived results of non-linear, static and dynamic, will increase. As in model, 2-storey, the push-over procedure represented approved results while the maximum difference is approx 5% at roof level.

In 5-storey structure the derived results of push-over procedure is compatible with those of non-linear dynamic analysis up to the third storey. But from there upward, the difference will be approx. 15%. In 10-storey structure the derived results of push-over procedure in the upper one-third of structure height is in fact upon low estimation and this difference will be approx 20% at roof level. Furthermore and finally, the uniform distribution (rectangular) among the different loading patterns submits more accurate responses in comparison with other patterns.

REFERENCES

1. A. K. Chopra and R. K. Goel, *Extension of Modal Pushover Analysis to Compute Member Forces*, Earthquake Spectra, 2005, Vol.21, No.1, pp. 125-139.
2. T. Tjhin and M. Aschheim and M. Hernandez and E. Montes, *Observation on the Reliability of Alternative Multiple-Mode Pushover Analysis Methods*, Journal of Structural Engineering, 2006, Vol.132, No.3, pp. 471-477.
3. M. J. N. Priestly, *Performance-Based Seismic Design*, Proceedings of the "12th World Conference on Earthquake Engineering", 2000.
4. FEMA, *Prestandard & Commentary for Rehabilitation of Buildings*, USA, 2000.
5. Applied Technology Council, *Seismic Evaluation and Retrofit of Concrete building*, USA, 1997.

A More Realistic Lateral Load Pattern for Design of Reinforced Concrete Buildings with Moment Frames and Shear Walls

Mahmood Hosseini[a] and Arash Khosahmadi[b]

[a] Associate Professor, Structural Engineering Research Center, International Institute of Earthquake Engineering and Seismology (IIEES), P.O. Box 19395/3913, Tehran, Iran, Email: hosseini@iiees.ac.ir

[b] Graduate Student, Earthquake Engineering Department, Engineering School, Science and Research Branch of the Islamic Azad University (IAU), P.O. Box 14155/4933, Tehran, Iran, Email: a_khoshamadi@yahoo.com

Abstract. In this research it has been tried to find a more realistic distribution pattern for the seismic load in reinforced concrete (R/C) buildings, having moment frames with shear walls as their lateral resisting system, by using Nonlinear Time History Analyses (NLTHA). Having shear wall as lateral load bearing system decreases the effect of infill walls in the seismic behavior of the building, and therefore the case of buildings with shear walls has been considered for this study as the first stage of the studies on lateral load patterns for R/C buildings. For this purpose, by assuming three different numbers of bays in each direction and also three different numbers of stories for the buildings, several R/C buildings, have been studied. At first, the buildings have been designed by the Iranian National Code for R/C Buildings. Then they have been analyzed by a NLTHA software using the accelerograms of some well-known earthquakes. The used accelerograms have been also scaled to various levels of peak ground acceleration (PGA) such as 0.35g, 0.50g, and 0.70g, to find out the effect of PGA in the seismic response. Numerical results have shown that firstly the values of natural period of the building and their shear force values, calculated by the code, are not appropriate in all cases. Secondly, it has been found out that the real lateral load pattern is quite different with the one suggested by the seismic code. Based on the NLTHA results a new lateral load pattern has been suggested for this kind of buildings, in the form of some story-dependent modification factors applied to the existing code formula. The effects of building's natural period, as well as its number of stories, are taken into account explicitly in the proposed new load pattern. The proposed load pattern has been employed to redesign the buildings and again by NLTHA the real lateral load distribution in each case has been obtained which has shown very good agreement with the proposed pattern.

Keywords: Nonlinear Time History Analysis, Story-dependent Response Modification Factor.

INTRODUCTION

It has been observed in some studies performed in almost a recent ago, on the seismic evaluation of existing buildings, either steel [1] or reinforced concrete (R/C) [2], that the distribution pattern of lateral loads along the building height is not close enough to the linear distribution, assumed by most codes. Therefore, some more realistic lateral load pattern seemed to be necessary. Following the studies, which had

CP1020, *2008 Seismic Engineering Conference Commemorating the 1908 Messina and Reggio Calabria Earthquake*, edited by A. Santini and N. Moraci
© 2008 American Institute of Physics 978-0-7354-0542-4/08/$23.00

dealt this fact indirectly, Hosseini and Motamedi [3] performed a study to approach this fact directly. That study showed that the actual lateral load distribution is not only far from the code load pattern, but also is dependent on various features of the building such as material, number of bays, number of stories, and the lateral load bearing system.

Later, it was tried by the first author and his graduate students to propose some new lateral load patterns for various types of buildings, including R/C buildings with moment frames and shear walls [4], steel frames with concentric X bracing [5], and steel frames with eccentric bracing [6-7], by introducing a somehow new concept, called "story-dependent response modification factor". Almost all of those studies showed that the lateral load pattern has an S-curve shape rather than a linear form with an additional value at the roof level, which is suggested by codes for high buildings. This paper presents the results of the study reported in reference [4].

THE BUILDINGS CONSIDERED FOR THIS STUDY

Nine R/C buildings with moment frames and shear walls as their load bearing systems were considered for the study. Three numbers of stories of 5, 10, and 15 were considered as well as three numbers of bays of 3, 4, and 5 in one direction with just 3 bays in the other direction. The span length was 4.0 meters in one direction and 5.0 meters in the other for all frames, and the story height was 3.0 meters in all cases. Therefore, the geometry of buildings was regular. The distribution of live and dead loads was also assumed uniform so that the buildings could be quite regular and symmetrical to avoid torsion effects. The lateral load bearing system was considered to be moment frames in one direction and combination of moment frames and shear walls in the other. Number of walls was two for 3-story buildings and four (one pair in each side) for 10- and 15-story buildings as shown in FIGURE 1.

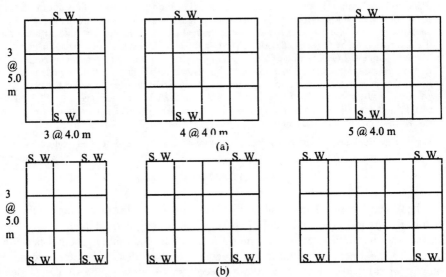

FIGURE 1. The buildings plans: (a) for 5-story buildings, (b) for 10- and 15-story buildings

956

For initial design of buildings the properties of construction materials, including concrete and steel were considered as f'_c = 250 kgf/cm^2, fy = 4000 kgf/cm^2 and 3000 kgf/cm^2 for longitudinal and lateral steel bars respectively. The minimum dimension for beam and columns' section was 30 cm and the ceiling system was joist and blocks. Is was tried to use the exact amount of required steel area in design of beams and columns instead of using a number of bars with known cross-sectional areas so that the amount of over-strength can be minimum. The earthquake loading was applied based on the Iranian Standard No. 2800, and considering the 0.35g as the design basis acceleration. The response modification values (R) were 8 for the direction with moment frames, and 9 for the direction with moment frames and shear walls. The calculation of base shear forces for the considered buildings for the direction with shear walls is shown in TABLE 1.

TABLE 1. Calculation of base shear forces of considered buildings in direction with shear walls

Code 2800	5 story			10 story			15 story		
	3 span	4 span	5 span	3 span	4 span	5 span	3 span	4 span	5 span
I	1	1	1	1	1	1	1	1	1
A	0.35	0.35	0.35	0.35	0.35	0.35	0.35	0.35	0.35
T	0.38	0.38	0.38	0.64	0.64	0.64	0.87	0.87	0.87
T_0	0.4	0.4	0.4	0.4	0.4	0.4	0.4	0.4	0.4
B	2.5	2.5	2.5	1.83	1.83	1.83	1.49	1.49	1.49
R	9	9	9	9	9	9	9	9	9
$C = ABI/R$	0.097	0.097	0.097	0.071	0.071	0.071	0.058	0.058	0.058
W (ton)	1071.2	1389.6	1694.6	2315.8	2904.7	3580.1	3607.7	4575.4	5543.2
$V = C*W$	103.9	134.8	164.4	164.4	206.2	254.2	209.2	265.4	321.5

It is seen in TABLE 1 that the values of building period (T) and base shear coefficient (C=ABI/R) do not depend on the number of spans of building frame. This is one of the shortcomings of most of seismic design codes, including Iranian Standard No. 2800, which will be discussed later in the paper.

NONLINEAR TIME HISTORY ANALYSES (NLTHA)

For performing the NLTHA the accelerograms of horizontal components of some well-known Iran and US earthquakes were used, including Naghan, Iran (1977), Tabas, Iran (1978), Manjil, Iran (2000), El Centro (1940) and San Fernando (1971). These earthquakes cover a wide range of frequency content and duration. The accelerograms were scaled to five various values of 0.3g, 0.4g, 0.5g, 0.6g, and 0.7g to find out the effect of peak ground acceleration (PGA) value on the base shear forces as well as lateral load distribution along the buildings' height. Regarding that each one of the nine considered buildings were analyzed subjected to five accelerograms, each one scaled to five various PGA values, in total 225 cases of NLTHA were performed for the study. The maximum values of base shear forces obtained form all of 225 analysis cases are shown in TABLE 2.

TABLE 2. The maximum shear forces obtained from 225 cases of NLTHA of nine studied buildings subjected to accelerograms of five selected earthquakes scaled to five PGA values of 0.3g to 0.7g

Base Shear	ton	5 story			10 story			15 story		
		3 span	4 span	5 span	3 span	4 span	5 span	3 span	4 span	5 span
Tabas	0.3	314	379	480	648	772	902	615	845	1020
	0.4	372	466	545	708	934	1060	731	˙030	1290
	0.5	422	629	786	897	˙070	1150	879	1110	1430
	0.6	508	783	808	1020	1200	1190	1050	1370	1520
	0.7	641	800	949	1070	1250	˙260	1110	1470	1620
Manjil	0.3	299	349	438	635	615	567	651	780	850
	0.4	373	452	573	644	648	817	719	861	908
	0.5	447	553	670	708	694	876	792	1020	1010
	0.6	486	532	564	845	777	915	897	1020	1200
	0.7	704	641	551	727	860	1000	1130	1070	1260
Naghan	0.3	341	405	439	575	622	632	862	959	874
	0.4	397	467	520	626	740	773	1060	983	905
	0.5	439	502	540	812	846	951	1130	991	1320
	0.6	472	700	545	685	935	1050	1140	1040	1370
	0.7	565	605	544	763	1010	1310	1080	1280	1570
El Centro	0.3	405	431	524	704	737	792	647	918	946
	0.4	447	523	599	746	823	791	871	1120	1110
	0.5	541	607	633	832	827	1040	1000	1180	1310
	0.6	581	655	713	896	862	1230	1120	1330	1600
	0.7	678	709	757	918	1140	1210	1360	1510	1700
San Fernando	0.3	280	341	302	432	524	668	570	806	822
	0.4	330	326	343	571	697	856	797	855	880
	0.5	365	377	410	685	837	1080	867	852	1050
	0.6	357	428	505	728	986	1280	913	917	˙280
	0.7	425	468	550	835	1070	1130	1010	1070	1380

It can be seen in TABLE 2 that maximum base shear values increase as number of spans and weight of the building increase, however, this increase is not proportional to the building weight, in spite of the code formulas implication. For example, in case of El Centro earthquake, scaled to PGA value of 0.3g, the maximum base shear value for 4-span frame, weighing 1389.6 tonf (see TABLE 1) is 431 tonf which is 1.064 times the maximum shear force of 3-span frame (405 tonf), weighing 1071.2 tonf, while the weights ratio is 1.297 in this case. Surprisingly, in case of San Fernando earthquake with PGA value of 0.3g the maximum base shear for the 5-span frame is even less than that of 4-span frame. It can be said that, looking overall, this inconsistency is higher for higher values of PGA. Furthermore, for some earthquakes the higher values of PGA have resulted in lower base shear forces. Such cases have been shown in gray in TABLE 2. This can be due to the higher energy absorption by plastic deformation of buildings in higher PGA values. The effect of number of spans, which is believed to be effective, to a great extent, in seismic response of buildings is not addressed in most of the existing seismic design codes.

As the main goal of this study was finding the lateral load pattern along the building height, the maximum lateral forces acting on each of the buildings floor levels during all time history analyses were calculated. For this purpose, the maximum story shear forces were obtained in each case from the time history, and then the maximum lateral forces were calculated by subtraction of maximum story

958

shear forces successively. The results of these calculations for all 225 cases of NLTHA, summarized in 45 sets of average values, are shown in FIGURE 2.

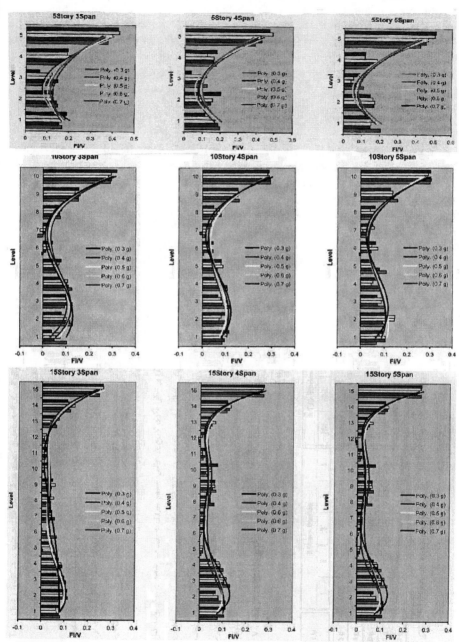

FIGURE 2. The effect of PGA values on the lateral load pattern of 5-, 10-, and 15-story buildings

In this figure each color horizontal bar in each floor level stands for the average of five maximum values, each obtained form one time history analysis. The fist thing which can be seen in this figure is the big difference between the obtained load patterns and the linear pattern suggested by most of seismic design codes. It can be also observed in this figure that although the values of average lateral load in some levels may be different drastically for various PGA values, in an overall look, the lateral load pattern is not so much sensitive to the variation of PGA value. It is worth mentioning that the drastic variations happen just in the levels which have little lateral forces, which are mostly the medium floors of buildings. It can be also observed that, in spite of the dependence of maximum base shear values on the number of spans or the associated weight of the building, the overall load pattern is not affected by the number of spans and the associated weight changes.

The other point realizable in FIGURE 2 is the similar pattern of lateral load in the three upper floors as well as a few lower floors of buildings regardless of their number of stories. Actually, the value of lateral load in the roof level is much more than the value given by code formula and can reach 2.5 times the code value in the case of 15-story buildings. The results also show that the values of lateral loads in lower floors are also higher than the corresponding code values, but in medium floors the values are less than the code values. Looking at the curves in FIGURE 2 it seems possible to fit some polynomials to them. Having this in mind it was tried to find some polynomials with appropriate order as shown in FIGURE 3.

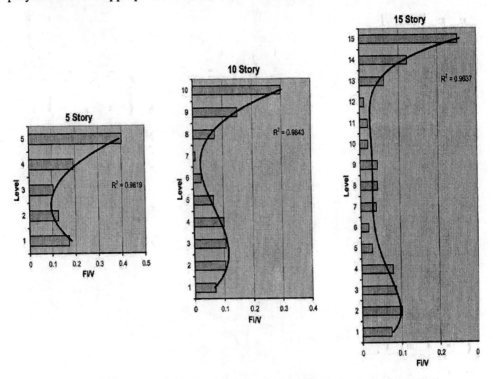

FIGURE 3. Polynomials fitted to the lateral load patterns by regression analysis

If FIGURE 3 the three mentioned parts in the load pattern along the building height, corresponding to the upper, the medium, and the lower floors can be realized easily, particularly for higher buildings. In this figure, the best polynomial curve for 5-story buildings is a second order curve, for the 10-story buildings a third order curve, and for 15-story buildings a fifth order curve. However, to make it easier for use as a code formula some simple modification factors can be introduced to be applied to existing lateral load distribution formula in the code. On this basis the following formula can be proposed.

$$F_i = \left\{ \begin{matrix} 2.6\,T \\ 1.2\,T \\ 0.6\,T \\ Ne^{-0.5i} \\ \cdot \\ \cdot \\ \cdot \\ Ne^{-0.5i} \end{matrix} \right\} \times \frac{w_i h_i}{\displaystyle\sum_{i=1}^{N} w_i h_i}\, V \qquad (1)$$

where T is the fundamental period of the building and N is the number of stories. In Eq. (1) the modification factors are shown inside { }, and they should be multiplied by the well-known code lateral load values shown as the second term in Eq. (1).

To find out the effect of using this formula in design of buildings the lateral load distribution were calculated for 4- to 18-story buildings of which some sample are shown in FIGURE 4.

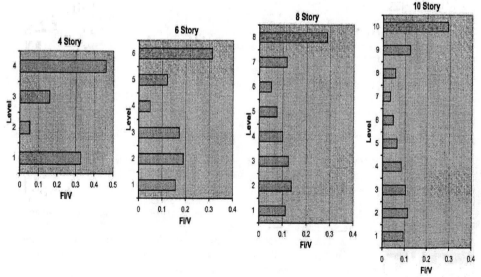

FIGURE 4. The lateral load distribution for 4- to 10-story buildings calculated by the proposed formula, given by Eq. (1)

Then two new 4-sapn buildings, one 10- and the other 15-story, were designed once by using Standard No. 2800 and once the proposed lateral load formula. The comparison between the lateral displacements of the two sets of buildings shows that the proposed formula results in less top displacement and more uniform drift values which is desired, particularly for taller buildings.

CONCLUSIONS

Based on the numerical results it can be concluded that:

- The number of bays, which has not been considered in the code formula, is effective in the natural period of the building, and therefore the maximum base shear force.
- The real lateral load pattern is quite different with the one suggested by most of the seismic codes.
- Based on the NLTHA results a new lateral load pattern can be suggested for this kind of buildings, in the form of some story-dependent modification factors applied to the existing code formula. The effects of building's natural period, as well as its number of stories, are taken into account explicitly in the proposed new load pattern.
- Using the proposed load pattern for design of R/C buildings with shear walls results in less top displacement and more uniform drifts, which is more desirable, particularly for tall buildings.

REFERENCES

1. K. Nasser Assadi and M. Hosseini, "A Study on the Ductility Factor of Common Steel Buildings (in Persian)", Proceedings of the 3rd International Conference on Seismology and Earthquake Engineering (SEE-3), IIEES, Tehran, Iran, May 1999.
2. M. Motamedi and M. Hosseini, "Using the Concept of Energy Dissipation and Distribution of Damage in Seismic Design of RC Buildings (in Persian)", Proceedings of the 3rd International Conference on Seismology and Earthquake Engineering (SEE-3), IIEES, Tehran, Iran, May 1999.
3. M. Hosseini and M. Motamedi, "A Study on the Distribution of Lateral Seismic Forces in the Height of R/C Buildings by Using Nonlinear Dynamic Analysis (in Persian)", Proceedings of the 1st Conference of Iranian Society of Civil Engineers (ISCE), Tehran, Iran, October 1999.
4. A. Khoshamadi, Identifying the Lateral Load Pattern for Reinforced Concrete Buildings with Moment Frames and Shear Walls Based on Nonlinear Dynamic Analysis, Master Thesis Under Supervision of Dr. Mahmood Hosseini Submitted to Earthquake Engineering Department, Engineering School, Science and Research Branch of the Islamic Azad University (IAU), Tehran, Iran, June 2004.
5. M. Hosseini, M. M. Yazdinejad and A. S. Moghadam, "Introducing a Story-Dependent Response Modification Factor for Steel Buildings with Concentrically Braced Frames", Proceedings of the 13th World Conference on Earthquake Engineering, Vancouver, Canada, August 2004.
6. M. Hosseini, and M. Esmaeili, "Story-dependent Response Modification Factors for Eccentrically Braced Frames: A New Lateral Load Pattern", Proceedings of the 8th US National Conference on Earthquake Engineering, San Francisco, USA, April 2006.
7. M. Hosseini and, M. R. Rezaee, "Proposing a New Lateral Load Pattern for Eccentrically Braced Steel Frames (in Persian)", Proceedings of the 7th International Congress on Civil Engineering, Tarbiat Modarres University, Tehran, Iran, May 2006.

The Seismic Reliability of Offshore Structures Based on Nonlinear Time History Analyses

Mahmood Hosseini[a], Somayyeh Karimiyani[b], Amin Ghafooripour[c] and Mohammad Javad Jabbarzadeh[d]

[a] Associate Professor, Civil Engineering Department, Graduate School, Tehran South Branch of the Islamic Azad University (IAU), Jamalzadeh St., Keshsavarz Blvd, Tehran 14187, Iran, Fax: (98) (21) 6643 7232, Email: Mahmood.hosseini@gmail.com and hosseini@iiees.ac.ir

[b] Graduate Student, Civil Engineering Department, Graduate School, Tehran South Branch of the Islamic Azad University (IAU), Jamalzadeh St., Keshsavarz Blvd, Tehran 14187, Iran, Fax: (98) (21) 6643 7232, Email: s_karimiyan@yahoo.com

[c] Assistant Professor, Civil Engineering Department, Tehran Central Branch of the Islamic Azad University (IAU), Tehran, Iran, Email: Amingh@ParsPadir.com

[d] PhD Student, Graduate Program, International Institute of Earthquake Engineering and Seismology (IIEES), P.O.Box 19395/3913, Tehran, Iran, Email: jabbarzadeh@iiees.ac.ir

Abstract. Regarding the past earthquakes damages to offshore structures, as vital structures in the oil and gas industries, it is important that their seismic design is performed by very high reliability. Accepting the Nonlinear Time History Analyses (NLTHA) as the most reliable seismic analysis method, in this paper an offshore platform of jacket type with the height of 304 feet, having a deck of 96 feet by 94 feet, and weighing 290 million pounds has been studied. At first, some Push-Over Analyses (POA) have been preformed to recognize the more critical members of the jacket, based on the range of their plastic deformations. Then NLTHA have been performed by using the 3-components accelerograms of 100 earthquakes, covering a wide range of frequency content, and normalized to three Peak Ground Acceleration (PGA) levels of 0.3g, 0.65g, and 1.0g. By using the results of NLTHA the damage and rupture probabilities of critical member have been studied to assess the reliability of the jacket structure. Regarding that different structural members of the jacket have different effects on the stability of the platform, an "importance factor" has been considered for each critical member based on its location and orientation in the structure, and then the reliability of the whole structure has been obtained by combining the reliability of the critical members, each having its specific importance factor.

Keywords: Jacket Type Structure, Push-Over and Nonlinear Time History Analyses, Critical Members, Importance Factor

INTRODUCTION

Offshore structures, particularly oil and gas platforms, are among vital structures all over the world, and many of them are located in seismic regions. Regarding the adverse effects of damage to these structures subjected to earthquake it is important that their seismic design is performed by very high reliability. Several studies have been performed on the reliability of offshore structures, of which some have dealt with

CP1020, 2008 Seismic Engineering Conference Commemorating the 1908 Messina and Reggio Calabria Earthquake, edited by A. Santini and N. Moraci
© 2008 American Institute of Physics 978-0-7354-0542-4/08/$23.00

seismic analysis, and a few of them are briefly reviewed here. Nadim and Gudmestad (1994) seems to be among the first researchers who have worked on reliability of an engineering system under a strong earthquake with application to offshore platforms [1]. Venkataramana and his colleagues (1998) have studied two structure models: (i) a jacket-type offshore structure and (ii) a tension-leg-platform, and have used the Kanai-Tajimi spectrum as the ground acceleration model in a frequency-domain random-vibration approach [2]. They have claimed that the response of tension-leg-platforms are highly dependent on the frequency parameter. Zhuang and his colleagues (1999) have also studied the seismic reliability offshore jacket platforms by means of nonlinear pushover failure analysis [3]. Jin and his colleagues (2002) have also worked on reliability of offshore jacket platforms subjected to seismic action by an stochastic approach [4]. They have reported that the responses of the platform considering the pile-soil-structure mutual interaction were less than those of a fixed platform with 6 times the diameter, and that the dynamic reliability index of offshore platforms was high and the failure probability was very low.

Recently, Finagenov and Glagovsky (2005) have performed a study on the assessment of reliability of offshore marine hydraulic structures under seismic impacts by both deterministic and stochastic approaches, in which the soil-structure-interaction has been taken into account with more precision [5]. It is seen that in spite of several studies on seismic reliability of offshore structures, the cases in which the sever earthquake and the nonlinear behavior of the structure have been taken into consideration are very few. It is believed that the most reliable kind of analysis for seismic design is Nonlinear Time History Analyses (NLTHA). On this basis, recently the authors has performed a study on the seismic reliability assessment of offshore structure, [6], whose results are briefly presented in this paper.

INTRODUCING THE JACKET AND ITS FEATURES

The studied offshore platform is of jacket type, and has been installed in Lavan oil field in Persian Gulf in 1970. It is 304 feet high, and has a deck of 96 feet by 94 feet, being carried on four inclined legs of 3 feet diameter. The total weight of jacket and deck is over 290 million pounds (more than 131,000 tonf), and its members are all of tubular section. FIGURE 1 shows the geometry of the jacket and platform.

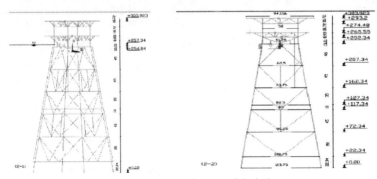

FIGURE 1. The geometric features of the jacket structure

964

The material of the jacket structure is high-strength steel with modulus of elasticity of 2.1E6 kgf/cm^2, and yielding stress of 3600 kgf/cm^2, giving a yielding strain of 0.171%. Based on the mentioned structural specifications of the jacket it modal properties up to 11 modes are given in TABLE 1, and the modal shapes of the first three modes are shown in FIGURE 2.

TABLE 1. Modal frequencies and periods of the jacket structure

Mode No.	1	2	3	4	5	6	7	8	9	10	11
Freq. (Hz)	0.414	0.428	0.749	1.314	1.315	1.429	1.435	1.492	1.512	1.654	1.655
Period (sec)	2.414	2.338	1.334	0.761	0.760	0.700	0.697	0.670	0.661	0.605	0.604

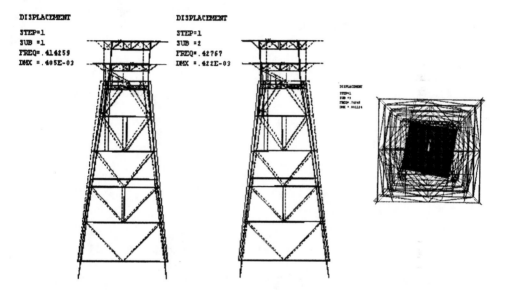

FIGURE 2. Modal shapes of the first three modes of the jacket structure

It can be seen in TABLE 1 that there is just slight difference between the frequencies of the 6th and upper modes up to the 11th. This means that the structure has several closely-spaced modes, and this make its modal analysis a very deliberate one, in which using the ordinary modal combination methods like SRSS is not adequate. Also it is seen in FIGURE 2 that although the first two modes are related to the lateral motions of the structure in the two main directions (X and Y) the third mode is a torsional mode.

Regarding the high number of elements in the jacket structure and the large volume of dynamic analyses outputs on the one hand, and the main goal of this study which is performing nonlinear time history analysis (NLTHA) for reliability assessment of the offshore structure, on the other, it was decided to perform at first a set of push-over analyses (POA) to find out the more critical members of the structure so that in the time history analyses only the results of the critical members are selected as outputs. The performed POA and their results are presented in the following section.

PUSH-OVER ANALYSES

The POA was preformed to recognize the more critical members of the jacket, based on the range of their plastic deformations. For this purpose a concentrated load was applied at the master joint of the top level of the upper platform once in one main direction (X) and once in the other main direction (Y). Since the jacket structure is a little asymmetric the POA were repeated for opposite directions (–X and –Y) as well. The more critical members of the structure identified based on the plastic deformations are shown in FIGURE 3.

FIGURE 3. The more critical members of the jacket structure identified by POA

Based on the POA the yielding forces and yielding displacements of the jacket structure were obtained as shown in TABLE 2.

TABLE 2. Yielding forces and yielding displacements of the jacket structure obtained from POA

	Push (X)	Push (-X)	Push (Y)	Push (-Y)
Yielding Force (lb)	88,499,185	92,158,817	84,579,280	86,261,183
Yielding Displ. (ft)	1.760	1.840	1.715	1.753
Stiffness (lb/ft)	50,278,384	50,076,738	49,316,160	49,213,052

The closeness of values of stiffness in directions X and –X, as well as Y and –Y directions show the satisfactory precision of POA. The close values of frequencies of the first two modes is also confirmed by the close values of stiffness in the two main directions of X and Y, which are around 50,175,000 lb/ft and 49,260,000 lb/ft respectively.

TIME HISTORY ANALYSES

Then NLTHA were performed by using the 3-components accelerograms of 100 earthquakes, covering a wide range of frequency content, all normalized to the same Peak Ground Acceleration (PGA) level. Three values of 0.3g, 0.65g, and 1.0g were used for the PGA level to find of the effect of earthquake intensity on the behavior of the jacket structure. In NLTHA the stress and strain values, particularly plastic strains in critical members, identified by POA, were of the main concern. To decrease the volume of NLTHA output the stress and strain values at only four locations in the section of critical members (say at 0, 90, 180, and 270 degrees in the tubular section) were calculated.

The number of locations in the sections of structural members, in which the strain value exceeded the elastic level in each time history, was chosen as the main damage criterion for nonlinear analyses. Since four locations in each section were considered to experience plastic deformation, and this could be the case at either end sections of each member the maximum number of locations with plastic deformation could be eight in each member. In some of these locations the strain value could exceed the rupture level (which was the strain value of 0.0034 based on Von Mises plasticity criterion). Assuming that these locations were just in the critical members, identified by POA, and shown in FIGURE 3, number of these locations were obtained for all 100 NLTHA cases with the PGA value of 1.0g, which showed that just 29 earthquakes were able to create plastic deformation or rupture cases in the jacket critical structural members, as shown in TABLE 3.

TABLE 3. The 29 more effective earthquakes and the number of locations with plastic deformation or rupture in the jacket structure obtained from NLTHA

No.	Earthquake Name	No. of plastic locations (N_p)	No. of rupture locations (N_r)	$N_p + N_r$	No.	Earthquake Name	No. of plastic locations (N_p)	No. of rupture locations (N_r)	$N_p + N_r$
1	Chi-chi, Taiwan 9	92	54	146	16	Birjand	56	14	70
2	Bajestan	77	40	117	17	Duzce, Turkey	55	21	76
3	Chi-chi, Taiwan 4	72	42	119	18	Imperial Valley 1	54	14	68
4	Boshrueh	70	38	108	19	Northridge 3	52	23	75
5	Turkey	68	29	97	20	Ferdows	52	21	73
6	Erzincan, Turkey	68	28	96	21	Gheshm	51	14	65
7	Northridge 2	64	38	102	22	Chi-chi, Taiwan 10	49	19	68
8	Sedeh 2	63	34	97	23	Tehran	49	12	61
9	Imperial Valley 2	60	32	92	24	Tehran 23	47	21	68
10	Bandarabbas	60	26	86	25	Abaregh	45	7	52
11	Imperial Valley	60	26	86	26	Deyhook	43	16	59
12	Chi-chi, Taiwan 2	60	26	86	27	Chi-chi, Taiwan 3	38	2	40
13	Khash	57	30	87	28	Rudbar	37	8	45
14	Rayen	57	27	84	29	Bandar Khamir	34	2	36
15	Sedeh	57	25	82	-	-	-	-	-

Considering the number of plastic locations as the damage criterion the first 15 earthquakes out of the 29 ones mentioned in TABLE 3 can be selected as the most damaging earthquakes for the jacket structure. On this basis the accelerograms of these earthquakes were scaled once to 0.65g and once to 0.30g for more NLTHA cases. On the other hand, paying attention to the results of NLTHA for various critical members, shown in FIGURE 3, it can be realized that in each case each of these elements experienced some different level of damage. On this basis, depending on the number of plastic locations (0 to 8) in each member a damage percent can be defined for it. Considering five levels of damage of: 1) 0%, namely elastic behavior, 2) less than 25%, 3) between 25% and 50%, 4) between 50% and 75%, and 5) more than 75% some damage probability density functions can be obtained as shown in FIGURE 4 for element No. 1 (in the lowest part of the jacket legs – see FIGURE 3) as the most critical element of the jacket structure.

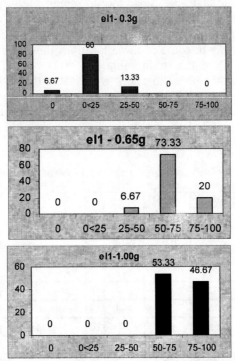

FIGURE 4. Damage percent of element No. 1 for various levels of PGA in NLTHA

It is seen in FIGURE 4 that the damage probability in element No. 1, which has almost normal distribution for PGA values of 0.3g and 0.65g, increases with increase in the PGA level. Similar graphs can be presented for other members. If the number of rupture cases is considered as the damage indicator, by using the results shown in TABLE 3 for rupture cases, and using the five aforementioned states, the rupture probability density functions can be obtained as shown in FIGURE 5 for again element No. 1.

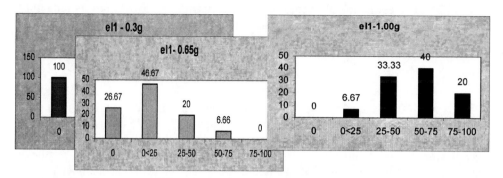

FIGURE 5. Rupture percent of element No. 1 for various levels of PGA in NLTHA

It is seen in FIGURE 5 that the rupture probability in element No. 1, which its statistical distribution is not far from normal, increases with increase in the PGA level. Similar results can be obtained for other members of the jacket structure, however, to assess the reliability of the jacket structure subjected to earthquakes, the damage or rupture probabilities of all members, or at least all critical members of the structure, should be combined by some logic. For this purpose an "importance factor" can be assigned to each structural member depending on its role in the whole seismic stability of the jacket structure. In this study the importance factor was considered to be 1.0 for legs members, 0.5 for diagonal members in vertical direction, and 0.25 for horizontal members. On this basis, some levels of reliability can be considered for the seismic reliability of the jacket structural members. Defining 10 reliability levels and using various colors for different levels figure similar to FIGURE 6 can be produced.

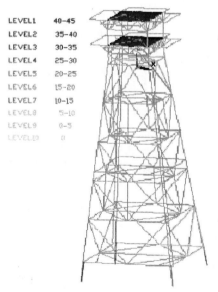

FIGURE 6. Reliability level of the jacket structural members based on their damage or rupture probabilities and importance factors

In FIGURE 6 as the member color changes form light yellow to red its reliability level decreases. Obviously, if importance factors other than those used in these calculations are used some other reliability levels would be obtained.

CONCLUSIONS

From the 100 earthquakes used in the study, covering a wide range of frequency content, duration and spectral characteristics, less than 30% could be damaging for the considered jacket structure, even by using a PGA value of 1.0g. This means that, in an overall view the reliability of the seismic design of the jacket structure is relatively high. However, the level of damage is not the same for different members. This implies that the reliability level of seismic design is not the same for all jacket structural members. The importance factor of members, which depends on its location, orientation, and load bearing situation, has also great effect on the reliability assessment. The NLTHA results also show that that none of the earthquake characteristics alone, can be the dominant factor. Instead, a combined factor in which various features of earthquake, including frequency content, energy, and spectral characteristics are taken into account can be suggested. Further research is needed to address this issue.

REFERENCES

1. F. Nadim, O. T. Gudmestad, Reliability of an engineering system under a strong earthquake with application to offshore platforms, *Structural Safety*, Vol. 14, No. 3, pp. 203-217 (1994).
2. K. Venkataramana, K. Kawano, T. Taniguchi, Earthquake response and reliability analysis of offshore structure, *Structural Safety and Reliability* (A. A. Balkema Uitgevers B. V. - Netherlands), Vol. 3, pp. 2029-2036 (1998).
3. Y. Zhuang, W. Jin, H. Li, Z. Song, X. Li, and D. Zou, A seismic reliability analysis approach on offshore jacket platforms, *Acta Oceanologica Sinica/Haiyang Xuebao*, Vol. 21, No. 5, pp. 129-136 (1999).
4. W. Jin, Z. Zheng, H. Li, Analysis of dynamic reliability of offshore jacket platforms subjected to seismic action, *Zhejiang Daxue Xuebao (Gongxue Ban)/Journal of Zhejiang University (Engineering Science)*, Vol. 36, No. 3, pp. 233-238 (May-June 2002).
5. O. M. Finagenov, V. B. Glagovsky, Assessment of Reliability of Offshore Marine Hydraulic Structures under Seismic Impacts, Proceedings of ISOPE-2005: Fifteenth International Offshore and Offshore and Polar Engineering Conference, Vol. 1-4, 2005.
6. S. Karimiyan, Seismic Reliability Analysis of Offshore Structures by Using Nonlinear Time History Analyses, Master Thesis under supervision of Dr. Mahmood Hosseini, Submitted to Civil Engineering Department, Graduate School, Tehran South Branch of the Islamic Azad University (IAU), Tehran, Iran, 2008.

The Effect of Corrosion on the Seismic Behavior of Buried Pipelines and a Remedy for Their Seismic Retrofit

Mahmood Hosseini[a], Shamila Salek[b] and Masoud Moradi[c]

[a] Associate Professor, Structural Engineering Research Center, International Institute of Earthquake Engineering and Seismology (IIEES), P.O. Box 19395/3913, Tehran, Iran, Email: hosseini@iiees.ac.ir

[b] Graduate Student, Civil Engineering Department, Graduate School, Tehran South Branch of the Islamic Azad University (IAU), Jamalzadeh St., Keshsavarz Blvd, Tehran 14187, Iran, Fax: (98) (21) 6643 7232, Email: shamila_salek@yahoo.com

[c] Graduate Student, Civil Engineering Department, Structrual Engineering Group, Khajeh Nasseeruddin Toosi University, Vali-Asr Ave., Tehran 15875-4416, Iran, Fax: (98) (21) 8877 9476, Email: masoud_moradi1360@yahoo.com

Abstract. The effect of corrosion phenomenon has been investigated by performing some sets of 3-Dimensional Nonlinear Time History Analysis (3-D NLTHA) in which soil structure interaction as well as wave propagation effects have been taken into consideration. The 3-D NLTHA has been performed by using a finite element computer program, and both states of overall and local corrosions have been considered for the study. The corrosion has been modeled in the computer program by introducing decreased values of either pipe wall thickness or modulus of elasticity and Poisson ratio. Three sets of 3-component accelerograms have been used in analyses, and some appropriate numbers of zeros have been added at the beginning of records to take into account the wave propagation in soil and its multi-support excitation effect. The soil has been modeled by nonlinear springs in longitudinal, lateral, and vertical directions. A relatively long segment of the pipeline has been considered for the study and the effect of end conditions has been investigated by assuming different kinds end supports for the segment. After studying the corroded pipeline, a remedy has been considered for the seismic retrofit of corroded pipe by using a kind of Fiber Reinforced Polymers (FRP) cover. The analyses have been repeated for the retrofitted pipeline to realize the adequacy of FRP cover. Numerical results show that if the length of the pipeline segment is large enough, comparing to the wave length of shear wave in the soil, the end conditions do not have any major effect on the maximum stress and strain values in the pipe. Results also show that corrosion can lead to the increase in plastic strain values in the pipe up to 4 times in the case of overall corrosion and up to 20 times in the case of local corrosion. The satisfactory effect of using FRP cover is also shown by the analyses results, which confirm the decrease of strain values to 1/3.

Keywords: Overall and local corrosion, 3-D nonlinear time history analysis, FRP cover

INTRODUCTION

There are several pipelines in the world which have been located in, or passing through corrosive environments. Although there are some provisions, enforced by the design codes, for prevention of corrosion, still this phenomenon does exist, and decreases the structural resistance of pipelines in several parts of the world. Apparently, in the case of old pipelines this issue is more crucial. It is clear that if the effect of this phenomenon is not taken into consideration in seismic evaluation of pipelines, the vulnerability results will not be reliable. However, there are very few cases of study on the effect of corrosion on the seismic behavior of pipelines. Isenburg (1978) has studied the role of corrosion in the seismic performance of buried steel pipelines in three United States earthquakes [1]. He has investigated the condition of underground water pipelines following earthquakes in Puget Sound, Santa Rosa, California, and San Fernando, California with emphasis on steel pipe damage in regions where the maximum ground displacements have been of the order of 10 cm. He has reported numerous leaks in the pipes due to local weaknesses caused by corrosion. He has also discussed corrosion control programs used by utilities including wrapping and cathodic protection, and replacement of corroded pipe with non-corroding asbestos-cement pipe. Eguchi and his colleagues (1995) have also studied pipeline replacement feasibility as a methodology to minimize seismic and corrosion risks to underground natural gas pipelines [2]. In a recent study, performed by the authors, the effect of corrosion phenomenon has been investigated by performing some sets of 3-Dimensional Nonlinear Time History Analysis (3-D NLTHA) in which soil structure interaction as well as wave propagation effects have been taken into consideration [3]. A retrofit remedy with Fiber reinforced Polymers (FRP) cover has been also proposed in that study. This paper is a brief format of that study.

MODELING AND ANALYSIS OF THE PIPE AND SOIL

To study the effect of corrosion on the seismic behavior of buried pipeline a two-stage modeling and analysis process was employed. At first the pipeline was modeled as a beam on elasto-plastic supports to find the most critical situation of the pipe based on the stress and strain values resulting form bending moment in combination with axial and shear forces. Then a very short segment of the pipe around the critical section was modeled by finite element method (FEM) to study deliberately the stress and strain condition in the pipe section. The same process was followed once for the intact pipe, once more for the corroded pipe and a third time for the retrofitted pipe. Both states of overall and local corrosions were considered in the study. The corrosion was modeled in the computer program by introducing decreased values of either modulus of elasticity and Poisson ratio of the pipe material or the pipe wall thickness, as suggested by other researchers, [ ], and described in detail hereinafter.

The pipeline segment considered for the study was of 400 meters length, 1.5 meters diameter, and 1.5 cm thickness. The soil was modeled by nonlinear springs in longitudinal, lateral, and vertical directions connected to nodes on the axis of the pipeline. The shear wave velocity was assumed to be 200 m/s, and regarding that the time step of digitized accelerograms was 0.02 seconds, the distance between the nodes on the pipeline axis for springs connections was considered as 4.00 meters to make it possible to take into account the wave propagation phenomenon in soil and the time difference between the seismic excitation of successive supports (multi-support excitation) by put some zeros at the beginning of the input records for various supports. FIGURE 1 shows the pipeline modeled as a beam on elsto-plastic supports.

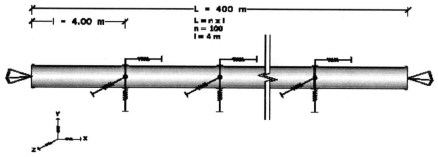

FIGURE 1. The pipeline modeled as a beam supported by elsto-plastic springs

It can be seen in FIGURE 1 that the studied pipe segment is consisted of 100 pipe elements, each having a length of 4.00 meters. To realize the effect of end conditions on the stress and strain values along the pipeline, four combinations of fixed-fixed, free-free, joint-joint, and fixed-free were considered. The conditions of the soil environment around the pipeline were assumed as follow:

- Soil type, dry sand with moderate compaction
- Soil density, $\gamma = 2000$ kgf/cm^3
- The angle of internal friction, $\varphi = 30$ degrees
- The coefficient of lateral soil pressure in rest, $k_0 = 0.5$
- The thickness of soil above the pipeline, $h = 1.0$ meter
- The burial depth, $H = 1.5$ meters
- The coefficient of friction between soil and pipe, $\mu = \tan \delta = 0.7$

The ultimate strengths of springs in longitudinal, lateral, and vertical directions were considered by using the following formulas [4].

$$t_u(x) = \frac{\pi D}{2} \gamma H (1 + k_0) \tan \delta \qquad (1)$$

$$p_u(y) = \gamma H N_{qh} D \qquad (2)$$

$$q_u(z) = H \gamma N_{qv} D \quad \text{(upward)} \qquad (3)$$

$$q_u(z) = c N_c D + \gamma' H N_q B + \frac{1}{2} \gamma D^2 N_\gamma \quad \text{(downward)} \qquad (4)$$

In these equations t_u, p_u, and q_u are respectively the ultimate strengths of soil springs in longitudinal (x), lateral (z), and vertical (y) directions, D is the external diameter of the pipe, H is the burial depth, c is the soil cohesion coefficient, γ' is the effective soil density, B is the effective breadth of the pipeline bed, and N_{qh}, N_{qv}, and N_γ are respectively the coefficients of soil load bearing capacities in difference situations. The calculated values of t_u, p_u, and q_u and their corresponding displacement values, x_y, z_y, and y_y, based on the conditions assumed for the study are respectively 3015 kgf/m, 15000 kgf/m, and 4500 kgf/m (upward) and 105000 kgf/m (downward), and 0.004 m, 0.080 m, and 0.023 m (upward) and 0.100 m (downward). The critical element in the pipeline segment was realized based on the maximum plastic axial strains. On this basis the force-displacement relations of soil spring in different directions are shown in FIGURE 2.

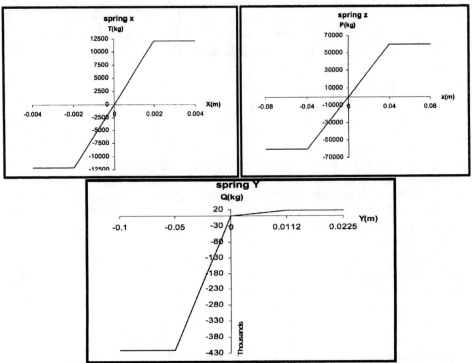

FIGURE 2. Force-displacement relations of soil spring in different directions

For 3-D NLTHA the displacement time histories of three sets of 3-component accelerograms of well known earthquakes, including Wittier Narrows and Northridge (recorded in two stations) earthquakes were used, whose specifications are shown in TABLE 1.

TABLE 1. The specifications of earthquakes whose displacement histories were used in the study

Earthquake	Station	Year	Duration (sec)	Component	PGA (g)	PGD (cm)
Whittier Narrows	Compton Castlegate	1987	31.18	E-W	0.332	5.04
				N-S	0.333	1.48
				UP	0.167	0.19
Northridge (I)	Canoga Park Topangacan	1994	24.99	E-W	0.420	20.17
				N-S	0.356	9.13
				UP	0.489	5.50
Northridge (II)	Jensen filter plant	1994	24.44	E-W (x)	0.424	43.06
				N-S (z)	0.593	24.00
				UP (y)	0.400	8.89

It is seen in TABLE 2 that the PGD value of Wittier Narrow earthquake as the weak earthquake is 5.04 cm, that of Northridge (I) as the moderate earthquake is 20.17 cm, and that of Northridge (II) as the strong earthquake is 43.06 cm. The strongest component of these three earthquakes, which is their E-W component in all cases, was assumed to act at a direction perpendicular to the longitudinal axis of pipeline to make the most severe lateral effect on the pipeline.

To model the overall corrosion, as mentioned before, two approaches of decreasing the thickness of pipe wall and decreasing the pipe material properties were employed. The values of either the pipe wall thickness or modulus of elasticity (E) and Poisson ratio (v) of the pipe material were decreased once 20% as the minor corrosion, once 30% as the moderate corrosion, and once 40% as the severe corrosion. For modeling the local corrosion just one case of 33% decrease in the pipe wall thickness in a small area of its wall was considered as shown in FIGURE 3.

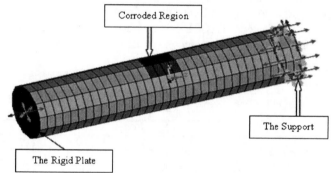

FIGURE 3. The FEM model of the pipe for studying the local corrosion

As it is seen in FIGURE 3 a rigid plate was used to apply the bending moment, and shear and axial forces to the short pipe segment. For retrofit of the pipe a kind of fiber reinforced polymer (FRP) of Aramid/Epoxy brand was used, and the RFP cover was modeled in the FEM program with a kind of shell element. The slide between FRP cover and the pipe wall was supposed not to happen. The specifications of the FRP cover are shown in TABLE 2.

TABLE 2. The specifications of FRP cover material

FRP Composite	Elastic Modulus (MPa)	Poisson Ratio	Shear Modulus (MPa)
Aramid/Epoxy	Ex = 13600	$v_{xy} = 0.32$	Gxy = 549.13
Aramid/Epoxy	Ey = 1482.1	$v_{yz} = 0.35$	Gyz = 547.00
Aramid/Epoxy	Ez = 1482.1	$v_{zx} = 0.35$	Gzx = 549.13

NUMERICAL RESULTS

Based on the performed 3-D NLTHA in both cases of beam on elasto-plastic supports model and the FEM model of the pipeline the numerical results were obtained with regard to the following issues:

- The effects of various end conditions on the location of the critical element in the pipeline and stress and strain values in it
- The effect of considering the time lag between excitations in various supports
- The effect of overall and local corrosions as well as the efficiency of the retrofit by FRP cover

The numerical results obtained with regard to the above issues indicated that:

- Various end conditions do not have much effect on the location of the critical section of the pipe, being somewhere around the midpoint of the pipeline length.
- The highest plastic deformation in the pipeline happens when the free-free end conditions is assumed.
- Considering the time lag between the excitations of various supports of the pipeline leads larger deformations in the pipeline, resulting in larger strain and stress values.
- The lateral movement of the pipeline follows to a great extent the lateral movement of the soil during earthquake.

With regard to the corrosion effects the maximum values of combined stresss, as Von Mises stress were used. TABLE 3 shows the maximum stresses and strains due to the overall corrosion modeled by decrease in E and v values.

TABLE 3. The maximum stress and strains in the pipe due to various levels of corrosion modeled by decrease in E and v values

Element No.	Max. Stress (Von Mises) kgf/m²	Max. Plastic Strain (Von Mises) for various percents of decrease in E and v values			
		0%	20%	30%	40%
32	4.7070E+07	6.1172E-02	1.1042E-01	1.6252E-01	2.1603E-01
33	4.7060E+07	5.0720E-02	9.6858E-02	1.1165E-01	1.9667E-01
37	4.7040E+07	3.4909E-02	6.7141E-02	7.4765E-02	1.4998E-01
38	4.7050E+07	3.6804E-02	5.0831E-02	7.9753E-02	1.4478E-01
39	4.7050E+07	3.5798E-02	5.0353E-02	7.7095E-02	1.2629E-01
43	4.7050E+07	4.2140E-02	7.1545E-02	9.0013E-02	1.3696E-01
44	4.7050E+07	4.0174E-02	7.0172E-02	8.5350E-02	1.3588E-01
49	4.7050E+07	3.8000E-02	6.1401E-02	8.1738E-02	1.1948E-01
50	4.7060E+07	4.2096E-02	6.5611E-02	8.9099E-02	1.2966E-01
61	4.7040E+07	3.6306E-02	5.5926E-02	7.8001E-02	1.1687E-01
62	4.7040E+07	3.7128E-02	5.7526E-02	7.9610E-02	1.1877E-01
67	4.7050E+07	3.2990E-02	5.1873E-02	7.0823E-02	1.1076E-01
68	4.7040E+07	3.5827E-02	5.5816E-02	7.6819E-02	1.1577E-01
74	4.7040E+07	3.3344E-02	5.2896E-02	7.1743E-02	1.1276E-01
80	4.7040E+07	3.2045E-02	5.0879E-02	6.9541E-02	1.0755E-01
The ratio of plastic strain of various corroded state to the non-corroded state		1.63	2.18	3.45	

It can be seen in TABLE 3 that the stress and strain values were exceeded the plastic limit in 15 elements out of the 100 elements of the pipe segment model (with element numbers of 32, 33, ..., 80). It is seen that the plastic strain is quite different for cases with different levels of corrosion, and can reach up to 3.45 times that of non-corroded state. The stress and plastic strain values of the overall corrosion modeled with 40% decrease in the pipe wall thickness are shown in TABLE 4.

TABLE 4. The maximum stress and plastic strains in the pipe in the case of overall corrosion modeled with 40% decrease in the pipe wall thickness

Element No.	Max. Stress (Von Mises) kgf/m²	Max. Plastic Strain (Von Mises)	Stress Ratio (Damaged Element/Undamaged Element)	Strain Ratio (Damaged Element/Undamaged Element)
32	5.48E+07	1.13E-01	1.17	1.85
33	5.56E+07	9.40E-02	1.18	1.85
37	5.37E+07	7.64E-02	1.14	2.19
38	5.77E+07	9.92E-02	1.23	2.70
39	5.84E+07	7.92E-02	1.24	2.21
43	5.95E+07	9.09E-02	1.27	2.16
44	5.75E+07	8.56E-02	1.22	2.13
49	5.23E+07	7.26E-02	1.11	1.91
50	5.43E+07	7.11E-02	1.15	1.69
61	5.68E+07	9.62E-02	1.21	2.65
62	5.59E+07	7.97E-02	1.19	2.15
67	5.86E+07	6.91E-02	1.25	2.09
68	5.87E+07	8.42E-02	1.25	2.35
74	5.85E+07	8.94E-02	1.24	2.68
80	5.83E+07	7.56E-02	1.24	2.36
The average ratios of stress and strain in damaged element to undamaged element			1.61	4.71

It is seen in TABLE 4 that the average ratios of stress and strain in damaged element to undamaged element are respectively 1.61 and 4.71. With regard to local element the stress and strain ratios of damaged element to undamaged element, obtained based on the results of FEM analysis are shown in TABLE 5.

TABLE 5. Stress and strain ratios of damaged element to undamaged element from FEM analysis

Element No.	Stress Ratio Damaged Element/Undamaged Element)	Strain Ratio Damaged Element/Undamaged Element)
397	2.53	12.20
398	1.90	9.07
399	1.86	9.41
400	2.40	11.14
401	3.4	18.48
402	1.97	9.26
403	2.03	10.09
404	2.2	10.46
405	1.67	8.59
406	2.01	11.60
407	2.14	12.40
408	1.93	10.55
Average	2.17	11.10

It is seen in TABLE 5 that although the average strain ratio due to local corrosion is around 11.10 this ratio can reach 20 in some cases, which means the local rupture of the pipe. With regard to the effect of FRP cover retrofit the stress and strain ratios of reinforced element to damaged element, obtained based on the results of FEM analysis are shown in TABLE 6 for the case of overall corrosion modeled by 20% decrease in E and v values as a sample, which shows that using the FRP cover can reduce the strain values to 1/3 of the damaged pipe.

TABLE 6. Stress and strain ratios of reinforced element to damaged element from FEM analysis

Element No.	Max. Stress (Von Mises) kgf/m²	Max. Plastic Strain (Von Mises)	Stress Ratio (Reinforced Element/Damaged Element)	Strain Ratio (Reinforced Element/Damaged Element)
32	3.7087E+07	3.2707E-02	0.79	0.30
33	3.0071E+07	3.5970E-02	0.64	0.37
37	3.7071E+07	1.9819E-02	0.79	0.30
38	3.7070E+07	6.8492E-03	0.79	0.13
39	3.7053E+07	8.8492E-03	0.79	0.18
43	3.7059E+07	2.1470E-02	0.79	0.30
44	3.2058E+07	3.1279E-02	0.68	0.45
49	3.5049E+07	9.2193E-03	0.74	0.15
50	3.7040E+07	4.7671E-02	0.79	0.73
61	3.7040E+07	2.1506E-02	0.79	0.38
62	3.6042E+07	2.0806E-02	0.77	0.36
67	3.7037E+07	2.7622E-02	0.79	0.53
68	3.7039E+07	2.1506E-02	0.79	0.39
74	3.5037E+07	1.2300E-02	0.74	0.23
80	3.2036E+07	9.8700E-03	0.68	0.19
Average Ratios			0.76	0.33

CONCLUSIONS

The numerical results show that if the length of the pipeline segment is large enough comparing to the wave length of shear wave in the soil the end conditions do not have any major effect on the maximum stress and strain values in the pipe. Results also show that corrosion can lead to the increase in stress and strain values in the pipe up to 4 times in the case of overall corrosion and up to 20 times in the case of local corrosion. The satisfactory effect of using FRP cover is also shown by the analyses results, which confirm the decrease of strain values to 1/3.

REFERENCES

1. J. Isenberg, The role of corrosion in the seismic performance of buried steel pipelines in three United States earthquakes, Weidlinger Associates, New York, 1978.
2. R. T. Eguchi, H. A. Seligson, and D. G. Honegger, Pipeline replacement feasibility study: A methodology for minimizing seismic and corrosion risks to underground natural gas pipelines, U.S. National Center for Earthquake Engineering Research (NCEER), March 1995.
3. S. Salek, Seismic Analysis of Buried Pipes with Corrosion and Proposing Their Retrofit Method, Master Thesis supervised by Dr. Mahmood Hosseini, submitted to Civil Engineering Department, Graduate School, Tehran South Branch of the Islamic Azad University (IAU), Tehran, Iran, 2008.
4. ALA, Guideline for the Design of Buried Steel Pipe, American Lifeline Alliance (ALA), American Society of Civil Engineers (ASCE), July 2001.

Nonlinear Behavior of Reinforced Concrete Shear Walls Using Macro Model

A. Jalali[a] and F. Dashti[b]

[a]Assistant Professor, Structural Eng. Dept., School of Civil Engineering, University of Tabriz 51664, Tabriz-Iran.
Tel: +98-411-3356024; Fax: +98-411-3344287; E-mail: jalali@tabrizu.ac.ir
[b]Graduate Student, Ditto

Abstract. A simple macro-model for reinforced concrete shear walls has been investigated. The proposed model consists of nonlinear spring elements representing flexural and shear behavior. The flexural behavior of the model is based on the uniaxial behavior of the vertical spring elements defined according to constitutive relations for materials and the tributary area assigned to each spring element which in turn leads to the integration of important material characteristics. The shear behavior is based on a trilinear force-displacement backbone curve assigned to each horizontal spring element. The model response has been predicted using nonlinear flexural and shear spring elements of the general purpose finite element program ABAQUS6.7. The analysis results show excellent agreement with experimental measurements of slender walls. The model turned out to be capable of simulating the nonlinear behavior of the selected test specimens at different stages of loading to a very good degree of accuracy within a few seconds of CPU time. The parametric studies conducted also show that the sensitivity of the model results to different modeling parameters is not significant.

In order to evaluate the advantages and deficiencies of the investigated model in comparison with the models based on the finite element approach, the nonlinear behavior of the selected test specimens has been predicted using microscopic models. Although the microscopic model could simulate some important aspects of wall behavior such as the interaction between shear and flexural response components observed even in relatively slender RC walls, distribution of cracks and stresses and local behavior, the CPU time was considerably greater than the one needed for the analysis of the investigated macro-model, and the response of the model was relatively sensitive to mesh size.

Taking the CPU time and simplicity into account, the observed agreement among three lateral load-displacement curves of experimental measurements, the macroscopic and microscopic models of test specimens indicates the efficiency of the investigated macro-model.

Keywords: Shear wall; Reinforced concrete; Macro-model; Flexural behavior; Shear behavior; Microscopic model

INTRODUCTION

Many structures of low and medium height consist of ductile moment- resisting frames. As the height of the structure increases, it is more efficient to provide the building with the required lateral strength and stiffness by means of a frame system interacting with structural walls. Considering the application of nonlinear static analysis in the new methods for seismic design and rehabilitation of structures, usually referred to as performance-based design, it is necessary to investigate simple and reasonably accurate models for predicting the nonlinear response of RC walls and wall

CP1020, 2008 Seismic Engineering Conference Commemorating the 1908 Messina and Reggio Calabria Earthquake,
edited by A. Santini and N. Moraci
© 2008 American Institute of Physics 978-0-7354-0542-4/08/$23.00

systems. The proposed procedures for analytical modeling of an RC shear wall are divided into two major groups, microscopic and macroscopic models. In this study the efficiency of a macroscopic model proposed by Orakcal and Wallace [1] in predicting the nonlinear behavior of selected slender RC walls subjected to monotonic loading has been evaluated, and its advantages and deficiencies in comparison with the microscopic models has been investigated following the 2D analysis of the selected test specimens using the finite element approach.

THE MACRO-MODEL DESCRIPTION

The macro-model adopted here is composed of several macro elements, the number of which depends on the expected accuracy and local behavior. Each macro element consists of vertical spring elements connected to rigid beams at the top and bottom levels representing the flexural response and a horizontal spring element, placed at the height ch, simulating the shear behavior of an RC wall (Fig. 1). As shown in Fig.1 two parallel spring elements representing the uniaxial behavior of concrete and steel are used to define the uniaxial behavior of the tributary area assigned to each couple of springs. The monotonic branch of the widely used hysteretic models of Menegotto and Pinto [2] and Mander et al.[3] is implemented for the reinforcing steel and concrete behavior, respectively. Since the connection points of steel spring elements and rigid beams are the same as the ones of concrete spring elements, the bond between concrete and the reinforcing steel has been taken into account and in order to model the contribution of cracked concrete to the tensile resistance of RC members, known as the effect of tension stiffening, the constitutive relations for materials have been modified according to the relations proposed by Belarbi and Hsu [4]. Since the efficiency of the model in predicting the response of slender RC walls was to be investigated the trilinear force-displacement backbone curve of the origin-oriented-hysteresis model (OOHM), proposed by Kabeyasawa et al. [5], in which the interaction between shear and flexural response components have not been taken into account, has been used to define the behavior of the horizontal spring elements. The value of $c = 0.4$ recommended by Vulcano et al. [6] based on comparison of the model response with experimental results has been used to define the location of the center of rotation along the height of each macro element.

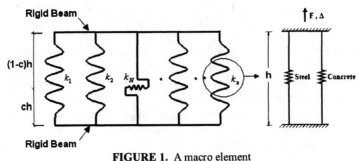

FIGURE 1. A macro element

EXPERIMENTAL CALIBRATION AND VERIFICATION

The wall specimens tested by Vallenas et al. [7] have been used to calibrate and assess the analytical model. The specimens selected include one framed wall where the boundary elements protrude from the surface of the wall (SW3) and one wall with rectangular cross section (SW5). The specimens were both subjected to monotonic loading. Figure 2 displays the simplest model configurations, with 3 macro elements stacked upon each other (1 macro element for each story), and with 4 uniaxial elements defined along the length of the wall for both specimens.

The analytical model was implemented in ABAQUS to allow comparison between experimental and analytical results. The model turned out to be capable of simulating the nonlinear behavior of the selected test specimens at different stages of loading to a very good degree of accuracy after 4 seconds of CPU time. Figures 3, 4 and 5 compare the measured and predicted lateral load-top displacement, lateral load – top flexural displacement and lateral load-top shear displacement responses for both specimens (SW3 and SW5). There are changes in the slope of the overall force-deformation diagrams at points A, B, C, D, E, F and G which are corresponding to changes in the behavior of the spring elements representing the axial response of the boundary elements and shear behavior of each story.

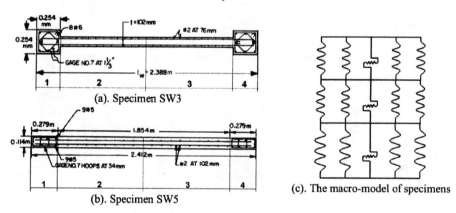

(a). Specimen SW3

(b). Specimen SW5

(c). The macro-model of specimens

FIGURE 2. Model discretization and tributary area assignment

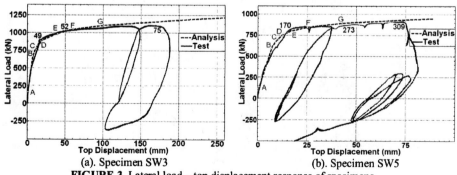

(a). Specimen SW3

(b). Specimen SW5

FIGURE 3. Lateral load – top displacement response of specimens

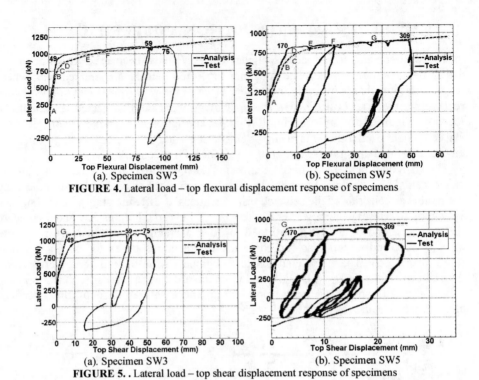

(a). Specimen SW3   (b). Specimen SW5

FIGURE 4. Lateral load – top flexural displacement response of specimens

(a). Specimen SW3   (b). Specimen SW5

FIGURE 5. . Lateral load – top shear displacement response of specimens

As shown in the figures above the analytical model provides a good prediction of the wall lateral overall and flexural displacement, and the discrepancy observed between analytical and experimental lateral load-top shear displacement of both specimens can be attributed to considering uncoupled shear and flexural responses in the analytical model. In other words, using the origin-oriented-hysteresis model to define the behavior of the horizontal spring elements, the interaction between shear and flexural response components observed even in relatively slender RC walls has not been considered in the analytical model. Deformation pattern of specimen SW3 before and after shear yielding is shown in Fig.6.

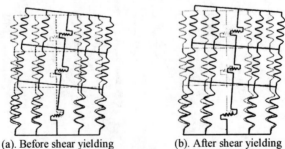

(a). Before shear yielding   (b). After shear yielding

FIGURE 6. Deformation pattern of the macroscopic model of specimen SW3 before and after shear yielding

PARAMETRIC STUDY

In order to evaluate the sensitivity of the macroscopic model to modeling parameters the lateral load top-displacement response of the models with different arrangement of elements has been compared with the one of the simplest possible model. The flexural deformation of the models with different arrangement of elements is shown in Fig.7 and the response of the models SW3B and SW3C (Fig.7) is compared with the one of the model SW3A in Fig.8. It has been observed that the sensitivity of the calculated global response (i.e., lateral load versus top displacement) to the selection of either the number of macro elements along the height of the first story of the wall (where the flexural and shear yielding occurs and is the most sensitive place to the number of macro elements) or the number of vertical springs along the length of the wall cross section is not significant. It should be mentioned that using more elements over the height or along the length of the wall allows for a more refined prediction of the wall local behavior, and has no influence on the CPU time needed for the analysis of the model.

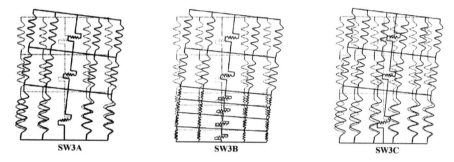

FIGURE 7. Sensitivity of the model flexural deformation to modeling parameters

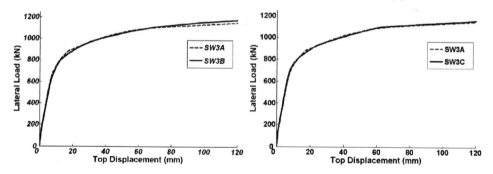

FIGURE 8. Sensitivity of the model response to modeling parameters

983

EVALUATION OF THE INVESTIGATED MACRO MODEL IN COMPARISON WITH MICROSCOPIC MODELS OF SPECIMENS

In order to evaluate the advantages and deficiencies of the investigated model in comparison with the models based on the finite element approach, the nonlinear behavior of the selected test specimens has been predicted using microscopic models. The specimens have been modeled using shell elements (S4R) and analyzed using the explicit solution method which is a true dynamic procedure originally developed to model high-speed impact events, and has proven valuable in solving static problems. ABAQUS/Explicit solves certain types of static problems more readily than ABAQUS/Standard does. Applying the explicit dynamic procedure to quasi-static problems requires some special considerations. Taking all these considerations into account, the microscopic model could simulate some important aspects of wall behavior such as the interaction between shear and flexural response components, distribution of cracks and stresses and local behavior in a CPU time (275 minutes for specimen SW3 and 366 minutes for specimen SW5) much greater than the one needed for the analysis of the investigated macro-model. The overall force-deformation diagram of a model of specimen SW3 with a relatively coarse mesh is indicated in Fig.9 for which sequence of plastic strain propagation can be observed in Fig.10.

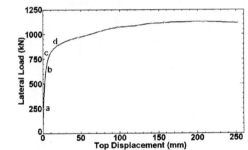

FIGURE 9. Force-deformation diagram of the microscopic model of specimen SW3 with a relatively coarse mesh

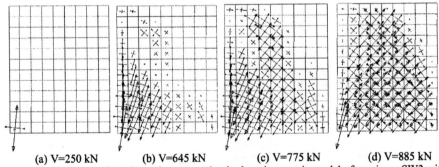

| (a) V=250 kN | (b) V=645 kN | (c) V=775 kN | (d) V=885 kN |

FIGURE 10. Sequence of plastic strain propagation in the microscopic model of specimen SW3 with a relatively coarse mesh

Cracking initiates at points where the tensile equivalent plastic strain is greater than zero, and the maximum principal plastic strain is positive (Fig.10). The direction of the vector normal to the crack plane is assumed to be parallel to the direction of the maximum principal plastic strain. The deformation pattern of the microscopic model of specimens with a relatively fine mesh is indicated in Fig.11. As shown in this figure the failure of specimen SW5 is accompanied by out of plane deformations which is a common phenomenon observed in plain rectangular walls, not having relatively stiff confined boundary elements. In fact such walls are prone to lateral buckling of the compression edge under large horizontal displacements, but since buckling of reinforcement and strength of concrete cover have not been taken into consideration in the microscopic model, failure was much more gradual in comparison with experimental results. Figure 12 displays the comparison between lateral load-displacement diagrams of the macroscopic and microscopic model of specimens. The descending branch of the load-displacement response of the microscopic model shows the capability of the model based on the finite element approach to predict strength degradation of structural walls under large horizontal displacements, which can not be simulated using the investigated macro-model. It should be mentioned that the response of the microscopic model was relatively sensitive to mesh refinement which in turn leads to the great increase in CPU time needed for the analysis of the model.

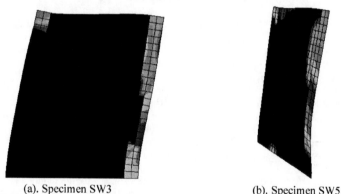

(a). Specimen SW3   (b). Specimen SW5

FIGURE 11. Ultimate deformation of the microscopic model of specimens

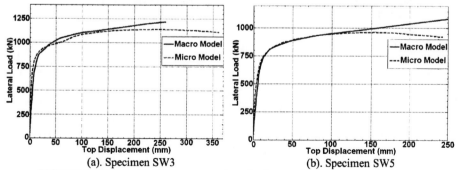

(a). Specimen SW3   (b). Specimen SW5

FIGURE 12. Response of the macroscopic and microscopic models of specimens

CONCLUSIONS

The investigated macro-model turned out to be capable of simulating the nonlinear behavior of the selected test specimens in different stages of loading to a very good degree of accuracy after a few seconds of CPU time.

Considering uncoupled shear and flexural responses in the model of slender walls, it was concluded it is not so significant to causes considerable discrepancy between analytical and experimental lateral load-top shear displacement response of such walls.

Although the microscopic model could simulate some important aspects of wall behavior such as the interaction between shear and flexural response components and distribution of cracks and stresses, taking the CPU time, simplicity, and parametric sensitivity into account, the observed agreement among three lateral load-displacement diagrams of experimental measurements, the macroscopic and microscopic models of test specimens indicates the efficiency of the investigated macro-model.

REFERENCES

1. Orakcal, K., and J. W. Wallace. 2004. Modeling of slender reinforced concrete walls. Proceedings, 13th World Conference on Earthquake Engineering. Vancouver, Canada
2. Menegotto, M., and E. Pinto. 1973. Method of analysis for cyclically loaded reinforced concrete plane frames including changes in geometry and non-elastic behavior of elements under combined normal force and bending, Proceedings, IABSE Symposium. Lisbon, Portugal.
3. Mander, J. B., M. J. N. Priestley, and R. Park. 1988a. Theoretical stress-strain model for confined concrete. ASCE Journal of Structural Engineering 114(8): 1804–1826.
4. Belarbi, H., and T. C. C. Hsu. 1994. Constitutive laws of concrete in tension and reinforcing bars stiffened by concrete. ACI Structural Journal 91(4): 465–474
5. Kabeyasawa, T., H. Shiohara, S. Otani, and H. Aoyama. 1983. Analysis of the full-scale seven-story reinforced concrete test structure. Journal of the Faculty of Engineering , The University of Tokyo 37(2): 431–478.
6. Vulcano, A., V. V. Bertero, and V. Colotti. 1988. Analytical modeling of RC structural walls. Proceedings, 9th World Conference on Earthquake Engineering (6). Tokyo-Kyoto, Japan.41–46.
7. Vallenas, J. M., Bertero, V. V. and Popov, E. P., 1979, Hysteretic behavior of reinforced concrete structural walls, Report No. UCB/EERC-79/20, EERC, University of California, Berkeley, California.1-268.

A Novel Approach for Modal Identification from Ambient Vibration

Azer A. Kasimzade and Sertac Tuhta

Ondokuz Mayis University, Department of Civil Engineering, Turkey
E-mail: azer@omu.edu.tr

Abstract. A novel approach of system characteristic matrix's correction in modal identification from ambient vibration is presented. As a result of this approach actual system characteristic matrices are determined more accurately with minimum error. It is reflected on to updating system parameters more reliable. In first approximation actual system characteristic matrices determined by singular value decomposition of block Hankel matrix which build from the response correlation matrix. In second approximation to make the system characteristic matrices optimal definite, for black-box modeling the input-output relation of the system used Kalman theory. Covariance of the non-measurable process noise and measurement noise matrixes are contained in Riccati equation are determined by expressing Hankel matrix's multiplicities from eigensolution of the system state matrix obtained in previous iteration. Another word process and measurement noises covariance matrixes indirectly is constructed only from measured output data. These iterations are repeated until satisfying estimated error. As a result of these iterations actual system characteristic matrices are determined more accurately with minimum error. Then from determined system characteristic matrices are extracted system modal parameters. These system modal parameters are used for the system modal updating for which direct and iterative methods are applied. Supporting to this algorithm realized code maybe interfaced with finite element codes.

Keywords: system identification, finite element modeling, updating, Kalman filter gain, operational modal analysis.

INTRODUCTION

When forced excitation tests are very difficult or only response data are measurable while the actual loading conditions are unknown, operation modal analysis (output-only modal identification techniques) remains the only technique for structural identification. The main advantage of this method is that no special, artificial type of excitation has to apply to the structure to determine its dynamic characteristics. Furthermore, if a structure has high period (more than one second) modes, it may be difficult to excite it with a shaker, whereas this is generally no problem for drop weight or ambient sources. However if mass-normalized mode shapes are required, ambient excitation cannot be used. Output -only modal identification techniques efficiently use with model updating tools to develop reliable finite element models of structures. Last year's Output – Only Model Identification studies of systems and

CP1020, *2008 Seismic Engineering Conference Commemorating the 1908 Messina and Reggio Calabria Earthquake*,
edited by A. Santini and N. Moraci
© 2008 American Institute of Physics 978-0-7354-0542-4/08/$23.00

results are given in appropriate references structural vibration solutions, partly in [1]. For the modal updating of the structure it is necessary to estimate sensitivity of reaction of examined system to change of parameters of a building [2, 3, 4].

System identification is the process of developing or improving a mathematical representation of a physical system using experimental data [5, 6, 7, 8, 9, 10, 11]. In engineering structures there are three types of identification: modal parameter identification; structural-modal parameter identification; control-model identification methods are used.

In the frequency domain the identification is based on the singular value decomposition of the spectral density matrix and it is denoted Frequency Domain Decomposition (FDD) and its further development Enhanced Frequency Domain Decomposition (EFDD).

In the time domain there are three different implementations of the Stochastic Subspace Identification (SSI) technique: Unweighted Principal Component (UPC); Principal component (PC); Canonical Variety Analysis (CVA) are used.

Below a novel approach of system characteristic matrix's correction in modal identification from ambient vibration is presented. In this algorithm the first approximation actual system characteristic matrices' determined by the data-driven stochastic subspace identification method. For an optimal estimation state vector another word to make the system characteristic matrices optimal definite in second approximation are obtained by applying the steady-state Kalman filter to the stochastic state-space model equation [12]. All off calculations are repeated until satisfying estimation error condition.

Determination of Approximate Values of System Characteristic Matrices

As mentioned above the purpose of operational modal analysis is to determine modal parameters of a real system, which is assumed to be linear time invariant, measuring only its response at specific locations, being the exciting load unknown.

In the vibration analysis identification of actual system (black-box model) supporting to experimental measurements determination approximate value of system characteristic matrices $\lfloor \hat{A} \rfloor \lfloor \hat{B} \rfloor \lfloor \hat{C} \rfloor$, instead of it appropriate unknown real matrices value are appeared first of all. Analysis sequences for evaluation this aim shown below.

The equation of motion of the continues system are arranged

$$[m]\{\ddot{u}(t)\} + [c]\{\dot{u}(t)\} + [k]\{u(t)\} = [d]\{f(t)\} \tag{1}$$

These are transformed to the state-space former of first order equations-i.e., a continuous-time state-space model of the system are evaluated as

$$\{\dot{z}(t)\} = [A_c]\{z(t)\} + [B_c]\{f(t)\} \tag{2a}$$

$$[A_c] = \begin{bmatrix} [0] & [I] \\ -[m]^{-1}[k] & -[m]^{-1}[c] \end{bmatrix} \tag{2b}$$

$$[B_c] = \begin{bmatrix} [0] \\ [m]^{-1}[d] \end{bmatrix}; \quad \{z(t)\} = \begin{bmatrix} u(t) \\ \dot{u}(t) \end{bmatrix} \tag{2c}$$

If the response of the dynamic system is measured by the m_1 output quantities in the output vector $\{y(t)\}$ using sensors (such as accelerometers, velocity, displacements, etc.,), for system model represented by the equations (2), appropriate measurement-output equation become as

$$\{y(t)\} = [C_a]\{\ddot{u}\} + [C_v]\{\dot{u}\} + [C_d]\{u\}$$
$$= [C]\{z(t)\} + [D]\{f(t)\} \tag{3a}$$

$$[C] = [[C_d] - [C_a][m]^{-1}[k] \quad [C_v] - [C_a][m]^{-1}[c]] \ (3b); \quad [D] = [C_a][m]^{-1}[d] \tag{3c}$$

Where $[m], [c], [k]$ mass ,damping, stiffness matrices of the structure are build by finite element method [13] ; $\{u\}$ is the vector of displacement; $[A_c]$, is an $n_1 (= 2n_2 ; n_2$ is the number of in depended coordinates) by n_1 state matrix ; $[d]$ is an n_2 by r_1 input influence matrix, characterizing the locations and type of known inputs $\{f(t)\}; [C_a], [C_v][C_d]$ are output influence matrices for acceleration, velocity, displacement for using sensors (such as accelerometers, tachometers, strain gages ,etc.,) respectively; $[C]$ is an $m_1 \times n_1$ output influence matrix for the state vector $\{z\}$ and displacement only; $[D]$ is an $m_1 \times r_1$ direct transmission matrix; r_1 is the number of inputs; m_1 is the number of outputs.

In the output –only modal analysis environment, the main assumption is that input force $\{F(t)\} = [d]\{f(t)\}$ comes from white noise or time impulse excitation. Under this hypothesis discrete-time stochastic state–space model may be written as:

$$\{z_{k+1}\} = [A]\{z_k\} + [B]\{f_k\} + \{w_k\} \tag{4}$$
$$\{y_k\} = [C]\{z_k\} + [D]\{f_k\} + \{v_k\} \tag{5a}$$

where $\{z_k\} = \{z(k\Delta t)\}$ is the discrete-time state vector;

$$[A] = e^{[A_c]\Delta t} = [I] + [A_c]\Delta t + (1/2!)([A_c]\Delta t)^2 + (1/3!)([A_c]\Delta t)^3 + \dots \tag{5b}$$

is the discrete-time system state matrix;

$$[B] = [B_c]\int_0^{\Delta t} e^{[A_c]\tau} d\tau = [B_c] [[I]\Delta t + (1/2!)[A_c](\Delta t)^2 + (1/3!)[A_c](\Delta t)^3 + \dots] \tag{5c}$$

is the discrete-time input matrix. If the real parts of all eigenvalues of $[A_c]$ are negative, then $[A], [B]$ series expansions converge. If none of eigenvalues of $[A_c]$ are zero, then $[B]$ may also by computed by $[B] = [[A] - [I]][A_c]^{-1}[B_c]$; $[I]$ is the identity matrix; $\{w_k\}$ is the process noise due to disturbance and modeling imperfections; $\{v_k\}$ is the measurement noise due to sensors' inaccuracies; $\{w_k\}, \{v_k\}$ vectors are non-measurable, but assumed that they are white noise with zero mean. If this white noise assumption is violated, in other words if the input contains also some dominant frequency components in addition to white noise , these frequency components cannot be separated from the eigen frequencies of the system and they will appear as poles of the system matrix $[A]$.

As shown from measurement-output equation (3) it indirectly is depend on system model (2) and contain appropriate system mass, damping, rigidity matrices $[m], [c], [k]$ respectively. For this reason to carry out measurement by the relation (5), it will be require known system model (2) with matrices $[m], [c], [k]$ previously for **zero** approximation. For **zero** approximation these known matrices are denoted as $[m], [c], [k]$ and include them $[A_c], [B_c], [A], [B], [C], [D]$ respectively.

In the real structures, exited by ambient vibration, the input $\{f(t)\}, \{f_k\}$ remains unmeasured and therefore it disappears from the equation (2, 3, 4, 5) respectively. Then to take into consideration this fact, the input is implicitly modeled by the noise terms $\{w_k\}, \{v_k\}$ and mentioned relation became as:

$$\{z_{k+1}\} = [A]\{z_k\} + \{w_k\} \tag{6.0}$$

$$\{y_k\} = [C]\{z_k\} + \{v_k\} \tag{7.0}$$

The main goal of the operational modal analysis, identification the modal parameters of the system from the output vector $\{y_k\}$, measured by the sensors located on the structure, whose dimensions are $n_{sensors} \times n_{samples}$. Where $n_{sensors}$ and $n_{samples}$ are the number of sensors and samples respectively. In the practical engineering of stochastic methods the signal $\{y_k\}$ (it may be displacement, velocity or acceleration) given by sensors can be sampled at discrete time intervals depending on the characteristic of the computer processing.

Measurements are realized by the formulas (6.0, 7.0) and then defined system matrices $[\hat{A}], [\hat{C}]$ by the next sequences. Measuring the output vector $\{y_k\}$ and $\{y_s\}_{ref}$ (it is a subset of the output vector $\{y_{k+s}\}$) by the sensors located on the $s + k = 1 \div (s_* + k_*)$ characteristic and $s = 1 \div s_*$ reference points appropriately of the building structure, then denoted the expected value operator as $E(...)$, the correlation function calculated as :

$$[R_k] = E(\{y_{k+s}\}\{y_s\}_{ref}^T) = \frac{1}{s_* - k} \sum_{s=0}^{s_*-k-1} \{y_{k+s}\}\{y_s\}_{ref}^T \tag{8}$$

From which the following block Hankel matrix build as:

$$[H_{p,q}] = \begin{bmatrix} [R_1] & [R_2] & \cdots & [R_q] \\ [R_2] & [R_3] & \cdots & [R_{q+1}] \\ \cdots & \cdots & \cdots & \cdots \\ [R_p] & [R_{p+1}] & \cdots & [R_{p+q-1}] \end{bmatrix} \tag{9}$$

The factorization of the block Hankel data matrix realized by using singular value decomposition,

$$[H(0)] = [U][\Sigma][V]^T \tag{10}$$

Removing eigen values nearly zero from matrix $[\Sigma]$, consequently defined its rank (n) and in accordance with matrices $[U_n], [\Sigma_n], [V_n]$ are evaluated respectively.

The **first** approximation of the system matrices $[\hat{A}], [\hat{B}], [\hat{C}]$ are calculated as:

$$\left|\hat{A}\right| = \left[\Sigma_n\right]^{-1/2}\left[U_n\right]^T\left[H(1)\right]\left[V_n\right]\left[\Sigma_n\right]^{-1/2} \tag{11}$$

$$\left[\hat{B}\right] = \left[\Sigma_n\right]^{1/2}\left[V_n\right]^T\left[E_r\right] = 0 \quad \text{for ambient vibration} \tag{12}$$

$$\left|\hat{C}\right| = \left[E_m\right]^T\left[U_n\right]\left[\Sigma_n\right]^{1/2} \tag{13}$$

$$\left[E_r\right]^T = \left(\left[I_{r_1}\right] \quad \left[0_{r_1}\right] \quad \cdots \quad \left[0_{r_1}\right]\right), \left[E_m\right]^T = \left(\left[I_{m_1}\right] \quad \left[0_{m_1}\right] \quad \cdots \quad \left[0_{m_1}\right]\right)$$

Where $[H(1)]$ a shifted is block Hankel matrix; r_1 and m_1 are the number of inputs and outputs respectively; $[I_i]$ is an identity matrix of order i; $[0_i]$ is a null matrix of order i.

To Make the System Characteristic Matrices Optimal Definite

Here, for reach theoretical target to find the best (or optimal) estimate $\{\hat{z}_k\}$ in the sense that the estimation error $\{e_k\} = \{z_k\} - \{\hat{z}_k\}$ is as small as possible, achievement of to make definite **first** approximate values of the building system characteristics $\left[\hat{A}\right], \left[\hat{C}\right]$, to the stochastic black–box models' is applied Kalman theory. Theoretically, the Kalman filter is very attractive, because it has a closed-form solution (i.e., Riccati equation) for its gain matrix. However, the Kalman filter requires information, including of covariance $E(\{w_k\}\{w_k\}^T) = [q\delta(k-j)] = [Q]$ of the non-measurable process noise $\{w(k)\}$ and covariance $E(\{v_k\}\{v_k\}^T) = [r\delta(k-j)] = [R]$ of the non-measurable measurement noise $\{v(k)\}$. For this reason, it is necessary to estimate this matrices $([Q],[R])$ indirectly from measured output data $\{y_k\}$ and $\{y_s\}_{ref} = \{y_{k+s}\}$. One of the crucial points [12] in presented method, determination these matrices at least approximately $[Q] \cong [\overline{Q}]$, $[R] \cong [\overline{R}] = [U_n]$ by expressing Hankel matrix's multiplicities from eigensolution of the system state matrix $\left|\hat{A}\right|$ as shown below (In references [7] by Observer/Kalman Filter Identification method to avoid this problem to make an effort estimate more approximately Kalman filter gain directly from experimental data (Markov parameters) without estimating the covariance (P) of the process and measurement noises and solving Riccati equation).

Shortly in this optimal make definition stage of matrices $\left[\hat{A}\right], \left[\hat{C}\right]$ for the **first** approximation, the system model represented by the relation (6.0, 7.0) become as

$$\{z_{k+1}\} = \left|\hat{A}\right|\{z_k\} + \{w_k\} \tag{6.1}$$

$$\{y_k\} = \left|\hat{C}\right|\{z_k\} + \{v_k\} \tag{7.1}$$

Under above assumptions and represented system model, the solution of this problem may be reached if the error covariance $[P] = E(\{e_k\}\{e_k\}^T)$ will satisfy discrete algebraic Riccati equation. Sequences of solution of formulated problem are given below.

Solving eigen problem of the matrix $\left|\hat{A}\right|$, consequently

$$\left([\Lambda],[\Psi]\right) = eig\left(\left|\hat{A}\right|\right) \tag{14}$$

And expressing it in the form $[\hat{A}] = [\Psi][\Lambda][\Psi]^T$, then the Hankel matrix's multiplicities are express as:

$$[H_0] = [[U_n][\Sigma_n]^{1/2}[\Psi]][[\Psi]^{-1}[\Sigma_n]^{1/2}[V_n]^T] \approx [\bar{P}][\bar{Q}] \qquad (15)$$

Where $[\bar{R}] = [U_n]$, $\quad [\bar{P}] = [[U_n][\Sigma_n]^{1/2}[\Psi]]$, $\quad [\bar{Q}] = [[\Psi]^{-1}[\Sigma_n]^{1/2}[V_n]^T]$

Supporting this result is evaluated Riccati equation:

$$[P] = [\hat{A}][P][\hat{A}]^T - [\hat{A}][P][\hat{C}]^T \left[[\bar{R}] + [\hat{C}][P][\hat{C}]^T\right]^{-1}[\hat{C}][P][\hat{A}]^T + [\bar{Q}] \qquad (16)$$

The existence of the Riccati equation solution is only possible, if the correlation function is positive definite. There are a few proposals in literature to guarantee a solution. But the existing experiences it remains an open problem in large scale stochastic realization theory. If a solution of the Riccati equation exists, after defining it $[P] = [\hat{P}]$, one can obtain Kalman gain of the building structure model:

$$[\hat{K}] = [\hat{A}][\hat{P}][\hat{C}]^T \left[[\bar{R}] + [\hat{C}][\hat{P}][\hat{C}]^T\right]^{-1} \qquad (17)$$

Kalman filter equation is evaluated

$$\{\hat{z}_{k+1}\} = [[\hat{A}] - [\hat{K}][\hat{C}]]\{\hat{z}_k\} + [\hat{K}]\{y_k\} = [\hat{A}]\{\hat{z}_k\} + [\hat{K}]\{\varepsilon_k\} \qquad (18)$$

With the output measurement $\{y_k\}$ satisfying

$$\{y_k\} = [\hat{C}]\{\hat{z}_k\} + \{\varepsilon_k\} \qquad (19)$$

The output residual $\{\varepsilon_k\}$ satisfies $\{\varepsilon_k\} = [\hat{C}]\{e_k\} + \{v_k\}$.

Comparing building system modeling by the Kalman filter equations (18, 19) with the first step system modeling equations, so

$$\{z_{k+1}\} = [\hat{A}]\{z_k\} + \{w_k\}; \qquad \{y_k\} = [\hat{C}]\{z_k\} + \{v_k\}$$

The error is evaluated as:
-the state estimation error

$$\{e_k\} = \{z_k\} - \{\hat{z}_k\}; \{E(e_k)\} = \{0\} \qquad (20)$$

-error dynamics (the state estimation error at steady-state reduces to)

$$\{e_{k+1}\} = [[\hat{A}] - [\hat{K}][\hat{C}]]\{e_k\} - [\hat{K}]\{v_k\} + \{w_k\} \qquad (21)$$

-output residual

$$\{\varepsilon_k\} = [\hat{C}]\{e_k\} + \{v_k\}; \{E(\varepsilon_k)\} = \{0\} \qquad (22)$$

If not satisfied condition (22), all off calculations are repeated until satisfying this condition. For example if not satisfied condition (22), in **second** approximation case relation (6.0, 7.0) became as

$$\{z_{k+1}\} = [\hat{A}]\{z_k\} + \{w_k\}; \quad \{y_k\} = [\hat{C}]\{z_k\} + \{v_k\} \qquad (23)$$

Measurements in **second** approximation case are realized by these formulas and then will defined system characteristic matrices $[\hat{A}], [\hat{C}]$, by the operations (8-13).

Then for the optimal make definition of the system matrices $[\hat{A}], [\hat{C}]$ relation (6.1, 7.1) became as

$$\{z_{k+1}\} = [\hat{A}]\{z_k\} + \{w_k\} \quad \{y_k\} = [\hat{C}]\{z_k\} + \{v_k\} \qquad (24)$$

992

In all off relation (14-22) instead of matrices $\left[\hat{A}\right], \left[\hat{C}\right]$, will take place matrices $\left[\tilde{A}\right], \left[\tilde{C}\right]$ respectively for this-**second** approximation case. As a result of this iterations (when satisfied condition (22)), are definite real building system parameters, another word, system identified completely and obtained system matrices will mark as $\left[\overline{A}\right], \left[\overline{C}\right]$ for the illustrating of operations in next section.

Computing of Identified System's Modal Parameters

After obtaining system parameters $\left[\overline{A}\right], \left[\overline{C}\right]$ as shown above, for computing system modal parameters (frequency, mode shapes and damping) in first as known, it is necessary transformation discrete time model (represented by the matrices $\left[\overline{A}\right], \left[\overline{C}\right]$) to the continuous time model (will be represented by the matrices $\left[\overline{A}_c\right], [C]$) by the inverse operation in relation $\left[\overline{A}_c\right] = \dfrac{1}{\Delta t}\left[\overline{A}\right]$. As a result continuous-time state-space model of the identified system became as

$$\{\dot{z}(t)\} = \left[\overline{A}_c\right]z(t) + \{w(t)\}; \quad \{y(t)\} = \left[\overline{C}\right]z(t) + \{v(t)\}$$

The matrices $\left[\overline{A}_c\right]$ and $\left[\overline{C}\right]$ are sufficient to compute the modal parameters. Computing eigen values $\left[\overline{\Lambda}\right]$ and eigenvectors $\left[\overline{\Psi}\right]$ of the matrix $\left[\overline{A}_c\right]$ $\left(\left[\overline{\Lambda}\right], \left[\overline{\psi}\right]\right) = eig\left[\overline{A}_c\right]$. The eigenvalue decomposition of $\left[\overline{A}_c\right]$ is given by $\left[\overline{A}_c\right] = \left[\overline{\Psi}\right]\left[\overline{\Lambda}\right]\left[\overline{\Psi}\right]^{-1}$. The continuous eigenvalues (or system poles) $\overline{\lambda}_k$ on the diagonal of $\left[\overline{\Lambda}\right]$ has representation $\overline{\lambda}_k = \overline{c}_k + i\overline{\omega}_k$ or $\overline{c}_k = \mathrm{Re}(\overline{\lambda}_k)$, $\overline{\omega}_k = \mathrm{Im}(\overline{\lambda}_k)$. Where $\overline{c}_k$ is the damping factor and $\overline{\omega}_k$ the damped natural frequency (related to natural cyclic frequency $\overline{f}_k = 1/\overline{T}_k$ and natural period $\overline{T}_k = 2\pi/\overline{\omega}_k$) of the k th mode. The damping ratio $\overline{\zeta}_k$ of the k th mode is given by $\overline{\zeta}_k = -\overline{c}_k/\sqrt{\overline{c}_k^2 + \overline{\omega}_k^2}$. The mode shape $\left[\overline{\phi}_k\right]$ of the k th mode at sensor locations are the observed parts of the system eigenvectors $\{\overline{\psi}_k\}$ of $\left[\overline{\psi}\right]$, given by the equation $\{\overline{\phi}_k\} = \left[\overline{C}\right]\{\overline{\Psi}_k\}$. The extracted mode shapes cannot be mass-normalized as this requires the measurement of the input force.

System Model Updating

Obtained modal parameters-damping $\overline{\zeta}_k$, period $\overline{T}_k = 2\pi/\overline{\omega}_k$ (which are contain the eigenvalue matrix $\left[\overline{\Lambda}\right]$), mode shape $\overline{\phi}_k$ using as a reference modal "experimental" data (in mean that they are defined supporting experimental result) the analytical (finite element model) stiffness, damping and mass matrices are corrected by the known direct [14] and iterative updating methods [15] using convergence criteria [16]. For speed up updating procedure, marked updating parameters previously are fuzzyfied in the user definable intervals these parameters using full factorial and orthogonal array testing on the base finite element method [2]. Some application result

partly presented in [4, 12]. Applied benchmark structure is shown in web side
www2.omu.edu.tr/akademik_birimler/muhendislik/insaat/akasimzade-engproje.htm

Conclusion

System characteristic matrix's correction in modal identification from ambient vibration is presented. In this algorithm the first approximation actual system characteristic matrices' determined by the data-driven stochastic subspace identification method. For an optimal estimation state vector another word to make the system characteristic matrices optimal definite in second approximation are obtained by applying the steady-state Kalman filter to the stochastic state-space model equation. All off calculations are repeated until satisfying estimation error condition.

As a result of this approach actual system characteristic matrices are determined more accurately with minimum error. It is reflected on to updating system parameters more reliable. For speed up updating procedure, marked updating parameters previously are fuzzyfied in the user definable intervals these parameters using full factorial and orthogonal array testing on the base finite element method. Supporting to this algorithm realized code maybe interfaced with finite element codes.

REFERENCES

1. Cunha, A., E. Caetano, From input-output to output-only modal Identification of Civil Engineering structures, First International Operational Modal Analysis Conference (R. Brincker, N. Moller. (Ed)), p. 11-29. Aalborg University, Copenhagen, Denmark, 2005.
2. Kasimzade, A.A., Bounds of the Structural Response imprecisely defined systems under Earthquake Action, P. Journal of Applied Sciences, Vol. 2, N. 10, 2002, p. 969-974.
3. Kasimzade, A.A., Tuhta S., Estimation of Sensitivity and Reliability Base Isolated Buildings under Earthquake Action, International Symposium on Network and Center–Based Research for Smart Structures Technologies and Earthquake Engineering (E. Tachibana, B.F. Spencer, Jr. and Y. Muakai (Ed)), p. 407-412. Osaka University, Osaka, Japan, 2004.
4. Kasimzade A.A., Particularities of Monitoring, Identification, Model Updating Hierarchy in Experimental Vibration Analysis of Structures, Experimental Vibration Analysis of Civil Engineering Structures, EVACES'07, October 24-26, 2007, Porto, Portugal
5. Bendat J.S., Nonlinear System Techniques and Applications, A Wiley-Interscience Publication, 1998.
6. Ibrahim, S.R., E.C. Miculcik, A Method for the Direct Identification of Vibration Parameters from the Free Response, The Shock and Vibration Bulletin, Vol. 47, N. 4, 1997, p.183-194.
7. Juang, J.N., Applied System Identification, Prentice Hall PTR, Englewood Cliffs, NS, 1994.
8. Peeters, B., System Identification and Damage Detection in Civil Engineering, Ph.D. Thesis, K.U., Leuven, Belgium, 2000.
9. ARTeMIS. Structural Vibration Solution, Aalborg, Denmark, 2003.
10. Ljung L., System Identification: Theory for the User, Prentice Hall, Second Edition, 1999.
11. Van Overschee P., De Moor B., Subspace Identification for Linear Systems: Theory-Implementation-Applictions, Kluver Academic Publications, Dordrecht, The Netherlands, 1996.
12. Kasimzade A. A., Coupling of the Control System and the System Identification Toolboxes with Application in Structural Dynamics, International Conference Control, Glasgow, Scotland, UK, 2006, 6p.
13. Kasimzade A.A., Fundamentals, Education and Application of FEM in Structural Mechanics, 9th US National Congress on Computational Mechanics, San Francisco, July 22-26, 2007.
14. Caesar B., Update and Identification of Dynamic Mathematical Models, 4th International Modal Analysis Conference, Los Angeles, 1986, p.394-401
15. Link M., Updating of Analytical Models Procedure and Experience, Conference on Modern Practice in Stress and Vibration Analysis, Sheffield, England, April 1993, p.35-52
16. Dascotte E and Vanhouckev P., Development of an Automatic Mathematical Model Updating Program Proceedings of the 7th International Modal Analysis Conference, IMAC, Las Vegas, Nevada, February 1989

Author Index

A

Abadi, M. M., 1837
Abate, G., 569
Abedi, K., 558
Abedi-Nik, F., 771
Ahlehagh, S., 1044
Ahmadi, A., 442
Alaggio, R., 76
Alam, M. J., 413, 419
Albanese, V., 1665
Al-Dabbeek, J. N., 191
Al Satari, M., 577, 1381
Amaddeo, C., 1312, 1819
Amidi, S., 1195
Amini, A., 585
Amini, F., 1245
Amini, O. N., 1775
Amiri, A., 642
Amiri, G. G., 270, 277, 719, 1195
Amiri, J. V., 1775
Amrei, S. A. R., 270, 277
Amrollahi, A., 1554
Antonucci, R., 1358
Apostol, B. F., 207
Arani, K. K., 1554
Asgarian, B., 1787
Ashtiany, M. G., 595
Asprone, D., 1677
Astaneh-Asl, A., 5, 1819, 1827
Astroza, R., 1795
Athanasopoulou, E., 1899
Ausilio, E., 199, 475
Avila, J. A., 779
Avilés, J., 674
Azad, A. K., 413
Azarbakht, A., 595

B

Bai, J.-W., 1685
Balan, S. F., 207
Balducci, F., 1358
Bandini, V., 485
Baratta, A., 216, 1573
Barbato, M., 1693

Barbosa, A., 930
Barszez, A.-M., 1581
Bartolomei, A., 1591
Baykal, G., 711
Bazzurro, P., 1607
Bekkouche, A., 493
Benadla, Z., 493
Benedetti, S., 1888
Benedettini, F., 76
Benzoni, G., 1312
Bergman, L. A., 1390
Betti, M., 787
Bian, H., 427
Bianco, A., 1320
Bilham, R., 224
Binda, L., 795
Biondi, G., 485, 501, 687
Blasone, E., 542
Bloch, J., 53
Bojorquez, E., 1599
Bontempi, F., 803
Bosi, A., 1803
Boumechra, N., 69
Brando, G., 19
Buffarini, G., 1366
Buonsanti, M., 1330, 1340, 1350
Bursi, O. S., 1093
Butti, F., 1505

C

Cacace, F., 1665
Cacciola, P., 1811
Caffrey, J. P., 121
Caglayan, B. O., 1701
Cairo, R., 602
Calderini, C., 816
Calderoni, B., 824
Caliò, I., 832
Calogero, C., 1657
Camelbeeck, T., 1581
Campione, G., 840
Candela, M., 1320
Cannizzaro, F., 832
Capilleri, P., 232
Carbonari, S., 610, 626

A6